THE COSMIC

PERSPECTIVE

Fourth Edition

About the Cover

The Earth view on the cover of The Cosmic Perspective, *Fourth Edition, was created by ARC Science Simulations using ARC's* **Face of the Earth**™ *computer model of the planet. New Milky Way image (top right corner) is courtesy of Dr. Mark A. Garlick.*

THE COSMIC PERSPECTIVE

Fourth Edition

Jeffrey Bennett
University of Colorado

Megan Donahue
Michigan State University

Nicholas Schneider
University of Colorado

Mark Voit
Michigan State University

PEARSON

Addison
Wesley

San Francisco   Boston   New York
Capetown   Hong Kong   London   Madrid   Mexico City
Montreal   Munich   Paris   Singapore   Sydney   Tokyo   Toronto

Editor-in-Chief: *Adam Black*
Senior Acquisitions Editor: *Lothlórien Homet*
Developmental Editor: *Elisa Adams*
Art Developmental Editor: *Jody Larson*
Assistant Editor: *Deb Greco*
Editorial Assistant: *Ashley Taylor Anderson*
Media Development Associate: *Brad Patterson*
Executive Marketing Manager: *Christy Lawrence*
Managing Media Producer: *Claire Masson*
Production Supervisor: *Shannon Tozier*
Production Management: *Elm Street Publishing Services, Inc.*
Composition: *Thompson Type*
Illustrators: *John & Judy Waller, Scientific Illustrators*
Manufacturing Manager: *Pam Augspurger*
Text Designer: *Side by Side Studios*
Cover Designer: *Jeff Puda*
Text Printer and Binder: *R. R. Donnelley, Willard*
Cover Printer: *Phoenix Color*

Library of Congress Cataloging-in-Publishing Data is on file at the Library of Congress.

ISBN 0-8053-9283-1 (The Cosmic Perspective, Student edition)
ISBN 0-8053-9294-7 (The Cosmic Perspective, Instructors edition)
ISBN 0-8053-9296-3 (The Solar System)
ISBN 0-8053-9210-6 (Stars, Galaxies, & Cosmology)

PEARSON

Addison
Wesley

1 2 3 4 5 6 7 8 9 10—DOW—08 07 06 05

www.aw-bc.com

Dedication

TO ALL WHO HAVE EVER WONDERED about the mysteries of the universe. We hope this book will answer some of your questions—and that it will also raise new questions in your mind that will keep you curious and interested in the ongoing human adventure of astronomy.

And, especially, to Michaela, Emily, Sebastian, Grant, Nathan, Brooke, and Angela. The study of the universe begins at birth, and we hope that you will grow up in a world with far less poverty, hatred, and war so that all people will have the opportunity to contemplate the mysteries of the universe into which they are born.

Brief Contents

Detailed Contents

PART III
LEARNING FROM OTHER WORLDS

PART VI
GALAXIES AND BEYOND

Preface

We humans have gazed into the sky for countless generations. We have wondered how our lives are connected to the Sun, Moon, planets, and stars that adorn the heavens. Today, through the science of astronomy, we know that these connections go far deeper than our ancestors ever imagined. This book tells the story of modern astronomy and the new perspective, *The Cosmic Perspective*, that astronomy gives us on ourselves and our planet.

This book grew out of our experience teaching astronomy to both college students and the general public over the past 25 years. During this time, a flood of new discoveries fueled a revolution in our understanding of the cosmos but had little impact on the basic organization and approach of most astronomy textbooks. We felt the time had come to rethink how to organize and teach the major concepts in astronomy to reflect this revolution in scientific understanding. This book is the result.

Who Is This Book For?

The Cosmic Perspective is designed as a textbook for college courses in introductory astronomy, but is suitable for anyone who is curious about the universe. We assume no prior knowledge of astronomy or physics, and the book is especially suited to students who do not intend to major in mathematics or science.

The Cosmic Perspective provides a comprehensive survey of modern astronomy, and it contains enough material for a two-semester introductory astronomy sequence. It may also be used for one-semester survey courses if professors choose their areas of emphasis; however, instructors of such courses may wish to consider the much shorter version of this book, titled *The Essential Cosmic Perspective*, that we designed specifically for one-semester survey courses.

About the Fourth Edition

Although the underlying philosophy and goals of *The Cosmic Perspective* remain the same as in past editions, this fourth edition differs from the third edition in many significant ways. Here, briefly, is a list of the major changes you'll find in this fourth edition:

- *New Chapters on Extrasolar Planets and Star Birth:* Reflecting recent developments in the science of astronomy, we have added two new chapters to the book: Chapter 13 on Other Planetary Systems and Chapter 16 on Star Birth. Note, however, that we have *not* increased the overall number of chapters; we made room for these new chapters by consolidating material from Third Edition Chapters 4 and 14 into other chapters.

- *New Pedagogy Centered on Learning Goals:* We have revised the motivational Learning Goals for each chapter to better reflect the types of questions that students are most likely to have even before reading this book. These new Learning Goals begin each chapter to focus students on the most important concepts ahead. Each section is carefully written to address the specific learning goal in its title. These learning goals are wrapped up at the end of each chapter in a visual summary that includes thumbnails of key figures. This approach helps keep students focused on the big picture and motivated through the reading. It also aids studying by making it easier for students to find topics in the book.

- *Streamlining and Updating of Presentation:* In the process of revising the book with this new pedagogy, we have carefully reviewed every sentence and every word to ensure that the book is as clear and streamlined as possible. We have also thoroughly updated the book with the many recent astronomical discoveries, including:

  - the discovery of an object larger than Pluto in our own solar system
  - the latest results from the rovers and orbiters studying Mars
  - the latest *Cassini* results from Saturn and *Huygens* results from Titan
  - the *Deep Impact* mission to Comet Tempel 1
  - supernovae observations providing stronger evidence for dark energy
  - recent results from the Spitzer Space Telescope
  - and much more

- *New Annotated Figures:* Key figures in each chapter now include the research-proven technique of "annotation"—carefully crafted text placed on the figure (in blue) to guide students through interpreting graphs, following process figures, and translating between different representations. In class tests, students have responded very positively to this new approach.

- *Expanded Exercise Sets:* We have significantly expanded the end-of-chapter exercise sets with additional multiple-choice and quantitative problems, in most cases providing at least twice as many exercises as we had in the third edition. Most of these are available online in MasteringAstronomy™ for automatic assignment, tutoring, and grading.

- *More Optional Mathematics:* As in prior editions, all mathematics in *The Cosmic Perspective* remains in optional Mathematical Insight boxes. We have expanded this feature by adding several new Mathematical Insights and by introducing an explicit and consistent "understand-solve-explain" problem-solving strategy to give students a consistent framework for solving homework problems.

- *MasteringAstronomy™ (www.masteringastronomy.com):* Users of prior editions of *The Cosmic Perspective* know of the wealth of interactive media that we have developed to support every chapter of our book. The Astronomy Place Web site (www.astronomyplace.com) proved so popular that more than 100,000 students have voluntarily registered to use the award-winning Interactive Tutorials, on-line quizzes, interactive figures, and other resources it provides. For the fourth edition, we have taken all these Web resources to a new level with **MasteringAstronomy**—the most advanced, educationally effective, and easy-to-use astronomy tutorial, assessment, and self-study system ever built. For students, MasteringAstronomy provides a wealth of tutorials and activities to build understanding, while quizzes and exercises allow them to test what they've learned. For instructors, MasteringAstronomy provides the unprecedented ability to quickly build, post, and automatically grade pre- and post-lecture diagnostic tests, weekly homework assignments, and exams of appropriate difficulty, duration, and content coverage. It also provides the ability to record detailed information on the step-by-step work of every student directly into a powerful and easy-to-use gradebook, and to evaluate results with a sophisticated suite of diagnostics. A 1-year subscription is included free with all new copies of *The Cosmic Perspective*, Fourth Edition.

Themes of *The Cosmic Perspective*

The Cosmic Perspective offers a broad survey of modern understanding of the cosmos and of how we have built that understanding. Such a survey can be presented in a number of different ways. We have chosen to interweave a few key themes throughout the book, each selected to help make the subject more appealing to students who may never have taken any formal science courses and who may begin the course with little understanding of how science works. We built our book around the following five key themes:

- *Theme 1: We are a part of the universe and thus can learn about our origins by studying the universe.* This is the overarching theme of *The Cosmic Perspective*, as we continually emphasize that learning about the universe helps us understand ourselves. Studying the intimate connections between human life and the cosmos gives students a reason to care about astronomy and also deepens their appreciation of the unique and fragile nature of our planet and its life.

- *Theme 2: The universe is comprehensible through scientific principles that anyone can understand.* The universe is comprehensible because the same physical laws appear to be at work in every aspect, on every scale, and in every age of the universe. Moreover, while professional scientists generally have discovered the laws, anyone can understand their fundamental features. Students can learn enough in one or two terms of astronomy to comprehend the basic reasons for many phenomena that they see around them—phenomena ranging from seasonal changes and phases of the Moon to the most esoteric astronomical images that appear in the news.

- *Theme 3: Science is not a body of facts but rather a process through which we seek to understand the world around us.* Many students assume that science is just a laundry list of facts. The long history of astronomy can show them that science is a process through which we learn about our universe—a process that is not always a straight line to the truth. That is why our ideas about the cosmos sometimes change as we learn more, as they did dramatically when we first recognized that Earth is a planet going around the Sun rather than the center of the universe. In this book, we continually emphasize the nature of science so that students can understand how and why modern theories have gained acceptance and why these theories may still change in the future.

- *Theme 4: A course in astronomy is the beginning of a lifelong learning experience.* Building upon the prior themes, we emphasize that what students learn in their astronomy course is not an end but a beginning. By remembering a few key physical principles and understanding the nature of science, students can follow astronomical developments for the rest of their lives. We therefore seek to motivate students enough that they will continue to participate in the ongoing human adventure of astronomical discovery.

- *Theme 5: Astronomy affects each of us personally with the new perspectives it offers.* We all conduct the daily business of our lives with reference to some "world view"— a set of personal beliefs about our place and purpose in the universe that we have developed through a combination of schooling, religious training, and personal thought. This world view shapes our beliefs and many of our actions. Although astronomy does not mandate a particular set of beliefs, it does provide perspectives on the architecture of the universe that can influence how we view ourselves and our world, and these perspectives can potentially affect our behavior. For example, someone who believes Earth to be at the center of the universe might treat our planet quite differently from someone who views it as a tiny and fragile world in the vast cosmos. In many respects, the role of astronomy in shaping world views may represent the deepest connection between the universe and the everyday lives of humans.

Pedagogical Principles of *The Cosmic Perspective*

No matter how an astronomy course is taught, it is very important to present material according to a clear set of pedagogical principles. The following list briefly summarizes the major pedagogical principles that we apply throughout this book. (The *Instructor's Guide* describes these principles in more detail.)

- *Stay focused on the big picture.* Astronomy is filled with interesting facts and details, but they are meaningless unless they fit into a big picture view of the universe. We therefore take care to stay focused on the big picture (essentially the themes discussed above) at all times. A major benefit of this approach is that although students may forget individual facts and details after the course is over, the big picture framework should stay with them for life.

- *Always provide context first.* We all learn new material more easily when we understand why we are learning it. In essence, this is simply the idea that it is easier to get somewhere when you know where you are going. We therefore begin the book (Chapter 1) with a broad overview of modern understanding of the cosmos, so that students can know what they will be studying in the rest of the book. We maintain this "context first" approach throughout the book by always telling students what they will be learning, and why, before diving into the details.

- *Make the material relevant.* It's human nature to be more interested in subjects that seem relevant to our lives. Fortunately, astronomy is filled with ideas that touch each of us personally. For example, the study of our solar system helps us better understand and appreciate our planet Earth, and the study of stars and galaxies helps us learn how we have come to exist. By emphasizing our personal connections to the cosmos, we make the material more meaningful, inspiring students to put in the effort necessary to learn it.

- *Emphasize conceptual understanding over "stamp collecting" of facts.* If we are not careful, astronomy can appear to be an overwhelming collection of facts that are easily forgotten when the course ends. We therefore emphasize a few key conceptual ideas that we use over and over again. For example, the laws of conservation of energy and conservation of angular momentum (introduced in Section 4.3) reappear throughout the book, and we find that the wide variety of features found on the terrestrial planets can be understood through just a few basic geological processes. Research shows that, long after the course is over, students are far more likely to retain such conceptual learning than individual facts or details.

- *Proceed from the more familiar and concrete to the less familiar and abstract.* It's well known that children learn best by starting with concrete ideas and then generalizing to abstractions later. The same is true for many adults. We therefore always try to "build bridges to the familiar"—that is, to begin with concrete or familiar ideas and then gradually draw more general principles from them.

- *Use plain language.* Surveys have found that the number of new terms in many introductory astronomy books is larger than the number of words taught in many first-year courses in foreign language. In essence, this means the books are teaching astronomy in what looks to students like a foreign language! Clearly, it is much easier for students to understand key astronomical concepts if they are explained in plain English without resorting to unnecessary jargon. We have gone to great lengths to eliminate jargon as much as possible or, at minimum, to replace standard jargon with terms that are easier to remember in the context of the subject matter.

- *Recognize and address student misconceptions.* Students do not arrive as blank slates. Most students enter our courses not only lacking the knowledge we hope to teach but often holding misconceptions about astronomical ideas. Therefore, to teach correct ideas, we must also help students recognize the paradoxes in their prior misconceptions. We address this issue in a number of ways, the most obvious being the presence of many Common Misconceptions boxes. These summarize commonly held misconceptions and explain why they cannot be correct.

The Topical (Part) Structure of *The Cosmic Perspective*

The Cosmic Perspective is organized into seven broad topical areas (the seven Parts in the table of contents), each approached in a distinctive way designed to help maintain the focus on the themes discussed earlier. Here, we summarize the guiding philosophy through which we have approached each topic. We also highlight a few of the major changes in the fourth edition and list a few of the key new figures that illustrate the general improvements we have made in this new edition.

Part I Developing Perspective (Chapters 1–3, S1)

Guiding Philosophy: Introduce the big picture, the process of science, and the historical context of astronomy.

The basic goal of these chapters is to give students a big picture overview and context for the rest of the book, and to help them develop an appreciation for the process of science and how science has developed through history. Chapter 1 offers an overview of our modern understanding of the cosmos, thereby giving students perspective on the entire universe. Chapter 2 provides an introduction to basic sky phenomena, including seasons and phases of the Moon, and a perspective on how phenomena we experience every day are tied to the broader cosmos. Chapter 3 discusses the nature of science, offering a historical perspective on the development of science and giving students perspective on how science works and how it differs from nonscience. The supplementary (optional) Chapter S1 goes into more detail about the sky, including celestial timekeeping and navigation.

New for the fourth edition As with the rest of the book, you will see our new pedagogy, enhanced exercise sets, and other general changes in these chapters. You might be particularly interested in the following changes for the fourth edition:

- We have significantly revised our discussion of the Nature of Science in Section 3.4, to make the distinction between science and nonscience even more clear to students.
- We have added new, optional Mathematical Insights to each of these chapters. In particular, Chapter 1 now contains four Mathematical Insights that, together, provide students with an overview of basic mathematical techniques that they can use throughout the book: an overview of problem solving, an introduction to unit analysis, scientific notation and order of magnitude estimation, and the use of simple formulas.

Key Figures The following figures will give you a sense of some of the revisions and enhancements we have made to the art program: 1.1, 1.2, 1.11, 1.17, 2.15, 2.24, 2.33, 3.13, 3.20, 3.24, S1.2, S1.13.

Part II Key Concepts for Astronomy (Chapters 4–6)

Guiding Philosophy: Bridges to the familiar.

These chapters lay the groundwork for understanding astronomy through what is sometimes called the "universality of physics"—the idea that a few key principles governing matter, energy, light, and motion explain both the phenomena of our daily lives and the mysteries of the cosmos. Each chapter begins with a section on science in everyday life in which we remind students how much they already know about scientific phenomena from their everyday experiences. We then build on this everyday knowledge to help students learn the formal principles of physics needed for the rest of their study of astronomy. Chapter 4 covers the laws of motion, the crucial conservation laws of angular momentum and energy, and the universal law of gravitation. Chapter 5 covers the nature of light and matter, the formation of spectra, and the Doppler effect. Chapter 6 covers telescopes and astronomical observing techniques.

New for the fourth edition Users of the third edition will want to be aware of the following organizational changes we have made to Part II for the fourth edition:

- We have eliminated the third edition's Chapter 4 (A Universe of Matter and Energy), instead integrating its content into the new Chapters 4 and 5.
- Section 4.3 on conservation laws in astronomy is new to this edition, introducing students to these two very important laws in a single, concise section.

Key Figures The following figures will give you a sense of some of the revisions and enhancements we have made to the art program: 4.5, 4.18, 4.23, 5.3, 5.7, 5.10, 5.14, 6.22, 6.29.

Part III Learning From Other Worlds (Chapters 7–13)

Guiding Philosophy: True Comparative Planetology.

This set of chapters begins in Chapter 7 with a broad overview of the solar system, including a 10-page tour that highlights some of the most important and interesting features of the Sun and each of the planets in our solar system. In the remaining chapters of this part, we seek to explain these features through a true comparative planetology approach, in which the discussion emphasizes the *processes* that shape the planets rather than the "stamp collecting" of facts about them. Chapter 8 uses the concrete features of the solar system presented in Chapter 7 to build student understanding of the current theory of solar system formation. Chapters 9 and 10 focus on the terrestrial planets, covering key ideas of geology and atmospheres, respectively. In both chapters, we start with examples from our own planet Earth to help students understand the types of features that are found throughout the terrestrial worlds and the fundamental processes that explain how these features came to be. We then complete each of these chapters by summarizing how the various processes have played out on each individual world. Chapter 11 covers the jovian planets and their moons and rings. Chapter 12 covers small bodies in the solar system—asteroids and comets, including the large comet-like objects of the Kuiper belt such as Pluto and the recently discovered "Planet X." It also covers cosmic collisions, including the impact linked to the extinction of the dinosaurs and a discussion of how seriously we should take the ongoing impact threat. Finally, Chapter 13 turns to the exciting new topic of other planetary systems that have been discovered in recent years. Note that Part III is essentially independent of Parts IV through VII, and thus can be covered either before or after them.

New for the fourth edition Users of the third edition will want to be aware of the following organizational changes we have made to Part III for the fourth edition:

- We have eliminated the third Chapter 14 (Planet Earth), instead integrating its content into the chapters on terrestrial geology and atmospheres (Chapters 9 and 10) and moving the material on the evolution of life on Earth to Chapter 24.
- Chapter 13 on Other Planetary Systems is entirely new.

Key Figures The following figures will give you a sense of some of the revisions and enhancements we have made to the art program: 7.1, 7.15, 8.12, 8.13, 9.15, 9.30, 10.6, 10.29, 11.9, 11.30, 12.12, 12.26, 13.6, 13.13.

Part IV A Deeper Look at Nature (Chapters S2–S4)

Guiding Philosophy: Ideas of relativity and quantum mechanics are accessible to anyone.

Nearly all students have at least heard of things like the prohibition on faster-than-light travel, curvature of space-time, and the uncertainty principle. But few (if any) students enter an introductory astronomy course with any idea of what these things mean, and they are naturally curious about them. Moreover, a basic understanding of ideas of relativity and quantum mechanics makes it possible to gain a much deeper appreciation of many of the most important and interesting topics in modern astronomy including black holes, gravitational lensing, and the overall geometry of the universe. The three chapters of Part IV cover special relativity (Chapter S2), general relativity (Chapter S3), and key astronomical ideas of quantum mechanics (Chapter S4). The main thrust throughout is to demystify relativity and quantum mechanics by convincing students that they are capable of understanding the key ideas despite the reputation of these subjects for being hard or counterintuitive. ***These chapters are labeled "supplementary" because coverage of them is optional.*** Covering them will give your students a deeper understanding of the topics that follow on stars, galaxies, and cosmology, but the later chapters are self-contained so that they may be covered without having covered Part IV at all.

New for the fourth edition The general flow of these chapters has not changed significantly, but we have made a few key enhancements to the pedagogy, particularly in the thought experiments of Chapters S2 and S3.

Key Figures The following figures will give you a sense of some of the revisions and enhancements we have made to the art program: S2.1, S2.4, S2.9, S3.1, S3.8, S3.12, S4.2, S4.11.

Part V Stars (Chapters 14–18)

Guiding Philosophy: We are intimately connected to the stars.

These are our chapters on stars and stellar life cycles. Chapter 14 covers the Sun in depth so that it can serve as our concrete model for building an understanding of other stars. Chapter 15 describes the general properties of other stars, how we measure these properties, and how we classify stars with the H-R diagram. The new Chapter 16 covers star birth, and the rest of stellar evolution is discussed in Chapter 17. Chapter 18 covers the end points of stellar evolution: white dwarfs, neutron stars, and black holes, along with a discussion of gamma-ray bursts.

New for the fourth edition Users of the third edition will want to be aware of the following organizational changes we have made to Part V for the fourth edition:

- We have greatly expanded our discussion of stellar evolution, splitting the third edition's single chapter on stellar evolution into two separate chapters: Chapter 16 on star birth and Chapter 17 on the life cycles of stars.
- Note that the material in Chapter 16 is almost entirely new, and includes discussion of the conditions under which star formation can occur, the first generation of stars, recent results from the Spitzer telescope, and much more.

Key Figures The following figures will give you a sense of some of the revisions and enhancements we have made to the art program: 14.8, 14.16, 14.23, 15.8, 16.4, 16.5, 16.8, 16.9, 16.11, 16.20, 17.19, 18.11, 18.19.

Part VI Galaxies and Beyond (Chapters 19–23)

Guiding Philosophy: Present galaxy evolution in a way that parallels the teaching of stellar evolution, integrating cosmological ideas in the places where they most naturally arise.

These chapters cover galaxies and cosmology. Chapter 19 presents the Milky Way as a paradigm for galaxies in much the same way that Chapter 14 uses the Sun as a paradigm for stars. Chapter 20 presents the variety of galaxies, and how we determine key parameters such as galactic distances and age, much as Chapter 15 presents the variety of stars and how we determine key stellar parameters. Chapter 21 discusses the current state of knowledge regarding galaxy evolution, much as Chapters 16 and 17 cover stellar evolution. Throughout these chapters, we integrate cosmological ideas as they arise. For example, we go into depth on Hubble's law in Chapter 20 because of its importance to the cosmic distance scale and to our understanding of what we see when we look at distant galaxies. This approach lays the groundwork for our discussion of dark matter and dark energy and their role in the fate of the universe in Chapter 22, and of the Big Bang in Chapter 23.

New for the fourth edition Users of the third edition will want to be aware of the following organizational changes we have made to Part V for the fourth edition:

- Chapter 19 on the Milky Way has been reorganized and now includes a discussion of the evolutionary history of our galaxy.
- We have moved the discussion of rotation curves from Chapter 19 to Chapter 22, although we still introduce the idea of dark matter in a general way when we discuss the rotation of the Milky Way Galaxy in Chapter 19.
- We have rewritten Chapter 22 to include much more discussion of dark energy and its possible relevance to the fate of the universe. In particular, please note our new Section 22.1, which explains the meaning of the

terms *dark matter* and *dark energy* so that students have the appropriate context for the subsequent discussion.

Key Figures The following figures will give you a sense of some of the revisions and enhancements we have made to the art program: 19.9, 19.17, 20.14, 20.16, 21.2, 21.23, 22.13, 22.14, 22.15, 22.17, 23.4, 23.13, 23.15.

Part VII Life on Earth and Beyond (Chapter 24)

Guiding Philosophy: The study of life on Earth helps us understand the search for life in the universe.

This part consists of a single chapter. It may be considered optional, to be used as time allows. Those who wish to teach a more detailed course on astrobiology may wish to consider the text *Life in the Universe*, by Bennett, Shostak, and Jakosky.

New for the fourth edition Users of the third edition will want to be aware of the following organizational changes we have made to Part VII for the fourth edition:

■ This chapter is quite different from the comparable chapter in the third edition, and in particular now includes a discussion of life on Earth, which formerly appeared in Chapter 14.

■ We have increased emphasis on how understanding astronomy gives us insight into the challenge of interstellar travel, largely designed so that students will be able to critically evaluate reports of UFOs.

Key Figures The following figures will give you a sense of some of the revisions and enhancements we have made to the art program: 24.3, 24.4, 24.7, 24.15.

Pedagogical Features of *The Cosmic Perspective*

Alongside the main narrative, *The Cosmic Perspective* includes a number of pedagogical devices designed to enhance student learning:

■ ENHANCED **Learning Goals** Presented as key questions, motivational learning goals begin every chapter, and every section of every chapter is carefully written to address the specific learning goal in the title. This helps students stay focused on the big picture and stay motivated by the understanding they will gain.

■ ENHANCED **Chapter Summary** The end-of-chapter summary offers a concise review of the learning goal questions, helping reinforce student understanding of key concepts from the chapter. Thumbnail figures are included to remind students of key illustrations and photos in the chapter.

■ NEW **Annotated Figures** Key figures in each chapter now include the research-proven technique of "annotation"—carefully crafted text placed on the figure (in

blue) to guide students through interpreting graphs, following process figures, and translating between different representations.

■ **Wavelength/Observatory Icons** For astronomical photographs (or astronomy art that may be confused with photographs), simple icons identify: the wavelength band; whether the image is a photo, artist's impression, or computer simulation; and whether the image came from ground-based or space-based observations.

■ NEW **Interactive Figures and Interactive Photos** In each chapter, key figures and photos include an icon directing the student to an interactive version available at www.masteringastronomy.com. These purpose-built interactive applets and animations target specific figures and photos that research shows students can better understand by manipulating time evolution, scale, geometry, or different representations. These dynamic tools allow students to actively engage with the figure or photograph, and quickly lead to deeper mastery of the concepts. They are also available for in-class presentation on the revised and expanded *Cosmic Lecture Launcher™ Instructor Resource Presentation Library* CD-ROM.

■ **MasteringAstronomy Interactive Tutorials** Lessons from within the highly acclaimed interactive tutorials on www.masteringastronomy.com are referenced above specific section titles to direct students to targeted, self-paced help.

■ **Think About It** This feature, which appears throughout the book as short questions integrated into the narrative, gives students the opportunity to reflect on important new concepts. It also serves as an excellent starting point for classroom discussions.

■ **Common Misconceptions** These boxes address popularly held but incorrect ideas related to the chapter material.

■ **Special Topic Boxes** These boxes contain supplementary discussion topics related to the chapter material but not prerequisite to the continuing discussion.

■ ENHANCED **Mathematical Insight Boxes** These boxes contain most of the mathematics used in the book and can be covered or skipped depending on the level of mathematics that you wish to include in your course. New to this edition, we have introduced a three-step problem-solving strategy—Understand, Solve, and Explain—in all the Mathematical Insight examples. This gives students a consistent and explicit structure for solving quantitative homework problems.

■ **The Big Picture** Every chapter narrative ends with this feature. It helps students put what they've learned in the chapter into the context of the overall goal of gaining a broader perspective on ourselves and our planet.

■ ENHANCED **End-of-Chapter Questions** Each chapter includes an extensive set of exercises that can be used for study, discussion, or assignment. NEW in this edition: All of the end-of-chapter exercises are organized into the following subsets:

● Review Questions: Questions that students should be able to answer from the reading alone.

- Does It Make Sense? (or similar title): A set of short statements for which students are expected to determine whether each statement makes sense, and to explain why or why not. These exercises are generally easy once students understand a particular concept, but very difficult otherwise; thus, they are an excellent probe of comprehension.
- Quick Quiz: New to this edition, this short multiple-choice quiz allows students to check their progress.
- Short-Answer/Essay Questions: Questions go beyond the Review Questions in asking for conceptual interpretation.
- Quantitative Problems: Problems that require some mathematics, usually based on topics covered in the Mathematical Insight boxes. This edition provides twice as many quantitative problems as previous editions.
- Discussion Questions: Open-ended questions for class discussions.

New to this edition, nearly all end-of-chapter questions are available at www.masteringastronomy.com for online homework assignment and automatic grading and diagnostics.

- **Media Explorations** Each chapter ends with a section or page of Media Explorations that highlight some of the many award-winning interactive tutorials available at www.masteringastronomy.com and other media available, such as the *Voyager: SkyGazer* CD-ROM, to aid students in studying the chapter material and provide suggestions for how to easily incorporate these into your course. Each activity targets specific learning objectives. These sections include suggested Web projects designed for independent research.
- **Cross-References** When a concept is covered in greater detail elsewhere in the book, we include a cross-reference in brackets to the relevant section (e.g., [Section 5.2]).
- **Glossary** A detailed glossary makes it easy for students to look up important terms.
- **Appendixes** The appendixes include a number of useful references and tables including key constants (Appendix A), key formulas (Appendix B), key mathematical skills (Appendix C), and numerous data tables and star charts.

MasteringAstronomy™—A New Paradigm in Astronomy Teaching

What is the single most important factor in student success in astronomy? Both research and common sense reveal the same answer: *study time*. No matter how good the teacher, or how good the textbook, students learn only when they spend adequate time studying. Unfortunately, limitations on resources for grading have prevented most instructors from assigning much homework despite its obvious benefits to student learning. And, limitations on help and office hours have made it difficult for students to make sure they use self-study time effectively. That, in a nutshell, is why we have created MasteringAstronomy. For students, it provides the first adaptive-learning, on-line system to coach them *individually*—responding to their errors with specific, targeted feedback, and providing hints for partial credit to help them when they get stuck. For professors, MasteringAstronomy provides unprecedented ability to automatically monitor and record students' step-by-step work and evaluate the effectiveness of assignments and exams. As a result, we believe that MasteringAstronomy will create a paradigm shift in the way astronomy courses are taught: For the first time, it will be possible, even in large classes, to ensure that each student spends his or her study time on optimal learning activities outside of class.

MasteringAstronomy provides students with a wealth of self-study resources including interactive tutorials targeting the most difficult concepts of the course, interactive versions of key figures and photos, and quizzes and other activities for self-assessment covering every chapter and every week. For professors, MasteringAstronomy provides the first library of tutoring activities and assessment problems pretested and informed by students nationally. You can choose from more than 4,500 activities and problems to automatically assign, grade, and track: pre- and post-lecture diagnostic quizzes, tutoring activities, end-of-chapter problems, even test bank questions. You can find a walk-through of major features of MasteringAstronomy in the front of this book, though of course the best way to become familiar with it is to spend some time on the Web site. We invite you to visit www.masteringastronomy.com to see it for yourself!

Finally, in a world where every publisher walks into a professor's office and claims that their Web site is better than anyone else's, we'd like to point out four reasons why you'll discover that MasteringAstronomy really does stand out from the crowd:

- MasteringAstronomy has been built specifically to support the structure and pedagogy of *The Cosmic Perspective*. You'll find the same concepts emphasized in the book and the Web site, using the same terminology and the same pedagogical approaches. This type of consistency ensures that students focus on the concepts, without the risk of becoming confused by different presentations.
- The major content of MasteringAstronomy has been developed through a close collaboration between the authors of *The Cosmic Perspective* and a team led by noted astronomy educators Tim Slater and Ed Prather. The high level of collaboration ensures the consistency and integration of text and media that our adopters expect. The Slater/Prather team bring tremendous expertise and insight from the latest research about how students can most effectively learn astronomy.
- The MasteringAstronomy platform uses the same unique student-driven engine as the highly successful MasteringPhysics product (the most widely adopted physics homework and tutorial system), developed by

a group led by MIT physicist David Pritchard. This robust platform gives instructors unprecedented power not only to tailor content to their own courses, but also to evaluate the effectiveness of assignments and exams. Using these unique diagnostics, instructors can now efficiently adapt their lectures, assignments, and tests to the students' needs as the course progresses.

■ Although the MasteringAstronomy platform is new to this fourth edition, it builds upon the proven success of The Astronomy Place—the site that accompanied prior editions of our book. You'll find the content from The Astronomy Place available on the new MasteringAstronomy site (along with much, much more)—content that has proved so valuable that more than 100,000 students have *voluntarily* registered to use the site as a core part of their studies.

As you might imagine, the time and expense that have gone into creating MasteringAstronomy are comparable to that of producing a textbook alone. For that reason, it is available as a subscription-based Web site. All new textbooks come with a free, 1-year subscription to MasteringAstronomy. Students buying a used book can purchase access online at the Web site.

Additional Supplements for *The Cosmic Perspective*

The Cosmic Perspective is much more than just a textbook. It is a complete package of teaching, learning, and assessment resources designed to help both teachers and students. In addition to MasteringAstronomy (described above), the following supplements are available with this book:

■ FREE with All New Books: *Voyager: SkyGazer, College Edition*™. Based on *Voyager III*, one of the world's most popular planetarium programs, *SkyGazer* makes it easy for students to learn constellations and explore the wonders of the sky through interactive exercises. The *SkyGazer* CD is packaged free with all new copies of this book. It comes preloaded with 75 demos, and suggested activities appear in the Media Explorations section at the end of each chapter in the book.

■ NEW *Starry Night*™: Now available as an additional option with *The Cosmic Perspective, Starry Night* has been acclaimed as the world's most realistic desktop planetarium software. This special version for college students has an easy-to-use point-and-click interface and is available as an additional bundle. Ask your Addison Wesley sales representative for details.

■ FREE with New Books: **The Addison-Wesley Astronomy Tutor Center**. This center provides one-on-one tutoring by qualified college instructors in any of four ways: via phone, fax, e-mail, and the Internet during evening and weekend hours. Tutor center instructors will answer questions and provide help with examples and exercises from the text. Tutor center registration is free with new books only when the professor orders

books with the special tutor center package. (Professors: Contact your local Addison Wesley sales representative if you wish to order this package.) Otherwise, it can be purchased separately. See www.aw-bc.com/tutorcenter for more information.

New! In-Class Active-Learning Tutorials (ISBN 0-8053-8296-8)
By Marvin De Jong
This new workbook provides a wealth of fifty 20-minute in-class tutorial activities to choose from. Designed for use in large lecture classes, these activities are also suitable for labs. These short, structured activities are designed for students to complete on their own or in peer-learning groups. Each activity targets specific learning objectives such as understanding Newton's laws, Mars's retrograde motion, tracking stars on the H-R diagram, or comparing the properties of planets.

New! Themes of the Times on Astronomy (ISBN 0-8053-9028-6)
A collection of 50 recent astronomy news articles from the *New York Times, Themes of the Times on Astronomy* brings the excitement of astronomical discovery to you and your students. Each article correlates to a chapter in *The Cosmic Perspective* and includes a series of follow-up questions for homework or class discussion.

Instructor-Only Supplements

Several additional supplements are available for instructors only. Contact your local Addison Wesley sales representative to find out more about the following supplements:

■ **The Cosmic Lecture Launcher v3.0 CD** (ISBN 0-8053-9285-8) This CD provides a wealth of presentation tools to help prepare course lectures. It includes a set of PowerPoint® slides for every chapter in the textbook; a comprehensive collection of more than 1,000 high-resolution figures from the book and other astronomical sources (such as NASA); a library of more than 300 interactive applets and animations, including new Interactive Figures and Interactive Photos; and even conceptual Clicker Quizzes (for Classroom Response Systems). All these materials were built specifically to enhance the same pedagogy presented in the text.

■ *Clickers in the Astronomy Classroom* (ISBN 0-8053-9616-0) This 100-page handbook by Douglas Duncan provides everything you need to know to successfully introduce or enhance your use of CRS (clicker) quizzing in your astronomy class—the research-proven benefits, common pitfalls to avoid, and a wealth of thought-provoking astronomy questions for every week of your course.

■ *Instructor's Guide* (ISBN 0-8053-9284-X) This guide contains a detailed overview of the text, sample syllabi for courses of different emphasis and duration, suggestions on teaching strategies, answers or discussion points for all Think About It questions in the text, solutions to end-of-chapter problems, and a detailed reference guide summarizing media resources available for every chapter and section in the book.

- **Carl Sagan's *Cosmos* (DVD or Video)** The *Best of Cosmos* and the complete, revised, enhanced, and updated *Cosmos* series are available free to qualified adopters of *The Cosmic Perspective*. A week-by-week guide of segments to include in your course is provided in the *Instructor's Guide*.

- **Test Bank** Available in both computerized (ISBN 0-8053-9286-6) and printed (ISBN 0-8053-9287-4) forms, the Test Bank contains a broad set of multiple-choice, true/false, and free-response questions for each chapter. The test bank is also available to instructors at www.masteringastronomy.com.

- **Transparency Acetates** (ISBN 0-8053-9289-0) For those who use overhead projectors in lectures, this set contains more than 100 images from the text.

Acknowledgments

A textbook may carry author names, but it is the result of hard work by a long list of committed individuals. We could not possibly list everyone who has helped, but we would like to call attention to a few people who have played particularly important roles. First, we thank our editors and friends at Addison Wesley, who have stuck with us through thick and thin, including Adam Black, Linda Davis, Stacie Kent, Christy Lawrence, Stacy Treco, Liana Allday, Deb Greco, Lothlorien Homet, and Shannon Tozier. Special thanks to our past and present production teams, especially Joan Marsh, Mary Douglas, and Brandi Nelson; our art and design team, Mark Ong, Judy Waller, John Waller, and Angel Chavez; our development team, including Elisa Adams and Jody Larson; our supplements team, including Lauren Jones and Ed Prather (for their work on the *Cosmic Lecture Launcher* CD), Jonathan Williams (for his work on the Test Bank), Douglas Duncan (for his work on Clicker Quizzes), Doug Lombardi (for his work on *Themes of the Times on Astronomy*), John Weiss (for writing solutions to end-of-chapter problems), and Michael LoPresto and Marvin De Jong (the workbook authors); our Web team, led by Claire Masson, Jim Dove, and Ian Shakeshaft; and our MasteringAstronomy team, led by Tim Slater, Ed Prather, and Brad Patterson.

We've also been fortunate to have an outstanding group of reviewers, whose extensive comments and suggestions helped us shape the book. We thank all those who have reviewed drafts of the book in various stages, including:

Christopher M. Anderson, *University of Wisconsin*

Peter S. Anderson, *Oakland Community College*

Keith Ashman, *University of Missouri, Kansas City*

Simon P. Balm, *Santa Monica College*

John Beaver, *University of Wisconsin at Fox Valley*

Peter A. Becker, *George Mason University*

Timothy C. Beers, *Michigan State University*

Jim Bell, *Cornell University*

Priscilla J. Benson, *Wellesley College*

Bernard W. Bopp, *University of Toledo*

Sukanta Bose, *Washington State University*

David Brain, *University of California Berkeley Space Sciences Laboratory*

David Branch, *University of Oklahoma*

James E. Brau, *University of Oregon*

Jean P. Brodie, *UCO/Lick Observatory, University of California, Santa Cruz*

James Brooks, *Florida State University*

Daniel Bruton, *Stephen F. Austin State University*

Amy Campbell, *Louisiana State University*

Eugene R. Capriotti, *Ohio State University*

Eric Carlson, *Wake Forest University*

David A. Cebula, *Pacific University*

Supriya Chakrabarti, *Boston University*

Dipak Chowdhury, *Indiana University Purdue University at Fort Wayne*

Chris Churchill, *New Mexico State University*

Josh Colwell, *University of Colorado*

Anita B. Corn, *Colorado School of Mines*

Philip E. Corn, *Red Rocks Community College*

Kevin Crosby, *Carthage College*

Christopher Crow, *Indiana University Purdue University at Fort Wayne*

Manfred Cuntz, *University of Texas at Arlington*

John M. Dickey, *University of Minnesota*

Robert Egler, *North Carolina State University at Raleigh*

Paul Eskridge, *Minnesota State University*

David Falk, *Los Angeles Valley College*

Robert A. Fesen, *Dartmouth College*

Tom Fleming, *University of Arizona*

Douglas Franklin, *Western Illinois University*

Sidney Freudenstein, *Metropolitan State College of Denver*

Martin Gaskell, *University of Nebraska*

Richard Gelderman, *Western Kentucky University*

Harold A. Geller, *George Mason University*

David Graff, *U.S. Merchant Marine Academy*

Richard Gray, *Appalachian State University*

Kevin Grazier, *Jet Propulsion Laboratory*

Robert Greeney, *Holyoke Community College*

Henry Greenside, *Duke University*

Alan Greer, *Gonzaga University*

David Griffiths, *Oregon State University*

David Grinspoon, *University of Colorado*

John Gris, *University of Delaware*

Bruce Gronich, *University of Texas, El Paso*

Jim Hamm, *Big Bend Community College*

Charles Hartley, *Hartwick College*

Joe Heafner, *Catawba Valley Community College*

Scott Hildreth, *Chabot College*

Mark Hollabaugh, *Normandale Community College*

Richard Holland, *Southern Illinois University, Carbondale*

Joseph Howard, *Salisbury University*

Richard Ignace, *University of Wisconsin*

James Imamura, *University of Oregon*

Douglas R. Ingram, *Texas Christian University*

Bruce Jakosky, *University of Colorado*

Adam G. Jensen, *University of Colorado*

Adam Johnston, *Weber State University*

Lauren Jones, *Gettysburg College*

William Keel, *University of Alabama*

Julia Kennefick, *University of Arkansas*

Steve Kipp, *University of Minnesota, Mankato*

Kurtis Koll, *Cameron University*

Ichishiro Konno, *University of Texas at San Antonio*

John Kormendy, *University of Texas, Austin*

Eric Korpela, *University of California, Berkeley*

Ted La Rosa, *Kennesaw State University*

Kristine Larsen, *Central Connecticut State University*

Ana Marie Larson, *University of Washington*

Larry Lebofsky, *University of Arizona*

Patrick Lestrade, *Mississippi State University*

Nancy Levenson, *University of Kentucky*

David M. Lind, *Florida State University*

Abraham Loeb, *Harvard University*

Michael LoPresto, *Henry Ford Community College*

William R. Luebke, *Modesto Junior College*

Darrell Jack MacConnell, *Community College of Baltimore City*

Marie Machacek, *Massachusetts Institute of Technology*

Loris Magnani, *University of Georgia*

Steven Majewski, *University of Virginia*

Phil Matheson, *Salt Lake Community College*

Marles McCurdy, *Tarrant County College*

Stacy McGaugh, *University of Maryland*

Barry Metz, *Delaware County Community College*

William Millar, *Grand Rapids Community College*

Dinah Moche, *Queensborough Community College of City University, New York*

Stephen Murray, *University of California, Santa Cruz*

Zdzislaw E. Musielak, *University of Texas, Arlington*

Charles Nelson, *Drake University*

Gerald H. Newsom, *Ohio State University*

Brian Oetiker, *Sam Houston State University*

John P. Oliver, *University of Florida*

Stacy Palen, *Weber State University*

Russell L. Palma, *Sam Houston State University*

Bryan Penprase, *Pomona College*

Eric S. Perlman, *University of Maryland, Baltimore County*

Peggy Perozzo, *Mary Baldwin College*

Charles Peterson, *University of Missouri, Columbia*

Cynthia W. Peterson, *University of Connecticut*

Jorge Piekarewicz, *Florida State University*

Lawrence Pinsky, *University of Houston*

Jascha Polet, *California State Polytechnic University, Pomona*

Matthew Price, *Oregon State University*

Harrison B. Prosper, *Florida State University*

Monica Ramirez, *Aims College, Colorado*

Christina Reeves-Shull, *Richland College*

Todd M. Rigg, *City College of San Francisco*

Elizabeth Roettger, *DePaul University*

Roy Rubins, *University of Texas, Arlington*

Carl Rutledge, *East Central University*

Rex Saffer, *Villanova University*

John Safko, *University of South Carolina*

James A. Scarborough, *Delta State University*

Britt Scharringhausen, *Ithaca College*

Ann Schmiedekamp, *Pennsylvania State University, Abington*

Joslyn Schoemer, *Denver Museum of Nature and Science*

James Schombert, *University of Oregon*

Gregory Seab, *University of New Orleans*

Larry Sessions, *Metropolitan State College of Denver*

Philip I. Siemens, *Oregon State University*

Caroline Simpson, *Florida International University*

Paul Sipiera, *William Harper Rainey College*

Earl F. Skelton, *George Washington University*

Michael Skrutskie, *University of Virginia*

Mark H. Slovak, *Louisiana State University*

Norma Small-Warren, *Howard University*

Dale Smith, *Bowling Green State University*

James R. Sowell, *Georgia Technical University*

John Spencer, *Lowell Observatory*

Darryl Stanford, *City College of San Francisco*

George R. Stanley, *San Antonio College*

John Stolar, *West Chester University*

Jack Sulentic, *University of Alabama*

Sean Sutton, *Mount Holyoke College*

Beverley A. P. Taylor, *Miami University*

Brett Taylor, *Radford University*

Donald M. Terndrup, *Ohio State University*

Frank Timmes, *School of Art Institute of Chicago*

David Trott, *Metro State College*

Trina Van Ausdal, *Salt Lake Community College*

Licia Verde, *University of Pennsylvania*

Nicole Vogt, *New Mexico State University*

Darryl Walke, *Rariton Valley Community College*

Fred Walter, *State University of New York, Stony Brook*

James Webb, *Florida International University*

Mark Whittle, *University of Virginia*

Paul J. Wiita, *Georgia State University*

Lisa M. Will, *Mesa Community College*

Jonathan Williams, *University of Florida*

Wayne Wooten, *Pensacola Junior College*

Scott Yager, *Brevard College*

Andrew Young, *Casper College*

Arthur Young, *San Diego State University*

Min S. Yun, *University of Massachusetts, Amherst*

Dennis Zaritsky, *University of California, Santa Cruz*

Robert L. Zimmerman, *University of Oregon*

Historical Accuracy Reviewer—Owen Gingerich, Harvard–Smithsonian

In addition, we thank the following colleagues who helped us clarify technical points or checked the accuracy of technical discussions in the book:

Phil Armitage, *University of Colorado*

Nahum Arav, *University of Colorado*

Thomas Ayres, *University of Colorado*

Cecilia Barnbaum, *Valdosta State University*

Rick Binzel, *Massachusetts Institute of Technology*

Howard Bond, *Space Telescope Science Institute*

David Brain, *University of California Berkeley Space Sciences Laboratory*

Humberto Campins, *University of Florida*

Robin Canup, *Southwest Research Institute*

Clark Chapman, *Southwest Research Institute*

Kelly Cline, *Carroll College*

Josh Colwell, *University of Colorado*

Mark Dickinson, *National Optical Astronomy Observatory*

Jim Dove, *Metropolitan State College of Denver*

Harry Ferguson, *Space Telescope Science Institute*

Andrew Hamilton, *University of Colorado*

Todd Henry, *Georgia State University*

Dennis Hibbert, *Everett Community College*

Dave Jewitt, *University of Hawaii*

Hal Levison, *Southwest Research Institute*

Mario Livio, *Space Telescope Science Institute*

McKim Malville, *University of Colorado*

Mark Marley, *New Mexico State University*

Linda Martel, *University of Hawaii*

Kevin McLin, *University of Colorado*

Michael Mendillo, *Boston University*

Rachel Osten, *National Radio Astronomy Observatory*

Bob Pappalardo, *University of Colorado*

Michael Shara, *American Museum of Natural History*

Bob Stein, *Michigan State University*

Glen Stewart, *University of Colorado*

John Stolar, *West Chester University*

Jeff Taylor, *University of Hawaii*

Dave Tholen, *University of Hawaii*

Nick Thomas, *MPI/Lindau (Germany)*

Dimitri Veras, *University of Colorado*

John Weiss, *University of Colorado*

Don Yeomans, *Jet Propulsion Laboratory*

Finally, we thank the many people who have greatly influenced our outlook on education and our perspective on the universe over the years, including Tom Ayres, Fran Bagenal, Forrest Boley, Robert A. Brown, George Dulk, Erica Ellingson, Katy Garmany, Jeff Goldstein, David Grinspoon, Robin Heyden, Don Hunten, Bruce Jakosky, Joan Marsh, Catherine McCord, Dick McCray, Dee Mook, Cheri Morrow, Charlie Pellerin, Carl Sagan, Mike Shull, John Spencer, and John Stocke.

Jeff Bennett
Megan Donahue
Nick Schneider
Mark Voit

About the Authors

JEFFREY BENNETT

Jeffrey Bennett holds a B.A. (1981) in biophysics from the University of California, San Diego, and an M.S. and Ph.D. (1987) in astrophysics from the University of Colorado, Boulder. He has taught at every level from preschool through graduate school, including more than fifty college classes in astronomy, physics, mathematics, and education. He served two years as a visiting senior scientist at NASA headquarters, where he created NASA's "IDEAS" grant program, started a program to fly teachers aboard NASA's airborne observatories (including the soon-to-be-flying SOFIA), and worked on numerous educational programs for the Hubble Space Telescope and other space science missions. He also proposed the idea for and helped develop the *Voyage* Scale Model Solar System, which opened in 2001 on the National Mall in Washington, D.C. (He is pictured here with the model Sun.) In addition to this astronomy textbook, he has written college-level textbooks in astrobiology, mathematics, and statistics, and a book for the general public, *On the Cosmic Horizon* (Addison-Wesley, 2001). He is also author of the award-winning children's books *Max Goes to the Moon* and *Max Goes to Mars.* When not working, he enjoys participating in masters swimming and in the daily adventures of life with his wife, Lisa; his children, Grant and Brooke; and his dog, Cosmo. You can read more about his projects on his personal Web site, www.jeffreybennett.com.

MEGAN DONAHUE

Megan Donahue is an associate professor in the Department of Physics and Astronomy at Michigan State University. Her current research is mainly on clusters of galaxies: their contents—dark matter, hot gas, galaxies, active galactic nuclei—and what they reveal about the contents of the universe and how galaxies form and evolve. She grew up on a farm in Nebraska and received a B.A. in physics from MIT, where she began her research career as an X-ray astronomer. She has a Ph.D. in astrophysics from the University of Colorado, for a thesis on theory and optical observations of intergalactic and intracluster gas. That thesis won the 1993 Trumpler Award from the Astronomical Society for the Pacific for an outstanding astrophysics doctoral dissertation in North America. She continued postdoctoral research in optical and X-ray observations as a Carnegie Fellow at Carnegie Observatories in Pasadena, California, and later as an STScI Institute Fellow at Space Telescope. Megan was a staff astronomer at the Space Telescope Science Institute until 2003, when she joined the MSU faculty. Megan is married to Mark Voit, and they collaborate on many projects, including this textbook and the raising of their children, Michaela, Sebastian, and Angela. Between the births of Sebastian and Angela, Megan qualified for and ran the 2000 Boston Marathon. These days, Megan runs, orienteers, and plays piano and bass guitar whenever her children allow it.

Nicholas Schneider is an associate professor in the Department of Astrophysical and Planetary Sciences at the University of Colorado and a researcher in the Laboratory for Atmospheric and Space Physics. He received his B.A. in physics and astronomy from Dartmouth College in 1979 and his Ph.D. in planetary science from the University of Arizona in 1988. In 1991, he received the National Science Foundation's Presidential Young Investigator Award. His research interests include planetary atmospheres and planetary astronomy, with a focus on the odd case of Jupiter's moon Io. He enjoys teaching at all levels and is active in efforts to improve undergraduate astronomy education. Off the job, he enjoys exploring the outdoors with his family and figuring out how things work.

Mark Voit is an associate professor in the Department of Physics and Astronomy at Michigan State University. He earned his B.A. in astrophysical sciences at Princeton University and his Ph.D. in astrophysics at the University of Colorado in 1990. He continued his studies at the California Institute of Technology, where he was a research fellow in theoretical astrophysics, and then moved on to Johns Hopkins University as a Hubble Fellow. Before going to Michigan State, Mark worked in the Office of Public Outreach at the Space Telescope, where he developed museum exhibitions about the Hubble Space Telescope and was the scientist behind NASA's HubbleSite. His research interests range from interstellar processes in our own galaxy to the clustering of galaxies in the early universe. He is married to coauthor Megan Donahue, and they try to play outdoors with their three children whenever possible, enjoying hiking, camping, running, and orienteering. Mark is also author of the popular book *Hubble Space Telescope: New Views of the Universe.*

How to Succeed in Your Astronomy Course

Using this Book

Each chapter in this book is designed to make it easy for you to study effectively and efficiently. To get the most out of each chapter, you might wish to use the following study plan:

■ A textbook is not a novel, and you'll learn best by reading the elements of this text in the following order:

1. Start by reading the Learning Goals and the introductory paragraphs at the beginning of the chapter so that you'll know what you are trying to learn.

2. Next, get an overview of key concepts by studying the illustrations and reading their captions. The illustrations highlight almost all of the major concepts, so this "illustrations first" strategy gives you an opportunity to survey the concepts before you read about them in depth. Watch for the Interactive Figure icons—when you see one, go to the Mastering-Astronomy Web site (www.masteringastronomy.com) to try the interactive version.

3. Read the chapter narrative, but save the boxed features (Common Misconceptions, Special Topics, Mathematical Insights) to read later. As you read, make notes on the pages to remind yourself of ideas you'll want to review later. Avoid using a highlight pen; underlining with pen or pencil is far more effective, because it forces you to take greater care and therefore helps keep you alert as you study. Be careful to underline selectively—it won't help you later if you've underlined everything.

4. After reading the chapter once, go back through and read the boxed material. You should read all of the Common Misconceptions and Special Topics boxes; whether you choose to read the Mathematical Insights is up to you and your instructor. Also watch for the MasteringAstronomy tutorial icons throughout the chapter; if a concept is giving you trouble, go to the Mastering Astronomy site to try the relevant tutorial.

5. Then turn your attention to the Chapter Summary. The best way to use the summary is to try to answer the Learning Goal questions for yourself before reading the short answers given in the summary.

■ After completing the reading as above, start testing your understanding with the end-of-chapter exercises. A good way to begin is to make sure you can answer all of the Review Questions; if you don't know an answer, look back through the chapter until you figure it out. Then test your understanding a little more deeply by trying the "Does It Make Sense?" and Quick Quiz questions.

■ You can further check your understanding and get feedback on difficulties by trying the online quizzes at www.masteringastronomy.com. Each chapter has two quizzes: a basic quiz and a conceptual quiz. Try the basic quiz first. Once you clear up any difficulties you have with it, try the conceptual quiz.

■ If your course has a quantitative emphasis, work through all of the examples in the Mathematical Insights before trying the quantitative problems for yourself. Remember that you should always try to answer questions qualitatively before you begin plugging numbers into a calculator. For example, make an order of magnitude estimate of what your answer should be so that you'll know your calculation is on the right track, and be sure that your answer makes sense and has the appropriate units.

■ If you have done all the above, you will have already made use of numerous resources on the Mastering-Astronomy Web site (www.masteringastronomy.com). Don't stop there; visit the site again and make use of other resources that will help you further build your understanding. These resources have been developed specifically to help you learn the most important ideas in your astronomy course, and they have been extensively tested to make sure they are effective. They really do work, and the only way you'll gain their benefits is by going on the Web site and using them.

The Key to Success: Study Time

The single most important key to success in any college course is to spend enough time studying. A general rule of thumb for college classes is that you should expect to study about 2 to 3 hours per week *outside* of class for each unit of credit. For example, based on this rule of thumb, a student taking 15 credit hours should expect to spend 30 to 45 hours each week studying outside of class. Combined with time in class, this works out to a total of 45 to 60 hours

If Your Course Is:	Time for Reading the Assigned Text (per week)	Time for Homework Assignments (per week)	Time for Review and Test Preparation (average per week)	Total Study Time (per week)
3 credits	2 to 4 hours	2 to 3 hours	2 hours	6 to 9 hours
4 credits	3 to 5 hours	2 to 4 hours	3 hours	8 to 12 hours
5 credits	3 to 5 hours	3 to 6 hours	4 hours	10 to 15 hours

spent on academic work—not much more than the time a typical job requires, and you get to choose your own hours. Of course, if you are working while you attend school, you will need to budget your time carefully.

As a rough guideline, your studying time in astronomy might be divided as shown in the table above. If you find that you are spending fewer hours than these guidelines suggest, you can probably improve your grade by studying longer. If you are spending more hours than these guidelines suggest, you may be studying inefficiently; in that case, you should talk to your instructor about how to study more effectively.

General Strategies for Studying

■ Don't miss class. Listening to lectures and participating in discussions is much more effective than reading someone else's notes. Active participation will help you retain what you are learning.

■ Take advantage of resources offered by your professor, whether it be e-mail, office hours, review sessions, online chats, or simply finding opportunities to talk to and get to know your professor. Most professors will go out of their way to help you learn in any way that they can.

■ Budget your time effectively. One or 2 hours each day is more effective, and far less painful, than studying all night before homework is due or before exams.

■ If a concept gives you trouble, do additional reading or studying beyond what has been assigned. And if you still have trouble, ask for help: You surely can find friends, peers, or teachers who will be glad to help you learn.

■ Working together with friends can be valuable in helping you understand difficult concepts. However, be sure

that you learn *with* your friends and do not become dependent on them.

■ Be sure that any work you turn in is of *collegiate quality*: neat and easy to read, well organized, and demonstrating mastery of the subject matter. Although it takes extra effort to make your work look this good, the effort will help you solidify your learning and is also good practice for the expectations that future professors and employers will have.

Preparing for Exams

■ Study the Review Questions, and rework problems and other assignments; try additional questions to be sure you understand the concepts. Study your performance on assignments, quizzes, or exams from earlier in the term.

■ Study the relevant online tutorials and chapter quizzes available at www.masteringastronomy.com.

■ Study your notes from lectures and discussions. Pay attention to what your instructor expects you to know for an exam.

■ Reread the relevant sections in the textbook, paying special attention to notes you have made on the pages.

■ Study individually *before* joining a study group with friends. Study groups are effective only if every individual comes prepared to contribute.

■ Don't stay up too late before an exam. Don't eat a big meal within an hour of the exam (thinking is more difficult when blood is being diverted to the digestive system).

■ Try to relax before and during the exam. If you have studied effectively, you are capable of doing well. Staying relaxed will help you think clearly.

THE COSMIC PERSPECTIVE

Fourth Edition

1

Our Place in the Universe

We shall not cease from exploration
And the end of all our exploring
Will be to arrive where we started
And know the place for the first time.

—**T.S. Eliot**

ar from city lights on a clear night, you can gaze upward at a sky filled with stars. Lie back and watch for a few hours, and you will observe the stars marching steadily across the sky. Confronted by the seemingly infinite heavens, you might wonder how Earth and the universe came to be. If you do, you will be sharing an experience common to humans around the world and in thousands of generations past.

Modern science offers answers to many of our fundamental questions about the universe and our place within it. We now know the basic content and scale of the universe. We know the age of Earth and the approximate age of the universe. And, although much remains to be discovered, we are rapidly learning how the simple ingredients of the early universe developed into the incredible diversity of life on Earth.

In this first chapter, we will survey the content and history of the universe, the scale of the universe, and the motions of Earth. We'll develop a "big picture" perspective on our place in the universe that will provide a base on which we can build a deeper understanding in the rest of the book.

1.1 Our Modern View of the Universe

If you observe the sky carefully, you can see why most of our ancestors believed that the heavens revolved about a stationary Earth. The Sun, Moon, planets, and stars appear to circle around our sky each day, and we cannot feel the constant motion of Earth as it rotates on its axis and orbits the Sun. Thus, it seems quite natural to assume that we live in an Earth-centered, or *geocentric*, universe.

Nevertheless, we now know that Earth is a planet orbiting a rather average star in a vast universe. The historical path to this knowledge was long and complex. In later chapters, we'll see that many ancient beliefs seemed to make a lot of sense in their day and changed only when people were confronted by strong evidence to the contrary. We'll also see how the process of science has enabled us to acquire this evidence and thereby discover that we are connected to the stars in ways our ancestors never imagined. First, however, it's useful to have at least a general picture of the universe as we know it today.

• What is our place in the universe?

We can describe our place in the universe with what we might call our "cosmic address," illustrated in Figure 1.1.

Earth is a planet in our **solar system**, which consists of the Sun and all the objects that orbit it: the planets and their moons, the chunks of rock we call *asteroids,* the balls of ice we call *comets,* and countless tiny particles of interplanetary dust.

Our Sun is a star, just like the stars we see in our night sky. The Sun and all the stars we can see with the naked eye make up only a small part of a huge, disk-shaped collection of stars called the **Milky Way Galaxy**. A **galaxy** is a great island of stars in space, containing from a few hundred million to a trillion or more stars. The Milky Way Galaxy is relatively large, containing more than 100 billion stars. Our solar system is located a little over halfway from the galactic center to the edge of the galactic disk.

Billions of other galaxies are scattered throughout space. Some galaxies are fairly isolated, but many others are found in groups. Our Milky Way, for example, is one of the two largest among about 40 galaxies in the **Local Group**. Groups of galaxies with more than a few dozen members are often called **galaxy clusters**.

On a very large scale, the universe appears weblike, with galaxies and galaxy clusters loosely arranged in giant chains and sheets. The galaxies and galaxy clusters are more tightly packed in some places than in others, forming giant structures called **superclusters**. The supercluster to which our Local Group belongs is called, not surprisingly, the **Local Supercluster**. Between the vast groupings of galaxies lie huge voids containing few, if any, galaxies. Finally, the **universe** is the sum total of all matter and energy, encompassing the superclusters and voids and everything within them.

THINK ABOUT IT

Some people think that our tiny physical size in the vast universe makes us insignificant. Others think that our ability to learn about the wonders of the universe gives us significance despite our size. What do *you* think?

• How did we come to be?

According to modern science, we humans are newcomers in an old universe. We'll devote much of the rest of this textbook to studying the scientific evidence that backs up this idea. To help prepare you for this study, let's look at a quick overview of the scientific history of the universe, as summarized in Figure 1.2 (page 5).

The Big Bang and the Expanding Universe As we'll discuss in Section 1.3, telescopic observations of distant galaxies show that the entire universe is *expanding*. That is, average distances between galaxies are increasing with time. If the universe is expanding, everything must have been closer together in the past. From the observed rate of expansion, astronomers estimate that the expansion started about 14 billion years ago. We call this beginning the **Big Bang**. The three cubes in the upper left corner of Figure 1.2 represent the expansion through time of a small piece of the entire universe.

Universe
approx. size: 10^{21} km

Local Supercluster

Local Group
approx. size: 3×10^{19} km

Milky Way Galaxy
approx. size: 10^{18} km

Solar System
(not to scale)
approx. size: 10^{10} km

Earth
approx. size: 10^4 km

Figure 1.1 Our cosmic address.

The universe as a whole has continued to expand ever since the Big Bang, but in many smaller regions, the force of gravity has drawn matter together. Structures such as galaxies and clusters of galaxies occupy regions where gravity has won out against the overall expansion. That is, while the universe as a whole continues to expand, individual galaxies and their contents do *not* expand. This idea is illustrated by the expanding cube in Figure 1.2. Notice that as the cube as a whole grows larger, the matter within it tends to clump into galaxies and galaxy clusters. Most galaxies, including our own Milky Way, probably formed within a few billion years after the Big Bang.

Stellar Lives and Galactic Recycling   Within galaxies, gravity drives the collapse of clouds of gas and dust to form stars and planets. Stars are not living organisms, but they nonetheless go through "life cycles." After their births in giant clouds of gas and dust, stars shine for millions or billions of years. The energy that makes them shine comes from **nuclear fusion**, the process in which lightweight atomic nuclei smash together and stick (or fuse) to make heavier nuclei. Nuclear fusion occurs deep in a star's core throughout its life. A star "dies" when it finally exhausts all its usable fuel for fusion.

In its final death throes, a star blows much of its content back out into space. In particular, massive stars die in titanic explosions called *supernovae*. The returned matter mixes with other matter floating between the stars in the galaxy, eventually becoming part of new clouds of gas and dust from which new generations of stars can be born. Thus, galaxies function as cosmic recycling plants, recycling material expelled from dying stars into new generations of stars and planets. This cycle is illustrated in the lower right of Figure 1.2. Our own solar system is a product of many generations of such recycling.

Basic Astronomical Objects, Units, and Motions

This box summarizes a few key astronomical definitions introduced in this chapter and used throughout the book.

Basic Astronomical Objects

star   Our Sun and other ordinary stars are large, glowing balls of gas that generate heat and light through nuclear fusion in their cores.

planet   A moderately large object that orbits a star. Planets may be rocky, icy, or gaseous in composition, and they shine primarily by reflecting light from their star.

moon (or satellite)   An object that orbits a planet. The term *satellite* is also used more generally to refer to any object orbiting another object.

asteroid   A relatively small and rocky object that orbits a star. Asteroids are sometimes called *minor planets*.

comet   A relatively small and ice-rich object that orbits a star.

Collections of Astronomical Objects

solar system   Our solar system consists of the Sun and all the material that orbits it, including the planets. The term *solar system* technically refers only to our own star system (because *solar* means "of the Sun"), but it is sometimes applied to other star systems.

star system   A star (sometimes more than one star) and any planets and other materials that orbit it. (Roughly half of all star systems contain two or more stars.)

galaxy   A great island of stars in space, containing from a few hundred million to a trillion or more stars, all held together by gravity and orbiting a common center.

cluster (or *group*) of galaxies   A collection of galaxies bound together by gravity. Small collections (up to a few dozen galaxies) are generally called *groups*, while larger collections are called *clusters*.

supercluster   A gigantic region of space where many individual galaxies and many groups and clusters of galaxies are packed more closely together than elsewhere in the universe.

universe (or cosmos)   The sum total of all matter and energy—that is, all galaxies and everything between them.

observable universe   The portion of the entire universe that can be seen from Earth, at least in principle. The observable universe is probably only a tiny portion of the entire universe.

Astronomical Distance Units

astronomical unit (AU)   The average distance between Earth and the Sun, which is about 150 million kilometers. (More technically, 1 AU is the length of the semimajor axis of Earth's orbit.)

light-year   The distance that light can travel in 1 year, which is about 9.46 trillion kilometers.

Terms Relating to Motion

rotation   The spinning of an object around its axis. For example, Earth rotates once each day around its axis, which is an imaginary line connecting the North Pole to the South Pole through the center of Earth.

orbit (revolution)   The orbital motion of one object around another. For example, Earth orbits around the Sun once each year.

expansion (of the universe)   We say that the universe is expanding because the average distance between galaxies is increasing with time. Note that while the universe as a whole is expanding, individual galaxies and their contents (as well as groups and clusters of galaxies) are *not* expanding.

Birth of the Universe. The expansion of the universe began with the hot and dense Big Bang. The cubes show how one region of the universe has expanded with time. The universe continues to expand, but on smaller scales gravity has pulled matter together to make galaxies.

Galaxies as Cosmic Recycling Plants. The early universe contained only two chemical elements: hydrogen and helium. All other elements were made by stars and recycled from one stellar generation to the next within galaxies like our Milky Way.

Life Cycles of Stars. Many generations of stars have lived and died in the Milky Way.

Stars are born in clouds of gas and dust; planets may form in surrounding disks.

Stars shine with energy released by nuclear fusion, which ultimately manufactures all elements heavier than hydrogen and helium.

Massive stars explode when they die, scattering the elements they've produced into space.

Earth and Life. By the time our solar system was born, about 4.6 billion years ago, about 2% of the original hydrogen and helium had been converted into heavier elements. Thus, we are "star stuff," because we and our planet are made from elements manufactured in stars that lived and died long ago.

Figure 1.2 Our cosmic origins.

Stars Manufacture the Elements of Earth and Life   The recycling of stellar material is connected to our existence in an even deeper way. By studying stars of different ages, we have learned that the early universe contained only the simplest chemical elements: hydrogen and helium (and a trace of lithium). We and Earth are made primarily of other elements, such as carbon, nitrogen, oxygen, and iron. Where did these other elements come from? Apparently, all the elements besides hydrogen and helium were manufactured by stars, either through the nuclear fusion that makes stars shine or through nuclear reactions accompanying the explosions that end stellar lives.

The recycling of matter in the Milky Way Galaxy had already been taking place for several billion years by the time our solar system formed, about 4.6 billion years ago. By that time, stars had converted about 2% of the original hydrogen and helium into heavier elements. Thus, the cloud that gave birth to our solar system was made of about 98% hydrogen and helium and 2% of everything else. This 2% may sound small, but it was more than enough to make the small rocky planets of our solar system, including Earth. On Earth, some of these elements became the raw ingredients of life, ultimately blossoming into the great diversity of life on Earth today.

In summary, most of the material from which we and our planet are made was created inside stars that died before the birth of our Sun. As astronomer Carl Sagan (1934–1996) used to say, we are "star stuff."

• How can we know what the universe was like in the past?

You may wonder how we can claim to know anything about what the universe was like in the distant past. It's

possible because we can actually see into the past by studying light from distant stars and galaxies.

Light travels extremely fast by earthly standards. The speed of light is 300,000 kilometers per second, a speed at which it would be possible to circle Earth nearly eight times in just 1 second. Nevertheless, even light takes a substantial amount of time to travel the vast distances in space. For example, light takes about 1 second to reach Earth from the Moon, and about 8 minutes to reach Earth from the Sun. Light from the stars takes many years to reach us, so we measure distances to the stars in units called **light-years**. One light-year is the distance that light can travel in 1 year—about 10 trillion kilometers, or 6 trillion miles. (You can calculate this distance by multiplying the speed of light by the number of seconds in 1 year; see Mathematical Insight 1.1.) Note that a light-year is a unit of *distance*, not time.

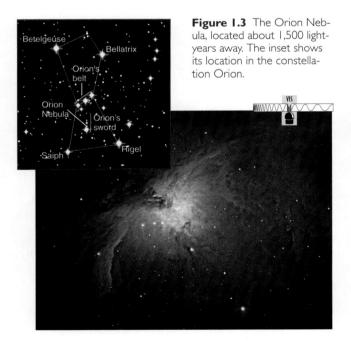

Figure 1.3 The Orion Nebula, located about 1,500 light-years away. The inset shows its location in the constellation Orion.

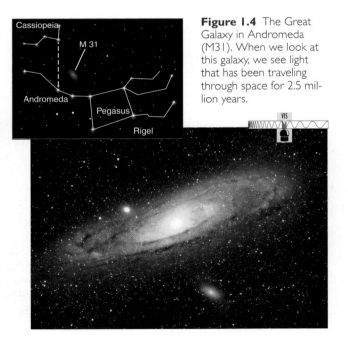

Figure 1.4 The Great Galaxy in Andromeda (M31). When we look at this galaxy, we see light that has been traveling through space for 2.5 million years.

Because light takes time to travel through space, we are led to a remarkable fact:

The farther away we look in distance, the further back we look in time.

For example, the brightest star in the night sky, Sirius, is about 8 light-years away, which means its light takes about 8 years to reach us. Thus, when we look at Sirius, we are seeing it as it was about 8 years ago.

The effect is more dramatic at greater distances. The Orion Nebula, a giant cloud in which stars and planets are forming, is located about 1,500 light-years from Earth (Figure 1.3). Thus, we see the Orion Nebula as it looked about 1,500 years ago—about the time of the fall of the Roman Empire. If any major events have occurred in the Orion Nebula since that time, we cannot yet know about them because the light from these events has not yet reached us. Figure 1.4 shows

an even more distant object: the Great Galaxy in Andromeda (also known as the Andromeda Galaxy and M 31), which lies about 2.5 million light-years from Earth. Its photograph is therefore a picture of this galaxy as it looked about 2.5 million years ago, when early humans were first walking on Earth.

It's also amazing to realize that any "snapshot" of a distant galaxy is a picture of both space and time. For example, because the Andromeda Galaxy is about 100,000 light-years in diameter, the light we see from the far side of the galaxy must have left on its journey to us some 100,000 years before the light from the near side. Thus, the single photograph in Figure 1.4 not only shows us how the Andromeda Galaxy looked more than 2 million years ago, but also shows different parts of the galaxy spread over a time period of 100,000 years. When we study the universe, it is impossible to separate space and time.

MATHEMATICAL INSIGHT **1.1**   **How Far Is a Light-Year? An Introduction to Astronomical Problem Solving**

The text states that a light-year is about 10 trillion kilometers, but how do we know? After all, this distance is far greater than any person or spacecraft has ever traveled, so we cannot possibly have measured it. Instead, we know the distance that light travels in a year through mathematics, which allows us to answer many questions that we cannot answer through direct measurement.

Although people often expect the mathematics needed in astronomy to be quite difficult (and it's true that professional astronomers use advanced mathematics in their work), many astronomical problems can be solved with little more than arithmetic or simple algebra; as we'll see below, answering the question "How far is a light-year?" requires only multiplication and division. Thus, the "trick" to solving this and other astronomical problems lies in finding an appropriate problem-solving process. Problem solving is more of an art than a science, but the following three-step process should help you whenever you have difficulty with a quantitative problem.

Step 1 U̲nderstand the problem: Ask yourself what the solution will look like (for example, What units will it have? Will it be big or small?) and what information you need to solve the problem. Draw a diagram or think of a simpler analogous problem to help you decide how to solve it.

Step 2 S̲olve the problem: Carry out the necessary calculations.

Step 3 E̲xplain your result: Be sure that your answer makes sense, and consider what you've learned by solving the problem.

You can remember this process by thinking of the first letters of each step: USE for Understand, Solve, and Explain. As you gain problem-solving experience, you may not always need to write out the three steps explicitly. However, they may help you whenever you get "stuck" while trying to solve a quantitative problem.

Example: How far is a light-year?

Solution: Let's go through the three-step process.

Step 1 Understand the problem: The question asks how *far,* so we are looking for a *distance.* By remembering the definition of a light-year, we know we are looking for the *distance that light*

can travel in 1 year. A simpler but analogous problem might therefore ask, say, "How far can your car travel in 2 hours?" By thinking about this simpler problem, you'll realize that all you need to know is the car's speed; for example, if you travel at a speed of 50 kilometers per hour, then you'll travel 100 kilometers in 2 hours. From there, you can realize that finding a distance from a speed just requires multiplying the speed by the time, that is: distance = speed × time. Since this problem asks how far light travels in a year, the speed is the speed of light and the time is 1 year.

Step 2 Solve the problem: From Step (1), we know that we need to multiply the speed of light by 1 year. The text states that the speed of light is 300,000 km/s. Because this speed is given *per second* and the time is 1 *year,* we'll need to convert a year to seconds. We can do this by knowing that there are 60 seconds in 1 minute, 60 minutes in 1 hour, 24 hours in 1 day, and 365 days in 1 year. We now carry out the calculations:

$$1 \text{ light-year} = (\text{speed of light}) \times (1 \text{ yr})$$

$$= \left(300{,}000\frac{\text{km}}{\cancel{s}}\right) \times \left(1 \cancel{\text{yr}} \times \frac{365 \cancel{\text{days}}}{1 \cancel{\text{yr}}}\right.$$

$$\left. \times \frac{24 \cancel{\text{hr}}}{1 \cancel{\text{day}}} \times \frac{60 \cancel{\text{min}}}{1 \cancel{\text{hr}}} \times \frac{60 \cancel{s}}{1 \cancel{\text{min}}}\right)$$

$$= 9{,}460{,}000{,}000{,}000 \text{ km (or 9.46 trillion km)}$$

Notice that the unit conversions are really just clever ways of multiplying by 1; see Appendix C.3 for a more detailed review of proper ways of working with units.

Step 3 Explain your result: In complete sentence form, our answer is: "One light-year is about 9.46 trillion kilometers." This answer makes sense: It has the expected units of distance (kilometers) and it is a long way, which we expect for the distance that light can travel in a year. We say "about" in the answer because we know it is not exact: For example, we know that a year is not exactly 365 days long. In fact, for most purposes, it makes more sense to approximate the answer further as: "One light-year is about 10 trillion kilometers."

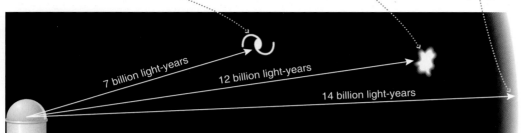

Far: *We see a galaxy 7 billion light-years away as it was 7 billion years ago–when the universe was about half its current age of 14 billion years.*

Farther: *We see a galaxy 12 billion light-years away as it was 12 billion years ago–when the universe was only about 2 billion years old.*

The limit of our observable universe: *Light from nearly 14 billion light-years away shows the universe as it looked shortly after the Big Bang, before galaxies existed.*

7 billion light-years

12 billion light-years

14 billion light-years

Beyond the observable universe: *We cannot see anything farther than 14 billion light-years away, because its light has not had enough time to reach us.*

Figure 1.5 The farther away we look in space, the further back we look in time. Thus, the age of the universe puts a limit on the size of the *observable* universe—the portion of the entire universe that we can observe, at least in principle.

THINK ABOUT IT

Suppose that, at this very moment, students are studying astronomy on a planet somewhere in the Andromeda Galaxy. What would they see as they look from afar at our Milky Way? Could they know that we exist here on Earth? Explain.

• Can we see the entire universe?

We see more distant galaxies as they were even further back into the past, a fact that allows us to observe the ways in which the universe has changed through time (Figure 1.5). Remember that we think the universe is about 14 billion years old. If we look at a galaxy that is 7 billion light-years away, we see it as it looked 7 billion years ago[*]—which means we see it as it was when the universe was only half its current age. If we look at a galaxy that is 12 billion light-years away, we see it as it was 12 billion years ago, when the universe was only 2 billion years old. Thus, by looking to great distances, we can see what the universe looked like when it was younger. The remarkable photo that opens this chapter (p. 1), taken by the Hubble Space Telescope, shows some galaxies that are more than 12 billion light-years away.

Even with the most powerful telescopes imaginable, there's a limit to how far we can see into space. This limit is imposed by the age of the universe. In a universe that is 14 billion years old, we cannot possibly see anything more than 14 billion light-years away. Beyond that—say, at a distance of 15 billion light-years—we'd be looking back to a time before the universe existed, which means there's nothing to see. Thus, our **observable universe**—the portion of the entire universe that we can potentially observe—consists only of objects that lie within 14 billion light-years of Earth. This fact does not put any limit on the size of the *entire* universe, which may be far larger than our observable universe. We simply have no hope of seeing or studying

anything beyond the bounds of our observable universe. Anything that exists beyond our observable universe is so far away that its light has not yet had time to reach us.

 Scale of the Universe Tutorial, Lessons 1–3

1.2 The Scale of the Universe

The numbers in our description of the size and age of the universe probably have little meaning for you—after all, they are literally astronomical. In this section, we will try to give those figures some meaning.

 Virtual Tour of the Solar System

• How big is Earth compared to our solar system?

One of the best ways to develop perspective on cosmic sizes and distances is to imagine our solar system shrunk down to a scale on which you could walk through it. The Voyage scale model solar system in Washington, D.C.,

Figure 1.6 This photo shows the pedestals housing the Sun (the gold sphere on the nearest pedestal) and the inner planets in the Voyage scale model solar system (Washington, D.C.). The building at the left is the National Air and Space Museum. (The model planets are encased in the sidewalk-facing disks that are visible at about eye level on the planet pedestals.)

---

[*]We will find in Chapter 20 that distances to very faraway galaxies are rather tricky to define in an expanding universe. In these examples we will simply define distances to galaxies in terms of how far their light has travelled.

makes such a walk possible (Figure 1.6). The Voyage model shows the Sun and the planets, and the distances between them, at *one ten-billionth* their actual sizes and distances. Figure 1.7a shows the planets at their scaled sizes, and 1.7b shows a map of the locations of the planets in the Voyage model. If you study the figures carefully, you'll notice several key features of our solar system:

● The Sun and planets are tiny compared to the distances between them. On the Voyage scale, the Sun is about the size of a large grapefruit, while the planets range in size from barely visible dots to marble-size Jupiter. Earth is about the size of the ball point in a pen and is located about 15 meters (16.5 yards) from the Sun.

● The planets fall into two clear groups by distance. The four inner planets—Mercury, Venus, Earth, and Mars—all lie within just a few steps of the Sun on this scale. The four large outer planets—Jupiter, Saturn, Uranus, and Neptune—are much more widely separated. For example, Jupiter is more than three times as far as Mars from the Sun.

● One of the most striking features of the solar system is its emptiness (that's why we call it *space!*). To show the complete orbits of the planets around the Sun, the Voyage model would require an area measuring over a kilometer (0.6 mile) on a side, which is larger than most college campuses. Spread over this large area, only the grapefruit-size Sun, the planets, and a few moons would be big enough to see with the naked eye.

THINK ABOUT IT

Earth is the only place in our solar system—and the only place we yet know of in the universe—with conditions suitable for human life. How does visualizing the Earth to scale affect your perspective on human existence? How does it affect your perspective on our planet? Explain.

Seeing our solar system to scale also helps us put space exploration into perspective. Although robotic spacecraft have visited all the planets through Neptune, humans have

---

SPECIAL TOPIC How Many Planets Are There In Our Solar System?

Everyone in college today grew up in an era when we were taught that there are nine planets in our solar system: Mercury, Venus, Earth, Mars, Jupiter, Saturn, Uranus, Neptune, and Pluto. That is why so many books, articles, web sites, and exhibits still refer to the "nine planets." For example, as you can see in Figure 1.7, the Voyage Scale Model Solar System counts Pluto as the ninth and last planet. There's even a well-known mnemonic for remembering the order of the planets that goes "My very excellent mother just sent us nine pizzas." Well, as you've probably heard by now, in 2005 scientists announced that someone ordered another pizza! More specifically, we learned that our solar system contains another object that is very much like Pluto, but slightly larger in size. (We'll discuss this object in Chapter 12.) Most news media immediately hailed the newly discovered object as the tenth planet, or "Planet X." However, it's not quite that simple, because scientists were already engaged in a debate about whether Pluto and other similarly sized objects should count as planets.

Pluto was discovered in 1930, and it was designated the ninth planet because early estimates of its size indicated that it was as large as some of the other planets in our solar system. But as our measurements became more precise, we learned that Pluto is actually far smaller and less massive than any of the first eight planets: Mercury, the smallest of these planets, is more than twice as large as Pluto in diameter and about 25 times as large in mass. Moreover, Pluto's orbit is much more eccentric ("stretched out") and more tilted to the ecliptic plane than that of any of the first eight planets, and its composition of ice and rock makes it much more similar to the thousands of known comets than to the first eight planets.

In the early 1990's, advances in technology allowed scientists to begin discovering other ice-rich objects orbiting in Pluto's region of the solar system. Because this region is formally called the Kuiper Belt (Kuiper rhymes with *piper*), these objects are often called "Kuiper Belt Objects" (or KBOs)—though in terms of composi-

tion, they are simply large comets. Scientists discovered dozens of new Kuiper Belt Objects with each passing year, and some were not much smaller in size than Pluto. As a result, some scientists began to argue that Pluto should be demoted from planet to Kuiper Belt Object, while others argued that some of the larger Kuiper Belt Objects should be upgraded to be called planets.

The decision on what objects count as planets rests with the International Astronomical Union (IAU). Prior to 2005, the IAU had rejected attempts to change the status of either Pluto or any other Kuiper Belt Object, essentially leaving intact the status quo from 1930. But the discovery of an object larger than Pluto has thrown this status quo into turmoil. After all, if Pluto counts as a planet, then surely an even larger object should also count.

Passions surrounding the definition of "planet" run so high that it may be years before the IAU makes an official decision regarding the status of Pluto and other large Kuiper Belt Objects, and even then the debate may continue. Meanwhile, we can identify at least three options in counting planets:

1. If we decide that objects like Pluto are too small to be planets, then the new object is also too small because it is so close in size to Pluto. Then there are 8 planets in our solar system.

2. If we decide that Pluto represents the minimum size cutoff, then the new object also makes it, and there are 10 planets in our solar system—and it's possible that other Pluto-size objects still remain to be discovered.

3. We could decide that the minimum size cutoff is somewhat smaller than Pluto, in which case the number of planets in our solar system could be 12, 15, 20, or many other numbers, depending on exactly where we place the cutoff.

Indeed, only one thing seems clear right now: If someone tries to tell you that there are nine planets in our solar system, they obviously haven't been staying up-to-date in astronomy.

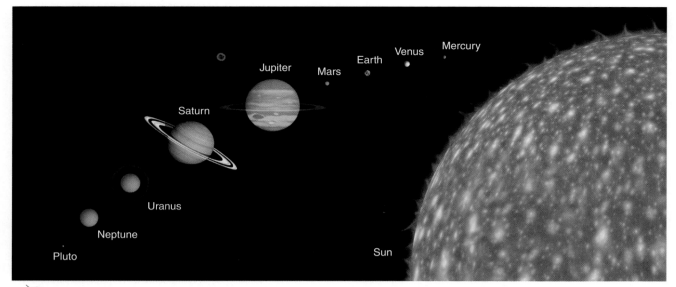

a 🎨 This painting shows the Sun and planets at one ten-billionth of their actual sizes. (Distances are not to scale in this painting.)

b This map shows the locations of the Sun and planets in the Voyage model.

Figure 1.7 The Voyage scale model represents the solar system at *one ten-billionth* of its actual size. Planets are shown in a line for easy comparison of distances; in reality, each planet orbits the Sun independently, and a perfect alignment essentially never occurs.

traveled only as far as the Moon. Twelve astronauts walked on the Moon between 1969 and 1972 as part of the Apollo program (Figure 1.8). On the 1-to-10-billion Voyage scale, the Moon is a barely visible speck about two thumb-widths (4 centimeters) from Earth (see Figure 2.21). Thus, if you place the model Earth in the center of your palm, the farthest place humans have ever reached also lies within your palm. Sending humans to Mars will require a journey some 200 times as far.

• How far away are the stars?

Imagine that you start at the Voyage model Sun in Washington, D.C. You walk the roughly 600-meter distance to Pluto (just over $\frac{1}{3}$ mile) and then decide to keep going to find the nearest star besides the Sun. How far would you have to go?

Amazingly, you would need to walk to California. That is, on the same scale that allows you to walk from the Sun to Pluto in minutes, even the nearest stars would be more

Figure 1.8 The Moon is the most distant place ever visited by humans, yet on the 1-to-10-billion scale of the Voyage model, it is only about 4 centimeters (1.5 inches) from Earth. This famous photograph from the first Moon landing (*Apollo 11* in July 1969) shows astronaut Buzz Aldrin, with Neil Armstrong reflected in his visor. Armstrong was the first to step onto the Moon's surface, saying, "That's one small step for [a] man, one giant leap for mankind."

Scaling is a great way to put space and time into perspective, and it usually requires nothing more than division. For example, in a 1-to-20 architectural scale model, a building that is actually 60 feet tall will be only 60 ÷ 20 = 3 feet tall in the model. The idea is the same for astronomical scaling, except that we must divide by much larger numbers. As a result, it's much easier to work with numbers expressed in *scientific notation;* that is, with the aid of powers of 10. The following examples show the idea. If you are unfamiliar with powers of 10 and scientific notation, or if you need a review, please see Appendices C.1 and C.2.

Example 1: How big is the Sun on the 1-to-10-billion scale described in the text?

Solution:
Step 1 Understand: We are looking for the *size* of the Sun on the scale, which means we need to know its actual size. From Appendix E.1, we find that the Sun's radius is 695,000 km, which we express in scientific notation as 6.95×10^5 km. To find the Sun's radius on the 1-to-10-billion scale, we need to divide its actual radius by 10 billion, or 10^{10}.

Step 2 Solve: Now we carry out the division:

$$\text{scaled radius of Sun} = \frac{\text{actual radius}}{10^{10}}$$
$$= \frac{6.95 \times 10^5 \text{ km}}{10^{10}}$$
$$= 6.95 \times 10^{(5-10)} \text{ km}$$
$$= 6.95 \times 10^{-5} \text{ km}$$

Notice that we used the rule that dividing powers of 10 means subtracting their exponents [Appendix C.1].

Step 3 Explain: We have found an answer, but it's hard to explain in its original form because most of us don't have a good sense of what 10^{-5} kilometer looks like. We need to go further and convert the answer into units that will be easier to interpret. We know there are 1,000 (or 10^3) meters in a kilometer and 100 (or 10^2) centimeters in a meter, so we can convert our answer to centimeters:

$$6.95 \times 10^{-5} \text{ km} \times \frac{10^3 \text{ m}}{1 \text{ km}} \times \frac{10^2 \text{ cm}}{1 \text{ m}} = 6.95 \text{ cm}$$

We can also convert this answer to inches by using data from Appendix Table C.3, which tell us that 1 cm = 0.3937 in.:

$$6.95 \text{ cm} \times 0.3937 \tfrac{\text{in.}}{\text{cm}} = 2.74 \text{ in.}$$

We've found that the scaled radius of the Sun on the 1-to-10-billion scale is just under 7 centimeters or $2\frac{3}{4}$ inches, which means a diameter of just under 14 centimeters or $5\frac{1}{2}$ inches. In other words, the Sun on a 1-to-10-billion scale is about the size of a large grapefruit.

Example 2: What scale do we need to use if we want the 100,000-light-year diameter of the Milky Way Galaxy to fit on a 100-meter-long football field?

Solution:
Step 1 Understand: This time we are looking to find *how many times larger* is the actual diameter of the galaxy than 100 meters, which means we need to divide the larger size by the smaller size. We can't do the division until we've put both sizes in the same units. We already know that a light-year is about 10^{13} kilometers (see Mathematical Insight 1.1), so we can use this fact and the fact that there are 10^3 meters in a kilometer to put both numbers in terms of meters.

Step 2 Solve: We now put the numbers in the same units and carry out the division; to make things easier to read, we write all the numbers as powers of 10, including the galaxy's diameter of 100,000 light-years, or 10^5 ly:

$$\frac{\text{galaxy diameter}}{\text{football field diameter}} = \frac{10^5 \text{ ly} \times \frac{10^{13} \text{ km}}{1 \text{ ly}} \times \frac{10^3 \text{ m}}{1 \text{ km}}}{10^2 \text{ m}}$$
$$= 10^{(5+13+3-2)} = 10^{19}$$

Notice how we've worked with powers of 10 in the calculation, and that the answer has no units because it simply tells us how many times larger one thing is than the other.

Step 3 Explain: We've found that we need a scale of 1 to 10^{19} to make the galaxy fit on a football field.

Example 3: The cosmic calendar in Figure 1.11 represents 14 billion years of real time. How much time does each second represent on the cosmic calendar?

Solution:
Step 1 Understand: We are looking for a real time that corresponds to the scaled time on the calendar.

Step 2 Solve: There are several ways to solve this problem, but it's particularly easy once we realize that the calendar scale is 1 to 14 billion, since 1 calendar year represents 14 billion real years. Thus, 1 second on the cosmic calendar represents 14 billion seconds of real time.

Step 3 Explain: Our initial answer is difficult to explain without further work because most people don't have a good sense of "14 billion seconds." We can make better sense of the answer by converting it to years:

$$1.4 \times 10^{10} \text{ s} \times \frac{1 \text{ min}}{60 \text{ s}} \times \frac{1 \text{ hr}}{60 \text{ min}} \times \frac{1 \text{ day}}{24 \text{ hr}} \times \frac{1 \text{ yr}}{365 \text{ day}} = 443.9 \text{ yr}$$

Recognizing that we've approximated the number of days in the year (and that we've used an approximate value of 14 billion years for the age of the universe), we'll say: "One second on the cosmic calendar represents about 440 years of real time." Thus, a tenth of a second represents about 44 years and a twentieth of a second (0.05 s) represents about 22 years—roughly the age of the average college student.

than 4,000 kilometers (2,500 miles) away. If this answer seems hard to believe, you can check it for yourself. A light-year is about 10 trillion kilometers, which becomes 1,000 kilometers on the 1-to-10-billion scale (because 10 trillion ÷ 10 billion = 1,000). The nearest star system to our own, a three-star system called Alpha Centauri (Figure 1.9), is about 4.4 light-years away. That distance becomes about 4,400 kilometers (2,700 miles) on the 1-to-10-billion scale, or roughly equivalent to the distance across the United States.

The tremendous distances to the stars give us some perspective on the technological challenge of astronomy. For example, because the largest star of the Alpha Centauri system is roughly the same size and brightness as our Sun, viewing it in the night sky is somewhat like being in Washington, D.C., and seeing a very bright grapefruit in San Francisco (neglecting the problems introduced by the curvature of the Earth). It may seem remarkable that we can see this star at all, but the blackness of the night sky allows the naked eye to see it as a faint dot of light. It looks much brighter through powerful telescopes, but we still cannot see any features of the star's surface.

Now, consider the difficulty of seeing *planets* orbiting nearby stars. It is equivalent to looking from Washington, D.C., and trying to see ball points or marbles orbiting grapefruits in California or beyond. You probably won't be surprised to learn that we cannot yet see any surface details on such planets. Indeed, the bigger surprise may be that we *have* discovered more than 150 planets around other stars, most through indirect techniques [Section 13.1].

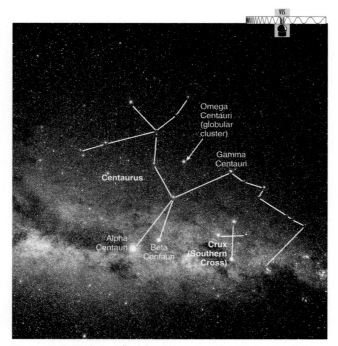

Figure 1.9 This photograph and diagram show the constellation Centaurus, which is visible from tropical and southern latitudes. Alpha Centauri's real distance is about 4.4 light-years, which becomes 4,400 kilometers (2,700 miles) on the 1-to-10-billion Voyage scale.

Our examination of stellar distances also offers a sobering lesson about the possibility of travel to the stars. Although science fiction shows like *Star Trek* and *Star Wars* make interstellar travel seem easy, the reality is far different. Consider the *Voyager 2* spacecraft. Launched in 1977, *Voyager 2* flew by Jupiter in 1979, Saturn in 1981, Uranus in 1986, and Neptune in 1989. (Its trajectory did not take it near Pluto.) *Voyager 2* is now bound for the stars at a speed of close to 50,000 kilometers per hour—about 100 times as fast as a speeding bullet. Even at this speed, *Voyager 2* would take about 100,000 years to reach Alpha Centauri if it were headed in that direction (which it's not). Convenient interstellar travel remains well beyond our present technology.

• How big is the Milky Way Galaxy?

The vast separation between our solar system and Alpha Centauri is typical of the separations among star systems here in the outskirts of the Milky Way Galaxy. Thus, the 1-to-10-billion scale is useless for thinking about distances beyond the nearest stars, because more distant stars would not fit on Earth with this scale. Visualizing the entire galaxy requires a new scale.

Let's further reduce our solar system scale by a factor of 1 billion (making it a scale of 1 to 10^{19}). On this new scale, each light-year becomes 1 millimeter, and the 100,000-light-year diameter of the Milky Way Galaxy becomes 100 meters, or about the length of a football field. Visualize a football field with a scale model of our galaxy centered over midfield. Our entire solar system is a microscopic dot located around the 20-yard line. The 4.4-light-year separation between our solar system and Alpha Centauri becomes just 4.4 millimeters on this scale—smaller than the width of your little finger. If you stood at the position of our solar system in this model, millions of star systems would lie within reach of your arms.

Another way to put the galaxy into perspective is to consider its number of stars—more than 100 billion. Imagine that tonight you are having difficulty falling asleep (perhaps because you are contemplating the scale of the universe). Instead of counting sheep, you decide to count stars. If you are able to count about one star each second, on average, how long would it take you to count 100 billion stars in the Milky Way? Clearly, the answer is 100 billion (10^{11}) seconds, but how long is that? Amazingly, 100 billion seconds turns out to be more than 3,000 years. (You can confirm this by dividing 100 billion by the number of seconds in 1 year.) Thus, you would need thousands of years just to *count* the stars in the Milky Way Galaxy, and this assumes you never take a break—no sleeping, no eating, and absolutely no dying!

• How big is the universe?

As incredible as the scale of our galaxy may seem, the Milky Way is only one of at least 100 billion galaxies in the observable universe. Just as it would take thousands of years

---

MATHEMATICAL INSIGHT **1.3** **Order of Magnitude Estimation**

We can often get a good feel for an idea by making an estimate rather than an exact calculation. In astronomy, the numbers are often so large that an estimate is useful even if it's good only to about the nearest power of 10. For example, the text states that there are about 10^{22} stars in the observable universe, a number found by estimating 100 billion galaxies with an average of 100 billion stars each. Because of the "ballpark" nature of the numbers we used, the real number of stars could easily be 10 times as large as our estimate (or 10^{23} stars), or 10 times smaller (10^{21} stars.) Nevertheless, the estimate of 10^{22} still gives us a good sense of the incredible number of stars in our universe. Estimates good to about the nearest power of 10 are usually called **order of magnitude estimates.**

Example 1: The text claims that "roughly speaking, the number of grains of sand [on all the beaches on Earth] is comparable to the number of stars in the observable universe." Check to see if this claim makes sense.

Solution:

Step 1 Understand: To verify the claim, we need to estimate the total number of grains of sand and see whether it is indeed close to the 10^{22} stars we've estimated to be in the observable universe. We find the total number of sand grains by dividing the total volume of sand on Earth's beaches by the average volume of an individual sand grain. The latter number is relatively easy to estimate: Find some sand and measure out a particular volume (such as 0.1 cm^3), then count the number of grains in this volume; dividing the volume by the number of grains will tell you the average volume of each grain. Some sand is more fine or coarse than other sand, but you'll typically find that the volume of individual sand grains is between about 0.1 and 1 cubic millimeter. For our order of magnitude estimate, let's call it 0.1 mm^3, which you should confirm to be equivalent to 10^{-10} m^3.

Now let's consider the total volume of sand on beaches. We can find a volume by multiplying a width, depth, and length. In this case, the depth and width are the average depth and width of dry sand on beaches that have sand, while the length is the total length of such sandy beaches along all the coastlines on Earth. The width and depth are not too hard to estimate. If you've ever dug down to dirt or wet sand on a beach, you'll know that the dry sand is certainly deeper than 0.1 meter on average but less deep than 10 meters. We can therefore use the intermediate value of 1 meter as a good order of magnitude estimate for beach depth. Similarly, the width of sandy beaches varies widely but

the average is certainly more than 1 meter and less than 100 meters, so we can choose the intermediate value of 10 meters as our order of magnitude estimate for beach width. The total length of sandy beach on Earth is a much more difficult number to estimate. A good strategy is to "solve" the problem without this number, then see what number it would have to be for the claim about the total number of grains to be true.

Step 2 Solve: As we've discussed, we have the following simple formula for the total number of grains of dry sand:

$$\text{total number of grains} = \frac{\text{beach depth} \times \text{beach width} \times \text{beach length}}{\text{average volume of 1 sand grain}}$$

Now, let's put in the estimates we've already made—1 m for beach depth, 10 m for beach width, and 10^{-10} m^3 for the volume of a sand grain:

$$\text{total number of grains} = \frac{1\text{m} \times 10\text{m} \times \text{beach length}}{10^{-10}\text{m}^3}$$
$$= \frac{10^{11}}{\text{m}} \times \text{beach length}$$

The claim we are trying to verify is that the total number of grains is roughly 10^{22}. For this to be the case, the beach length would have to be 10^{11} meters, since that would give us $10^{11} \times 10^{11} = 10^{22}$ on the right side of the equation above. (Notice how the units work out correctly.)

Step 3 Explain: We've found that the claim of 10^{22} grains of sand is reasonable *if* the total length of a sandy beach is close to 10^{11} meters, or 10^8 kilometers (which is 100 million kilometers). It's not easy to estimate the total length of a sandy beach, since beaches do not follow straight lines and it's hard to account for all the islands and inlets on Earth. Nevertheless, if you spend some time studying a globe or atlas, you can probably convince yourself that 100 million kilometers is an overestimate of the total length of beach. In fact, it's probably an overestimate by at least a factor of 10. Thus, the actual number of grains of sand may be closer to 10^{20} or 10^{21} than to 10^{22}. However, this does not mean the claim was wrong. When you consider all the order of magnitude estimates we've made—which include the number of galaxies, the number of stars per galaxy, the volume of beaches, and the volume of individual sand grains—getting to within a factor of 10 or even 100 is pretty good. Thus, it's reasonable to say that the number of stars in the observable universe is comparable to the number of grains of sand on all the beaches on Earth.

to count the stars in the Milky Way, it would take thousands of years to count all the galaxies.

Think for a moment about the total number of stars in all these galaxies. If we assume 100 billion stars per galaxy, the total number of stars in the observable universe is roughly 100 billion × 100 billion, or 10,000,000,000,000,000,000,000 (10^{22}).

How big is this number? Visit a beach. Run your hands through the fine-grained sand. Imagine counting each tiny grain of sand as it slips through your fingers. Then imagine counting every grain of sand on the beach and continuing to count *every* grain of dry sand on *every* beach on Earth. If you could actually complete this task, you would find that, roughly speaking, the number of grains of sand is comparable to the number of stars in the observable universe (Figure 1.10).

Contemplate the fact that there may be as many stars in the observable universe as grains of sand on all the beaches on Earth and that each star is a potential sun for a system of planets. With so many possible homes for life, do you think it is conceivable that life exists only on Earth? Why or why not?

• How do our lifetimes compare to the age of the universe?

Now that we have developed some perspective on the scale of space, we can do the same for the scale of time. Imagine the entire history of the universe, from the Big Bang to the present, compressed into a single year. We can represent this history with a *cosmic calendar,* on which the Big Bang

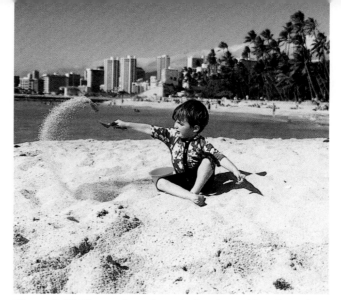

Figure 1.10 The number of stars in the observable universe is comparable to the number of grains of dry sand on all the beaches on Earth.

takes place at the first instant of January 1 and the present day is just before the stroke of midnight on December 31 (Figure 1.11). For a universe that is about 14 billion years old, each month on the cosmic calendar represents a little more than 1 billion years.

On this time scale, the Milky Way Galaxy probably formed sometime in February. Many generations of stars lived and died in the subsequent cosmic months, enriching the galaxy with the "star stuff" from which we and our planet are made.

Our solar system and our planet did not form until early September on this scale, or 4.6 billion years ago in

Figure 1.11 The cosmic calendar compresses the 14-billion-year history of the universe into 1 year, so that each month represents a little more than 1 billion years (more precisely, 1.17 billion years). This cosmic calendar is adapted from a version created by Carl Sagan.

THE HISTORY OF THE UNIVERSE IN 1 YEAR

| January 1: The Big Bang | February: The Milky Way forms | September 3: The Earth forms | September 22: Early life on earth | December 17: Cambrian explosion | December 26: Rise of the dinosaurs | December 30: Extinction of the dinosaurs |

real time. By late September, life on Earth was flourishing. However, for most of Earth's history, living organisms remained relatively primitive and microscopic in size. On the scale of the cosmic calendar, recognizable animals became prominent only in mid-December. Early dinosaurs appeared on the day after Christmas. Then, in a cosmic instant, the dinosaurs disappeared forever—probably due to the impact of an asteroid or a comet [Section 12.4]. In real time, the death of the dinosaurs occurred some 65 million years ago, but on the cosmic calendar it was only yesterday. With the dinosaurs gone, small furry mammals inherited Earth. Some 60 million years later, or around 9 P.M. on December 31 of the cosmic calendar, early hominids (human ancestors) began to walk upright.

Perhaps the most astonishing thing about the cosmic calendar is that the entire history of human civilization falls into just the last half-minute. The ancient Egyptians built the pyramids only about 11 seconds ago on this scale. About 1 second ago, Kepler and Galileo proved that Earth orbits the Sun rather than vice versa. The average college student was born about 0.05 second ago, around 11:59:59.95 P.M. on the cosmic calendar. On the scale of cosmic time, the human species is the youngest of infants, and a human lifetime is a mere blink of an eye.

1.3 Spaceship Earth

Wherever you are as you read this book, you probably have the feeling that you're "just sitting here." Nothing could be further from the truth. In fact, you are being spun in circles as Earth rotates, you are racing around the Sun in Earth's orbit, and you are careening through the cosmos in the Milky Way Galaxy. In the words of noted inventor and philosopher R. Buckminster Fuller (1895–1983), you are a traveler on *spaceship Earth*. In this section, we'll take a brief look at the motion of spaceship Earth through the universe.

• How is Earth moving in our solar system?

The most basic motions of Earth are its **rotation** (spin) and its **orbit** (sometimes called *revolution*) around the Sun.

Earth rotates once each day around its axis, which is the imaginary line connecting the North Pole to the South Pole through the center of Earth. Earth rotates from west to east—counterclockwise as viewed from above the North Pole—which is why the Sun and stars appear to rise in the east and set in the west each day.

Although we do not feel any obvious effects from Earth's rotation, the speed of rotation is substantial (Figure 1.12). Unless you live very near the North or South Poles, you

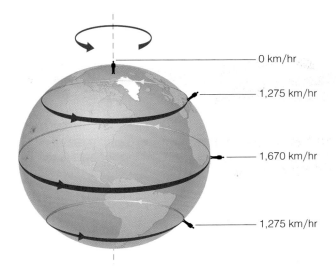

Figure 1.12 As Earth rotates, your speed around Earth's axis depends on your location: The closer you are to the equator, the faster you are traveling with rotation. Notice that Earth rotates from west to east—which is why the Sun appears to rise in the east and set in the west.

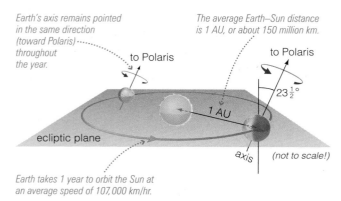

Earth's axis remains pointed in the same direction (toward Polaris) throughout the year.

to Polaris

The average Earth–Sun distance is 1 AU, or about 150 million km.

to Polaris

$23\frac{1}{2}°$

1 AU

ecliptic plane

axis

(not to scale!)

Earth takes 1 year to orbit the Sun at an average speed of 107,000 km/hr.

Figure 1.13 Earth takes a year to complete an orbit of the Sun, but its orbital speed is still surprisingly high. Notice that Earth both rotates and orbits counterclockwise as viewed from above the North Pole.

are whirling around Earth's axis at a speed of more than 1,000 kilometers per hour (600 miles per hour)—faster than most airplanes travel.

At the same time Earth is rotating, it is also orbiting the Sun, completing one orbit each year (Figure 1.13). Earth's average orbital distance is called an **astronomical unit**, or **AU**, which is equivalent to about 150 million kilometers (93 million miles). Again, even though we don't feel the effects of this motion, the speed is quite impressive. The orbit is not quite a perfect circle, which makes Earth's orbital speed vary slightly [Section 3.3], but at all times we are racing around the Sun at a speed in excess of 100,000 kilometers per hour (60,000 miles per hour). This is about 100 times as fast as a speeding bullet and faster than the launch speed of any spacecraft.

As you study Figure 1.13, notice that Earth's orbital path defines a flat plane that we call the **ecliptic plane**. Earth's

axis is tilted by $23\frac{1}{2}°$ from a line *perpendicular* to the ecliptic plane. This **axis tilt** happens to be oriented so that it points almost directly at a star called *Polaris*, or the *North Star*. Keep in mind that the idea of axis tilt makes sense only in relation to the ecliptic plane. That is, the idea of "tilt" by itself has no meaning in space, where there is no absolute up or down. In space, "up" and "down" mean only "away from the center of Earth (or another planet)" and "toward the center of Earth," respectively.

If there is no up or down in space, why do you think that nearly all globes have the North Pole on top and the South Pole on the bottom? Would it be equally correct to have the South Pole on top or to turn the globe sideways? Explain.

Notice also that Earth orbits the Sun in the same direction that it rotates on its axis: counterclockwise as viewed from above the North Pole. This is not a coincidence but a consequence of the way our planet was born. As we'll discuss in Chapter 8, Earth and the other planets were born in a spinning disk of gas that surrounded our Sun when it was young, and Earth rotates and orbits in the same direction as the disk was spinning. Notice also that we give a special name to the average distance between the Sun and Earth. This distance is called an **astronomical unit**, or **AU**, and it is about 150 million kilometers (93 million miles).

• How is our solar system moving in the Milky Way Galaxy?

Rotation and orbit are only a small part of the travels of spaceship Earth. In fact, our entire solar system is on a great journey within the Milky Way Galaxy (Figure 1.14).

Figure 1.14 This painting illustrates the motion of our solar system within the Milky Way Galaxy. The "zoom in" box shows that stars in our local solar neighborhood move essentially at random relative to one another. But the entire galaxy is rotating at the same time, so that all these stars orbit around the center of the galaxy.

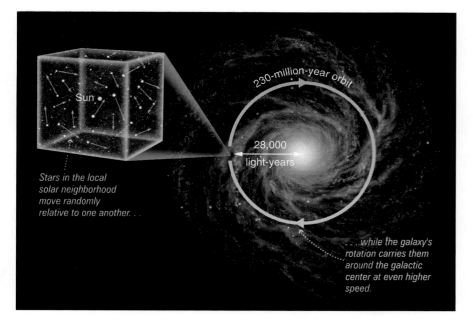

Sun

Stars in the local solar neighborhood move randomly relative to one another. . .

230-million-year orbit

28,000 light-years

. . .while the galaxy's rotation carries them around the galactic center at even higher speed.

Our Local Solar Neighborhood Let's begin with the motion of our solar system relative to nearby stars in what we call our *local solar neighborhood,* the region of the Sun and nearby stars. As you can see in Figure 1.14, this neighborhood is tiny compared to the entire galaxy. It also offers an important reminder of the incredible scale of the galaxy. Imagine drawing the tiniest dot that you can make on this painting of the galaxy. Your dot will probably be about 10,000 times smaller than the picture of the galaxy as a whole, which means the dot will cover a region representing more than 10 million stars! (The entire galaxy contains more than 100 billion stars, and 100 billion ÷ 10,000 = 10 million.) We usually think of our local solar neighborhood as an even smaller region of the galaxy, containing just a few thousand to a few million of the nearest stars.

The box in Figure 1.14 shows that stars within the local solar neighborhood move essentially at random relative to one another. These stars (or the stars of any other small region of the galaxy) also generally move quite fast relative to one another. For example, we are moving relative to nearby stars at an average speed of about 70,000 kilometers per hour (40,000 miles per hour), about three times as fast as the Space Station orbits Earth. Given these high speeds, why don't we see nearby stars racing around our sky?

The answer lies in their vast distances from us. You've probably noticed that a distant airplane appears to move through your sky more slowly than one flying close overhead. If we extend this idea to the stars, we find that even at speeds of 70,000 kilometers per hour, stellar motions would be noticeable to the naked eye only if we watched them for thousands of years. That is why the patterns in the constellations seem to remain fixed. Nevertheless, in 10,000 years the constellations will be noticeably different from those we see today. In 500,000 years they will be unrecognizable. If you

---

MATHEMATICAL INSIGHT 1.4 **Speeds of Rotation and Orbit**

In the first three Mathematical Insights, we investigated general problem-solving techniques, the use of units, the use of scientific notation, and estimation. Now we will see how simple formulas can expand the range of astronomical problems we can solve. In this case, we'll use the simple formula for the circumference of a circle, $C = 2 \times \pi \times$ radius (or $C = 2\pi r$), to confirm the speeds of rotation and orbit given in the text.

Example 1: How fast is a person on Earth's equator moving with Earth's rotation?

Solution:

Step 1 Understand: The question *how fast* tells us we are looking for a *speed* as the answer. If you remember that highway speeds are posted in miles per hour, you'll always remember that we find speed by dividing a distance (such as miles) by a time (such as hours). In this case, the distance is the circumference of Earth around the equator, because that is how far a person travels with each rotation of our planet (see Figure 1.12). The time is 24 hours, because that is how long each rotation takes.

Step 2 Solve: From Appendix E.1, Earth's equatorial radius is 6,378 km, so its circumference is $2 \times \pi \times 6,378$ km = 40,074 km. We divide this distance by the time of 24 hours:

$$\text{rotation speed at equator} = \frac{\text{equatorial circumference}}{\text{length of day}}$$
$$= \frac{40,074 \text{ km}}{24 \text{ hr}}$$
$$= 1,670 \text{ km/hr}$$

We can convert the speed to miles per hour by using the conversion from kilometers to miles given in Appendix Table C.3: 1,670 km × 0.6214 mi/km = 1,038 mi/hr.

Step 3 Explain: A person at Earth's equator is being carried around with Earth's rotation at a speed of about 1,670 kilometers per hour, or a little over 1,000 miles per hour. If you know

that commercial jets typically travel at about 500 miles per hour, you'll see that Earth's equatorial rotation speed is twice as fast. (See Problem 52 at the end of the chapter to learn how to find the speed at other latitudes.)

Example 2: How fast is Earth orbiting the Sun?

Solution:

Step 1 Understand: We are again looking for a speed, so we need to divide a distance by a time. In this case the distance is the circumference of Earth's orbit, because that is how far Earth travels each time it goes around the Sun. (Earth's orbit is not a perfect circle, but it's close enough for a circle to be a good approximation.) The time is 1 year, because that is how long each orbit takes.

Step 2 Solve: The text tells us that Earth's average distance from the Sun (1 AU) is about 149.6 million (1.496×10^8) km, so the orbit circumference is about $2 \times \pi \times 1.496 \times 10^8$ km $\approx 9.40 \times 10^8$ km. We find the orbital speed by dividing this orbital distance by 1 year, which we convert to hours so that our speed will come out in kilometers per hour rather than kilometers per year:

$$\text{orbital speed} = \frac{\text{orbital circumference}}{1 \text{ yr}}$$
$$= \frac{9.40 \times 10^8 \text{ km}}{1 \text{ yr} \times \frac{365 \text{ day}}{\text{yr}} \times \frac{24 \text{ hr}}{\text{day}}}$$
$$\approx 107,000 \text{ km/hr}$$

Step 3 Explain: Earth's average speed as it orbits the Sun is about 107,000 km/hr. (You should confirm that this is about 66,000 mi/hr.) On the Web, you can look up the fact that a typical speeding bullet travels between about 800 and 1,800 km/hr. Thus, roughly speaking, Earth's orbital speed is about 100 times as fast as a speeding bullet.

could watch a time-lapse movie made over millions of years, you *would* see stars racing across our sky.

Despite the chaos of motion in the local solar neighborhood over millions and billions of years, collisions between star systems are extremely rare. Explain why. (*Hint:* Consider the sizes of star systems, such as the solar system, relative to the distances between them.)

Galactic Rotation If you look closely at leaves floating in a stream, their motions relative to one another might appear random, just like the motions of stars in the local solar neighborhood. As you widen your view, you see that all the leaves are being carried in the same general direction by the downstream current. In the same way, as we widen our view beyond the local solar neighborhood, the seemingly random motions of its stars give way to a simpler and even faster motion: The entire Milky Way Galaxy is rotating. Our solar system, located about 28,000 light-years from the galactic center, completes one orbit of the galaxy in about 230 million years. Even if you could watch from outside our galaxy, this motion would be unnoticeable to your naked eye. However, if you calculate the speed of our solar system as we orbit the center of the galaxy, you will find that it is close to 800,000 kilometers (500,000 miles) per hour.

Careful study of the galaxy's rotation reveals one of the greatest mysteries in science—one that we will study in depth in Chapter 22. Stars at different distances from the galactic center orbit at different speeds, and we can learn how mass is distributed in the galaxy by measuring these speeds. Such studies have turned up an enormous surprise: It seems that the stars in the disk of the galaxy represent only the "tip of the iceberg" compared to the mass of the entire galaxy (Figure 1.15). That is, most of the mass of the galaxy seems to be located outside the visible disk, in what we call the galaxy's *halo*. We don't know the nature of this mass. Because we have not detected any light coming from it, we call it **dark matter**. Studies of other galaxies suggest that they also are made mostly of dark matter. In fact, most of the mass in the universe seems to be made of this mysterious dark matter, but we do not yet know what it is.

• How do galaxies move within the universe?

The billions of galaxies in the universe also move relative to one another. Within the Local Group (see Figure 1.1), some of the galaxies move toward us, some move away from us, and at least two small galaxies (known as the Large and Small Magellanic Clouds) apparently orbit our Milky Way Galaxy. Again, the speeds are enormous by earthly standards. For example, the Milky Way is moving toward the Great Galaxy in Andromeda at about 300,000 kilometers per hour (180,000 miles per hour). Despite this high speed, we needn't worry about a collision anytime soon. Even if the

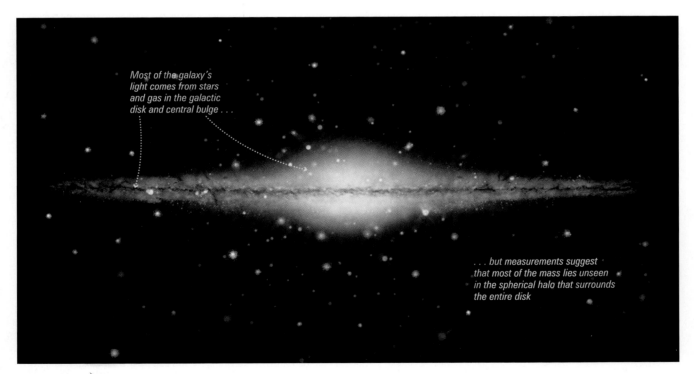

Most of the galaxy's light comes from stars and gas in the galactic disk and central bulge . . .

. . . but measurements suggest that most of the mass lies unseen in the spherical halo that surrounds the entire disk

Figure 1.15 This painting shows an edge-on view of the Milky Way Galaxy. Most visible stars reside within the galaxy's central bulge and thin *disk*, which run horizontally across the page in this figure. However, careful study of galactic rotation suggests that most of the mass lies in the galactic *halo*—a large, spherical region that surrounds and encompasses the disk. Because this mass emits no light that we have detected, we call it *dark matter*.

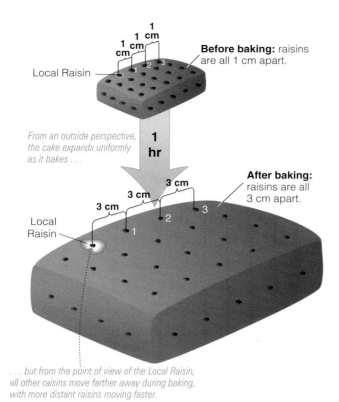

Before baking: raisins are all 1 cm apart.

Local Raisin

1 cm
1 cm
1 cm

From an outside perspective, the cake expands uniformly as it bakes . . .

1 hr

After baking: raisins are all 3 cm apart.

Local Raisin

3 cm
3 cm
3 cm

. . . but from the point of view of the Local Raisin, all other raisins move farther away during baking, with more distant raisins moving faster.

Distances and Speeds as Seen from the Local Raisin

Raisin Number	Distance Before Baking	Distance After Baking (1 hour later)	Speed
1	1 cm	3 cm	2 cm/hr
2	2 cm	6 cm	4 cm/hr
3	3 cm	9 cm	6 cm/hr
⋮	⋮	⋮	⋮

Figure 1.16 Interactive Figure An expanding raisin cake offers an analogy to the expanding universe. Someone living in one of the raisins inside the cake could figure out that the cake is expanding by noticing that all other raisins are moving away, with more distant raisins moving away faster. In the same way, we know that we live in an expanding universe because all galaxies outside our Local Group are moving away from us, with more distant ones moving faster.

that Raisin 1 starts out at a distance of 1 centimeter before baking and ends up at a distance of 3 centimeters after baking, which means it moves a distance of 2 centimeters away from the Local Raisin during the hour of baking. Hence, its speed as seen from the Local Raisin is 2 centimeters per hour. Raisin 2 moves from a distance of 2 centimeters before baking to a distance of 6 centimeters after baking, which means it moves a distance of 4 centimeters away from the Local Raisin during the hour. Hence, its speed is 4 centimeters per hour, or twice as fast as the speed of Raisin 1. Generalizing, the fact that the cake is expanding means that all raisins are moving away from the Local Raisin, with more distant raisins moving away faster.

Hubble's discovery that galaxies are moving in much the same way as the raisins in the cake, with most moving away from us and more distant ones moving away faster, implies that the universe in which we live is expanding much like the raisin cake. If you now imagine the Local Raisin as representing our Local Group of galaxies and the other raisins as representing more distant galaxies or clusters of galaxies, you have a basic picture of the expansion of the universe. Like the expanding dough between the raisins in the cake, *space* itself is growing between galaxies. More distant galaxies move away from us faster because they are carried along with this expansion like the raisins in the expanding cake. Many billions of light-years away, we see galaxies moving away from us at speeds approaching the speed of light.

There's one important distinction between the raisin cake and the universe: A cake has a center and edges, but we do not think the same is true of the entire universe. Anyone living in any galaxy in an expanding universe sees just what we see—other galaxies moving away, with more distant ones moving away faster. Because the view from each point in the universe is about the same, no place can claim to be any more "central" than any other place. The idea that the universe has no center or edges may sound strange, and it is not something that we can easily visualize. However, as we'll discuss in Chapters S2 and S3, space and time don't necessarily obey our commonsense notions

Milky Way and Andromeda Galaxies are approaching each other head-on (which they might not be), it will be billions of years before any collision begins.

When we look outside the Local Group, however, we find two astonishing facts recognized in the 1920s by Edwin Hubble, for whom the Hubble Space Telescope was named:

1. Virtually every galaxy outside the Local Group is moving *away* from us.
2. The more distant the galaxy, the faster it appears to be racing away.

These facts might make it sound like we suffer from a cosmic case of chicken pox, but there is a much more natural explanation: *The entire universe is expanding.* We'll save details about this expansion and how Hubble discovered it for Chapter 20, but you can understand the basic idea by thinking about a raisin cake baking in an oven.

Imagine that you make a raisin cake in which the distance between adjacent raisins is 1 centimeter. You place the cake into the oven, where it expands as it bakes. After 1 hour, you remove the cake, which has expanded so that the distance between adjacent raisins has increased to 3 centimeters (Figure 1.16). The expansion of the cake seems fairly obvious. But what would you see if you lived *in* the cake, as we live in the universe?

Pick any raisin (it doesn't matter which one), call it the Local Raisin, and identify it in the pictures of the cake both before and after baking. Figure 1.16 shows one possible choice for the Local Raisin, with three nearby raisins labeled. The accompanying table summarizes what you would see if you lived within the Local Raisin. Notice, for example, about them. Einstein showed that the three dimensions of space—length, width, and depth—are intertwined with a fourth dimension that we perceive as time. Once you learn how to think about the universe in four dimensions, the idea that the universe has no center or edges won't seem so strange.

It's also important to realize that, unlike the case with a raisin cake, Hubble couldn't actually *see* galaxies moving apart with time—the distances are far too vast for any motion to be noticeable on the time scale of a human life. Instead, he measured the speeds of galaxies by spreading their light into spectra and observing what we call *Doppler shifts* [Section 5.5]. We still use the same technique today, which illustrates how modern astronomy depends both on making careful observations and on using current understanding of the laws of nature to explain what we see.

• Are we ever sitting still?

As we have seen, we are never truly sitting still. Figure 1.17 summarizes the motions we have discussed. We spin around Earth's axis at more than 1,000 km/hr, while our planet orbits the Sun at more than 100,000 km/hr. Our solar system moves among the stars of the local solar neighborhood at typical speeds of 70,000 km/hr, while also orbiting the center of the Milky Way Galaxy at a speed of about 800,000 km/hr. Our galaxy moves among the other galaxies of the Local Group, while all other galaxies move away from us at speeds that grow greater with distance in our expanding universe. Spaceship Earth is carrying us on a remarkable journey!

1.4 The Human Adventure of Astronomy

In relatively few pages, we've laid out a fairly complete overview of our modern scientific ideas about the universe. But our goal in this book is not for you simply to be able to recite these ideas. Rather, it is to help you understand the evidence supporting them and the extraordinary story of how they developed.

• How has the study of astronomy affected human history?

Astronomy is a human adventure in the sense that it affects virtually everyone—even those who have never looked at the sky—because the development of astronomy has been so deeply intertwined with the development of civilization as a whole. Revolutions in astronomy have gone hand in hand with the revolutions in science and technology that have shaped modern life.

Witness the repercussions of the Copernican revolution, which showed us that Earth is not the center of the universe but rather just one planet orbiting the Sun. This revolution, which we will discuss further in Chapter 3, began when Copernicus published his idea of a Sun-centered solar system in 1543. Three later figures—Tycho Brahe, Johannes Kepler, and Galileo—provided the key evidence that eventually led to wide acceptance of the Copernican idea. The revolution culminated with Isaac Newton's uncovering of the laws of motion and gravity. Newton's work, in turn,

Earth rotates around its axis once each day, carrying people in most parts of the world around the axis at more than 1,000 km/hr.

Earth orbits the Sun once each year, moving at more than 100,000 km/hr.

The Solar System moves relative to nearby stars, typically at a speed of 70,000 km/hr.

The Milky Way Galaxy rotates, carrying our Sun around its center once every 230 million years, at a speed of about 800,000 km/hr.

Our galaxy moves relative to others in the Local Group; we are traveling toward the Andromeda Galaxy at about 300,000 km/hr.

The universe expands. The more distant an object, the faster it moves away from us; the most distant galaxies are receding from us at speeds close to the speed of light.

Figure 1.17 This figure summarizes the basic motions of Earth in the universe, along with their associated speeds.

became the foundation of physics that helped fuel the industrial revolution.

More recently, the development of space travel and the computer revolution have helped fuel tremendous progress in astronomy. We've learned a lot about our solar system by sending probes to the planets, and many of our most powerful observatories, including the Hubble Space Telescope, reside in space. On the ground, computer design and control have led to tremendous growth in the size and power of telescopes, particularly in the past decade.

Many of these efforts, and the achievements they spawned, led to profound social change. The most famous example is the fate of Galileo, whom the Vatican put under house arrest in 1633 for his claims that Earth orbits the Sun. Although the Church soon recognized that Galileo was right, he was formally vindicated only in 1992 with a statement by Pope John Paul II. In the meantime, his case spurred great debate in religious circles and profoundly influenced both theological and scientific thinking.

As you progress through this book and learn about astronomical discovery, try to keep in mind the context of the human adventure. You will then be learning not just about a science, but also about one of the great forces that has shaped our modern world. This context will also lead you to think about how the many astronomical mysteries that remain—such as the makeup of dark matter, the events in the first instant of the Big Bang, and the possibility of life beyond Earth—may influence our future.

What would it mean to us if we were ever to learn the complete story of our cosmic origins? How would our view of Earth be changed if we were to learn that Earth-like planets are common, or exceedingly rare? Only time may answer these questions, but the chapters ahead will give you the foundation you need to understand how we changed from a primitive people looking at patterns in the night sky to a civilization capable of asking deep questions about our existence.

Putting Chapter 1 into Context

In this first chapter, we developed a broad overview of our place in the universe. As we consider the universe in more depth in the rest of the book, remember the following "big picture" ideas:

- Earth is not the center of the universe but is instead a planet orbiting a rather ordinary star in the Milky Way Galaxy. The Milky Way Galaxy, in turn, is one of billions of galaxies in our observable universe.

- We are "star stuff." The atoms from which we are made began as hydrogen and helium in the Big Bang and were later fused into heavier elements by massive stars. When these stars died, they released their atoms into space, where our galaxy recycled them into new stars and planets. Our solar system formed from such recycled matter some 4.6 billion years ago.

- Cosmic distances are literally astronomical, but we can put them into perspective with the aid of scale models and other scaling techniques. When you think about these enormous scales, don't forget that every star is a sun and that every planet is a unique world.

- We are latecomers on the scale of cosmic time. The universe was already more than half its current age when our solar system formed, and several billion more years passed before humans arrived on the scene.

- All of us are being carried through the cosmos on spaceship Earth. Although we cannot feel this motion in our everyday lives, the associated speeds are surprisingly high. Learning about the motions of spaceship Earth gives us a new perspective on the cosmos and helps us understand its nature and history.

1.1 Our Modern View of the Universe

- **What is our place in the universe?** Earth is a planet orbiting the Sun. Our Sun is one of more than 100 billion stars in the **Milky Way Galaxy.** Our galaxy is one of about 40 galaxies in the **Local Group.** The Local Group is one small part of the **Local Supercluster,** which is one small part of the **universe.**

- **How did we come to be?** The universe began in the **Big Bang** and has been expanding ever since, except in localized regions where gravity has caused matter to collapse into galaxies and stars. The Big Bang essentially

 produced only two chemical elements: hydrogen and helium. The rest have been produced by stars and recycled within galaxies from one generation of stars to the next, which is why we are "star stuff."

- **How can we know what the universe was like in the past?** We can actually see into the past by studying light from distant stars and galaxies. Light takes time to travel through space, so the farther away we look in distance, the further back we look in time. Thus, when we look billions of **light-years** away, we see pieces of the universe as they were billions of years ago.

- **Can we see the entire universe?** No. The age

of the universe limits the extent of our **observable universe.** For a universe that is 14 billion years old, our observable universe extends to a distance of 14 billion light-years. If we tried to look beyond that distance, we'd be trying to look to a time before the universe existed.

1.2 The Scale of the Universe

- **How big is Earth compared to our solar system?** On a scale of 1 to 10 billion, the Sun is about

the size of a grapefruit. Planets are much smaller, with Earth the size of a ball point and Jupiter the size of a marble on this scale. The distances between planets are huge compared to their sizes, with Earth orbiting 15 meters from the Sun on this scale.

- **How far away are the stars?** On the 1-to-10-billion scale, it is possible to walk from the Sun to Pluto in just a few minutes. On the same scale, the nearest stars besides the Sun are thousands of kilometers away.

- **How big is the Milky Way Galaxy?** Using a scale on which the Milky Way galaxy is the size of a football field, the distance to the nearest star would be only about 4 millimeters. There are so many stars in our galaxy that it would take thousands of years just to count them.

- **How big is the universe?** The observable universe contains at least 100 billion galaxies. The total number of stars in the observable universe is comparable to the number of grains of dry sand on all the beaches on Earth.

- **How do our lifetimes compare to the age of the universe?** On a cosmic calendar that

compresses the history of the universe into 1 year, human civilization is just a few seconds old, and a human lifetime lasts only a fraction of a second.

1.3 Spaceship Earth

- **How is Earth moving in our solar system?**

Earth **rotates** on its axis once each day and **orbits** the Sun once each year. Earth orbits at an average distance from the Sun of 1 **AU** and with an **axis tilt** of $23\frac{1}{2}°$ to a line perpendicular to the **ecliptic plane.**

- **How is our solar system moving in the Milky Way Galaxy?** We move seemingly randomly relative

to other stars in our local solar neighborhood. The speeds are substantial by earthly standards, but the stars are so far away that their motions are undetectable to the naked eye. Our Sun and other stars in our neighborhood orbit the center of the galaxy every 230 million years, because the entire galaxy is rotating.

- **How do galaxies move within the universe?** Galaxies move essentially at random within the Local Group, but all galaxies beyond the Local Group are moving away from us. More distant galaxies are moving faster, which tells us that we live in an expanding universe.

- **Are we ever sitting still?** We are never truly sitting

still. We spin around Earth's axis and orbit the Sun. Our solar system moves among the stars of the local solar neighborhood while orbiting the center of the Milky Way Galaxy. Our galaxy moves among the other galaxies of the Local Group, while all other galaxies move away from us in our expanding universe.

1.4 The Human Adventure of Astronomy

- **How has the study of astronomy affected human history?** Throughout history, astronomy has developed hand in hand with social and technological development. Astronomy thereby touches each one of us and is a human adventure that all can enjoy.

EXERCISES AND PROBLEMS

For instructor-assigned homework go to MasteringASTRONOMY www.masteringastronomy.com

Review Questions

Short-Answer Questions Based on the Reading

1. What do we mean by a *geocentric* universe? In broad terms, contrast a geocentric view of the universe with our modern view of the universe.
2. Briefly describe the major levels of structure (such as planet, star, galaxy) in the universe.
3. What do we mean when we say that the universe is *expanding?* How does expansion lead to the idea of the *Big Bang?*
4. What did Carl Sagan mean when he said that we are "star stuff"?
5. How fast does light travel? What is a *light-year?*
6. Explain the statement *The farther away we look in distance, the further back we look in time.*
7. What do we mean by the *observable universe?* Is it the same thing as the entire universe?
8. Describe the solar system as it looks on the 1-to-10-billion scale used in the text. How far away are other stars on this same scale?

9. Describe at least one way to put the scale of the Milky Way Galaxy into perspective and at least one way to put the size of the observable universe into perspective.

10. Imagine describing the cosmic calendar to a friend. In your own words, give your friend a feel for how the human race fits into the scale of time.

11. Define *astronomical unit, ecliptic plane,* and *axis tilt.* Explain how each is related to Earth's rotation and/or orbit.

12. What is the shape of the Milky Way Galaxy? Where is our solar system located within the galaxy? How does our solar system move within the galaxy?

13. Distinguish between our galaxy's *disk* and *halo.* Where do most visible stars reside? Where does the mysterious *dark matter* seem to reside?

14. What key observations by Edwin Hubble lead us to conclude that the universe is expanding? Use the raisin cake model to explain how these observations imply expansion.

Test Your Understanding

Does It Make Sense?

Decide whether the statement makes sense (or is clearly true) or does not make sense (or is clearly false). Explain your reasoning.

Example: I walked east from our base camp at the North Pole.

Solution: The statement does not make sense because *east* has no meaning at the North Pole—all directions are south from the North Pole.

15. Our solar system is bigger than some galaxies.

16. The universe is billions of light-years in age.

17. It will take me light-years to complete this homework assignment!

18. Someday we may build spaceships capable of traveling a light-year in only a decade.

19. Astronomers recently discovered a moon that does not orbit a planet.

20. NASA plans soon to launch a spaceship that will photograph our Milky Way Galaxy from beyond its halo.

21. The observable universe is the same size today as it was a few billion years ago.

22. Photographs of distant galaxies show them as they were when they were much younger than they are today.

23. At a nearby park, I built a scale model of our solar system in which I used a basketball to represent Earth.

24. Because nearly all galaxies are moving away from us, we must be located at the center of the universe.

Quick Quiz

Choose the best answer to each of the following. Explain your reasoning with one or more complete sentences.

25. Which of the following correctly lists our "cosmic address" from small to large? (a) Earth, solar system, Milky Way Galaxy, Local Group, Local Supercluster, universe; (b) Earth, solar system, Local Group, Local Supercluster, Milky Way Galaxy, universe; (c) Earth, Milky Way Galaxy, solar system, Local Group, Local Supercluster, universe.

26. When we say the universe is *expanding,* we mean that: (a) Everything in the universe is growing in size. (b) The average distance between galaxies is growing with time. (c) The universe is getting older.

27. If stars existed but galaxies did not: (a) We would probably still exist anyway. (b) We would not exist because life on Earth depends on the light produced by galaxies. (c) We would not exist because we are made of material that required recycling in galaxies.

28. Could we see a galaxy that is 20 billion light-years away? (a) Yes, if we had a big enough telescope. (b) No, because it would be beyond the bounds of our observable universe. (c) No, because a galaxy could not possibly be that far away.

29. The star Betelgeuse is about 425 light-years away. If it explodes tonight: (a) We'll know because it will be brighter than the full Moon in the sky. (b) We'll know because debris from the explosion will rain down on us from space. (c) We won't know about it until 425 years from now.

30. If we represent the solar system on a scale that allows us to walk from the Sun to Pluto in a few minutes, then: (a) The planets would be the size of basketballs and the nearest stars would be a few miles away. (b) The planets would all be marble size or smaller and the nearest stars would be thousands of miles away. (c) The planets would be microscopic and the stars would be light-years away.

31. The total number of stars in the observable universe is roughly equivalent to: (a) the number of grains of sand on all the beaches on Earth. (b) the number of grains of sand on Miami Beach. (c) infinity.

32. The age of our solar system is about: (a) one-third of the age of the universe. (b) three-fourths of the age of the universe. (c) two billion years less than the age of the universe.

33. An astronomical unit is: (a) any planet's average distance from the Sun. (b) Earth's average distance from the Sun. (c) any large astronomical distance.

34. The fact that nearly all galaxies are moving away from us, with more distant ones moving faster, helped us to conclude that: (a) The universe is expanding. (b) Galaxies repel each other like magnets. (c) Our galaxy lies near the center of the universe.

Investigate Further

In-Depth Questions to Increase Your Understanding

Short-Answer/Essay Questions

35. *Our Cosmic Origins.* Write one to three paragraphs summarizing why we could not be here if the universe did not contain both stars and galaxies.

36. *Alien Technology.* Some people believe that Earth is regularly visited by aliens who travel here from other star systems. For this to be true, how much more advanced than our own technology would the space travel technology of the aliens have to be? Write one to two paragraphs to give a sense of the technological difference. (*Hint:* The ideas of scale in this chapter can help you contrast the distance the aliens would have to travel easily with the distances we are now capable of traveling.)

37. *Stellar Collisions.* Is there any danger that another star will come crashing through our solar system in the near future? Explain.

38. *Raisin Cake Universe.* Suppose that all the raisins in a cake are 1 centimeter apart before baking and 4 centimeters apart after baking.

   a. Draw diagrams to represent the cake before and after baking.

   b. Identify one raisin as the Local Raisin on your diagrams. Construct a table showing the distances and speeds of other raisins as seen from the Local Raisin.

   c. Briefly explain how your expanding cake is similar to the expansion of the universe.

39. *Scaling the Local Group of Galaxies.* Both the Milky Way Galaxy and the Great Galaxy in Andromeda (M 31) have a diameter of about 100,000 light-years. The distance between the two galaxies is about 2.5 million light-years.
    a. Using a scale on which 1 centimeter represents 100,000 light-years, draw a sketch showing both galaxies and the distance between them to scale.
    b. How does the separation between galaxies compare to the separation between stars? Based on your answer, discuss the likelihood of galactic collisions in comparison to the likelihood of stellar collisions.
40. *The Cosmic Perspective.* Write a short essay describing how the ideas presented in this chapter affect your perspectives on your own life and on human civilization.

Quantitative Problems
Be sure to show all calculations clearly and state your final answers in complete sentences.

41. *Distances by Light.* Just as a light-year is the distance that light can travel in 1 year, we define a light-second as the distance that light can travel in 1 second, a light-minute as the distance that light can travel in 1 minute, and so on. Calculate the distance in both kilometers and miles represented by each of the following:
    a. 1 light-second;   c. 1 light-hour;
    b. 1 light-minute;   d. 1 light-day.
42. *Moonlight and Sunlight.* How long does it take light to travel from:
    a. the Moon to Earth?   b. the Sun to Earth?
43. *Communication with Mars.* We use radio waves, which travel at the speed of light, to communicate with robotic spacecraft. How long does it take a message to travel from Earth to a spacecraft on Mars when:
    a. Mars it at its closest to Earth?
    b. Mars is at its farthest from Earth? (*Data:* The distance from Earth to Mars ranges between about 56 and 400 million kilometers.)
44. *Communication with Pluto.* About how long does it take light to travel from Earth to Pluto? You can assume each planet is at its average distance from the Sun.
45. *Moon to Stars.* How many times greater is the distance from Earth to Alpha Centauri (4.4 light-years) than the distance from Earth to the Moon?
46. *Saturn vs. the Milky Way.* Photos of Saturn and photos of galaxies can look so similar that children often think the photos show similar objects. (After all, like Saturn, at least some galaxies have a large, round center surrounded by a disk that looks like rings.) In reality, a galaxy is far larger than any planet. About how many times larger is the diameter of the Milky Way Galaxy than the diameter of Saturn's rings? (*Data:* Saturn's rings are about 270,000 km in diameter; the Milky Way is 100,000 light-years in diameter.)
47. *Galaxy Scale.* Consider the 1-to-10^{19} scale on which the disk of the Milky Way Galaxy fits on a football field. On this scale, how far is it from the Sun to Alpha Centauri (real distance: 4.4 light-years)? How big is the Sun itself on this scale? Compare the Sun's size on this scale to the size of a typical atom (real diameter: about 10^{-10} meter).

48. *Universal Scale.* Suppose we wanted to make a scale model of the Local Group of Galaxies, in which the Milky Way Galaxy was the size of a marble (about 1 cm in diameter).
    a. How far from the Milky Way Galaxy would the Andromeda Galaxy be on this scale?
    b. How far would the Sun be from Alpha Centauri on this scale?
    c. How far would it be on this scale from the Milky Way Galaxy to the most distant galaxies in the observable universe?
49. *Driving Trips.* Imagine that you could drive your car at a constant speed of 100 km/hr (62 mi/hr), even across oceans and in space. (In reality, the law of gravity would make driving through space at a constant speed all but impossible.) How long would it take to drive:
    a. around Earth's equator?
    b. from the Sun to Earth?
    c. from the Sun to Pluto?
    d. to Alpha Centauri (4.4 light-years away)?
50. *Faster Trip.* Suppose you wanted to reach Alpha Centauri in 100 years.
    a. How fast would you have to go, in km/hr?
    b. How many times faster is the speed you found in (a) than the speeds of our fastest current spacecraft (around 50,000 km/hr)?
51. *Galactic Rotation Speed.* We are located about 28,000 light-years from the galactic center and we orbit the center once every 230 million years. How fast are we traveling around the galaxy? Give your answer in both km/hr and mi/hr.
52. *Earth Rotation Speed.* Mathematical Insight 1.3 shows how to find Earth's equatorial rotation speed. To find the rotation speed at any other latitude, you need the following fact: The radial distance from Earth's axis at any latitude is equal to the equatorial radius times the cosine of the latitude. Use this fact to find the rotation speed for a person at latitude:
    a. 30°N;
    b. 60°N;
    c. your latitude. (*Hint:* When using the cosine (cos) function on your calculator, be sure your calculator is set to recognize angles in degree mode, as opposed to radian or gradient mode.)

Discussion Questions

53. *Eliot Quote.* Think carefully about the chapter-opening quotation from T.S. Eliot. What do you think he means? Explain clearly.
54. *Infant Species.* In the last few tenths of a second before midnight on December 31 of the cosmic calendar, we have developed an incredible civilization and learned a great deal about the universe, but we also have developed technology with which we could destroy ourselves. The midnight bell is striking, and the choice for the future is ours. How far into the next cosmic year do you think our civilization will survive? Defend your opinion.
55. *A Human Adventure.* Astronomical discoveries clearly are important to science, but are they also important to our personal lives? Defend your opinion.

Interactive Tutorials

Tutorial Review of Key Concepts

Use the interactive **Tutorial** at **www.mastering astronomy.com** to review key concepts from this chapter.

Scale of the Universe Tutorial
Lesson 1 Distance Scales: The Solar System
Lesson 2 Distance Scales: Stars and Galaxies
Lesson 3 Powers of 10

Supplementary Tutorial Exercises

Use the interactive **Tutorial Lessons** to explore the following questions.

Scale of the Universe Tutorial, Lesson 2
1. Could we use spaceships like those that we use to explore our solar system to explore planets around other stars? Why or why not?
2. How does the distance between galaxies in the Local Group compare to the sizes of the galaxies?
3. Why are collisions between galaxies more likely than collisions between stars within galaxies?

Scale of the Universe Tutorial, Lesson 3
1. Why are powers of 10 useful for describing and comparing distances in the universe?
2. If you begin from the scale of the observable universe, by how many powers of 10 must you zoom in to see the orbits of the planets in our solar system?
3. How many powers of 10 separate the scale on which you can see all the planetary orbits from the scale on which you can see individual people?

 Exploring the Sky and Solar System

Of the many activities available on the *Voyager: SkyGazer* **CD-ROM** accompanying your book, use the following files to observe key phenomena covered in this chapter.

Go to the **File: Basics** folder for the following demonstrations.
1. Chicago 10000AD
2. Dragging the Sky

Go to the **Explore** menu for the following demonstrations.
1. Solar Neighborhood
2. Paths of the Planets

 Movies

Check out the following narrated and animated short documentaries available on **www.masteringastronomy.com** for a helpful review of key ideas covered in this chapter.

From the Big Bang to Galaxies Movie

Web Projects

1. *Astronomy on the Web.* The Web contains a vast amount of astronomical information. Starting from the links on the textbook Web site, spend at least an hour exploring astronomy on the Web. Write two or three paragraphs summarizing what you learned from your Web surfing. What was your favorite astronomical Web site, and why?
2. *Tour Report.* Take the virtual tour of the Voyage scale model solar system on **www.masteringastronomy.com**. After completing it, imagine that a friend asks you the following questions. Answer each question in one paragraph.
   a. Is the Sun really much bigger than Earth?
   b. Is it true that the Sun uses nuclear energy?
   c. Would it be much harder to send humans to Mars than to the Moon?
   d. In elementary school, I heard that Neptune is farther from the Sun than Pluto. Is this true?
   e. I read that Pluto is not really a planet. What's the story?
   f. Why didn't they have any stars besides the Sun in the scale model?
   g. What was the most interesting thing you learned during your tour?
3. *NASA Missions.* Visit the NASA Web site to learn about upcoming astronomy missions. Write a one-page summary of the mission you feel is most likely to give us new astronomical information during the time you are enrolled in your astronomy course.
4. *The Hubble Ultra Deep Field.* The photo that opens this chapter is called the Hubble Ultra Deep Field. Find the photo on the Hubble Space Telescope Web site, and learn how it was taken, what it shows, and what we've learned from it. Write a short report on your findings.

2

Discovering the Universe for Yourself

LEARNING GOALS

2.1 Patterns in the Night Sky
- What does the universe look like from Earth?
- Why do stars rise and set?
- Why do the constellations we see depend on latitude and time of year?

2.2 The Reason for Seasons
- What causes the seasons?
- How do we mark the progression of the seasons?
- How does the orientation of Earth's axis change with time?

2.3 The Moon, Our Constant Companion
- Why do we see phases of the Moon?
- What causes eclipses?

2.4 The Ancient Mystery of the Planets
- What was once so mysterious about planetary motion in our sky?
- Why did the ancient Greeks reject the real explanation for planetary motion?

> *We had the sky, up there, all speckled with stars, and we used to lay on our backs and look up at them, and discuss about whether they was made, or only just happened.*
>
> **Mark Twain, *Huckleberry Finn***

This is an exciting time in the history of astronomy. A new generation of telescopes is probing the depths of the universe. Increasingly sophisticated space probes are collecting new data about the planets and other objects in our solar system. Rapid advances in computing technology are allowing scientists to analyze the vast amount of new data and to model the processes that occur in planets, stars, galaxies, and the universe.

A goal of this book is to help you share in the ongoing adventure of astronomical discovery. One of the best ways to become a part of this adventure is to do what other humans have done for thousands of generations: Go outside, observe the sky around you, and contemplate the awe-inspiring universe of which you are a part. In this chapter, we'll discuss a few key ideas that will help you understand what you see in the sky.

2.1 Patterns in the Night Sky

Today we take for granted that we live on a small planet orbiting an ordinary star in one of many galaxies in the universe. But this fact is not obvious from a casual glance at the night sky, and we've learned about our place in the cosmos only through a long history of careful observations of patterns in the sky. In this section, we'll discuss the major features of the night sky, considering both how they appear to our eyes and how we understand them in light of our current knowledge of the universe.

• What does the universe look like from Earth?

Shortly after sunset, as daylight fades to darkness, the sky appears to fill slowly with stars. On clear, moonless nights far from city lights, more than 2,000 stars may be visible to your naked eye, along with the whitish band of light that we call the *Milky Way* (Figure 2.1). As you look at the stars, your mind may group them into patterns that look like familiar shapes or objects. If you observe the sky night after night or year after year, you will recognize the same patterns of stars. Indeed, the patterns we see today are virtually the same ones that generations of our ancestors saw over the last several thousand years.

Constellations   People of nearly every culture gave names to patterns they saw in the sky. The pattern that the Greeks named Orion, the hunter (see Figure 1.3), was seen by the ancient Chinese as a supreme warrior called *Shen*. Hindus in ancient India also saw a warrior, called *Skanda*, who rode

Figure 2.1   A "fish-eye" photograph of the Australian night sky. The Milky Way is the prominent whitish band with dark lanes running through it. A comet is also visible near the upper left of this photo. (It is Comet Hyakutake, which shined in Earth's sky for a few months in 1996.)

a peacock and was the general of a great celestial army. Aborigines of northern Australia saw the three stars of Orion's belt as three fishermen in a canoe. In southern California, these three stars climb almost straight up into the sky as they rise in the east, which may explain why the Chemehuevi Indians of the California desert saw them as a line of three sure-footed mountain sheep.* These are but a few of the many names, each accompanied by a rich folklore, that have been given to the pattern of stars we call Orion.

We usually refer to Orion and other patterns of stars as constellations, but to astronomers the term has a more precise meaning: A **constellation** is a *region* of the sky with well-defined borders. Just as every spot of land in the United States is part of some state, every point in the sky belongs to some constellation. For example, Figure 2.2 shows the borders of the constellation Orion and several of its neighbors. The familiar patterns of stars merely help us locate these constellations.

The official borders of the constellations were set in 1928 by members of the International Astronomical Union (IAU), an association of astronomers from around the world. The IAU divided the sky into 88 constellations (see Appendix I), choosing constellation names that were famil-

---

*These and other constellation stories are found in E. C. Krupp, *Beyond the Blue Horizon* (Oxford University Press, 1991).

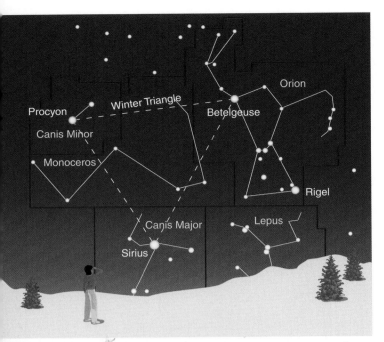

Figure 2.2 Red lines mark official borders of several constellations near Orion. Yellow lines connect recognizable patterns of stars within constellations. Sirius, Procyon, and Betelgeuse form a pattern that spans several constellations and is called the *Winter Triangle*. It is easy to find on clear winter evenings.

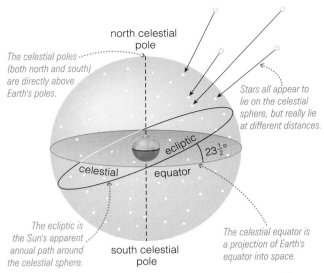

Figure 2.3 Stars appear to lie on a great celestial sphere that surrounds the Earth.

iar to European and American astronomers. Most constellations visible in the Northern Hemisphere have names that can be traced back to civilizations of the ancient Middle East. Many Southern Hemisphere constellations carry official names that originated with seventeenth-century European explorers.

Learning your way around the constellations is no more difficult than learning your way around your own neighborhood, and recognizing the patterns of just 20–40 constellations is enough to make the entire sky seem familiar. The best way to learn the constellations is to go out and view them, guided by a few visits to a planetarium and star charts like the ones in the back of this book (Appendix J). The *SkyGazer* software that comes with this book can also help you learn constellations.

The Celestial Sphere The stars in a particular constellation appear to lie close to one another but may be quite far apart in reality, because they may lie at very different distances from Earth. This illusion occurs because we lack depth perception when we look into space, a consequence of the fact that the stars are so far away [Section 1.2]. The ancient Greeks mistook this illusion for reality, imagining the stars to lie on a great **celestial sphere** that surrounds the Earth (Figure 2.3).

We now know that Earth does not really lie in the center of a giant ball of stars, but we can still use the idea of a celestial sphere to help us understand and map the sky. For reference, we identify four special points and circles.

- The **north celestial pole** is the point directly over Earth's North Pole.
- The **south celestial pole** is the point directly over Earth's South Pole.
- The **celestial equator**, which is a projection of Earth's equator into space, makes a complete circle around the celestial sphere.
- The **ecliptic** is the path the Sun follows as it appears to circle around the celestial sphere once each year.

Notice that the ecliptic crosses the celestial equator at a $23\frac{1}{2}°$ angle, because that is the tilt of Earth's axis. In fact, the ecliptic on the celestial sphere is the projection into space of Earth's orbital plane (called the *ecliptic plane* [Section 1.3]).

Just as a globe shows us the layout of Earth's surface, a model of the celestial sphere shows how the stars appear to be arranged in the sky. A typical model of the celestial sphere (Figure 2.4) shows patterns of bright stars, borders of the 88 official constellations, and the celestial poles, celestial equator, and ecliptic.

The Milky Way As your eyes adapt to darkness at a dark site, you'll begin to see the whitish band of light called the *Milky Way* (see Figure 2.1). Our Milky Way Galaxy gets its name from this band of light. You can see only part of the Milky Way at any particular time, but it circles all the way around the celestial sphere, passing through more than a dozen constellations. The widest and brightest parts of the Milky Way are most easily seen from the Southern Hemisphere, which probably explains why the Aborigines of Australia gave names to the patterns they saw within the Milky Way in the same way other cultures named patterns of stars.

The band of light called the Milky Way bears an important relationship to the Milky Way Galaxy: *It traces our galaxy's disk of stars—the galactic plane—as it appears from our location in the outskirts of the galaxy.* Figure 2.5 shows the idea. The Milky Way Galaxy is shaped like a thin pan-

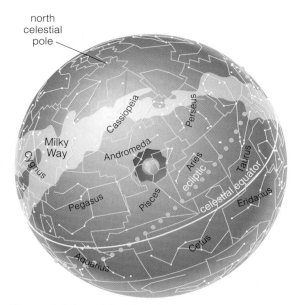

Figure 2.4 A model of the celestial sphere showing bright stars and the borders of constellations. Remember that the celestial sphere represents the view from Earth, so we imagine Earth to reside in the center of the sphere.

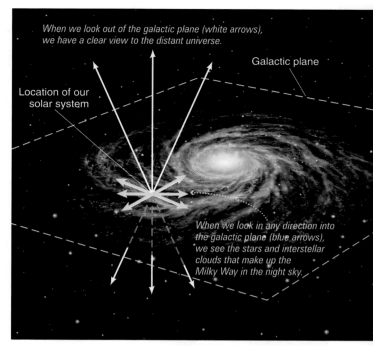

Figure 2.5 This painting shows how our galaxy's structure affects our view from Earth. Remember that all the stars visible to the naked eye lie quite close to us compared to the size of the galaxy.

cake with a bulge in the middle. We view the universe from our location a little more than halfway out from the center of this "pancake." No matter what direction we look *within* the disk of the galaxy, our view takes in countless stars and vast clouds of interstellar gas and dust. These stars and clouds form the Milky Way in our night sky. We see far fewer stars and relatively little gas and dust in our own galaxy when we look in directions pointing *away* from the galactic plane. Thus, as shown by the white arrows in Figure 2.5, these directions offer a clear view to the far reaches of the universe, as long as we look with sufficiently powerful telescopes.

If you examine the Milky Way closely, you'll notice that it varies in width and has dark lanes running through it. The width depends on the direction we are looking into the galactic disk: It is widest in the direction of the constellation Sagittarius, because that is the direction in which we are looking toward the galaxy's wide, central bulge. The dark lanes appear in regions where particularly dense interstellar clouds obscure our view of stars behind them. The Milky Way's abundant gas and dust prevents us from seeing more than a few thousand light-years into our galaxy's disk.* This limitation kept much of our own galaxy hidden from view until just a few decades ago, when new technologies allowed us to peer through the clouds by observing forms of light that are invisible to our eyes (such as radio waves and X-rays [Section 5.2]).

THINK ABOUT IT

Consider a distant galaxy located in the same direction from Earth as the center of our own galaxy (but much farther away). Could we see it with our eyes? Explain.

---

*There are a few small "windows" through which we can see much farther; see Figure 15.4 for an example.

The Local Sky   The celestial sphere, banded by the Milky Way, is a useful way of thinking about the appearance of the universe from Earth. But it is not what we actually see when we go outside. Picture yourself standing in a flat, open field. The sky appears to take the shape of a dome, making it easy to understand why people of many ancient cultures imagined that we lived on a flat Earth under a great dome encompassing the world. We see only half of the celestial sphere at any particular moment from any particular location, while the other half is blocked from view by the ground. The half of the celestial sphere that you see at any time represents what we call your **local sky**—the sky as seen from wherever you happen to be standing.

We use several key features of the local sky to serve as references for locating objects (Figure 2.6). The boundary between Earth and sky defines the **horizon**. The point directly overhead is the **zenith**. The **meridian** is an imaginary half-circle stretching from the horizon due south, through the zenith, to the horizon due north.

We can pinpoint the position of any object in the local sky by stating its **direction** along the horizon and its **altitude** above the horizon. For example, Figure 2.6 shows a person pointing to a star located in the direction of southeast at an altitude of 60°. (The zenith has an altitude of 90° but no direction, because it is straight overhead.)

In addition to stating an object's location in our sky, we often want to describe its size or how far it is from other objects in the sky. For example, to describe the pattern of stars in a particular constellation, we might wish to say how large the pattern appears in the sky and where individual stars are located relative to others. It's easy to do this, as long as we remember that we can describe only *angular*

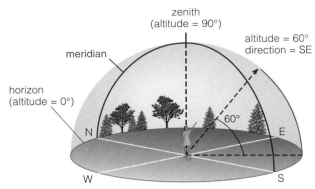

Figure 2.6 From any place on Earth, the local sky looks like a dome (hemisphere). This diagram shows key reference points in the local sky. It also shows how we can describe any position in the local sky by its altitude and direction.

The Moon Illusion

You've probably noticed that the full moon appears to be larger when it is near the horizon than when it is high in your sky. However, this apparent size change is an illusion. If you measure the angular size of the full moon on a particular night, you'll find that it is about the same whether the Moon is near the horizon or high in the sky. The Moon's angular size in the sky depends only on its true size and its distance from Earth. Although this distance varies over the course of the Moon's monthly orbit, it does not change enough to cause a noticeable effect on a single night. You can confirm that the Moon's angular size remains the same by measuring it. You may also be able to make the illusion go away by viewing the Moon upside down between your legs when it is on the horizon.

sizes and distances in the sky. Our lack of depth perception on the celestial sphere means that we have no way to judge the true sizes or separations of the objects we see.

The **angular size** of an object is the angle it appears to span in your field of view. For example, the angular sizes of the Sun and Moon are each about $\frac{1}{2}°$ (Figure 2.7a). Notice that angular size does not by itself tell us much about an object's true size, because angular size also depends on distance: The farther away an object is, the smaller it will look in angular size. For example, the Sun is about 400 times larger in diameter than the Moon, but it has the same angular size in our sky because it is also about 400 times farther away.

The **angular distance** between a pair of objects in the sky is simply the angle that appears to separate them. For example, the angular distance between the "pointer stars" at the end of the Big Dipper's bowl is about 5° (Figure 2.7b). You can use your outstretched hand to make rough estimates of angles in the sky (Figure 2.7c).

For more precise astronomical measurements, we subdivide each degree into 60 **arcminutes** and subdivide each arcminute into 60 **arcseconds** (Figure 2.8). We abbreviate

arcminutes with the symbol ′ and arcseconds with the symbol ″. For example, we read 35°27′15″ as "35 degrees, 27 arcminutes, 15 arcseconds."

THINK ABOUT IT

Children often try to describe the sizes of objects in the sky (such as the Moon or an airplane) in inches or miles, or by holding their fingers apart and saying "it was THIS big." Can we really describe objects in the sky in this way? Why or why not?

• Why do stars rise and set?

If you spend a few hours out under a starry sky, you'll soon notice that the universe seems to be circling around us. Stars move gradually across the sky from east to west, and

Figure 2.7 We measure *angular sizes* or *angular distances*, rather than actual sizes or distances, when we look at objects in the sky.

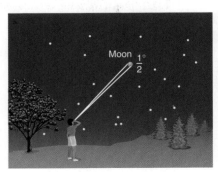

a The angular size of the Moon is about 1/2° (which is also the angular size of the Sun).

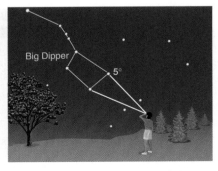

b The angular distance between the two "pointer stars" of the Big Dipper (which point to the North Star, Polaris; see Figure 2.13a) is about 5°.

Stretch out your arm as shown here.

c You can estimate angular sizes or distances with your outstretched hand.

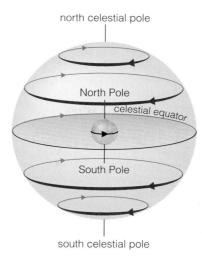

Figure 2.8 We subdivide each degree into 60 arcminutes and each arcminute into 60 arcseconds. (The figure is not drawn to scale.)

Figure 2.9 Earth rotates from west to east (black arrow), making the celestial sphere *appear* to rotate around us from east to west (red arrows).

you'll always see some stars rising in the east while others are setting in the west. Many ancient people took this appearance of movement at face value, concluding that we lie in the center of a universe that rotates around us each day. Today we know that the ancients had it backward: It is Earth that rotates each day, not the rest of the universe. The stars, along with the Sun, Moon, and planets, all appear to rise in the east and set in the west because Earth rotates in the opposite direction, from west to east.

THINK ABOUT IT

Do distant galaxies also rise and set like the stars in our sky? Why or why not?

Because the stars all appear to reside on a celestial sphere, we can picture the movement of the sky by showing the celestial sphere rotating around Earth (Figure 2.9). From this perspective it is very easy to see how the universe seems to turn around us: Every object on the celestial sphere appears to make a simple daily circle around Earth. However, the motion can look a little more complex in the local sky.

In the local sky, the horizon cuts the celestial sphere in half. Figure 2.10 shows the idea for a typical location in the Northern Hemisphere. As Earth rotates—making the celestial sphere appear to circle around us—we see stars and other objects on the celestial sphere move across our local sky. Different stars follow different paths through the sky, depending on where they are located on the celestial sphere. If you study Figure 2.10 carefully, you'll notice the following key facts about the paths of stars (and other celestial objects) through the sky:

● Stars relatively near the north celestial pole remain perpetually above the horizon. They never rise or set but instead make daily counterclockwise circles around the north celestial pole. We say that such stars are **circumpolar**.
● Stars relatively near the south celestial pole never rise above the horizon at all.
● All other stars have daily circles that are partly above the horizon and partly below it. Because Earth rotates from

west to east (counterclockwise as viewed from above the North Pole), the stars appear to rise in the east and set in the west.

The photo that opens this chapter (p. 26) shows a beautiful time-exposure photograph that helps us see the daily paths of stars. It was taken at Arches National Park in Utah. Paths of circumpolar stars are visible within the arch; notice that the complete daily circles for these stars are above the horizon (although the photo shows only a portion of each circle). The north celestial pole lies at the center of these circles. The circles grow larger for stars farther from the north celestial pole. If they are large enough, the circles cross the horizon, so that the stars rise in the east and set in the west. The same ideas apply in the Southern Hemisphere, except that circumpolar stars are those near the south celestial pole.

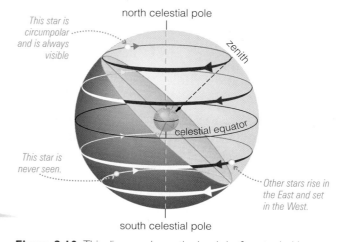

Figure 2.10 This diagram shows the local sky for a typical location in the Northern Hemisphere. The local horizon slices through half the celestial sphere, which is why the local sky looks like a dome. Because the slice is at an angle to the equator, the simple daily circles of stars are tilted in the local sky. Note: It may be easier to follow the star paths in the local sky if you rotate the page so that the zenith points up.

If you hold a quarter in front of your eye, it can block your entire field of view. But as you move it farther way, it appears to get smaller and it blocks less of your view. Figure (a) summarizes the idea by showing the quarter in cross section, so we can see how its angular diameter decreases with distance.

It's useful to have a formula telling us how an object's angular size depends on its physical size and distance, and we can find the formula with a little mathematical trick that works when the angular size is small. In Figure (b), we've made the quarter from Figure (a) look like a tiny piece of a circle going all the way around your eye. The radius of the circle is the *distance* from your eye to the quarter, the angle from your eye is the quarter's *angular size*, and we've labeled the quarter's actual diameter as its *physical size*. Now, notice that as long as the angular size is relatively small (less than a few degrees), we can pretend that the quarter's physical size (diameter) is a small piece of the circle we've drawn. This is the trick we needed: The quarter's angular size is now the *same fraction* of the full 360° circle as its physical size is of the circle's physical circumference. Since the circumference of this circle is $2\pi \times (distance)$, we can write what we've found as:

$$\frac{\text{angular size}}{360°} = \frac{\text{physical size}}{2\pi \times \text{distance}}$$

Multiplying both sides by 360° and rearranging a bit, we have a formula that allows us to determine angular size when we know physical size and distance:

$$\text{angular size} = \text{physical size} \times \frac{360°}{2\pi \times \text{distance}}$$

(This formula is sometimes called the *small-angle formula*, since it is valid only when the angular size is small.) In astronomy, we generally measure an object's angular size and often have a way of determining its distance (we'll discuss distance measurement techniques in later chapters). Thus, it's useful to rearrange the formula so we can calculate physical size when we know angular size and distance. You should confirm that a little algebra tells us that:

$$\text{physical size} = \text{angular size} \times \frac{2\pi \times \text{distance}}{360°}$$

The context of a problem will tell you which form of the formula to choose.

Example 1: The angular diameter of the Moon is about 0.5° and the Moon is about 380,000 km away. Estimate the Moon's actual diameter.

Solution:

Step 1 Understand: We are asked to find the Moon's actual diameter given its angular diameter and distance. The formula below tells us how to find physical *size* from angular size and distance; we can use this formula once we realize that, in this case, the "size" is a diameter.

Step 2 Solve: We now use the formula to calculate the Moon's physical size (diameter) from the given values of its angular size and distance:

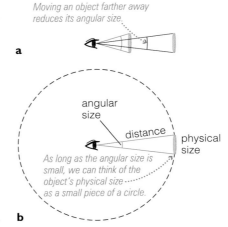

Moving an object farther away reduces its angular size.

a

angular size

distance ⟶ physical size

As long as the angular size is small, we can think of the object's physical size as a small piece of a circle.

b

Angular size depends on physical size and distance.

$$\text{physical size} = \text{angular size} \times \frac{2\pi \times \text{distance}}{360°}$$
$$= 0.5° \times \frac{2\pi(380,000 \text{ km})}{360°}$$
$$\approx 3,300 \text{ km}$$

Step 3 Explain: We have found that the Moon's diameter is about 3,300 kilometers. We can check that our answer makes sense by comparing it to the value for the Moon's diameter given in Appendix E. Our estimate of 3,300 kilometers is fairly close to the Moon's actual diameter of 3,476 kilometers, which we could find by using more precise values for the Moon's angular diameter and distance.

Example 2: Suppose the two headlights on a car are separated by 1.5 meters and you are looking at the car from a distance of 500 meters. What is the angular separation of the headlights?

Solution:

Step 1 Understand: In this case we are asked about "separation" between two lights rather than size, but the idea is the same. We simply replace size with separation in the below formulas and we have all the information we need to solve the problem.

Step 2 Solve: Because we know the physical separation and distance in this case, we use the formula in the first form that we found above.

$$\text{angular separation} = \text{physical separation} \times \frac{360°}{2\pi \times \text{distance}}$$
$$= 1.5 \text{ m} \times \frac{360°}{2\pi(500 \text{ m})} = 0.17°$$

Step 3 Explain: We have found that the angular separation of the two headlights is 0.17°. However, remember that it is more common to express fractions of a degree in arcminutes or arcseconds. Remembering that there are 60 arcminutes in 1°, our answer of 0.17° is equivalent to 0.17° × 60 arcmin/1° = 10.2 arcminutes. In other words, the angular separation of the headlights is about 10 arcminutes, which is about $\frac{1}{3}$ of the 30 arcminute (0.5°) angular diameter of the full moon.

• Why do the constellations we see depend on latitude and time of year?

If you stay in one place, the basic patterns of motion in the sky will stay the same from one night to the next. However, you'll see different constellations at different times of year. For example, the constellation Orion is prominent in the February evening sky but visible only shortly before dawn in September. In addition, you'll see a different set of constellations from those you see at home if you travel far to the north or south. For example, you'll never see the constellation Centaurus (see Figure 1.9) from the northern United States, but it will be circumpolar in your sky if you travel to southern Chile. Let's explore why the constellations depend on location and time of year.

Variation with Latitude   To understand why the visible constellations vary with northward or southward travel, we must first review how we locate points on Earth (Figure 2.11a). **Latitude** measures north-south position on Earth and is defined to be 0° at the equator and increases northward and southward. The North Pole and South Pole have latitude 90°N and 90°S, respectively. Note that "lines of latitude" are imaginary *circles* running parallel to the equator. **Longitude** measures east-west position, so "lines of longitude" are semicircles (half-circles) extending from the North Pole to the South Pole. By international treaty, the line of longitude passing through Greenwich, England, is defined to be longitude 0°. This line is called the **prime meridian** (Figure 2.11b). Stating a latitude and a longitude pinpoints a location on Earth. For example, Rome lies at about 42°N latitude and 12°E longitude and Miami lies at about 26°N latitude and 80°W longitude.

You can see why visible constellations vary with latitude by looking again at Figure 2.10. If we move the person

standing on Earth to a different latitude, it shifts the position of her horizon and zenith on the celestial sphere. For example, Figure 2.12 shows how the local sky appears relative to the celestial sphere for the latitudes of the North Pole (90°N) and Sydney, Australia (34°S).

Note that although the sky varies with latitude, it does *not* vary with longitude. For example, Charleston (South Carolina) and San Diego are at about the same latitude, so people in both cities see the same set of constellations at night. (However, because they are in different time zones,

Figure 2.11 We can locate any place on Earth's surface by its latitude and longitude.

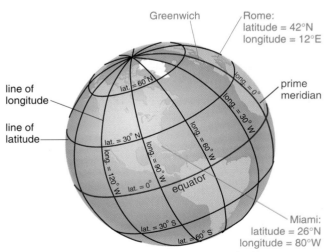

a Latitude measures angular distance north or south of the equator. Longitude measures angular distance east or west of the prime meridian, which passes through Greenwich, England.

b The entrance to the Old Royal Greenwich Observatory, near London. The line emerging from the door marks the prime meridian.

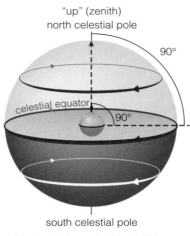

Figure 2.12 Interactive Figure ⟋ The sky varies with latitude. Notice that the altitude of the celestial pole that is visible in your sky is always equal to your latitude.

a The local sky at the North Pole (latitude 90°N).

b The local sky at latitude 34°S.

night comes to Charleston about 3 hours before it comes to San Diego.)

If you study diagrams like those in Figures 2.10 and 2.12 carefully, you can discover many interesting facts about the sky. For example, notice that all stars visible at the North Pole are circumpolar, and you can see only the stars that lie on the northern half of the celestial sphere. These facts explain why the Sun remains above the horizon for 6 months at the North Pole: The Sun lies on the northern half of the celestial sphere for half of each year (see the yellow dots representing the Sun in Figure 2.4), so during these 6 months it must circle the North polar sky daily just like a circumpolar star.

The diagrams also show a fact that is very important to navigation:

The altitude of the celestial pole in your sky is equal to your latitude.

For example, if the north celestial pole appears in your sky at an altitude of 40° above your north horizon, your latitude is 40°N. Similarly, if the south celestial pole appears in your sky at an altitude of 34° above your south horizon, your latitude is 34°S. Thus, you can determine your latitude just by finding the celestial pole in your sky. Finding the north celestial pole is fairly easy, because it lies very close to the star Polaris, also known as the North Star (Figure 2.13a).

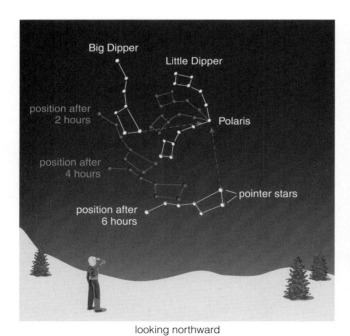

looking northward

a In the Northern Hemisphere, the pointer stars of the Big Dipper point to the North Star, Polaris, which lies within 1° of the north celestial pole. Note that the sky appears to turn counterclockwise around the north celestial pole.

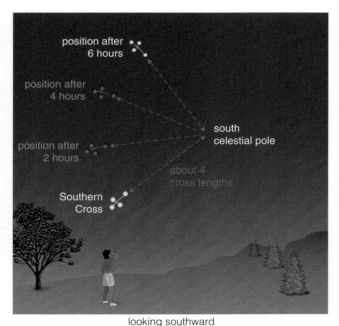

looking southward

b In the Southern Hemisphere, the Southern Cross points to the south celestial pole, which is not marked by any bright star. The sky appears to turn clockwise around the south celestial pole.

Figure 2.13 Interactive Figure ⟋ You can determine your latitude by measuring the altitude of the celestial pole in your sky.

In the Southern Hemisphere, you can find the south celestial pole with the aid of the Southern Cross (Figure 2.13b). We'll discuss celestial navigation and how the sky varies with latitude in more detail in Chapter S1.

THINK ABOUT IT

What is your latitude? Based on your latitude, where is the north (or south) celestial pole in your sky? Where should you look to see circumpolar stars? What portion of the celestial sphere is never visible in your sky?

Variation with Time of Year   The night sky changes throughout the year because of Earth's changing position in its orbit around the Sun. Figure 2.14 shows how this works. As we orbit the Sun over the course of a year, the Sun *appears* to move against the background of the distant stars in the constellations. We don't see the Sun and the stars at the same time, but if we could we'd notice the Sun gradually moving eastward along the ecliptic, completing one circuit each year. The constellations along the ecliptic are called the constellations of the **zodiac**. (Tradition places 12 constellations along the zodiac, but the official borders include a wide swath of a thirteenth constellation, Ophiuchus.)

The Sun's apparent location along the ecliptic determines which constellations we see at night. For example, Figure 2.14 shows that the Sun appears to be in Leo in late August. We therefore cannot see Leo in late August, because it moves with the Sun through the daytime sky. However,

COMMON MISCONCEPTIONS

What Makes the North Star Special?

Most people are aware that the North Star, Polaris, is a special star. Contrary to a relatively common belief, however, it is *not* the brightest star in the sky. More than 50 other stars are either considerably brighter or comparable in brightness. Polaris is special because it is so close to the north celestial pole. This position makes it very useful in navigation, because it closely marks the direction of due north and because its altitude in your sky is nearly equal to your latitude.

we can see Aquarius all night long, since it is opposite Leo on the celestial sphere. Six months later, in February, we see Leo at night while Aquarius is above the horizon only in the daytime.

THINK ABOUT IT

Based on Figure 2.14 and today's date, in what constellation does the Sun currently appear? What constellation of the zodiac will be on your meridian at midnight? What constellation of the zodiac will you see in the west shortly after sunset? Go outside at night to confirm your answers.

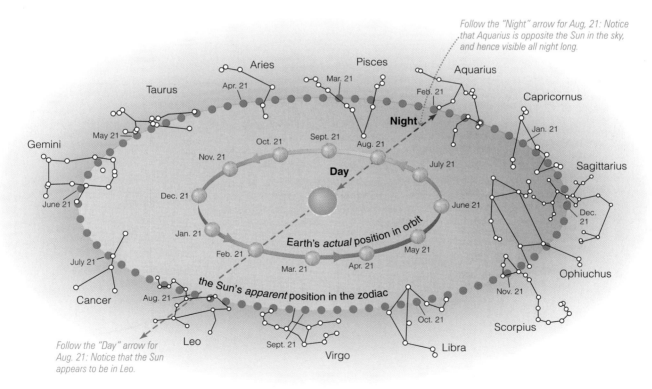

Figure 2.14   Interactive Figure   This diagram shows why the Sun appears to move steadily eastward along the ecliptic, through the constellations of the zodiac. As Earth orbits the Sun, we see the Sun against the background of different zodiac constellations at different times of year. For example, on August 21 the Sun appears to be in Leo, because it is between us and the much more distant stars that make up Leo.

2.2 The Reason for Seasons

We have seen how Earth's rotation makes the sky appear to circle us daily and how the night sky changes as Earth orbits the Sun each year. The combination of Earth's rotation and its orbit also leads to the progression of the seasons. In this section, we'll explore the reason for seasons.

• What causes the seasons?

You know that we have seasonal changes, such as longer and warmer days in summer and shorter and cooler days in winter. But why do the seasons occur? The answer is that the tilt of Earth's axis causes sunlight to fall differently on Earth at different times of year. Figure 2.15 shows the idea.

Notice that Earth's axis remains pointed in the same direction in space (toward Polaris) throughout the year [Section 1.3]. However, because Earth orbits the Sun, the orientation of the axis *relative to the Sun* changes over the course of a year. For example, look at the left side of Figure 2.15. The Northern Hemisphere is tipped toward the Sun, mak-

ing it summer there, while the Southern Hemisphere is tipped away from the Sun, making it winter there. Half an orbit later (the right side of Figure 2.15), the situation is reversed. The Northern Hemisphere is in winter because it is tipped away from the Sun, while the Southern Hemisphere is in summer because it is tipped toward the Sun. Thus, the two hemispheres experience opposite seasons.

We can also see why the days are longer and warmer in summer and shorter and cooler in winter. Sunlight strikes the summer hemisphere at a steeper angle than it strikes the winter hemisphere. The steeper angle means sunlight is more concentrated, which is why summer tends to be warmer than winter. The steeper angle also means that the Sun follows a longer and higher path through the summer sky, which is why the days are long and midday shadows are short. You can see the annual change in how high the Sun rises in your sky by observing the Sun's position at the same time each day (Figure 2.16).

Notice that the seasons on Earth are caused only by the axis tilt and *not* by any change in Earth's distance from the Sun. Although Earth's orbital distance varies over the course of each year, the variation is fairly small: Earth is only about 3% farther from the Sun at its farthest point than at

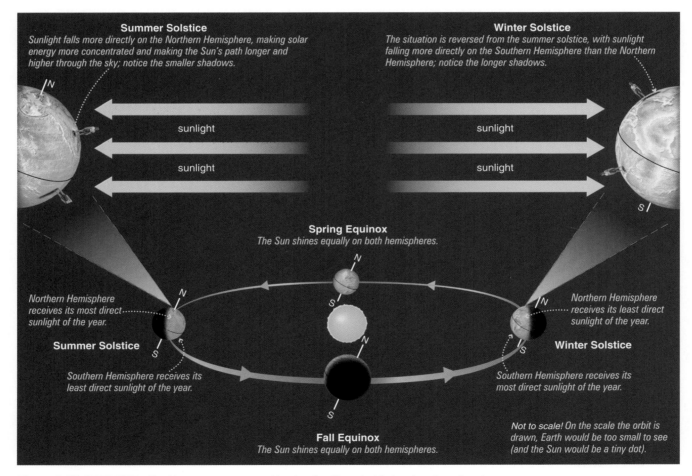

Summer Solstice
Sunlight falls more directly on the Northern Hemisphere, making solar energy more concentrated and making the Sun's path longer and higher through the sky; notice the smaller shadows.

Winter Solstice
The situation is reversed from the summer solstice, with sunlight falling more directly on the Southern Hemisphere than the Northern Hemisphere; notice the longer shadows.

sunlight

sunlight

sunlight

sunlight

Spring Equinox
The Sun shines equally on both hemispheres.

Northern Hemisphere receives its most direct sunlight of the year.

Northern Hemisphere receives its least direct sunlight of the year.

Summer Solstice

Southern Hemisphere receives its least direct sunlight of the year.

Winter Solstice

Southern Hemisphere receives its most direct sunlight of the year.

Not to scale! On the scale the orbit is drawn, Earth would be too small to see (and the Sun would be a tiny dot).

Fall Equinox
The Sun shines equally on both hemispheres.

Figure 2.15 Interactive Figure Earth's seasons are caused by the tilt of its axis. Notice that the axis points in the same direction (toward Polaris) throughout the year, which means the Northern Hemisphere is tipped toward the Sun on one side of the orbit and away from the Sun on the other side of the orbit. The same is true for the Southern Hemisphere, but on opposite sides of the orbit.

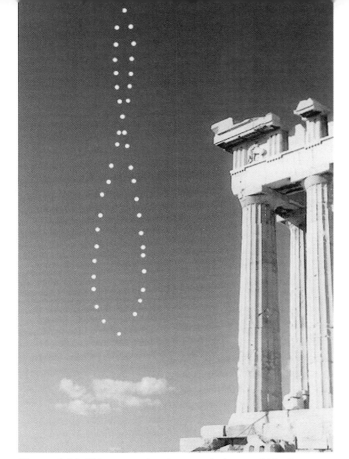

Figure 2.16 This composite photograph shows midday images of the Sun at 7- to 11-day intervals over the course of a year, always from the same spot (the Parthenon in Athens, Greece) and at the same time of day (technically, at the same "mean solar time"). Notice the dramatic change in the Sun's midday altitude over the course of the year. We'll discuss the reasons for the "figure 8" (called an *analemma*) in Chapter S1.

its nearest. The difference in the strength of sunlight due to this small change in distance is easily overwhelmed by the effects caused by the $23\frac{1}{2}°$ axis tilt. (*Note:* Some people also misinterpret diagrams like Figure 2.15, getting the mistaken impression that orbital position can make one hemisphere closer to the Sun. The mistake happens because figures like this are not drawn to scale. If the figure were drawn to scale, Earth would be microscopic, showing that there can be no practical difference in the distances of the two hemispheres from the Sun.)

Jupiter has an axis tilt of about 3°, small enough to be insignificant. Saturn has an axis tilt of about 27°, or slightly greater than that of Earth. Both planets have nearly circular orbits around the Sun. Do you expect Jupiter to have seasons? Do you expect Saturn to have seasons? Explain.

• How do we mark the progression of the seasons?

Today, we use a calendar to mark the progression of the seasons. In ancient times, however, people tracked the seasons

The Cause of Seasons

When asked what causes the seasons, many people mistakenly answer that the seasons are caused by variations in Earth's distance from the Sun. By recognizing that the Northern and Southern Hemispheres experience opposite seasons, you'll see that Earth's varying distance from the Sun *cannot* be the cause of the seasons. If that were the case, then both hemispheres would have summer at the same time. The real cause of the seasons is Earth's axis tilt, which causes the Northern and Southern Hemispheres to alternately receive more or less direct sunlight.

by observing the Sun's changing position in the sky. You can see these changes for yourself. For example, in summer the Sun traces a long, high arc across the sky, while in winter its path is short and low. Let's explore how the Sun's changing path allows us to mark the changing of the seasons.

Solstices and Equinoxes To help us mark the changing seasons, we define four special moments in the year, each of which corresponds to one of the four special positions in Earth's orbit shown in Figure 2.15.

- The **summer solstice**, which occurs around June 21 each year, is the moment when the Northern Hemisphere receives its most direct sunlight (and the Southern Hemisphere gets its least direct sunlight).
- The **winter solstice**, which occurs around December 21, is the moment when the Northern Hemisphere receives its least direct sunlight.
- The **spring equinox** (or *vernal equinox*) occurs around March 21, at the moment when the Northern Hemisphere goes from being tipped slightly away from the Sun to being tipped slightly toward the Sun.
- The **fall equinox** (or *autumnal equinox*) occurs around September 22 and marks the opposite change, when the Northern Hemisphere first starts to be tipped away from the Sun.

The exact dates and times of the solstices and equinoxes vary slightly from year to year, but stay within a couple of days of the dates given here. In fact, our modern calendar was designed specifically to ensure that the solstices and equinoxes occur around the same dates each year. As we'll discuss in Chapter S1, the calendar keeps this correspondence between dates and seasons by including a leap year in almost but not quite every four-year period. The pattern, set in the late 1500s, makes the average length of the calendar year almost precisely equal to the time between one spring equinox and the next.

Identifying the Solstices and Equinoxes Ancient people recognized the days on which the solstices and equinoxes occurred by observing the Sun in the sky. Many ancient structures were used for this purpose, including Stonehenge in England and the Sun Dagger in New Mexico [Section 3.1].

The equinoxes occur on the only two days of the year on which the Sun rises precisely due east and sets precisely due west (Figure 2.17). These are also the only two days when sunlight falls equally on both hemispheres. The summer solstice occurs on the day that the Sun follows its longest and highest path through the Northern Hemisphere sky (and its shortest and lowest path through the Southern Hemisphere sky). It is therefore the day that the Sun rises and sets farther to the north than on any other day of the year, and on which the noon Sun reaches its highest point in the Northern Hemisphere sky. The opposite is true on the day of the winter solstice, when the Sun rises and sets farthest to the south and the noon Sun is lower in the Northern Hemisphere sky than on any other day of the year.

First Days of Seasons   We usually say that each equinox and solstice marks the first day of a season. For example, the day of the summer solstice is usually said to be the "first day of summer." Notice, however, that the summer solstice occurs when the Northern Hemisphere has its *maximum* tilt toward the Sun. Thus, you might wonder why we consider the summer solstice to be the beginning rather than the midpoint of summer.

In part, the choice of the summer solstice as the "first" day of summer is somewhat arbitrary. However, the choice makes sense in at least two ways. First, it was much easier for ancient people to identify the days on which the Sun reached extreme positions in the sky—such as when it reached its highest point on the summer solstice—than other days in between. Second, we usually think of the seasons in terms of weather, and the solstices and equinoxes

correspond quite well with the beginnings of seasonal weather patterns. For example, although the Sun's path through the Northern Hemisphere sky is longest and highest around the time of the summer solstice, the warmest days tend to come one to two months later. To understand why, think about what happens when you heat a pot of cold soup. Even though you may have the stove turned on high from the start, it takes a while for the soup to warm up. In the same way, it takes some time for sunlight to heat the ground and oceans from the cold of winter to the warmth of summer. Thus, "midsummer" in terms of weather comes in late July and early August, which makes the summer solstice a pretty good choice for the "first day of summer." For similar reasons, the winter solstice is a good choice for the first day of winter, and the spring and fall equinoxes are good choices for the first days of those seasons.

Seasons Around the World   The names of the solstices and equinoxes reflect the northern seasons, which can make things sound strange when we talk about seasons in the Southern Hemisphere. For example, on the *summer* solstice it is *winter* in the Southern Hemisphere. This apparent injustice to people in the Southern Hemisphere arose because the solstices and equinoxes were named long ago by people living in the Northern Hemisphere. A similar injustice is inflicted on people living in equatorial regions. If you study Figure 2.14 carefully, you'll see that Earth's equator gets its most direct sunlight on the two equinoxes and its least direct sunlight on the solstices. Thus, people living near the equator don't experience four seasons in the same way as people living at mid-latitudes. Instead, equatorial regions generally have rainy and dry seasons, with the rainy seasons coming when the Sun is higher in the sky.

In addition, seasonal variations around the times of the solstices are more extreme at high latitudes. For example, Alaska has much longer summer days and much longer winter nights than Florida. In fact, the Sun becomes circumpolar at very high latitudes (within the *Arctic* and *Antarctic Circles*) in the summer. The Sun never sets during these summer days in what we call the *land of the midnight Sun*

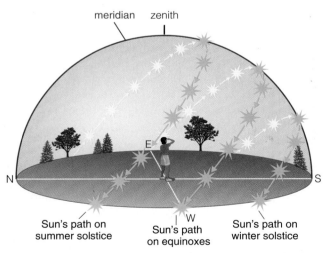

Figure 2.17 This diagram shows the Sun's path on the solstices and equinoxes for the Northern Hemisphere sky (latitude 40°N). Notice that the Sun rises exactly due east and sets exactly due west only on the equinoxes. The summer solstice occurs on the day that the Sun rises and sets farthest to the north and reaches its highest point in the sky. The winter solstice occurs on the day that the Sun rises and sets farthest to the south and traces its lowest path through the sky.

High Noon

When is the Sun directly overhead in your sky? Many people answer "at noon." It's true that the Sun reaches its *highest* point each day when it crosses the meridian, giving us the term "high noon" (though the meridian crossing is rarely at precisely 12:00; see Section S1.2). However, unless you live in the Tropics (between latitudes 23.5°S and 23.5°N), the Sun is *never* directly overhead. In fact, any time you can see the Sun as you walk around, you can be sure it is *not* at your zenith. Unless you are lying down, seeing an object at the zenith requires tilting your head back into a very uncomfortable position.

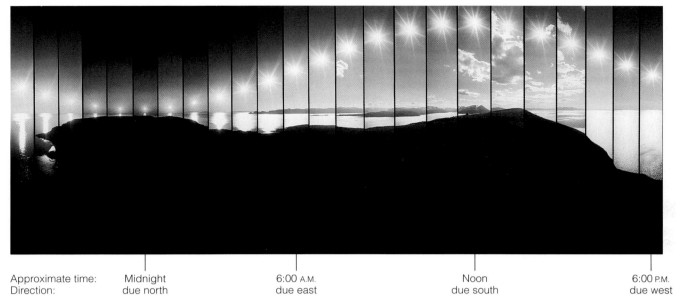

| Approximate time: | Midnight | 6:00 A.M. | Noon | 6:00 P.M. |
| Direction: | due north | due east | due south | due west |

Figure 2.18 This sequence of photos shows the progression of the Sun all the way around the horizon on the summer solstice at the Arctic Circle. Notice that the Sun does not set but instead skims the northern horizon at midnight. It then gradually rises higher, reaching its highest point at noon, when it appears due south.

(Figure 2.18). Of course, the name "land of noon darkness" would be more appropriate in the winter, when the Sun never rises above the horizon at these high latitudes.

Why Orbital Distance Doesn't Affect Our Seasons As we've discussed, Earth's distance from the Sun varies slightly over the course of each year, but not enough to be a factor in causing the seasons. Still, we might expect the varying orbital distance to make the seasons slightly more extreme in one or the other hemisphere. For example, Earth is closest to the Sun (perihelion) in early January and farthest from the Sun (aphelion) in early July, which means that the Northern Hemisphere receives slightly more intense winter sunlight and less intense summer sunlight than does the Southern Hemisphere. This "extra" winter sunlight should tend to make the Northern Hemisphere winter less cold, while the reduced summer sunlight should tend to make the Northern Hemisphere summer less hot. Thus, we would expect the Northern Hemisphere to have slightly more moderate seasons than the Southern Hemisphere. However, weather records show that the opposite is true: Northern Hemisphere seasons are slightly more extreme than those of the Southern Hemisphere.

The main reason for this surprising fact becomes obvious when you look at a map of Earth (Figure 2.19). Most of Earth's land lies in the Northern Hemisphere, with far more ocean in the Southern Hemisphere. As you can see at any beach, lake, or pool, water takes longer to heat or cool than soil or rock (largely because sunlight heats bodies of water to a depth of many meters while heating only the very top layer of land). The water temperature therefore remains fairly steady both day and night, while the ground can heat up and cool down dramatically. The Southern Hemisphere's larger amount of ocean moderates its climate.

The Northern Hemisphere, with more land and less ocean, heats up and cools down more easily, explaining why it has the more extreme seasons.

A secondary reason for the Northern Hemisphere's more extreme seasons is that Earth moves slightly faster in its orbit when it is closer to the Sun and slightly slower when it is farther from the Sun (a fact embodied in Kepler's second law [Section 3.3]). Thus, Earth moves slightly slower during the Northern Hemisphere summer (when Earth is farther from the Sun), which makes the northern summer last about 2–3 days longer than the southern summer. A couple of extra days of more direct sunlight means more time to heat up.

In summary, Earth's variation in distance from the Sun makes virtually no difference at all to the seasons, because other effects are far more important. However, the same is not necessarily true for other planets, especially if they

Figure 2.19 A map of Earth shows that most land lies in the Northern Hemisphere while most ocean lies in the Southern Hemisphere. The climate-moderating effects of water therefore make Southern Hemisphere seasons less extreme than Northern Hemisphere seasons.

have significantly greater distance variations. For example, Mars varies in distance from the Sun much more than does Earth. As a result, Mars has significantly more extreme seasons in one hemisphere than the other, even though it has nearly the same axis tilt as Earth [Section 10.4].

• How does the orientation of Earth's axis change with time?

We have now discussed both daily and seasonal changes in the sky. These changes are all you are likely to notice in your daily life. However, if you could observe the sky over many centuries, you would begin to notice other changes in its appearance. One of the most important of these slow changes is called **precession**, which gradually changes the orientation of Earth's axis in space.

Precession occurs with many rotating objects. You can see it easily by spinning a top (Figure 2.20a). As the top spins rapidly, you'll notice that its axis sweeps out a circle at a somewhat slower rate. We say that the top's axis *precesses*. Earth's axis precesses in much the same way, but far more slowly (Figure 2.20b). Each cycle of Earth's precession takes about 26,000 years. Today, the axis points toward Polaris, making it our North Star. However, because of precession, Polaris has been a good North Star for only a few centuries, and a few centuries from now it will no longer be a good marker for due north. Some 13,000 years from now, the axis will point toward the star Vega (within a few degrees), making Vega the North Star at that time. In 26,000 years, the axis will again point the way it does now. At most other times, the axis will not point near any bright star.

Notice that while precession affects the orientation of Earth's axis, it does not affect the *amount* of the axis tilt and therefore does not affect the pattern of the seasons.

However, because the solstices and equinoxes correspond to points in Earth's orbit that depend on the direction the axis points in space, their positions in the orbit gradually shift with the cycle of precession. As a result, the constellations associated with the solstices and equinoxes gradually change. For example, a couple thousand years ago the Sun appeared in the constellation Cancer on the day of the summer solstice, but it now appears in Gemini on that day. That explains something you can see on any world map: The latitude at which the Sun is directly overhead on the summer solstice ($23\frac{1}{2}°$ N) is called the *Tropic of Cancer*, telling us that it got its name back when the Sun used to appear in Cancer on the summer solstice.

THINK ABOUT IT

What constellation will the Sun be in on the summer solstice about 2,000 years from now? (*Hint:* Figure 2.14 shows the names and order of the zodiac constellations.)

Why does precession occur? It is caused by gravity's effect on a tilted, rotating object that is *not* a perfect sphere. You have probably seen how gravity affects a top. If you try to balance a nonspinning top on its point, it will fall over almost immediately. This happens because a top that is not spherical will inevitably be leaning a little to one side. No matter how slight this lean, gravity will quickly tip the nonspinning top over. However, if you spin the top rapidly, it does not fall over so easily. The spinning top stays upright because rotating objects tend to keep spinning around the same rotation axis (a consequence of something known as the *law of conservation of angular momentum* [Section 4.3]). This tendency prevents gravity from immediately pulling the spinning top over, since falling over would mean a change

Figure 2.20 Interactive Figure ↘ Precession affects the orientation of a spinning object's axis but not the amount of its tilt.

The top spins rapidly around its axis . . .

precession of axis

rotation

rotation

. . . while its axis more slowly sweeps out a circle of precession.

a A spinning top wobbles, or precesses, more slowly than it spins.

Polaris

Vega

Earth's axis today

precession of axis

Earth's axis in A.D. 15,000

Earth rotates around its axis every 24 hours . . .

rotation

rotation

. . . while its axis sweeps out a circle of precession every 26,000 years.

Earth's orbit

b Earth's axis also precesses. Each precession cycle takes about 26,000 years.

Sun Signs

You probably know your astrological "Sun sign." When astrology began a few thousand years ago, your Sun sign was supposed to represent the constellation in which the Sun appeared on your birth date. However, because of precession, this is no longer the case for most people. For example, if your birthday is the spring equinox, March 21, a newspaper horoscope will show that your Sun sign is Aries but the Sun actually appears in Pisces on that date. The astrological Sun signs are based on the positions of the Sun among the stars described by the Greek scientist Ptolemy in his book *Tetrabiblos* [Section 3.5], which was written in about A.D. 150. Because almost 2,000 years have passed since that time, Earth's axis has moved about $\frac{1}{13}$ of the way through its 26,000-year precession cycle since the Sun signs got their names and dates. As a result, the astrological Sun signs are off by nearly a month from the actual positions of the Sun among the constellations today.

in the spin axis from near-vertical to horizontal. Instead, gravity succeeds only in making the axis trace circles of precession. As friction slows the top's spin, the circles of precession get wider and wider, and ultimately the top falls over. If there were no friction to slow its spin, the top would spin and precess forever.

The rotating Earth precesses because of gravitational tugs from the Sun and Moon. Earth is not quite a perfect sphere, instead bulging slightly at its equator. Because the equator is tilted $23\frac{1}{2}°$ to the ecliptic plane, the gravitational attractions of the Sun and Moon try to pull the equatorial bulge into the ecliptic plane, effectively trying to "straighten out" Earth's axis tilt. However, like the spinning top, Earth tends to keep rotating around the same axis. Gravity therefore does not succeed in changing Earth's axis tilt and instead only makes the axis precess. If you want to gain a better understanding of precession and how it works, you might wish to purchase a simple toy *gyroscope*. Gyroscopes are essentially rotating wheels mounted in a way that allows them to move freely, which makes it easy to see how their spin rate affects their motion. (The fact that gyroscopes tend to keep the same rotation axis makes them very useful in aircraft and spacecraft navigation.)

Phases of the Moon Tutorial, Lessons 1–3

2.3 The Moon, Our Constant Companion

Aside from the Sun, the Moon is the brightest and most noticeable object in our sky. But the Moon behaves in some ways that may at first seem rather strange. Sometimes it is visible in the daytime and sometimes only at night. Its appearance also changes, so that sometimes we see less of it and sometimes we see more of it. Occasionally, the Moon's position is just right to cause a lunar or solar eclipse. The Moon behaves in these special ways because it is our constant companion in space, orbiting Earth as we orbit the Sun. In this section, we'll discuss the how and why of the Moon's appearance in our sky.

Before we get into the details, it's helpful to know a few basic facts about the Moon. The Moon orbits Earth about once every $27\frac{1}{3}$ days. Its average distance from Earth is about 380,000 kilometers (240,000 miles), but the precise distance varies because the orbit is not quite a perfect circle. Figure 2.21 shows the Moon's orbit on the same scale we used for the model solar system in Section 1.2. Remember that on this scale, the Sun is about the size of a large grapefruit and is located about 15 meters from Earth. Thus, the entire orbit of the Moon would fit easily inside the Sun, and for practical purposes we can consider the Earth and Moon to share the same orbit around the Sun.

Like all objects in space, the Moon appears to reside on the celestial sphere. Earth's daily rotation makes the Moon appear to rise in the east and set in the west each day. In addition, because of its orbit around Earth, the Moon appears to move eastward from night to night through the constellations of the zodiac. Each circuit through the constellations takes the same $27\frac{1}{3}$ days that the Moon takes to orbit Earth. If you do the math, you'll see that this means the Moon moves relative to the stars by about $\frac{1}{2}°$—its own

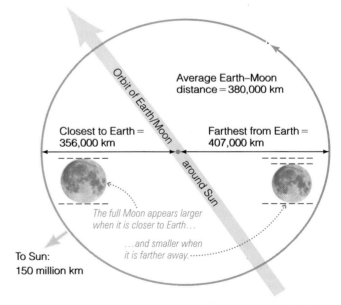

Figure 2.21 The Moon's orbit around Earth, shown on the 1-to-10-billion scale used in Section 1.2 (see Figure 1.6). The segment shown of our orbit around the Sun looks nearly straight because the distance to the Sun is so great in comparison to the size of the Moon's orbit. The inset photos contrast the relative angular size of the full moon in our sky when the Moon is at the near and far points of its orbit; of course, full moon occurs only when the Moon is opposite the Sun as seen from Earth.

angular size—each hour. You can notice this gradual motion in just a few hours by checking the Moon's position in comparison to bright stars near it in the sky.

• Why do we see phases of the Moon?

As the Moon moves through the sky, both its appearance and the times at which it rises and sets change with the cycle of **lunar phases**. The phase of the Moon on any given day depends on its position relative to the Sun as it orbits Earth.

The easiest way to understand the lunar phases is with the simple demonstration illustrated in Figure 2.22. Take a ball outside on a sunny day. (If it's dark or cloudy, you can use a flashlight instead of the Sun; put the flashlight on a table a few meters away and shine it toward you.) Hold the ball at arm's length to represent the Moon while your head represents Earth. Slowly spin around (counterclockwise), so that the ball goes around you just like the Moon orbits Earth. As you turn, you'll see the ball go through phases just like the Moon. If you think about what's happening, you'll realize that the phases of the ball result from just two basic facts:

1. Half the ball always faces the Sun (or flashlight) and therefore is bright, while the other half faces away from the Sun and therefore is dark.
2. As you look at the ball at different positions in its "orbit" around your head, you see different combinations of its bright and dark faces.

For example, when you hold the ball directly opposite the Sun, you see only the bright portion of the ball, which represents the "full" phase. When you hold the ball at its "first-quarter" position, half the face you see is dark and the other half is bright.

We see lunar phases for the same reason. Half the Moon is always illuminated by the Sun, but the amount of this illuminated half that we see from Earth depends on the Moon's position in its orbit. The photographs in Figure 2.22 show how the phases look. Each complete cycle of phases, from one new moon to the next, takes about $29\frac{1}{2}$ days—hence the origin of the word *month* (think "moonth"). This is slightly longer than the Moon's actual orbital period because of Earth's motion around the Sun during the time the Moon is orbiting around Earth [see Figure S1.3].

The different phases not only look different but also rise, reach their highest points, and set at different times. For example, because full moon occurs when the Moon is

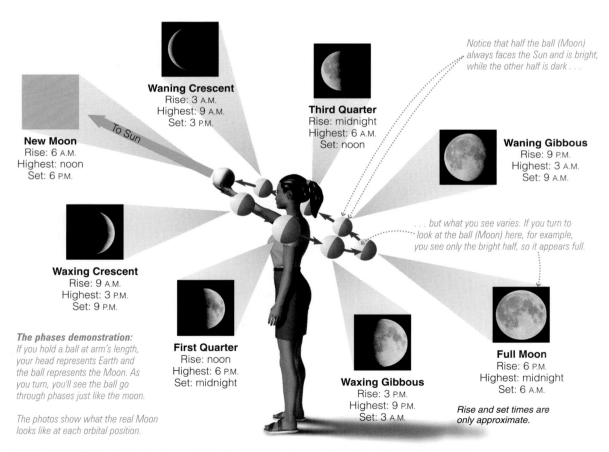

Figure 2.22 Interactive Figure ▸ A simple demonstration illustrates the phases of the Moon. The half of the ball (Moon) facing the Sun is always illuminated while the half facing away is always dark, but you see the ball go through phases as it orbits around your head (Earth). (The new moon photo shows blue sky, because a new moon is always close to the Sun in the sky and hence hidden from view by the bright light of the Sun.)

opposite the Sun in the sky, the full moon must rise around sunset, reach its highest point in the sky at midnight, and set around sunrise. Similarly, because first-quarter moon occurs when the Moon is about 90° east of the Sun in our sky, it must rise around noon, reach its highest point around 6 P.M., and set around midnight. Figure 2.22 lists the approximate rise, highest point, and set times for each phase. (The exact times vary with latitude, time of year, and other factors.)

THINK ABOUT IT

Suppose you go outside in the morning and notice that the visible face of the Moon is half-light and half-dark. Is this a first-quarter or third-quarter moon? How do you know?

As you study the phases in Figure 2.22, it's worth noting at least three interesting things about their names. First, notice that the phases from new to full are said to be *waxing*, which means "increasing"; the phases from full to new are said to be *waning*, which means "decreasing." Second, notice that there's no phase called a "half moon." Instead, what most people think of as a "half moon" is either a first-quarter or third-quarter phase. These phases get their names because the cycle of phases is usually taken to begin with new moon. Thus, a first-quarter moon occurs when the Moon is one-quarter of the way through its cycle, and a third-quarter moon occurs at the three-quarters point of the cycle. Third, while nearly everyone is familiar with the term *crescent* for the waxing and waning crescent phases, few people know the name *gibbous* (pronounced with a hard *g* as in "gift") for the waxing and waning gibbous phases. Notice that a gibbous moon is essentially the opposite of a crescent moon—a crescent moon has a small sliver of light while a gibbous moon has a small sliver of dark. The term *gibbous* literally means "hump-backed," so you can probably see how the gibbous moon gets its name. We see the two gibbous phases just as often as we see the crescent phases, so it's worth knowing their names.

The Moon's Synchronous Rotation   Although we see many *phases* of the Moon, we do not see many *faces*. In fact, from Earth we always see the same face of the Moon.* This fact tells us that the Moon must rotate once on its axis in the same time that it makes a single orbit of Earth, a trait that we call **synchronous rotation** (because the periods of rotation and orbit are the same, or synchronized).

To see why the Moon must rotate synchronously to keep the same face toward Earth, try the following demonstration. Place a ball on a table to represent Earth while you represent the Moon. Start by facing the ball. If you do not rotate as you walk around the ball, you'll be looking away from it by the time you are halfway around your orbit (Figure 2.23a). The only way you can face the ball at all times is by completing exactly one rotation while you complete one orbit (Figure 2.23b). (We'll learn in Chapter 5 *why* the Moon's periods of rotation and orbit are the same.)

The View from the Moon   A good way to solidify your understanding of the lunar phases is to imagine that you live on the side of the Moon that faces Earth. For example, ask yourself what you would see as you looked at Earth at a time when people on Earth saw a new moon. By remembering that a new moon occurs when the Moon is between the Sun and Earth, you'll realize that from the Moon you'd be looking at Earth's daytime side and hence would see a *full earth*. Similarly, at full moon you would be facing the night side of Earth and would see a *new earth*. In general, you'd always see Earth in a phase opposite the phase of the Moon being seen by people on Earth at the same time. Moreover, because the Moon always shows nearly the same face to Earth, Earth would appear to hang nearly stationary in your sky as it went through its cycle of phases.

THINK ABOUT IT

How long would each day and night last if you lived on the Moon? Explain.

---

*Because the Moon's orbital speed varies while its rotation rate is steady, the visible face appears to wobble slightly back and forth as it orbits Earth. This effect, called *libration*, allows us to see a total of about 59% of the Moon's surface over the course of a month, even though we see only 50% of the Moon at any single time.

a If you do not rotate while walking around the model, you will not always face it.

b You will face the model at all times only if you rotate exactly once during each orbit.

Figure 2.23 The fact that we always see the same face of the Moon means that the Moon must rotate once in the same amount of time that it takes to orbit Earth once. You can see why by walking around a model of Earth while imagining that you are the Moon.

Moon in the Daytime and Stars on the Moon

In traditions and stories, night is so closely associated with the Moon that many people mistakenly believe that the Moon is visible only in the nighttime sky. In fact, the Moon is above the horizon as often in the daytime as at night, though it is easily visible only when its light is not drowned out by sunlight. For example, a first-quarter moon is easy to spot in the late afternoon as it rises through the eastern sky, and a third-quarter moon is visible in the morning as it heads toward the western horizon.

 Another misconception appears in illustrations that show a star in the dark portion of the crescent moon. The star in the dark portion appears to be in front of the Moon, which is impossible because the Moon is much closer to us than is any star.

The "Dark Side" of the Moon

The term *dark side of the Moon* really should be used to mean the night side—that is, the side facing away from the Sun. Unfortunately, *dark side* traditionally meant what would better be called the *far side*—the face that never can be seen from Earth. Many people still refer to the far side as the "dark side," even though this side is not necessarily dark. For example, during new moon the far side faces the Sun and hence is completely sunlit (see Figure 2.22). The only time the far side is completely dark is at full moon, when it faces away from both the Sun and Earth.

Thinking about the view from the Moon clarifies another interesting feature of the lunar phases: The dark portion of the lunar face is not *totally* dark. Imagine that you are standing on the Moon when it is in a crescent phase. Because it's nearly new moon as seen from Earth, you would see nearly full earth in your sky. Just as we can see at night by the light of the Moon, the light of Earth would illuminate your night moonscape. (In fact, because Earth is much larger than the Moon, the full earth is much bigger and brighter in the lunar sky than the full moon is in Earth's sky.) This faint light illuminating the "dark" portion of the Moon's face is often called the *ashen light* or *earthshine*. This light enables us to see the outline of the full face of the Moon even when the Moon is not full.

 Eclipses Tutorial, Lessons 1–3

• What causes eclipses?

The phases of the Moon are the most obvious effect of the Moon's orbit around Earth. Occasionally, however, we see something much more dramatic. The Moon and Earth cast shadows in sunlight, and these shadows can create **eclipses** when the Sun, Earth, and Moon fall into a straight line. There are two basic types of eclipse:

● A **lunar eclipse** occurs when Earth lies directly between the Sun and Moon, so that Earth's shadow falls on the Moon.
● A **solar eclipse** occurs when the Moon lies directly between the Sun and Earth, so that the Moon's shadow falls on the Earth. People living within the area covered

by the Moon's shadow will see the Sun blocked or partially blocked from view.

To understand when and why eclipses occur, we must investigate the Moon's orbit of Earth a little more deeply.

Conditions for Eclipses Look once more at Figure 2.22. The figure makes it look like the Sun, Earth, and Moon line up with every new and full moon. Thus, if this figure told the whole story of the Moon's orbit, we would have both a lunar and a solar eclipse every month—but we don't.

The missing piece of the story in Figure 2.22 is that the Moon's orbit is slightly inclined (by about 5°) to the ecliptic plane (the plane of Earth's orbit around the Sun). To visualize this inclination, imagine the ecliptic plane as the surface of a pond, as shown in Figure 2.24. Because of the inclination of its orbit, the Moon spends most of its time either above or below this surface. It crosses *through* this surface only twice during each orbit: once coming out and once going back in. The two points in each orbit at which the Moon crosses the surface are called the **nodes** of the Moon's orbit.

As you study Figure 2.24, notice that the nodes are aligned approximately the same way (diagonally on the page) throughout the year. As a result, the nodes lie in a straight line with the Sun and Earth only about twice each year. (For reasons we'll discuss shortly, it is not *exactly* twice each year.) Thus, we find the following two conditions that must be met for an eclipse to occur:

1. The phase of the Moon must be full (for a lunar eclipse) or new (for a solar eclipse), since those are the only phases at which the Sun, Earth, and Moon can lie in a straight line.
2. The new or full moon must occur during one of the periods when the nodes of the Moon's orbit are aligned with the Sun and Earth, since those are the only times when the new and full moons lie in the ecliptic plane.

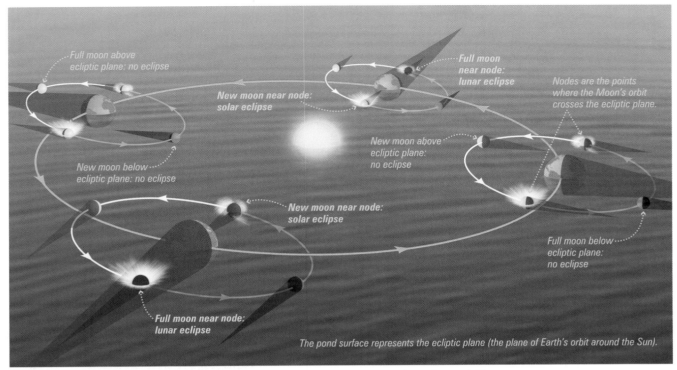

Figure 2.24 Interactive Figure. This illustration represents the ecliptic plane as the surface of a pond. The Moon's orbit is tilted by about 5° to the ecliptic plane, so the Moon spends half of each orbit above the plane (the pond surface) and half below it. Eclipses occur only when the Moon is at both a node (passing through the pond surface) *and* a phase of either new moon (for a solar eclipse) or full moon (for a lunar eclipse)—as is the case with the lower left and top right orbits shown. (Not to scale.)

Although there are two basic types of eclipse—lunar and solar—each of these types can look different depending on precisely how the shadows fall. The shadow of the Moon or Earth consists of two distinct regions: a central **umbra**, where sunlight is completely blocked, and a surrounding **penumbra**, where sunlight is only partially blocked (Figure 2.25). Thus, an umbral shadow is totally dark, while a penumbral shadow is only slightly darker than no shadow. Let's see how this affects eclipses.

Lunar Eclipses A lunar eclipse begins at the moment when the Moon's orbit first carries it into Earth's penumbra. After that, we will see one of three types of lunar eclipse (Figure 2.26). If the Sun, Earth, and Moon are nearly perfectly aligned, the Moon will pass through Earth's umbra, and we will see a **total lunar eclipse**. If the alignment is somewhat less perfect, only part of the full moon will pass through the umbra (with the rest in the penumbra), and we will see a **partial lunar eclipse**. If the Moon passes through Earth's penumbra only, we will see a **penumbral lunar eclipse**.

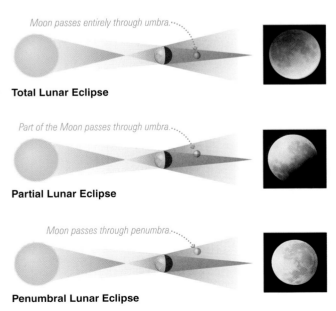

Moon passes entirely through umbra.
Total Lunar Eclipse

Part of the Moon passes through umbra.
Partial Lunar Eclipse

Moon passes through penumbra.
Penumbral Lunar Eclipse

Figure 2.26 Interactive Figure. The three types of lunar eclipse.

penumbra
umbra

Figure 2.25 The shadow cast by an object in sunlight. Sunlight is fully blocked in the umbra and partially blocked in the penumbra.

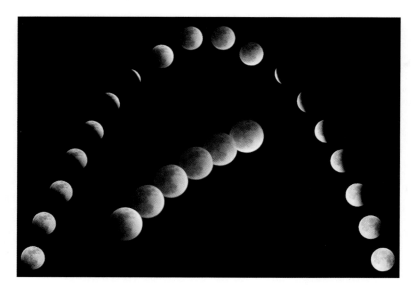

Figure 2.27 This sequence of photos, taken at 5-minute intervals, shows the progression of a total lunar eclipse; the central photos show more detail during totality. Notice Earth's curved shadow advancing across the Moon during the partial phases, and the redness of the full moon during totality. The arc-shaped arrangement was chosen for effect and has no inherent meaning.

Penumbral eclipses are the most common type of lunar eclipse, but they are the least visually impressive because the full moon darkens only slightly. Earth's umbral shadow clearly darkens part of the Moon's face during a partial lunar eclipse, and the curvature of this shadow demonstrates that Earth is round. A total lunar eclipse is particularly spectacular because the Moon becomes dark and eerily red during **totality**—the time during which the Moon is entirely engulfed in the umbra (Figure 2.27). Totality typically lasts about an hour. The Moon becomes dark because it is in shadow, and red because Earth's atmosphere scatters some of the red light from the Sun enough so that it is directed toward the Moon.

Solar Eclipses We can also see three types of solar eclipse (Figure 2.28). If a solar eclipse occurs when the Moon is relatively close to Earth in its orbit, the Moon's umbra touches a small area of Earth's surface (no more than about 270 kilometers in diameter). Anyone within this area will see a **total solar eclipse**. Surrounding the region of totality is a much larger area (typically about 7,000 kilometers in diameter) that falls within the Moon's penumbral shadow.

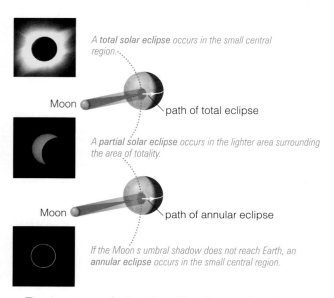

A **total solar eclipse** occurs in the small central region.

Moon — path of total eclipse

A **partial solar eclipse** occurs in the lighter area surrounding the area of totality.

Moon — path of annular eclipse

If the Moon's umbral shadow does not reach Earth, an **annular eclipse** occurs in the small central region.

a The three types of solar eclipse. The diagrams show the Moon's shadow falling on Earth; note the dark central umbra surrounded by the much lighter penumbra.

b This photo from Earth orbit shows the Moon's shadow (umbra) on Earth during a total solar eclipse. Notice that only a small region of Earth experiences totality at any one time.

Figure 2.28 Interactive Figure During a solar eclipse, the Moon's small shadow moves rapidly across the face of the Earth.

Figure 2.29 This multiple-exposure photograph shows the progression of a total solar eclipse. Totality (central image) lasts only a few minutes, during which time we can see the faint corona around the outline of the Sun. This photo was taken July 22, 1990, in La Paz, Mexico. The foreground church was photographed at a different time of day.

Anyone within this region will see a **partial solar eclipse**, in which only part of the Sun is blocked from view.

If the eclipse occurs when the Moon is relatively far from Earth in its orbit, the Moon's slightly smaller angular size means its umbral shadow may not quite reach Earth. In that case, anyone in the small region of Earth directly behind the umbra will see an **annular eclipse**, in which a ring of sunlight surrounds the disk of the Moon. Again, anyone in the surrounding penumbral shadow will see a partial solar eclipse.

The combination of Earth's rotation and the orbital motion of the Moon causes the Moon's umbral and penumbral shadows to race across the face of Earth at a typical speed of about 1,700 kilometers per hour. As a result, the umbral shadow traces a narrow path across Earth, and totality never lasts more than a few minutes in any particular place.

A total solar eclipse is a spectacular sight. It begins when the disk of the Moon first appears to touch the Sun.

Over the next couple of hours, the Moon appears to take a larger and larger "bite" out of the Sun. As totality approaches, the sky darkens and temperatures fall. Birds head back to their nests, and crickets begin their nighttime chirping. During the few minutes of totality, the Moon completely blocks the normally visible disk of the Sun, allowing the faint *corona* to be seen (Figure 2.29). The surrounding sky takes on a twilight glow, and planets and bright stars become visible in the daytime. As totality ends, the Sun slowly emerges from behind the Moon over the next couple of hours. However, because your eyes have adapted to the darkness, totality appears to end far more abruptly than it began.

Predicting Eclipses Few phenomena have so inspired and humbled humans throughout the ages as eclipses. For many cultures, eclipses were mystical events associated with fate or the gods, and countless stories and legends surround

---

SPECIAL TOPIC Does the Moon Influence Human Behavior?

From myths of werewolves to stories of romance under the full moon, human culture is filled with claims that our behavior is influenced by the phase of the Moon. Can we say anything scientific about such claims?

The Moon clearly has important influences on Earth. For example, the Moon is primarily responsible for the tides [Section 4.5]. However, the Moon's tidal force cannot directly affect objects as small as people.

If a physical force from the Moon cannot affect human behavior, could we be influenced in other ways? Certainly, anyone who lives near the oceans is influenced by the rising and falling of the tides. For example, fishermen and boaters must follow the tides. Thus, although the Moon does not influence their behavior directly, it does so indirectly through its effect on the oceans.

Physiological patterns in many species appear to follow the lunar phases; for example, some crabs and turtles lay eggs only at full moon. No human trait is so closely linked to lunar phases, but the average human menstrual cycle is so close in length to a lunar month that it is difficult to believe the similarity is mere coincidence. Nevertheless, aside from the physiological cycles and the influence of tides on people who live near the oceans, claims that the lunar phase affects human behavior are difficult to verify scientifically. For example, although it is possible that the full moon brings out certain behaviors, it may also simply be that some behaviors are easier to exhibit when the sky is bright. A beautiful full moon may bring out your desire to walk on the beach under the moonlight, but there is no scientific evidence to suggest that the full moon would affect you the same way if you were confined to a deep cave.

them. One legend holds that the Greek philosopher Thales (c. 624–546 B.C.) successfully predicted the year (but presumably not the precise time) that a total eclipse of the Sun would be visible in the area where he lived, which is now part of Turkey. Coincidentally, the eclipse occurred as two opposing armies (the Medes and the Lydians) were massing for battle. The eclipse so frightened the armies that they put down their weapons, signed a treaty, and returned home. Because modern research shows that the only eclipse visible in that part of the world at about that time occurred on May 28, 585 B.C., we know the precise date on which the treaty was signed—the earliest historical event that can be dated precisely.

Much of the mystery of eclipses probably stems from the relative difficulty of predicting them. Look again at Figure 2.24. The two periods each year when the nodes of the Moon's orbit are nearly aligned with the Sun are called eclipse seasons. Each eclipse season lasts a few weeks, so some type of lunar eclipse occurs during each eclipse season's full moon, and some type of solar eclipse occurs during its new moon.

If Figure 2.24 told the whole story, eclipse seasons would occur every 6 months and predicting eclipses would be easy. For example, if eclipse seasons always occurred in January and July, eclipses would always occur on the dates of new and full moons in those months. But the figure does not show one important thing about the Moon's orbit: The nodes slowly move around the orbit. As a result, eclipse seasons occur slightly less than 6 months apart (about 173 days apart) and do not recur in the same months year after year.

The combination of the changing dates of eclipse seasons and the $29\frac{1}{2}$-day cycle of lunar phases makes eclipses recur in a cycle of about 18 years $11\frac{1}{3}$ days. This cycle is called the saros cycle. Astronomers in many ancient cultures identified the saros cycle and thus could predict *when* eclipses would occur. However, the saros cycle does not account for all the complications involved in predicting eclipses. If a solar eclipse occurred today, the one that would occur 18 years $11\frac{1}{3}$ days from now would not be visible from the same places on Earth and might not be of the same type. For example, one might be total and the other only partial. No ancient culture achieved the ability to predict eclipses in every detail.

Today, we can predict eclipses because we know the precise details of the orbits of Earth and the Moon. Many astronomical software packages, including the *SkyGazer* package that comes with this book, do the necessary calculations automatically. Table 2.1 lists upcoming lunar eclipses of all types; notice that, as we expect, eclipses generally come a little less than 6 months apart. Figure 2.30 shows paths of totality for total solar eclipses (but not for partial or annular eclipses) over the next couple of decades, using color coding to show eclipses that repeat with the saros cycle.

THINK ABOUT IT

In Table 2.1, notice that there's one exception to the "rule" of eclipses coming a little less than 6 months apart: the 2009 lunar eclipses of July 7 and August 6. How can eclipses occur just a month apart like this? Should you be surprised that both these lunar eclipses are penumbral? Explain.

2.4 The Ancient Mystery of the Planets

We've now covered the appearance and motion of the stars, Sun, and Moon in the sky. That leaves us with the planets yet to discuss. As we'll soon see, planetary motion posed an ancient mystery that played a critical role in the development of modern civilization.

Table 2.1 Lunar Eclipses 2006–2010*

Date	Type	Where You Can See It	Date	Type	Where You Can See It
Mar. 14, 2006	penumbral	Americas, Europe, Africa, Asia	Feb. 9, 2009	penumbral	eastern Europe, Asia, Australia, Pacific, western North America
Sep. 7, 2006	partial	Europe, Africa, Asia, Australia	Jul. 7, 2009	penumbral	Australia, Pacific, Americas
Mar. 3, 2007	total	Americas, Europe, Africa, Asia	Aug. 6, 2009	penumbral	Americas, Europe, Africa, western Asia
Aug. 28, 2007	total	eastern Asia, Australia, Pacific, Americas			
Feb. 21, 2008	total	central Pacific, Americas, Europe, Africa	Dec. 31, 2009	partial	Europe, Africa, Asia, Australia
			Jun. 26, 2010	partial	eastern Asia, Australia, Pacific, western Americas
Aug. 16, 2008	partial	South America, Europe, Africa, Asia, Australia	Dec. 21, 2010	total	eastern Asia, Australia, Pacific, Americas, Europe

*Dates are based on Universal Time and hence are those in Greenwich, England, at the time of the eclipse; to see an eclipse, check a news or Web source for the exact local time and date. Data from NASA's Eclipse Home Page maintained by Fred Espanak.

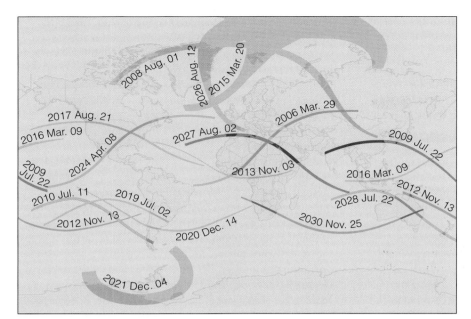

Five planets are easy to find with the naked eye: Mercury, Venus, Mars, Jupiter, and Saturn. Mercury is visible only infrequently, and then only just after sunset or just before sunrise because it is so close to the Sun. Venus often shines brightly in the early evening in the west or before dawn in the east. If you see a very bright "star" in the early evening or early morning, it is probably Venus. Jupiter, when it is visible at night, is the brightest object in the sky besides the Moon and Venus. Mars is often recognizable by its reddish color, though you should check a star chart to make sure you aren't looking at a bright red star. Saturn is also easy to see with the naked eye, but because many stars are just as bright as Saturn, it helps to know where to look. (It also helps to know that planets tend not to twinkle as much as stars.) Sometimes several planets may appear close together in the sky, offering a particularly beautiful sight (Figure 2.31).

• What was once so mysterious about planetary motion in our sky?

Ancient people carefully observed the movements of the planets among the stars, and you can observe the same motions for yourself. On any particular night, the planets move through the sky just like stars, rising in the east and setting in the west. However, if you observe planetary positions among the stars over a period of weeks or months, you'll notice that the planets wander slowly through the constellations of the zodiac. (The word *planet* comes from the Greek for "wandering star.") By itself, this wandering might not be too surprising. After all, the Sun and Moon also move among the constellations. But the planets move in what seems to be a very strange way.

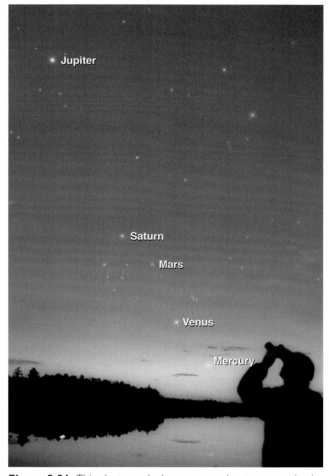

Figure 2.31 This photograph shows a rare planetary grouping in which all five planets that are easily visible to the naked eye appeared close together in the sky. It was taken near Chatsworth, New Jersey, just after sunset on April 23, 2002. The next such close grouping of these five planets in our sky will not occur until September 2040.

Unlike the Sun and Moon, which move steadily eastward relative to the stars, the planets vary substantially in both speed and brightness as they move among the stars. Moreover, while the planets *usually* move eastward relative to the stars, they occasionally reverse course completely, moving westward rather than eastward through the zodiac. A period during which a planet appears to move westward relative to the stars is called a period of **apparent retrograde motion** (*retrograde* means "backward"). These periods may last from a few weeks to a few months, depending on the planet (Figure 2.32).

Ancient astronomers could easily "explain" the daily paths of the stars through the sky by imagining that the celestial sphere was real and that it really rotated around Earth each day. But the apparent retrograde motion of the planets posed a far greater mystery: What could cause the planets sometimes to go backward? As we'll discuss in Chapter 3, the ancient Greeks came up with some very clever ways to explain the occasional backward motion of the planets, despite being wedded to the incorrect idea of an Earth-centered universe. However, the Greek explanation was quite complex, and ultimately wrong.

In contrast, apparent retrograde motion has a simple explanation in a Sun-centered solar system. You can demonstrate it for yourself with the help of a friend (Figure 2.33a). Pick a spot in an open field to represent the Sun. You can represent Earth, walking counterclockwise around the Sun, while your friend represents a more distant planet (such as Mars, Jupiter, or Saturn) by walking counterclockwise around the Sun at a greater distance. Your friend should walk more slowly than you, because more distant planets orbit the Sun more slowly. As you walk,

watch how your friend appears to move relative to buildings or trees in the distance. Although both of you always walk the same way around the Sun, your friend will appear to move backward against the background during the part of your "orbit" at which you catch up to and pass him or her. To understand the apparent retrograde motions of Mercury and Venus, which are closer to the Sun than is Earth, simply switch places with your friend and repeat the demonstration.

This demonstration applies directly to the planets. For example, because Mars takes about 2 years to orbit the Sun (actually 1.88 years), it covers about half its orbit during the 1 year in which Earth makes a complete orbit. If you trace lines of sight from Earth to Mars from different points in their orbits, you will see that the line of sight usually moves eastward relative to the stars but moves westward during the time when Earth is passing Mars in its orbit (Figure 2.33b). Like your friend in the demonstration, Mars never actually changes direction. It only *appears* to change direction from our perspective on Earth.

• Why did the ancient Greeks reject the real explanation for planetary motion?

If the apparent retrograde motion of the planets is so readily explained by recognizing that Earth orbits the Sun, why wasn't this idea accepted in ancient times? In fact, the idea that Earth goes around the Sun was suggested as early as 260 B.C. by the Greek astronomer Aristarchus. No one knows why Aristarchus proposed a Sun-centered solar system, but the fact that it explains planetary motion so naturally

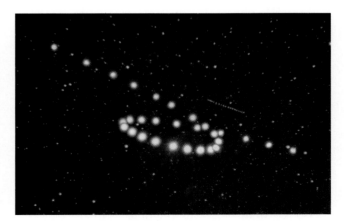

a This diagram shows Jupiter's approximate position among the stars in our sky from late 2006 to early 2008. Notice the backward path between May and September 2007.

Figure 2.32 Apparent retrograde motion.

b This composite photograph shows Mars between early June and late November 2003. The progression goes from right to left, with apparent retrograde motion in the middle. Notice that Mars is biggest and brightest in the middle of the retrograde loop, because that is where it is closest to Earth in its orbit. The 29 individual photos were taken at 5- to 8-day intervals. (The white dots in a line just right of center are the planet Uranus, which by coincidence was in the same part of the sky.)

Apparent retrograde motion occurs between positions 3 and 5, as the inner person (planet) passes the outer person (planet).

Follow the lines of sight from inner person (planet) to outer person (planet) to see where the outer one appears against the background.

a The retrograde motion demonstration: Watch how your friend (in red) usually appears to you (in blue) to move forward against the background of the building in the distance but appears to move backward as you catch up to and pass him or her in your "orbit."

b This diagram shows how the idea from the demonstration applies to a planet. Follow the lines of sight from Earth to Mars in numerical order. Notice that Mars appears to move westward relative to the distant stars as Earth passes it by in its orbit (roughly from points 3 to 5 in the diagram).

Figure 2.33 Interactive Figure⮌ Apparent retrograde motion—the occasional "backward" motion of the planets relative to the stars—has a simple explanation in a Sun-centered solar system.

probably played a role (see box, p. 52). Nevertheless, Aristarchus's contemporaries rejected his idea, and the Sun-centered solar system did not gain wide acceptance until almost 2,000 years later.

Although there were many reasons why the Greeks were reluctant to abandon the idea of an Earth-centered universe, one of the most important was their inability to detect something called **stellar parallax.** Extend your arm and hold up one finger. If you keep your finger still and alternately close your left eye and right eye, your finger will appear to jump back and forth against the background. This apparent shifting, called *parallax,* occurs because your two eyes view your finger from opposite sides of your nose. If you move your finger closer to your face, the parallax increases. If you look at a distant tree or flagpole instead of your finger, you may not notice any parallax at all. Thus, parallax depends on distance, with nearer objects exhibiting greater parallax than more distant objects.

If you now imagine that your two eyes represent Earth at opposite sides of its orbit around the Sun and that your finger represents a relatively nearby star, you have the idea of stellar parallax. That is, because we view the stars from different places in our orbit at different times of year, nearby stars should *appear* to shift back and forth against the background of more distant stars (Figure 2.34).

Because the Greeks believed that all stars lie on the same celestial sphere, they actually expected to see stellar parallax

in a slightly different way. If Earth orbited the Sun, they reasoned that at different times of year we would be closer to different parts of the celestial sphere and thus would notice changes in the angular separations of stars. However, no matter how hard they searched, they could find no sign

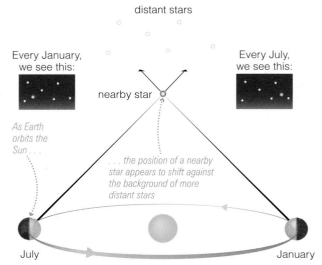

distant stars

Every January, we see this:

nearby star

Every July, we see this:

As Earth orbits the Sun . . .

. . . the position of a nearby star appears to shift against the background of more distant stars

July   January

Figure 2.34 Stellar parallax is an apparent shift in the position of a nearby star as we look at it from different places in Earth's orbit. This figure is greatly exaggerated; in reality, the amount of shift is far too small to detect with the naked eye.

You've probably heard of Copernicus, whose work in the mid-1500s started the revolution that ultimately overturned the ancient belief in an Earth-centered universe [Section 3.3]. However, Copernicus was *not* the first person to propose the idea that Earth goes around the Sun. By the time it was adopted by Copernicus, the Sun-centered idea had already been around for nearly 2,000 years.

As far as we know, the Greek scientist Aristarchus (c. 310–230 B.C.) was the first person to suggest that Earth goes around the Sun rather than vice versa. He proposed the idea in about 260 B.C. Little of Aristarchus's work survives to the present day, so we cannot know what motivated him to suggest an idea so contrary to the prevailing wisdom of an Earth-centered universe. However, it's quite likely that he was motivated by the fact that a Sun-centered system offers a much more natural explanation for the apparent retrograde motion of the planets. To account for the lack of detectable stellar parallax, Aristarchus suggested that the stars were extremely far away.

Aristarchus further strengthened his argument by estimating the sizes of the Moon and the Sun. By observing the shadow of Earth on the Moon during a lunar eclipse, he estimated the Moon's diameter to be about one-third of Earth's diameter—only slightly higher than the actual value. He then used a geometric argument, based on measuring the angle between the Moon and the Sun at first- and third-quarter phases, to conclude that the Sun must be

larger than Earth. (Aristarchus's measurements were imprecise, so he estimated the Sun's diameter to be about seven times Earth's rather than the correct value of about 100 times.) His conclusion that the Sun is larger than Earth may have been another reason why he felt that Earth should orbit the Sun, rather than vice versa.

Although he was probably the first to suggest that Earth orbits the Sun, Aristarchus's ideas were in large part built upon the work of earlier scholars. For example, Heracleides (c. 388–315 B.C.) had previously suggested that Earth rotates. Aristarchus may have drawn on this idea to explain the apparent daily rotation of the stars in our sky. Heracleides also suggested that not all heavenly bodies circle Earth: Based on the fact that Mercury and Venus always stay fairly close to the Sun in the sky, he argued that these two planets must orbit the Sun. Thus, in suggesting that *all* the planets orbit the Sun, Aristarchus was extending the ideas of Heracleides and others before him.

Aristarchus gained little support among his contemporaries, but his ideas never died. They were kept alive by scholars throughout history, and Copernicus was well aware of Aristarchus's claims when he proposed his own version of the Sun-centered system. Thus, our modern understanding of the universe owes at least some debt to the remarkable vision of a man who lived more than 2,200 years ago.

---

of stellar parallax. They concluded that one of the following must be true:

1. Earth orbits the Sun but the stars are so far away that stellar parallax is undetectable to the naked eye.
2. There is no stellar parallax because Earth remains stationary at the center of the universe.

Unfortunately, with notable exceptions such as Aristarchus, the Greeks rejected the correct answer (the first one) because they could not imagine that the stars could be *that* far away. Today, we can detect stellar parallax with the aid of telescopes, providing direct proof that Earth really does orbit the Sun. Careful measurements of stellar parallax also provide the most reliable means of measuring distances to nearby stars [Section 15.1].

THINK ABOUT IT

How far apart are opposite sides of Earth's orbit? How far away are the nearest stars? Describe the challenge of detecting stellar parallax. It may help to visualize Earth's orbit and the distance to the stars on the 1-to-10-billion scale used in Chapter 1.

The ancient mystery of the planets drove much of the historical debate over Earth's place in the universe. In many ways, the modern technological society we take for granted today can be traced directly to the scientific revolution that began in the quest to explain the strange wandering of the

planets among the stars in our sky. We will turn our attention to this revolution in the next chapter.

THE BIG PICTURE

Putting Chapter 2 into Context

In this chapter, we surveyed the phenomena of our sky. Keep the following "big picture" ideas in mind as you continue your study of astronomy:

● You can enhance your enjoyment of learning astronomy by observing the sky. The more you learn about the appearance and apparent motions of the sky, the more you will appreciate what you can see in the universe.

● From our vantage point on Earth, it is convenient to imagine that we are at the center of a great celestial sphere—even though we really are on a planet orbiting a star in a vast universe. We can then understand what we see in the local sky by thinking about how the celestial sphere appears from our latitude.

● Most of the phenomena of the sky are relatively easy to observe and understand. The more complex phenomena—particularly eclipses and apparent retrograde motion of the planets—challenged our ancestors for thousands of years. The desire to understand these phenomena helped drive the development of science and technology.

2.1 Patterns in the Night Sky

- **What does the universe look like from Earth?**
Stars and other celestial objects appear to lie on a great **celestial sphere** surrounding Earth. We divide the celestial sphere into **constellations** with well-defined borders. From any location on Earth, we see half the celestial sphere at any one time as the dome of our **local sky**, in which the **horizon** is the boundary between Earth and sky, the **zenith** is the point directly overhead, and the **meridian** runs from due south to due north through the zenith.

- **Why do stars rise and set?** Earth's rotation makes stars appear to circle around Earth each day. A star whose complete circle lies above our horizon is said to be **circumpolar**. Other stars have circles that cross the horizon, making them rise in the east and set in the west each day.

- **Why do the constellations we see depend on latitude and time of year?** The visible constellations vary with time of year because our night sky lies in different directions in space as we orbit the Sun. The constellations vary with **latitude** because your latitude determines the orientation of your horizon relative to the celestial sphere. The sky does not vary with **longitude.**

2.2 The Reason for Seasons

- **What causes the seasons?** The tilt of Earth's axis causes the seasons. The axis points in the same direction (toward Polaris) throughout the year. Thus, as Earth orbits the Sun, sunlight hits different parts of Earth more directly at different times of the year.

- **How do we mark the progression of the seasons?** The **summer** and **winter solstices** are the times during the year when the Northern Hemisphere gets its most and least direct sunlight, respectively. The **spring** and **fall equinoxes** are the two times when both hemispheres get equally direct sunlight.

- **How does the orientation of Earth's axis change with time?** Earth's 26,000-year cycle of **precession** changes the orientation of the axis in space, although the tilt remains about $23\frac{1}{2}°$. The changing orientation of the axis does not affect the pattern of seasons, but it changes the identity of any "north star" and shifts the locations of the solstices and equinoxes in Earth's orbit.

2.3 The Moon, Our Constant Companion

- **Why do we see phases of the Moon?** The **phase** of the Moon depends on its position relative to the Sun as it orbits Earth. The half of the Moon facing the Sun is always illuminated while the other half is dark, but from Earth we see varying combinations of the illuminated and dark faces.

- **What causes eclipses?** We see a **lunar eclipse** when Earth's shadow falls on the Moon, and a **solar eclipse** when the Moon blocks our view of the Sun. We do not see an eclipse at every new and full moon because the Moon's orbit is slightly inclined to the ecliptic plane. Eclipses come in different types, depending on where the dark **umbral** and lighter **penumbral** shadows fall.

2.4 The Ancient Mystery of the Planets

- **What was once so mysterious about planetary motion in our sky?** Planets generally drift eastward relative to the stars over the course of the year, but for weeks or months they reverse course in their periods of **apparent retrograde motion.** This motion occurs when Earth passes by (or is passed by) another planet in its orbit, but it posed a major mystery to ancient people who assumed Earth to be at the center of the universe.

- **Why did the ancient Greeks reject the real explanation for planetary motion?** The Greeks rejected the idea that Earth goes around the Sun in part because they could not detect **stellar parallax**— slight apparent shifts in stellar positions over the course of the year. To most Greeks, it seemed unlikely that the stars could be so away as to make parallax undetectable to the naked eye, even though that is, in fact, the case.

Review Questions

Short-Answer Questions Based on the Reading

1. What are *constellations?* How did they get their names?
2. Suppose you were making a model of the celestial sphere with a ball. Briefly describe all the things you would need to mark on your celestial sphere.
3. On a clear, dark night, the sky may appear to be "full" of stars. Does this appearance accurately reflect the way stars are distributed in space? Explain.
4. Why does the *local sky* look like a dome? Define *horizon, zenith,* and *meridian.* How do we describe the location of an object in the local sky?
5. Explain why we can measure only *angular sizes* and *angular distances* for objects in the sky. What are *arcminutes* and *arcseconds?*
6. What are *circumpolar stars?* Are more stars circumpolar at the North Pole or in the United States? Explain.
7. What are *latitude* and *longitude?* Does the sky vary with latitude? Does it vary with longitude? Explain.
8. What is the *zodiac,* and why do we see different parts of it at different times of year?
9. Suppose Earth's axis had no tilt. Would we still have seasons? Why or why not?
10. Briefly describe what is special about the summer and winter solstices and the spring and fall equinoxes.
11. What is *precession,* and how does it affect the sky that we see from Earth?
12. Briefly describe the Moon's cycle of *phases.* Can you ever see a full moon at noon? Explain.
13. What do we mean when we say that the Moon exhibits *synchronous rotation?* What does this tell us about the Moon's periods of rotation and orbit?
14. Why don't we see an *eclipse* at every new and full moon? Describe the conditions that must be met for us to see a *solar* or *lunar eclipse.*
15. What do we mean by the *apparent retrograde motion* of the planets? Why was it difficult for ancient astronomers to explain but easy for us to explain?
16. What is *stellar parallax?* Briefly describe the role it played in making ancient astronomers believe in an Earth-centered universe.

Test Your Understanding

Does It Make Sense?

Decide whether the statement makes sense (or is clearly true) or does not make sense (or is clearly false). Explain your reasoning. (For an example, see Chapter 1, "Does It Make Sense?")

17. The constellation Orion didn't exist when my grandfather was a child.
18. When I looked into the dark lanes of the Milky Way with my binoculars, I saw what must have been a cluster of distant galaxies.
19. Last night the Moon was so big that it stretched for a mile across the sky.
20. I live in the United States, and during my first trip to Argentina I saw many constellations that I'd never seen before.

21. Last night I saw Jupiter right in the middle of the Big Dipper. (*Hint:* Is the Big Dipper part of the zodiac?)
22. Last night I saw Mars move westward through the sky in its apparent retrograde motion.
23. Although all the known stars appear to rise in the east and set in the west, we might someday discover a star that will appear to rise in the west and set in the east.
24. If Earth's orbit were a perfect circle, we would not have seasons.
25. Because of precession, someday it will be summer everywhere on Earth at the same time.
26. This morning I saw the full moon setting at about the same time the Sun was rising.

Quick Quiz

Choose the best answer to each of the following. Explain your reasoning with one or more complete sentences.

27. Two stars that are in the same constellation: (a) must both be part of the same cluster of stars in space. (b) must both have been discovered at about the same time. (c) may actually be very far away from each other.
28. The North Celestial Pole is 35° above your northern horizon. This tells you that: (a) you are at latitude 35°N. (b) you are at longitude 35°E. (c) it is the winter solstice.
29. Beijing and Philadelphia have about the same latitude but very different longitudes. Therefore, tonight's night sky in these two places: (a) will look about the same. (b) will have completely different sets of constellations. (c) will have partially different sets of constellations.
30. In winter, Earth's axis points toward the star Polaris. In spring: (a) the axis also points toward Polaris. (b) the axis points toward Vega. (c) the axis points toward the Sun.
31. When it is summer in Australia, it is: (a) winter in the United States. (b) summer in the United States. (c) spring in the United States.
32. If the Sun rises precisely due east: (a) you must be located at Earth's equator. (b) it must be the day of either the spring or fall equinox. (c) it must be the day of the summer solstice.
33. A week after full moon, the Moon's phase is: (a) first quarter. (b) third quarter. (c) new.
34. The fact that we always see the same face of the Moon tell us that: (a) the Moon does not rotate. (b) the Moon's rotation period is the same as its orbital period. (c) the Moon looks the same on both sides.
35. If there is going to be a total lunar eclipse tonight, then you know that: (a) the Moon's phase is full. (b) the Moon's phase is new. (c) the Moon is unusually close to Earth.
36. When we see Saturn going through a period of apparent retrograde motion, it means: (a) Saturn is temporarily moving backward in its orbit of the Sun. (b) Earth is passing Saturn in its orbit, with both planets on the same side of the Sun. (c) Saturn and Earth must be on opposite sides of the Sun.

Investigate Further

In-Depth Questions to Increase Your Understanding

Short-Answer/Essay Questions

37. *New Planet.* Suppose we discover a planet in another solar system that has a circular orbit and an axis tilt of 35°. Would

you expect this planet to have seasons? If so, would you expect them to be more extreme than the seasons on Earth? If not, why not?

38. *Your View.*
    a. Find your latitude and longitude, and state the source of your information.
    b. Describe the altitude and direction in your sky at which the north or south celestial pole appears.
    c. Is Polaris a circumpolar star in your sky? Explain.

39. *View from the Moon.* Assume you live on the Moon near the center of the face that looks toward Earth.
    a. Suppose you see a full Earth in your sky. What phase of the Moon would people on Earth see? Explain.
    b. Suppose people on Earth see a full moon. What phase would you see for Earth? Explain.
    c. Suppose people on Earth see a waxing gibbous moon. What phase would you see for Earth? Explain.
    d. Suppose people on Earth are viewing a total lunar eclipse. What would you see from your home on the Moon? Explain.

40. *View from the Sun.* Suppose you lived on the Sun (and could ignore the heat). Would you still see the Moon go through phases as it orbits Earth? Why or why not?

41. *A Farther Moon.* Suppose the distance to the Moon were twice its actual value. Would it still be possible to have a total solar eclipse? Why or why not?

42. *A Smaller Earth.* Suppose Earth were smaller. Would solar eclipses be any different? If so, how? What about lunar eclipses? Explain.

43. *Observing Planetary Motion.* Find out what planets are currently visible in your evening sky. At least once a week, observe the planets and draw a diagram showing the position of each visible planet relative to stars in a zodiac constellation. From week to week, note how the planets are moving relative to the stars. Can you see any of the apparently wandering features of planetary motion? Explain.

44. *A Connecticut Yankee.* Find the book *A Connecticut Yankee in King Arthur's Court* by Mark Twain. Read the portion that deals with the Connecticut Yankee's prediction of an eclipse (or read the entire book). In a one- to two-page essay, summarize the episode and explain how it helped the Connecticut Yankee gain power.

Quantitative Problems

Be sure to show all calculations clearly and state your final answers in complete sentences.

45. *Arcminutes and Arcseconds.* There are 360° in a full circle.
    a. How many arcminutes are in a full circle?
    b. How many arcseconds are in a full circle?
    c. The Moon's angular size is about $\frac{1}{2}$°. What is this in arcminutes? In arcseconds?

46. *Latitude Distance.* Earth's radius is approximately 6,370 km.
    a. What is Earth's circumference?
    b. What distance is represented by each degree of latitude?
    c. What distance is represented by each arcminute of latitude?
    d. Can you give similar answers for the distances represented by a degree or arcminute of longitude? Why or why not?

47. *Angular Conversions I.* The following angles are given in degrees and fractions of degrees. Rewrite them in degrees, arcminutes, and arcseconds.
    a. 24.3° d. 0.01°
    b. 1.59° e. 0.001°
    c. 0.1°

48. *Angular Conversions II.* The following angles are given in degrees, arcminutes, and arcseconds. Rewrite them in degrees and fractions of degrees.
    a. 7°38′42″ d. 1′
    b. 12′54″ e. 1″
    c. 1°59′59″

49. *Moon Speed.* The Moon takes about $27\frac{1}{3}$ days to complete each orbit of Earth. About how fast is the Moon going as it orbits Earth? Give your answer in km/hr.

50. *Scale of the Moon.* The Moon's diameter is about 3,500 km and its average distance from Earth is about 380,000 km. How big and how far from Earth is the Moon on the 1-to-10-billion scale used in Chapter 1? Compare the size of the Moon's orbit to the size of the Sun on this scale.

51. *Angular Size of Your Finger.* Measure the width of your index finger and the length of your arm. Based on your measurements, calculate the angular width of your index finger at arm's length. Does your result agree with the approximations shown in Figure 2.7c? Explain.

52. *Find the Sun's Diameter.* The Sun has an angular diameter of about 0.5° and an average distance of about 150 million km. What is the Sun's approximate physical diameter? Compare your answer to the actual value of 1,390,000 km.

53. *Find a Star's Diameter.* The supergiant star Betelgeuse (in the constellation Orion) has a measured angular diameter of 0.044 arcsecond. Its distance has been measured to be 427 light-years. What is the actual diameter of Betelgeuse? Compare your answer to the size of our Sun and the Earth-Sun distance.

54. *Eclipse Conditions.* The Moon's precise equatorial diameter is 3,476 km, and its orbital distance from Earth varies between 356,400 km and 406,700 km. The Sun's diameter is 1,390,000 km and its distance from Earth ranges between 147.5 and 152.6 million km.
    a. Find the Moon's angular size at its minimum and maximum distances from Earth.
    b. Find the Sun's angular size at its minimum and maximum distances from Earth.
    c. Based on your answers to (a) and (b), is it possible to have a total solar eclipse when the Moon and Sun are both at their maximum distances? Explain.

Discussion Questions

55. *Earth-Centered Language.* Many common phrases reflect the ancient Earth-centered view of our universe. For example, the phrase "the Sun rises each day" implies that the Sun is really moving over Earth. We know that the Sun only *appears* to rise as the rotation of Earth carries us to a place where we can see the Sun in our sky. Identify other common phrases that imply an Earth-centered viewpoint.

56. *Flat Earth Society.* Believe it or not, there is an organization called the Flat Earth Society. Its members hold that Earth is flat and that all indications to the contrary (such as pictures of Earth from space) are fabrications made as part of a conspiracy to hide the truth from the public. Discuss the evidence for a round Earth and how you can check it for yourself. In light of the evidence, is it possible that the Flat Earth Society is correct? Defend your opinion.

Interactive Tutorials

Tutorial Review of Key Concepts

Use the interactive **Tutorials** at **www.mastering astronomy.com** to review key concepts from this chapter.

Seasons Tutorial

Lesson 1 Factors Affecting Seasonal Changes
Lesson 2 The Solstices and Equinoxes
Lesson 3 The Sun's Position in the Sky

Phases of the Moon Tutorial

Lesson 1 The Causes of Lunar Phases
Lesson 2 Time of Day and Horizons
Lesson 3 When the Moon Rises and Sets

Eclipses Tutorial

Lesson 1 Why and When Do Eclipses Occur?
Lesson 2 Types of Solar Eclipses
Lesson 3 Lunar Eclipses

Supplementary Tutorial Exercises

Use the interactive **Tutorial Lessons** to explore the following questions.

Seasons Tutorial, Lesson 1

1. What factors affect a planet's surface temperature?
2. Why does axis tilt have a greater effect than orbital distance on Earth's seasons?
3. Why is the equator always hot and why are the poles always cold despite the passage of the seasons?

Phases of the Moon Tutorial, Lesson 3

1. When does a new moon rise and set?
2. In which phase does the Moon set just after the Sun?
3. What factors influence our view of the Moon from Earth?

Eclipses Tutorial, Lesson 1

1. What happens during a solar eclipse?
2. What happens during a lunar eclipse?
3. What role does the Moon's orbit play in the appearance of a solar eclipse?

 Exploring the Sky and Solar System

Of the many activities available on the *Voyager: SkyGazer* **CD-ROM** accompanying your book, use the following files to observe key phenomena covered in this chapter.

Go to the **File: Basics** folder for the following demonstrations.

1. Wide Field Milky Way
2. Eclipse 1991–1992 Views
3. Winter Sky

Go to the **File: Demo** folder for the following demonstrations.

1. Russian Midnight Sun
2. Earth Orbiting the Moon
3. Mars in Retrograde

Go to the **Explore** menu for the following demonstrations.

1. Shadows on Earth
2. Phases of the Planets

Web Projects

1. Sky Information. Search the Web for sources of daily information about sky phenomena (such as lunar phases, times of sunrise and sunset, or dates of equinoxes and solstices). Identify and briefly describe your favorite source.
2. Constellations. Search the Web for information about the constellations and their mythology. Write a short report about one or more constellations.
3. Upcoming Eclipse. Find information about an upcoming solar or lunar eclipse that you might have a chance to witness. Write a short report about how you could best witness the eclipse, including any necessary travel to a viewing site, and what you can expect to see. Bonus: Describe how you could photograph the eclipse.

The Science of Astronomy

We especially need imagination in science. It is not all mathematics, nor all logic, but is somewhat beauty and poetry.

Maria Mitchell (1818–1889), astronomer and the first woman elected to the American Academy of Arts and Sciences

Today we know that Earth is a planet orbiting a rather ordinary star, in a galaxy of more than a hundred billion stars, in an incredibly vast universe. We know that Earth, along with the entire cosmos, is in constant motion. We know that, on the scale of cosmic time, human civilization has existed for only the briefest moment. How did we manage to learn these things?

It wasn't easy. Astronomy is the oldest of the sciences, with roots extending as far back as recorded history allows us to see. But the most impressive advances in knowledge have occurred in just the past few centuries.

In this chapter, we will trace how modern astronomy grew from its roots in ancient observations, including those of the Greeks. We'll pay special attention to the unfolding of the Copernican revolution, which overturned the ancient belief in an Earth-centered universe and laid the foundation for nearly all of modern science. Finally, we'll explore the nature of modern science and the scientific method.

3.1 The Ancient Roots of Science

A common stereotype holds that scientists walk around in white lab coats and somehow think differently than other people. In reality, scientific thinking is a fundamental part of human nature. In this section, we will trace the roots of science to experiences common to nearly all people and nearly all cultures.

• In what ways do all humans employ scientific thinking?

Scientific thinking comes naturally to us. By about a year of age, a baby notices that objects fall to the ground when she drops them. She lets go of a ball—it falls. She pushes a plate of food from her high chair—it falls too. She continues to drop all kinds of objects, and they all plummet to Earth. Through powers of observation, the baby learns about the physical world, finding that things fall when they are unsupported. Eventually, she becomes so certain of this fact that, to her parents' delight, she no longer needs to test it continually.

One day somebody gives the baby a helium balloon. She releases it, and to her surprise it rises to the ceiling! Her understanding of nature must be revised. She now knows that the principle "all things fall" does not represent the whole truth, although it still serves her quite well in most

situations. It will be years before she learns enough about the atmosphere, the force of gravity, and the concept of density to understand *why* the balloon rises when most other objects fall. For now, she is delighted to observe something new and unexpected.

The baby's experiences with falling objects and rising balloons exemplify scientific thinking. In essence, it is a way of learning about nature through careful observation and trial-and-error experiments. Rather than thinking differently than other people, modern scientists simply are trained to organize everyday thinking in a way that makes it easier for them to share their discoveries and employ their collective wisdom. This type of clear and organized thinking is at the heart of science.

THINK ABOUT IT

Describe a few cases where you have learned by trial and error while cooking, participating in sports, fixing something, on the job, or in any other situation.

Just as learning to communicate through language, art, or music is a gradual process for a child, the development of science has been a gradual process for humanity. Science in its modern form requires painstaking attention to detail, relentless testing of each piece of information to ensure its reliability, and a willingness to give up old beliefs that are not consistent with observed facts about the physical world. For professional scientists, these demands are the "hard work" part of the job. At heart, professional scientists are like the baby with the balloon, delighted by the unexpected and motivated by those rare moments when they—and all of us—learn something new about the universe.

• How did astronomical observations benefit ancient societies?

We will discuss modern science shortly, but first we will explore how it arose from the observations of ancient peoples. Our exploration begins in central Africa, where people of many indigenous societies learned to predict the weather with reasonable accuracy by making careful observations of the Moon. Remember that the Moon begins its monthly cycle as a crescent in the western sky just after sunset. Through long traditions of sky watching, central African societies discovered that the orientation of the crescent "horns" relative to the horizon is closely tied to local rainfall patterns (Figure 3.1).

No one knows when central Africans first developed the ability to predict weather using the lunar crescent, but we do know that people began paying attention to the sky long ago. The earliest known written astronomical record comes from central Africa near the border between modern-day Congo and Uganda. It consists of an animal bone (known as the *Ishango* bone) etched with patterns that appear to be part of a lunar calendar, probably carved around 6500 B.C.

Figure 3.1 The roots of science lie in careful observation of the world around us. This diagram shows how central Africans used the orientation of the waxing crescent moon to predict the rainfall. The graph depicts the annual rainfall pattern in central Nigeria, and the Moon diagrams show the varying angle of the "horns" of a waxing crescent moon relative to the western horizon. The technique works because the orientation depends on the relative positions of the Sun and Moon along the ecliptic, which varies with the time of year, and because the equatorial climate features a rainy season and a dry season rather than the four seasons we experience at temperate latitudes. (Adapted from *Ancient Astronomers* by Anthony F. Aveni.)

Why did ancient people make such careful and detailed observations of the sky? In part, it was probably to satisfy their inherent curiosity. In the daytime, they surely recognized the importance of the Sun to their lives. At night, without electric light, they were much more aware of the starry sky than we are today. Thus, it's not surprising that they paid attention to patterns of motion in the sky and developed ideas and stories to explain what they saw.

Astronomy also played a practical role in ancient times by enabling people to keep track of time and seasons, a crucial skill for anyone who depends on agriculture for survival. This ability may seem quaint today, when digital watches tell us the precise time and date, but it required considerable knowledge and skill in ancient times, when the only clocks and calendars were in the sky.

Modern measures of time come directly from ancient observations of motion in the sky. The length of our day is the time it takes the Sun to make one full circuit of the sky. The length of a month comes from the Moon's cycle of phases [Section 2.3], and our year is based on the cycle of the seasons [Section 2.2]. The seven days of the week were named after the seven naked-eye objects that appear to move among the constellations: the Sun, the Moon, and the five planets* recognized in ancient times (Table 3.1).

• What did ancient civilizations achieve in astronomy?

Nearly all ancient civilizations practiced astronomy at some level. Many built remarkable structures for observing

*The word *planet*, which means "wanderer," originally referred to the Sun and the Moon as well as the five visible planets. Earth was not considered a planet in ancient times, since it was assumed to be stationary at the center of the universe.

the sky. Let's explore a few of the many ways that ancient societies studied the sky.

Determining the Time of Day In the daytime, ancient peoples could tell time by observing the Sun's path through the sky. Many cultures probably used the shadows cast by sticks as simple sundials [Section S1.2]. The ancient Egyptians built huge obelisks, often inscribed or decorated in homage to the Sun, that probably also served as clocks (Figure 3.2).

At night, ancient people could estimate the time from the position and phase of the Moon by using the ideas we discussed in Chapter 2 (see Figure 2.22). For example, recall that a first-quarter moon sets around midnight, so it is not yet midnight if the first-quarter moon is still above your western horizon. The positions of the stars can also

Table 3.1 The Seven Days of the Week and the Astronomical Objects They Honor

In English, the correspondence between celestial objects and days is clear only if we look at the names of the objects used by the Teutonic tribes that lived in the region of modern-day Germany. French and Spanish are included because they make some of the correspondences easier to see.

Object	Teutonic Name	English	French	Spanish
Sun	Sun	Sunday	dimanche	domingo
Moon	Moon	Monday	lundi	lunes
Mars	Tiw	Tuesday	mardi	martes
Mercury	Woden	Wednesday	mercredi	miércoles
Jupiter	Thor	Thursday	jeudi	jueves
Venus	Fria	Friday	vendredi	viernes
Saturn	Saturn	Saturday	samedi	sábado

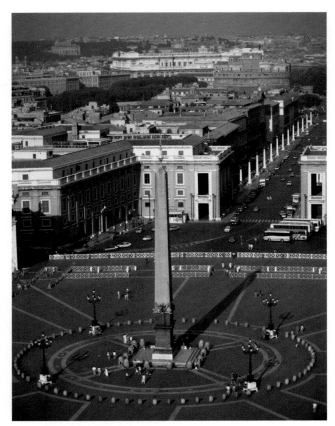

Figure 3.2 This ancient Egyptian obelisk, which stands 83 feet tall and weighs 331 tons, resides in St. Peter's Square at the Vatican in Rome. It is one of 21 surviving obelisks from ancient Egypt, most of which are now scattered around the world. Shadows cast by the obelisks may have been used to tell time.

indicate the time, as long as you know the approximate date. For example, in December the constellation Orion rises around sunset, reaches the meridian around midnight, and sets around sunrise. Hence, if it is winter and Orion is setting, dawn must be approaching. Most ancient peoples probably were adept at estimating the time of night, although written evidence is sparse.

We can trace the origins of our modern clock to ancient Egypt, some 4,000 years ago. The Egyptians divided the daylight into 12 equal parts, and we still break the 24-hour day into 12 hours each of A.M. and P.M. (The abbreviations A.M. and P.M. stand for the Latin terms *ante meridiem* and *post meridiem*, respectively, which mean "before the middle of the day" and "after the middle of the day.") However, unlike our modern hours, the Egyptian "hours" varied in length because the amount of daylight varies during the year. For example, the daylight "summer hours" were longer than the daylight "winter hours," because one-twelfth of the daylight lasts longer in summer than in winter. Only much later in history did the hour become a fixed amount of time, subdivided into 60 equal minutes each consisting of 60 equal seconds.

The Egyptians also divided the night into 12 equal parts. Egyptian *star clocks*, often found painted on the coffin lids of Egyptian pharaohs, essentially cataloged where particu-

lar stars appear in the sky at particular times of night and particular times of year. By knowing the date from their calendar and observing the positions of particular stars in the sky, the Egyptians could use the star clocks to estimate the time of night.

By about 1500 B.C., Egyptians had abandoned star clocks in favor of clocks that measure time by the flow of water through an opening of a particular size, just as hourglasses measure time by the flow of sand through a narrow neck.* These *water clocks* had the advantage of being useful even when the sky was cloudy. They eventually became the primary timekeeping instruments for many cultures, including the Greeks, Romans, and Chinese. Water clocks, in turn, were replaced by mechanical clocks in the 1600s and by electronic clocks in the twentieth century. Despite the availability of other types of clocks, sundials were common throughout ancient times and remain popular today both for their decorative value and as reminders that the Sun and stars once were our only guides to time.

Marking the Seasons   Many ancient cultures built structures to help them mark the seasons. One of the oldest standing human-made structures served such a purpose: Stonehenge in southern England, which was constructed in stages from about 3100 B.C. to about 1550 B.C. (Figure 3.3). Stonehenge also served as a social gathering place and probably as a religious site. No one knows whether its original purpose was social or astronomical. Perhaps it was built for both— in ancient times, social rituals and practical astronomy probably were deeply intertwined.

Among the most spectacular structures used to mark the seasons was the Templo Mayor in the Aztec city of Tenochtitlán, located on the site of modern-day Mexico City (Figure 3.4). Twin temples stood on a flat-topped, 150-foot-high pyramid. From the vantage point of a royal observer watching from the opposite side of the plaza, the Sun rose through the notch between the temples on the equinoxes. Like Stonehenge, the Templo Mayor served important social and religious functions in addition to its astronomical role. Before the Conquistadors destroyed it, other Spanish visitors reported elaborate rituals there, sometimes including human sacrifice, at times determined by astronomical observations. After its destruction, stones from the Templo Mayor were used to build the nearby cathedral in the great plaza of Mexico City.

Ancient buildings and cities were often aligned with the cardinal directions (north, south, east, and west), enabling people to work and live in an environment that essentially represented a miniature version of the heavens. This type of alignment is common to ancient architecture around the world, and is found at such diverse sites as the Egyptian pyramids, the Forbidden City in China, and ceremonial kivas built by the Anasazi people of the American southwest

---

*Hourglasses using sand were not invented until about the eighth century A.D., long after the advent of water clocks. Natural sand grains vary in size, so making accurate hourglasses required technology for making uniform grains of sand.

a The remains of Stonehenge today.

Figure 3.3 Stonehenge, in southern England.

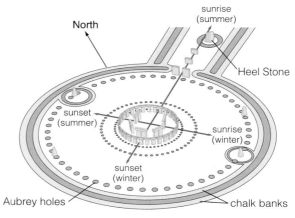

b This sketch shows how archaeologists believe Stonehenge looked upon its completion in about 1550 B.C. Several astronomical alignments are shown as they appear from the center. For example, the Sun rises directly over the Heel Stone on the summer solstice.

(Figure 3.5). Many modern cities retain this layout, which is why you'll find so many streets that run directly north-south or east-west.

Other structures were used to mark the Sun's position on special dates such as the winter or summer solstice. For example, someone among the ancient Anasazi people carved a 19-turn spiral—often called the *Sun dagger*—on a vertical cliff face near the top of Fajada butte in Chaco Canyon, New Mexico (Figure 3.6). Three large slabs of rock in front of the spiral produce different patterns of light and shadow throughout the year. The slabs form a dagger of sunlight that pierces the center of the carved spiral only once each year—at noon on the summer solstice. Two daggers of light bracket the spiral at winter solstice, and a single dagger cuts through the center of a smaller spiral on each equinox.

Lunar Calendars Many ancient civilizations paid particular attention to lunar phases, often using them as the basis for calendars. The months on a lunar calendar generally have

either 29 or 30 days, chosen to make the average agree with the approximately $29\frac{1}{2}$-day lunar cycle. Thus, a 12-month lunar calendar has only 354 or 355 days, or about 11 days fewer than a calendar based on the Sun. Such a calendar is still used in the Muslim religion. That is why the month-long fast of Ramadan (the ninth month) begins about 11 days earlier with each subsequent year.

Some lunar calendars remain roughly synchronized with solar calendars by taking advantage of an interesting coincidence: On a solar calendar, the lunar phases repeat on the same dates about every 19 years. For example, there was a full moon on October 17, 2005, and there will be a full moon 19 years later, on October 17, 2024. This 19-year cycle on which the dates of lunar phases repeat is called the *Metonic cycle*, because the Greek astronomer Meton recognized it in 432 B.C. (However, Babylonian astronomers almost certainly knew of the cycle centuries earlier.)

Lunar phases repeat with the 19-year Metonic cycle because 19 years on a solar calendar is almost precisely

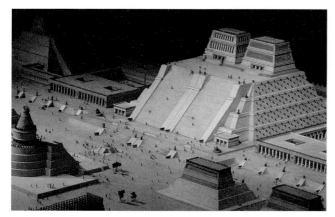

Figure 3.4 This scale model shows the Templo Mayor and the surrounding plaza as they are thought to have looked before the Spanish Conquistadors destroyed Aztec civilization. The structure was used to help mark the seasons.

Figure 3.5 This large structure, more than 20 meters in diameter, is an Anasazi kiva in Chaco Canyon, New Mexico. It was built approximately 1,000 years ago. Its main axis is aligned almost precisely north-south.

a A single dagger of sunlight pierces the center of the carved spiral only at noon on the summer solstice. Other special patterns occur on the winter solstice and the equinoxes, and the Sun Dagger can also be used to observe an 18.6-year cycle of the moon.

b This photo shows Fajada butte in Chaco Canyon, New Mexico, where the Sun dagger is located on a vertical cliff near the top.

Figure 3.6 The Sun Dagger. Three large slabs of rock lie in front of a 19-turn carved spiral, creating patterns of light and shadow that vary throughout the year.

235 months on a lunar calendar. (See Problem 47 at the end of this chapter to verify this fact.) Thus, a lunar calendar will remain roughly in sync with a solar calendar if it has 235 months in each 19-year period. Because 19 years with 12 months each would mean only $19 \times 12 = 228$ months, or 7 short of the needed 235 months, lunar calendars that follow the Metonic cycle add a thirteenth month to 7 of every 19 years. For example, the Jewish calendar has a thirteenth month in the third, sixth, eighth, eleventh, fourteenth, seventeenth, and nineteenth years of each 19-year cycle. This scheme keeps the dates of Jewish holidays within about a 1-month range on a solar calendar, with precise dates repeating every 19 years. It also explains why the date of Easter changes from year to year: The New Testament ties the date of Easter to the Jewish festival of Passover. In a slight modification of the original scheme, most Western Christians now celebrate Easter on *the first Sunday after the first full moon after March 21*. If the full moon falls on Sunday, Easter is the following Sunday. (Eastern Orthodox churches calculate the date of Easter differently, because they base the date on the Julian rather than the Gregorian calendar [Section S1.1].)

In addition to following the lunar phases, some ancient cultures discovered other lunar cycles. In the Middle East more than 2,500 years ago, the ancient Babylonians achieved remarkable success in predicting eclipses, thanks to their recognition of the approximately 18-year saros cycle [Section 2.3]. The Mayans of Central America also appear to have been experts at eclipse prediction. The Mayan calendar featured a sacred cycle that almost certainly was related to eclipses. This Mayan cycle, called the *sacred round,* lasted 260 days—almost exactly $1\frac{1}{2}$ times the 173.32 days between successive eclipse seasons. Unfortunately, we know little more about the extent of Mayan knowledge because the Spanish Conquistadors burned most Mayan writings.

The Anasazi people also appear to have understood lunar cycles quite well. The rise and set positions of the full moon vary in an 18.6-year cycle (due to precession of the Moon's orbit), so that the full moon rises at its most southerly point along the eastern horizon only once every 18.6 years. At this time, known as a "major lunar standstill," the shadow of the full moon passes through the slabs of rock to lie tangent to the edge of the spiral in the Sun dagger (see Figure 3.6); 9.3 years later the lunar shadow cuts through the center of the spiral. The major lunar standstill was apparently also observed by the Anasazi at nearby Chimney Rock and at Mesa Verde in Colorado.

Ancient Structures and Archaeoastronomy   Many ancient cultures apparently also made careful observations of planets and stars. The Chinese, for example, began keeping remarkably detailed records of astronomical observations at least 5,000 years ago. Their careful record keeping allowed Chinese astronomers to make many important discoveries. By the 1400s, the Chinese had built a great observatory in Beijing, which still stands today (Figure 3.7).

It's easy to establish the astronomical intentions of ancient cultures that left extensive written records, such as the Chinese and the Egyptians. In other cases, however, claims that ancient structures served astronomical purposes can be much more difficult to evaluate.

Figure 3.7 This photo shows a model of the celestial sphere and other instruments on the roof of the ancient astronomical observatory in Beijing. The observatory was built in the 1400s; the instruments shown here were built later and show a European influence brought by Jesuit missionaries.

Figure 3.8 The Big Horn Medicine Wheel in Wyoming. A study once claimed that its "spokes" have astronomically significant alignments, but later research showed the claim was in error—making this a good example of how science adapts as new data come to light.

Figure 3.9 Hundreds of lines and patterns are etched in the sand of the Nazca desert in Peru. This aerial photo shows a large etched figure of a hummingbird.

The study of ancient structures in search of astronomical connections is called *archaeoastronomy*, a word that combines archaeology and astronomy. Scientists engaged in archaeoastronomy usually start by evaluating an ancient structure to see whether it shows any particular astronomical alignments. For example, they may check to see whether an observer in a central location would see particular stars rise above specially marked stones, or whether sunlight enters through a window only on special days like the solstices or equinoxes. However, the mere existence of astronomical alignments is not enough to establish that a structure had an astronomical purpose; the alignments may be coincidental. Moreover, many claimed alignments have later proven to be in error.

Native American Medicine Wheels offer a good example of the pitfalls that can occur in trying to establish the intentions of ancient builders. Medicine wheels are stone circles found throughout the northern plains of the United States. In the 1970s, a study of the Big Horn Medicine Wheel in Wyoming (Figure 3.8) seemed to indicate that its 28 "spokes" were aligned with the rise and set of particular stars. However, later research showed that the original study had failed to take into account the motion of stars as they rise above the horizon and the way the atmosphere affects the visibility of stars at the latitude of Big Horn. In reality the spokes do *not* show any special alignments with bright stars. Moreover, if Medicine Wheels really did serve an astronomical purpose, we'd expect all of them to have been built with consistent alignments—but that is not the case. Nevertheless, many popular books and articles still report the original claims regarding astronomical purposes for Medicine Wheels, even though there is no evidence to support these claims.*

Other cases are more ambiguous. For example, ancient people in what is now Peru etched hundreds of lines and

patterns in the sand of the Nazca desert. Many of the lines point to places where bright stars or the Sun rise at particular times of year, but that doesn't prove anything: With hundreds of lines, random chance ensures that many will have astronomical alignments no matter how or why they were made. The patterns, many of which are large figures of animals (Figure 3.9), have evoked even more debate. Some people think they may be representations of constellations recognized by the people who lived in the region, but we really do not know for sure.

Animal figures like that in Figure 3.9 show up clearly only when seen from above. As a result, some UFO enthusiasts argue that the patterns must have been created by aliens. What do you think of this argument? Defend your opinion.

In some cases, scientists can use other clues to establish the intentions of ancient builders. For example, lodges built by the Pawnee people in Kansas feature strategically placed holes for observing the passage of constellations that figure prominently in Pawnee folklore. In this case, the correspondence between the folklore and the structural features provides a strong case for deliberate intent rather than coincidence. Similar evidence suggests that people of the Inca Empire of South America built cities and structures with deliberate astronomical alignments. Inca traditions held that its rulers were descendents of the Sun and therefore demanded that movements of the Sun be watched closely. Many of the cities and ceremonial centers of the Inca, such as the World Heritage Site of Machu Picchu, appear to have been designed so that particular arrangements of roads, temples, and pillars on the horizon marked the positions where the Sun rose or set on special dates like the solstices (Figure 3.10). Other Inca cities and ceremonial centers show similar alignments for the Sun, as well as alignments with the rise and set positions of bright stars such as the Pleiades.

A different type of evidence makes a convincing case for the astronomical sophistication of ancient Polynesians.

---

*Textbooks are not immune to this problem; early editions of this textbook also reported the original research without the later corrections.

Figure 3.10 The World Heritage Site of Machu Picchu has structures aligned with sunrise at the winter and summer solstices.

The Polynesians colonized and traveled among many islands of the mid- and South Pacific, even though the next island in a journey usually was too distant to be seen. Their success clearly depended upon superb navigational skills, including skills of celestial navigation. Indeed, anthropological studies have shown that the most esteemed position in Polynesian culture was that of the Navigator, a person who had acquired the knowledge necessary to navigate great distances among the islands. The Navigators employed a combination of detailed knowledge of astronomy and equally impressive knowledge of the patterns of waves and swells around different islands (Figure 3.11). The stars provided the broad navigational sense, pointing the Navigators in the general direction of their destination. As they neared a destination, the wave and swell patterns guided them to their precise landing point. A Navigator memorized all his skills and passed them to the next generation through a well-developed program for training future Navigators. Unfortunately, with the advent of modern navigational technology, many of the skills of the Navigators have been lost.

Figure 3.11 A Micronesian stick chart, an instrument used by Polynesian Navigators to represent swell patterns around islands.

From Observations to Science Before a structure such as Stonehenge or the Templo Mayor could be built, careful observations had to be made and repeated over and over to ensure their validity. Some long-term cycles became clear only when generations of observers recorded and preserved their discoveries. Carefully recorded, repeatable observations also underlie modern science. Thus, at least some elements of modern science were present in many early human cultures.

The extent to which scientific ideas developed in different societies depended on practical needs, social and political customs, and interactions with other cultures. For example, despite their "head start" in astronomical record keeping, Chinese science and technology had clearly fallen behind that of Europe by the end of the 1600s. Some historians argue that a primary reason for this decline was that the Chinese tended to regard their science and technology as state secrets. This secrecy may have slowed Chinese scientific development by preventing the broad-based collaborative science that fueled the European advance beginning in the Renaissance.

In Central America, the ancient Mayans also were ahead of their time in many ways. For example, their system of numbers and mathematics looks distinctly modern. They invented the concept of zero some 500 years before its introduction in the Eurasian world (by Hindu mathematicians, around A.D. 600). The Aztecs, Incas, Anasazi, and other ancient peoples of the Americas may have been quite advanced in many other areas as well, but few written records survive to tell the tale.

If the circumstances of history had been different, any one of these or many other cultures might have been the first to develop what we consider to be modern science. In the end, however, history takes only one of countless possible paths. The path that led to modern science emerged from the ancient civilizations of the Mediterranean and the Middle East—and especially from ancient Greece.

3.2 Ancient Greek Science

By 3000 B.C., civilization was well established in two major regions of the Middle East: Egypt and Mesopotamia (Figure 3.12). Their geographical locations placed these civilizations at a crossroads for travelers, merchants, and armies of Europe, Asia, and Africa. This mixing of cultures fostered creativity, and the broad interactions among peoples ensured that new ideas spread throughout the region.

Over the next 2,500 years, numerous great cultures arose. The ancient Egyptians built the Great Pyramids between 2700 and 2100 B.C., using their astronomical knowledge to orient the Pyramids with the cardinal directions. They also invented papyrus scrolls and ink-based writing. The Babylonians invented a method of writing on clay tablets and developed arithmetic to serve in commerce and later in astronomical calculations. Many more of our mod-

Figure 3.12 The highlighted region on this map shows the approximate extent of the ancient civilizations of Egypt and Mesopotamia. The map also identifies a few other places discussed in this text. The overlying white borders show modern country borders.

ern principles of commerce, law, and religion originated with the cultures of Egypt and Mesopotamia. But much of what we now call *science* came from ancient Greece, which rose as a power in the Middle East around 500 B.C.

• Why does modern science trace its roots to the Greeks?

The ancient Greeks stand out as the first people who tried to explain the motions of astronomical objects by using logic and geometry, without resorting to the supernatural. In this sense, they were the first people to seek knowledge in a way that looked something like what we now call science.

Greek philosophers sought to explain the architecture of the universe by constructing **models** of nature. More than two thousand years later, this idea of modeling nature is central to modern science.

Scientific models differ somewhat from the models you may be familiar with in everyday life. In our daily lives, we tend to think of models as miniature physical representations, such as model cars or airplanes. In contrast, a scientific model is a conceptual representation whose purpose is to explain and predict observed phenomena. For example, a model of Earth's climate uses logic and mathematics to represent what we know about how climate works. Its purpose is to explain and predict climate changes, such as the changes that may occur with global warming. Just as a model airplane does not faithfully represent every aspect of a real airplane, a scientific model may not fully explain all our observations of nature. Nevertheless, even the failings of a scientific model can be useful, because they often point the way toward building a better model.

The Greek models of nature sought to explain things such as the properties of matter and the motions of the stars, Sun, Moon, and planets. Many of the scientific ideas discussed in this book originated with the Greeks. (Figure 3.13 shows a time line of the most influential Greek thinkers about astronomy.) Although the Greek models may seem primitive from our modern perspective, they were an enormous step forward in scientific thinking. As an example of the modeling process, we'll look next at how the Greeks developed models to explain celestial motion.

• How did the Greeks explain planetary motion?

As we discussed in Section 2.4, most ancient Greek philosophers assumed that Earth must reside at the center of the universe. The Greeks therefore developed a sophisticated **geocentric model** of the cosmos—so named because it placed a spherical Earth at the center of the universe—that they used to explain the motions of the Sun, Moon, planets, and stars.

Early Development of the Geocentric Model We generally trace the origin of Greek science to the philosopher Thales (c. 624–546 B.C.; pronounced *thay-lees*). We encountered Thales earlier because of his legendary prediction of a solar eclipse [Section 2.3]. Thales was the first person known to have addressed the question, "What is the universe made of?" without resorting to supernatural explanations. His own guess—that the universe fundamentally consists of water and that Earth is a flat disk floating in an infinite ocean—was not widely accepted even in his own time. Nevertheless, just by asking the question he suggested that the world is inherently understandable and thus inspired others to come up with better models for the structure of the universe.

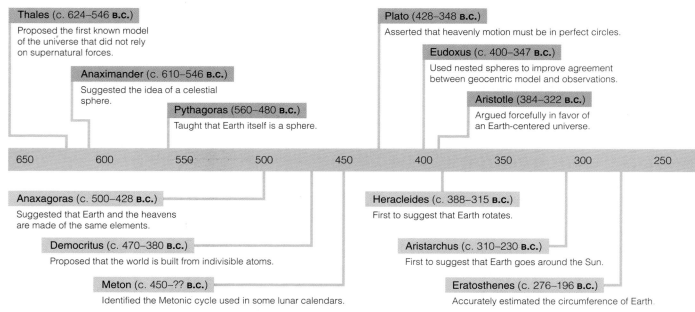

Thales (c. 624–546 **B.C.**)
Proposed the first known model of the universe that did not rely on supernatural forces.

Anaximander (c. 610–546 **B.C.**)
Suggested the idea of a celestial sphere.

Pythagoras (560–480 **B.C.**)
Taught that Earth itself is a sphere.

Plato (428–348 **B.C.**)
Asserted that heavenly motion must be in perfect circles.

Eudoxus (c. 400–347 **B.C.**)
Used nested spheres to improve agreement between geocentric model and observations.

Aristotle (384–322 **B.C.**)
Argued forcefully in favor of an Earth-centered universe.

| 650 | 600 | 550 | 500 | 450 | 400 | 350 | 300 | 250 |

Anaxagoras (c. 500–428 **B.C.**)
Suggested that Earth and the heavens are made of the same elements.

Democritus (c. 470–380 **B.C.**)
Proposed that the world is built from indivisible atoms.

Meton (c. 450–?? **B.C.**)
Identified the Metonic cycle used in some lunar calendars.

Heracleides (c. 388–315 **B.C.**)
First to suggest that Earth rotates.

Aristarchus (c. 310–230 **B.C.**)
First to suggest that Earth goes around the Sun.

Eratosthenes (c. 276–196 **B.C.**)
Accurately estimated the circumference of Earth.

Figure 3.13 Time line for major Greek figures in the development of astronomy. (All these individuals are discussed in this book but not necessarily in this chapter.)

A more sophisticated idea followed soon after, proposed by a student of Thales named Anaximander (c. 610–546 B.C.). Anaximander suggested that Earth floats in empty space surrounded by a sphere of stars and two separate rings along which the Sun and Moon travel. Thus, we credit him with inventing the idea of a celestial sphere [Section 2.1]. Interestingly, Anaximander imagined Earth itself to be cylindrical rather than spherical in shape. He probably chose this shape because he knew Earth had to be curved in a north-south direction to explain changes in the constellations with latitude. Because the visible constellations do not change with longitude, he apparently saw no need for curvature in the east-west direction.

We do not know precisely when the Greeks first thought that Earth is round, but this idea was already being taught by about 500 B.C. by the famous mathematician Pythagoras (c. 560–480 B.C.). He and his followers imagined Earth as a sphere floating at the center of the celestial sphere. Much of their motivation for adopting a spherical Earth probably was philosophical: The Pythagoreans had a mystical interest in mathematical perfection, and they considered a sphere to be geometrically perfect. More than a century later, Aristotle cited observations of Earth's curved shadow on the Moon during lunar eclipses as evidence for a spherical Earth.

The supposed heavenly perfection of spheres influenced models of the cosmos for many centuries. Plato (428–348 B.C.), whose philosophy was based much more on pure thought than on observations, asserted that all heavenly objects move in perfect circles at constant speeds and therefore must reside on huge spheres encircling Earth (Figure 3.14). Greeks who took observations more seriously found this model problematic: The apparent retrograde motion of the planets [Section 2.4], already well known by

that time, clearly showed that planets do not move at constant speeds around Earth.

An ingenious solution came from one of Plato's colleagues, Eudoxus (c. 400–347 B.C.). He created a model in which the Sun, the Moon, and the planets each had its own sphere, which in turn was nested within several other spheres. Individually, the nested spheres turned in perfect circles. By carefully choosing the sizes, rotation axes, and rotation

Figure 3.14 This model represents the Greek idea of the heavenly spheres (c. 400 B.C.). Earth is a sphere that rests in the center. The Moon, the Sun, and the planets each have their own spheres. The outermost sphere holds the stars.

Apollonius (c. 240–190 **B.C.**)
Introduced circles upon circles to explain retrograde motion.

Hipparchus (c. 190–120 **B.C.**)
Developed many of the ideas of the Ptolemaic model, discovered precession, invented the magnitude system for describing stellar brightness.

Ptolemy (c. **A.D.** 100–170)
His Earth-centered model of the universe remained in use for some 1,500 years.

B.C. A.D.

200   150   100   50   0   50   100   150

Major steps in the development of the geocentric model

Other milestones of Greek astronomy

speeds for the invisible spheres, Eudoxus was able to make them work together in a way that reproduced many of the observed motions of the Sun, Moon, and planets in our sky. Other Greeks refined the model by comparing its predictions to observations and adding more spheres to improve the agreement.

This is how things stood when Aristotle (384–322 B.C.) arrived on the scene. Whether Eudoxus and his followers thought of the nested spheres as real physical objects is not clear, but Aristotle certainly did. In the cosmic model of Aristotle, all the spheres responsible for celestial motion were transparent and interconnected like the gears of a giant machine. Furthermore, Earth's position at the center of everything—as well as its composition and shape—was explained as a natural consequence of gravity. Aristotle argued that gravity pulled heavy things toward the center of the universe and allowed lighter things to float toward the heavens. Thus, all the dirt, rock, and water of the universe collected at the center, forming a spherical Earth. We now know that Aristotle was wrong about both gravity and Earth's location. However, largely because of his persuasive arguments for an Earth-centered universe, the geocentric view dominated Western thought for almost 2,000 years.

SPECIAL TOPIC Eratosthenes Measures Earth

In a remarkable feat, the Greek astronomer and geographer Eratosthenes estimated the size of Earth in about 240 B.C. He did it by comparing the altitude of the Sun on the summer solstice in the Egyptian cities of Syene (modern-day Aswan) and Alexandria.

Eratosthenes knew that the Sun passed directly overhead in Syene on the summer solstice. He also knew that in the city of Alexandria to the north, the Sun came within only 7° of the zenith on the summer solstice. He therefore reasoned that Alexandria must be 7° of latitude to the north of Syene (see figure). Because 7° is $\frac{7}{360}$ of a circle, he concluded that the north-south distance between Alexandria and Syene must be $\frac{7}{360}$ of the circumference of Earth.

Eratosthenes estimated the north-south distance between Syene and Alexandria to be 5,000 stadia (the *stadium* was a Greek unit of distance). Thus, he concluded that:

$$\frac{7}{360} \times \text{circumference of Earth} = 5000 \text{ stadia}$$

From this he found Earth's circumference to be about 250,000 stadia.

Today, we don't know exactly what distance a stadium meant to Eratosthenes. Based on the actual sizes of Greek stadiums, it must have been about $\frac{1}{6}$ kilometer. Thus, Eratosthenes estimated the circumference of the Earth to be about $\frac{250,000}{6} = 42,000$ kilometers—remarkably close to the real value of just over 40,000 kilometers.

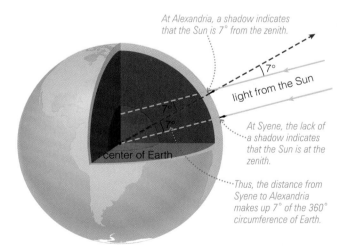

At Alexandria, a shadow indicates that the Sun is 7° from the zenith.

light from the Sun

At Syene, the lack of a shadow indicates that the Sun is at the zenith.

Thus, the distance from Syene to Alexandria makes up 7° of the 360° circumference of Earth.

Center of Earth

At noon on the summer solstice, the Sun appears at the zenith in Syene but 7° shy of the zenith in Alexandria. Thus, 7° of latitude, which corresponds to a distance of $\frac{7}{360}$ of Earth's circumference, must separate the two cities.

Ptolemy's Synthesis of the Geocentric Model Greek modeling of the cosmos became much more quantitative after Aristotle, culminating in the work of Claudius Ptolemy (c. A.D. 100–170; pronounced *TOL-e-mee*). Ptolemy's model still placed Earth at the center of the universe, but it differed in significant ways from the nested spheres of Eudoxus and Aristotle. We refer to Ptolemy's model as the **Ptolemaic model** to distinguish it from earlier geocentric models.

To explain the apparent retrograde motion of the planets, the Ptolemaic model applied an idea first suggested by Apollonius (c. 240–190 B.C.). This idea held that each planet moves around Earth on a small circle that turns upon a larger circle (Figure 3.15). (The small circle is sometimes called an *epicycle,* and the larger circle is called a *deferent.*) A planet following this circle-upon-circle motion traces a loop as seen from Earth, with the backward portion of the loop mimicking apparent retrograde motion.

Ptolemy also relied heavily on the work of Hipparchus (c. 190–120 B.C.), considered one of the greatest of the Greek astronomers. Among his many accomplishments, Hipparchus developed the circle-upon-circle idea of Apollonius into a model that could predict planetary positions. To do this, Hipparchus had to add several features to the basic idea; for example, he included even smaller circles that moved upon the original set of small circles, and he positioned the large circles slightly off-center from Earth.

Ptolemy's great accomplishment was to adapt and synthesize these earlier ideas into a single system that agreed quite well with the astronomical observations available at the time. In the end, he created and published a model that could correctly forecast future planetary positions to within a few degrees of arc—roughly equivalent to the size

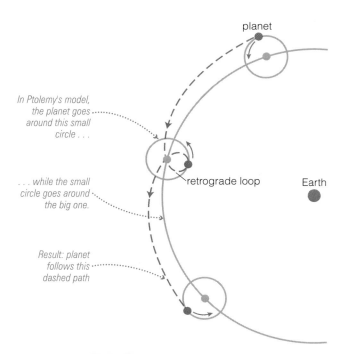

planet

In Ptolemy's model, the planet goes around this small circle . . .

. . . while the small circle goes around the big one.

retrograde loop

Earth

Result: planet follows this dashed path

Figure 3.15 Interactive Figure. This diagram shows how the Ptolemaic model accounted for apparent retrograde motion. Each planet is assumed to move around a small circle that turns upon a larger circle. The resulting path (dashed) includes a loop in which the planet goes backward as seen from Earth.

of your hand viewed at arm's length against the sky. His model generally worked so well that it remained in use for the next 1,500 years. When his book describing the model was translated by Arabic scholars around A.D. 800, they gave it the title *Almagest,* derived from words meaning "the greatest compilation."

• How was Greek knowledge preserved through history?

One reason Greek thought gained such broad influence was that the Greeks proved as adept at politics and war as they were at philosophy. In about 330 B.C., Alexander the Great (356–323 B.C.) began a series of conquests that expanded the Greek Empire throughout the Middle East, absorbing all the former empires of Egypt and Mesopotamia. Alexander was more than just a military leader. He was also keenly interested in science and education. As a teenager, he had been tutored by none other than Aristotle.

Alexander encouraged the pursuit of knowledge and respect for foreign cultures. On the Nile delta in Egypt, he founded the city of Alexandria, which soon became a center of world culture. Not long after his death, the city of Alexandria commenced work on a great library and research center: the Library of Alexandria, which opened in about 300 B.C. (Figure 3.16). It remained the world's preeminent center of research for some 700 years, surviving well past the fall of the classical Greek Empire and into the heyday of the Roman Empire.

a This rendering shows an artist's reconstruction, based on scholarly research, of how the Great Hall may have looked in the ancient Library of Alexandria.

b A similar rendering to (a), this time showing a scroll room in the ancient library.

c The New Library of Alexandria, in Egypt, which opened in 2003.

Figure 3.16 The ancient Library of Alexandria thrived for some 700 years, starting in about 300 B.C.

Destruction of the Library of Alexandria   Much of the Library of Alexandria's long history remains unknown today, in part because the books that recorded this history were deliberately destroyed along with the Library. Nevertheless, historians are confident that the Library's demise was intertwined with the life and death of one of the few prominent female scholars of the ancient world: a woman named Hypatia (A.D. 370–415). Hypatia was one of the last resident scholars of the Library, as well as the director of the observatory in Alexandria and one of the leading mathematicians and astronomers of her time. Unfortunately, she became a scapegoat during a time of rising sentiment against free inquiry and was murdered by an anti-intellectual mob in A.D. 415. The final destruction of the Library took place not long after her death.

At its peak, the Library of Alexandria may have held more than a half million books, handwritten on papyrus scrolls. Most of these scrolls were ultimately burned, their contents lost forever. In some cases, we were left with small fragments or titles or descriptions from other works, teasing us with intriguing subject matter about which we will never read. In commemoration of the ancient Library, the country of Egypt recently built a New Library of Alexandria (more commonly called the *Bibliotheca Alexandrina*). Egypt hopes that this new library, which opened in 2003, will once again make Alexandria a global center for research and scientific excellence.

THINK ABOUT IT

Estimate the number of books you're likely to read in your lifetime, and compare this number to the half million books once housed in the Library of Alexandria. Can you think of other ways to put into perspective the loss of ancient wisdom resulting from the destruction of the Library of Alexandria?

Islamic Preservation of Science   Much of Greek knowledge was lost with the destruction of the Library of Alexandria. That which survived was preserved primarily thanks to the rise of a new center of intellectual inquiry in Baghdad (in present-day Iraq). While European civilization fell into the period of intellectual decline known as the Dark Ages, scholars of the new religion of Islam sought knowledge of mathematics and astronomy in hopes of better understanding the wisdom of Allah. During the eighth and ninth centuries A.D., scholars working in the Muslim empire centered in Baghdad translated and thereby saved many ancient Greek works.

Around A.D. 800, the Islamic leader Al-Mamun (A.D. 786–833) established a "House of Wisdom" in Baghdad with a mission much like that of the destroyed Library of Alexandria. Founded in a spirit of openness and tolerance, the House of Wisdom employed Jews, Christians, and Muslims, all working together in scholarly pursuits. Using the translated Greek scientific manuscripts as building blocks, these scholars developed the mathematics of algebra and many new instruments and techniques for astronomical observation. Most of the official names of constellations and stars come from Arabic because of the work of the scholars at Baghdad. If you look at a star chart, you will see that the names of many bright stars begin with *al* (for example, Aldebaran, Algol), which simply means "the" in Arabic.

Toward a Scientific Renaissance   The Islamic world of the Middle Ages was in frequent contact with Hindu scholars from India, who in turn brought knowledge of ideas and discoveries from China. Hence, the intellectual center in Baghdad achieved a synthesis of the surviving work of the ancient Greeks and that of the Indians and the Chinese. The accumulated knowledge of the Arabs spread throughout the Byzantine Empire (the eastern part of the former Roman Empire). When the Byzantine capital of Constantinople (modern-day Istanbul) fell to the Turks in 1453, many Eastern scholars headed west to Europe, carrying with them the knowledge that helped ignite the European Renaissance. It is to this story that we turn next in our quest to understand the origins of modern science.

3.3 The Copernican Revolution

The Greeks and other ancient peoples developed many important ideas of science, but what we now think of as science arose during the European Renaissance. Within a half century after the fall of Constantinople, Polish scientist Nicholas Copernicus began the work that ultimately overturned the Earth-centered Ptolemaic model. Over the next century and a half, philosophers and scientists (who were often one and the same) debated and tested his radical view of the cosmos. Ultimately, the new ideas introduced by Copernicus fundamentally changed the way we perceive our place in the universe. This dramatic change, known as the **Copernican revolution**, spurred the development of virtually all modern science and technology.

• How did Copernicus, Tycho, and Kepler challenge the Earth-centered idea?

Recall that Aristarchus suggested the idea of a Sun-centered solar system more than 1,700 years before Copernicus was born [Section 2.4]. However, before the time of the Copernican revolution, few people took the idea seriously . Let's examine how three key individuals—Nicholas Copernicus, Tycho Brahe, and Johannes Kepler—challenged the prevailing dogma that our planet must be the center of the universe.

Copernicus   Nicholas Copernicus was born in Toruń, Poland, on February 19, 1473. His family was wealthy and he received a first-class education in mathematics, medicine, and law. He began studying astronomy in his late teens. By that time, tables of planetary motion based on the Ptolemaic model were noticeably inaccurate. But few people were willing to undertake the difficult calculations required to revise the tables. Indeed, the best tables available had been compiled some two centuries earlier under the guidance of Spanish monarch Alphonso X (1221–1284). Commenting on the tedious nature of the work required to

Copernicus (1473–1543)

make these *Alfonsine Tables,* the monarch is said to have complained, "If I had been present at the creation, I would have recommended a simpler design for the universe."

In his quest for a better way to predict planetary positions, Copernicus adopted Aristarchus's Sun-centered idea. He was probably motivated in large part by the much simpler explanation for apparent retrograde motion offered by a Sun-centered system (see Figure 2.33). As he worked out the mathematical details of his model, Copernicus also discovered simple geometric relationships that allowed him to calculate each planet's orbital period around the Sun and its relative distance from the Sun in terms of the Earth-Sun distance (see Mathematical Insight S1.1, p. 94). The success of his model in providing a geometric layout for the solar system further convinced him that the Sun-centered idea must be correct.

Copernicus was hesitant to publish his work, fearing that his suggestion that Earth moved would be considered absurd. Nevertheless, he discussed his system with other scholars, generating great interest. At the urging of some of these scholars, including some high-ranking officials of the Catholic Church, he finally agreed to publish his work. Copernicus saw the first printed copy of his book, *De Revolutionibus Orbium Caelestium* ("Concerning the Revolutions of the Heavenly Spheres"), on the day he died—May 24, 1543.

Early supporters of Copernicus were drawn to the aesthetic advantages of his model. However, it did not make substantially better predictions than Ptolemy's model, largely because Copernicus still believed that heavenly

---

SPECIAL TOPIC Aristotle

Aristotle (384–322 B.C.) is among the best-known philosophers of the ancient world. Both his parents died when he was a child, and he was raised by a family friend. In his 20s and 30s, he studied under Plato (428–348 B.C.) at Plato's Academy. He later founded his own school, called the Lyceum, where he studied and lectured on virtually every subject. Historical records tell us that his lectures were collected and published in 150 volumes. About 50 of these volumes survive to the present day.

Many of Aristotle's scientific discoveries were about the nature of plants and animals. He studied more than 500 animal species in detail, dissecting specimens of nearly 50 species, and came up with a strikingly modern classification system. For example, he was the first person to recognize that dolphins should be classified with

land mammals rather than with fish. In mathematics, he is known for laying the foundations of mathematical logic. Unfortunately, he was far less successful in physics and astronomy, areas in which many of his claims turned out to be wrong.

Despite his wide-ranging discoveries and writings, Aristotle's philosophies were not particularly influential until many centuries after his death. His books were preserved and valued by Islamic scholars but were unknown in Europe until they were translated into Latin in the twelfth and thirteenth centuries. Aristotle achieved his near-reverential status only after St. Thomas Aquinas (1225–1274) integrated Aristotle's philosophy into Christian theology. In the ancient world, Aristotle's greatest influence came indirectly, through his role as the tutor of Alexander the Great.

---

motion must occur in perfect circles. Because the true orbits of the planets are *not* circles, Copernicus found it necessary to add circles upon circles to his system, just as Ptolemy had done. As a result, his complete model was no more accurate and no less complex than the Ptolemaic model, and the Sun-centered idea won relatively few converts in the 50 years after it was published. After all, why trade thousands of years of tradition for a new model that worked equally poorly?

Tycho Part of the difficulty faced by astronomers who sought to improve either the Ptolemaic or the Copernican system was a lack of quality data. The telescope had not yet been invented, and existing naked-eye observations were not very accurate. In the late 1500s, Danish nobleman Tycho Brahe (1546–1601), usually known simply as Tycho (pronounced *tie-koe*), set about correcting this problem.

When Tycho was a young boy, his family discouraged his interest in astronomy. He therefore followed his passion in secret, learning the constellations from a miniature model of a celestial sphere that he kept hidden. As he grew older, Tycho was often arrogant about both his noble birth and his intellectual abilities. At age 20, he fought a duel with another student over which of them was the better mathematician. Part of Tycho's nose was cut off, and he designed a replacement piece made of silver and gold.

In 1563, Tycho decided to observe a widely anticipated alignment of Jupiter and Saturn. To his surprise, the alignment occurred nearly two days later than Copernicus had predicted. Resolving to improve the state of astronomical prediction, he set about compiling careful observations of stellar and planetary positions in the sky.

Tycho's fame grew after he observed what he called a *nova*, meaning "new star," in 1572. By measuring its parallax and comparing it to the parallax of the Moon, he proved that the nova was much farther away than the Moon. (Today, we know that Tycho saw a *supernova*—the explosion of a distant star [Section 17.3].) In 1577, Tycho made similar observations of a comet and proved that it too lay in the realm of the heavens. Others, including Aristotle, had argued that comets were phenomena of Earth's atmosphere. King Frederick II of Denmark decided to sponsor Tycho's ongoing work, providing him with money to build an unparalleled observatory for naked-eye observations (Figure 3.17). After Frederick II died in 1588, Tycho moved to Prague, where German emperor Rudolf II supported his work.

Over a period of three decades, Tycho

Tycho Brahe (1546–1601)

and his assistants compiled naked-eye observations accurate to better than 1 arcminute—less than the thickness of a fingernail viewed at arm's length. Because the telescope was invented shortly after his death, Tycho's data remain the best set of naked-eye observations ever made. Despite the quality of his observations, Tycho never succeeded in coming up with a satisfying explanation for planetary motion. He was convinced that the *planets* must orbit the Sun, but his inability to detect stellar parallax [Section 2.4] led him to conclude that Earth must remain stationary. Thus, he advocated a model in which the Sun orbits Earth while all other planets orbit the Sun. Few people took this model seriously.

Kepler Tycho failed to explain the motions of the planets satisfactorily, but he succeeded in finding someone who could: In 1600, he hired the young German astronomer Johannes Kepler (1571–1630). Kepler and Tycho had a strained relationship, but Tycho recognized the talent of his young apprentice. In 1601, as he lay on his deathbed, Tycho begged Kepler to find a system that would make sense

Figure 3.17 Tycho Brahe in his naked-eye observatory, which worked much like a giant protractor. He could sit and observe a planet through the rectangular hole in the wall as an assistant used a sliding marker to measure the angle on the protractor.

Johannes Kepler (1571–1630)

of the observations so "that it may not appear I have lived in vain."*

Kepler was deeply religious and believed that understanding the geometry of the heavens would bring him closer to God. Like Copernicus, he believed that Earth and the other planets traveled around the Sun in circular orbits. He worked diligently to match circular motions to Tycho's data.

Kepler labored with particular intensity to find an orbit for Mars, which posed the greatest difficulties in matching the data to a circular orbit. After years of calculation, Kepler found a circular orbit that matched all of Tycho's observations of Mars's position along the ecliptic (east-west) to within 2 arcminutes. However, the model did not correctly predict Mars's positions north or south of the ecliptic. Because Kepler sought a physically realistic orbit for Mars, he could not (as Ptolemy and Copernicus had done) tolerate one model for the east-west positions and another for the north-south positions. He attempted to find a unified model with a circular orbit. In doing so, he found that some of his predictions differed from Tycho's observations by as much as 8 arcminutes.

Kepler surely was tempted to ignore these discrepancies and attribute them to errors by Tycho. After all, 8 arcmin-

---

*For a particularly moving version of the story of Tycho and Kepler, see Episode 3 of Carl Sagan's *Cosmos* video series.

utes is barely one-fourth the angular diameter of the full moon. But Kepler trusted Tycho's careful work. The small discrepancies finally led Kepler to abandon the idea of circular orbits—and to find the correct solution to the ancient riddle of planetary motion. About this event, Kepler wrote:

> If I had believed that we could ignore these eight minutes [of arc], I would have patched up my hypothesis accordingly. But, since it was not permissible to ignore, those eight minutes pointed the road to a complete reformation in astronomy.

Kepler's key discovery was that planetary orbits are not circles but instead are a special type of oval called an **ellipse**. You probably know how to draw a circle by putting a pencil on the end of a string, tacking the string to a board, and pulling the pencil around (Figure 3.18a). Drawing an ellipse is similar, except that you must stretch the string around *two* tacks (Figure 3.18b). The locations of the two tacks are called the **foci** (singular, **focus**) of the ellipse. The long axis of the ellipse is called its *major axis*, each half of which is called a **semimajor axis**; as we'll see shortly, the length of the semimajor axis is particularly important in astronomy. The short axis is called the *minor axis*. By altering the distance between the two foci while keeping the length of string the same, you can draw ellipses of varying **eccentricity**, a quantity that describes how much an ellipse deviates from a perfect circle (Figure 3.18c). A circle is an ellipse with zero eccentricity, and greater eccentricity means a more elongated ellipse.

Kepler's decision to trust the data over his preconceived beliefs marked an important transition point in the history of science. Once he abandoned perfect circles in favor of ellipses, Kepler soon came up with a model that could predict planetary positions with far greater accuracy than

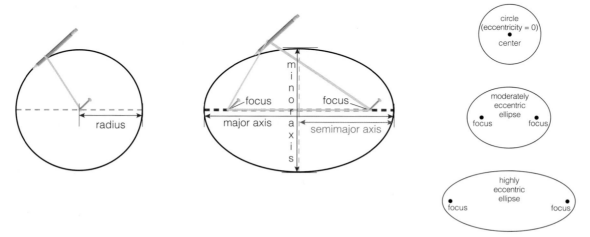

a Drawing a circle with a string of fixed length.

b Drawing an ellipse with a string of fixed length.

c *Eccentricity* describes how much an ellipse deviates from a perfect circle.

Figure 3.18 Interactive Figure An ellipse is a special type of oval. These diagrams show how an ellipse differs from a circle and how different ellipses vary in their eccentricity.

Ptolemy's Earth-centered model. Kepler's model withstood the test of time and became accepted not only as a model of nature but as a deep, underlying truth about planetary motion.

 Orbits and Kepler's Laws Tutorial, Lessons 2–4

• What are Kepler's three laws of planetary motion?

Kepler summarized his discoveries in three simple laws that we now call **Kepler's laws of planetary motion**. He published the first two laws in 1610 and the third in 1618.

Kepler's First Law Kepler's first law tells us that *the orbit of each planet around the Sun is an ellipse with the Sun at one focus* (Figure 3.19). (Nothing is at the other focus.) In essence, this law tells us that a planet's distance from the Sun varies during its orbit. It is closest at the point called **perihelion** (from the Greek for "near the Sun") and farthest at the point called **aphelion** (from the Greek for "away from the Sun"). The *average* of a planet's perihelion and aphelion distances is the length of its *semimajor axis*. We will refer to this simply as the planet's average distance from the Sun.

Kepler's Second Law Kepler's second law states that *as a planet moves around its orbit, it sweeps out equal areas in equal times*. As shown in Figure 3.20, this means the planet moves a greater distance when it is near perihelion than it does in the same amount of time near aphelion. That is, the planet travels faster when it is nearer to the Sun and slower when it is farther from the Sun.

Kepler's Third Law Kepler's third law tells us that *more distant planets orbit the Sun at slower average speeds, obeying*

MATHEMATICAL INSIGHT **3.1** **Eccentricity and Planetary Orbits**

We describe how much a planet's orbit differs from a perfect circle by stating its orbital *eccentricity*. There are several equivalent ways to define the eccentricity of an ellipse, but the simplest is shown in Figure 1. We define c to be the distance from each focus to the center of the ellipse and a to be the length of the semimajor axis. The eccentricity is defined to be

$$e = \frac{c}{a}$$

Notice that $c = 0$ in a perfect circle, because a circle is an ellipse with both foci *in* the center, so this formula gives an eccentricity of 0 for a perfect circle, just as we expect.

You can find the eccentricity of planetary orbits in many tables of planetary data, such as Table E.2 in Appendix E of this book. Once you know the eccentricity, the following formulas allow you to calculate the perihelion and aphelion distances of the planet (Figure 2):

$$\text{perihelion distance} = a(1 - e)$$
$$\text{aphelion distance} = a(1 + e)$$

Example: What are Earth's perihelion and aphelion distances?

Solution:

Step 1 Understand: From the given formulas, we see that calculating Earth's perihelion and aphelion distances requires knowing Earth's orbital eccentricity and the length of its semimajor axis, which is its average distance from the Sun. From Table E.2, Earth's orbital eccentricity is $e = 0.017$ and its average distance is $a = 149.6$ million km.

Step 2 Solve: We use the values we've found for e and a in the above equations:

$$
\begin{aligned}
\text{Earth's perihelion distance} &= a(1 - e) \\
&= (149.6 \times 10^6 \text{ km})(1 - 0.017) \\
&= 147.1 \times 10^6 \text{ km} \\
\text{Earth's aphelion distance} &= a(1 + e) \\
&= (149.6 \times 10^6 \text{ km})(1 + 0.017) \\
&= 152.1 \times 10^6 \text{ km}
\end{aligned}
$$

Step 3 Explain: We have found that Earth's perihelion distance is 147.1 million kilometers and its aphelion distance is 152.1 million kilometers. Because perihelion is Earth's closest point to the Sun and aphelion is its farthest point, we've learned that Earth's distance from the Sun varies between 147.1 and 152.1 million kilometers over the course of each orbit.

Figure 1

Figure 2

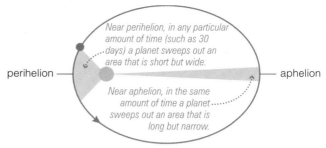

Figure 3.19 *Interactive Figure* Kepler's first law: The orbit of each planet about the Sun is an ellipse with the Sun at one focus. (The eccentricity shown here is exaggerated compared to the actual eccentricities of the planets.)

Figure 3.20 *Interactive Figure* Kepler's second law: As a planet moves around its orbit, it sweeps out equal areas (the shaded regions) in equal times.

a precise mathematical relationship. The relationship is written

$$p^2 = a^3$$

where p is the planet's orbital period in years and a is its average distance from the Sun in astronomical units. Figure 3.21a shows the $p^2 = a^3$ law graphically. Notice that the square of each planet's orbital period (p^2) is indeed equal to the cube of its average distance from the Sun (a^3). Because Kepler's third law relates a planet's orbital distance to its orbital time (period), we can use the law to calculate a planet's average orbital speed.* Figure 3.21b shows the result, confirming that more distant planets orbit the Sun more slowly.

The fact that more distant planets move more slowly led Kepler to suggest that planetary motion might be the result of a force from the Sun. He even speculated about the nature of this force, guessing that it might be related to magnetism. (This idea, shared by Galileo, was first suggested by William Gilbert [1544–1603], an early believer in the Copernican system.) Kepler was right about the existence of a force, but wrong in his guess of magnetism. More than a half century later, Isaac Newton explained planetary motion as a consequence of the gravitational force attracting planets to the Sun [Section 4.4].

---

*To calculate orbital speed from Kepler's third law: Remember that speed = distance/time. For a complete planetary orbit, the distance is orbital circumference, which is $2\pi a$ (where a is the semimajor axis, roughly the "radius" of the orbit); the time is the orbital period, p. Thus, the orbital speed is $(2\pi a)/p$. From Kepler's third law, $p = a^{3/2}$. Plugging this value for p into the orbital speed equation, we find that a planet's orbital speed is $2\pi/\sqrt{a}$; the graph of this equation is the curve in Figure 3.21b.

THINK ABOUT IT

Suppose a comet has an orbit that brings it quite close to the Sun at its perihelion and beyond Mars at its aphelion, but with an average distance (semimajor axis) of 1 AU. According to Kepler's laws, how long would the comet take to complete each orbit of the Sun? Would it spend most of its time close to the Sun, far from the Sun, or somewhere in between? Explain.

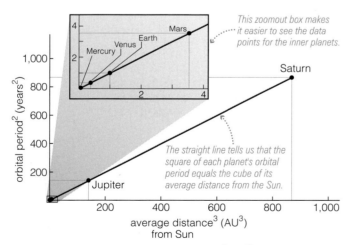

a This graph shows that Kepler's third law ($p^2 = a^3$) does indeed hold true; for simplicity, the graph shows only the planets known in Kepler's time.

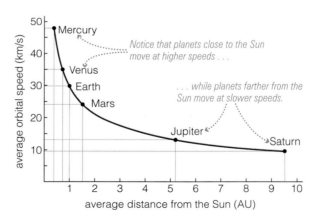

b This graph shows how the orbital speeds of the planets depend on their distances from the Sun: More distant planets orbit the Sun more slowly. (Kepler knew the form of the relationship but could not determine speeds in km/s because the numerical value of the astronomical unit was not yet known.)

Figure 3.21 Graphs based on Kepler's third law.

• How did Galileo solidify the Copernican revolution?

The success of Kepler's laws in matching Tycho's data provided strong evidence in favor of Copernicus's placement of the Sun, rather than Earth, at the center of the solar system. Nevertheless, many scientists still voiced reasonable objections to the Copernican view. There were three basic objections, all rooted in the 2,000-year-old beliefs of Aristotle and other ancient Greeks.

- First, Aristotle had held that Earth could not be moving because, if it were, objects such as birds, falling stones, and clouds would be left behind as Earth moved along its way.
- Second, the idea of noncircular orbits contradicted Aristotle's claim that the heavens—the realm of the Sun, Moon, planets, and stars—must be perfect and unchanging.
- Third, as the ancients had argued, stellar parallax ought to be detectable if Earth orbits the Sun [Section 2.4].

Galileo (1564–1642)

Galileo Galilei (1564–1642), a contemporary and correspondent of Kepler, answered all three objections.

Galileo (nearly always known by only his first name) defused the first objection with experiments that almost single-handedly overturned the Aristotelian view of physics. In particular, he used experiments with rolling balls to demonstrate that a moving object remains in motion *unless* a force acts to stop it (an idea now codified in Newton's first law of motion [Section 4.2]). This contradicted Aristotle's claim that the natural tendency of any

---

MATHEMATICAL INSIGHT 3.2 Kepler's Third Law

When Kepler discovered his third law ($p^2 = a^3$), he knew only that it applied to the orbits of planets around the Sun. However, it turns out to apply much more generally; for example, it works for asteroids and comets as well as planets. We'll see its most general form in Mathematical Insight 4.3, but even in its original form we can use Kepler's third law under the following conditions:

1. In its original form $p^2 = a^3$, Kepler's third law applies only to objects *orbiting the Sun* or another star of precisely the same mass as the Sun.
2. We can use Kepler's third law in its original form only if we measure orbital *periods in years* and orbital *distances in AU*.

Example 1: The largest asteroid, Ceres, orbits the Sun at an average distance (semimajor axis) of 2.77 AU. What is its orbital period?

Solution:

Step 1 Understand: We know that Kepler's third law will allow us to find an orbital period (p) from an average distance (a), but only if the problem meets the two conditions described above. The first condition is met because the asteroid orbits the Sun. The second is met because we are given the orbital distance in AU. Thus, we can use Kepler's third law in its original form.

Step 2 Solve: Notice that Kepler's law is written in a way that gives us p^2, but we are looking for the orbital period, p. Thus, we first solve Kepler's third law for p and then substitute the given value of $a = 2.77$ AU.

$$p^2 = a^3 \implies p = \sqrt{a^3} = \sqrt{2.77^3} = 4.6$$

Notice that because of the special conditions attached to the use of Kepler's third law in its original form, we do *not* include units

when working with it; we know we'll get a period in years as long as we start with a distance in AU.

Step 3 Explain: We have found that Ceres has an orbital period of 4.6 years, meaning that it takes 4.6 years to complete each orbit around the Sun.

Example 2: A new planet is discovered to be orbiting a star with the same mass as our Sun. The planet orbits the star every 3 months. What is its average distance from its star?

Solution:

Step 1 Understand: This time we are given an orbital period and want to find the orbital distance. Again, we can use Kepler's third law in its original form only if the problem meets the two conditions above. The first condition is met because the planet is orbiting a star with the same mass as our Sun. The second is not met, because the orbital period is given in months rather than years. However, we can easily convert it to years: Three months is $\frac{1}{4}$ year, so we have $p = 0.25$ year.

Step 2 Solve: Kepler's law is written in a way that gives us a^2, but we are looking for the average distance, a. We therefore solve Kepler's third law for a and then substitute the orbital period of $p = 0.25$ year:

$$p^2 = a^3 \implies a = \sqrt[3]{p^2} = \sqrt[3]{0.25^2} = 0.40$$

Step 3 Explain: We have found that the planet orbits its star at an average distance of 0.4 AU. By comparing this result to the distances of planets in our own solar system given in Table E.2, we find that this planet's average orbital distance is just slightly larger than that of the planet Mercury in our own solar system.

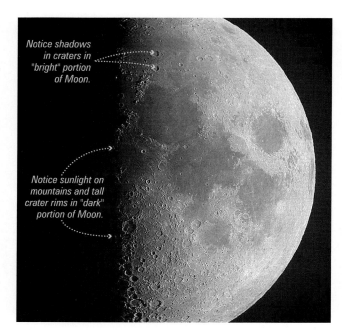

Figure 3.22 Interactive Photo The shadows cast by mountains and crater rims near the dividing line between the light and dark portions of the lunar face prove that the Moon's surface is not perfectly smooth.

Figure 3.23 A page from Galileo's notebook written in 1610. His sketches show four "stars" near Jupiter (the circle) but in different positions at different times (and sometimes hidden from view). Galileo soon realized that the "stars" were actually moons orbiting Jupiter.

moving object is to come to rest. Galileo concluded that objects such as birds, falling stones, and clouds that are moving with Earth should stay with Earth unless some force knocks them away. This same idea explains why passengers in an airplane keep moving with the airplane even when they leave their seats.

Tycho's supernova and comet observations already had challenged the validity of the second objection by showing that the heavens could change. Galileo shattered the idea of heavenly perfection after he built a telescope in late 1609. (Galileo did *not* invent the telescope; Hans Lippershey invented it in 1608. However, Galileo took what was little more than a toy and turned it into a scientific instrument.) Through his telescope, Galileo saw sunspots on the Sun, which were considered "imperfections" at the time. By noticing the shadows cast near the dividing line between the light and dark portions of the lunar face, he was also able to use his telescope to prove that the Moon has mountains and valleys like the "imperfect" Earth (Figure 3.22). If the heavens were in fact not perfect, then the idea of elliptical orbits (as opposed to "perfect" circles) was not so objectionable.

The third objection—the absence of observable stellar parallax—had been of particular concern to Tycho. Based on his estimates of the distances of stars, Tycho believed that his naked-eye observations were sufficiently precise to detect stellar parallax if Earth did in fact orbit the Sun. Refuting Tycho's argument required showing that the stars were more distant than Tycho had thought and therefore too distant for him to have observed stellar parallax. Al-

though Galileo didn't actually prove this fact, he provided strong evidence in its favor. For example, his telescope allowed him to discover that the Milky Way is made up of countless individual stars. This discovery helped him argue that the stars were far more numerous and more distant than Tycho had imagined.

In hindsight, the final nails in the coffin of the Earth-centered universe came with two of Galileo's earliest discoveries through the telescope. First, he observed four moons clearly orbiting Jupiter, *not* Earth (Figure 3.23). (By itself, this observation still did not rule out a stationary, central Earth. However, it showed that moons can orbit a moving planet like Jupiter, which overcame some critics' complaints that the Moon could not stay with a moving Earth.) Soon thereafter, he observed that Venus goes through phases in a way that makes sense only if it orbits the Sun and not Earth (Figure 3.24).

With Earth clearly removed from its position at the center of the universe, the scientific debate turned to the question of whether Kepler's laws were the correct model for our solar system. The most convincing evidence came in 1631, when astronomers observed a transit of Mercury across the Sun's face. Kepler's laws had predicted the transit with overwhelmingly better success than any competing model.

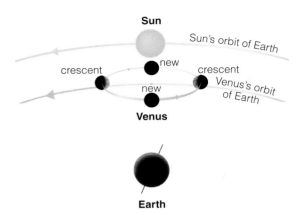

Ptolemaic View of Venus

Copernican View of Venus

a In the Ptolemaic system, Venus orbits Earth, moving around a small circle on its larger orbital circle; the center of the small circle lies on the Earth-Sun line. Thus, if this view were correct, Venus's phases would range only from new to crescent.

b In reality, Venus orbits the Sun, so from Earth we can see it in many different phases. This is just what Galileo observed, allowing him to prove that Venus really does orbit the Sun.

Figure 3.24 Interactive Figure Galileo's telescopic observations of Venus proved that it orbits the Sun rather than Earth.

Although we now recognize that Galileo won the day, the story was more complex in his own time, when Catholic Church doctrine still held Earth to be the center of the universe. On June 22, 1633, Galileo was brought before a Church inquisition in Rome and ordered to recant his claim that Earth orbits the Sun. Nearly 70 years old and fearing for his life, Galileo did as ordered and his life was spared. However, legend has it that as he rose from his knees, he whispered under his breath, *Eppur si muove*—Italian for "And yet it moves." (Given the likely consequences if Church officials had heard him say this, most historians doubt the legend.)

The Church did not formally vindicate Galileo until 1992, but Church officials had given up the argument long before that. Galileo's book, *Dialogue Concerning the Two Chief World Systems*, was removed from the Church's index of banned books in 1824. Today, Catholic scientists are at the forefront of much astronomical research, and official Church teachings are compatible not only with Earth's planetary status but also with the theories of the Big Bang and the subsequent evolution of the cosmos and of life.

3.4 The Nature of Science

The story of how our ancestors gradually figured out the basic architecture of the cosmos exhibits many features of what we now consider "good science." For example, we have seen how models were formulated and tested against observations, and modified or replaced when they failed those tests. The story also illustrates some classic mistakes, such as the apparent failure of anyone before Kepler to

question the belief that orbits must be circles. The ultimate success of the Copernican revolution led scientists, philosophers, and theologians to reassess the various modes of thinking that played a role in the 2,000-year process of discovering Earth's place in the universe. Let's examine how the principles of modern science emerged from the lessons learned in the Copernican revolution.

• How can we distinguish science from nonscience?

Perhaps surprisingly, it turns out to be quite difficult to define the term *science* precisely. The word comes from the Latin *scientia*, meaning "knowledge," but not all knowledge is science. For example, you may know what music you like best, but your musical taste is not a result of scientific study.

The Idealized Scientific Method One reason science is difficult to define is that not all science works in the same way. It is often said that science follows the "scientific method," which means we propose an idea and then perform experiments or observations to test it. As an example of the scientific method, consider what you would do if your flashlight suddenly stopped working. You might wonder why it has stopped working, and you might *hypothesize* that the batteries have died. In other words, you've created a tentative explanation, or **hypothesis**, for the flashlight's failure. A hypothesis is sometimes called an *educated guess*—in this case, it is "educated" because you already know that flashlights need batteries. Your hypothesis then allows you to make a simple prediction: If you replace the batteries with new ones, the flashlight should work. You can test this prediction by replacing the batteries. If the flashlight now works, you've confirmed your hypothesis. If it doesn't, you must revise or discard your hypothesis, usually in favor of some other one that you can also test (such as that the bulb is dead). Figure 3.25 illustrates the basic flow of this process.

However, while the scientific method can be a useful idealization, science rarely progresses in such an orderly way. Science can also progress when someone goes out and looks at nature in a general way in hopes of learning something new and unexpected, rather than conducting a careful set of experiments. For example, Galileo wasn't looking for anything in particular when he pointed his telescope at the sky and made his first startling discoveries. Furthermore, scientists are human beings, and their intuition and personal beliefs inevitably influence their work. Copernicus, for example, adopted the idea that Earth orbits the Sun not because he had carefully tested it but because he believed it made more sense than the prevailing view of an Earth-centered universe. While his intuition guided him to the right general idea, he erred in the specifics because he still clung to Plato's ancient belief that heavenly motion must be in perfect circles.

Given that the idealized scientific method is an overly simplistic characterization of science, how can we tell what is science and what is not? To answer this question, we must look a little deeper at the distinguishing characteristics of scientific thinking.

SPECIAL TOPIC And Yet It Moves

The case of Galileo is often portrayed as a simple example of conflict between science and religion, but the reality was much more complex, with deep divisions inside the Church hierarchy. Perhaps the clearest evidence for a more open-minded Church comes from the case of Copernicus, whose revolutionary work was strongly supported by many Church officials. A less-well-known and even earlier example concerns Nicholas of Cusa (1401–1464), who published a book arguing for a Sun-centered solar system in 1440, more than a century before Copernicus. Nicholas was ordained a priest in the same year that his book was published, and he was later elevated to Cardinal. Clearly, his views caused no problems for Church officials of the time. (Copernicus probably was not aware of this earlier work by Nicholas of Cusa.)

Many other scientists received similar support from within the Church. Indeed, for most of his life, Galileo counted Cardinals (and even the Pope who later excommunicated him) among his friends. Some historians suspect that Galileo got into trouble less for his views than for the way in which he portrayed them. For example, in 1632—just a year before his famous trial—he published a book in which two fictional characters debated the geocentric and Sun-centered views. He named the character taking the geocentric position Simplicio—essentially "simple-minded"—and someone apparently convinced the Pope that the character was meant to represent him. If it was personality rather than belief that got Galileo into trouble, he was not the only one. Another early supporter of Copernicus, Giordano Bruno (1548–1600), drew the wrath of the Church after essentially writing that no rational person could disagree with him—even though the evidence for the Copernican system was still quite weak at the time, as Kepler had not yet discovered his laws of planetary motion. Bruno was branded a heretic and burned at the stake.

The evidence supporting the idea that Earth rotates and orbits the Sun was quite strong by the time of Galileo's trial in 1633, but it was still indirect. Thus, conservatives within the Church could at least make some case against Galileo. Today, we have much more direct proof that Galileo was correct when he supposedly whispered of Earth *Eppur si muove*—"And yet it moves."

French physicist Jean Foucault provided the first direct proof of *rotation* in 1851. Foucault built a large pendulum that he carefully started swinging. Any pendulum tends to swing always in the same plane, but Earth's rotation made Foucault's pendulum appear to twist slowly in a circle. Today, *Foucault pendulums* are a popular attraction at many science centers and museums. A second direct proof that Earth rotates is provided by the *Coriolis effect*, first described by French physicist Gustave Coriolis (1792–1843). The Coriolis effect [Section 10.2], which would not occur if Earth were not rotating, is responsible for things such as the swirling of hurricanes and the fact that missiles that travel great distances on Earth deviate from straight-line paths.

The first direct proof that Earth orbits the Sun came from English astronomer James Bradley (1693–1762). To understand Bradley's proof, imagine that starlight is like rain, falling straight down. If you are standing still you should hold your umbrella straight over your head, but if you are walking through the rain, you should tilt your umbrella forward, because your motion makes the rain appear to be coming down at an angle. Bradley discovered that observing light from stars requires that telescopes be tilted slightly in the direction of Earth's motion—just like the umbrella. This effect is called the *aberration of starlight*. Stellar parallax also provides direct proof that Earth orbits the Sun, and it was first measured in 1838 by German astronomer Friedrich Bessel.

Foucault pendulums are popular attractions at many science centers, museums, and other public places.

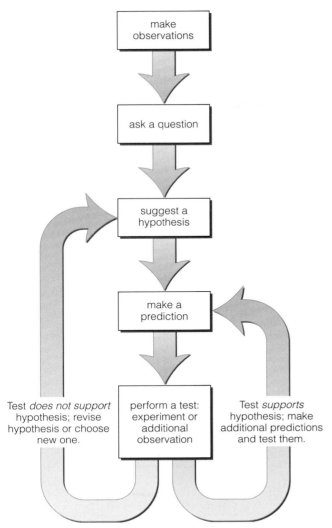

Figure 3.25 This diagram illustrates what we often call the *scientific method*.

Boxes within Figure 3.25:
- make observations
- ask a question
- suggest a hypothesis
- make a prediction
- perform a test: experiment or additional observation
- Test *does not support* hypothesis; revise hypothesis or choose new one.
- Test *supports* hypothesis; make additional predictions and test them.

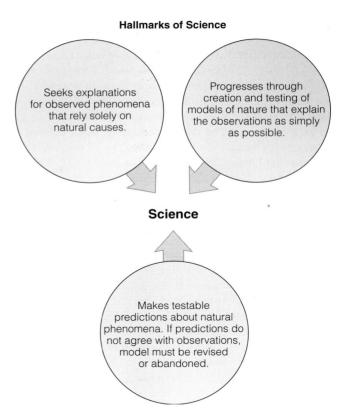

Hallmarks of Science

- Seeks explanations for observed phenomena that rely solely on natural causes.
- Progresses through creation and testing of models of nature that explain the observations as simply as possible.

Science

- Makes testable predictions about natural phenomena. If predictions do not agree with observations, model must be revised or abandoned.

Figure 3.26 Hallmarks of science.

Hallmarks of Science   One way to define scientific thinking is to list the criteria that scientists use when they judge competing models of nature. Historians and philosophers of science have examined (and continue to examine) this issue in great depth, and different experts express somewhat different viewpoints on the details. Nevertheless, everything we now consider to be science shares the following three basic characteristics, which we will refer to as the "hallmarks" of science (Figure 3.26):

- Modern science seeks explanations for observed phenomena that rely solely on natural causes.
- Science progresses through the creation and testing of models of nature that explain the observations as simply as possible.
- A scientific model must make testable predictions about natural phenomena that would force us to revise or abandon the model if the predictions do not agree with observations.

Each of these hallmarks is evident in the story of the Copernican revolution. The first shows up in the way Tycho's exceptionally careful measurements of planetary motion motivated Kepler to come up with a better explanation for those motions. The second is evident in the way several competing models were compared and tested, most notably those of Ptolemy, Copernicus, and Kepler. We see the third in the fact that each model could make precise predictions about the future motions of the Sun, Moon, planets, and stars in our sky. When a model's predictions failed, the model was modified or ultimately discarded. Kepler's model gained acceptance in large part because its predictions matched Tycho's observations much better than Ptolemy's model.

Occam's Razor   The criterion of simplicity in the second hallmark deserves further explanation. Remember that the original model of Copernicus did *not* match the data noticeably better than Ptolemy's model. Thus, if scientists had judged Copernicus's model solely on the accuracy of its predictions, they might have rejected it immediately. However, many scientists found elements of the Copernican model appealing, such as the simplicity of its explanation for apparent retrograde motion. They therefore kept the model alive until Kepler found a way to make it work.

In fact, if agreement with data were the sole criterion for judgment, we could imagine a modern-day Ptolemy adding millions or billions of additional circles to the geocentric model in an effort to improve its agreement with observations. A sufficiently complex geocentric model could in principle reproduce the observations with almost perfect accuracy—but it still would not convince us that Earth is the center of the universe. We would still choose the Copernican view over the geocentric view because its predictions would be just as accurate yet would follow from a much simpler model of nature. The idea that scientists should prefer the simpler of two models that agree equally well with observations is called *Occam's razor*, after the medieval scholar William of Occam (1285–1349).

Verifiable Observations   The third hallmark of science asks the question of what counts as an "observation" against which a prediction can be tested. Consider the claim that aliens are visiting Earth in UFOs. Proponents of this claim say that many thousands of eyewitness observations of UFO encounters provide evidence that it is true. But should these personal testimonials count as *scientific* evidence? On the surface, the answer may not be obvious, because all scientific studies involve eyewitness accounts on some level. For example, only a handful of scientists have personally made detailed tests of Einstein's theory of relativity, and it is their personal reports of the results that have convinced other scientists of the theory's validity. However, there's a very important difference between personal testimony about a scientific test and an observation of a UFO: The first is at least in principle verifiable by anyone, while the second is not.

Understanding this difference is crucial to understanding what counts as science and what does not. Even though you may never have conducted a test of Einstein's theory of relativity yourself, there's nothing stopping you from doing so. It might require several years of study before you have the necessary background to conduct the test, but you could then confirm the results reported by other scientists. In other words, while you may currently be trusting the eyewitness testimony of scientists, you always have the option of verifying their testimony for yourself.

In contrast, there is no way for you to verify someone's eyewitness account of a UFO. Without hard evidence such as photographs or pieces of the UFO, there is nothing that you could evaluate for yourself, even in principle. (And in those cases where "hard evidence" for UFO sightings has been presented, scientific study has never yet found the evidence to be strong enough to support the claim of alien spacecraft [Section 24.5].) Moreover, scientific studies of eyewitness testimony show it to be notoriously unreliable. For example, different eyewitnesses often disagree on what they saw even immediately after an event has occurred. As time passes, memories of the event may change further. In some cases in which memory has been checked against reality, people have reported vivid memories of events that never happened at all. This explains something that virtually all of us have experienced: disagreements with a friend about who did what and when. Since both people cannot

be right in such cases, at least one person must have a memory that differs from reality.

Because of its demonstrated unreliability, eyewitness testimony alone should *never* be used as evidence in science, no matter who reports it or how many people offer similar testimony. It can be used in support of a scientific model only when it is backed up by independently verifiable evidence that anyone could in principle check for him- or herself. (For the same reason, eyewitness testimony alone is usually insufficient for a conviction in criminal court; at least some other evidence, such as motive, is required.)

Science and Pseudoscience   It's important to realize that science is not the only valid way of seeking knowledge. For example, suppose you are shopping for a car, learning to play drums, or pondering the meaning of life. In each case, you might make observations, exercise logic, and test hypotheses. Yet these pursuits clearly are not science, because they are not directed at developing testable explanations for observed natural phenomena. As long as nonscientific searches for knowledge make no claims about how the natural world works, they do not conflict with science.

However, you will often hear claims about the natural world that seem to be based on observational evidence but do not treat evidence in a truly scientific way. Such claims are often called **pseudoscience**, which literally means "false science." To distinguish real science from pseudoscience,

---

COMMON MISCONCEPTIONS

Eggs on the Equinox

One of the hallmarks of science holds that you needn't take scientific claims on faith. In principle, at least, you can always test them for yourself. Consider the claim, repeated in news reports every year, that the spring equinox is the only day on which you can balance an egg on its end. Many people believe this claim, but you'll be immediately skeptical if you think about the nature of the spring equinox. The equinox is merely a point in time at which sunlight strikes both hemispheres equally (see Figure 2.15). It's difficult to see how sunlight could affect an attempt to balance eggs (especially if the eggs are indoors), and there is no difference in the strength of either Earth's gravity or the Sun's gravity on that day compared to any other day.

More important, you can test this claim directly. It's not easy to balance an egg on its end, but with practice you'll find that you can do it on any day of the year, not just on the spring equinox. Not all scientific claims are so easy to test for yourself, but the basic lesson should be clear: Before you accept any scientific claim, you should demand at least a reasonable explanation of the evidence that backs it up.

a good first step is to check whether a particular claim exhibits all three hallmarks of science. Consider the example of people who claim a psychic ability to "see" the future. If they truly can see the future, psychics should be able to make specific, testable predictions. In this sense, "seeing" the future sounds scientific, since we can test it. However, numerous studies have tested the predictions of "seers" and have found that their predictions come true no more often than would be expected by pure chance. If the "seers" were scientific, they would admit that this evidence undercuts their claim of psychic abilities. Instead, they generally make excuses. For example, they might say that the predictions didn't come true because of some type of "psychic interference." Making testable claims but then ignoring the results of the tests marks as pseudoscience the claimed ability to see the future.

Objectivity in Science The idea that science is objective, meaning that all people should be able to find the same results, is very important to the validity of science as a means of seeking knowledge. However, there is a difference between the overall objectivity of science and the objectivity of individual scientists. In particular, because science is practiced by human beings, individual scientists bring their personal biases and beliefs to their scientific work.

Personal bias can influence the way a scientist proposes or tests a model. For example, most scientists choose their research projects based on personal interests rather than on some objective formula. In some extreme cases, scientists have even been known to cheat—either deliberately or subconsciously—to obtain a result they desire. For example, in the late nineteenth and early twentieth centuries, some astronomers (including, most notably, Percival Lowell) claimed to see artificial canals in their blurry telescopic images of Mars [Section 9.4]. They hypothesized that Mars was home to a dying civilization that used the canals to transport water to thirsty cities (leading H. G. Wells to write his novel *The War of the Worlds*). However, no such canals actually exist. These astronomers apparently allowed their beliefs about extraterrestrial life to influence the way they interpreted blurry images—in essence, a form of cheating, though certainly not intentional.

Bias can sometimes show up even in the thinking of the scientific community as a whole. Some valid ideas may not be considered by any scientist because the ideas fall too far outside the general patterns of thought, or **paradigm**, of the time. Einstein's theory of relativity provides an example. Many scientists in the decades before Einstein had gleaned hints of the theory but did not investigate them, at least in part because the ideas seemed too outlandish. (Chapters S2 and S3 discuss Einstein's special and general theories of relativity.)

The beauty of science is that it encourages continued testing by many people. Even if personal biases affect

---

SPECIAL TOPIC Logic and Science

In science, we attempt to acquire new knowledge through logical reasoning. We carry out the process of reasoning by constructing arguments in which we begin with a set of premises and try to draw appropriate conclusions. This concept of a logical argument differs somewhat from the definition of argument in everyday life. In particular, a logical argument need not imply any animosity or dissension.

There are two basic types of argument: *deductive* and *inductive*. Both are important in science.

The following is a simple example of a deductive argument:

PREMISE: All planets orbit the Sun in ellipses with the Sun at one focus.

PREMISE: Earth is a planet.

CONCLUSION: Earth orbits the Sun in an ellipse with the Sun at one focus.

As long as the two premises are true, the conclusion must also be true—this is the essence of deduction. Note the construction of a deductive argument: The first premise is a general statement that applies to all planets, and the conclusion is a specific statement that applies only to Earth. As this example suggests, we often use deduction to *deduce* a specific prediction from a more general theory. If the specific prediction proves to be false, then something must be wrong with the premises from which it was deduced. If it proves true, then we've acquired a piece of evidence in support of the premises.

Contrast the deductive argument above with the following example of an *inductive argument*:

PREMISE: Birds fly up into the air but eventually come back down.

PREMISE: People who jump into the air fall back down.

PREMISE: Rocks thrown into the air come back down.

PREMISE: Balls thrown into the air come back down.

CONCLUSION: What goes up must come down.

Because each premise supports the conclusion, you might at first agree that this is a strong inductive argument and that its conclusion probably is true. However, no matter how many more examples you consider of objects that go up and come down, you could never *prove* that the conclusion is true—only that it seems likely to be true. Moreover, a single counterexample can prove the conclusion of an inductive argument to be false. For example, in this case we know that the conclusion is false because the spacecraft *Voyager 2* was launched from Earth and never fell back down.

Inductive arguments *generalize* from specific facts to a broader model or theory. Thus, inductive arguments are the arguments used to build scientific theories, because we use them to *infer* general principles from observations and experiments. That is why theories can never be proved true beyond all doubt—they can only be shown to be consistent with ever larger bodies of evidence. Theories *can* be proved false, however, if they fail to account for observed or experimental facts.

some results, tests by others will eventually uncover the mistakes. Similarly, if a new idea is correct but falls outside the accepted paradigm, sufficient testing and verification of the idea will eventually force a change in the paradigm. Thus, although individual scientists rarely follow the idealized scientific method, the collective action of many scientists over many years ensures that science as a whole remains objective.

• What is a scientific theory?

The most successful scientific models explain a wide variety of observations in terms of just a few general principles. When a powerful yet simple model makes predictions that survive repeated and varied testing, scientists elevate its status and call it a **theory**. Some famous examples are Isaac Newton's theory of gravity, Charles Darwin's theory of evolution, and Albert Einstein's theory of relativity.

Note that the scientific use of the word *theory* is quite different from its everyday use, in which we equate a theory more closely with speculation or a hypothesis. In everyday life, someone might get a new idea and say, for example, "I have a new theory about why people enjoy the beach." Without the support of a broad range of evidence that others have tested and confirmed, this idea is really only a hypothesis. In contrast, Newton's theory of gravity qualifies as a scientific theory because it uses simple physical principles to explain a great many observations and experiments.

Despite its success in explaining observed phenomena, a scientific theory can never be proved true beyond all doubt, because ever more sophisticated observations may eventually disagree with its predictions. However, anything that qualifies as a scientific theory must be supported by a large, compelling body of evidence.

In this sense, a scientific theory is not at all like a hypothesis or any other type of guess. We are free to change a hypothesis at any time, because it has not yet been carefully tested. In contrast, we can discard or replace a scientific theory only if we have an alternate way of explaining the evidence that supports it.

Again, the theories of Newton and Einstein offer good examples. A vast body of evidence supports Newton's theory of gravity, but by the late 1800s scientists had begun to discover cases where its predictions did not perfectly match observations. These discrepancies were explained only when Einstein developed his general theory of relativity in the early twentieth century, which was able to match the observations. Still, the many successes of Newton's theory could not be ignored, and Einstein's theory would not have gained acceptance if it had not been able to explain these successes equally well. It did, and that is why we now view Einstein's theory as a broader theory of gravity than Newton's theory. Some scientists today are seeking a theory of gravity that will go beyond Einstein's. If any new theory ever gains acceptance, it will have to match all the successes of Einstein's theory as well as work in new realms where Einstein's theory does not.

THINK ABOUT IT

When people claim that something is "only a theory," what do you think they mean? Does this meaning of "theory" agree with the definition of a theory in science? Do scientists always use the word *theory* in its "scientific" sense? Explain.

3.5 Astrology

We have discussed the development of astronomy and the nature of science in some depth. Now let's talk a little about a subject often confused with the science of astronomy: the similarly named subject of *astrology*. Today, astronomy and astrology are very different practices. In ancient times, however, astrology and astronomy often went hand in hand, and astrology played an important role in the historical development of astronomy.

• How is astrology different from astronomy?

The basic tenet of astrology is that the apparent positions of the Sun, Moon, and planets among the stars in our sky influence human events. The origins of this idea are easy to understand. After all, the position of the Sun in the sky certainly influences our lives, since it determines the seasons and the times of daylight and darkness. Similarly, the Moon directly affects us by creating the tides and sometimes providing moonlight to see by at night. Because the planets also appear to move among the stars, it seemed reasonable to ancient people that planets also influence our lives, even if these influences were much more subtle and difficult to discover.

Ancient astrologers hoped that they might learn *how* the positions of the Sun, Moon, and planets influence our lives. They charted the skies, seeking correlations with events on Earth. For example, if an earthquake occurred when Saturn was entering the constellation Leo, might Saturn's position have caused the earthquake? If the king became ill when Mars appeared in the constellation Gemini and the first-quarter moon appeared in Scorpio, might it mean another tragedy for the king when this particular alignment of the Moon and Mars next recurred? Surely, the ancient astrologers thought, the patterns of influence would eventually become clear. The astrologers hoped that they might someday learn to forecast human events with the same reliability with which astronomical observations of the Sun could forecast the coming of spring.

Because forecasts of the seasons and forecasts of human events were imagined to be closely related, astrologers and astronomers usually were one and the same in the ancient world. For example, in addition to his books on astronomy, Ptolemy published a treatise on astrology called *Tetrabiblios* that remains the foundation for much of astrology today. But Ptolemy himself recognized that astrology stood upon a far shakier foundation than astronomy. In the introduction to *Tetrabiblios*, Ptolemy compared astronomical and astrological predictions:

> [Astronomy], which is first both in order and effectiveness, is that whereby we apprehend the aspects of the movements of sun, moon, and stars in relation to each other and to the earth. . . . I shall now give an account of the second and less sufficient method [of prediction (astrology)] in a proper philosophical way, so that one whose aim is the truth might never compare its perceptions with the sureness of the first, unvarying science. . . .

Other ancient scientists probably also recognized that their astrological predictions were far less reliable than their astronomical ones. Nevertheless, if there were even a slight possibility that astrologers could forecast the future, no king or political leader would dare to be without one. Astrologers held esteemed positions as political advisers in the ancient world and were provided with the resources they needed to continue charting the heavens and human history. Indeed, wealthy political leaders' support of astrology made possible much of the development of ancient astronomy.

Throughout the Middle Ages and into the Renaissance, many astronomers continued to practice astrology. For example, Kepler cast numerous *horoscopes*—the predictive charts of astrology (Figure 3.27)—even as he was discovering the laws of planetary motion. However, given Kepler's later description of astrology as "the foolish stepdaughter of astronomy" and "a dreadful superstition," he may have cast the horoscopes solely as a source of much-needed income. Modern-day astrologers also claim Galileo as one of their own, in part for his having cast a horoscope for the

Grand Duke of Tuscany. However, Galileo's astronomical discoveries changed human history, while the horoscope he cast was just plain wrong: The Duke died just a few weeks after Galileo's horoscope predicted that he would have a long and fruitful life.

The scientific triumph of Kepler and Galileo in showing Earth to be a planet orbiting the Sun heralded the end of the linkage between astronomy and astrology. Astronomy has since gained status as a successful science that helps us understand our universe, while astrology no longer has any connection to the modern science of astronomy.

• Does astrology have any scientific validity?

Although astronomers gave up on it centuries ago, astrology remains popular with the general public. Many people read their daily horoscopes in newspapers, and some pay significant fees to have personal horoscopes cast by "professional" astrologers. In fact, more people earn incomes by casting horoscopes than through astronomical research, and books and articles on astrology often outsell all but the most popular books on astronomy. With so many people giving credence to astrology, is it possible that it has some scientific validity after all?

Testing Astrology   The question of scientific validity can be more difficult to answer than it sounds, because there's no general agreement among astrologers even on such basic things as what astrology is or what it can predict. For example, "Western astrology" is quite different in nature from the astrology practiced in India and China. Some astrologers no longer make testable predictions at all; rather, they give vague guidance about how to live one's life. Most newspaper horoscopes fall into this category. Although your horoscope may seem to ring true at first, a careful read will usually show it to be so vague as to be untestable. For example, a horoscope that says, "It is a good day to spend time with your friends" may be good advice but doesn't offer much to test. If you read the horoscopes for all 12 astrological signs, you'll probably find that several of them apply equally well to you. When astrology offers only vague advice rather than testable predictions, the question of scientific validity does not apply.

THINK ABOUT IT

Look in a local newspaper for today's weather forecast and for your horoscope. Contrast the nature of their predictions. By the end of the day, you will know if the weather forecast was accurate. Will you know whether your horoscope was accurate? Explain.

Nevertheless, most professional astrologers still earn their livings by casting horoscopes that either predict future events in an individual's life or describe characteristics of the person's personality and life. If the horoscope predicts

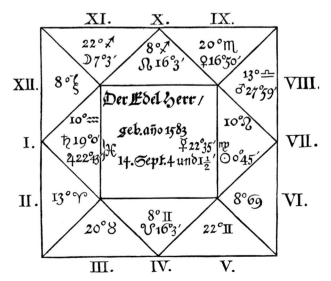

Figure 3.27 This chart is an example of a horoscope, in this case one cast by Kepler, who called astrology "a dreadful superstition."

future events, we can check to see whether the predictions come true. If it describes a person's personality and life, the description can be checked for accuracy. A scientific test of astrology requires evaluating many horoscopes and comparing their accuracies to what would be expected by pure chance. For example, suppose a horoscope states that a person's best friend is female. Because roughly half the population of the United States is female, an astrologer who casts 100 such horoscopes would be expected by pure chance to be right about 50 times. Thus, we would be impressed with the predictive ability of the astrologer only if he or she were right much more often than 50 times out of 100.

In hundreds of scientific tests, astrological predictions have never proved to be accurate by a substantially greater margin than expected from pure chance. Similarly, in tests in which astrologers are asked to cast horoscopes for people they have never met, the horoscopes fail to match actual personality profiles more often than expected by chance. The verdict is clear: The methods of astrology are useless for predicting the past, the present, or the future.

Examining the Underpinnings of Astrology   In science, observations and experiments are the ultimate judge of any idea. No matter how outlandish an idea might appear, it cannot be dismissed if it successfully meets observational or experimental tests. The idea that Earth rotates and orbits the Sun at one time seemed outlandish, yet today it is so strongly supported by the evidence that we consider it a fact. The idea that the positions of the Sun, Moon, and planets among the stars influence our lives might sound outlandish today, but if astrology were to make predictions that came true, adherence to the principles of science would force us to take astrology seriously. However, given that scientific tests of astrology have never found any evidence that its predictive methods work, it is worth looking at its premises to see whether they make sense. Might there be a few kernels of wisdom buried within the lore of astrology?

Let's begin with one of the key premises of astrology: There is special meaning in the patterns of the stars in the constellations. This idea may have seemed quite reasonable in ancient times, when the stars were assumed to be fixed on an unchanging celestial sphere. Today we know that the patterns of the stars in the constellations are accidents of the moment. Long ago the constellations did not look the same, and they will look still different far in the future [Section 1.3]. Moreover, the stars in a constellation don't necessarily have any *physical* association (see Figure 2.3). Because stars vary in distance, two stars that appear on opposite sides of our sky might well be closer together than two stars in the same constellation. Constellations are only *apparent* associations of stars, with no more physical reality than the water in a desert mirage.

Astrology also places great importance on the positions of the planets among the constellations. Again, this idea might have seemed quite reasonable in ancient times, when it was thought that the planets truly wandered among the stars. Today we know that the planets only *appear* to wander among the stars. In reality, the planets are in our own solar system, while the stars are vastly farther away. It is difficult to see how mere appearances could have profound effects on our lives.

Many other ideas at the heart of astrology are equally suspect. For example, most astrologers claim that a proper horoscope must account for the positions of *all* the planets. Does that mean that all horoscopes cast before the discoveries of Neptune in 1846 and Pluto in 1930 were invalid? If so, why didn't astrologers notice that something was wrong with their horoscopes and predict the existence of these planets? (Astronomers *did* predict the existence of Neptune by using Newton's universal law of gravitation; see p. 130.) In recent years, astronomers have discovered many Pluto-like objects orbiting in the outer region of the solar system [Section 12.3]. Given their similarity to Pluto, shouldn't these objects also be included when casting horoscopes, and does that mean that horoscopes cannot be accurate until we've discovered every single one of these objects? What about planets orbiting other stars, which far outnumber the planets of our own solar system? Given seemingly unanswerable questions like these, there seems little hope that astrology will ever meet its ancient goal of forecasting human events.

THE BIG PICTURE

Putting Chapter 3 into Context

In this chapter, we focused on the scientific principles through which we have learned so much about the universe. Key "big picture" concepts from this chapter include the following:

- The basic ingredients of scientific thinking—careful observation and trial-and-error testing—are a part of everyone's experience. Modern science simply provides a way of organizing this everyday thinking to facilitate the learning and sharing of new knowledge.

- Although our understanding of the universe is growing rapidly today, each new piece of knowledge rests on ideas that came before.

- The Copernican revolution, which overthrew the ancient Greek belief in an Earth-centered universe, did not occur instantaneously. It unfolded over a period of more than a century, during which many of the characteristics of modern science first appeared.

- Several key hallmarks distinguish science from other ways of gathering knowledge: Science seeks natural explanations for observed phenomena, it progresses through the creation and testing of models that explain the observations as simply as possible, and it revises or abandons models whose predictions disagree with observations.

- Astronomy and astrology once developed hand in hand, but today they represent very different things.

3.1 The Ancient Roots of Science

- **In what ways do all humans employ scientific thinking?** Scientific thinking relies on the same type of trial-and-error thinking that we use in our everyday lives, but in a carefully organized way.

- **How did astronomical observations benefit ancient societies?** Ancient cultures used astronomical observations to help them keep track of time and the seasons, crucial skills for people who depended on agriculture for survival.

- **What did ancient civilizations achieve in astronomy?** Ancient astronomers were accomplished observers who learned to tell the time of day and the time of year, to track cycles of the Moon, and to observe planets and stars. Many ancient structures aided in astronomical observations.

3.2 Ancient Greek Science

- **Why does modern science trace its roots to the Greeks?** The Greeks developed **models** of nature and emphasized the importance of having the predictions of those models agree with observations of nature.

- **How did the Greeks explain planetary motion?** The Greek **geocentric model** reached its culmination with the model of Ptolemy. The **Ptolemaic model** explained apparent retrograde motion by having each planet move on a small circle whose center moves around Earth on a larger circle.

  retrograde loop

- **How was Greek knowledge preserved through history?** While Europe was in its Dark Ages, Islamic scholars in the Middle East preserved and extended ancient Greek knowledge. After the fall of Constantinople, scholars moved west to Europe, where their knowledge helped ignite the European Renaissance.

3.3 The Copernican Revolution

- **How did Copernicus, Tycho, and Kepler challenge the Earth-centered idea?** Copernicus created a Sun-centered model of the solar system designed to replace the Ptolemaic model, but it was no more accurate than Ptolemy's because Copernicus still used perfect circles. Tycho's accurate, naked-eye observations provided the data needed to improve on Copernicus's model. Kepler found a model of planetary motion that fit Tycho's data.

- **What are Kepler's three laws of planetary motion?** (1) The orbit of each planet is an ellipse with the Sun at one focus. (2) As a planet moves around its orbit, it sweeps out equal areas in equal times. (3) More distant planets orbit the Sun at slower average speeds, obeying the precise mathematical relationship $p^2 = a^3$.

- **How did Galileo solidify the Copernican revolution?** Galileo's experiments and telescopic observations overcame remaining objections to the Copernican idea of Earth as a planet orbiting the Sun. Although not everyone accepted his results immediately, in hindsight we see that Galileo sealed the case for the Sun-centered solar system.

  Venus

  Earth

3.4 The Nature of Science

- **How can we distinguish science from nonscience?** Science generally exhibits these three hallmarks: (1) Modern science seeks explanations for observed phenomena that rely solely on natural causes. (2) Science progresses through the creation and testing of models of nature that explain the observations as simply as possible. (3) A scientific model must make testable predictions about natural phenomena that would force us to revise or abandon the model if the predictions do not agree with observations.

- **What is a scientific theory?** A scientific **theory** is a simple yet powerful model that explains a wide variety of observations in terms of just a few general principles, and has attained the status of a theory by surviving repeated and varied testing.

3.5 Astrology

- **How is astrology different from astronomy?** Astronomy and astrology share common roots in ancient observations of the sky but they are very different today. Astronomy is a modern science that has taught us much about the universe. Astrology is a search for hidden influences on human lives based on the positions of planets and stars in the sky.

- **Does astrology have any scientific validity?** Scientific tests have shown that astrological predictions do not prove to be accurate more than we can expect by pure chance, showing that the predictions have no scientific validity.

Review Questions

Short-Answer Questions Based on the Reading

1. In what way is scientific thinking natural to all of us? How does modern science differ from this everyday type of thinking?
2. Why did ancient peoples study astronomy? Describe the astronomical achievements of at least four ancient cultures.
3. How are the names of the seven days of the week related to astronomical objects?
4. Describe at least three ways that ancient people determined either the time of day or the time of year.
5. What is a lunar calendar? What is the Metonic cycle? Explain why the dates of Ramadan cycle through our solar calendar while the dates of Jewish holidays and Easter remain within about a 1-month period.
6. What do we mean by a *model* in science?
7. Summarize the development of the Greek *geocentric model*.
8. Who was Ptolemy? How did the Ptolemaic model account for the apparent retrograde motion of planets in our sky?
9. What was the *Copernican revolution*, and how did it change the human view of the universe?
10. Why wasn't the Copernican model immediately accepted? Describe the roles of Tycho, Kepler, and Galileo in the eventual triumph of the Sun-centered model.
11. What is an *ellipse*? Define the *focus* and the *eccentricity* of an ellipse. Why are ellipses important in astronomy?
12. Clearly state each of *Kepler's laws of planetary motion*. For each law, describe in your own words what it means in a way that almost anyone could understand.
13. What is the difference between a *hypothesis* and a *theory* in science?
14. Describe each of the three hallmarks of science and give an example of how we can see each one in the unfolding of the Copernican revolution. What is *Occam's razor*? Why doesn't science accept personal testimony as evidence?
15. What do we mean by *pseudoscience*? How is it different from other types of nonscience?
16. What is the basic idea behind *astrology*? Explain why this idea seemed reasonable in ancient times but is no longer given credence by scientists.

Test Your Understanding

Does It Make Sense?

Decide whether the statement makes sense (or is clearly true) or does not make sense (or is clearly false). Explain your reasoning. (For an example, see Chapter 1, "Does It Make Sense?")

17. Ancient astronomers failed to realize that Earth goes around the Sun because they just weren't as smart as people today.
18. In ancient Egypt, children whose parents gave them "1 hour" to play got to play longer in the summer than in the winter.
19. If the planet Uranus had been identified as a planet in ancient times, we'd probably have eight days in a week.
20. The date of Christmas (December 25) is set each year according to a lunar calendar.

21. When navigating in the South Pacific, the Polynesians found their latitude with the aid of the pointer stars of the Big Dipper.
22. The Ptolemaic model reproduced apparent retrograde motion by having planets move sometimes counterclockwise and sometimes clockwise in their circles.
23. According to Kepler's laws, Earth would take longer to orbit the Sun if it had a larger mass.
24. In science, saying that something is a theory means that it is really just a guess.
25. A scientific theory should never gain acceptance until it has been proved true beyond all doubt.
26. Ancient astronomers were convinced of the validity of astrology as a tool for predicting the future.

Quick Quiz

Choose the best answer to each of the following. Explain your reasoning with one or more complete sentences.

27. Stonehenge was useful for: (a) telling the time of day; (b) determining the season; (c) predicting lunar eclipses.
28. With each 19-year Metonic cycle: (a) the lunar phases repeat on the same dates of the year; (b) solar eclipses repeat at the same times and places; (c) Ramadan occurs on the same dates of the year.
29. In the Greek geocentric model, the retrograde motion of a planet occurs when: (a) Earth is about to pass the planet in its orbit around the Sun; (b) The planet actually goes backward in its orbit around Earth; (c) The planet is aligned with the Moon in our sky.
30. Which of the following was *not* a major advantage of Copernicus's Sun-centered model over the Ptolemaic model? (a) It made significantly better predictions of planetary positions in our sky. (b) It offered a more natural explanation for the apparent retrograde motion of planets in our sky. (c) It allowed calculation of the orbital periods and distances of the planets.
31. When we say that a planet has a highly eccentric orbit, we mean that: (a) it is spiraling in toward the Sun; (b) its orbit is an ellipse with the Sun at one focus; (c) in some parts of its orbit it is much closer to the Sun than in other parts.
32. Earth is closer to the Sun in January than in July. Therefore: (a) Earth travels faster in its orbit around the Sun in July than in January. (b) Earth travels faster in its orbit around the Sun in January than in July. (c) It is summer in January and winter in July.
33. According to Kepler's third law: (a) Mercury travels fastest in the part of its orbit in which it is closest to the Sun. (b) Jupiter orbits the Sun at a faster speed than Saturn. (c) Pluto has a highly eccentric orbit.
34. Tycho Brahe's contribution to astronomy included: (a) inventing the telescope; (b) proving that Earth orbits the Sun; (c) collecting data that enabled Kepler to discover the laws of planetary motion.
35. Galileo's contribution to astronomy included: (a) discovering the laws of planetary motion; (b) discovering the law of

gravity; (c) making observations and conducting experiments that dispelled scientific objections to the Sun-centered model.

36. Which of the following is not true about scientific progress? (a) Science progresses through the creation and testing of models of nature. (b) Science advances through the scientific method. (c) Science avoids explanations that invoke the supernatural.

Investigate Further
In-Depth Questions to Increase Your Understanding

Short-Answer/Essay Questions

37. *Earth's Shape.* It took thousands of years for humans to deduce that Earth is spherical. For each of the following alternative models of Earth's shape, identify one or more observations that you could make for yourself and that would invalidate the model.
    a. A flat Earth.
    b. A cylindrical Earth (shaped like a fallen log), like that proposed by Anaximander.
    c. A football-shaped Earth.

38. *Lunar Calendars.*
    a. Find the dates of the Jewish festival of Chanukah (also spelled "Hanukah") for this year and the next three years. Based on what you have learned in this chapter, explain why the dates change as they do.
    b. Find the dates of the Muslim fast for Ramadan for this year and the next three years. Based on what you have learned in this chapter, explain why the dates change as they do.

39. *Copernican Players.* Using a bulleted list format, make a one-page "executive summary" of the major roles that Copernicus, Tycho, Kepler, and Galileo played in overturning the ancient belief in an Earth-centered universe.

40. *What Makes It Science?* Choose a single idea in the modern view of the cosmos as discussed in Chapter 1, such as "The universe is expanding," "The universe began with a Big Bang," "We are made from elements manufactured by stars," or "The Sun orbits the center of the Milky Way Galaxy once every 230 million years."
    a. Briefly describe how the idea you have chosen is rooted in each of the three hallmarks of science discussed in this chapter. (That is, explain how it is based on observations, how our understanding of it depends on a model, and how the model is testable.)
    b. No matter how strongly the evidence may support a scientific idea, we can never be certain beyond all doubt that the idea is true. For the idea you have chosen, describe an observation that might cause us to call the idea into question. Then briefly discuss whether you think that, overall, the idea is likely or unlikely to hold up to future observations. Defend your opinion.

41. *Influence on History.* Based on what you have learned about the Copernican revolution, write a one- to two-page essay about how you believe it altered the course of human history.

42. *Scientific Test of Astrology.* Find out about at least one scientific test that has been conducted to test the validity of astrology. Write a short summary of how the test was conducted and what conclusions were reached.

43. *Your Own Astrological Test.* Devise your own scientific test of astrology. Clearly define the methods you will use in your test and how you will evaluate the results. Then carry out the test and write a report on your methods and results.

44. *Cultural Astronomy.* Choose a particular culture of interest to you, and research the astronomical knowledge and accomplishments of that culture. Write a two- to three-page summary of your findings.

45. *Astronomical Structures.* Choose an ancient astronomical structure of interest to you (e.g., Stonehenge, Templo Mayor, Pawnee lodges) and research its history. Write a two- to three-page summary of your findings. If possible, also build a scale model of the structure or create detailed diagrams to illustrate how the structure was used.

46. *Venus and the Mayans.* The planet Venus apparently played a particularly important role in Mayan society. Research the evidence and write a one- to two-page summary of current knowledge about the role of Venus in Mayan society.

Quantitative Problems
Be sure to show all calculations clearly and state your final answers in complete sentences.

47. *The Metonic Cycle.* The length of our calendar year is 365.2422 days, and the Moon's monthly cycle of phases averages 29.5306 days in length. By calculating the number of days in each, confirm that 19 solar years is almost precisely equal to 235 cycles of the lunar phases. Show your work clearly, then write a few sentences explaining how this fact can be used to keep a lunar calendar roughly synchronized with a solar calendar.

48. *Chinese Calendar.* The traditional Chinese lunar calendar has 12 months in most years but adds a thirteenth month to 22 of every 60 years. How many days does this give the Chinese calendar in each 60-year period? How does this compare to the number of days in 60 years on a solar calendar? Based on your answers, explain how this scheme is similar to the scheme used by lunar calendars that follow the Metonic cycle. (*Hint:* You'll need the data given in Problem 47.)

49. *Method of Eratosthenes I.* You are an astronomer on planet Nearth, which orbits a distant star. It has recently been accepted that Nearth is spherical in shape, though no one knows its size. One day, while studying in the library of Alectown, you learn that on the equinox your sun is directly overhead in the city of Nyene, located 1,000 kilometers due north of you. On the equinox, you go outside in Alectown and observe that the altitude of your sun is 80°. What is the circumference of Nearth? (*Hint:* Apply the technique used by Eratosthenes to measure Earth's circumference.)

50. *Method of Eratosthenes II.* You are an astronomer on planet Tirth, which orbits a distant star. It has recently been accepted that Tirth is spherical in shape, though no one knows its size. One day, you learn that on the equinox your sun is directly overhead in the city of Tyene, located 400 kilometers due north of you. On the equinox, you go outside in Alectown and observe that the altitude of your sun is 86°. What is the circumference of Tirth? (*Hint:* Apply the technique used by Eratosthenes to measure Earth's circumference.)

51. *Mars Orbit.* Find the perihelion and aphelion distances of Mars. (*Hint:* You'll need data from Appendix E.)

52. *Most Eccentric Orbit.* Which planet in our solar system has the most eccentric orbit? Find the planet's perihelion and aphelion distances.

53. *Least Eccentric Orbit.* Which planet in our solar system has the least eccentric orbit? Find the planet's perihelion and aphelion distances.

54. *Sedna Orbit.* The recently discovered object Sedna orbits our Sun beyond Pluto, at an average distance (semimajor axis) of 509 AU. What is its orbital period?

55. *New Planet Orbit.* A newly discovered planet orbits a distant star with the same mass as the Sun at an average distance of 112 million kilometers. Its orbital eccentricity is 0.3. Find the planet's orbital period and its nearest and farthest orbital distances from its star.

56. *Halley Orbit.* Halley's comet orbits the Sun every 76.0 years and has an orbital eccentricity of 0.97.
   a. Find its semimajor axis distance.
   b. Find its perihelion and aphelion distances.
   c. Does Halley's comet spend most of its time near its perihelion distance, its aphelion distance, or halfway in between? Explain.

Discussion Questions

57. *The Impact of Science.* The modern world is filled with ideas, knowledge, and technology that developed through science and application of the scientific method. Discuss some of these things and how they affect our lives. Which of these impacts do you think are positive? Which are negative? Overall, do you think science has benefited the human race? Defend your opinion.

58. *The Importance of Ancient Astronomy.* Why was astronomy important to people in ancient times? Discuss both the practical importance of astronomy and the importance it may have had for religious or other traditions. Which do you think was more important in the development of ancient astronomy, its practical or its philosophical role? Defend your opinion.

59. *Astronomy and Astrology.* Why do you think astrology remains so popular around the world even though it has failed all scientific tests of its validity? Do you think the popularity of astrology has any positive or negative social consequences? Defend your opinions.

MEDIA EXPLORATIONS

For self-study activities go to **Mastering**ASTRONOMY www.masteringastronomy.com

Interactive Tutorials

Tutorial Review of Key Concepts

Use the interactive **Tutorial** at **www.mastering astronomy.com** to review key concepts from this chapter.

Orbits and Kepler's Laws Tutorial
Lesson 2 Kepler's First Law
Lesson 3 Kepler's Second Law
Lesson 4 Kepler's Third Law

Supplementary Tutorial Exercises

Use the interactive **Tutorial Lesson** to explore the following questions.

Orbits and Kepler's Laws Tutorial, Lesson 2
1. When is an ellipse a circle?
2. Use the ellipse tool in Lesson 2 to show that the semimajor axis of an ellipse is a planet's average orbital radius:
   a. Orbital radius when the planet is closest to the Sun: perihelion = ___ AU
   b. Orbital radius when the planet is farthest from the Sun: aphelion = ___ AU

   c. Average of perihelion and aphelion = ___ AU
   d. Length of the major axis of the ellipse = ___ AU
   e. Length of the semimajor axis of the ellipse = ___ AU
   Are the average orbital radius (c) and the semimajor axis (e) equal?

 Exploring the Sky and Solar System

Of the many activities available on the *Voyager: SkyGazer* **CD-ROM** accompanying your book, use the following files to observe key phenomena covered in this chapter.

Go to the **File: Basics** folder for the following demonstrations.
1. Ptolemy on Venus
2. Pluto's Orbit

Go to the **File: Demo** folder for the following demonstrations.
1. Hale-Bopp Path
2. Venus-Earth-Moon

Go to the **Explore** menu for the following demonstrations.
1. Paths of the Planets

Web Projects

1. *Easter.* Find out when different sects of Christianity celebrate Easter. Then research how and why different sects set different dates for Easter. Summarize your findings in a one- to two-page report.
2. *Greek Astronomers.* Many ancient Greek scientists had ideas that, in retrospect, seem well ahead of their time. Choose one or more of the following ancient Greek scientists, and

learn enough about their work in science and astronomy to write a one- to two-page "scientific biography."

Thales	Anaximander	Pythagoras
Anaxagoras	Empedocles	Democritus
Meton	Plato	Eudoxus
Aristotle	Callipus	Aristarchus
Archimedes	Eratosthenes	Apollonius
Hipparchus	Seleucus	Ptolemy
Hypatia		

3. *The Ptolemaic Model.* This chapter gives only a very brief description of Ptolemy's model of the universe. Investigate the model in greater depth. Using diagrams and text as needed, give a two- to three-page description of the model.

4. *The Galileo Affair.* In recent years, the Roman Catholic Church has devoted a lot of resources to learning more about the trial of Galileo and to understanding past actions of the Church in the Galilean case. Learn more about such studies, and write a short report about the current Vatican view of the case.

5. *Science or Pseudoscience.* Choose some pseudoscientific claim that has been in the news recently, and learn more about it and how scientists have "debunked" it. Write a short summary of your findings.

Celestial Timekeeping and Navigation

Supplementary Chapter

In ancient times, the practical need for timekeeping and navigation was one of the primary reasons for the study of astronomy. The celestial origins of timekeeping and navigation are still evident. The time of day comes from the location of the Sun in the local sky, the month comes from the Moon's cycle of phases, and the year comes from the Sun's annual path along the ecliptic. The very name of the "North Star" tells us how it can be an aid to navigation.

Today, we can tell the time by glancing at an inexpensive electronic watch and navigate with handheld devices that receive signals from satellites of the global positioning system (GPS). But knowing the celestial basis of timekeeping and navigation can still be useful, particularly for understanding the rich history of astronomical discovery. In this chapter, we will explore the apparent motions of the Sun, Moon, and planets in greater detail, which will allow us to study the principles of celestial timekeeping and navigation.

S1.1 Astronomical Time Periods

Today, our clocks and calendars are beautifully synchronized to the rhythms of the heavens. Precision measurements allow us to ensure that our clocks keep pace with the Sun's daily trek across our sky, while our calendar holds the dates of the equinoxes and solstices as steady as possible.

In earlier chapters, we saw how this synchronicity took root in ancient observations of the sky. However, it is only much more recently that we have come to understand timekeeping in all its details. In this section, we will look in more detail at basic measures of time and our modern, international system of timekeeping.

• How do we define the day, month, year, and planetary periods?

By now you know that the length of the day corresponds to Earth's rotation, the length of the month to the cycle of lunar phases, and the length of the year to our orbit around the Sun. However, when we look carefully at each case, we find that the correspondence is not quite as simple as we might at first guess. Instead, we are forced to define two different types of day, month, and year. We also define planetary periods in two different ways. Let's take a look at how and why we make these distinctions.

The Length of the Day   We usually think of a day as the time it takes for Earth to rotate once, but if you measure this time period you'll find that it is *not* exactly 24 hours. Instead, Earth's rotation period is about 4 minutes short of 24 hours. What's going on?

We can understand the answer by thinking about the movement of the stars and Sun across our sky. Remember that the daily circling of the stars in our sky is an illusion created by Earth's rotation (see Figure 2.9). Thus, you can measure Earth's rotation period by measuring how long it takes for any star to go from its highest point in the sky one day to its highest point the next day (Figure S1.1a). This time period, which we call a **sidereal day**, is about 23 hours 56 minutes (more precisely, $23^h56^m4.09^s$). *Sidereal* (pronounced *sy-DEAR-ee-al*) means "related to the stars"; note that you'll measure the same time no matter what star you choose. For practical purposes, the sidereal day is Earth's precise rotation period.

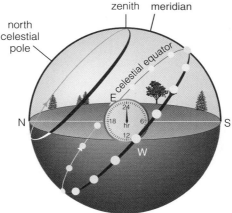

Figure S1.1 Using the sky to measure the length of a day.

a A sidereal day is the time it takes any star to make a circuit of the local sky. It is about 23 hours 56 minutes.

b A solar day is measured similarly, but by timing the Sun rather than a distant star. The length of the solar day varies over the course of the year but averages 24 hours.

Our 24-hour day, which we call a **solar day**, is based on the time it takes for the *Sun* to make one circuit around the local sky. You can measure this time period by measuring how long it takes the Sun to go from its highest point in the sky one day to its highest point the next day (Figure S1.1b). The solar day is indeed 24 hours on average, although it varies slightly (up to 25 seconds longer or shorter than 24 hours) over the course of a year.

A simple demonstration shows why the solar day is about 4 minutes longer than the sidereal day. Set an object to represent the Sun on a table, and stand a few steps away to represent Earth. Point at the Sun and imagine that you also happen to be pointing toward a distant star that lies in the same direction. If you rotate (counterclockwise) while standing in place, you'll again be pointing at both the Sun and the star after one full rotation (Figure S1.2a). However, because Earth rotates at the same time it orbits the Sun, you can make the demonstration more realistic by taking a couple of steps around the Sun (counterclockwise) while you rotate (Figure S1.2b). After one full rotation, you will again be pointing in the direction of the distant star, so this represents a sidereal day. But it does not represent a solar day, because you will not yet be pointing back at the Sun. If you wish to point again at the Sun, you need to make up for your orbital motion by making slightly more than one full rotation. This "extra" bit of rotation makes a solar day longer than a sidereal day.

The only problem with this demonstration is that it exaggerates Earth's daily orbital motion. Because Earth takes about 365 days (1 year) to make a full 360° orbit around the Sun, Earth actually moves only about 1° per day around its orbit. Thus, a solar day represents about 361° of rotation, rather than the 360° for a sidereal day (Figure S1.2c). The extra 1° rotation takes about $\frac{1}{360}$ of Earth's rotation period, which is about 4 minutes.

The Length of the Month As we discussed in Chapter 2, our month comes from the Moon's $29\frac{1}{2}$-day cycle of phases (think "moonth"). More technically, the $29\frac{1}{2}$-day period required for each cycle of phases is called a **synodic month**. The word *synodic* comes from the Latin *synod*, which means "meeting." A synodic month gets its name because the Sun and the Moon "meet" in the sky with every new moon.

Just as a solar day is not Earth's true rotation period, a synodic month is not the Moon's true orbital period. Earth's motion around the Sun means that the Moon must complete more than one full orbit of Earth from one new moon to the next (Figure S1.3). The Moon's true orbital period, or **sidereal month**, is only about $27\frac{1}{3}$ days. Like the sidereal day, the sidereal month gets its name because it describes how long it takes the Moon to complete an orbit relative to the positions of distant stars.

The Length of the Year A year is related to Earth's orbital period, but again there are two slightly different definitions for the length of the year. The time it takes for Earth to complete one orbit relative to the stars is called a **sidereal year**. But our calendar is based on the cycle of the seasons, which we measure as the time from the spring equinox one year to the spring equinox the next year. This time period, called a **tropical year**, is about 20 minutes shorter than the sidereal year. A 20-minute difference might not seem like much, but it would make a calendar based on the sidereal year get out of sync with the seasons by 1 day every 72 years—a difference that would add up over centuries.

The difference between the sidereal year and the tropical year arises from Earth's 26,000-year cycle of axis precession [Section 2.3]. Precession not only changes the orientation of the axis in space but also changes the locations in Earth's orbit at which the seasons occur. Each year, the location of the equinoxes and solstices among the stars shifts about

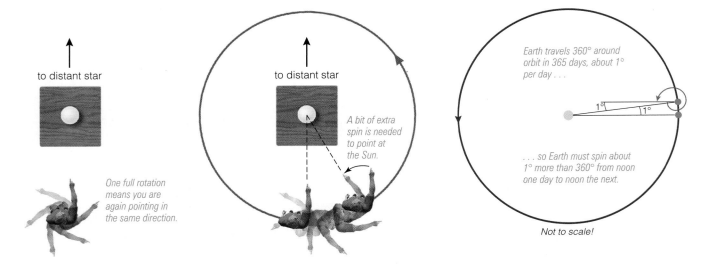

a One full rotation represents a sidereal day and means that you are again pointing to the same distant star.

b While you are "orbiting" the Sun, one rotation still returns you to pointing at a distant star, but you need slightly more than one full rotation to return to pointing at the Sun.

c Earth travels about 1° per day around its orbit, so a solar day requires about 361° of rotation.

Figure S1.2 A demonstration showing why a solar day is slightly longer than a sidereal day.

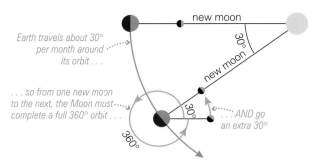

new moon

30°

new moon

30°

Earth travels about 30° per month around its orbit . . .

. . . so from one new moon to the next, the Moon must complete a full 360° orbit . . .

. . . AND go an extra 30°

360°

Figure S1.3 Interactive Figure ↘ The Moon completes one 360° orbit in about $27\frac{1}{3}$ days (a sidereal month), but the time from new moon to new moon is about $29\frac{1}{2}$ days (a synodic month).

$\frac{1}{26,000}$ of the way around the orbit. If you do the math, you'll find that $\frac{1}{26,000}$ of a year is about 20 minutes, which explains the 20-minute difference between the tropical year and the sidereal year.

Planetary Periods   Although planetary periods are not used in our modern timekeeping, they were important to many ancient cultures. For example, the Mayan calendar was based in part on the apparent motions of Venus. In addition, Copernicus's ability to determine orbital periods of planets with his Sun-centered model played an important role in keeping the model alive long enough for its ultimate acceptance [Section 3.3].

A planet's **sidereal period** is the time the planet takes to orbit the Sun. (As usual, it has the name *sidereal* because it is measured relative to distant stars.) For example, Jupiter's sidereal period is 11.86 years, so it takes about 12 years for Jupiter to make a complete circuit around the constellations of the zodiac. Thus, Jupiter appears to move through roughly one zodiac constellation each year. If Jupiter is currently in Libra (as it is for much of 2006), it will be in Scorpius at this time next year and Sagittarius the following year, returning to Libra in 12 years.

A planet's **synodic period** is the time between when it is lined up with the Sun in our sky one time and the next similar alignment. (As with the Moon, the term *synodic* refers to the planet's "meeting" the Sun in the sky.) Figure S1.4 shows that the situation is somewhat different for

planets nearer the Sun than Earth (that is, Mercury and Venus) and planets farther away (all the rest of the planets).

Look first at the situation for the more distant planet in Figure S1.4. As seen from Earth, this planet will sometimes line up with the Sun in what we call a **conjunction**. At other special times, it will appear exactly opposite the Sun in our sky, or at **opposition**. We cannot see the planet during conjunction with the Sun because it is hidden by the Sun's glare and rises and sets with the Sun in our sky. At opposition, the planet moves through the sky like the full moon, rising at sunset, reaching the meridian at midnight, and setting at dawn. Note that the planet is closest to Earth at opposition and hence appears brightest in our sky at this time.

Now look at the planet that is *nearer* than Earth to the Sun in Figure S1.4. This planet never has an opposition but instead has two conjunctions—an "inferior conjunction" between Earth and the Sun and a "superior conjunction" when the planet appears behind the Sun as seen from Earth. Two other points are important for the inner planets: their points of **greatest elongation**, when they appear farthest from the Sun in our sky. At its greatest eastern elongation, Venus appears about 46° east of the Sun in our sky, which means it shines brightly in the evening sky. Similarly, at its greatest western elongation, Venus appears about 46° west of the Sun and shines brightly in the predawn sky. In between the times when Venus appears in the morning sky and the times when it appears in the evening sky, Venus disappears from view for a few weeks with each conjunction. Mercury's pattern is similar, but because it is closer to the Sun, it never appears more than about 28° from the Sun in our sky. That makes Mercury difficult to see, because it is almost always obscured by the glare of the Sun.

THINK ABOUT IT

Do we ever see Mercury or Venus at midnight? Explain.

As you study Figure S1.4, you might wonder whether Mercury or Venus ever falls directly in front of the Sun at inferior conjunction, creating mini-eclipses as they block a little of the Sun's light. They do, but only rarely, because

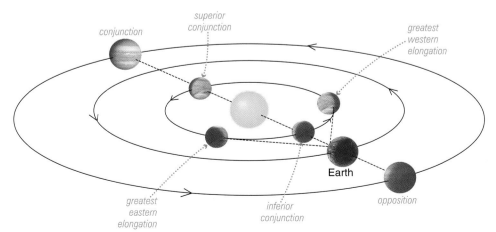

superior conjunction

conjunction

greatest western elongation

greatest eastern elongation

inferior conjunction

Earth

opposition

Figure S1.4 This diagram shows important positions of planets relative to Earth and the Sun. For a planet farther from the Sun than Earth (such as Jupiter), conjunction occurs when the planet appears aligned with the Sun in the sky, and opposition occurs when the planet appears on our meridian at midnight. Planets nearer the Sun (such as Venus) have two conjunctions and never get farther from the Sun in our sky than at their greatest elongations. (Adapted from *Advanced Skywatching*, by Burnham *et al.*)

their orbital planes are slightly tilted compared to Earth's orbital plane (the ecliptic plane). As a result, Mercury and Venus usually appear slightly above or below the Sun at inferior conjunction. But on rare occasions, we do indeed see Mercury or Venus appear to pass directly across the face of the Sun during inferior conjunction (Figure S1.5). Such events are called **transits**. Mercury transits occur an average of a dozen times per century, with the first two of this century on November 8, 2006, and May 9, 2016. Venus transits come in pairs 8 years apart, with more than a century between the second of one pair and the first of the next. We are currently between the two transits of a pair: The first occurred on June 8, 2004, and the second will occur on June 6, 2012. After that, it will be 105 years until the first of the next pair of Venus transits, which will occur in 2117 and 2125.

• How do we tell the time of day?

Telling time seems simple, but in fact there are several different ways to define the time of day, even after we agree that time should be based on the 24-hour solar day. Let's explore some of the ways of telling time, and see how they ultimately led to our modern system in which we can synchronize clocks anywhere in the world.

Apparent Solar Time If we base time on the Sun's *actual* position in the local sky, as is the case when we use a sundial (Figure S1.6), we are measuring **apparent solar time**. Noon is the precise moment when the Sun is highest in the sky (on the meridian) and the sundial casts its shortest shadow. Before noon, when the Sun is rising upward through the sky, the apparent solar time is *ante meridiem* ("before the

---

MATHEMATICAL INSIGHT **S1.1** **The Copernican Layout of the Solar System**

As discussed in Chapter 3, Copernicus favored the Sun-centered model partly because it allowed him to calculate orbital periods and distances for the planets. Let's see how the Copernican system allows us to determine orbital (sidereal) periods of the planets.

We cannot directly measure orbital periods, because our own movement around the Sun means that we look at the planets from different points in our orbit at different times. However, we can measure synodic periods simply by seeing how much time passes between one particular alignment (such as opposition or inferior conjunction) and the next. Figure 1 shows the geometry for a planet *farther from the Sun than Earth* (such as Jupiter), under the assumption of circular orbits (which is what Copernicus assumed). Study the figure carefully to notice the following key facts:

● The dashed brown curve shows the planet's complete orbit. The time the planet requires for one complete orbit is its orbital (sidereal) period, P_{orb}.

● The solid brown arrow shows how far the planet travels along its orbit from one opposition to the next. The time between oppositions is defined as its synodic period, P_{syn}.

● The dashed blue curve shows Earth's complete orbit; Earth takes $P_{Earth} = 1$ yr to complete an orbit.

● The solid red curve (and red arrow) shows how far Earth goes during the planet's synodic period; it is *more* than one complete orbit because Earth must travel a little "extra" to catch back up with the planet. The time it takes Earth to travel the "extra" distance (the thick part of the red curve) must be the planet's synodic period minus 1 year, or $P_{syn} - 1$ yr.

● The angle that the planet sweeps out during its synodic period is equal to the angle that Earth sweeps out as it travels the "extra" distance. Thus, the *ratio* of the planet's complete orbital period (P_{orb}) to its synodic period (P_{syn}) must be equal to the *ratio* of Earth's orbital period (1 yr) to the time required for the "extra" distance (see Appendix C.5 for a review of ratios). Since we already found that the time required for this extra distance is $P_{syn} - 1$ yr, we write:

$$\frac{P_{orb}}{P_{syn}} = \frac{1 \text{ yr}}{(P_{syn} - 1 \text{ yr})}$$

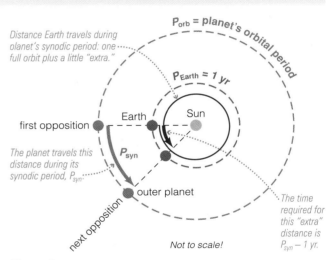

Figure 1

Multiplying both sides by P_{syn} gives us the final equation:

$$P_{orb} = P_{syn} \times \frac{1 \text{ yr}}{(P_{syn} - 1 \text{ yr})}$$
[for planets farther from the Sun than Earth]

The geometry is slightly different when Earth is the outer planet, as shown in Figure 2. You should convince yourself that in this case the two equal ratios are $1 \text{ yr}/P_{syn} = P_{orb}/(P_{syn} - P_{orb})$. With a little algebra, you can solve this equation for P_{orb}:

$$P_{orb} = P_{syn} \times \frac{1 \text{ yr}}{(P_{syn} + 1 \text{ yr})}$$
[for planets closer to the Sun than Earth]

Copernicus knew the synodic periods of the planets and was therefore able to use the above equations (in a slightly different form) to calculate their true orbital periods. From there, he was able to use the geometry of planetary alignments to compute the

---

Figure S1.5 This photo was taken in Florida during the transit of Venus on June 8, 2004. Venus is the small black dot near the right edge of the Sun's face.

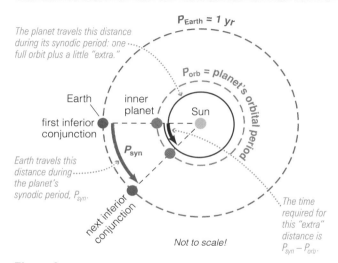

The planet travels this distance during its synodic period: one full orbit plus a little "extra."

$P_{Earth} = 1\ yr$

P_{orb} = planet's orbital period

Earth   inner planet   Sun

first inferior conjunction

P_{syn}

Earth travels this distance during the planet's synodic period, P_{syn}.

next inferior conjunction

The time required for this "extra" distance is $P_{syn} - P_{orb}$.

Not to scale!

Figure 2

distances to the planets in terms of the Earth-Sun distance. (That is, he calculated distances in AU, but he did not have a method for determining how far the Earth is from the Sun.) His results, which were quite close to modern values, made so much sense to him that he felt he must have uncovered some deep truth about nature.

Example 1: Jupiter's synodic period is 398.9 days, or 1.092 years. What is its actual orbital period?

Solution:

Step 1 Understand: We are given Jupiter's synodic period (P_{syn} in the above equations) and are asked to find its orbital period (P_{orb}). Because Jupiter is farther from the Sun than Earth, we can use the first equation.

Step 2 Solve: We use the equation for a planet farther from the Sun than Earth with $P_{syn} = 1.092$ yr (Jupiter's synodic period):

$$P_{orb} = P_{syn} \times \frac{1\ yr}{(P_{syn} - 1\ yr)}$$
$$= 1.092\ yr \times \frac{1\ yr}{(1.092\ yr - 1\ yr)}$$
$$= 11.87\ yr$$

Step 3 Explain: We have found that Jupiter's orbital period is 11.87 years. In other words, simply by measuring the time that passes from when Jupiter is opposite the Sun in one year to when it is again opposite the next year (which is Jupiter's synodic period), we have learned that Jupiter takes a little less than 12 years to orbit the Sun. Notice that the answer makes sense in that it is longer than Earth's orbital period of 1 year, just as we expect for a planet that is farther than Earth from the Sun.

Example 2: Venus's synodic period is 583.9 days. What is its actual orbital period?

Solution:

Step 1 Understand: We are given Venus's synodic period, $P_{syn} = 583.9$ days, and are asked to find its orbital period. Because Venus is closer to the Sun than Earth, we need the second equation. Also, because we are given Venus's synodic period in days, for unit consistency we need to convert it to years; you should confirm that 583.9 days = 1.599 yr.

Step 2 Solve: We simply plug the value of P_{syn} into the equation for a planet closer to the Sun than Earth:

$$P_{orb} = P_{syn} \times \frac{1\ yr}{(P_{syn} + 1\ yr)}$$
$$= 1.599\ yr \times \frac{1\ yr}{(1.599\ yr + 1\ yr)}$$
$$= 0.6152\ yr$$

Setp 3 Explain: We have found that Venus takes 0.6152 year to orbit the Sun. This number is easier to interpret if we convert it to days or months; you should confirm that it is equivalent to 224.7 days, or about $7\frac{1}{2}$ months. Notice that the answer makes sense in that it is shorter than Earth's orbital period of 1 year, just as we expect for a planet that is closer than Earth to the Sun.

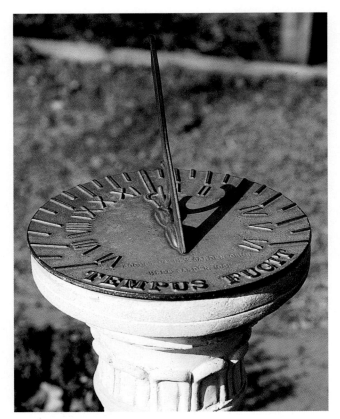

Figure S1.6 A basic sundial consists of a dial marked by numerals, and a stick, or *gnomon*, that casts a shadow. Here, the shadow is on the Roman numeral III, indicating that the apparent solar time is 3:00 P.M. (The portion of the dial without numerals represents nighttime hours.) Because the Sun's path across the local sky depends on latitude, a particular sundial will be accurate only for a particular latitude.

middle of the day"), or A.M. For example, if the Sun will reach the meridian 2 hours from now, the apparent solar time is 10 A.M. After noon, the apparent solar time is *post meridiem* ("after the middle of the day"), or P.M. If the Sun crossed the meridian 3 hours ago, the apparent solar time is 3 P.M. Note that, technically, noon and midnight are *neither* A.M. nor P.M. However, by convention we usually say that noon is 12 P.M. and midnight is 12 A.M.

THINK ABOUT IT

Is it daytime or nighttime at 12:01 A.M.? 12:01 P.M.? Explain.

Mean Solar Time Suppose you set a clock to precisely 12:00 when a sundial shows noon today. If every solar day were precisely 24 hours, your clock would always remain synchronized with the sundial. However, while 24 hours is the *average* length of the solar day, the actual length of the solar day varies throughout the year. As a result, your clock will not remain perfectly synchronized with the sundial. For example, your clock is likely to read a few seconds before or after 12:00 when the sundial reads noon tomorrow, and within a few weeks your clock time may differ from the apparent solar time by several minutes. Your clock (assuming

it is accurate) will again be synchronized with the Sun on the same date next year, since it keeps track of the average length of the solar day.

If we average the differences between the time a clock would read and the time a sundial would read, we can define **mean solar time** (*mean* is another word for *average*). A clock set to mean solar time reads 12:00 each day at the time that the Sun crosses the meridian *on average*. The actual mean solar time at which the Sun crosses the meridian varies over the course of the year in a fairly complex way (see "Solar Days and the Analemma," p. 98). The result is that, on any given day, a clock set to mean solar time may read anywhere from about 17 minutes before noon to 15 minutes after noon (that is, from 11:43 A.M. to 12:15 P.M.) when a sundial indicates noon.

Although the lack of perfect synchronization with the Sun might at first sound like a drawback, mean solar time is actually more convenient than apparent solar time (the sundial time)—as long as you have access to a mechanical or electronic clock. Once set, a reliable mechanical or electronic clock can always tell you the mean solar time. In contrast, precisely measuring apparent solar time requires a sundial, which is useless at night or when it is cloudy.

Like apparent solar time, mean solar time is a *local* measure of time. That is, it varies with longitude because of Earth's west-to-east rotation. For example, clocks in New York are set 3 hours ahead of clocks in Los Angeles. If clocks were set precisely to local mean solar time, they would vary even over relatively short east-west distances. For example, mean solar clocks in central Los Angeles would be about 2 minutes behind mean solar clocks in Pasadena, because Pasadena is slightly farther east.

Standard, Daylight, and Universal Time Clocks displaying mean solar time were common during the early history of the United States. However, by the late 1800s, the growth of railroad travel made the use of mean solar time increasingly problematic. Some states had dozens of different "official" times, usually corresponding to mean solar time in dozens of different cities, and each railroad company made schedules according to its own "railroad time." The many time systems made it difficult for passengers to follow the scheduling of trains.

On November 18, 1883, the railroad companies agreed to a new system that divided the United States into four time zones, setting all clocks within each zone to the same time. That was the birth of **standard time**, which today divides the world into time zones (Figure S1.7). Depending on where you live within a time zone, your standard time may vary somewhat from your mean solar time. In principle, the standard time in a particular time zone is the mean solar time in the *center* of the time zone, so that local mean solar time within a 1-hour-wide time zone would never differ by more than a half-hour from standard time. However, time zones often have unusual shapes to conform to social, economic, and political realities, so larger variations between standard time and mean solar time sometimes occur.

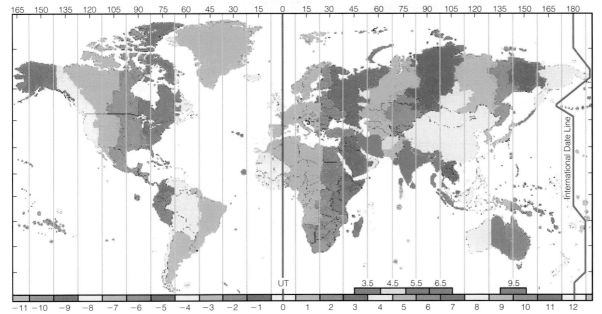

Figure S1.7 Time zones around the world. The numerical scale at the bottom shows hours ahead of (positive numbers) or behind (negative numbers) the time in Greenwich, England; the scale at the top is longitude. The vertical lines show standard time zones as they would be in the absence of political considerations. The color-coded regions show the actual time zones. Note, for example, that all of China uses the same standard time, even though the country is wide enough to span several time zones, and that the time zones of Russia and other countries of the former USSR are all shifted by about an hour. Note also that a few countries use time zones centered on a half-hour (the upper set of four colored bars), rather than an hour, relative to Greenwich time.

In most parts of the United States, clocks are set to standard time for only part of the year. Between the first Sunday in April and the last Sunday in October,* most of the United States changes to **daylight saving time**, which is 1 hour ahead of standard time. Because of the 1-hour advance with daylight saving time, clocks read around 1 P.M. (rather than around noon) when the Sun is on the meridian.

For purposes of navigation and astronomy, it is useful to have a single time for the entire Earth. For historical reasons, this "world" time was chosen to be the mean solar time in Greenwich, England—the place that also defines longitude 0° (see Figure 2.11). Today, this *Greenwich mean time* (*GMT*) is often called **universal time** (**UT**). (Outside astronomy, it is more commonly called *universal coordinated time* [UTC]. Many airlines and weather services call it "Zulu time," because Greenwich's time zone is designated "Z" and "zulu" is a common way of phonetically identifying the letter *Z*.)

Seasons Tutorial, Lesson 2

• When and why do we have leap years?

Our modern calendar is based on the length of the tropical year, which is the amount of time from one spring equinox to the next. Thus, the calendar should stay synchronized

---

*Under legislation passed in 2005, beginning in 2007 daylight saving time will extend four weeks longer in the United States: It will start on the second Sunday in March and end on the first Sunday in November.

with the seasons. Getting this synchronization just right was a long process in human history.

The origins of our calendar go back to ancient Egypt. By 4200 B.C., the Egyptians were using a calendar that counted 365 days in a year. However, because the length of a year is about $365\frac{1}{4}$ days (rather than exactly 365 days), the Egyptian calendar drifted out of phase with the seasons by about 1 day every 4 years. For example, if the spring equinox occurred on March 21 one year, 4 years later it occurred on March 22, 4 years after that on March 23, and so on. Over many centuries, the spring equinox moved through many different months. To keep the seasons and the calendar synchronized, Julius Caesar decreed the adoption of a new calendar in 46 B.C. This *Julian calendar* introduced the concept of **leap year:** Every fourth year has 366 days, rather than 365, so that the average length of the calendar year is $365\frac{1}{4}$ days.

The Julian calendar originally had the spring equinox falling around March 24. If it had been perfectly synchronized with the tropical year, this calendar would have ensured that the spring equinox occurred on the same date every 4 years (that is, every leap-year cycle). It didn't work perfectly, however, because a tropical year is actually about 11 minutes short of $365\frac{1}{4}$ days. Thus, the moment of the spring equinox slowly advanced by an average of 11 minutes per year. By the late 1500s, the spring equinox was occurring on March 11.

Concerned by this drift in the date of the spring equinox, Pope Gregory XIII introduced a new calendar in 1582. This *Gregorian calendar* was much like the Julian calendar, with two important adjustments. First, Pope Gregory decreed that the day in 1582 following October 4 would be October 15. By eliminating the 10 dates from October 5 through October 14, 1582, he pushed the date of the spring equinox in 1583 from March 11 to March 21. (He chose March 21 because it was the date of the spring equinox in A.D. 325, which was the time of the Council of Nicaea, the first ecumenical council of the Christian church.) Second, the Gregorian calendar added an exception to the rule of having leap year every 4 years: Leap year is skipped when a century changes (for example, in years 1700, 1800, 1900) *unless* the century year is divisible by 400. Thus, 2000 was a leap year because it is divisible by 400 (2,000 ÷ 400 = 5), but 2100 will *not* be a leap year. These adjustments make

---

SPECIAL TOPIC Solar Days and the Analemma

The average length of a solar day is 24 hours, but the precise length varies over the course of the year. Two effects contribute to this variation.

The first effect arises from Earth's varying orbital speed. Recall that, in accord with Kepler's second law, Earth moves slightly faster when it is closer to the Sun in its orbit and slightly slower when it is farther from the Sun. Thus, Earth moves slightly farther along its orbit each day when it is closer to the Sun. This means that the solar day requires more than the average amount of "extra" rotation (see Figure S1.2) during these periods—making these solar days longer than average. Similarly, the solar day requires less than the average amount of "extra" rotation when it is in the portion of its orbit farther from the Sun—making these solar days shorter than average.

The second effect arises from the tilt of Earth's axis, which causes the ecliptic to be inclined by $23\frac{1}{2}°$ to the celestial equator on the celestial sphere. Because the length of a solar day depends on the Sun's apparent *eastward* motion along the ecliptic, the inclination would cause solar days to vary in length even if Earth's orbit were perfectly circular. To see why, suppose the Sun appeared to move exactly 1° per day along the ecliptic. Around the times of the sol-stices, this motion would be entirely eastward, making the solar day slightly longer than average. Around the times of the equinoxes, when the motion along the ecliptic has a significant northward or southward component, the solar day would be slightly shorter than average.

Together, the two effects make the actual length of solar days vary by up to about 25 seconds (either way) from the 24-hour average. Because the effects accumulate at particular times of year, the apparent solar time can differ by as much as 17 minutes from the mean solar time. The net result is often depicted visually by an **analemma** (Figure 1), which looks much like a figure 8. You'll find an analemma printed on many globes (Figure 2.16 shows a photographic version).

By using the horizontal scale on the analemma (Figure 1), you can convert between mean and apparent solar time for any date. (The vertical scale shows the declination of the Sun, which is discussed in Section S1.2.) For example, the dashed line shows that on November 10, a mean solar clock is about 17 minutes "behind the Sun," or behind apparent solar time. Thus, if the apparent solar time is 6:00 P.M. on November 10, the mean solar time is only 5:43 P.M. The discrepancy between mean and apparent solar times is called the **equation of time**. It is often plotted as a graph (Figure 2), which gives the same results as reading from the analemma.

The discrepancy between mean and apparent solar time also explains why the times of sunrise and sunset don't follow seasonal patterns perfectly. For example, the winter solstice around December 21 has the shortest daylight hours (in the Northern Hemisphere), but the earliest sunset occurs around December 7, when the Sun is still well "behind" mean solar time.

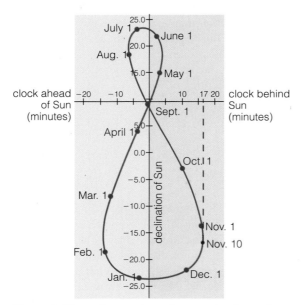

Figure 1 The analemma shows the annual pattern of discrepancies between apparent and mean solar time. For example, the dashed red line shows that on November 10, a mean solar clock reads 17 minutes behind (earlier than) apparent solar time.

Figure 2 The discrepancies can also be plotted on a graph as the equation of time.

the average length of the Gregorian calendar year almost exactly the same as the actual length of a tropical year, which ensures that the spring equinox will occur on March 21 every fourth year for thousands of years to come.

Today, the Gregorian calendar is used worldwide for international communication and commerce. (Many countries still use traditional calendars, such as the Chinese, Islamic, and Jewish calendars, for cultural purposes.) However, as you might guess, the Pope's decree was not immediately accepted in regions not bound to the Catholic Church. For example, the Gregorian calendar was not adopted in England or in the American colonies until 1752, and it was not adopted in China until 1912 or in Russia until 1919.

S1.2 Celestial Coordinates and Motion in the Sky

We are now ready to turn our attention from timekeeping to navigation. The goal of celestial navigation is to use the Sun and the stars to find our position on Earth. Before we can do that, we need to understand the apparent motions of the sky in more detail than we covered in Chapter 2. We'll begin in this section by discussing how we locate objects on the celestial sphere, which will then allow us to explore how positions on the celestial sphere determine motion in the local sky. With this background, we'll be ready to explore the principles of celestial navigation in the final section of this chapter.

• How do we locate objects on the celestial sphere?

Recall from Chapter 2 that the celestial sphere is an illusion, but one that is quite useful when looking at the sky. We can make the celestial sphere even more useful by giving it a set of **celestial coordinates** that function much like the coordinates of latitude and longitude on Earth. Just as we can locate a city on Earth by its latitude and longitude, we will use an object's celestial coordinates to describe its precise location on the celestial sphere.

We have already discussed the basic features of the celestial sphere that will serve as starting points for our coordinate system: the north and south celestial poles, the celestial equator, and the ecliptic. Figure S1.8 shows these locations on a schematic diagram. The arrow along the ecliptic indicates the direction in which the Sun appears to move over the course of each year. It is much easier to visualize the celestial sphere if you make a model with a simple plastic ball. Use a felt-tip pen to mark the north and south celestial poles on your ball, and then add the celestial equator and the ecliptic. Note that the ecliptic crosses the celestial equator on opposite sides of the celestial sphere at an angle of $23\frac{1}{2}°$ (because of the tilt of Earth's axis).

Equinoxes and Solstices Remember that the equinoxes and solstices are special moments in the year that help define

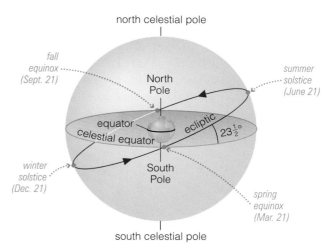

Figure S1.8 Schematic diagram of the celestial sphere, shown without stars. As a study aid, you should use a plastic ball as a model of the celestial sphere, marking it with the same special locations.

the seasons [Section 2.2]. For example, the *spring equinox*, which occurs around March 21 each year, is the moment when spring begins for the Northern Hemisphere and fall begins for the Southern Hemisphere. These moments correspond to positions in Earth's orbit (see Figure 2.15) and hence to apparent locations of the Sun along the ecliptic. As shown in Figure S1.8, the spring equinox occurs when the Sun is on the ecliptic at the point where it crosses from south of the celestial equator to north of the celestial equator. This point is also called the spring equinox. Thus, the term *spring equinox* has a dual meaning: It is the *moment* when spring begins and also the *point* on the ecliptic at which the Sun appears to be located at that moment.

Figure S1.8 also shows the points marking the summer solstice, fall equinox, and winter solstice, with the dates on which the Sun appears to be located at each point. Remember that the dates are approximate because of the leap-year cycle and because a tropical year is not exactly $365\frac{1}{4}$ days. (For example, the spring equinox may occur anytime between about March 20 and March 23.)

Although no bright stars mark the locations of the equinoxes or solstices among the constellations, you can find them with the aid of nearby bright stars (Figure S1.9). For example, the spring equinox is located in the constellation Pisces and can be found with the aid of the four bright stars in the Great Square of Pegasus. Of course, when the Sun is located at this point around March 21, we cannot see Pisces or Pegasus because they are close to the Sun in our daytime sky.

<hr>

THINK ABOUT IT

Using your plastic ball as a model of the celestial sphere (which you have already marked with the celestial poles, equator, and ecliptic), mark the locations and approximate dates of the equinoxes and solstices. Based on the dates for these points, approximately where along the ecliptic is the Sun on April 21? On November 21? How do you know?

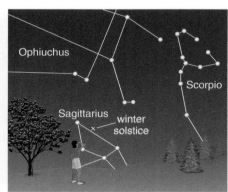

Figure S1.9 These diagrams show the locations among the constellations of the equinoxes and solstices. No bright stars mark any of these points, so you must find them by studying their positions relative to recognizable patterns. The time of day and night at which each point is above the horizon depends on the time of year.

Celestial Coordinates We can now add our system of celestial coordinates to the celestial sphere. Because this will be the third coordinate system we've used in this book, it's easier to understand if we first review the other two. Figure S1.10a shows the coordinates of *altitude* and *direction* (or *azimuth**) we use in the local sky. Figure S1.10b shows the coordinates of *latitude* and *longitude* we use on Earth's surface. Finally, Figure S1.10c shows our new system of celestial coordinates. As you can see in the figure, these coordinates are called **declination (dec)** and **right ascension (RA)**.

---

*Azimuth is usually measured clockwise around the horizon from due north. By this definition, the azimuth of due north is 0°, of due east is 90°, of due south is 180°, and of due west is 270°.

If you compare Figures S1.10b and c, you'll see that declination on the celestial sphere is very similar to latitude on Earth, while right ascension is very similar to longitude. Let's start by looking at declination; notice the following key points:

- Just as lines of latitude are parallel to Earth's equator, lines of declination are parallel to the celestial equator.
- Just as Earth's equator has lat = 0°, the celestial equator has dec = 0°.
- Latitude is labeled *north* or *south* relative to the equator, while declination is labeled *positive* or *negative*. For example, the North Pole has lat = 90°N, while the north celestial pole has dec = +90°; the South Pole has lat = 90°S, while the south celestial pole has dec = −90°.

Figure S1.10 Celestial coordinate systems.

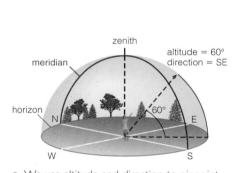

a We use altitude and direction to pinpoint locations in the local sky.

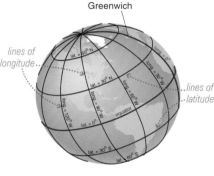

b We use latitude and longitude to pinpoint locations on Earth.

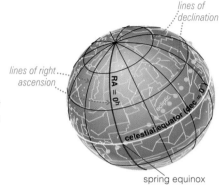

c We use declination and right ascension to pinpoint locations on the celestial sphere.

Next, notice the close correspondence between right ascension and longitude:

- Just as lines of longitude extend from the North Pole to the South Pole, lines of right ascension extend from the north celestial pole to the south celestial pole.
- Just as there is no natural starting point for longitude, there is no natural starting point for right ascension. By international treaty, longitude zero (the prime meridian) is the line of longitude that runs through Greenwich, England. By convention, right ascension zero is the line of right ascension that runs through the spring equinox.
- Longitude is measured in *degrees* east or west of Greenwich, while right ascension is measured in *hours* (and minutes and seconds) east of the spring equinox. A full 360° circle around the celestial equator goes through 24 hours of right ascension, so each hour of right ascension represents an angle of 360° ÷ 24 = 15°.

As an example of how we use celestial coordinates to locate objects on the celestial sphere, consider the bright star Vega. Its coordinates are dec = +38°44′ and RA = 18^h35^m (Figure S1.11). The positive declination tells us that Vega is 38°44′ *north* of the celestial equator. The right ascension tells us that Vega is 18 hours 35 minutes east of the spring equinox. Translating the right ascension from hours to angular degrees, we find that Vega is about 279° east of the spring equinox (because 18 hours represents 18 × 15° = 270° and 35 minutes represents $\frac{35}{60}$ × 15° ≈ 9°).

On your plastic ball model of the celestial sphere, add a scale for right ascension along the celestial equator and also add a few circles of declination, such as declination 0°, ±30°, ±60°, and ±90°. Where is Vega on your model?

The Vega example also shows why right ascension is measured in units of time: It is because time units make it easier to track the daily motions of objects through the local sky. All objects with a particular right ascension cross the meridian at the same time, which means that all stars with RA = 0^h cross the meridian at the same time that the spring equinox crosses the meridian. Thus, when we use time units, the right ascension of any object tells us how long *after* the spring equinox the object crosses the meridian. For example, Vega's right ascension, 18^h35^m, tells us that on any particular day, Vega crosses the meridian about 18 hours 35 minutes after the spring equinox. (This is 18 hours 35 minutes of *sidereal time* later, which is not exactly the same as 18 hours 35 minutes of solar time; see Mathematical Insight S1.2.)

Stars are so far away that they take thousands of years or more to move noticeably on the celestial sphere. Nevertheless, the celestial coordinates of stars are not quite constant, because they are tied to the celestial equator and the celestial equator gradually moves relative to the constellations with Earth's 26,000-year cycle of axis precession [Section 2.2]. (Axis precession does not affect Earth's orbit, so it does not affect the location of the ecliptic among the constellations.) Even over just a few decades, the coordinate changes are significant enough to make a difference in precise astronomical work—such as aiming a telescope at a particular object. As a result, careful observations require almost constant updating of celestial coordinates. Star catalogs therefore always state the year for which coordinates are given (for example, "epoch 2000"). Astronomical software can automatically calculate day-to-day celestial coordinates for the Sun, Moon, and planets as they wander among the constellations.

Celestial Coordinates of the Sun Unlike stars, which remain fixed in the patterns of the constellations on the celestial sphere, the Sun moves gradually along the ecliptic. It takes a year for the Sun to make a full circuit of the ecliptic, which means it moves through all 24 hours of right ascension over the course of the year. Each month, the Sun moves approximately one-twelfth of the way around the ecliptic, meaning that its right ascension changes by about 24 ÷ 12 = 2 hours per month. Figure S1.12 shows the ecliptic marked with

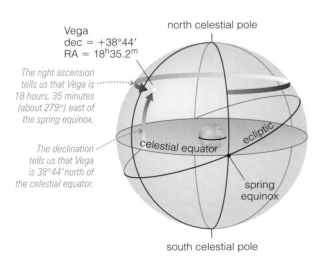

Figure S1.11 This diagram shows how we interpret the celestial coordinates of Vega.

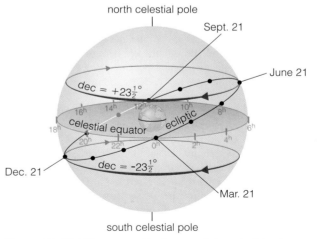

Figure S1.12 We can use this diagram of the celestial sphere to determine the Sun's right ascension and declination at monthly intervals.

the Sun's monthly position and a scale of celestial coordinates. From this figure, we can create a table of the Sun's month-by-month celestial coordinates.

Table S1.1 starts from the spring equinox, when the Sun has declination 0° and right ascension 0^h. You can see in the shaded areas of the table that while RA advances steadily through the year, the Sun's declination changes much more rapidly around the equinoxes than around the solstices. For example, the Sun's declination changes from −12° on February 21 to 12° on April 21, a difference of 24° in just 2 months. In contrast, during the 2 months around the summer solstice (that is, between May 21 and July 21), the declination varies only between +20° and its maximum of +23$\frac{1}{2}$°. This behavior explains why the number of daylight

---

MATHEMATICAL INSIGHT S1.2 Time by the Stars

The clocks we use in daily life are set to solar time, ticking through 24 hours for each day of mean solar time. In astronomy, it is also useful to have clocks that tell time by the stars, or **sidereal time**. Just as we define *solar time* according to the Sun's position relative to the meridian, *sidereal time* is based on the positions of stars relative to the meridian. We define the **hour angle (HA)** of any object on the celestial sphere to be the time since it last crossed the meridian. (For a circumpolar star, hour angle is measured from the *higher* of the two points at which it crosses the meridian each day.) For example:

- If a star is crossing the meridian now, its hour angle is 0^h.
- If a star crossed the meridian 3 hours ago, its hour angle is 3^h.
- If a star will cross the meridian 1 hour from now, its hour angle is −1^h or, equivalently, 23^h.

By convention, time by the stars is based on the hour angle of the spring equinox. That is, the **local sidereal time (LST)** is

$$LST = HA_{\text{spring equinox}}$$

For example, the local sidereal time is 00:00 when the spring equinox is *on* the meridian. Three hours later, when the spring equinox is 3 hours west of the meridian, the local sidereal time is 03:00.

Note that, because right ascension tells us how long after the spring equinox an object reaches the meridian, the local sidereal time is also equal to the right ascension (RA) of objects currently crossing your meridian. For example, if your local sidereal time is 04:30, stars with RA = 4^{h}30^m are currently crossing your meridian. This idea leads to an important relationship among any object's current hour angle, the current local sidereal time, and the object's right ascension:

$$HA_{\text{object}} = LST − RA_{\text{object}}$$

This formula will make sense to you if you recognize that an object's right ascension tells us the time by which it trails the spring equinox on its daily trek through the sky. Because the local sidereal time tells us how long it has been since the spring equinox was on the meridian, the difference LST − RA$_{\text{object}}$ must tell us the position of the object relative to the meridian.

Sidereal time has one important subtlety: Because the stars (and the celestial sphere) appear to rotate around us in one sidereal day (23^{h}56^m), sidereal clocks must tick through 24 hours of sidereal time in 23 hours 56 minutes of solar time. That is, a sidereal clock gains about 4 minutes per day over a solar clock. As a result, you cannot immediately infer the local sidereal time from the local solar time, or vice versa, without either doing some calculations or consulting an astronomical table. Of course, the easiest way to determine the local sidereal time is with a clock that ticks at the sidereal rate. Astronomical observatories always have sidereal clocks, and you can buy moderately priced telescopes that come with sidereal clocks.

Example 1: Suppose the local apparent solar time is 9:00 P.M. on the spring equinox (March 21). What is the local sidereal time?

Solution:

Step 1 Understand: The trick to this problem is understanding exactly what we are looking for. We are asked to find the local sidereal time, which is defined as the hour angle of the spring equinox in the local sky. Thus, we need to know where the spring equinox is located in the local sky. We are given the key clue: The date is the day of the spring equinox, the one day of the year on which the Sun is located in the same position as the spring equinox in the sky.

Step 2 Solve: We can now find the hour angle of the spring equinox from the hour angle of the Sun. We are told that the local apparent solar time is 9:00 P.M., which means that the Sun is 9 hours past the meridian and thus has an hour angle of 9 hours. Because the spring equinox and the Sun are located in the same place on this one date of the year, the hour angle of the spring equinox is also 9 hours.

Step 3 Explain: The hour angle of the spring equinox is 9 hours, which means the local sidereal time is LST = 09:00.

Example 2: Suppose the local sidereal time is LST = 04:00. When will Vega cross the meridian?

Solution:

Step 1 Understand: We are given the local sidereal time, which tells us the hour angle of the spring equinox in the local sky. To determine when Vega will cross the meridian, we need to know its hour angle, which we can calculate from its right ascension and the formula given above. Figure S1.11 shows us that Vega has RA = 18^{h}35^m, so we have all the information we need. (Appendix F also gives RA and dec for Vega and other stars.)

Step 2 Solve: We now use the formula to find Vega's hour angle from the local sidereal time and Vega's right ascension:

$$HA_{\text{Vega}} = LST − RA_{\text{Vega}} = 4:00 − 18:35 = −14:35$$

Step 3 Explain: Vega's hour angle is −14 hours 35 minutes, which means it will cross your meridian 14 hours and 35 minutes from now. This also means that Vega crossed your meridian 9 hours and 25 minutes ago (because 14^{h}35^m + 9^{h}25^m = 24^h). (Note that these are intervals of sidereal time.)

Table S1.1 The Sun's Approximate Celestial Coordinates at 1-Month Intervals

Approximate Date	RA	Dec
Mar. 21 (spring equinox)	0 hr	0°
Apr. 21	2 hr	+12°
May 21	4 hr	+20°
June 21 (summer solstice)	6 hr	$+23\frac{1}{2}°$
July 21	8 hr	+20°
Aug. 21	10 hr	+12°
Sept. 21 (fall equinox)	12 hr	0°
Oct. 21	14 hr	−12°
Nov. 21	16 hr	−20°
Dec. 21 (winter solstice)	18 hr	$-23\frac{1}{2}°$
Jan. 21	20 hr	−20°
Feb. 21	22 hr	−12°

hours increases rapidly in spring and decreases rapidly in fall, while the number of daylight hours remains long and nearly constant for a couple of months around the summer solstice and short and nearly constant for a couple of months around the winter solstice.

THINK ABOUT IT

On your plastic ball model of the celestial sphere, add dots along the ecliptic to show the Sun's monthly positions. Based on your model, what are the Sun's approximate celestial coordinates on your birthday?

Seasons Tutorial, Lesson 3

How do stars move through the local sky?

We can use our system of celestial coordinates to gain a better understanding of the way stars move through the local sky. Earth's rotation makes all celestial objects appear to circle around Earth each day (see Figure 2.9), but what we actually see in the local sky is more complex because we see only half the celestial sphere at one time (the ground blocks our view of the other half). Let's explore the appearance of the local sky. As we'll see, the path of any star through your local sky depends on only two things: (1) your latitude; and (2) the declination of the star whose path you want to know.

The Sky at the North Pole   The daily paths of stars are easiest to understand for the local sky at the North Pole, so let's begin there before moving on to other latitudes. Figure S1.13a shows the rotating celestial sphere and your orientation relative to it when you are standing at the North Pole. Your "up" points toward the north celestial pole, which therefore marks your zenith. Earth blocks your view of anything south of the celestial equator, which therefore runs along your horizon. To make it easier for you to visualize the local sky, Figure S1.13b shows your horizon extending to the celestial sphere. The horizon is marked with directions, but remember that all directions are south from the North Pole. We therefore cannot define a meridian for the North Pole, since a meridian would have to run from the north to the south points on the horizon and there are no such unique points at the North Pole.

Notice that the daily circles of the stars keep them at constant altitudes above or below the North Polar horizon.

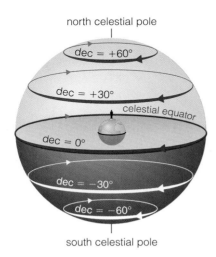

a The orientation of the local sky, relative to the celestial sphere, for an observer at the North Pole.

Figure S1.13   Interactive Figure   The sky at the North Pole.

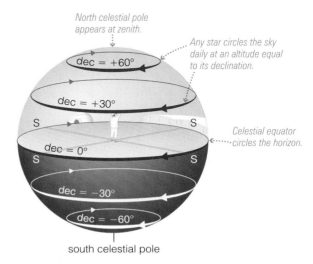

b Extending the horizon to the celestial sphere makes it easier to visualize the local sky at the North Pole. (Note: To understand the extension, think of Earth in part (a) as being extremely small compared to the celestial sphere.)

Moreover, the altitude of any star is equal to its declination. For example, a star with declination +60° circles the sky at an altitude of 60°, and a star with declination −30° remains 30° below your horizon at all times. As a result, all stars north of the celestial equator are circumpolar at the North Pole, meaning that they never fall below the horizon. Stars south of the celestial equator can never be seen at the North Pole. If you are having difficulty visualizing the star paths, it may help you to watch star paths as you rotate your plastic ball model of the celestial sphere.

You should also notice that right ascension does not affect a star's path at all: The path depends only on declination. As we'll see shortly, this rule holds for all latitudes. Right ascension affects only the *time* of day and year at which a star is found in a particular position in your sky.

The Sky at the Equator After the Poles, the equatorial sky is the next easiest case to understand. Imagine that you are standing somewhere on Earth's equator (lat = 0°), such as in Ecuador, in Kenya, or on the island of Borneo. Figure S1.14a shows that "up" points directly away from (perpendicular to) Earth's rotation axis. Figure S1.14b shows the local sky more clearly by extending the horizon to the celestial sphere and rotating the diagram so the zenith is up. As everywhere except at the poles, the meridian extends from the horizon due south, through the zenith, to the horizon due north.

Look carefully at how the celestial sphere appears to rotate in the local sky. The north celestial pole remains stationary on your horizon due north. As we should expect, its altitude of 0° is equal to the equator's latitude [Section 2.1]. Similarly, the south celestial pole remains stationary on your horizon due south. At any particular time, half the celestial equator is visible, extending from the horizon due east, through the zenith, to the horizon due west. The other half lies below the horizon. As the equatorial sky appears to turn, all star paths rise straight out of the eastern horizon

and set straight into the western horizon, with the following features:

- **Stars with dec = 0°** lie *on* the celestial equator and therefore rise due east, cross the meridian at the zenith, and set due west.
- **Stars with dec > 0°** rise north of due east, reach their highest point on the meridian in the north, and set north of due west. Their rise, set, and highest point depend on their declination. For example, a star with dec = +30° rises 30° north of due east, crosses the meridian 30° to the north of the zenith—that is, at an *altitude* of 90° − 30° = 60° in the north—and sets 30° north of due west.
- **Stars with dec < 0°** rise south of due east, reach their highest point on the meridian in the south, and set south of due west. For example, a star with dec = −50° rises 50° south of due east, crosses the meridian 50° to the south of the zenith—that is, at an *altitude* of 90° − 50° = 40° in the south—and sets 50° south of due west.

Notice that exactly half of any star's daily circle lies above the horizon, which means that every star at the equator is above the horizon for exactly half of each sidereal day, or just under 12 hours, and below the horizon for the other half of the sidereal day. Also, notice again that right ascension does not affect star paths, although it does affect the time of day and year at which a star will be in a particular place along its path.

THINK ABOUT IT

Are any stars circumpolar at the equator? Are there stars that never rise above the horizon at the equator? Explain.

Skies at Other Latitudes Star tracks may at first seem more complex at other latitudes, with their mixtures of circum-

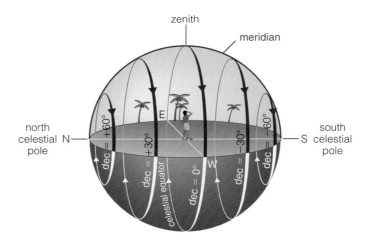

a The orientation of the local sky, relative to the celestial sphere, for an observer at Earth's equator.

b Extending the horizon and rotating the diagram make it easier to visualize the local sky at the equator.

Figure S1.14 Interactive Figure The sky at the equator.

polar stars and stars that rise and set. However, they are easy to understand if we apply the same basic strategy we've used for the North Pole and equator. Let's consider latitude 40°N, such as in Denver, Indianapolis, Philadelphia, or Beijing. First, as shown in Figure S1.15a, imagine standing at this latitude on a basic diagram of the rotating celestial sphere. Note that "up" points to a location on the celestial sphere with declination +40°. To make it easier to visualize the local sky, we next extend the horizon and rotate the diagram so the zenith is up (Figure S1.15b).

As we would expect, the north celestial pole appears 40° above the horizon due north, since its altitude in the local sky is always equal to the latitude. Half the celestial equator is visible. It extends from the horizon due east, to the meridian at an altitude of 50° in the south, to the horizon due west. By comparing this diagram to that of the local sky for the equator, you can probably notice a general rule for the celestial equator at any latitude except the poles:

The celestial equator always extends from due east on your horizon to due west on your horizon, crossing the meridian at an altitude of 90° *minus* your latitude.

The celestial equator crosses the meridian south of the zenith for locations in the Northern Hemisphere and north of the zenith for locations in the Southern Hemisphere.

If you study Figure S1.15b carefully (or, better yet, rotate your plastic ball model of the celestial sphere with the same orientation), you'll notice the following features of the sky for latitude 40°N:

- **Stars with dec = 0°** lie *on* the celestial equator and therefore follow the path of the celestial equator through the local sky. That is, for latitude 40°N, these stars rise due east, cross the meridian at altitude 90° − 40° = 50° in the south, and set due west.
- **Stars with dec > (90° − lat)** are circumpolar. Thus, for latitude 40°N, stars with declination greater than

90° − 40° = 50° are circumpolar, because they lie *within* 40° of the north celestial pole.

- **Stars with dec > 0° but which are not circumpolar** follow paths parallel to but north of the celestial equator: They rise north of due east and set north of due west, and cross the meridian to the north of the place where the celestial equator crosses it by an amount equal to their declination. For example, because the celestial equator at latitude 40° crosses the meridian at altitude 50° in the south, a star with dec = +30° crosses the meridian at altitude 50° + 30° = 80° in the south. Similarly, a star with dec = +60° crosses the meridian 60° farther north than the celestial equator, which means at altitude 70° in the north. (To calculate this result, note that the sum 50° + 60° = 110° goes 20° past the zenith altitude of 90°, making it equivalent to 90° − 20° = 70°.)
- **Stars with dec < (−90° + lat)** never rise above the horizon. Thus, for latitude 40°N, stars with declination less than −90° + 40° = −50° never rise above the horizon, because they lie within 40° of the south celestial pole.
- **Stars with dec < 0° but which are sometimes visible** follow paths parallel to but south of the celestial equator: They rise south of due east and set south of due west, and cross the meridian south of the place where the celestial equator crosses it by an amount equal to their declination. For example, a star with dec = −30° crosses the meridian at altitude 50° − 30° = 20° in the south.

You should also notice that the fraction of any star's daily circle that is above the horizon—and hence the amount of time it is above the horizon each day—depends on its declination. Because exactly half the celestial equator is above the horizon, stars on the celestial equator (dec = 0°) are

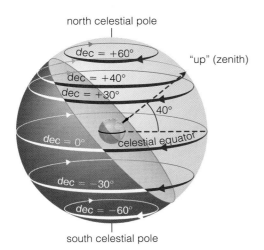

a The orientation of the local sky, relative to the celestial sphere, for an observer at latitude 40°N. Because latitude is the angle to Earth's equator, "up" points to the circle on the celestial sphere with declination +40°.

Figure S1.15 Interactive Figure. The sky at 40°N latitude.

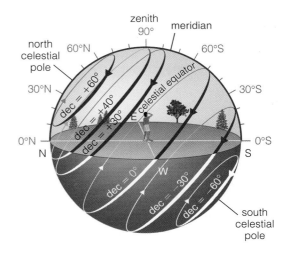

b Extending the horizon and rotating the diagram so the zenith is up make it easier to visualize the local sky. The blue scale along the meridian shows altitudes and directions in the local sky.

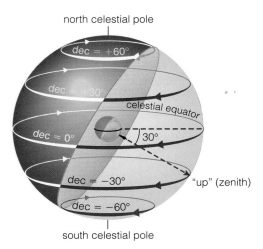

a The orientation of the local sky for an observer at latitude 30°S, relative to the celestial sphere. "Up" points to the circle on the celestial sphere with declination −30°.

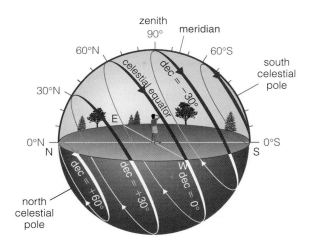

b Extending the horizon and rotating the diagram so the zenith is up make it easier to visualize the local sky. Note that the south celestial pole is visible at altitude 30° in the south, while the celestial equator stretches across the northern half of the sky.

Figure S1.16 Interactive Figure The sky at 30°S latitude.

above the horizon for about 12 hours per day. For northern latitudes like 40°N, stars with positive declinations have more than half their daily circles above the horizon and hence are above the horizon for more than 12 hours each day (with the range extending to 24 hours a day for the circumpolar stars). Stars with negative declinations have less than half their daily circles above the horizon and hence are above the horizon for less than 12 hours each day (with the range going to zero for stars that are never above the horizon).

We can apply the same strategy we used in Figure S1.15 to find star paths for other latitudes. Figure S1.16 shows the process for latitude 30°S. Note that the south celestial pole is visible to the south and that the celestial equator passes through the northern half of the sky. If you study the diagram carefully, you can see how star tracks depend on declination.

THINK ABOUT IT

Study Figure S1.16 for latitude 30°S. Describe the path of the celestial equator. Does it obey the 90° − lat rule given earlier? Describe how star tracks differ for stars with positive and negative declinations. What declination must a star have to be circumpolar at this latitude?

• How does the Sun move through the local sky?

Like the stars and other objects on the celestial sphere, the Sun's path depends only on its declination and your latitude. However, because the Sun's declination changes over the course of the year, the Sun's path also changes.

Figure S1.17 shows the Sun's path on the equinoxes and solstices for latitude 40°N. On the equinoxes, the Sun is on the celestial equator (dec = 0°) and therefore follows the

celestial equator's path: It rises due east, crosses the meridian at altitude 50° in the south, and sets due west. Like any object on the celestial equator, it is above the horizon for 12 hours. On the summer solstice, when the Sun has dec = $+23\frac{1}{2}°$ (see Table S1.1), the Sun rises well north of due east,[*] reaches an altitude of $50° + 23\frac{1}{2}° = 73\frac{1}{2}°$ when it crosses the meridian in the south, and sets well north of due west. The daylight hours are long because much more than half the Sun's path is above the horizon. On the winter solstice, when the Sun has dec = $-23\frac{1}{2}°$, the Sun rises well south of due east, reaches an altitude of only $50° - 23\frac{1}{2}° = 26\frac{1}{2}°$ when it crosses the meridian in the south, and sets well

---

[*]Calculating exactly how far north of due east the Sun rises is beyond the scope of this book, but *SkyGazer, Starry Night,* and other astronomical software packages can tell you exactly where (and at what time) the Sun rises and sets each day.

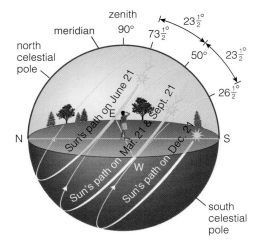

Figure S1.17 Interactive Figure The Sun's daily paths for the equinoxes and solstices at latitude 40°N.

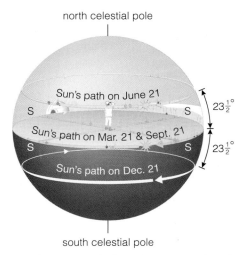

Figure S1.18 Special latitudes defined by the Sun's path through the sky.

Figure S1.19 Interactive Figure Daily paths of the Sun for the equinoxes and solstices at the North Pole.

south of due west. The daylight hours are short because much less than half the Sun's path is above the horizon.

We could make a similar diagram to show the Sun's path on various dates for any latitude. However, the $23\frac{1}{2}°$ tilt of Earth's axis makes the Sun's path particularly interesting at the special latitudes shown in Figure S1.18. Let's investigate.

The Sun at the North and South Poles Remember that the celestial equator circles the horizon at the North Pole. Figure S1.19 shows how we use this fact to find the Sun's path in the North Polar sky. Because the Sun appears *on* the celestial equator on the day of the spring equinox, the Sun circles the North Polar sky *on the horizon* on March 21, completing a full circle in 24 hours (1 solar day). Over the next 3 months, the Sun continues to circle the horizon each day, circling at gradually higher altitudes as its declination increases. It reaches its highest point on the summer solstice, when its declination of $+23\frac{1}{2}°$ means that it circles the North Polar sky at an altitude of $23\frac{1}{2}°$. After the summer solstice, the daily circles gradually fall lower over the next 3 months, reaching the horizon on the fall equinox. Then, because the Sun's declination is negative for the next 6 months (until the following spring equinox), it remains below the North Polar horizon. Thus, the North Pole essentially has 6 months of daylight and 6 months of darkness, with an extended twilight that lasts a few weeks beyond the fall equinox and an extended dawn that begins a few weeks before the spring equinox.

The situation is the opposite at the South Pole. Here the Sun's daily circle first reaches the horizon on the fall equinox. The daily circles then rise gradually higher, reaching a maximum altitude of $23\frac{1}{2}°$ on the *winter* solstice, and then slowly fall back to the horizon on the spring equinox. Thus, the South Pole has the Sun above the horizon during the 6 months it is below the north polar horizon.

Although we've correctly described the Sun's true position in the polar skies over the course of the year, two effects complicate what we actually see at the poles around the times of the equinoxes. First, the atmosphere bends

light enough so that the Sun *appears* to be slightly above the horizon even when it is actually slightly below it. Near the horizon, this bending makes the Sun appear about 1° higher than it would in the absence of an atmosphere. Second, the Sun's angular size of about $\frac{1}{2}°$ means that it does not fall below the horizon at a single moment but instead sets gradually. Together, these effects mean that the Sun appears above each polar horizon for slightly longer (by several days) than 6 months each year.

The Sun at the Equator At the equator, the celestial equator extends from the horizon due east, through the zenith, to the horizon due west. The Sun therefore follows this path on each equinox, reaching the zenith at local noon (Figure S1.20). Following the spring equinox, the Sun's increasing declination means that it follows a daily track that takes it gradually northward in the sky. It is farthest north on the summer solstice, when it rises $23\frac{1}{2}°$ north of due east,

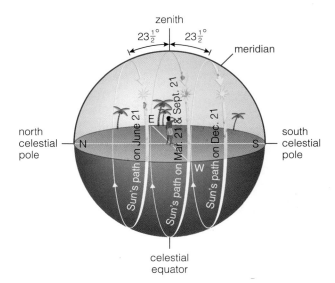

Figure S1.20 Interactive Figure Daily paths of the Sun for the equinoxes and solstices at the equator.

crosses the meridian at altitude $90° - 23\frac{1}{2}° = 66\frac{1}{2}°$ in the north, and sets $23\frac{1}{2}°$ north of due west. Over the next 6 months, it gradually tracks southward until the winter solstice, when its path is the mirror image (across the celestial equator) of its summer solstice path.

Like all objects in the equatorial sky, the Sun is always above the horizon for half a day and below it for half a day. Moreover, the Sun's track is highest in the sky on the equinoxes and lowest on the summer and winter solstices. That is why equatorial regions do not have four seasons like temperate regions [Section 2.2]. The Sun's path in the equatorial sky also makes it rise and set perpendicular to the horizon every day of the year, making for a more rapid dawn and a briefer twilight than at other latitudes.

The Sun at the Tropics   The circles of latitude 23.5°N and 23.5°S are called the **tropic of Cancer** and the **tropic of Capricorn**, respectively (see Figure S1.18). The region between these two circles, generally called the *tropics*, represents the parts of Earth where the Sun can sometimes reach the zenith at noon.

Figure S1.21 shows why the tropic of Cancer is special. The celestial equator extends from due east on the horizon to due west on the horizon, crossing the meridian in the south at an altitude of $90° - 23\frac{1}{2}°$ (the latitude) $= 66\frac{1}{2}°$. The Sun follows this path on the equinoxes (March 21 and September 21). As a result, the Sun's path on the summer solstice, when it crosses the meridian $23\frac{1}{2}°$ northward of the celestial equator, takes it to the zenith at local noon. Because the Sun has its maximum declination on the summer solstice, the tropic of Cancer marks the northernmost latitude at which the Sun ever reaches the zenith. Similarly, at the tropic of Capricorn the Sun reaches the zenith at local noon on the winter solstice, making this the southernmost latitude at which the Sun ever reaches the zenith. Between the two tropic circles, the Sun passes through the zenith twice a year; the precise dates vary with latitude.

The Sun at the Arctic and Antarctic Circles   At the equator, the Sun is above the horizon for 12 hours each day year-round. At latitudes progressively farther from the equator, the daily time that the Sun is above the horizon varies progressively more with the seasons. The special latitudes at which the Sun remains continuously above the horizon for a full day each year are the polar circles: the **Arctic Circle** at latitude 66.5°N and the **Antarctic Circle** at latitude 66.5°S (see Figure S1.18). Poleward of these circles, the length of continuous daylight (or darkness) increases beyond 24 hours, reaching the extreme of 6 months at the North and South Poles.

Figure S1.22 shows why the Arctic Circle is special. The celestial equator extends from due east on the horizon to due west on the horizon, crossing the meridian in the south at an altitude of $90° - 66\frac{1}{2}°$ (the latitude) $= 23\frac{1}{2}°$. As a result, the Sun's path is circumpolar on the summer solstice: It skims the northern horizon at midnight, rises through the eastern sky to a noon maximum altitude of 47° in the south (which is the celestial equator's maximum altitude of $23\frac{1}{2}°$ plus the Sun's summer solstice declination of $23\frac{1}{2}°$), and then gradually falls through the western sky until it is back on the horizon at midnight (see the photograph of this path in Figure 2.18). At the Antarctic Circle, the Sun follows the same basic pattern on the winter solstice, except that it skims the horizon in the south and rises to a noon maximum altitude of 47° in the north.

However, as at the North and South Poles, what we actually see at the polar circles is slightly different from this idealization. Again, the bending of light by Earth's atmosphere and the Sun's angular size of about $\frac{1}{2}°$ make the Sun *appear* to be slightly above the horizon even when it is slightly below it. Thus, the Sun seems not to set for several days, rather than for a single day, around the summer solstice at the Arctic Circle (the winter solstice at the Antarctic Circle). Similarly, the Sun appears to peek above the horizon momentarily, rather than not at all, around the winter solstice at the Arctic Circle (the summer solstice at the Antarctic Circle).

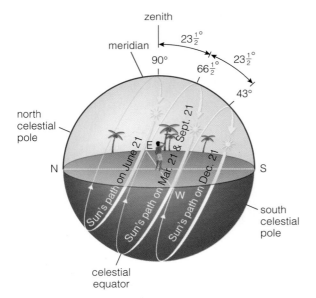

Figure S1.21 Interactive Figure Daily paths of the Sun for the equinoxes and solstices at the tropic of Cancer.

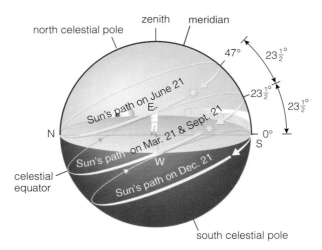

Figure S1.22 Interactive Figure Daily paths of the Sun for the equinoxes and solstices at the Arctic Circle.

S1.3 Principles of Celestial Navigation

We now have all the background we need to cover the basic principles of celestial navigation. Imagine that you're on a ship at sea, far from any landmarks. How can you figure out where you are? It's easy if you understand the apparent motions of the sky that we have already discussed in this chapter.

• How can you determine your latitude?

Determining latitude is particularly easy if you can find the north or south celestial pole: Your latitude is equal to the altitude of the celestial pole in your sky. In the Northern Hemisphere at night, you can determine your approximate latitude by measuring the altitude of Polaris. Because Polaris has a declination within 1° of the north celestial pole, its altitude is within 1° of your latitude. For example, if Polaris has altitude 17°, your latitude is between 16°N and 18°N.

If you want to be more precise, you can determine your latitude from the altitude of *any* star as it crosses your meridian. For example, suppose Vega happens to be crossing your meridian right now and it appears in your southern sky at altitude 78°44'. Because Vega has dec = +38°44' (see Figure S1.11), it crosses your meridian 38°44' north of the celestial equator. As shown in Figure S1.23a, you can conclude that the celestial equator crosses your meridian at an altitude of precisely 40° in the south. Your latitude must therefore be 50°N because the celestial equator always

crosses the meridian at an altitude of 90° minus the latitude. You know you are in the Northern Hemisphere because the celestial equator crosses the meridian in the south.

In the daytime, you can find your latitude from the Sun's altitude on your meridian if you know the date and have a table that tells you the Sun's declination on that date. For example, suppose the date is March 21 and the Sun crosses your meridian at altitude 70° in the north (Figure S1.23b). Because the Sun has dec = 0° on March 21, you can conclude that the celestial equator also crosses your meridian in the north at altitude 70°. You must be in the Southern Hemisphere, because the celestial equator crosses the meridian in the north. From the rule that the celestial equator crosses the meridian at an altitude of 90° minus the latitude, you can conclude that you are at latitude 20°S.

• How can you determine your longitude?

You can determine your longitude by comparing the current position of an object in your sky with its position as seen from some known longitude. As a simple example, suppose you use a sundial to determine that the apparent solar time is 1:00 P.M., which means the Sun passed the meridian 1 hour ago. You immediately call a friend in England and learn that it is 3:00 P.M. in Greenwich (or you carry a clock that keeps Greenwich time). You now know that your local time is 2 hours earlier than the local time in Greenwich, which means you are 2 hours west of Greenwich. (An earlier time means that you are *west* of Greenwich, because Earth rotates from west to east.) Each hour corresponds to 15° of longitude, so "2 hours west of Greenwich" means longitude 30°W.

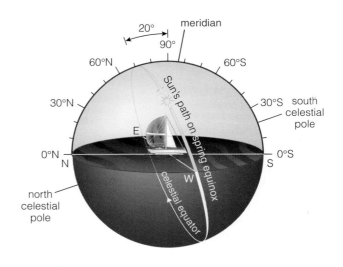

a Because Vega has dec = +38°44', it crosses the meridian 38°44' north of the celestial equator. Because Vega crosses the meridian at altitude 78°44' in the south, the celestial equator must cross the meridian at altitude 40° in the south. Thus, the latitude must be 50°N.

b To determine latitude from the Sun's meridian crossing, you must know the Sun's declination, which you can determine from the date. The case shown is for the spring equinox, when the Sun's declination is 0° and hence follows the path of the celestial equator through the local sky. Because the celestial equator crosses the meridian at 70° in the north, the latitude must be 20°S.

Figure S1.23 Determining latitude from a star and from the Sun.

At night, you can find your longitude by comparing the positions of stars in your local sky and at some known longitude. For example, suppose Vega is on your meridian and a call to your friend reveals that it won't cross the meridian in Greenwich until 6 hours from now. In this case, your local time is 6 hours later than the local time in Greenwich. Thus, you are 6 hours east of Greenwich, or at longitude 90°E (because $6 \times 15° = 90°$).

Celestial Navigation in Practice Although celestial navigation is easy in principle, at least three practical considerations make it more difficult in practice. First, finding either latitude or longitude requires a tool for measuring angles in the sky. One such device, called an *astrolabe,* was invented by the ancient Greeks and significantly improved by Islamic scholars during the Middle Ages. The astrolabe's faceplate (Figure S1.24a) could be used to tell time, because it consisted of a rotating star map and horizon plates for specific latitudes. Today you can buy similar rotatable star maps, called *planispheres.* Most astrolabes contained a sighting

stick on the back that allowed users to measure the altitudes of bright stars in the sky. These measurements could then be correlated against special markings under the faceplate (Figure S1.24b). Astrolabes were effective but difficult and expensive to make. As a result, medieval sailors often measured angles with a simple pair of calibrated perpendicular sticks, called a *cross-staff* or *Jacob's staff* (Figure S1.24c). A more modern device called a *sextant* allows much more precise angle determinations by incorporating a small telescope for sightings (Figure S1.24d). Sextants are still used for celestial navigation on many ships. If you want to practice celestial navigation yourself, you can buy an inexpensive plastic sextant at many science-oriented stores.

A second practical consideration is knowing the celestial coordinates of stars and the Sun so that you can determine their paths through the local sky. At night, you can use a table listing the celestial coordinates of bright stars. In addition to knowing the celestial coordinates, you must either know the constellations and bright stars extremely well or

Figure S1.24 Navigational instruments.

a The faceplate of an astrolabe; many astrolabes had sighting sticks on the back for measuring positions of bright stars.

b A copper engraving of Italian explorer Amerigo Vespucci (for whom America was named) using an astrolabe to sight the Southern Cross. The engraving by Philip Galle, from the book *Nova Reperta,* was based on an original by Joannes Stradanus in the early 1580s.

d A sextant.

c A woodcutting of Ptolemy holding a cross-staff (artist unknown).

carry star charts to help you identify them. For navigating by the Sun in the daytime, you'll need a table listing the Sun's celestial coordinates on each day of the year.

The third practical consideration applies to determining longitude: You need to know the current position of the Sun (or a particular star) in a known location, such as Greenwich, England. Although you could find this out by calling a friend who lives there, it's more practical to carry a clock set to universal time (that is, Greenwich mean time). In the daytime, the clock makes it easy to determine your longitude. If apparent solar time is 1:00 P.M. in your location and the clock tells you that it is 3:00 P.M. in Greenwich, then you are 2 hours west of Greenwich, or at longitude 30°W. The task is more difficult at night, because you must compare the position of a *star* in your sky to its current position in Greenwich. You can do this with the aid of detailed astronomical tables that allow you to determine the current position of any star in the Greenwich sky from the date and the universal time.

Historically, this third consideration created enormous problems for navigation. Before the invention of accurate clocks, sailors could easily determine their latitude but not their longitude. Indeed, most of the European voyages of discovery beginning in the 1400s relied on little more than guesswork about longitude, although some sailors learned complex mathematical techniques for estimating longitude through observations of the lunar phases. More accurate longitude determination, upon which the development of extensive ocean commerce and travel depended, required the invention of a clock that would remain accurate on a ship rocking in the ocean swells. By the early 1700s, solving this problem was considered so important that the British government offered a substantial monetary prize for the solution. John Harrison claimed the prize in 1761, with a clock that lost only 5 seconds during a 9-week voyage to Jamaica.[*]

The Global Positioning System In the past couple of decades, a new type of celestial navigation has supplanted traditional methods. It finds positions relative to a set of satellites in Earth orbit. These satellites of the **global positioning system** (**GPS**) in essence function like artificial stars. The satellites' positions at any moment are known precisely from their orbital characteristics. The GPS currently uses about two dozen satellites orbiting Earth at an altitude of 20,000 kilometers. Each satellite transmits a radio signal that a small radio receiver can receive—rain or shine, day or night. Each GPS receiver has a built-in computer that calculates your precise position on Earth by comparing the signals received from several GPS satellites.

The United States originally built the GPS in the late 1970s for military use. Today, the many applications of the

---

[*]The story of the difficulties surrounding the measurement of longitude at sea and how Harrison finally solved the problem is chronicled in *Longitude*, by Dava Sobel (Walker and Company, 1995).

COMMON MISCONCEPTIONS

Compass Directions

Most people determine direction with the aid of a compass rather than the stars. However, a compass needle doesn't actually point to true geographic north. Instead, the compass needle responds to Earth's magnetic field and points to *magnetic* north, which can be substantially different from true north. If you want to navigate precisely with a compass, you need a special map that takes into account local variations in Earth's magnetic field. Such maps are available at most camping stores. They are not perfectly reliable, however, because the magnetic field also varies with time. In general, celestial navigation is much more reliable for determining direction than using a compass.

GPS include automobile navigation systems as well as systems for helping airplanes land safely, guiding the blind around town, and helping lost hikers find their way. Geologists have used the GPS to measure *millimeter*-scale changes in Earth's crust.

With rapid growth in the use of GPS navigation, the ancient practice of celestial navigation is in danger of becoming a lost art. Fortunately, many amateur clubs and societies are keeping the skills of celestial navigation alive.

THE BIG PICTURE

Putting Chapter S1 into Context

In this chapter, we built upon concepts from the first three chapters to form a more detailed understanding of celestial timekeeping and navigation. We also learned how to determine paths for the Sun and the stars in the local sky. As you look back at what you've learned, keep in mind the following "big picture" ideas:

● Our modern systems of timekeeping are rooted in the apparent motions of the Sun through the sky. Although it's easy to forget these roots when you look at a clock or a calendar, the sky was the only guide to time for most of human history.

● The term *celestial navigation* sounds a bit mysterious, but it refers to simple principles that allow you to determine your location on Earth. Even if you're never lost at sea, you may find the basic techniques of celestial navigation useful to orient yourself at night (for example, on your next camping trip).

● If you understand the apparent motions of the sky discussed in this chapter and also learn the constellations and bright stars, you'll feel very much "at home" under the stars at night.

S1.1 Astronomical Time Periods

- **How do we define the day, month, year, and planetary periods?** Each of these is defined in two ways. A **sidereal day** is Earth's rotation period, which is about 4 minutes shorter than the 24-hour **solar day** from noon one day to noon the next day. A **sidereal month** is the Moon's orbital period of about $27\frac{1}{3}$ days; a **synodic month** is the $29\frac{1}{2}$ days required for the Moon's cycle of phases. A **sidereal year** is Earth's orbital period, which is about 20 minutes longer than the **tropical year** from one spring equinox to the next. A planet's **sidereal period** is its orbital period, and its **synodic period** is the time from one opposition or conjunction to the next.

- **How do we tell the time of day?** There are several time measurement systems. **Apparent solar time** is based on the Sun's position in the local sky. **Mean solar time** is also local, but it averages the changes in the Sun's rate of motion over the year. **Standard time** and **daylight saving time** divide the world into time zones. **Universal time** is the mean solar time in Greenwich, England.

- **When and why do we have leap years?** We usually have a **leap year** every 4 years because the length of the year is $365\frac{1}{4}$ days. However, it is not exactly $365\frac{1}{4}$ days, so our calendar skips a leap year in century years not divisible by 400.

S1.2 Celestial Coordinates and Motion in the Sky

- **How do we locate objects on the celestial sphere?** **Declination** is given as an angle describing an object's position north or south of the celestial equator. **Right ascension**, usually measured in hours (and minutes and seconds), tells us how far east an object is located relative to the spring equinox.

- **How do stars move through the local sky?** A star's path through the local sky depends on its declination and your latitude. Latitude tells you the orientation of your sky relative to the celestial sphere, while declination tells you how a particular star's path compares to the path of the celestial equator through your sky.

- **How does the Sun move through the local sky?** The Sun's path also depends on its declination and your latitude, but it varies throughout the year because its declination varies. The Sun's varying path helps define special latitudes, including the **tropics of Cancer and Capricorn** and the **Arctic and Antarctic Circles**.

S1.3 Principles of Celestial Navigation

- **How can you determine your latitude?** You can determine your latitude from the altitude of the celestial pole in your sky or by measuring the altitude and knowing the declination of a star (or the Sun) as it crosses your meridian.

- **How can you determine your longitude?** To determine longitude you must know the position of the Sun or a star in your sky and its position at the same time in the sky of Greenwich, England (or some other specific location). This is most easily done if you have a clock that tells universal time.

For instructor-assigned homework go to **Mastering**ASTRONOMY www.masteringastronomy.com

Review Questions

Short-Answer Questions Based on the Reading

1. Why is a *sidereal day* shorter than a *solar day*?
2. What is the difference between a *sidereal month* and a *synodic month*? Between a *sidereal year* and a *tropical year*? Between a planet's *sidereal period* and its *synodic period*?
3. What do we mean by *opposition, conjunction,* and *greatest elongation* for planets? Explain both for planets closer than Earth to the Sun and for planets farther than Eartth from the Sun.
4. Under what circumstances do we see a *transit* of a planet across the Sun?

5. What is *apparent solar time*? Why is it different from *mean solar time*? How are *standard time, daylight saving time*, and *universal time* related to mean solar time?

6. Describe the origins of the Julian and Gregorian calendars. Which one do we use today?

7. What do we mean when we describe the equinoxes and solstices as points on the celestial sphere? How are these points related to the times of year that we call the equinoxes and solstices?

8. What are *declination* and *right ascension*? How are these celestial coordinates similar to latitude and longitude on Earth? How are they different?

9. How and why do the Sun's celestial coordinates change over the course of each year?

10. Suppose you are standing at the North Pole. Where is the celestial equator in your sky? Where is the north celestial pole? Describe the daily motion of the sky. Do the same for the sky at the equator and at latitude 40°N.

11. Describe the Sun's paths through the local sky on the equinoxes and on the solstices for latitude 40°N. Do the same for the North Pole, South Pole, and equator.

12. What is special about the tropics of Cancer and Capricorn? Describe the Sun's path on the solstices at these latitudes. Do the same for the Arctic and Antarctic Circles.

13. Briefly describe how you can use the Sun or stars to determine your latitude and longitude.

14. What is the global positioning system?

Test Your Understanding

Does It Make Sense?

Decide whether the statement makes sense (or is clearly true) or does not make sense (or is clearly false). Explain your reasoning. (For an example, see Chapter 1, "Does It Make Sense?")

(*Hint:* For statements that involve coordinates—such as altitude, longitude, or declination—check whether the correct coordinates are used for the situation. For example, it does not make sense to describe a location on Earth by an altitude since altitude makes sense only for positions of objects in the local sky.)

15. Last night I saw Venus shining brightly on the meridian at midnight.

16. The apparent solar time was noon, but the Sun was just setting.

17. My mean solar clock said it was 2:00 P.M., but a friend who lives east of here had a mean solar clock that said it was 2:11 P.M.

18. When the standard time is 3:00 P.M. in Baltimore, it is 3:15 P.M. in Washington, D.C.

19. Last night around 8:00 P.M. I saw Jupiter at an altitude of 45° in the south.

20. The latitude of the stars in Orion's belt is about 5°N.

21. Today the Sun is at an altitude of 10° on the celestial sphere.

22. Los Angeles is west of New York by about 3 hours of right ascension.

23. The summer solstice is east of the vernal equinox by 6 hours of right ascension.

24. Even though my UT clock had stopped, I was able to find my longitude by measuring the altitudes of 14 different stars in my local sky.

Quick Quiz

Choose the best answer to each of the following. Explain your reasoning with one or more complete sentences.

25. The time from one spring equinox to the next is the: (a) sidereal day; (b) tropical year; (c) synodic month.

26. Jupiter is brightest when it is: (a) at opposition; (b) at conjunction; (c) closest to the Sun in its orbit.

27. Venus is easiest to see in the evening when it is: (a) at superior conjunction; (b) at inferior conjunction; (c) at greatest eastern elongation.

28. In the winter, your wristwatch tells: (a) apparent solar time; (b) standard time; (c) universal time.

29. A star that is located 30° north of the celestial equator has: (a) declination = 30°; (b) right ascension = 30°; (c) latitude = 30°.

30. A star's path through your sky depends on your latitude and the star's: (a) declination; (b) right ascension; (c) both declination and right ascension.

31. At latitude 50°N, the celestial equator crosses the meridian at altitude: (a) 50° in the south; (b) 50° in the north; (c) 40° in the south.

32. At the North Pole on the summer solstice, the Sun: (a) remains stationary in the sky; (b) reaches the zenith at noon; (c) circles the horizon at altitude $23\frac{1}{2}°$.

33. If you know a star's declination, you can determine your latitude if you also: (a) measure its altitude when it crosses the meridian; (b) measure its right ascension; (c) know the universal time.

34. If you measure the Sun's position in your local sky, you can determine your longitude if you also: (a) measure its altitude when it crosses the meridian; (b) know its right ascension and declination; (c) know the universal time.

Investigate Further

In-Depth Questions to Increase Your Understanding

Short-Answer/Essay Questions

35. *Opposite Rotation.* Suppose Earth rotated in the opposite direction from its revolution; that is, suppose it rotated clockwise (as seen from above the North Pole) while still orbiting counterclockwise around the Sun each year. Would the solar day still be longer than the sidereal day? Explain.

36. *No Precession.* Suppose Earth's axis did *not* precess. Would the sidereal year still be different from the tropical year? Explain.

37. *Fundamentals of Your Local Sky.* Answer each of the following for *your* latitude.
    a. Where is the north (or south) celestial pole in your sky?
    b. Describe the location of the meridian in your sky. Specify its shape and at least three distinct points along it (such as the points at which it meets your horizon and its highest point).
    c. Describe the location of the celestial equator in your sky. Specify its shape and at least three distinct points along it (such as the points at which it meets your horizon and crosses your meridian).
    d. Does the Sun ever appear at your zenith? If so, when? If not, why not?
    e. What range of declinations makes a star circumpolar in your sky? Explain.
    f. What is the range of declinations for stars that you can never see in your sky? Explain.

38. *Sydney Sky.* Repeat Problem 37 for the local sky in Sydney, Australia (latitude 34°S).

39. *Path of the Sun in Your Sky.* Describe the path of the Sun through your local sky for each of the following days.
    a. The spring and fall equinoxes.
    b. The summer solstice.
    c. The winter solstice.
    d. Today. (*Hint:* Estimate the right ascension and declination of the Sun for today's date by using the data in Table S1.1).

40. *Sydney Sun.* Repeat Problem 39 for the local sky in Sydney, Australia (latitude 34°S).

41. *Lost at Sea I.* During an upcoming vacation, you decide to take a solo boat trip. While contemplating the universe, you lose track of your location. Fortunately, you have some astronomical tables and instruments, as well as a UT clock. You thereby put together the following description of your situation:
    ● It is the spring equinox.
    ● The Sun is on your meridian at altitude 75° in the south.
    ● The UT clock reads 22:00.
    a. What is your latitude? How do you know?
    b. What is your longitude? How do you know?
    c. Consult a map. Based on your position, where is the nearest land? Which way should you sail to reach it?

42. *Lost at Sea II.* Repeat Problem 41, based on the following description of your situation:
    ● It is the day of the summer solstice.
    ● The Sun is on your meridian at altitude $67\frac{1}{2}°$ in the north.
    ● The UT clock reads 06:00.

43. *Lost at Sea III.* Repeat Problem 41, based on the following description of your situation:
    ● Your local time is midnight.
    ● Polaris appears at altitude 67° in the north.
    ● The UT clock reads 01:00.

44. *Lost at Sea IV.* Repeat Problem 41, based on the following description of your situation:
    ● Your local time is 6 A.M.
    ● From the position of the Southern Cross, you estimate that the south celestial pole is at altitude 33° in the south.
    ● The UT clock reads 11:00.

45. *Powering Spirit.* Suppose that it is currently northern summer on Mars, and that the Mars Exploration Rover *Spirit* is in Gusev Crater near 15° north latitude. Spirit's operators have discovered that its solar panels need to receive just a few more minutes of sunlight each day to power the rover through the Martian night. What should they do? Explain.

46. *The Sun from Mars.* Mars has an axis tilt of 25.2°, only slightly larger than that of Earth. Compared to Earth, is the range of latitudes on Mars for which the Sun can reach the zenith larger or smaller? Is the range of latitudes for which the Sun is circumpolar larger or smaller? Make a sketch of Mars similar to the one for Earth in Figure S1.18.

Quantitative Problems

Be sure to show all calculations clearly and state your final answers in complete sentences.

47. *Solar and Sidereal Days.* Suppose Earth orbited the Sun in 6 months rather than 1 year but had the same rotation period. How much longer would a solar day be than a sidereal day? Explain.

48. *Saturn's Orbital Period.* Saturn's synodic period is 378.1 days. What is its actual orbital period?

49. *Mercury's Orbital Period.* Mercury's synodic period is 115.9 days. What is its actual orbital period?

50. *New Asteroid.* You discover an asteroid with a synodic period of 429 days. What is its actual orbital period?

51. *Using the Analemma I.* It's February 15 and your sundial tells you the apparent solar time is 18 minutes until noon. What is the mean solar time?

52. *Using the Analemma II.* It's July 1 and your sundial tells you that the apparent solar time is 3:30 P.M. What is the mean solar time?

53. *Find the Sidereal Time.* It is 4 P.M. on the spring equinox. What is the local sidereal time?

54. *Where's Vega?* The local sidereal time is 19:30. When will Vega cross your meridian?

55. *Find Right Ascension.* You observe a star that has an hour angle of 13 hours (13^h) when the local sidereal time is 8:15. What is the star's right ascension?

56. *Where's Orion?* The Orion Nebula has declination of about $-5.5°$ and right ascension of 5^h25^m. If you are at latitude 40°N and the local sidereal time is 7:00, approximately where does the Orion Nebula appear in your sky?

57. *Meridian Crossings of the Moon and Phobos.* Estimate the time between meridian crossings of the Moon for a person standing on Earth. Repeat your calculation for meridian crossings of the Martian moon Phobos. Use the Appendices in the back of the book if necessary.

58. *Mercury's Rotation Period.* Mercury's sidereal day is approximately $\frac{2}{3}$ of its orbital period, or about 58.6 days. Estimate the length of Mercury's solar day. Compare to Mercury's orbital period of about 88 days.

Discussion Questions

59. *Northern Chauvinism.* Why is the solstice in June called the *summer solstice,* when it marks winter for places like Australia, New Zealand, and South Africa? Why is the writing on maps and globes usually oriented so that the Northern Hemisphere is at the top, even though there is no up or down in space? Discuss.

60. *Celestial Navigation.* Briefly discuss how you think the benefits and problems of celestial navigation might have affected ancient sailors. For example, how did they benefit from using the north celestial pole to tell directions, and what problems did they experience because of the difficulty in determining longitude? Can you explain why ancient sailors generally hugged coastlines as much as possible on their voyages? What dangers did this type of sailing pose? Why did the Polynesians become the best navigators of their time?

Interactive Tutorials

Tutorial Review of Key Concepts

Use the interactive **Tutorials** at **www.mastering astronomy.com** to review key concepts from this chapter.

Seasons Tutorial
Lesson 2 The Solstices and Equinoxes
Lesson 3 The Sun's Position in the Sky

Supplementary Tutorial Exercises

Use the interactive **Tutorial Lessons** to explore the following questions.

Seasons Tutorial, Lesson 2
1. How are the solstices and equinoxes related to Earth's orbital position around the Sun?
2. What are day and night like at the North Pole?
3. How does the length of the day vary with the seasons at the Antarctic Circle?

Seasons Tutorial, Lesson 3
1. When can you see the Sun directly over the equator?
2. Where can you see the Sun directly overhead on the winter solstice?
3. When is the Sun seen directly overhead at your latitude?

 Exploring the Sky and Solar System

Of the many activities available on the *Voyager: SkyGazer* **CD-ROM** accompanying your book, use the following files to observe key phenomena covered in this chapter.

Go to the **File: Basics** folder for the following demonstrations.
1. Analemma
2. Rubber Horizon
3. Three Cities

Go to the **File: Demo** folder for the following demonstrations.
1. Venus Transit of 1769
2. Celestial Poles
3. Russian Midnight Sun

 Movies

Check out the following narrated and animated short documentaries available on **www.masteringastronomy.com** for a helpful review of key ideas covered in this chapter.

The Celestial Sphere Movie

Time and Seasons Movie

Web Projects

1. *Sundials.* Although they are no longer necessary for timekeeping, sundials remain popular for their cultural and artistic value. Search the Web for pictures and information about interesting sundials around the world. Write a short report about at least three sundials that you find particularly interesting.
2. *The Analemma.* Learn more about the analemma and its uses from information available on the Web. Write a short report on your findings.
3. *Calendar History.* Investigate the history of the Julian or Gregorian calendar in greater detail. Write a short summary of some interesting aspect of the history you learn from your Web research. (For example, why did Julius Caesar allow one year to have 445 days? How did our months end up with 28, 30, or 31 days?)
4. *Global Positioning System.* Learn more about the global positioning system and its uses. Write a short report summarizing how you think the growing availability of GPS will affect our lives over the next 10 years.

4

Making Sense of the Universe
Understanding Motion, Energy, and Gravity

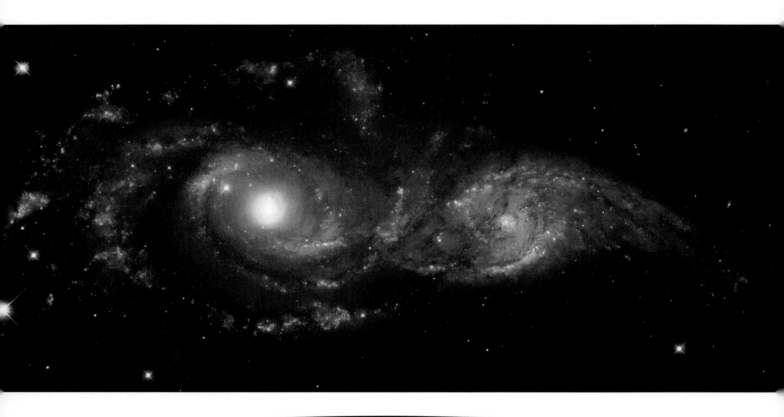

LEARNING GOALS

4.1 Describing Motion: Examples from Daily Life
- How do we describe motion?
- How is mass different from weight?

4.2 Newton's Laws of Motion
- How did Newton change our view of the universe?
- What are Newton's three laws of motion?

4.3 Conservation Laws in Astronomy
- Why do objects move at constant velocity if no force acts on them?
- What keeps a planet rotating and orbiting the Sun?
- Where do objects get their energy?

4.4 The Universal Law of Gravitation
- What determines the strength of gravity?
- How does Newton's law of gravity extend Kepler's laws?

4.5 Orbits, Tides, and the Acceleration of Gravity
- How do gravity and energy together allow us to understand orbits?
- How does gravity cause tides?
- Why do all objects fall at the same rate?

If I have seen farther than others, it is because I have stood on the shoulders of giants.

Isaac Newton

The history of the universe is essentially a story about the interplay between matter and energy since the beginning of time. Interactions between matter and energy began in the Big Bang and continue today in everything from the microscopic jiggling of atoms to gargantuan collisions of galaxies. Understanding the universe therefore depends on becoming familiar with how matter responds to the ebb and flow of energy.

You might guess that it would be difficult to understand the many interactions that shape the universe, because they occur on so many different size scales. However, we now know that just a few physical laws govern the movements of everything from atoms to galaxies. The Copernican revolution spurred the discovery of these laws, and Galileo deduced some of them from his experiments. But it was Sir Isaac Newton, one of the most influential human beings of all time, who put all of the pieces together into a simple system of laws describing both motion and gravity.

In this chapter, we'll discuss the laws that describe motion and energy in the universe, including Newton's laws of motion, the laws of conservation of angular momentum and of energy, and the universal law of gravitation. By understanding these laws, you will be able to make sense of the wide-ranging phenomena you will encounter as you continue your study of astronomy.

4.1 Describing Motion: Examples from Daily Life

Think about what happens when you throw a ball to a dog. The ball leaves your hand, traveling in some particular direction at some particular speed. During its flight, the ball is pulled toward Earth by gravity, slowed by air resistance, and pushed by gusts of wind. Despite the complexity of the ball's motion, the dog still catches it.

We humans can perform an even better trick: We have learned how to figure out where the ball will land even before throwing it. In fact, we can use the same basic trick to predict the motions of objects throughout the universe, and we can perform it with such extraordinary precision that we can land a spaceship on target on Mars after sending it on a journey of hundreds of millions of kilometers.

Our primary goal in this chapter is to understand how humans have learned to make sense of motion in the universe. We all have a great deal of experience with motion and a natural intuition about how motion works, but in science we need to define our ideas and terms precisely. In this section, we'll use examples from everyday life to explore some of the fundamental ideas that underlie our understanding of motion. After this introduction to the basic ideas, we'll go on to explore the laws that govern all motion.

• How do we describe motion?

You are probably familiar with the terms used to describe motion in science—terms such as *velocity, acceleration,* and *momentum.* However, their scientific definitions may differ subtly from those you use in casual conversation. Let's investigate the precise meanings of these terms.

Speed, Velocity, and Acceleration   We use three basic terms to describe how an object is moving. These terms are all familiar to you from driving a car:

- The **speed** of the car tells us how far it will go in a certain amount of time. For example, 100 km/hr (about 60 mi/hr) is a speed, and it tells us that the car will cover a distance of 100 kilometers if it is driven at this speed for an hour.
- The **velocity** of the car tells us both its speed and direction. For example, "100 km/hr going due north" describes a velocity.
- The car has an **acceleration** if its velocity is changing in any way, whether in speed or direction or both.

You are undoubtedly familiar with the term *acceleration* as it applies to increasing speed. In science, we also say that you are accelerating when you slow down or turn (Figure 4.1). Slowing occurs when acceleration is in a direction opposite to the motion. In this case, we say that your acceleration is negative, causing your velocity to decrease. Turning changes your velocity because it changes the direction in which you are moving. Thus, turning is a form of acceleration even if your speed remains constant.

You can often feel the effects of acceleration. For example, as you speed up in a car you feel yourself being pushed back into your seat. As you slow down you feel yourself being pulled forward from the seat. As you drive around a curve you feel yourself being pushed away from the

30 km/hr   60 km/hr

This car is accelerating because its velocity is increasing.

60 km/hr   60 km/hr

This car is accelerating because its direction is changing, even though its speed stays constant.

60 km/hr   30 km/hr   0 km/hr

This car is accelerating because its velocity is decreasing. (negative acceleration).

Figure 4.1 Speeding up, turning, and slowing down are all examples of acceleration.

direction of your turn. In contrast, you don't feel such effects when moving at *constant velocity*. That is why you don't feel any sensation of motion when you're traveling in an airplane on a smooth flight.

The Acceleration of Gravity   One of the most important types of acceleration is that caused by gravity, which makes objects accelerate as they fall. In a legendary experiment in which he supposedly dropped weights from the Leaning Tower of Pisa, Galileo demonstrated that gravity accelerates all objects by the same amount, regardless of their mass. This fact may be surprising because it seems to contradict everyday experience: A feather floats gently to the ground, while a rock plummets. However, air resistance causes this difference in acceleration. If you dropped a feather and a rock on the Moon, where there is no air, both would fall at exactly the same rate.

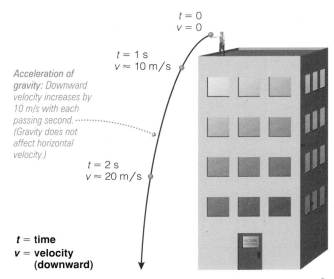

Figure 4.2 On Earth, the acceleration of gravity is about 10 m/s² (more precisely, 9.8 m/s²), which means an unsupported object's downward velocity increases by about 10 m/s with each passing second (assuming we neglect air resistance).

THINK ABOUT IT

Find a piece of paper and a small rock. Hold both at the same height, one in each hand, and let them go at the same instant. The rock, of course, hits the ground first. Next, crumple the paper into a small ball and repeat the experiment. What happens? Explain how this experiment suggests that gravity accelerates all objects by the same amount.

The acceleration of a falling object is called the **acceleration of gravity**, abbreviated *g*. On Earth, the acceleration of gravity causes falling objects to fall faster by 9.8 meters per second (m/s), or about 10 m/s, with each passing second. For example, suppose you drop a rock from a tall building. At the moment you let it go, its speed is 0 m/s. After 1 second, the rock will be falling downward at about 10 m/s. After 2 seconds, it will be falling at about 20 m/s. In the absence of air resistance, its speed would continue to increase by about 10 m/s each second until it hits the ground (Figure 4.2). We therefore say that the acceleration of gravity is about 10 *meters per second per second,* or 10 *meters per second squared,* which we write as 10 m/s². (More precisely, $g = 9.8$ m/s².)

Momentum and Force   The concepts of speed, velocity, and acceleration describe how an individual object moves, but most of the interesting phenomena we see in the universe result from interactions between objects. We need two additional concepts to describe these interactions:

● An object's **momentum** is its combination of mass and velocity. (Mathematically, we express this idea as momentum = mass × velocity.)
● The only way to change an object's momentum is to apply a **force** to it.

We can understand these concepts by considering the effects of collisions. Imagine that you're innocently stopped at a red light in your car when a bug flying at a velocity of 30 km/hr due south slams into your windshield. What will

happen to your car? Not much, except perhaps a bit of a mess on your windshield. Next imagine that a 2-ton truck runs the red light and hits you head-on with the same velocity as the bug. Clearly, the truck will cause far more damage. To see why, let's consider the momentum and force in each collision.

Before the collisions, the truck's much greater mass means that it has far more momentum than the bug, even though both the truck and the bug are moving with the same velocity. During the collisions, the bug and the truck each transfer some of their momentum to your car. The bug has very little momentum to give to your car, so it does not exert much of a force. In contrast, the truck imparts enough of its momentum to cause a dramatic and sudden change in your car's momentum. You feel this sudden change in momentum as a force, and it can do great damage to you and your car.

The mere presence of a force does not always cause a change in momentum. For example, a moving car is always affected by forces of air resistance and friction with the road—forces that will slow your car if you take your foot off the gas pedal. However, you can maintain a constant velocity, and hence constant momentum, if you step on the gas pedal hard enough to overcome the slowing effects of these forces.

In fact, forces of some kind are always present. For example, on Earth we always feel the force of gravity and the electromagnetic forces acting between atoms. The **net force** (or *overall force*) acting on an object represents the combined effect of all the individual forces put together. There is no net force on your car when you are driving at constant velocity, because the force generated by the engine to turn the wheels precisely offsets the forces of air resistance and road friction. A change in momentum occurs only when the net force is not zero.

Changing an object's momentum means changing its velocity, as long as its mass remains constant. Thus, the presence of a net force causes an object to accelerate. Moreover, anytime an object accelerates, a net force must be causing the acceleration. That is why you feel forces (pushing you forward, backward, or to the side) when you accelerate in your car. We can use the same ideas to understand many astronomical processes. For example, planets are always accelerating as they orbit the Sun, because their direction of travel constantly changes as they go around their orbits. We can therefore conclude that some force must be causing this acceleration. As we'll discuss shortly, Isaac Newton identified this force as gravity.

Moving in Circles   Think about an ice skater spinning in place. She isn't going anywhere, so she has no overall velocity and hence no overall momentum. Nevertheless, every part of her body is moving in a circle as she spins, so these parts have momentum even though her overall momentum is zero. Is there a way to describe the total momentum from each part of her body as she spins? Yes—we say that her spin gives her **angular momentum**, which you can also think of as "circling momentum" or "turning momentum." (The term *angular* arises because a complete circle turns through an *angle* of 360°.)

Any object that is either spinning or moving along a curved path has angular momentum, which makes angular momentum very important in astronomy. For example, Earth has angular momentum due both to its rotation (its *rotational angular momentum*) and to its orbit around the Sun (its *orbital angular momentum*). In Section 4.3, we'll use the idea of angular momentum to explain why Earth and other planets don't need any fuel to keep rotating and orbiting the Sun.

Because angular momentum is a special type of momentum, an object's angular momentum can change only when a special type of force is applied to it. To see why, consider what happens when you try to open a swinging door. Opening the door means making it rotate on its hinges, which means giving the door some angular momentum. If you push directly on the hinges, it will have no effect on the door, even if you push with a very strong force. However, even a light force can make the door rotate if you push on the part of the door that is farthest from the hinges. The type of force that can change an object's angular momentum is called a **torque**, which you can think of as a "twisting force." As the door example shows, the amount of torque depends not only on how much force you use to push on the door, but also on where you push.

Changing a tire offers another familiar example of torque. Turning the bolts on a tire means making them rotate, which requires giving them some angular momentum. A longer wrench means you can push from much farther out than you can with a short wrench, so you can turn the bolts with less strength. We will see many more applications of angular momentum in astronomy throughout the rest of the book.

• How is mass different from weight?

In discussing momentum and force, we've referred several times to the idea of *mass*. In daily life, we usually think of mass as something you can measure with a bathroom scale, but technically the scale measures your weight, not your mass. The distinction between mass and weight rarely matters when we are talking about objects on Earth, but it is very important in astronomy:

● Your **mass** is the amount of matter in your body.
● Your **weight** describes a *force* that is acting on your mass. In other words, it is what a scale measures when you stand on it.*

To understand the difference between mass and weight, imagine standing on a scale in an elevator (Figure 4.3). Your mass will be the same no matter how the elevator is moving, but your weight can vary. When the elevator is stationary or moving at constant velocity, the scale reads your "normal" weight. When the elevator accelerates upward, the floor exerts a greater force than it does when you are at rest. You feel heavier, and the scale verifies your greater weight. When the elevator accelerates downward, the floor and the scale exert a weaker force on you, so the scale registers less weight. Be sure to note that the scale shows a weight different from your "normal" weight only when the elevator is *accelerating*, not when it is going up or down at constant speed.

<hr>

THINK ABOUT IT

Find a small bathroom scale and take it with you on an elevator ride. How does your weight change when the elevator accelerates upward or downward? Does it change when the elevator is moving at constant speed? Explain your observations.

<hr>

Thus, your mass depends only on the amount of matter in your body and is the same anywhere, but your weight can vary because the forces acting on you can vary. For example, your mass would be the same on the Moon as on Earth, but you would weigh less on the Moon because of its weaker gravity.

Free-Fall and Weightlessness   Now consider what happens if the elevator cable breaks (see the last frame in Figure 4.3). The elevator and you are suddenly both in **free-fall**—falling without any resistance to slow you down. The floor drops away at the same rate that you fall, allowing you to "float" freely above it, and the scale reads zero because you are no longer held to it. In other words, your free-fall has made you **weightless**.

In fact, you are in free-fall whenever there's nothing to *prevent* you from falling. For example, you are in free-fall

<hr>

*Some physics texts distinguish between your "true weight" measured at rest and your "apparent weight" measured when you stand on an accelerating scale (as in an elevator). We will not make this distinction in this book.

Figure 4.3
Interactive Figure

Mass is not the same as weight. The man's mass never changes, but his weight is different when the elevator accelerates.

elevator stationary or moving at constant velocity	elevator **accelerating** upward	elevator **accelerating** downward	elevator in **free-fall**
Normal weight	Heavier-than-normal weight	Lighter-than-normal weight	Weightless

when you jump off a chair or spring from a diving board or trampoline. Surprising as it may seem, you have therefore experienced weightlessness many times in your life. You can experience it right now simply by jumping off your chair. Of course, your weightlessness lasts for only the very short time before you hit the ground.

Weightlessness in Space   You've probably seen video of astronauts floating weightlessly in the Space Shuttle or the Space Station. But why are they weightless? Many people guess that there's no gravity in space, but that's not true. After all, it is gravity that makes the Space Shuttle and Space Station orbit Earth. Astronauts are weightless for the same reason you are weightless when you jump off a chair: They are in free-fall.

More specifically, astronauts are weightless the entire time they are orbiting Earth because they are in a *constant state of free-fall*. The idea that an orbiting object is in a "constant state of free-fall" may sound a bit strange, but it's easy to understand. Imagine a tower that goes all the way to the Space Station's orbit, about 350 kilometers above Earth. Figure 4.4 shows the idea. If you stepped off the tower, you would fall downward, remaining weightless until you hit the ground (or until air resistance had a noticeable effect on you). Now, imagine that instead of stepping off the tower, you ran and jumped out of the tower. You'd still fall to the ground, but because of your forward motion you wouldn't fall straight down. You'd land a short distance away from the base of the tower.

The faster you ran out of the tower, the farther you'd go before landing. If you could somehow use a rocket to run fast enough—about 28,000 km/hr (17,000 mi/hr) at the orbital altitude of the Space Station—a very interesting thing would happen: By the time gravity had pulled you downward as far as the length of the tower, you'd already have moved far enough around Earth that you'd no longer be going down at all. Instead, you'd be just as high above Earth's surface as you'd been all along, but a good portion of the way around the world. In other words, you'd be orbiting Earth.

Thus, the Space Shuttle, the Space Station, and all other orbiting objects stay in orbit because they are constantly "falling around" the Earth. Their constant state of free-fall makes these spacecraft and everything in them weightless.

THINK ABOUT IT

In the *Hitchhiker's Guide to the Galaxy* books, author Douglas Adams says that the trick to flying is to throw yourself at the ground and miss. Although this phrase does not really explain flying, which involves lift from air, it does describe an orbit fairly well. Explain.

 Motion and Gravity Tutorial, Lesson 1

4.2 Newton's Laws of Motion

The human "trick" of being able to figure out where a ball will land before it is thrown or to predict how other

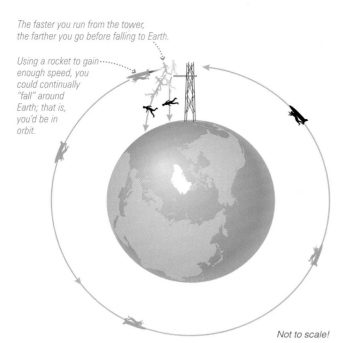

The faster you run from the tower, the farther you go before falling to Earth.

Using a rocket to gain enough speed, you could continually "fall" around Earth; that is, you'd be in orbit.

Not to scale!

Figure 4.4 Interactive Figure This figure explains why astronauts are weightless and float freely in space. It shows that if you could leap from a tall tower with enough speed (perhaps with the aid of a rocket) you could travel forward so fast that you'd orbit Earth. You'd then be in a constant state of free-fall, which means you'd be weightless. *Note:* On the scale shown here, the tower extends far higher than the Space Station's orbit; the rocket orientation assumes that it rotates once with each orbit, as is the case for the Space Shuttle. (Adapted from *Space Station Science* by Marianne Dyson.)

motions will unfold requires understanding precisely how forces affect objects in motion. The complexity of motion in daily life might lead you to guess that the laws governing motion would also be complex. For example, if you watch a falling piece of paper waft lazily to the ground, you'll see it rock irregularly back and forth in a seemingly unpredictable pattern. However, the complexity of this motion arises because the paper is affected by a variety of forces, including gravity and the changing forces caused by air currents. If you could analyze the forces individually, you'd find that each force affects the paper's motion in a simple, predictable way. Sir Isaac Newton (1642–1727) discovered the remarkably simple system of laws that govern all motion.

• How did Newton change our view of the universe?

Newton was born in Lincolnshire, England, on Christmas Day in 1642. His father, a farmer who never learned to read or write, died 3 months before his premature birth. Newton had a difficult childhood and showed few signs of unusual talent. He attended Trinity College at Cambridge, where he earned his keep by performing menial labor, such as cleaning the boots and bathrooms of wealthier students and waiting on their tables.

The plague hit Cambridge shortly after Newton graduated, and he returned home. By his own account, he experienced a moment of inspiration in 1666 when he saw an apple fall to the ground. He suddenly realized that the gravity making the apple fall was the same force that held the Moon in orbit around Earth. In that moment, Newton shattered the remaining vestiges of the Aristotelian view of the world, which for centuries in Europe had been taken as unquestioned truth.

Aristotle (384–322 B.C.) had made many claims about the physics of motion, using his ideas to support his belief in an Earth-centered cosmos. He had also maintained that the heavens were totally distinct from Earth, so that physical laws on Earth did not apply to heavenly motion. By the time Newton saw the apple fall, the Copernican revolution had displaced Earth from a central position, and Galileo's experiments had shown that the laws of physics were not what Aristotle had believed.

Sir Isaac Newton (1642–1727)

Newton's sudden insight delivered the final blow to Aristotle's physics. When Newton realized that gravity operated in the heavens as well as on Earth, he eliminated Aristotle's distinction between the two realms. For the first time in history, the heavens and Earth were brought together as one *universe.* Newton's insight also heralded the birth of the modern science of *astrophysics* (although the term wasn't coined until much later). Astrophysics applies physical laws discovered on Earth to phenomena throughout the cosmos.

Over the next 20 years, Newton's work completely revolutionized mathematics and science. He quantified the laws of motion and gravity, conducted crucial experiments regarding the nature of light, built the first reflecting telescopes, and invented the mathematics of calculus. The compendium of Newton's discoveries is so tremendous that it would take a complete book just to describe them, and many more books to describe their influence on civilization. When Newton died in 1727, at age 84, English poet Alexander Pope composed the following epitaph:

> *Nature, and Nature's laws lay hid in the Night.*
> *God said,* Let Newton be! and *all was Light.*

We'll discuss Newton's laws of motion in the rest of this section, and later in the chapter we'll turn our attention to Newton's discoveries about gravity.

• What are Newton's three laws of motion?

Newton published the laws of motion and gravity in 1687, in a book usually called *Principia,* short for *Philosophiae Naturalis Principia Mathematica* ("Mathematical Principles of Natural Philosophy"). He enumerated three laws that apply to all motion, so we now call them **Newton's laws of motion.** These laws govern the motion of everything from our daily movements here on Earth to the movements of planets, stars, and galaxies throughout the universe. Figure 4.5 summarizes the three laws, which we now discuss in a little more detail.

Newton's First Law Newton's first law of motion essentially restates Galileo's discovery that objects will remain in motion unless a force acts to stop them. It can be stated as follows:

> *Newton's first law: An object moves at constant velocity if there is no net force acting upon it.*

Thus, objects at rest (velocity = 0) tend to remain at rest, and objects in motion tend to remain in motion with no change in either their speed or their direction.

The idea that an object at rest should remain at rest is rather obvious: A car parked on a flat street won't suddenly start moving for no reason. But what if the car is traveling along a flat, straight road? Newton's first law says that the car should keep going at the same speed forever *unless* a force acts to slow it down. You know that the car eventually will come to a stop if you take your foot off the gas pedal, so we must conclude that one or more forces are stopping the car—in this case forces arising from friction and air resistance. If the car were in space, and therefore unaffected by friction or air, it would keep moving forever (though gravity would eventually alter its speed and direction). That is why interplanetary spacecraft need no fuel to keep going after they are launched into space, and why astronomical objects don't need fuel to travel through the universe.

Newton's first law also explains why you don't feel any sensation of motion when you're traveling in an airplane on a smooth flight. As long as the plane is traveling at constant velocity, no net force is acting on it or on you. Therefore, you feel no different from the way you would feel at rest. You can walk around the cabin, play catch with a person a few rows forward, or relax and go to sleep just as though you were "at rest" on the ground.

Newton's Second Law Newton's second law of motion tells us what happens to an object when a net force *is* present. We have already seen that a net force will change an object's momentum, accelerating it in the direction of the force. Newton's second law quantifies this relationship, telling us that the amount of the acceleration depends on the object's

Newton's first law of motion:
An object moves at constant velocity unless a net force acts to change its speed or direction.

Example: *A spaceship needs no fuel to keep moving in space.*

Newton's second law of motion:
Force = mass × acceleration

Example: *A baseball accelerates as the pitcher applies a force by moving his arm. (Once the ball is released, the force from the pitcher's arm ceases, and the ball's path changes only because of the forces of gravity and air resistance.)*

Newton's third law of motion:
For any force, there is always an equal and opposite reaction force.

Example: *A rocket is propelled upward by a force equal and opposite to the force with which gas is expelled out its back.*

Figure 4.5 Newton's three laws of motion.

mass and the strength of the net force. We usually write this law as an equation:

Newton's second law: **Force = mass × acceleration (or F = ma).**

(The law is sometimes instead written as force = rate of change in momentum.)

This law explains why you can throw a baseball farther than you can throw a shot-put. The force your arm delivers to both the baseball and the shot-put equals the product of mass and acceleration. Because the mass of the shot-put is greater than that of the baseball, the same force from your arm gives the shot-put a smaller acceleration. Because of its smaller acceleration, the shot-put leaves your hand with less speed than the baseball and thus travels a shorter distance before hitting the ground. Astronomically, Newton's second law explains why a large planet such as Jupiter has a greater effect on asteroids and comets than a small planet such as Earth [Section 12.4]. Jupiter exerts a stronger gravitational force on passing asteroids than does Earth, so Newton's second law tells us that Jupiter sends the asteroids scattering with a greater acceleration.

We can also use Newton's second law of motion to understand acceleration around curves. Suppose you swing a ball on a string around your head (Figure 4.6a). The ball is accelerating even if it has a steady speed, because it is constantly changing direction and therefore has a changing velocity. What makes it accelerate? According to Newton's second law, the taut string must be applying a force to the ball. We can understand this force by considering what happens when the string breaks and the force disappears (Figure 4.6b). In that case, the ball simply flies off in a straight line. Thus, when the string is intact, the force must be pulling the ball *inward* to keep it from flying off. Because acceleration must be in the same direction as the force, we conclude that the ball has an inward acceleration as it moves around the circle.

The same idea helps us understand the force on a car moving around a curve or a planet orbiting around the Sun. In the case of the car, the force comes from friction between the tires and the road. The tighter the curve or the faster the car is going, the greater the force needed to keep the car moving around it. If the force is not great enough, the car skids outward. For a planet moving around the Sun, the force pulling inward is gravity, which we'll discuss shortly. Thus, an orbiting planet is always accelerating toward the Sun. Indeed, it was Newton's discovery of the precise nature of this acceleration that helped him deduce the law of gravity.

Newton's Third Law Think for a moment about standing still on the ground. Your weight exerts a downward force, so if this force were acting alone, Newton's second law would demand that you be accelerating downward. The fact that you are not falling means that the ground must be pushing back up on you with exactly the right amount of force to offset the force you exert downward on the floor. This fact is embodied in Newton's third law of motion, which tells us that any force is always opposed by an equal and opposite reaction force.

Newton's third law: For any force, there is always an equal and opposite reaction force.

This law is very important in astronomy, because it tells us that objects always attract *each other* through gravity. For example, your body always exerts a gravitational force on the Earth identical to the force that Earth exerts on you, except that it acts in the opposite direction. Of course, the same force means a much greater acceleration for you than for Earth (because your mass is so much smaller than Earth's), which is why we see you falling toward Earth when you jump off a chair, rather than Earth falling toward you.

Newton's third law also explains how a rocket works: A rocket engine generates a force that drives hot gas out the back, which creates an equal and opposite force that propels the rocket forward.

The inward force along the string keeps the ball moving in a circle.

F = ma

v

If the string breaks, the inward force is gone...

string breaks

v ...so the ball moves with constant velocity from the point of the break.

a When you swing a ball on a string, the string exerts a force that pulls the ball inward.

b If the string breaks, the ball flies off in a straight line at constant velocity.

Figure 4.6 Newton's second law of motion tells us that an object going around a curve has an acceleration pointing toward the inside of the curve.

What Makes a Rocket Launch?

If you've ever watched a rocket launch, it's easy to see why many people believe that the rocket "pushes off" the ground. In fact, the ground has nothing to do with the rocket launch. The rocket's launch is explained by Newton's third law of motion. To balance the force driving gas out the back of the rocket, an equal and opposite force must propel the rocket forward. Thus, rockets can be launched horizontally as well as vertically, and a rocket can be "launched" in space (for example, from a space station) with no need for any nearby solid ground.

4.3 Conservation Laws in Astronomy

Newton's laws of motion are easy to state, but they may seem a bit arbitrary. Why, for example, should every force be opposed by an equal and opposite reaction force? In the centuries since Newton first stated his laws, we have learned that they are not arbitrary at all, but instead reflect deeper aspects of nature known as *conservation laws*. In this section, we'll explore three of the most important conservation laws for astronomy: *conservation of momentum, conservation of angular momentum,* and *conservation of energy.* We'll see some immediate examples of how they apply to astronomy, and then use these laws over and over throughout the rest of the book.

• Why do objects move at constant velocity if no force acts on them?

The first of our conservation laws, the law of **conservation of momentum**, states that the total momentum of inter-

acting objects cannot change as long as no external force is acting on them; that is, their total momentum is *conserved.* An individual object can gain or lose momentum only if some other object's momentum changes by a precisely opposite amount.

The law of conservation of momentum is implicit in Newton's laws. To see why, watch a game of pool. Newton's second law tells us that when one pool ball strikes another, it exerts a force that changes the momentum of the second ball. At the same time, Newton's third law tells us that the second ball exerts an equal and opposite force on the first one—which means that the first ball's momentum changes by precisely the same amount as the second ball's momentum, but in the opposite direction. Thus, the total combined momentum of the two balls remains the same both before and after the collision (Figure 4.7). Note that no *external* forces are accelerating the balls.

Rockets offer another good example of conservation of momentum in action. When you fire a rocket engine, the total momentum of the rocket and the hot gases it shoots out the back must stay the same. In other words, the amount of forward momentum the rocket gains is equal to the amount of backward momentum in the gas that shoots out the back. That is why forces between the rocket and the gases are always equal and opposite.

From the perspective of conservation of momentum, Newton's first law makes perfect sense. When no net force acts on an object, there is no way for the object to transfer any momentum to or from any other object. Thus, in the absence of a net force, an object's momentum must remain unchanged—which means the object must continue to move exactly as it has been moving.

According to current understanding of the universe, conservation of momentum is an absolute law that always holds true. For example, it holds even when you jump up into the air. So you may wonder, Where do I get the momentum that carries me upward? The answer is that

---

MATHEMATICAL INSIGHT 4.1 **Units of Force, Mass, and Weight**

Newton's second law, $F = ma$, shows that the unit of force is equal to a unit of mass multiplied by a unit of acceleration. For example, if a mass of 1 kilogram accelerates at 10 m/s^2, the magnitude of the responsible force is:

$$\text{force} = \text{mass} \times \text{acceleration}$$
$$= 1 \text{ kg} \times 10\frac{\text{m}}{\text{s}^2} = 10\frac{\text{kg} \times \text{m}}{\text{s}^2}$$
$$= 10 \text{ newtons}$$

The standard unit of force is the *kilogram-meter per second squared,* called the **newton** for short.

We can now further clarify the difference between mass and weight. When you stand on a scale, it records the downward force that you exert on it, which is equal and opposite to the upward force it exerts on you. If the scale were suddenly pulled out from under your feet, you would begin accelerating downward with the

acceleration of gravity. Thus, when you are standing still, the scale must support you with a force equal to your mass times the acceleration of gravity. Your weight must also equal this force (but in an opposite direction):

$$\text{weight} = \text{mass} \times \text{acceleration of gravity}$$

Your apparent weight may differ from this value if forces besides gravity are affecting you.

Like any force, weight has units of mass times acceleration. Thus, although we commonly speak of weights in *kilograms,* this usage is not technically correct: Kilograms are a unit of mass, not of force. You may safely ignore this technicality as long as you are dealing with objects on Earth that are not accelerating. In space or on other planets, the distinction between mass and weight is important and cannot be ignored.

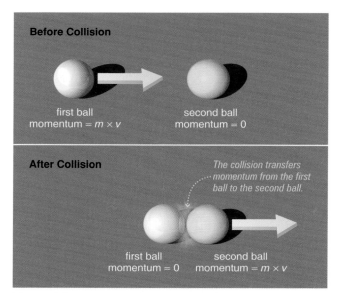

Before Collision

first ball
momentum = $m \times v$

second ball
momentum = 0

After Collision

The collision transfers momentum from the first ball to the second ball.

first ball
momentum = 0

second ball
momentum = $m \times v$

Figure 4.7 Conservation of momentum demonstrated on a pool table.

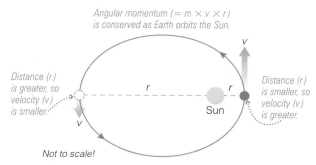

Angular momentum (= $m \times v \times r$) is conserved as Earth orbits the Sun.

Distance (r) is greater, so velocity (v) is smaller.

Distance (r) is smaller, so velocity (v) is greater.

Sun

Not to scale!

Figure 4.8 Earth's angular momentum always stays the same as it orbits the Sun, so it moves faster when it is closer to the Sun and slower when it is farther from the Sun. It needs no fuel to keep orbiting because no forces are acting in a way that could change its angular momentum.

as your legs propel you skyward, they are actually pushing Earth in the other direction, giving Earth's momentum an equal and opposite kick. However, Earth's huge mass renders its acceleration undetectable. During your brief flight, the gravitational force between you and Earth pulls you back down, transferring your momentum back to Earth. The total momentum of you and the rest of Earth remains the same at all times.

• What keeps a planet rotating and orbiting the Sun?

Perhaps you've wondered how Earth manages to keep rotating and going around the Sun day after day and year after year. The answer comes from our second conservation law: the law of **conservation of angular momentum**. Recall that rotating or orbiting objects have angular momentum because they are moving in circles or going around curves, and that angular momentum can be changed only by a "twisting force," or *torque*. The law of conservation of angular momentum states that as long as there is no external torque, the total angular momentum of a set of interacting objects cannot change. An individual object can change its angular momentum only by transferring some angular momentum to or from another object. Because astronomical objects can have angular momentum due to both their rotation and orbit, let's consider both cases.

Orbital Angular Momentum   Consider Earth's orbit around the Sun. A simple formula tells us Earth's angular momentum at any point in its orbit:

$$\text{angular momentum} = m \times v \times r$$

where m is Earth's mass, v is its speed (or velocity) around the orbit, and r is the "radius" of the orbit, by which we mean its distance from the Sun (Figure 4.8). Because there

are no objects around to give or take angular momentum from Earth as it orbits the Sun, Earth's orbital angular momentum must always stay the same. This explains two key facts about Earth's orbit:

1. Earth needs no fuel or push of any kind to keep orbiting the Sun—it will keep orbiting as long as nothing comes along to take angular momentum away.
2. Because Earth's angular momentum at any point in its orbit depends on the product of its speed and orbital radius (distance from the Sun), Earth's orbital speed must be faster when it is nearer to the Sun (and the radius is shorter) and slower when it is farther from the Sun (and the radius is longer).

The second fact is just what Kepler's second law of planetary motion tells us [Section 3.3]. Thus, the law of conservation of angular momentum tells us *why* Kepler's law is true.

Rotational Angular Momentum   The same idea explains why Earth keeps rotating. As long as Earth isn't transferring any of the angular momentum of its rotation to another object, it keeps rotating at the same rate. (In fact, Earth is very gradually transferring some of its rotational angular momentum to the Moon, and as a result, Earth's rotation is gradually slowing down; see Section 4.5.)

Conservation of angular momentum also explains why we see so many spinning disks in the universe, such as the disks of galaxies like the Milky Way and disks of material orbiting young stars. The idea is easy to illustrate with an ice skater spinning in place (Figure 4.9). Because there is so little friction on ice, the angular momentum of the ice skater remains essentially constant. When she pulls in her extended arms, she decreases her radius—which means her velocity of rotation must increase. Stars and galaxies are both born from clouds of gas that start out much larger in size. These clouds almost inevitably have some small net rotation, though it may be imperceptible. However, like the spinning skater as she pulls in her arms, they must spin faster as gravity makes them shrink in size. (We'll discuss why the clouds also flatten into disks in Chapter 8.)

In the product m × v × r, extended arms mean larger radius and smaller velocity of rotation.

Bringing in her arms decreases her radius and therefore increases her rotational velocity.

Figure 4.9 A spinning skater conserves angular momentum.

THINK ABOUT IT

How does conservation of angular momentum explain the spiraling of water going down a drain?

Energy Tutorial, Lesson 1

• Where do objects get their energy?

Our third crucial conservation law for astronomy is the law of **conservation of energy**. This law tells us that, like momentum and angular momentum, energy cannot appear out of nowhere or disappear into nothingness. Thus, objects can gain or lose energy only by exchanging energy with other objects. Because of this law, the story of the universe is a story of the interplay of energy and matter: All actions in the universe involve exchanges of energy or the conversion of energy from one form to another.

Throughout the rest of this book, we'll see numerous cases in which we can understand astronomical processes simply by tracing the ways in which energy is transformed and exchanged. For example, we'll see that planetary interiors cool with time only because they radiate energy into space, and that the Sun became hot because of energy released by the gas that formed it. It's no exaggeration to say that, by applying the laws of conservation of momentum, conservation of angular momentum, and conservation of energy, we can understand almost every major process that occurs in the universe.

Basic Types of Energy   Before we can fully understand the law of conservation of energy, we need to know what energy is.

In essence, energy is what makes matter move. Because this statement is so broad, we often distinguish between many different types of energy. For example, we talk about the energy we get from the food we eat, the energy that makes our cars go, and the energy a light bulb emits. Fortunately, we can classify all these various types of energy into just three major categories (Figure 4.10):

● Energy of motion, or **kinetic energy** (*kinetic* comes from a Greek word meaning "motion"). Falling rocks, orbiting planets, and the molecules moving in the air around us are all examples of objects with kinetic energy.

● Energy carried by light, or **radiative energy** (the word *radiation* is often used as a synonym for *light*). All light carries energy, which is why light can cause changes in matter. For example, light can alter molecules in our eyes—thereby allowing us to see—or warm the surface of a planet.

● Stored energy, or **potential energy**, which might later be converted into kinetic or radiative energy. For example, a rock perched on a ledge has *gravitational* potential energy because it will fall if it slips off the edge, and gasoline contains *chemical* potential energy that can be converted into the kinetic energy of a moving car.

Regardless of which type of energy we are dealing with, we can measure the amount of energy with the same standard units. For Americans, the most familiar units of energy are *Calories*, which are shown on food labels to tell us how much energy our bodies can draw from the food. A

Energy can be converted from one form to another.

kinetic energy
(energy of motion)

radiative energy
(energy of light)

potential energy
(stored energy)

Figure 4.10 The three basic categories of energy. Energy can be converted from one form to another, but it can never be created or destroyed, an idea embodied in the law of conservation of energy.

typical adult needs about 2,500 Calories of energy from food each day. In science, the standard unit of energy is the **joule**. One food Calorie is equivalent to about 4,184 joules, so the 2,500 Calories used daily by a typical adult is equivalent to about 10 million joules. Table 4.1 compares various energies in joules.

Thermal Energy—The Kinetic Energy of Many Particles Although there are only three major categories of energy, we sometimes divide them into various subcategories. In astronomy, the most important subcategory of kinetic energy is **thermal energy**, which represents the collective kinetic energy of the many individual particles (atoms and molecules) moving randomly within a substance like a rock or the air or the gas within a distant star. In such cases, it is much easier to talk about the thermal energy of the object rather than about the kinetic energies of its billions upon billions of individual particles. Note that all objects contain thermal energy even when they are sitting still, because the particles within them are always jiggling about randomly. These random motions can contain substantial energy— the thermal energy of a parked car due to the random motion of its atoms is much greater than the kinetic energy of the car moving at highway speed.

Thermal energy gets its name because it is related to temperature, but temperature and thermal energy are not quite the same thing. Thermal energy measures the *total* kinetic energy of all the randomly moving particles in a substance, while **temperature** measures the *average* kinetic energy of the particles. Thus, for a particular object, a higher

lower temperature   higher temperature

These particles are moving relatively slowly, which means low temperature . . .

. . . And now the same particles are moving faster, which means higher temperature.

Figure 4.11 Temperature is a measure of the average kinetic energy of the particles (atoms and molecules) in a substance. (Longer arrows represent higher speeds.)

temperature simply means that the particles on average have more kinetic energy and hence are moving faster (Figure 4.11). You're probably familiar with temperatures measured in *Fahrenheit* or *Celsius,* but in science we often use the **Kelvin** temperature scale (Figure 4.12). The Kelvin scale never has negative temperatures, because it starts from the coldest possible temperature, known as *absolute zero* (0 K).

Thermal energy depends on temperature, because a higher average kinetic energy for the particles in a substance must also lead to a higher total energy. However, thermal energy also depends on the number or density of the particles. You can see why by considering what would happen if you quickly thrust your arm into and out of a hot oven and a pot of boiling water.

The air in a hot oven is much hotter in temperature than the water boiling in a pot (typically 400°F for the oven versus 212°F for boiling water). However, the boiling water would scald your arm almost instantly, while you can safely put your arm into the oven air for a few seconds. The reason for this difference is density (Figure 4.13). If air or water is hotter than your body, molecules striking your skin transfer thermal energy to molecules in your arm. The higher temperature in the oven means that the air molecules strike

Table 4.1 Energy Comparisons

Item	Energy (joules)
Energy of sunlight at Earth (per m² per second)	1.3×10^3
Energy from metabolism of a candy bar	1×10^6
Energy needed to walk for 1 hour	1×10^6
Kinetic energy of a car going 60 mi/hr	1×10^6
Daily food energy need of average adult	1×10^7
Energy released by burning 1 liter of oil	1.2×10^7
Thermal energy of parked car	1×10^8
Energy released by fission of 1 kilogram of uranium-235	5.6×10^{13}
Energy released by fusion of hydrogen in 1 liter of water	7×10^{13}
Energy released by 1-megaton H-bomb	5×10^{15}
Energy released by major earthquake (magnitude 8.0)	2.5×10^{16}
Annual U.S. energy consumption	10^{20}
Annual energy generation of Sun	10^{34}
Energy released by a supernova	$10^{44}-10^{46}$

373.15 K	100°C	212°F	water boils
273.15 K	0°C	32°F	water freezes
0 K	−273.15°C	−459.67°F	absolute zero
Kelvin	Celsius	Fahrenheit	

Figure 4.12 Three common temperature scales: Kelvin, Celsius, and Fahrenheit. Scientists generally prefer the Kelvin scale. (The degree symbol ° is not usually used with the Kelvin scale.)

The air in the oven is hotter (400°F) than the boiling water (212°F) in the pot . . .

. . . but the water in the pot contains more thermal energy because of its much higher density.

212°F   400°F

Figure 4.13 Thermal energy depends on both the temperature and density of particles in a substance.

your skin harder, on average, than the molecules in the boiling water. However, because the *density* of water is so much higher than the density of air (meaning water has far more molecules in the same amount of space), many more molecules strike your skin each second in the water. Thus, while each individual molecule that strikes your skin transfers a little less energy in the boiling water than in the oven, the sheer number of molecules hitting you in the water means that more thermal energy is transferred to your arm. That is why the water causes a rapid burn.

THINK ABOUT IT

In air or water that is colder than your body temperature, thermal energy is transferred from you to the surrounding cold air or water. Use this fact to explain why falling into a 32°F (0°C) lake is much more dangerous than standing naked outside on a 32°F day.

The environment in space provides another example of the difference between temperature and heat. Surprisingly, the temperature in low Earth orbit can be several thousand degrees. However, astronauts working in Earth orbit (outside the Space Shuttle or Station) can get very cold and sometimes need heated space suits and gloves. The astronauts can get cold despite the high temperature, because the extremely low density of space means that relatively few particles are available to transfer thermal energy to them. (You may wonder how the astronauts become cold given that the low density also means the astronauts cannot transfer much of their own thermal energy to the particles in space. It turns out that they lose their body heat by emitting *thermal radiation,* which we will discuss in Section 5.3.)

Potential Energy in Astronomy   Many types of potential energy are important in astronomy, but two are particularly important: gravitational potential energy and the potential energy of mass itself, or mass-energy. Let's examine each of these two types of potential energy in a little more depth.

An object's **gravitational potential energy** depend on its mass and how far it can fall as a result of gravity. An object has more gravitational potential energy when it is higher and less when it is lower. For example, if you throw a ball up into the air, it has more potential energy when it is high up than it does near the ground. Because energy must be conserved during the ball's flight, the ball's kinetic energy must increase when its gravitational potential energy decreases, and vice versa (Figure 4.14a). That is why the ball travels fastest (has the most kinetic energy) when it is closest to the ground, where it has the least gravitational potential energy. The higher it is, the more gravitational potential energy it has and thus the slower the ball travels (less kinetic energy).

The same general idea explains how stars become hot (Figure 4.14b). Before a star forms, its matter is spread out

The total energy (kinetic + potential) is the same at all points in the ball's flight.

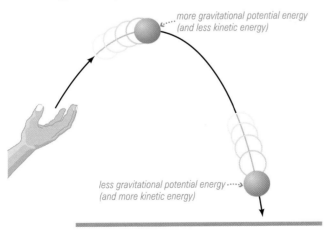

more gravitational potential energy (and less kinetic energy)

less gravitational potential energy (and more kinetic energy)

a The ball has more gravitational potential energy when it is high up than when it is near the ground.

Figure 4.14 Two examples of gravitational potential energy.

Energy is conserved: As the cloud contracts, gravitational potential energy is converted to thermal energy (and some of this energy is converted to radiation).

less gravitational potential energy (and more thermal energy)

more gravitational potential energy (and less thermal energy)

b A cloud of interstellar gas can contract due to its own gravity. It has more gravitational potential energy when it is spread out than when it shrinks in size.

in a large, cold cloud of gas. Most of the individual gas particles are far from the center of this large cloud and therefore have a lot of gravitational potential energy. The particles lose gravitational potential energy as the cloud contracts under its own gravity, and this "lost" potential energy ultimately gets converted into thermal energy, making the center of the cloud hot.

The second particularly important form of potential energy in astronomy is the energy contained in mass itself, often called **mass-energy**. Einstein discovered that mass is a form of potential energy, and he expressed the idea that mass can be converted to other forms of energy with his famous formula

$$E = mc^2$$

where E is the amount of potential energy, m is the mass of the object, and c is the speed of light.

A small amount of mass contains a huge amount of energy. For example, the energy released by a 1-megaton H-bomb—enough to destroy a major city—comes from converting only about 0.1 kilogram of mass (about 3 ounces—a quarter of a can of soda) into energy (Figure 4.15). The Sun generates energy by converting a tiny fraction of its mass into energy through a similar process of nuclear fusion [Section 14.2].

Just as Einstein's formula tells us that mass can be converted into other forms of energy, it also tells us that energy can be transformed into mass. This process is especially important to understanding what we think happened dur-

Figure 4.15 The energy released by this H-bomb comes from the conversion of only about 0.1 kilogram of mass into energy in accordance with the formula $E = mc^2$.

ing the early moments in the history of the universe, when some of the energy of the Big Bang turned into the mass from which we are now made [Section 23.1]. Scientists also use this idea to search for undiscovered particles of matter,

---

MATHEMATICAL INSIGHT **4.2** **Mass-Energy**

It's easy to calculate mass-energies with Einstein's formula $E = mc^2$. Once we calculate an energy, we can compare it to other known energies.

Example: Suppose a 1-kilogram rock were completely converted to energy. How much energy would it release? Compare this to the energy released by burning 1 liter of oil.

Solution:

Step 1 Understand: We are asked how much energy would be released by converting all the mass of a 1-kilogram rock to energy. Thus, we need to know the total mass-energy of the rock, which we can then compare to the energy released by burning a liter of oil.

Step 2 Solve: The total mass-energy of the rock is given by $E = mc^2$, where m is the 1-kilogram mass and $c = 3 \times 10^8$ m/s:

$$E = mc^2 = 1 \text{ kg} \times \left(3 \times 10^8 \, \frac{\text{m}}{\text{s}} \right)^2$$

$$= 1 \text{ kg} \times \left(9 \times 10^{16} \, \frac{\text{m}^2}{\text{s}^2} \right)$$

$$= 9 \times 10^{16} \, \frac{\text{kg} \times \text{m}^2}{\text{s}^2}$$

$$= 9 \times 10^{16} \text{ joules}$$

We now compare this energy to the energy released by burning 1 liter of oil, which is 12 million joules (see Table 4.1). Dividing the mass-energy of the rock by the energy released by burning 1 liter of oil, we find:

$$\frac{9 \times 10^{16} \text{ joules}}{1.2 \times 10^7 \text{ joules}} = 7.5 \times 10^9$$

Step 3 Explain: We have found that converting a 1-kilogram rock completely to energy would release 9×10^{16} joules of energy, which is about 7.5 billion times as much energy as we get from burning 1 liter of oil. Moreover, by looking it up, we can find that 7.5 billion liters of oil is roughly the amount of oil that *all* cars in the United States use in a week. Thus, complete conversion of the mass of a single 1-kilogram rock to energy could yield enough energy to power all the cars in the United States for a week. Unfortunately, no technology available now or in the foreseeable future can release all the mass-energy of a rock.

---

by using large machines called *particle accelerators* to create subatomic particles from energy.

Conservation of Energy We have seen that energy comes in three basic categories—kinetic, radiative, and potential—and explored several subcategories that are especially important in astronomy: thermal energy, gravitational potential energy, and mass-energy. Now we are ready to return to the question of "Where do objects get their energy?" Because energy cannot be created or destroyed, objects always get their energy from other objects. Ultimately, we can always trace an object's energy back to the Big Bang [Section 1.1], the beginning of the universe in which all matter and energy is thought to have come into existence.

For example, imagine that you've thrown a baseball. It is moving, so it has kinetic energy. Where did this kinetic energy come from? The baseball got its kinetic energy from the motion of your arm as you threw it. Your arm, in turn, got its kinetic energy from the release of chemical potential energy stored in your muscle tissues. Your muscles got this energy from the chemical potential energy stored in the foods you ate. The energy stored in the foods came from sunlight, which plants convert into chemical potential energy through photosynthesis. The radiative energy of the Sun was generated through the process of nuclear fusion, which releases some of the mass-energy stored in the Sun's supply of hydrogen. The mass-energy stored in the hydrogen came from the birth of the universe in the Big Bang. After you throw the ball, its kinetic energy will ultimately end up as thermal energy of the air or ground. According to present understanding, the total energy content of the universe was determined in the Big Bang. It remains the same today and will stay the same forever into the future.

4.4 The Universal Law of Gravitation

Newton's three laws of motion describe how objects in the universe move in response to forces. The laws of conservation of momentum, angular momentum, and energy offer an alternative and often simpler way of thinking about what happens when a force causes some change in the motion of one or more objects. Either way, however, we cannot fully understand motion unless we also understand the forces that lead to changes in motion. In astronomy, the most important force is gravity, which governs virtually all large-scale motion in the universe.

 Motion and Gravity Tutorial, Lesson 2

• What determines the strength of gravity?

Isaac Newton discovered the basic law that describes how gravity works. Newton expressed the force of gravity math-

*The **universal law of gravitation** tells us the strength of the gravitational attraction between the two objects.*

$$F_g = G\frac{M_1 M_2}{d^2}$$

M_1 and M_2 are the masses of the two objects.

d is the distance between the centers of the two objects.

Figure 4.16 The universal law of gravitation is an *inverse square law*, which means that the force of gravity declines with the *square* of the distance d between two objects.

ematically with his **universal law of gravitation**. Three simple statements summarize this law:

● Every mass attracts every other mass through the force called *gravity.*

● The strength of the gravitational force attracting any two objects is *directly proportional* to the product of their masses. For example, doubling the mass of *one* object doubles the force of gravity between the two objects.

● The strength of gravity between two objects decreases with the *square* of the distance between their centers. That is, the gravitational force follows an **inverse square law** with distance. For example, doubling the distance between two objects weakens the force of gravity by a factor of 2^2, or 4.

These three statements tell us everything we need to know about Newton's universal law of gravitation. Mathematically, all three statements can be combined into a single equation, usually written like this:

$$F_g = G\frac{M_1 M_2}{d^2}$$

where F_g is the force of gravitational attraction, M_1 and M_2 are the masses of the two objects, and d is the distance between their centers (Figure 4.16). The symbol G is a constant called the **gravitational constant**, and its numerical value has been measured to be $G = 6.67 \times 10^{-11} \text{ m}^3/(\text{kg} \times \text{s}^2)$.

THINK ABOUT IT

How does the gravitational force between two objects change if the distance between them triples? If the distance between them drops by half?

 Orbits and Kepler's Law Tutorial, Lessons 1–4

• How does Newton's law of gravity extend Kepler's laws?

By the time Newton published *Principia* in 1687, Kepler's three laws of planetary motion [Section 3.3] had already been known and tested for some 70 years. Kepler's laws had proven so successful at predicting planetary positions that there was little doubt about their validity. However, there was great debate among scientists about *why* Kepler's laws hold true.

Newton solved the mystery by showing that Kepler's laws are consequences of the laws of motion and the universal law of gravitation. For example, we've already seen how we can understand Kepler's second law, that a planet moves faster when it is closer to the Sun, by thinking about conservation of angular momentum. Kepler's third law, that average orbital speed is higher for planets with smaller average orbital distance, comes directly from the fact that planets closer to the Sun have a stronger gravitational attraction to the Sun. In essence, Newton used the mathematical expressions of his laws of gravity and motion to show that a planet orbiting the Sun should automatically have an elliptical orbit that obeys Kepler's laws.

Newton's explanation of Kepler's laws sealed the triumph of the Copernican revolution. Prior to Newton, it was still possible to see Kepler's model of planetary motion as "just" another model, albeit one that fit the observational data far better than any previous model. By explaining Kepler's laws in terms of basic laws of physics, Newton removed virtually all remaining doubt about the legitimacy of the Sun-centered solar system.

In addition, Newton found that Kepler's laws were only part of the story of how objects move in response to gravity. Remember that Kepler had discovered his laws by analyzing the orbits of planets around the Sun, and he therefore had no reason to think that his laws would apply to any other cases, such as moons orbiting planets or comets orbiting the Sun. However, when Newton analyzed his equations for gravity and motion, he discovered that Kepler's laws are just one special case of a more general set of rules about orbiting objects. These more general rules explain the motion of objects throughout the universe; as we will see throughout this book, they are a crucial part of the foundations of modern astronomy. Let's explore four ways in which Newton extended Kepler's laws.

Planets Are Not the Only Objects with Elliptical Orbits   Kepler wrote his first two laws for planets orbiting the Sun, but Newton showed that any object going around another object will obey these two laws. For example, the orbits of a satellite around Earth, of a moon around a planet, and of an asteroid around the Sun are all ellipses in which the orbiting object moves faster at the nearer points in its orbit and slower at the farther points.

Ellipses Are Not the Only Possible Orbital Paths   Kepler was right when he found that ellipses (which include circles) are the only possible shapes for **bound orbits**—orbits in which an object goes around another object over and over again. (The term *bound orbit* comes from the idea that gravity creates a *bond* that holds the objects together.) However, Newton discovered that objects can also follow **unbound orbits**—paths that bring an object close to another object just once. For example, some comets that

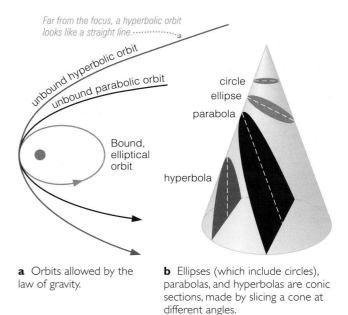

a Orbits allowed by the law of gravity.

b Ellipses (which include circles), parabolas, and hyperbolas are conic sections, made by slicing a cone at different angles.

Figure 4.17

enter the inner solar system follow unbound orbits. They come in from afar just once, loop around the Sun, and never return.

More specifically, Newton showed that the allowed orbital paths are ellipses, parabolas, and hyperbolas (Figure 4.17a). Bound orbits are ellipses, while unbound orbits can be either parabolas or hyperbolas. Together, these shapes are known in mathematics as the *conic sections*, because they can be made by slicing through a cone at different angles (Figure 4.17b). Note that objects on unbound orbits still obey the basic principle of Kepler's second law: They move faster when they are closer to the object they are orbiting, and slower when they are farther away.

Objects Orbit Their Center of Mass   We usually think of one object orbiting another object, like a planet orbiting the Sun or the Moon orbiting Earth. However, Newton showed that two objects attracted by gravity actually *both* orbit around their common **center of mass**—the point at which the two objects would balance if they were somehow connected (Figure 4.18). For example, in a binary star system in which both stars have the same mass, we would see both stars tracing ellipses around a point halfway between them. When one object is more massive than the other, the center of mass lies closer to the more massive object.

The idea that objects orbit their common center of mass holds even for the Sun and planets. However, the Sun is so much more massive than the planets that the center of mass between the Sun and any planet actually lies either inside or nearly inside the Sun, making it difficult for us to notice the Sun's motion about this center. Nevertheless, with precision measurements we can indeed detect the Sun's slight motion around this center of mass. As we will see in Chapter 13, similar measurements of other stars are

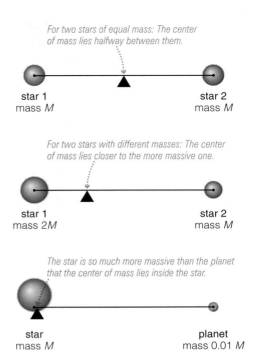

For two stars of equal mass: The center of mass lies halfway between them.

star 1
mass *M*

star 2
mass *M*

For two stars with different masses: The center of mass lies closer to the more massive one.

star 1
mass 2*M*

star 2
mass *M*

The star is so much more massive than the planet that the center of mass lies inside the star.

star
mass *M*

planet
mass 0.01 *M*

Figure 4.18 Interactive Figure Two objects attracted by gravity orbit their common *center of mass*—the point at which they would balance if they were somehow connected.

Figure 4.19 Newton's version of Kepler's third law enables us to measure masses throughout the universe. In addition, it shows that when one object orbits a much more massive object, the orbital period depends only on its average orbital distance. Thus, the astronaut and the Space Shuttle share the same orbit and stay together—even as both orbit Earth at a speed of some 25,000 km/hr.

the primary method astronomers have used to discover planets around those stars.

Orbital Characteristics Tell Us the Masses of Distant Objects   Recall that the precise statement of Kepler's third law is $p^2 = a^3$, where *p* is a planet's orbital period in years and *a* is the planet's average distance from the Sun in AU. Newton found that this statement is actually a special case of a more general equation that we call **Newton's version of Kepler's third law** (see Mathematical Insight 4.3). This more general equation allows us to measure orbital period and distance in any units we wish (rather than only in years and AU) and also shows that the relationship between the orbital period and the average distance depends on the masses of the orbiting objects. In particular, when an object is much less massive than the object it orbits, we can calculate the mass of the central object from the orbital period and average distance of the orbiting object.

Newton's version of Kepler's third law is the primary means by which we determine masses throughout the universe. For example, it allows us to calculate the mass of the Sun from Earth's orbital period (1 year) and its average distance from the Sun. Similarly, measuring the orbital period and average distance of one of Jupiter's moons allows us to calculate Jupiter's mass, and measuring the orbital periods and average distances of stars in binary star systems allows us to determine their masses.

Newton's version of Kepler's third law also explains another important characteristic of orbital motion. It shows that the orbital period of a *small* object orbiting a much more massive object depends only on its orbital distance, not on its mass. That is why an astronaut does not need a

tether to stay close to the Space Shuttle or the Space Station during a space walk (Figure 4.19). The spacecraft and the astronaut are both much smaller than Earth and thus stay together because they have the same orbital distance and hence the same orbital period.

4.5 Orbits, Tides, and the Acceleration of Gravity

Newton's universal law of gravitation has applications that go far beyond explaining Kepler's laws. In this final section, we'll explore three important concepts that we can understand with the help of the universal law of gravitation: how orbits can change, how gravity produces tides, and the acceleration of gravity.

• How do gravity and energy together allow us to understand orbits?

We've seen that Newton's law of universal gravitation explains Kepler's laws of planetary motion, which describe the simple and stable orbits of the planets. By extending Kepler's laws, Newton also explained many other stable orbits, such as the orbit of a satellite around Earth or of a moon around a planet. But orbits do not always stay the same. For example, you've probably heard of satellites crashing to Earth from orbit, proving that orbits can sometimes change dramatically. To understand how and why orbits sometimes change, we need to consider the role of energy in orbits.

Orbital Energy Consider the orbit of a planet around the Sun. An orbiting planet has both kinetic energy (because it is moving around the Sun) and gravitational potential energy (because it would fall toward the Sun if it stopped orbiting). The planet's kinetic energy depends on its orbital speed, and its gravitational potential energy depends on its distance from the Sun. Because the planet's distance and speed both vary as it orbits the Sun, the amounts of

---

MATHEMATICAL INSIGHT **4.3** **Newton's Version of Kepler's Third Law**

Newton's version of Kepler's third law relates the orbital periods, distances, and masses of any pair of orbiting objects, such as the Sun and a planet, a planet and a moon, or two stars in a binary star system. Mathematically, we write it as follows:

$$p^2 = \frac{4\pi^2}{G(M_1 + M_2)}a^3$$

where M_1 and M_2 are the masses of the two objects, p is their orbital period, and a is the distance between their centers. The term $4\pi^2$ is simply a number ($4\pi^2 \approx 4 \times 3.14^2 = 39.44$), and G is the gravitational constant. Any units for p and a can be used in this equation, as long as the units are consistent with the units used for the gravitational constant G.

Newton's version of Kepler's third law gives us the power to measure the masses of distant objects. Any time we measure the orbital period and distance of an orbiting object, we can use Newton's equation to calculate the sum $M_1 + M_2$ of the two objects involved in the orbit. If one object is much more massive than the other, we essentially learn the mass of the massive object. For example, in the case of a planet orbiting the Sun, the sum $M_{Sun} + M_{planet}$ is pretty much just M_{Sun} because the Sun is so much more massive than any of the planets. Thus, knowing any planet's orbital period and distance from the Sun allows us to calculate the mass of the Sun.

Historically, the value of G wasn't determined experimentally until more than 100 years after Newton published *Principia*. Only then did it become possible to use Newton's version of Kepler's third law to measure masses in absolute units. The following examples show some of the remarkable power of Newton's version of Kepler's third law.

Example 1: Use the fact that Earth orbits the Sun in 1 year at an average distance of 150 million kilometers (1 AU) to calculate the mass of the Sun.

Solution:

Step 1 Understand: We know that we can calculate the mass with Newton's version of Kepler's third law. We start by realizing that for Earth orbiting the Sun, this law takes the form:

$$(p_{Earth})^2 = \frac{4\pi^2}{G(M_{Sun} + M_{Earth})}(a_{Earth})^3$$

Because the Sun is so much more massive than Earth, the sum of their masses is approximately the mass of the Sun alone: $M_{Sun} + M_{Earth} \approx M_{Sun}$. Using this approximation, we can write the equation as:

$$(p_{Earth})^2 \approx \frac{4\pi^2}{G \times M_{Sun}}(a_{Earth})^3$$

Step 2 Solve: Because we already know Earth's orbital period (p_{Earth}) and distance (a_{Earth}), the above equation contains only

one unknown: the Sun's mass (M_{Sun}). We can therefore find the Sun's mass by solving the equation for this one unknown. We multiply both sides by M_{Sun} and divide both sides by $(p_{Earth})^2$:

$$M_{Sun} \approx \frac{4\pi^2}{G}\frac{(a_{Earth})^3}{(p_{Earth})^2}$$

We now plug in the known values. Earth's orbital period is $p_{Earth} = 1$ yr, which is the same as 3.15×10^7 seconds. Its average distance from the Sun is $a_{Earth} \approx 150$ million km, or 1.5×10^{11} m. Using these values and the experimentally measured value $G = 6.67 \times 10^{-11}$ m^3/(kg $\times$ s^2), we find:

$$M_{Sun} \approx \frac{4\pi^2}{\left(6.67 \times 10^{-11}\frac{m^3}{kg \times s^2}\right)}\frac{(1.5 \times 10^{11}\ m)^3}{(3.15 \times 10^7\ s)^2}$$
$$= 2 \times 10^{30}\ kg$$

Step 3 Explain: Simply by knowing Earth's orbital period and distance from the Sun and the gravitational constant G, we have used Newton's version of Kepler's third law to find that the Sun's mass is about 2×10^{30} kilograms.

Example 2: A *geosynchronous satellite* orbits Earth in the same amount of time that Earth rotates: 1 sidereal day. If a geosynchronous satellite is also in an equatorial orbit, it is said to be *geostationary* because it remains fixed in the sky (it maintains a constant altitude and direction) as seen from the ground (see Problem 47). Calculate the orbital distance of a geosynchronous satellite.

Solution:

Step 1 Understand: Again, we start by putting Newton's version of Kepler's third law into a form we can apply to this problem. A satellite is much less massive than Earth, so $M_{Earth} + M_{satellite} \approx M_{Earth}$, so we can use the law in this form:

$$(p_{satellite})^2 \approx \frac{4\pi^2}{G \times M_{Earth}}(a_{satellite})^3$$

Step 2 Solve: Because we want to know the satellite's distance, we solve for $a_{satellite}$ by multiplying both sides of the equation by $(G \times M_{Earth})$, dividing both sides by $4\pi^2$, and then taking the cube root of both sides:

$$a_{satellite} \approx \sqrt[3]{\frac{G \times M_{Earth}}{4\pi^2}(p_{satellite})^2}$$

We know that $p_{satellite} = 1$ sidereal day $\approx 86,164$ seconds. You should confirm that substituting this value and the mass of Earth yields $a_{satellite} \approx 42,000$ km.

Step 3 Explain: We have found that a geosynchronous satellite orbits at a distance of 42,000 kilometers above the *center* of the Earth, which is about 35,600 kilometers above Earth's surface.

Total orbital energy = gravitational potential energy + kinetic energy

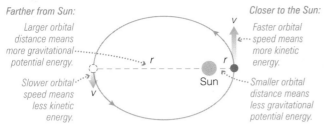

Farther from Sun:
Larger orbital distance means more gravitational potential energy.

Slower orbital speed means less kinetic energy.

Closer to the Sun:
Faster orbital speed means more kinetic energy.

Smaller orbital distance means less gravitational potential energy.

Figure 4.20 The total orbital energy of a planet stays constant throughout its orbit, because its gravitational potential increases when its kinetic energy decreases, and vice versa.

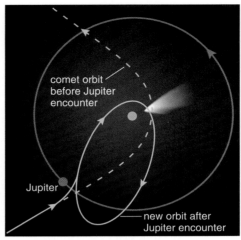

comet orbit before Jupiter encounter

Jupiter

new orbit after Jupiter encounter

Figure 4.21 Depiction of a comet in an unbound orbit of the Sun that happens to pass near Jupiter. The comet loses orbital energy to Jupiter, changing its unbound orbit to a bound orbit around the Sun.

its gravitational potential energy and kinetic energy also vary (Figure 4.20). However, the planet's total **orbital energy**—the sum of its kinetic and gravitational potential energies—always stays the same. This fact is a consequence of the law of conservation of energy. As long as no other object causes the planet to gain or lose orbital energy, its orbital energy cannot change and its orbit must remain the same.

Generalizing from planets to other objects leads us to a very important idea about motion throughout the cosmos: *Orbits cannot change spontaneously.* Left undisturbed, planets would forever keep the same orbits around the Sun, moons would keep the same orbits around their planets, and stars would keep the same orbits around their galaxies.

Gravitational Encounters Although orbits cannot change spontaneously, they can change through exchanges of energy. One way that two objects can exchange orbital energy is through a **gravitational encounter**, in which they pass near enough so that each can feel the effects of the other's gravity. For example, in the rare cases in which a comet happens to pass near a planet, the comet's orbit can change dramatically. Figure 4.21 shows a comet headed toward the Sun on an unbound orbit. The comet's close passage by Jupiter allows the comet and Jupiter to exchange energy. In this case, the comet loses so much orbital energy that its orbit changes from unbound to a bound, elliptical orbit. Jupiter gains exactly as much energy as the comet loses, but the effect on Jupiter is unnoticeable because of its much greater mass.

Spacecraft engineers can use the same basic idea in reverse. The two *Voyager* spacecraft that visited the outer planets in the 1980s and the *Cassini* spacecraft now orbiting Saturn were all deliberately sent past Jupiter on paths that caused them to gain orbital energy at Jupiter's expense. This extra orbital energy helped the spacecraft reach more distant planets. Of course, the effect of the tiny spacecraft on Jupiter was unnoticeable.

A similar dynamic sometimes occurs naturally and may explain why most comets orbit so far from the Sun. Astronomers think that most comets once orbited in the same region of the solar system as the large outer planets [Section 12.2]. Gravitational encounters with Jupiter or the other large planets then caused some of these comets to

be "kicked" into much more distant orbits around the Sun. In some cases, the comets may have been ejected from the solar system completely.

Atmospheric Drag Friction can cause objects to lose orbital energy. For example, consider a satellite orbiting Earth. If the orbit is fairly low—say, just a few hundred kilometers above Earth's surface—the satellite experiences a bit of drag from Earth's thin upper atmosphere. This drag gradually causes the satellite to lose orbital energy until it finally plummets to Earth. The satellite's lost orbital energy is converted to thermal energy in the atmosphere, which is why a falling satellite usually burns up.

Friction may also have played a role in shaping the current orbits of some moons and planets. For example, some of the small moons of Jupiter and the other outer planets may once have orbited the Sun independently. Their orbits could not have changed spontaneously. However, the outer planets probably once were surrounded by clouds of gas [Section 8.4], and friction would have slowed objects passing through this gas. Some of these small objects probably lost just enough energy to friction to allow them to be "captured" as moons.

Escape Velocity An object that gains orbital energy moves into an orbit with a higher average altitude. For example, if we want to boost the orbital altitude of a spacecraft, we can give it more orbital energy by firing a rocket. The chemical potential energy released by the rocket fuel is converted to orbital energy for the spacecraft. The Hubble Space Telescope is still in orbit only because it has gotten periodic boosts during visits by the Space Shuttle. Without these boosts, atmospheric drag would by now have caused the telescope to fall to Earth.

If we give a spacecraft enough orbital energy, it may end up in an unbound orbit that allows it to *escape* Earth completely (Figure 4.22). For example, if we want to send a

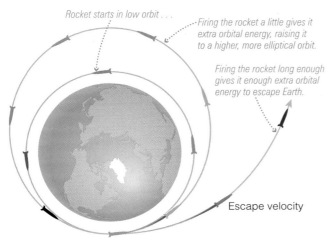

Rocket starts in low orbit . . .

Firing the rocket a little gives it extra orbital energy, raising it to a higher, more elliptical orbit.

Firing the rocket long enough gives it enough extra orbital energy to escape Earth.

Escape velocity

Figure 4.22 Interactive Figure ⟶ An object with escape velocity has enough orbital energy to escape Earth completely.

space probe to Mars, we must use a large rocket that gives the probe enough energy to leave Earth orbit. Although it would probably make more sense to say that the probe

achieves "escape energy," we instead say that it achieves **escape velocity**. The escape velocity from Earth's surface is about 40,000 km/hr, or 11 km/s, meaning that this is the minimum velocity required to escape Earth's gravity if a spacecraft starts near the surface.

Notice that the escape velocity does not depend on the mass of the escaping object—*any* object must travel at a velocity of 11 km/s to escape from Earth, whether it is an individual atom or molecule escaping from the atmosphere, a spacecraft being launched into deep space, or a rock blasted into the sky by a large impact. Escape velocity *does* depend on whether the object starts from the surface or from someplace high above the surface. Because gravity weakens with distance, it takes less energy—and hence a lower escape velocity—to escape from a point high above Earth than from Earth's surface.

• How does gravity cause tides?

If you've spent time near an ocean, you're probably aware of the rising and falling of the tides. In most places, tides rise and fall twice each day. We can understand the basic

---

MATHEMATICAL INSIGHT **4.4** **Escape Velocity**

A simple formula allows us to calculate the escape velocity from any planet, moon, or star:

$$v_{escape} = \sqrt{2 \times G \times \frac{M}{R}}$$

where M is the object's mass, R is the starting distance above the object's center, and G is the gravitational constant. If you use this formula to calculate the escape velocity from an object's surface, replace R with the object's radius.

Example 1: Calculate the escape velocity from the Moon. Compare it to that from Earth.

Solution:

Step 1 Understand: The text has already told us that the escape velocity from Earth's surface is 11 km/s. We can find the Moon's escape velocity with the formula given above. We can find the Moon's mass and radius in Appendix E; they are, respectively, $M = 7.4 \times 10^{22}$ kg and $R = 1.7 \times 10^6$ m.

Step 2 Solve: We substitute the Moon's mass and radius into the escape velocity formula:

$$v_{escape} = \sqrt{2 \times G \times \frac{M_{Moon}}{R_{Moon}}}$$

$$= \sqrt{\frac{2 \times \left(6.67 \times 10^{-11} \frac{m^3}{kg \times s^2} \right) \times (7.4 \times 10^{22} \text{ kg})}{1.7 \times 10^6 \text{ m}}}$$

$$\approx 2,400 \text{ m/s} = 2.4 \text{ km/s}$$

Step 3 Explain: We have found that the escape velocity from the Moon is 2.4 km/s, which is less than one-fourth the 11-km/s escape velocity from Earth.

Example 2: Suppose a future space station orbits Earth in geosynchronous orbit, which is 42,000 kilometers above the center of Earth (see Mathematical Insight 4.3). At what velocity must a spacecraft be launched from the station to escape Earth? Is there any advantage to launching from the space station instead of from Earth's surface?

Solution:

Step 1 Understand: In this case, we want to find the escape velocity from a satellite orbiting 42,000 kilometers above the center of Earth. We therefore will use the escape velocity formula with the mass of Earth ($M_{Earth} = 6.0 \times 10^{24}$ kg) and the satellite's distance above the center of Earth ($R = 42,000$ km $= 4.2 \times 10^7$ m).

Step 2 Solve: We substitute the needed values into the escape velocity formula:

$$v_{escape} = \sqrt{2 \times G \times \frac{M_{Earth}}{R_{orbit}}}$$

$$= \sqrt{\frac{2 \times \left(6.67 \times 10^{-11} \frac{m^3}{kg \times s^2} \right) \times (6.0 \times 10^{24} \text{ kg})}{4.2 \times 10^7 \text{ m}}}$$

$$= 4,400 \text{ m/s} = 4.4 \text{ km/s}$$

Step 3 Explain: The escape velocity from geosynchronous orbit is 4.4 km/s—considerably lower than the 11-km/s escape velocity from Earth's surface. Thus, it requires substantially less fuel to launch the spacecraft from the space station than from Earth. (In addition, the spacecraft would already have the orbital velocity of the space station.) Of course, this assumes that the space station is already in place and that the spacecraft is assembled at the space station.

cause of tides by examining the gravitational attraction between Earth and the Moon. We'll then see how the same ideas explain many other phenomena that we can observe throughout the universe, including the synchronous rotation of our own Moon and many other worlds.

The Moon's Tidal Force Gravity attracts Earth and the Moon toward each other (with the Moon staying in orbit as it "falls around" Earth), but it affects different parts of Earth slightly differently: Because the strength of gravity declines with distance, the gravitational attraction of each part of Earth to the Moon becomes weaker as we go from the side of Earth facing the Moon to the side facing away from the Moon. This difference in attraction creates a "stretching force," or **tidal force**, that stretches the entire Earth to create two tidal bulges—one facing the Moon and one opposite the Moon (Figure 4.23). If you are still unclear about why there are *two* tidal bulges, think about a rubber band: If you pull on a rubber band it will stretch in both directions relative to its center, even if you pull on only one side. In the same way, Earth stretches on both sides even though the Moon is tugging harder on only one side.

Tides affect both land and ocean, but we generally notice only the ocean tides because water flows much more readily than land. Earth's rotation carries any location through each of the two bulges each day, creating two high tides. Low tides occur when the location is at the points halfway between the two tidal bulges. The two "daily" high tides actually come slightly more than 12 hours apart. Because of its orbital motion around Earth, the Moon reaches its highest point in the sky at any location about every 24 hours 50 minutes, rather than every 24 hours. Thus, the tidal cycle

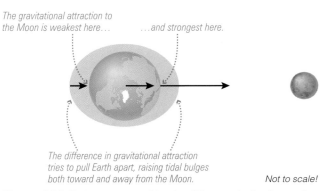

The gravitational attraction to the Moon is weakest here...   *...and strongest here.*

The difference in gravitational attraction tries to pull Earth apart, raising tidal bulges both toward and away from the Moon.   *Not to scale!*

Figure 4.23 Tides are created by the difference in the force of attraction between different parts of Earth and the Moon. There are two daily high tides as any location on Earth rotates through the two tidal bulges. (The diagram highly exaggerates the tidal bulges, which raise the oceans only about 2 meters and the land only about a centimeter.)

of two high tides and two low tides actually takes about 24 hours 50 minutes, with each high tide occurring about 12 hours 25 minutes after the previous one.

The height and timing of tides can vary considerably from place to place around the Earth, depending on factors such as latitude, the orientation of the coastline (such as whether it is north-facing or west-facing), and the depth and shape of any channel through which the rising tide must flow. For example, while the tide rises gradually in most locations, the incoming tide near the famous abbey on Mont-Saint-Michel, France, moves much faster than a person can swim (Figure 4.24). In centuries past, the Mont was an island twice a day at high tide but was connected

Figure 4.24 Photographs of high and low tide at the abbey at Mont-Saint-Michel, France, one of the world's most popular tourist destinations. Here the tide rushes in much faster than a person can swim. Before a causeway was built (visible to the left, with cars on it), the Mont was accessible by land only at low tide. At high tide, it became an island.

The Origin of Tides

Many people believe that tides arise because the Moon pulls Earth's oceans toward it. But if that were the whole story, there would be a bulge only on the side of Earth facing the Moon, and hence only one high tide each day. The correct explanation for tides must account for why Earth has *two* tidal bulges.

Only one explanation works: Earth must be stretching from its center in both directions (toward and away from the Moon). The stretching force, or tidal force, arises from the *difference* between the force of gravity attracting different parts of Earth to the Moon. In fact, stretching due to tides affects many objects, not just Earth. Many moons are stretched into slightly oblong shapes by tidal forces caused by their parent planets, and mutual tidal forces stretch close binary stars into teardrop shapes. In regions where gravity is extremely strong, such as near a black hole, tides could even stretch spaceships or people.

to the mainland at low tide. Many pilgrims drowned when they were caught unprepared by the tide rushing in. (Today, a human-made land bridge keeps the island connected to the mainland.) Another unusual tidal pattern occurs in coastal states along the northern shore of the Gulf of Mexico, where topography and other factors combine to make only one noticeable high tide and low tide each day.

The Tidal Effect of the Sun   The Sun also exerts a tidal force on Earth, causing Earth to stretch along the Sun-Earth line. You might at first guess that the Sun's tidal force would be more than the Moon's, since the Sun's mass is more than a million times that of the Moon. Indeed, the *gravitational* force between Earth and the Sun is much greater than that between Earth and the Moon, which is why Earth orbits the Sun. However, the much greater distance to the Sun (than to the Moon) means that the *difference* in the Sun's pull on the near and far sides of Earth is relatively small.

The overall tidal force caused by the Sun is about one-third that caused by the Moon (Figure 4.25). When the tidal forces of the Sun and the Moon work together, as is the case at both new moon and full moon, we get the especially pronounced *spring tides* (so named because the water tends to "spring up" from the Earth). When the tidal forces of the Sun and the Moon counteract each other, as is the case at first- and third-quarter moon, we get the relatively small tides known as *neap tides*.

Explain why any tidal effects on Earth caused by the other planets would be extremely small (in fact, so small as to be unnoticeable).

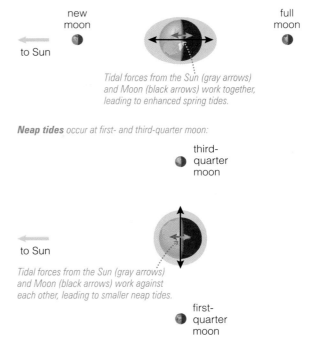

Spring tides occur at new moon and full moon:

Tidal forces from the Sun (gray arrows) and Moon (black arrows) work together, leading to enhanced spring tides.

Neap tides occur at first- and third-quarter moon:

Tidal forces from the Sun (gray arrows) and Moon (black arrows) work against each other, leading to smaller neap tides.

Figure 4.25   Interactive Figure   The Sun exerts a tidal force on Earth about one-third as strong as that from the Moon. When the tidal forces from the Sun and Moon work together at new moon and full moon, we get enhanced *spring tides*. When they work against each other, at first- and third-quarter moons, we get smaller *neap tides*.

Tidal Friction   So far, we have talked as if Earth slides smoothly through the tidal bulges as it rotates. But because tidal forces stretch the Earth itself, the process involves some friction, called **tidal friction**. Figure 4.26 shows the effects of this friction. In essence, the Moon's gravity tries to keep the tidal bulges on the Earth-Moon line, while Earth's rotation tries to pull the bulges around with it. The resulting "compromise" keeps the bulges just ahead of the Earth-Moon line at all times.

The slight misalignment of the tidal bulges with the Earth-Moon line causes two important effects. First, because the Moon's gravity is always pulling back on the bulges, Earth's rotation slows very gradually. Second, the gravity of the bulges pulls the Moon slightly ahead in its orbit, thereby adding to the Moon's orbital energy and causing the Moon to move gradually farther from Earth. These effects are barely noticeable on human time scales. For example, tidal friction increases the length of a day by only about 1 second every 50,000 years.* But the effects add up over billions of years. Early in Earth's history, a day may have been only 5 or 6 hours long and the Moon may have been one-tenth or less its current distance from Earth. These changes also provide a great example of conservation of angular momentum: The Moon's growing orbit gains the angular momentum that Earth loses as its rotation slows.

---

*This effect is overwhelmed on short time scales by effects due to slight changes in Earth's internal mass distribution; these changes can alter Earth's rotation period by a second or more per year, which is why "leap seconds" are occasionally added to or subtracted from the year.

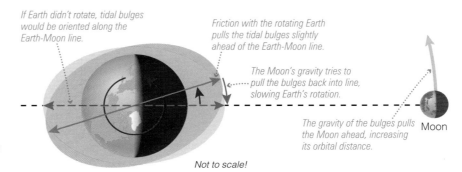

Figure 4.26 Earth's rotation pulls its tidal bulges slightly ahead of the Earth-Moon line, leading to gravitational effects that gradually slow Earth's rotation and increase the Moon's orbital energy and distance.

If Earth didn't rotate, tidal bulges would be oriented along the Earth-Moon line.

Friction with the rotating Earth pulls the tidal bulges slightly ahead of the Earth-Moon line.

The Moon's gravity tries to pull the bulges back into line, slowing Earth's rotation.

The gravity of the bulges pulls the Moon ahead, increasing its orbital distance.

Moon

Not to scale!

The Moon's Synchronous Rotation Recall that the Moon always shows (nearly) the same face to Earth, a trait called **synchronous rotation** (see Figure 2.23). Synchronous rotation may seem like an extraordinary coincidence, but it is a natural consequence of tidal friction.

Because Earth is more massive than the Moon, Earth exerts a greater tidal force on the Moon than the Moon does on Earth. This tidal force gives the Moon two tidal bulges along the Earth-Moon line, much like the two tidal bulges that the Moon creates on Earth. (The Moon does not have visible tidal bulges, but it does indeed have excess mass along the Earth-Moon line.) Thus, if the Moon were rotating through its tidal bulges in the same way that Earth rotates through its tidal bulges, the resulting friction would cause the Moon's rotation to slow down. This is exactly what we think happened long ago.

The Moon probably once rotated much faster than it does today. As a result, it *did* rotate through its tidal bulges, and its rotation gradually slowed. Once the Moon's rotation slowed to the point at which the Moon and its bulges rotated at the same rate—that is, synchronously with the orbital period—there was no further source for tidal friction. Thus, the Moon's synchronous rotation was a natural outcome of Earth's tidal effects on the Moon. In fact, if the Moon and Earth stay together long enough (for another few hundred billion years), the gradual slowing of Earth's rotation will eventually make Earth keep the same face to the Moon as well.

Tidal Effects on Other Worlds Tidal forces affect not only Earth and the Moon, but also many other objects. Synchronous rotation caused by tidal friction is especially common. For example, Jupiter's four large moons (Io, Europa, Ganymede, and Callisto) keep nearly the same face toward Jupiter at all times, as do many other moons. Pluto and its moon Charon *both* rotate synchronously: Like two dancers, they always keep the same face toward each other. Many binary star systems also rotate in this way. Some moons and planets exhibit variations on synchronous rotation. For example, Mercury rotates exactly three times for every two orbits of the Sun. This pattern ensures that Mercury's tidal bulge always aligns with the Sun at perihelion, where the Sun exerts its strongest tidal force.

Tidal forces play other roles in the cosmos as well. They can alter the shapes of objects by stretching them along the line of tidal bulges. More dramatically, in Chapter 11 we'll see how tidal forces have led to the astonishing volcanic activity of Jupiter's moon Io and the possibility of a subsurface ocean on its moon Europa. As you study astronomy, you'll encounter many more cases where tides and tidal friction play important roles.

• Why do all objects fall at the same rate?

Throughout the remainder of the text, we will see many more applications of the universal law of gravitation. For now, let's look at just one more: Galileo's discovery that the acceleration of a falling object is independent of its mass.

If you drop a rock, the force acting on the rock is the force of gravity. The two masses involved are the mass of Earth and the mass of the rock, which we'll denote M_{Earth} and M_{rock}, respectively. The distance between their *centers* is the distance from the *center of Earth* to the center of the rock. If the rock isn't too far above Earth's surface, this distance is approximately the radius of Earth, R_{Earth} (about 6,400 kilometers). That is, $d \approx R_{Earth}$. Thus, the force of gravity acting on the rock is:

$$F_g = G\frac{M_{Earth}M_{rock}}{d^2} \approx G\frac{M_{Earth}M_{rock}}{(R_{Earth})^2}$$

According to Newton's second law of motion ($F = ma$), this force is equal to the product of the mass and the acceleration of the rock. That is:

$$G\frac{M_{Earth}\cancel{M_{rock}}}{(R_{Earth})^2} = \cancel{M_{rock}}a_{rock}$$

Note that M_{rock} "cancels" because it appears on both sides of the equation (as a multiplier), giving Galileo's result that the acceleration of the rock—or of any falling object—does not depend on the object's mass.

The fact that objects of different masses fall with the same acceleration struck Newton as an astounding coincidence, even though his own equations showed it to be so. For the next 240 years, this seemingly odd coincidence remained just that—a coincidence—in the minds of scientists. However, in 1915 Einstein discovered that it is not a coincidence at all. Rather, it reveals something deeper about the nature of gravity and of the universe. Einstein described the new insights in his *general theory of relativity* (the topic of Chapter S3).

The text shows that the acceleration of a falling rock near the surface of Earth is:

$$a_{\text{rock}} = G \times \frac{M_{\text{Earth}}}{(R_{\text{Earth}})^2}$$

Because this formula applies to *any* falling object on Earth, it is the *acceleration of gravity, g*. Calculating *g* is easy. Simply look up Earth's mass (6.0×10^{24} kg) and radius (6.4×10^6 m) in Appendix E, and then "plug in":

$$g = G \times \frac{M_{\text{Earth}}}{(R_{\text{Earth}})^2}$$
$$= \left(6.67 \times 10^{-11} \frac{\text{m}^3}{\text{kg} \times \text{s}^2}\right) \times \frac{6.0 \times 10^{24} \text{ kg}}{(6.4 \times 10^6 \text{ m})^2}$$
$$= 9.8 \frac{\text{m}}{\text{s}^2}$$

We can find the acceleration of gravity on the surface of any other world by using the same formula, using the other world's mass and radius instead of Earth's.

Example 1: What is the acceleration of gravity on the surface of the Moon?

Solution:

Step 1 Understand: We want to find the acceleration of gravity on the Moon's surface, so we can simply use the formula given above. We look up the values for the Moon's mass (7.4×10^{22} kg) and radius (1.7×10^6 m).

Step 2 Solve: We substitute the values of the Moon's mass and radius into the formula for the acceleration of gravity:

$$g_{\text{Moon}} = G \times \frac{M_{\text{Moon}}}{(R_{\text{Moon}})^2}$$
$$= \left(6.67 \times 10^{-11} \frac{\text{m}^3}{\text{kg} \times \text{s}^2}\right) \times \frac{7.4 \times 10^{22} \text{ kg}}{(1.7 \times 10^6 \text{ m})^2}$$
$$= 1.7 \frac{\text{m}}{\text{s}^2}$$

Step 3 Explain: The acceleration of gravity on the Moon is 1.7 m/s², or about one-sixth that on Earth. Thus, objects on the Moon weigh about one-sixth of what they would weigh on Earth. If you can lift a 50-kilogram barbell on Earth, you'll be able to lift a 300-kilogram barbell on the Moon.

Example 2: The Space Station orbits at an altitude roughly 300 kilometers above Earth's surface. What is the acceleration of gravity at this altitude?

Solution:

Step 1 Understand: Because the Space Station is significantly above Earth's surface, we cannot use the approximation $d \approx R_{\text{Earth}}$ that we used in the text. Instead, we must go back to Newton's second law and set the gravitational force on the Space Station equal to its mass times acceleration. The acceleration in this equation will be the acceleration of gravity at the Space Station's altitude.

Step 2 Solve: We write Newton's second law with the force being the force of gravity acting between Earth and the Space Station, which we set equal to the Space Station's mass times its acceleration:

$$G \times \frac{M_{\text{Earth}} \cancel{M_{\text{station}}}}{d^2} = \cancel{M_{\text{station}}} \times a_{\text{station}}$$

You should confirm that when we solve this equation for the acceleration of gravity, we find:

$$a_{\text{station}} = G \times \frac{M_{\text{Earth}}}{d^2}$$

In this case, the distance *d* is the 6,400-kilometer radius of Earth *plus* the 300-kilometer altitude of the Station, or $d = 6{,}700$ km $= 6.7 \times 10^6$ m. Thus, the gravitational acceleration of the Space Station when orbiting Earth is:

$$a_{\text{station}} = G \times \frac{M_{\text{Earth}}}{d^2}$$
$$= \left(6.67 \times 10^{-11} \frac{\text{m}^3}{\text{kg} \times \text{s}^2}\right) \times \frac{6.0 \times 10^{24} \text{ kg}}{(6.7 \times 10^6 \text{ m})^2}$$
$$= 8.9 \frac{\text{m}}{\text{s}^2}$$

Step 3 Explain: The acceleration of gravity in low-Earth orbit is 8.9 m/s². Notice that this is only about 10% less than the 9.8 m/s² acceleration of gravity at Earth's surface. Thus, we see again that lack of gravity cannot be the reason why astronauts are weightless in orbit; rather, they are weightless because they are in free-fall.

THE BIG PICTURE

Putting Chapter 4 into Context

We've covered a lot of ground in this chapter, from the scientific terminology of motion to the story of how Newton explained universal motion. Be sure you grasp the following "big picture" ideas:

- Understanding the universe requires understanding motion. Motion may seem complex, but it can be understood simply through Newton's three laws of motion.

- Today, we comprehend Newton's laws of motion through deeper principles, including the laws of conservation of angular momentum and of energy. We can understand many processes in the universe by following how energy is exchanged between different objects.

- Newton also discovered the universal law of gravitation, which explains how gravity holds planets in their orbits and much more—including how satellites can reach and stay in orbit, the nature of tides, and why the Moon rotates synchronously with Earth.

- Perhaps most important, Newton's discoveries showed that the same physical laws we observe on Earth apply throughout the universe. The universality of physics opens up the entire cosmos as a possible realm of human study.

4.1 Describing Motion: Examples from Daily Life

- **How do we describe motion?** **Speed** is the rate at which an object is moving. **Velocity** is speed in a certain direction. **Acceleration** is a change in velocity, meaning a change in either speed or direction. **Momentum** is mass × velocity. A **force** can change an object's momentum, causing it to accelerate.

- **How is mass different from weight?** An object's mass is the same no matter where it is located, but its weight varies with the strength of gravity or other forces acting on the object. An object becomes **weightless** when it is in **free-fall**, even though its mass is unchanged.

4.2 Newton's Laws of Motion

- **How did Newton change our view of the universe?** Newton showed that the same physical laws that operate on Earth also operate in the heavens, making it possible for us to learn about the universe by studying physical laws here on Earth.

- **What are Newton's three laws of motion?** (1) An object moves at constant velocity if there is no net force acting upon it. (2) Force = mass × acceleration ($F = ma$). (3) For any force, there is always an equal and opposite reaction force.

4.3 Conservation Laws in Astronomy

- **Why do objects move at constant velocity if no force acts on them?** **Conservation of momentum** means that an object's momentum cannot change unless it transfers momentum to or from other objects. When no force is present, no momentum can be transferred so an object must maintain its speed and direction.

- **What keeps a planet rotating and orbiting the Sun?** **Conservation of angular momentum** means that a planet's rotation and orbit cannot change unless it transfers angular momentum to another object. The planets in our solar system do not exchange substantial angular momentum with each other or anything else, so their orbits and rotation rates remain quite steady.

- **Where do objects get their energy?** Energy is always conserved—it can be neither created nor destroyed. Thus, objects must have gotten whatever energy they now have from exchanges of energy with other objects. Energy comes in three basic categories—**kinetic**, **radiative**, and **potential**— though sometimes we use subcategories for convenience.

kinetic energy

radiative energy   potential energy

4.4 The Universal Law of Gravitation

- **What determines the strength of gravity?** According to the **universal law of gravitation**, every object attracts every other object with a gravitational force that is directly proportional to the product of the objects' masses and declines with the square of the distance between their centers:

$$F_g = G \frac{M_1 M_2}{d^2}$$

- **How does Newton's law of gravity extend Kepler's laws?** Newton extended Kepler's laws in four main ways: (1) He showed that any object going around another object will obey Kepler's first two laws. (2) He showed that **bound orbits** in ellipses (which include circles) are not the only possible orbital shape—orbits can also be **unbound** parabolas or hyperbolas. (3) He showed that two objects actually orbit their common **center of mass**. (4) **Newton's version of Kepler's third law** allows us to calculate the masses of orbiting objects from their orbital periods and distances.

4.5 Orbits, Tides, and the Acceleration of Gravity

- **How do gravity and energy together allow us to understand orbits?** Gravity determines orbits, and an object cannot change its orbit unless it gains or loses **orbital energy**—the sum of its kinetic and gravitational potential energies—through energy transfer with other objects. If an object gains enough orbital energy, it may achieve **escape velocity** and leave the gravitational influence of the object it was orbiting.

- **How does gravity cause tides?** The Moon's gravity creates a **tidal force** that stretches Earth along the Earth-Moon line, causing the tidal bulges responsible for our daily tides. Tidal forces can also lead to **tidal friction**, which is gradually slowing Earth's rotation and explains the synchronous rotation of the Moon.

- **Why do all objects fall at the same rate?** Newton's equations show that the acceleration of gravity is independent of the mass of a falling object, so all objects fall at the same rate.

Review Questions

Short-Answer Questions Based on the Reading

1. How does *speed* differ from *velocity?* Give an example in which you can be traveling at constant speed but not at constant velocity.

2. What do we mean by *acceleration?* What is the *acceleration of gravity?* Explain what we mean when we state an acceleration in units of m/s².

3. What is *momentum?* How can momentum be affected by a *force?* What do we mean when we say that momentum will be changed only by a *net force?*

4. What is *free-fall,* and why does it make you *weightless?* Briefly describe why astronauts are weightless in the Space Station.

5. State each of *Newton's three laws of motion.* For each law, give an example of its application.

6. What are the laws of *conservation of momentum, conservation of angular momentum,* and *conservation of energy?* For each, give an example of how it is important in astronomy.

7. Define *kinetic energy, radiative energy,* and *potential energy.* For each type of energy, give at least two examples of objects that either have it or use it.

8. Define *temperature* and *thermal energy.* How are they related? How are they different?

9. Which has more gravitational potential energy: a rock on the ground or a rock that you hold out the window of a 10-story building? Explain.

10. What do we mean by *mass-energy?* Is it a form of kinetic, radiative, or potential energy? How is the idea of mass-energy related to the formula $E = mc^2$?

11. Summarize the *universal law of gravitation* in words. Then state the law mathematically, explaining the meaning of each symbol in the equation.

12. What is the difference between bound and unbound orbit? What orbital shapes are possible?

13. What do we need to know if we want to measure an object's mass with *Newton's version of Kepler's third law?* Explain.

14. Explain why orbits cannot change spontaneously. How can atmospheric drag affect an orbit? How can a *gravitational encounter* cause an orbit to change? How can an object achieve *escape velocity?*

15. Explain how the Moon creates tides on Earth. Why do we have two high and low tides each day?

16. How do the tides vary with the phase of the Moon? Why?

17. What is *tidal friction?* What effects does it have on Earth? How does it explain the Moon's synchronous rotation?

18. Would you fall at the same rate on the Moon as on Earth? Explain.

Test Your Understanding

Does It Make Sense?

Decide whether the statement makes sense (or is clearly true) or does not make sense (or is clearly false). Explain your reasoning. (For an example, see Chapter 1, "Does It Make Sense?")

19. If you could go shopping on the Moon to buy a pound of chocolate, you'd get a lot more chocolate than if you bought a pound on Earth. (*Hint:* Pounds are a unit of weight, not mass.)

20. Suppose you could enter a vacuum chamber (on Earth), that is, a chamber with no air in it. Inside this chamber, if you dropped a hammer and a feather from the same height at the same time, both would hit the bottom at the same time.

21. When an astronaut goes on a space walk outside the Space Station, she will quickly float away from the station unless she has a tether holding her to the station or constantly fires thrusters on her space suit.

22. Newton's version of Kepler's third law allows us to calculate the mass of Saturn from orbital characteristics of its moon Titan.

23. If we could somehow replace the Sun with a giant rock that has precisely the same mass, Earth's orbit would not change.

24. The fact that the Moon rotates once in precisely the time it takes to orbit Earth once is such an astonishing coincidence that scientists probably never will be able to explain it.

25. Venus has no oceans, so it could not have tides even if it had a moon (which it doesn't).

26. If an asteroid passed by Earth at just the right distance, it would be captured by Earth's gravity and become our second moon.

27. When I drive my car at 30 miles per hour, it has more kinetic energy than it does at 10 miles per hour.

28. Someday soon, scientists are likely to build an engine that produces more energy than it consumes.

Quick Quiz

Choose the best answer to each of the following. Explain your reasoning with one or more complete sentences.

29. Which one of the following describes an object that is accelerating? (a) A car traveling on a straight, flat road at 50 miles per hour. (b) A car traveling on a straight uphill road at 30 miles per hour. (c) A car going around a circular track at a steady 100 miles per hour.

30. Suppose you visit another planet: (a) Your mass and weight would be the same as they are on Earth. (b) Your mass would be the same as on Earth, but your weight would be different. (c) Your weight would be the same as on Earth, but your mass would be different.

31. Which person is weightless? (a) A child in the air as she plays on a trampoline. (b) A scuba diver exploring a deep-sea wreck. (c) An astronaut on the Moon.

32. Consider the statement "There's no gravity in space." This statement is: (a) Completely false. (b) False if you are close to a planet or moon, but true in between the planets. (c) Completely true.

33. If you want to make a rocket turn left, you need to: (a) Fire an engine that shoots out gas to the left. (b) Fire an engine that shoots out gas to the right. (c) Spin the rocket counterclockwise.

34. Earth's angular momentum when it is at perihelion (nearest to the Sun) in its orbit is: (a) Greater than its angular momentum at aphelion. (b) Less than its angular momentum at aphelion. (c) Exactly the same as its angular momentum at aphelion.

35. As an interstellar gas cloud shrinks in size, its gravitational potential energy: (a) stays the same at all times. (b) gradually is transformed into other forms of energy. (c) gradually grows larger.

36. If Earth were twice as far from the Sun, the force of gravity attracting Earth to the Sun would be: (a) twice as strong. (b) half as strong. (c) one-quarter as strong.

37. According to the law of universal gravitation, what would happen to Earth if the Sun were somehow replaced by a black hole of the same mass? (a) Earth would be quickly sucked into the black hole. (b) Earth would slowly spiral in to the black hole. (c) Earth's orbit would not change.

38. If the Moon were closer to Earth, high tides would: (a) be higher than they are now. (b) be lower than they are now. (c) occur three or more times a day rather than twice a day.

Investigate Further

In-Depth Questions to Increase Your Understanding

Short-Answer/Essay Questions

39. *Units of Acceleration.*
   a. If you drop a rock from a very tall building, how fast will it be going after 4 seconds? Explain.
   b. As you sled down a steep, slick street, you accelerate at a rate of 4 m/s^2. How fast will you be going after 5 seconds? Explain.
   c. You are driving along the highway at a speed of 60 miles per hour when you slam on the brakes. If your acceleration is at an average rate of −20 miles per hour per second, how long will it take to come to a stop?

40. *Weightlessness.* Astronauts are weightless when in orbit in the Space Shuttle. Are they also weightless during the Shuttle's launch? How about during its return to Earth? Explain.

41. *Gravitational Potential Energy.*
   a. Why does a bowling ball perched on a cliff ledge have more gravitational potential energy than a baseball perched on the same ledge?
   b. Why does a diver on a 10-meter platform have more gravitational potential energy than a diver on a 3-meter diving board?
   c. Why does a 100-kilogram satellite orbiting Jupiter have more gravitational potential energy than a 100-kilogram satellite orbiting Earth, assuming both satellites orbit at the same distance from their planets' centers?

42. *Einstein's Famous Formula.*
   a. What is the meaning of the formula $E = mc^2$? Be sure to define each variable.
   b. How does this formula explain the generation of energy by the Sun?
   c. How does this formula explain the destructive power of nuclear bombs?

43. *The Gravitational Law.*
   a. How does quadrupling the distance between two objects affect the gravitational force between them?
   b. Suppose the Sun were somehow replaced by a star with twice as much mass. What would happen to the gravitational force between Earth and the Sun?
   c. Suppose Earth were moved to one-third of its current distance from the Sun. What would happen to the gravitational force between Earth and the Sun?

44. *Allowable Orbits?*
   a. Suppose the Sun were replaced by a star with twice as much mass. Could Earth's orbit stay the same? Why or why not?
   b. Suppose Earth doubled in mass (but the Sun stayed the same as it is now). Could Earth's orbit stay the same? Why or why not?

45. *Head-to-Foot Tides.* You and Earth attract each other gravitationally, so you should also be subject to a tidal force resulting from the difference between the gravitational attraction felt by your feet and that felt by your head (at least when you are standing). Explain why you can't feel this tidal force.

46. *Synchronous Rotation.* Suppose the Moon had rotated *more slowly* when it formed than it does now. Would it still have ended up in synchronous rotation? Why or why not?

47. *Geostationary Orbit.* A satellite in geostationary orbit appears to remain stationary in the sky as seen from any particular location on Earth.
   a. Briefly explain why a geostationary satellite must orbit Earth in 1 *sidereal* day, rather than 1 solar day.
   b. Explain why a geostationary satellite must orbit around Earth's equator, rather than in some other orbit (such as around the poles).
   c. Home satellite dishes (such as those used for television) receive signals from communications satellites. Explain why these satellites must be in geostationary orbit.

48. *Elevator to Orbit.* Some people have proposed building a giant elevator from Earth's surface to geosynchronous orbit. The top of the elevator would then have the same orbital distance and period as any satellite in geosynchronous orbit.
   a. Suppose you were to let go of an object at the top of the elevator. Would the object fall? Would it orbit Earth? Explain.
   b. Briefly explain why (not counting the huge costs for construction) the elevator would make it much cheaper and easier to put satellites in orbit or to launch spacecraft into deep space.

Quantitative Problems

Be sure to show all calculations clearly and state your final answers in complete sentences.

49. *Energy Comparisons.* Use the data in Table 4.1 to answer each of the following questions.
   a. Compare the energy of a 1-megaton hydrogen bomb to the energy released by a major earthquake.
   b. If the United States obtained all its energy from oil, how much oil would be needed each year?
   c. Compare the Sun's annual energy output to the energy released by a supernova.

50. *Moving Candy Bar.* We can calculate the kinetic energy of any moving object with a very simple formula: kinetic energy $= \frac{1}{2}mv^2$, where m is the object's mass and v is its velocity or speed. Table 4.1 shows that metabolizing a candy bar releases about 10^6 joules. How fast must the candy bar travel to have the same 10^6 joules in the form of kinetic energy? (Assume the candy bar's mass is 0.2 kilogram.) Is your answer faster or slower than you expected?

51. *Spontaneous Human Combustion.* Suppose that all the mass in your body were suddenly converted into energy according to the formula $E = mc^2$. How much energy would be released? Compare this to the energy released by a 1-megaton hydrogen bomb (see Table 4.1). What effect would your disappearance have on your surroundings?

52. *Fusion Power.* No one has yet succeeded in creating a commercially viable way to produce energy through nuclear fusion. However, suppose we could build fusion power plants

using the hydrogen in water as a fuel. Based on the data in Table 4.1, how much water would we need each minute in order to meet U.S. energy needs? Could such a reactor power the entire United States with the water flowing from your kitchen sink? Explain. (*Hint:* Use the annual U.S. energy consumption to find the energy consumption per minute, and then divide by the energy yield from fusing 1 liter of water to figure out how many liters would be needed each minute.)

53. *Understanding Newton's Version of Kepler's Third Law I.* Imagine another solar system, with a star of the same mass as the Sun. Suppose there is a planet in that solar system with a mass twice that of Earth orbiting at a distance of 1 AU from the star. What is the orbital period of this planet? Explain. (*Hint:* The calculations for this problem are so simple that you will not need a calculator.)

54. *Understanding Newton's Version of Kepler's Third Law II.* Suppose a solar system has a star that is four times as massive as our Sun. If that solar system has a planet the same size as Earth orbiting at a distance of 1 AU, what is the orbital period of the planet? Explain. (*Hint:* The calculations for this problem are so simple that you will not need a calculator.)

55. *Using Newton's Version of Kepler's Third Law I.*
   a. The Moon orbits Earth in an average time of 27.3 days at an average distance of 384,000 kilometers. Use these facts to determine the mass of Earth. (*Hint:* You may neglect the mass of the Moon, since its mass is only about $\frac{1}{80}$ of Earth's.)
   b. Jupiter's moon Io orbits Jupiter every 42.5 hours at an average distance of 422,000 kilometers from the center of Jupiter. Calculate the mass of Jupiter. (*Hint:* Io's mass is very small compared to Jupiter's.)
   c. You discover a planet orbiting a distant star that has about the same mass as the Sun. Your observations show that the planet orbits the star every 63 days. What is its orbital distance?

56. *Using Newton's Version of Kepler's Third Law II.*
   a. Pluto's moon Charon orbits Pluto every 6.4 days with a semimajor axis of 19,700 kilometers. Calculate the *combined* mass of Pluto and Charon. Compare this combined mass to the mass of Earth, which is about 6×10^{24} kilograms.
   b. Calculate the orbital period of the Space Shuttle in an orbit 300 kilometers above Earth's surface.
   c. The Sun orbits the center of the Milky Way Galaxy every 230 million years at a distance of 28,000 light-years. Use these facts to determine the mass of the galaxy. (As we'll discuss in Chapter 22, this calculation actually tells us only the mass of the galaxy *within* the Sun's orbit.)

57. *Escape Velocity.* Calculate the escape velocity from each of the following.
   a. The surface of Mars (mass = $0.11 M_{Earth}$, radius = $0.53 R_{Earth}$).
   b. The surface of Mars's moon Phobos (mass = 1.1×10^{16} kg, radius = 12 km).
   c. The cloud tops of Jupiter (mass = $317.8 M_{Earth}$, radius = $11.2 R_{Earth}$).
   d. Our solar system, starting from Earth's orbit. (*Hint:* Most of the mass of our solar system is in the Sun; $M_{Sun} = 2.0 \times 10^{30}$ kg.)
   e. Our solar system, starting from Saturn's orbit.

58. *Weights on Other Worlds.* Calculate the acceleration of gravity on the surface of each of the following worlds. How much would *you* weigh, in pounds, on each of these worlds?
   a. Mars (mass = $0.11 M_{Earth}$, radius = $0.53 R_{Earth}$).
   b. Venus (mass = $0.82 M_{Earth}$, radius = $0.95 R_{Earth}$).
   c. Jupiter (mass = $317.8 M_{Earth}$, radius = $11.2 R_{Earth}$). Bonus: Given that Jupiter has no solid surface, how could you weigh yourself on Jupiter?
   d. Jupiter's moon Europa (mass = $0.008 M_{Earth}$, radius = $0.25 R_{Earth}$).
   e. Mars's moon Phobos (mass = 1.1×10^{16} kg, radius = 12 km).

59. *Gees.* Acceleration is sometimes measured in *gees*, or multiples of the acceleration of gravity: 1 gee (1*g*) means $1 \times g$, or 9.8 m/s^2; 2 gees (2*g*) means $2 \times g$, or 2×9.8 m/s^2 = 19.6 m/s^2; and so on. Suppose you experience 6 gees of acceleration in a rocket.
   a. What is your acceleration in meters per second squared?
   b. You will feel a compression force from the acceleration. How does this force compare to your normal weight?
   c. Do you think you could survive this acceleration for long? Explain.

60. *Earth's 2nd Moon.* Suppose Earth had a second moon, called Swisscheese, with an average orbital distance double the Moon's and a mass about the same as the Moon's.
   a. Is Swisscheese's orbital period longer or shorter than the Moon's? Explain.
   b. The Moon's orbital period is about one month. Apply Kepler's 3rd law to find the approximate orbital period of Swisscheese. (*Hint:* If you form the ratio of the orbital distances of Swisscheese and the Moon, you can solve this problem with Kepler's original version of his third law rather than looking up all the numbers you'd need to apply Newton's version of Kepler's third law.)
   c. In words, describe how tides would differ due to the presence of this second moon. Consider the cases when the two moons are on the same side of Earth, on opposite sides of Earth, and 90° apart in their orbits.

Discussion Questions

61. *Knowledge of Mass-Energy.* Einstein's discovery that energy and mass are equivalent has led to technological developments that are both beneficial and dangerous. Discuss some of these developments. Overall, do you think the human race would be better or worse off if we had never discovered that mass is a form of energy? Defend your opinion.

62. *Perpetual Motion Machines.* Every so often, someone claims to have built a machine that can generate energy perpetually from nothing. Why isn't this possible according to the known laws of nature? Why do you think claims of perpetual motion machines sometimes receive substantial media attention?

63. *Tidal Complications.* The ocean tides on Earth are much more complicated than they might at first seem from the simple physics that underlies tides. Discuss some of the factors that make the real tides so complicated and how these factors affect the tides. Consider the following factors: the distribution of land and oceans; the Moon's varying distance from Earth in its orbit; and the fact that the Moon's orbital plane is not perfectly aligned with the ecliptic and that neither the Moon's orbit nor the ecliptic is aligned with Earth's equator.

Interactive Tutorials

Tutorial Review of Key Concepts
Use the interactive **Tutorial** at **www.mastering astronomy.com** to review key concepts from this chapter.

Motion and Gravity Tutorial
Lesson 1, Lesson 2

Energy Tutorial
Lesson 1 Kinetic Energy and Gravitational Potential Energy

Orbits and Kepler's Laws Tutorial
Lessons 1–4

Supplementary Tutorial Exercises
Use the interactive **Tutorial Lessons** to explore the following questions.

Motion and Gravity, Lesson 2
1. Suppose Earth were twice as large in radius but still had the same mass. How much would you weigh? Explain.
2. Does the force of gravity between Earth and a spaceship launched from Earth ever become zero? Explain.
3. If a spaceship somehow came to a stop precisely $\frac{3}{4}$ of the way from Earth to the Moon, would it fall toward Earth, toward the Moon, or stay where it is? Explain.

Energy, Lesson 1
1. What properties of an object determine its kinetic energy? Its gravitational potential energy?
2. What condition must be met for an object launched from Earth to escape Earth's gravity?

Orbits and Kepler's Laws Tutorial, Lesson 4
1. Use the tool in Lesson 4 to predict the orbital period of an asteroid that has an orbital radius of 4 AU ($p =$ _____).
2. Use Kepler's third law, $p^2 = a^3$, to check your prediction ($a^3 =$ _____). Now take the square root of the result

($p =$ _____). Is your orbital period the same as the answer you calculated for question 1?
3. Use the tool to predict the orbital radius of an asteroid that has a period of 2 years. Does your answer agree with what you would expect from Kepler's third law?
4. Does the eccentricity of the orbit affect your answers to the last three questions?

Exploring the Sky and Solar System

Of the many activities available on the *Voyager: SkyGazer* **CD-ROM** accompanying your book, use the following files to observe key phenomena covered in this chapter.

Go to the **File: Basics** folder for the following demonstrations.
1. Planet Paths   2. Planet Orrery   3. Follow a Planet

Go to the **File: Demo** folder for the following demonstrations.
1. Earth and Venus   2. Hyakutake at Perihelion
3. Pluto's Orbit

Go to the **Explore** menu for the following demonstrations.
1. Solar System   2. Paths of the Planets

Movies

Check out the following narrated and animated short documentary available on **www.masteringastronomy.com** for a helpful review of key ideas covered in this chapter.

Orbits in the Solar System

Web Projects

1. *Space Station.* Visit a NASA site with pictures from the Space Station. Choose two photos that illustrate some facet of Newton's laws of motion or gravity. Explain how what is going on is related to Newton's laws.
2. *Energy Comparisons.* Using information from the Energy Information Administration Web site, choose some aspect of U.S. or world energy use that interests you. Write a short report on this issue.
3. *Tide Tables.* Find a tide table or tide chart for a beach town that you'd like to visit. Explain how to read the table and discuss any differences between the actual tidal pattern and the idealized tidal pattern described in this chapter.
4. *Space Elevator.* Read more about space elevators (see Problem 48) and how they might make it easier and cheaper to get to Earth orbit or beyond. Write a short report about the feasibility of building a space elevator, and briefly discuss the pros and cons of such a project.

5

Light and Matter
Reading Messages from the Cosmos

May the warp be the white light of morning,
May the weft be the red light of evening,
May the fringes be the falling rain,
May the border be the standing rainbow.
Thus weave for us a garment of brightness.

Song of the Sky Loom (Native American)

Ancient observers could discern only the most basic features of the light that they saw—features such as color and brightness. Over the past several hundred years, we have discovered that light carries far more information. Today, we can analyze the light of distant objects to learn what they are made of, how hot they are, how fast they are moving, and much more.

We are lucky that light carries so much information. Our present spacecraft can reach only objects within our own solar system, and except for an occasional meteorite falling from the sky, the cosmos does not come to us. In contrast to the limited reach of spacecraft, light travels throughout the universe, carrying its treasury of information wherever it goes. Light is truly the cosmic messenger, bringing the stories of distant objects to our home here on Earth.

In this chapter, we will focus our attention on learning how to read the messages carried to us by light from the cosmos. We'll begin with a brief look at the basic interactions of light and matter that create those messages, and then study the properties of light and matter individually and in some detail. With that background, we'll be ready to explore the formation of light spectra, so that we can understand how light can encode so much information about distant objects.

 Light and Spectroscopy Tutorial, Lesson 1

5.1 Light in Everyday Life

Look around you right now. What do you see? You may be tempted to list nearby objects, but all you're really seeing is light that has interacted with those objects. Through intuition and experience, you're able to interpret the colors and patterns of the light and turn them into information about the objects and substances that surround you.

Astronomers do much the same thing when studying the universe with telescopes. Telescopes collect the light of distant objects, and astronomers then extract information from the light. Knowing how to extract the maximum amount of information from light requires a deeper understanding of what light is and how it interacts with matter. We'll start developing that understanding by taking a closer look at our everyday experience with light.

• How do we experience light?

Even without opening your eyes, you can deduce that light is a form of energy. Outside on a hot, sunny day you can feel your skin warm as it absorbs sunlight. If you remember that greater warmth means more molecular motion, you'll realize that sunlight must be transferring its energy to the molecules in your skin. The energy that light carries is called *radiative energy*; recall that it is one of the three basic categories of energy, along with kinetic and potential energy (see Figure 4.10). But how do we measure the amount of energy that light carries?

Energy and Power We measure energy in units of joules [Section 4.3]. However, in astronomy we are often more interested in the *rate* at which light carries energy toward or away from us than in the total amount of energy it carries. After all, because light always travels through space at the speed of light, we cannot hold light in our hands in the same way that we can hold a hot potato, which has thermal energy, or a rock, which has gravitational potential energy. In science, the rate of energy flow is called **power**. The standard units of power are called **watts**. A power of 1 watt means an energy flow of 1 joule per second:

$$1 \text{ watt} = 1 \text{ joule/s}$$

Thus, for every second that you leave a 100-watt light bulb turned on, you will have to pay the utility company for 100 joules of energy. Interestingly, the power requirement of an average human—about 10 million joules per day—is about the same as that of a 100-watt light bulb.

Light and Color Everyday experience tells us that light comes in different forms that we call *colors*. You've probably seen a prism split light into the rainbow of light called a **spectrum** (Figure 5.1). The basic colors in a rainbow-like spectrum are red, orange, yellow, green, blue, and violet. We see *white* when the basic colors are mixed in roughly equal proportions. Light from the Sun or a light bulb is often called *white light*, because it contains all the colors of

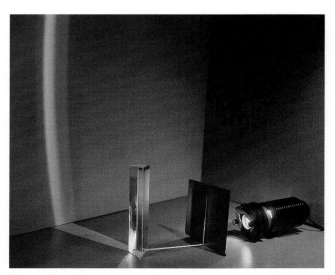

Figure 5.1 When we pass white light through a prism, it disperses into a rainbow of color that we call a *spectrum*.

Reflection (mirror): angle of incidence = angle of reflection.

50°

50°

Scattering: The screen scatters light from the projector in many directions...

screen

projection booth

...so that every person in the audience sees light from all parts of the screen.

a A mirror reflects light along a simple path: The angle at which the light strikes the mirror is the same angle at which it is reflected.

b A movie screen scatters light in many different directions, so that each member of the audience can watch the movie. The pages in a book do the same thing, which is why you can read them from different angles and distances.

Figure 5.2
Reflection and scattering.

the rainbow. *Black* is what we perceive when there is no light and hence no color.

The wide variety of all possible colors comes from mixtures of just a few colors in varying proportions. Your television takes advantage of this fact to simulate a huge range of colors by combining only red, green, and blue light; these three colors are often called the primary colors of vision, because they are the colors directly detected by cells in your eyes. Colors tend to look different on paper, so artists generally work with an alternate set of primary colors: red, yellow, and blue. If you do any graphic design work, you may be familiar with the "CMYK" process that mixes the four colors cyan, magenta, yellow, and black to produce a great variety of color; the four-color CMYK process was used to print this book.

THINK ABOUT IT

If you have a magnifying glass handy, hold it close to your TV set to see the individual red, blue, and green dots. If you don't have a magnifying glass, try splashing a few droplets of water onto your TV screen (carefully!). What do you see? What are the drops of water doing?

You can produce a spectrum with either a prism or a **diffraction grating**, which is a piece of plastic or glass etched with many closely spaced lines. If you have a compact disc or DVD handy, you can make a spectrum for yourself. The bottom of a CD or DVD is etched with many closely spaced circles and therefore acts like diffraction grating. That's why you see rainbows of color on the bottom of the disc when you hold it up to the light.

• How do light and matter interact?

If you think about the interactions between light and matter that you see in everyday life, you'll realize that light can interact with matter in four basic ways:

- **Emission:** When you turn on a lamp, electric current heats the filament of the light bulb to a point at which it *emits* visible light.
- **Absorption:** If you place your hand near a lit light bulb, your hand *absorbs* some of the light, and this absorbed energy warms your hand.
- **Transmission:** Some forms of matter, such as glass or air, *transmit* light. That is, they allow light to pass through.
- **Reflection/scattering:** Light can bounce off matter, leading to what we call *reflection* (when the bouncing is all in the same general direction) or *scattering* (when the bouncing is more random). Figure 5.2 contrasts reflection and scattering.

Materials that transmit light are said to be *transparent*, and materials that absorb light are called *opaque.* Many materials are neither perfectly transparent nor perfectly opaque. For example, dark sunglasses and clear eyeglasses are both at least partially transparent, but the dark glasses absorb more and transmit less light. In addition, many materials affect different colors of light differently. For example, red glass transmits red light but absorbs other colors, while a green lawn reflects (scatters) green light but absorbs all other colors.

Let's put these ideas together to understand what happens when you walk into a room and turn on the light switch (Figure 5.3). The light bulb begins to emit white light, which is a mix of all the colors in the spectrum. Some of this light exits the room, transmitted through the windows. The rest of the light strikes the surfaces of objects inside the room, and the material properties of each object determine the colors it absorbs or reflects. The light coming from each object therefore carries an enormous amount of information about the object's location, shape and structure, and composition. You acquire this information when light enters your eyes, where special cells in your retina absorb it and send signals to your brain. Your brain interprets the messages that light carries, recognizing materials and objects in the process we call vision.

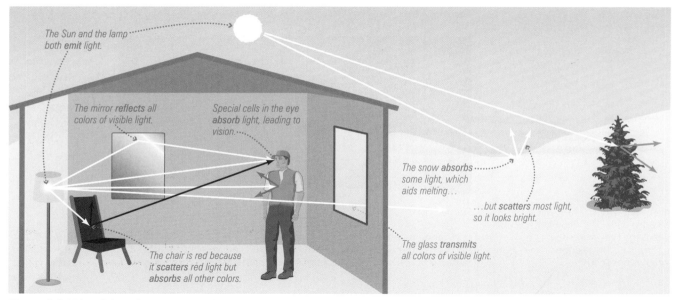

Figure 5.3 When light strikes any piece of matter in the universe, that matter reacts in one or a combination of four ways: emission, absorption, transmission, and reflection (or scattering).

In the figure:

- The Sun and the lamp both *emit* light.
- The mirror **reflects** all colors of visible light.
- Special cells in the eye **absorb** light, leading to vision...
- The snow **absorbs** some light, which aids melting...
- ...but *scatters* most light, so it looks bright.
- The glass *transmits* all colors of visible light.
- The chair is red because it *scatters* red light but **absorbs** all other colors.

All the information that light brings to Earth from the universe was encoded by one of the four basic interactions between light and matter common to our everyday experience. However, our eyes perceive only a tiny fraction of all the information contained in light. Modern instruments can break light into a much wider variety of colors and can analyze those colors in far greater detail. In order to understand how to decode that information, we need to examine the nature of light and matter more closely.

5.2 Properties of Light

Light has always been familiar to people, but its actual nature remained a mystery for most of human history. The first real insights into the nature of light came with experiments that Isaac Newton performed in the 1660s. It was already well known that passing light through a prism produced a rainbow of colors, but most people assumed that the colors were a property of the prism rather than of the light itself. Newton dispelled this belief by placing a second prism in front of the light of just one color, such as red, from the first prism. If the rainbow of color came from the prism itself, the second prism would have produced a rainbow just like the first. But it did not: When only red light entered the second prism, only red light emerged, proving that the color was a property of the light and not of the prism.

Newton's work tells us something about the nature of color, but it still does not tell us exactly what light *is*. In this section, we'll investigate modern understanding of the nature of light.

• What is light?

Newton himself guessed that light is made up of countless tiny particles. However, other scientists soon conducted experiments that demonstrated that light behaves like waves. Thus began one of the most important debates in scientific history: Is light a wave or a particle? To understand this question, and our modern answer to it, we must first understand the differences between particles and waves.

Particles and Waves in Everyday Life   Marbles, baseballs, and individual atoms and molecules are all examples of *particles*. A particle of matter can sit still or it can move from one place to another. If you throw a baseball at a wall, it moves from your hand to the wall.

Now think about what happens when you toss a pebble into a pond (Figure 5.4). The ripples moving out from the place where the pebble lands are *waves* consisting of *peaks*, where the water is higher than average, and *troughs*, where the water is lower than average. If you watch as the waves pass by a floating leaf, you'll see the leaf rise up with the peak and drop down with the trough, but the leaf itself does *not* travel across the pond's surface with the wave. This observation tells us that the particles (molecules) of water are moving up and down but are not flowing outward. That is, the waves carry *energy* outward from the place where the pebble landed but do not carry matter along with them. In essence, a particle is a *thing*, while a wave is a *pattern* revealed by its interaction with particles.

For our purposes in this book, we can characterize the waves on the pond by three basic properties: wavelength, frequency, and speed.* Their **wavelength** is the distance from one peak to the next (or one trough to the next). Their **frequency** is the number of peaks passing by any point each second. For example, if we see the leaf bobbing up and down three times each second, we know that three peaks

---

*A fourth characteristic of a wave is its amplitude, defined as half the height from trough to peak. Amplitude is related to the brightness of light, but we will not use it in our study of light in this book.

Wavelength is the distance from one peak to the next (or one trough to the next).

pebble thrown into pond

trough

peak

speed of wave moving outward

Leaf bobs up and down with frequency of wave.

Figure 5.4 Interactive Figure↘ Tossing a pebble into a pond generates waves. The waves carry energy outward, but matter does *not* travel outward. Matter, such as a floating leaf or the water itself, bobs up and down and sloshes back and forth as the waves pass by.

must be passing by the leaf each second. We then say that the waves have a frequency of three **cycles per second** (referring to the up-and-down "cycles" of the passing waves). Cycles per second often are called **hertz (Hz)**, so we can also describe this frequency as 3 Hz. The **speed** of the waves tells us how fast their peaks travel across the pond. Because the waves carry energy, the speed essentially tells us how fast the energy travels from one place to another.

A simple formula relates the wavelength, frequency, and speed of any wave. Suppose a wave has a wavelength of 1 centimeter and a frequency of 3 hertz. The wavelength tells us that each time a peak passes by, the wave peak has traveled 1 centimeter. The frequency tells us that three peaks pass by each second. Thus, the speed of the wave must be 3 centimeters per second. If you try a few more similar examples, you'll find the general rule:

$$\text{wavelength} \times \text{frequency} = \text{speed}$$

Light as an Electromagnetic Wave You've probably heard that light is a wave, but it isn't quite like the waves we see in everyday life. More familiar waves always move through some form of matter. For example, the waves on the pond move through the water, causing particles (molecules) of water to vibrate up and down and slosh back and forth, while sound waves move through air and cause air molecules to vibrate back and forth. The vibrations of matter allow the waves to transmit energy from one place to another, even though particles of matter do not travel along with the waves. In contrast to these everyday examples of waves, we do not see anything move up and down when light travels through space. So what, exactly, is "waving" when a light wave passes by?

The answer is what scientists call electric and magnetic fields. The concept of a **field** is a bit abstract. Fields associated with forces, such as electric and magnetic fields, describe the strength of the force that any particle would experience at any point in space. For example, recall that the force Earth exerts on an orbiting satellite depends on the mass of Earth, the mass of the satellite, and the distance of the satellite from Earth's center [Section 4.4]. We can therefore characterize the force that *any* object would experience at the satellite's distance by saying that Earth creates a *gravitational field* equal to the force on the satellite divided by the satellite's mass. In other words, Earth's gravitational field at any point in space depends only on Earth's mass and the distance of that point from Earth's center. Likewise, we can describe the forces that charged particles exert on one another in terms of *electric fields* and *magnetic fields*.

Light waves are vibrations of both electric and magnetic fields caused by the motions of charged particles. We therefore say that light is an **electromagnetic wave.** Just as the ripples on a pond will cause a leaf to bob up and down, the vibrations of the electric field in an electromagnetic wave will cause an electron to bob up and down. If you could set up electrons in a row, they would wriggle like a snake as light passed by (Figure 5.5a). The distance between peaks in this row of electrons would tell us the wavelength of the light wave, while the number of times each electron bobs up and down would tell us the frequency (Figure 5.5b).

All light travels through empty space at the same speed— the **speed of light** (often represented by the letter *c*), which is about 300,000 kilometers per second. Because the speed of any wave is its wavelength times its frequency, we find a very important relationship between wavelength and frequency for light: *The longer the wavelength, the lower the frequency, and vice versa.* For example, light waves with a wavelength of 1 centimeter must have half the frequency

If you could line up electrons, they would bob up and down with the vibrating electric field of a passing light wave.

a Electrons move when light passes by, showing that light carries a vibrating electric field.

Wavelength is the distance between adjacent peaks of the electric (and magnetic) field . . .

. . . while frequency is the number of times each second that the electric (and magnetic) field vibrates up and down (or side to side) at any point.

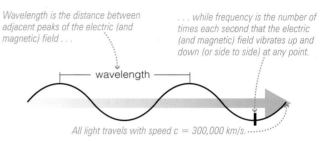

wavelength

All light travels with speed c = 300,000 km/s.

b The vibrations of the electric field determine the wavelength and frequency of a light wave. Light also carries a magnetic field (not shown) that vibrates perpendicular to the direction of the electric field vibrations.

Figure 5.5 Interactive Figure↘ Light is an electromagnetic wave.

If you go to the store to buy a pair of sunglasses, you'll face a dizzying array of choices. Sunglasses come in different styles, different tints, and with different efficiencies in blocking ultraviolet and infrared light. Most of these choices should at least make sense to you (well, perhaps not all of the styles), but one option may be less familiar: The labels on some sunglasses say that they are "polarized." What does this mean? The term comes from a property of light, called *polarization*, that has to do with the direction in which a light wave vibrates and how those vibrations change when light bounces off or passes through matter. Polarization turns out to be important not only to your sunglasses but also to astronomy.

It's easiest to understand the idea of polarization if you think about how waves move on a string when you shake one end of it. The string vibrates either up and down or back and forth while the wave itself moves along the string, perpendicular to the direction of vibration. Light waves move in a similar way, with the electric and magnetic fields vibrating either up and down or side to side compared with the direction of travel. For example, the wave shown in Figure 5.5b is moving to the right on the page while its electric field vibrates up and down on the page.

The direction of vibration affects the way light interacts with matter. As Figure 5.5a indicates, an electric field that vibrates up and down will make electrons move up and down as the wave passes by. Thus, the direction in which the electric field vibrates determines the direction in which charged particles will vibrate as the wave passes by. Because the direction of wave vibration matters, we give it a name: the *polarization* of the wave. In other words, an individual wave moving toward you can be polarized either with its vibrations up and down or side to side, or some direction in between those two. (In all cases, the polarization is perpendicular to the direction of travel.)

We rarely notice polarization in everyday life because our eyes are not equipped to detect this property of light. Each of the light waves entering our eyes (or, more technically, each individual photon) has a certain direction of polarization. If all the waves taken together have no preferential direction of vibration, we say that the light is *unpolarized*. However, some physical processes can produce waves with a particular direction of polarization—which is where your sunglasses and astronomy come in.

When light reflects off a flat, horizontal surface like the ground or a lake, the surface tends to reflect light in which the electric field is vibrating along (parallel to) the surface and to transmit or absorb light with electric fields vibrating in other directions. As a result, the reflected light becomes polarized, with its electric field vibrating horizontally. Polarized sunglasses are designed to block out horizontally polarized light, which means they block the light reflected from horizontal surfaces—which is otherwise known as "glare." Of course, the polarized glasses work only if you are wearing them horizontally as well; if you turn a pair of polarized sunglasses so that the two lenses are no longer horizontal to the ground, they will not block glare effectively.

In astronomy, we aren't worried about glare from distant objects, but if we learn that a light source is producing polarized light, it tells us something about the nature of the source. For example, light that passes through clouds of interstellar dust tends to be polarized, telling us that the dust grains in the cloud must be preferentially absorbing light with electric fields vibrating in a particular direction. More detailed analysis has taught us that the polarization arises because the microscopic dust grains have an elongated shape and all tend to be aligned in the same way due to magnetic fields within the clouds. Polarization arises in many other astronomical contexts as well, including the study of the leftover radiation from the Big Bang. However, while polarization has provided important insights into many astronomical processes, its analysis can be fairly technical, and we will not discuss it further in this book.

---

of light waves with a wavelength of $\frac{1}{2}$ centimeter, and one-fourth the frequency of light waves with a wavelength of $\frac{1}{4}$ centimeter (Figure 5.6).

Photons: "Particles" of Light In our everyday lives, waves and particles appear distinctly different: No one would confuse the ripples on a pond with a baseball. However, experiments have shown that light behaves as *both* a wave and a particle. We therefore say that light comes in individual "pieces," called **photons**, that have properties of both particles and waves. Like baseballs, photons of light can be counted individually and can hit a wall one at a time. Like waves, each photon is characterized by a wavelength and a frequency. The idea that light can be both a wave and a particle may seem quite strange, but it is fundamental to our modern understanding of physics. (We will discuss some of the implications of this "wave-particle duality" in Chapter S4.)

Just as a moving baseball carries a specific amount of kinetic energy, each photon of light carries a specific amount of radiative energy. The shorter the wavelength of the light (or, equivalently, the higher its frequency), the higher the energy of the photons.

To sum up, our modern understanding maintains that light is both a particle and a wave, an idea we describe by saying that light consists of individual photons character-

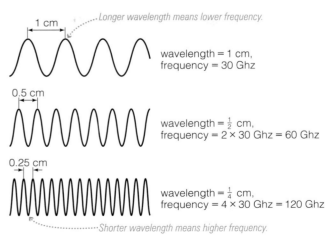

Figure 5.6 Because all light travels through space at the same speed, light of longer wavelength must have lower frequency, and vice versa. (GHz stands for gigahertz, or 10^9 hertz.)

ized by wavelength, frequency, and energy. The wavelength, frequency, and energy are related in a very specific way so that all photons of light travel through space at the same speed—the speed of light.

THINK ABOUT IT

If you assume that each of the three waves shown in Figure 5.6 represents a photon of light, which one has the most energy? Which one has the least energy? Explain.

• What is the electromagnetic spectrum?

Newton's experiments proved that white light is actually a mix of all the colors in the rainbow. Later scientists found that there is light "beyond the rainbow" as well. Just as there are sounds that our ears cannot hear (such as the sound of a dog whistle), there is light that our eyes cannot see. In fact, the spectrum of **visible light** that splits into the rainbow of color is only a tiny part of the complete range of light's wavelengths. Visible light differs from other forms of light only in the wavelength and frequency of the photons.

Because we sometimes describe light as an electromagnetic wave, the complete spectrum of light is usually called the **electromagnetic spectrum**. Light itself is often called **electromagnetic radiation**. Figure 5.7 shows the complete electromagnetic spectrum.

Notice that we give names to various portions of the electromagnetic spectrum. Visible light has wavelengths ranging from about 400 nanometers at the blue or violet end of the rainbow to about 700 nanometers at the red end. (A nanometer [nm] is a billionth of a meter.) Light with wavelengths somewhat longer than red light is called **infrared**, because it lies beyond the red end of the rainbow. **Radio waves** are the longest-wavelength light. Thus, radio waves are a form of light, *not* a form of sound. The region near the border between infrared and radio waves, where

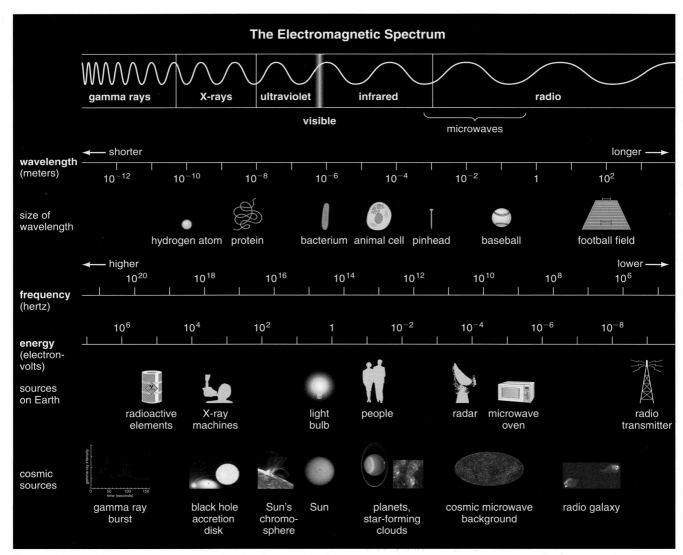

Figure 5.7 The electromagnetic spectrum. Notice that wavelength increases as we go from gamma rays to radio waves, while frequency and energy increase in the opposite direction.

wavelengths range from micrometers to millimeters, is sometimes given the name **microwaves**.

On the other side of the spectrum, light with wavelengths somewhat shorter than blue light is called **ultraviolet**, because it lies beyond the blue (or violet) end of the rainbow. Light with even shorter wavelengths is called **X rays**, and

the shortest-wavelength light is called **gamma rays**. Notice that visible light is an extremely small part of the entire electromagnetic spectrum: The reddest red that our eyes can see has only about twice the wavelength of the bluest blue, but the radio waves from your favorite radio station are a billion times longer than the X rays used in a doctor's office.

MATHEMATICAL INSIGHT **5.1** **Wavelength, Frequency, and Energy**

The speed of any wave is the product of its wavelength and its frequency. Because all forms of light travel at the same speed (in a vacuum) of $c = 3 \times 10^8$ m/s, we can write:

$$\lambda \times f = c$$

where λ (the Greek letter *lambda*) stands for wavelength and f stands for frequency. This formula is simple but revealing: Because the speed c is constant, the formula tells us that frequency must go up when wavelength goes down, and vice versa.

The formula for the radiative energy (E) carried by a photon of light is:

$$E = h \times f$$

where h is *Planck's constant* ($h = 6.626 \times 10^{-34}$ joule $\times$ s). Thus, energy increases in proportion to the frequency of the photon.

Example 1: The numbers on a radio dial for FM radio stations are the frequencies of radio waves in megahertz (MHz), or millions of hertz. If your favorite radio station is "93.3 on your dial," it broadcasts radio waves with a frequency of 93.3 million cycles per second. What is the wavelength of these radio waves?

Solution:

Step 1 Understand: Radio waves are a form of light, so we know they obey the relationship

$$\lambda \times f = c$$

The speed of light is always the same and, in this case, we are given the frequency ($f = 93.3$ MHz). Thus we simply solve for the wavelength (λ).

Step 2 Solve: We solve the above equation for wavelength by dividing both sides by the frequency (f), which gives us:

$$\lambda = \frac{c}{f}$$

Now we plug in the speed of light (3×10^8 m/s) and the frequency. Remember that 1 MHz = 10^6 Hz, so the frequency in this case is 93.3×10^6 Hz. Also remember that hertz (Hz) is an abbreviation for "cycles per second"; the "cycles" have no units, so the actual units of hertz are simply "per second," equivalent to 1/s. Thus, we find:

$$\lambda = \frac{c}{f} = \frac{3 \times 10^8 \frac{m}{s}}{93.3 \times 10^6 \frac{1}{s}} = 3.2 \text{ m}$$

Step 3 Explain: We have found that radio waves with a frequency of 93.3 MHz have a wavelength of 3.2 meters. That is why radio towers are so large—they need to be longer than the size of the waves they are transmitting.

Example 2: The wavelength of green visible light (in the middle of the visible spectrum) is about 550 nanometers (1 nm = 10^{-9} m). What is the frequency of this light?

Solution:

Step 1 Understand: Again, we know that visible light obeys the relation $\lambda \times f = c$. This time we are given the wavelength ($\lambda = 550$ nm $= 550 \times 10^{-9}$ m), so we need to solve the equation for the frequency.

Step 2 Solve: We solve the equation ($\lambda \times f = c$) for frequency by dividing both sides by the wavelength (λ), which gives us:

$$f = \frac{c}{\lambda}$$

Now we plug in the speed of light (3×10^8 m/s) and the wavelength ($\lambda = 550 \times 10^{-9}$ m) to find:

$$f = \frac{c}{\lambda} = \frac{3 \times 10^8 \frac{m}{s}}{550 \times 10^{-9} \text{ m}} = 5.45 \times 10^{14} \frac{1}{s}$$

Step 3 Explain: Our answer is about 5.5×10^{14} 1/s; remembering that the units 1/s are called hertz, this answer is equivalent to 5.5×10^{14} Hz, or 550 trillion Hz. Thus, green visible light has a frequency of about 550 trillion Hz. Notice that this frequency is extremely high, which explains in part why it took so long for humans to realize that light behaves like a wave.

Example 3: What is the energy of a visible-light photon with wavelength 550 nanometers?

Solution:

Step 1 Understand: The energy of a photon is given by the formula $E = h \times f$. From Example 2, we also know that $f = c/\lambda$. Thus, we can write the energy formula as:

$$E = h \times f = h \times \frac{c}{\lambda}$$

Step 2 Solve: We plug in the wavelength ($\lambda = 550 \times 10^{-9}$ m) and Planck's constant ($h = 6.626 \times 10^{-34}$ joule $\times$ s) to find:

$$E = h \times \frac{c}{\lambda} = (6.626 \times 10^{-34} \text{ joule} \times \text{s}) \times \frac{3 \times 10^8 \frac{m}{s}}{550 \times 10^{-9} \text{ m}}$$
$$= 3.6 \times 10^{-19} \text{ joule}$$

Step 3 Explain: The energy of a single visible-light photon is about 3.6×10^{-19} joule. Note that this energy for a single photon is extremely small compared to, say, the energy of 100 joules used each second by a 100-watt light bulb.

Can You Hear Radio Waves?

Most people associate the term *radio* with sound, but radio waves are a form of *light* with long wavelengths—too long for our eyes to see. Radio stations encode sounds (such as voices and music) as electrical signals and broadcast the signals as radio waves. What we call "a radio" in daily life is an electronic device that receives these radio waves and decodes them to re-create the sounds played at the radio station. Televisions, cell phones, and other wireless devices also work by encoding and decoding information in the form of light called radio waves. Moreover, there's nothing special about encoding sounds with radio waves—any form of light could work in principle. Indeed, fiber-optic cables transmit sounds and data by encoding them into infrared or visible light. Thus, we cannot actually hear radio waves; we hear only the decoded signals that they carry.

What Could You See with X-Ray Vision?

The idea of having "X-ray vision" that would allow you to see through walls is part of our Hollywood culture, and dates back to old Superman comics. But is it really possible? At first, the idea might seem reasonable; after all, you've probably seen a doctor or dentist holding up "X rays" that allow you to see into your bones or teeth. However, the "X rays" that the doctor or dentist holds up are not a form of light; they are just pieces of film. The images of your bones and teeth are made with the help of a special machine that works somewhat like the flash on an ordinary camera but emits X rays instead of visible light. This machine flashes the X rays at you, and a piece of photographic film (or an electronic detector) records the X rays that are transmitted through your body. Thus, what you see is the image left by these X rays; you cannot see the X rays themselves, because they have wavelengths far too short for human eyes to see. If you think about this process, you'll realize that X-ray vision would do you no good, even if it were possible. People, walls, and other ordinary objects do not emit any X rays of their own, so there'd be nothing to see with your X-ray vision.

The different energies of different forms of light explain many familiar effects in everyday life. Radio waves carry so little energy that they have no noticeable effect on our bodies. However, radio waves can make electrons move up and down in an antenna, which is how the antenna of your car radio receives the radio waves coming from a radio station. Molecules moving around in a warm object emit infrared light, which is why we sometimes associate infrared light with heat. Receptors in our eyes respond to visible-light photons, making vision possible. Ultraviolet photons carry enough energy to harm cells in our skin, causing sunburn or skin cancer. X-ray photons have enough energy to penetrate through skin and muscle but can be blocked by bones or teeth. That is why doctors and dentists can see our underlying bone structures on photographs taken with X-ray light.

Just as different colors of visible light may be absorbed or reflected differently by the objects we see (see Figure 5.3), the various portions of the electromagnetic spectrum may interact in very different ways with matter. For example, a brick wall is opaque to visible light but transmits radio waves, which is why we can listen to radios inside buildings. Similarly, glass that is transparent to visible light may be opaque to ultraviolet light. In general, certain types of matter tend to interact more strongly with certain types of light, so each type of light carries different information about distant objects in the universe. Astronomers therefore seek to observe light of all wavelengths, using telescopes adapted to detecting each form of light, from radio to gamma rays [Section 6.4].

5.3 Properties of Matter

Light carries information about matter across the universe, but we are usually more interested in the matter with which the light interacts than we are in the light itself. Planets, stars, and galaxies are made of matter, and we must understand the nature of matter if we are to decode the messages we receive in light.

• What is the structure of matter?

Like the nature of light, the nature of matter remained mysterious for most of human history. Nevertheless, ancient philosophers came up with some ideas that are still with us today.

The ancient Greek philosopher Democritus (c. 470–380 B.C.) wondered what would happen if we broke a piece of matter, such as a rock, into ever smaller pieces. Democritus believed that the rock would eventually break into particles so small that nothing smaller could be possible. He called these particles *atoms*, a Greek term meaning "indivisible." Building upon the beliefs of earlier Greek philosophers, Democritus thought that all materials were composed of just four basic *elements:* fire, water, earth, and air. He proposed that the different properties of the elements could be explained by the physical characteristics of their atoms. For example, Democritus suggested that atoms of water were smooth and round, so water flowed and had no fixed shape, while burns were painful because atoms of fire were thorny. He also imagined atoms of earth to be rough and jagged, so that they could fit together like pieces of a three-dimensional jigsaw puzzle, and he used this idea to suggest that the universe began as a chaotic mix of atoms that slowly clumped together to form Earth.

Is Radiation Dangerous?

Many people associate the word *radiation* with danger. However, the word *radiate* simply means "to spread out from a center" (note the similarity between *radiation* and *radius* [of a circle]). Radiation is energy being carried through space. If energy is being carried by particles of matter, such as protons or neutrons, we call it *particle radiation*. If energy is being carried by light, we call it *electromagnetic radiation*.

High-energy forms of radiation, such as particles from radioactive materials or X rays, are dangerous because they can penetrate body tissues and cause cell damage. Low-energy forms of radiation, such as radio waves, are usually harmless. The visible-light radiation from the Sun is necessary to life on Earth. Thus, while some forms of radiation are dangerous, others are harmless or beneficial.

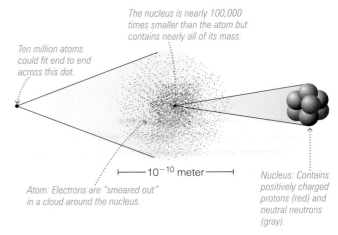

The nucleus is nearly 100,000 times smaller than the atom but contains nearly all of its mass.

Ten million atoms could fit end to end across this dot.

|—— 10^{-10} meter ——|

Atom: Electrons are "smeared out" in a cloud around the nucleus.

Nucleus: Contains positively charged protons (red) and neutral neutrons (gray).

Figure 5.8 The structure of a typical atom. Notice that atoms are extremely tiny: The atom shown in the middle is magnified to about 1 billion times its actual size, and the nucleus on the right is magnified to about 100 trillion times its actual size.

Although Democritus was wrong about there being only four types of atoms and about their specific properties, he was on the right track. All ordinary matter is indeed composed of **atoms,** and the properties of ordinary matter depend on the physical characteristics of their atoms. However, by modern definition, atoms are *not* indivisible because they are composed of even smaller particles.

Atoms come in different types, and each type corresponds to a different chemical **element.** Today, we have identified more than 100 chemical elements, and fire, water, earth, and air are *not* among them. Some of the most familiar chemical elements are hydrogen, helium, carbon, oxygen, silicon, iron, gold, silver, lead, and uranium. Appendix D gives the periodic table of all the elements.

Atomic Structure   Each chemical element consists of a different type of atom, and atoms are in turn made of particles that we call **protons, neutrons,** and **electrons** (Figure 5.8). Protons and neutrons are found in the tiny **nucleus** at the center of the atom. The rest of the atom's volume contains electrons, which surround the nucleus. Although the nucleus is very small compared to the atom as a whole, it contains most of the atom's mass, because protons and neutrons are each about 2,000 times as massive as an electron. Note that atoms are incredibly small: Millions could fit end to end across the period at the end of this sentence. The number of atoms in a single drop of water (typically, 10^{22} to 10^{23} atoms) may exceed the number of stars in the observable universe.

The properties of an atom depend mainly on the **electrical charge** in its nucleus; an object's electric charge is a measure of how strongly it will interact with electromagnetic fields. We define the electrical charge of a proton as the basic unit of positive charge, which we write as $+1$. The electron

has an electrical charge that is precisely opposite that of a proton, so we say it has negative charge (-1). Neutrons are electrically neutral, meaning that they have no charge.

Oppositely charged particles attract one another, and similarly charged particles repel. An atom is held together by the attraction between the positively charged protons in its nucleus and the negatively charged electrons that surround the nucleus. Ordinary atoms have identical numbers of electrons and protons, making them electrically neutral overall.*

Although we can think of electrons as tiny particles, they are not quite like tiny grains of sand and they don't orbit the nucleus the way planets orbit the Sun. Instead, the electrons in an atom form a kind of "smeared out" cloud that surrounds the nucleus and gives the atom its apparent size. The electrons aren't really cloudy, but it is impossible to pinpoint their positions in the atom.

In Figure 5.8, you can see that the electrons give the atom a size far larger than its nucleus even though they represent only a tiny portion of the atom's mass. If you imagine an atom on a scale that makes its nucleus the size of your fist, its electron cloud would be many miles wide.

Atomic Terminology   Astronomical objects contain many different types of atoms, and it's important to have a way to describe the atoms. You've probably learned the basic terminology of atoms in past science classes, but let's review it just to be sure.

Each different chemical element contains a different number of protons in its nucleus. This number is its **atomic number.** For example, a hydrogen nucleus contains just one proton, so its atomic number is 1. A helium nucleus contains two protons, so its atomic number is 2.

---

*You may wonder why electrical repulsion doesn't cause the positively charged protons in a nucleus to fly apart from one another. The answer is that an even stronger force, called the *strong force*, overcomes electrical repulsion and holds the nucleus together [Section S4.2].

The Illusion of Solidity

Bang your hand on a table. Although the table feels solid, it is made almost entirely of empty space! Nearly all the mass of the table is contained in the nuclei of its atoms. But the volume of an atom is more than a trillion times the volume of its nucleus, so relatively speaking, the nuclei of adjacent atoms are nowhere near to touching one another. The solidity of the table comes from a combination of electrical interactions between the charged particles in its atoms and the strange quantum laws governing the behavior of electrons. If we could somehow pack all the table's nuclei together, the table's mass would fit into a microscopic speck. Although we cannot pack matter together in this way, nature can and does—in *neutron stars*, which we will study in Chapter 18.

The *combined* number of protons and neutrons in an atom is called its **atomic mass number.** The atomic mass number of ordinary hydrogen is 1 because its nucleus is just a single proton. Helium usually has two neutrons in addition to its two protons, giving it an atomic mass number of 4. Carbon usually has six protons and six neutrons, giving it an atomic mass number of 12.

Every atom of a given element contains exactly the same number of protons, but the number of neutrons can vary. For example, all carbon atoms have six protons, but they may have six, seven, or eight neutrons. Versions of an element with different numbers of neutrons are called **isotopes** of the element. Isotopes are named by listing their element name and atomic mass number. For example, the most common isotope of carbon has 6 protons and 6 neutrons, giving it atomic mass number $6 + 6 = 12$, so we call it carbon-12. The other isotopes of carbon are carbon-13 (six protons and seven neutrons give it atomic mass number 13) and carbon-14 (six protons and eight neutrons give it atomic mass number 14). We can also write the atomic mass number of an isotope as a superscript to the left of the element symbol: ^{12}C, ^{13}C, ^{14}C. We read ^{12}C as "carbon-12." Figure 5.9 summarizes basic atomic terminology.

The symbol ^{4}He represents helium with an atomic mass number of 4. ^{4}He is the most common form of helium, containing two protons and two neutrons. What does the symbol ^{3}He represent?

Molecules The number of different material substances is far greater than the number of chemical elements because atoms can combine to form **molecules.** Some molecules consist of two or more atoms of the same element. For example, we breathe O_2, oxygen molecules made of two oxy-

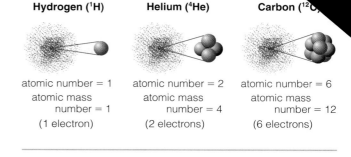

atomic number = number of protons
atomic mass number = number of protons + neutrons
(A neutral atom has the same number of electrons as protons)

Hydrogen (^{1}H)   **Helium (^{4}He)**   **Carbon (^{12}C)**

atomic number = 1
atomic mass number = 1
(1 electron)

atomic number = 2
atomic mass number = 4
(2 electrons)

atomic number = 6
atomic mass number = 12
(6 electrons)

*Different **isotopes** of a given element contain the same number of protons, but different numbers of neutrons.*

Isotopes of Carbon

carbon-12   carbon-13   carbon-14

^{12}C
(6 protons
+ 6 neutrons)

^{13}C
(6 protons
+ 7 neutrons)

^{14}C
(6 protons
+ 8 neutrons)

Figure 5.9 Terminology of atoms.

gen atoms. Other molecules, such as water, are made up of atoms of two or more different elements. The symbol H_2O tells us that a water molecule contains two hydrogen atoms and one oxygen atom. Substances composed of molecules with two or more different types of atoms are called **compounds.** Thus, water is a compound.

The chemical properties of a molecule are different from those of its individual atoms. For example, molecular oxygen (O_2) behaves very differently from atomic oxygen (O), and water behaves very differently from pure hydrogen or pure oxygen.

• What are the phases of matter?

Interactions between light and matter also depend on the physical state of the matter. Everyday experience tells us that a substance can behave dramatically differently in different *phases*, even though it is still made of the same atoms or molecules. For example, molecules of H_2O can exist in three familiar phases: as **solid** ice, as **liquid** water, and as the **gas** we call water vapor. How can the same molecules look and act so differently in these different phases?

You are probably familiar with the idea of a **chemical bond**, the name we give to the interactions between electrons that hold the atoms in a molecule together. For example, we say that chemical bonds hold the hydrogen and oxygen atoms together in a molecule of H_2O. Similar but much weaker interactions among electrons hold together the many water molecules in a block of ice or a pool of water. Thus, we can think of the interactions that keep

...molecules close together as other

...of bonds, the phases of solid, liquid, ...rength of the bonds between neigh- ...ecules. Phase changes occur when ...oken and replaced by another. Changes ...emperature (or both) can cause phase ...r to think first about temperature: As a substance ...ted, the average kinetic energy of its particles increases, enabling the particles to break the bonds holding them to their neighbors.

Phase Changes in Water Water is the only familiar substance that we see in all three phases (solid, liquid, gas) in our everyday lives. Let's therefore take water as an example of what happens as we heat a substance and observe the phase changes.

At low temperatures (below 0°C at sea level on Earth), water molecules have a relatively low average kinetic energy. Each molecule is therefore bound tightly to its neighbors, making the *solid* structure of ice. As long as the temperature remains below freezing, the water molecules in ice remain rigidly held together. However, the molecules within this rigid structure are always vibrating, and higher temperature means greater vibrations. Thus, if we start with ice at a very low temperature, the molecular vibrations will grow gradually stronger as the temperature rises toward the melting point of 0°C.

The melting point (0°C at ordinary pressure) is the temperature at which the molecules finally have enough energy to break the solid bonds of ice. The molecules can then move much more freely among one another, allowing the water to flow as a *liquid*. However, the molecules in liquid water are not completely free of one another, as we can tell from the fact that droplets of water can stay intact. Thus, adjacent molecules in liquid water are still held together by a type of bond, though a much looser bond than the one that holds them together in solid ice.

If we continue to heat the water, the increasing kinetic energy of the molecules will ultimately break the bonds between neighboring molecules altogether. The molecules will then be able to move freely, and freely moving particles constitute what we call a *gas*. Above the boiling point (100°C at sea level), all the bonds between adjacent molecules are broken so that the water can exist only as a gas.

We see ice melting into liquid water and liquid water boiling into gas so often that it's tempting to think that's the end of the story. However, a little thought should convince you that the reality has to be more complex. For example, you know that Earth's atmosphere contains water vapor that condenses to form clouds and rain. But Earth's temperature is well below the boiling point of water (luckily for us!), so how is it that our atmosphere can contain water in the gas phase?

You'll understand the answer if you remember that temperature is a measure of the *average* kinetic energy of the particles in a substance [Section 4.3]; individual particles may have substantially lower or higher energies than the average. Thus, even at the low temperatures at which most water molecules are bound together as ice or liquid, a few molecules will always have enough energy to break free of their neighbors and enter the gas phase. In other words, some gas (water vapor) is always present along with solid ice or liquid water. The process by which molecules escape from a solid is called **sublimation**, and the process by which molecules escape from a liquid is called **evaporation**. Higher temperatures lead to higher rates of sublimation or evaporation.

THINK ABOUT IT

Global warming is expected to increase Earth's average temperature by up to a few degrees over the coming century. Based on what you've learned about phase changes, how would you expect global warming to affect the total amount of cloud cover on Earth? Explain.

Molecular Dissociation and Ionization The state of matter changes even further at higher temperatures. To see what happens, let's return to the example of heating the water. Above the boiling point, all the water will have entered the gas phase. What happens if we continue to raise the temperature?

The molecules in a gas move freely, but they often collide with one another. As the temperature rises, the molecules move faster and the collisions become more violent. At high enough temperatures, the collisions will be so violent that they can break the chemical bonds holding individual water molecules together. The molecules then split into pieces, a process we call **molecular dissociation**. (In the case of water, molecular dissociation usually breaks one hydrogen atom away from a remaining molecule that

---

COMMON MISCONCEPTIONS

One Phase at a Time?

In daily life, we usually think of H_2O as being in the phase of either solid ice, liquid water, or water vapor, with the phase depending on the temperature. In reality, two or even all three phases can exist at the same time. In particular, some sublimation *always* occurs over solid ice, and some evaporation *always* occurs over liquid water. You can tell that evaporation always occurs, because an uncovered glass of water will gradually empty as the liquid evaporates into gas. You can see that sublimation occurs by watching the snow pack after a winter storm: Even if the snow doesn't melt into liquid, it will gradually disappear as some of the ice sublimates into water vapor.

consists of one hydrogen atom and one oxygen atom [OH]; at even higher temperatures, the OH dissociates into individual atoms.)

At still higher temperatures, collisions can break the bonds holding electrons around the nuclei of individual atoms, allowing the electrons to go free. The loss of one or more negatively charged electrons leaves the remaining atom with a net positive charge. Such charged atoms are called **ions**. The process of stripping electrons from atoms is called **ionization**.

Thus, at temperatures of several thousand degrees, what once was water becomes a hot gas consisting of freely moving electrons and positively charged ions of hydrogen and oxygen. This type of hot gas, in which atoms have become ionized, is called a **plasma**. Because a plasma contains many charged particles, its interactions with light are different from those of a gas consisting of neutral atoms, which is one reason that plasma is sometimes referred to as "the fourth phase of matter." However, because the electrons and ions are not bound to one another, it is also legitimate to call plasma a gas. That is why we sometimes say that the Sun is made of hot gas and sometimes say that it is made of plasma; both statements are correct.

The degree of ionization in a plasma depends on its temperature and composition. A neutral hydrogen atom contains only one electron, so hydrogen can be ionized only once; the remaining hydrogen ion, designated H^+, is simply a proton. Oxygen, with atomic number 8, has eight electrons when it is neutral, so it can be ionized multiple times. *Singly ionized* oxygen is missing one electron, so it has a charge of +1 and is designated O^+. *Doubly ionized* oxygen, or O^{++}, is missing two electrons; *triply ionized* oxygen, or O^{+3}, is missing three electrons; and so on. At temperatures of several million degrees, oxygen can be *fully ionized*, in which case all eight electrons are stripped away and the remaining ion has a charge of +8.

Figure 5.10 summarizes the changes that occur as we heat water from ice to a fully ionized plasma. Other chemical substances go through similar phase changes, but the changes generally occur at different temperatures for different substances.

Phases and Pressure Temperature is the primary factor determining the phase of a substance and the ways in which light interacts with it, but pressure also plays a role. You're undoubtedly familiar with the idea of pressure in an everyday sense: For example, you can put more pressure on your arm by squeezing it. In science, we use a more precise definition: **pressure** is the *force per unit area* pushing on an object's surface. You feel more pressure when you squeeze your arm because squeezing increases the force on each square centimeter of your arm's surface. Similarly, piling rocks on a table increases the weight (force) on the table, which therefore increases the pressure on the surface of the table; if the pressure becomes too great, the table breaks. The gas in an atmosphere also creates pressure, because the

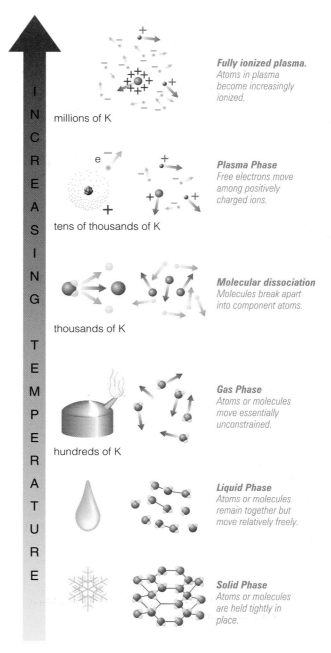

Figure 5.10 Interactive Figure. The general progression of phase changes in water.

weight of the gas bears down on everything beneath it. For example, at sea level on Earth, the weight of the atmosphere creates a pressure of about 14.7 pounds per square inch. That is, the total weight of all the air above each square inch of Earth's surface is about 14.7 pounds [Section 10.1].

Pressure can affect phases in a variety of ways. For example, deep inside Earth, the pressure is so high that Earth's inner metal core remains solid, even though the temperature is high enough that the metal would melt into liquid under less extreme pressure conditions [Section 9.1]. On

Atmosphere

The pressure of water vapor determines how frequently water molecules return to the ocean by colliding with its surface.

At the same time, some water molecules are always evaporating from the ocean.

Ocean

Figure 5.11 Evaporation of water molecules from the ocean is balanced in part by molecules of water vapor in Earth's atmosphere returning to the ocean. The rate at which these molecules return is directly related to the pressure created by water vapor in the atmosphere.

a planetary surface, atmospheric pressure can determine whether water is stable in liquid form.

Remember that liquid water is always evaporating (or ice sublimating) at a low level, because a few molecules randomly get enough energy to break the bonds holding them to their neighbors. On Earth, enough liquid water has evaporated from the oceans to make water vapor an important ingredient of our atmosphere. Some of these atmospheric water vapor molecules collide with the ocean surface, where they can "stick" and rejoin the ocean—essentially the opposite of evaporation (Figure 5.11). The greater the pressure created by water vapor molecules in our atmosphere,* the higher the rate at which water molecules return to the ocean. This direct return of water vapor molecules from the atmosphere helps keep the total amount of water in Earth's oceans fairly stable.** On the Moon, where the lack of atmosphere means no pressure from water vapor at all, liquid water would evaporate quite quickly (as long as the temperature were high enough that it did not freeze first). The same is true on Mars, because the atmosphere lacks enough water vapor to balance the rate of evaporation.

High pressure can also cause gases to dissolve in liquid water. For example, sodas are made by putting water in contact with high-pressure carbon dioxide gas. Because of the high pressure, many more carbon dioxide molecules enter the water than are released, so the water becomes "carbonated"—that is, it has a lot of dissolved carbon dioxide. When you open a bottle of carbonated water, exposing it to air with ordinary pressure, the dissolved carbon dioxide quickly bubbles up and escapes.

---

*Technically, this is known as the *vapor pressure* of water in the atmosphere. We can also measure vapor pressure for other atmospheric constituents, and the total gas pressure is the sum of all the individual vapor pressures.
**Rain and snow also contribute, of course; however, even if Earth's temperature rose enough so that raindrops and snowflakes could no longer form, only a small fraction of Earth's ocean water would evaporate before the return rate of water vapor molecules balanced the evaporation rate.

• How is energy stored in atoms?

Now that we have reviewed the structure of matter and how its phase depends on temperature and pressure, it is time to return to the primary goal of this chapter: understanding how we learn about distant objects by studying their light. To produce light, these objects must somehow transform energy contained in matter into the vibrations of electric and magnetic fields that we call light. We therefore need to focus on the charged particles within atoms, particularly the electrons, because only particles that have charge can interact with light.

Atoms actually contain energy in three different ways. First, by virtue of their mass, they possess mass-energy in the amount mc^2. Second, they possess kinetic energy by virtue of their motion. Third, and most important to reading the messages encoded in light, atoms contain *electrical potential energy* that depends on the arrangement of their electrons around their nuclei. To interpret the messages carried by light, we must understand how electrons store and release their electrical potential energy.

Energy Levels in Atoms The energy stored by electrons in atoms has a very strange but important property: The electrons can have only particular amounts of energy, and not other energies in between. As an analogy, suppose you're washing windows on a building. If you use an adjustable platform to reach high windows, you can stop the platform at any height above the ground. But if you use a ladder, you can stand only at *particular* heights—the heights of the rungs of the ladder—and not at any height in between. The possible energies of electrons in atoms are like the possible heights on a ladder. Only a few particular energies are possible, and energies between these special few are not possible. The possible energies are known as the **energy levels** of an atom.

Let's consider the energy levels of hydrogen, shown in Figure 5.12. Hydrogen has the simplest set of energy levels of any atom, because it has only one electron. Notice that, on the far right of the diagram, the various energy levels are labeled with energies in units of **electron-volts**, or **eV** for short. ($1 \text{ eV} = 1.60 \times 10^{-19}$ joule.)

The lowest possible energy level is called level 1 or the **ground state**. It represents the minimum amount of energy that the electron can have when it is part of a hydrogen atom, so its energy is labeled 0 eV. Each of the other possible energy levels represents an **excited state**, so named because an electron in any of these levels is "excited" and will soon (usually in a tiny fraction of a second) "calm down" and return to the ground state. Each excited state, from level 2 on upward, is also labeled with an energy in electron-volts; this energy represents the "extra" energy of an electron in that level compared to the ground state.

Energy Level Transitions An electron can rise from a low energy level to a higher one or fall from a high level to a lower one; such changes are usually called **energy level**

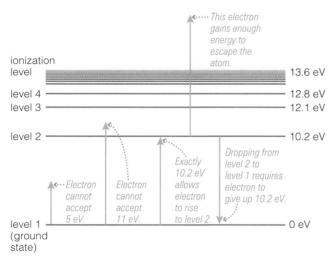

ionization level — 13.6 eV

level 4 — 12.8 eV
level 3 — 12.1 eV

level 2 — 10.2 eV

This electron gains enough energy to escape the atom.

Electron cannot accept 5 eV.

Electron cannot accept 11 eV.

Exactly 10.2 eV allows electron to rise to level 2.

Dropping from level 2 to level 1 requires electron to give up 10.2 eV.

level 1 (ground state) — 0 eV

Figure 5.12 Energy levels for the electron in a hydrogen atom. The electron can change energy levels only if it gains or loses the amount of energy separating the levels. If the electron gains enough energy to reach the ionization level, it can escape from the atom, leaving behind a positively charged ion. (The many levels between level 4 and the ionization level are not labeled.)

transitions. Because energy must be conserved, energy level transitions can occur only when an electron gains or loses the specific amount of energy separating two levels. For example, an electron in level 1 can rise to level 2 only if it gains 10.2 eV of energy. If you try to give the electron 5 eV of energy, it won't accept it because that is not enough energy to reach level 2. Similarly, if you try to give it 11 eV, it won't accept it because it is too much for level 2 but not enough to reach level 3. Once in level 2, the electron can return to level 1 by giving up 10.2 eV of energy. Figure 5.12 shows several examples of allowed and disallowed energy level transitions.

Notice that the amount of energy separating the various levels gets smaller near the top. For example, it takes more energy to raise the electron from level 1 to level 2 than from level 2 to level 3, which in turn takes more energy than the transition from level 3 to level 4. If the electron gains enough energy to reach the *ionization level,* it escapes the atom completely, thereby ionizing the atom. Any excess energy beyond that needed for ionization becomes kinetic energy of the free-moving electron.

THINK ABOUT IT

Are there any circumstances under which an electron in a hydrogen atom can lose 2.6 eV of energy? Explain.

Quantum Physics If you think about it, the idea that electrons in atoms are restricted to particular energy levels is quite bizarre. It is as if you had a car that could go around a track only at particular speeds and not at speeds in between. How strange it would seem if your car suddenly changed its speed from 5 miles per hour to 20 miles per hour without first passing through a speed of 10 miles per

hour! In scientific terminology, the electron's energy levels in an atom are said to be *quantized,* and the study of the energy levels of electrons (and other particles) is called *quantum physics* (or *quantum mechanics*). We will explore more of the astonishing implications of quantum physics in Chapter S4.

Electrons have quantized energy levels in all atoms, not just in hydrogen. Moreover, the allowed energy levels differ from element to element and from one ion of an element to another ion of the same element. Even molecules have quantized energy levels. As we will see shortly, the different energy levels of different atoms and molecules allow light to carry "fingerprints" that can tell us the chemical composition of distant objects.

5.4 Learning from Light

Matter leaves its fingerprints whenever it interacts with light. Examining the color of an object is a crude way of studying the clues left by the matter it contains. For example, a red shirt absorbs all visible photons except those in the red part of the spectrum, so we know that it must contain a dye with these special light-absorbing characteristics. If we take light and disperse it into a spectrum, we can see the spectral fingerprints more clearly. For example, the photograph that opens this chapter (p. 145) shows the Sun's visible-light spectrum in great detail, with the rainbow of color stretching in horizontal rows from the upper left to the lower right of the photograph. We see similar dark or bright lines when we look at almost any spectrum, whether it is the spectrum of the flame from the gas grill in someone's backyard or the spectrum of a distant galaxy whose light we collect with a gigantic telescope. As long as we collect enough light to see details in the spectrum, we can learn many fundamental properties of the object we are viewing, no matter how far away the object is located.

The process of obtaining a spectrum and reading the information it contains is called **spectroscopy.** Spectra often look much like rainbows, at least if they are made only from visible light. However, for the purposes of examining spectra in greater detail, it's often more useful to display them as graphs that show the amount of radiation, or **intensity**, at different wavelengths. The graphs of intensity in this book simply show the relative amount of energy received from an object in each wavelength of light. In other words, the intensity at a given wavelength is proportional to the number of photons observed at that wavelength times the energy of those photons.

For example, Figure 5.13 shows a schematic spectrum of light from a celestial body such as a planet. Notice that this spectrum extends all the way from the ultraviolet on the left to the infrared on the right, which means the graph is showing us details that would be invisible to our eyes in a photograph. At wavelengths where a lot of light is coming from the celestial body, the intensity is high, while at wavelengths where there is little light, the intensity is low.

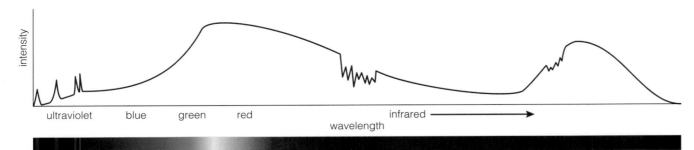

Figure 5.13 A schematic spectrum obtained from the light of a distant object. The "rainbow" at bottom shows how the light would appear when viewed through a prism or diffraction grating; of course, our eyes cannot see the ultraviolet or infrared light. The graph shows the corresponding intensity of the light at each wavelength. The intensity is high where the rainbow is bright and low where it is dim (such as in places where the rainbow shows dark lines).

Our goal in this section is to learn how to interpret astronomical spectra. The bumps and wiggles in Figure 5.13 arise from several different processes, making it a good case study for spectral interpretation. We'll consider these processes one at a time, then we'll return to interpret Figure 5.13 at the end of this section.

 Light and Spectroscopy Tutorial, Lessons 2–4

• What are the three basic types of spectra?

Let's begin our study of spectra by learning how to classify them. Laboratory studies show that spectra come in three basic types: continuous, emission line, and absorption line. Figure 5.14 shows an example of each type, along with the conditions that produce it. (The rules that specify these conditions are often called *Kirchhoff's laws*.) Study the figure carefully to notice the following key ideas:

1. The spectrum of a common (incandescent) light bulb is a rainbow of color. Because the rainbow spans a broad range of wavelengths without interruption, we call it a **continuous spectrum**.
2. A thin or low-density cloud of gas does not produce a continuous spectrum. Instead, it emits light only at specific wavelengths that depend on its composition and temperature. The spectrum therefore consists of bright **emission lines** that we see against a black background, making what we call an **emission line spectrum**.
3. If the cloud of gas lies between us and a light bulb, we still see most of the continuous light emitted by the light bulb. However, the cloud absorbs light of specific wavelengths, so that the spectrum shows dark **absorption lines** over the background rainbow from the light bulb.* We call this an **absorption line spectrum**.

---

*More technically, we'll see an absorption line spectrum as long as the cloud is cooler in temperature than the source of illumination, which in this case means the light bulb filament.

Each of the three spectra is shown both as a band of light (essentially as it would be recorded on a photograph) and as a graph of intensity versus wavelength. Notice that absorption lines appear as dips on a background of relatively high-intensity light, while emission lines look like spikes on a background with little or no intensity.

We can apply some of these ideas to the solar spectrum that opens this chapter. Notice that it shows numerous absorption lines over a background rainbow of color. This tells us that we are essentially looking at a hot light source through gas that is absorbing some of the colors, much as we see when looking through the cloud of gas to the light bulb in Figure 5.14c. For the solar spectrum, the hot light source is the hot interior of the Sun, while the "cloud" is the relatively cool and low-density layer of gas that makes up the Sun's visible surface.

• How does light tell us what things are made of?

We have just seen *how* different viewing conditions lead us to see different types of spectra, so we are now ready to discuss *why*. Let's start with emission and absorption line spectra. Emission and absorption lines form as a direct consequence of the fact that each type of atom, ion, or molecule possesses a unique set of energy levels. This fact allows us to learn the compositions of distant objects in the universe, as we can see by considering what happens in a cloud of gas consisting solely of hydrogen atoms.

Emission Line Spectra The atoms in any cloud of gas are constantly colliding with one another, exchanging energy in each collision. Most of the collisions simply send the atoms careening off in new directions. However, a few of the collisions transfer the right amount of energy to bump an electron from a low energy level to a higher energy level.

Electrons can't stay in higher energy levels for long. They always fall back down to the ground state, level 1, usually in a tiny fraction of a second. The energy the electron loses when it falls to the lower energy level must go somewhere,

The light bulb produces light of all visible wavelengths (colors).

hot light source

prism

Continuous Spectrum

The spectrum shows a smooth, continuous rainbow of light.

A graph of the spectrum is also continuous; notice that intensity varies slightly at different wavelengths.

intensity

wavelength

a

The atoms in a warm gas cloud emit light only at specific wavelengths (colors) determined by the cloud's composition and temperature.

cloud of gas

prism

Emission Line Spectrum

We see bright emission lines at specific wavelengths (colors), but no other light.

The graph shows an upward spike at the wavelength of each emission line.

intensity

wavelength

b

If light from a hot source passes through a cooler gas cloud, atoms in the cloud absorb light at wavelengths determined by the cloud's composition and temperature.

hot light source

cloud of gas

prism

Absorption Line Spectrum

We see dark absorption lines where the cloud has absorbed light of specific wavelengths (colors).

The graph shows a dip in intensity at the wavelength of each absorption line.

intensity

wavelength

c

Figure 5.14 Interactive Figure These diagrams show examples of the conditions under which we see the three basic types of spectra.

and often it goes to *emitting* a photon of light. The emitted photon must have the same amount of energy that the electron loses, which means that it has a specific wavelength (and frequency). Figure 5.15a shows the energy levels in hydrogen that we saw in Figure 5.12. This time, however, it is also labeled with the wavelengths of the photons emitted by various downward transitions of an electron from a higher energy level to a lower one. For example, the transition from level 2 to level 1 emits an ultraviolet photon of wavelength 121.6 nm, and the transition from level 3 to level 2 emits a red visible-light photon of wavelength 656.3 nm.*

Although electrons that rise to higher energy levels in a gas quickly return to level 1, new collisions can raise other electrons into the higher levels. Thus, as long as the

gas remains even moderately warm, collisions are always bumping some electrons into levels from which they fall back down and emit photons with some of the wavelengths shown in Figure 5.15a. The gas therefore emits light with these specific wavelengths. That is why a warm gas cloud produces an emission line spectrum, as shown in Figure 5.15b. The bright emission lines appear at the wavelengths that correspond to downward transitions of electrons, and the rest of the spectrum is dark (black). The specific set of lines that we see depends on the cloud's temperature as well as its composition: At higher temperatures, electrons are more likely to be bumped to higher energy levels.

THINK ABOUT IT

If nothing continues to heat the hydrogen gas, all the electrons eventually will end up in the lowest energy level (the ground state, or level 1). Use this fact to explain why we should *not* expect to see an emission line spectrum from a very cold cloud of hydrogen gas.

---

*Astronomers call transitions between level 1 and other levels the *Lyman* series of transitions. The transition between level 1 and level 2 is Lyman α, between level 1 and level 3 Lyman β, and so on. Similarly, transitions between level 2 and higher levels are called *Balmer* transitions. Other sets of transitions also have names.

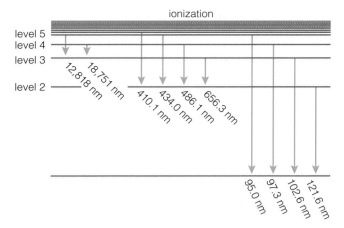

ionization

level 5
level 4
level 3

level 2

12,818 nm
18,751 nm
410.1 nm
434.0 nm
486.1 nm
656.3 nm
95.0 nm
97.3 nm
102.6 nm
121.6 nm

a Energy level transitions in hydrogen correspond to photons with specific wavelengths. Only a few of the many possible transitions are labeled.

410.1 nm 434.0 nm 486.1 nm 656.3 nm

b This spectrum shows emission lines produced by downward transitions between higher levels and level 2 in hydrogen.

410.1 nm 434.0 nm 486.1 nm 656.3 nm

c This spectrum shows absorption lines produced by upward transitions between level 2 and higher levels in hydrogen.

Figure 5.15 Interactive Figure An atom emits or absorbs light only at specific wavelengths that correspond to changes in the atom's energy as an electron undergoes transitions between its allowed energy levels.

Absorption Line Spectra Now, suppose a light bulb illuminates the hydrogen gas from behind (as in Figure 5.14c). The light bulb emits light of all wavelengths, producing a spectrum that looks like a rainbow of color. However, the hydrogen atoms can absorb those photons that have the right amount of energy needed to raise an electron from a low energy level to a higher one. Figure 5.15c shows the result. It is an absorption line spectrum, because the light bulb produces a continuous rainbow of color while the hydrogen atoms remove (absorb) light at specific wavelengths.

Given that electrons in high energy levels quickly return to lower levels, you might wonder why photons emitted in downward transitions don't cancel out the effects of those absorbed in upward transitions. Finding the answer requires looking more deeply at what happens to the absorbed photons. Two things can happen after an electron absorbs a photon and rises to a higher energy level. The first is that the electron quickly returns to its original level, emitting a photon of the same energy that it absorbed.

However, the emitted photon can be going in any random direction, which means that we still see an absorption line because photons that were coming toward us are redirected away from our line of sight. Alternatively, the electron can lose its energy in some other way, either by dropping back down to a different energy level or by transferring its energy to another particle in a subsequent collision. Again, we are left with an absorption line because photons of a specific wavelength have been removed from the spectrum of light that's coming toward us.

You can now see why the dark absorption lines in Figure 5.15c occur at the same wavelengths as the emission lines in Figure 5.15b: Both types of lines represent the same energy level transitions, except in opposite directions. For example, electrons moving downward from level 3 to level 2 in hydrogen can emit photons of wavelength 656.3 nm (producing an emission line at this wavelength), while electrons absorbing photons with this wavelength can rise up from level 2 to level 3 (producing an absorption line at this wavelength).

Chemical Fingerprints The fact that hydrogen emits and absorbs light at specific wavelengths makes it possible to detect its presence in distant objects. For example, imagine that you look through a telescope at an interstellar gas cloud, and its spectrum looks like that shown in Figure 5.15b. Because only hydrogen produces this particular set of lines, you can conclude that the cloud is made of hydrogen. In essence, the spectrum contains a "fingerprint" left by hydrogen atoms.

Real interstellar clouds are not made solely of hydrogen. However, the other chemical constituents in the cloud leave fingerprints on the spectrum in much the same way. Every type of atom has its own unique spectral fingerprint, because it has its own unique set of energy levels. For example, Figure 5.16 shows emission line spectra for helium, sodium, and neon.

Not only does each chemical element produce a unique spectral fingerprint, but *ions* of a particular element (atoms that are missing one or more electrons) produce fingerprints different from those of neutral atoms (Figure 5.17). For example, the spectrum of doubly ionized neon (Ne^{++}) is different from that of singly ionized neon (Ne^{+}), which in turn is different from that of neutral neon (Ne). These differences can help us determine the temperature of a hot

helium

sodium

neon

Figure 5.16 Visible-light emission line spectra for helium, sodium, and neon. The patterns and wavelengths of lines are different for each element, giving each a unique spectral fingerprint.

hydrogen lines

Figure 5.17 The emission line spectrum of the Orion Nebula in a portion of the ultraviolet (about 350–400 nm). The lines are labeled with the chemical elements or ions that produce them (He = helium; O = oxygen; Ne = neon). The many hydrogen lines are all transitions from high levels to level 2.

a We can think of a two-atom molecule as two balls connected by a spring. Although this model is simplistic, it illustrates how molecules can rotate and vibrate. The rotations and vibrations can have only particular amounts of energy and therefore produce unique spectral fingerprints.

b This spectrum of molecular hydrogen (H_2) shows that molecular spectra consist of lines bunched into broad molecular bands.

Figure 5.18 Like atoms and ions, molecules also emit or absorb light at specific wavelengths.

gas or plasma, and they come in handy when we're trying to measure the surface temperatures of stars [Section 15.1]. At higher temperatures, more highly charged ions will be present, so we can estimate the temperature by identifying the ions that are creating spectral lines.

Molecules also produce spectral fingerprints. Like atoms, molecules can produce spectral lines when their electrons change energy levels. But molecules can also produce spectral lines in two other ways. Because they are made of two or more atoms bound together, molecules can vibrate and rotate (Figure 5.18a). Vibration and rotation also require energy, and the possible energies of rotation and vibration in molecules are quantized much like electron energy levels in atoms. Thus, a molecule can absorb or emit a photon when it changes its rate of vibration or rotation. The energy changes in molecules are usually smaller than those in atoms and therefore produce lower-energy photons, and the energy levels also tend to be bunched more closely together than in atoms. Molecules therefore produce spectra with many sets of tightly bunched lines, called **molecular bands** (Figure 5.18b), that are usually found in the infrared portion of the electromagnetic spectrum. That is one reason why infrared telescopes and instruments are so important to astronomers.

Over the past century, scientists have conducted laboratory experiments to identify the spectral lines of every chemical element and of many ions and molecules. Thus, when we see lines in the spectrum of a distant object, we can usually determine what chemicals produced them. For example, if we see spectral lines of hydrogen, helium, and carbon in the spectrum of a distant star, we know that all three elements are present in the star. Moreover, with detailed analysis, we can determine the relative proportions of the various elements. That is how we have learned the chemical compositions of objects throughout the universe.

• How does light tell us the temperatures of planets and stars?

We have seen how emission and absorption line spectra form, and how we can use them to determine the composition of a cloud of gas. Now we are ready to turn our atten-

tion to continuous spectra. Although continuous spectra can be produced in more than one way, light bulbs, planets, and stars produce them in a way that can help us determine their temperatures.

Thermal Radiation: Every Body Does It In a cloud of gas that produces a simple emission or absorption line spectrum, the individual atoms or molecules are essentially independent of one another. Most photons pass easily through such a gas, except those that cause energy level transitions in the atoms or molecules of the gas. However, the atoms and molecules within most of the objects we encounter in everyday life, such as rocks, light bulb filaments, and people, cannot be considered independent. Thus, their energy levels are far more complex. They tend to absorb light across a broad range of wavelengths, meaning that any light striking them cannot easily pass through and that any light emitted inside them cannot easily escape. In fact, the same is true of almost any large or dense object, including planets and stars.

In order to understand the spectra of such objects, let's consider an idealized case, in which an object absorbs all the photons that strike it and does not allow photons inside it to escape easily. Photons trying to escape are quickly absorbed by an atom or molecule, which quickly reemits the photon—but often with a slightly different wavelength and in a different direction.* In effect, the emitted photons bounce randomly around inside the object, constantly exchanging energy with the object's atoms or molecules. By the time the photons finally escape the object and fly off

---

*One of the reasons that photons are reemitted with a slightly different wavelength has to do with the Doppler effect (see Section 5.5). If the absorbing atom (or molecule) is moving, there will be a shift in wavelength that depends on the atom's velocity and the direction in which the photon escapes.

into space, their radiative energies have become randomized so that they are spread over a wide range of wavelengths. The wide wavelength range of the photons explains why the spectrum of light from such an object is smooth or *continuous,* like a pure rainbow without any absorption or emission lines.

Most important, the spectrum from such an object depends on only one thing: the object's *temperature.* To understand why, remember that temperature represents the average kinetic energy of the atoms or molecules in an object [Section 4.3]. Because the randomly bouncing photons interact so many times with those atoms or molecules, they end up with energies that match the kinetic energies of the object's atoms or molecules—which means the photon energies depend only on the object's temperature, regardless of what the object is made of. The temperature dependence of this light explains why we call it **thermal radiation** (sometimes known as *blackbody* radiation), and why its spectrum is called a **thermal radiation spectrum.** Thermal radiation spectra are the most common type of continuous spectra.

No real object emits a perfect thermal radiation spectrum, but almost all familiar objects—including the Sun, the planets, rocks, and even you—emit light that approximates thermal radiation. Figure 5.19 shows graphs of the idealized thermal radiation spectra of three stars and a human, each with its temperature given on the Kelvin scale (see Figure 4.12). Be sure to notice that these spectra show the intensity of light emitted from each square meter of an object's surface, not the total amount of light emitted by the object; that is, they tell us the emitted intensity *per unit area.* Thus, for example, a very large 3,000 K star can emit more total

light than a small 15,000 K star, even though the hotter star emits much more light per unit area of its surface.

The Two Laws of Thermal Radiation   If you compare the spectra in Figure 5.19, you'll see that temperature has two effects on the spectrum; the quantitative descriptions of these effects (see Mathematical Insight 5.2) are often called the *laws* of thermal radiation:

- Law 1 (Stefan-Boltzmann law): *Each square meter of a hotter object's surface emits more light at all wavelengths.* For example, each square meter on the surface of the 15,000 K star emits a lot more light at every wavelength than each square meter of the 3,000 K star, and the hotter star emits light at some ultraviolet wavelengths that the cooler star does not emit at all.
- Law 2 (Wien's law ["Wien" is pronounced *veen*]): *Hotter objects emit photons with a higher average energy,* which means a shorter average wavelength. That is why the "humps" of the spectra are at shorter wavelengths for hotter objects. For example, the hump for the 15,000 K star is in ultraviolet light, the hump for the 5,800 K Sun is in visible light, and the hump for the 3,000 K star is in the infrared.

You can see these properties in action by playing with a light that has a dimmer switch. The first property is apparent as you turn the dimmer switch up: The filament in the light bulb gets hotter and the light brightens. When you turn the switch back down, the filament gets cooler and the light dims. (You can verify the changing temperature by placing your hand near the bulb.) The color of the light illustrates the second property. Before you turn on the switch, the filament is too cool to emit visible light at all, instead emitting only infrared radiation that your eyes cannot see. When you first turn the switch on low, the filament begins to glow red or orange—the longest wavelength of visible light. As you turn the switch higher and the filament gets hotter, the average wavelength of the emitted photons moves toward the blue end of the visible spectrum. The mix of colors emitted at this higher temperature makes the filament and its light look white to your eyes.

Because thermal radiation spectra depend only on temperature, we can use them to measure the temperatures of distant objects. In many cases we can estimate temperatures simply from the object's colors. Notice that while hotter objects emit more light at *all* wavelengths, the biggest difference appears at the shortest wavelengths. An object with a temperature of 310 K, which is about human body temperature, emits mostly in the infrared and emits no visible light at all—which explains why we don't glow in the dark! A relatively cool star, with a 3,000 K surface temperature, emits mostly red light. That is why some bright stars in our sky, such as Betelgeuse (in Orion) and Antares (in Scorpius), appear reddish in color. The Sun's 5,800 K surface emits most strongly in green light (around 500 nm), but the Sun looks yellow or white to our eyes because it also emits other colors throughout the visible spectrum. Hotter stars emit

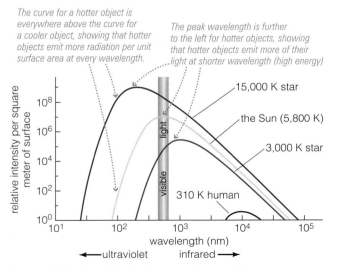

The curve for a hotter object is everywhere above the curve for a cooler object, showing that hotter objects emit more radiation per unit surface area at every wavelength.

The peak wavelength is further to the left for hotter objects, showing that hotter objects emit more of their light at shorter wavelength (high energy)

Figure 5.19 Interactive Figure Graphs of idealized thermal radiation spectra demonstrate the two laws of thermal radiation: (1) Each square meter of a *hotter object's surface emits more light at all wavelengths;* (2) hotter objects emit photons with a higher average energy. Notice that the graph uses power-of-10 scales on both axes, so that we can see all the curves even though the differences between them are quite large.

mostly in the ultraviolet, but because our eyes cannot see ultraviolet, they appear blue-white in color. If an object were heated to a temperature of millions of degrees, it would radiate mostly X rays. Some astronomical objects are indeed hot enough to emit X rays, such as disks of gas encircling exotic objects like neutron stars and black holes (see Chapter 18).

• How do we interpret an actual spectrum?

We've covered the processes that form the three idealized types of spectra (continuous or thermal, absorption line, and emission line). The spectra of real astronomical objects are usually combinations of these three idealized types. In addition, real spectra often show features produced by one other process familiar from everyday life: reflection (or scattering). Let's consider how reflection affects a spectrum, and then we'll be ready to return to the interpretation of the spectrum we began with in Figure 5.13.

Reflected Light Spectra   Most of the light we see in daily life is either reflected or scattered. The source of the light is thermal radiation from the Sun or a lamp. After this light strikes the ground, clouds, people, or other objects, we see only the wavelengths of light that are reflected (or scattered). For example, a red sweatshirt absorbs blue light and reflects red light, so its visible spectrum looks like the thermal ra-diation spectrum of its light source—the Sun—but with blue light missing.

In the same way that we can distinguish lemons from limes, we can use color information in reflected light to learn about celestial objects. Different fruits, different rocks, and even different atmospheric gases reflect and absorb light at different wavelengths. Although the absorption features that show up in spectra of reflected light are not as distinct as the emission and absorption lines for thin gases, they still provide useful information. For example, the surface materials of a planet determine how much light of different colors is reflected or absorbed. The reflected light gives the planet its color, while the absorbed light heats the surface and helps determine its temperature.

Putting It All Together   Figure 5.20 shows the same spectrum we began with in Figure 5.13, but this time with labels indicating the processes responsible for its various features. What can we say about this object from its spectrum? The hump of thermal emission peaking in the infrared shows that this object has a surface temperature of about 225 K, well below the 273 K freezing point of water. The absorption bands in the infrared come mainly from carbon dioxide, which tells us that the object has a carbon dioxide atmosphere. The emission lines in the ultraviolet come from hot gas in a high, thin layer of the object's atmosphere. The reflected light looks like the Sun's 5,800 K thermal

---

MATHEMATICAL INSIGHT **5.2 Laws of Thermal Radiation**

The two rules of thermal radiation have simple mathematical formulas. Law 1, called the *Stefan-Boltzmann law* (named after its discoverers), is expressed as:

$$\text{emitted power (per square meter of surface)} = \sigma T^4$$

where σ (Greek letter *sigma*) is a constant with a measured value $\sigma = 5.7 \times 10^{-8}$ watt/($m^2 \times$ Kelvin^4). Note that this equation is for power *per square meter of surface*. Finding the total power that an object radiates requires multiplying this power by the object's surface area.

Law 2, called *Wien's law*, is expressed approximately as:

$$\lambda_{max} \approx \frac{2{,}900{,}000}{T\,(\text{Kelvin})}\,\text{nm}$$

where λ_{max} (read as *lambda-max*) is the wavelength (in nanometers) of maximum intensity, which is the peak of the hump in a thermal radiation spectrum.

Example: Consider a 15,000 K object that emits thermal radiation. How much power does it emit per square meter? What is its wavelength of peak intensity?

Solution:

Step 1 Understand: We can calculate the total emitted power per square meter from the Stefan-Boltzmann law and the wavelength of maximum intensity from Wien's law.

Step 2 Solve: First we use the Stefan-Boltzmann law to calculate the emitted power per square meter. We plug in the object's temperature of T = 15,000 K:

$$\sigma T^4 = 5.7 \times 10^{-8}\,\frac{\text{watt}}{m^2 \times K^4} \times (15{,}000\ K)^4$$
$$= 2.9 \times 10^9\ \text{watt}/m^2$$

We find the wavelength of maximum intensity with Wien's law:

$$\lambda_{max} \approx \frac{2{,}900{,}000}{15{,}000\ (\text{Kelvin})}\,\text{nm} \approx 190\ \text{nm}$$

Step 3 Explain: A 15,000 K object emits a total power of 2.9 billion watts per square meter. Its wavelength of maximum intensity is 190 nm, which is in the ultraviolet portion of the electromagnetic spectrum. By using these facts in reverse, we can learn about astronomical objects. For example, if an object has a thermal radiation spectrum that peaks at a wavelength of 190 nm, we know that its surface temperature is about 15,000 K. Furthermore, knowing the temperature of the object tells us how much power it emits per square meter of surface, allowing us to calculate its total size from its total radiated power.

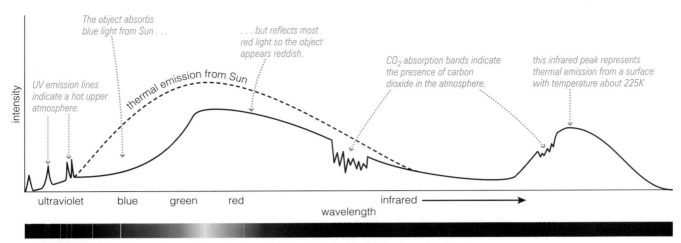

Figure 5.20 The spectrum of Figure 5.13, with interpretation. We conclude that the object looks red in color because it absorbs more blue light than red light from the Sun. The absorption lines tell us that the object has a carbon dioxide atmosphere, and the emission lines tell us that its upper atmosphere is hot. The hump in the infrared (near the right of the diagram) tells us that the object has a surface temperature of about 225 K. It is an idealized spectrum of the planet Mars.

radiation except that much of the blue light is missing, so the object must be reflecting sunlight and must look red in color. Perhaps by now you have guessed that this figure represents the spectrum of the planet Mars.

 The Doppler Effect Tutorial, Lessons 1–2

5.5 The Doppler Shift

You're probably already amazed at the volume of information contained in light, but there is still more. In particular, we can learn about the motion of a distant object (relative to us) from changes in its spectrum caused by the **Doppler effect**.

• How does light tell us the speed of a distant object?

You've probably noticed the Doppler effect on the *sound* of a train whistle near train tracks. If the train is stationary, the pitch of its whistle sounds the same no matter where you stand (Figure 5.21a). But if the train is moving, the pitch will sound higher when the train is coming toward you and lower when it's moving away from you. Just as the train passes by, you can hear the dramatic change from high to low pitch—a sort of "weeeeeeee–oooooooooh" sound. To understand why, we have to think about what happens to the sound waves coming from the train (Figure 5.21b). When the train is moving toward you, each pulse of a

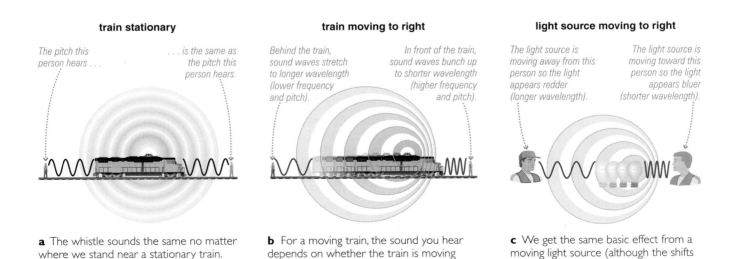

train stationary

The pitch this person hears . . .   . . . is the same as the pitch this person hears.

a The whistle sounds the same no matter where we stand near a stationary train.

train moving to right

Behind the train, sound waves stretch to longer wavelength (lower frequency and pitch).   In front of the train, sound waves bunch up to shorter wavelength (higher frequency and pitch).

b For a moving train, the sound you hear depends on whether the train is moving toward you or away from you.

light source moving to right

The light source is moving away from this person so the light appears redder (longer wavelength).   The light source is moving toward this person so the light appears bluer (shorter wavelength).

c We get the same basic effect from a moving light source (although the shifts are usually too small to notice by eye).

Figure 5.21 The Doppler effect. Each circle represents the crests of sound (or light) waves going in all directions from the source. For example, the circles from the train might represent waves emitted 0.001 second apart.

sound wave is emitted a little closer to you. The result is that waves are bunched up between you and the train, giving them a shorter wavelength and higher frequency (pitch). After the train passes you by, each pulse comes from farther away, stretching out the wavelengths and giving the sound a lower frequency.

The Doppler effect causes similar shifts in the wavelengths of light (Figure 5.21c). If an object is moving toward us, the light waves bunch up between us and the object, so that its entire spectrum is shifted to shorter wavelengths. Because shorter wavelengths of visible light are bluer, the Doppler shift of an object coming toward us is called a **blueshift**. If an object is moving away from us, its light is shifted to longer wavelengths. We call this a **redshift** because longer wavelengths of visible light are redder. For convenience, astronomers use the terms *blueshift* and *redshift* even when they aren't talking about visible light.

Spectral lines provide the reference points we use to identify and measure Doppler shifts (Figure 5.22). For example, suppose we recognize the pattern of hydrogen lines in the spectrum of a distant object. We know the **rest wavelengths** of the hydrogen lines—that is, their wavelengths in stationary clouds of hydrogen gas—from laboratory experiments in which a tube of hydrogen gas is heated so that the wavelengths of the spectral lines can be measured. If the hydrogen lines from the object appear at longer wavelengths, then we know they are redshifted and the object is moving away from us. The larger the shift, the faster the object is moving. If the lines appear at shorter wavelengths, then we know they are blueshifted and the object is moving toward us.

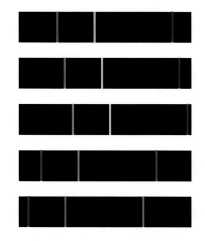

Laboratory spectrum
Lines at rest wavelengths.

Object 1 *Lines redshifted:*
Object moving away from us.

Object 2 *Greater redshift:*
Object moving away faster
than Object 1.

Object 3 *Lines blueshifted:*
Object moving toward us.

Object 4 *Greater blueshift:*
Object moving toward us
faster than Object 3.

Figure 5.22 Interactive Figure Spectral lines provide the crucial reference points for measuring Doppler shifts.

THINK ABOUT IT

Suppose the hydrogen emission line with a rest wavelength of 121.6 nm (the transition from level 2 to level 1) appears at a wavelength of 120.5 nm in the spectrum of a particular star. Given that these wavelengths are in the ultraviolet, is the shifted wavelength closer to or farther from blue visible light? Why, then, do we say that this spectral line is *blueshifted*?

Notice that the Doppler shift tells us only the part of an object's full motion that is directed toward or away from us (the object's *radial* component of motion). Doppler shifts do not give us any information about how fast an object is

---

MATHEMATICAL INSIGHT **5.3** **The Doppler Shift**

As long as an object's radial velocity is small compared to the speed of light (that is, less than a few percent of c), we can use a simple formula to calculate the radial velocity (toward or away from us) of an object from its Doppler shift:

$$\frac{v_{rad}}{c} = \frac{\lambda_{shift} - \lambda_{rest}}{\lambda_{rest}}$$

where v_{rad} is the radial velocity of the object, λ_{rest} is the rest wavelength of a particular spectral line, and λ_{shift} is the shifted wavelength of the same line. (And, as always, c is the speed of light.)

If the result is positive, the object has a redshift and is moving away from us. A negative result means that the object has a blueshift and is moving toward us.

Example: The rest wavelength of one of the visible lines of hydrogen is 656.285 nm. This line is easily identifiable in the spectrum of the bright star Vega, but it appears at a wavelength of 656.255 nm. What is the radial velocity of Vega?

Solution:

Step 1 Understand: We notice that the line's wavelength in Vega's spectrum is slightly shorter than its rest wavelength. Thus, the line is blueshifted and Vega's radial motion is *toward* us. We can calculate the radial velocity from the given formula.

Step 2 Solve: In the radial velocity formula, the rest wavelength is $\lambda_{rest} = 656.285$ nm. The shifted wavelength is the wavelength in Vega's spectrum, $\lambda_{shift} = 656.255$ nm. We find:

$$\frac{v_{rad}}{c} = \frac{\lambda_{shift} - \lambda_{rest}}{\lambda_{rest}}$$

$$= \frac{656.255 \text{ nm} - 656.285 \text{ nm}}{656.285 \text{ nm}}$$

$$= -4.5712 \times 10^{-5}$$

Step 3 Explain: The answer we have found tells us Vega's radial velocity as a fraction of the speed of light; it is negative because Vega is moving toward us. To convert our answer to a speed in km/s, we multiply by the speed of light:

$$v_{rad} = -4.5712 \times 10^{-5} \times c$$

$$= -4.5712 \times 10^{-5} \times (3 \times 10^5 \text{ km/s})$$

$$= -13.7 \text{ km/s}$$

Vega is moving *toward* us at 13.7 km/s. This speed is typical of stars in our neighborhood of the galaxy.

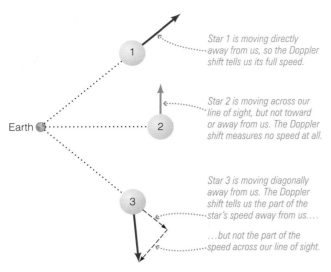

Star 1 is moving directly away from us, so the Doppler shift tells us its full speed.

Star 2 is moving across our line of sight, but not toward or away from us. The Doppler shift measures no speed at all.

Star 3 is moving diagonally away from us. The Doppler shift tells us the part of the star's speed away from us....

...but not the part of the speed across our line of sight.

Figure 5.23 Interactive Figure The Doppler shift tells us only the portion of an object's speed that is directed toward or away from us. It does not give us any information about how fast an object is moving across our line of sight.

moving across our line of sight (the object's *tangential* component of motion). For example, consider three stars all moving at the same speed, with one moving directly away from us, one moving across our line of sight, and one moving diagonally away from us (Figure 5.23). The Doppler shift will tell us the full speed only of the first star. The

Doppler shift will not measure any speed for the second star, because none of its motion is directed toward or away from us. For the third star, the Doppler shift will tell us only the part of the star's speed that is directed away from us. To measure how fast an object is moving across our line of sight, we must observe it long enough to notice how its position gradually shifts across our sky.

• How does light tell us the rotation rate of a distant object?

The Doppler effect not only tells us how fast a distant object is moving toward or away from us but also can reveal information about motion *within* the object. For example, suppose we look at spectral lines of a planet or star that happens to be rotating (Figure 5.24).

As the object rotates, light from the part of the object rotating toward us will be blueshifted, light from the part rotating away from us will be redshifted, and light from the center of the object won't be shifted at all. The net effect, if we look at the whole object at once, is to make each spectral line appear *wider* than it would if the object were not rotating. The faster the object is rotating, the broader in wavelength the spectral lines become. Thus, we can determine the rotation rate of distant objects by measuring the width of their spectral lines. We will see later that the Doppler effect on the spectra of celestial objects reveals even more information.

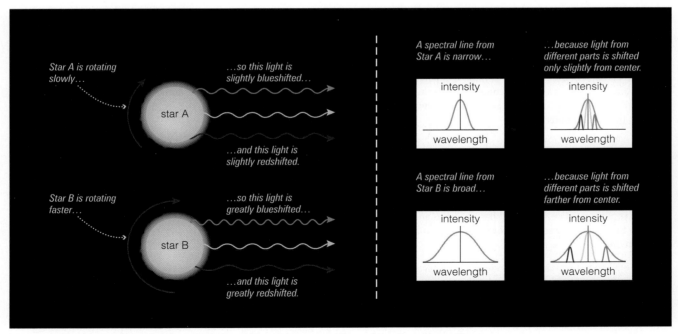

Figure 5.24 The Doppler effect broadens the widths of the spectral lines of rotating objects. One side of a rotating object is moving toward us, creating a blueshift, while the other side is rotating away from us, creating a redshift. The faster the rotation, the greater the spread in wavelength between the light from the two sides. When we look at a spectral line for the object as a whole, we see light from all parts of the object—that is, the parts with blueshifts as well as the parts with redshifts—at the same time. Thus, the greater Doppler shifts in a fast-rotating object make the overall spectral line wider because the light is spread over a greater range of wavelengths.

Putting Chapter 5 into Context

This chapter was devoted to one essential purpose: understanding how we learn about the universe by observing the light of distant objects. "Big picture" ideas that will help you keep your understanding in perspective include the following:

● Light and matter interact in ways that allow matter to leave "fingerprints" on light. Thus, we can learn a great deal about the objects we observe by carefully analyzing their light. Most of what we know about

the universe comes from information that we receive from light.

● The visible light that our eyes can see is only a small portion of the complete electromagnetic spectrum. Different portions of the spectrum may contain different pieces of the story of a distant object, so it is important to study all forms of light.

● There is far more to light than meets the eye. By dispersing the light of a distant object into a spectrum, we can determine the object's composition, surface temperature, motion toward or away from us, rotation rate, and more.

5.1 Light in Everyday Life

● **How do we experience light?** Light carries radiative energy that it can exchange with matter. **Power** is the *rate* of energy transfer, measured in **watts**: 1 watt = 1 joule/s. The colors of light contain a great deal of information about the matter with which it has interacted.

● **How do light and matter interact?** Matter can emit light, absorb light, transmit light, or reflect (or scatter) light.

5.2 Properties of Light

● **What is light?** Light is an **electromagnetic wave**, but also comes in individual "pieces" called **photons**. Each photon has a precise wavelength, frequency, and energy: The shorter the wavelength, the higher the frequency and energy.

● **What is the electromagnetic spectrum?** In order of decreasing wavelength (increasing frequency and energy), the forms of light are **radio waves**, **microwaves**, **infrared**, **visible light**, **ultraviolet**, **X rays**, and **gamma rays**.

5.3 Properties of Matter

● **What is the structure of matter?** Ordinary matter is made of **atoms**, which are made of **protons**, **neutrons**, and **electrons**. Atoms of different **chemical elements** have different numbers of protons. **Isotopes** of a particular chemical element all have the same number of protons but different numbers of neutrons. **Molecules** are made from two or more atoms.

● **What are the phases of matter?** The appearance of matter depends on its phase: **solid**, **liquid**, or **gas**. Gas is always present along with solid or liquid phases; solids **sublimate** into gas and liquids **evaporate** into gas. At very high temperatures, molecular dissociation breaks up molecules and **ionization** strips electrons from atoms; an ionized gas is called a **plasma**.

● **How is energy stored in atoms?** Electrons can exist only in particular **energy levels** within an atom. **Energy level transitions**, in which an electron moves from one energy level to another, can occur only when the electron gains or loses just the right amount of energy.

5.4 Learning from Light

● **What are the three basic types of spectra?** There are three basic types of spectra: a **continuous** spectrum, which looks like a rainbow of light; an **absorption line** spectrum, in which specific colors are missing from the rainbow; and an **emission line** spectrum, in which we see light only with specific colors against a black background.

● **How does light tell us what things are made of?** Emission or absorption lines occur only at specific wavelengths that correspond to particular energy level transitions in atoms or molecules. Every kind of atom, ion, and molecule produces a unique set of spectral lines, so we can determine composition by identifying these lines.

● **How does light tell us the temperatures of planets and stars?** Objects such as planets and stars produce **thermal radiation** spectra, the most common type of continuous spectra. We can determine temperature from these spectra because hotter objects emit more total radiation per unit area and emit photons with a higher average energy.

- **How do we interpret an actual spectrum?**
The spectrum of a real object generally shows a combination of features of emission line, absorption line, and thermal radiation spectra, as well as features produced by reflection. By carefully studying the spectral features, we can learn a great deal about the object that produced them.

5.5 The Doppler Shift

- **How does light tell us the speed of a distant object?** The **Doppler effect** tells us how fast an object is moving toward or away from us. Spectral lines are

shifted to shorter wavelengths (a **blueshift**) in objects moving toward us and to longer wavelengths (a **redshift**) in objects moving away from us.

- **How does light tell us the rotation rate of a distant object?** The Doppler effect broadens spectral lines of rotating objects. The faster the rotation, the wider the spectral line.

EXERCISES AND PROBLEMS

For instructor-assigned homework go to MasteringASTRONOMY   www.masteringastronomy.com

Review Questions

Short-Answer Questions Based on the Reading

1. What is the difference between *energy* and *power*? What units do we use to measure power?
2. What is a *spectrum*, and how do we see one?
3. Give an example from everyday life of each of the four major types of interaction between matter and light.
4. How does a particle differ from a wave in everyday life? Define *wavelength, frequency,* and *speed* for a wave.
5. What do we mean when we say that light is an *electromagnetic wave*? How is wavelength related to frequency for electromagnetic waves? Explain.
6. What is a *photon*? In what way is a photon like a particle? In what way is it like a wave?
7. List the different forms of light in order from lowest to highest energy. Would the list be different if you went in order from lowest to highest frequency? From shortest to longest wavelength? Explain.
8. Briefly describe the structure of an atom. How big is an atom? How big is the *nucleus* in comparison to the entire atom?
9. What determines an atom's *atomic number*? What determines its *atomic mass number*? Under what conditions are two atoms different *isotopes* of the same element? What is a *molecule*?
10. What is *electrical charge*? Will an electron and a proton attract or repel one another? Will two electrons attract or repel one another? Explain.
11. Describe the phase changes of water as you heat it starting from its solid phase, ice. What happens at very high temperatures? What is a *plasma*?
12. What do we mean when we say that *energy levels* are quantized in atoms? Under what circumstances can *energy level transitions* occur?
13. How do we convert a spectrum shown as a band of light (like a rainbow) into a graph of the spectrum?
14. Describe the conditions that would cause us to see each of the three basic types of spectra. What do we see in the Sun's spectrum shown on the opening page of this chapter?
15. How can we use emission or absorption lines to determine the chemical composition of a distant object?
16. Describe two ways in which the thermal radiation spectrum of an 8,000 K star would differ from that of a 4,000 K star.

17. Describe each of the key features of the spectrum in Figure 5.20 and explain what it tells us about the object.
18. Describe the *Doppler effect* for light and what we can learn from it. What does it mean to say that radio waves are *blueshifted*? Why does the Doppler effect widen the spectral lines of rotating objects?

Test Your Understanding

Does It Make Sense?

Decide whether the statement makes sense (or is clearly true) or does not make sense (or is clearly false). Explain your reasoning. (For an example, see Chapter 1, "Does It Make Sense?")

19. The walls of my room are transparent to radio waves.
20. Because of their higher frequencies, X rays must travel through space faster than radio waves.
21. If you could see infrared light, you would see a glow from the backs of your eyelids when you closed your eyes.
22. If you had X-ray vision, you could read this entire book without turning any pages.
23. Two isotopes of the element rubidium differ not only in their number of neutrons, but also in their number of protons.
24. A "white hot" object is hotter than a "red hot" object.
25. If the Sun's surface became much hotter (while the Sun's size remained the same), the Sun would emit more ultraviolet light but less visible light than it currently emits.
26. If you could view a spectrum of light reflecting off a blue sweatshirt, you'd find the entire rainbow of color (looking the same as a spectrum of white light).
27. Nearly all galaxies show redshifts, so they must all be red in color.
28. If a distant galaxy has a substantial redshift (as viewed from our galaxy), then anyone living in that galaxy would see a substantial redshift in a spectrum of the Milky Way Galaxy.

Quick Quiz

Choose the best answer to each of the following. Explain your reasoning with one or more complete sentences.

29. Why is a sunflower yellow? (a) It emits yellow light. (b) It absorbs yellow light. (c) It reflects yellow light.

30. Blue light has higher frequency than red light. Thus, blue light has: (a) higher energy and shorter wavelength than red light; (b) higher energy and longer wavelength than red light; (c) lower energy and shorter wavelength than red light.

31. Radio waves are: (a) a form of sound; (b) a form of light; (c) a type of spectrum.

32. Compared to an atom as a whole, an atomic nucleus: (a) is very tiny but has most of the mass; (b) is quite large and has most of the mass; (c) is very tiny and has very little mass.

33. Some nitrogen atoms have seven neutrons and some have eight neutrons. These two forms of nitrogen are: (a) ions of each other; (b) phases of each other; (c) isotopes of each other.

34. Sublimation is the process by which: (a) solid material enters the gas phase; (b) liquid material enters the gas phase; (c) solid material becomes a liquid.

35. If you heat a rock until it glows, its spectrum will be: (a) a thermal radiation spectrum; (b) an absorption line spectrum; (c) an emission line spectrum.

36. The set of spectral lines that we see in a star's spectrum depends on the star's: (a) interior temperature; (b) chemical composition; (c) rotation rate.

37. A star whose spectrum peaks in the infrared is: (a) cooler than our Sun; (b) hotter than our Sun; (c) larger than our Sun.

38. A spectral line that appears at a wavelength of 321 nm in the laboratory appears at a wavelength of 328 nm in the spectrum of a distant object. We say that the object's spectrum is: (a) redshifted; (b) blueshifted; (c) skewed.

Investigate Further
In-Depth Questions to Increase Your Understanding

Short-Answer/Essay Questions

39. *Atomic Terminology Practice I.*
    a. The most common form of iron has 26 protons and 30 neutrons in its nucleus. State its atomic number, atomic mass number, and number of electrons if it is electrically neutral.
    b. Consider the following three atoms: Atom 1 has 7 protons and 8 neutrons; atom 2 has 8 protons and 7 neutrons; atom 3 has 8 protons and 8 neutrons. Which two are *isotopes* of the same element?
    c. Oxygen has atomic number 8. How many times must an oxygen atom be ionized to create an O^{+5} ion? How many electrons are in an O^{+5} ion?

40. *Atomic Terminology Practice II.*
    a. Consider fluorine atoms with nine protons and 10 neutrons. What are the atomic number and atomic mass number of this fluorine? Suppose we could add a proton to this fluorine nucleus. Would the result still be fluorine? Explain. What if we added a neutron to the fluorine nucleus?
    b. The most common isotope of gold has atomic number 79 and atomic mass number 197. How many protons and neutrons does the gold nucleus contain? If it is electrically neutral, how many electrons does it have? If it is triply ionized, how many electrons does it have?
    c. The most common isotope of uranium is ^{238}U, but the form used in nuclear bombs and nuclear power plants is ^{235}U. Given that uranium has atomic number 92, how many neutrons are in each of these two isotopes of uranium?

41. *The Fourth Phase of Matter.*
    a. Explain why nearly all the matter in the Sun is in the plasma phase.
    b. Based on your answer to part (a), explain why plasma is the most common phase of matter in the universe.
    c. If plasma is the most common phase of matter in the universe, why is it so rare on Earth?

42. *Energy Level Transitions.* The following labeled transitions represent an electron moving between energy levels in hydrogen. Answer each of the following questions and explain your answers.

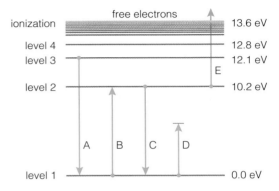

    a. Which transition could represent an atom that *absorbs* a photon with 10.2 eV of energy?
    b. Which transition could represent an atom that *emits* a photon with 10.2 eV of energy?
    c. Which transition represents an electron that is breaking free of the atom?
    d. Which transition, as shown, is *not* possible?
    e. Would transition A represent emission or absorption of light? How would the wavelength of the emitted or absorbed photon compare to that of the photon involved in transition C? Explain.

43. *Spectral Summary.* Clearly explain how studying an object's spectrum can allow us to determine each of the following properties of the object.
    a. The object's surface chemical composition.
    b. The object's surface temperature.
    c. Whether the object is a low-density cloud of gas or something more substantial.
    d. Whether the object has a hot upper atmosphere.
    e. Whether the object is reflecting blue light from a star.
    f. The speed at which the object is moving toward or away from us.
    g. The object's rotation rate.

44. *Orion Nebula.* To the eye (through a telescope), much of the Orion Nebula looks like a glowing cloud of gas. What type of spectrum would you expect to see from the glowing parts of the nebula? Why?

45. *Neptune's Spectrum.* The planet Neptune is colder than Mars and it appears blue in color. (a) Make a sketch similar to Figure 5.20 for Mars, but instead showing the spectrum you'd expect to see from Neptune. Label the axes clearly, and briefly describe each of the features shown in your spectrum in much the same way that Figure 5.20 describes the features in Mars's spectrum. (b) Suppose a very large asteroid crashed into Neptune, causing its atmosphere to become 10K warmer for a short time. List two ways in which the spectrum you drew in part (a) would differ when the atmosphere became

warmer. (c) Suppose Neptune rotated much faster. How would you expect its spectral lines to change?

46. *The Doppler Effect.* In hydrogen, the transition from level 2 to level 1 has a rest wavelength of 121.6 nm. Suppose you see this line at a wavelength of 120.5 nm in Star A, at 121.2 nm in Star B, at 121.9 nm in Star C, and at 122.9 nm in Star D. Which stars are coming toward us? Which are moving away? Which star is moving fastest relative to us (either toward or away from)? Explain your answers without doing any calculations.

Quantitative Problems

Be sure to show all calculations clearly and state your final answers in complete sentences.

47. *Human Wattage.* A typical adult uses about 2,500 Calories of energy each day. Use this fact to calculate the typical adult's average *power* requirement, in watts. (*Hint:* 1 Calorie = 4,184 joules.)

48. *Electric Bill.* Your electric utility bill probably shows your energy use for the month in units of *kilowatt-hours*. A kilowatt-hour is defined as the energy used in 1 hour at a rate of 1 kilowatt (1,000 watts); that is, 1 kilowatt-hour = 1 kilowatt × 1 hour. Use this fact to convert 1 kilowatt-hour into joules. If your bill says you used 900 kilowatt-hours, how much energy did you use in joules?

49. *Radio Station.* What is the wavelength of a radio photon from an "AM" radio station that broadcasts at 1,120 kilohertz? What is its energy?

50. *UV Photon.* What is the energy (in joules) of an ultraviolet photon with wavelength 120 nm? What is its frequency?

51. *X-Ray Photon.* What is the wavelength of an X-ray photon with energy 10 keV (10,000 eV)? What is its frequency? (*Hint:* 1 eV = 1.60×10^{-19} joule.)

52. *How Many Photons?* Suppose that all the energy from a 100-watt light bulb came in the form of photons with wavelength 600 nm. (This is not quite realistic; see Problem 59.)
   a. Calculate the energy of a *single* photon with wavelength 600 nm.
   b. How many 600-nm photons must be emitted each second to account for all the light from this 100-watt light bulb?
   c. Based on your answer to part (b), explain why we don't notice the particle nature of light in our everyday lives.

53. *Thermal Radiation Laws I.* Consider a 3,000 K object that emits thermal radiation. How much power does it emit per square meter? What is its wavelength of peak intensity?

54. *Thermal Radiation Laws II.* Consider a 50,000 K object that emits thermal radiation. How much power does it emit per square meter? What is its wavelength of peak intensity?

55. *Hotter Sun.* Suppose the surface temperature of the Sun were about 12,000 K, rather than 6,000 K.
   a. How much more thermal radiation would the Sun emit?
   b. What would happen to the Sun's wavelength of peak emission?
   c. Do you think it would still be possible to have life on Earth? Explain.

56. *Taking the Sun's Temperature.* The Sun radiates a total power of about 4×10^{26} watts into space. The Sun's radius is about 7×10^8 meters.
   a. Calculate the average power radiated by each square meter of the Sun's surface. (*Hint:* The formula for the surface area of a sphere is $A = 4\pi r^2$.)

   b. Using your answer from part (a) and the Stefan-Boltzmann law, calculate the average surface temperature of the Sun. (*Note:* The temperature calculated this way is called the Sun's *effective temperature.*)

57. *Doppler Calculations I.* In hydrogen, the transition from level 2 to level 1 has a rest wavelength of 121.6 nm. Suppose you see this line at a wavelength of 120.5 nm in Star A and at 121.2 nm in Star B. Calculate each star's speed, and be sure to state whether it is moving toward or away from us.

58. *Doppler Calculations II.* In hydrogen, the transition from level 2 to level 1 has a rest wavelength of 121.6 nm. Suppose you see this line at a wavelength of 121.9 nm in Star C and at 122.9 nm in Star D. Calculate each star's speed, and be sure to state whether it is moving toward or away from us.

59. *Understanding Light Bulbs.* A standard (incandescent) light bulb uses a hot tungsten coil to produce a thermal radiation spectrum. The temperature of this coil is typically about 3,000 K.
   a. What is the wavelength of maximum intensity for a standard light bulb? Compare this to the 500-nm wavelength of maximum intensity for the Sun.
   b. Overall, do you expect the light from a standard bulb to be the same as, redder than, or bluer than light from the Sun? Why? Use your answer to explain why professional photographers use a different type of film for indoor photography than for outdoor photography.
   c. Do standard light bulbs emit all their energy as visible light? Use your answer to explain why light bulbs are usually hot to touch.
   d. *Fluorescent* light bulbs primarily produce emission line spectra rather than thermal radiation spectra. Explain why, if the emission lines are in the visible part of the spectrum, a fluorescent bulb can emit more visible light than a standard bulb of the same wattage.
   e. *Compact fluorescent* light bulbs are designed to produce so many emission lines in the visible part of the spectrum that their light looks very similar to the light of standard bulbs. However, they are much more energy efficient: A 15-watt compact fluorescent bulb typically emits as much visible light as a standard 75-watt bulb. Although compact fluorescent bulbs generally cost more than standard bulbs, is it possible that they could save you money? Besides initial cost and energy efficiency, what other factors must be considered?

Discussion Questions

60. *The Changing Limitations of Science.* In 1835, French philosopher Auguste Comte stated that science would never allow us to learn the composition of stars. Although spectral lines had been seen in the Sun's spectrum by that time, not until the mid-1800s did scientists recognize that spectral lines give clear information about chemical composition (primarily through the work of Foucault and Kirchhoff). Why might our present knowledge have seemed unattainable in 1835? Discuss how new discoveries can change the apparent limitations of science. Today, other questions seem beyond the reach of science, such as the question of how life began on Earth. Do you think such questions will ever be answerable through science? Defend your opinion.

61. *Your Microwave Oven.* A *microwave oven* emits microwaves that have just the right wavelength needed to cause energy level changes in water molecules. Use this fact to explain how

a microwave oven cooks your food. Why doesn't a microwave oven make a plastic dish get hot? Why do some clay dishes get hot in the microwave? Why do dishes that aren't themselves heated by the microwave oven sometimes still get hot when you heat food on them?

62. *Democritus and the Path of History*. Besides his belief in atoms, Democritus held several other strikingly modern notions. For example, he maintained that the Moon was a world with mountains and valleys and that the Milky Way was composed of countless individual stars—ideas that weren't generally accepted until the time of Galileo, more than 2,000 years later. Unfortunately, we know of Democritus's work only secondhand because none of the 72 books he is said to have written survived the destruction of the Library of Alexandria. Do you think history might have been different if the work of Democritus had not been lost? Defend your opinion.

MEDIA EXPLORATIONS

For self-study activities go to **Mastering**ASTRONOMY www.masteringastronomy.com

 Interactive Tutorials

Tutorial Review of Key Concepts

Use the interactive **Tutorials** at **www.mastering astronomy.com** to review key concepts from this chapter.

Light and Spectroscopy Tutorial
Lesson 1 Radiation, Light, and Waves
Lesson 2 Spectroscopy
Lesson 3 Atomic Spectra—Emission and Absorption Lines
Lesson 4 Thermal Radiation

Doppler Effect Tutorial
Lesson 1 Understanding the Doppler Shift
Lesson 2 Using Emission and Absorption Lines to Measure the Doppler Shift

Supplementary Tutorial Exercises

Use the interactive **Tutorial Lessons** to explore the following questions.

Light and Spectroscopy Tutorial, Lesson 3
1. Why do neon lights come in so many different colors?
2. How can we tell the difference between lithium and carbon in the spectrum of a distant object?
3. How will an interstellar cloud affect the spectrum of light that we see from a star that lies behind it?

Light and Spectroscopy Tutorial, Lesson 4
1. How would an object's spectrum influence the choice of telescope used to observe it?
2. Consider the spectra of hot stars as compared to those of cooler stars. Which stars would be best observed with an ultraviolet telescope? Explain.

Doppler Effect Tutorial, Lesson 1
1. As the source of a sound passes you, which two properties will you hear change? Which change is due to the Doppler effect?
2. Which property of a sound wave is responsible for the pitch that your ears hear? How does the Doppler effect change the sound you hear?
3. How can the Doppler effect for light be used to measure the speed of a car?

Doppler Effect Tutorial, Lesson 2
1. How can we use precise wavelengths of spectral lines to measure the speed of a distant object?
2. Does it matter whether lines are in emission or absorption for the Doppler effect? Explain.

Web Projects

1. *Kids and Light*. Visit one of the many Web sites designed to teach middle and high school students about light. Read the content and try the activities. If you were a teacher, would you find the site useful for your students? Why or why not? Write a one-page summary of your conclusions.
2. *Light Bulbs*. Learn about alternatives to standard incandescent light bulbs, such as fluorescent lights, compact fluorescent lights, and halogen lights. Write a short report summarizing some of the advantages and disadvantages of each technology.
3. *Medical Imaging*. Learn about CAT scans or other technologies for medical imaging of the human body. How do they work? How are such technologies similar to those used by astronomers to learn about the universe? Write a short report summarizing your findings.

6

Telescopes
Portals of Discovery

All of this has been discovered and observed these last days thanks to the telescope that I have [built], after having been enlightened by divine grace.

Galileo

We are in the midst of a great revolution in human understanding of the universe. Astonishing new discoveries about the early history of the universe, about the lives of galaxies and stars, and about planets orbiting other stars are frequent features of the daily news.

The primary fuel for this astronomical revolution has been recent and significant advances in telescope technology. A growing number of ever larger telescopes are operating around the world, and new technologies have vastly improved the quality of data that we can obtain with telescopes on the ground. Meanwhile, telescopes lofted into space are offering views of the heavens unobstructed by Earth's atmosphere while also allowing us to study wavelengths of light that do not penetrate to the ground. By studying light from across the entire spectrum, we can learn far more about the universe than we can by studying visible light alone.

Because telescopes are the portals through which we study the universe, understanding them can help us understand both the triumphs and the limitations of modern astronomy. In this chapter, we will explore the basic principles by which telescopes work, along with some of the technological advances fueling the current revolution in astronomy.

6.1 Eyes and Cameras: Everyday Light Sensors

We observe the world around us with the five basic senses: touch, taste, smell, hearing, and sight. We learn about the world by using our brains to analyze and interpret the data that our senses record. The science of astronomy progresses similarly. We collect data about the universe, and then we analyze and interpret the data. Within our solar system, we can analyze some matter directly. We can sample Earth's surface and study meteorites that fall to Earth. Occasionally, we send spacecraft to sample the surfaces or atmospheres of other worlds. Aside from these few samples collected in our own solar system, nearly all other data about the universe come to us in the form of light.

Astronomers collect light with telescopes and record light with cameras and other instruments. Thus, a good way to begin our study of telescopes is to examine the principles of eyes and cameras, our everyday light sensors.

• How does your eye form an image?

You've already seen many astronomical images in this book. Although most of these images have been made with

Figure 6.1 A simplified diagram of the human eye.

powerful telescopes, the basic processes of image making are the same as those that occur in our eyes.

The eye is a remarkably complex organ, but its basic components are a *lens*, a *pupil*, and a *retina* (Figure 6.1). The retina contains light-sensitive cells (called *cones* and *rods*) that, when triggered by light, send signals to the brain via the optic nerve.

The lens of your eye creates an image on your retina because it bends light much like a simple glass lens.* To understand how that works, imagine a light wave coming toward you from a large distance. In that case, we can represent the wave's path as a simple *ray* of light traveling in a straight line toward you. The peaks and troughs of the light wave's electric and magnetic fields are perpendicular to the path of the ray. The wave slows down when it hits your eye because light generally travels more slowly through denser matter than through air. This change in speed causes the wave to bend, a phenomenon known as *refraction*. If the wave hits at an angle, the side of the wave nearest your eye slows down first, allowing the far side to catch up (Figure 6.2a). The end result is a change in the direction the light is traveling. Figure 6.2b shows an example of how the bending of light entering Earth's atmosphere from space distorts the Sun's image at sunset.

The bending of light that happens in your eye is precisely tuned to make vision possible. Light rays that enter the lens farther from the center are bent more, and rays that pass directly through the center are not bent at all. In this way, parallel rays of light, such as the light from a distant star, converge to a point called the **focus**. Figure 6.3 shows the idea for both a glass lense and an eye. The fact that parallel rays of light converge to a sharp focus explains why distant stars appear as *points* of light to our eyes or on photographs.

Light rays that are *not* parallel, such as those from a nearby object, enter a lens from different directions. These rays do not all converge at the focus, but they still follow precise rules as they are bent through the lens; some of

---

*Actually, both the lens and cornea of your eye are responsible for focusing an image. The cornea does about two-thirds of the work and the lens does the remaining third. For simplicity, the eye diagrams in this book show all the bending occurring at the lens.

a Light that hits glass at an angle bends as it enters the glass. The wide yellow ribbons in this figure represent light waves. The darker bands on those ribbons (perpendicular to the direction of light travel) represent the positions of wave peaks.

b Earth's atmosphere also bends light slightly. The Sun looks slightly squashed at sunset because light from the lower portion of the Sun passes through more atmosphere and therefore bends slightly more than light from the upper portion.

Figure 6.2 Light bends when it moves from one substance to another, a phenomenon called *refraction*.

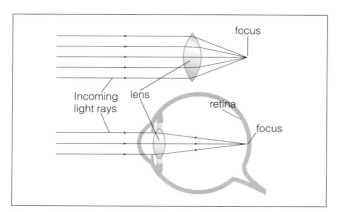

Figure 6.3 A glass lens bends parallel rays of light to a point called the *focus* of the lens. In an eye with perfect vision, rays of light are bent to a focus on the retina.

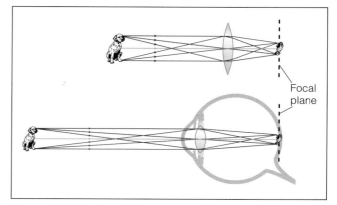

Figure 6.4 Light from different parts of an object focuses at different points to make an (upside-down) image of the object.

these rules are illustrated by the ray paths in Figure 6.4. The end result is that the rays are bent to form an **image** of the original object. The place where the image appears in focus is called the **focal plane** of the lens. In an eye with perfect vision, the focal plane is on your retina. (The retina actually is curved, rather than a flat plane, but we will ignore this detail.) The image formed by a lens is upside down but is flipped right side up by your brain, where the true miracle of vision occurs.

The pupil of the eye does not play a direct role in image formation, but it controls the amount of light that enters the eye by adjusting the size of its opening. The pupil dilates (opens wider) in low light levels, allowing your eye to gather as many photons of light as possible. It constricts in bright light so that your eye is not overloaded with light.

THINK ABOUT IT

Cover one eye and look toward a bright light. Then look immediately into a mirror and compare the openings of your two pupils. Which one is wider, and why? Why do eye doctors dilate your pupils during eye exams?

• How do we record images?

If we want to remember an image or study it in more detail, it's useful to record it with a camera. The basic operation of a camera is quite similar to that of an eye (Figure 6.5), except that the camera makes a permanent record of the image on film or by using some type of electronic detector. To keep our discussion simple, we'll use the term **detector** to mean any device that records light, including photographic

detector (CCD)

shutter

lens

Figure 6.5 A camera works much like an eye. When the shutter is open, light passes through the lens to form an image on the detector (which may be film or an electronic device).

film and various types of digital-imaging chips. The camera lens plays the role of the lens of the eye and the detector plays the role of the retina. The camera shutter is analogous to an eyelid: Light can reach the detector only when the shutter is open. Fancier cameras even have an adjustable circular opening that controls the amount of light entering the camera just as the pupil controls the amount of light entering the eye. (The circular opening through which light enters is called the camera's *aperture*, so cameras in which the opening size is adjustable are said to have "aperture control.")

Recording an image with a camera offers at least two important advantages over simply looking at or drawing it. First, a recorded image is much more reliable and detailed than a drawing. Second, with a camera we can control the **exposure time**, the amount of time during which light collects on the detector. A longer exposure time means that more photons reach the detector, allowing the detector to record details that would be too faint to see with our eyes alone.

Thanks to the advantages of cameras over the human eye, the advent of photography in the mid-1800s spurred a leap in astronomical data collection. Until recently, these images were recorded on photographic film or glass plates coated with light-sensitive chemicals. These chemicals change in response to light, thereby recording an image.

Today, most images are recorded digitally with electronic detectors called *charge-coupled devices*, or **CCDs**. A CCD is a chip of silicon engineered to be extraordinarily sensitive to photons. The chip is physically divided into a grid of squares called *picture elements*, or **pixels** for short. When a photon strikes a pixel, it causes a bit of electric charge to accumulate. Each subsequent photon striking the same pixel adds to this accumulated electric charge. After an exposure is complete, a computer measures the total electric charge in each pixel, thus determining how many photons have struck each one. The overall image is stored on a memory chip as an array of numbers representing the results from each pixel. Many consumer camera chips now have 6 million or more pixels, and professional cameras can have significantly more.

Find a digital camera (either your own or a friend's). Where is its lens? Where is its detector? Can you control its exposure time manually? How many pixels does its CCD have?

CCDs (and other types of electronic detectors) have three major advantages over photographic film. First, they are much more sensitive to light. Most photons striking film have no effect at all: Fewer than 10% of the visible photons reaching the film cause a change in the light-sensitive chemicals that coat it. In contrast, CCDs can accurately record 90% or more of the photons that strike them. Second, CCDs have a much wider *dynamic range* than photographic film, meaning that they can more easily record both dim and bright light at the same time. For example, if you try to record an image of a galaxy on film, the bright galactic center will be overexposed by the time you capture enough light to see details in the outskirts of the galaxy. A CCD may still be able to show details in both the dim and bright regions of the galaxy. Third, a digital image can be manipulated through techniques of *image processing* to bring out details that otherwise might be missed in analyzing the image.

 Telescopes Tutorial, Lessons 1–2

6.2 Telescopes: Giant Eyes

Telescopes are essentially giant eyes that can collect far more light than our naked eyes can. They allow us to see objects that are far too faint for our eyes to see and with far more detail than our eyes can notice. In addition, we can connect scientific instruments to telescopes, allowing us to record and analyze the light they collect. For example, sophisticated cameras can make high-quality images of light collected by a telescope, and spectrographs can disperse the light into spectra that can reveal an object's chemical composition, temperature, speed, and more. In this section, we'll investigate the basic principles of telescopes.

• What are the two most important properties of a telescope?

Telescopes can differ in a number of ways, but the two most fundamental properties of any telescope are its *light-collecting area* and its *angular resolution*. Let's investigate each of these properties.

A telescope's **light-collecting area** tells us how much total light the telescope can collect at one time. Telescopes are generally round, so we usually characterize a telescope's "size" by stating the *diameter* of its light-collecting area. For example, a "10-meter telescope" has a light-collecting area that is 10 meters in diameter. Such a telescope has a light-collecting area more than a million times that of the human eye. Because area is proportional to the *square* of a telescope's diameter, a relatively small increase in diameter can

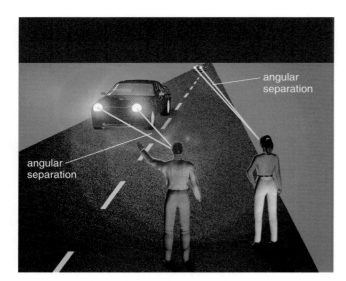

Figure 6.6 Interactive Figure This diagram shows how angular separation depends on distance. The headlights on the car have the same physical separation in both cases, but their angular separation is larger when the car is closer. Similarly, two stars separated by a particular distance will have a larger angular separation if they are nearby than if they are farther away.

Magnification and Telescopes

You are probably familiar with the way binoculars and telephoto camera lenses make objects appear larger in size—the phenomenon we call *magnification*. It's useful to know the magnification of binoculars and telephoto lenses, because it tells you what you can expect to see when you look through them. However, magnification is not a very useful way to describe a telescope's power, because magnification won't do you any good if the telescope collects too little light or has poor angular resolution. No matter how much you magnify the image, you cannot see details if the telescope does not collect enough light to show them, and you can never see details smaller than the angular resolution of the telescope. That is why magnifying an image too much just makes it look blurry, without revealing any new details, and why a telescope's light-collecting area and angular resolution are much more important than its magnification.

mean a big increase in light-collecting area. A 10-meter telescope has five times the diameter of a 2-meter telescope, so its light-collecting area is $5^2 = 25$ times as great.

THINK ABOUT IT

What is the population of your hometown? Suppose everyone in your hometown looked at the dark sky at the same time. How would the total amount of light entering everyone's eyes compare with the light collected by a 10-meter telescope? Explain.

Angular resolution gets its name because it is the smallest angle over which we can tell that two dots—or two stars—are distinct. For example, the human eye has an angular resolution of about 1 arcminute $\left(\frac{1}{60}^\circ\right)$, which means that two stars will appear distinct only if they have at least this much angular separation in the sky. If the stars are separated by less than 1 arcminute, our eyes will not be able to distinguish them individually and they will appear as a single star. The angular separation between two points of light depends both on their actual separation and on their distance from us [Section 2.1]; Figure 6.6 shows the idea.

Large telescopes can have amazing angular resolution. For example, the 2.4-meter Hubble Space Telescope has an angular resolution of about 0.05 arcsecond (for visible light), which would allow you to read this book from a distance of almost 1 kilometer. In principle, larger telescopes should have even better (smaller) angular resolution. However, all larger telescopes today are located on the ground, where effects of Earth's atmosphere can limit their angular resolution.

The ultimate limit to a telescope's resolving power comes from the properties of light. Because light is an electromagnetic wave [Section 5.2], beams of light can interfere with one another like overlapping sets of ripples on a pond (Figure 6.7). This *interference* causes a blurring of images that limits a telescope's angular resolution even when all

Figure 6.7 This computer-generated image shows how overlapping sets of ripples on a pond interfere with one another. The effects of the two sets of ripples add in some places, making the water rise extra high or fall extra low, and cancel in other places, making the water surface flat. Light waves also exhibit interference. (The colors in this image are for visual effect only.)

Angular resolution is the smallest angle over which we can tell that two dots are distinct. If we want to know whether two objects can be resolved, we simply need to determine their angular separation: If their angular separation is less than the angular resolution, then we will not be able to see the objects individually (they will be blurred together), and if it is more, then we will be able to tell them apart. Recall from Mathematical Insight 2.1 that we can calculate angular separation in degrees with this formula:

$$\text{angular separation} = \text{physical separation} \times \frac{360°}{2\pi \times \text{distance}}$$

With telescopes, angular resolution is usually measured in arcseconds or fractions of arcseconds. Thus, it's useful to revise the above formula to give an answer in arcseconds rather than degrees. Because there are 3,600 arcseconds in 1 degree (see Figure 2.8), we convert the units by multiplying by $\frac{3600''}{1°}$:

$$\frac{\text{angular}}{\text{separation}} = \text{physical separation} \times \frac{360°}{2\pi \times \text{distance}} \times \frac{3600''}{1°}$$

We can simplify this formula by multiplying out the numbers; you should confirm that $360 \times 3600 \div 2\pi \approx 206{,}265$. The formula now reads:

$$\text{angular separation} = 206{,}265'' \times \frac{\text{physical separation}}{\text{distance}}$$

Remember that the physical separation and distance must both have the same units (such as kilometers or miles) in order for the angular separation to come out in arcseconds.

Example 1: Two stars in a binary star system are separated by 200 million kilometers. The system is located 20 light-years away. Can we study the stars individually with the Hubble Space Telescope's angular resolution of 0.05 arcsecond?

Solution:

Step 1 Understand: We are asked whether we can see two stars individually with the Hubble Space Telescope, which means we need to determine whether their angular separation is greater than or less than the telescope's angular resolution. The telescope can see them individually only if the angular resolution is smaller than the angular separation of the stars.

Step 2 Solve: We are given the telescope angular resolution of 0.05 arcsecond, so we need to calculate the angular separation of the two stars and see whether it is smaller or larger than 0.05 arcsecond. Note, however, that we are given the separation of the stars in kilometers and their distance in light-years. Therefore we must first put both numbers into the same units before we can use the formula we found above. Remembering that 1 light-year $\approx 10^{13}$ km (see Mathematical Insight 1.1), we find that 20 light-years $\approx 20 \times 10^{13}$ km $= 2 \times 10^{14}$ km. Using scientific notation, we write the physical separation of the two stars as 200 million km $= 2 \times 10^8$ km. We can now use our formula to find the angular separation of the two stars:

$$\text{angular separation} = 206{,}265'' \times \frac{\text{physical separation}}{\text{distance}}$$
$$= 206{,}265'' \times \frac{2 \times 10^8 \text{ km}}{2 \times 10^{14} \text{ km}}$$
$$= 0.2''$$

Step 3 Explain: We have found that the angular separation of the two stars is 0.2 arcsecond, which is greater than the telescope's angular resolution of 0.05 arcsecond. We can therefore see and study the stars individually.

Example 2: If you looked at this book with a telescope that has Hubble's angular resolution of 0.05 arcsecond, how far away could you place the book and still be able to read it?

Solution:

Step 1 Understand: In this case, we can read the book if we can resolve its individual letters. So we must decide what we mean by the "angular separation" in the case of reading letters. There's a number of ways to think about this question, but here's one simple approach: You know that, in principle, we could make the letters out of tiny dots (just as they are made on a computer screen). Let's say that to read them clearly, you'd want the letters to be at least 10 dots tall (and 10 dots wide). If you measure, you'll find that the letters in this book are about 2 millimeters tall, so the dots would be separated by 0.2 millimeter. We can use this value as the physical separation of the dots. The telescope will be able to resolve the dots as long as their angular separation is at least as great as the telescope resolution of 0.05 arcsecond; we'll therefore use this value as the angular separation. We now need to calculate the distance at which the dots will have this angular separation.

Step 2 Solve: Solving the above formula for the distance, we find (you should confirm the algebra for yourself):

$$\text{distance} = 206{,}265'' \times \frac{\text{physical separation}}{\text{angular separation}}$$

We now plug in the angular separation of 0.05 arcsecond and the physical separation of the dots of 0.2 millimeter:

$$\text{distance} = 206{,}265'' \times \frac{0.2 \text{ mm}}{0.05''} \approx 825{,}000 \text{ mm}$$

Step 3 Explain: We've found the distance at which we could read the book, but the answer is difficult to interpret when written as nearly a million millimeters. To make the answer more meaningful, we convert it to meters:

$$825{,}000 \text{ mm} \times \frac{1 \text{ m}}{1{,}000 \text{ mm}} = 825 \text{ m}$$

A telescope with an angular resolution of 0.05 arcsecond should be able resolve the letters in this book as long as the book is within about 800 meters, or a little less than 1 kilometer.

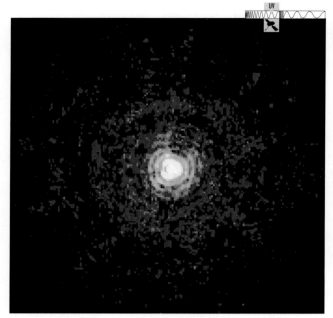

Figure 6.8 When examined in detail, a Hubble Space Telescope image of a star has rings (represented as green and purple in the figure) resulting from the wave properties of light. With higher angular resolution, the rings would be smaller.

Figure 6.9 Refracting telescopes.

other conditions are perfect. That is why even a high-quality telescope in space, such as the Hubble Space Telescope, cannot have perfect angular resolution* (Figure 6.8).

The angular resolution that a telescope could achieve if it were limited only by the interference of light waves is called its **diffraction limit**. (*Diffraction* is a technical term for the specific effects of interference that limit telescope resolution.) The diffraction limit depends on both the diameter of the telescope's primary mirror and the wavelength of the light being observed (see Mathematical Insight 6.2, p. 185). For any particular wavelength of light, a larger telescope has a smaller diffraction limit, meaning it can achieve a better (smaller) angular resolution. For any particular telescope, the diffraction limit is larger (poorer angular resolution) for longer-wavelength light. That is why, for example, a radio telescope must be far larger than a visible-light telescope to achieve the same angular resolution.

• What are the two basic designs of telescopes?

Telescopes come in two basic designs: *refracting* and *reflecting*. A **refracting telescope** operates much like an eye, using transparent glass lenses to focus the light from distant objects (Figure 6.9). The earliest telescopes, including those that Galileo built, were refracting telescopes. The world's largest refracting telescope, completed in 1897, has a lens

_____

*In fact, the Hubble Space Telescope's primary mirror was made with the wrong shape, which further limited its angular resolution until corrective optics were installed in 1993. The corrective optics consist of small mirrors that correct for most of the blurring created by the primary mirror.

that is 1 meter (40 inches) in diameter and a telescope tube that is 19.5 meters (64 feet) long.

A **reflecting telescope** uses a precisely curved *primary mirror* to gather light (Figure 6.10). This mirror reflects the gathered light to a *secondary mirror* that lies in front of it. The secondary mirror then reflects the light to a focus at a place where the eye or instruments can observe it—sometimes through a hole in the primary mirror, and sometimes out the side of the telescope (perhaps with the aid of additional small mirrors). The fact that the secondary mirror prevents some light from reaching the primary mirror might seem like a drawback to reflecting telescopes, but in practice it is not a problem because only a small fraction of the incoming light is blocked.

Nearly all telescopes used in current astronomical research are reflectors, mainly for two practical reasons. First, because light passes *through* the lens of a refracting telescope, lenses must be made from clear, high-quality glass with precisely shaped surfaces on both sides. In contrast, only the reflecting surface of a mirror must be precisely shaped, and the quality of the underlying glass is not a factor. Second, large glass lenses are extremely heavy and can be held in place only by their edges. Since the large lens is at the top of a refracting telescope, it is difficult to stabilize refracting telescopes and to prevent large lenses from deforming. The primary mirror of a reflecting telescope is mounted at the bottom, where its weight presents a far less serious problem. (A third problematic feature of lenses, called *chromatic aberration,* occurs because a lens brings different colors of light into focus at slightly different places. This problem can be minimized by using combinations of lenses.)

For a long time, the main factor limiting the size of reflecting telescopes was the sheer weight of the glass needed

Cassegrain Focus   Newtonian Focus   Nasmyth/Coudé Focus

a Three variations on the basic design of a reflecting telescope. In all cases, a reflecting telescope collects light with a precisely curved primary mirror that reflects light back upward to the secondary mirror. In the *Cassegrain* design, the secondary mirror reflects the light through a hole in the primary mirror, so that the light can be observed with cameras or instruments beneath the telescope. In the *Newtonian* design, the secondary mirror reflects the light out to the side of the telescope. In the *Nasmyth* and *Coudé* designs, a third mirror is used to reflect light out the side but lower down than in the Newtonian design.

b The Gemini North telescope, located on the summit of Mauna Kea, Hawaii, is a reflecting telescope with the Cassegrain design. The primary mirror, visible at the bottom of the large lattice tube, is 8 meters in diameter. The secondary mirror, located in the smaller central lattice, reflects light back down through the hole visible in the center of the primary mirror.

Figure 6.10   Interactive Figure   Reflecting telescopes.

for their primary mirrors. Recent technological innovations have made it possible to build lighter-weight mirrors, such as the one in the Gemini telescope shown in Figure 6.10b, or to make many small mirrors work together as one large one. For example, Figure 6.11 shows the primary mirror

of one of the 10-meter Keck telescopes, which consists of 36 smaller mirrors that function together as one.

These new mirror-building technologies are fueling a revolution in the building of large telescopes. Before the 1990s, the 5-meter Hale telescope on Mount Palomar

Figure 6.11 This photo on the right shows the primary mirror of one of the Keck telescopes on Mauna Kea, with a man in the center for scale. The mirror is made up of 36 smaller, hexagonal mirrors, arranged in a honeycomb pattern. The photo on the left shows two Keck telescopes, with the primary mirrors visible inside the domes.

Table 6.1 Largest Optical (Visible-Light) Telescopes

Size	Name	Sponsor	Location	Opened	Special Features
10.4 m	Gran Telescopio Canarias	Spain, Mexico, U. Florida	Canary Islands	2005	Segmented primary mirror based on mirrors for Keck telescopes
10 m	Keck I and Keck II	Cal Tech, U. California, NASA	Mauna Kea, HI	1993/1996	Two identical 10-m telescopes, each with a primary mirror consisting of 36 1.8-m hexagonal segments
9.2 m	Hobby-Eberly	U. Texas, Penn State, Stanford, Germany	Mt. Locke, TX	1997	Consists of 91 1-m segments, for a total diameter of 11 m, but only 9.2 m can be used at a time; designed primarily for spectroscopy
9.2 m	South African Large Telescope	South Africa, Rutgers, UW–Madison, UNC–Chapel Hill, Dartmouth, Carnegie-Mellon, 5 others	South Africa	2004	Based on design of Hobby-Eberly telescope
2 × 8.4 m	Large Binocular Telescope	U. Arizona, Ohio State U., Italy, Germany	Mt. Graham, AZ	2006	Two 8.4-m mirrors on a common mount, giving light-collecting area of 11.8-m telescope
4 × 8.2 m	Very Large Telescope	European Southern Observatory	Cerro Paranal, Chile	2000	Four separate 8-m telescopes designed to work individually or together as the equivalent of a 16-m telescope
8.3 m	Subaru	Japan	Mauna Kea, HI	1999	Japan's first large telescope project
8 m	Gemini North and South	U.S., U.K., Canada, Chile, Brazil, Argentina	Mauna Kea, HI (North); Cerro Pachon, Chile (South)	1999	Twin telescopes, one in each hemisphere
6.5 m	Magellan I and II	Carnegie Institute, U. Arizona, Harvard, U. Michigan, MIT	Las Campanas, Chile	2000/2002	Twin 6.5-m telescopes, known respectively as the Walter Baade and Landon Clay telescopes
6.5 m	MMT	Smithsonian Institution, U. Arizona	Mt. Hopkins, AZ	2000	Replaced an older telescope in the same observatory

(outside San Diego) reigned for more than 40 years as the most powerful telescope in the world. Today, it does not even make the top-10 list for telescope size (Table 6.1). Astronomers are currently working on designs for even larger telescopes, and it's quite likely that we'll see telescopes with diameters of 30 meters or more within the next couple of decades.

• What do astronomers do with telescopes?

Cartoon images of astronomers usually show them looking through a telescope by eye (Figure 6.12). In fact, astronomers rarely have the chance to look through large telescopes by eye, because it is much more important to record observations with cameras or other equipment.

Every astronomical observation is unique, and astronomers use many different kinds of instruments and detectors to extract the information contained in the light collected by a telescope. Nevertheless, most observations fall into one of three basic categories: *imaging*, which yields photographs (images) of astronomical objects; *spectroscopy*, in which astronomers obtain and study spectra; and *timing*, which tracks how an object changes with time.* Let's look at each category in a little more detail.

Imaging At its most basic, an imaging instrument is simply a camera. Astronomers sometimes place *filters* in front of the camera that allow only particular colors or wavelengths of light to pass through. In fact, most of the richly hued

---

*Some astronomers include a fourth general category called *photometry*, which is the accurate measurement of light intensity from a particular object at a particular time. We do not list this as a separate category because today's detectors can generally perform photometry at the same time that they are being used for imaging, spectroscopy, or timing.

Figure 6.12 This isn't really the way professional astronomers work.

The actual light collected . . .

blue filter

green filter

red filter

. . . is combined to show a full-color image.

VIS

Figure 6.13 Interactive Photo In astronomy, color images are usually constructed by combining several images taken through different filters.

images in this and other astronomy books are made by combining images recorded through different filters (Figure 6.13).

Today, many astronomical observatories record images made from nonvisible light. For example, Figure 6.14 is an "X-ray image." What exactly does this mean, given that X rays are invisible to the eye? The easiest way to understand the idea is to think about X rays at a doctor's office. When the doctor makes an "X ray of your arm," he or she uses a machine that sends X rays through your arm. The X rays that pass through are recorded on a piece of X-ray sensitive film. We cannot see the X rays themselves, but we can see the image left behind on the film. Astronomical images work the same way. Whether a camera records visible or nonvisible light, it makes an image showing the intensity of the light.

The simplest way to display a nonvisible image is in black and white, with brighter regions of the image corresponding to brighter light coming from the object. (A doctor's X ray is actually a negative, making bones look bright because few X rays pass through them.) In astronomy, nonvisible images are often color-coded. In Figure 6.14, for example, the colors correspond to X rays of different energy. (Remember that higher energy means higher frequency and shorter wavelength.) All images recorded in nonvisible light must use some kind of color or gray-scale coding. Sometimes the colors correspond to different wavelengths (as in Figure 6.14), sometimes to different intensities of light, and sometimes to physical properties.

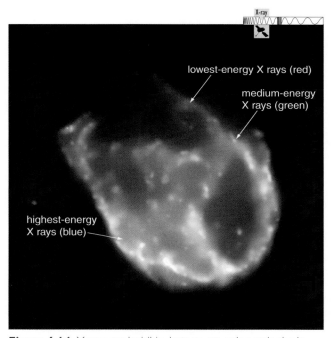

X-ray

lowest-energy X rays (red)

medium-energy X rays (green)

highest-energy X rays (blue)

Figure 6.14 X rays are invisible, but we can color-code the information recorded by an X-ray detector to make an image of the object as it would appear in X rays. This image, from NASA's Chandra X-Ray Observatory, shows X-ray emission from the debris of a stellar explosion (a supernova remnant named N132D). The image is color-coded, with different colors representing X rays of different energy.

Color-coded images are common even outside astronomy. For example, medical images from CAT scans and MRIs are usually displayed in color, even though neither type of imaging uses visible light. What do you think the colors mean in CAT scans and MRIs? How are the colors useful to doctors?

Spectroscopy Instruments called **spectrographs** use diffraction gratings (or other devices) to separate the various colors of light into spectra, which are then recorded with a detector such as a CCD (Figure 6.15).

As we discussed in Chapter 5, a spectrum can reveal a wealth of information about an object, including its chemical composition, temperature, and rotation rate. However, just as the amount of information we can glean from an image depends on the angular resolution, the information we can glean from a spectrum depends on the **spectral resolution**: The higher the spectral resolution, the more detail we can see (Figure 6.16).

In principle, astronomers would always like the highest possible spectral resolution. However, higher spectral resolution comes at a price. A telescope collects only so much light in a given amount of time, and the spectral resolution depends on how widely the spectrograph spreads out this light. If the spectrograph spreads out the light too much, the spectrum may become so dim that nothing can be

recorded without a very long exposure time. Thus, for the same telescope, making a spectrum of an object requires a longer exposure time than making an image, and high-resolution spectra take longer than low-resolution spectra.

Timing Many astronomical objects vary with time. For example, some stars undergo sudden outbursts caused by tremendous explosions. Other stars, including our own Sun, vary much less and much more slowly as sunspots or starspots cover more or less of their surfaces. Some objects vary periodically; looking for small, periodic changes in brightness is an important planet-hunting strategy [Section 13.1].

For a slowly varying object, a timing experiment may be as simple as comparing a set of images or spectra obtained on different nights. For more rapidly varying sources, specially designed instruments essentially make rapid multiple exposures, in some cases recording the arrival time of every individual photon.

The results of timing experiments are often shown as **light curves**: graphs that show how an object's intensity varies with time. For example, Figure 6.17 shows a light curve for the star Mira. Notice that the light output of Mira varies by more than a factor of 100 as it rises and falls with a period of a little less than a year.

An Astronomer's Job Using imaging, spectroscopy, and timing, astronomers can put together detailed pictures of the objects they study. However, a professional astronomer typically spends no more than a few nights a year gathering data at a large telescope. Some astronomers make no obser-

Figure 6.15 The basic design of a spectrograph. In this diagram, the spectrograph is attached to the bottom of a reflecting telescope, with light entering the spectrograph through a hole in the primary mirror. A narrow slit (or small hole) at the entrance to the spectrograph allows only light from the object of interest to pass through. This light bounces from the collimating mirror to the diffraction grating, which disperses the light into a spectrum. The dispersed light is then focused by the camera mirror and recorded by a CCD, giving an image of the spectrum.

Figure 6.16 Higher spectral resolution means that we can see more details in the spectrum. The two ultraviolet spectra above both show the same object in the same wavelength band. However, we see far more detail with higher spectral resolution, including individual spectral lines that appear merged together at lower spectral resolution. (The spectrum shows absorption lines created when interstellar gas absorbs light from a more distant star.)

vations at all, instead focusing on the development of models to explain astronomical observations made by others.

The day-to-day life of an observational astronomer goes something like this: After identifying an important unanswered question, the astronomer proposes a set of observations to an organization that owns a large telescope. The astronomer must write the proposal clearly and eloquently, explaining exactly how she or he will carry out the observations and why these observations would be a good use of telescope time. Often, several astronomers with similar interests collaborate on a proposal, all planning to work together to make the observations and analyze the data. A committee of other astronomers then evaluates all the proposals that have been submitted, deciding which ones are worthy of receiving telescope time and which ones are not.

In most cases, the amount of telescope time requested in worthy proposals is much larger than the actual amount of time that is available. For example, if 100 worthy proposals each require 10 nights of observing time in the next year, there's only enough time for about one-third of them (or even fewer, since the weather isn't always perfect!). Thus,

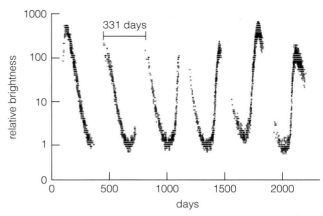

Figure 6.17 This graph shows a light curve for the variable star Mira (in the constellation Cetus), with data spanning several years. Centuries of observation show that Mira's brightness varies with an average period of 331 days.

the various individuals and groups who write proposals essentially compete against one another, and only a fraction of them are awarded time on the telescope.

MATHEMATICAL INSIGHT **6.2** **The Diffraction Limit**

A simple formula gives the diffraction limit of a telescope in arcseconds:

$$\text{diffraction limit} \atop \text{(arcseconds)} \approx 2.5 \times 10^5 \times \left(\frac{\text{wavelength of light}}{\text{diameter of telescope}} \right)$$

The wavelength of light and the diameter of the telescope must be in the same units.

Example 1: What is the diffraction limit of the 2.4-meter Hubble Space Telescope for visible light with a wavelength of 500 nanometers?

Solution:

Step 1 Understand: We are asked for the diffraction limit, which is the best angular resolution the telescope can achieve in principle. Above, we are given a formula for calculating the diffraction limit, and we already have all the information we need to use the formula.

Step 2 Solve: For light with a wavelength of 500 nanometers (500×10^{-9} meters), the diffraction limit of the Hubble Space Telescope is approximately:

$$\text{diffraction limit} \approx 2.5 \times 10^5 \times \left(\frac{\text{wavelength}}{\text{telescope diameter}} \right)$$
$$= 2.5 \times 10^5 \times \frac{500 \times 10^{-9} \text{ m}}{2.4 \text{ m}}$$
$$= 0.05 \text{ arcsecond}$$

Step 3 Explain: The Hubble Space Telescope has a diffraction limited angular resolution of 0.05 arcsecond for visible light with wavelength 500 nanometers. Thus, if it actually achieves its diffraction limit, the telescope can resolve objects that are

separated by more than 0.05 arcsecond. Objects separated by less will be blurred together.

Example 2: Suppose you wanted to achieve a diffraction limit of 0.001 arcsecond for visible light of wavelength 500 nanometers. How large a telescope would you need?

Solution:

Step 1 Understand: In this case we are looking for the size of the telescope with a particular diffraction limit. We have a formula for calculating the diffraction limit from the telescope size, so we need to rearrange this formula to tell us telescope size when we are given the diffraction limit. You should confirm that when we multiply both sides of the given formula by the telescope diameter and divide both sides by the diffraction limit, we get the following new formula:

$$\text{telescope diameter} \approx 2.5 \times 10^5 \times \left(\frac{\text{wavelength}}{\text{diffraction limit}} \right)$$

Step 2 Solve: Now we substitute the given values for the wavelength and the diffraction limit:

$$\text{telescope diameter} \approx 2.5 \times 10^5 \times \frac{500 \times 10^{-9} \text{ m}}{0.001 \text{ arcsecond}}$$
$$= 125 \text{ m}$$

Step 3 Explain: An optical telescope would need a mirror diameter (or separation between mirrors, for an interferometer) of 125 meters to achieve an angular resolution of 0.001 arcsecond. Note that this diameter is longer than a football field! Note also that it makes sense that we would need a telescope 50 times as large as the Hubble Space Telescope to achieve an angular resolution 50 times better than Hubble's.

Because the telescope time is so precious, it's crucial to use it efficiently. Thus, the real work begins after an observing proposal is accepted. The astronomers who wrote the proposal must prepare carefully to make sure they use the telescope time well. Then, after completing their observations, they spend even more time analyzing and interpreting their data.

6.3 Telescopes and the Atmosphere

From the time Galileo first turned his telescope to the heavens in 1609 [Section 3.3] until the dawn of the space age just a few decades ago, all astronomical telescopes were located on the ground. Even today, the vast majority of observatories are ground based, and that will probably hold true long into the future. Telescopes on the ground are much less expensive to build, operate, and maintain than telescopes in space. Nevertheless, Earth's surface is far from ideal as an observing site. In this section, we'll explore some of the problems that Earth's atmosphere poses for astronomical observations and learn why, despite the higher costs, dozens of telescopes have been lofted into Earth orbit or beyond.

• How does Earth's atmosphere affect ground-based observations?

The most obvious problems with observing from the ground are daylight and weather. Our daytime sky is bright because the atmosphere scatters sunlight, and this brightness drowns out the dim light of most astronomical objects. Thus, most astronomical observations are practical only at night. Even then, we can observe only when the sky is clear rather than cloudy. (The atmosphere does not scatter most radio waves, so radio telescopes can operate day and night and under cloudy skies.)

The constraints of daylight and weather affect the timing of observations, but by themselves do not hinder observations on clear nights. However, our atmosphere creates three other problems that inevitably affect astronomical observations: the scattering of human-made light, the blurring of images by atmospheric motion, and the fact that most forms of light cannot reach the ground at all.

Light Pollution Just as our atmosphere scatters sunlight in the daytime, it also scatters the bright lights of cities at night (Figure 6.18). That is why you cannot see as many stars when you look at the sky from a big city as you can see

---

SPECIAL TOPIC Would You Like Your Own Telescope?

Just a few years ago, a decent telescope for personal use would have set you back a few thousand dollars and taken weeks of practice to learn to use effectively. Today, you can get a good-quality telescope for just a few hundred dollars and built-in computer drives can make it very easy to use.

Before you start thinking about what telescope to buy, it's important to understand what a personal telescope can and cannot do. A telescope will allow you to look for yourself at light that has traveled vast distances through space to reach your eyes. It can be a rewarding experience, but the images in your telescope will *not* look like the beautiful photographs in this book—those are obtained with much larger telescopes and sophisticated cameras. In addition, while your telescope can in principle let you see many distant objects, including star clusters, nebulae, and galaxies, it won't allow you to find anything unless you first set it up properly. Even computer-driven telescopes, usually called "go to" telescopes, typically take 15 minutes to a half-hour to set up for each use (and longer when you are first learning).

If your goal is just to see the Moon and a few other objects with relatively little effort, you should probably go for a good pair of binoculars rather than a telescope. Indeed, it's generally a good idea for everyone to buy binoculars before buying a telescope, since binoculars will help you learn about viewing the sky and are a lot less expensive than good telescopes. Binoculars are generally described by two numbers, such as 7 × 35 or 12 × 50. The first number is the magnification; for example "7×" means that objects will look seven times closer through the binoculars than to your eye. The second number is the diameter of each lens in millimeters. As with telescopes, larger lenses mean more light and better views. However, larger lenses also tend to be heavier and more difficult to hold

steady. If you buy a large pair of binoculars, you should also get a tripod to help hold it steady.

If you decide to go ahead with a telescope, the first rule to remember is that magnification is *not* the key factor to consider in telescope selection. Telescopes that are advertised by their magnification, such as "650 power," are usually junk. Forget them, and instead focus on three factors when choosing your telescope:

1. *The light-collecting area* (also called *aperture*). Most personal telescopes are reflectors, so a "6-inch" telescope means a primary mirror 6 inches in diameter.
2. *Optical quality*. A poorly made telescope won't do you much good. Your best bet if you cannot do side-by-side comparisons is to stick with a major telescope manufacturer (such as Meade, Celestron, or Orion).
3. *Portability*. A large, bulky telescope can be great if you plan to keep it on your roof, but it won't be much fun on camping trips. Depending on how you plan to use your telescope, you'll need to make trade-offs between size and portability.

Most important, remember that a telescope is an investment: It will cost real money to buy, and you will probably keep it for many years. As you should with any investment, learn all you can before you settle on a particular model. Read reviews of telescopes in magazines such as *Astronomy*, *Mercury*, and *Sky and Telescope*. Talk to knowledgeable salespeople at stores that specialize in telescopes. And find a nearby astronomy club that holds observing sessions at which you can try out some telescopes and learn from experienced telescope users. Astronomy clubs can be found almost everywhere, and you can usually find them by typing "astronomy club" and your city into a Web search engine.

Figure 6.18 Earth at night: It's pretty, but to astronomers it's light pollution. This image, a composite made from hundreds of satellite photos, shows the bright lights of cities around the world as they appear from Earth orbit at night.

from an unpopulated area. The same problem of scattered light can seriously hinder astronomical observations. For example, the 2.5-meter telescope at Mount Wilson, the world's largest when it was built in 1917, would be much more useful today if it weren't located so close to the lights of what was once the small town of Los Angeles. The problem of scattered light at night is known as **light pollution,** because it is caused by human-made light that obscures our view of the sky.

Similar but less serious light pollution hinders many other telescopes, including the Mount Palomar telescopes near San Diego and the telescopes of the National Optical Astronomy Observatory on Kitt Peak near Tucson. Fortunately, many communities are working to reduce light pollution. Placing reflective covers on the tops of streetlights directs more light toward the ground, rather than toward the sky. Using "low-pressure" (sodium) lights also helps. Because these lights shine most brightly in just a few wavelengths of light, special filters can be used to absorb these wavelengths while transmitting most of the light from the astronomical objects under study. An extra benefit is that both reflective covers and low-pressure lights offer significant energy savings to communities that use them.

Twinkling and Atmospheric Turbulence The second major problem is somewhat less obvious but equally serious: distortion of light by the atmosphere. Winds and other air currents ensure that air in our atmosphere is constantly moving and mixing around, a phenomenon that we call **turbulence.** The ever changing motions of air due to turbulence constantly change the atmosphere's light-bending properties, so that light rays passing through the atmosphere are constantly being bent by slightly different amounts. As a result, our view of things outside Earth's atmosphere appears to jiggle around, rather like your view of things outside the water when you look at them from the bottom of a swim-

ming pool. In the sky, we see this jiggling as the familiar twinkling of stars. Twinkling may be beautiful to the naked eye, but it causes problems for astronomers because it blurs astronomical images.

THINK ABOUT IT

Put a coin at the bottom of a cup of water. If you stir the water, the coin will appear to move around, even if it remains stationary on the bottom. What makes the coin appear to move? How is this similar to the way that our atmosphere makes stars appear to twinkle?

The blurring of images by turbulence places a severe limit on the angular resolution that ground-based telescopes can achieve. Even at the best observing sites, atmospheric turbulence typically limits the angular resolution to no better than about 0.5 arcsecond. Thus, most large telescopes cannot come anywhere near achieving the angular resolution that their size would offer in the absence of an atmosphere. For example, if you calculate its diffraction limit (see Mathematical Insight 6.2), you'll find that a 2.5-meter telescope would have angular resolution of 0.05 arcsecond in the absence of atmospheric effects—10 times better than what the atmosphere actually permits.

Astronomers can partially mitigate effects of weather, light pollution, and atmospheric distortion by choosing appropriate sites for observatories. The key criteria are that the sites be dark (limiting light pollution), dry (limiting rain and clouds), calm (limiting turbulence), and high (placing them above at least part of the atmosphere). Islands are often ideal, and the 4,300-meter (14,000-foot) summit of Mauna Kea on the Big Island of Hawaii is home to many of the world's best observatories (Figure 6.19).

In addition, a remarkable new technology called **adaptive optics** can eliminate much of the blurring caused by

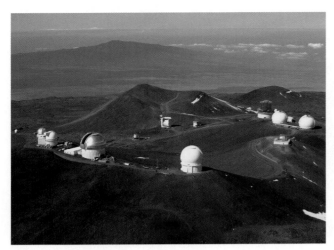

Figure 6.19 Observatories on the summit of Mauna Kea in Hawaii. Mauna Kea meets all the key criteria for an observing site: It is far from big-city lights, high in altitude, and in an area where the air tends to be calm and dry.

our atmosphere. Atmospheric turbulence causes blurring because it makes the light of a star seem to dance around when viewed through a telescope. Adaptive optics essentially makes the telescope's mirrors do an opposite dance, canceling out the atmospheric distortions (Figure 6.20). The shape of the mirror (often the secondary or even a third or fourth mirror) is changed slightly many times each second to compensate for the rapidly changing atmospheric distortions. A computer calculates the necessary changes by monitoring distortions in the image of a bright star near the object under study. In some cases, if there is no bright star near the object of interest, the observatory shines a laser into the sky to create an *artificial star* (a point of light in Earth's atmosphere) that it can monitor for distortions.

a Atmospheric distortion makes this ground-based image of a double star look like a single star.

b When the same telescope is used with adaptive optics, the two stars are clearly distinguished. The angular separation between the two stars is 0.38 arcsecond.

Figure 6.20 The technology of adaptive optics can enable a ground-based telescope to overcome most of the blurring caused by Earth's atmosphere. (Both these images were taken in near-infrared light with the Canada-France-Hawaii telescope; colors represent infrared intensity.)

Why do we put telescopes into space?

The ultimate solution to the problems faced by ground-based observatories is to put telescopes into space, where they are above the atmosphere and thereby unaffected by daylight, weather, light pollution, and atmospheric turbulence. That is one reason why the Hubble Space Telescope (Figure 6.21) was built and why it has been so successful despite the relatively small size of its primary mirror. Notice that Hubble's 2.4-meter mirror is much smaller than the mirror sizes of the ground-based telescopes listed in Table 6.1.

However, while Hubble's visible-light images were far superior to anything possible from the ground when it was launched in 1990, the new technology of adaptive optics is helping ground-based observatories catch up rapidly. At least in some cases, ground-based observatories are now able to equal or better the image quality of the Hubble Space Telescope. So does that mean that the Hubble Space Telescope and other space observatories are becoming obsolete? No, and here's why: Earth's atmosphere poses one major problem that no Earth-bound technology can overcome— *Our atmosphere prevents most forms of light from reaching the ground at all.*

If we studied only visible light, we'd be missing much of the story that light brings to us from the cosmos. Planets are relatively cool and emit primarily infrared light. The hot upper layers of stars like the Sun emit ultraviolet and X-ray light. Some cosmic events are so violent that they can produce gamma rays that travel through space to Earth. Indeed, most astronomical objects emit light over a broad

Figure 6.21 The Hubble Space Telescope orbits Earth. Its position above the atmosphere allows it to get undistorted views of the heavens. Hubble can observe infrared and ultraviolet light as well as visible light.

range of wavelengths. If we want to understand the universe, we must observe light all across the electromagnetic spectrum. We simply can't do this from the ground.

Figure 6.22 shows the depths to which different forms of light penetrate Earth's atmosphere. Notice that only radio waves, visible light, and small parts of the infrared spectrum can be observed from the ground. Thus, the most important reason for putting telescopes into space is to allow us to observe light that does not penetrate Earth's atmosphere. Indeed, the Hubble Space Telescope often observes in ultraviolet or infrared wavelengths that do not

reach the ground, which is why it would remain a valuable observatory even if ground-based telescopes were some day to match all its visible-light capabilities.

The Hubble Space Telescope is the most famous observatory in space, but there are many others. Most of these observe parts of the electromagnetic spectrum that do not reach the ground. For example, the Chandra X-Ray Observatory is in space because an X-ray telescope would be completely useless on the ground. Table 6.2 lists some of the most important existing and planned telescopes in space.

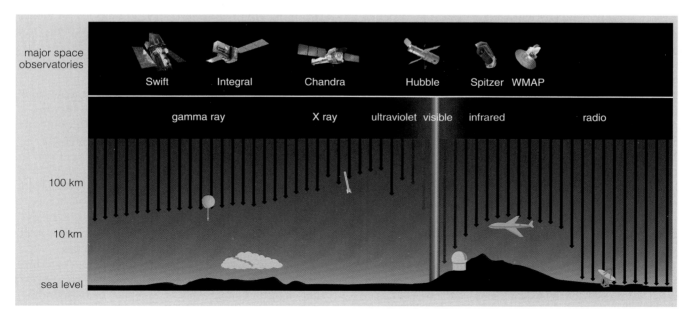

Figure 6.22 This diagram shows the approximate depths to which different wavelengths of light penetrate Earth's atmosphere. Note that most of the electromagnetic spectrum—except for visible light, a small portion of the infrared, and radio waves—can be observed only from very high altitudes or from space.

Table 6.2 Selected Present and Future Major Observatories in Space

Name	Launch Year	Lead Space Agency	Special Features
Hubble Space Telescope	1990	NASA	Optical, infrared, and ultraviolet observations
Far Ultraviolet Spectroscopic Explorer (FUSE)	1999	NASA	Ultraviolet spectroscopy
Chandra X-Ray Observatory	1999	NASA	X-ray imaging and spectroscopy
XMM–Newton	1999	ESA*	European-led mission for X-ray spectroscopy
Wilkinson Microwave Anisotropy Probe (WMAP)	2001	NASA	Study of the cosmic microwave background
International Gamma-Ray Astrophysics Laboratory (INTEGRAL)	2002	ESA*	Gamma-ray imaging, spectroscopy, and timing
Spitzer Infrared Telescope	2003	NASA	Infrared observations of the cosmos
Swift	2004	NASA	Study of gamma-ray bursts
Planck	2007[†]	ESA*	Transit search for extrasolar Earth-like planets
Kepler	2008[†]	NASA	Study of the cosmic microwave background
Space Interferometry Mission (SIM)	2011[†]	NASA	First mission for optical interferometry in space
James Webb Space Telescope (JWST)	2012[†]	NASA	Follow-on to Hubble Space Telescope
Terrestrial Planet Finder (TPF)	2014[†]	NASA	Search for Earth-like planets around other stars

*European Space Agency.

[†] Tentative launch year.

6.4 Telescopes and Technology

Astronomers today are making new discoveries at an astonishing rate, driven largely by the availability of more and larger telescopes, including space telescopes that can observe previously inaccessible portions of the electromagnetic spectrum. However, larger telescopes are not the only fuel for the current astronomical revolution.

Some new technologies make it possible to obtain better images or spectra with existing telescopes. For example, electronic detectors are constantly improving. As a result, a relatively small telescope equipped with the latest camera technology can record images as good as those that could be captured only by much larger telescopes in the past. In addition, adaptive optics allows telescopes to overcome problems of atmospheric distortion.

Other technologies make it possible to record and analyze data more efficiently. For example, obtaining spectra of distant galaxies used to be a very time-consuming and labor-intensive task. Today, astronomers can sometimes obtain hundreds of spectra simultaneously in a single telescopic observation. Once astronomers have these spectra, they use computers to analyze the vast amount of data the spectra contain.

In this section, we'll focus our attention on two important areas of modern telescope technology that we have not discussed previously: the technology used to observe different parts of the electromagnetic spectrum and a remarkable technology that allows many small telescopes to work together to obtain higher-quality images.

• How do we observe nonvisible light?

The basic idea behind all telescopes is the same: to collect as much light as possible with as much resolution as possible. In addition, nearly all telescopes used by professional astronomers (except some telescopes for gamma-ray wavelengths) are essentially reflecting telescopes, using mirrors to bring light to a focus. Nevertheless, telescopes for most nonvisible wavelengths require some variations on the basic design used for visible-light telescopes. Let's investigate, going in order of decreasing wavelength.

Radio Telescopes Although it may take a little thought to realize it, a specialized kind of radio telescope is now the most common type of telescope in the world, so common that you'll see dozens of them on a drive through almost any neighborhood and you may already own one yourself. Why? Because every satellite dish is essentially a small radio telescope designed to collect radio waves from a satellite in Earth orbit. Just by looking at a satellite dish, you can see that it operates by the same basic principles as a reflecting telescope (Figure 6.23). The metal dish is the mirror, and it is shaped to bring the radio waves to a focus in front of the dish; that's where you see the receiver, which functions like the secondary mirror. The receiver collects the radio waves reflected by the primary mirror and sends them to the television (or some other communication device).

The only major differences between satellite dishes and astronomical radio telescopes are where they look in the

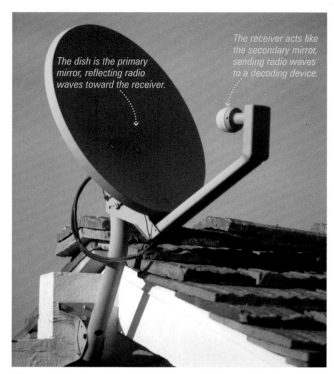

The dish is the primary mirror, reflecting radio waves toward the receiver.

The receiver acts like the secondary mirror, sending radio waves to a decoding device.

Figure 6.23 A satellite TV dish is essentially a small radio telescope.

Figure 6.24 The Arecibo radio telescope stretches across a natural valley in Puerto Rico. At 305 meters across, it is the world's largest single telescope.

sky, their sizes, and their purposes. Communication satellites have geostationary orbits in which they orbit above Earth's equator in exactly the same amount of time that Earth rotates each day (see Mathematical Insight 4.3); thus, a dish aimed at a particular satellite always points to the same spot in the local sky. Astronomical radio telescopes point toward cosmic radio sources that rise and set just like stars in the sky, so long exposures require tracking the sources across the sky. Astronomical radio telescopes are larger than satellite dishes both because cosmic radio sources are fainter than satellite signals (and thus require a larger light-collecting area) and because they are used to make images of the cosmic sources (and thus require decent angular resolution). Angular resolution is unimportant for satellite dishes, because they are not used to make images of the satellites in space; radio and television signals are encoded in the radio waves themselves, so the dish needs only to collect the radio waves and send them to a decoding device like a television.

The long wavelengths of radio waves mean that very large telescopes are necessary to achieve reasonable angular resolution. That is why radio telescopes are so much larger than visible-light telescopes. For example, the world's largest single radio telescope, the Arecibo radio dish, stretches 305 meters (1,000 feet) across a natural valley in Puerto Rico (Figure 6.24). Despite its large size, Arecibo's angular resolution is only about 1 arcminute at commonly observed radio wavelengths (for example, 21 centimeters [Section 19.2])—a few hundred times worse than the visible-light resolution of the Hubble Space Telescope. Fortu-

nately, through an amazing technique that we'll discuss shortly (interferometry), radio telescopes can work together to achieve much better angular resolution.

If you look again at Figure 6.22, you'll see that radio waves are the only form of light besides visible light that we can observe easily from the ground. Moreover, because the atmosphere does not distort radio waves the way it distorts visible light, there's no inherent advantage to observing from space. However, "radio-wave pollution" is an even more serious impediment to radio astronomy than light pollution is to visible-light astronomy. Humans use many portions of the radio spectrum so heavily that radio signals from cosmic sources are almost completely drowned out. Astronomers therefore hope someday to put radio telescopes into deep space or on the far side of the Moon, where human signals will be blocked. In addition, because radio telescopes can be made to work together, putting them into space can in principle allow them to be spread out over a much greater distance.

Infrared Telescopes Most of the infrared portion of the spectrum is close enough in wavelength to visible light to behave quite similarly. Thus, infrared telescopes generally look much the same as visible-light telescopes. In fact, visible-light telescopes can in principle collect and focus much of the infrared spectrum; the practical limitation comes from Earth's atmosphere.

As you can see in Figure 6.22, infrared light from space generally does not reach the ground. A few portions of the infrared spectrum can be observed from high mountaintops, and many mountaintop observatories do indeed observe in the infrared. The higher you go in the atmosphere, the more infrared light becomes accessible. NASA has recently started flying an airborne infrared telescope called SOFIA (Stratospheric Observatory for Infrared Astronomy). SOFIA carries a 2.5-meter infrared telescope that looks out

Figure 6.25 This photograph shows NASA's airborne observatory, SOFIA, during a test flight prior to installation of its 2.5-meter infrared telescope. The black rectangle marks the location where the telescope has since been installed into the fuselage.

Figure 6.26 🎨 This painting shows the Spitzer Infrared Telescope, which trails millions of kilometers behind Earth in the same orbit around the Sun. The background is an artistic rendition of infrared emission from a star-forming cloud.

through a large hole cut in the fuselage of a Boeing 747 airplane (Figure 6.25). Advanced technologies make it possible for the airplane to fly smoothly despite its large hole and to keep the telescope pointed accurately at observing targets during flight.

Extreme infrared light (the longest wavelengths of infrared) poses more difficult observing challenges. Remember that all objects emit thermal radiation characteristic of their temperatures [Section 5.4]. At ordinary temperatures, the ground, an airplane, and even a telescope itself emit enough long-wavelength infrared light to interfere with any attempt to observe these wavelengths from the cosmos. The only solution to this problem is to put the telescopes into space so they get away from Earth's own heat and to cool the telescopes so they emit less and lower-energy infrared light. NASA's Spitzer Infrared Telescope (Figure 6.26), launched in 2003, is cooled with liquid helium to just a few degrees above absolute zero. Spitzer is kept far from Earth's own heat with a unique orbit: Instead of orbiting Earth like the Hubble Space Telescope, Spitzer shares Earth's orbit of the Sun but trails millions of kilometers behind our planet. A Sun shield blocks out light from the Sun.

Ultraviolet Telescopes   Like infrared light, much of the ultraviolet spectrum is close enough in wavelength to visible light to behave similarly. (Very short-wavelength ultraviolet light, sometimes called *extreme ultraviolet*, behaves like X rays, which we'll discuss below.) Thus, visible-light mirrors can in principle collect and focus this ultraviolet light. However, Earth's atmosphere almost completely absorbs ultraviolet light, making most ultraviolet observations impossible from the ground.

At present, there are three major ultraviolet observatories in space: The Far Ultraviolet Spectroscopic Explorer (FUSE), designed exclusively for ultraviolet spectroscopy; the Galaxy Evolution Explorer (GALEX), a relatively small

telescope that is conducting an ultraviolet survey of the entire sky; and the Hubble Space Telescope, which is capable of high-resolution imaging and spectroscopy in ultraviolet light. Hubble's continued operation depends on periodic Space Shuttle missions to repair it, update it, and boost it back up as atmospheric drag lowers its orbit. The last servicing mission was in 2002, and the observatory is unlikely to survive past about 2007 without another servicing mission. Hubble therefore faces an uncertain future, because NASA cancelled future servicing missions in the aftermath of the 2003 Columbia Space Shuttle accident. However, NASA has since agreed to reconsider this decision, and Hubble may yet be serviced in time to keep it alive for at least several more years.

X-Ray Telescopes   Besides needing to be in space, X-ray observations pose another challenge: X rays have sufficient energy to penetrate many materials, including living tissue and ordinary mirrors. While this property makes X rays very useful to medical doctors, it gives astronomers headaches.

Trying to focus X rays is somewhat like trying to focus a stream of bullets. If the bullets are fired directly at a metal sheet, they will puncture or damage the sheet. However, if the metal sheet is angled so that the bullets barely graze its surface, then it will slightly deflect the bullets. Specially designed mirrors can deflect X rays in much the same way. Such mirrors are called **grazing incidence** mirrors because X rays merely graze their surfaces as they are deflected toward the focal plane. X-ray telescopes, such as NASA's Chandra X-Ray Observatory, generally consist of several nested grazing incidence mirrors (Figure 6.27).

Chandra offers the best angular resolution of any X-ray telescope yet built, but a European X-ray telescope called XMM–Newton has a larger light-collecting area. Astronomers therefore use the two observatories in the way best suited to science goals. For example, Chandra is better for

a 🔊 Artist illustration of the Chandra X-ray Observatory, which orbits Earth.

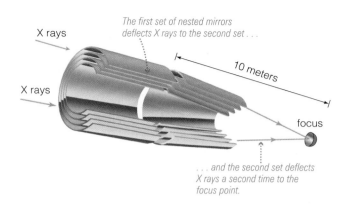

X rays

X rays

The first set of nested mirrors deflects X rays to the second set . . .

10 meters

focus

. . . and the second set deflects X rays a second time to the focus point.

b This diagram shows the arrangement of Chandra's nested, cylindrical X ray mirrors. Each mirror is 0.8 meter long and between 0.6 and 1.2 meters in diameter.

Figure 6.27 The Chandra X-Ray Observatory focuses X rays that enter the front of the telescope by deflecting them twice so that they end up focused at the back of the telescope.

making images of X-ray sources, while XMM–Newton's larger light-collecting area allows it to obtain more detailed X-ray spectra.

<hr>

THINK ABOUT IT

If you look straight down at your desktop, you probably cannot see your reflection. But if you glance along the desktop surface (or another smooth surface, such as that of a book), you should see reflections of objects in front of you. Explain how these reflections represent *grazing incidence* for visible light.

Gamma-Ray Telescopes Gamma rays can penetrate even grazing incidence mirrors and therefore cannot be focused in the traditional sense. Capturing such high-energy light at all requires detectors so massive that the photons cannot simply pass through them. The largest gamma-ray observatory to date was the 17-ton Compton Gamma Ray Observatory (Figure 6.28), which was launched in 1991 and destroyed in a controlled crash to Earth in 2000.

Gamma rays come from a number of different types of astronomical objects, but the most mysterious sources produce short bursts of gamma rays that quickly fade away [Section 18.4]. Because gamma rays are difficult to focus, Compton and other past gamma-ray telescopes were unable to pinpoint the precise locations of the sources of these gamma-ray bursts. They located the sources well enough for us to know they are in distant galaxies, but not well enough for us to identify the objects that produced them. In 2004, NASA launched a new gamma-ray observatory named Swift. In addition to its gamma-ray telescope, the Swift satellite carries X-ray and visible-light telescopes. As soon as Swift detects a burst of gamma rays, these other telescopes rapidly swing toward the source to look for corresponding bursts in X-rays and visible light. Swift thereby

gets much more accurate positions for the sources of the gamma-ray bursts.

Looking Beyond Light We have learned virtually everything we know about distant objects by observing light. However, light is not the only form of information that travels through the universe, and astronomers have begun to build and use telescopes designed to observe at least three other types of "cosmic messengers." First, there's an extremely light-weight type of subatomic particle known as the *neutrino* [Sections S4.2, 14.2] that is produced by nuclear reactions,

Figure 6.28 This photograph shows the Compton Gamma Ray Observatory being deployed from the Space Shuttle in 1991. Compton revolutionized our understanding of the gamma-ray universe before its demise (in a controlled reentry into the atmosphere).

including nuclear fusion in the Sun and the reactions that accompany the explosions of distant stars. Astronomers have already had some success with "neutrino telescopes"—typically located in deep mines or under water or ice—that have provided valuable insights about the Sun and stellar explosions. Second, Earth is continually bombarded by very high-energy subatomic particles from space known as *cosmic rays* [Section 19.2]. We still know little about their origin (including how they are accelerated to such high speeds) and chemical composition, but astronomers are now using both satellites and ground-based detectors to catch and study them. Third, Einstein's general theory of relativity predicts the existence of something called *gravitational waves* [Section S3.4] that are very different in nature from light but travel at the speed of light. For decades, we've had indirect evidence that gravitational waves really exist, but until recently, direct detection of them was beyond our technological capabilities. Today, the first gravitational wave telescopes are up and running in Washington and Louisiana (as well as Italy and Germany), and astronomers hope they will eventually be able to detect gravitational waves from exotic objects like orbiting neutron stars and black holes.

• How can multiple telescopes work together?

Individual telescopes always face limits on their capabilities. Even in space, the diffraction limit places a fundamental constraint on the angular resolution of a telescope of any particular size. In addition, while astronomers would always like larger telescopes, the current state of technology and budgetary considerations create practical limits on telescope size.

Figure 6.30 The Very Large Array (VLA) in New Mexico consists of 27 telescopes that can be moved along train tracks. The telescopes work together through interferometry and can achieve an angular resolution equivalent to that of a single radio telescope almost 40 kilometers across.

These constraints ultimately limit the amount of light that we can collect with telescopes. Even if we put a bunch of telescopes together, there's no getting around the fact that their total light-collecting area is simply the sum of their individual areas. However, remember that the two key properties of a telescope are light-collecting area and angular resolution. Amazingly, astronomers have found a way to make the angular resolution of a group of telescopes far better than that of any individual telescope.

Interferometry In the 1950s, radio astronomers developed an ingenious technique for improving the angular resolution of radio telescopes: They learned to link two or more individual telescopes to achieve the angular resolution of a much larger telescope (Figure 6.29). This technique is called **interferometry** because it works by taking advantage of the wavelike properties of light that cause interference (see Figure 6.7). The procedure relies on precisely timing when radio waves reach each dish and on using computers to analyze the resulting interference patterns.

One famous example of radio interferometry, the Very Large Array (VLA) near Socorro, New Mexico, consists of 27 individual radio dishes that can be moved along railroad tracks laid down in the shape of a Y (Figure 6.30). The light-gathering capability of the VLA's 27 dishes is simply equal to their combined area, equivalent to that of a single telescope 130 meters across. But its angular resolution is equivalent to that of a much larger telescope: When the 27 dishes are spaced as widely as possible, the VLA can achieve an angular resolution that otherwise would require a single radio telescope with a diameter of almost 40 kilometers. Today, astronomers can achieve even higher

Interferometry allows these two small radio telescopes to obtain the angular resolution of a much larger telescope.

Figure 6.29 This diagram shows the basic idea behind interferometry: Smaller telescopes work together to obtain the angular resolution of a much larger telescope. Note that interferometry improves angular resolution but does *not* affect the total light-collecting area, which is simply the sum of the light-collecting areas of the individual telescopes.

Figure 6.31 🎨 An artist's conception of a possible future lunar observatory. The Moon has no atmosphere to create distortion and has a solid surface that makes it ideal as a home for astronomical observatories.

angular resolution by linking radio telescopes around the world.

Interferometry is more difficult for shorter-wavelength (higher-frequency) light, but astronomers are rapidly learning to use the technique beyond the radio portion of the spectrum. Some success has already been achieved at infrared and visible wavelengths. Indeed, new telescopes are now often built in pairs (such as the Keck and Magellan telescope pairs) or with more than one telescope on a common mount (such as the Large Binocular Telescope) so that they can be used for infrared and visible-light interferometry. In addition, astronomers are testing technologies that may allow interferometry to be extended all the way to X rays.

The Future of Astronomy in Space   It will always be cheaper and easier to do astronomy from ground-based observatories, but space is likely to play an ever increasing role in astronomy. NASA is already at work on follow-ons to the Hubble Space Telescope, such as the James Webb Space Telescope (JWST), which may launch as early as 2012. Later telescopes are likely to be interferometers, using several telescopes together to achieve the angular resolution of far larger telescopes. The potential value of such interferometers is enormous. For example, within the next two decades astronomers hope to launch a suite of small space telescopes that could work together as an interferometer to achieve the angular resolution of a single telescope hundreds of meters across [Section 13.4].

For the more distant future, many astronomers dream of an observatory on the Moon (Figure 6.31). Because the Moon has no atmosphere, it offers all the advantages of telescopes in space while also offering the ease of operating on a solid surface. Some day, astronomers may use telescopes in space or on the Moon as giant interferometers, offering views of distant objects that may be as detailed in comparison to Hubble Space Telescope images as Hubble's images are in comparison to the naked eye's.

THE BIG PICTURE

Putting Chapter 6 into Context

In this chapter, we've focused on the technological side of astronomy: the telescopes that we use to learn about the universe. Keep in mind the following "big picture" ideas as you continue to learn about astronomy:

● Technology drives astronomical discovery. Every time we build a bigger telescope, develop a more sensitive detector, or open up a new wavelength region to study, we learn more about the universe than was previously possible.

● Telescopes work much like giant eyes, enabling us to see the universe in great detail. New technologies for making larger telescopes, along with advances in adaptive optics and interferometry, are making ground-based telescopes more powerful than ever.

● For the ultimate in observing the universe, space is the place! Telescopes in space allow us to detect light from across the entire spectrum while also avoiding the distortion caused by Earth's atmosphere.

6.1 Eyes and Cameras: Everyday Light Sensors

- **How does your eye form an image?** Your eye brings parallel rays of light to a **focus** on your retina. Glass lenses work similarly, so distant objects form an **image** that is in focus on the **focal plane**.

- **How do we record images?** We record images with a **detector** such as photographic film or a **CCD**. Recorded images are more reliable than just seeing by eye and can record more light if we use a longer **exposure time**.

6.2 Telescopes: Giant Eyes

- **What are the two most important properties of a telescope?** A telescope's most important properties are its **light-collecting area**, which determines how much light it gathers, and its **angular resolution**, which determines how much detail we can see in its images.

- **What are the two basic designs of telescopes?** A **refracting telescope** forms an image by bending light through a lens. A **reflecting telescope** forms an image by focusing light with mirrors.

- **What do astronomers do with telescopes?** The three primary uses of telescopes are **imaging** to create pictures of distant objects, **spectroscopy** to study the spectra of distant objects, and **timing** to study how a distant object's brightness changes with time.

6.3 Telescopes and the Atmosphere

- **How does Earth's atmosphere affect ground-based observations?** Earth's atmosphere limits visible-light observations to nighttime and clear weather. **Light pollution** can lessen the quality of observations, and atmospheric **turbulence** makes stars twinkle, blurring their images. The technology of **adaptive optics** can overcome some of the blurring due to turbulence.

- **Why do we put telescopes into space?** Telescopes in space are above Earth's atmosphere and the problems it causes for observations. Most important, telescopes in space can observe all wavelengths of light, while telescopes on the ground can observe only visible light, radio waves, and small portions of the infrared.

6.4 Telescopes and Technology

- **How do we observe nonvisible light?** Telescopes for nonvisible light often use variations on the basic design of a reflecting telescope. Radio telescopes use large metal dishes as their primary mirrors. Infrared telescopes are sometimes cooled to very low temperature. X-ray telescopes use **grazing incidence** reflections rather than direct reflections.

- **How can multiple telescopes work together?** The technique of **interferometry** allows multiple telescopes to be linked in a way that allows them to obtain the angular resolution of a much larger telescope.

EXERCISES AND PROBLEMS

For instructor-assigned homework go to MasteringASTRONOMY www.masteringastronomy.com

Review Questions

Short-Answer Questions Based on the Reading

1. How does your eye focus light? How is a glass lens similar? What do we mean by the *focal plane* of a lens?
2. For purposes of astronomy, what advantages does a camera have over the human eye? What advantages do *CCDs* have over photographic film?

3. What are the two key properties of a telescope, and why is each important?
4. What is the *diffraction limit*, and how does it depend on a telescope's size and the wavelength of light being observed?
5. How do *reflecting telescopes* differ from *refracting telescopes*? Which type is more commonly used by professional astronomers, and why?

6. What are the three basic categories of astronomical observation, and how is each conducted?

7. What do we mean when we speak of images made from nonvisible light, such as X-ray or infrared images? What do the colors in these images mean?

8. What do we mean by *spectral resolution*? Why is higher spectral resolution more difficult to achieve?

9. List at least three ways in which Earth's atmosphere can hinder astronomical observations, and explain why putting a telescope into space helps in each case. What problem can *adaptive optics* help with?

10. Study Figure 6.22 and describe how deeply each portion of the electromagnetic spectrum penetrates Earth's atmosphere. Based on your answers, why is space astronomy so important to our understanding of the universe?

11. How do telescopes for nonvisible wavelengths differ from those for visible light? Answer for each major wavelength band and give examples of important observatories in those bands.

12. What is interferometry, and how can it improve astronomical observations?

Test Your Understanding

Does It Make Sense?

Decide whether the statement makes sense (or is clearly true) or does not make sense (or is clearly false). Explain your reasoning. (For an example, see Chapter 1, "Does It Make Sense?")

13. The image was blurry because the photographic film was not placed at the focal plane.

14. By using a CCD, I can photograph the Andromeda Galaxy with a shorter exposure time than I would need with photographic film.

15. I have a reflecting telescope in which the secondary mirror is bigger than the primary mirror.

16. The photograph shows what appears to be just two distinct stars, but each of those stars is actually a binary star system.

17. I built my own 14-inch telescope that has a lower diffraction limit than most large professional telescopes.

18. Now that I've bought a spectrograph, I can use my home telescope for spectroscopy as well as imaging.

19. If you lived on the Moon, you'd never see stars twinkle.

20. New technologies will soon allow astronomers to use X-ray telescopes on Earth's surface.

21. Thanks to adaptive optics, the telescope on Mount Wilson can now make ultraviolet images of the cosmos.

22. Thanks to interferometry, a properly spaced set of 10-meter radio telescopes can achieve the angular resolution of a single 100-kilometer radio telescope.

Quick Quiz

Choose the best answer to each of the following. Explain your reasoning with one or more complete sentences.

23. How much greater is the light-collecting area of a 6-meter telescope than a 3-meter telescope? (a) two times (b) four times (c) six times

24. Suppose that two stars are separated in the sky by 0.1 arcsecond. If you look at them with a telescope that has an angular resolution of 0.5 arcsecond, what will you see? (a) Two distinct stars. (b) One point of light that is the blurred image of both stars. (c) Nothing at all.

25. The diffraction limit is a limit on: (a) A telescope's size. (b) A telescope's angular resolution. (c) A telescope's spectral resolution.

26. The Hubble Space Telescope obtains higher-resolution images than most ground-based telescopes because it is: (a) larger. (b) closer to the stars. (c) above Earth's atmosphere.

27. If you see the color red in an X-ray image from the Chandra X-Ray Observatory, it means: (a) the object really is red in color. (b) the red parts are hotter than the bluer parts. (c) it depends: the colors are chosen arbitrarily to represent something about the X rays recorded by the telescope.

28. The twinkling of stars is caused by: (a) variations in stellar brightness with time. (b) light pollution. (c) motion of air in our atmosphere.

29. If you wanted a radio telescope to achieve the same angular resolution as a visible-light telescope, it would need to be: (a) much larger. (b) slightly larger. (c) in space.

30. Where should you put a telescope designed for ultraviolet observations? (a) in Earth orbit. (b) on an airplane. (c) on a high mountaintop.

31. Which technology can allow a single ground-based telescope to achieve images as sharp as those from the Hubble Space Telescope? (a) adaptive optics (b) grazing incidence (c) interferometry

32. Interferometry uses two or more telescopes to achieve: (a) a light-collecting area equivalent to that of a much larger telescope. (b) an angular resolution equivalent to that of a much larger telescope. (c) both the light-collecting area and angular resolution of a much larger telescope.

Investigate Further

In-Depth Questions to Increase Your Understanding

Short-Answer/Essay Questions

33. *Image Resolution.* What happens if you take a photograph from a newspaper, magazine, or book and blow it up to a larger size? Can you see any more detail than you could see before? Explain clearly, and relate your answer to the concepts of magnification and angular resolution in astronomical observations.

34. *Type of Observation.* For each of the following, decide what type of observation (imaging, spectroscopy, timing) you would need to make. Explain clearly.
    a. Studying how a star's hot upper atmosphere changes with time.
    b. Learning the composition of a distant star.
    c. Determining how fast a distant galaxy is moving away from Earth.

35. *Telescope Location.* Where do you live? In light of the problems faced by ground-based observatories, is your location a good place for an astronomical observatory? Why or why not?

36. *Telescope Technology.* Suppose you were building a space-based observatory consisting of five individual telescopes. Which would be the best way to use these telescopes: as five individual telescopes with adaptive optics, or as five telescopes without adaptive optics linked together for interferometry? Explain your reasoning clearly.

37. *Filters.* Describe what an American flag would look like if you viewed it through a filter that transmits only red light. What would it look like through a filter that transmits only blue light?

38. *Megapixels.* Digital cameras whose detectors have more than 6 million pixels are advertised to have over 6 megapixels, and that number will surely get greater in the future. Astronomical instruments are also boasting ever greater numbers of pixels. Is it better to use those extra pixels to observe larger portions of the sky or to observe smaller details in the sky? Explain how the answer to this question depends on the type of telescope being used and recent advances in telescope technology.

39. *Lunar Observatory.* Some astronomers dream of someday having an observatory on the Moon. What advantages would a lunar observatory have over ground-based observatories on Earth? What advantages would it have over space-based observatories? What would the disadvantages be?

40. *Project: Twinkling Stars.* Using a star chart, identify 5–10 bright stars that should be visible in the early evening. On a clear night, observe each of these stars for a few minutes. Note the date and time, and for each star record the following information: approximate altitude and direction in your sky, brightness compared to other stars, color, and how much the star twinkles compared to other stars. Study your record. Can you draw any conclusions about how brightness and position in your sky affect twinkling? Explain.

41. *Project: Radio Pollution.* The rapid growth in wireless communications has vastly increased the amount of radio "noise" in the vicinity of radio telescopes. However, a few sections of the radio spectrum are protected for the sake of radio astronomy. Find out what those regions are and explain why they were chosen for special protection.

42. *Project: Personal Telescope Review.* By going to stores or looking in online or printed catalogs, find three telescopes that you could buy for under $1,000. Evaluate each one against the following criteria: light-collecting area, angular resolution, construction quality, and portability. Give each telescope a rating of 1 to 4 stars (4 is best) and state which one you would recommend for purchase to someone planning to buy a telescope in this price range.

Quantitative Problems

Be sure to show all calculations clearly and state your final answers in complete sentences.

43. *Light-Collecting Area.*
    a. How much greater is the light-collecting area of one of the 10-meter Keck telescopes than that of the 5-meter Hale telescope?
    b. Suppose astronomers built a 100-meter telescope. How much greater would its light-collecting area be than that of the 10-meter Keck telescope?

44. *Close Binary System.* Suppose that two stars in a binary star system are separated by a distance of 100 million kilometers and are located at a distance of 100 light-years from Earth. What is the angular separation of the two stars? Give your answer in both degrees and arcseconds. Can the Hubble Space Telescope resolve the two stars?

45. *Finding Planets.* Suppose you were looking at our own solar system from a distance of 10 light-years.
    a. What angular resolution would you need to see the Sun and Jupiter as distinct points of light?
    b. What angular resolution would you need to see the Sun and Earth as distinct points of light?

c. How do the angular resolutions you found in parts (a) and (b) compare to the angular resolution of the Hubble Space Telescope? Comment on the challenge of making images of planets around other stars.

46. *Diffraction Limit of the Eye.*
    a. Calculate the diffraction limit of the human eye, assuming a wide-open pupil so that your eye acts like a lens with diameter 0.8 centimeter, for visible light of 500-nanometer wavelength. How does this compare to the diffraction limit of a 10-meter telescope?
    b. Now remember that humans have two eyes that are approximately 7 centimeters apart. Estimate the diffraction limit for human vision, assuming that your "optical interferometer" is just as good as one eyeball as large as the separation of two regular eyeballs.

47. *The Size of Radio Telescopes.* What is the diffraction limit of a 100-meter radio telescope observing radio waves with a wavelength of 21 centimeters? Compare this to the diffraction limit of the 2.4-meter Hubble Space Telescope for visible light. Use your results to explain why, to be useful, radio telescopes must be much larger than optical telescopes.

48. *Your Satellite Dish.* Suppose you have a satellite dish that is 0.5 meter in diameter and you want to use it as a radio telescope. What is the diffraction limit on the angular resolution of your dish, assuming that you want to observe radio waves with a wavelength of 21 centimeters? Would it be very useful as an astronomical radio telescope?

49. *Hubble's Field of View.* Large telescopes often have small fields of view. For example, the Hubble Space Telescope's (HST) new advanced camera has a field of view that is roughly square and about 0.06° on a side.
    a. Calculate the angular area of the HST's field of view in square degrees.
    b. The angular area of the entire sky is about 41,250 square degrees. How many pictures would the HST have to take with its camera to obtain a complete picture of the entire sky?

50. *Hubble Sky Survey?* In Problem 49, you found out how many pictures HST would require to photograph the entire sky. If you assume that it would take 1 hour for each picture, how many years would HST need to obtain photos of the entire sky? Use your answer to explain why astronomers would like to have more than one large telescope in space.

51. *Visible-Light Interferometry.* Technological advances are now making it possible to link visible-light telescopes so that they can achieve the same angular resolution as a single telescope over 300 meters in size. What is the angular resolution (diffraction limit) of such a system of telescopes for observations at a wavelength of 500 nanometers?

52. *The Size of an X-Ray Observatory.* Spacecraft designed to observe X rays from space need to be quite long because X rays are so hard to focus. Photons entering on parallel paths get deflected by the telescope's mirrors onto paths that eventually converge at a focus, but they are deflected by only about 2 degrees from their original direction. On the Chandra Observatory the largest X-ray deflecting mirrors are about 4 meters in diameter. Based on this information, estimate the length of the Chandra X-Ray Observatory.

53. *Science and Technology Funding.* Technological innovation clearly drives scientific discovery in astronomy, but the reverse is also true. For example, Newton made his discoveries in part because he wanted to explain the motions of the planets, but his discoveries have had far-reaching effects on our civilization. Congress often must make decisions between funding programs with purely scientific purposes ("basic research") and programs designed to develop new technologies. If you were a member of Congress, how would you try to allocate spending between basic research and technology? Why?

54. *A Lunar Observatory.* Do the potential benefits of building an astronomical observatory on the Moon justify its costs at the present time? If it were up to you, would you recommend that Congress begin funding such an observatory? Defend your opinions.

MEDIA EXPLORATIONS

For self-study activities go to **Mastering ASTRONOMY** www.masteringastronomy.com

 Interactive Tutorials

Tutorial Review of Key Concepts
Use the interactive **Tutorials** at **www.mastering astronomy.com** to review key concepts from this chapter.

Light and Spectroscopy Tutorial
Lesson 2 Spectroscopy
Lesson 3 Atomic Spectra—Emission and Absorption Lines
Lesson 4 Thermal Radiation

Telescopes Tutorial
Lesson 1 Optics and Light Gathering Power
Lesson 2 Angular Resolution

Supplementary Tutorial Exercises
Use the interactive **Tutorial Lessons** to explore the following questions.

Light and Spectroscopy Tutorial, Lessons 2–4
1. How would an object's spectrum influence the choice of telescope used to observe it?
2. Consider the spectra of hot stars as compared to those of cooler stars. Which stars would be best observed with an ultraviolet telescope? Explain.
3. Are there spectral features that can be observed only with telescopes in space? Explain.

Telescopes Tutorial, Lesson 1
1. To your eye, a star looks like a point of light in the sky, but the diagrams in this tutorial lesson show that the light from a single star enters a telescope as a series of parallel light rays. Why?
2. What is the purpose of the secondary mirror in a reflecting telescope? Why doesn't the fact that it blocks some starlight create a major problem for observations?
3. Describe how the light-collecting area of a reflecting telescope depends on its diameter.

Telescopes Tutorial, Lesson 2
1. Why do many stars that appear as single stars to the naked eye appear as pairs of stars through a telescope?
2. Describe how the angular resolution of a telescope depends on the diameter of the primary mirror.
3. Which telescope has better angular resolution: a 1-meter telescope observing ultraviolet light of wavelength 2×10^{-7} m, or a 10-meter telescope observing infrared light with wavelength 1×10^{-5} m? Explain.

Web Projects

1. *Major Ground-Based Observatories.* Take a virtual tour of one of the world's major astronomical observatories. Write a short report on why the observatory is useful to astronomy.
2. *Space Observatory.* Visit the Web site of a major space observatory, either existing or under development. Write a short report about the observatory, including its purpose, its orbit, and how it operates.
3. *Really Big Telescopes.* Several studies are under way in hopes of building telescopes far larger than any now in operation. Learn about one or more of these projects (such as OWL, the Swedish 50-m Optical Telescope, or the Thirty Meter Telescope [TMT]), and write a short report about the telescope's prospects and potential capabilities.

7

Our Planetary System

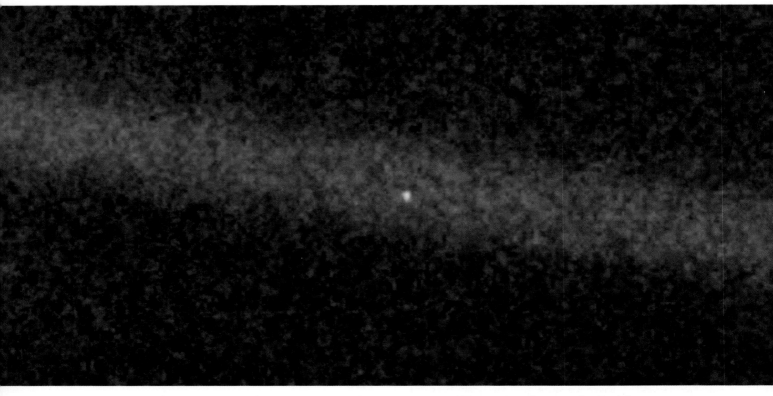

Earth, photographed from the outskirts of our solar system by the Voyager spacecraft. The "sunbeam" surrounding Earth is an artifact of light scattering in the camera.

LEARNING GOALS

7.1 Studying the Solar System
- **What does the solar system look like?**
- **What can we learn by comparing the planets to one another?**
- **What are the major features of the Sun and planets?**

7.2 Patterns in the Solar System
- **What features of our solar system provide clues to how it formed?**

7.3 Spacecraft Exploration of the Solar System
- **How do robotic spacecraft work?**

We succeeded in taking that picture [left], and, if you look at it, you see a dot. That's here. That's home. That's us. On it, everyone you ever heard of, every human being who ever lived, lived out their lives . . . on a mote of dust, suspended in a sunbeam.

—Carl Sagan

Our ancestors long ago recognized the motions of the planets through the sky, but it has been only a few hundred years since we learned that Earth is also a planet that orbits the Sun. Even then, we knew little about the other planets until the advent of large telescopes. More recently, the dawn of space exploration has brought us far greater understanding of other worlds. We've lived in this solar system all along, but only now are we getting to know it.

In this chapter, we'll explore our solar system like newcomers to the neighborhood. We'll begin by discussing what we hope to learn by studying the solar system, and in the process we'll take a brief tour of major features of the Sun and planets. We'll then consider some of the overall patterns we observe in the solar system—patterns that we will seek to explain in subsequent chapters. Finally, we'll discuss the use of spacecraft to explore the solar system, examining how we are coming to learn so much more about our neighbors.

7.1 Studying the Solar System

Galileo's telescopic observations began a new era in astronomy in which the Sun, Moon, and planets could be studied for the first time as *worlds*, rather than as mere lights in the sky. Since that time, people have studied these worlds in different ways. Sometimes we study them individually—for example, when we seek to map the geography of Mars or to probe the atmospheric structure of Jupiter. Other times we compare the worlds to one another, seeking to understand their similarities and differences. This latter approach is called **comparative planetology**. Astronomers use the term *planetology* broadly to include moons, asteroids, and comets as well as planets.

We will use the comparative planetology approach for most of our study of the solar system in this book. Before we can compare the planets, however, we must have a general idea of the nature of our solar system and of the characteristics of individual worlds.

Scale of the Universe Tutorial, Lesson 1

• What does the solar system look like?

The first step in getting to know our solar system is to visualize what it looks like as a whole. Imagine having the perspective of an alien spacecraft making its first scientific survey of our solar system. What would we see as we viewed the solar system from beyond the orbits of the planets?

Without a telescope, the answer would be "not much." Remember that the Sun and planets are all quite small compared to the distances between them [Section 1.2]—so small that if we viewed them from the outskirts of our solar system, the planets would be only pinpoints of light, and even the Sun would be just a small bright dot in the sky. But if we magnify the sizes of the planets by about a million times compared to their distances from the Sun, and show their orbital paths, we'd get the picture in Figure 7.1 (on the next page). Note that, even with this magnification, we do not see the many asteroids and comets that orbit the Sun, although the figure shows the location of the asteroid belt where many asteroids orbit. The figure does not show moons, either, although a few of the largest moons would be barely visible on the magnified scale.

As you study Figure 7.1, you'll quickly see that our solar system is *not* a random collection of worlds. For example, all the planets orbit the Sun in the same direction and in nearly the same plane, and the four inner planets are much smaller and closer together than the next four planets.

Additional patterns become clear when we look at the solar system from a viewpoint high above the ecliptic plane, as in Figure 7.2. This perspective shows us that most of the planets have nearly circular orbits, and the small white circles around all the planets except Mercury and Venus (which have no moons) show that most major moons orbit their planets in the same direction that their planets rotate. Again, the planet sizes in Figure 7.2 are magnified about a million times compared to the scale used for the orbits. The planets would be far too small to see if we tried to show them on the same scale as the orbits in this figure.

THINK ABOUT IT

Do you see any other patterns in Figures 7.1 and 7.2? For example, do you see any patterns to rotation or axis tilt? Are there exceptions to any of the general patterns you see?

The patterns revealed in Figures 7.1 and 7.2 suggest an organization to the solar system that is unlikely to have arisen by random processes. In science, we must therefore seek an explanation for the existence of these patterns. Indeed, the development of a theory that seems to explain most of the observed patterns has been one of the major triumphs of the scientific study of the solar system. We'll discuss this theory in Chapter 8, after we complete our overview of the solar system and search for additional patterns and features that a successful theory must explain.

Formation of the Solar System Tutorial, Lesson 1

• What can we learn by comparing the planets to one another?

We've stated that we'll take a comparative approach to our study of the worlds in our solar system, but what exactly

Figure 7.1 Interactive Figure The layout of our solar system as it would appear from beyond Pluto, if we could magnify the sizes of the planets by about a million times. Notice that all the planets orbit the Sun in the same direction. The tilt of each planet's rotation axis is also shown, with a circling arrow to indicate the direction of rotation. The dots between Mars and Jupiter represent the asteroid belt.

Close-up of
Inner Solar System

The Solar System

Mars

Earth

Mercury

Sun

Venus

Jupiter

Saturn

Uranus

Neptune

Pluto

*Orbits to scale; planet sizes
exaggerated about 1 million
times relative to orbits.
Sun not to scale.*

does this mean? The essence of comparative planetology lies in the idea that we can learn more about an individual world, including our own Earth, by studying it in the context of other objects in our solar system. It is much like learning more about a person by getting to know his or her family, friends, and culture.

While we still can learn much by studying planets individually, the comparative planetology approach has demonstrated its value in at least three key ways:

- Comparative study has revealed similarities and differences among the planets that have helped guide the development of our theory of solar system formation, thereby giving us a better understanding of how we came to exist here on Earth.
- Comparative study has given us deep new insights into the physical processes that have shaped Earth and other worlds—insights that can help us better understand and manage our own planet.
- Comparative study has allowed us to apply lessons from our solar system to the study of the many recently discovered planetary systems around other stars. These lessons help us understand both the general principles that govern star systems and the specific circumstances under which Earth-like planets—and possibly life—might exist elsewhere.

The comparative planetology approach should also benefit you as a student by helping you stay focused on *processes* rather than on a collection of facts. After more than four decades of planetary exploration, the "facts" can easily prove overwhelming; for example, you can find atlases with names and descriptions for hundreds of mountains, craters, and valleys on Mars alone. Your goal should not be to memorize all these features of individual worlds—though it will be useful to know names for a handful of the most significant features—but rather to think about how these features came to exist and why the features on one world are similar to or different from those on other worlds. By staying focused on this goal, you'll gain a far deeper understanding of the history of Earth and the rest of our solar system.

• What are the major features of the Sun and planets?

We cannot achieve our goal of comparing the planets unless we first know a little bit about them. The following 10 pages offer a brief tour through our solar system, beginning at the Sun and continuing to each of the planets. The tour highlights a few of the most important features of each world we visit—just enough information so that you'll be ready for the comparative study we'll undertake in later chapters. The side of each page shows the planets to scale, using the 1-to-10-billion scale introduced in Chapter 1. The map along the bottom of each page shows the locations of the Sun and each of the planets in the Voy-

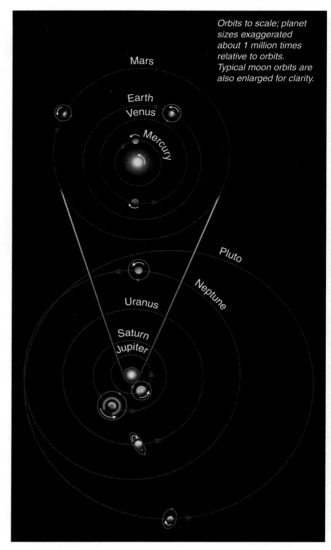

Figure 7.2 Interactive Figure View of the solar system from high above Earth's North Pole. Notice that most orbits are nearly circular. The short, curved arrow for each planet indicates the planet's direction of rotation. The small circles around the planets indicate the orbital direction of major moons. (Mercury and Venus have no moons.) Notice that most major moons orbit in the same direction that their planet rotates.

age scale model solar system in Washington, DC (see Figures 1.6 and 1.7), so that you can see relative distances from the Sun. Table 7.1, which follows the tour, summarizes key planetary data.

THINK ABOUT IT

As you read through the tour pages, make a brief list of the similarities and differences you notice among the planets. In addition, try to answer the following questions as you read: (1) Are all planets made of the same ingredients? (2) Are all planets "Earth-like" with solid surfaces? (3) Can you organize the planets into groups with common characteristics?

Figure 7.3 The Sun contains more than 99.9% of the total mass in our solar system.

Earth shown for size comparison

a A visible-light photograph of the Sun's surface. The dark splotches are sunspots—each large enough to swallow several Earths.

b This ultraviolet photograph, from the SOHO spacecraft, shows a huge streamer of hot gas on the Sun. The image of Earth was added for size comparison.

• The Sun

- Radius: 695,000 km = 108 R_{Earth}
- Mass: 333,000 M_{Earth}
- Composition (by mass): 98% hydrogen and helium, 2% other elements

The Sun is by far the largest and brightest object in our solar system. It contains more than 99.9% of the solar system's total mass, making it more than a thousand times as massive as everything else in the solar system combined.

The Sun's surface looks solid in photographs (Figure 7.3), but it is actually a roiling sea of hot (about 5,800 K, or 6,100°C or 11,000°F) hydrogen and helium gas. The surface is speckled with sunspots that appear dark in photographs only because they are slightly cooler than their surroundings. Solar storms sometimes send streamers of hot gas soaring far above the surface.

The Sun is gaseous throughout. If you could plunge into the Sun, you'd find ever higher temperatures as you went deeper. The source of the Sun's energy lies deep in its core, where the temperatures and pressures are so high that the Sun is a nuclear fusion power plant. Each sec-

ond, fusion transforms about 600 million tons of the Sun's hydrogen into 596 million tons of helium. The "missing" 4 million tons becomes energy in accord with Einstein's famous formula $E = mc^2$ [Section 4.3]. Despite losing 4 million tons of mass each second, the Sun contains so much hydrogen that it has already shone steadily for almost 5 billion years and will continue to shine for some 5 billion years more.

The Sun is certainly the most influential object in our solar system. Its gravity governs the orbits of the planets. Its heat is the primary influence on the temperatures of planetary surfaces and atmospheres. And it is the source of virtually all the visible light in our solar system—the Moon and planets shine only by virtue of the sunlight they reflect. In addition, charged particles flowing outward from the Sun (the solar wind) help shape planetary magnetic fields and can influence planetary atmospheres. Nevertheless, we can understand almost all the present characteristics of the planets without knowing much more about the Sun than what we have just discussed. We'll therefore save more detailed study of the Sun for Chapter 14, where we will study it as our prototype for understanding other stars.

The Voyage scale model solar system represents sizes and distances in our solar system at one ten-billionth of their actual values (see Figure 1.7). The strip along the side of the page shows the sizes of the Sun and planets on this scale, and the map above shows their locations in the Voyage model on the National Mall in Washington, D.C. The Sun is about the size of a large grapefruit on this scale.

Figure 7.4 This image shows what it would look like to be orbiting a few hundred kilometers above Mercury's surface with your back toward the Sun. The view was created using imagery from NASA's *Mariner 10* spacecraft but with computer manipulation to provide color and the orbital viewpoint. The inset (right) is a composite photograph of the full disk of Mercury as seen by *Mariner 10*; the blank strip at the upper right was not photographed. (Image at left from the Voyage scale model solar system, developed by the Challenger Center for Space Science Education, the Smithsonian Institution, and NASA. Image created by ARC Science Simulations © 2001.)

• Mercury

- Average distance from the Sun: 0.39 AU
- Radius: 2,440 km = 0.38 R_{Earth}
- Mass: 0.055 M_{Earth}
- Average density: 5.43 g/cm^3
- Composition: rocks, metals
- Average surface temperature: 700 K (day), 100 K (night)
- Moons: 0

Mercury is the innermost planet of our solar system, and the smallest of the first eight planets. It is a desolate, cratered world with no active volcanoes, no wind, no rain, and no life. Because there is virtually no air to scatter sunlight or color the sky, you could see stars even in the daytime if you stood on Mercury with your back toward the Sun.

You might expect Mercury to be very hot because of its closeness to the Sun, but in fact it is a world of both hot and cold extremes. Tidal forces from the Sun have forced Mercury into an unusual rotation pattern [Section 4.5]: Its 58.6-day rotation period means it rotates exactly three times for every two of its 87.9-day orbits of the Sun. This combination of rotation and orbit gives Mercury days and nights that last about three Earth months each. Daytime temperatures reach 425°C—nearly as hot as hot coals. At

night or in shadow, the temperature falls below −150°C —far colder than Antarctica in winter.

Mercury is the least studied of the inner planets. Its proximity to the Sun makes it difficult to observe through telescopes, and it has been visited by just one spacecraft: *Mariner 10*, which obtained images of one hemisphere (Figure 7.4) during three rapid flybys of Mercury in 1974–1975. The images show vast numbers of craters, ancient lava flows, and tall, steep cliffs that run hundreds of kilometers in length. As we'll discuss in Chapter 9, the cliffs may be wrinkles from an episode of "planetary shrinking" early in Mercury's history.

The influence of Mercury's gravity on *Mariner 10*'s orbit allowed scientists to determine Mercury's mass and to make inferences about its interior composition and structure. Mercury appears to be made mostly of iron, making it the most metal-rich of the planets. Scientists hope to learn much more about Mercury soon. A NASA spacecraft called *MESSENGER*, launched in 2004, will observe Mercury during three flybys in 2008 and 2009 and then begin orbiting Mercury in 2011.

Figure 7.5 The image above shows an artistic rendition of the surface of Venus as scientists think it would appear to our eyes. The surface topography is based on data from NASA's *Magellan* spacecraft. The inset (left) shows the full disk of Venus photographed by NASA's *Pioneer Venus Orbiter* with cameras sensitive to ultraviolet light. With visible light, cloud features cannot be distinguished from the general haze. (Image above from the Voyage scale model solar system, developed by the Challenger Center for Space Science Education, the Smithsonian Institution, and NASA. Image by David P. Anderson, Southern Methodist University © 2001.)

• Venus

- Average distance from the Sun: 0.72 AU
- Radius: 6,051 km = 0.95 R_{Earth}
- Mass: 0.82 M_{Earth}
- Average density: 5.24 g/cm^3
- Composition: rocks, metals
- Average surface temperature: 740 K
- Moons: 0

Venus, the second planet from the Sun, is nearly identical in size to Earth. Prior to the era of spacecraft visits, Venus stood out largely for its strange rotation: It rotates on its axis very slowly and in the opposite direction of Earth, so that days and nights are very long and the Sun rises in the west and sets in the east instead of rising in the east and setting in the west. Its surface is completely hidden from view by dense clouds, so we knew little about it until a few decades ago, when spacecraft carrying cloud-penetrating radar finally allowed us to study Venus in detail (Figure 7.5). Because we knew so little about it, some science fiction writers used its Earth-like size, thick atmosphere, and closer distance to the Sun to speculate that it might be a lush, tropical paradise—a "sister planet" to Earth.

The reality is far different. We now know that an extreme *greenhouse effect* bakes Venus's surface to an incredible 470°C (about 880°F), trapping heat so effectively that nighttime offers no relief. Day and night, Venus is hotter than a pizza oven. All the while, the thick atmosphere bears down on the surface with a pressure equivalent to what exists nearly a kilometer (0.6 mile) beneath the ocean's surface on Earth. Even if we could somehow survive these hazards, we'd find no oxygen to breathe and no water to drink. Far from being a beautiful sister planet to Earth, Venus resembles a traditional view of hell.

The surface of Venus has mountains, valleys, and craters and shows many signs of past (and possibly present) volcanic activity. Aside from these superficial similarities, however, the geology of Venus's surface appears to be quite different from that of Earth. We will devote much of Chapters 9 and 10 to understanding how and why Venus became so unlike Earth, both in its surface geology and in its atmosphere.

a This image (left), computer generated from satellite data, shows the striking contrast between the daylight and nighttime hemispheres of Earth. The day side reveals little evidence of human presence, but at night our presence is revealed by the lights of human activity (mostly from cities as well as agricultural, oil, and gas fires). (From the Voyage scale model solar system, developed by the Challenger Center for Space Science Education, the Smithsonian Institution, and NASA. Image created by ARC Science Stimulations © 2001.)

b Earth and the Moon, shown to scale. The Moon is about one-fourth as large as Earth in diameter, while its mass is about 1/80 Earth's mass. If you wanted to show the distance between Earth and Moon on the same scale, you'd need to hold these two photographs about 1 meter (3 feet) apart.

Figure 7.6 Earth, our home planet.

• Earth

- Average distance from the Sun: 1.00 AU
- Radius: 6,378 km $= 1$ R_{Earth}
- Mass: 1.00 M_{Earth}
- Average density: 5.52 g/cm^3
- Composition: rocks, metals
- Average surface temperature: 290 K
- Moons: 1

Beyond Venus, we next encounter our home planet. Although Earth is a barely visible speck when viewed on the scale of our entire solar system, it is the only known oasis of life. Even if we someday discover microscopic life on other worlds in our solar system, Earth is the only world on which humans could survive without a protective enclosure. It is the only planet with oxygen for us to breathe and ozone to shield us from deadly solar radiation. It is the only planet with abundant surface water to nurture life. Temperatures are pleasant for us because Earth's atmosphere contains just enough carbon dioxide and water vapor to maintain a moderate greenhouse effect.

Despite Earth's small size, its beauty is striking (Figure 7.6a). Blue oceans cover nearly three-fourths of the surface, broken by the continental land masses and scattered islands. The polar caps are white with snow and ice, and white clouds are scattered above the surface. At night, the glow of artificial lights clearly reveals the presence of an intelligent civilization.

Earth is the first planet on our tour with a moon. Moreover, our Moon is surprisingly large in comparison to Earth (Figure 7.6b); although it is not the largest moon in the solar system, almost all other moons are much smaller in comparison to the planets they orbit. The question of why Earth has such a large moon has long been one of the major mysteries of our solar system.

Figure 7.7 The image below shows the surface of Mars photographed by NASA's *Opportunity* rover in 2004. Notice the tracks leading away from the casing that held the rover as it descended to the surface. The globe shows the full disk of Mars; the horizontal "gash" across the center is the giant canyon Valles Marineris.

Mars

- Average distance from the Sun: 1.52 AU
- Radius: $3,397$ km $= 0.53\ R_{Earth}$
- Mass: $0.11\ M_{Earth}$
- Average density: 3.93 g/cm^3
- Composition: rocks, metals
- Average surface temperature: 220 K
- Moons: 2 (very small)

The next planet on our tour is Mars, the last of the four inner planets of our solar system (Figure 7.7). Mars is larger than Mercury and the Moon but smaller than Venus and Earth. About half Earth's size in diameter, Mars has a mass about 10% that of Earth. Mars has two tiny moons, Phobos and Deimos, that look much like typical asteroids and may once have roamed freely in the asteroid belt. These moons are too small to influence Mars in any of the important ways that our Moon influences Earth (such as tides).

Mars is a world of wonders, with ancient volcanoes that dwarf the largest mountains on Earth, a great canyon that runs nearly one-fifth of the way around the planet, and polar caps made of frozen carbon dioxide ("dry

ice") and water ice. Although Mars is frozen today, the presence of dried-up riverbeds, rock-strewn floodplains, and minerals that form in water offers clear evidence that Mars had at least some warm and wet periods in the distant past. Thus, Mars may once have been hospitable to life, though major flows of liquid water probably ceased at least 2–3 billion years ago.

Mars looks almost Earth-like in photographs taken by spacecraft on its surface, but you wouldn't want to visit without a space suit. The air pressure is far less than that on top of Mount Everest, the temperature is usually well below freezing, the trace amounts of oxygen would not be nearly enough to breathe, and the lack of atmospheric ozone would leave you exposed to deadly ultraviolet radiation from the Sun.

Mars is the most studied planet besides Earth. More than a dozen spacecraft have flown past, orbited, or landed on Mars, and plans are in the works for many more missions to Mars. We may even send humans to Mars within the next few decades. By overturning rocks in ancient riverbeds or chipping away at ice in the polar caps, explorers will help us learn whether Mars has ever been home to life.

Figure 7.8 This image shows what it would look like to be orbiting near Jupiter's moon Io as Jupiter comes into view. Notice the Great Red Spot to the left of Jupiter's center. The extraordinarily dark rings discovered during the *Voyager* missions are exaggerated to make them visible. This computer visualization was created using data from both NASA's *Voyager* and *Galileo* missions. (From the Voyage scale model solar system, developed by the Challenger Center for Space Science Education, the Smithsonian Institution, and NASA. Image created by ARC Science Simulations © 2001.)

• Jupiter

- Average distance from the Sun: 5.20 AU
- Radius: 71,492 km = 11.2 R_{Earth}
- Mass: 318 M_{Earth}
- Average density: 1.33 g/cm^3
- Composition: mostly hydrogen and helium
- Cloud-top temperature: 125 K
- Moons: at least 63

To reach the orbit of Jupiter from Mars, we must traverse a distance that is more than double the total distance from the Sun to Mars, passing through the asteroid belt along the way. Upon our arrival, we find a planet much larger than any we have seen so far (Figure 7.8).

Jupiter is so different from the planets of the inner solar system that we must adopt an entirely new mental image of the term *planet*. Its mass is more than 300 times that of Earth, and its volume is more than 1,000 times that of Earth. Its most famous feature—a long-lived storm called the Great Red Spot—is itself large enough to swallow two or three Earths. Like the Sun, Jupiter is made primarily of hydrogen and helium and has no solid surface. If we plunged deep into Jupiter, the increasing gas pressure would crush us long before we ever reached its core.

Jupiter reigns over dozens of moons and a thin set of rings (too faint to be seen in most photographs). Most of the moons are very small, but four are large enough that we'd consider them planets if they orbited the Sun. These four moons—Io, Europa, Ganymede, and Callisto—are often called the *Galilean moons,* because Galileo discovered them shortly after he first turned his telescope toward the heavens [Section 3.3]. They are also planetlike in having varied and interesting geology. Io is the most volcanically active place in the solar system. Europa has an icy crust that may hide a subsurface ocean of liquid water, making it a promising place to search for life. Ganymede and Callisto may also have subsurface oceans, and their surfaces have many features that remain mysterious.

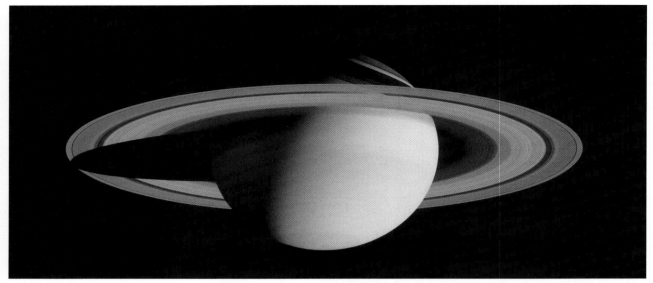

Figure 7.9 *Cassini*'s view of Saturn. We see the shadow of the rings on Saturn's sunlit face, and the rings become lost in Saturn's shadow on the night side.

• Saturn

- Average distance from the Sun: 9.54 AU
- Radius: 60,268 km $= 9.4\ R_{Earth}$
- Mass: 95.2 M_{Earth}
- Average density: 0.70 g/cm^3
- Composition: mostly hydrogen and helium
- Cloud-top temperature: 95 K
- Moons: at least 47

The journey from Jupiter to Saturn is a long one: Saturn orbits nearly twice as far from the Sun as Jupiter. Saturn, the second-largest planet in our solar system, is only slightly smaller than Jupiter in diameter. However, its lower density makes it considerably less massive than Jupiter (about one-third Jupiter's mass). Like Jupiter, Saturn is made mostly of hydrogen and helium and has no solid surface.

Saturn is famous for its spectacular rings (Figure 7.9). Although all four of the giant outer planets have rings, only Saturn's rings can be seen easily through a small telescope. The rings may look solid from a distance, but in reality they are made of countless small particles, each of which orbits Saturn like a tiny moon. If you could wander into the rings, you'd find yourself surrounded by chunks of rock and ice that range in size from dust grains to city blocks.

Like Jupiter, Saturn is orbited by many moons. Most are the size of small asteroids or comets, but a few are much larger. Perhaps the most fascinating moon is Titan, which is larger than the planet Mercury and blanketed by a thick atmosphere. On Titan's surface, you'd find an atmospheric pressure even greater than on Earth, and you could inhale air with roughly the same nitrogen content as air on Earth. However, you'd need to bring your own oxygen, and you'd certainly want a warm space suit for protection against the frigid outside temperatures. NASA's *Cassini* spacecraft, designed to explore Saturn and its rings and moons, carried a probe called *Huygens* that landed on Titan in January 2005. There, it found a bizarre landscape carved by erosion despite a temperature of −180°C.

THINK ABOUT IT

Find the current status of the *Cassini* mission (from news reports or the mission Web site). Is the spacecraft still operating? What is its next target of study? What recent discoveries has it helped us make?

Figure 7.10 This image shows a view from a vantage point high above Uranus's moon Ariel. The ring system is shown, although it would actually be too dark to see from this vantage point. This computer simulation is based on data from NASA's *Voyager 2* mission. (From the Voyage scale model solar system, developed by the Challenger Center for Space Science Education, the Smithsonian Institution, and NASA. Image created by ARC Science Simulations © 2001.)

• Uranus

- Average distance from the Sun: 19.2 AU
- Radius: 25,559 km $= 4.0$ R$_{\text{Earth}}$
- Mass: 14.5 M$_{\text{Earth}}$
- Average density: 1.32 g/cm^3
- Composition: hydrogen, helium, hydrogen compounds
- Cloud-top temperature: 60 K
- Moons: at least 27

It's another long journey to our next stop on the tour, as Uranus lies twice as far from the Sun as Saturn. Uranus (normally pronounced *YUR-uh-nus*) is much smaller than either Jupiter or Saturn but still much larger than Earth. It is made largely of hydrogen, helium, and *hydrogen compounds* such as water (H_2O), ammonia (NH_3), and methane (CH_4). Methane gas gives Uranus its pale blue-green color (Figure 7.10). Like the other giants of the outer solar system, Uranus lacks a solid surface. More than two dozen moons orbit Uranus, along with a set of rings somewhat similar to those of Saturn but much darker and more difficult to see.

The entire Uranus system—planet, rings, and moon orbits—is tipped on its side compared to the rest of the planets. This unusual orientation may be the result of a cataclysmic collision that Uranus suffered as it was forming some 4.6 billion years ago. It also makes for the most extreme pattern of seasons on any planet. If you lived on a platform floating in Uranus's atmosphere near its north pole, you'd have continuous daylight for half of each orbit, or 42 years. Then, after a very gradual sunset, you'd enter into a 42-year-long night.

Only one spacecraft has visited Uranus. *Voyager 2* flew past all four of the giant outer planets before heading out of the solar system. Much of our current understanding of Uranus comes from that mission, though powerful new telescopes are also capable of studying this planet. Scientists would love an opportunity to study Uranus and its interesting rings and moons in much greater detail, but no missions to Uranus are currently under development.

Figure 7.11 This image shows what it would look like to be orbiting Neptune's moon Triton as Neptune itself comes into view. The dark rings are exaggerated to make them visible in this computer simulation using data from NASA's *Voyager 2* mission. (From the Voyage scale model solar system, developed by the Challenger Center for Space Science Education, the Smithsonian Institution, and NASA. Image created by ARC Science Simulations © 2001.)

• Neptune

- Average distance from the Sun: 30.1 AU
- Radius: 24,764 km = 3.9 R_{Earth}
- Mass: 17.1 M_{Earth}
- Average density: 1.64 g/cm^3
- Composition: hydrogen, helium, hydrogen compounds
- Cloud-top temperature: 60 K
- Moons: at least 13

The journey from the orbit of Uranus to the orbit of Neptune is the longest yet in our tour, calling attention to the vast emptiness of the outer solar system. Nevertheless, Neptune looks nearly like a twin of Uranus, although it is more strikingly blue (Figure 7.11). It is slightly smaller than Uranus in size, but a higher density makes it slightly more massive even though the two planets share very similar compositions. Like Uranus, only the *Voyager 2* spacecraft has visited Neptune. No further missions are currently planned.

Neptune has rings and numerous moons. Its largest moon, Triton, is larger than Pluto and is one of the most fascinating moons in the solar system. Its icy surface has features that appear to be somewhat like geysers, although they spew nitrogen gas rather than water into the sky [Section 11.2]. Even more surprisingly, Triton is the only large moon in the solar system that orbits its planet "backward"—that is, in a direction opposite to the direction in which Neptune rotates. The attempt to understand how such a large moon ended up orbiting Neptune backward has given us new insights into the history of the outer solar system.

Figure 7.12 Pluto and Charon, as photographed by the Hubble Space Telescope. Because no spacecraft has been to Pluto, we do not yet have clear pictures of this tiny planet or its moon.

• Pluto (and Planet X)

- Average distance from the Sun: 39.5 AU
- Radius: 1,160 km = 0.18 R_{Earth}
- Mass: 0.0022 M_{Earth}
- Average density: 2.0 g/cm^3
- Composition: ices, rock
- Average surface temperature: 40 K
- Moons: 3

We conclude our tour at Pluto, which reigned for nearly 75 years as the ninth and last planet in our solar system. Pluto's average distance from the Sun lies as far beyond Neptune as Neptune lies beyond Uranus. Its great distance makes Pluto cold and dark. From Pluto, the Sun would be little more than a bright light among the stars. Pluto has a moon, Charon, that is locked together with it in synchronous rotation [Section 4.5]: From one side of Pluto, Charon would dominate the sky, while from the other side Charon would never be seen.

We've known for a few decades that Pluto is very different in character from the first eight planets—it is much smaller in size and has an orbit that is much more eccentric and more inclined to the ecliptic plane. Its composition of ice and rock is also quite different from that of any of the first eight planets, although it is virtually identical to that of many known comets. By the 1990s, these facts had led to a debate over whether Pluto should be called a "planet" at all.

The debate remains unresolved, but it has taken on new urgency since the 2005 announcement that astronomers had found another distant ball of ice and rock that is slightly larger than Pluto (see Special Topic box, p. 9). The object, which many people refer to as "Planet X",* currently lies about twice as far as Pluto from the Sun. However, its orbit is even more eccentric and more inclined to the ecliptic plane than Pluto's, so that it comes within Pluto's orbit every 560 years.

Although Planet X is the only icy object so far known to be larger than Pluto, we may yet discover others. Moreover, hundreds of other large iceballs have previously been found in the same general region of the solar system, and some of these objects are not that much smaller than Pluto [Section 12.3]. Thus, regardless of whether we choose to call Pluto a "planet," it's now clear that it represents a different type of object than the first eight planets: Pluto is one of hundreds of large iceballs—essentially large comets—that orbit the Sun near and beyond Neptune, in the region of the solar system known as the *Kuiper belt.*

The great distances and small sizes of Pluto and other similar worlds makes them difficult to study. As you can see in Figure 7.12, even the best telescopic views of Pluto and Charon reveal little detail. Better information should be coming soon, however. A spacecraft called *New Horizons* is scheduled to launch toward Pluto in 2006. It will fly past Pluto sometime after mid-2015, and will then visit other objects of the Kuiper belt as it continues its journey outward through our solar system.

---

*The object does not yet have an official name as this book goes to press. It bears the temporary designation 2003 UB313, but its discoverer, Michael Brown, refers to it by the code name Xena, after the TV warrior princess. A few people have jokingly suggested calling it Goofy, so that Pluto would not be the only dog in the outer solar system.

Table 7.1 Planetary Data[*]

Photo	Planet	Relative Size	Average Distance from Sun (AU)	Average Equatorial Radius (km)	Mass (Earth = 1)	Average Density (g/cm³)	Orbital Period	Rotation Period	Axis Tilt	Average Surface (or Cloud Tops) Temperature[†]	Composition	Known Moons (2005)	Rings?
	Mercury	·	0.387	2,440	0.055	5.43	87.9 days	58.6 days	0.0°	700 K (day) 100 K (night)	Rocks, metals	0	No
	Venus	•	0.723	6,051	0.82	5.24	225 days	243 days	177.3°	740 K	Rocks, metals	0	No
	Earth	•	1.00	6,378	1.00	5.52	1.00 year	23.93 hours	23.5°	290 K	Rocks, metals	1	No
	Mars	·	1.52	3,397	0.11	3.93	1.88 years	24.6 hours	25.2°	220 K	Rocks, metals	2	No
	Jupiter	●	5.20	71,492	318	1.33	11.9 years	9.93 hours	3.1°	125 K	H, He, hydrogen compounds[§]	63	Yes
	Saturn	●	9.54	60,268	95.2	0.70	29.4 years	10.6 hours	26.7°	95 K	H, He, hydrogen compounds[§]	47	Yes
	Uranus	•	19.2	25,559	14.5	1.32	83.8 years	17.2 hours	97.9°	60 K	H, He, hydrogen compounds[§]	27	Yes
	Neptune	•	30.1	24,764	17.1	1.64	165 years	16.1 hours	29.6°	60 K	H, He, hydrogen compounds[§]	13	Yes
	Pluto	·	39.5	1,160	0.0022	2.0	248 years	6.39 days	112.5°	40 K	Ices, rock	1	No
	Planet X[‡]	·	67.9	1,430	?	?	560 years	?	?	?	Ices, rock	3	?

§Includes water (H_2O), methane (CH_4), and ammonia (NH_3).

7.2 Patterns in the Solar System

Take a little time to think about our general picture of the solar system so far. Look back at the patterns of motion you see in Figures 7.1 and 7.2. Review the tour we've taken of individual worlds and consider patterns you may have noticed about size, distance, composition, and number of moons. Identify some of these patterns in Table 7.1, along with any other patterns you might notice. Clearly, as we've said from the beginning of this chapter, our solar system is not a random collection of worlds. Rather, it is a family of worlds exhibiting many patterns and traits that would be difficult to attribute to coincidence.

• What features of our solar system provide clues to how it formed?

In science, we assume that the patterns and traits we observe among the planets are consequences of the way our solar system was born and has evolved through time. Thus, when we seek to understand the origin of our solar system, we are looking for a way to explain these patterns and traits. We can summarize them by listing four major features of our solar system:

1. **Patterns of motion among large bodies.** The Sun, planets, and large moons generally orbit and rotate in a very organized way.
2. **Two major types of planets.** The first eight planets from the Sun divide clearly into two groups: the small, rocky planets that are close together and close to the Sun, and the large, gas-rich planets that are farther apart and farther out.
3. **Asteroids and comets.** Between and beyond the planets, huge numbers of asteroids and comets orbit the Sun. The locations, orbits, and compositions of these asteroids and comets follow distinct patterns.
4. **Exceptions to the rules.** The generally orderly solar system also has some notable "exceptions to the rules." For example, Earth is unique among the inner planets in having a large moon, and Uranus has an odd, sideways tilt. We will consider these exceptions as a feature in themselves, because a successful theory of our solar system must make allowances for such exceptions even as it explains the general rules.

Because these four features will be so important to our study of the solar system in later chapters, let's investigate each of them in a little more detail.

Feature 1: Patterns of Motion Among Large Bodies   Figures 7.1 and 7.2 reveal numerous patterns of motion among the large bodies of our solar system. (In this context, a "body" is simply an individual object such as the Sun, a planet, or a moon.) For example:

- All planetary orbits are nearly circular and lie nearly in the same plane.
- All planets orbit the Sun in the same direction: counterclockwise as viewed from high above Earth's North Pole.
- Most planets rotate in the same direction in which they orbit (counterclockwise as viewed from above the North Pole), with fairly small axis tilts. The Sun also rotates in this same direction.
- Most of the solar system's large moons exhibit similar properties in their orbits around their planets—for example, orbiting in their planet's equatorial plane in the same direction that the planet rotates.

We consider these orderly patterns together as the first major feature of our solar system. As we'll see in Chapter 8, our theory of solar system formation explains all these patterns as consequences of processes that occurred during the very first stages of the birth of our solar system.

Feature 2: The Existence of Two Types of Planets   Our brief planetary tour showed that the four inner planets are quite different from the four large outer planets. We say that these two groups represent two distinct planetary classes: *terrestrial* and *jovian*.

The **terrestrial planets** are the four planets of the inner solar system: Mercury, Venus, Earth, and Mars. (*Terrestrial* means "Earth-like.") The four terrestrial planets are relatively small and dense, with rocky surfaces and an abundance of metals deep in their interiors. They have few moons, if any, and none have rings. We often count our Moon as a fifth terrestrial world, because its history has been shaped by the same processes that have shaped the terrestrial planets.

The **jovian planets** are the four large planets of the outer solar system: Jupiter, Saturn, Uranus, and Neptune. (*Jovian* means "Jupiter-like.") The jovian planets are much larger in size and lower in average density than the terrestrial planets. They have rings and numerous moons. Their compositions are also quite different from those of the terrestrial worlds. They are made mostly of hydrogen, helium, and **hydrogen compounds**—compounds containing hydrogen, such as water (H_2O), ammonia (NH_3), and methane (CH_4).

Because the jovian planets are made mostly of substances that are gases under earthly conditions, they are sometimes called "gas giants." They do not have solid surfaces and look like balls of gas from the outside. However, the intense pressures and temperatures of the jovian planet interiors transform these gases into forms unlike anything we ordinarily see on Earth. Table 7.2 contrasts the general traits of the terrestrial and jovian planets.

Notice that Pluto is left out in the cold, both literally and figuratively, because it is much smaller and farther from the Sun (on average) than any of the terrestrial or jovian planets. As we discussed in our tour (p. 213) and will discuss further in Chapter 12, recent discoveries suggest that Pluto is just one of many similar objects that roam the solar system beyond the orbit of Neptune.

Table 7.2 Comparison of Terrestrial and Jovian Planets

Terrestrial Planets	Jovian Planets
Smaller size and mass	Larger size and mass
Higher density	Lower density
Made mostly of rock and metal	Made mostly of hydrogen, helium, and hydrogen compounds
Solid surface	No solid surface
Few (if any) moons and no rings	Rings and many moons
Closer to the Sun (and closer together), with warmer surfaces	Farther from the Sun (and farther apart), with cool temperatures at cloud tops

Feature 3: Asteroids and Comets The third major feature of our solar system is the existence of vast numbers of small objects orbiting the Sun. These objects fall into two major groups: asteroids and comets.

Asteroids are small, rocky bodies that orbit the Sun much like planets, but they are much smaller than planets (Figure 7.13). Even the largest of the asteroids have radii of only a few hundred kilometers, which means they are dwarfed by our Moon. Most asteroids are found within the **asteroid belt** between the orbits of Mars and Jupiter (see Figure 7.1).

Comets are also small objects that orbit the Sun, but they are made largely of ices (such as water ice, ammonia ice, and methane ice) mixed with rock. You are probably familiar with the occasional appearance of comets in the inner solar system, where they may become visible to the naked eye with long, beautiful tails (Figure 7.14). These visitors,

which may delight sky watchers for a few weeks or months, are actually quite rare among comets. The vast majority of comets never visit the inner solar system, instead orbiting the Sun in the extreme outer reaches of the solar system where they remain frozen and lack tails. As we'll discuss further in Chapter 8, there are two major groupings of comets in our solar system: One group orbits the Sun in the same direction and nearly the same plane as the planets, occupying a broad region beyond the orbit of Neptune that we call the *Kuiper belt* (Kuiper rhymes with *piper*); Pluto and the recently discovered "Planet X" are just the largest known of the icy objects in the Kuiper belt. The other group of comets does not show any simple pattern to its orbits, and resides much farther from the Sun in a region called the *Oort cloud* (*Oort* rhymes with *court*).

Feature 4: Exceptions to the Rules The fourth key feature of our solar system is that the general rules have a few notable exceptions. For example, while most of the planets rotate in the same direction as they orbit, Uranus rotates nearly on its side, and Venus rotates "backward"—clockwise, rather than counterclockwise, as viewed from high above Earth's North Pole. Similarly, while most large moons orbit their planets in the same direction as their planets rotate, many small moons have much more unusual orbits.

One of the most interesting exceptions concerns our own Moon. While the other terrestrial planets have either no moons (Mercury and Venus) or very tiny moons (Mars, with two moons), Earth has one of the largest moons in the solar system.

Summary of the Four Features Figure 7.15 summarizes the four general features of the solar system that we have discussed. In Chapter 8, we will turn our attention to the task of finding a theory that can explain all these features.

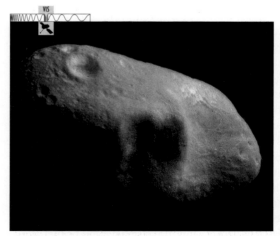

Figure 7.13 The asteroid Eros (photographed from the *NEAR* spacecraft). Its appearance is probably typical of most asteroids. Eros is about 40 kilometers in length. Like other small objects in the solar system, it is not spherical.

Figure 7.14 Comet Hale-Bopp, photographed over Boulder, Colorado, during its appearance in 1997.

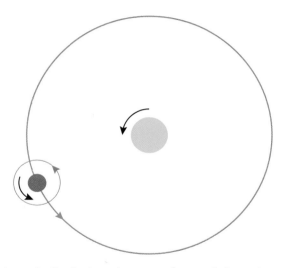

1. Large bodies in the solar system have orderly motions. All planets and most satellites have nearly circular orbits going in the same direction in nearly the same plane. The Sun and most of the planets rotate in this same direction as well.

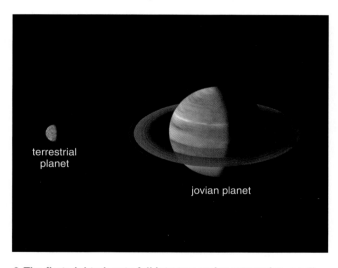

2. The first eight planets fall into two major categories: small, rocky terrestrial planets near the Sun and large, hydrogen-rich jovian planets farther out. The jovian planets have many moons and rings made of rock and ice.

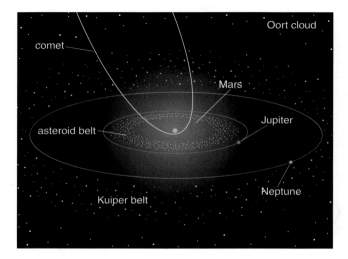

3. Swarms of asteroids and comets populate the solar system. Asteroids are concentrated in the asteroid belt, and comets populate the regions known as the Kuiper belt and the Oort cloud.

Figure 7.15 Four major features of the solar system.

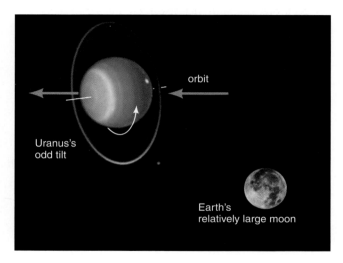

4. Several notable exceptions to these general trends stand out, such as planets with unusual axis tilts or surprisingly large moons, and moons with unusual orbits.

7.3 Spacecraft Exploration of the Solar System

How have we learned so much about the solar system? Much of our knowledge comes from telescopic observations, using both ground-based telescopes and telescopes in Earth orbit such as the Hubble Space Telescope. In one case—our Moon—we have learned a lot by sending astronauts to explore the terrain and bring back rocks for laboratory study. In a few other cases, we have samples of distant worlds that have come to us as meteorites. But most of the data fueling the recent revolution in our understanding of the solar system have come from robotic spacecraft. To date, we have sent robotic spacecraft to all the terrestrial and jovian planets, as well as to many moons, asteroids, and comets. In this section, we'll briefly investigate how we use robotic spacecraft to explore the solar system.

• How do robotic spacecraft work?

The spacecraft we send to explore the planets are robots suited for long space journeys. They carry specialized equipment for scientific study. All spacecraft have computers used to control their major components, power sources such as solar cells, propulsion systems, and devices to point cameras and other instruments precisely at their targets. Robotic spacecraft operate primarily with preprogrammed instructions. They carry radios for communication, which allow them to receive additional instructions from Earth and to send home the data they collect. Most robotic spacecraft make one-way trips from Earth, never physically returning but sending their data back from space in the same way we send radio and television signals around the world.

Broadly speaking, the robotic missions we send to explore other worlds fall into four major categories:

- **Flyby:** A spacecraft on a flyby goes past a world just once and then continues on its way.
- **Orbiter:** An orbiter is a spacecraft that orbits the world it is studying, allowing longer-term study during its repeated orbits.

- **Lander** or **probe:** These spacecraft are designed to land on a planet's surface or probe a planet's atmosphere by flying through it. Some landers carry rovers to explore wider regions.
- **Sample return mission:** A sample return mission requires a spacecraft designed to return to Earth carrying a sample of the world it has studied.

The choice of spacecraft type depends on both scientific objectives and cost. In general, a flyby is the lowest-cost way to visit another planet, and some flybys gain more "bang for the buck" by visiting multiple planets. For example, *Voyager 2* flew past Jupiter, Saturn, Uranus, and Neptune before continuing on its way out of our solar system (Figure 7.16).

SPECIAL TOPIC How Did We Learn the Scale of the Solar System?

In this chapter, we have presented the layout of the solar system as we know it today, when we have very precise measurements of planetary sizes and distances. But how did we learn the scale of the solar system?

By the mid-1600s, Kepler's laws had given us precise values of planetary distances in astronomical units (AU)—that is, distances *relative* to the Earth-Sun distance [Section 3.3]. However, no one knew the actual value of the AU in absolute units like miles or kilometers. A number of seventeenth-century astronomers proposed ideas for measuring the Earth-Sun distance, but none were practical. Then, in 1716, Edmond Halley (best-known today for having his name on Halley's comet) hit upon the idea that would ultimately solve the problem: He realized that during a planetary *transit*, when a planet appears to pass across the face of the Sun as seen from Earth [Section S1.1], observers in different locations on Earth would see the planet trace slightly different paths across the Sun. In essence, careful observations of the transit from different places allows precise measurement of the planet's parallax, from which we can use geometry to measure its distance (see figure). Because scientists in Halley's time already knew planetary distances in AU, knowing just one planet's actual distance from parallax would allow them to calculate the actual value of the AU and hence the actual distances of all the other planets.

Only Mercury and Venus can produce transits visible from Earth. Halley realized that although we see transits of Mercury far more often than transits of Venus, the measurements would be much easier with Venus because its closer distance to Earth during a transit would give it a larger parallax. Thus, Halley advocated sending expeditions to different parts of the world to observe transits of Venus. Unfortunately, Venus transits are rare, occurring in pairs 8 years apart about every 120 years. Halley did not live to see a Venus transit, but later astronomers followed his advice, mounting expeditions to observe transits that occurred in 1761 and 1769.

The transit observations turned out to be quite difficult in practice, partly due to the inherent challenge that long expeditions posed at that time and partly because getting the geometry right required very precise timing of the beginning of Venus's transit across the Sun. Astronomers discovered that this timing was more difficult than Halley had guessed, because of optical effects that occur during a transit. Nevertheless, astronomers kept studying the data from the 1761 and 1769 transits for many decades, and by the mid-1800s these data allowed them to pin down the value of the astronomical

unit to within about 5% of its modern value of 149.6 million kilometers. The next Venus transits occurred in 1874 and 1882. Photography had been invented by then, making observations more reliable, so in principle those transits could have allowed further refinement of the AU. However, by that time photography and better telescopes also made it possible to observe parallax of planets against stars, and by 1877 such observations had given us the value of the AU to within 0.2% of its modern value. More recently, a Venus transit occured in 2004 (see Figure S1.5) and another will occur in 2012.

Today, we measure the distance to Venus very precisely by bouncing radio waves off its surface with radar, a technique known as *radar ranging*. Because we know the speed of light, measuring the time it takes for the radio waves to make the round trip to Venus tells us the precise round-trip distance. We then use this distance and Venus's known distance in AU to calculate the actual value of the AU. Once we know the value of the AU, we can determine the actual distances of all the planets from the Sun, and we can determine their actual sizes from their angular sizes and distances. Indeed, we now know the layout of the solar system so well that we can launch spacecraft from Earth and send them to precise places on or around distant worlds.

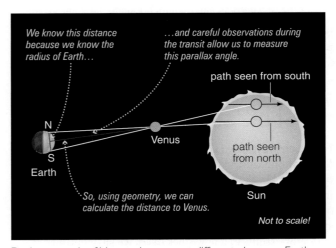

During a transit of Venus, observers at different places on Earth will see it trace slightly different paths across the Sun. The precise geometry of these events therefore allows computation of Venus's true distance, which in turn allows computation of the Sun's distance. *Adapted from Sky and Telescope, Feb. 2004.*

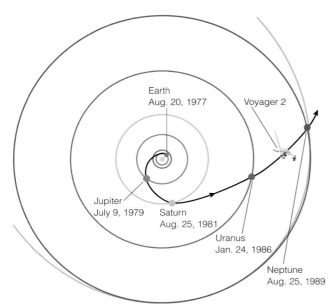

Figure 7.16 The trajectory of *Voyager 2*, which made flybys of each of the four jovian planets in our solar system.

Flybys   Flybys tend to be cheaper than other missions because they are generally less expensive to launch into space. Launch costs depend largely on weight, and onboard fuel is a significant part of the weight of a spacecraft heading to another planet. Once a spacecraft is on its way, the lack of friction or air drag in space means that it can maintain its orbital trajectory through the solar system without using any fuel at all. Fuel is needed only when the spacecraft needs to change from one trajectory (orbit) to another. Moreover, with careful planning, some trajectory changes can be made by taking advantage of the gravity of other planets. If ou look closely at the *Voyager 2* trajectory in Figure 7.16, you'll see that the spacecraft made significant trajectory changes as it passed by Jupiter and Saturn. In effect, it made these changes for free by using gravity to bend its path rather than by burning fuel. (This technique is known as a "gravitational slingshot," and it essentially speeds up the spacecraft significantly while slowing the planet by an unnoticeable amount.)

THINK ABOUT IT

Study the *Voyager 2* trajectory in Figure 7.16. Given that Saturn orbits the Sun every 29 years, Uranus orbits the Sun every 84 years, and Neptune orbits the Sun every 165 years, would it be possible to send another flyby mission to all four jovian planets if we launched it now? Explain.

Although a flyby offers only a relatively short period of close-up study, it can provide valuable scientific information. Flybys generally carry small telescopes, cameras, and spectrographs. Because these instruments are brought within a few tens of thousands of kilometers of other worlds (or closer), they can obtain much higher-resolution images and spectra than even the largest current telescopes viewing these worlds from Earth. In addition, flybys sometimes give us information that would be very difficult to obtain from Earth. For example, *Voyager 2* helped us discover Jupiter's rings and learn about the rings of Saturn, Uranus, and Neptune through views in which the rings were backlit by the Sun. Such views are possible only from beyond each planet's orbit.

Flybys may also carry instruments to measure local magnetic field strength or to sample interplanetary dust. The gravitational effects of the planets and their moons on the spacecraft itself provide information about object masses and densities. Like the backlit views of the rings, these types of data cannot be gathered from Earth. Indeed, most of what we know about the masses and compositions of moons comes from data gathered by spacecraft that have flown past them.

Orbiters   An orbiter can study another world for a much longer period of time than a flyby. Like the spacecraft used for flybys, orbiters often carry cameras, spectrographs, and instruments for measuring the strength of magnetic fields. Missions to Venus, Mars, and Saturn's moon Titan have also used radar to make precise altitude measurements of surface features. In the cases of Venus and Titan, radar observations give us our only clear surface views, because we cannot otherwise see through the thick cloud cover on these worlds.

An orbiter is generally more expensive than a flyby for an equivalent weight of scientific instruments, primarily because it must carry added fuel to change from an interplanetary trajectory to a path that puts it into orbit around another world. Careful planning can minimize the added expense. For example, recent Mars orbiters have saved on fuel costs by carrying only enough fuel to enter highly elliptical orbits around Mars. The spacecraft then settled into the smaller, more circular orbits needed for scientific observations by skimming the Martian atmosphere at the low point of every elliptical orbit. Atmospheric drag slowed the spacecraft with each orbit and, over several months, circularized the spacecraft orbit. (This technique of using the atmosphere to slow the spacecraft and change its orbit is called *aerobraking*.) We have sent orbiters to the Moon and to the planets Venus, Mars, Jupiter, and Saturn, and to the asteroid Eros.

Landers or Probes   The most "up close and personal" study of other worlds comes from spacecraft that send probes into the atmospheres or landers to the surfaces. For example, in 1995, the *Galileo* spacecraft dropped a probe into Jupiter's atmosphere [Section 11.1]. The probe collected temperature, pressure, composition, and radiation measurements for about an hour as it descended; it was then destroyed by Jupiter's high interior pressures and temperatures.

On planets with solid surfaces, a lander can offer close-up surface views, local weather monitoring, and the ability to carry out automated experiments. Some landers carry robotic rovers able to venture across the surface, such as the *Spirit* and *Opportunity* rovers that landed on Mars in

The aeroshell protects the rover from fiery temperatures as it enters the Martian atmosphere. Six minutes to landing.

With the parachute deployed three retrorockets fire their engines, suspending the lander 30–50 feet above the Martian surface.

Protected by large airbags, the lander falls away from the parachute, landing safely on Mars. The lander bounces for several minutes, traveling hundreds of meters.

After bouncing to a stop, the airbags are retracted and the lander's petals deploy, creating a ramp for the rover to descend.

The rover deploys the solar arrays, wheels, cameras, and other instruments and begins its exploration of the Martian surface.

Figure 7.17 The Mars Exploration Rovers used parachutes and air bags to land safely on Mars.

2004 and operated long beyond their planned lifetimes. Landers typically require fuel to slow their descent to a planetary surface, but clever techniques can reduce cost. For example, the recent Mars landers hit the surface of Mars at crash-landing speed but were protected by cocoons of air bags deployed on the way down. These air bags allowed the landers to bounce along the surface of Mars before finally coming to rest (Figure 7.17).

Sample Return Missions While probes and landers can carry out experiments on surface rock or atmospheric samples, the experiments must be designed in advance and must fit inside the spacecraft. These limitations make scientists long for missions that will scoop up samples from other worlds and return them to Earth for more detailed study. To date, the only sample return missions have been to the Moon.

Astronauts collected samples during the *Apollo* missions, and the then–Soviet Union sent robotic spacecraft to collect rocks from the Moon in the early 1970s. Many scientists are working toward a sample return mission to Mars, and they hope to launch such a mission within the next decade or so. (A slight variation on the theme of a sample return mission is the *Stardust* mission, which collected comet dust on a flyby and will return to Earth in 2006.)

Combination Spacecraft Many missions combine more than one type of spacecraft. For example, the *Galileo* mission to Jupiter included an orbiter that studied Jupiter and its moons as well as the probe that entered Jupiter's atmosphere. The *Cassini* spacecraft, which has been orbiting Saturn since 2004, included flybys of Venus, Earth, and Jupiter during its 7-year trip to Saturn (Figure 7.18a). The spacecraft

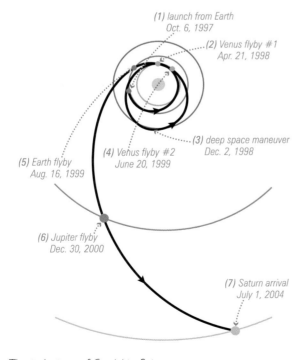

(1) launch from Earth
Oct. 6, 1997

(2) Venus flyby #1
Apr. 21, 1998

(3) deep space maneuver
Dec. 2, 1998

(4) Venus flyby #2
June 20, 1999

(5) Earth flyby
Aug. 16, 1999

(6) Jupiter flyby
Dec. 30, 2000

(7) Saturn arrival
July 1, 2004

a The trajectory of *Cassini* to Saturn

Figure 7.18 The *Cassini* mission to Saturn.

b Artist's conception of the *Huygens* probe landing on Titan. From upper right to lower left: As the *Cassini* mothership passes in the distance, the probe enters the atmosphere, parachutes down, and finally lands on the surface. *Huygens* landed in January 2005, collecting data and transmitting images for several hours before its batteries ran out. (See Figure 11.29 for actual images of Titan taken by the *Huygens* probe.)

Table 7.3 Selected Robotic Missions to Other Worlds

Name	Type	Destination	Lead Space Agency	Arrival Year	Mission Highlights
New Horizons	Flyby	Pluto	NASA	2015[†]	First mission to Pluto, scheduled for 2006 launch
Rosetta	Flyby and lander	Comet 67P/ Churyumov-Gerasimenko	ESA[*]	2014[†]	Launched in 2004, carries lander and orbiter for comet
MESSENGER	Orbiter	Mercury	NASA	2011[†]	Studies Mercury's surface, atmosphere, and interior
Hayabusa	Lander and sample return	Asteroid Itokawa	JAXA[**]	2007[†]	Collected asteroid sample in 2005, to return to Earth in 2007
Stardust	Flyby and sample return	Comet Wild 2	NASA	2006[†]	Collected dust from comet in 2004, to return to Earth in 2006
Mars Reconnaissance Orbiter	Orbiter	Mars	NASA	2006[†]	High-resolution mapper will identify future landing sites
Deep Impact	Flyby and "lander"	Comet Tempel 1	NASA	2005	"Lander" to hit at 10 km/sec, flyby spacecraft to study the impact
Cassini	Orbiter	Saturn	NASA	2004	Included probe (*Huygens*) that landed on Titan
Mars Exploration Rovers	Lander	Mars	NASA	2004	Rovers *Spirit* and *Opportunity* explored surface
Mars Express	Orbiter	Mars	ESA[*]	2004	Climate studies of Mars
Venus Express	Orbiter	Venus	ESA[*]	2006	Low-cost Mars Express spinoff focused on atmosphere, surface
Mars Odyssey 2001	Orbiter	Mars	NASA	2001	Detailed study of Martian surface features and composition
Near Earth Asteroid Rendezvous (NEAR)	Orbiter	Eros (asteroid)	NASA	2000	First spacecraft dedicated to in-depth study of an asteroid
Mars Global Surveyor	Orbiter	Mars	NASA	1997	Detailed imaging of surface from orbit
Mars Pathfinder	Lander	Mars	NASA	1997	Carried *Sojourner*, the first robotic rover on Mars
Galileo	Orbiter	Jupiter	NASA	1995	Dropped probe into Jupiter; close-up study of moons
Magellan	Orbiter	Venus	NASA	1990	Detailed radar mapping of surface of Venus
Voyager 1	Flyby	Jupiter, Saturn	NASA	1979, 1980	Unprecedented views of Jupiter and Saturn; continuing out of solar system
Voyager 2	Flyby	Jupiter, Saturn, Uranus, Neptune	NASA	1979, 1981, 1986, 1989	Only mission to Uranus and Neptune; continuing out of solar system
Vikings 1 and 2	Orbiter and lander	Mars	NASA	1976	First landers on Mars

[*]European Space Agency.   [**]Japan Aerospace Exploration Agency.   [†]Scheduled arrival year.

itself is an orbiter that is studying Saturn and its moons, but it also carried the *Huygens* probe that descended through the atmosphere and landed on the surface of Saturn's moon Titan (Figure 7.18b).

Exploration Past, Present, and Future   Over the past several decades, studies using both telescopes on Earth and robotic spacecraft have allowed us to learn the general characteristics of all the major planets and moons in our solar system as well as the general characteristics of asteroids and comets. Telescopes will continue to play an important role in future observations, but for detailed study we will probably continue to depend on spacecraft.

Table 7.3 (p. 221) lists some significant robotic missions of the past and present. The next few years promise many new discoveries as missions arrive at their destinations. Over the longer term, all the world's major space agencies have high hopes of launching numerous and diverse missions to study our solar system. For example, nearly a dozen more missions are on the drawing board for Mars alone. If these missions come to fruition, the next couple of decades will see spacecraft that answer many specific questions about the nature of our solar system and its numerous worlds.

Putting Chapter 7 into Perspective

In this chapter, we've introduced the major features of our solar system and discussed some important patterns and trends that provide clues to its formation. As you continue your study of the solar system, keep in mind the following "big picture" ideas:

- Our solar system is not a random collection of objects moving in random directions. Rather, it is highly organized, with clear patterns of motion and with most objects falling into just a few basic categories.

- Each planet has its own unique and interesting features. Becoming familiar with the planets is an important first step in understanding the root causes of their similarities and differences.

- Much of what we now know about the solar system comes from spacecraft exploration. Choosing the type of mission to send to a planet involves many considerations, from the scientific to the purely political. Many missions are currently under way, offering us hope of learning much more in the near future.

SUMMARY OF KEY CONCEPTS

7.1 Studying the Solar System

- **What does the solar system look like?**   The

planets are tiny compared to the distances between them. Our solar system consists of the Sun, the planets and their moons, and vast numbers of asteroids and comets. Each world has its own unique character, but there are many clear patterns among the worlds.

- **What can we learn by comparing the planets to one another?**   Comparative studies reveal the similarities and differences that give clues to solar system formation, and highlight the underlying processes that give each planet its unique appearance.

- **What are the major features of the Sun and planets?**   The Sun is the dominant object in our solar

system. Each of the planets has unique features, but we also observe important similarities among the terrestrial worlds and among the jovian worlds, similarities that give us key clues to the processes that govern the evolution of worlds.

7.2 Patterns in the Solar System

- **What features of our solar system provide clues to how it formed?**   Four major features

provide clues: (1) The Sun, planets, and large moons generally rotate and orbit in a very organized way. (2) The first eight planets from the Sun divide clearly into two groups: terrestrial and jovian. (3) The solar system contains huge numbers of asteroids and comets. (4) There are some notable exceptions to these general patterns.

7.3 Spacecraft Exploration of the Solar System

- **How do robotic spacecraft work?**   Spacecraft

can be categorized as flyby, orbiter, lander/probe, or sample return missions. In all cases, robotic spacecraft carry their own propulsion, power, and communication systems, and can operate under preprogrammed control or with updated instructions from ground controllers.

Review Questions

Short-Answer Questions Based on the Reading

1. What do we mean by *comparative planetology*? Does it apply only to planets?
2. What would the solar system look like to your naked eye if you could view it from beyond the orbit of Pluto?
3. Briefly describe the overall layout of the solar system as it is shown in Figures 7.1 and 7.2.
4. For the Sun and each of the planets in our solar system, describe at least two features that you find interesting.
5. Describe at least four orderly patterns of motion that we observe in our solar system.
6. What are the basic differences between the *terrestrial* and *jovian* planets? Which planets in our solar system fall into each group?
7. What do we mean by *hydrogen compounds*? In what kinds of planets or small bodies are they major ingredients?
8. In what ways does Pluto resemble a terrestrial planet? In what ways does it resemble a jovian planet? In what ways does it resemble neither?
9. What are *asteroids*? Where do we find most asteroids in our solar system?
10. What are *comets*? How do they differ from asteroids?
11. What is the *Kuiper belt*? What is the *Oort cloud*? How do the orbits of comets differ in the two regions?
12. Describe at least two "exceptions to the rules" that we find in our solar system.
13. Describe and distinguish between space missions that are *flybys, orbiters, landers* or *probes*, and *sample return missions*. What are the advantages and disadvantages of each mission type?
14. For a few of the most important past, present, or future robotic missions to the planets, describe their targets, types, and mission highlights.

Test Your Understanding

Does It Make Sense?

Decide whether the statement makes sense (or is clearly true) or does not make sense (or is clearly false). Explain your reasoning. (For an example, see Chapter 1, "Does It Make Sense?")

15. Pluto orbits the Sun in the opposite direction of all the other planets.
16. If Pluto were as large as the planet Mercury, we would classify it as a terrestrial planet.
17. Comets in the Kuiper belt and Oort cloud have long, beautiful tails that we can see when we look through telescopes.
18. Our Moon is about the same size as moons of the other terrestrial planets.
19. The mass of the Sun compared to the mass of all the planets combined is like the mass of an elephant compared to the mass of a cat.
20. On average, Venus is the hottest planet in the solar system—even hotter than Mercury.
21. The weather conditions on Mars today are much different than they were in the distant past.
22. Moons cannot have atmospheres, active volcanoes, or liquid water.

23. Saturn is the only planet in the solar system with rings.
24. We could probably learn more about Mars by sending a new spacecraft on a flyby than by any other method of studying the planet.

Quick Quiz

Choose the best answer to each of the following. Explain your reasoning with one or more complete sentences.

25. The largest terrestrial planet and jovian planet are, respectively: (a)Venus and Jupiter. (b) Earth and Jupiter. (c) Earth and Pluto.
26. Which terrestrial planets have had volcanic activity at some point in their histories? (a) only Earth (b) Earth and Mars (c) all of them
27. Large moons orbit their planets in the same direction the planet rotates . . . (a) rarely. (b) half of the time. (c) most of the time.
28. Which of the following three kinds of objects resides closer to the Sun on average? (a) comets (b) asteroids (c) jovian planets
29. What's unusual about our Moon? (a) It's the only moon that orbits a terrestrial planet. (b) It's by far the largest moon in the solar system. (c) It's not the largest moon, but it's surprisingly large relative to the planet it orbits.
30. Planetary orbits are: (a) very eccentric (stretched-out) ellipses and in the same plane. (b) fairly circular and in the same plane. (c) fairly circular but oriented in every direction.
31. Which have more moons on average? (a) jovian planets (b) terrestrial planets (c) Terrestrial and jovian planets both have about the same number of moons.
32. The most abundant ingredient of the Sun and Jupiter is: (a) ionized metal. (b) hydrogen. (c) ammonia.
33. Are there any exceptions to the rule that planets rotate with small axis tilts and in the same direction that they orbit the Sun? (a) no (b) Venus is the only exception. (c) Venus, Uranus, and Pluto are exceptions.
34. The *Cassini* spacecraft: (a) flew past Pluto. (b) landed on Mars. (c) is orbiting Saturn.

Investigate Further

In-Depth Questions to Increase Your Understanding

Short-Answer/Essay Questions

35. *Planetary Tour.* Based on the brief planetary tour in this chapter, which planet besides Earth do you think is the most interesting, and why? Defend your opinion clearly in two or three paragraphs.
36. *Patterns of Motion.* In one or two paragraphs, explain why the existence of orderly patterns of motion in our solar system should suggest that the Sun and the planets all formed at one time from one cloud of gas, rather than as individual objects at different times.
37. *Two Classes of Planets.* In terms a friend or roommate would understand, write one or two paragraphs explaining why we say that the planets fall into two major categories and what those categories are. Do all planets fit in the two groups? Explain.
38. *Solar System Trends.* Study the planetary data in Table 7.1 to answer each of the following.

a. Notice the relationship between distance from the Sun and surface temperature. Describe the trend, explain why it exists, and explain any notable exceptions to the trend.

b. The text says that planets can be classified as either terrestrial or jovian, with Pluto as a misfit. Describe in general how the columns for density, composition, and distance from the Sun support this classification.

c. Describe the trend you see in orbital periods and explain the trend in terms of Kepler's third law.

39. *Comparing Planetary Conditions.* Use the planetary data in Table 7.1 and Appendix E to answer each of the following.

a. Which column of data would you use to find out which planet has the shortest days? Do you see any notable differences in the length of a day for the different types of planets? Explain.

b. Which planets would you expect not to have seasons? Why?

c. Which column tells you how much a planet's orbit deviates from a perfect circle? Based on that column, are there any planets for which you would expect the surface temperature to vary significantly over the course of each orbit? Explain.

Quantitative Problems

Be sure to show all calculations clearly and state your final answers in complete sentences.

40. *Size Comparisons.* How many Earths could fit inside Jupiter (assuming you could fill up all the volume)? How many Jupiters could fit inside the Sun? The equation for the volume of a sphere is $V = (\frac{4}{3})\pi r^3$.

41. *Asteroid Orbit.* Ceres, the largest asteroid, has an orbital semimajor axis of 2.77 AU. Use Kepler's third law to find its orbital period. Compare your answer with Table 7.1 and name the planets that orbit just inside and outside Ceres's orbit.

42. *Density Classification.* Imagine that a new planet is discovered in our solar system with a mass of 5.97×10^{25} kilograms and a radius of 12,800 kilometers . Based just on its density, would we consider it the largest terrestrial planet or the

smallest jovian planet? Explain. (*Hint:* Be careful to convert your density to units of grams per cubic centimeter in order to compare with the terrestrial and jovian planet data in this chapter.)

43. *Escape Velocity.* By studying the data in Appendix E, briefly describe how escape velocity is related to mass and radius. Is the trend what you expect based on what you learned about escape velocity in Chapter 5?

44. *Comparative Weight.* Suppose you weigh 100 pounds. How much you would weigh on each of the other planets in our solar system? Assume you can stand either on the surface or in an airplane in the planet's atmosphere. (*Hint:* Recall from Chapter 5 that weight is mass times the acceleration of gravity. The surface gravity column in Appendix E tells you how the acceleration of gravity on other planets compares to that on Earth.)

45. *Mission to Pluto.* The *New Horizons* spacecraft will take about 9 years to travel from Earth orbit to Pluto. About how fast will it be traveling on average? Assume that its trajectory is close to a straight line. Give your answer in AU/year and in km/hr.

46. *Empty Solar System.* If you could line up the Sun and all the planets end-to-end, how far would they stretch? Compare your answer to the distance between the Sun and Pluto. What does this tell you about the "emptiness" of the solar system?

47. *Planetary Parallax.* Suppose observers at Earth's North Pole and South Pole use a transit of the Sun by Venus to discover that the angular size of Earth as viewed from Venus would be 62.8 arcseconds. Earth's radius is 6378 km. Estimate the distance between Venus and Earth in km and AU. Compare your answer with information from the chapter.

Discussion Questions

48. *Where Would You Go?* Suppose you could visit any one of the planets or moons in our solar system for 1 week. Which object would you choose to visit, and why?

49. *Planetary Priorities.* Suppose you were in charge of developing and prioritizing future planetary missions for NASA. What would you choose as your first priority for a new mission, and why?

MEDIA EXPLORATIONS

For self-study activities go to MasteringASTRONOMY www.masteringastronomy.com

 Interactive Tutorials

Tutorial Review of Key Concepts

Use the interactive **Tutorial** at **www.mastering astronomy.com** to review key concepts from this chapter.

Scale of the Universe Tutorial
Lesson 1 Distances of Scale: Our Solar System

Formation of the Solar System Tutorial
Lesson 1 Comparative Planetology

Orbits and Kepler's Laws Tutorial
Lesson 2 Kepler's First Law
Lesson 3 Kepler's Second Law
Lesson 4 Kepler's Third Law

Supplementary Tutorial Exercises
Use the interactive **Tutorial Lessons** to explore the following questions.

Scale of the Universe Tutorial, Lesson 1
1. How big is an astronomical unit (AU) in comparison to more familiar units?
2. Why do we use AUs to describe distances in our solar system? Is the AU a good unit for describing distances to stars? Explain.
3. How big is the Sun relative to the Moon's orbit of Earth?
4. Explore the layout of the solar system, focusing on the scale of the various planetary orbits. Which of the key patterns discussed in this chapter are evident from the scaled view? Which require deeper study to uncover?

Formation of the Solar System Tutorial, Lesson 1
1. What features distinguish the terrestrial planets from the jovian planets?
2. What is the difference between axis tilt, orbital eccentricity, and orbital inclination? What patterns do we see in these properties among the planets?
3. Which solar system bodies constitute "exceptions to the rules," and in what ways?

 Exploring the Sky and Solar System

Of the many activities available on the *Voyager: SkyGazer* CD-ROM accompanying your book, use the following files to observe key phenomena covered in this chapter.

Go to the **File: Basics** folder for the following demonstrations.
1. Saturn's Phases
2. Tracking Venus
3. Planet Panel

Go to the **File: Demo** folder for the following demonstrations.
1. Earth and Venus
2. Trailing Saturn
3. Triple Conjunction of 7 BC

Go to the **Explore** menu for the following demonstrations.
1. Solar System
2. Paths of Planets

 Movies

Check out the following narrated and animated short documentaries available on **www.masteringastronomy.com** for a helpful review of key ideas covered in this chapter.

Orbits in the Solar System

History of the Solar System Movie

Web Projects

1. *Current Mission.* Visit the Web page for one of the current missions listed in Table 7.3. Write a one- to two-page summary of the mission's basic design, goals, and current status.
2. *Mars Missions.* Go to the home page for NASA's Mars Exploration Program. Write a one- to two-page summary of the plans for future exploration of Mars.

8

Formation of the Solar System

The evolution of the world may be compared to a display of fireworks that has just ended: some few red wisps, ashes and smoke. Standing on a cooled cinder, we see the slow fading of the suns, and we try to recall the vanished brilliance of the origin of the worlds.

G. Lemaître (1894–1966),
astronomer and Catholic priest

How did Earth come to be? How old is it? Is it unique? Our ancestors could do little more than guess at the answers to these questions, and they imagined Earth to be fundamentally different from objects in the heavens. Today we know Earth as just one of many worlds in our solar system, and our solar system as just one of many planetary systems in the universe.

We seek to understand the origin of our planet from the clues we discover as we study the rest of our solar system. These clues point to a common origin for the Sun and all the planets, moons, asteroids, and comets that orbit it. By studying the clues in depth, scientists have put together a theory that explains most of the observed features of our solar system. We'll explore the theory in this chapter, using it as a case study in the way that science seeks to answer fundamental questions about our origins.

Our theory of solar system formation is important not only because it helps us understand our cosmic origins, but also because it holds the key to understanding the nature of planets. If the planets in our solar system all formed together, then their differences must be attributable to physical processes that occurred during the birth and subsequent evolution of the solar system. Thus, our study of the solar system's birth will form the basis for our comparative study of the planets in subsequent chapters. It will also help us understand and interpret recent discoveries of planets around other stars, a topic we will study in depth in Chapter 13.

8.1 The Search for Origins

The development of any scientific theory is an interplay between observations and attempts to explain those observations [Section 3.4]. Hypotheses that seem to make sense at one time might later be dismissed because they fail to explain new data. For example, ancient Greek ideas about Earth's origins probably seemed quite reasonable when people assumed that Earth was the center of the universe, but they no longer made sense after Kepler and Galileo proved that Earth is a planet orbiting the Sun.

By the end of the seventeenth century, the Copernican revolution [Section 3.3] and Newton's discovery of the universal law of gravitation [Section 4.4] had given us a basic understanding of the layout and motion of the planets and moons in our solar system. It was only natural that scientists would begin to speculate about how this system came to be. We'll spend most of this chapter discussing our mod-

ern theory of the solar system's origins. First, however, let's explore the origins of the theory and the way in which we evaluate it.

• What properties of our solar system must a formation theory explain?

Remember that a hypothesis can rise to the status of a scientific theory only if it offers a detailed, physical model that explains a broad range of observed facts. In Chapter 7, we discussed four major structural features of our solar system (see Figure 7.15). These features hold the key to testing any hypothesis that claims to explain the origin of the solar system. If the hypothesis fails to explain even one of the four features, then it cannot be correct. If it successfully explains all four, then we might reasonably assume it is on the right track. Thus, we have the following four criteria for the success of a solar system formation theory:

1. It must explain the patterns of motion we discussed in Chapter 7.
2. It must explain why planets fall into the two major categories—the small, rocky *terrestrial* planets near the Sun and the large, hydrogen-rich *jovian* planets farther out.
3. It must explain the existence of huge numbers of asteroids and comets and why these objects reside primarily in the regions we call the *asteroid belt*, the *Kuiper belt*, and the *Oort cloud*.
4. It must explain all the general patterns while at the same time making allowances for exceptions to the general rules, such as the odd axis tilt of Uranus and the existence of Earth's large Moon.

The theory can gain even further support if it makes predictions that are borne out by new observations or experiments. For example, a theory that explains the origin of *our* solar system ought to be able to make predictions about *other* solar systems. Today, we can test such predictions with the rapidly growing body of observations of young stars that appear to be in the early stages of planet formation and of actual planets discovered around other stars. As we will see in both this chapter and Chapter 13, our modern theory of solar system formation has required only minor modifications to account for these new observations, a fact that has put it on a foundation that is stronger today than ever before.

• What theory best explains the features of our solar system?

After the Copernican revolution, many scientists speculated about the origin of the solar system. However, we generally credit two eighteenth-century scientists with proposing the hypothesis that ultimately blossomed into our modern scientific theory of the origin of the solar system. Around 1755, German philosopher Immanuel Kant proposed that our solar system formed from the gravitational collapse of an interstellar cloud of gas. About 40 years later, French mathematician Pierre-Simon Laplace put forth the same idea independently (he apparently was unaware of Kant's proposal). Because an interstellar cloud is usually called a *nebula* (Latin for "cloud"), their idea became known as the *nebular hypothesis.*

The nebular hypothesis remained popular throughout the nineteenth century. By the early twentieth century, however, scientists had found a few aspects of our solar system that the nebular hypothesis did not seem to explain well—at least in its original form as described by Kant and Laplace. While some scientists sought to modify the nebular hypothesis, others looked for entirely different ideas about how the solar system might have formed.

During much of the first half of the twentieth century, the nebular hypothesis faced stiff competition from a hypothesis proposing that the planets represent debris from a near-collision between the Sun and another star. According to this *close encounter hypothesis,* the planets formed from blobs of gas that had been gravitationally pulled out of the Sun during the near-collision.

Today, the close encounter hypothesis has been discarded. Calculations showed that it could not account for either the observed orbital motions of the planets or the neat division of the planets into two major categories (terrestrial and jovian). Historically, these failures alone should have been enough to doom the hypothesis, but it lost favor for another reason as well: It required a highly improbable event. Given the vast separation between star systems in our region of the galaxy, the chance that any two stars would pass close enough to cause a substantial gravitational disruption is so small that it would be difficult to imagine it happening even in the one case needed to make our own solar system. While low probability alone could not rule out the close encounter hypothesis, it certainly did not help the case for a hypothesis that also failed on other grounds. Moreover, the low probability of forming planets in collisions leads to a testable prediction that we now know to be wrong: If planets are born only in very rare events, then planets should be very rare around other stars. The fact that we now know planets to be quite common [Section 13.1] therefore counts as an added strike against the close encounter hypothesis.

While the close encounter hypothesis was losing favor, new ideas about the physics of planet formation led to modifications of the nebular hypothesis. As scientists investigated the implications of the idea that our solar system formed from a large cloud of gas, they developed new models that led to natural explanations for all four general features of our solar system. By the latter decades of the twentieth century, so much evidence had accumulated in favor of the nebular hypothesis that it achieved the status of a scientific theory—the **nebular theory** of our solar system's birth. Moreover, as we noted earlier, the theory has gained further support by successfully predicting much of what we have discovered about other planetary systems. While recent discoveries of planets around other stars have revealed a few surprises, overall they have been consistent with what we expect if the nebular theory is correct.

 Formation of the Solar System Tutorial, Lessons 1–2

8.2 The Birth of the Solar System

We are now ready to examine the nebular theory of solar system formation in more depth. In this section, we'll begin at the beginning, examining the origins of the cloud of gas that gave birth to our solar system and explaining how the gravitational collapse of this cloud led to orderly patterns of motion. Remember that the same ideas are likely to apply to the births of other star systems as well.

• Where did the solar system come from?

The nebular theory begins with the idea that our solar system was born from a cloud of gas that collapsed under its own gravity. We usually refer to this cloud as the **solar nebula**. But where did this gas come from?

Galactic Recycling According to the evidence as we understand it today, the universe was born in the Big Bang [Section 1.1]. Hydrogen and helium (and a trace of lithium) were the only chemical elements present when the universe was young. All the heavier elements have been produced since that time by stars, through nuclear fusion in their cores or the explosions that accompany their deaths. When stars die, they release much of their content back into space, including newly produced heavy elements. This material can then be recycled into new generations of stars (Figure 8.1). (We'll study the galactic recycling process in more detail in Chapter 19.)

Although this process of creating heavy elements in stars and recycling them within the galaxy has probably gone on for most of the 14-billion-year history of our universe, only a small fraction of the original hydrogen and helium has been converted into heavier elements. By the time our solar system formed about 4.6 billion years ago, about 2% (by mass) of the original hydrogen and helium in the galaxy had been converted to heavier elements; the other 98% remained hydrogen and helium. The Sun, which represents nearly all the mass in our solar system, still has this basic composition today.

Thus, the gas that made up the solar nebula was the result of billions of years of galactic recycling that occurred before our solar system was born. The Sun and planets were born from this gas, and Earth and the other terrestrial worlds ended up being made primarily from the heavier elements mixed within it. As we saw in Chapter 1, we are therefore "star stuff," because we and our planet are made of elements forged in stars that lived and died long ago.

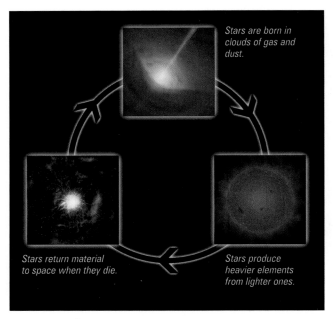

Figure 8.1 🎨 This figure, which is a portion of Figure 1.3 earlier in the book, summarizes the galactic recycling process.

Stars are born in clouds of gas and dust.

Stars return material to space when they die.

Stars produce heavier elements from lighter ones.

Figure 8.2 Interactive Photo This photograph shows the Orion Nebula, an interstellar cloud in which new star systems are forming. Over the next few million years, thousands of stars will be born in this gas cloud. Some of these stars may end up with their own planetary systems.

THINK ABOUT IT

Could a solar system like ours have formed with the first generation of stars after the Big Bang? Explain.

Evidence from Other Gas Clouds Strong observational evidence supports the idea that our solar system was born from an interstellar cloud of gas, because stars that appear to be in the process of formation today are always found within interstellar clouds [Section 16.1].

Figure 8.2 shows the Orion Nebula, one of the most famous star-forming clouds. Some stars have already been born within this vast cloud of gas, and thousands more will be born there over the next few million years, some perhaps with planets. Moreover, each star we observe in the process of formation is surrounded by relatively dense gas that appears to be collapsing under its own gravity, suggesting that it is forming within its own "solar nebula," just as we think our own solar system formed some 4.6 billion years ago.

• What caused the orderly patterns of motion in our solar system?

The solar nebula probably began as a large, roughly spherical cloud of very cold and very low-density gas. Initially, this gas was probably so spread out—perhaps over a region a few light-years in diameter—that gravity alone may not have been able to pull it together to start its collapse. Instead, the collapse may have been triggered by a cataclysmic event, such as the impact of a shock wave from the explosion of a nearby star (a supernova).

Once the collapse started, the law of gravity ensured that it would continue. Remember that the strength of gravity follows an inverse square law with distance [Section 4.4]. Because the mass of the cloud remained the same as it shrank, the strength of gravity increased as the diameter of the cloud decreased. For example, when the diameter decreased by half, the force of gravity increased by a factor of four.

Because gravity pulls inward in all directions, you might at first guess that the solar nebula would have remained spherical as it shrank. Indeed, the idea that gravity pulls in all directions explains why the Sun and the planets are spherical. However, gravity is not the only physical law that affects the collapse of a cloud of gas. When we consider everything that should have occurred in the collapsing cloud, we find that the solar nebula should have become a spinning, flattened disk surrounding a central star.

Heating, Spinning, and Flattening As the solar nebula shrank in size, three important processes altered its density, temperature, and shape, changing it from a large, diffuse (spread-out) cloud to a much smaller spinning disk (Figure 8.3):

- *Heating.* The temperature of the solar nebula increased as it collapsed. Such heating represents energy conservation in action [Section 4.3]. As the cloud shrank, its gravitational potential energy was converted to the kinetic energy of individual gas particles falling inward. These particles crashed into one another, converting the kinetic energy of their inward fall to the random motions of thermal energy (see Figure 4.12b). The Sun formed in the center, where temperatures and densities were highest.
- *Spinning.* Like an ice skater pulling in her arms as she spins, the solar nebula rotated faster and faster as it

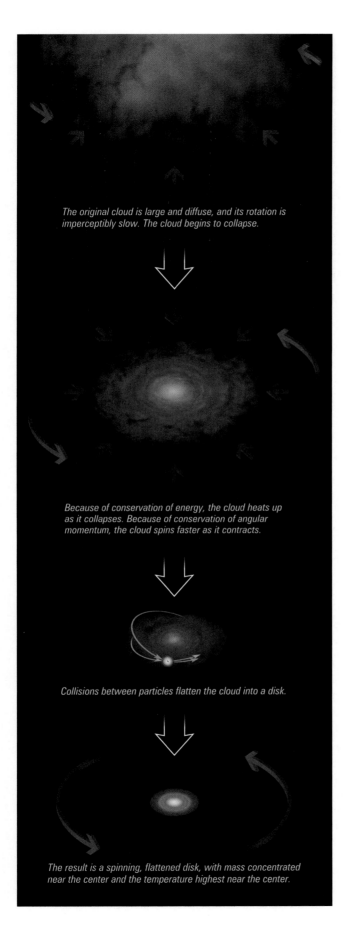

The original cloud is large and diffuse, and its rotation is imperceptibly slow. The cloud begins to collapse.

Because of conservation of energy, the cloud heats up as it collapses. Because of conservation of angular momentum, the cloud spins faster as it contracts.

Collisions between particles flatten the cloud into a disk.

The result is a spinning, flattened disk, with mass concentrated near the center and the temperature highest near the center.

shrank in radius. This increase in rotation rate represents conservation of angular momentum in action [Section 4.3]. The rotation of the cloud may have been imperceptibly slow before its collapse began, but the cloud's shrinkage made fast rotation inevitable. The rapid rotation helped ensure that not all the material in the solar nebula collapsed into the center: The greater the angular momentum of a rotating cloud, the more spread out it will be.

- *Flattening.* The solar nebula flattened into a disk. This flattening is a natural consequence of collisions between particles in a spinning cloud. A cloud may start with any size or shape, and different clumps of gas within the cloud may be moving in random directions at random speeds. These clumps collide and merge as the cloud collapses, and each new clump has the average velocity of the clumps that formed it. Thus, the random motions of the original cloud become more orderly as the cloud collapses, changing the cloud's original lumpy shape into a rotating, flattened disk. Similarly, collisions between clumps of material in highly elliptical orbits reduce their eccentricities, making their orbits more circular.

The formation of the spinning disk explains the orderly motions of our solar system today. The planets all orbit the Sun in nearly the same plane because they formed in the flat disk. The direction in which the disk was spinning became the direction of the Sun's rotation and the orbits of the planets. Computer models show that the planets would also have tended to rotate in this same direction as they formed—which is why most planets rotate the same way today—though the small sizes of planets compared to the entire disk allowed some exceptions to arise. The fact that collisions in the disk tended to make orbits more circular explains why most planets in our solar system have nearly circular orbits.

THINK ABOUT IT

You can create a simple analogy to the organized motions of the solar system by sprinkling pepper into a bowl of water and stirring it quickly in random directions. Because the water molecules are always colliding with one another, the motion of the pepper grains will tend to settle down into a slow rotation representing the average of the original, random velocities. Try the experiment several times, stirring the water differently each time. Do the random motions ever exactly cancel out, resulting in no rotation at all? Describe what occurs, and explain how it is similar to what took place in the solar nebula.

Testing the Model If we are correct about the processes that occurred in the solar nebula, the same processes ought to occur in other clouds of gas that collapse to form stars.

a We see this disk edge-on around the star AU Microscopii, confirming its flattened shape.

b This photo shows a disk around the star HD 141569A. The colors are not real; a black-and-white image has been tinted red to bring out faint detail.

Figure 8.4 These Hubble Space Telescope photos show flattened, spinning disks of material around other stars.

Thus, the formation of a disk in which planets can form should be a natural part of the star formation process, and we can test whether our model makes sense by searching for such disks elsewhere.

Observational evidence does indeed support the idea that our solar nebula collapsed in the way envisioned by the nebular model. The heating that occurs in a collapsing cloud of gas means that the gas should emit thermal radiation [Section 5.4], primarily in the infrared. We've detected infrared radiation from many nebulae where star systems appear to be forming today. We've even seen structures around other stars that appear to be flattened, spinning disks (Figure 8.4).

Other support for the model comes from computer simulations of the formation process. A simulation begins with a set of data representing the conditions we observe in interstellar clouds. Then, with the aid of a computer, we apply the laws of physics to predict the changes that should occur over time. These computer simulations successfully reproduce most of the general characteristics of motion in our solar system, suggesting that the nebular theory is on the right track.

Finally, evidence that our ideas about the formation of flattened disks are correct comes from many other structures in the universe. We expect flattening to occur anywhere that orbiting particles can collide, which explains why we find so many cases of flat disks, including the disks of spiral galaxies like the Milky Way, the disks of planetary rings, and the *accretion disks* surrounding many neutron stars and black holes [Section 18.3].

8.3 The Formation of Planets

The planets began to form after the solar nebula had collapsed into a flattened disk of perhaps 200 AU in diameter

(about twice the present-day diameter of Pluto's orbit). In this section, we'll discuss how the planets formed within the spinning disk of gas. Along the way, we'll see how the process of planet formation led to the existence of the two basic types of planets.

• Why are there two major types of planets?

The churning and mixing of gas in the solar nebula should have ensured that the nebula had the same composition throughout: 98% hydrogen and helium plus 2% heavier elements. Yet the solar system ended up with two very different major types of planets: the small, rocky terrestrial planets and the large, gaseous jovian planets. How did the compositions of the planets end up being so different, when they were all born in the same cloud of gas? The locations of these two major classes of planets provide the key to understanding how they formed. The terrestrial planets formed in the warm, inner regions of the swirling disk and the jovian planets formed in the colder, outer regions. Let's explore how temperature differences led directly to different types of planets.

Condensation: Sowing the Seeds of Planets   In the center of the collapsing solar nebula, gravity drew together enough material to form the Sun. In the surrounding disk, however, the gaseous material was too spread out for gravity alone to clump it up. Instead, material had to begin clumping in some other way and then grow in size until gravity could start pulling it together into planets. In essence, planet formation required the presence of "seeds"—solid bits of matter from which gravity could ultimately build planets.

The basic process of seed formation was probably much like the formation of snowflakes in clouds on Earth: When the temperature is low enough, some atoms or molecules in a gas may bond and solidify. (Pressures in the solar nebula were generally too low to allow the condensation of liquid droplets.) The general process in which solid (or liquid) particles form in a gas is called **condensation**—we say that the particles *condense* out of the gas. The particles that initially condense from a gas are microscopic in size, but they can grow larger with time.

Different materials condense at different temperatures. On the basis of their condensation properties, the ingredients of the solar nebula fell into four major categories, summarized in Table 8.1:

- **Hydrogen and helium gas (98% of the solar nebula).** These gases never condense under the conditions present in a nebula.
- **Hydrogen compounds (1.4% of the solar nebula).** Materials such as water (H_2O), methane (CH_4), and ammonia (NH_3) can solidify into **ices** at low temperatures (below about 150 K under the low pressure of the solar nebula).
- **Rock (0.4% of the solar nebula).** Rocky material is gaseous at very high temperatures, but condenses into solid bits of mineral at temperatures between about 500 K and 1,300 K, depending on the type of rock. (A *mineral* is a piece of rock with a particular chemical composition and structure.)
- **Metal (0.2% of the solar nebula).** Metals such as iron, nickel, and aluminum are also gaseous at very high temperatures, but condense into solid form at higher temperatures than rock—typically in the range of 1,000 K to 1,600 K.

Because hydrogen and helium gas made up 98% of the solar nebula's mass and did not condense, the vast majority of the nebula remained gaseous at all times. However, hydrogen compounds, rock, and metal could condense into solid form wherever the temperature allowed (Figure 8.5). In the innermost regions of the nebula near the forming Sun, where the temperature was above 1,600 K, it was too hot for any material to condense. Near what is now Mercury's orbit, the temperature was low enough for metals and some types of rock to condense into tiny, solid particles, but other types of rock and all the hydrogen compounds remained gaseous. More types of rock could condense, along with the metals, at the distances from the Sun where Venus, Earth, and Mars would form. In the region where the asteroid belt would eventually be located, temperatures were low enough to allow dark, carbon-rich minerals to condense, along with minerals containing small amounts of water. Hydrogen compounds could condense into ices only beyond the **frost line,** which lay between the present-day orbits of Mars and Jupiter.

The frost line marked the key transition between the warm inner regions of the solar system where terrestrial planets formed and the cool outer regions where jovian planets formed. Inside the frost line, only metal and rock could condense into solid "seeds," which is why the terres-

Table 8.1 Materials in the Solar Nebula

A summary of the four types of materials present in the solar nebula. The squares represent the relative proportions of each type (by mass).

	Examples	Typical Condensation Temperature	Relative Abundance (by mass)
Hydrogen and Helium Gas	hydrogen, helium in nebula	do not condense	98%
Hydrogen Compounds	water (H_2O) methane (CH_4) ammonia (NH_3)	<150 K	1.4%
Rock	various minerals	500– 1,300 K	0.4%
Metals	iron, nickel, aluminum	1,000– 1,600 K	0.2%

trial planets ended up being made of metal and rock. Beyond the frost line, where it was cold enough for hydrogen compounds to condense into ices, the solid seeds were built of ice along with metal and rock. Moreover, because hydrogen compounds were nearly three times as abundant in the nebula as metal and rock combined (see Table 8.1), the total amount of solid material was far greater beyond the frost line than within it. The stage was set for the birth of two types of planets: planets born from seeds of metal and rock in the inner solar system and planets born from seeds of ice (as well as metal and rock) in the outer solar system.

THINK ABOUT IT

Consider a region of the solar nebula in which the temperature was about 1,300 K. Based on the data in Table 8.1, what fraction of the material in this region was gaseous? What were the solid particles in this region made of? After you have answered these questions, do the same for a region with a temperature of 100 K. Would the 100 K region be closer to or farther from the Sun? Explain.

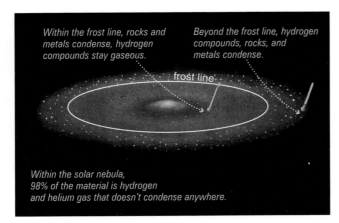

Within the frost line, rocks and metals condense, hydrogen compounds stay gaseous.

Beyond the frost line, hydrogen compounds, rocks, and metals condense.

frost line

Within the solar nebula, 98% of the material is hydrogen and helium gas that doesn't condense anywhere.

Figure 8.5 Interactive Figure, Temperature differences in the solar nebula led to different kinds of condensed materials, sowing the seeds for two different kinds of planets.

• How did the terrestrial planets form?

We have seen that inner regions of the solar nebula contained tiny seeds of solid metal and rock, and we know that the inner solar system ended up with terrestrial planets made of metal and rock. Thus, the basic story of the terrestrial planets seems fairly clear: The solid seeds of metal and rock gradually grew into the full-fledged planets we see today. Because rock and metal made up such a small amount of the material in the solar nebula, the planets they built ended up relatively small in size. All that's left for us to understand are the details of how the seeds grew in size.

The process by which small "seeds" grew into planets is called **accretion** (Figure 8.6). Accretion began with the microscopic solid particles that condensed from the gas of the solar nebula. These particles orbited the forming Sun with the same orderly, circular paths as the gas from which they condensed. Individual particles therefore moved at nearly

the same speed as neighboring particles, so "collisions" were more like gentle touches. Although the particles were far too small to attract each other gravitationally at this point, they were able to stick together through electrostatic forces—the same "static electricity" that makes hair stick to a comb. Small particles thereby began to combine into larger ones. As the particles grew in mass, gravity began to aid the process of their sticking together, accelerating their growth into boulders large enough to count as **planetesimals**, which means "pieces of planets."

The planetesimals grew rapidly at first. As a planetesimal grew larger, it would have both more surface area to make contact with other planetesimals and more gravity to attract them. Some planetesimals probably grew to hundreds of kilometers in size in only a few million years—a long time in human terms, but only about 1/1,000 the present age of the solar system. However, once the planetesimals reached these relatively large sizes, further growth became more difficult.

Gravitational encounters [Section 4.5] between planetesimals tended to alter their orbits, particularly those of the smaller planetesimals. With different orbits crossing each other, collisions between planetesimals tended to occur at higher speeds and hence became more destructive. Such collisions tended to produce fragmentation more often than accretion. Only the largest planetesimals avoided being shattered and thus were able to grow into full-fledged, terrestrial planets.

Theoretical evidence in support of this model comes from computer simulations of the accretion process, which support the scenario described above. Observational evidence comes from meteorites, many of which appear to be surviving fragments from the early period of condensation in our solar system [Section 9.1]. Meteorites often contain metallic grains embedded in a variety of rocky minerals (Figure 8.7), just as we expect for the planetesimals of the inner solar system. Meteorites thought to come from the

Early in the accretion process, there are many relatively large planetesimals on crisscrossing orbits.

As time passes, a few planetesimals grow larger by accreting smaller ones, while others shatter in collisions.

Ultimately, only the largest planetesimals avoid shattering and grow into full-fledged planets.

Not to scale!

Figure 8.6 Interactive Figure, These diagrams show how planetesimals gradually accrete into terrestrial planets.

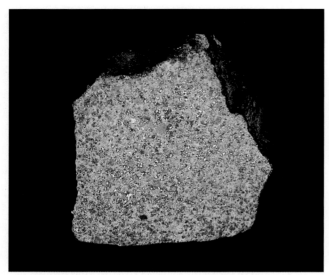

Figure 8.7 Shiny flakes of metal are clearly visible in this slice through a meteorite (a few centimeters across), mixed in among the rocky material. Such metallic flakes are just what we would expect to find if condensation really occurred in the solar nebula as described by the nebular theory.

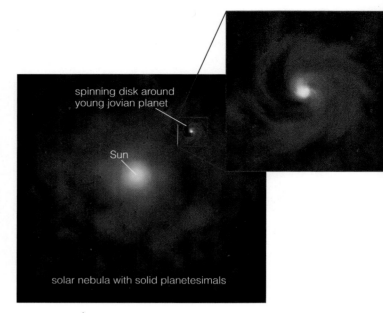

spinning disk around young jovian planet

Sun

solar nebula with solid planetesimals

Figure 8.8 The young jovian planets were surrounded by disks of gas, much like the disk of the entire solar nebula but smaller in size. According to the leading model, the planets grew as large ice-rich planetesimals captured hydrogen and helium gas from the solar nebula. This painting shows the gas and planetesimals surrounding one jovian planet in the larger solar nebula.

outskirts of the asteroid belt contain abundant carbon-rich materials, and some contain water—again as we expect for planetesimals that formed in that region.

• How did the jovian planets form?

Accretion should have occurred similarly in the outer solar system, but condensation of ices meant that there was more solid material. Thus, the planetesimals that accreted in the outer solar system contained large amounts of ice in addition to metal and rock. The solid objects that reside in the outer solar system today, such as comets and the moons of the jovian planets, still show this ice-rich composition. However, the growth of icy planetesimals cannot be the whole story of jovian planet formation, because the jovian planets themselves are *not* made mostly of ice. Instead, they contain large amounts of hydrogen and helium gas.

The precise way in which the jovian planets formed remains a topic of scientific debate, but in the leading model they acquired their gas after beginning existence as large, ice-rich planetesimals. According to this model, some of the abundant, icy planetesimals of the outer solar system grew to masses many times that of Earth. With these large masses, their gravity became strong enough to capture and hold some of the hydrogen and helium gas that made up the vast majority of the surrounding solar nebula. As the growing planets accumulated gas, their gravity grew stronger still, allowing them to capture even more gas. Ultimately, the jovian planets grew so much that they bore little resemblance to the icy seeds from which they started, instead ending up with large abundances of hydrogen and helium gas.

This model also explains the many large moons of the jovian planets. The same processes of heating, spinning, and flattening that made the disk of the solar nebula should also have affected the gas drawn by gravity to the young

jovian planets. Thus, each jovian planet probably was surrounded by its own disk of gas, spinning in the same direction that the planet rotated (Figure 8.8). Moons accreted from ice-rich planetesimals within these disks, which is why these moons tend to orbit in the same direction as their planet rotates, with nearly circular orbits lying close to the equatorial plane of their parent planet.

This model of accretion followed by gas capture explains the observed features of the jovian planets quite well. Nevertheless, some scientists have recently advanced a competing model in which disturbances (technically, instabilities) within the disk of the solar nebula could have led clumps of gas to collapse and form jovian planets without first forming an icy planetesimal. Because this model is relatively new, scientists have not yet worked out enough details to know whether it can explain all the characteristics of the jovian planets. Most planetary scientists therefore still favor the accretion and gas capture model, but good practice of science demands that we remain open to considering new ideas.

• What ended the era of planet formation?

The vast majority of the hydrogen and helium gas in the solar nebula never became part of any planet. So what became of it? Apparently, it was swept into interstellar space by the **solar wind**, which consists of charged particles (such as protons and electrons) that are continually blown off the surface of the Sun in all directions. Although the solar wind is fairly weak today, observations of other stars show that winds tend to be much stronger in young stars. Thus, the young Sun should have had a strong solar wind—strong enough to have swept huge quantities of gas out of the solar system. The radiation from the young Sun also played a role in clearing out the gas.

Sealing the Fates of Planets   The clearing of the gas sealed the compositional fate of the planets. If the gas had not been cleared soon after the planets formed, it might have continued to cool until hydrogen compounds could have condensed into ices even in the inner solar system. In that case, the terrestrial planets might have accreted abundant ice, and perhaps hydrogen and helium gas as well, changing their basic nature. At the other extreme, if the gas had been blown out too early, the raw materials of the planets might have been swept away before the planets could fully form. Although these extreme scenarios did not occur in our solar system, studies of other planetary systems suggest that they may sometimes occur around other stars [Section 13.3].

Explaining the Sun's Rotation   The young Sun's strong solar wind also helps explain what was once considered a surprising aspect of the Sun's rotation. According to the law of conservation of angular momentum, the spinning disk of the solar nebula should have spun fastest near its center, where most of the mass became concentrated. Thus, the young Sun should have been rotating very fast. Today the Sun rotates quite slowly, with each full rotation taking about a month. If the young Sun really did rotate fast, as our theory seems to demand, how did its rotation slow down?

Angular momentum cannot simply disappear, but it is possible to transfer angular momentum from one object to another—and then get rid of the second object. A spinning skater can slow her spin by grabbing her partner and then pushing him away. In the 1950s, scientists realized that the young Sun's rapid rotation would have generated a magnetic field far stronger than that of the Sun today. This strong magnetic field actually helped create the strong solar wind. More generally, the strong magnetic field would have made the young Sun much more active on its surface than it is today [Section 14.3]. For example, large "solar storms"

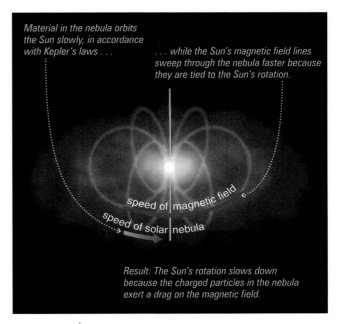

Material in the nebula orbits the Sun slowly, in accordance with Kepler's laws . . .

. . . while the Sun's magnetic field lines sweep through the nebula faster because they are tied to the Sun's rotation.

speed of magnetic field

speed of solar nebula

Result: The Sun's rotation slows down because the charged particles in the nebula exert a drag on the magnetic field.

Figure 8.9   The young Sun should have rotated rapidly, but today the Sun rotates quite slowly. As shown in this diagram, the Sun probably lost angular momentum because of drag between slow-moving charged particles in the solar nebula and the Sun's rotating magnetic field (represented by purple loops).

and sunspots would have been much more common, causing the emission of much more ultraviolet and X-ray light. This high-energy radiation from the young Sun ionized gas in the solar nebula, creating many charged particles.

As we will discuss in more detail in Chapter 14, charged particles and magnetic fields tend to stick together. As the Sun rotated, its magnetic field dragged the charged particles along faster than the rest of the nebula, adding to their angular momentum. As the particles were gaining angular momentum, the Sun was losing it. The young Sun's strong solar wind then blew these particles into interstellar space, leaving the Sun with the greatly diminished angular momentum and much slower rotation that we see today (Figure 8.9).

Although we cannot prove that the young Sun really did lose angular momentum in this way, support for the idea comes from observations of other stars. When we look at young stars that have recently formed in interstellar clouds, we find that nearly all of them rotate rapidly and have strong magnetic fields and strong winds [Section 16.2]. Older stars, in contrast, almost invariably rotate slowly, like our Sun. This suggests that nearly all stars have their original rapid rotations slowed by transferring angular momentum to charged particles in their disks—particles that are later swept away—just as our theory suggests happened with the Sun.

8.4 The Aftermath of Planet Formation

So far, we have seen how the nebular theory explains our solar system's patterns of motion and its two major types of planets. We still have two features of our solar system left

---

to explain: the existence, composition, and orbital properties of the myriad asteroids and comets, and the exceptions to the general rules. These two features turn out to be closely related—both involve the planetesimals that remained among the forming planets in the early solar system.

• Where did asteroids and comets come from?

The strong wind from the young Sun cleared excess gas from the solar nebula, but many planetesimals remained scattered between the newly formed planets. These "leftovers" became asteroids and comets. We'll wait until Chapter 12 to discuss the details of why most asteroids ended up grouped in the asteroid belt while most comets ended up split between two regions, the Kuiper belt and the Oort cloud. However, we can understand the basic ideas here.

Asteroids are the rocky leftover planetesimals of the inner solar system. The four inner planets must have swept up most of the planetesimals near their orbits, leaving relatively few leftovers between them. However, no planet formed between Mars and Jupiter (a fact attributable to effects from Jupiter's strong gravity [Section 12.1]), so this region—the asteroid belt—must once have been dense with rocky planetesimals. In fact, the asteroid belt probably once contained enough planetesimals to form another terrestrial planet, but most of them ultimately crashed into the inner planets or were ejected from the solar system. Today, the total mass of the asteroids in the asteroid belt is only a tiny fraction of the mass of any terrestrial planet.

Comets are the ice-rich leftover planetesimals of the outer solar system. Comets must have initially populated the entire outer solar system, presumably out to some distance beyond Neptune. Those that orbited beyond Neptune should have been able to grow fairly large without disruption by the jovian planets. Moreover, they would have remained on their original orbits in the same plane as the orbits of the planets. These large comets thereby became the comets of the Kuiper belt, which follow the same basic orbital patterns as the planets. Pluto, which seemed a misfit when compared to the first eight planets, seems much more at home when considered as a large comet of the Kuiper belt [Section 12.3].

In contrast to the stable orbital paths for comets beyond Neptune, the comets that roamed between the jovian planets almost inevitably suffered close gravitational encounters with these planets. Ultimately, these encounters caused most of these comets to be "kicked" into orbits that carried them far from the Sun. Because the "kicks" could have sent these comets flying in random directions, their new orbits no longer displayed the original orderly motion of the disk in which the planets formed. That is why these comets now populate the Oort cloud, with randomly oriented orbits very far from the Sun.

Evidence that asteroids and comets really are leftover planetesimals comes from analysis of meteorites, spacecraft visits to comets and asteroids, and computer simulations of solar system formation. In fact, the nebular theory actu-

ally *predicts* the existence of both the Oort cloud and the Kuiper belt—a prediction first made in the 1950s. Thus, the discoveries, beginning in the 1990s, of numerous objects orbiting in the Kuiper belt represent a triumph for the nebular theory. Comets in the distant Oort cloud remain undetectable with present technology, but there is little doubt that they exist, because their presence is the only way to explain the comets that enter the inner solar system from afar.

• How do we explain "exceptions to the rules"?

We have now explained all the major features of our solar system except for the "exceptions to the rules," such as our surprisingly large Moon and the odd rotation of Uranus. Today, we think that most of these exceptions arose from collisions or close encounters with leftover planetesimals.

The Heavy Bombardment The asteroids and comets that exist today probably represent only a small fraction of the vast numbers of leftover planetesimals that roamed the young solar system. The rest are now gone. Some of these "lost" planetesimals may have been flung into deep space, but many others must have collided with the planets. The vast majority of these collisions occurred in the first few hundred million years of our solar system's history, during the period we call the **heavy bombardment.**

Every world in our solar system must have been pelted by impacts during the heavy bombardment (Figure 8.10). Some of these impacts left scars that we can still see today as impact craters. These impacts did more than just batter

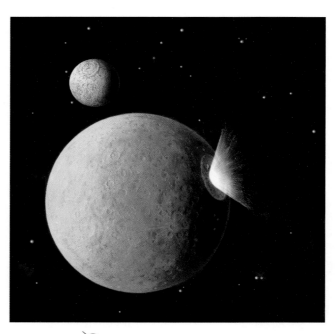

Figure 8.10 🛰 Around 4 billion years ago, Earth, its Moon, and the other planets were heavily bombarded by leftover planetesimals. This painting shows the young Earth and Moon, with an impact in progress on Earth.

the planets. They also brought materials from other regions of the solar system—a fact that is critical to our existence on Earth today.

Remember that the terrestrial planets were built from planetesimals made of metal and rock. These planetesimals probably contained no water or other hydrogen compounds at all, because it was too hot for these compounds to condense in our region of the solar nebula. How, then, did Earth come to have the water that makes up our oceans? The answer is that water, along with other hydrogen compounds, must have been brought to Earth and the other terrestrial planets by the impact of planetesimals that formed farther from the Sun. We don't yet know whether these planetesimals came primarily from the outer asteroid belt, where rocky planetesimals contained small amounts of water and other hydrogen compounds, or whether they were comets containing huge amounts of ice. In either case, the water we drink and the air we breathe probably once were part of planetesimals floating beyond the orbit of Mars.

<hr>
THINK ABOUT IT
<hr>

Jupiter's gravity played a major role in flinging asteroids and comets into the inner solar system. Given this fact, what can you say about Jupiter's role in bringing water to Earth? How might Earth be different if Jupiter had never formed?

Captured Moons   We can easily explain the orbits of most large jovian planet moons by their formation in a disk that swirled around the forming planet. However, some moons have unusual orbits—orbits in the "wrong" direction (opposite the rotation of their planet) or with large inclinations to the planet's equator. These unusual moons are probably leftover planetesimals that were captured into orbit around a planet.

It's not easy for a planet to capture a moon. An object cannot switch from an unbound orbit (for example, an asteroid whizzing by Jupiter) to a bound orbit (for example, a moon orbiting Jupiter) unless it somehow loses orbital energy [Section 4.5]. For the jovian planets, captures probably occurred when passing planetesimals lost energy to drag in the extended and relatively dense gas that surrounded these planets as they formed. The planetesimals would have been slowed by friction with the gas, just as artificial satellites are slowed by drag in encounters with Earth's atmosphere. If friction reduced a passing planetesimal's orbital energy enough, it could have become an orbiting moon. Because of the random nature of the capture process, the captured moons would not necessarily orbit in the same direction as their planet or in its equatorial plane. Computer models suggest that this capture process would have worked only on objects of a few kilometers in size. Most of the small moons of the jovian planets are a few kilometers across, supporting the idea that they were captured in this way. Mars may have captured its two small moons, Phobos and Deimos, in a similar way at a time when the planet had a much more extended atmosphere than it does today (Figure 8.11).

a Phobos   **b** Deimos

Figure 8.11 The two moons of Mars, shown here in photos taken by the *Viking* spacecraft, are probably captured asteroids. Phobos is only about 13 kilometers across and Deimos is only about 8 kilometers across—making each of these two moons small enough to fit within the boundaries of a typical large city.

• How do we explain the existence of our Moon?

Capture processes explain the origin of many moons, but they cannot explain our own Moon. Our Moon is much too large to have been captured by a small planet like Earth. We can also rule out the possibility that our Moon formed simultaneously with Earth. If the Moon and Earth had formed together, both should have accreted from planetesimals of the same type and should therefore have approximately the same composition and density. That is not the case. The Moon's density is considerably lower than Earth's, indicating that it has a very different average composition and could not have formed in the same way or at the same time as our planet. We must look for another way to explain the origin of our surprisingly large Moon.

Giant Impacts and the Formation of Our Moon   The largest leftover planetesimals may have been huge, perhaps the size of Mars. When one of these planet-size planetesimals collided with a planet, the spectacle would have been awesome. Such a **giant impact** could have significantly altered a planet's fate.

What would have happened if a Mars-size object had collided with the young Earth? Depending on exactly where and how fast the object struck Earth, the blow might have tilted Earth's axis, changed its rotation rate, or completely shattered our planet. The most interesting case arises when we consider what would have happened if the impact had blasted away rock from Earth's outer layers (mantle and crust) and sent this material into orbit around our planet. According to computer simulations, this orbiting material could have reaccreted to form our Moon (Figure 8.12).

Today, such a giant impact is the leading hypothesis for explaining the origin of our Moon. Strong support for this hypothesis comes from two features of the Moon's composition. First, the Moon's overall composition is quite similar to that of Earth's outer layers—just as we should expect if it were made from material blasted away from those layers. Second, the Moon has a much smaller proportion of easily vaporized ingredients (such as water) than Earth. This fact

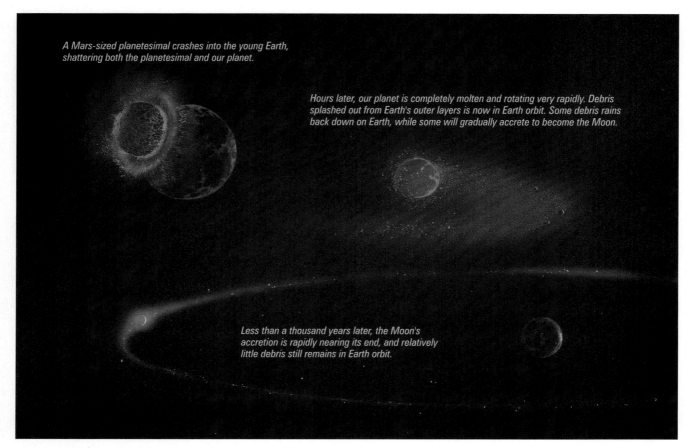

A Mars-sized planetesimal crashes into the young Earth, shattering both the planetesimal and our planet.

Hours later, our planet is completely molten and rotating very rapidly. Debris splashed out from Earth's outer layers is now in Earth orbit. Some debris rains back down on Earth, while some will gradually accrete to become the Moon.

Less than a thousand years later, the Moon's accretion is rapidly nearing its end, and relatively little debris still remains in Earth orbit.

Figure 8.12 Artist's conception of the Moon's formation by a giant impact. The fact that ejected material came mostly from Earth's outer rocky layers explains why the Moon contains very little metal. The impact must have occurred more than 4.4 billion years ago, since that is the age of the oldest Moon rocks. As shown, the Moon formed quite close to a rapidly rotating Earth, but over billions of years tidal forces have slowed Earth's rotation and moved the Moon's orbit outward (see Figure 4.26).

supports the hypothesis because the heat of the impact would have vaporized these ingredients. As gases, they would not have participated in the process of reaccretion that formed the Moon.

Other Giant Impacts Giant impacts may also explain many of the other exceptions to the general trends. For example, giant impacts may have been responsible for tilting the axes of many planets (including Earth) and perhaps for tipping Uranus on its side. Pluto's moon Charon may have formed in a giant impact similar to the one that formed our Moon. Mercury's surprisingly high density may be the result of a giant impact that blasted away its rocky outer layers, leaving behind a planet made almost entirely of the metal that was once in its core.

Unfortunately, we can do little to test whether a particular giant impact really occurred billions of years ago. The difficulty in proving the giant impact hypothesis makes the idea controversial, and most planetary scientists didn't take the idea seriously when it was first proposed. But no other idea so effectively explains the formation of our Moon and other "exceptions" we've discussed. Moreover, numerous giant impacts certainly *should* have occurred, given the number of large leftover planetesimals predicted

by the nebular theory. Random giant impacts are therefore considered the most likely explanation for many of the exceptions to our general planetary rules.

• Was our solar system destined to be?

We've now seen that the nebular theory accounts for all the major features of our solar system. Of course, that does not mean that it explains *everything* about the formation of our solar system. As we'll discuss in Chapter 13, discoveries of other planetary systems have already forced us to revise parts of the theory to allow for a wider range of possibilities than we observe in our own solar system. Nevertheless, the basic theory is widely accepted today because it is so strongly supported by the available evidence. Figure 8.13 summarizes our current understanding of how our solar system formed. The entire process of planet formation probably took no more than a few tens of millions of years, about 1% of the current age of the solar system.

Assuming the nebular theory is correct, was our solar nebula "destined" to form the solar system we see today? Probably not. The first stages of planet formation were orderly and inevitable according to the nebular theory.

Figure 8.13 A summary of the process by which our solar system formed, according to the nebular theory.

A large, diffuse interstellar gas cloud (solar nebula) contracts due to gravity.

Contraction of Solar Nebula: *As it contracts, the cloud heats, flattens, and spins faster, becoming a spinning disk of dust and gas.*

The Sun will be born in the center.

Planets will form in the disk.

Warm temperatures allow only metal/rock "seeds" to condense in inner solar system.

Condensation of Solid Particles: *Hydrogen and helium remain gaseous, but other materials can condense into solid "seeds" for building planets.*

Cold temperatures allow "seeds" to contain abundant ice in the outer solar system.

Terrestrial planets are built from metal and rock.

Accretion of Planetesimals: *Solid "seeds" collide and stick together. Larger ones attract others with their gravity, growing bigger still.*

The seeds of jovian planets grow large enough to attract hydrogen and helium gas, making them into giant, mostly gaseous planets; moons form in disks of dust and gas that surround the planets.

Clearing the Nebula: *The solar wind blows remaining gas into interstellar space.*

Terrestrial planets remain in the inner solar system.

Jovian planets remain in the outer solar system.

"Leftovers" from the formation process become asteroids (metal/rock) and comets (mostly ice).

Not to scale

The creation of a spinning disk, condensation within that disk, and the first stages of accretion were relatively gradual processes that probably would happen all over again if we turned back the clock. However, the final stages of accretion, and giant impacts in particular, are inherently random in nature and probably would not happen again in just the same way. A larger or smaller planet might form at Earth's location and might suffer from a larger giant impact or from no giant impact at all. However, we don't yet know whether these differences would fundamentally alter the solar system or simply change a few "minor" details—such as the possibility of life on Earth. What we do know is that, among the billions of other star systems in our galaxy, it's unlikely that any other is exactly like ours.

8.5 The Age of the Solar System

The nebular theory seems to give us a very good understanding of how our solar system was born. But *when* was it born, and how do we know? The answer is that the planets began to form through accretion about 4.6 billion years ago, a fact we learn by determining the age of the oldest rocks in the solar system.

• How does radioactivity reveal an object's age?

The first step in understanding how we learn the age of our solar system is to understand exactly what we mean by the age of a rock. A rock is a collection of a great many atoms held together in solid form. The atoms must be older than Earth, having been forged in the Big Bang or in stars that lived long ago. We cannot determine the ages of the individual atoms, because old atoms are indistinguishable from young ones. However, some atoms undergo changes with time that allow us to determine how long they have been held in place within the rock's solid structure. Thus, by the age of a rock we mean the time since its atoms became locked together in their present arrangement, which in most cases means the time *since the rock last solidified.*

The method by which we measure the age of a rock is known as **radiometric dating**. This method relies on careful measurement of the proportions of various atoms and isotopes in the rock. Remember that each chemical element is uniquely characterized by the number of protons in its nucleus. Different *isotopes* of the same element differ only in their number of neutrons [Section 5.3].

A **radioactive isotope** has a nucleus that can undergo spontaneous change, or **radioactive decay**, such as breaking apart or having one of its protons turn into a neutron. Decay can change one element into an entirely different one, with different chemical properties. For example, potassium-40 is a radioactive isotope with nuclei that decay when a proton turns into a neutron, changing the

potassium-40 into argon-40. We say that potassium-40 is the *parent isotope*, because it is the original isotope before the decay, and argon-40 is the *daughter isotope* left behind by the decay process.

For any single nucleus, radioactive decay is an instantaneous event. However, in a large collection of atoms, individual nuclei will decay at different times. Thus, if we start with a collection of atoms of a particular parent isotope, we will find that it gradually transforms itself into a collection of atoms of the daughter isotope. The rate at which this transformation occurs is characterized by the parent isotope's **half-life**, the time it would take for half of the parent nuclei in the collection to decay. For example, laboratory measurements show the half-life of potassium-40 to be 1.25 billion years, which means that if we started with any particular amount of potassium-40, it would take 1.25 billion years for half of it to decay into argon-40. Every radioactive isotope has its own unique half-life, which may be anywhere from a fraction of a second to many billions of years. (We can measure half-lives even when they are billions of years because we only need to observe the decay of a tiny fraction of the nuclei to calculate how long it would take half of them to decay.)

Because we can measure decay rates so precisely, in principle it is easy to determine the age of a rock. Consider a small piece of rock that contained 1 microgram of potassium-40 and no argon-40 when it formed (solidified) long ago. Because potassium-40 has a half-life of 1.25 billion years, half the original potassium-40 would have decayed into argon-40 by the time the rock was 1.25 billion years old. Thus, after 1.25 billion years, the rock contained $\frac{1}{2}$ microgram of potassium-40 and $\frac{1}{2}$ microgram of argon-40. Half again would have decayed by the end of the next 1.25 billion years, so after 2.5 billion years the rock contained $\frac{1}{4}$ microgram of potassium-40 and $\frac{3}{4}$ microgram of argon-40. After three half-lives, or 3.75 billion years, only $\frac{1}{8}$ microgram of potassium-40 remained, while $\frac{7}{8}$ microgram had become argon-40. Figure 8.14 summarizes the gradual decrease in the amount of potassium-40 and the corresponding rise in the amount of argon-40.

We can now see the essence of radiometric dating. Suppose you find a rock that contains equal numbers of atoms of potassium-40 and argon-40. If you *assume* that all the argon came from potassium decay, then it must have taken precisely one half-life for the rock to end up with equal amounts of the two isotopes. You could therefore conclude that the rock is 1.25 billion years old (see Figure 8.14). The only question is whether you are right in assuming that the rock lacked argon-40 when it formed. In this case, knowing a bit of "rock chemistry" helps. Potassium-40 is a natural ingredient of many minerals in rocks, but argon-40 is a gas that never combines with other elements and did not condense in the solar nebula. Therefore, if you find argon-40 gas trapped inside minerals, you can be sure that it came from radioactive decay of potassium-40. (Another assumption is that no argon-40 gas has escaped from the rock. Rock chemistry tells us that as long as the rock has not been sig-

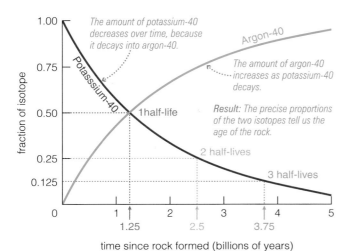

Figure 8.14 Potassium-40 is radioactive, decaying into argon-40 with a half-life of 1.25 billion years. The red line shows the decreasing amount of potassium-40, and the blue line shows the increasing amount of argon-40. The remaining amount of potassium-40 drops by half with each successive half-life.

nificantly heated since the time it formed, atoms of gases such as argon-40 will remain trapped within it.)

Rocks from the Moon offer another example of radiometric dating. Rocks that the *Apollo* astronauts brought

back from the ancient lunar highlands [Section 9.3] contain minerals with a very small amount of uranium-238, which decays (in several steps) into lead-206 with a half-life of about 4.5 billion years. Lead and uranium have very different chemical behaviors, and some minerals start with virtually no lead. Laboratory analysis of such minerals in lunar rocks shows that they now contain almost equal proportions of atoms of uranium-238 and lead-206. We conclude that half the original uranium-238 has decayed, turning into the same number of lead-206 atoms. The lunar rock therefore must be about one half-life old, or about 4.5 billion years old. More precise work shows these lunar rocks to be about 4.4 billion years old.

THINK ABOUT IT

If future scientists examine the same lunar rocks 4.5 billion years from now, what proportions of uranium-238 and lead-206 will they find? Explain.

Radiometric dating is possible with many other radioactive isotopes as well. In many cases, we can date a rock that contains more than one radioactive isotope, so agreement between the ages calculated from the different isotopes gives us confidence that we have dated the rock correctly. Moreover, we can often check the results we obtain from radio-

---

MATHEMATICAL INSIGHT **8.1** **Radiometric Dating**

The amount of a radioactive substance decays by half with each half-life, so we can express the decay process with a simple formula relating the current amount of a radioactive substance in a rock to the original amount:

$$\frac{\text{current amount}}{\text{original amount}} = \left(\frac{1}{2}\right)^{t/t_{\text{half}}}$$

where t is the time since the rock formed and t_{half} is the half-life of the radioactive material. We can solve this equation for the age t by taking the logarithm of both sides and rearranging the terms. The resulting general equation for the age is:

$$t = t_{\text{half}} \times \frac{\log_{10}\left(\frac{\text{current amount}}{\text{original amount}}\right)}{\log_{10}\left(\frac{1}{2}\right)}$$

Example: You heat and chemically analyze a small sample of a meteorite. Potassium-40 and argon-40 are present in a ratio of approximately 0.85 unit of potassium-40 atoms to 9.15 units of gaseous argon-40 atoms. (The units are unimportant, because only the relative amounts of the parent and daughter materials matter.) How old is the meteorite?

Solution:

Step 1 Understand: The formula given above allows us to find the age of a rock from the current and original amounts of a

parent isotope. In this case, we are told that the current amount of the parent isotope potassium-40 is 0.85 unit. We are not told the original amount, but we can figure it out from the fact that the sample now has 9.15 units of the daughter isotope argon-40. As discussed in the text, the fact that argon-40 is gaseous makes it reasonable to assume that *all* the trapped argon-40 in the meteorite is a decay product of potassium-40. Thus, the 9.15 units of argon must originally have been potassium-40, which means the original amount of potassium-40 was $0.85 + 9.15 = 10$ units.

Step 2 Solve: We use the formula given above, with the current amount = 0.85 unit and the original amount = 10 units:

$$t = 1.25 \text{ billion yr} \times \frac{\log_{10}\left(\frac{0.85}{10}\right)}{\log_{10}\left(\frac{1}{2}\right)}$$

$$= 1.25 \text{ billion yr} \times \left(\frac{-1.07}{20.301}\right)$$

$$= 4.45 \text{ billion yr}$$

Step 3 Explain: We have found an age of 4.45 billion years. In other words, radiometric dating of this meteorite tells us that it solidified about 4.45 billion years ago. (*Note:* This example captures the essence of radiometric dating, but real cases generally require more detailed analysis.)

Recall that the solar nebula probably could not have started collapsing on its own, because gravity was too weak when the gas was still spread out over a large region of space. We therefore suspect that some cataclysmic event triggered the collapse. But what type of event? Although we may never know for sure, radioactive elements in meteorites give us some clues. Radioactive elements are made only deep inside stars or in violent stellar explosions (supernovae). Thus, the presence of these elements in meteorites and on Earth underscores the fact that our solar system is made from the remnants of past generations of stars.

We have learned even more from the discovery that some meteorites contain the rare isotope xenon-129. Xenon is gaseous even at extremely low temperatures, so it could not have condensed and become trapped in the meteorites as they condensed in the solar nebula. Instead, the xenon in meteorites must be a product of radioactive decay; some other isotope must have condensed in the solar nebula, then decayed into the gaseous xenon after it was already trapped in the meteorite. Laboratory studies show that xenon-129 is a decay product of iodine-129, so we conclude that it was iodine-129 rather than xenon-129 that was present in the solar nebula. Moreover, iodine-129 has a relatively short half-life of 17 million years. Thus, no more than a few tens of millions of years could have passed between the time the iodine was produced in a stellar explosion and the time that it condensed in the solar nebula. Otherwise, too much of the iodine-129 would have decayed for significant amounts to have condensed and later decayed into xenon.

Summing up, we conclude that iodine (and presumably other materials) produced by a nearby stellar explosion ended up as part of the gas in the solar nebula, and that this explosion occurred no more than a few tens of millions of years before the collapse of the solar nebula began. This timing suggests that the shock wave from the exploding star may have been the trigger for the collapse. Once this shock wave got the collapse started, gravity took over and made the rest of the collapse inevitable.

metric dating against other methods of measuring ages, giving us even more confidence that the technique works. For example, with many fairly recent archaeological artifacts, we can confirm the ages obtained from radiometric dating by measuring ages from tree ring data. Similarly, astronomers have discovered completely independent methods for estimating the Sun's age. Although none of these methods give an age as precise as the age we obtain from radiometric dating, they all confirm that the Sun's age is in the range of 4 to 5 billion years. Overall, the technique of radiometric dating has been checked in so many ways and relies on such basic scientific principles that there is no longer any serious scientific debate about its validity.

• When did the planets form?

Radiometric dating tells us how long it has been since a rock solidified, which is not the same as the age of a planet as a whole. For example, we find rocks of many different ages on Earth. Some rocks are quite young because they formed recently from molten lava; others are much older. The oldest Earth rocks are about 4 billion years old, but even these are not as old as Earth itself, because Earth's entire surface has been reshaped through time.

Moon rocks brought back by the *Apollo* astronauts date to as far back as 4.4 billion years ago. Although they are older than Earth rocks, these Moon rocks must still be younger than the Moon itself. The ages of these rocks also tell us that the giant impact thought to have created the Moon must have occurred more than 4.4 billion years ago.

To go all the way back to the origin of the solar system, we must find rocks that have not melted or vaporized since they first condensed in the solar nebula. Meteorites that have fallen to Earth are our source of such rocks. Many meteorites appear to have remained unchanged since they condensed and accreted in the early solar system. Careful analysis of radioactive isotopes in meteorites shows that the oldest ones formed about 4.55 billion years ago, so this time must mark the beginning of accretion in the solar nebula. Because the planets apparently accreted within a few tens of millions of years, Earth and the other planets formed about 4.5 billion years ago. Thus, the age of our solar system is only about a third of the 14-billion-year age of our universe.

THE BIG PICTURE

Putting Chapter 8 into Perspective
In this chapter, we've described the current scientific theory of our solar system's formation and we've seen how this theory explains the major features we observe. As you continue your study of the solar system, keep in mind the following "big picture" ideas:

- The nebular theory of solar system formation gained wide acceptance because of its success in explaining the major characteristics of our solar system.

- Most of the general features of the solar system were determined by processes that occurred very early in the solar system's history.

- Chance events may have played a large role in determining how individual planets turned out. No one knows how different our solar system might be if it started over.

- We have learned the age of our solar system—about 4.6 billion years—from radiometric dating of the oldest meteorites. This age agrees with the ages we can estimate through a variety of other techniques, making it quite clear that we ourselves are recent arrivals on a very old planet.

8.1 The Search for Origins

- **What properties of our solar system must a formation theory explain?** A successful theory must explain patterns of motion, the differences between terrestrial and jovian planets, the presence of asteroids and comets, and exceptions to the rules.

- **What theory best explains the features of our solar system?** The **nebular theory**, which holds that the solar system formed from the gravitational collapse of a great cloud of gas and dust, successfully explains all the major features of our solar system.

8.2 The Birth of the Solar System

- **Where did the solar system come from?** The cloud that gave birth to our solar system was the product of recycling of gas through many generations of stars within our galaxy. This material consisted of 98% hydrogen and helium and 2% everything else combined.

- **What caused the orderly patterns of motion in our solar system?** A collapsing gas cloud naturally tends to heat up, spin faster, and flatten out as it shrinks in size. Thus, our solar system began as a spinning disk of gas and dust. The orderly motions we observe today all came from the orderly motion of this spinning disk.

8.3 The Formation of Planets

- **Why are there two major types of planets?** The two types of planets formed from two different types of solid "seeds" that condensed from gas in the solar nebula. Within the **frost line**, temperatures were so high that only metal and rock could condense; beyond the frost line, cooler temperatures also allowed more abundant **hydrogen compounds** to condense into ice.

- **How did the terrestrial planets form?** Terrestrial planets formed inside the frost line, where **accretion** allowed tiny, solid grains of metal and rock to grow into **planetesimals** that ultimately merged to make the planets we see today.

- **How did the jovian planets form?** Accretion built ice-rich planetesimals in the outer solar system, and some of these icy planetesimals grew large enough for their gravity to draw in hydrogen and helium gas, building the jovian planets.

- **What ended the era of planet formation?** The young Sun had a strong solar wind, which ultimately swept away material not yet accreted onto the planets. Once the solar nebula was cleared away, planet formation ceased because there was little material left to accrete.

8.4 The Aftermath of Planet Formation

- **Where did asteroids and comets come from?** Asteroids are the rocky leftover planetesimals of the inner solar system, and comets are the ice-rich leftover planetesimals of the outer solar system.

- **How do we explain "exceptions to the rules"?** Most of the exceptions probably arose from collisions or close encounters with leftover planetesimals, especially during the **heavy bombardment** that occurred early in the solar system's history.

- **How do we explain the existence of our Moon?** Our Moon is probably the result of a **giant** **impact** between a Mars-size planetesimal and the young Earth. The impact blasted material from Earth's outer layers into orbit, where it reaccreted to form the Moon.

- **Was our solar system destined to be?** A solar system generally like ours was probably destined to form from the collapse of the solar nebula, but the nature of individual planets in our solar system results from significant random events that could have happened differently or not at all.

8.5 The Age of the Solar System

- **How does radioactivity reveal an object's age?** **Radiometric dating** is based on carefully measuring the proportions of radioactive isotopes and their decay products within rocks. The ratio of the two changes with time and provides a reliable measure of the rock's age.

- **When did the planets form?** The planets began to accrete in the solar nebula about 4.6 billion years ago, a fact we determine from radiometric dating of the oldest meteorites.

Review Questions

Short-Answer Questions Based on the Reading

1. What are the four major features of our solar system that provide clues to how it formed? Describe each one briefly.
2. What is the *nebular theory,* and why is it widely accepted by scientists today?
3. What do we mean by the *solar nebula?* What was it made of, and where did it come from?
4. Describe each of the three key processes that led the solar nebula to take the form of a spinning disk. What observational evidence supports this scenario?
5. List the four categories of materials in the solar nebula by their condensation properties and abundance. Which ingredients are present in terrestrial planets? In jovian planets? In comets and asteroids? Explain why.
6. What was the *frost line* in the solar nebula? Explain how temperature differences led to the formation of two distinct types of planets.
7. Briefly describe the process by which terrestrial planets are thought to have formed.
8. How was the formation of jovian planets similar to that of the terrestrial planets? How was it different? Why did the jovian planets end up with so many moons?
9. What is the *solar wind,* and what roles did it play in the early solar system?
10. In the context of planet formation, what are asteroids and comets? Briefly explain why we find asteroids in the asteroid belt and comets in the Kuiper belt and Oort cloud.
11. What was the *heavy bombardment,* and when did it occur?
12. How do we think the Moon formed, and what evidence supports this hypothesis? Should we be surprised that a *giant impact* could have affected our planet?
13. Describe the technique of *radiometric dating.* What is a half-life?
14. How old is the solar system, and how do we know?

Test Your Understanding

Surprising Discoveries? ❓

Suppose we found a solar system with the property described. (These are *not* real discoveries.) In light of what you've learned about the formation of our own solar system, decide whether the discovery should be considered reasonable or surprising. Explain.

15. A solar system has five terrestrial planets in its inner solar system and three jovian planets in its outer solar system.
16. A solar system has four large jovian planets in its inner solar system and seven small planets made of rock and metal in its outer solar system.
17. A solar system has 10 planets that all orbit the star in approximately the same plane. However, five planets orbit in one direction (e.g., counterclockwise), while the other five orbit in the opposite direction (e.g., clockwise).
18. A solar system has 12 planets that all orbit the star in the same direction and in nearly the same plane. The 15 largest moons in this solar system orbit their planets in nearly the same direction and plane as well. However, several smaller moons have highly inclined orbits around their planets.

19. A solar system has six terrestrial planets and four jovian planets. Each of the six terrestrial planets has at least five moons, while the jovian planets have no moons at all.
20. A solar system has four Earth-size terrestrial planets. Each of the four planets has a single moon that is nearly identical in size to Earth's Moon.
21. A solar system has many rocky asteroids and many icy comets. However, most of the comets orbit in the inner solar system, while the asteroids orbit in far-flung regions much like the Kuiper belt and Oort cloud of our solar system.
22. A solar system has several planets similar in composition to the jovian planets of our solar system but similar in mass to the terrestrial planets of our solar system.
23. A solar system has several terrestrial planets and several larger planets made mostly of ice. (*Hint:* What would happen if the solar wind started earlier or later than in our solar system?)
24. Radiometric dating of meteorites from another solar system shows that they are a billion years younger than rocks from the terrestrial planets of the same system.

Quick Quiz

Choose the best answer to each of the following. Explain your reasoning with one or more complete sentences.

25. How many of the planets orbit the Sun in the same direction that Earth does? (a) a few (b) most (c) all
26. What fraction of the moons of the planets orbit in the same direction that their planets rotate? (a) some (b) most (c) all
27. The solar nebula was 98% . . . ? (a) rock and metal. (b) hydrogen compounds. (c) hydrogen and helium.
28. Which of the following did *not* occur during the collapse of the solar nebula? (a) spinning faster (b) heating up (c) concentrating denser materials nearer the Sun
29. What is Jupiter's main ingredient? (a) rock and metal (b) hydrogen compounds (c) hydrogen and helium
30. Which lists the major steps of solar system formation in the correct order? (a) collapse, accretion, condensation (b) collapse, condensation, accretion (c) accretion, condensation, collapse
31. Leftover ice-rich planetesimals are called (a) comets. (b) asteroids. (c) neither of the above: there were no leftover ice-rich planetesimals.
32. Why didn't a terrestrial planet form at the location of the asteroid belt? (a) There was never enough material in that part of the solar nebula. (b) The solar wind cleared away nebular material there. (c) Jupiter's gravity kept planetesimals from accreting.
33. What's the leading theory for the origin of the Moon? (a) It formed in place along with the Earth. (b) It formed from the material ejected in a giant impact. (c) It split out of a rapidly rotating Earth.
34. About how old is the solar system? (a) 4.5 million years (b) 4.5 billion years (c) 4.5 trillion years

Investigate Further

In-Depth Questions to Increase Your Understanding

Short-Answer/Essay Questions

35. *Patterns of Motion.* In one or two paragraphs, explain why the existence of orderly patterns of motion in our solar system should suggest that the Sun and the planets all formed

at one time from one cloud of gas, rather than as individual objects at different times.

36. *A Cold Solar Nebula.* Suppose the entire solar nebula had cooled to 50 K before the solar wind cleared it away. How would the composition and sizes of the planets of the inner solar system be different from what we see today? Explain your answer in a few sentences.

37. *An Early Solar Wind.* Suppose the solar wind had cleared away the solar nebula before the seeds of the jovian planets could gravitationally draw in hydrogen and helium gas. How would the planets of the outer solar system be different? Would they still have many moons? Explain your answer in a few sentences.

38. *Angular Momentum.* Suppose our solar nebula had begun with much more angular momentum than it did. Do you think planets could still have formed? Why or why not? What if the solar nebula had started with zero angular momentum? Explain your answers in one or two paragraphs.

39. *Two Kinds of Planets.* The jovian planets differ from the terrestrial planets in a variety of ways. Using phrases or sentences that members of your family would understand, explain why the jovian planets differ from the terrestrial planets in each of the following: composition, size, density, distance from the Sun, and number of satellites.

40. *Two Kinds of Planetesimals.* Why are there two kinds of "leftovers" from solar system formation? List as many differences between comets and asteroids as you can.

41. *History of the Elements.* Our bodies (and most living things) are made mostly of water: H_2O. Summarize the "history" of a typical hydrogen atom from its creation to the formation of the Earth. Do the same for a typical oxygen atom. (*Hint:* Which elements were created in the Big Bang, and where were the others created?)

42. *Our Messy Solar System.* How would Earth and other planets be different if all the "wandering planetesimals" had stayed in their location of the solar system instead of colliding or passing near other planets? Describe at least three major ways in which our solar system would differ.

43. *Pluto.* How does the nebular theory explain the origin of objects like Pluto? How was its formation similar to jovian and terrestrial planets, and how was it different?

44. *Understanding Radiometric Dating.* Imagine you had the good fortune to find a rocky meteorite in your backyard. Qualitatively, how would you expect its ratio of potassium-40 and argon-40 to be different from other rocks in your yard? Explain why, in a few sentences.

Quantitative Problems

Be sure to show all calculations clearly and state your final answers in complete sentences.

45. *Radiometric Dating.* You are analyzing rocks that contain small amounts of potassium-40 and argon-40. The half-life of potassium-40 is 1.25 billion years.
    a. You find a rock that contains equal amounts of potassium-40 and argon-40. How old is it? Explain.
    b. You find a rock that contains three times as much argon-40 as potassium-40. How old is it? Explain.

46. *Dating Lunar Rocks.* You are analyzing Moon rocks that contain small amounts of uranium-238, which decays into lead with a half-life of about 4.5 billion years.
    a. In a rock from the lunar highlands, you determine that 55% of the original uranium-238 remains, while the other 45% has decayed into lead. How old is the rock?

b. In a rock from the lunar maria, you find that 63% of the original uranium-238 remains, while the other 37% has decayed into lead. Is this rock older or younger than the highlands rock? By how much?

47. *Carbon-14 Dating.* The half-life of carbon-14 is about 5,700 years.
    a. You find a piece of cloth painted with organic dye. By analyzing the dye, you find that only 77% of the carbon-14 originally in the dye remains. When was the cloth painted?
    b. A well-preserved piece of wood found at an archaeological site has 6.2% of the carbon-14 it must have had when it was living. Estimate when the wood was cut.
    c. Is carbon-14 useful for establishing Earth's age? Why or why not?

48. *Martian Meteorite.* Some unusual meteorites thought to be chips from Mars contain small amounts of radioactive thorium-232 and its decay product, lead-208. The half-life for this decay process is 14 billion years. Analysis of one such meteorite shows that 94% of the original thorium remains. How old is this meteorite?

49. *Collapsing Cloud.* The time it takes for a cloud 100,000 AU in diameter to collapse to form a new star turns out to be about *half* the time it would take an object to orbit the star on an elliptical orbit with a semimajor axis of 50,000 AU (half the 100,000 AU diameter). Use Kepler's third law to find the collapse time, assuming the star has the same mass as the Sun.

50. *Monster Iceballs.* The ice-rich planetesimals that formed the cores of the jovian planets were about 10 times more massive than Earth. Assuming such an iceball has a density of about 2 g/cm^3, what would its radius be? How does this compare to Earth's radius?

51. *Icy Earth.* How massive would Earth have been if it had accreted hydrogen compounds in addition to rock and metal? Assume the same proportion of the ingredients as listed in Table 8.1. What would Earth's mass be, if it had been possible to capture hydrogen and helium gas in the proportion listed in the table?

52. *Big Moons.* Use the data in Appendix E to compare Earth's Moon, Jupiter's moon Ganymede, Saturn's moon Titan, and Pluto's moon Charon. Rank them in order of their actual size (radius, in kilometers). Then rank them in order of the ratio of their size to the size of the planet they orbit. What can you conclude from the differences in the two rankings?

53. *What Are the Odds?* The fact that all the planets orbit the Sun in the same direction is cited as support for the nebular hypothesis. Imagine that there's a different hypothesis in which planets can be created orbiting the Sun in either direction. Under this hypothesis, what is the probability that nine planets would end up traveling the same direction? (*Hint:* It's the same probability as that of flipping a coin nine times and getting nine heads or nine tails.)

54. *Spinning Up the Solar Nebula.* The orbital speed of the solar nebula at Pluto's average distance from the Sun is about 5 km/sec. When the solar nebula was about 40,000 AU in radius, how fast would the same material have orbited? (Remember that when the cloud was larger, motions were more random, so we can calculate only the average.) Use the law of conservation of angular momentum (see Section 4.3).

Discussion Questions

55. *Theory and Observation.* Discuss the interplay between theory and observation that has led to our modern theory of solar system formation. What role does technology play in allowing us to test this theory?

56. *Lucky to Be Here?* Considering the overall process of solar system formation, do you think it was likely for a planet like Earth to have formed? Could random events in the early history of the solar system have prevented our being here today? What implications do your answers have for the possibility of Earth-like planets around other stars? Defend your opinions.

57. *Rocks From Other Solar Systems.* The chapter states that some of the "leftovers" from planetary formation were likely ejected from our solar system, and the same has presumably happened in other star systems. Given that fact, should we expect to find rocks in our solar system that actually formed in other star systems? How rare or common would you expect such rocks to be? As you consider your answer, be sure to consider the distance between stars. Suppose that we *did* find a meteorite identified as a leftover from another stellar system. What do you think we would learn from it?

MEDIA EXPLORATIONS

For self-study activities go to MasteringASTRONOMY   www.masteringastronomy.com

 Interactive Tutorials

Tutorial Review of Key Concepts
Use the interactive **Tutorial** at **www.mastering astronomy.com** to review key concepts from this chapter.

Formation of the Solar System Tutorial
Lesson 1 Comparative Planetology
Lesson 2 Formation of the Protoplanetary Disk
Lesson 3 Formation of Planets

Supplementary Tutorial Exercises
Use the interactive **Tutorial Lessons** to explore the following questions.

Formation of the Solar System Tutorial, Lesson 2
1. Study the animations of the formation of the solar nebula. Why does the nebula flatten as it collapses?
2. Use the tool in the tutorial to make a graph of temperatures in the solar nebula. How does the temperature vary with the distance from the center?
3. Why didn't any planets form within 0.3 AU of the Sun?

Formation of the Solar System Tutorial, Lesson 3
1. What is the composition of planetesimals that form outside the frost line?
2. Why didn't the terrestrial planets capture hydrogen and helium gas?
3. What is Pluto made of, and why?

 Exploring the Sky and Solar System

Of the many activities available on the *Voyager: SkyGazer* **CD-ROM** accompanying your book, use the following files to observe key phenomena covered in this chapter.

Go to the **File: Basics** folder for the following demonstrations.
1. Saturn's Phases
2. Tracking Venus
3. Planet Panel

Go to the **File: Demo** folder for the following demonstrations.
1. Earth and Venus
2. Trailing Saturn
3. Triple Conjunction of 7 BC

Go to the **Explore** menu for the following demonstrations.
1. Solar System
2. Paths of Planets

 Movies

Check out the following narrated and animated short documentaries available on **www.masteringastronomy.com** for a helpful review of key ideas covered in this chapter.

Orbits in the Solar System   *History of the Solar System*

Web Projects

1. *Spitzer Space Telescope.* NASA's latest "Great Observatory" operates at the infrared wavelengths especially useful for studying star and planet formation. Visit the Spitzer Web site to see if recent discoveries are confirming the nebular theory for planet formation or are requiring us to broaden our understanding of the process. Summarize your findings in a one- to two-page report.

2. *Other Solar Nebulas.* New technology enables better and better imaging of possible planet-forming regions around other stars. Search the Internet for recent images from the Hubble Space Telescope and major ground-based observatories listed in Table 6.1. Write a one- to two-page report, including pictures, describing how the new images support or expand our ideas about our own solar nebula.

9

Planetary Geology

Earth and the Other Terrestrial Worlds

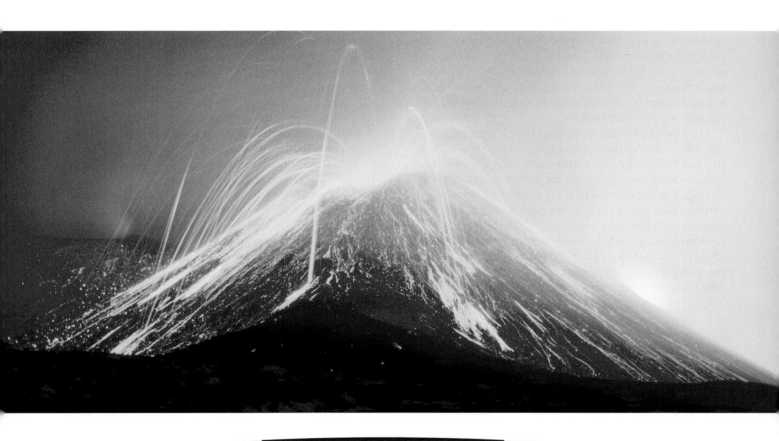

LEARNING GOALS

Nothing is rich but the inexhaustible wealth of nature. She shows us only surfaces, but she is a million fathoms deep.

Ralph Waldo Emerson

It's easy to take for granted the qualities that make Earth so suitable for human life: a temperature neither boiling nor freezing, abundant water, a protective atmosphere, and a relatively stable environment. But we need look only as far as our neighboring terrestrial worlds—the three other inner planets and the Moon—to see how fortunate we are. As we learned in Chapter 7, the Moon, likely made of debris from Earth, is airless and barren. Mercury is much the same. Venus is a searing hothouse, while Mars has an atmosphere so thin and cold that liquid water cannot last on its surface today.

How did the terrestrial worlds come to be so different, when all were made similarly from metal and rock that had condensed in the solar nebula? Why did Earth alone develop the conditions that permit abundant life? We can find the answers to these questions only with careful, comparative study of the planets. We'll begin the quest in this chapter by focusing on the geology of the terrestrial worlds, and then explore their atmospheres in Chapter 10. In each case, we'll start by exploring the general processes that shape the planets and then examine how these processes have played out on the individual

worlds. By the end, you'll see that the histories of the terrestrial worlds were largely determined by the properties with which they were endowed at birth.

 Formation of the Solar System Tutorial, Lesson 1

9.1 Connecting Planetary Interiors and Surfaces

Earth's surface seems solid and steady, but every so often it offers us a reminder that nothing about it is permanent. If you live in Alaska or California, you've probably felt the ground shift beneath you in an earthquake. In Washington State, you may have seen the recent rumblings of Mount St. Helens. In Hawaii, a visit to the still-active Kilauea volcano will remind you that you are standing on mountains of volcanic rock protruding from the ocean floor.

Volcanoes and earthquakes are not the only processes acting to reshape Earth's surface. They are not even the most dramatic: Far greater change can occur on the rare occasions when an asteroid or a comet slams into Earth. Much more gradual processes can also have spectacular effects. The Colorado River causes only small changes in the landscape from year to year, but its unrelenting flow over the past few million years carved the Grand Canyon. The Rocky Mountains were once twice as tall as they are today and have been cut down in size through tens of millions of years

Venus

Earth

Mercury

Mercury is heavily cratered, but also has long, steep cliffs—one is visible here as the long curve that passes through the center of the image.

The central structure is a tall, twin-peaked volcano on Venus. Both images are based on radar data from the Magellan space craft, because Venus's thick clouds prevent us from seeing the surface in visible light.

Earth has a variety of geological features visible in this photo from orbit.

of erosion by wind, rain, and ice. Entire continents even move slowly about, completely rearranging the map of Earth every few hundred million years.

Earth is not alone in having undergone tremendous change since its birth. The surfaces of all five terrestrial worlds—Mercury, Venus, Earth, the Moon, and Mars—must have looked quite similar when they were young. All five were made of rocky material that condensed in the solar nebula, and all five were subjected early on to the impacts of the heavy bombardment [Section 8.4]. The great differences in their present-day appearances must therefore be the result of changes that have occurred through time. Ultimately, these changes must be traceable to fundamental properties of the planets.

Figure 9.1 shows global views of the terrestrial surfaces to scale, along with sample close-up views from orbit. Profound differences between these worlds are immediately obvious. Mercury and the Moon show the scars of their battering during the heavy bombardment: They are densely covered by craters except in areas that appear to be volcanic plains. Bizarre bulges and odd volcanoes dot the surface of Venus. Mars, despite its middling size, has the solar system's largest volcanoes and a huge canyon cutting across its surface, along with numerous features that appear to have been shaped by running water. Earth has surface features similar to all those on the other terrestrial worlds, and more—including a unique layer of living organisms that covers almost the entire surface of the planet.

Our purpose in this chapter is to understand how these profound differences among the terrestrial planets came to be. We will be discussing what is often called **planetary geology**; because geology is the study of Earth (*geo* means "Earth"), planetary geology is the extension of this science to other worlds. Note that we use the term *planetary geol-ogy* for the study of any solid world, even if it is a moon rather than a planet. Although we can see geological features only on planetary surfaces, most of these features are shaped by processes that take place deep beneath the surface. Thus, the first step in understanding planetary geology is to understand what planets are like on the inside.

• What are terrestrial planets like on the inside?

We cannot see inside Earth or any other planet, but we still know quite a bit about what planetary interiors are like. Our most detailed information about Earth's interior comes from **seismic waves**, vibrations that travel both through Earth's interior and along its surface after an earthquake (*seismic* comes from the Greek word for "shake"). In much the same way that shaking a gift offers clues about what's inside, the vibrations that follow earthquakes can tell us about the inside of Earth. By combining our understanding of Earth's interior with a variety of other clues, we have developed a fairly clear picture of the interiors of each of the terrestrial worlds (see box, p. 250).

Layering by Density   All the terrestrial worlds have layered interiors. You are probably familiar with the idea of dividing the interior into three basic layers by density:

● **Core:** The highest-density material, consisting primarily of metals such as nickel and iron, resides in the central core.

● **Mantle:** Rocky material of moderate density—mostly minerals that contain silicon, oxygen, and other elements—forms the thick mantle that surrounds the core.

Mars

Figure 9.1 Global views to scale, along with sample close-ups viewed from orbit, of each of the five terrestrial worlds. All the photos were taken with visible light, except the Venus photos, which are based on radar data.

Earth's Moon

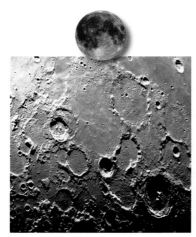

The Moon's surface is heavily cratered in most places.

Mars has impact craters like the one near the upper right, but it also has features that look much like dried up riverbeds.

- **Crust:** The lowest-density rock, such as granite and basalt (a common form of volcanic rock), forms the thin crust, essentially representing the world's outer skin.

Figure 9.2 shows these layers for the five terrestrial worlds. Although it is not shown in the figure, seismic studies indicate that Earth's metallic core consists of two distinct regions: a solid *inner core* and a molten (liquid) *outer core*. Because temperature increases with depth in a planet, it may seem surprising that Earth's inner core is solid while the outer core is molten. The explanation is that the inner core is kept solid by the higher pressure at this depth, even

though the temperature is also higher. Venus may have a similar core structure, but without seismic data we cannot be sure.

The distinct layering of the terrestrial worlds occurs for the same reason that liquids separate by density. For example, in a mixture of oil and water, gravity pulls the denser water to the bottom, driving the less dense oil to the top. The process by which gravity separates materials by density is called **differentiation** (because it results in layers made of *different* materials). The layered interiors of the terrestrial worlds tell us that all the worlds underwent differentiation at some time in the past, which means all these worlds

---

SPECIAL TOPIC How Do We Know What's Inside a Planet?

Our deepest drills have barely pricked Earth's surface, penetrating much less than 1% of the way into the interior. On other worlds, we've done nothing more than examine surface rocks. How, then, can we claim to know what Earth or any other planet is like on the inside?

On Earth, much of our information about the interior comes from *seismic waves*, vibrations created by earthquakes. Seismic waves come in two basic types that are analogous to the two ways that you can generate waves in a Slinky (Figure 1). Pushing and pulling on one end of a Slinky (while someone holds the other end still) generates a wave in which the Slinky is bunched up in some places and stretched out in others. Waves like this in rock are called P waves. The *P* stands for *primary,* because these waves travel fastest and are the first to arrive after an earthquake, but it is easier to think of it as meaning *pressure* or *pushing.* P waves can travel through almost any material—whether solid, liquid, or gas—because molecules can always push on their neighbors no matter how weakly they are bound together. (Sound travels as a pressure wave and thus is quite similar to P waves.)

The liquid outer core bends P waves . . .

. . . but stops S waves.

Figure 2 Because S waves do not reach the side of Earth opposite an earthquake, we infer that part of Earth's core is liquid.

Shaking a Slinky slightly up and down generates an up-and-down motion all along its length. Such up-and-down (or side-to-side) waves in rock are called S waves. The *S* stands for *secondary* but is easier to remember as meaning *shear* or *side-to-side.* S waves travel only through solids, because the bonds between neighboring molecules in a liquid or gas are too weak to transmit up-and-down or sideways forces.

The precise speeds and directions in which seismic waves travel through Earth depend on the composition, density, pressure, temperature, and phase (solid or liquid) of the material they pass through. For example, P waves reach the side of the world opposite an earthquake, but S waves do not. This tells us that a liquid layer has stopped the S waves, which is how we know that Earth has a liquid layer in its outer core (Figure 2). More careful analysis of seismic waves has allowed geologists to develop a detailed picture of what Earth looks like on the inside.

We have also used seismic waves to study the Moon's interior, thanks to monitoring stations left on the Moon by the *Apollo* astronauts. To learn about the interiors of other worlds, we use less direct clues. One key piece of information is a planet's average density, which we can calculate simply by dividing its mass by its volume. We can find a planet's mass by applying Newton's version of Kepler's third law to the orbital properties of its moons or of spacecraft that have visited it [Section 4.4] and its size through careful study of telescopic or spacecraft images. If we find that the density of the planet's surface rock is much less than its overall average density, then we know it must contain much more dense rock or metal inside. We can also learn about a planet's interior from precise measurements of its gravity, which tell us how mass is distributed within the planet; from studies of its magnetic field, which is generated deep inside the planet; and from observations of surface rocks that appear to have emerged from deep within the interior.

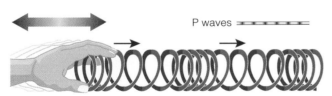

P waves ⬛⬛⬛⬛⬛⬛⬛

P waves result from compression and stretching in the direction of travel.

S waves 〜〜〜〜〜〜

S waves vibrate up and down or side to side perpendicular to the direction of travel.

Figure 1 Slinky examples demonstrating P and S waves.

Earth **Venus** **Mars** **Mercury** **Moon**

rocky crust
(lower density)

mantle
(medium density)

metal core
(highest density)

rigid lithosphere
(crust and
part of mantle)

lithosphere

lithosphere lithosphere lithosphere

Key: crust mantle ⊥ lithosphere core

Figure 9.2 Interior structures of the terrestrial worlds, shown to scale and in order of decreasing size. The color-coded regions show the core-mantle-crust layering, which is based on density. Dashed lines indicate the lithosphere, which encompasses the crust and part of the mantle on each world and is defined by the strength of the rock rather than the density. (The thicknesses of the crust and lithosphere on Venus and Earth are exaggerated to make them visible in this figure.)

must once have been hot enough inside for their interior rock and metal to melt and separate by density. Dense metals like iron sank toward the center, driving less dense rocky material toward the surface.

Layering by Strength: The Lithosphere   Although the activity of volcanoes may give the impression that Earth is molten inside, in most cases the lava actually rises upward from a fairly narrow layer of rock in the upper mantle that is only partially molten. The vast majority of Earth's interior is solid rock, and only the outer core is fully molten. However, the interior rock is not equally strong throughout, and differences in rock strength are often more important than differences in density for understanding many geological processes.

The idea that rock can vary in strength may sound strange, since we often think of rock as the very definition of strength. However, remember that rocks feel solid only because of the electrical bonds between their molecules [Section 5.3]. Although these bonds are strong, they can still break and re-form when subjected to sustained stress. Thus, over millions and billions of years, rocky material can slowly deform and flow. The long-term behavior of rock is much like that of the popular toy Silly Putty, which breaks like a brittle solid when you pull it sharply but deforms and stretches when you pull it slowly (Figure 9.3). Also like Silly Putty, rock becomes softer and easier to deform when it is warmer.

The fact that rock can deform and flow explains why small moons and asteroids are "potato-shaped" while larger worlds are spherical. The weak gravity of a small object is unable to overcome the rigidity of its rocky material, so the object retains the shape it had when it was born. For a larger world, gravity can overcome the strength of solid rock, slowly deforming and molding it into a spherical shape. Gravity will make any rocky object more than about 500 kilometers in diameter into a sphere within about 1 billion years.

Larger worlds become spherical more quickly, especially if they are molten (or gaseous) at some point in their history.

In terms of rock strength, a planet's outer layer consists of relatively cool and rigid rock, called the **lithosphere** (*lithos* is Greek for "stone"), that "floats" on warmer, softer rock beneath. The lithosphere generally encompasses the

Figure 9.3 Silly Putty stretches when pulled slowly but breaks cleanly when pulled rapidly. Rock behaves just the same, but on a longer timescale.

Earth Is Not Full of Molten Lava

Many people guess that Earth is full of molten lava (more technically known as *magma*). This misconception may arise partly because we see molten lava emerging from inside Earth when a volcano erupts. However, Earth's mantle and crust are almost entirely solid. The lava that erupts from volcanoes comes only from a narrow region of partially molten material beneath the lithosphere. Indeed, the only part of Earth's interior that is fully molten is the outer core, which lies so deep within the planet that core material never erupts directly to the surface.

crust and part of the mantle, as shown by the dashed circles in Figure 9.2. Beneath the lithosphere, in the lower part of the mantle, the higher temperatures allow rock to deform and to flow much more easily. The thickness of the lithosphere is very important to geology. A thin lithosphere is brittle and can crack easily. A thick lithosphere is much stronger and can prevent volcanic eruptions and the creation of mountain ranges.

Comparing Planetary Interiors If you study the interior structures in Figure 9.2, you'll see that the basic interior structure is the same for each of the terrestrial worlds: a very thin crust, a much thicker mantle, and a central core. However, there are also some important differences, particularly in core size and in the extent to which the lithospheres reach into the mantles. With a little thought, you'll realize that these differences are linked closely with planetary size.

Notice that smaller planets generally have smaller cores, just as we would expect for worlds that were all made from metal and rock that condensed in the inner regions of the solar nebula. However, two worlds seem to deviate from the expected pattern. Mercury has a surprisingly large core, even when we account for the fact that it formed in the region of the solar nebula where the planetesimals should have been most metal-rich. That is the primary reason why scientists suspect that Mercury once suffered a giant impact [Section 8.4] that blasted away much of its original mantle and crust. The Moon's core deviates from the pattern in the opposite way, as it seems surprisingly small in comparison to the Moon's overall size. Again, the explanation may lie with a giant impact; in this case, we expect a small core if the Moon really did form from debris blasted out of Earth's outer layers (after Earth had undergone differentiation), since that debris would have contained very little high-density metal.

Lithospheric thickness is also closely related to planetary size: smaller planets tend to have thicker lithospheres. The two largest terrestrial planets, Earth and Venus, have very thin lithospheres that extend only a short way into their upper mantles. The smaller worlds—Mars, Mercury, and

the Moon—have thick lithospheres that extend nearly to their cores.

• What causes geological activity?

The most profound differences between the geologies of the terrestrial worlds occur in their relative amounts of **geological activity**, a term used to describe how much ongoing change occurs on their surfaces. For example, we say that Earth is geologically active, because volcanic eruptions, earthquakes, erosion, and other geological processes continually reshape its surface. In contrast, the Moon and Mercury have virtually no geological activity, which is why their surfaces today look essentially the same as they did a billion or more years ago.

Interior heat is the primary driver of geological activity. Volcanoes can erupt on Earth because Earth is quite hot inside—hot enough that, in at least some cases, rock melts into molten lava. In contrast, the Moon has no active volcanoes, because its interior is too cool to melt rock and push it to the surface. But what makes some planetary interiors hotter than others? To find the answer, we must investigate how interiors heat up and cool off. As we'll see, we can ultimately trace a planet's internal heat, and hence its geological activity, back to its size.

How Interiors Get Hot A hot interior contains a lot of thermal energy, and the law of conservation of energy tells us that this energy must have come from somewhere [Section 4.3]. Although you might first guess that the Sun would be the heat source, this is not the case: Sunlight is the primary heat source for the *surfaces* of the terrestrial planets, but virtually none of this solar energy penetrates more than a few meters into the ground. Internal heat is a product of the planets themselves, not of the Sun. In fact, three sources of energy explain nearly all the interior heat of the terrestrial worlds (Figure 9.4):

- **Heat of accretion:** Accretion deposits energy brought in from afar by colliding planetesimals. An incoming planetesimal has a lot of gravitational potential energy when it is far away. As it approaches a forming planet, its gravitational potential energy is converted to kinetic energy, causing it to accelerate. Upon impact, much of the kinetic energy is converted to heat, adding to the thermal energy of the planet.

- **Heat from differentiation:** Differentiation also converts gravitational potential energy into thermal energy. Remember that differentiation occurs as denser materials sink and less dense materials rise in a molten interior. This process adds mass to the planet's core and reduces the mass of the outer layers. Thus, much of the planet's mass effectively moves inward, losing gravitational potential energy. The friction generated as the mass separates by density converts the lost gravitational potential energy to thermal energy, heating the interior. The same thing happens when you drop a brick into a pool: As the brick sinks to the bottom, friction with the surround-

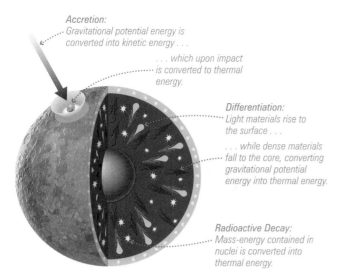

Accretion:
Gravitational potential energy is
converted into kinetic energy . . .

. . . which upon impact
is converted to thermal
energy.

Differentiation:
Light materials rise to
the surface . . .

. . . while dense materials
fall to the core, converting
gravitational potential
energy into thermal energy.

Radioactive Decay:
Mass-energy contained in
nuclei is converted into
thermal energy.

Figure 9.4 The three main heat sources for terrestrial planet interiors are accretion, differentiation, and radioactive decay. Only the latter is still a major heat source today.

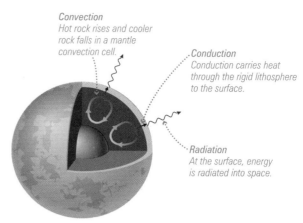

Convection
Hot rock rises and cooler
rock falls in a mantle
convection cell.

Conduction
Conduction carries heat
through the rigid lithosphere
to the surface.

Radiation
At the surface, energy
is radiated into space.

Figure 9.5 The three main cooling processes in planets. Convection can occur only in a planet that is still hot inside.

ing water heats the pool—though the amount of heat from a single brick is too small to notice.

- **Heat from radioactive decay:** Radioactive decay affects the terrestrial worlds because the rock and metal planetesimals that built them contained radioactive isotopes of elements such as uranium, potassium, and thorium. When radioactive nuclei decay, subatomic particles fly off at high speeds, colliding with neighboring atoms and heating them. In essence, this transfers some of the mass-energy ($E = mc^2$) of the radioactive nucleus to the thermal energy of the planetary interior.

Note that while accretion and differentiation deposited heat into planetary interiors only when the planets were very young, radioactive decay continues to heat the terrestrial planets to this day. Indeed, over billions of years, the total amount of heat deposited by radioactive decay has been several times as great as the amount that was deposited initially by accretion and differentiation. Nevertheless, the rate of radioactive decay declines with time, so it was a more significant heat source when the planets were young.

The combination of the three heat sources explains how the terrestrial interiors ended up with their core-mantle-crust structures. The many violent impacts that occurred during the latter stages of accretion deposited so much energy that the outer layers of the young planets began to melt. This started the process of differentiation, and the process of differentiation itself then began to release additional heat. Along with the substantial heat from early radioactive decay, the interiors became hot enough to melt and differentiate throughout.

How Interiors Cool Off   Cooling a planetary interior requires transporting heat outward. Just as there are three basic heating processes for planetary interiors, there are also three basic cooling processes (Figure 9.5):

- **Convection:** Convection transports heat upward when hot material expands and rises while cooler material contracts and falls. It can occur whenever there is strong heating from below. For example, you can see convection in a pot of soup on a hot burner. (You may also be familiar with convection in weather—warm air near the ground tends to rise while cool air above tends to fall.)
- **Conduction:** Conduction is the process in which hot material transfers heat to cooler material through contact, just as a hot potato transfers its heat to your cooler hand when you pick it up. Conduction occurs through the microscopic collisions of many individual atoms or molecules. Molecules of materials in close contact are constantly colliding with one another, so the faster-moving molecules in hot material tend to transfer some of their energy to the slower-moving molecules of cooler material.
- **Radiation:** An object can also lose heat through radiation; that is, by radiating energy (light) into space. Remember that all objects emit *thermal radiation* characteristic of their temperatures [Section 5.4] and therefore are always losing some energy to space. Note that while convection and conduction can transport interior heat toward a planet's surface, the planet ultimately loses this heat to space only through radiation.

Let's start with Earth as an example of how heat moves from the deep interior to space. Convection is the most important process in Earth's deep interior. Hot rock from deep in the mantle gradually rises, slowly cooling as it makes its way upward. By the time it reaches the top of the mantle, the rock has transferred all its excess heat to its surroundings, so it is now cool and it begins to fall. The ongoing process of convection creates individual **convection cells** within the mantle. Figure 9.5 shows a few mantle convection cells as small circles, with arrows indicating the direction of flow. Keep in mind that mantle convection primarily involves solid rock, not molten rock. Because solid rock flows quite slowly, mantle convection is a very slow process. At the typical rate of mantle convection on Earth—about 1 centimeter per year—it would take about

Pressure and Temperature

Some people think that Earth's interior should be hot just because the internal pressure is so high. After all, if we compress a gas from low pressure to high pressure, it heats up. The same is not necessarily true of rock. High pressure compresses rock only slightly, so the compression causes little increase in temperature. Thus, while high pressures and temperatures sometimes go together in planets, they don't have to. In fact, after all the radioactive elements decay (billions of years from now), Earth's deep interior will become quite cool even though the pressure will be the same as it is today. The temperatures inside Earth and the other planets can remain high only if there is an internal source of heat, such as accretion, differentiation, or radioactive decay.

100 million years for a piece of rock to be carried from the base of the mantle to the top.

Mantle convection stops at the base of the lithosphere, where the rock is too strong to flow as readily as it does lower down. Thus, from the base of the lithosphere to the surface, heat continues upward primarily through conduction. (Some heat also reaches the surface through volcanic eruptions that directly carry hot rock upward.) When the heat finally reaches Earth's surface, it is radiated away into space.

The Role of Planetary Size Size is the single most important factor in planetary cooling: Just as a hot potato remains hot inside much longer than a hot pea, a large planet can stay hot inside much longer than a small one. You can see why size is the critical factor by picturing a large planet as a smaller planet wrapped in extra layers of rock. The extra rock acts as insulation, so it takes much longer for interior heat to reach the surface.

Thus, we can trace interior heat, heat transport, and geological activity all back to planetary size. The relatively small sizes of the Moon and Mercury probably allowed their interiors to cool within a billion years or so after they formed. As they cooled, their lithospheres thickened and mantle convection was confined to deeper and deeper layers. Ultimately, the mantle convection probably stopped altogether. With insufficient internal heat to drive any further movement of interior rock, the Moon and Mercury are now geologically "dead," meaning they have no more heat-driven geological activity.

In contrast, the much larger size of Earth has allowed our planet to stay quite hot inside. Mantle convection keeps interior rock in motion and the heat keeps the lithosphere thin, which is why geological activity can continually reshape the surface. Venus probably remains nearly as active as Earth, thanks to its very similar size. Mars, with a size in between that of the other terrestrial worlds, probably rep-

resents an intermediate case: It has certainly cooled significantly during its history, but it probably still retains enough internal heat for at least some geological activity. No one knows whether Mars still has any convection in its lower mantle.

In general, what kind of planet would you expect to have the *thickest* lithosphere: the largest planet, the smallest planet, the planet closest to the Sun, or the planet farthest from the Sun? Why? (Assume the planet has the same composition in all four cases.)

• Why do some planetary interiors create magnetic fields?

A planet's interior structure also determines another important planetary property: the presence or absence of a **magnetic field** that can affect charged particles or magnetized objects in its vicinity. If you've ever used a compass, you know that Earth has a magnetic field that determines the direction in which the compass needle points. The global extent of Earth's magnetic field gives us a strong clue that the field is generated *inside* our planet. By examining how a planetary interior can generate a magnetic field, we'll see why Earth has a relatively strong magnetic field and the rest of the terrestrial planets do not.

You are probably familiar with the general pattern of the magnetic field created by an iron bar (Figure 9.6a). Earth's magnetic field is generated by a process more similar to that of an *electromagnet,* in which the magnetic field arises as a battery forces charged particles (electrons) to move along a coiled wire (Figure 9.6b). Earth does not contain a battery, but charged particles move with the molten metals in its liquid outer core (Figure 9.6c). Internal heat causes the liquid metals to rise and fall (convection), while Earth's rotation twists and distorts the convection pattern of these molten metals. The result is that electrons in the molten metals move within Earth's outer core in much the same way they move in an electromagnet, generating Earth's magnetic field.

We can generalize these ideas to other worlds. There are three basic requirements for a global magnetic field:

1. An interior region of electrically conducting fluid (liquid or gas), such as molten metal.
2. Convection in that layer of fluid.
3. At least moderately rapid rotation.

Earth is unique among the terrestrial worlds in meeting all three requirements, which is why it is the only terrestrial world with a strong magnetic field. The Moon has no magnetic field, either because it lacks a metallic core altogether or because its core has long since solidified and ceased convecting. Mars likewise has virtually no magnetic field today, probably because of core solidification. Venus probably has a molten core layer much like that of Earth, but either its convection or its 243-day rotation period is too slow to

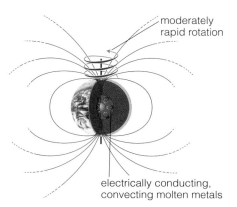

moderately
rapid rotation

electrically conducting,
convecting molten metals

a This photo shows how a bar magnet influences iron filings (small black specks) around it. The *magnetic field lines* (red) represent this influence graphically.

b A similar magnetic field is created by an electromagnet, which is essentially a wire wrapped around a bar and attached to a battery. The field is created by the battery-forced motion of charged particles (electrons) along the wire.

c Earth's magnetic field also arises from the motion of charged particles. The charged particles move within Earth's liquid outer core, which is made of electrically conducting, convecting molten metals.

Figure 9.6 Sources of magnetic fields.

MATHEMATICAL INSIGHT **9.1** **The Surface Area-to-Volume Ratio**

Mathematically, we can see why large planets cool more slowly than smaller ones by thinking about their relative surface areas and volumes. The total amount of heat contained in a planet depends on the planet's total *volume*. However, this heat can escape directly to space only from the planet's surface, so the key factor in the planet's heat-loss rate is its *surface area*. As heat escapes from the surface, more heat flows upward from the interior to replace it. This process will continue until the interior is no hotter than the surface. Thus, the total time it takes for a planet to lose its internal heat must be closely related to the ratio of the surface area through which it loses heat to the volume that contains heat, which is what we call the *surface area-to-volume ratio*.

$$\text{surface area-to-volume ratio} = \frac{\text{surface area}}{\text{volume}}$$

To see exactly how the surface area-to-volume ratio affects heat loss, consider a spherical object like a planet. The surface area of a sphere is given by the formula $4\pi r^2$ and the volume is given by the formula $\frac{4}{3}\pi r^3$, where r is the radius of the sphere. Thus, for a sphere:

$$\underset{\text{(for a sphere)}}{\text{surface area-to-volume ratio}} = \frac{4\pi r^2}{\frac{4}{3}\pi r^3} = \frac{3}{r}$$

Because the radius r appears in the denominator, we conclude that *larger objects have smaller surface area-to-volume ratios*. This idea holds true not just for spheres, but for objects of any shape. Thus, if two objects start with the same internal temperature, the larger one will take longer to cool off. This principle explains many everyday phenomena. For example, crushing an ice cube does not change the total volume of the ice, but the many smaller pieces have a greater combined surface area than the original ice cube. Thus, crushing ice increases its surface area-to-volume ratio, putting more area in contact with the surrounding liquid or air. That is why crushed ice will cool a drink more quickly than an equal volume larger ice cubes.

Example 1: Compare the surface area-to-volume ratios of the Moon and Earth. How does your answer help explain why Earth's interior is much hotter than the Moon's?

Solution:

Step 1 Understand: Comparing the two surface area-to-volume ratios will tell us the relative rates at which the two worlds would lose heat if they both started with the same internal temperature. "Comparing" in this context means seeing how much larger one ratio is than the other. We know that the smaller Moon should have the larger ratio, so we'll find our answer by dividing the Moon's surface area-to-volume ratio by the Earth's surface area-to-volume ratio.

Step 2 Solve: We've already found that the surface area-to-volume ratio for a spherical world is $3/r$. Dividing the surface area-to-volume ratios for the Moon and Earth, we find:

$$\frac{\text{surface area-to-volume ratio (Moon)}}{\text{surface area-to-volume ratio (Earth)}} = \frac{3/r_{\text{Moon}}}{3/r_{\text{Earth}}} = \frac{r_{\text{Earth}}}{r_{\text{Moon}}}$$

From Appendix E, the radii of the Moon and Earth are $r_{\text{Moon}} = 1{,}738$ km and $r_{\text{Earth}} = 6{,}378$ km. Substituting these values into our equation we find:

$$\frac{\text{surface area-to-volume ratio (Moon)}}{\text{surface area-to-volume ratio (Earth)}} = \frac{6{,}378 \text{ km}}{1{,}738 \text{ km}} = 3.7$$

Step 3 Explain: We've found that the Moon's surface area-to-volume ratio is nearly four times as large as Earth's, which means the Moon would cool four times faster if both worlds started with the same temperature. However, Earth's larger size also gave it much more internal heat to begin with, which amplifies the difference in heat loss found by the surface area-to-volume ratio alone. Overall, we conclude that the Moon should have cooled many times faster than Earth, which explains why the Moon's interior is so much cooler than Earth's interior today.

generate a magnetic field. Mercury remains an enigma: It possesses a measurable magnetic field despite its small size and slow, 59-day rotation. The reason may be Mercury's huge metal core, which may still be partly molten and convecting.

The same three requirements for a magnetic field also apply to jovian planets and stars. For example, Jupiter's strong magnetic field comes from its rapid rotation and its layer of convecting, metallic hydrogen that conducts electricity [Section 11.1]. The Sun's magnetic field is generated by the combination of convection of ionized gas (plasma) in its interior and rotation.

THINK ABOUT IT

Recall that the Sun had a strong solar wind when it was young (which cleared out the remaining gas of the solar nebula [Section 8.3]) because its magnetic field was much stronger than it is today. What single factor explains why the young Sun had a much stronger magnetic field than the Sun today?

Magnetic fields can be very useful. On Earth, the magnetic field helps us navigate. Elsewhere, magnetic fields provide us with important information about planetary interiors. However, magnetic fields have virtually no effect on the structure of a planet itself. From the standpoint of planetary science, their most important role is that they can form a protective *magnetosphere* surrounding a planet, with potentially profound effects on a planet's atmosphere and any inhabitants [Section 10.1].

 Shaping Planetary Surfaces Tutorial, Lessons 1–3

9.2 Shaping Planetary Surfaces

Now that we have discussed how Earth and other terrestrial worlds work on the inside, we are ready to turn to their surfaces. The surfaces of the planets are remarkably flat. For example, the tallest mountains on Earth rise only about 10 kilometers above sea level, less than $\frac{1}{600}$ of Earth's radius, and are small enough that they could be represented by grains of sand on a typical globe. Nevertheless, surface features tell us a great deal about the histories of the planets.

THINK ABOUT IT

Find a globe of Earth that shows mountains in raised relief (that is, a globe that has bumps to represent mountains). Are the heights of the mountains correctly scaled relative to Earth's size? Explain.

• What processes shape planetary surfaces?

Earth offers a huge variety of geological surface features, and the variety only increases when we survey other worlds. But on closer examination, geologists have found that almost all surface features can be explained by just four major geological processes:

- **Impact cratering:** the blasting of bowl-shaped *impact craters* by asteroids or comets striking a planet's surface.
- **Volcanism:** the eruption of molten rock, or *lava*, from a planet's interior onto its surface.
- **Tectonics:** the disruption of a planet's surface by internal stresses.
- **Erosion:** the wearing down or building up of geological features by wind, water, ice, and other phenomena of planetary weather.

Let's examine each of these processes in a little more detail.

Impact Cratering   The scarred faces of the Moon and Mercury (see Figure 9.1) attest to the battering that the terres-

Figure 9.7   Interactive Figure   Artist's conception of the impact process.

a Meteor Crater in Arizona was created about 50,000 years ago by the impact of an asteroid about 50 meters across. The crater is more than a kilometer across and almost 200 meters deep.

b This photo shows a crater, named Tycho, on the Moon. Note the classic shape and central peak.

Figure 9.8 Impact craters.

trial worlds have taken from leftover planetesimals, such as comets and asteroids. They also immediately reveal an important feature of impact cratering: Small craters far outnumber large ones, confirming that many more small asteroids and comets orbit the Sun than large ones. While the Moon and Mercury bear the most obvious scars, all the terrestrial worlds have suffered similarly.

An impact crater forms when an asteroid or comet slams into a world with a solid surface (Figure 9.7). Impacting objects typically hit the surface at a speed between about 40,000 and 250,000 km/hr (10–70 km/s). At such a tremendous speed, the impact releases enough energy to vaporize solid rock and excavate a crater (the Greek word for "cup"). Debris from the blast shoots high above the surface and then rains down over a large area. If the impact is large enough, some of the ejected material can completely escape from the planet.

Craters generally end up being circular because an impact blasts out material in all directions, no matter the direction of the incoming object. Laboratory experiments show that craters are generally about 10 times as wide as the objects that create them and about 10–20% as deep as they are wide. For example, an asteroid 1 kilometer in diameter will blast out a crater about 10 kilometers wide and 1–2 kilometers deep. A large crater may have a central peak, which forms when the center rebounds after impact in much the same way that water rebounds after you drop a pebble into it. Figure 9.8 shows two impact craters with classic shapes, one on Earth and one on the Moon.

Although impact craters are almost always circular, details of their shapes can provide important information about geological conditions on a planetary surface. To illustrate how, Figure 9.9 contrasts three craters on Mars. The crater in Figure 9.9a has a fairly simple bowl shape,

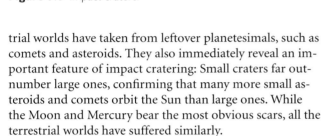

A simple bowl-shaped crater, showing a sharp rim . . .

. . . and a ring of ejected debris.

Unusual ridges suggest the impact debris was muddy.

This crater rim looks like it was eroded by rainfall.

a A crater with a "standard" shape.

b This crater was probably made by an impact into icy ground.

c This crater shows evidence of erosion; see Figure 9.29b for a close-up view of this crater's floor.

Figure 9.9 Crater shapes on Mars tell us about Martian geology. (Photos from *Mars Global Surveyor* (a) and *Viking* spacecraft (b and c).)

suggesting that it formed just as we expect from the basic cratering process. The crater in Figure 9.9b also has a fairly "standard" shape, but notice the extra large bump in its center and the way it is surrounded by a pattern of what looks like mud flows. This suggests that the surface had underground water or ice that melted or vaporized on impact; the muddy debris then flowed across the surface before hardening into the pattern we see today. The crater in Figure 9.9c shows obvious signs of erosion: It lacks a sharp rim and its floor no longer has the classic bowl shape. Again, this shape tells us about Martian geological history, in this case suggesting that ancient rainfall eroded the crater and that the crater bottom was once a lake. Studies of crater shapes on other worlds provide similar clues to their surface conditions and history.

Volcanism The second major geological process is volcanism. On Earth, we generally associate volcanism with the mountains we call volcanoes. In the broader context of planetary geology, we use the term *volcanism* to refer to any eruption of molten lava onto the surface, whether it comes from a tall volcano or simply rises through a crack in a planet's lithosphere.

Volcanism occurs when underground molten rock, technically called *magma,* finds a path through the lithosphere to the surface (Figure 9.10). Molten rock tends to rise for three main reasons. First, molten rock is generally less dense than solid rock, and lower-density materials tend to rise when surrounded by higher-density materials. Second, remember that most of Earth's interior is not molten. Thus, a "chamber" of molten rock (a *magma chamber*) may be squeezed by the surrounding solid rock, which drives the molten rock upward under pressure. Third, molten rock often contains trapped gases that expand as it rises, which can make it rise much faster and lead to dramatic eruptions. The same molten rock that is called magma when it is underground is called *lava* once it erupts onto the surface.

The result of an eruption depends on how easily the molten lava flows across the surface. Lava that is "runny"

molten rock in upper mantle

Figure 9.10 Interactive Figure Volcanism. This photo shows the eruption of an active volcano on the flanks of Kilauea on the Big Island of Hawaii. The inset shows the underlying process: Molten rock collects in a "magma chamber" and can erupt upward.

can flow far before cooling and solidifying, while "thick" lava tends to collect in one place.* Broadly speaking, lava can shape three different types of volcanic features:

- The runniest lavas flow far and flatten out before solidifying, creating vast **volcanic plains** (Figure 9.11a).
- Lavas that are somewhat thicker can solidify before they can completely spread out, creating what we call **shield**

---

*You may hear the "thickness" of a liquid described by the technical term *viscosity.* For example, liquid water has a low viscosity that allows it to flow quickly and easily, while thick and slow-flowing honey and molasses have much higher viscosities.

Lava plains (maria) on the Moon

10 km

VIS

a Very runny lava makes flat lava plains like these on the moon. (The long, winding channel near the upper left was made by a river of molten lava.)

Olympus Mons (Mars)

20 km

VIS

b Slightly thicker lava makes shallow-sloped shield volcanoes, such as Olympus Mons on Mars.

Mount Hood (Earth)

c The thickest lavas make steep-sloped stratovolcanoes like Oregon's Mount Hood.

Figure 9.11 The shapes of volcanic structures depend on whether the molten lava is runny or thick.

volcanoes (so named because of their shape). Shield volcanoes can be very tall but are not very steep. Examples include the mountains of the Hawaiian islands on Earth and Olympus Mons on Mars (Figure 9.11b).

- The thickest lavas can't flow very far before solidifying and therefore build up tall, steep **stratovolcanoes** (Figure 9.11c). Examples of stratovolcanoes on Earth include Mount Fuji in Japan, Mount Kilimanjaro in Tanzania, Mount Shasta in California, Mount Hood in Oregon, and Mount Rainier and Mount St. Helens in Washington State.

Both lava plains and shield volcanoes are made of **basalt**, a mixture of many different minerals that erupts from volcanoes as a high-density but fairly runny lava. All the terrestrial worlds and some jovian moons show evidence of volcanic plains and shield volcanoes, telling us that basalt is common throughout the solar system. Stratovolcanoes are made by lower-density volcanic rock that erupts as a thicker lava; this type of lava is common on Earth but apparently much rarer in the rest of the solar system.

Volcanic structures are the most obvious result of volcanism, but from our standpoint as living beings, volcanism has had a far more important effect on our planet: It explains the existence of our atmosphere and oceans. Recall that Earth and the other terrestrial planets accreted from rocky and metallic planetesimals; the water and gas in their interiors presumably came from icy planetesimals that originated in more distant reaches of the solar system and crashed into the growing terrestrial worlds [Section 8.4]. This water and gas became trapped in the interiors of the planets in much the same way the gas in a carbonated beverage is trapped in a pressurized bottle. When molten rock erupts onto the surface as lava, the release of pressure expels the trapped gases in a process we call **outgassing**. Outgassing can be either quite dramatic, as during a volcanic eruption (Figure 9.12a), or more gradual, as when gas escapes from volcanic vents (Figure 9.12b). Virtually all the gas that made the atmospheres of Venus, Earth, and Mars—and the water vapor that rained down to form Earth's

b This photo shows more gradual outgassing from a volcanic vent in Volcanoes National Park, Hawaii.

a This photo shows the eruption of Mount St. Helens (Washington State) on May 18, 1980. Note the tremendous outgassing that accompanies the eruption.

Figure 9.12 Examples of outgassing, which released the gases that ultimately made the atmospheres of Venus, Earth, and Mars.

oceans—originally was released from the planetary interiors by outgassing.

Tectonics The third major geological process, tectonics, gets its name from the Greek word for "builder," *tekton*; you'll notice the same root in the word *architect*, which means "master builder." In geology, tectonics refers to "building" performed on planetary surfaces—that is, any surface reshaping that results from stretching, compression, or other forces acting on the lithosphere.

Tectonic features can arise in a variety of ways. For example, the weight of a newly formed volcano can bend or crack the lithosphere beneath it, or a plume of hot material rising from a planet's deep interior can push up on the lithosphere to create a bulge. However, most tectonic features form as a direct or indirect result of mantle convection (Figure 9.13). For example, we see compression features in places where adjacent convection cells push rock together. Such crustal compression helped create the Appalachian Mountains of the eastern United States. In other places, mantle convection may tend to pull apart the crust, making

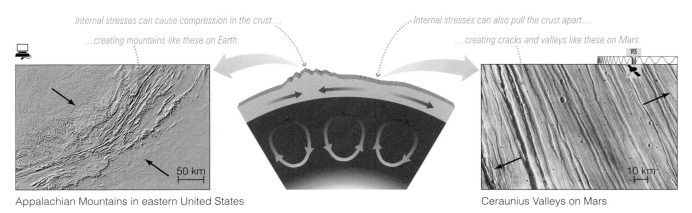

Internal stresses can cause compression in the crust...

...creating mountains like these on Earth.

Internal stresses can also pull the crust apart...

...creating cracks and valleys like these on Mars.

Appalachian Mountains in eastern United States

50 km

Ceraunius Valleys on Mars

10 km

VIS

Figure 9.13 Interactive Figure Tectonic forces can produce a wide variety of features. Mountains and fractured plains are among the most common. The Mars image is a visible-light photo from orbit. The Earth image was computer generated from topographical data.

cracks and valleys. Examples of such cracks and valleys include the Guinevere Plains on Venus, the Ceraunius Valleys on Mars, and New Mexico's Rio Grande Valley. (The river named the Rio Grande came *after* the valley formed from tectonic processes.)

On Earth, the ongoing stress of mantle convection ultimately fractured Earth's lithosphere into more than a dozen pieces, or *plates*. These plates move over, under, and around each other in a process we call **plate tectonics**. The movements of the plates explain nearly all Earth's major geological features, including the arrangement of the continents, the nature of the seafloor, and the origin of earthquakes. Because plate tectonics appears to be unique to Earth among the terrestrial worlds in our solar system, we'll save further discussion of it for Section 9.6.

Erosion The last of our four major geological processes is erosion. *Erosion* is a blanket term for a variety of processes that break down or transport rock through the action of ice, liquid, or gas. The shaping of valleys by glaciers (ice), the carving of canyons by rivers (liquid), and the shifting of sand dunes by wind (gas) are all examples of erosion.

We often associate erosion with the breakdown of existing features, but erosion also builds things. Sand dunes, river deltas, and lake-bed deposits are all examples of features built by erosion. Indeed, erosion built much of the surface rock on Earth. Over long periods of time, erosion

has piled sediments into layers on the floors of oceans and seas, forming what we call **sedimentary rock**. The gorgeous layered rock of the Grand Canyon is sedimentary rock that was built by erosion. Figure 9.14 shows several examples of erosion on Earth.

• Why do the terrestrial planets have different geological histories?

The terrestrial worlds show great differences in the extent to which the four geological processes have operated. The surfaces of the Moon and Mercury are dominated by impact craters, but show virtually no features of erosion. Volcanic and tectonic features are found on all the worlds, but we find tall volcanoes only on Earth, Venus, and Mars. And Earth is the only terrestrial world on which features of erosion dominate the landscape. Why do the processes operate so differently on different worlds? Ultimately, we can trace the answer back to fundamental planetary properties like size, distance from the Sun, and rate of rotation.

Figure 9.15 summarizes the key connections between a planet's fundamental properties and the geological processes of volcanism, tectonics, and erosion. Let's explore these connections in a little more detail. We'll then be ready to explore the connection between planetary properties and impact cratering, the only one of the four geological processes with an external cause.

Figure 9.14 A few examples of erosion on Earth.

a The Colorado River continues to carve the Grand Canyon after millions of years.

b Glaciers created Yosemite Valley during ice ages.

c Wind erosion wears away rocks and builds up sand dunes, such as these in a California desert.

d Erosional debris also creates geological features, as seen in this river delta.

The Role of Planetary Size

Small Terrestrial Planets

Interior cools rapidly...

...so that tectonic and volcanic activity cease after a billion years or so. Many ancient craters therefore remain.

Lack of volcanism means little outgassing, and low gravity allows gas to escape more easily; no atmosphere means no erosion.

Large Terrestrial Planets

Warm interior causes mantle convection...

...leading to ongoing tectonic and volcanic activity; most ancient craters have been erased.

Outgassing produces an atmosphere and strong gravity holds it, so that erosion is possible.

Core may be molten, producing a magnetic field if rotation is fast enough.

The Role of Distance from the Sun

Planets Close to the Sun

Surface is too hot for rain, snow, or ice, so little erosion occurs.

High atmospheric temperature allows gas to escape more easily.

Sun

Planets at Intermediate Distances from the Sun

Moderate surface temperatures can allow for oceans, rain, snow, and ice, leading to substantial erosion.

Gravity can more easily hold atmospheric gases.

Planets Far from the Sun

Low surface temperatures can allow for ice and snow, but no rain or oceans, limiting erosion.

Atmosphere may exist, but gases can more easily condense to make surface ice.

The Role of Planetary Rotation

Slow Rotation

Less wind and weather means less erosion, even with a substantial atmosphere.

Slow rotation means weak magnetic field, even with a molten core.

Rapid Rotation

More wind and weather means more erosion.

Rapid rotation is necessary for a global magnetic field.

Figure 9.15 A planet's fundamental properties of size, distance from the Sun, and rotation rate are responsible for its geological history. This illustration shows the role of each key property separately. A planet's overall geological evolution depends on the combination of all the effects shown above.

Planetary Properties Controlling Volcanism and Tectonics Volcanism and tectonics both require internal heat, which is why they tend to go hand in hand: More volcanism generally also means more tectonics. The connection with internal heat also allows us to tie these processes back to planetary size. As shown in Figure 9.15, larger planets have more internal heat and hence more volcanic and tectonic activity.

All the terrestrial worlds probably had some degree of volcanism and tectonics when their interiors were young and hot. As an interior cools, however, volcanic and tectonic activity subsides. The Moon and Mercury lack ongoing volcanism and tectonics because their small size allowed their interiors to cool long ago. Earth has active volcanism and tectonics because it is large enough to still have a hot interior. Venus, nearly the same size as Earth, must still be hot inside and probably also has active volcanism and tectonics, though we lack conclusive evidence of volcanic eruptions in the past few tens of millions of years. Mars, with its smaller size, certainly has much less volcanism and tectonic activity today than it did in the distant past.

Planetary Properties Controlling Erosion The link between planetary properties and erosion is a little more complex, because erosion can occur only on a planet with a significant atmosphere. Otherwise, there can be no surface liquids, no wind, and no climate change to cause glacial movements. An atmosphere is not the only requirement for erosion, however. A stationary atmosphere that simply bears down without stirring the surface would not cause any erosion. Erosion comes from processes such as wind and rain, which we generally refer to as "weather." Thus, the basic requirements for erosion are the existence of an atmosphere and significant weather affecting the surface.

We can therefore trace a planet's level of erosion back to all three fundamental planetary properties shown in Figure 9.15. The planet's size is important because a terrestrial world can have an atmosphere only if it has had significant outgassing, and outgassing is part of volcanism. In addition, size determines the strength of gravity, and atmospheric gases are more easily lost to space on worlds with weaker gravity. The planet's distance from the Sun affects its temperature; if all else is equal (such as planetary size), the higher temperatures on a world closer to the Sun will make it much easier for atmospheric gases to escape to space, while the colder temperatures on a world farther from the Sun may cause atmospheric gases to freeze out. Distance is also important because water erosion is much more effective with liquid water than with water vapor or ice, and therefore is strongest when moderate temperature allows water vapor to condense into liquid form. Planetary rotation is important to erosion because rotation is the primary driver of winds and other weather [Section 10.2]: Faster rotation means stronger winds and storms.

The Moon and Mercury lack significant atmospheres and erosion because they lack outgassing today, and any atmospheric gases they had in the distant past have been lost to space. Mars has limited erosion; it has a very thin atmosphere because much of the water vapor and carbon dioxide out-gassed in its past either escaped to space or lies frozen in its polar caps or beneath the surface. Venus and Earth probably had similar amounts of outgassing, but cooler temperatures on Earth led to condensation and the ultimate removal of most of this gas from Earth's atmosphere, though still leaving enough for plenty of erosion. Most of Venus's gas remained in its atmosphere, making its atmosphere much thicker than Earth's, but Venus has little erosion because its slow rotation gives it virtually no wind on its surface.

• How does a planet's surface reveal its geological age?

We've focused on the four geological processes and their causes individually, but they also interact with one another to shape planetary surfaces. For example, volcanism produces lava flows that can erase old impact craters, and outgassing can create atmospheres that may lead to erosion. These interactions lead to one of the most useful insights in planetary geology: We can learn about a planet's overall level of geological activity and the "geological age" of its surface simply by looking at how closely impact craters are crowded together. We consider surfaces with ongoing volcanism, tectonics, or erosion to be geologically young, because they are constantly being reshaped by these processes. In contrast, a surface that has not changed in billions of years is considered geologically old. Remember that we are speaking only of the age *of the surface as it now appears,* not the age of the planet itself: All the planets are about 4.5 billion years old [Section 8.5].

To understand how crater crowding reveals surface ages, remember that all the terrestrial worlds should have been battered by impacts during the heavy bombardment. The number of impacts on a surface of any given size should have been quite similar, regardless of a planet's size, distance from the Sun, or rotation rate. Why, then, do we see such different numbers of craters on the different worlds? Careful study of the Moon reveals the key clues. The degree of crowding among craters varies greatly from place to place on the Moon (Figure 9.16). In the regions on the Moon known as the *lunar highlands*, craters are so crowded that we see craters on top of other craters. In the regions known as the *lunar maria*, we see only a few craters on top of generally smooth volcanic plains.

The sheer number of impact craters in the lunar highlands, especially the very common small ones, tells us that no part of the Moon could have escaped heavy cratering during the early history of the solar system. Thus, the relative lack of craters in the maria tells us that some process must have *erased* these craters from the maria after the heavy bombardment ended. (We'll discuss the precise process in Section 9.3). Radiometric dating of rocks brought back from the Moon by the *Apollo* astronauts confirms this idea. Rocks from the heavily cratered lunar highlands have ages as old as 4.4 billion years. Rocks from the maria are hundreds of millions of years younger, with ages between 3.0–3.9 billion years. Because the maria contain only about 3% as many craters as the highlands even though they are just a few

Lunar maria are huge impact basins that were flooded by lava. Only a few small craters appear on the maria.

Lunar highlands are ancient and heavily cratered.

a This *Apollo* photograph of the Moon shows some areas are much more heavily cratered than others. (This view of the Moon is *not* the one we see from Earth.)

b This map shows the entire surface of the Moon in the same way a flat map of Earth represents the entire globe. Radiometric dating shows the heavily cratered lunar highlands to be a half-billion or more years older than the darkly colored lunar maria, telling us that the impact rate dropped dramatically after the end of the heavy bombardment.

Figure 9.16 These photos of the Moon show us that crater crowding is closely related to surface age.

hundred million years younger, the heavy bombardment must have subsided quite early in the solar system's history. Compared to the period of heavy bombardment, relatively few impacts have occurred in the last 4 billion years.

More precise study of lunar cratering has allowed scientists to reconstruct the rate of impacts during much of the Moon's history. Because impacts are essentially random events, the same changes in impact rate over time must apply to all the terrestrial worlds. The degree of crater crowding therefore provides a good measure of the geological age of a planetary surface. A surface that is crowded with craters must have remained essentially undisturbed since the end of the heavy bombardment. A surface with few craters must have had its ancient craters erased by more recent geological processes.

Thus, the key factor determining the number of craters on a planetary surface today is its degree of past geological activity in the form of volcanism, tectonics, and erosion. All the terrestrial worlds must have been covered with impact craters in their early histories. Those that still have lots of craters today, like the Moon and Mercury, are those that have had little geological activity. Those with few craters today, like Venus and Earth, must have had extensive geological activity that has erased essentially all their ancient craters. The only craters we see on Venus and Earth are those few that have formed relatively recently (within the last few hundred million years).

The fact that more geological activity means fewer craters allows us to tie impact cratering primarily to one planetary property: size. Because the interiors of smaller planets cool more quickly, smaller worlds have had less

geological activity and therefore retain more of their ancient craters. Larger worlds had at least as many impact craters when they were young as small worlds do today—in fact, larger worlds probably suffered more impacts because their stronger gravity tended to draw in and accelerate planetesimals—but their hotter interiors have supported enough geological activity to erase most of their ancient craters.

THINK ABOUT IT

Mercury is heavily cratered, much like the lunar highlands. Venus has relatively few craters. Which planet has the older surface? Is there any difference between the ages of the two planets as a whole? Explain how the surfaces of Mercury and Venus can have different ages while the planets themselves are the same age.

9.3 Geology of the Moon and Mercury

In the rest of this chapter we will investigate the geological histories of the terrestrial worlds, with the goal of explaining why each ended up as it did. Because smaller worlds have simpler geological histories, we'll start with the Moon and Mercury.

Look back at Figure 9.1, and you'll see that both these worlds are considerably smaller than Venus, Earth, or Mars. Their small size means that they long ago lost most of their internal heat, leaving them without any energy source to power ongoing geological activity. Small size also explains

a The Moon   **b** Mercury

Figure 9.17 Similar views of the Moon and Mercury, shown to scale. The Mercury photo is a composite of photos from the *Mariner 10* spacecraft [Section 7.3], which photographed only parts of the surface.

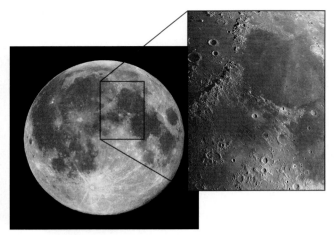

Figure 9.18 The familiar face of the Moonshows numerous dark, smooth maria.

their lack of significant atmospheres: Their gravity is too weak to hold gas for long periods of time, and without ongoing volcanism they lack the outgassing needed to replenish gas lost in the past.

Figure 9.17 shows the similar-looking faces of the Moon and Mercury. Their most obvious surface features are their numerous impact craters, showing that impact cratering has been by far the most important geological process on both worlds. However, closer examination shows a few features that are volcanic or tectonic in origin. Apparently, the Moon and Mercury had geological activity in the past. This should make sense: Long ago, before they had a chance to cool, these worlds were hot enough inside for at least some volcanism and tectonics.

• What geological processes shaped our Moon?

We've already discussed the fact that different parts of the Moon's surface have different geological ages. Now we are ready to examine in more detail the processes that have shaped the varying features of the Moon's surface.

Impact Cratering on the Moon   We've seen that the familiar face of the full moon shows abundant cratering, but not all regions of the surface look the same. The brighter regions, called the *lunar highlands,* are heavily cratered. Other regions, known as the **lunar maria**, look smoother and darker (Figure 9.18). Indeed, the maria got their name because they look much like oceans when seen from afar; *maria* (singular, *mare*) is Latin for "seas." But the entire surface of the Moon should have been packed with craters during the heavy bombardment. So what happened to the craters that must once have been located in the regions where we now see maria?

Volcanism and Tectonics in the Lunar Maria   The smooth and dark appearance of the lunar maria suggests that they were made by a flood of molten lava, and studies of moon rocks confirm this suggestion. Indeed, the maria are the Moon's most noticeable volcanic features.

Figure 9.19 shows how we think the maria formed. During the heavy bombardment, craters covered the Moon's entire surface. The largest impacts of the heavy bombardment violently fractured the Moon's lithosphere beneath the huge craters they created. However, the Moon's interior had already cooled since its formation, and there was no molten rock to flood these craters immediately. Instead, the lava floods came hundreds of millions of years later, thanks to heat released by the decay of radioactive elements in the Moon's interior. This heat gradually built up during the Moon's early history, until mantle material melted about 3 to 4 billion years ago. The molten rock then welled up through the cracks in the lithosphere, flooding the largest impact craters with lava. The maria are generally circular because they are essentially flooded craters (and craters are almost always round). Their dark color comes from the dense, iron-rich rock (basalt) that rose up from the lunar mantle as molten lava.

The flat surfaces of the maria tell us that the lunar lava spread easily and far, which means that it must have been among the runniest lava in the solar system. Its runniness can probably be traced to the Moon's formation. Recall that we think the Moon formed as the result of a giant impact of a Mars-sized body with Earth [Section 8.4]. This impact would have vaporized water and released other trapped gases, so these materials would not have become part of the Moon when it reaccreted from the giant impact debris. Water and trapped gases tend to increase the "thickness" of erupting lava in the same way that tiny bubbles increase the "thickness" of whipped cream: When you buy whipping cream in a carton, it is liquid and too thin to shape into anything, but the whipping process adds tiny air bubbles that make it thick enough so that you can build whipped cream "mountains." With no water or trapped gases, the Moon's lava would have lacked the gas bubbles

a This illustration shows the Mare Humorum region as it probably looked about 4 billion years ago, when it would have been completely covered in craters.

b Around that time, a huge impact excavated the crater that would later become Mare Humorum. The impact fractured the Moon's lithosphere and erased the many craters that had existed earlier.

c A few hundred million years later, enough heat from radioactive decay built up to melt the Moon's upper mantle. Molten lava welled up through the lithospheric cracks, flooding the impact crater.

d This photo shows Mare Humorum as it appears today, and the inset shows its location on the Moon.

Figure 9.19 The lunar maria formed between 3 and 4 billion years ago, when molten lava flooded large craters that had formed hundreds of millions of years earlier. This sequence of diagrams represents the formation of Mare Humorum.

that make lava "thick" on Earth and other worlds. The lunar lava was so runny that in places it flowed like rivers of molten rock, carving out long, winding channels (see the channel in Figure 9.11a).

The Moon's relatively few tectonic features are also found within the maria. These features, which look much like surface wrinkles, probably were created by small-scale tectonic stresses that arose when the lava that flooded the maria cooled and contracted (Figure 9.20).

Surprisingly, the Moon's far side—which we never see from Earth—has very few maria like the ones that we see on its near side (see Figure 9.16). Instead, when spacecraft first photographed the far side in the 1960s, they revealed a landscape looking almost entirely like the heavily cratered highlands of the near side. We now know that the entire far side essentially has a higher altitude than the near side. No one knows what caused this difference in altitude between the two sides, but it apparently allowed lava to well up and fill crater basins only on the near side, thus explaining why the near side has maria.

The Moon Today: Rare Impacts and Little More The Moon's era of geological activity is long gone. Today, the Moon is a desolate and nearly unchanging place. Rare impacts may occur in the future, but we are unlikely ever to witness a major one. Little happens on the Moon, aside from the occasional visit of robotic spacecraft or astronauts from Earth (Figure 9.21a).

The only ongoing geological change on the Moon is a very slow "sandblasting" of the surface by *micrometeorites*, sand-size particles from space. These tiny particles burn up as meteors in the atmospheres of Earth, Venus, and Mars but rain directly onto the surface of the airless Moon. The micrometeorites gradually pulverize the surface rock, which explains why the lunar surface is covered by a thin layer of powdery "soil." The *Apollo* astronauts left their footprints in this powdery surface (Figure 9.21b). Pulverization by micrometeorites is a very slow process, and the astronauts' footprints will last millions of years before they are finally erased. Over billions of years, the micrometeorite impacts can gradually smooth out rough crater rims.

Aside from the micrometeorite sandblasting and rare larger impacts, the Moon has been "geologically dead" since the maria formed more than 3 billion years ago. Nevertheless, the Moon remains a prime target of future exploration. Further studies of lunar geology will help us understand both the Moon's history and the history of our solar system. The Moon may also offer an excellent location for astronomical observatories, since it has no air to obstruct our view but a stable ground on which to build large telescopes. Someday, perhaps in the not-too-distant future,

Figure 9.20 The features that look like wrinkles within maria like this one (Mare Imbrium) arose from tectonic stresses associated with the cooling and contracting of the lava that flooded the maria between 3 and 4 billion years ago.

a Astronaut Gene Cernan takes the Lunar Roving Vehicle for a spin during the final *Apollo* mission to the Moon (*Apollo 17*, December 1972).

b The *Apollo* astronauts left clear footprints, like this one, in the Moon's powdery "soil." The powder is the result of gradual pulverization of surface rock by micrometeorites. Micrometeorites will eventually erase the astronauts' footprints, but not for millions of years.

Figure 9.21 The Moon today is geologically dead, but it is nevertheless a world from which we can learn a lot about the history of our solar system.

humans will build a lunar colony, with living quarters underground for protection against dangerous solar and cosmic radiation.

• What geological processes shaped Mercury?

Mercury looks so much like the Moon that it's often difficult at first to tell which world you are looking at in surface photos. The similarities extend to all the geological processes, though there are also a few differences.

Impact craters are visible almost everywhere on Mercury, indicating an ancient surface. However, Mercury's craters are less crowded together than the craters in the most ancient regions of the Moon, suggesting that molten lava later covered up some of the craters that formed on Mercury during the heavy bombardment (Figure 9.22a). As on the Moon, these lava flows probably occurred when heat from radioactive decay accumulated enough to melt part of the mantle. Although we have not found evidence of lava flows as large as those that created the lunar maria, the lesser crater crowding and the many smaller lava plains suggest that Mercury had at least as much volcanism as the Moon.

The largest single surface feature on Mercury is a huge impact crater called the *Caloris Basin* (Figure 9.22b). The Caloris Basin spans more than half of Mercury's radius, and its multiple rings bear witness to the violent impact that created it. The impact must have reverberated throughout the planet—we see evidence of violent surface shaking on the precise opposite side of Mercury from the Caloris Basin (Figure 9.22c). The Caloris Basin has few craters within it, indicating that it must have formed at a time when the heavy bombardment was already subsiding.

The most surprising features of Mercury are its many tremendous cliffs—evidence of a type of past tectonics quite different from anything we have found on any other terrestrial world (Figure 9.23). Mercury's cliffs have vertical faces up to 3 or more kilometers high and typically run for hundreds of kilometers across the surface. They probably formed when tectonic forces compressed the crust, causing the surface to crumple. Because crumpling would have shrunk the portions of the surface it affected, Mercury as a whole could not have stayed the same size unless other parts of the surface expanded. However, we find no evidence of "stretch marks" on Mercury. Can it be that the whole planet simply shrank?

Apparently so. Early in its history, Mercury's larger size and greater iron content allowed it to gain more internal heat from accretion and differentiation than did the Moon. This heat swelled the size of the large iron core. Later, as the core cooled, it contracted by perhaps as much as 20 kilometers in radius. The mantle and lithosphere must have contracted along with the core, generating the tectonic

Figure 9.22 Features of impact cratering and volcanism on Mercury. (Photos from *Mariner 10.*)

a Close-up view of Mercury's surface, showing impact craters and smooth regions where lava apparently covered up craters.

b Part of the Caloris Basin (outlined by the dashed circular ring), a large impact crater on Mercury.

c The surface on the precise opposite side of Mercury from the Caloris Basin. Seismic energy from the impact focused here and jumbled the crust.

Mercury's core and mantle shrank . . .

Some portions of the crust were forced to slide under others.

Today we see long, steep cliffs created by this crustal movement.

. . . causing Mercury's crust to contract.

Shrinkage not to scale!

VIS

a Mercury's long cliffs probably formed when its core cooled and contracted, causing the mantle and lithosphere to shrink. This diagram shows how the cliffs probably formed as the surface crumpled.

b This cliff extends about 100 kilometers in length, and its vertical face is as much as 2 kilometers tall. (Photo from *Mariner 10*.)

Figure 9.23 Long cliffs on Mercury offer evidence that the entire planet shrank early in its history, perhaps by as much as 20 kilometers in radius.

stresses that created the great cliffs. The contraction probably also closed off any remaining volcanic vents, ending Mercury's period of volcanism.

Because Mercury is larger than the Moon, it probably retained its internal heat a bit longer. Nevertheless, it is still small enough that its interior should have cooled relatively quickly. Crater counts on Mercury suggest that all its volcanic and tectonic activity ceased within the first billion years after its formation. Like the Moon, Mercury has been geologically dead for most of its history.

9.4 Geology of Mars

Mars is the next stop on our geological tour. Because Mars is much larger than the Moon or Mercury (see Figure 9.1), we expect it to have had a more interesting and varied geological history. However, it is much smaller than Venus or Earth—its radius is about half that of Earth and its mass is only about 10% that of Earth—so we expect it to be less geologically active than our own world. Observations confirm this basic picture, though Mars's greater distance from the Sun also plays a role in its geology; Mars is about 50% farther than Earth from the Sun.

• How did Martians invade popular culture?

Have you ever noticed that people often speak of Martians but rarely of, say, Venusians or Jupiterians? It's no accident, but rather the result of misconceptions that arose from early telescopic observations of Mars.

The story begins in the late eighteenth century with the brother and sister astronomers William and Caroline Herschel. Although they are most famous for discovering the

planet Uranus, they also made many observations of Mars. Their observations revealed several uncanny resemblances to Earth, including a very similar axis tilt, a day just slightly longer than 24 hours, the presence of polar caps, and seasonal variations in appearance over the course of the Martian year (about 1.9 Earth years). In 1784, William Herschel gave a talk in which he spoke with confidence about "inhabitants" of Mars.

The hypothetical Martians got a bigger break about a century later. In 1879, Italian astronomer Giovanni Schiaparelli reported seeing a network of 79 linear features on Mars through his telescope. He named these features *canali*, by which he meant the Italian word for "channels," but it was frequently translated as "canals." Inspired by what sounded like evidence of intelligent life, American astronomer Percival Lowell commissioned the building of an observatory in Flagstaff, Arizona, for the study of Mars.

The Lowell Observatory opened in 1894. Barely a year later, Lowell published detailed maps of the Martian canals and the first of three books in which he argued that the canals were the work of an advanced civilization. He suggested that Mars was falling victim to unfavorable climate changes and that the canals had been built to carry water from the poles to thirsty cities elsewhere. Lowell's work drove rampant speculation about Martians and fueled science fiction fantasies such as H. G. Wells's *The War of the Worlds* (published in 1898). The public mania drowned out the skepticism of astronomers who saw no canals through their telescopes or in their photographs (Figure 9.24).

The debate over the existence of Martian canals and cities was not entirely put to rest until 1965, when NASA's *Mariner 4* spacecraft flew past Mars and sent back photos of a barren, cratered surface. Lowell's canals were nowhere to be seen, confirming that they had been the work of a vivid imagination looking at blurred telescopic images.

The lack of ancient cities may have dampened the enthusiasm of science fiction writers, but scientists find Mars as intriguing as ever. More spacecraft have visited Mars than any other planet, and plans call for future spacecraft visits to Mars about every 26 months to coincide with optimal orbital alignments of Earth and Mars. Indeed, while we

VIS

Figure 9.24 The image at left is a telescopic photo of Mars. The image at right is a drawing of Mars made by Percival Lowell. As you can see, Lowell had a vivid imagination. Nevertheless, if you blur your eyes while looking at the photo, you might see how some of the features resemble what Lowell thought he saw.

now know there have never been Martian cities, it's possible that Mars has harbored more primitive life. The search for past or present life on Mars is one of the major goals of Mars exploration.

• What are the major geological features of Mars?

Mars's total surface area is only about one-fourth that of Earth, but both planets have nearly the same amount of *land* area because Earth's surface is about three-fourths covered by water. Thus, there's plenty of territory to explore on Mars. For context, Figure 9.25 shows the full surface of Mars and the locations of features we will discuss in this book. As we'll see, close study of Mars reveals extensive evidence of all four geological processes.

Impact Cratering on Mars Aside from the polar caps, the most striking feature of Figure 9.25 is the dramatic difference in terrain around different parts of Mars. Much of the southern hemisphere has relatively high elevation and is scarred by numerous large impact craters, including the very large crater known as the Hellas Basin. In contrast, the northern plains show few impact craters and tend to

be below the average Martian surface level. The differences in crater crowding clearly show that the southern highlands are a much older surface than the northern plains, which must have had their early craters erased by other geological processes.

THINK ABOUT IT

Which fundamental planetary property (size, distance from the Sun, or rotation rate) explains why Earth does not have any terrain that is as heavily cratered as the southern highlands of Mars?

Volcanism on Mars Further study suggests that volcanism was the most important of the processes responsible for erasing ancient craters on Mars, although tectonics and erosion also played a part. However, no one knows why volcanism affected the northern plains so much more than the southern highlands or why the two regions differ so much in elevation.

More dramatic evidence of volcanism on Mars comes from several towering shield volcanoes. One of these, Olympus Mons, is the tallest known volcano in the solar system (Figure 9.26; see also Figure 9.12b). Its base is some 600 kilo-

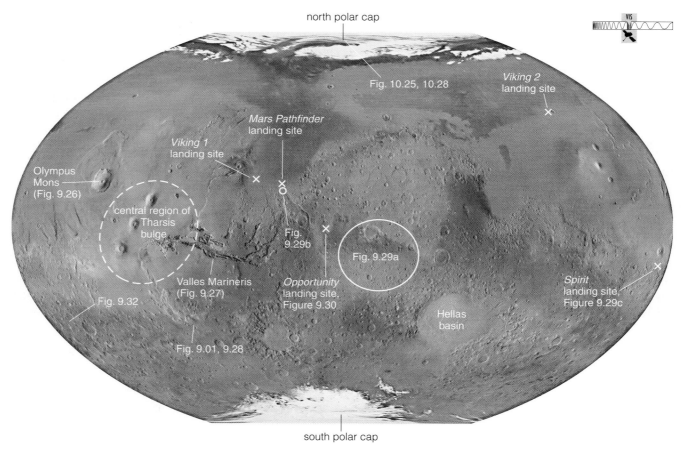

Figure 9.25 Interactive Photo This image showing the full surface of Mars is a composite made by combining more than 1,000 images with more than 200 million altitude measurements from the *Mars Global Surveyor* mission. Several key geological features are labeled, and the locations of features shown in close-up photos elsewhere in this book are marked.

Figure 9.26 Olympus Mons, the tallest volcano in the solar system, covers an area the size of Arizona and rises higher than Mount Everest on Earth. Note the tall cliff around its rim and the central volcanic crater from which lava erupted.

meters across, large enough to cover an area the size of Arizona. Its peak stands about 26 kilometers above the average Martian surface level, or some three times as high as Mount Everest stands above sea level on Earth. Much of Olympus Mons is rimmed by a cliff that in places is 6 kilometers high.

Olympus Mons and several other large volcanoes are concentrated on or near the continent-size *Tharsis Bulge* (see Figure 9.25). Tharsis, as it is usually called, is some 4,000 kilometers across, and most of it rises several kilometers above the average Martian surface level. It was probably created by a long-lived plume of rising mantle material that bulged the surface upward and provided the molten rock for the eruptions that built the giant volcanoes.

Could volcanoes still be active on Mars? Until recently, we didn't think so. We expect Mars to be much less volcanically active than Earth, because its smaller size has allowed its interior to cool much more, and Martian volcanoes show enough impact craters on their slopes to suggest that they have been inactive for at least tens of millions of years and possibly much longer. However, geologically speaking, tens

of millions of years is not that long. In addition, radiometric dating of meteorites that appear to have come from Mars (so-called Martian meteorites [Section 24.3]) shows some of them to be made of volcanic rock that solidified from molten lava as little as 180 million years ago—quite recent in the 4.5-billion-year history of the solar system. Given this evidence of geologically recent volcanic eruptions, it is likely that Martian volcanoes will erupt again someday, though not necessarily in our lifetimes. Nevertheless, the Martian interior is presumably cooling and its lithosphere thickening, so that Mars will become as geologically dead as the Moon and Mercury within a few billion years.

Tectonics and Valles Marineris Mars also has tectonic features, though none on a global scale like the plate tectonics of Earth. The most prominent tectonic feature is the long, deep system of valleys called *Valles Marineris* (Figure 9.27). Named for the *Mariner 9* spacecraft that first imaged it, Valles Marineris extends almost a fifth of the way along the planet's equator. It is as long as the United States is wide and almost four times as deep as Earth's Grand Canyon.

No one knows exactly how Valles Marineris formed. Parts of the canyon are completely enclosed by high cliffs on all sides, so neither flowing lava nor water could have been responsible. Extensive cracks on its western end run up against the Tharsis Bulge (see Figure 9.25), suggesting a connection between the two features. Valles Marineris may have formed through tectonic stresses accompanying the uplift of material that created the Tharsis Bulge, cracking the surface and leaving the tall cliff walls of the valleys.

• What geological evidence tells us that water once flowed on Mars?

Impacts, volcanism, and tectonics explain most of the major geological features of Mars, but closer examination shows ample evidence of features of erosion. For example, Figure 9.28 looks much like dry riverbeds on Earth seen from above. These channels appear to have been carved by running water, though no one knows whether the water came from runoff after rainfall, from erosion by water-rich debris flows, or from an underground source. Regardless of the specific mechanism, water was almost certainly responsible, because it is the only substance that could have been liquid under past Martian conditions and that is sufficiently abundant to have created such extensive erosion features.

Figure 9.27 Valles Marineris is a huge system of valleys on Mars created in part by tectonic stresses.

Figure 9.28 This photo, taken by the *Viking Orbiter*, shows what appear to be dried-up riverbeds. Toward the top of the image we see many individual tributaries, which merge into the larger "river" near the lower right. Counts of craters near the channels indicate that they formed more than 3 billion years ago.

If you were to visit Mars, however, the idea that parts of the surface were shaped by flowing water might seem quite strange. No liquid water exists anywhere on the surface of Mars today. We know this not only because we've studied most of the surface in reasonable detail, but also because the surface conditions would not allow liquid water to be present as lakes, rivers, or even puddles. In most places and at most times, Mars is so cold that any liquid water would immediately freeze into ice. Even when the temperature rises above freezing, as it often does at midday near the equator, the air pressure is so low that liquid water would quickly evaporate [Section 5.3]. In other words, liquid water is *unstable* on Mars today: If you donned a spacesuit and took a cup

of water outside your pressurized spaceship, the water would either freeze or boil away (or both) almost immediately.

Evidence for Ancient Water Flows The fact that we see water erosion features on Mars but that liquid water is absent today tells us that Mars must once have had very different surface conditions—conditions such as warmer temperatures and greater air pressure that would have allowed water to flow and rain to fall. This warmer and wetter period must have ended long ago, as you can see by looking again at Figure 9.28. Notice that a few impact craters lie on top of the channels. From counts of the craters in and near them, it appears that these channels are at least 2–3 billion years old, meaning that water has not flowed through them since that time.

Other orbital evidence also argues that Mars had rain and surface water in the distant past. Figure 9.29a shows a broad region of the ancient, heavily cratered southern highlands. Notice the indistinct rims of many large craters and the relative lack of small craters. Both facts argue for ancient rainfall, which would have eroded crater rims and erased small craters altogether. Figure 9.29b shows a close-up of a crater floor with sculpted patterns that suggest it was once the site of a lake. Ancient rains may have filled the crater, allowing sediments to build up from material that settled to the bottom. The sculpted patterns in the crater bottom may have been created as erosion by wind exposed layer upon layer of sedimentary rock, much as erosion by the water in the Colorado River exposed the layers visible in the walls of the Grand Canyon on Earth. Figure 9.29c shows a three-dimensional perspective of the surface that suggests water once flowed between two ancient crater lakes.

Figure 9.29 More orbital evidence of past water on Mars.

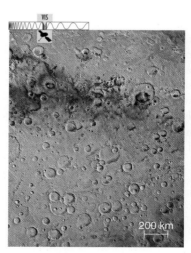

a This photo shows a broad region of the southern highlands on Mars. The eroded rims of large craters and the lack of many small craters suggest erosion by rainfall.

b This close-up view shows the floor of the crater shown earlier in Figure 9.9c. The sculpted patterns appear to be layers of sedimentary rock that were laid down at a time when the crater was filled with water.

c This computer-generated perspective view shows how a Martian valley forms a natural passage between two possible ancient lakes (shaded blue). Vertical relief is exaggerated 14 times to reveal the topography. In 2004, the *Spirit* rover landed in the crater (called Gusev) at the bottom of this picture.

Surface studies have provided additional evidence for past water on Mars. In 2004, two robotic rovers named *Spirit* and *Opportunity* landed on nearly opposite sides of Mars. *Spirit* landed in Gusev Crater, the site of the possible ancient lake shown in Figure 9.29c. *Opportunity* landed in the Meridiani Plains, where orbital spacecraft had detected spectroscopic hints of minerals that form in water. The twin rovers carried dozens of cameras, several instruments to identify rock composition, and a grinder to expose fresh rock for analysis. The rovers, designed for just 3 months of operation, were still going strong as this book was being written, more than a year and a half after their arrival.

Both rovers found solid evidence that liquid water was once plentiful on Mars. Rocks at the *Opportunity* landing site contain tiny spheres—nicknamed "blueberries" although they're neither blue nor as large as the berries we find in stores—and odd indentations suggesting that they formed in standing water, or possibly by groundwater percolating through rocks (Figure 9.30). Compositional analysis showed that the abundant "blueberries" contained the iron-rich mineral hematite, and other rocks contained the sulfur-rich mineral jarosite; both minerals form in water. Detailed chemical analysis supports the case for formation in an acidic, salty environment—possibly a sea or ocean. Moreover, a close look at the layering of the sedimentary rocks suggests a changing environment of waves and/or wind. Taken together, the orbital and surface studies provide overwhelming evidence for abundant water in Mars's past.

Martian Water Today   If water once flowed over large portions of Mars, where did it all go? As we'll discuss in Chapter 10, much of the water was probably lost to space forever. However, significant amounts of water apparently still remain, frozen at the polar caps and in the top meter or so of the surface soil around much of the rest of the planet (Figure 9.31).

If water ice exists on Mars this close to the surface, even more water probably lies deeper underground. If there is still volcanic heat on Mars, this water may sometimes melt and flow. Although we have found no geological evidence to suggest that any large-scale water flows have occurred on Mars in the past billion years, orbital photographs offer tantalizing hints of smaller-scale water flows in much more recent times.

The strongest evidence for liquid water in recent times comes from photos of gullies on crater and channel walls. In Figure 9.32, for example, note the striking similarity to the gullies we see on almost any eroded slope on Earth.

a This sequence zooms in on a rock outcropping near the rover's landing site. The outcrop stands about knee-high. The close-up shows a piece of the rock about 3 centimeters across (about the width of your big toe). The layered structure, the odd indentations, and the small sphere all support the idea that the rock formed from sediments in standing water.

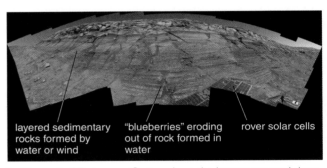

layered sedimentary rocks formed by water or wind   "blueberries" eroding out of rock formed in water   rover solar cells

b The wall of Endurance Crater, where the impact exposed deep bedrock layers profoundly altered by water. Close-ups show the durable hematite "blueberries" eroding out of the rock and tumbling downhill in piles. Changing tilts of the rock layers hint at changing wind or waves during formation.

Figure 9.30 These photos were taken by the *Opportunity* rover, which landed on Mars in 2004.

Figure 9.31 This map, made with data from *Mars Odyssey*, represents the hydrogen content of the Martian surface soil. The blue areas contain the most hydrogen, probably because they represent regions in which the top meter or so of surface soil contains water ice.

Figure 9.32 This photograph from the *Mars Global Surveyor* shows gullies on a crater wall. The gullies may have been formed by water melting under the protective cover of snowpack.

One hypothesis suggests that the gullies form when snow accumulates on the crater walls in winter and melts away in spring. Because the gullies are relatively small (note the scale bar in Figure 9.32), they should be gradually covered over by blowing sand during Martian dust storms. Gullies that are still clearly visible must be no more than a few million years old. Geologically speaking, this time is short enough to make it quite likely that water flows are still forming gullies today. Thus, while Mars's era of abundant surface water is long gone, small amounts of liquid water may still flow on occasion.

Even if water sometimes does still flow, Mars clearly was much warmer and wetter at times in the past than it is today. Ironically, Percival Lowell's supposition that Mars was drying up has turned out to be basically correct, although in a very different way than he imagined. We'll discuss the reasons for Mars's dramatic climate change in Chapter 10.

9.5 Geology of Venus

The surface of Venus is searing hot with brutal pressure [Section 7.2], making it seem quite unlike the "sister planet" to Earth it is sometimes called. However, beneath the surface, Venus must be quite similar to Earth. Venus is nearly the same size as Earth (see Figure 9.1), with a radius only about 5% smaller than Earth's. Venus and Earth are also quite close together relative to the overall size of the solar system, so both should have been built from the same kinds of planetesimals. The very similar densities of Venus and Earth (see Table 7.1) offer further evidence that both planets have the same overall composition.

With the same basic size and the same basic composition, we expect the interiors of Venus and Earth to be quite similar in structure and to retain about the same level of internal heat today. Nevertheless, we see ample evidence of "skin-deep" differences in surface geology.

• What are the major geological features of Venus?

Venus's thick cloud cover prevents us from seeing through to its surface, but we can study its geological features with radar because radio waves can pass through clouds. **Radar mapping** bounces radio waves off the surface and uses the reflections to create three-dimensional images of the surface. From 1990 to 1993, the *Magellan* spacecraft used radar to map the surface of Venus, discerning features as small as 100 meters across. Scientists have named almost all the geological features on Venus for goddesses and famous women.

Figure 9.33a shows a global map of Venus based on the *Magellan* radar observations. Three large, elevated "continents" (the three regions labeled "Terra" in Figure 9.33a) are the biggest features on the surface.

Venus has some impact craters (Figure 9.33b), but far fewer than Mercury, the Moon, or Mars. Moreover, the relatively few impact craters are distributed fairly uniformly over the entire planet, suggesting that the surface is about the same age everywhere. Precise crater counts suggest that the surface of Venus is about 750 million years old. Thus, no craters from the heavy bombardment remain, and the only craters we see on Venus are those created by impacts during the past 750 million years. Older craters must have been erased by other geological processes. Geologists have also noted an absence of small craters on Venus—the very type of crater that is most common on other worlds. Apparently, Venus lacks small craters because small impactors completely burn up as they enter Venus's thick atmosphere. Other terrestrial worlds have smaller craters (as well as large ones), because their thinner atmospheres allow smaller objects to survive the plunge to the surface.

Volcanism is clearly important on Venus. The surface shows abundant evidence of lava plains and many volcanic mountains. Some mountains are shield volcanoes (Figure 9.33c), indicating eruptions in which the lava was about as runny as the lava that formed the Hawaiian Islands on Earth. A few volcanoes have steeper sides, indicating eruptions of a higher-viscosity lava (Figure 9.33d). Volcanoes almost undoubtedly remain active on Venus, though we have not observed any eruptions. However, the composition of Venus's clouds suggests that volcanoes must still be active on geological timescales (erupting within the past 100 million years). The clouds contain sulfuric acid, which is made from sulfur dioxide (SO_2) and water. Sulfur dioxide enters the atmosphere through volcanic outgassing, but once in the atmosphere it is steadily removed by chemical reactions with surface rocks. Thus, the fact that sulfuric acid clouds still exist means that outgassing must continue to supply sulfur dioxide to the atmosphere.

The most remarkable features on Venus are tectonic in origin. Its crust is quite contorted. In some regions, the surface appears to be fractured in a regular pattern (Figure 9.33e). Other tectonic features, called *coronae* (Figure 9.33f), point directly to mantle convection on Venus. *Coronae* (Latin for "crowns") were probably formed by hot, rising plumes in the mantle. The plumes push up on the crust, forming concentric tectonic stretch marks on the surface. The plumes also force lava to the surface, explaining why numerous volcanoes are found near coronae.

The biggest difference between the geology of Venus and that of Earth is the lack of erosion on Venus. We might

c *These two volcanic peaks are probably much like the shield volcanoes that make up the Hawaiian Islands on Earth.*

100 km

Ishtar Terra

d *The round blobs are steep stratovolcanoes, apparently built from a "thick" lava.*

10 km

10 km

Aphrodite Terra

e *Tectonic forces have fractured and twisted the crust.*

Lada Terra

a *This central image shows the full surface of Venus, 98% of which was mapped by Magellan. Notice the three large, elevated "continents" called Ishtar Terra, Lada Terra, and Aphrodite Terra.*

10 km

b *Impact craters, like this one, are relatively rare on Venus and are distributed uniformly over the surface.*

50 km

f *This round corona was probably made by a mantle plume. It is dotted with small volcanoes (the round dots) and surrounded by tectonic stress marks.*

Figure 9.33 These images show the surface of Venus as revealed by radar observations from the *Magellan* spacecraft. Bright regions in the radar images represent rough areas or higher altitudes.

RADIO

naively expect Venus's thick atmosphere to produce strong erosion, but the view both from orbit and on the surface suggests that erosion is only a minor process. The then–Soviet Union landed two probes on Venus's surface in 1975. Before the intense surface heat destroyed them, the probes returned images of a bleak, volcanic landscape with little evidence of erosion (Figure 9.34). We can trace the lack of erosion on Venus to two simple facts. First, Venus is far too hot for any type of rain or snow on its surface. Second, Venus has virtually no wind or weather because of its slow rotation. Venus rotates so slowly—once every 243 days—that its atmosphere barely stirs the surface.

THINK ABOUT IT

Suppose Venus rotated as fast as Earth. How would this change its relative levels of volcanism, tectonics, and erosion? Explain.

• Does Venus have plate tectonics?

While we can easily explain the lack of erosion on Venus, another "missing feature" in Venus's geology is very surprising: Venus shows no evidence at all of Earth-like plate tectonics. Almost all the major features of Earth's geology are attributable to plate tectonics, including the ridges that

extend thousands of kilometers along the seafloor, the deep ocean trenches found near continental boundaries, and long mountain ranges like the Rockies and Himalayas. Venus does not have any similar features.

VIS

10 cm

Figure 9.34 This photo from the former Soviet Union's *Venera* lander shows the surface around the base of the spacecraft (visible at lower right). This region of the planet is covered in volcanic plains formed some 750 million years ago. Many volcanic rocks are visible, hardly affected by erosion despite their age.

Moreover, Venus shows evidence of a very different type of global geological change. On Earth, plate tectonics resculpts the surface gradually, so that different regions have different ages. But, as we've already discussed, the uniform distribution of impact craters on Venus tells us that the entire surface is about 750 million years old. Apparently, Venus's surface was somehow "repaved" at that time.

We do not know how much of the repaving was due to tectonic processes and how much was due to volcanism, but both probably were important. It is even possible that plate tectonics played a role before and during the repaving, only to stop for some reason after the repaving episode was over. Regardless of whether Venus may have had plate tectonics in the past, its lack of plate tectonics today poses a major mystery.

Earth's lithosphere fractured into plates because of forces arising from the underlying mantle convection. The lack of plate tectonics on Venus suggests either that it has weaker mantle convection or that its lithosphere somehow resists fracturing. The first possibility seems unlikely: Because Venus is so similar in size to Earth, it should have a similar level of mantle convection. Most scientists therefore suspect that Venus's lithosphere resists fracturing into plates because it is thicker and stronger than Earth's lithosphere, though we have no direct evidence to support this hypothesis.

Even if a thicker and stronger lithosphere explains the lack of plate tectonics on Venus, we are still left with the question of why the lithospheres of Venus and Earth should differ. One possible answer is Venus's high surface temperature. Venus is so hot that any water in its crust and mantle has probably been baked out over time. Water tends to soften and lubricate rock, so its loss would have tended to thicken and strengthen Venus's lithosphere. If this idea is correct, then Venus might have had plate tectonics if it had not become so hot in the first place.

The mystery of plate tectonics is just one of many reasons why scientists would like to explore Venus in greater detail. Unfortunately, it is probably the most difficult to study of the terrestrial worlds. Its thick cloud cover always hides the surface from orbiting cameras, so only improved radar mapping will tell us more about its global geology. The harsh surface conditions make it difficult for robotic landers to survive, and sending astronauts to Venus seems almost completely out of the question. Still, we are hoping to learn more from a European mission called *Venus Express*; if all goes well, it will reach Venus in mid-2006.

THINK ABOUT IT

What is the current status of the *Venus Express* mission? If it has reached Venus, has it made any important new discoveries?

9.6 The Unique Geology of Earth

Most aspects of Earth's geology make sense in light of what we know about other terrestrial worlds. Earth's size— largest of the five terrestrial worlds—explains why our

planet retains internal heat and hence remains volcanically and tectonically active. Earth's rampant erosion by water and wind is explained by the combination of our planet's size, distance from the Sun, and rotation rate: Earth's size leads to active volcanism and hence outgassing, its distance from the Sun keeps temperatures in a range in which outgassed water can be liquid, and its moderately rapid rotation drives weather that produces erosion.

The only feature of our planet's geology that stands clearly apart from the geology we see on other worlds is its unique style of tectonics. As we've seen, even similar-size Venus lacks Earth-like plate tectonics. Moreover, the importance of plate tectonics goes far beyond basic geology; as we'll discuss in Chapter 10, plate tectonics plays a crucial role in the stability of Earth's climate. Indeed, our own existence represents the culmination of a 4-billion-year chain of biological evolution that probably could not have occurred without the climate stability that we owe to plate tectonics.

• How do we know that Earth's surface is in motion?

The term *plate tectonics* refers to the scientific theory that explains much of Earth's surface geology as a result of the slow motion of *plates*—fractured pieces of the lithosphere. According to the theory, the lithosphere fractured because of stresses generated by underlying mantle convection. The plates essentially "float" over the mantle, gradually moving over, under, and around each other as convection moves Earth's interior rock. (Because the term *plate tectonics* refers to the theory, it is generally considered to be singular rather than plural, despite the -*s* on *tectonics*.)

Earth's lithosphere is broken into more than a dozen plates, with known boundaries shown in Figure 9.35. The motions of the plates are barely noticeable on human timescales—a few centimeters per year, which is about the rate at which your fingernails grow. However, geologists can now measure plate motions by comparing readings taken with the global positioning system (GPS) [Section S1.3] on either side of plate boundaries.

The GPS measurements offer the most direct evidence of plate motion. However, the overall theory of plate tectonics rests on three more significant lines of evidence: evidence of past continental motion, evidence that plates spread apart on seafloors, and a difference between the nature of Earth's crust on the seafloors and the continents. Let's investigate each of these.

Continental Motion   We usually trace the origin of the theory of plate tectonics to a slightly different idea proposed in 1912 by German meteorologist and geologist Alfred Wegener: *continental drift*, the idea that continents gradually drift across the surface of the Earth. Wegener got his idea in part from the puzzlelike fit of continents such as South America and Africa (Figure 9.36). He also noted that similar types of distinctive rocks and rare fossils were found in eastern

Figure 9.35 This relief map shows plate boundaries (solid yellow lines), with arrows to represent directions of plate motion. Color represents elevation, progressing from blue (lowest) to red (highest). Labels identify some of the geological features discussed in the text.

South America and western Africa, suggesting that these two regions once had been close together.

Despite these strong hints, no one at the time knew of a mechanism that could allow the continents to push their way through the solid rock beneath them. Wegener suggested that Earth's gravity and tidal forces from the Sun and Moon were responsible, but other scientists quickly showed that these forces were too weak to move continents around. As a result, Wegener's idea of continental drift was widely rejected by geologists for decades after he proposed it, even though his idea of a "continental fit" for Africa and South America ultimately gained acceptance. Today, far more extensive fossil evidence makes it very clear that the continents really were arranged differently in the past.

Seafloor Spreading   In the mid-1950s, scientists began to observe geological features that suggested a mechanism for continental motion. In particular, they discovered *mid-ocean ridges* (such as the mid-Atlantic Ridge shown in Figure 9.35) along which mantle material erupts onto the ocean floor, pushing apart the existing seafloor on either side. This **seafloor spreading** helped explain how the continents could move apart with time.

Today, we understand that continents move in concert with underlying mantle convection, driven by the heat released from Earth's interior. Because this idea is quite different from Wegener's original notion of continents plowing through the solid rock beneath them, geologists no longer use the term *continental drift* and instead consider continental motion within the context of plate tectonics.

Seafloor Crust and Continental Crust   The third line of evidence for plate tectonics comes from the fact that Earth's surface has two very different types of crust: one type found on seafloors and the other on continents (Figure 9.37). **Seafloor crust** is thinner, denser, and younger in age than **continental crust**. No such compositional difference has been found on other planets.

Seafloor crust forms a layer over the mantle that is typically only 5–10 kilometers thick. It is made primarily of basalt; recall that basalt is a mixture of many different minerals that tends to have relatively high density but is fairly runny when melted. The runniness explains why molten basalt spreads outward from volcanoes along mid-ocean ridges. Radiometric dating shows that seafloor basalt is quite young—usually less than 200 million years old—indicating that it erupted to the surface relatively recently in geological history. Further evidence of the young age of seafloors comes from studies of impact craters. Large impacts should occur more or less uniformly over Earth's surface, and the oceans won't stop a large asteroid or comet from making a huge crater. However, we find far

Figure 9.36 The puzzle-like fit of South America and Africa.

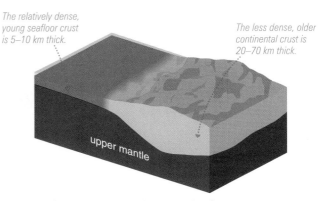

The relatively dense, young seafloor crust is 5–10 km thick.

The less dense, older continental crust is 20–70 km thick.

Figure 9.37 Earth has two distinct kinds of crust.

fewer large craters on the seafloor than on the continents, which means that seafloor crust must have been created more recently.

Continental crust forms a much thicker layer over the mantle than seafloor crust. It is typically between 20 and 70 kilometers thick, although it sticks up only slightly higher than seafloor crust because its sheer weight presses it down farther into the mantle below. Most of the rock found in continental crust is lower in density than the basaltic rock of the seafloor; for example, the most common type of rock on the continents, granite, is significantly less dense than basalt. Moreover, while seafloor crust is uniformly young, continental crust spans a wide range of ages. Some rocks in the continental crust date to as old as 4.0 billion years.

The two types of crust make it clear that Earth's surface must undergo continual change. New seafloor crust continually emerges at sites of seafloor spreading, while the wide age range of continental crust tells us that the continents have gradually built up over time.

• How is Earth's surface shaped by plate tectonics?

The theory of plate tectonics explains all three lines of evidence (continental motion, seafloor spreading, and two types of crust) as direct results of the way plates move about on Earth. Over millions of years, the movements involved in plate tectonics act like a giant conveyor belt for Earth's lithosphere (Figure 9.38).

Recycling of the Seafloor   Mid-ocean ridges occur at places where mantle material rises upward, creating new seafloor crust and pushing plates apart. The newly formed basaltic crust cools and contracts as it spreads sideways from the central ridge, giving seafloor spreading regions their characteristic ridged shape (see Figure 9.35). Worldwide along the mid-ocean ridges, new crust covers an area of about 2 square kilometers every year, enough to replace the entire

seafloor within about 200 million years—and thus explaining the less-than-200-million-year age of seafloor rocks.

Over tens of millions of years, any piece of seafloor crust gradually makes its way across the ocean bottom, then finally gets recycled into the mantle in the process we call **subduction**. Subduction occurs where a seafloor plate meets a continental plate, which is generally somewhat offshore at the edge of a sloping continental shelf. As the dense seafloor crust of one plate pushes under the less dense continental crust of another plate, it can pull the entire surface downward to form a deep *ocean trench*. At some trenches, the ocean depth is more than 8 kilometers (5 miles).

Beneath a subduction zone, the descending seafloor crust heats up and may begin to melt as it moves deeper into the mantle. If enough melting occurs, the molten rock may erupt upward. As you can see in Figure 9.38, the process of subduction tends to make the melting occur under the edges of the continents, which is why so many active volcanoes tend to be found along those edges. Moreover, the lowest-density material tends to melt first, which is why the continental crust emerging from these landlocked volcanoes is lower in density than seafloor crust. This fact explains why steep-sided stratovolcanoes, made from low-density but thick lava (see Figure 9.11c), are common on Earth's continents but rare everywhere else in the solar system: Other worlds lack anything like Earth's crustal recycling with plate tectonics and therefore generally have only volcanic plains and shield volcanoes made from basalt. (Venus may have a few stratovolcanoes, but far fewer than Earth.)

The conveyor-like process of plate tectonics undoubtedly is driven by the heat flow from mantle convection, although the precise relationship between the convection cells and the plates remains an active topic of research. For example, it is not clear whether mantle convection simply pushes plates apart at the sites of seafloor spreading or whether the plates are denser than the underlying material and pull themselves down at subduction zones.

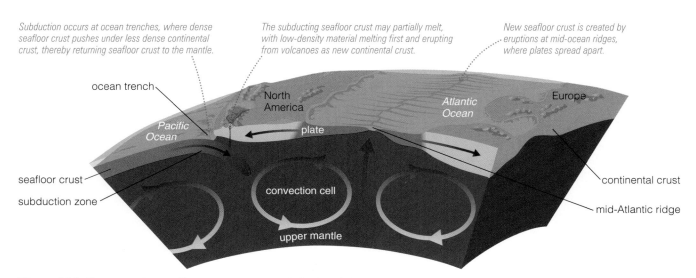

Subduction occurs at ocean trenches, where dense seafloor crust pushes under less dense continental crust, thereby returning seafloor crust to the mantle.

The subducting seafloor crust may partially melt, with low-density material melting first and erupting from volcanoes as new continental crust.

New seafloor crust is created by eruptions at mid-ocean ridges, where plates spread apart.

ocean trench
North America
plate
Atlantic Ocean
Europe
Pacific Ocean
seafloor crust
subduction zone
convection cell
upper mantle
continental crust
mid-Atlantic ridge

Figure 9.38 Plate tectonics acts like a giant conveyor belt for Earth's lithosphere.

Building Up Continents   Note that, unlike seafloor crust, continental crust does not get recycled back into the mantle. As a result, the continents are gradually growing with time. Billions of years ago, Earth was even more of an ocean planet than it is today.

The present-day continents have been built up over billions of years, but their histories are not just a simple "piling up" of continental crust. Instead, the continents are continually reshaped by volcanism and stresses associated with plate tectonics, as well as by erosion. Figure 9.39 shows some of the complex geological history of North America, which we can use as an example of how the continents have been built.

The west coast regions of Alaska, British Columbia, Washington, Oregon, and most of California began as numerous volcanic islands in the Pacific. Long ago, plate motion carried these islands from their ocean birthplaces toward the North American mainland, where subduction continues to take the seafloor crust downward into the mantle (see Figure 9.38). Because the islands are made of lower-density continental crust, they remained on the surface and attached themselves to the edge of North America, even as the seafloor crust beneath them slid back into the mantle. Similar processes are affecting other islands today. Alaska's Aleutian Islands, a string of volcanoes located over a region where one seafloor plate is subducting under another (see the plate boundaries in Figure 9.35), are gradually growing and will someday merge together. Japan and the Philippines represent a later stage in this process; each of them once contained many small islands that merged into the fewer islands we see today. As these islands continue to grow and merge, they may eventually create a new continent or merge with an existing one.

Other portions of North America have been shaped by erosion. The Great Plains and the Midwest once were ancient seas. Erosion gradually filled these seas with sediment that hardened into sedimentary rock. The deep south formed from the buildup of sediments that were carried to these regions after eroding from other parts of the continent. Northeastern Canada features some of the oldest continental crust on Earth, worn down over time by erosion.

Tectonic processes are responsible for the mountain ranges of North America and other continents. The Sierra Nevada range in California, which lies over the region where the Pacific plate subducts under North America, formed when partially molten granite rose up from the mantle and pushed the surface rock higher. The original surface rock was sedimentary, but erosion gradually wore it away and left the granite exposed, which is why the Sierra Nevada mountains are made largely of granite.

Other mountain ranges were created by collisions between two continent-bearing plates. The Himalayas, already the tallest mountains on Earth, offer the most famous example of this process (Figure 9.40). They are still growing taller as the plate carrying the Indian subcontinent rams into the plate carrying Eurasia. Because both colliding plates are made of low-density continental crust, neither plate can subduct under the other. Instead, they push into each other and the resulting compression pushes the land upward to make mountains.

The Appalachian range in the eastern United States was also built by colliding continental plates, but in this case the collisions occurred several times. Over a period of a few hundred million years, North America apparently collided twice with South America and then with western Africa.

volcanic and tectonic mountains made over subduction zones

sediments deposited in ancient seas, later turned to rock

ancient continental crust, heavily eroded

Figure 9.39 The major geological features of North America record the complex history of plate tectonics. Only the basic processes behind the largest features are shown here.

former Pacific islands scraped onto continent during subduction

deep sedimentary layers formed by erosion of continent

mountains formed by repeated collisions with other continents

Figure 9.40 This satellite photo shows the Himalayas, which are still slowly growing as the plate carrying India pushes into the Eurasian plate. Arrows indicate the directions of plate motion.

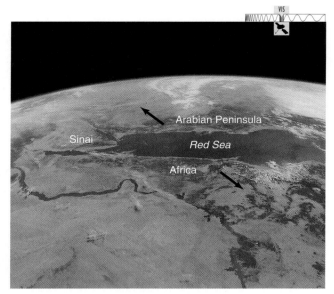

Figure 9.41 When continental plates pull apart, the crust thins and deep rift valleys form. This process tore the Arabian Peninsula from Africa, forming the Red Sea. Arrows indicate the directions of plate motion.

The Appalachians probably were once as tall as the Himalayas are now, but tens of millions of years of erosion transformed them into the fairly modest mountain range we see today. Similar processes contributed to the formation of the Rocky Mountains in the United States and Canada.

Rifts, Faults, and Earthquakes So far we have discussed features formed in places where plates are colliding, either with a seafloor plate subducting under a continental plate or with two continental plates pushing into each other. Different types of features occur in places where plates pull apart or slide sideways relative to one another.

In places where continental plates are pulling apart, the crust thins and can create a large *rift valley*. The East African rift zone is an example (see Figure 9.35). This rift is slowly growing and will eventually tear the African continent apart. At that point, rock rising upward with mantle convection will begin to erupt from the valley floor, creating a new zone of seafloor spreading. A similar process tore the Arabian Peninsula from Africa, creating the Red Sea (Figure 9.41).

Places where plates slip sideways relative to each other are marked by what we call **faults**—fractures in the lithosphere. The San Andreas Fault in California marks a line where the Pacific plate is moving northward relative to the continental plate of North America (Figure 9.42). In about 20 million years, this motion will bring Los Angeles and San Francisco together. The two plates do not slip smoothly against each other. Instead, their rough surfaces catch against each other. Stress builds up until it is so great that it forces a rapid and violent shift, causing an earthquake. In contrast to the usual motion of plates, which proceeds at a few centimeters per year, an earthquake can move plates by several *meters* in a few seconds. The movement can raise mountains, level cities, set off destructive tsunamis, and make the whole planet vibrate with seismic waves (Figure 9.43).

THINK ABOUT IT

By studying the plate boundaries in Figure 9.35, explain why the west coast states of California, Oregon, and Washington are prone to more earthquakes and volcanoes than other parts of the United States. Find the locations of recent earthquakes and volcanic eruptions worldwide. Do the locations fit the pattern you expect?

Hot Spots Not all volcanoes occur near plate boundaries. Sometimes, a plume of hot mantle material may rise in what we call a **hot spot**. The Hawaiian Islands are the result

Figure 9.42 California's San Andreas Fault marks a boundary where plates are sliding sideways as shown by the white arrows; asterisks indicate recent earthquakes. The inset photo shows a place along the San Andreas Fault where the painted lines in a road allows us to see how far the two sides of the fault moved in an earthquake.

Figure 9.43 These photos show the ancient Iranian city of Bam before and after a major earthquake in 2004. The quake killed more than 40,000 people.

of a hot spot that has been erupting basaltic lava for tens of millions of years. Plate tectonics gradually carries the Pacific plate over the hot spot, forming a chain of volcanic islands as different parts of the plate lie directly above the hot spot at different times (Figure 9.44). Today, most of the lava erupts on the Big Island of Hawaii, giving much of this island a young, rocky surface. About a million years ago, the Pacific plate lay farther to the southeast (relative to its current location), and the hot spot built the island of Maui. Before that, the hot spot created other islands, including Oahu (3 million years ago), Kauai (5 million years ago), and Midway (27 million years ago). The older islands are more heavily eroded. Midway has eroded so much that it barely rises above sea level. If plate tectonics were not moving the plate relative to the hot spot, a single huge volcano would have formed—perhaps looking somewhat like Olympus Mons on Mars (see Figure 9.26).

The movement of the plate over the hot spot continues today, building underwater volcanoes that eventually will rise above sea level to become new Hawaiian islands. The growth of a future island, named Loihi, is already well under way—prime beach real estate should be available there in about a million years or so. Hot spots can also occur beneath continental crust. For example, a hot spot is responsible for the geysers and hot springs of Yellowstone National Park. The continental plate is moving in a southwestern direction over this hot spot.

THINK ABOUT IT

Find your hometown or current home on the map in Figure 9.35 or Figure 9.39 (or both). Based on what you have learned about Earth's geology, describe the changes your home has undergone over millions (or billions) of years.

Plate Tectonics Through Time We can use the current motions of the plates to project the arrangement of continents millions of years into the past or the future. For example,

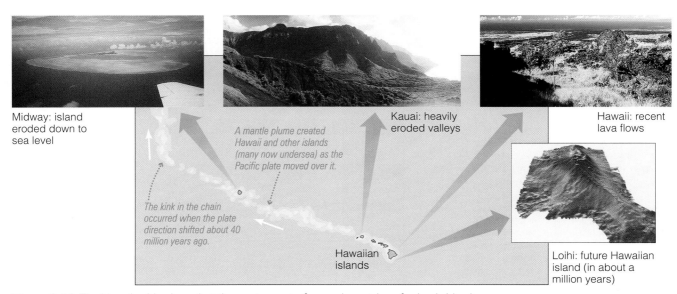

Midway: island eroded down to sea level

A mantle plume created Hawaii and other islands (many now undersea) as the Pacific plate moved over it.

The kink in the chain occurred when the plate direction shifted about 40 million years ago.

Hawaiian islands

Kauai: heavily eroded valleys

Hawaii: recent lava flows

Loihi: future Hawaiian island (in about a million years)

Figure 9.44 The Hawaiian Islands are just the most recent of a very long string of volcanic islands made by a mantle hot spot. The image of Loihi (lower right) was obtained by sonar, as it is still entirely underwater.

at a speed of 2 centimeters per year, a plate will travel 2,000 kilometers in 100 million years. Figure 9.45 shows several past arrangements of the continents, along with one future arrangement. Note that about 200 million years ago, the present-day continents were together in a single "supercontinent," sometimes called *Pangaea* (which means "all lands").

Mapping the sizes and locations of continents at even earlier times is more difficult. However, studies of magnetized rocks (which can record the orientation of ancient magnetic fields) and comparisons of fossils found in different places around the world have allowed geologists to map the movement of the continents much further into the past. It seems that, over the past billion years or more, the continents have slammed together, pulled apart, spun around, and changed places on the globe. Central Africa once lay at Earth's South Pole, and Antarctica once was near the equator. The continents continue to move, and their current arrangement is no more permanent than any past arrangement.

• Was Earth's geology destined from birth?

Now that we have completed our geological tour of the terrestrial worlds, it's interesting to ask whether fundamental planetary properties shape all geological destiny. If so, then we should be able to predict the geological features of terrestrial worlds we may someday find around other stars. If not, then we still have more to learn before we'll know whether other worlds could have Earth-like geology.

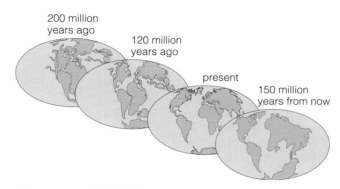

Figure 9.45 Interactive Figure. Past, present, and future arrangements of Earth's continents. The present continents were all combined into a single "supercontinent" about 200 million years ago.

Figure 9.46 summarizes the key trends we've seen among the terrestrial worlds with respect to impact cratering, volcanism, and tectonics. All the worlds were affected similarly by impact cratering during the heavy bombardment. Volcanism and tectonics have operated to very different extents on the different worlds, but we can trace the major differences to differences in planetary size. The interiors of the smallest worlds cooled quickly, so they have not had volcanism or tectonics for billions of years. The interiors of the larger worlds have stayed hot longer, allowing volcanism and tectonics to continue for much longer time periods.

The major remaining question concerns whether fundamental properties can also explain Earth's unique plate tectonics. The lack of plate tectonics on Venus tells us that

Figure 9.46 This diagram summarizes the geological histories of the terrestrial worlds. The brackets along the top indicate that impact cratering has affected all worlds similarly. The arrows represent volcanic and tectonic activity. A thicker and darker arrow means more volcanic/tectonic activity, and the arrow length tells us how long this activity persisted. Notice that the trend follows the order of planetary size: Earth remains active to this day; Venus has also been quite active, though we are uncertain whether it remains so; Mars has had an intermediate level of activity and might still have low-level volcanism; Mercury and the Moon have had very little volcanic/tectonic activity. Erosion is not shown, because it has played a significant role only on Earth (ongoing) and on Mars (where it was quite significant in the past and continues at low levels today).

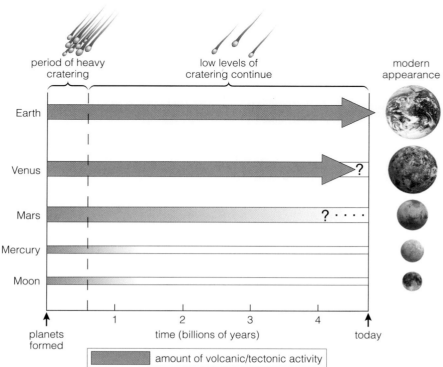

we cannot attribute plate tectonics solely to size, since Venus is so similar in size to Earth. However, if we are correct in guessing that Venus lacks plate tectonics because its high surface temperature led to loss of water and lithospheric thickening, then a lower surface temperature might have led to plate tectonics on Venus as well as Earth. As we'll discuss in detail in the next chapter, we can ultimately trace Venus's high surface temperature to its nearness to the Sun. Thus, while plenty of uncertainty still remains, it seems quite likely that the broad geological histories of Earth and the other terrestrial worlds were indeed destined from birth by the fundamental properties of size, distance from the Sun, and rate of rotation.

THE BIG PICTURE

Putting Chapter 9 into Perspective

In this chapter, we have explored the geology of the terrestrial worlds. As you continue your study of the solar system, keep the following "big picture" ideas in mind:

- The terrestrial worlds all looked much the same when they were born, so their present-day differences are a result of geological processes that occurred in the ensuing 4.5 billion years.

- The extent to which different geological processes operate on the different worlds depends largely on their fundamental properties, especially their size.

- A planet's geology is largely destined from its birth— which means we should be able to predict the geology of as-yet-undiscovered planets once we know their fundamental properties.

- Earth has been affected by the same geological processes affecting the other terrestrial worlds. However, erosion is far more important on Earth than on any other terrestrial world, and Earth's unique plate tectonics may be very important to our existence.

SUMMARY OF KEY CONCEPTS

9.1 Connecting Planetary Interiors and Surfaces

- **What are terrestrial planets like on the inside?** In order of decreasing density and depth, the interior structure consists of **core, mantle,** and **crust.** The crust and part of the mantle together make up the rigid **lithosphere.** In general, a thinner lithosphere allows more geological activity.

- **What causes geological activity?** Interior heat drives geological activity, by causing mantle convection, keeping the lithosphere thin, and keeping the interior partially molten. **Radioactive decay** is the main source of internal heat in terrestrial planets today. Larger planets retain internal heat longer.

- **Why do some planetary interiors create magnetic fields?** A planetary **magnetic field** requires three things: an interior layer of electrically conducting fluid, convection of that fluid, and rapid rotation. Among the terrestrial planets, only Earth has all three characteristics.

9.2 Shaping Planetary Surfaces

- **What processes shape planetary surfaces?** The four major geological processes are **impact cratering, volcanism, tectonics,** and **erosion.**

- **Why do the terrestrial planets have different geological histories?** Fundamental planetary properties, especially size, determine a planet's geological history. Larger planets have more volcanism and tectonics, while erosion depends on a planet's size, distance from the Sun, and rotation rate.

- **How does a planet's surface reveal its geological age?** Surfaces crowded with impact craters are ancient, implying that no volcanism, tectonics, or erosion has occurred recently. "Younger" surfaces with few craters have undergone some volcanic, tectonic, and/or erosional processes. Surfaces with no craters are very young, and must have ongoing geological activity.

9.3 Geology of the Moon and Mercury

- **What geological processes shaped our Moon?** The lunar surface is a combination of ancient, heavily cratered terrain and vast lava plains called the lunar **maria.** Some small tectonic features are also present. The Moon lacks erosion because of its lack of atmosphere.

- **What geological processes shaped Mercury?** Mercury resembles the Moon in being dominated by impact cratering and volcanism. It also has tremendous tectonic cliffs that probably formed when the whole planet cooled and contracted in size.

9.4 Geology of Mars

- **How did Martians invade popular culture?**
Superficial similarities between the planets led to
speculation about civilized Martians. Astronomer Percival
Lowell thought he saw canals built by an advanced
civilization, but the canals do not really exist.

- **What are the major geological features of
Mars?** Mars shows evidence of all four geological

 processes in its past. It has
the tallest volcano and the
biggest canyon in the solar
system, evidence of a period
of great volcanic and tectonic
activity. It also has abundant
craters and evidence of
erosion by wind and flowing
water.

- **What geological evidence tells us that water
once flowed on Mars?** Images of dry river channels

 and eroded craters, along with
chemical analysis of Martian
rocks, all show that water once
flowed on Mars. Any periods
of rainfall seem to have ended
at least 2–3 billion years ago.
Mars today still has water ice
underground and in its polar
caps, and could possibly have
pockets of underground liquid
water.

9.5 Geology of Venus

- **What are the major geological features of
Venus?** Venus's surface shows evidence of major volcanic

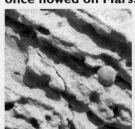 and tectonic activity in the
past billion years, as expected
for a planet nearly as large as
Earth. This activity explains
the relative lack of craters.
Erosion is only a minor
process despite the thick
atmosphere, because of the
high temperature (no rainfall)

and slow rotation (little wind). Venus almost certainly
remains geologically active today.

- **Does Venus have plate tectonics?** Venus may
undergo occasional planetwide resurfacing, but we do not
see any evidence of plate tectonics of the type we see on
Earth. The lack of plate tectonics may mean that Venus
has a stiffer and stronger lithosphere than Earth, perhaps
because the high surface temperature has baked out any
water that might have softened the lithospheric rock.

9.6 The Unique Geology of Earth

- **How do we know that Earth's surface is in
motion?** The evidence for **plate tectonics** includes

 the way continents "fit"
together, which tells us that
they were close together in
the past; the creation of new
crust from seafloor spreading;
and the compositional
differences between seafloor
and continental crust. Today,
we can measure plate motion
directly with GPS technology.

- **How is Earth's surface shaped by plate
tectonics?** Plate tectonics has led to many unique

 features of our geology,
especially **seafloor spreading**
zones and the building of
continents along **subduction
zones**. The shifting of plates has completely changed Earth's
geological appearance many times in the past billion years
alone.

- **Was Earth's geology destined from birth?** It
seems likely that the geological histories of Earth and the
other terrestrial worlds were indeed destined from birth by
the fundamental properties of size, distance from the Sun,
and rate of rotation.

EXERCISES AND PROBLEMS

For instructor-assigned homework go to **Mastering**ASTRONOMY www.masteringastronomy.com

Review Questions

Short-Answer Questions Based on the Reading

1. What do we mean by *planetary geology?* Briefly summarize
the different geological appearances of the five terrestrial
worlds.

2. What is differentiation, and how did it lead to the core-
mantle-crust structures of the terrestrial worlds?

3. What do we mean by the *lithosphere,* and why isn't it listed
as one of the three layers by composition? How does lithospheric thickness vary among the five terrestrial worlds?

4. Summarize the processes by which planetary interiors get hot and cool off. Why do large planets retain internal heat longer than smaller planets?

5. Why does Earth have a global *magnetic field?* Why don't the other terrestrial worlds have similarly strong magnetic fields?

6. Define each of the four major geological processes, and give examples of features shaped by each process.

7. What is *outgassing,* and why is it so important to our existence?

8. How do crater counts tell us the age of a planetary surface? Briefly explain why the Moon is so much more heavily cratered than Earth.

9. Summarize the ways in which a terrestrial world's size, distance from the Sun, and rotation rate each affect its relative level of impact cratering, volcanism, tectonics, and erosion.

10. Briefly summarize the geological history of the Moon. Be sure to explain the origin of the lunar *maria.*

11. Briefly summarize the geological history of Mercury. Be sure to explain how Mercury's great cliffs are thought to have formed.

12. Choose five features visible on the global map of Mars (Figure 9.25) and explain what we think each feature is and how it formed.

13. Explain why liquid water is not stable on Mars today, but why we nonetheless think it flowed in the distant past on Mars. Could there still be liquid water anywhere on Mars today? Explain.

14. Choose at least three major geological features of Venus and explain how we think each one formed.

15. What evidence tells us that Venus was "repaved" about 750 million years ago? What difference between the lithospheres of Venus and Earth might explain the lack of plate tectonics on Venus?

16. Describe the conveyor-like action of plate tectonics on Earth. How does this action explain the differences between seafloor crust and continental crust?

17. Briefly explain how each of the following geological features of Earth are formed: seafloors, continents, islands, mountain ranges, rift valleys, and faults.

18. To what extent do we think the geologies of the terrestrial worlds were destined from their births?

Test Your Understanding

Surprising Discoveries?

Suppose we were to make the following discoveries. (These are *not* real discoveries.) In light of your understanding of planetary geology, decide whether the discovery should be considered reasonable or surprising. Explain your answer, if possible tracing your logic back to the terrestrial world's fundamental properties of size, distance from the Sun, and rotation rate. (In some cases, both views can be defended.)

19. The next mission to Mercury photographs part of the surface never seen before and detects vast fields of sand dunes.

20. Seismographs placed on the surface of Mercury record frequent and violent earthquakes.

21. A future orbiter observes a volcanic eruption on Venus.

22. A Venus radar mapper discovers extensive regions of layered sedimentary rocks, similar to those found on Earth.

23. Radiometric dating of rocks brought back from one lunar crater shows that the crater was formed only a few tens of millions of years ago.

24. New orbital photographs of craters on Mars that have gullies also show pools of liquid water to be common on the crater bottoms.

25. Drilling into the Martian surface, a robotic spacecraft discovers liquid water deep beneath the slopes of a Martian volcano.

26. Clear-cutting in the Amazon rain forest on Earth exposes vast regions of ancient terrain that is as heavily cratered as the lunar highlands.

27. Seismic studies on Earth reveal a "lost continent" that held great human cities just a few thousand years ago but that is now buried deep underground off the western coast of Europe.

28. We find a planet in another solar system that orbits at the same distance from a Sun-like star as Earth orbits the Sun and has Earth-like plate tectonics, but it is only the size of the Moon.

29. We find a planet in another solar system that is as large as Earth but as heavily cratered as the Moon.

30. We find a planet in another solar system with Earth-like seafloor crust and continental crust but that apparently lacks plate tectonics or any other kind of crustal motion.

Quick Quiz

Choose the best answer to each of the following. Explain your reasoning with one or more complete sentences.

31. What is the most important long-lasting internal heat source responsible for geological activity? (a) accretion (b) radioactive decay (c) sunlight

32. In general, what kind of planet would you expect to have the *thickest* lithosphere? (a) the largest planet (b) the smallest planet (c) the planet farthest from the Sun

33. Which of a planet's fundamental properties has the greatest effect on its level of volcanic and tectonic activity? (a) size (b) distance from the Sun (c) rotation rate

34. What is the name of the outer rigid layer of a planet? (a) crust (b) mantle (c) lithosphere

35. Which describes our understanding of flowing water on Mars? (a) It was never important. (b) It was important once, but no longer. (c) It is a major process on the Martian surface today.

36. What do we conclude if a planet has few impact craters of any size? (a) The planet was never bombarded by asteroids or comets. (b) Its atmosphere stopped impactors of all sizes. (c) Other geological processes have wiped out craters.

37. How many of the five terrestrial worlds have surfaces being constantly reshaped by plate tectonics? (a) one (b) two (c) three or more (Be sure that you explain *why* plate tectonics occurs on this many worlds and not more.)

38. On how many of the five terrestrial worlds has erosion been an important process? (a) one (b) two (c) three or more (Be sure that you explain *why* erosion is important on this many worlds and not more.)

39. How many of the terrestrial worlds have lava plains or shield volcanoes? (a) one (b) three (c) five (Be sure to explain *why* and when these volcanic features formed.)

40. How many of the five terrestrial worlds are considered "geologically dead"? (a) none (b) two (c) four (Be sure to explain *why* these worlds became geologically dead.)

Investigate Further
In-Depth Questions to Increase Your Understanding

Short-Answer/Essay Questions

41. *Dating Planetary Surfaces.* We have discussed two basic techniques for determining the age of a planetary surface: studying the abundance of impact craters and radiometric dating of surface rocks. Which technique seems more reliable? Which technique is more practical? Explain.

42. *Comparative Erosion.* Consider erosion on Mercury, Venus, the Moon, and Mars. Which of these four worlds has the greatest erosion? Why? For each world, write a paragraph explaining its level of erosion, tracing it back to fundamental properties.

43. *Miniature Mars.* Suppose Mars had turned out to be significantly smaller than its current size—say, the size of our Moon. How would this have affected the number of geological features due to each of the four major geological processes? Do you think Mars would be a better or worse candidate for harboring extraterrestrial life? Summarize your answers in two or three paragraphs.

44. *Civilization on Mars.* Based on what we can see on the surface of Mars, does it seem possible that Mars once had a civilization with cities on the surface but that the evidence has now been erased or buried underground? Explain.

45. *Change in Fundamental Properties.* Consider Earth's size and distance from the Sun. Choose one property and suppose that it had been different (for example, smaller size or greater distance). Describe how this change might have affected Earth's subsequent geological history and the possibility of our existence today on Earth.

46. *Predictive Geology.* Suppose another star system has a rocky terrestrial planet twice as large as Earth but at the same distance from its star (which is just like our Sun) and with a similar rotation rate. In one or two paragraphs, describe the type of geology you would expect it to have.

47. *Mystery Planet.* It's the year 2098, and you are designing a robotic mission to a newly discovered planet around a nearby star that is nearly identical to our Sun. The planet is as large in radius as Venus, rotates with the same daily period as Mars, and lies 1.2 AU from its star. Your spacecraft will orbit but not land on the planet.
    a. Some of your colleagues believe that the planet has no metallic core. How could you support or refute their hypothesis?
    b. Other colleagues suspect that the planet has no atmosphere, but the instruments designed to study the planet's atmosphere fail due to a software error. However, the spacecraft can still photograph geological features. How could you use the spacecraft's photos of geological features to determine whether a significant atmosphere is (or was) present on this planet?

48. *Experiment: Geological Properties of Silly Putty.* Roll room temperature Silly Putty into a ball and measure its diameter. Place the ball on a table and gently place one end of a heavy book on it. After 5 seconds, measure the height of the squashed ball. Repeat the experiment two more times, the first time warming the Silly Putty in hot water before you start and the second time cooling it in ice water before you start. How do the different temperatures affect the rate of "squashing"? How does the experiment relate to planetary geology? Explain.

49. *Experiment: Planetary Cooling in a Freezer.* To simulate the cooling of planetary bodies of different sizes, use a freezer and two small plastic containers of similar shape but different sizes. Fill each container with cold water and put both into the freezer at the same time. Checking every hour or so, record the time and your estimate of the thickness of the "lithosphere" (the frozen layer) in the two tubs. How long does it take the water in each tub to freeze completely? Describe in a few sentences the relevance of your experiment to planetary geology. Extra credit: Plot your results on a graph with time on the *x*-axis and lithospheric thickness on the *y*-axis. What is the ratio of the two freezing times?

50. *Amateur Astronomy: Observing the Moon.* Any amateur telescope has resolution adequate to identify geological features on the Moon. The light highlands and dark maria should be evident, and shadowing is visible near the line between night and day. Try to observe the Moon near the first- or third-quarter phase. Sketch or photograph the Moon at low magnification, and then zoom in on a region of interest. Again sketch or photograph your field of view, label its features, and identify the geological process that created them. Look for craters, volcanic plains, and tectonic features. Estimate the size of each feature by comparing it to the size of the whole Moon (radius = 1,738 km).

Quantitative Problems
Be sure to show all calculations clearly and state your final answers in complete sentences.

51. *Moon and Mars.* Compare the surface area-to-volume ratios of the Moon and Mars. Based on your answer, what should you expect about interior heat on Mars compared to that on the Moon?

52. *Earth and Venus.* Compare the surface area-to-volume ratios of Venus and Earth. Based on your answer, how should you expect the amount of interior heat to compare on these two worlds?

53. *Doubling Your Size.* Just as the surface area-to-volume ratio depends on size, so can other properties. To see how, suppose that your size suddenly doubled; that is, your height, width, and depth all doubled. (For example, if you were 5 feet tall before, you now are 10 feet tall.)
    a. By what factor has your waist size increased?
    b. How much more material will be required for your clothes? (*Hint:* Clothes cover the surface *area* of your body.)
    c. By what factor has your weight increased? (*Hint:* Weight depends on the *volume* of your body.)
    d. The pressure on your weight-bearing joints depends on how much *weight* is supported by the *surface area* of each joint. How has the pressure on your weight-bearing joints changed?

54. *Lunar Footprints.* Micrometeorites will eventually erase the footprints left on the Moon by the *Apollo* astronauts. Assume that the Moon is hit by about 25 million micrometeorite impacts each day (this number comes from observations of meteors in Earth's atmosphere) and that these impacts strike randomly around the Moon's surface. Also assume that it will take about 20 such impacts to destroy a footprint. Using these assumptions, estimate how long it will take before a footprint is erased. (*Hints:* Use the Moon's surface area to determine the impact rate per square centimeter, and estimate the size of footprint.)

55. *Impact Energies.* An asteroid of 1-kilometer diameter will make a crater of about 10-kilometer diameter. How much kinetic energy does the asteroid have, if it strikes the surface at 20 km/sec? Convert your answer to megatons of TNT: 1 megaton is about 4×10^{15} joules. Compare the impactor to the largest nuclear weapons, currently less than 100 megatons of energy.

56. *Internal vs. External Heating.* In daylight, the Earth's surface absorbs about 400 watts per square meter. Earth's internal radioactivity produces a total of 3 trillion watts that leak out through our planet's entire surface. Calculate the amount of heat from radioactive decay that flows outward through each square meter of Earth's surface (your answer should have units of watts per square meter). Compare quantitatively to solar heating, and comment on why internal heating drives geological activity.

57. *Plate Tectonics.* Typical motions of one plate relative to another are 1 centimeter per year. At this rate, how long would it take for two continents 3,000 kilometers apart to collide? What are the global consequences of motions like this?

58. *More Plate Tectonics.* Consider a seafloor spreading zone creating 1 centimeter of new crust over its entire 2,000-kilometer length every year. How many square kilometers of surface will this create in 100 million years? What fraction of Earth's surface does this constitute?

59. *Planet Berth.* Imagine a planet, which we'll call *Berth*, orbiting a star identical to the Sun at a distance of 1 AU. Assume that Berth has eight times as much mass as Earth, and is twice as large as Earth in diameter.
   a. How does Berth's density compare to Earth's?
   b. How does Berth's surface area compare to Earth's?
   c. Based on your answers to (a) and (b), discuss how Berth's geological history is likely to have differed from Earth's.

Discussion Questions

60. *What Is Predictable?* We've found that much of a planet's geological history is destined from its birth. Briefly explain why, and discuss the level of detail that is predictable. For example, was Mars's general level of volcanism predictable? Could we have predicted a mountain as tall as Olympus Mons or a canyon as long as Valles Marineris? Explain.

61. *Worth the Effort?* Politicians often argue over whether planetary missions are worth the expense involved. Based on what we have learned by comparing the geologies of the terrestrial worlds, do you think the missions that have given us this knowledge have been worth their expense? Defend your opinion.

62. *Evidence of Our Civilization.* Discuss how the geological processes will affect the evidence of our current civilization in the distant future. For example, what evidence of our current civilization will survive in 100,000 years? in 100 million years? Do you think that future archaeologists or alien visitors will be able to know that we existed here on Earth?

MEDIA EXPLORATIONS

For self-study activities go to Mastering**ASTRONOMY** www.masteringastronomy.com

 Interactive Tutorials

Tutorial Review of Key Concepts
Use the following interactive **Tutorials** at **www.mastering astronomy.com** to review key concepts from this chapter.

Shaping Planetary Surfaces Tutorial
Lesson 1 The Four Geological Processes
Lesson 2 What Do Geological Processes Depend On?
Lesson 3 Planet Surface Evolution

Supplementary Tutorial Exercises
Use the interactive **Tutorial Lessons** to explore the following questions.

Shaping Planetary Surfaces Tutorial, Lesson 1
   1. Watch the impact cratering animation. How does the crater size depend on the size of the impactor?

2. Watch the tectonics animation. What happens in the mantle to cause compression and mountain building?
3. What happens in the mantle to cause tectonic stretching?

Shaping Planetary Surfaces Tutorial, Lesson 2
   1. How has the rate of crater formation in the solar system changed over time?
   2. Why do we see so many more impact craters on the Moon and Mercury than on Earth?
   3. Why is erosion so much more dominant on Earth than on any of the other terrestrial worlds?

Shaping Planetary Surfaces Tutorial, Lesson 3
   1. How has Earth's geology differed from that of the other terrestrial planets and why?
   2. Suppose Earth had been born at half its current size. How would you expect it to have been different? Would life still be possible? Explain.
   3. Suppose Earth had been born 10% larger than its current size. How would you expect it to have been different? Would life still be possible? Explain.

 Exploring the Sky and Solar System

Of the many activities available on the *Voyager: SkyGazer* CD-ROM accompanying your book, use the following files to observe key phenomena covered in this chapter.

Go to the **File: Demo** folder for the following demonstrations.
1. Earth and Venus
2. Venus-Earth-Moon

Go to the **Explore** menu for the following demonstrations.
1. Solar System
2. Paths of the Planets

 Movies

Check out the following narrated and animated short documentary available on **www.masteringastronomy.com** for a helpful review of key ideas covered in this chapter.

History of the Solar System

Web Projects

1. *"Coolest" Surface Photo.* Visit the Astronomy Picture of the Day Web site, and search for past images of the terrestrial worlds. Look at many of them, and choose the one you think is the "coolest." Make a printout, write a short description of what it shows, and explain what you like about it.

2. *Water on Mars.* Go to the home page for NASA's Mars Exploration Program, and look for the latest evidence concerning recent water flows on Mars. Write a few paragraphs describing the new evidence and what it tells us. How will future missions help resolve the questions?

3. *Planetary Geology.* Visit the Nine Planets Web site, which has summaries of features on each of the planets. Choose one of the terrestrial worlds, and learn about at least two geological features on it that we did not highlight in this chapter. Are the features consistent with what we expect based on the ideas of "geological destiny" discussed in this chapter? Explain.

4. *Volcanoes and Earthquakes.* Learn about one major earthquake or volcanic eruption that occurred during the past decade. Report on the geological conditions that led to the event, as well as on its geological and biological consequences.

10

Planetary Atmospheres
Earth and the Other Terrestrial Worlds

LEARNING GOALS

10.1 Atmospheric Basics
- What is an atmosphere?
- How does the greenhouse effect warm a planet?
- Why do atmospheric properties vary with altitude?

10.2 Weather and Climate
- What creates wind and weather?
- What factors can cause long-term climate change?
- How does a planet gain or lose atmospheric gases?

10.3 Atmospheres of the Moon and Mercury
- Do the Moon and Mercury have any atmosphere at all?

10.4 The Atmospheric History of Mars
- What is Mars like today?
- Why did Mars change?

10.5 The Atmospheric History of Venus
- What is Venus like today?
- How did Venus get so hot?

10.6 Earth's Unique Atmosphere
- How did Earth's atmosphere end up so different?
- Why does Earth's climate stay relatively stable?
- How might human activity change our planet?

For the first time in my life, I saw the horizon as a curved line. It was accentuated by a thin seam of dark blue light—our atmosphere. Obviously this was not the ocean of air I had been told it was so many times in my life. I was terrified by its fragile appearance.

Ulf Merbold, astronaut (Germany)

Life as we know it would be impossible on Earth without our atmosphere. This layer of gas surrounding the planet supplies the oxygen we breathe, shields us from harmful ultraviolet and X-ray radiation from the Sun, protects us from continual bombardment by micrometeorites, generates rain-giving clouds, and traps just enough heat to keep Earth habitable. We couldn't have designed a better atmosphere if we'd tried.

Despite forming under similar conditions, our neighboring planets have ended up with atmospheres utterly inhospitable to us. No other planet in our solar system has air that we could breathe or temperature and pressure conditions in which we could survive without a space suit. Mars, only slightly farther from the Sun than Earth, has an atmosphere that leaves its surface cold, dry, and barren. Venus, only slightly closer to the Sun, has an atmosphere that makes its surface conditions resemble a classical view of hell.

How did Earth end up with conditions so fortunate for life as we know it? In this chapter, we'll investigate the processes that govern planetary atmospheres and the reasons why the terrestrial worlds ended up so different. We'll also see how Earth's climate remains relatively stable, and how human activity may be threatening the very stability that we depend on.

10.1 Atmospheric Basics

All the planets of our solar system have an atmosphere to at least some degree, as do several of the solar system's larger moons. The jovian planets are essentially atmosphere throughout, because they are made largely of gaseous material. In contrast, atmospheric gases around solid bodies—including the terrestrial worlds, jovian moons with atmospheres, and Pluto—make up only a minuscule fraction of a world's total mass. Nevertheless, we can use the same principles to understand virtually all planetary atmospheres. We'll study these principles in this chapter, exploring how they explain the atmospheres of the terrestrial worlds. Later, we will see that these principles also apply to the jovian planets and other worlds with atmospheres.

The terrestrial atmospheres are even more varied than the terrestrial geologies. Figure 10.1 shows global and surface views of each of the five terrestrial worlds. Table 10.1 lists general characteristics of their atmospheres. Even a quick scan of the photos and the table reveals vast differences between the worlds.

The Moon and Mercury have so little atmosphere that it's quite reasonable to call them "airless" worlds. These worlds therefore have no wind or weather of any kind. At the other extreme, a thick atmosphere composed almost entirely of carbon dioxide completely shrouds Venus, which has surface conditions so hot and harsh that not even robotic space probes can survive there for long. Mars also has a carbon dioxide atmosphere, but its air is so thin that your body tissues would bulge painfully if you stood on the surface without wearing a full space suit.

Earth stands out in having an atmosphere that is hospitable to surface life, including human life. The composition of our atmosphere—mostly nitrogen and oxygen—

Figure 10.1 Views of the terrestrial worlds and their atmospheres from orbit and from the surface. The surface views for Mercury and Venus are artists' conceptions; the others are photos. The global views are visible-light photos taken from spacecraft. (Venus appears in gibbous phase as it was seen by the *Galileo* spacecraft during its Venus flyby en route to Jupiter.)

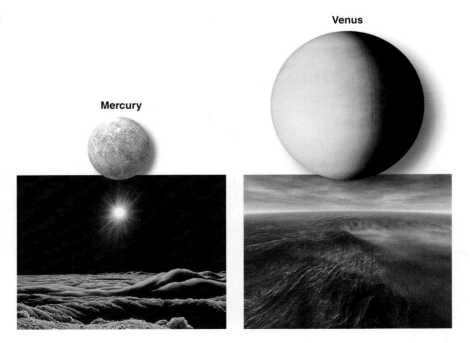

Table 10.1 Atmospheres of the Terrestrial Worlds

World	Composition of Atmosphere	Surface Pressure*	Average Surface Temperature	Winds, Weather Patterns	Clouds, Hazes
Mercury	helium, sodium, oxygen	10^{-14} bar	day: 425°C (797°F); night: −175°C (−283°F)	none: too little atmosphere	none
Venus	96% carbon dioxide (CO_2) 3.5% nitrogen (N_2)	90 bars	470°C (878°F)	slow winds, no violent storms, acid rain	sulfuric acid clouds
Earth	77% nitrogen (N_2) 21% oxygen (O_2) 1% argon H_2O (variable)	1 bar	15°C (59°F)	winds, hurricanes, rain, snow	H_2O clouds, pollution
Moon	helium, sodium, argon	10^{-14} bar	day: 125°C (257°F); night: −175°C (−283°F)	none: too little atmosphere	none
Mars	95% carbon dioxide (CO_2) 2.7% nitrogen (N_2) 1.6% argon	0.007 bar	−50°C (−58°F)	winds, dust storms	H_2O and CO_2 clouds, dust

* 1 bar = the pressure at sea level on Earth.

has no counterpart on any of the other terrestrial worlds. Earth's surface temperature and pressure are just right to allow for the liquid water that is so important to life. Earth's weather is far more dynamic and varied than the weather on Venus or Mars.

Indeed, Earth's atmosphere is so different from the atmospheres of the other terrestrial worlds that you might at first guess that it was made in a completely different way. However, the same basic processes are at work in all planetary atmospheres. Our primary goal in this chapter is to understand how these processes led to such profound differences among the planets. Let's begin by discussing a few basic properties of atmospheres.

• What is an atmosphere?

An *atmosphere* is a layer of gas that surrounds a world. In most cases, it is a surprisingly thin layer. On Earth, for example, about two-thirds of the air in the atmosphere lies within 10 kilometers of Earth's surface (Figure 10.2). You could represent this air on a standard globe with a layer only as thick as a dollar bill.

Earth

Earth's Moon

Mars

Figure 10.2 Earth's atmosphere, visible in this photograph from the Space Shuttle, makes a very thin layer over Earth's surface. Most of the air is in the lowest 10 kilometers of the atmosphere, visible along the edge of the planet.

The air that makes up any atmosphere is a mixture of many different gases that may consist either of individual atoms or of molecules. In general, temperatures in the terrestrial atmospheres are low enough (even on Venus) for atoms to combine into molecules. For example, the air we breathe consists of *molecular* nitrogen (N_2) and oxygen (O_2), as opposed to individual atoms (N or O). Other common molecules in terrestrial atmospheres include water (H_2O) and carbon dioxide (CO_2).

Atmospheric Pressure Collisions of individual atoms or molecules in an atmosphere create *pressure* [Section 5.3] that pushes in all directions. On Earth, for example, the nitrogen and oxygen molecules in the air fly around at average

speeds of about 500 meters per second—fast enough to cross your bedroom a hundred times in 1 second. Given that a single breath of air contains more than a billion trillion molecules, you can imagine how frequently molecules collide. On average, each molecule in the air around you will suffer a million collisions in the time it takes to read this paragraph. These collisions create the pressure that pushes in all directions, and that in essence holds up the atmosphere so that it does not collapse under its own weight.

A balloon offers a good example of how pressure works in a gas. The air molecules inside a balloon exert pressure pushing outward as they constantly collide with the balloon's inside surface. At the same time, air molecules outside the balloon collide with the balloon from the other side, exerting pressure that by itself would make the balloon collapse. A balloon stays inflated when the inward pressure and the outward pressure are balanced (Figure 10.3a). (We are ignoring the tension in the rubber of the balloon walls.) Imagine that you blow more air into the balloon (Figure 10.3b). The extra molecules inside mean more collisions with the balloon wall, momentarily making the pressure inside greater than the pressure outside. The balloon therefore expands until the inward and outward pressures are again in balance.

If you heat the balloon (Figure 10.3c), the gas molecules begin moving faster and therefore collide harder and more frequently with the inside of the balloon, which also momentarily increases the inside pressure until the balloon expands. As it expands, the pressure inside it decreases and the balloon comes back into pressure balance. Conversely, cooling a balloon makes it contract, because the outside pressure momentarily exceeds the inside pressure. You can watch this contraction happen by putting a balloon in a refrigerator or freezer.

We can understand **atmospheric pressure** by applying similar principles. Gas in an atmosphere is held down by gravity. The atmosphere above any given altitude therefore has some weight that presses downward, tending to compress the atmosphere beneath it. At the same time, the fast-

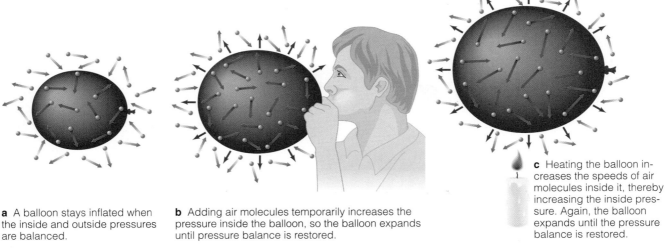

a A balloon stays inflated when the inside and outside pressures are balanced.

b Adding air molecules temporarily increases the pressure inside the balloon, so the balloon expands until pressure balance is restored.

c Heating the balloon increases the speeds of air molecules inside it, thereby increasing the inside pressure. Again, the balloon expands until the pressure balance is restored.

Figure 10.3 Gas pressure in a balloon depends on both density and temperature.

moving molecules exert pressure in all directions, including upward, which tends to make the atmosphere expand. Planetary atmospheres exist in a perpetual balance between the downward weight of their gases and the upward push of their gas pressure.

The higher you go in an atmosphere, the less the weight of the gas above you. Thus, the pressure must also become less as you go upward, which explains why the pressure decreases as you climb a mountain or ascend in an airplane. You can visualize what happens by imagining the atmosphere as a very big stack of pillows (Figure 10.4). The pillows at the bottom are highly compressed because of the weight of all the pillows above. Going upward, the pillows are less and less compressed because less weight lies on top of them.

In planetary science, we usually measure atmospheric pressure in a unit called the **bar** (as in *barometer*). One bar is roughly equal to Earth's atmospheric pressure at sea level. It is also equivalent to 1.03 kilograms per square centimeter or 14.7 pounds per square inch. In other words, if you gathered up all the air directly above any 1 square inch of Earth's surface, you would find that it weighs about 14.7 pounds.

THINK ABOUT IT

Suppose Mars had exactly the same total amount of air above each square inch of surface as does Earth. Would the atmospheric pressure be higher, lower, or the same as on Earth? Explain. (*Hint:* Remember that weight depends on the strength of gravity [Section 4.1], and that surface gravity is weaker on Mars than Earth.)

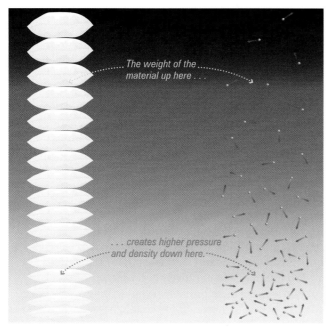

Figure 10.4 Atmospheric density and pressure decrease with altitude for the same reason that a giant stack of pillows would be more compressed at the bottom than at the top. The pressure in each layer is enough to hold up the weight of layers above, counteracting the force of gravity.

You might wonder why you don't feel Earth's atmospheric weight bearing down on you. After all, if we placed 14.7 pounds of weights on every square inch of your shoulders, you'd certainly feel the downward pressure. You don't notice the atmospheric pressure for two reasons: First, the pressure pushes in all directions, so it pushes upward and inward on you as well as downward. Second, the fluids in your body push outward with an equivalent pressure, so there is no net pressure trying to compress or expand your body. You can tell that the pressure is there, however, because you'll quickly notice any pressure changes. For example, even slight changes in pressure as you go up or down in altitude can cause your ears to "pop." More extreme changes, such as those that affect deep-sea divers when they rise too rapidly, can be deadly. You'd face even greater pressure differences if you visited other planets; as shown in Table 10.1, atmospheric pressure varies by a factor of 10^{16} among the terrestrial worlds.

Where Does an Atmosphere End? If you study Figure 10.4, you'll see that there is no clear boundary between the atmosphere and space above. Instead, the atmosphere gradually fades away as the pressure and density decrease with increasing altitude. At some point, the density becomes so low that we can't really think of the gas as "air" anymore. Collisions between atoms or molecules are rare at these altitudes, and the gas is so thin that it would look and feel as if you had already entered space. On Earth, for example, the gas is so thin above an altitude of about 60 kilometers that the sky is black even in the daytime, much like the sky on the Moon. For this reason, an altitude of about 60 kilometers is often described as "the edge of space."

However, some gas is still present even above 60 kilometers. Earth's tenuous upper atmosphere extends for several hundred kilometers above Earth's surface. The Space Shuttle, the Space Station, and many other satellites orbit Earth within these outer reaches of the atmosphere (Figure 10.5). The low-density gas may be barely noticeable under most conditions, but it still exerts drag on orbiting spacecraft. That is why satellites in low-Earth orbit slowly spiral downward, eventually burning up as they reenter the denser layers of the atmosphere [Section 4.5]. The Space Station would suffer the same fate without periodic boosts from the Space Shuttle or other rockets.

How Atmospheres Affect Planets It may be fairly obvious that Earth's atmosphere is important to our existence, but the full range of atmospheric effects is more than most people realize. To provide context for our study of planetary atmospheres, here is a brief list of the key effects of atmospheres that we will study in more detail in the rest of this chapter.

- As we've already seen, atmospheres create pressure that can in turn determine whether water can exist in liquid form on the surface.
- Atmospheres absorb and scatter light. Scattering explains why daytime skies are bright on worlds with

Figure 10.5 This photograph shows the visible glow of gas around the tail of an orbiting Space Shuttle. Some low-density gas is present even at the Shuttle's altitude of a few hundred kilometers, though most of Earth's atmospheric gas lies within 10 kilometers of the surface and altitudes above 60 kilometers are generally considered to be "space." The aurora visible in the background is additional proof of the great vertical extent of the atmosphere, since the aurora is created by interactions between charged particles and atmospheric gas at altitudes above 100 kilometers.

atmospheres and black on airless worlds like the Moon and Mercury. Absorption can keep the surface safe from dangerous radiation; for example, ozone in Earth's atmosphere prevents most solar ultraviolet radiation from reaching the ground.

- Atmospheres can create wind and weather and play a major role in long-term climate change.
- Interactions between atmospheric gases and the solar wind can create a protective *magnetosphere* around planets with strong magnetic fields.
- In some cases, atmospheres contain gases that can make planetary surfaces warmer than they would be otherwise, through what we call the **greenhouse effect**.

The greenhouse effect is arguably the most important effect that an atmosphere can have on its planet, and we'll therefore focus our attention on it now.

• How does the greenhouse effect warm a planet?

You've probably heard of the greenhouse effect, because it is an important part of the environmental problem known as *global warming* [Section 10.6]. But you may be surprised to learn that the greenhouse effect is also critical to the existence of life on Earth. Without the greenhouse effect, Earth's surface would be too cold for liquid water to flow and for life to flourish. Let's explore how the greenhouse

effect works and what it does to the surface temperatures of the terrestrial worlds.

How the Greenhouse Effect Works The basic idea behind the greenhouse effect is quite simple (Figure 10.6). Some of the visible light from the Sun is absorbed by the ground (the rest is reflected). The ground returns the absorbed energy back toward space, but in the form of infrared rather than visible light. The reason planets emit infrared light is that their temperatures are too low to emit visible light. Remember that all objects emit thermal radiation by virtue of their temperatures [Section 5.4]. Planetary temperatures are in a range in which nearly all emitted energy is in the infrared.

The greenhouse effect works by "trapping" some of the infrared light that a planet emits, slowing its return to space. This trapping occurs when atmospheric gases absorb the infrared light. Gases that are particularly good at absorbing infrared light are called **greenhouse gases**. The most important greenhouse gases are water vapor (H_2O), carbon dioxide (CO_2), and methane (CH_4). These gases absorb infrared light effectively because their molecular structures make them prone to begin rotating or vibrating when struck by a photon of infrared light (see Figure 5.18).

After a greenhouse gas molecule absorbs a photon of infrared light, it reemits a photon of infrared light in some random direction. This photon is usually absorbed by another greenhouse molecule, which does the same thing. The net result is that greenhouse gases tend to slow the

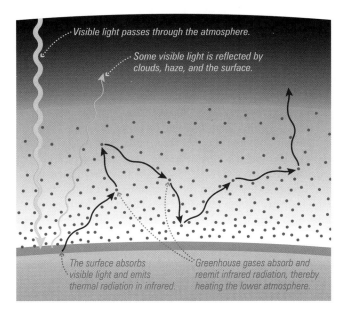

- Visible light passes through the atmosphere.
- Some visible light is reflected by clouds, haze, and the surface.

The surface absorbs visible light and emits thermal radiation in infrared.

Greenhouse gases absorb and reemit infrared radiation, thereby heating the lower atmosphere.

Figure 10.6 Interactive Figure. The greenhouse effect. The lower atmosphere becomes warmer than it would be if it had no greenhouse gases such as water vapor, carbon dioxide, and methane.

escape of infrared radiation from the lower atmosphere, while their molecular motions heat the surrounding air. In this way, the greenhouse effect makes the surface and the lower atmosphere warmer than they would be from sunlight alone. The more greenhouse gases present, the greater the degree of surface warming.

It's important to realize that the greenhouse effect does not alter a planet's overall energy balance. The total amount of energy that a planet receives from the Sun must be precisely balanced with the amount of energy it returns to space through reflection and radiation. If it weren't balanced, the planet would rapidly heat up (if it received more energy than it returned) or cool down (if it returned more energy than it received). All the energy "trapped" by the greenhouse effect still escapes to space, just not as directly as it would otherwise. A blanket offers a good analogy: You stay warmer under a blanket not because the blanket itself provides any heat, but rather because it slows the escape of your body heat into the cold outside air. In essence, the greenhouse effect makes the total amount of energy in the lower atmosphere greater than it would be otherwise—making it warmer—while leaving the total amount of energy that escapes to space unchanged.

THINK ABOUT IT

Clouds on Earth are made of water (H_2O), which acts as a very effective greenhouse gas even when the water vapor condenses into droplets in clouds. Use this fact to explain why clear nights tend to be colder than cloudy nights.

Incidentally, the name *greenhouse effect* is a bit of a misnomer. The name comes from botanical greenhouses, usually built mostly of glass, which keep the air inside them warmer than it would be otherwise. These greenhouses

trap heat simply by not letting hot air rise, which is very different from the way infrared absorption traps heat and leads to the "greenhouse effect" in planetary atmospheres.

"No Greenhouse" Planetary Temperatures We can better appreciate the importance of the greenhouse effect by considering planetary temperatures in its absence. Remember that a terrestrial planet's interior heat has very little effect on its surface temperature; sunlight is the only significant energy source for the surface [Section 9.1]. Without the greenhouse effect, a planet's average surface temperature would depend only on how much sunlight it absorbed. This absorbed sunlight would be radiated directly back to space as infrared light. Moreover, the amount of absorbed sunlight—and hence the average surface temperature—would depend on only two things:

- *The planet's distance from the Sun,* which determines the amount of energy incoming as sunlight. The closer a planet is to the Sun, the greater the intensity of the incoming sunlight.
- *The planet's overall reflectivity,* which determines the relative proportions of incoming sunlight that the planet reflects and absorbs. (You may sometimes see reflectivity described by the term *albedo.*) The higher the reflectivity, the less light absorbed and the cooler the planet.

A planet's reflectivity depends on its surface composition and color. The darker the surface, the more light it absorbs and the less it reflects. For example, clouds, snow, and ice reflect 70% or more of the light that hits them, while rocks typically reflect only about 20% of the light that hits them and absorb the other 80%.

THINK ABOUT IT

Who gets hotter in the summer sun, you wearing a white shirt or a person in black next to you? Explain.

COMMON MISCONCEPTIONS

The Greenhouse Effect Is Bad

The greenhouse effect is often in the news, usually in discussions about environmental problems, but in itself the greenhouse effect is not a bad thing. In fact, we could not exist without it, since it is responsible for keeping our planet warm enough for liquid water to flow in the oceans and on the surface. The "no greenhouse" temperature of Earth is well below freezing. Why, then, is the greenhouse effect discussed as an environmental problem? The reason is that human activity is adding more greenhouse gases to the atmosphere—which might change Earth's climate [Section 10.6]. While the greenhouse effect makes Earth livable, it is also responsible for the searing 470°C temperature of Venus—proving that it's possible to have too much of a good thing.

Both distance from the Sun and reflectivity have been measured for all the terrestrial worlds. With a little mathematics (see Mathematical Insight 10.1), it's possible to use these measurements to calculate the temperature that any planet would have without greenhouse gases. For example, such calculations show that Earth's **global average temperature**—that is, the average temperature for the entire planet—would be a chilly $-16°C$ ($+3°F$) without the greenhouse effect. Because this "no greenhouse" temperature is well below the freezing point of water, we conclude that our planet would be frozen over in the absence of a greenhouse effect. Clearly, the greenhouse effect is very important to life on Earth.

Greenhouse Warming of the Terrestrial Worlds   Table 10.2 lists the distance and reflectivity for each terrestrial world and compares the world's calculated "no greenhouse" temperatures to actual surface temperatures. The "no greenhouse" temperatures for Mercury and the Moon lie between their actual day and night temperatures, just as we should expect on worlds with very little atmosphere and hence no greenhouse effect. For the other planets, however, the greenhouse effect plays an important role.

The greenhouse effect is fairly weak on Mars, where the actual average surface temperature is only 6°C higher than the "no greenhouse" temperature. Earth has a moderate greenhouse effect that makes the actual global average temperature about 15°C (59°F), which is 31°C warmer than the "no greenhouse" temperature. Venus has an extreme greenhouse effect that bakes the surface to a temperature more than 500°C hotter than it would be otherwise.

If you think about it, this greenhouse warming is quite remarkable. Earth's atmosphere is made almost entirely of nitrogen and oxygen, which have no effect on infrared light and do not contribute to the greenhouse effect. (Molecules with only two atoms, especially those with two of the same kind of atom, such as N_2 and O_2, are poor infrared absorbers because they have very few ways to vibrate and rotate.) Greenhouse gases are only trace constituents of Earth's atmosphere, yet they warm our planet enough to raise the global average temperature from well below freezing to well above it.

THINK ABOUT IT

Suppose nitrogen and oxygen were greenhouse gases. Assuming our atmosphere still had the same composition and density, how would Earth be different?

MATHEMATICAL INSIGHT 10.1   "No Greenhouse" Temperatures

As discussed in the text, the "no greenhouse" temperature of a planet in our solar system depends only on the planet's distance from the Sun and its overall reflectivity. From those two factors and the Sun's brightness, we can derive the following formula for a planet's "no greenhouse" temperature:

$$T = 280 \text{ K} \times \sqrt[4]{\frac{(1 - \text{reflectivity})}{d^2}}$$

where d is the distance from the Sun in AU and the reflectivity is stated as a fraction between 0 and 1. For example, if a planet reflects 20% of the incoming sunlight, its reflectivity would be stated as 0.2. The symbol $\sqrt[4]{}$ means the fourth root, or $\frac{1}{4}$ power.

We will not derive the formula here, but you can see how it makes sense. The term $(1 - \text{reflectivity})$ is the proportion of sunlight that the planet absorbs, which is the light that heats its surface. This term is divided by d^2 because the amount of energy in sunlight (per unit area) declines with the square of distance from the Sun, just as gravity declines with the square of distance. Thus, the full term $(1 - \text{reflectivity})/d^2$ represents the total amount of energy that the planet absorbs from sunlight (per unit area) each second. This energy warms the surface, and the surface returns the energy to space by thermal radiation [Section 5.4]. The energy emitted by thermal radiation depends on temperature raised to the fourth power (see Mathematical Insight 5.2). Calculating the planet's temperature therefore requires taking the fourth root of the absorbed energy, or the fourth root of $(1 - \text{reflectivity})/d^2$. The 280 K in the formula happens to be the temperature of a perfectly black planet at 1 AU from the Sun.

Example: Calculate the "no greenhouse" temperature of Mercury.

Solution:

Step 1 Understand: We can use the above formula to find a "no greenhouse" temperature by knowing a planet's distance from the Sun in AU and its reflectivity as a decimal fraction. Table 10.2 gives us both values for Mercury, except that it gives the reflectivity as a percentage. Of course, we can easily convert 12% to the decimal fraction 0.12.

Step 2 Solve: We now plug in Mercury's distance from the Sun, $d = 0.387$ AU, and its reflectivity of 0.12 to find its "no greenhouse" temperature:

$$T = 280 \text{ K} \times \sqrt[4]{\frac{(1 - \text{reflectivity})}{d^2}}$$

$$= 280 \text{ K} \times \sqrt[4]{\frac{(1 - 0.12)}{0.387^2}} = 436 \text{ K}$$

Step 3 Explain: We have found that Mercury's "no greenhouse" temperature is 436 K. We can check this value against the value shown in Table 10.2 by converting it to Celsius. The only difference between the Celsius and Kelvin scales is their starting point: 0 Kelvin is equivalent to $-273°C$ (see Figure 4.12). Thus, we convert the Kelvin temperature to Celsius by subtracting 273, finding that 436 K is equivalent to $436 - 273 = 163°C$. Notice that this value agrees with the "no greenhouse" temperature for Mercury shown in the table. It also makes sense: Because Mercury has no greenhouse effect, its "no greenhouse" temperature should be midway between its day and night temperatures. The precise halfway point of the day and night temperatures shown in Table 10.2 is 125°C; although this is not an exact match for the 163°C that we calculated, it is close enough for us to be confident that we have the right general idea.

Table 10.2 The Greenhouse Effect on the Terrestrial Worlds

World	Average Distance from Sun (AU)	Reflectivity	"No Greenhouse" Average Surface Temperature*	Actual Average Surface Temperature	Greenhouse Warming (actual temperature minus "no greenhouse" temperature)
Mercury	0.387	12%	163°C	425°C (day), −175°C (night)	—
Venus	0.723	75%	−40°C	470°C	510°C
Earth	1.00	29%	−16°C	15°C	31°C
Moon	1.00	12%	−2°C	125°C (day), −175°C (night)	—
Mars	1.524	16%	−56°C	−50°C	6°C

*The "no greenhouse" temperature is calculated by assuming no change to the atmosphere other than lack of greenhouse warming. Thus, for example, Venus ends up with a lower "no greenhouse" temperature than Earth even though it is closer to the Sun, because the high reflectivity of its bright clouds means that it absorbs less sunlight than Earth.

• Why do atmospheric properties vary with altitude?

The greenhouse effect can warm a planet's surface and lower atmosphere, but other processes affect the temperature of the atmosphere at higher altitudes. The way in which temperature varies with altitude determines what is often called the **atmospheric structure**. Earth's atmospheric structure has four basic layers (Figure 10.7), each of which affects the planet in a distinct way:

- The **troposphere** is the lowest layer of the atmosphere. The temperature drops with altitude in the troposphere, something you've probably noticed if you've ever climbed a mountain.
- The **stratosphere** begins where the temperature stops dropping and instead begins to rise with altitude. High in the stratosphere, the temperature falls again.*
- The **thermosphere** begins at a high altitude where the temperature again starts to rise.
- The **exosphere** is the uppermost region in which the atmosphere gradually fades away into space.

The layering is shaped by the way atmospheric gas interacts with sunlight. As shown in Figure 10.7, visible light warms the surface so that the ground emits infrared light that heats the troposphere. The absorption of ultraviolet light from the Sun warms the stratosphere, while solar X rays and some solar ultraviolet light heat the thermosphere and exosphere. Let's now look at these ideas in more detail, so that we'll understand why atmospheric properties vary with altitude. Along the way, we'll also see why storms occur only in the troposphere, why ozone is so important to life on Earth, and why the sky is blue.

Interactions Between Light and Atmospheric Gases The key to understanding atmospheric structure lies in interactions between atmospheric gases and energy from the Sun. Essentially all the energy in terrestrial atmospheres comes from sunlight.

Most light coming from the Sun is visible light, but the Sun also emits significant amounts of ultraviolet light and

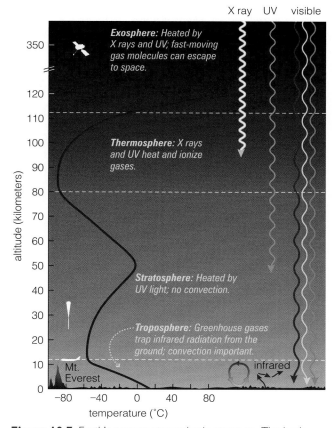

Figure 10.7 Earth's average atmospheric structure. The background color represents the air density and pressure, both of which decline rapidly with altitude. The curve shows average temperatures for each altitude. The squiggly arrows to the right represent photons and show where they are typically absorbed in the atmosphere.

*Technically, the region where the temperature falls again is called the *mesosphere,* but we will not make this distinction in this book.

X rays. The light emitted from a planet is almost entirely in the infrared. Atmospheric gases interact with each of these forms of light in different ways that are summarized in Figure 10.8:

- X rays have enough energy to knock electrons free from almost any atom or molecule; that is, they ionize [Section 5.3] the atoms or molecules they strike. Thus, virtually all atmospheric gases absorb X rays and become ionized in the process.
- Ultraviolet photons can break apart some molecules. For example, a molecule of water vapor (H_2O) can split apart if it absorbs an ultraviolet photon. Weakly bonded molecules, such as ozone (O_3), are particularly good at absorbing ultraviolet light, splitting apart in the process.
- Visible-light photons usually pass through atmospheric gases without being absorbed, but some are *scattered* so that their direction changes.
- As we discussed earlier, infrared light can be absorbed by greenhouse gases—molecules such as water vapor (H_2O), carbon dioxide (CO_2), and methane (CH_4) that easily begin rotating and vibrating.

With these ideas in mind, let's examine the reasons for Earth's atmospheric structure, working our way from the ground up.

Visible Light: Warming the Surface and Coloring the Sky The transparency of atmospheric gases to visible light explains why visible light reaches the ground and warms the surface. However, the small amount of visible-light scatter-ing that occurs in the atmosphere has two very important effects.

First, scattering explains why the daytime sky is bright and we cannot see the stars. Without scattering, our sky would look like the daytime lunar sky, with the Sun just a very bright circle set against a black, star-studded background. Scattering also prevents shadows on Earth from being pitch black. On the Moon, shadows receive little scattered sunlight and are extremely cold and dark.

Second, the scattering of visible light explains why our sky is blue. Visible light consists of all the colors of the rainbow, but not all the colors are scattered equally. Gas molecules scatter blue light (higher energy) much more effectively than red light (lower energy). The difference in scattering is so great that, for practical purposes, we can imagine that only the blue light gets scattered. When the Sun is overhead, this scattered blue light reaches our eyes from all directions and the sky appears blue (Figure 10.9). At sunset or sunrise, the sunlight must pass through a greater amount of atmosphere on its way to us. Most of the blue light is scattered away, leaving only red light to color the sky.

Infrared Light and the Troposphere The troposphere gets its structure because of the way atmospheric gases interact with infrared light radiated upward by the ground. That is, the greenhouse effect warms the troposphere. Because the infrared light absorbed in the troposphere comes from the surface, more is absorbed closer to the ground than at higher altitudes. That is why the temperature drops with altitude in the troposphere. (The relatively small amount of infrared light coming from the Sun is mostly absorbed by atmospheric molecules before it reaches the ground and thus has relatively little effect on the atmosphere.)

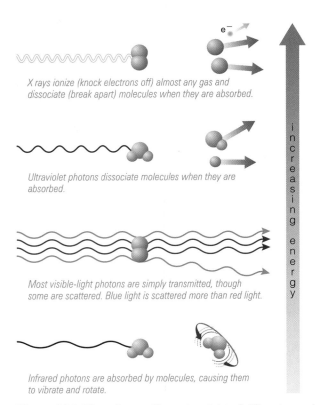

X rays ionize (knock electrons off) almost any gas and dissociate (break apart) molecules when they are absorbed.

Ultraviolet photons dissociate molecules when they are absorbed.

Most visible-light photons are simply transmitted, though some are scattered. Blue light is scattered more than red light.

Infrared photons are absorbed by molecules, causing them to vibrate and rotate.

Figure 10.8 The primary effects when light of different energies strikes common atmospheric gases.

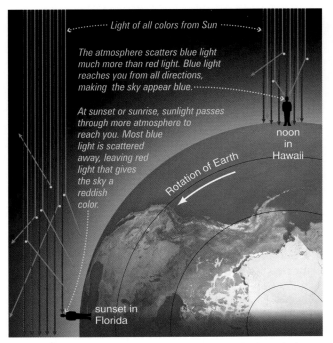

Figure 10.9 This diagram summarizes why the daytime sky is blue and sunsets (and sunrises) are red.

Why Is the Sky Blue?

If you ask around, you'll find a wide variety of misconceptions about why the sky is blue. For example, some people think the sky is blue because of light reflecting from the oceans. Anyone who lives in Nebraska will immediately know that this must be wrong—Nebraska is too far from any ocean for its sky to be influenced by such reflection. Other people claim that "air is blue." It's not obvious what someone might mean by this statement, but it's clearly wrong. If air molecules emitted blue light, then air would glow blue even in the dark. If air molecules were blue because they reflected blue light and absorbed red light, then no red light could reach us at sunset. The real explanation for the blue sky is light scattering, as shown in Figure 10.9. This explanation not only makes sense for our blue sky but also explains our red sunsets.

The drop in temperature with altitude, combined with the relatively high density of air in the troposphere, explains why the troposphere is the only layer of the atmosphere with storms. The primary cause of storms is the churning of air by convection, in which warm air rises and cool air falls. Remember that convection occurs only when there is strong heating from below [Section 9.1]. In the troposphere, the heating from the ground can drive convection. Indeed, the troposphere gets its name from this convection, since *tropos* is Greek for "turning."

Ultraviolet Light and the Stratosphere   The primary source of heat in the troposphere—the absorption of infrared radiation from the ground—is not important in higher layers of the atmosphere. Once infrared radiation from the ground reaches the top of the troposphere, it travels essentially unhindered to space. Thus, we need to consider only the effects of sunlight in investigating atmospheric structure above the troposphere.

As we move upward through Earth's stratosphere, the absorption of solar ultraviolet light by ozone becomes the primary source of atmospheric heating. This heating tends to be stronger at higher altitudes, because almost all the ultraviolet light from the Sun gets absorbed before it reaches the bottom of the stratosphere. As a result, the lower stratosphere gets warmer with increasing altitude—the opposite of the situation in the troposphere.

Convection cannot occur in the lower stratosphere because heat cannot rise if the air is even hotter higher up. The lack of convection makes the air relatively stagnant and *stratified* (or layered)—with layers of warm air overlying cooler air—which is how the stratosphere gets its name. The lack of convection also means that the stratosphere essentially has no weather and no rain. Pollutants that reach the stratosphere—including the ozone-destroying

chemicals known as chlorofluorocarbons (CFCs)—remain there for decades.

Note that a planet can have a stratosphere *only* if its atmosphere contains molecules that are particularly good at absorbing ultraviolet photons. Ozone (O_3) plays this role on Earth, but the lack of oxygen in the atmospheres of the other terrestrial worlds means that they also lack ozone. As a result, Earth is the only terrestrial world with a stratosphere, at least in our solar system. (The jovian planets have stratospheres due to other ultraviolet-absorbing molecules [Section 11.1].)

X Rays and the Thermosphere   Nearly all gases are good X-ray absorbers, so X rays from the Sun are absorbed by the first gases they encounter as they enter the atmosphere. The density of gas in the exosphere is too low to absorb significant amounts of the incoming X rays, leaving most X rays to be absorbed in the thermosphere. Temperatures rise quite high in the thermosphere (*thermos* is Greek for "hot"), but you wouldn't feel hot there because the density and pressure are so low [Section 4.3]. Virtually no X rays penetrate beneath the thermosphere, which is why X-ray telescopes are useful only on very high-flying balloons, rockets, and spacecraft.

In addition to heating the thermosphere, X rays ionize a small but important fraction of its gas. The portion of the thermosphere that contains most of the ionized gas is called the *ionosphere.* Thus, the ionosphere is part of the thermosphere, not a separate layer. The ionosphere is very important to radio communications on Earth. The ionosphere reflects most radio broadcasts back to Earth's surface—almost as though Earth were wrapped in aluminum foil. Without reflection by the ionosphere, radio communication would work only between locations in sight of each other.

The Exosphere   The exosphere is the extremely low-density gas that forms the gradual and fuzzy boundary between the atmosphere and space (*exo* means "outermost" or "outside"). The gas density in the exosphere is so low that collisions between atoms or molecules are very rare, although the high temperature means that gas particles move quite rapidly. Lightweight gas molecules sometimes reach escape velocity [Section 4.5] and fly off into space.

Higher Altitudes Are Always Colder

Many people think that the low temperatures in the mountains are just the result of lower pressures, but Figure 10.7 shows that it's not that simple. The higher temperatures near sea level on Earth are a result of the greenhouse effect, which traps more heat at lower altitudes. If Earth had no greenhouse gases, mountaintops wouldn't be so cold—or, more accurately, sea level wouldn't be so warm.

Comparative Structures of Terrestrial Atmospheres We've discussed everything we need to understand the structures of all the terrestrial atmospheres. The Moon and Mercury have so little gas that they essentially contain only an exosphere and have no structure to speak of. Figure 10.10 contrasts the structures of the lower atmospheres of Venus, Earth, and Mars. The following key ideas should make sense to you:

- All three planets have a warm troposphere at the bottom of the atmosphere, created by the greenhouse effect, and a warm thermosphere at the top, where solar X rays are absorbed.
- Only Earth has the extra "bump" of a stratosphere, because only Earth has a layer of ultraviolet-absorbing gas (ozone). Without the warming in our unique layer of stratospheric ozone, the middle altitudes of Earth's atmosphere would be almost as cold as those on Mars.
- All three planets are warmer at the surface than they would be without the greenhouse effect, although the amount of warming is far greater for Venus than for Earth or Mars.

Would astronauts on the Moon need protection from solar X rays? What about astronauts on Mars? Explain.

Magnetospheres and the Solar Wind We have yet to consider one more type of energy coming from the Sun: the low-density breeze of subatomic charged particles that we call the *solar wind* [Section 8.3]. The solar wind does not significantly affect atmospheric structure, but it can have other important effects. The effects depend largely on whether

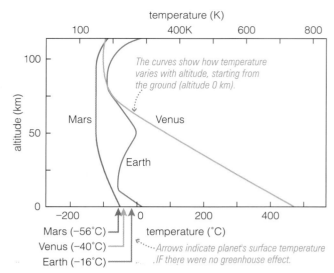

Figure 10.10 Temperature profiles for Venus, Earth, and Mars. Note that Venus and Earth are considerably warmer than they would be without the greenhouse effect. (The thermospheres and exospheres, which extend above the top of the graph, are qualitatively alike.)

a planet has a strong enough magnetic field [Section 9.1] to divert the charged particles of the solar wind.

On worlds that do not have strong magnetic fields, solar wind particles can directly impact the surface or atmosphere. For example, solar wind particles continually hit the upper atmospheres of Venus and Mars and the surfaces of Mercury and the Moon. In contrast, Earth's strong magnetic field creates a **magnetosphere** that acts like a protective bubble surrounding our planet (Figure 10.11a). The

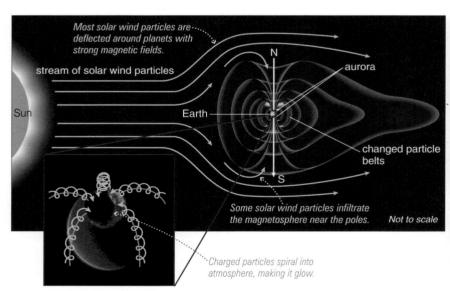

a This diagram shows how Earth's magnetosphere deflects solar wind particles. Some particles accumulate in charged particle belts encircling our planet. The inset is a photo of a ring of auroras around the North Pole; the bright crescent at its left is part of the day side of Earth.

b This photograph shows the aurora near Yellowknife, Northwest Territories, Canada. In a video, you would see these lights dancing about in the sky.

Figure 10.11 Interactive Figure Earth's magnetosphere acts like a protective bubble, shielding our planet from the charged particles of the solar wind.

magnetosphere deflects most of the charged particles from the Sun around our planet.

The magnetosphere still allows a few solar wind particles to get through, especially near the magnetic poles. Once inside the magnetosphere, these particles move along magnetic field lines, collecting in **charged particle belts** that encircle our planet. (These belts are often called the *Van Allen belts* after their discoverer.) The high energies of the particles in charged particle belts can be hazardous to spacecraft and astronauts passing through them.

Charged particles trapped in the magnetosphere also create the beautiful spectacle of light we call the **aurora** (Figure 10.11b). Variations in the solar wind can buffet the magnetosphere and give energy to particles trapped there. If a trapped particle gains enough energy, it can follow the magnetic field all the way down to Earth's atmosphere, where it collides with atmospheric atoms and molecules. These collisions cause the atoms and molecules to radiate and produce the moving lights of the aurora. Because the charged particles follow the magnetic field, auroras are most common near the magnetic poles and are best viewed at high latitudes. In the Northern Hemisphere, the aurora is often called the *aurora borealis,* or northern lights. In the Southern Hemisphere, it is called the *aurora australis,* or southern lights. The aurora can also be seen from space, where the lights can look much like surf in the upper atmosphere (see Figure 10.5).

Surface Temperature of Terrestrial Planets Tutorial, Lesson 1

10.2 Weather and Climate

So far, we've talked mostly about *average* conditions in planetary atmospheres, such as the average temperature on the surface and at each altitude above the surface. From our experience on Earth, however, we know that surface and atmospheric conditions constantly change. These changes are generally described as *weather* or *climate.*

Weather and climate are closely related, but they are not quite the same thing. In any particular location, some days may be hotter or cooler, clearer or cloudier, or calmer or stormier than others. The ever-varying combination of winds, clouds, temperature, and pressure is what we call **weather**.

Climate is the long-term average of weather, which means it can change only on much longer time scales. For example, Antarctic deserts have a cold, dry climate, while tropical rain forests have a hot, wet climate. Geological records show that climates can change, but most of the time the changes occur gradually over decades, centuries, or millennia.

Weather and climate can be hard to distinguish on a human time scale. For example, a drought lasting a few years may be the result of either random weather fluctuations or the beginning of a gradual change to a drier climate. The difficulty in distinguishing between random weather

---

SPECIAL TOPIC Weather and Chaos

Scientists today have a very good understanding of the physical laws and mathematical equations that govern the behavior and motion of atoms in the air, oceans, and land. Why, then, do we have so much trouble predicting the weather? To understand why weather is so unpredictable, we must look at the nature of scientific prediction.

Let's say you want to predict the location of a car on a road 1 minute from now. You need two basic pieces of information: where the car is now, and how it is moving. If the car is now passing Smith Road and heading north at 1 mile per minute, it will be 1 mile north of Smith Road in 1 minute.

Now suppose you want to predict the weather. Again, you need two basic types of information: (1) the current weather and (2) how weather changes from one moment to the next.

You could attempt to predict the weather by creating a "model world." For example, you could overlay a globe of Earth with graph paper and then specify the current temperature, pressure, cloud cover, and wind within each square. These are your starting points, or initial conditions. Next, you could input all the initial conditions into a computer, along with a set of equations (physical laws) that describe the processes that can change weather from one moment to the next.

Now you run your computer model to predict the weather for the next month in New York City. The model might tell you that tomorrow will be warm and sunny, with cooling during the next week and a major storm passing through a month from now. But suppose you run the model again, making one minor change in the initial conditions—say, a small change in the wind speed some-

where over Brazil. This slightly different initial condition will not change the weather prediction for tomorrow in New York City. For next week's weather, the new model may yield a slightly different prediction. For next month's weather, however, the two predictions may not agree at all!

The disagreement between the two predictions arises because the laws governing weather can cause very tiny changes in initial conditions to be greatly magnified over time. This extreme sensitivity to initial conditions is sometimes called the *butterfly effect:* If initial conditions change by as much as the flap of a butterfly's wings, the resulting prediction may be very different.

The butterfly effect is a hallmark of *chaotic systems.* Simple systems are described by linear equations in which, for example, increasing a cause produces a proportional increase in an effect. In contrast, chaotic systems are described by nonlinear equations, which allow for subtler and more intricate interactions. For example, the economy is nonlinear because a rise in interest rates does not automatically produce a corresponding change in consumer spending. Weather is nonlinear because a change in the wind speed in one location does not automatically produce a corresponding change in another location.

Despite their name, chaotic systems are not necessarily random. In fact, many chaotic systems have a kind of underlying order that explains the general features of their behavior even while details at any particular moment remain unpredictable. In a sense, many chaotic systems—like the weather—are "predictably unpredictable."

and real climate trends is an important part of the debate about human influences on the climate. We'll return to this topic in Section 10.6; for now, let's focus on understanding weather and climate generally.

• What creates wind and weather?

Wind, rain, and other weather phenomena are all driven by energy in the atmosphere, which means only planets with atmospheres can have weather. Even then, weather varies dramatically among different worlds. Earth has the most diverse weather of the terrestrial planets, so we'll use it as our example of how weather works.

The complexity of weather makes it difficult to predict, and at best the local weather can be predicted only a week or so in advance (see Special Topic box, p. 299). Nevertheless, when we look at Earth as a whole, we can identify certain general characteristics of weather.

Global Wind Patterns The wind's direction and strength can change rapidly in any particular place. On a more global scale, however, winds blow in distinctive patterns. For example, winds cause storms moving in from the Pacific to hit the west coast of the United States first and then make their way eastward across the Rocky Mountains and the Great Plains, heading to the east coast. This and other general wind patterns make up what we call Earth's **global wind patterns** (or *global circulation*).

Figure 10.12 shows Earth's global wind patterns. Notice how the wind direction changes back and forth with latitude: Equatorial winds blow from east to west, midlatitude winds blow from west to east, and high-latitude winds blow like equatorial winds from east to west. What causes this pattern of winds? Two factors explain it: atmospheric heating and planetary rotation. Let's examine each in turn.

THINK ABOUT IT

Hurricanes usually form over warm, equatorial waters. Based on the wind patterns shown in Figure 10.12, which direction will a hurricane move after forming in the eastern Atlantic Ocean? Explain why these wind patterns make the southeastern United States and the Caribbean especially vulnerable to hurricanes.

Atmospheric Heating and Circulation Cells The first major factor affecting global wind patterns is atmospheric heating. Earth's equatorial regions receive more heat from the Sun than do polar regions. The excess heat makes the atmosphere expand above the equator so that it rises upward and flows toward the poles. Near the poles, cool air descends and flows toward the equator. If Earth's rotation did not influence this process, the result would be two huge **circulation cells**, one in each hemisphere (Figure 10.13). We'll see shortly that rotation causes these huge cells to split into somewhat smaller cells, but this fact doesn't change the general picture of circulation cells carrying air between the equator and the poles. (Circulation cells are often called *Hadley cells* after the man who first suggested their existence in 1735.)

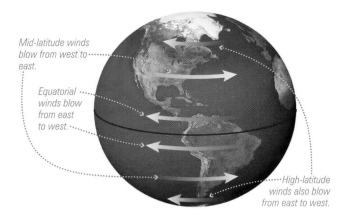

Figure 10.12 Schematic of Earth's global wind patterns. Notice that midlatitude winds blow from west to east, explaining why storms generally move across the United States from the west coast toward the east coast.

The circulation cells transport heat both from lower to higher altitudes and from the equator to the poles. They therefore make Earth's polar regions much warmer than they would be in the absence of circulation. The same idea applies to different extents on Venus and Mars. On Venus, the dense atmosphere allows the circulation cells to transport so much thermal energy that temperatures are nearly the same at the equator and the poles. On Mars, the circulation cells transport very little heat because the atmosphere is so thin, so the poles remain much colder than the equator.

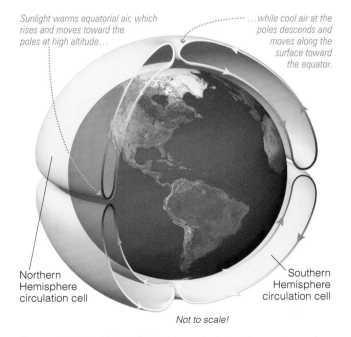

Figure 10.13 Interactive Figure Atmospheric heating creates a circulation cell in each hemisphere that carries warm equatorial air toward the poles while cool polar air moves toward the equator. This diagram shows how the circulation cells would work if our planet's rotation didn't affect them. The vertical extent of the circulation cells is greatly exaggerated; in reality, the tops of the cells are only a few kilometers above Earth's surface.

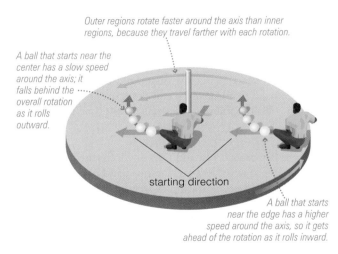

Outer regions rotate faster around the axis than inner regions, because they travel farther with each rotation.

A ball that starts near the center has a slow speed around the axis; it falls behind the overall rotation as it rolls outward.

starting direction

A ball that starts near the edge has a higher speed around the axis, so it gets ahead of the rotation as it rolls inward.

Figure 10.14 Interactive Figure ↖ The Coriolis effect on a merry-go-round rotating counterclockwise. Notice that the ball's path deviates to the *right* regardless of whether it is rolled inward or outward. If the merry-go-round were rotating clockwise, both balls would veer to the left.

Rotation and the Coriolis Effect Planetary rotation is the second major factor in global wind patterns, because rotation can alter wind patterns through what we call the **Coriolis effect** (named for the French physicist who first explained it).

It's easy to understand the Coriolis effect on a spinning merry-go-round (Figure 10.14). The outer parts of the merry-go-round move at a faster speed than the inner parts, because they have a greater distance to travel around the axis with each rotation. If you sit near the edge and roll a ball toward the center, the ball begins with your relatively high speed around the axis. As it rolls inward, the ball's

high speed makes it move ahead of the slower-moving inner regions. On a merry-go-round rotating counterclockwise, the ball therefore deviates *to the right* instead of heading straight inward. The Coriolis effect also makes the ball deviate to the right if you roll it outward from a position near the center. In that case, the ball starts with your slower speed around the center and lags behind as it rolls outward to faster-moving regions. Again, it deviates to the right. (If the merry-go-round rotates clockwise rather than counterclockwise, the deviations go to the left instead of the right.)

The Coriolis effect alters the path of air on the rotating Earth in much the same way. Earth's equatorial regions travel faster than polar regions that are closer to the rotation axis (see Figure 1.12). Air moving away from the equator therefore has "extra" speed that causes it to move ahead of Earth's rotation to the east, while air moving toward the equator lags behind Earth's rotation to the west. In either case, moving air turns to the *right* in the Northern Hemisphere and to the *left* in the Southern Hemisphere, which explains why storms circulate in opposite directions in the two hemispheres (Figure 10.15). Uneven heating and cooling of the Earth's surface creates regions of slightly higher pressure ("H" on weather maps) or lower pressure ("L" on weather maps) than average. Storms generally occur around low-pressure regions, which draw air inward from surrounding regions. The Coriolis effect makes the inward-flowing air rotate counterclockwise in the Northern Hemisphere and clockwise in the Southern Hemisphere. Thus, instead of evening out the pressure quickly, the inflowing winds end up moving in a circular pattern around the low-pressure zone. That is why storms can persist for days or weeks while being carried across the globe.

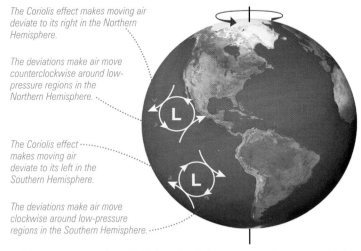

The Coriolis effect makes moving air deviate to its right in the Northern Hemisphere.

The deviations make air move counterclockwise around low-pressure regions in the Northern Hemisphere.

The Coriolis effect makes moving air deviate to its left in the Southern Hemisphere.

The deviations make air move clockwise around low-pressure regions in the Southern Hemisphere.

a Low-pressure regions ("L") draw in air from surrounding areas, and the Coriolis effect causes this air to circulate counterclockwise in the Northern Hemisphere and clockwise in the Southern Hemisphere.

Notice the opposite directions of storm circulation in the Northern and Southern Hemispheres.

b This photograph shows the opposite directions of storm circulation in the two hemispheres. Notice the counterclockwise swirling of the Northern Hemisphere storm near the upper left and the clockwise swirls in the Southern Hemisphere.

Figure 10.15 Interactive Figure ↖ The Coriolis effect works on the rotating Earth much as it does on a merry-go-round, because regions near the equator move at faster speed around the axis than regions near the poles.

The Coriolis effect also is important for long-range missiles. Suppose you are in the Northern Hemisphere on Earth and are trying to send a missile to hit a target that lies 5,000 kilometers due north of you. Should you aim directly at the target, somewhat to the left, or somewhat to the right? Explain.

The Coriolis effect plays an even more important role in shaping Earth's global wind patterns: It splits each of the two huge circulation cells shown in Figure 10.13 into three smaller circulation cells (Figure 10.16). You can understand the basic reason by considering the case of air flowing southward along the surface from the North Pole. Without rotation, this air would travel 10,000 kilometers due south to the equator. But on the rotating Earth, the Coriolis effect diverts this air to the right (westward) well before it reaches the equator. This diversion also explains the global wind directions: Surface air moves toward the equator in the cells near the equator and near the poles, so the Coriolis effect diverts this air into westward winds (see Figure 10.12). In contrast, surface air moves toward the poles in the midlatitude cells, so the Coriolis effect diverts it into winds that blow eastward. In essence, the Coriolis effect on a rotating planet tends to divert air moving north or south into east-west winds.

The Coriolis effect operates to some extent on all planets. Its strength depends on both a planet's size and rotation rate: Larger size and faster rotation both contribute to a stronger Coriolis effect. Among the terrestrial planets,

Earth is the only one with a Coriolis effect strong enough to split the two large circulation cells. Venus has a very weak Coriolis effect because of its slow rotation, while Mars has a weak Coriolis effect because of its small size. The Coriolis effect is much stronger on the large and fast-rotating jovian planets, and it can therefore split their circulation cells into many more smaller cells than on Earth [Section 11.1].

Clouds and Precipitation In addition to winds, the other key components of weather are rain, snow, and hail, which together are called **precipitation** in weather reports. Precipitation requires clouds. We may think of clouds as imperfections on a sunny day, but they have profound effects on Earth and other planets. Besides being the source of precipitation, they can alter a planet's energy balance. Clouds reflect sunlight back to space, thereby reducing the amount of sunlight that warms a planet's surface, but they also tend to be made from greenhouse gases that contribute to planetary warming. Despite their significant effects, clouds generally are made from minor ingredients of the atmosphere (see Table 10.1).

On Earth, clouds are made from tiny droplets of liquid water or flakes of ice, a fact that you can feel if you walk through a cloud on a mountaintop. Thus, clouds are produced by condensation of water vapor (Figure 10.17). The water vapor enters the atmosphere through evaporation of surface water (or sublimation of ice and snow). Convection then carries the water vapor to high, cold regions of the troposphere, where it can condense to form clouds.

The Coriolis effect diverts air flowing north-south into east-west winds, which causes the single circulation cell in each hemisphere to become three cells.

Northern Hemisphere circulation cells

Southern Hemisphere circulation cells

Not to scale!

Figure 10.16 Interactive Figure On Earth, the Coriolis effect causes each of the two large circulation cells that would be present without rotation (see Figure 10.13) to split into three smaller cells.

Toilets in the Southern Hemisphere

A common myth holds that water circulates "backward" in the Southern Hemisphere—that toilets flush the opposite way, water spirals down sink drains the opposite way, and so on. If you visit an equatorial town, you may even find charlatans who, for a small fee, will demonstrate how to change the direction of water spiraling down a drain simply by stepping across the equator. This myth sounds similar to the Coriolis effect—which really does make hurricanes spiral in opposite directions in the two hemispheres—but it is completely untrue.

The Coriolis effect is noticeable only when material moves significantly closer to or farther from Earth's rotation axis, which means distances of hundreds of kilometers. It is completely negligible for objects as small as toilets or drains, or even for tornadoes or dust devils; the direction of swirl in these cases is determined by small-scale effects such as the shape of the sink or the initial direction of rotation. Thus, you can find water swirling down drains in either direction, as well as tornadoes and dust devils twisting in either direction, no matter where you are located on Earth.

Figure 10.17 The cycle of water on Earth's surface and in the atmosphere.

3. Water vapor condenses into droplets or flakes, forming clouds.

4. Drops and flakes grow larger.

2. Convection carries vapor higher, to cooler regions.

5. Rain and snow fall to surface.

1. Water evaporates into atmosphere

Clouds can also form as winds blow over mountains, since the mountains physically push the air high enough to allow condensation. The condensed droplets or ice flakes start out very small, but gradually grow larger. If they get large enough so that the upward convection currents can no longer hold them aloft, then they begin to fall toward the surface as rain, snow, or hail.

Stronger convection means more clouds and precipitation. That is why thunderstorms are common on summer afternoons, when the sunlight-warmed surface drives strong convection. The linkage between clouds and convection also explains why Earth has lush jungles near the equator and deserts at latitudes of 20°–30° north or south. Equatorial regions experience high rainfall because they receive more sunlight, which drives strong convection. This high rainfall depletes the air of moisture as Earth's circulation cells carry it away from the equator (see Figure 10.16), leaving little moisture to fall as rain at the latitudes of the deserts.

• What factors can cause long-term climate change?

Long-term climate changes are more difficult to notice than day-to-day variations in the weather, but geological evidence proves that they occur. On Earth, we are well aware of ice ages and warm periods in our planet's past. Other planets have gone through even more dramatic changes. For example, our geological tour of Mars showed that it must have been warm and wet in the distant past but now is very cold and dry [Section 9.4].

Four major factors can lead to long-term climate change on the terrestrial worlds:

- **Solar brightening:** The Sun has grown gradually brighter with time, increasing the amount of warming sunlight that reaches the planets.
- **Changes in axis tilt:** The tilt of Earth and the other planets can change over long periods of time.
- **Changes in reflectivity:** An increase in a planet's reflectivity means a decrease in the amount of sunlight that it absorbs, and vice versa.
- **Changes in greenhouse gas abundance:** More abundant greenhouse gases result in greater warming by the atmosphere.

Figure 10.18 summarizes these four factors, which we now investigate in a little more detail. Keep in mind that more than one factor may be acting at any one time.

Solar brightening: As the Sun brightens with time, the increasing sunlight tends to warm the planets.

Changes in reflectivity: Higher reflectivity tends to cool a planet, while lower reflectivity leads to warming.

Changes in axis tilt: Greater tilt makes more extreme seasons, while smaller tilt keeps polar regions colder.

Changes in greenhouse gas abundance: An increase in greenhouse gases slows escape of infrared radiation, warming the planet, while a decrease leads to cooling.

Figure 10.18 Four major factors affecting long-term climate change.

Solar Brightening Both theoretical models of the Sun and observations of other Sun-like stars tell us that the Sun has gradually brightened with age [Section 14.2]: The Sun today is thought to be about 30% brighter than it was when our solar system was young. This brightening tends to warm climates with time, though we'd expect the effects to be noticeable only over periods of tens of millions of years or longer. Moreover, the brightening is so gradual that other climate change factors can easily overwhelm it. For example, while solar brightening alone would mean that all planets should be warmer today than they were in the past, both Mars and Earth are cooler than they were at times in their early histories.

Changes in Axis Tilt Small gravitational tugs from moons, other planets, and the Sun can change a planet's axis tilt over thousands or millions of years. For example, while Earth's current axis tilt currently is about $23\frac{1}{2}°$, the tilt varies over tens of thousands of years between about 22° and 25° (Figure 10.19). These small changes affect the climate by making seasons more or less extreme. Greater tilt means more extreme seasons, with warmer summers and colder winters. The extra summer warmth tends to prevent ice from building up, making the whole planet warmer on average. In contrast, smaller tilts mean less sunlight falls in polar regions in summer, making these regions colder.

Earth's past periods of smaller axis tilt correlate well with colder climate and ice ages, especially when considered along with other small changes in Earth's rotation and orbit. Thus, they are thought to be a primary factor in climate changes on Earth. (The cyclical changes in Earth's axis tilt and orbit are often called *Milankovitch cycles*, after the Serbian scientist who suggested their role in climate change.) As we'll discuss shortly, Mars probably experiences much more extreme changes in axis tilt and climate than Earth, and similar climate cycles may occur on some of the icy moons of the outer solar system.

Changes in Reflectivity Changes in reflectivity affect climate because they change the proportion of sunlight absorbed and reflected. If a planet reflects more sunlight, it must absorb less, which can lead to planetwide cooling. Microscopic dust particles (sometimes called *aerosols*) released by volcanic eruptions can reflect sunlight and thus cool a planet. Small but measurable planetwide cooling has been detected on Earth following a major volcanic eruption. The cooling can continue for years if the dust particles are blasted all the way into the stratosphere.

Human activity is currently altering Earth's reflectivity, although we are not sure in which direction or by precisely how much. Smog particles can act like volcanic dust, reflecting sunlight before it reaches the ground. Deforestation also increases reflectivity because it removes sunlight-absorbing plants. On the other hand, paving a road with blacktop decreases reflectivity, because black roads absorb most of the sunlight that hits them. With different human activities altering Earth's reflectivity in different directions, the overall effects are not easy to gauge.

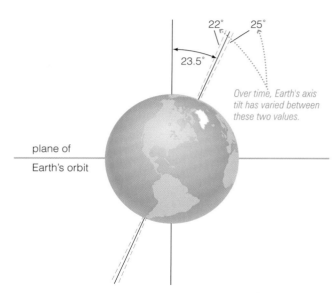

Figure 10.19 Earth's axis tilt is currently about $23\frac{1}{2}°$, but it varies over time between about 22° and 25°.

Changes in Greenhouse Gas Abundance Perhaps the most important factor in long-term climate change is a change in the abundance of greenhouse gases, which strengthens or weakens the greenhouse effect. If the abundance of greenhouse gases increases, the planet generally will warm. If the planet warms enough, increased evaporation and sublimation may add substantial amounts of gas to the planet's atmosphere, leading to an increase in atmospheric pressure. Conversely, if the abundance of greenhouse gases decreases, the planet generally will cool, and atmospheric pressure may decrease as gases freeze.

• How does a planet gain or lose atmospheric gases?

Of the factors that affect planetary climate, changes in greenhouse gas concentration appear to have had the greatest effect on the long-term climates of Venus, Earth, and Mars. But how can the greenhouse gas concentration change? In fact, such changes generally occur as part of more general changes in the abundances of atmospheric gases. Thus, before we can understand the specific reasons for long-term climate changes on the terrestrial worlds, we must investigate how atmospheres gain and lose gas.

Sources of Atmospheric Gas During their births some four and a half billion years ago, the terrestrial worlds were too small to capture significant amounts of hydrogen and helium gas before the solar wind cleared away the solar nebula [Section 8.3]. Thus, they were essentially born without atmospheres. So where did the gas in terrestrial atmospheres come from? In general, three different processes can provide gas to terrestrial atmospheres (Figure 10.20).

As we discussed in Chapter 9, volcanic *outgassing* has been the primary source of gases for the terrestrial atmospheres. Recall that the terrestrial worlds were built primarily of metal and rock, but impacts of ice-rich planetesimals

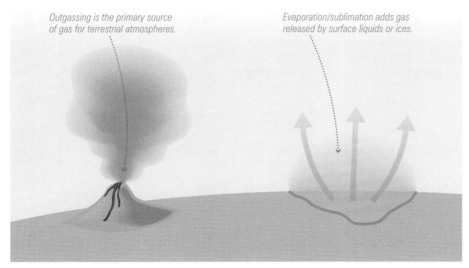

Outgassing is the primary source of gas for terrestrial atmospheres.

Evaporation/sublimation adds gas released by surface liquids or ices.

Impacts of particles and photons on worlds without atmospheres can eject small amounts of gas from surface.

Figure 10.20 Three processes that can provide gas to terrestrial atmospheres.

(from beyond the frost line [Section 8.3]) brought in water and gas that became trapped in their interiors. Studies of volcanic eruptions show that the most common gases released by outgassing are water (H_2O), carbon dioxide (CO_2), nitrogen (N_2), and sulfur-bearing gases (H_2S and SO_2).*

Some worlds have a secondary source of atmospheric gas: evaporation of surface liquids or sublimation [Section 5.3] of surface ices into gas. After outgassing creates an atmosphere, some atmospheric gases may condense to become surface liquids or ices. For example, Earth's oceans and polar caps were made from water vapor that condensed as rain or snow, while Mars's polar caps contain both frozen carbon dioxide and water ice. Surface liquids or ices continually exchange gas with the atmosphere, and overall climate changes can alter the balance of this exchange. If a planet warms, the rates of evaporation and sublimation can increase, thereby adding gas to the atmosphere.

Outgassing and evaporation/sublimation are the only important sources of gas on planets that have significant atmospheres, such as Venus, Earth, and Mars. However, a third source can produce very small amounts of gas on worlds that are essentially unprotected by an atmosphere: Micrometeorites, solar wind particles, and high-energy solar photons can knock individual atoms or molecules free from the surface. In essence, the particles and photons affect the surface much like sandblasting, but on a much smaller scale. As a result, an "airless" world like the Moon or Mercury can end up with an extremely low-density atmosphere consisting of both the atoms and molecules released from the surface and the particles that blasted them away. This "surface ejection" process does not affect planets with more substantial atmospheres, because the atmo-

spheres prevent small particles and high-energy solar photons from reaching the surface.

Losses of Atmospheric Gas Planets can also lose atmospheric gas through five different processes, all summarized in Figure 10.21. Two of these processes simply recycle gas from the atmosphere to the planets' surface or interior. The first, *condensation*, is the reverse of the release of gas through evaporation or sublimation; gases in the atmosphere condense and fall as rain, hail, or snow. On Mars, for example, it is cold enough for carbon dioxide to condense into solid ice, especially at the poles. That is why the Martian polar caps contain large amounts of frozen carbon dioxide (also known as "dry ice"). The second process removes gas from the atmosphere through *chemical reactions* that incorporate the gas into surface metal or rock. Rusting is a familiar example of this process: Iron rusts when it reacts with oxygen, thereby removing the oxygen from the atmosphere and incorporating it into the metal. This process is so efficient that it could remove all the oxygen in Earth's atmosphere in just a few million years; Earth retains atmospheric oxygen only because oxygen is continually resupplied by plants and other photosynthetic organisms.

The removal of gas by condensation or chemical reactions is not necessarily permanent, since it leaves open the possibility of the gas returning to the atmosphere at a later time. In contrast, the other three loss processes shown in Figure 10.21 are always permanent, because the gravity of the terrestrial planets is too weak to recapture lost gases.

The first of these permanent loss processes is the loss of gas through impacts: The same impacts that create large craters may blast significant amounts of atmospheric gas into space. All the terrestrial worlds probably lost gas in this way during the heavy bombardment; Mars probably suffered the greatest loss, because its small size compared to Venus and Earth allowed gas to escape more easily.

---

*Among these materials, only water existed in modest quantities in the solar nebula. The others were created by chemical reactions that occurred *inside* the planets after gas became trapped.

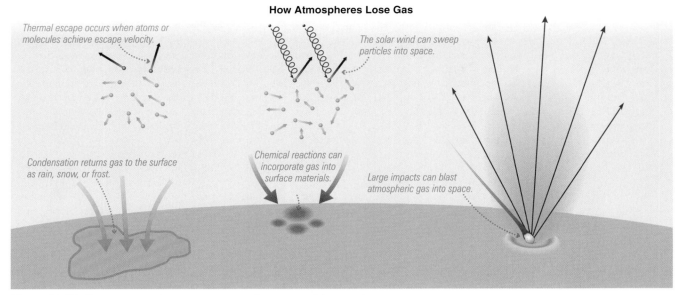

Thermal escape occurs when atoms or molecules achieve escape velocity.

The solar wind can sweep particles into space.

Condensation returns gas to the surface as rain, snow, or frost.

Chemical reactions can incorporate gas into surface materials.

Large impacts can blast atmospheric gas into space.

Figure 10.21 Five processes that can remove gas from terrestrial atmospheres.

The second permanent loss process primarily affects planets that lack a protective magnetosphere. Without a magnetosphere, particles from the solar wind constantly pelt the exosphere and sometimes strip ionized gas particles into space. In essence, the solar wind sweeps away some of a planet's atmospheric gas. This process has had little effect on Earth, because Earth's magnetic field creates a magnetosphere that protects our atmosphere. In contrast, as we'll discuss in Section 10.4, Mars may have lost much of its early atmosphere to stripping by the solar wind.

The third permanent loss process, called **thermal escape**, occurs when an atom or a molecule of gas in a planet's exosphere achieves escape velocity [Section 4.5] and flies off into space. In principle, thermal escape can affect any planetary atmosphere. However, its relative importance on any world depends on three distinct factors:

1. The planet's escape velocity: In general, more massive planets have higher escape velocities, making it more difficult for gases to escape.
2. The atmospheric temperature: Individual particles in a gas move at a wide range of random speeds [Section 4.3], but the higher the temperature, the faster their *average* speed and the more likely that some particles will reach escape velocity.
3. The mass of the gas particles: Because all gas particles have the same average kinetic energy at a particular temperature, lightweight gases such as hydrogen or helium move at faster average speeds than heavier ones such as oxygen or carbon dioxide (because kinetic energy depends on both mass and velocity [Section 4.3]). Lightweight gases are therefore more likely to achieve escape velocity.

All the terrestrial planets are small enough and warm enough for hydrogen and helium to escape. Thus, they were unable to hold for long any hydrogen or helium gas that

they may have captured from the solar nebula when they were very young. Venus, Earth, and Mars were able to retain heavier gases in their atmospheres, which is why they have substantial atmospheres today.

10.3 Atmospheres of the Moon and Mercury

We have now covered all the basic ideas that we need to understand the atmospheric histories of the terrestrial worlds. In the rest of this chapter, we'll use the atmospheric source and loss processes to learn how and why each world ended up with its current atmosphere. We'll use the ideas we've studied about weather to understand the conditions we find on planets with atmospheres. And we'll use the four factors that affect long-term climate change to discuss how and why the climates of Venus, Earth, and Mars have followed such different paths through time. As we did in Chapter 9, we'll proceed through our tour of the terrestrial atmospheres in order of planetary size. We therefore begin with the two smallest terrestrial worlds, the Moon and Mercury.

• Do the Moon and Mercury have any atmosphere at all?

We usually don't think of the Moon and Mercury as having atmospheres, but they are not totally devoid of gas. However, their gas densities are far too low to scatter sunlight as our atmosphere does. Even in broad daylight, you would see a pitch-black sky surrounding the bright Sun. Moreover, their atmospheres are too thin to absorb much solar radiation or to have any kind of atmospheric structure such as a troposphere, stratosphere, or thermosphere. In

The amount of gas that an atmosphere loses to thermal escape depends on the speeds of the gas particles (atoms or molecules). In general, gas particles of a particular type move at a wide range of speeds. For example, Figure 1 shows the range of speeds of sodium atoms at the Moon's daytime temperature ($T = 400$ K). The peak in the figure represents the most common speed of the sodium atoms. This speed, called the *peak thermal velocity* of the sodium atoms, is given by the following formula:

$$v_{\text{thermal}} = \sqrt{\frac{2kT}{m}}$$

where m is the mass of a single atom, T is the temperature in Kelvin, and $k = 1.38 \times 10^{-23}$ joule/Kelvin is *Boltzmann's constant*. The T in the formula tells us that the peak thermal velocity increases with temperature; thus, particle speeds are generally lower for low temperatures and higher for high temperatures. The m in the formula tells us that at any particular temperature, particles of lighter gases (smaller m) move at faster speeds than particles of heavier gases.

If the peak thermal velocity of a particular type of gas is greater than the escape velocity, then most of the gas particles will quickly escape to space. However, gas can escape even if the peak thermal velocity is much less than the escape velocity, because the wide range of gas particle speeds ensures that at least some particles are moving much faster than the peak velocity. For example, Figure 1 shows that while the peak thermal velocity of sodium atoms on the Moon (about 0.5 km/s) is much less than the Moon's escape velocity (about 2.4 km/s), a small fraction of the sodium atoms are moving faster than the escape velocity. As these atoms escape, collisions between the remaining gas atoms (or between gas atoms and the Moon's surface) continually ensure that a few atoms always attain escape velocity.

In principle, at any moment, a small fraction of the particles in any gas will be moving with a speed above the escape velocity. Some of these particles will also be moving in the right direction for them to escape (and are unlikely to suffer a collision since they are in an exosphere), which means that the gas could slowly leak away into space. However, this loss may be insignificant if the peak thermal velocity is far below the escape velocity. As a rule of thumb, we can expect thermal loss to be significant if the peak thermal velocity is more than a few percent of the escape velocity; if the peak thermal velocity of a particular type of gas particle is above about 20% of the escape velocity, then this gas will be completely lost to space in less than a few billion years (that is, in a time less than the age of our solar system).

Example: The Moon has a very thin exosphere of sodium atoms but retains virtually no hydrogen gas at all. Explain why. Useful data: The Moon's daytime temperature is $T = 400$ K; the Moon's escape velocity is about 2.4 km/s; the mass of a hydrogen atom is 1.67×10^{-27} kg; the mass of a sodium atom is 3.84×10^{-26} kg (about 23 times the mass of a hydrogen atom).

Solution:

Step 1 Understand: If we assume that the Moon loses gas through thermal escape, then the rate of escape for any particular type of gas depends on how its peak thermal velocity compares to the Moon's escape velocity. From Figure 1, we already know that the peak thermal velocity for sodium atoms is about 0.5 km/s. We don't yet know the peak thermal velocity for hydrogen, so we need to calculate it.

Step 2 Solve: From the formula and the given data, the peak thermal velocity for hydrogen atoms at the Moon's daytime temperature is:

$$v_{\text{thermal (H atoms)}} = \sqrt{\frac{2kT}{m_{H\,atom}}}$$

$$= \sqrt{\frac{2 \times \left(1.38 \times 10^{-23} \frac{\text{joule}}{\text{K}}\right) \times (400\ \text{K})}{1.67 \times 10^{-27}\ \text{kg}}}$$

$$\approx 2{,}600\ \text{m/s} = 2.6\ \text{km/s}$$

Step 3 Explain: The peak thermal velocity of the hydrogen atoms is 2.6 km/s, which is slightly greater than the Moon's escape velocity of about 2.4 km/s. Thus, hydrogen atoms quickly escape, which is why the Moon cannot retain a hydrogen atmosphere. In contrast, the peak thermal velocity for sodium atoms (about 0.5 km/s) is only about 20% of the Moon's escape velocity. Thus, while the Moon slowly loses sodium atoms to space, the rate of escape is quite slow. Moreover, micrometeorites, solar wind particles, and high-energy photons constantly blast new sodium atoms from the Moon's surface into its atmosphere, so the Moon retains a thin sodium exosphere.

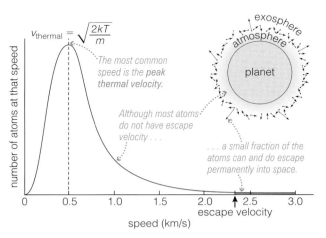

Figure 1 Interactive Figure In a gas at a given temperature, different atoms are always moving at different speeds. This plot shows the range of speeds of sodium atoms in the lunar atmosphere.

a The Moon's exosphere.   **b** Mercury's exosphere.

Figure 10.22 These "images" of the atmospheres of the Moon and Mercury—which are essentially exospheres only—are based on data collected with instruments sensitive to emission lines from sodium atoms (the same emission lines that give sodium street lights their yellow color). The colors shown here are not real, but rather represent the varying gas densities from highest (purple) to lowest (black). The inset photos of the Moon and Mercury are on the same scale as the images, so you can see that their exospheres extend quite far above their surfaces.

essence, the Moon and Mercury have only thin exospheres, without any other atmospheric layers.

The total amount of gas in the exospheres of the Moon and Mercury is very small. If you could condense the entire atmosphere of either the Moon or Mercury into solid form, you would have so little material that you could almost store it in a dorm room. The low density of the gas means that collisions between atoms or molecules are rare. The gas particles therefore can rise as high as their speeds allow—sometimes even escaping to space. As a result, the exospheres of the Moon and Mercury extend thousands of kilometers into space (Figure 10.22).

Source and Loss Processes on the Moon and Mercury   The Moon and Mercury may once have had some gas released by volcanic outgassing, but any such gas is long gone. Recall that volcanic activity on these worlds ceased billions of years ago, so there is no ongoing source of outgassing. Some of the gas released long ago was probably lost through impacts or stripping by the solar wind, but it would have been lost to thermal escape anyway. Mercury cannot hold much of an atmosphere because its small size and high daytime temperature mean that nearly all gas particles eventually achieve escape velocity. Because it is smaller than Mercury, gases escape about as easily from the Moon even though it is also somewhat cooler than Mercury.

The only ongoing source of gas on the Moon and Mercury is the "surface ejection" that occurs when micrometeorites, solar wind particles, or high-energy solar photons knock free surface atoms and molecules. This gas never accumulates because it is lost as quickly as it is gained. Some of the relatively few particles released from the surface are blasted upward fast enough to achieve escape velocity and therefore escape directly to space. The rest bounce around like tiny rubber balls, arcing hundreds of kilometers into the sky before crashing back down to the surface. Each gas particle typically bounces a few dozen times before being absorbed back into the surface.

Ice in Polar Craters   In general, we do not expect to find either liquid water or solid water ice on the Moon or Mercury. The lack of significant atmospheric pressure means that water cannot remain stable in liquid form, and daytime temperatures are high enough that any solid ice would long ago have sublimated into gas and escaped to space. However, recent observations strongly suggest that both the Moon and Mercury have some water ice at the bottoms of craters near their poles.

If you think about it, the existence of this polar ice should not be too surprising. Both the Moon and Mercury have certainly been struck many times by ice-rich comets. The ice would vaporize upon impact, and any water molecules that did not achieve escape velocity would bounce randomly around the surface until they settled down. Some molecules would have settled into craters near the poles, where they condensed into ice. Because the bottoms of these polar craters lie in nearly perpetual shadow, it is possible that water ice could remain perpetually frozen. Spacecraft observations of the Moon suggest that this is indeed the case, and radar observations of Mercury suggest that it also has ice in polar craters. Although the polar ice probably has little effect on either of these worlds, it could prove very important for future exploration. In particular, the ice in polar craters offers a potential source of water for future human colonies on the Moon.

10.4 The Atmospheric History of Mars

We began this chapter with the goal of understanding how and why the terrestrial atmospheres came to differ so profoundly. The cases of the Moon and Mercury have proved to be quite easy to understand. However, because outgassing supplied substantial early atmospheres to the three larger worlds—Mars, Venus, and Earth—their very different present atmospheres must reflect atmospheric gains and losses that have occurred through time. Again, planetary size is the most important influence on these gains and losses: Size ultimately determines the level of volcanism and outgassing on each planet and, as we'll see, size has also played a crucial role in determining which loss processes have occurred. We therefore continue our tour in size order, which brings us next to Mars.

Mars is only 40% larger in radius than Mercury, but its surface reveals a much more fascinating and complex atmospheric history. As we discussed in Chapter 9, orbital photographs and studies by robotic landers tell us that water once flowed on the Martian surface. In the distant past, Mars must have been very different from its current "freeze-dried" state.

• What is Mars like today?

The present-day surface of Mars looks much like some deserts or volcanic plains on Earth (see Figures 7.7, 9.30, and 10.1). However, its thin atmosphere makes it unlike

Seasons on Mars

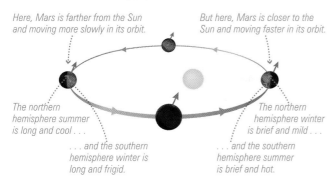

Here, Mars is farther from the Sun and moving more slowly in its orbit.

But here, Mars is closer to the Sun and moving faster in its orbit.

The northern hemisphere summer is long and cool . . .

The northern hemisphere winter is brief and mild . . .

. . . and the southern hemisphere winter is long and frigid.

. . . and the southern hemisphere summer is brief and hot.

Figure 10.23 The ellipticity of Mars's orbit makes seasons more extreme (hotter summers and colder winters) in the southern hemisphere than in the northern hemisphere.

any place on Earth. The low atmospheric pressure—less than 1% of that on Earth's surface—explains why liquid water is unstable on the Martian surface [Section 9.4]: Any liquid water would rapidly disappear, with some evaporating and some freezing into ice.

The atmosphere is made mostly of carbon dioxide, but the total amount of gas is so small that it creates only a weak greenhouse effect (see Table 10.2). The temperature is usually well below freezing, with a global average of about −50°C (−58°F). The lack of oxygen means that Mars lacks an ozone layer, so much of the Sun's damaging ultraviolet radiation passes unhindered to the surface.

Martian Seasons and Winds Mars has seasons much like those on Earth because of its similar axis tilt, but they last about twice as long (because a Martian year is almost twice as long as an Earth year). On Mars, however, the shape of its orbit introduces a second important effect. Mars's more

elliptical orbit puts it significantly closer to the Sun during the southern hemisphere summer and farther from the Sun during the southern hemisphere winter (Figure 10.23). Mars therefore has more extreme seasons in its southern hemisphere—that is, shorter, hotter summers and longer, colder winters—than in its northern hemisphere.

The seasonal changes create the major feature of Mars's weather: winds blowing from the summer pole to the winter pole. Polar temperatures at the winter pole drop so low (about −130°C) that carbon dioxide condenses into "dry ice" at the polar cap. At the same time, frozen carbon dioxide at the summer pole sublimates into carbon dioxide gas (Figure 10.24). During the peak of summer, nearly all the carbon dioxide may sublime from the summer pole, leaving only a residual polar cap of water ice (Figure 10.25). The atmospheric pressure therefore increases at the summer pole and decreases at the winter pole, driving strong pole-to-pole winds. Overall, as much as one-third of the total carbon dioxide of the Martian atmosphere moves seasonally between the north and south polar caps.

The direction of the pole-to-pole winds on Mars changes with the alternating seasons. Sometimes these winds initiate huge dust storms, particularly when the more extreme summer approaches in the southern hemisphere (Figure 10.26). At times, airborne dust shrouds the surface so much that no surface markings can be distinguished and large dunes form and shift with the strong winds. As the dust settles out onto the surface, it can change the color or reflectivity over vast areas, creating seasonal changes in appearance that fooled some astronomers in the late 1800s and early 1900s into thinking they saw changes in vegetation.

The Martian winds often spawn *dust devils*, swirling winds that you may have seen over desert sands or dry dirt on Earth. Dust devils look much like miniature tornadoes, but they rise up from the ground rather than coming down

October 1996: *Late winter, with polar cap near maximum size.*

January 1997: *Midspring, notice shrinkage of polar cap.*

March 1997: *Early Summer, with polar cap near minimum size.*

Figure 10.24 These images from the Hubble Space Telescope show the region around Mars's north pole, which is at the center of each image. Notice how the polar cap shrinks as summer approaches, due to sublimation of carbon dioxide ice. This sublimation at the summer pole occurs at the same time that carbon dioxide condenses at the winter pole. These changes drive winds from the summer pole to the winter pole.

Figure 10.25 This image, from *Mars Global Surveyor*, shows the residual north polar cap during northern hemisphere summer. The white material is primarily water ice, with dust mixed in.

from the sky. The air in dust devils is heated from below by the sunlight-warmed ground, and swirls because of the way it interacts with prevailing winds. Dust devils on Mars are especially common during summer in either hemisphere and can be far larger than their counterparts on Earth. The Mars rovers have even photographed dust devils (Figure 10.27), and some have apparently passed directly over the rovers without causing damage. In one case, a dust devil apparently "cleaned" dust off the solar panels of the rover Spirit, restoring power that had been lost as dust accumulated.

Color of the Martian Sky Martian winds and dust storms leave Mars with perpetually dusty air, which helps explain the colors of the Martian sky. The air on Mars scatters visible sunlight much like the air on Earth; that is, it scatters blue light more than red light. However, because there is much less air on Mars than Earth, it scatters less light. By itself this scattering would make the Martian sky a deep blue (almost black).

The dust in the Martian atmosphere has an opposite effect. Dust tends to absorb blue light, so by itself it would leave the sky with a brownish-pink color. The actual color of the Martian sky comes from the combination of the effects of air and dust. The result depends on the time of day, the season, and how much dust is in the air. Most of the time, the daytime Martian sky ends up yellow-brown. However, it sometimes ends up with other colors, such as blue or even green, particularly in the mornings and evenings.

Mars Climate and Axis Tilt The weather on Mars does not change much from one year to the next. However, changes in its axis tilt probably cause Mars to undergo longer-term

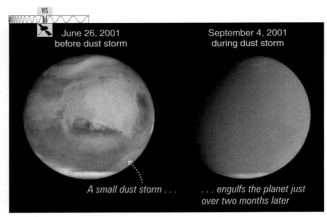

June 26, 2001
before dust storm

September 4, 2001
during dust storm

A small dust storm . . .

. . . engulfs the planet just over two months later

Figure 10.26 These two Hubble Space Telescope photos contrast the appearance of the same face of Mars in the absence (left) and presence (right) of a global dust storm.

cycles of climate change. Theoretical calculations suggest that Mars's axis tilt varies far more than that of Earth—from as little as 0° to 60° or more on time scales of hundreds of thousands to millions of years. This extreme variation arises for two reasons. First, Jupiter's gravity has a greater effect on the axis of Mars than on Earth, because Mars's orbit is closer to Jupiter's orbit. Second, Earth's axis is stabilized by the gravity of our relatively large Moon. Mars's two tiny moons, Phobos and Deimos (see Figure 8.11), are far too small to offer any stabilizing influence on its axis.

As we discussed earlier, changes in axis tilt affect both the severity of the seasons and the global average temperature. When Mars's axis tilt is small, the poles may stay in a perpetual deep freeze for tens of thousands of years. With more carbon dioxide frozen at the poles, the atmosphere becomes thinner, lowering the pressure and weakening the greenhouse effect. When the axis is highly tilted, the summer pole becomes quite warm, allowing substantial amounts of water ice to sublime, along with carbon dioxide, into the atmosphere. The pressure therefore increases, and Mars becomes warmer as the greenhouse effect strengthens—

Figure 10.27 This photograph shows a dust devil on Mars, photographed by the *Spirit* rover in early 2005.

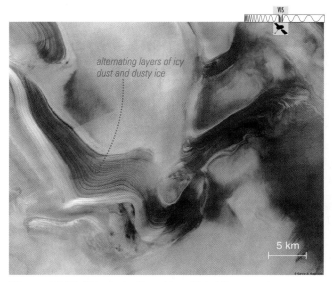

alternating layers of icy dust and dusty ice

5 km

Figure 10.28 This photo from *Mars Global Surveyor* shows a region about 40 kilometers wide at the edge of a Martian polar cap. Note the alternating layers of ice and dust, which are probably the result of climate changes caused by the changing axis tilt of Mars.

although probably not by enough to allow liquid water to become stable at the surface. The Martian polar regions show layering of dust and ice that probably reflects changes in climate due to the changing axis tilt (Figure 10.28).

• Why did Mars change?

Climate change due to Mars's varying axis tilt may have had important effects, but it does not explain the most dramatic climate change in Martian history: the change from the world of flowing water that existed long ago to the cold and dry world we see today.

The geological evidence we discussed in Chapter 9 leaves little room for doubt that Mars had wetter and warmer periods, probably with rainfall, that came to an end more than 3 billion years ago. The full extent of these periods is a topic of considerable scientific debate. Some scientists think that Mars may have been continuously warm and wet for much of its first billion years of existence. Others think that Mars may have had only intermittent periods of rainfall, perhaps triggered by the heat of large impacts, and that ancient lakes, ponds, or oceans may have been completely ice-covered.

Either way, Mars must have had a much denser and warmer atmosphere to allow rainfall of any kind. Warmer temperatures would have been needed to keep water from freezing, and higher density would have been needed to make the atmospheric pressure great enough to allow for stable liquid water. Computer simulations show that Mars today could have liquid water if it had a carbon dioxide atmosphere about 400 times as dense as the current Martian atmosphere; this atmosphere would give Mars a surface pressure about three times that on Earth today. However, because the Sun was dimmer in the distant past (see Fig-

ure 10.18), even more greenhouse warming would have been needed to allow for liquid water when Mars was young. Current models are unable to account for the necessary additional warming with carbon dioxide gas alone, but additional greenhouse warming may have been provided by carbon dioxide ice clouds or atmospheric methane.

The idea that Mars once had a much thicker atmosphere and stronger greenhouse effect makes sense. Mars has plenty of ancient volcanoes, and calculations show that these volcanoes could have supplied the atmosphere with 400 times as much carbon dioxide as it has today. Moreover, if Martian volcanoes outgassed carbon dioxide and water in the same proportions as do volcanoes on Earth, Mars would have had enough water to fill oceans tens or even hundreds of meters deep.

The bigger question is not whether Mars once had a denser atmosphere but where all the atmospheric gas went. In particular, Mars must somehow have lost most of its carbon dioxide gas. This loss would have weakened the greenhouse effect until the planet essentially froze over. Some of the carbon dioxide condensed to make the polar caps and some may be chemically bound to surface rock. However, the bulk of the gas was probably lost to space.

The precise way in which Mars lost its carbon dioxide gas is not clear, but recent data suggest a close link to a change in Mars's magnetic field (Figure 10.29). Early in its history, Mars probably had molten metals in its core, much like Earth today. Circulating with convection and Mars's rotation, these metals should have produced a magnetic field and a protective magnetosphere around Mars. However, the magnetic field weakened as the small planet cooled and the core solidified, leaving the atmosphere vulnerable to solar wind particles. These solar wind particles could have stripped gas out of the Martian atmosphere and into space. Had Mars kept its magnetic field, it might have retained much of its atmosphere and might have a more moderate climate today.

Much of the water once present on Mars is also probably gone for good. Like the carbon dioxide, some water vapor may have been stripped away by the solar wind. However, Mars also lost water in another way. Because Mars lacks an ultraviolet-absorbing stratosphere, atmospheric water molecules would have been easily broken apart by ultraviolet photons. The hydrogen atoms that broke away from the water molecules would have been lost rapidly to space through thermal escape. With these hydrogen atoms gone, the water molecules could not be made whole again. Initially, oxygen from the water molecules would have remained in the atmosphere, but over time this oxygen was lost too. The solar wind probably stripped some of the oxygen away from the atmosphere, and the rest probably was drawn out of the atmosphere through chemical reactions with surface rock. This oxygen literally rusted the Martian rocks, giving the "red planet" its distinctive tint.

In summary, Mars changed primarily because of its relatively small size. It was big enough for volcanism and outgassing to release plenty of water and atmospheric gas early

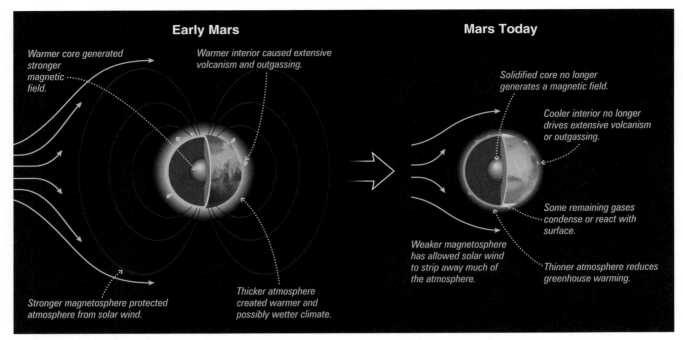

Figure 10.29 Some 3 billion years ago, Mars underwent dramatic climate change, ensuring that rain could never fall again.

in its history, but too small to maintain the internal heat needed to prevent the loss of this water and gas. As its interior cooled, its volcanoes quieted and stopped releasing gas into the atmosphere, while its relatively weak gravity and the loss of its magnetic field allowed existing gas to be stripped away to space. If Mars had been as large as Earth, so that it could still have outgassing and a global magnetic field, it might still have a moderate climate today. Mars's distance from the Sun also helped seal its fate: Even with its small size, Mars might still have some flowing water if it were significantly closer to the Sun, where the extra warmth could melt the water that remains frozen underground and at the polar caps.

The history of the Martian atmosphere holds important lessons for us on Earth. Mars apparently was once a world with pleasant temperatures and with streams, rain, glaciers, lakes, and possibly oceans. It had all the necessities for life as we know it. This once-hospitable planet turned into a frozen and barren desert at least 3 billion years ago, and it is unlikely that Mars will ever again be warm enough for its frozen water to flow. Any life that may have existed on Mars is either extinct or hidden away in a few choice locations, such as in underground water near not-quite-dormant volcanoes. As we consider the possibility of future climate change on Earth, Mars presents us with an ominous example of how drastically things can change.

> **THINK ABOUT IT**
>
> Some people have proposed "terraforming" Mars—that is, making it more Earth-like—by finding a way to release all the carbon dioxide frozen in its polar caps into its atmosphere. If Mars once had oceans in the distant past, could the release of gas from the polar caps allow it to have oceans again? Why or why not?

10.5 The Atmospheric History of Venus

Venus presents a stark contrast to Mars. Its thick carbon dioxide atmosphere creates an extreme greenhouse effect that bakes the surface to unbearable temperatures. It is fairly easy to understand why Venus is so different from Mars: Its larger size allowed it to retain more interior heat, leading to greater volcanism [Section 9.5]. The associated outgassing released the vast quantities of carbon dioxide that create Venus's strong greenhouse effect.

Venus becomes more mysterious when we compare it to Earth. Because Venus and Earth are so similar in size, we might naively expect both planets to have had similar atmospheric histories. Clearly, this is not the case. In this section, we'll explore how Venus's atmosphere ended up so different from Earth's, and in the process we'll learn important lessons about the habitability of our own planet.

• What is Venus like today?

If you stood on the surface of Venus, you'd feel a searing temperature higher than that of a self-cleaning oven and a tremendous pressure 90 times greater than that on Earth. A deep-sea diver would have to go nearly 1 kilometer (0.6 mile) beneath the ocean surface on Earth to feel comparable pressure. Venus's atmosphere consists almost entirely of carbon dioxide (CO_2). It has virtually no molecular oxygen (O_2), so you could not breathe the air even if you cooled it to a comfortable temperature.

Moving through the thick air near Venus's surface would feel like a cross between swimming and flying: Its density is about 10% that of water. Looking upward, you'd see a perpetually overcast sky, with only weak sunlight filtering

Figure 10.30 Clouds on Venus are all that can be seen in this ultraviolet image taken by the *Pioneer Venus Orbiter*.

through the thick clouds above. Because the thick atmosphere scatters nearly all the blue light away, the dimly lit sky appears reddish-orange in color.

The weather forecast for the surface of Venus today and every day is dull, dull, dull. Venus's slow rotation (243 Earth days) means a very weak Coriolis effect. As a result, Venus has little wind on its surface and never has hurricane-like storms. The top wind speeds measured by the Soviet Union's *Venera* landers were only about 6 kilometers per hour. No rain falls, because any gases that condense in the cool upper atmosphere evaporate long before they reach the ground. The weak Coriolis effect also means that Venus's atmosphere has just two large circulation cells, much like what Earth would have if rotation didn't split the cells (see Figure 10.13). The thick atmosphere makes the circulation so efficient at transporting heat from the equator to the poles that the surface temperature is virtually the same everywhere. That is, the poles are no cooler than the equator and night is just as searingly hot as day. Moreover, Venus has no seasons because it has virtually no axis tilt.* Thus, temperatures are the same year-round.

While Venus has very dull surface weather, more interesting things occur higher up in its atmosphere. Because its high surface temperature drives strong convection, Venus is covered with bright, reflective clouds (Figure 10.30). The clouds are made of droplets containing sulfuric acid (H_2SO_4), which condenses high in the troposphere where temperatures are 400°C cooler than on the surface (see Figure 10.10). The droplets sometimes fall through the upper troposphere as sulfuric acid rain, but they evaporate at least 30 kilometers above the surface because of the higher temperatures closer to the ground. The upper atmosphere of Venus also has some fast winds, though no one knows precisely why.

*In tables (such as Table 7.1), Venus's axis tilt is usually written as 177.4°. This may sound large, but notice that it is nearly 180°—the same tilt as 0° but "upside down." It is written this way because Venus rotates backward compared to its orbit, and backward rotation is equivalent to forward rotation that is upside down.

• How did Venus get so hot?

It's tempting to attribute Venus's high surface temperature solely to the fact that it is closer than Earth to the Sun. However, as we discussed earlier, Venus's cloud cover would actually make the planet quite cold without its strong greenhouse effect (see Table 10.2). Thus, the real question is why Venus has such a strong greenhouse effect.

On a simple level, the answer is Venus's thick carbon dioxide atmosphere; the total amount of carbon dioxide in Venus's atmosphere is nearly 200,000 times as great as the amount in Earth's atmosphere. However, a deeper question still remains. Given their similar sizes and compositions, we expect Venus and Earth to have had similar levels of volcanic outgassing—and the released gas ought to have had about the same composition on both worlds. Why, then, is Venus's atmosphere so different from Earth's?

The Fate of Outgassed Water and Carbon Dioxide   Our understanding of planetary geology suggests that huge amounts of water and carbon dioxide should have been outgassed into the atmospheres of both Venus and Earth. Venus's atmosphere does indeed have huge amounts of carbon dioxide, but it has virtually no water. Earth's atmosphere has very little of either gas. What happened to all the outgassed water on Venus, and to the outgassed water and carbon dioxide on Earth?

We can easily account for both missing gases on Earth. The huge amounts of water vapor released into our atmosphere condensed into rain, forming our oceans. Thus, the water is still here, but mostly in liquid rather than gaseous form. The huge amount of carbon dioxide released into our atmosphere is also still here, but in solid form: Carbon dioxide dissolves in water, where it can undergo chemical reactions to make **carbonate rocks** (rocks rich in carbon and oxygen) such as limestone. Earth has about 170,000 times as much carbon dioxide locked up in rocks as in its atmosphere—which means that Earth does indeed have almost as much total carbon dioxide as Venus. Of course, the fact that Earth's carbon dioxide is mostly in rocks rather than in the atmosphere makes all the difference in the world: If this carbon dioxide were in our atmosphere, our planet would be nearly as hot as Venus and certainly uninhabitable.

Having explained what happened to the water and carbon dioxide outgassed on Earth, and knowing that Venus's carbon dioxide still remains in its atmosphere, we are left with the question of what happened to Venus's water. Venus today is incredibly dry, with a surface far too hot to have either liquid water or ice. It is even too hot for water to be chemically bound in surface rock, and any water deeper in its crust or mantle was probably baked out long ago. Measurements also show very little water in the atmosphere. Overall, the total amount of water on Venus is about 10,000 times smaller than the total amount on Earth. Moreover, the absence of water explains why Venus retains so much carbon dioxide in its atmosphere: Without oceans, carbon dioxide cannot dissolve or become locked away in carbonate rocks. Thus, if it is true that a huge amount of water was outgassed on Venus, it has somehow disappeared.

The leading hypothesis for the disappearance of Venus's water invokes some of the same processes thought to have removed water from Mars. Ultraviolet light from the Sun breaks apart water molecules in Venus's atmosphere. The hydrogen atoms then escape to space (through thermal escape), ensuring that the water molecules can never re-form. The oxygen from the water molecules was lost to a combination of chemical reactions with surface rocks and stripping by the solar wind; Venus's lack of magnetic field leaves its atmosphere vulnerable to the solar wind.

Acting over billions of years, the breakdown of water molecules and the escape of hydrogen can easily explain the loss of an ocean's worth of water from Venus. Moreover, careful study of Venus's atmospheric composition offers evidence that it did indeed lose a great deal of water. Recall that most hydrogen nuclei contain just a single proton, but a tiny fraction of all hydrogen atoms (about 1 in 50,000 on Earth) contain a neutron in addition to the proton, making the isotope of hydrogen that we call *deuterium*. Water molecules that contain an atom of deuterium (called "heavy water") behave chemically just like ordinary water and can be broken apart by ultraviolet light just as easily. However, a deuterium atom is twice as heavy as an ordinary hydrogen atom and thus does not escape to space as easily when the water molecule is broken apart. If Venus lost a huge amount of hydrogen from water molecules to space, the rare deuterium atoms would have been more likely to remain behind than the ordinary hydrogen atoms. Measurements show that this is the case. The fraction of deuterium among hydrogen atoms is a hundred times higher on Venus than on Earth, suggesting that a substantial amount of water must have been broken apart and its hydrogen lost to space. We cannot determine exactly how much water Venus has lost, but it certainly seems plausible that it really did outgas as much water as Earth and then lost virtually all of it.

To summarize our discussion so far, we can successfully explain the lack of water on Venus today *if* we make a crucial assumption about water in Venus's past: The water must have all been in the atmosphere as water vapor, where ultraviolet light could break it down, rather than in liquid oceans like the water on Earth. Thus, our quest to understand Venus's high temperature leads to one more question: Why didn't Venus get oceans like Earth, which would have al-

lowed its carbon dioxide to be trapped in carbonate rocks and prevented its water from being lost to space?

The Runaway Greenhouse Effect To understand why Venus does not have oceans, let's consider what would happen if we could somehow move Earth to the orbit of Venus.

The greater intensity of sunlight would almost immediately raise Earth's global average temperature by about 30°C, from its current 15°C to about 45°C (113°F). Although this is still well below the boiling point of water, the higher temperature would lead to increased evaporation of water from the oceans. The higher temperature would also allow the atmosphere to hold more water vapor before the vapor condensed to make rain.

The combination of more evaporation and greater atmospheric capacity for water vapor would substantially increase the total amount of water vapor in Earth's atmosphere. Because water vapor is a greenhouse gas, this added water vapor would strengthen the greenhouse effect, driving temperatures a little higher. The higher temperatures, in turn, would lead to even more ocean evaporation and more water vapor in the atmosphere—strengthening the greenhouse effect even further. In other words, we'd have a self-sustaining process in which each little bit of additional water vapor in the atmosphere would mean higher temperature and even more water vapor.

The process would rapidly spin out of control, resulting in a **runaway greenhouse effect** (Figure 10.31). Our planet would heat up until the oceans were completely evaporated and the carbonate rocks had released all their carbon dioxide back into the atmosphere. By the time the runaway process was complete, temperatures on our "moved Earth" would be even higher than they are on Venus today, thanks to the combined greenhouse effects of carbon dioxide and water vapor in the atmosphere. The water vapor would then gradually disappear, as ultraviolet light broke water molecules apart and the hydrogen escaped to space. In short, moving Earth to Venus's orbit would essentially turn our planet into another Venus.

This scenario suggests a simple explanation for why Venus is so much hotter than Earth. Even though Venus's distance from the Sun is only about 30% less than Earth's distance, the difference in distance was critical. On Earth,

Figure 10.31 This diagram shows how, if Earth were placed at Venus's distance from the Sun, the runaway greenhouse effect would cause the oceans to evaporate completely.

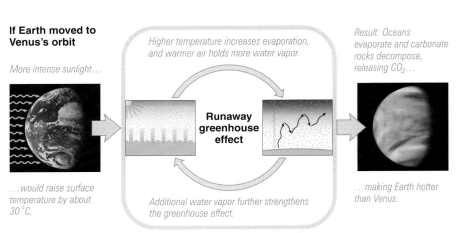

If Earth moved to Venus's orbit

More intense sunlight…

…would raise surface temperature by about 30°C.

Higher temperature increases evaporation, and warmer air holds more water vapor.

Runaway greenhouse effect

Additional water vapor further strengthens the greenhouse effect.

Result: Oceans evaporate and carbonate rocks decompose, releasing CO$_2$…

…making Earth hotter than Venus.

it was cool enough for water to rain down to make oceans. The oceans then dissolved carbon dioxide and chemical reactions locked it away in carbonate rocks, leaving our atmosphere with only enough greenhouse gases to make our planet pleasantly warm. On Venus, the greater intensity of sunlight made it just enough warmer that oceans either never formed or soon evaporated. Without oceans to dissolve carbon dioxide and make carbonate rock, carbon dioxide accumulated in the atmosphere, leading to a runaway greenhouse effect.

The next time you see Venus shining brightly as the morning or evening "star," consider the radically different path it has taken from that taken by Earth—and thank your lucky star. If Earth had formed a bit closer to the Sun or if the Sun had been slightly hotter, our planet might have suffered the same greenhouse-baked fate.

<hr>THINK ABOUT IT<hr>

We've seen that moving Earth to Venus's orbit would cause our planet to become Venus-like. If we could somehow move Venus to Earth's orbit, would it become Earth-like? Why or why not?

A Pleasant Early Venus? Venus's closeness to the Sun may have sealed its ultimate fate, but it's possible that Venus might have been more moderate in its early history. Remember that the Sun has gradually brightened with age. Thus, sunlight was less intense on all the planets when they were young. Some 4 billion years ago, the intensity of sunlight at Venus was not much greater than it is at Earth today. Rain might well have fallen on the young planet Venus, and oceans could have formed. It's even conceivable that life could have arisen on the young Venus.

As the Sun gradually brightened, however, any liquid water or life on Venus was doomed. The runaway greenhouse effect raised the temperature so high that all the water evaporated. In the upper atmosphere, ultraviolet light broke apart the water molecules, and the hydrogen escaped to space. If Venus had oceans in its youth, the water is now gone forever.

We'll probably never know for sure whether Venus ever had oceans. The global "repaving" of Venus's surface by tectonics and volcanism [Section 9.5] would long ago have covered up any shorelines or other geological evidence of past oceans, and the high surface temperatures would have baked out any gases that might once have been incorporated into surface rock. If the climate once was pleasant, it's unlikely that any evidence survives to tell the tale.

10.6 Earth's Unique Atmosphere

Earth's atmosphere, composed mostly of nitrogen and oxygen, is the only one under which human life is possible outside an artificial environment. This atmosphere provides our planet with just enough warmth and pressure to enable water to cycle between all three phases (solid ice, liquid water, and gaseous water vapor). It protects us from harmful solar radiation and produces the weather patterns that variously bring us days of sunshine, clouds, and rain or snow. In this section, we'll discuss how and why our atmosphere is so hospitable, and we'll conclude by considering how we humans may be altering the very balances upon which we depend for survival.

• How did Earth's atmosphere end up so different?

If you think about our comparative study so far, you'll realize that most of the major physical differences in the atmospheres of the terrestrial planets are easy to understand. For example, the Moon and Mercury lack substantial atmospheres because of their small sizes, while a very strong greenhouse effect causes the high surface temperature on Venus.

Explaining differing atmospheric compositions is more challenging, especially when we consider the cases of Venus, Earth, and Mars. In all three cases, outgassing should have released the same gases. How, then, did Earth's atmosphere end up so different? We can break down this general question into four separate questions:

1. Why did Earth retain most of its outgassed water—enough to form vast oceans—while Venus and Mars lost theirs?
2. Why does Earth have so little carbon dioxide (CO_2) in its atmosphere compared to Venus, when Earth should have outgassed about as much of it as Venus?
3. Why is Earth's atmosphere composed primarily of nitrogen (N_2) and oxygen (O_2), when these gases are only trace constituents in the atmospheres of Venus and Mars?
4. Why does Earth have an ultraviolet-absorbing stratosphere, while Venus and Mars do not?

Water and Carbon Dioxide We have already answered the first two questions in our discussions of the atmospheres of Mars and Venus. On Mars, some of the outgassed water was lost after solar ultraviolet light broke water vapor molecules apart, and the rest froze and may remain in the polar caps or underground. On Venus, it was too hot for water vapor to condense, so virtually all the water molecules were ultimately broken apart, allowing the hydrogen atoms to escape to space. Earth retained its outgassed water because temperatures were low enough for water vapor to condense into rain and thereby form oceans. Evidence from tiny mineral grains (known as zircons) suggests that Earth had oceans as early as 4.3–4.4 billion years ago.

The presence of oceans also explains the low level of carbon dioxide in our atmosphere. Most of the carbon dioxide outgassed by volcanism on Earth dissolved in the oceans, where chemical reactions ultimately turned it into carbonate rocks. Even today, about 60 times as much carbon dioxide is dissolved in the oceans as is present in the atmosphere, and carbonate rocks contain some 170,000 times as much CO_2 as the atmosphere.

Nitrogen, Oxygen, and Ozone   Turning our attention to the third question, it's relatively easy to explain the substantial nitrogen content (77%) of our atmosphere. Although water vapor and carbon dioxide are the two most common gases released by outgassing, the third most common gas is nitrogen. Because most of Earth's water ended up in the oceans and most of the carbon dioxide ended up in rocks, our atmosphere was left with nitrogen as its dominant ingredient. The oxygen content (21%) of our atmosphere is a little more mysterious.

Molecular oxygen (O_2) is not a product of outgassing or any other geological process. Moreover, oxygen is a highly reactive chemical that would disappear from the atmosphere in just a few million years if it were not continuously resupplied. Fire, rust, and the discoloration of freshly cut fruits and vegetables are everyday examples of chemical reactions that remove oxygen from the atmosphere (often called *oxidation reactions*). Similar reactions between oxygen and surface materials (especially iron-bearing minerals) give rise to the reddish appearance of much of Earth's rock and clay, including the beautiful reds of Arizona's Grand Canyon. Thus, we must explain not only how oxygen got into Earth's atmosphere in the first place, but also how the amount of oxygen remains relatively steady even while chemical reactions can remove it rapidly from the atmosphere.

The answer to the oxygen mystery is *life*. Plants and many microorganisms release oxygen through photosynthesis. Photosynthesis takes in CO_2 and, through a fairly complex chain of chemical reactions, releases O_2. Because chemical reactions can remove oxygen from the atmosphere, it took a long time for oxygen to accumulate in Earth's atmosphere. According to present evidence, it took at least a billion years of photosynthesis before the buildup of atmospheric oxygen began, and Earth's atmosphere probably has had enough oxygen for us to breathe only for the past few hundred million years [Section 24.1]. If you had a time machine and spun the dial to arrive randomly at any point in Earth's past, there's only about a 1 in 10 chance that you would arrive at a time when you'd be able to breathe the outside air. Today, plants and single-celled photosynthetic organisms return oxygen to the atmosphere in approximate balance with the rate at which animals and chemical reactions consume oxygen, keeping the oxygen content relatively steady.

THINK ABOUT IT

Suppose that, somehow, all photosynthetic life on Earth died out. What would happen to the oxygen in our atmosphere? Could animals, including us, still survive?

Life and oxygen also explain the presence of Earth's ultraviolet-absorbing stratosphere. In the upper atmosphere, chemical reactions involving solar ultraviolet light transform some of the O_2 into molecules of O_3, or ozone (Figure 10.32). The O_3 molecule is more weakly bound than the O_2 molecule, which allows it to absorb solar ultraviolet

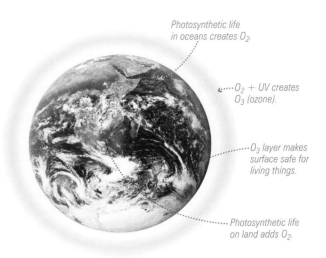

Photosynthetic life in oceans creates O_2.

O_2 + UV creates O_3 (ozone).

O_3 layer makes surface safe for living things.

Photosynthetic life on land adds O_2.

Figure 10.32 The origin of oxygen and ozone in Earth's atmosphere.

energy even better. The absorption of solar energy by ozone heats the upper atmosphere, creating the layer we call the stratosphere. This ozone layer prevents harmful ultraviolet radiation from reaching the surface. Mars and Venus lack photosynthetic life and therefore have too little O_2 and consequently too little ozone to form a stratosphere.

Maintaining Balance   We have answered all four basic questions about Earth's atmosphere. However, a deeper look at what we have learned still leaves us with a major mystery. Our oceans exist because Earth has a greenhouse effect that is "just right" to keep them from either freezing or boiling away, and the presence of liquid water allowed life to arise and produce oxygen. But *why* does the amount of carbon

COMMON MISCONCEPTIONS

Ozone—Good or Bad?

Ozone often generates confusion, because in human terms it is sometimes good and sometimes bad. In the stratosphere, ozone acts as a protective shield from the Sun's ultraviolet radiation. Thus, the fact that human activity (especially the release of the chemicals known as CFCs) has been reducing stratospheric ozone is a significant environmental problem. At the same time, ozone is poisonous to most living creatures and therefore is a bad thing when it is found near Earth's surface.

In fact, ozone is one of the main ingredients in urban air pollution. It is produced as a by-product of automobiles and industry. Some people wonder whether we might be able to transport this ozone to the stratosphere and thereby alleviate the effects of ozone depletion in the stratosphere. Unfortunately, this plan won't work. Even if we could find a way to transport this ozone, all the ozone ever produced in urban pollution would barely make a dent in the amount lost from the stratosphere.

dioxide in our atmosphere stay "just right"? For example, why didn't the chemical reactions either remove *all* the carbon dioxide, or leave so much of it that the oceans would boil away? More generally, the long-term existence of Earth's oceans tells us that our planet has enjoyed remarkable climate stability—presenting a stark contrast to the dramatic climate changes that apparently occurred on Mars and Venus.

• Why does Earth's climate stay relatively stable?

Earth's long-term climate stability has clearly been important to the ongoing evolution of life—and hence to our own relatively recent arrival as a species (see Figure 1.11). Had our planet undergone a runaway greenhouse effect like Venus, life would certainly have been extinguished. If Earth had suffered loss of atmosphere and a global freezing like Mars, any surviving life would have been driven to hide in underground pockets of liquid water.

Earth's climate is not perfectly stable—our planet has endured numerous ice ages and warm periods in the past. Nevertheless, even in the deepest ice ages and warmest warm periods, Earth's temperature has remained in a range in which ocean water has remained at least partially liquid and hence could harbor life. This long-term climate stability is even more remarkable when you remember that the Sun has brightened substantially over the past 4 billion years, yet Earth's temperature has managed to stay in nearly the same range throughout this time. Apparently, the strength of the greenhouse effect self-adjusts to keep the climate stable. How does it do this?

The Carbon Dioxide Cycle   The mechanism by which Earth self-regulates its temperature is called the **carbon dioxide cycle**, or the **CO_2 cycle** for short. Let's follow the cycle as illustrated in Figure 10.33, starting with the carbon dioxide in the atmosphere:

- Atmospheric carbon dioxide dissolves in rainwater, creating a mild acid.
- The mildly acidic rainfall erodes rocks on Earth's continents, and rivers carry the broken-down minerals to the oceans.
- In the oceans, calcium from the broken-down minerals combines with dissolved carbon dioxide and falls to the ocean floor, making carbonate rocks such as limestone.*
- Over millions of years, the conveyor belt of plate tectonics carries the carbonate rocks to subduction zones, and subduction carries them down into the mantle.
- As they are pushed deeper into the mantle, some of the subducted carbonate rock melts and releases its carbon dioxide, which then outgasses back into the atmosphere through volcanoes.

---

*During the past half billion years or so, the carbonate minerals have been made by shell-forming sea animals, falling to the bottom in the seashells left after the animals die. Without the presence of animals, chemical reactions would do the same thing—and apparently did for most of Earth's history.

The CO_2 cycle acts as a thermostat for Earth, because of the way that changes in temperature feed back into the cycle. You are probably familiar with what we generally call **feedback processes**—processes in which a change in one property amplifies (positive feedback) or counteracts (negative feedback) the behavior of the rest of the system. For example, if someone brings a microphone too close to a loudspeaker, it picks up and amplifies small sounds from the speaker. These amplified sounds are again picked up by the microphone and further amplified, causing a loud screech. This sound feedback is an example of *positive feedback*, because it automatically amplifies itself. The screech usually leads to a form of *negative feedback*: The embarrassed person holding the microphone moves away from the loudspeaker, thereby stopping the positive sound feedback.

> **THINK ABOUT IT**
>
> Think of at least one other everyday example each of positive feedback and negative feedback.

The CO_2 cycle has a built-in form of negative feedback that returns Earth's temperature toward "normal" whenever it warms up or cools down (Figure 10.34). The negative feedback occurs because the overall rate at which carbon dioxide is pulled from the atmosphere is very sensitive to temperature: the higher the temperature, the higher the rate at which carbon dioxide is removed.

Consider first what happens if Earth warms up a bit. The warmer temperature means more evaporation and rainfall, pulling more CO_2 out of the atmosphere. The

Atmospheric CO_2 dissolves in rainwater.

Rainfall erodes rock on land; rivers carry broken-down minerals to the sea.

Volcanoes outgas CO_2.

Broken-down minerals react with dissolved CO_2 to form carbonate rocks.

Carbonate rocks subduct and melt, releasing CO_2.

Figure 10.33 This diagram shows how the CO_2 cycle continually moves carbon dioxide from the atmosphere to the ocean to rock and back to the atmosphere. Note that plate tectonics (subduction in particular) plays a crucial role in the cycle.

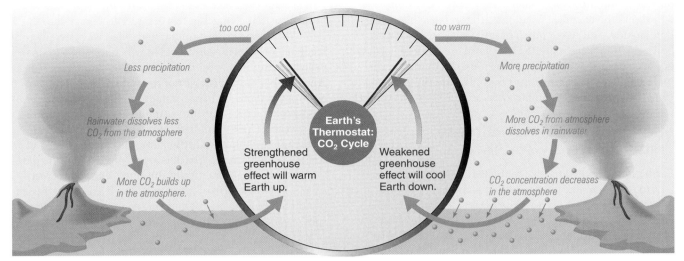

Figure 10.34 The carbon dioxide cycle acts as a thermostat for Earth through negative feedback processes. Cool temperatures cause atmospheric CO_2 to increase, and warm temperatures cause atmospheric CO_2 to decline.

reduced atmospheric CO_2 concentration leads to a weakened greenhouse effect that counteracts the initial warming and cools the planet back down. Similarly, if Earth cools a bit, precipitation decreases and less CO_2 is dissolved in rainwater, allowing the CO_2 released by volcanism to build back up in the atmosphere. The increased CO_2 concentration strengthens the greenhouse effect and warms the planet back up. Overall, the natural thermostat of the carbon dioxide cycle has allowed the greenhouse effect to strengthen or weaken just enough to keep Earth's climate fairly stable, regardless of what other changes have occurred on our planet.

Ice Ages and Other Long-Term Climate Change   While Earth's climate has remained stable enough for the oceans to remain at least partly liquid throughout history, the climate has not been perfectly steady. Numerous warmer periods and numerous ice ages have occurred. Such variations are possible because the CO_2 cycle does not act instantly. When something begins to change the climate, it takes time for the feedback mechanisms of the CO_2 cycle to come into play because of their dependence on the gradual action of mineral formation in the oceans and of plate tectonics. Calculations show that the time to stabilize atmospheric CO_2 through the CO_2 cycle is about 400,000 years. That is, if the amount of CO_2 in the atmosphere were to rise due to, say, increased volcanism, it would take some 400,000 years for the CO_2 cycle to restore temperatures to their current values.*

Ice ages occur when the global average temperature drops by a few degrees. The slightly lower temperatures lead to increased snowfall that may cover continents with

ice down to fairly low latitudes. For example, the northern United States was entirely covered with glaciers during the peak of the most recent ice age, which ended only about 10,000 years ago. The causes of ice ages are complex and not fully understood. Over periods of tens or hundreds of millions of years, the Sun's gradual brightening and the changing arrangement of the continents around the globe have at least in part influenced the climate. During the past few million years—a period too short for solar changes or continental motion to have a significant effect—the ice ages appear to have been strongly influenced by small changes in Earth's axis tilt and other characteristics of Earth's rotation and orbit (the *Milankovitch cycles* that we noted earlier).

Geologists have recently discovered evidence of several particularly long and deep ice ages between about 750 and 580 million years ago. During these periods, glaciers appear to have advanced all the way to the equator. If so, the global temperature must have dropped low enough for the oceans to begin to freeze worldwide. This ocean freezing would have set up a positive feedback process that would have cooled Earth even further, because ice reflects much more sunlight than water. (About 90% of incoming sunlight is reflected by ice, compared to about 5% by water.) With less sunlight being absorbed, the surface would cool further, which in turn would allow the oceans to freeze further. Geologists suspect that in this way, our planet may have entered the periods we now call **snowball Earth** (Figure 10.35). We do not know precisely how extreme the cold became. Some models suggest that at the peak of a snowball Earth period, the global average temperature may have been as low as $-50°C$ ($-58°F$) and the oceans may have been frozen to a depth of 1 kilometer or more.

How did Earth recover from a snowball phase? Even if Earth's surface got cold enough for the ocean surface to freeze completely, the interior still remained hot. As a result, volcanism would have continued to add CO_2 to the

---

*This time scale applies to ocean/atmosphere equilibrium only. The time scale for crust recycling is much longer, while shorter-term climate variations in atmospheric CO_2 concentration can occur through factors besides the inorganic CO_2 cycle, such as cycling of carbon dioxide by life.

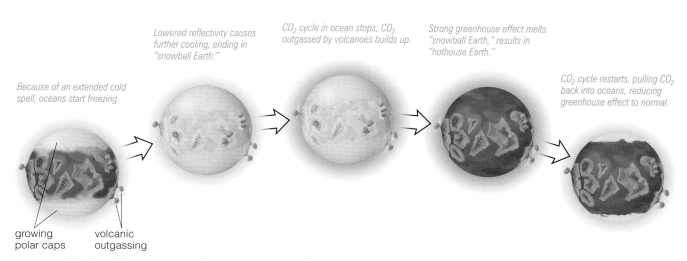

Because of an extended cold spell, oceans start freezing.

Lowered reflectivity causes further cooling, ending in "snowball Earth."

CO_2 cycle in ocean stops; CO_2 outgassed by volcanoes builds up.

Strong greenhouse effect melts "snowball Earth," results in "hothouse Earth."

CO_2 cycle restarts, pulling CO_2 back into oceans, reducing greenhouse effect to normal.

growing polar caps

volcanic outgassing

Figure 10.35 The CO_2 cycle rescues Earth from a "snowball" phase.

atmosphere. Oceans covered by ice would have been unable to absorb this CO_2 gas, and the CO_2 content of the atmosphere would have gradually built up and strengthened the greenhouse effect. As long as the ocean surfaces remained frozen, the CO_2 buildup would have continued. Scientists suspect that, over a period of 10 million years or so, the CO_2 content of our atmosphere may have increased 1,000-fold. Eventually, the strengthening greenhouse effect would have warmed Earth enough to start melting the ocean surface ice. The feedback processes that started the snowball Earth episode then moved quickly in reverse.

As the ocean surface melted, more sunlight would have been absorbed (again, because liquid water absorbs more and reflects less sunlight than ice), warming the planet further. Moreover, because the CO_2 concentration was so high, the warming would have continued well past current temperatures—perhaps taking the global average temperature to higher than 50°C (122°F). Thus, in just a few centuries, Earth would have emerged from a snowball phase into a hothouse phase. Geological evidence supports the occurrence of dramatic increases in temperature at the end of each snowball Earth episode. Earth then slowly recovered from the hothouse phase as the CO_2 cycle removed carbon dioxide from the atmosphere. This recovery would have taken the 400,000 years required for stabilization of atmospheric CO_2.

THINK ABOUT IT

Suppose Earth did not have plate tectonics. Could the planet ever recover from a "snowball" phase? Explain.

The snowball Earth episodes must have had severe consequences for any life on Earth at the time. Indeed, the end of the snowball Earth episodes roughly coincides with a dramatic increase in the diversity of life on Earth (the Cambrian explosion [Section 24.1]). Some scientists suspect that the environmental pressures caused by the snowball Earth periods may have led to a burst of evolution. If so,

we might not be here today if not for the dramatic climate changes of the snowball Earth episodes.

Earth's Long-Term Future Climate Despite going through periods of ice ages, snowball Earth, and unusual warmth, Earth's climate has remained suitable for life for some 4 billion years. However, the continuing brightening of the Sun ultimately will overheat our planet.

The numerous feedback processes that affect climate make it difficult to make precise predictions about future climate. According to some climate models, the warming Sun could cause Earth to begin losing its water as soon as a billion or so years from now. If so, then life on Earth has already completed about 75% of its history on this planet. However, uncertainties in the models make it quite possible that they are wrong. The CO_2 cycle may well keep the climate steady for much more than the next billion years. If so, water and life could persist for as long as 3–4 billion years to come.

About 3–4 billion years from now, however, the Sun will have grown so warm that sunlight on Earth will be as intense as it is on Venus today. The effect will be the same as if we moved Earth to Venus's orbit today (see Figure 10.31)—a runaway greenhouse effect. The rising temperature on Earth will cause increased ocean evaporation (if the oceans still exist), and the water vapor in the atmosphere will further strengthen the greenhouse effect. The positive feedback won't stop until the oceans have evaporated away and all the carbon dioxide has been released from carbonate rocks. Our planet will become a Venus-like hothouse, with temperatures far too high for liquid water to exist.

Thus, Earth's habitability for life will come to an end between 1 and 4 billion years from now. Although this may seem depressing, remember that a billion years is a very long time—equivalent to some 10 million human lifetimes and far longer than humans have existed so far. If you want to lose sleep worrying about the future, there are far more immediate threats, including those that we will discuss next.

• How might human activity change our planet?

We humans are well adapted to the unique, present-day conditions on our planet. The amount of oxygen in our atmosphere, the average temperature of our planet, and the ultraviolet-absorbing ozone layer are just what we need to survive. We have seen that these "ideal" conditions are no accident, but instead consequences of our planet's unique geology and biology.

Nevertheless, the stories of the dramatic and permanent climate changes that occurred on Venus and Mars should teach us to take nothing for granted. Our planet may regulate its own temperature quite effectively, and life may keep the level of oxygen and ozone in the atmosphere quite steady, but it is possible for humans to interfere with these natural mechanisms.

Some of the dangers posed by human activity are already clear. For example, we learned in the 1970s that human-made chemicals known as CFCs can rise high into the atmosphere, where they destroy some of the ozone that shields our planet from dangerous ultraviolet light. In the 1980s, we found a large and ominous "hole" forming in the ozone layer over Antarctica each spring. This clear evidence of a human role in altering a crucial aspect of our planet's habitability led to international action and treaties intended to stop ozone destruction.

Another well-recognized danger comes from extinctions. According to some estimates, human activity is driving species to extinction so rapidly that up to half of today's plant and animal species may be gone by the end of this century. This rate of extinction is comparable with that in past *mass extinctions,* such as the one that wiped out the dinosaurs some 65 million years ago [Section 12.4]. Prior mass extinctions generally led to the loss of the planet's dominant species at the time—and we are the dominant species today. Although we cannot predict the consequences of current extinctions with any certainty, it is at least conceivable that we could be unwittingly clearing the way for a new set of dominant species.

The threat that causes the most debate is **global warming**—an apparent increase in Earth's global average temperature over the past century or so. The debate centers on the question of whether the warming has been caused by human input of carbon dioxide and other greenhouse gases (such as methane) into the atmosphere. The carbon dioxide cycle operates too slowly to absorb the excess greenhouse gases released by human activity, so it is reasonable to think that the warming may be occurring because these gases strengthen the greenhouse effect.

The Global Warming Debate   The potential threat of global warming is a hot political issue because alleviating the threat would require finding new energy sources and making other changes that would dramatically affect the world's economy. However, a major research effort has gradually added to our understanding of the potential threat, particularly in the past decade. Although some controversy still exists about how serious the problem of global warming might be, at least three facts are now clear:

1. Measurements show that Earth has indeed warmed up over the past 50 years, by about 0.5°C. Although this may sound small, it is a significant increase in such a short time period. The warming trend has continued and gotten stronger in recent years.
2. The burning of fossil fuels and other human activity is clearly increasing the amounts of greenhouse gases in the atmosphere. The current concentration of carbon dioxide in Earth's atmosphere is significantly higher than it has been at any time during the past 400,000 years, and the concentration is rising rapidly (Figure 10.36).
3. Because we understand the basic mechanism of the greenhouse effect (see Figure 10.6), we can have no doubt that a continually rising concentration of greenhouse gases would eventually make our planet warm up.

However, while continued input of greenhouse gases would eventually cause global warming, it's difficult to prove that human activity has caused the global warming observed to date. The problem is that numerous feedback mechanisms can counter or enhance greenhouse warming on timescales of years, decades, and even centuries, and we don't yet understand all these mechanisms. For example, increased evaporation of ocean water could enhance the greenhouse effect because water vapor is also a greenhouse gas, but it could counter global warming if it caused the formation of more clouds that prevented sunlight from reaching the ground. Melting of polar ice, on the other hand, makes global warming even worse, because it changes part of Earth's surface from high reflectivity to low, leading to more absorption of solar energy.

Scientists seek to understand the problem better by creating sophisticated computer models of the climate and comparing the model predictions to actual observations. Current models suggest that global temperatures are increasing in a manner consistent with what we expect if human activity is the cause (Figure 10.37). Thus, while plenty of uncertainties still remain, most scientists are now convinced that human activity is indeed causing global warming.

Potential Consequences of Global Warming   Clearly, global warming would mean a higher global average surface temperature for Earth. Beyond that simple statement, however, the issue of consequences becomes much more complicated and the uncertainties involved in prediction even greater.

Changing weather patterns would ensure that different regions of Earth experience different degrees of warming. Some regions would even become colder. Other regions might experience more rainfall or might become deserts. The greater overall warmth of the atmosphere would tend to mean more evaporation from the oceans, leading to more numerous and more intense storms.

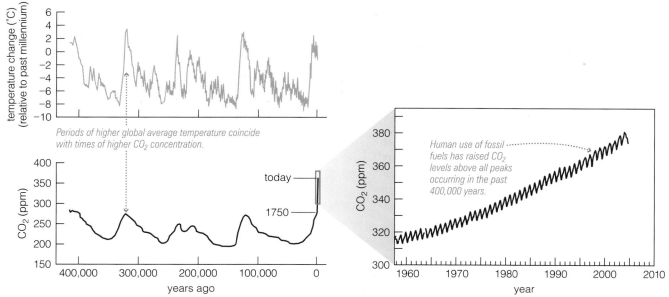

Figure 10.36 This diagram shows the atmospheric concentration of carbon dioxide and global average temperature over the past 400,000 years. Notice that both vary together: more carbon dioxide goes with higher temperature, and vice versa. The CO_2 data for the past half-century come from direct measurements (made at Mauna Loa); most of the earlier data come from studies of air bubbles trapped in Antarctic ice (ice core samples). The concentration is measured in parts per million (ppm), which is the number of CO_2 molecules among every 1 million air molecules.

Another potential threat comes from rising sea level. Water expands very slightly as it warms—so slightly that we don't notice the change in a glass of water, but enough to cause changes in sea level. Sea level has already risen some 20 centimeters in the past hundred years, and could rise another meter as the oceans warm. Such increases in sea level would threaten many coastal regions with more severe flooding. A much greater increase in sea level may occur if polar ice and glaciers melt. Such melting appears to be occurring to some extent already. For example, the famous "snows" (glaciers) of Mount Kilimanjaro are already in retreat and may be gone within the next decade or so.

Secondary effects, such as those arising from changes to ocean currents or ecological changes, pose an even more intractable problem. For example, it is difficult to know how forests and other ecosystems will respond to climate changes induced by global warming. As a result, we cannot easily predict the impact of such climate changes on food production, fresh water availability, or other issues critical to the well-being of human populations.

Given the current uncertainties, no one can predict the precise impact of global warming over the next century. However, the lesson from the solar system is clear: Dramatic and deadly change can occur unexpectedly, and we do not know how our tampering might affect the finely balanced mechanisms that control Earth's climate.

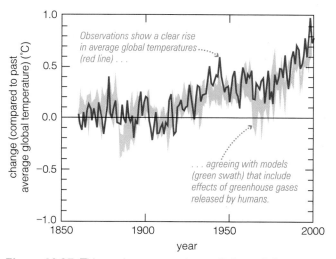

Figure 10.37 This graph compares the predictions of climate models (green swath) with observed temperature changes (red line) since about 1860. The thickness of the green swath for the models represents the range of variation between several different models. These models take into account both natural factors (such as slight changes in the Sun's brightness and the effects of volcanic eruptions) and emission of greenhouse gases by humans. The agreement is not perfect—telling us we still have much to learn—but it is good enough to give us confidence that greenhouse gases are, indeed, causing global warming.

THINK ABOUT IT

If you were a political leader, how would *you* deal with the uncertain threat of global warming?

Putting Chapter 10 into Perspective

With what we have learned in this chapter and the previous chapter, we now have a complete "big picture" view of how the terrestrial worlds started out so similar yet ended up so different. As you continue your studies, keep in mind the following important ideas:

● Atmospheres affect planets in a variety of ways. For example, they absorb and scatter light, distribute heat, and create the weather that can lead to erosion. Perhaps most important, the greenhouse effect allows atmospheres to make a planet warmer than it would be otherwise.

● Atmospheric properties differ widely among the terrestrial worlds, but we can trace these differences to root causes. For example, only the larger worlds have significant atmospheres. The smaller worlds lack the internal heat needed for volcanism and the associated outgassing that releases atmospheric gases, and they also lack the gravity necessary to retain atmospheric gases.

● The climate histories of Venus and Mars suggest that major climate change should be the rule, not the exception. Mars was once warm and wet, but its small size and lack of a magnetic field caused it to lose gas and freeze over some 3 billion or more years ago. Venus may once have had oceans, but its proximity to the Sun doomed it to a runaway greenhouse effect.

● We are here to talk about these things today only because Earth somehow has managed to be the exception, a planet on which climate has remained relatively consistent. We humans are ideally adapted to Earth today, but we have no guarantee that Earth will remain as hospitable in the future, especially as we tinker with the balance of greenhouse gases that has kept our climate stable.

SUMMARY OF KEY CONCEPTS

10.1 Atmospheric Basics

● **What is an atmosphere?** An atmosphere is a

layer of gas that surrounds a world and that can have profound effects on a planet's surface, including creating pressure; absorbing and scattering sunlight; creating wind and weather; interacting with the solar wind to create a magnetosphere; and creating a **greenhouse effect** that can make a planet's surface warmer than it would be otherwise.

● **How does the greenhouse effect warm a planet?** **Greenhouse gases** such as carbon dioxide,

methane, and water vapor absorb infrared light emitted from a planet's surface. The absorbed photons are quickly reemitted, but in random directions. The result acts much like a blanket, warming the planet's surface.

● **Why do atmospheric properties vary with altitude?** **Atmospheric structure** is determined by the

way atmospheric gases interact with sunlight. On Earth, the basic structure consists of the **troposphere**, where most greenhouse warming occurs; the **stratosphere**, where **ozone** absorbs ultraviolet light from the Sun; the **thermosphere**, where solar X rays are absorbed; and the **exosphere**, the extremely low-density outer layer of the atmosphere.

10.2 Weather and Climate

● **What creates wind and weather?** **Global wind patterns** are shaped by atmospheric heating and the **Coriolis effect** caused by a planet's rotation. Convection in the troposphere can lead to the formation of clouds and rain, hail, or snow.

● **What factors can cause long-term climate change?** The Sun has been gradually brightening since its birth more than 4 billion years ago, which tends to warm the planets. But other effects can be even more important in climate change, including changes in axis tilt, changes in reflectivity, and changes in greenhouse gas abundance.

● **How does a planet gain or lose atmospheric gases?** The three basic sources of atmospheric gas

are **outgassing**, evaporation or sublimation, and the "sandblasting" of the surface on worlds with very little atmospheric gas. The five basic loss processes are condensation, chemical reactions with surface materials, the blasting away of gas by impacts, the stripping of gas by the solar wind, and **thermal escape**.

10.3 Atmospheres of the Moon and Mercury

● **Do the Moon and Mercury have any atmosphere at all?** The Moon and Mercury have only very thin exospheres consisting of gas particles released by the "sandblasting" of the surface by micrometeorites, solar wind particles, and high-energy solar photons.

10.4 The Atmospheric History of Mars

- **What is Mars like today?** Mars is cold and dry, with an atmospheric pressure so low that water is unstable. Martian weather is driven largely by seasonal changes that cause carbon dioxide alternately to condense and sublime at the poles, creating pole-to-pole winds and sometimes leading to huge dust storms.

- **Why did Mars change?** Mars's atmosphere must once have been much thicker with a much stronger greenhouse effect, so change must have occurred due to loss of atmospheric gas. Much of the lost gas probably was stripped away by the solar wind, which was able to reach the atmosphere as Mars cooled and lost its magnetic field and protective magnetosphere. Water was probably also lost because ultraviolet light could break apart water molecules in the atmosphere, and the lightweight hydrogen then escaped to space.

10.5 The Atmospheric History of Venus

- **What is Venus like today?** Venus has a thick atmosphere that creates its strong greenhouse effect, explaining why it is so hot. It rotates slowly and therefore has a weak Coriolis effect and weak winds, and is far too hot for rain to fall. Its atmospheric circulation keeps temperatures about the same day and night, and its lack of axis tilt means no seasonal changes.

- **How did Venus get so hot?** Venus's distance from the Sun ultimately led to a **runaway greenhouse effect**: Venus became too hot to develop liquid oceans like those on Earth. Without oceans to dissolve outgassed carbon dioxide and lock it away in carbonate rocks, all of Venus's

carbon dioxide remained in its atmosphere, creating its intense greenhouse effect.

10.6 Earth's Unique Atmosphere

- **How did Earth's atmosphere end up so different?** Temperatures on Earth were just right for outgassed water vapor to condense and form oceans. The oceans dissolve carbon dioxide and ultimately lock it away in carbonate rocks, thereby keeping the greenhouse effect moderate. Nitrogen from outgassing remained in the atmosphere. Oxygen and ozone were produced by photosynthesis, which was possible because the moderate conditions allowed the origin and evolution of abundant life.

- **Why does Earth's climate stay relatively stable?** Earth's long-term climate is remarkably stable due to feedback processes that tend to counter any warming or cooling that occurs. The most important feedback process is the **carbon dioxide cycle**, which naturally regulates the strength of Earth's greenhouse effect.

- **How might human activity change our planet?** Ozone depletion can leave surface life more vulnerable to dangerous solar ultraviolet radiation, and the high rate of extinctions could have unknown consequences. The human release of greenhouse gases into the atmosphere may already be causing global warming and certainly would affect the climate if it continues.

CO_2 concentration over the past 400,000 years

For instructor-assigned homework go to MasteringASTRONOMY www.masteringastronomy.com

Review Questions
Short-Answer Questions Based on the Reading

1. Briefly summarize the different atmospheric properties of the five terrestrial worlds. How do they differ in surface temperature and pressure? What are the different atmospheres made of?

2. Using a balloon as an analogy, explain the origin of gas pressure. What do we mean by *atmospheric pressure*? Why does pressure decrease with altitude? What is 1 *bar* of pressure?

3. Is there any atmosphere at the orbital altitude of the Space Station above Earth? Explain.

4. What does the *greenhouse effect* do to a planet? Explain the role of greenhouse gases and describe the basic mechanism of the greenhouse effect.

5. If there were no greenhouse effect, what factors would determine a planet's surface temperature? How do the "no green-

house" temperatures of the terrestrial planets compare to their actual temperatures, and why?

6. Describe Earth's basic atmospheric structure, from the ground up. How do interactions of sunlight and gases explain the existence of each of the atmospheric layers?

7. Why is the sky blue? Why are sunrises and sunsets red?

8. Why does convection occur in the troposphere, leading to active weather, but not in the stratosphere?

9. What is ozone? How does the absence of ozone on Venus and Mars explain why these planets lack a stratosphere?

10. What is a *magnetosphere*? Briefly explain how a magnetosphere can protect a planet from the solar wind and lead to the creation of auroras.

11. What is the difference between weather and climate?

12. What are Earth's global wind patterns? Explain why our planet has circulation cells that transport warm air toward

the poles and cool air toward the equator. How does our planet's rotation affect these cells?

13. What are clouds made of? How does rain or snow form?

14. Describe each of the four factors that can lead to long-term climate change, and the effects of each factor.

15. Describe each of the processes by which a planet can gain or lose atmospheric gas. What factors determine whether gases are lost through thermal escape?

16. Why do the Moon and Mercury have so little atmospheric gas? How is it possible that they might nonetheless have water ice in polar craters?

17. Why does Mars have seasons? How do seasons on Mars differ from seasons on Earth?

18. How do we think that Mars lost atmospheric gas? What basic planetary property (such as size, distance from the Sun, or rotation rate) would have had to be different for Mars to have retained a thicker atmosphere?

19. What do we mean by a *runaway greenhouse effect?* Explain why this process occurred on Venus but not on Earth.

20. Describe several ways in which Earth's atmosphere is unique among the terrestrial worlds, and how each unique feature is important to our existence.

21. What is the *carbon dioxide cycle,* and why is it so crucial to life on Earth?

22. Briefly summarize the debate over global warming and its potential consequences.

Test Your Understanding

Does It Make Sense? ❓

Decide whether the statement makes sense (or is clearly true) or does not make sense (or is clearly false). Explain your reasoning. (For an example, see Chapter 1, "Does It Make Sense?")

23. If Earth's atmosphere did not contain molecular nitrogen, X rays from the Sun would reach the surface.

24. If the molecular oxygen content of Earth's atmosphere increases, it will cause our planet to warm up.

25. Earth's oceans must have formed at a time when no greenhouse effect operated on Earth.

26. In the distant past, when Mars had a thicker atmosphere, it also had a stratosphere.

27. If Earth rotated faster, hurricanes would be more common and more severe.

28. Mars would still have seasons even if its orbit around the Sun were perfectly circular rather than elliptical.

29. Mars once may have been warmer than it is today, but it could never have been warmer than Earth because it is farther from the Sun than is Earth.

30. If the solar wind were much stronger, Mercury might develop a carbon dioxide atmosphere.

31. If Earth had as much carbon dioxide in its atmosphere as Venus, our planet would be too hot for liquid water to exist on the surface.

32. A planet in another solar system has an Earth-like atmosphere with plentiful oxygen but no life of any kind.

Quick Quiz

Choose the best answer to each of the following. Explain your reasoning with one or more complete sentences.

33. Which terrestrial world has the most atmosphere? (a) Venus (b) Earth (c) Mars

34. The greenhouse effect happens in which layer of an atmosphere? (a) the troposphere (b) the stratosphere (c) the lithosphere

35. What kind of light warms the stratosphere? (a) infrared (b) visible (c) ultraviolet

36. Which of the following is a strong greenhouse gas? (a) nitrogen (b) water vapor (c) oxygen

37. In which direction do hurricanes in the Southern Hemisphere rotate? (a) clockwise (b) counterclockwise (c) equally probable in either direction

38. What is the leading theory for Venus's complete lack of water in any form? (a) It formed closer to the Sun and accreted very little water. (b) It is locked away in the crust. (c) It was broken apart into H and O, which were lost from the atmosphere.

39. What kind of gas is most affected by thermal escape? (a) greenhouse gases (b) light gases (c) all gases equally

40. Which gas is the suspected cause of global warming? (a) ozone (b) water vapor (c) CO_2

41. What causes the release of oxygen gas into Earth's atmosphere? (a) outgassing (b) evaporation/sublimation (c) life

42. Where is the CO_2 that has outgassed from Earth's volcanoes? (a) in the atmosphere (b) escaped to space (c) locked up in rocks

Investigate Further

In-Depth Questions to Increase Your Understanding

Short-Answer/Essay Questions

43. *Clouds of Venus.* Table 10.2 shows that Venus's surface temperature in the absence of the greenhouse effect is lower than Earth's, even though it is closer to the Sun.
    a. Explain this unexpected result in one or two sentences.
    b. Suppose Venus had neither clouds nor greenhouse gases. What do you think would happen to the surface temperature of Venus? Why?
    c. How are clouds and volcanoes linked on Venus? What change in volcanism might result in the disappearance of clouds? Explain.

44. *Atmospheric Structure.* Study the curve showing Earth's average atmospheric structure in Figure 10.7. Sketch similar curves, and explain each curve in words, to show how the atmospheric structure would be different in each of the following cases.
    a. Suppose Earth had no greenhouse gases.
    b. Suppose the Sun emitted no ultraviolet light.
    c. Suppose the Sun had a higher output of X rays.

45. *Magic Mercury.* Suppose we could magically give Mercury the same atmosphere as Earth. Assuming this magical intervention happened only once, would Mercury be able to keep its new atmosphere? Explain.

46. *A Swiftly Rotating Venus.* Suppose Venus had rotated as rapidly as Earth throughout its history. Briefly explain how and why you would expect it to be different in terms of each of the following: geological processes, atmospheric circulation, magnetic field, and climate history. Write a few sentences about each.

47. *Coastal Winds.* During the daytime, heat from the Sun tends to make the air temperature warmer over land near the coast than over the water offshore. At night, when land cools off faster than the sea, the temperatures tend to be cooler over land than over the sea. Use these facts to predict the directions in which winds generally blow during the day and at night

in coastal regions. For example, do the winds blow out to sea or in toward land? (*Hint:* How do these conditions resemble those that cause planetary circulation cells?) Explain your reasoning in a few sentences. (Diagrams might help.)

48. *Sources and Losses.* Choose one process by which atmospheres can gain gas and one by which they can lose gas. For each process, write a few sentences that describe it and how it depends on each of the following fundamental planetary properties: size, distance from the Sun, and rotation rate.

49. *Project: Atmospheric Science in the Kitchen.*
    a. Find an empty plastic bottle, such as a water bottle with a screwtop that makes a good seal. Warm the air inside by filling the bottle partway with hot water and then shaking and emptying the bottle. Seal the bottle and place it into the refrigerator or freezer. What happens after 15 minutes? Explain why this happens, imagining that you can see the individual air molecules.
    b. You may have noticed that loose ice cubes in the freezer gradually shrink away to nothing or that frost sometimes builds up in old freezers. What technical terms used in this chapter describe these phenomena? On what terrestrial planet do these same processes play a major role in changing the atmospheric pressure?

50. *Two Paths Diverged.* By looking back to fundamental properties such as size and distance from the Sun, explain why Earth has oceans and very little atmospheric carbon dioxide, while similar-size Venus has a thick carbon dioxide atmosphere.

51. *Change in Fundamental Properties.* Consider Earth's size and distance from the Sun. Choose one property and suppose that it had been different (for example, smaller size or greater distance). Describe how this change might have affected Earth's subsequent history and the possibility of life on Earth.

52. *Feedback Processes in the Atmosphere.* As the Sun gradually brightens in the future, how can the CO_2 cycle respond to reduce the warming effect? Which parts of the cycle will be affected? Is this an example of positive or negative feedback?

53. *Earth to Mars.* Section 10.5 discusses what might happen to Earth if it were suddenly moved to the orbit of Venus. What do you think would happen to Earth if it were suddenly moved to the orbital distance of Mars? Write a few sentences explaining your answer.

54. *Global Warming.* What, if anything, should we be doing to alleviate the threat of global warming that we are not doing already? Write a one-page editorial summarizing and defending your opinion.

Quantitative Problems
Be sure to show all calculations clearly and state your final answers in complete sentences.

55. *The Mass of an Atmosphere.* What is the total mass of Earth's atmosphere? You may use the fact that 1 bar is the pressure exerted by 10,000 kilograms pushing down on a square meter in Earth's gravity. Remember that every square meter of Earth experiences this pressure from the atmosphere above it. Alternatively, you can start with the English units value for pressure of 14.7 pounds per square inch and convert to kilograms for your final answer. Remember that the surface area of a sphere of radius r is $4\pi r^2$.

56. *An Early Steam Atmosphere?* Many planetary scientists consider it likely that Earth's oceans were temporarily "boiled away" during the heavy bombardment period shortly after formation. What atmospheric pressure would result if today's

oceans were converted to steam? Today's oceans cover 75% of Earth's surface to an average depth of 3.5 kilometers, and the density of water is 1,000 kilograms per cubic meter. (*Hint:* The previous problem shows how to convert a surface pressure to the total mass of an atmosphere, and this problem does the reverse.)

57. *The Role of Reflectivity.* Using Mathematical Insight 10.1, find out the maximum possible "no greenhouse" temperature for a completely black planet at 1 AU. What temperature do you find for a perfectly white planet? Explain. Finally, find out what reflectivity would be necessary to keep the average temperature exactly at the freezing point, and compare this to the Earth's actual reflectivity in Table 10.2.

58. *The Cooling Clouds of Venus.* Table 10.2 shows that Venus's temperature in the absence of the greenhouse effect is lower than Earth's, even though Venus is closer to the Sun. What would the Venus "no greenhouse" temperature be if its clouds were more transparent, giving a reflectivity the same as Earth's? What would the surface temperature be in this example if the greenhouse effect simply increased the surface temperature by the same number of degrees that it does today?

59. *Mars's Elliptical Orbit.* Mars's distance from the Sun varies from 1.38 AU to 1.66 AU. How much would this change its surface temperature? (*Hint:* Find the difference of the "no greenhouse" temperature at the two extremes.) Comment on how this affects Mars's seasons, connecting your answer to Figure 10.23.

60. *Escape from Venus.*
    a. Calculate the escape velocity from Venus's exosphere, which begins about 200 kilometers above the surface. (*Hint:* You'll need the radius of Venus and the formula in Mathematical Insight 5.3.)
    b. Calculate and compare the thermal speeds of hydrogen and deuterium atoms at the exospheric temperature of 350 K. The mass of a hydrogen atom is 1.67×10^{-27} kilograms, and the mass of a deuterium atom is about twice the mass of a hydrogen atom.
    c. In a few sentences, comment on the relevance of these calculations to the question of whether Venus has lost large quantities of water.

61. *Escape from Jupiter.* The atmospheric escape processes of this chapter also apply to the outer solar system. Jupiter's exosphere is even hotter than the Venus case of the previous problem. Will this be enough to cause even faster escape of hydrogen?
    a. Calculate the escape velocity from Jupiter's exosphere, which begins about 1,000 kilometers above the cloud tops. (*Hint:* You'll need the radius of Jupiter and the formula in Mathematical Insight 4.4.)
    b. Calculate and compare the thermal speeds of hydrogen atoms at the exospheric temperature of 800 K. The mass of a hydrogen atom is 1.67×10^{-27} kilograms.
    c. In a few sentences, state whether or not you consider the escape of hydrogen from Jupiter's hot exosphere to be significant, and why.

62. *Abundance versus Importance.* Make a table summarizing Earth's main atmospheric gases using those in Table 10.1, plus CO_2.
    a. Across the top of the table, list the five main gases in the atmosphere from most abundant to least abundant. In the next row, state the fraction of the atmosphere in percent. Find the percent values for those gases listed in

Table 10.1. For water, which can vary from place to place and time to time, use 1%. For CO_2, find the current fraction from Figure 10.35 measured in ppm. Convert ppm (part per million) to percent, remembering that 1% is "one part per hundred."

b. For each gas, decide on a "rank" in importance as a greenhouse gas. The most important gas gets a "1," etc. For those that are not greenhouse gases, put an "X." Below your table, comment on the relationship between abundance and importance as a greenhouse gas, if any.

c. Add another row of notes for each gas listing why it is important in our atmosphere, based on the content of this chapter. If its importance has not been discussed, put "X." For some gases there is more than one important aspect. Below the table, comment on whether the most abundant gases have the most important effects, in your opinion.

Discussion Questions

63. *Lucky Earth.* The climate histories of Venus and Mars make it clear that it's not "easy" to get a pleasant climate like that of Earth. How does this affect your opinion about whether Earth-like planets might exist around other stars? Explain.

64. *Terraforming Mars.* Some people have suggested that we might be able to carry out planetwide engineering of Mars that would cause its climate to warm and its atmosphere to thicken. This type of planet engineering is called *terraforming*, because its objective is to make a planet more Earth-like and thus easier for humans to live on. Discuss possible ways to terraform Mars, at least in principle. Do any of these ideas seem practical? Does it seem like a good idea? Defend your opinions.

65. *Terraforming Venus.* As a follow-up to the previous question, think of any ways in which it would be possible to terraform Venus. Again, discuss the possibilities in principle as well as their practicality.

66. *Don't Worry, Be Happy.* There are many skeptics of global warming. Research their opinions in newspapers, in magazines, and on the Internet. Do they disagree with the data, saying that Earth is not getting warmer? Do they disagree with the conclusion that humans are the primary cause? Do they disagree with the idea that action must be taken? Defend or refute their findings based on your own studies.

MEDIA EXPLORATIONS

For self-study activities go to **Mastering**ASTRONOMY   www.masteringastronomy.com

Interactive Tutorials

Surface Temperature of Terrestrial Planets Tutorial
Lesson 1 Energy Balance
Lesson 2 Role of Planet's Distance from the Sun
Lesson 3 Role of Planet's Reflectivity
Lesson 4 Role of Planet's Atmosphere

Supplementary Tutorial Exercises
Use the interactive **Tutorial Lessons** to explore the following questions.

Surface Temperature of Terrestrial Planets Tutorial, Lesson 1
1. How does the rate at which a planet emits radiation depend on the amount of sunlight it receives?
2. Does a planet's size affect its temperature? Explain.
3. How does a planet's rotation rate affect its day and night temperatures? Does rotation rate affect the average temperature? Explain.

Surface Temperature of Terrestrial Planets Tutorial, Lesson 2
1. Use the tool in Lesson 2 to predict Mercury's surface temperature. How does this compare to Mercury's actual day and night temperatures? Explain.

2. Use the tool in Lesson 2 to predict Earth's surface temperature. Why isn't this Earth's actual surface temperature? Explain.

Surface Temperature of Terrestrial Planets Tutorial, Lesson 3
1. Use the tool in Lesson 3 to predict the range of possible "no greenhouse" temperatures for Earth by moving the reflectivity slider. If Earth were black, would it be as warm as the observed temperature of +15°C?
2. Using the tool in Lesson 3, determine which causes a bigger temperature difference: changing the reflectivity from 0.0 to 0.1, or changing it from 0.9 to 1.0. Explain.

Surface Temperature of Terrestrial Planets Tutorial, Lesson 4
1. How would Earth's temperature be different if the carbon dioxide concentration doubled but all other gas concentrations remained the same?
2. A small warming could lead to increased evaporation of water. What would happen to the temperature if, along with the doubling of carbon dioxide as in question 1, the atmospheric water vapor concentration increased by 10% (to 1.1 times its current value)?
3. How might increased cloud cover offset the warming due to the greenhouse effect?
4. Increased cloud cover also means increased atmospheric water vapor. Use the tools to explore the balance between warming due to more water vapor and cooling due to more cloud cover. If we continued to add greenhouse gases to the atmosphere for centuries, what would ultimately happen to our planet's temperature? Explain.

Exploring the Sky and Solar System

Of the many activities available on the *Voyager: SkyGazer* CD-ROM accompanying your book, use the following files to observe key phenomena covered in this chapter.

Go to the **File: Basics** folder for the following demonstrations.
1. Planet Paths
2. Planet Orrery
3. Follow a Planet

Go to the **File: Demo** folder for the following demonstrations.
1. Earth and Venus
2. Hyakutake at Perihelion
3. Pluto's Orbit

Go to the **Explore** menu for the following demonstrations.
1. Solar System
2. Paths of the Planets

 Movies

Check out the following narrated and animated short documentary available on **www.masteringastronomy.com** for a helpful review of key ideas covered in this chapter.

History of the Solar System

Web Projects

1. *Human Threats to Earth.* Write an in-depth research report, three to five pages in length, about current understanding and controversy regarding one of the following issues: global warming, ozone depletion, or the loss of species on Earth due to human activity. Be sure to address both the latest knowledge about the issue and proposals for alleviating any dangers associated with it. End your report by making your own recommendations about what, if anything, needs to be done to prevent damage to Earth.
2. *Spacecraft Study of Atmospheres.* Learn about a current or planned mission to study the atmosphere of one of the terrestrial worlds (possibly including Earth). Write a one- to two-page essay describing the mission and what we hope to learn from it.
3. *Martian Weather.* Find the latest weather report for Mars from spacecraft and other satellites. What season is it in the northern hemisphere? When was the most recent dust storm? What surface temperature was most recently reported from Mars's surface, and where was the lander located? Summarize your findings by writing a 1-minute script for a television news update on Martian weather.

11

Jovian Planet Systems

Do there exist many worlds, or is there but a single world? This is one of the most noble and exalted questions in the study of Nature.

St. Albertus Magnus (1206–1280)

In Roman mythology, the namesakes of the jovian planets are rulers among gods: Jupiter is the king of the gods, Saturn is Jupiter's father, Uranus is the lord of the sky, and Neptune rules the sea. However, our ancestors could not have foreseen the true majesty of the four jovian planets. The smallest, Neptune, is large enough to contain the volume of more than 50 Earths. The largest, Jupiter, has a volume some 1,400 times that of Earth. These worlds are totally unlike the terrestrial planets. They are essentially giant balls of gas, with no solid surface on which to stand.

Why should we care about a set of worlds so different from our own? Apart from satisfying natural curiosity, studies of the jovian planets and their moons help us understand the birth and evolution of our solar system—which in turn helps us understand our own planet Earth. In addition, the jovian planets provide stepping stones to understanding the more than 150 planets so far discovered around other stars, because these planets are probably jovian in nature. In this chapter, we'll explore the jovian planet systems, first focusing on the planets themselves, then on their many moons, and finally on their beautifully complex rings.

 Formation of the Solar System Tutorial, Lesson 1

11.1 A Different Kind of Planet

We toured the four jovian planets of our solar system briefly in Section 7.1. We saw that they are radically different from the terrestrial worlds in size, density, and composition. Now we are ready to explore the jovian planets in a little more depth.

In the context of human history, it's only quite recently that we recognized the vast differences between the jovian and terrestrial worlds. Jupiter and Saturn are easily visible to the naked eye, but their appearance in the night sky doesn't tell us how they differ in character from other visible planets. Uranus and Neptune were discovered only in 1781 and 1846, respectively (see box, p. 331), so they were unknown even during the Copernican revolution.

Astronomers first recognized the immense sizes of Jupiter and Saturn about 250 years ago. Recall that we need to know both angular size and distance to calculate an object's true size (see Mathematical Insight 2.1). Although Copernicus had figured out the *relative* distances to the known planets in astronomical units (AU), scientists did not establish the absolute scale of the solar system until the 1760s, when they were able to measure the true length of

an astronomical unit with data from a transit of Venus (see Special Topic, p. 218). Only then could scientists calculate the true sizes of Jupiter and Saturn from their distances and angular sizes (measured through telescopes). Knowing the scale of the solar system also told scientists the distances of orbiting moons; because the orbital periods of the moons were easy to measure, the orbital distances allowed scientists to use Newton's version of Kepler's third law [Section 4.4] to calculate the masses of the jovian planets. Together, the measurements of size and mass revealed the low densities of the jovian planets, proving that these worlds are very different in nature from Earth.

The real revolution in understanding jovian planet systems began with the first spacecraft sent to visit them: *Pioneer 10* and *Pioneer 11*, which flew past Jupiter and Saturn in the early 1970s. The *Voyager* missions followed soon thereafter; *Voyager 1* flew past Jupiter and Saturn, and *Voyager 2* flew past all four jovian planets. More recently, we've learned much more about Jupiter from the *Galileo* mission, which orbited Jupiter from 1995 to 2003, and about Saturn from the *Cassini* mission, which entered Saturn orbit in 2004. The discovery of more than 150 presumably jovian planets around other stars [Section 13.2] has also heightened interest in our own solar system's jovian planets.

• Are jovian planets all alike?

Figure 11.1 shows a montage of the jovian planets compiled by the *Voyager* spacecraft, along with basic data and Earth included for scale. All four share the properties of immense size and a low-density composition of hydrogen, helium, and hydrogen compounds—just as we expect for worlds that formed beyond the frost line of the solar nebula, in the cold, outer regions of our solar system [Section 8.3]. However, a close look at the images and data in Figure 11.1 shows that the jovian planets differ substantially from one another in size, mass, density, composition, and even color. If you look closely, you'll even see subtle differences in shape, with Saturn much less like a perfect sphere than the others. As we'll see, these differences can also be traced to the birth of these planets.

Jovian Planet Composition Jupiter and Saturn are made almost entirely of hydrogen and helium, with just a few percent of their masses coming from hydrogen compounds and even smaller amounts of rock and metal. Thus, their overall compositions are much more similar to the composition of the Sun than to the compositions of the terrestrial planets. Some people have gone as far as to call Jupiter a "failed star" because it has a starlike composition but lacks the nuclear fusion needed to make it shine. Its lack of fusion is due to its mass: Although Jupiter is large for a planet, it is much less massive than any star. As a result, its gravity is too weak to compress its interior to the extreme temperatures and densities needed for nuclear fusion. Jupiter would have needed to grow to about 80 times its current

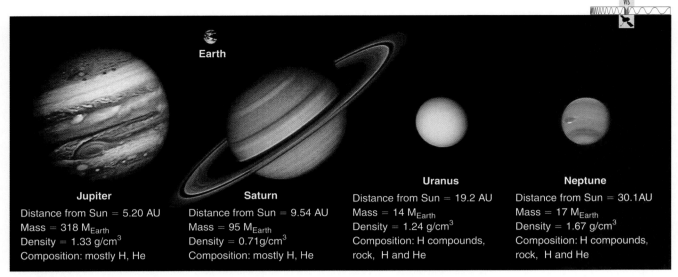

Earth

Jupiter
Distance from Sun = 5.20 AU
Mass = 318 M_{Earth}
Density = 1.33 g/cm^3
Composition: mostly H, He

Saturn
Distance from Sun = 9.54 AU
Mass = 95 M_{Earth}
Density = 0.71g/cm^3
Composition: mostly H, He

Uranus
Distance from Sun = 19.2 AU
Mass = 14 M_{Earth}
Density = 1.24 g/cm^3
Composition: H compounds, rock, H and He

Neptune
Distance from Sun = 30.1AU
Mass = 17 M_{Earth}
Density = 1.67 g/cm^3
Composition: H compounds, rock, H and He

Figure 11.1 Jupiter, Saturn, Uranus, and Neptune, shown to scale with Earth for comparison.

mass to have become a star. Of course, where some people see a failed star, others see an extremely successful planet.

Uranus and Neptune are not only much smaller than Jupiter and Saturn overall, but also contain proportionally much smaller amounts of hydrogen and helium. These gases make up much less than half their total masses. Rather than being made mostly of hydrogen and helium, Uranus and Neptune are made primarily of hydrogen compounds such as water (H_2O), methane (CH_4), and ammonia (NH_3), along with smaller amounts of metal and rock.

The differences in composition among the jovian planets can probably be traced to their origins. Recall that the jovian planets are thought to have formed in a very different way from the terrestrial planets [Section 8.3]. The terrestrial planets grew only through accretion of solid planetesimals containing rock and metal. Because rock and metal were quite rare in the solar nebula, the terrestrial planets never grew massive enough for their gravity to hold much of the abundant hydrogen and helium gas that made up most of the nebula.

The jovian planets formed in the outer solar system (beyond the frost line), where it was cold enough for hydrogen compounds to condense into ices (see Figure 8.5). Because hydrogen compounds were so much more abundant than metal and rock, some of the ice-rich planetesimals of the outer solar system grew to great size. Once these planetesimals became sufficiently massive, their gravity allowed them to draw in the hydrogen and helium gas that surrounded them. All four jovian planets are thought to have grown from ice-rich planetesimals of about the same mass—roughly 10 times the mass of Earth. Thus, their differences in composition stem from the amount of hydrogen and helium gas that they captured.

Jupiter and Saturn captured so much hydrogen and helium gas that these gases now make up the vast majority of their masses. The ice-rich planetesimals from which they grew now represent only a small fraction of their overall masses—about 3% in Jupiter's case and about 10% in Saturn's case.

Uranus and Neptune pulled in much less hydrogen and helium gas. For example, notice in Figure 11.1 that Uranus's mass is about 14 times Earth's mass. Since it is thought to have grown around an ice-rich planetesimal that was about 10 times Earth's mass, the drawn-in hydrogen and helium gas must make up only about a third of Uranus's total mass. The bulk of its mass consists of material from the original ice-rich planetesimal: hydrogen compounds mixed with smaller amounts of rock and metal. The same is true for Neptune, though its higher density suggests that it may have formed around a slightly more massive ice-rich planetesimal.

We can understand why the jovian planets captured different amounts of gas by considering their distances from the Sun. As the solar system formed, the solid particles that condensed far from the Sun must have been much more widely spread out than particles that condensed nearer to the Sun. Thus, at greater distances from the Sun, it took longer for small particles to accrete into large, icy planetesimals with gravity strong enough to pull in gas from the surrounding nebula. Jupiter would have been the first jovian planet to get a planetesimal large enough for its gravity to start drawing in gas, followed by Saturn, Uranus, and Neptune. Because all the planets stopped accreting gas at the same time—when the solar wind blew all the remaining gas into interstellar space—the more distant planets had less time to capture gas and thus ended up smaller in size.

Density Differences   Notice in Figure 11.1 that Saturn is considerably less dense than Uranus or Neptune. This should make sense when you compare compositions. After all, the hydrogen compounds, rock, and metal that make up Uranus and Neptune are normally much more dense than hydrogen or helium gas. By the same logic, we'd expect Jupiter to be even less dense than Saturn—but it's not. To understand Jupiter's surprisingly high density, we need to think about how massive planets are affected by their own gravity.

Saturn's average density of 0.71 g/cm^3 is less than that of water. As a result, it is sometimes said that Saturn could float on a giant ocean. Suppose there really were a gigantic planet with a gigantic ocean and we put Saturn on the ocean's surface. Would it float? If not, what would happen?

Building a planet of hydrogen and helium is a bit like making one out of very fluffy pillows. Imagine assembling a planet pillow by pillow. As each new pillow is added, those on the bottom are compressed more by those above. As the lower layers are forced closer together, their mutual gravitational attraction increases, compressing them even further. At first the stack grows substantially with each pillow, but eventually the growth slows until adding pillows barely increases the height of the stack (Figure 11.2a).

This analogy explains why Jupiter is only slightly larger than Saturn in radius even though it is more than three times as massive. The extra mass of Jupiter compresses its interior to a much higher density. More precise calculations show that Jupiter's radius is almost the maximum possible radius for a jovian planet. If much more gas were added to Jupiter, its weight would actually compress the interior

---

SPECIAL TOPIC   How Were Uranus, Neptune, and Pluto Discovered?

The planets Mercury, Venus, Mars, Jupiter, and Saturn were all known to ancient people. Each can be seen with the naked eye and each clearly wanders among the fixed stars of the constellations. In contrast, the planets (and other objects) beyond Saturn are relatively recent discoveries in human history.

Uranus was the first "discovered" planet. Although faintly visible to the naked eye, Uranus is so faint and moves so slowly in its 84-year orbit of the Sun that ancient people did not recognize it as a planet. It was even recorded as a star on some detailed early sky charts. Uranus was recognized as a planet in 1781 by English astronomers William and Caroline Herschel. William and Caroline were brother and sister, and they worked together both in building their telescopes and in much of their observing. William became the more famous of the pair by far, partly because Caroline's gender made many scientists of the time unwilling to take her seriously, but also because he was nearly 12 years older than his sister and had been the one who drew her into astronomical research. William Herschel originally suggested naming the new planet Georgium Sidus, Latin for George's star, in honor of his patron, King George III. Fortunately, the idea of "Planet George" never caught on. Instead, many eighteenth- and nineteenth-century astronomers referred to the new planet as Herschel. The modern name Uranus—after the mythological father of Saturn—was first suggested by one of Herschel's contemporaries, astronomer Johann Bode. This name was generally accepted by the mid-nineteenth century.

Neptune's discovery followed next, and it represented an important triumph for Newton's law of universal gravitation and the young science of astrophysics. By the mid-1800s, careful observations of Uranus had shown its orbit to be slightly inconsistent with that predicted by Newton's law of gravity—at least if it were being influenced only by the Sun and the other known planets. In the early 1840s in England, a student named John Adams suggested that the inconsistency could be explained if there were an unseen "eighth planet" orbiting the Sun beyond Uranus. According to the official story, he used Newton's theory to predict the location of the planet but was unable to convince British astronomers to carry out a telescopic search. However, documents from the time suggest his prediction may not have been as precise as the official history claims. Meanwhile, in the summer of 1846, French astronomer Urbain Leverrier independently made very precise calculations. He sent a letter to Johann Galle of the Berlin Observatory, suggesting a search for the eighth planet. On the night of September 23, 1846, Galle pointed his telescope to the position suggested by Leverrier. There, within 1° of its predicted position, he saw the planet Neptune. Hence, Neptune's discovery truly was made by mathematics and physics and was only confirmed with a telescope.

As a side note to this story, Leverrier had such faith in Newton's universal law of gravitation that he also suggested a second unseen planet, this one orbiting closer to the Sun than Mercury. He got this idea because other astronomers had identified slight discrepancies between Mercury's actual orbit and the orbit predicted by Newton's theory. He assumed that no one had yet seen the planet, which he called Vulcan, because it was so close to the Sun. Leverrier died in 1877, still believing that Vulcan would someday be discovered. In fact, Vulcan does not exist. Mercury's actual orbit does not match the orbit predicted by Newton's law of gravity because Newton's theory is not the whole story of gravity. About 40 years after Leverrier's death, Einstein showed that Newton's theory is only an approximation of a broader theory of gravity, known today as Einstein's *general theory of relativity*. Einstein's theory predicts an orbit for Mercury that matches its actual orbit. This match was one of the first key pieces of evidence in favor of Einstein's theory [Section S3.4].

Pluto was discovered in 1930 by American astronomer Clyde Tombaugh, culminating a search that began when astronomers analyzed the orbit of Neptune. The story of Pluto's discovery at first seemed much the same as that of Neptune's. Just as discrepancies between the predicted orbit and the actual orbit of Uranus led to the prediction that Neptune must exist, lingering discrepancies between the updated prediction for Uranus's orbit and its actual orbit suggested the existence of an even more distant "ninth planet." Tombaugh found Pluto just 6° from the position in the sky where this ninth planet had been predicted to lie, so it seemed that the search had been successful. However, while initial estimates suggested that Pluto was much larger than Earth, we now know that Pluto has a radius of only 1,160 kilometers and a mass of just 0.002 Earth mass—making it far too small to affect the orbit of Neptune. In retrospect, the supposed orbital irregularities of Neptune appear to have been errors in measurement. No "ninth planet" is needed to explain the orbit of Neptune. The discovery of Pluto was just a happy coincidence.

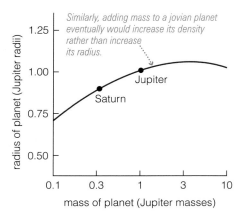

Adding pillows may increase height . . .

. . . but eventually compresses the stack, making density greater.

Similarly, adding mass to a jovian planet eventually would increase its density rather than increase its radius.

radius of planet (Jupiter radii)

mass of planet (Jupiter masses)

Figure 11.2 The relationship between mass and radius for a planet made of hydrogen and helium.

a A tall enough stack of pillows offers an analogy to massive hydrogen/helium planets: adding more pillows compresses the stack rather than increasing its height.

b This graph shows how radius depends on mass for a hydrogen/helium planet. Notice that Jupiter is only slightly larger in radius than Saturn, despite being three times as massive. Gravitational compression of a planet much more massive than Jupiter would actually make it smaller in size.

enough to make the planet *smaller* rather than larger (Figure 11.2b). Thus, some extrasolar planets that are larger in mass than Jupiter are probably smaller in size. In fact, the smallest stars are significantly smaller in radius than Jupiter, even though they are 80 times as massive.

Jovian Planet Rotation and Shape   The jovian planets rotate much more rapidly than any of the terrestrial worlds. However, precisely defining their rotation rates can be difficult because they lack solid surfaces. We can measure a terrestrial rotation period simply by watching the apparent movement of a mountain or crater as a planet rotates. On jovian planets, we can observe only the movements of clouds. Cloud movements can be deceptive, because their apparent speeds are a combination of winds and planetary rotation. In addition, observations of clouds at different latitudes suggest that the jovian planets do not rotate like solid balls. Instead, their rotation rates vary with latitude: Equatorial regions complete each rotation in less time than polar regions. (The Sun also has a rotation rate that differs with latitude.)

We can measure the rotation rates of the jovian interiors by tracking emissions from charged particles trapped in their magnetospheres. This technique tells us the rotation period of the magnetosphere, which should be the same as the rotation period deep in the interior, where the magnetic field is generated. Measurements show that the jovian "day" ranges from about 10 hours on Jupiter and Saturn to 16–17 hours on Uranus and Neptune.

The rapid rotation rates of the jovian planets affect their shapes (Figure 11.3). Gravity alone would make the jovian planets into perfect spheres. Rotation makes them less like perfect spheres because it makes material bulge outward. Material near the equator, where speeds around the rotation axis are highest, is flung outward in the same way you

feel yourself flung outward when you ride on a merry-go-round. The size of the equatorial bulge depends on the balance between the strength of gravity (which pulls material inward) and the rate of rotation (which pushes material outward). The balance tips most strongly toward flattening on Saturn: With its rapid 10-hour rotation period and relatively weak surface gravity, Saturn is about 10% wider at its equator than at its poles. In addition to altering a planet's appearance, the equatorial bulge exerts an extra gravitational pull that helps keep moons and rings aligned with the equator.

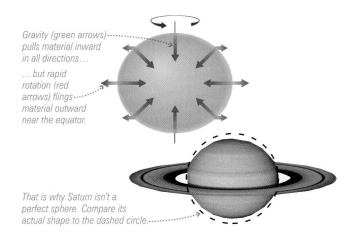

Gravity (green arrows) pulls material inward in all directions . . .

. . . but rapid rotation (red arrows) flings material outward near the equator.

That is why Saturn isn't a perfect sphere. Compare its actual shape to the dashed circle.

Figure 11.3   Interactive Figure   Because of their rapid rotation, the jovian planets are not quite spherical. Saturn shows the biggest difference between its actual shape and a perfect sphere.

• What are jovian planets like on the inside?

The jovian planets are often called "gas giants," making it sound as if they are entirely gaseous like air on Earth. Based on their compositions, Jupiter and Saturn may seem to deserve this name; after all, they became giants primarily by capturing so much hydrogen and helium gas. The name may seem less fitting for Uranus and Neptune, since they are made mostly of materials besides pure hydrogen and helium. However, closer inspection shows that the name is a little misleading even for Jupiter and Saturn, because their strong gravity compresses most of the "gas" into forms of matter quite unlike anything we are familiar with in everyday life on Earth.

You may wonder how we can claim to know what the jovian planets are like on the inside. The answer is a combination of theoretical modeling and laboratory experiments. Just as with the terrestrial planets, detailed observations of the strength of a planet's gravity and of its magnetic field can help us put together a model of the planet's interior structure. We then use the model to predict observable phenomena, such as a planet's average density, and modify it until the predictions match the data. Today, advanced computer models successfully explain the observed sizes, densities, atmospheric compositions, and precise shapes of the jovian planets. Thus, we are confident that the interior structure found by the models is fairly close to reality. Laboratory studies also provide important data; for example, laboratory experiments have shown us how hydrogen and helium act under the tremendous temperatures and pressures that exist deep beneath the jovian cloud tops.

Inside Jupiter   Let's begin discussion of the jovian interiors by using Jupiter as a prototype. Jupiter's lack of a solid surface makes it tempting to think of the planet as "all atmosphere," but you could not fly through Jupiter's interior in the way airplanes fly through air. A spacecraft plunging into Jupiter would find increasingly higher temperatures and pressures as it descended. The *Galileo* spacecraft dropped a scientific probe into Jupiter in 1995, and it collected measurements for about an hour before the ever-increasing pressures and temperatures destroyed it. *Galileo* provided valuable data about Jupiter's atmosphere, but didn't last long enough to sample the interior: It survived only to a depth of about 200 kilometers, or about 0.3% of Jupiter's radius.

In fact, while Jupiter has no solid surface, computer models tell us that it still has fairly distinct interior layers (Figure 11.4). The layers do not differ much in composition—all except the core are mostly hydrogen and helium; instead they differ in the phase (such as liquid or gas) of their hydrogen. To get a better sense of this layering, imagine plunging head-on into Jupiter in a futuristic space suit that allows you to survive the extreme interior conditions.

Near the cloud tops, you'll find the temperature to be a brisk 125 K ($-148°C$), the density to be a low 0.0002 g/cm^3, and the atmospheric pressure to be about 1 bar (the same as the pressure at sea level on Earth [Section 10.1]). As you plunge downward, conditions quickly become more extreme.

By a depth of 7,000 kilometers—about 10% of the way to the center—you'll find that the temperature has increased to a scorching 2,000 K and the pressure has reached 500,000 bars. The density is about 0.5 g/cm^3, or about half that of water. Under these conditions, hydrogen acts more like a liquid than a gas, which is why the layer that begins at this depth is labeled *liquid hydrogen* in Figure 11.4. Of course, like the rest of Jupiter, the layer also contains helium and hydrogen compounds.

At a depth of 14,000 kilometers, or about 20% of the way to the center, the density has become about as high as that of water (1.0 g/cm^3). The temperature is near 5,000 K, almost as hot as the surface (but not the interior) of the Sun. The pressure has reached 2 million bars. This extreme pressure forces hydrogen into a compact, metallic form. Just as is the case with everyday metals, electrons are free to move around in *metallic hydrogen*, so it conducts electricity quite effectively. This layer extends through most of the rest of Jupiter's interior and, as we'll see shortly, it is where Jupiter's magnetic field is generated.

You'll finally reach Jupiter's core at a depth of 60,000 kilometers, or about 10,000 kilometers from the center. The core temperature is some 20,000 K and the pressure is about 100 million bars. The density of the core is about 25 g/cm^3, much denser than any material you'll find on Earth's surface. The core is a mix of hydrogen compounds, rock, and metal, but these materials bear little resemblance to familiar solids or liquids because of the combination of high temperature and extreme pressure. Moreover, while the rock and metal of the terrestrial planets have separated into

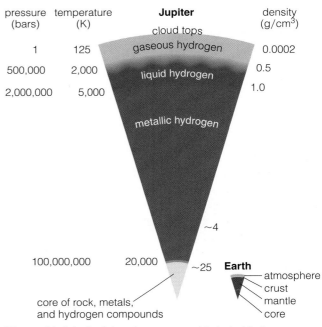

Figure 11.4 Jupiter's interior structure, labeled with the pressure, temperature, and density at various depths. Earth's interior structure is shown to scale for comparison. Note that Jupiter's core is only slightly larger than Earth but is about 10 times as massive.

layers by composition (that is, into the layers we call the core, mantle, and crust), Jupiter's core materials are probably all mixed together. The core contains about 10 times as much mass as the entire Earth, but it is only about the same size as Earth because it is compressed to such high density.

Comparing Jovian Interiors Computer models of the jovian interiors yield the somewhat surprising result that the cores of all four jovian planets are quite similar in composition and mass. Thus, the jovian planet interiors differ mainly in the hydrogen/helium layers that surround their cores. Figure 11.5 contrasts the four jovian interiors. Remember that while the outer layers are named for the phase of their hydrogen, they also contain helium and hydrogen compounds.

Saturn is the most similar to Jupiter, just as we should expect given its similar size and composition. It has the same set of four layers as Jupiter, and the layers differ from those of Jupiter only because of Saturn's lower mass and weaker gravity. The lower mass makes the weight of the overlying layers less on Saturn than on Jupiter, so you must travel deeper into Saturn to find the layer where pressure changes hydrogen from one form to another. That is why Saturn has a thicker layer of gaseous hydrogen and a much thinner and more deeply buried layer of metallic hydrogen.

Pressures within Uranus and Neptune are not high enough to form liquid or metallic hydrogen at all. Each of these two planets has only a thick layer of gaseous hydrogen surrounding its core of hydrogen compounds, rock, and metal. This core material may be liquid, making for very odd "oceans" buried deep inside Uranus and Neptune. The cores of Uranus and Neptune are larger in radius than the cores of Jupiter and Saturn, even though they have about the same mass, because they are less compressed by the lower weight of their overlying layers. The less extreme interior conditions also allowed Uranus's and Neptune's cores to differentiate, so that hydrogen compounds reside in a layer above a central layer of rock and metal.

Internal Heat When we studied the terrestrial worlds, we found that internal heat was the major driver of surface geology. The jovian planets do not have any geology, because they have no solid surfaces, but their internal heat is still important for atmospheric processes. Again, let's start by examining Jupiter's internal heat, then extend the idea to the other jovian planets.

Jupiter has a tremendous amount of internal heat, and like any hot object it gradually loses this heat to space by emitting thermal radiation [Section 5.4]. In fact, Jupiter emits almost twice as much energy as it receives from the Sun. This heat contributes significant energy to Jupiter's upper atmosphere. (For comparison, Earth's internal heat contributes only 0.005% as much energy to the surface as does sunlight.)

What keeps Jupiter so hot inside? Jupiter's large size means it loses internal heat very slowly (see Mathematical Insight 9.1), but calculations show that the remaining heat of accretion and differentiation is not enough to explain Jupiter's present-day heat loss. Radioactive decay adds further heat, but still not enough to account for all the internal heat. The most likely explanation for Jupiter's excess heat is that the planet is still slowly contracting, as if it has not quite finished forming. Contraction converts gravitational potential energy to thermal energy, so ongoing contraction represents an ongoing source of internal heat. Although we have not measured any such contraction, theoretical considerations show that it is probably occurring. Moreover, calculations tell us that contraction could easily explain Jupiter's internal heat, even if the contraction is so gradual that we have little hope of ever measuring it directly.

Regardless of the specific mechanism, Jupiter has undoubtedly lost substantial heat during the 4.5 billion years since its formation. Jupiter's interior must have been much warmer in the distant past, and this heat would have "puffed up" its atmosphere. Thus, Jupiter was larger and reflected more sunlight in the distant past, and it must have been even more prominent in Earth's sky than it is today.

Like Jupiter, Saturn emits nearly twice as much energy as it receives from the Sun, suggesting that it, too, must have some ongoing source of heat. However, Saturn's mass is too small for it to be generating this heat by contracting

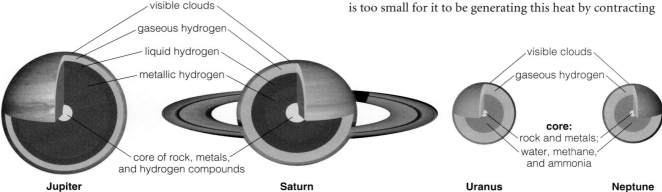

Figure 11.5 These diagrams compare the interior structures of the jovian planets, shown approximately to scale. All four planets have cores of rock, metal, and hydrogen compounds, with masses about 10 times the mass of Earth's core. They differ primarily in the depth of the hydrogen/helium layers that surround their cores. (The cores of Uranus and Neptune are differentiated into separate layers of rock/metal and hydrogen compounds.)

like Jupiter. Instead, Saturn's pressure and its lower interior temperatures may allow helium to condense into liquid form at relatively high levels within the interior. The helium droplets slowly rain down to the deeper interior. This gradual helium rain represents a sort of ongoing *differentiation* [Section 9.1], because it means that higher-density material (the helium droplets) is still sinking inside the planet. *Voyager* and *Cassini* observations have confirmed that Saturn's atmosphere is somewhat depleted of helium (compared to Jupiter's atmosphere), just as we should expect if helium has been raining down into Saturn's interior for billions of years.

Neither Uranus nor Neptune has internal conditions that would allow helium rain to form, and most of their original heat from accretion should have escaped long ago. This explains why Uranus emits virtually no excess internal energy. Neptune, however, is much more mysterious. Like Jupiter and Saturn, Neptune emits nearly twice as much energy as it receives from the Sun. The only reasonable explanation for Neptune's internal heat is that the planet is somehow still contracting, somewhat like Jupiter, thereby converting gravitational potential energy into thermal energy. However, we do not yet know why a planet of Neptune's size would still be contracting more than 4 billion years after its formation.

• What is the weather like on jovian planets?

Jovian atmospheres, by which we mean the portions of the planets that are visible to telescopes and spacecraft, have dynamic winds and weather, with colorful, cloudy skies and enormous storms. These atmospheres are made mostly of hydrogen and helium gas, mixed with small amounts of various hydrogen compounds. Because oxygen, carbon, and nitrogen are among the most common elements in the universe besides hydrogen and helium, the most common hydrogen compounds are methane (CH_4), ammonia (NH_3), and water (H_2O). Spectroscopy also reveals the presence of small amounts of more complex hydrogen compounds, such as acetylene (C_2H_2), ethane (C_2H_6), and propane (C_3H_8). Although all these hydrogen compounds together make up only a minuscule fraction of the jovian planet atmospheres, they are responsible for virtually all aspects of their appearances. Some of these compounds condense to form the clouds that are so prominent in telescope and spacecraft images. Others are responsible for the great variety of colors we see among the jovian planets. Without these compounds in their atmospheres, the jovian planets would be uniform, colorless balls of gas.

THINK ABOUT IT

Several gases in Jupiter's atmosphere—including methane, propane, and acetylene—are used as highly flammable fuels here on Earth. Jupiter also has plenty of lightning to provide sparks. Why don't these gases ignite in Jupiter's atmosphere? (*Hint:* What's missing from Jupiter's atmosphere that's necessary for ordinary fire?)

We can understand the jovian atmospheres using the same basic principles that we used with the terrestrial planet atmospheres. Of course, the different compositions, colder temperatures, and more rapid rotation rates of the jovian planets lead to some fascinating differences between terrestrial and jovian atmospheres. As we did for the jovian planet interiors, let's examine different aspects of the jovian atmospheres by starting with Jupiter as the prototype for each feature. We'll then use the general differences between the jovian planets to understand differences in their weather.

Atmospheric Structure and Clouds Telescopic observations of Jupiter, along with data returned from the *Galileo* probe during its 1995 plunge into Jupiter, tell us that the temperature structure of Jupiter's atmosphere is very similar to that of Earth (see Figure 10.7). High above the cloud tops, Jupiter's *thermosphere* consists of very-low-density gas heated to perhaps 1,000 K by solar X rays and by energetic particles from Jupiter's magnetosphere. Below the thermosphere but still above the clouds, we find Jupiter's *stratosphere*. Recall that a planet can have a stratosphere only if it has a gas that can absorb ultraviolet light from the Sun. Ozone plays this role on Earth. Jupiter lacks molecular oxygen and ozone, but has a few minor atmospheric ingredients that absorb solar ultraviolet photons. This absorption gives the stratosphere a peak temperature of around 200 K. Chemical reactions driven by the solar ultraviolet photons also create a smoglike haze that masks the color and sharpness of the clouds below. Below the stratosphere lies Jupiter's *troposphere*, where the temperature rises with depth because greenhouse gases trap both solar heat and Jupiter's own internal heat. Figure 11.6 shows how temperature varies with altitude in Jupiter's atmosphere.

In the troposphere, the warmer temperatures below drive strong convection. This convection is responsible for the thick clouds that enshroud Jupiter. Recall that clouds form when a gas condenses to make tiny liquid droplets or solid flakes. Water vapor is the only gas that can condense in Earth's atmosphere, which is why clouds on Earth are made of water droplets or ice flakes that can produce rain or snow. In contrast, Jupiter's atmosphere has several gases that can condense to form clouds. Each of these gases condenses at a different temperature, leading to distinctive cloud layers at different altitudes.

Jupiter has three primary cloud layers, which are also shown in Figure 11.6. We can understand them by imagining that we could watch gases rising through the troposphere with convection. Deep in the troposphere, the gases include three ingredients that will condense when temperatures are low enough: water (H_2O), ammonium hydrosulfide (NH_4SH), and ammonia (NH_3). The rising gas first encounters temperatures cool enough for water vapor to condense into liquid water but not cool enough for the other gases to condense. Thus, the lowest layer of clouds contains water droplets. As the remaining gas continues its rise, it next reaches an altitude at which ammonium hydrosulfide condenses to make the second cloud layer. Finally, after rising another 50 kilometers, the gas reaches an altitude

Figure 11.6 This graph shows the temperature structure of Jupiter's atmosphere. Jupiter has at least three distinct cloud layers because different atmospheric gases condense at different temperatures and hence at different altitudes.

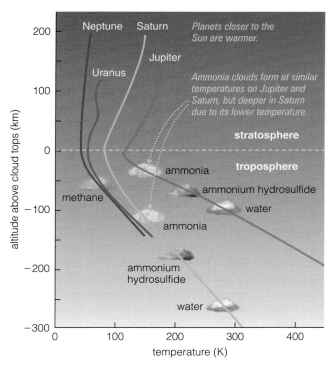

Figure 11.7 The figure contrasts the atmospheric structures and cloud layers of the four jovian planets. Note that the altitude scale is different on this figure than on Figure 11.6, allowing us to show the tropospheres and stratospheres only.

at which ammonia (NH_3) condenses to make the upper cloud layer.

Data from telescopic and spacecraft observations tell us that the atmospheric structures of the other three jovian planets are quite similar to that of Jupiter (Figure 11.7). The primary difference among them is that the atmospheres get progressively cooler with increasing distance from the Sun, just as we would expect. These temperature differences lead the planets to have their cloud layers at different altitudes.

Saturn has the same set of three cloud layers as Jupiter, but the lower overall temperatures causes these layers to lie deeper in Saturn's atmosphere. For example, to find the relatively warm temperatures at which water vapor can condense to form water clouds, we must look about 200 kilometers deeper into Saturn than into Jupiter. Saturn's cloud layers are also separated by greater vertical distances than Jupiter's, because Saturn's weaker gravity causes less atmospheric compression.

Uranus and Neptune are so cold that any cloud layers similar to those of Jupiter or Saturn would be buried too deep in their atmospheres for us to see. Thus, we do not know whether such clouds exist on these two planets. However, high in the atmospheres of Uranus and Neptune, we can see clouds made from flakes of methane snow. Methane can condense in the very cold upper tropospheres of Uranus and Neptune but not in the warmer tropospheres of Jupiter and Saturn.

Clouds and Colors The spectacular colors of the jovian planets are probably the first thing that jumps out at you when

you look at the photos in Figure 11.1. Many mysteries remain about precisely why the jovian planets are so colorful, but clouds are responsible for many of the color features. Different types of clouds apparently reflect light of different colors. For example, Earth's clouds look white from space because they are made of water that reflects the white light of the Sun.

Water clouds on Jupiter and Saturn probably also reflect white light, as do the upper layer clouds made of ammonia. In contrast, the mid-layer clouds of ammonium hydrosulfide reflect brown and red light, although no one knows exactly why. Most likely, the colors in these clouds come from as-yet-unidentified compounds that are produced by chemical reactions deeper in the atmosphere and are then carried upward by convection. The fact that Saturn's clouds lie deeper in its atmosphere than Jupiter's probably explains Saturn's more subdued colors: Less light penetrates to the depths at which Saturn's clouds are found, and the light they reflect is more obscured by the atmosphere above them.

The blue colors of Uranus and Neptune come from methane gas, which is at least 20 times more abundant (by percentage) on these planets than on Jupiter or Saturn. Methane gas in the upper atmospheres of Uranus and Neptune absorbs red light, allowing only blue light to penetrate to the level at which the methane clouds exist. The methane clouds reflect this blue light upward, giving the planets their blue colors (Figure 11.8). Uranus has a lighter blue color than Neptune, probably because it has more smog-like haze to scatter sunlight before it reaches the level of the methane clouds. The extra haze is probably a result

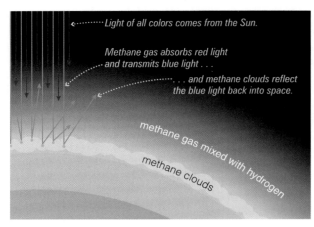

Light of all colors comes from the Sun.

Methane gas absorbs red light
and transmits blue light . . .

. . . and methane clouds reflect
the blue light back into space.

methane gas mixed with hydrogen

methane clouds

Figure 11.8 Neptune and Uranus look blue because methane gas absorbs red light but transmits blue light. Clouds of methane snowflakes reflect the transmitted blue light back to space.

of Uranus's "sideways" axis tilt, which leads to extreme seasons. With one hemisphere remaining sunlit for decades, gases have plenty of time to interact with solar ultraviolet light and to make the chemical ingredients of the haze. Continuous sunlight may also explain why Uranus has a surprisingly hot thermosphere that extends thousands of kilometers above its cloud tops. As we'll discuss later, Uranus's thermosphere has an important influence on its rings.

Global Winds and Storms on Jupiter Jupiter has dynamic weather, with strong winds and powerful storms. To understand Jupiter's weather, we must think about how its rapid rotation affects its global winds.

Jupiter has planetwide circulation cells similar to those on Earth (see Figure 10.16). As on Earth, solar heat causes equatorial air to expand and spill toward the poles, while cooler polar air flows toward the equator. Also as on Earth, the Coriolis effect splits the large equator-to-pole circulation cells into smaller cells. However, Jupiter's greater size and faster rotation make the Coriolis effect much stronger on Jupiter than on Earth. Instead of being split into just three smaller cells encircling each hemisphere, as on Earth, Jupiter's circulation cells split into many alternating bands of rising and falling air. These bands are visible as the "stripes" of alternating color in photographs of Jupiter.

The alternating colors of the stripes come from differences in the clouds that form in rising and falling air. As shown in Figure 11.9, the entire planet is blanketed with the reddish-brown, ammonium hydrosulfide clouds that make up Jupiter's middle cloud layer. In contrast, the upper layer of white ammonia clouds forms only in places where rising air carries ammonia gas to altitudes high enough and cold enough for the ammonia to condense. Thus, the white bands (sometimes called *zones*) represent regions of rising air with cold ammonia clouds, while the darker bands

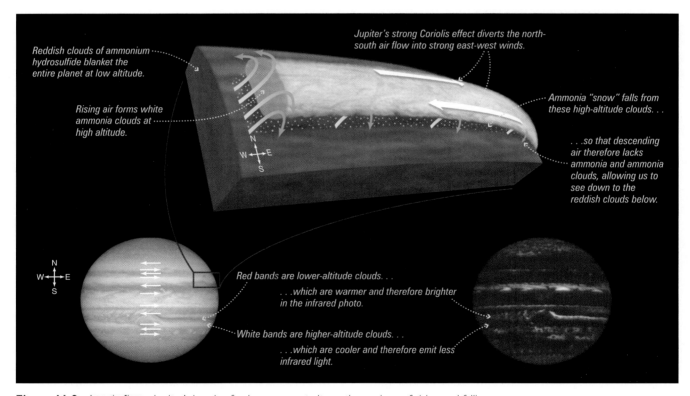

Reddish clouds of ammonium hydrosulfide blanket the entire planet at low altitude.

Jupiter's strong Coriolis effect diverts the north-south air flow into strong east-west winds.

Rising air forms white ammonia clouds at high altitude.

Ammonia "snow" falls from these high-altitude clouds. . .

. . .so that descending air therefore lacks ammonia and ammonia clouds, allowing us to see down to the reddish clouds below.

Red bands are lower-altitude clouds. . .
. . .which are warmer and therefore brighter in the infrared photo.

White bands are higher-altitude clouds. . .
. . .which are cooler and therefore emit less infrared light.

Figure 11.9 Interactive Figure Jupiter's bands of color represent alternating regions of rising and falling air: We see white ammonia clouds in regions of rising air, and see down to the underlying layer of reddish ammonium hydrosulfide clouds in regions of falling air. The white arrows on the visible-light photo indicate wind directions. The infrared photo at the right, taken nearly simultaneously with the visible-light photo, confirms that the red bands represent warmer, deeper clouds and the white bands represent cooler, higher-altitude clouds.

(sometimes called *belts*) represent regions of falling air in which we can see down to a lower, warmer cloud deck. This distinction between Jupiter's bright and dark bands is analogous to the difference on Earth between cloudy, rainy equatorial regions (regions of generally rising air) and the clear desert skies found at latitudes roughly 20°–30° north and south of the equator (regions of descending air). Notice how the infrared photo of Jupiter in Figure 11.9 confirms this idea: Warmer air emits more infrared light, so the warmer reddish bands appear brighter in the infrared photograph than the cooler white bands.

You might wonder why ammonia clouds can form only in the rising air and not in the falling air. The answer goes back to the clouds themselves. Jupiter's lower atmosphere has ammonia throughout, so there's always ammonia in the air rising upward. However, once the ammonia condenses to form clouds of ammonia ice, these ice flakes can fall back downward against the air currents. In essence, the ammonia rises up and forms clouds, then falls straight back down as ammonia "snow." Thus, the air that continues to rise above the clouds has very little ammonia left in it. This ammonia-depleted air then spills to the north and south and returns downward in the bands of falling air, explaining why these bands lack the ammonia needed to form clouds. It's analogous to the way that air descending over Earth's desert latitudes has been depleted in water by precipitation over the equator, which explains why desert regions are so dry.

The alternating bands of rising and falling air shape Jupiter's global wind patterns. As on Earth, the rising and falling air drives slow winds that are directed north or south, but Jupiter's strong Coriolis effect diverts these winds into fast east or west winds. The east-west winds have peak speeds above 400 km/hr, making hurricane winds on Earth seem mild by comparison. The winds are generally strongest at the equator and at the boundaries between bands of rising and falling air.

Jupiter's global wind patterns are sometimes interrupted by powerful storms, much as storms can interrupt Earth's global wind patterns. Of course, Jupiter's storms dwarf those that we see on Earth. Indeed, Jupiter's most famous feature—the **Great Red Spot**—is a giant storm more than twice as wide as the entire planet Earth. It is somewhat like a hurricane on Earth, except that its winds circulate around a high-pressure region rather than a low-pressure region (Figure 11.10). It is also extremely long-lived compared to storms on Earth: Astronomers have seen it throughout the three centuries during which telescopes have been powerful enough to detect it.

Other, smaller storms are constantly brewing in Jupiter's atmosphere. They are small only in comparison to the Great Red Spot. Brown ovals are low-pressure storms with their cloud tops deeper in Jupiter's atmosphere, and white ovals are high-pressure storms topped with ammonia clouds.

No one knows what drives Jupiter's storms, why Jupiter has only one Great Red Spot, or why the Great Red Spot has persisted so much longer than storms on Earth. Storms on Earth lose their strength when they pass over land, so per-

Figure 11.10 This photograph shows Jupiter's Great Red Spot, a huge high-pressure storm that is large enough to swallow two or three Earths. The overlaid diagram shows a weather map of the region.

haps Jupiter's biggest storms last for centuries simply because no solid surface is present to sap their energy. The bright colors of the Great Red Spot also pose a mystery: We might expect its high-altitude clouds to be white like the high-altitude ammonia clouds elsewhere, but instead, of course, they are red. The colors may be the result of chemicals formed by interactions between the storm's high-altitude gas and solar ultraviolet light, but no one really knows for sure.

Jupiter does not have seasons, because it has no appreciable axis tilt. In fact, Jupiter's polar temperatures are quite similar to its equatorial temperatures, presumably because heat from Jupiter's interior keeps the planet uniformly warm.

Global Winds and Storms on Saturn, Uranus, and Neptune The other jovian planets also have dramatic weather patterns (Figure 11.11). As on Jupiter, Saturn's rapid rotation creates alternating bands of rising and falling air, along with rapid east-west winds. In fact, Saturn's winds are even faster than Jupiter's—a surprise that scientists have yet to explain. We might expect seasons on Saturn, because it has an axis tilt similar to that of Earth. Some seasonal weather changes have been observed, but Saturn's internal heat keeps temperatures about the same year-round and planetwide.

Neptune's atmosphere is also banded, and we have seen a high-pressure storm, called the Great Dark Spot, similar to Jupiter's Great Red Spot. However, the Great Dark Spot did not last as long, having disappeared from view just 6 years after its discovery. Like Saturn, Neptune has an axis tilt similar to Earth's, but it has relatively little seasonal change because of its internal heat.

The greatest surprise in jovian weather comes from Uranus. When *Voyager 2* flew past Uranus in 1986, photographs revealed virtually no clouds and no banded structure like those found on the other jovian planets. Scientists attributed the lack of weather to the lower internal heat of Uranus. However, more recent observations from the Hubble Space Telescope and groundbased adaptive optics telescopes show storms raging in Uranus's atmosphere. The storms may be brewing because of the changing seasons:

a A close-up of Jupiter's atmosphere in its southern hemisphere, with the Great Red Spot visible near the top. The white oval is a low-pressure storm.

b A close-up of Saturn's atmosphere. Its banded appearance is very similar to that of Jupiter, but it has even faster winds.

c This infrared photograph from the Keck telescope shows several storms (the bright blotches) brewing on Uranus. Uranus's ring shows up in red.

d Neptune's atmosphere has bands much like Jupiter and Saturn, and occasional strong storms. The large storm to the left of center is known as the Great Dark Spot.

Figure 11.11 Selected views of weather patterns on the four jovian planets.

Thanks to Uranus's extreme axis tilt and 84-year orbit of the Sun [Section 7.1], its northern hemisphere is just beginning to see sunlight for the first time in decades.

• Do jovian planets have magnetospheres like Earth's?

Recall that Earth has a global magnetic field generated by the movements of charged particles in our planet's metallic outer core (see Figure 9.6). The jovian planets also have global magnetic fields generated by motions of charged particles deep in their interiors. Just as with Earth, these magnetic fields create bubble-like *magnetospheres* that surround the planets and shield them from the solar wind.

Jupiter's Magnetosphere Jupiter's magnetic field is by far the strongest of the four jovian planets, and it is some 20,000 times stronger than Earth's magnetic field. As we discussed in Chapter 9, a planet can have a global magnetic field if it has: (1) an interior region of electrically conducting fluid; (2) convection in that layer of fluid; and (3) at least moderately rapid rotation. In Jupiter's case, the electrically conducting fluid region is its thick layer of metallic hydrogen (see Figure 11.4). The great extent of this region, combined with Jupiter's rapid rotation, explains Jupiter's strong magnetic field.

Jupiter's strong magnetic field also gives it an enormous magnetosphere (Figure 11.12). It begins to deflect the solar wind some 3 million kilometers (about 40 Jupiter radii)

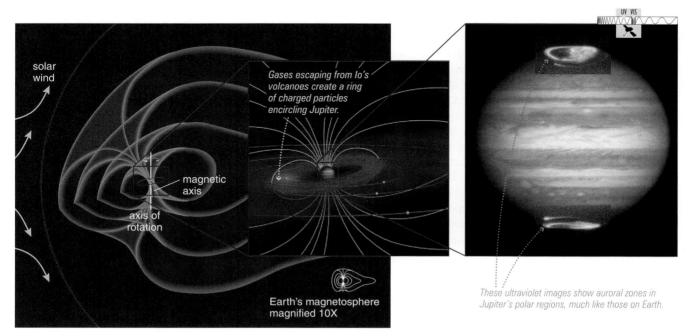

Figure 11.12 Jupiter's strong magnetic field gives it an enormous magnetosphere. Gases escaping from Io feed the donut-shaped Io torus, and particles entering Jupiter's atmosphere near its magnetic poles contribute to auroras on Jupiter. The image at the right is a composite of ultraviolet images of the polar regions overlaid on a visible image of the whole planet, all taken by the Hubble Space Telescope.

before the solar wind even reaches Jupiter. If our eyes could see this part of Jupiter's magnetosphere, it would be larger than the full moon in our sky.

Jupiter's magnetosphere traps far more charged particles than Earth's magnetosphere, largely because it has a source of particles that Earth's lacks. Nearly all the charged particles in Earth's magnetosphere come from the solar wind, but in Jupiter's case its volcanically active moon, Io, contributes many additional particles. These particles help create auroras on Jupiter. They also create belts of very intense radiation around Jupiter, which can cause damage to orbiting spacecraft.

Jupiter's magnetosphere, in turn, has important effects on Io and the other moons of Jupiter. The charged particles bombard the surfaces of Jupiter's icy moons, with each particle blasting away a few atoms or molecules. This process alters the surface materials and can even generate thin atmospheres, much like the thin atmospheres of Mercury and the Moon [Section 10.3]. On Io, bombardment by charged particles leads to the continuous escape of gases released by volcanic outgassing. The escaping gases (sulfur, oxygen, and a hint of sodium) are ionized and feed a donut-shaped charged particle belt, called the *Io torus*, that approximately traces Io's orbit (the red "donut" in Figure 11.12).

Comparing Jovian Magnetospheres   The other jovian planets also have magnetic fields and magnetospheres, but theirs are much weaker than Jupiter's (although still much stronger than Earth's). The strength of each planet's magnetic field depends primarily on the size of the electrically conducting layer buried in its interior. Saturn's magnetic field is weaker than Jupiter's because it has a thinner layer of electrically conducting metallic hydrogen. Uranus and Neptune, smaller still, have no metallic hydrogen at all. Their relatively weak magnetic fields must be generated in their core "oceans" of hydrogen compounds, rock, and metal.

The size of a planet's magnetosphere depends not only on the magnetic field strength, but also on the pressure of the solar wind against it. The pressure of the solar wind is weaker at greater distances from the Sun, so the magnetospheric "bubbles" surrounding more distant planets are larger than they would be if these planets were closer to the Sun. Thus, Uranus and Neptune have moderate-size magnetospheres despite their weak magnetic fields (Figure 11.13). No other magnetosphere is as full of charged particles as Jupiter's, primarily because no other jovian planet has a satellite like Io. Because trapped particles generate auroras, Jupiter has the brightest auroras, while those of the more distant jovian planets are progressively weaker.

We still have much to learn about the jovian magnetic fields and magnetospheres. For example, we generally expect magnetic fields to be closely aligned with planetary rotation, because the magnetic fields are generated within the rotating interiors of planets. However, while this is the case for Jupiter and Saturn (whose magnetic fields are inclined to their rotation axes by 10° and 0°, respectively), it is not the case for Uranus and Neptune. *Voyager* observations showed that the magnetic field of Uranus is tipped

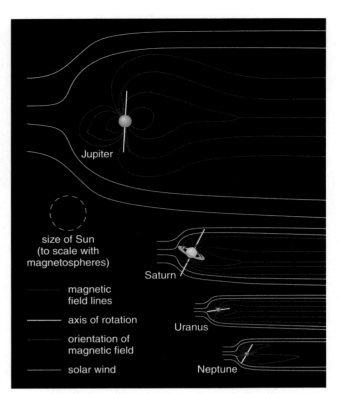

Figure 11.13 Comparison of jovian planet magnetospheres. The size of Jupiter's magnetosphere is particularly impressive in light of the greater pressure from the solar wind nearer the Sun. Note the significant tilts of the magnetic fields of Uranus and Neptune compared to their rotation axes. (Planets enlarged for clarity.)

by a whopping 60° relative to its rotation axis, and the magnetic field's center is also significantly offset from the planet's center. Neptune's magnetic field is inclined by 46° to its rotation axis. No one has yet explained these surprising observations.

11.2 A Wealth of Worlds: Satellites of Ice and Rock

The jovian planets are majestic and fascinating, but they are only the beginning of our exploration of jovian planet *systems*. Each of the four jovian systems includes numerous moons and a set of rings. The total mass of all the moons and rings put together is minuscule compared to any one of the jovian planets, but the remarkable diversity of these satellites makes up for their lack of size. In this section, we'll explore a few of the most interesting aspects of the jovian moons. We'll then discuss rings in the following section.

• What kinds of moons orbit the jovian planets?

We now know of more than 100 moons orbiting the jovian planets. Jupiter has the most, with more than 60 moons known to date. Because there are so many moons, it's helpful to organize them into three groups by size: small moons less than about 300 kilometers in diameter, medium-size moons ranging from about 300 to 1,500 kilometers in diameter, and large moons more than 1,500 kilometers in diameter. These categories are useful because size relates to geological activity. The large moons show evidence of active and ongoing geology, and most of the medium-size moons seem to have had geological activity in the past.

Figure 11.14 shows a montage of all the medium-size and large moons. These moons resemble the terrestrial planets in many ways. Each is spherical with a solid surface and its own unique geology. Some possess atmospheres, hot interiors, and even magnetic fields. The two largest—Jupiter's moon Ganymede and Saturn's moon Titan—are larger than the planet Mercury. Four others are larger than Pluto: Jupiter's moons Io, Europa, and Callisto and Neptune's moon Triton. However, they differ from terrestrial worlds in their compositions: Because they formed in the cold outer solar system, most of these worlds contain substantial amounts of ice in addition to metal and rock. As we will see, the fact that ice can undergo geological change at much lower temperatures than rock helps explain the surprising geological activity of many moderately sized moons.

Most of the medium-size and large moons probably formed by accretion within the disks of gas surrounding individual jovian planets [Section 8.3]. That explains why their orbits are almost circular and lie close to the equatorial plane of their parent planet, and also why these moons orbit in the same direction in which their planet rotates. Nearly all these moons also share an uncanny trait: They always keep the same face turned toward their planet, just as our Moon always shows the same face to Earth. This synchronous rotation arose from the strong tidal forces [Section 4.5] exerted by the jovian planets, which caused each moon to end up with equal periods of rotation and orbit regardless of how fast the moon rotated when it formed.

The medium-size and large moons tend to be the most interesting, but small moons are far more numerous. Many of the small moons are probably captured asteroids or comets, and as a result they do not follow any particular orbital patterns. Dozens of the smallest moons have been discovered only within the past few years, and many more may yet be discovered. (See Table E.3 in Appendix E for a complete list of the known moons and their general properties.)

The small moons' shapes are irregular, like potatoes (Figure 11.15), because their gravities are too weak to force their rigid material into spheres. We have not studied these moons in depth, but we expect their small sizes to allow for little if any geological activity. For the most part, the small moons are just chunks of ice and rock held captive by the gravity of a massive jovian planet.

In case you are wondering, scientists originally named Jupiter's moons for the mythical lovers of the Roman god Jupiter. However, the planet Jupiter has so many moons that more recent discoveries are named for its namesake's more distant mythological relations. Saturn's moon Titan is named for the Greek gods called the Titans, who ruled before Zeus (Jupiter); other moons of Saturn are named for individual Titans. Moons of Uranus all take their names from characters in the works of William Shakespeare and Alexander Pope, while moons of Neptune are all characters related to the sea in Greek and Roman mythology.

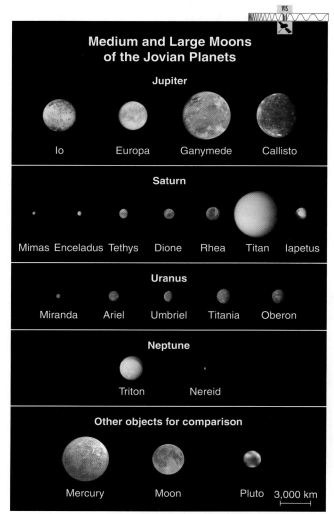

Figure 11.14 The medium-size and large moons of the jovian planets, with sizes (but not distances) shown to scale.

Figure 11.15 These photos from the *Voyager* spacecraft show five of Saturn's smaller moons. All are much smaller than the smallest moons shown in Figure 11.14. Their irregular shapes are due to their small size, which makes their gravities too weak to force them into spheres.

1000 km

| Io | Europa | Ganymede | Callisto |

Figure 11.16 This set of photos, taken by the *Galileo* spacecraft orbiting Jupiter, shows global views of the four Galilean moons as we know them today. Sizes are shown to scale. (Io is about the size of Earth's Moon.)

• Why are Jupiter's Galilean moons so geologically active?

We are now ready to embark on a brief tour of the most interesting moons of the jovian planets. Our first stop is Jupiter. Jupiter's four largest moons, known as the *Galilean moons* because Galileo discovered them [Section 3.3], are all large enough that they would count as planets if they orbited the Sun (Figure 11.16).

Io: The Most Volcanically Active World in the Solar System For anyone who thinks of moons as barren, geologically dead places like our own Moon, Io shatters the stereotype. When the *Voyager* spacecraft first photographed Io up close more than two decades ago, we discovered a world with a surface so young that not a single impact crater has survived from past impacts. Moreover, *Voyager* cameras recorded volcanic eruptions in progress as the spacecraft passed by. We now know that Io is by far the most volcanically active world in our solar system. Large volcanoes pockmark its entire surface (Figure 11.17), and eruptions are so frequent that they have buried virtually every impact crater. Io probably also has tectonic activity, because tectonics and volcanism generally go hand in hand. However, debris from volcanic eruptions has probably buried most tectonic features.

Many of Io's frequent eruptions are surprisingly similar to those of volcanoes on Earth. Some of the taller volcanoes appear to be shield volcanoes built from flows of basalt lava, and the eruptions are accompanied by outgassing. The primary gases released from Io's volcanoes are sulfur and sulfur dioxide (SO_2). Some of this gas escapes into space, where it supplies ionized gas to the Io torus and Jupiter's magnetosphere, and some of it gives Io a very thin atmosphere. Much of the gas condenses and falls back to the surface, where the sulfur gives Io its distinctive red and orange colors and the sulfur dioxide makes a white frost. As hot lava flows across the surface (left photo in Figure 11.17), it can re-vaporize the sulfur dioxide surface ice in much the same way that lava flowing into the ocean vaporizes water on Earth. Io's low gravity and thin atmosphere allow tall plumes of this vaporized sulfur dioxide to rise upward to altitudes of hundreds of kilometers (right photo in Figure 11.17). Over a period of a few months, fallout from these tall volcanic plumes can blanket an area the size of Arizona (Figure 11.18).

Io's active volcanoes tell us that it must be quite hot inside. However, Io is only about the size of our geologically dead Moon, so it should have long ago lost any heat from its birth and it is too small for radioactivity to provide much ongoing heat. How, then, can Io be so hot inside? The only

Figure 11.17 Interactive Photo Io is the most volcanically active body in the solar system. Most of the black, brown, and red spots on the surface are recently active volcanic features. White and yellow areas are sulfur dioxide and sulfur deposits, respectively, from volcanic gases. (Photographs from the *Galileo* spacecraft; some colors slightly enhanced or altered.)

20 km

This close-up shows the glow of intensely hot lava from a volcanic eruption.

jets of SO_2 gas

SO_2 frost

Basalt lava flow

10 km

This 80-km-high gas plume was created when hot lava flowed over sulfur dioxide frost, causing it to sublimate explosively into gas.

Some sulfur-rich volcanoes deposit red, orange, or yellow fallout.

An eruption during the Galileo mission produced dusty fallout that covered an area larger than Arizona in a matter of months.

200 km

Figure 11.18 This enhanced-color photo shows how the fallout from volcanic eruptions can cover large areas of the surface in a matter of months or years.

possible answer is that some other ongoing process must be heating Io's interior. Scientists have identified this process and call it **tidal heating**, because it arises from effects of tidal forces exerted by Jupiter.

Just as Earth exerts a tidal force that causes the Moon to keep the same face toward us at all times [Section 4.5], a tidal force from Jupiter makes Io keep the same face toward Jupiter as it orbits. But Jupiter's mass make this tidal force far larger than that which Earth exerts on the Moon. Moreover, Io's orbit is slightly elliptical, so its orbital speed and distance from Jupiter vary. This variation means that the strength and direction of the tidal force change slightly as Io moves through each orbit, which in turn changes the size and orientation of Io's tidal bulges (Figure 11.19a). The result is that Io is continuously being flexed in different directions, which generates friction inside it. The flexing heats the interior in the same way that flexing warms Silly Putty. Tidal heating generates tremendous heat on Io— more than 200 times as much heat (per gram of mass) as the radioactive heat driving much of Earth's geology. This

heat explains Io's volcanic activity. The energy for tidal heating ultimately comes from Jupiter's rotation, which is gradually slowing, although at a rate too small to be observed.

However, we are still left with a deeper question: Why is Io's orbit slightly elliptical, when almost all other large satellites have nearly circular orbits? The answer lies in an interesting dance executed by Io and its neighboring moons (Figure 11.19b). During the time Ganymede takes to complete one orbit of Jupiter (about 7 days), Europa completes exactly two orbits and Io completes exactly four orbits. The three moons therefore line up periodically, and the gravitational tugs they exert on one another add up over time.* Because the tugs are always in the same direction with each alignment, they tend to stretch out the orbits, making them slightly elliptical. The effect is much like that of pushing a child on a swing. If timed properly, a series of small pushes can add up to a *resonance* that causes the child to swing quite high. For the three moons, the resonance that makes their orbits elliptical comes from the small gravitational tugs that repeat at each alignment.

The phenomenon of orbital periods falling into a simple mathematical relationship is called an **orbital resonance**. As we will discuss later in this and the next two chapters, resonances are quite common in planetary systems. For example, they affect not only the Galilean moons but also planetary rings, the asteroid belt, and the Kuiper belt in our solar system, and we have even discovered resonances among the known systems of planets around other stars.

---

*You may wonder why periodic alignments occur. Like synchronous rotation, they are not a coincidence but rather a consequence of feedback from the tides that the moons raise on Jupiter. The periodic tugs the moons exert on one another actually work to sustain the recurring alignments.

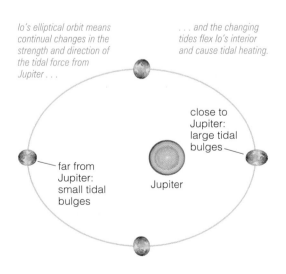

Io's elliptical orbit means continual changes in the strength and direction of the tidal force from Jupiter . . .

. . . and the changing tides flex Io's interior and cause tidal heating.

close to Jupiter: large tidal bulges

far from Jupiter: small tidal bulges

Jupiter

a Tidal heating arises because Io's elliptical orbit (exaggerated in this diagram) causes varying tides.

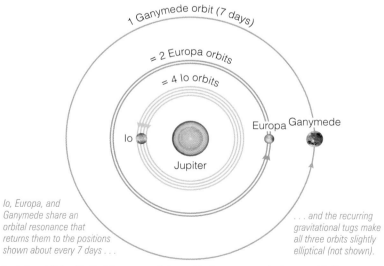

1 Ganymede orbit (7 days)

= 2 Europa orbits

= 4 Io orbits

Io

Jupiter

Europa   Ganymede

Io, Europa, and Ganymede share an orbital resonance that returns them to the positions shown about every 7 days . . .

. . . and the recurring gravitational tugs make all three orbits slightly elliptical (not shown).

b Io's orbit is elliptical because of the orbital resonance it shares with Europa and Ganymede.

Figure 11.19 These diagrams explain the cause of tidal heating on Io. Tidal heating has a weaker effect on Europa and Ganymede, because they are farther from Jupiter and tidal forces weaken with distance.

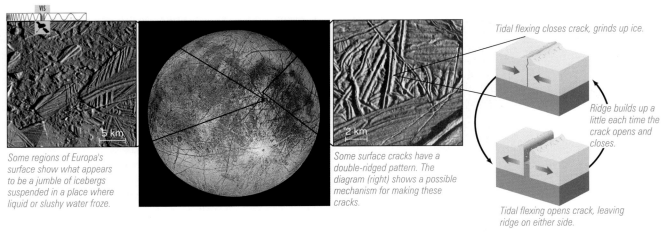

Some regions of Europa's surface show what appears to be a jumble of icebergs suspended in a place where liquid or slushy water froze.

Some surface cracks have a double-ridged pattern. The diagram (right) shows a possible mechanism for making these cracks.

Tidal flexing closes crack, grinds up ice.

Ridge builds up a little each time the crack opens and closes.

Tidal flexing opens crack, leaving ridge on either side.

Figure 11.20 Europa's icy crust may hide a deep, liquid water ocean beneath its surface. These photos are from the *Galileo* spacecraft; colors are enhanced in the global view.

Europa: The Water World? Europa offers a stark contrast to Io. Instead of active volcanoes dotting its surface, Europa is covered by water ice (Figure 11.20). Nevertheless, its fractured, frozen surface must hide an interior made hot by the same type of tidal heating that powers Io's volcanoes—but tidal heating is weaker on Europa because it lies farther from Jupiter. Europa has only a handful of impact craters, suggesting a surface no more than a few tens of millions of years old. Clearly, active geology has covered up older craters on Europa. But what could be doing the covering up?

Scientists suspect that water (H_2O) is the agent of change on Europa's surface: either liquid water rising up from beneath the icy crust, or interior ice that is just warm enough to undergo convection that allows some of it to rise up and flow across the surface. Close-up photos of the surface support this idea. For example, some photos show what appear to be jumbled icebergs suspended in a region where liquid or slushy water has frozen (left side of Figure 11.20). Other photos show double-ridged cracks on the surface (right side of Figure 11.20) that may be created when tidal flexing allows water to well up and build up ridges.

Models of Europa's interior give us further insight into its geology. Detailed measurements of the strength of gravity over different parts of Europa from the *Galileo* spacecraft suggest that Europa has a metallic core and a rocky mantle. Above the rocky mantle, Europa has enough water to make a layer of ice about 100 kilometers thick (Figure 11.21). The upper 5 to 25 kilometers should be solidly frozen, but beneath this icy crust, tidal heating provides enough warmth to create a layer of either liquid water or relatively warm, convecting ice. If it is liquid water, then Europa's ocean may contain more than twice as much liquid water as all of Earth's oceans combined.

We do not yet have enough data to be certain that Europa's subsurface layer consists of liquid water rather than convecting ice. However, magnetic field data from the *Galileo* spacecraft strongly argue for liquid water. Europa is one of only a few moons in the solar system that have a magnetic field at all. Moreover, Europa's magnetic field changes as Jupiter rotates, suggesting that it is created (or "induced") in response to the rotation of Jupiter's strong magnetic field. In order to respond in this way to Jupiter's magnetic field, Europa would have to have a *liquid* layer of electrically conducting material. Thus, a salty ocean would fit the bill, but convecting ice would not. The data further suggest that Europa's liquid ocean must be global in extent, and that it is about as salty as Earth's oceans. Additional support for this idea comes from another of *Galileo*'s instruments, which found evidence for salty compounds on Europa's surface—possible seepage from a briny deep.

Taken together, the photographic and magnetic-field evidence makes a strong case for a deep ocean of liquid water on Europa. Perhaps as on Earth's seafloor, lava erupts from vents on Europa's seafloor, sometimes violently enough to jumble the icy crust above (Figure 11.22). And, knowing that primitive life thrives near seafloor vents on Earth, we can wonder whether Europa might also be a home to life—a possibility we will explore further in Chapter 24.

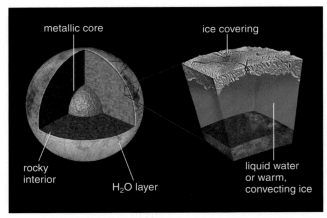

metallic core

ice covering

rocky interior

H_2O layer

liquid water or warm, convecting ice

Figure 11.21 This diagram shows one model of Europa's interior structure. There is little doubt that the H_2O layer is real, but we are not yet certain that the material beneath the icy crust is actually liquid water rather than just relatively warm, convecting ice.

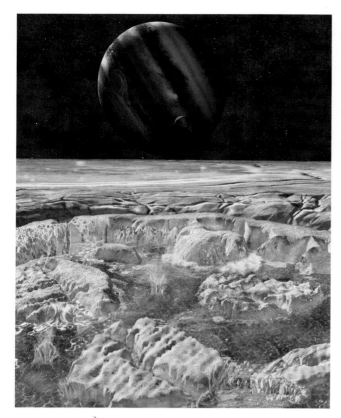

Figure 11.22 🎨 Tidal heating may give Europa a subsurface ocean beneath its icy crust. This artist's conception imagines a region where an undersea volcano has disrupted the crust.

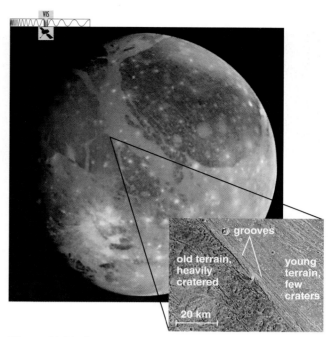

Figure 11.23 Ganymede, the largest moon in the solar system, has both old and young regions on its surface of water ice. The dark regions are heavily cratered and must be billions of years old, while the light regions are younger landscapes where eruptions of water have presumably erased ancient craters; the long grooves in the light regions were probably formed by water erupting along surface cracks. Notice that the boundary between the two types of terrain can be quite sharp.

Ganymede: King of the Moons   Ganymede is the largest moon in the solar system, and its surface tells of a complex geological history. Like Europa, Ganymede has a surface of water ice. However, while Europa's surface appears relatively young everywhere, Ganymede's surface appears to have a dual personality (Figure 11.23). Some regions are dark and densely cratered, suggesting that they look much the same today as they did billions of years ago. Other regions are light-colored with very few craters, suggesting that liquid water has recently erupted and refrozen. In some cases, fairly sharp boundaries separate the two types of terrain.

The young terrain argues for occasional upwelling of liquid water or icy slush to the surface. This material would cover craters before refreezing, explaining why there are so few craters in this terrain. The long grooves are probably made by the eruption of water or slush along a crack in the surface. As the water in the crack freezes, it expands and pushes outward, creating the groove.

If liquid water occasionally wells up to the surface, could it mean that Ganymede has a subsurface ocean like that thought to exist on Europa? The case for an ocean on Ganymede is less strong than the case for an ocean on Europa but is also bolstered by magnetic-field measurements. Like Europa, Ganymede has a magnetic field that varies with Jupiter's rotation, suggesting the presence of a salty ocean beneath the surface.

One difficulty with the idea of an ocean on Ganymede is figuring out where the heat to melt subsurface ice might come from. Because Ganymede is farther from Jupiter than Europa or Io, its tidal heating is weaker and could not by itself supply enough heat to melt ice today. However, Ganymede's larger size means it should retain more heat from radioactive decay. Perhaps tidal heating and radioactive decay together provide enough heat to make a liquid layer beneath the icy surface. If so, the lesser heating on Ganymede means that its ocean probably lies at least 150 kilometers beneath the surface—much deeper than Europa's potential ocean.

Callisto: Last of the Galilean Moons   The outermost Galilean moon, Callisto, looks most like what scientists originally expected for an outer solar system satellite: a heavily cratered iceball (Figure 11.24). The bright, circular patches on its surface are impact craters. They are bright presumably because large impacts blast out "clean" ice from deep underground, and this ice reflects more light than the dirtier ice that has been on the surface longer.

Craters make sense on an old surface like Callisto's, but other features are more difficult to interpret. For example, close-up photos show a dark, powdery substance concentrated in low-lying areas, leaving ridges and crests bright white (small photo in Figure 11.24). The dark powder may be debris left behind when ice sublimates into gas from Callisto's surface, much as dark material is left behind on a comet's nucleus when ice sublimates away.

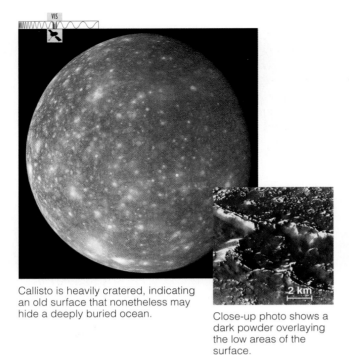

Callisto is heavily cratered, indicating an old surface that nonetheless may hide a deeply buried ocean.

2 km

Close-up photo shows a dark powder overlaying the low areas of the surface.

Figure 11.24 Callisto, the outermost of the four Galilean moons, has a heavily cratered, icy surface.

Figure 11.25 Titan, as photographed by the *Cassini* spacecraft, is enshrouded by a methane gas and tan haze. The inset shows the haze layers extending a hundred kilometers above a portion of the moon. *Cassini* was outfitted with filters designed to peer through the atmosphere at the specific near-infrared wavelengths of light that are least affected by the atmosphere or haze.

Despite its relatively large size (the third-largest moon in the solar system), Callisto lacks volcanic and tectonic features. This tells us that it lacks any significant source of internal heat. In fact, gravity measurements by the *Galileo* spacecraft showed that Callisto never underwent differentiation: Dense rock and lighter ice are thoroughly mixed throughout most of its interior, which means that its interior has never warmed significantly. The lack of heat is not surprising: Callisto has no tidal heating because it does not participate in the orbital resonances that affect the other three Galilean moons, and we do not expect its icy interior to contain enough radioactive material to supply much heat through radioactive decay.

Nevertheless, it is possible that Callisto, too, may hide a subsurface ocean. Like Europa and Ganymede, Callisto has a magnetic field that suggests the presence of a salty interior ocean. Perhaps Callisto's surface provides just enough insulation for heat from radioactive decay to melt water beneath the thick, icy crust. If so, we face the intriguing possibility that there could be three "water worlds" orbiting Jupiter—with far more total ocean than we find here on Earth.

• What is special about Titan and other major moons of the outer solar system?

Leaving Jupiter behind, we will continue our tour of moons by discussing the major moons of Saturn, Uranus, and Neptune. We'll go in planetary order, first focusing special attention on Saturn's enigmatic Titan, the second-largest moon in the solar system after Ganymede.

Saturn's Moon Titan: Piercing the Veil   Titan is unique among the moons of our solar system in having a thick atmosphere—so thick that it hides the surface from view, except at a few specific wavelengths of light (Figure 11.25). Titan's reddish color comes from chemicals in its atmosphere, much like those that make smog over cities on Earth. The atmosphere is about 90% molecular nitrogen (N_2), not that different from the 77% nitrogen content of Earth's atmosphere. However, on Earth the rest of the atmosphere is mostly oxygen, while the rest of Titan's atmosphere consists of argon, methane (CH_4), ethane (C_2H_6), and other hydrogen compounds.

Titan's atmospheric composition is different from that of any other world in our solar system, which is not too surprising given our general understanding of atmospheric source and loss processes [Section 10.2]. Titan's icy composition supplies methane and ammonia gas through evaporation and sublimation. Solar ultraviolet light breaks down some of those molecules, releasing hydrogen atoms and leaving highly reactive compounds containing carbon and nitrogen. The hydrogen atoms can leave Titan forever by thermal escape, while the remaining molecular fragments react to make the other ingredients of Titan's atmosphere. For example, the abundant molecular nitrogen is made after ultraviolet light breaks down ammonia (NH_3) molecules, and ethane is made from methane.

The methane and ethane in Titan's atmosphere are both greenhouse gases, and therefore give Titan an appreciable greenhouse effect [Section 10.1] that makes it warmer than it would be otherwise. Still, because of its great distance from the Sun, its surface temperature is a frigid 93 K ($-180°C$). The surface pressure on Titan is about 1.5 times the sea level pressure on Earth, which would be fairly comfortable if not for the lack of oxygen and the cold temperatures.

A moon with a thick atmosphere would be intriguing enough, but we have at least two other reasons for special interest in Titan. First, its complex atmospheric chemistry probably produces numerous organic chemicals—the chemicals that are the basis of life. Few scientists think there could be actual life on Titan, due to the cold temperatures, but many hope that further study of Titan will teach us about the chemistry that may have occurred on Earth before life actually arose. Second, although it is far too cold for liquid water to exist on Titan, conditions are right for methane or ethane rain, perhaps making rivers feeding into lakes or oceans.

NASA and the European Space Agency (ESA) found Titan so intriguing that they combined forces to explore it. NASA built the *Cassini* "mother ship" that carried an ESA-built probe called *Huygens* designed to parachute to the surface for a soft landing. The probe landed successfully on January 14, 2005 (Figure 11.26). During its descent it photographed river valleys merging together, flowing down to what looks like a shoreline. On impact, instruments on the probe discovered that the landing was not so hard: The surface has a hard crust but is a bit squishy below, perhaps like sand with liquid mixed in. The view from the surface shows "ice boulders" rounded by erosion. All these results support the idea of a wet climate—but wet with liquid methane rather than liquid water.

The analysis of the data from Titan is still in its early stages, but we have already learned a great deal. The brighter regions in the photographs are icy hills, perhaps made by ice volcanoes. The dark valleys are probably created when methane rain carries down "smog particles" that concentrate on river bottoms. The vast plains into which the valleys appear to empty are indeed the low regions, but they do not appear to be liquid. They too are probably made dark from smog particles carried there by the rivers. All

in all, conditions on Titan appear to be analogous to the desert southwest of the United States, where infrequent rainfall carves valleys and creates vast dry lakes called "playas" where the water evaporates or soaks into the ground.

Perhaps the most astonishing result from the *Huygens* mission is how familiar the landscape looks in this alien environment with these unfamiliar materials. Instead of liquid water, Titan has liquid methane. Instead of rock, Titan has ice. Instead of molten lava, Titan has a slush of water ice mixed with ammonia. Instead of surface dirt, Titan's surface has smog particles that rain out of the sky. Evidently, the similarities between the physical processes that occur on Titan and Earth are far more important in shaping the landscapes than the fact that the two worlds have very different compositions and temperatures. The *Cassini* mission should continue to operate in Saturn orbit until at least 2008, and scientists hope that additional radar mapping of Titan will teach us more about the surface, especially in areas that may be very different from the *Huygens* landing site.

THINK ABOUT IT

What other geological features might you expect on Titan? Remember than the *Huygens* probe saw just one location, and its view might be as limited as a single landing spot on Earth.

Saturn's Medium-Size Moons   The *Cassini* mission is also teaching us more about Saturn's other moons, especially its six medium-size moons (Figure 11.27). The photographs suggest that these moons have had a complex history.

Only Mimas, the smallest of these six moons, shows little evidence of past volcanism or tectonics. It is essentially a heavily cratered iceball. One huge crater is sometimes called "Darth Crater," because of Mimas's resemblance to

Figure 11.26 This sequence zooms in on the *Huygens* landing site on Titan. Left: a global view taken by the Cassini orbiter. Center: an aerial mosaic of images taken by the probe during descent. Right: a surface view taken by the probe after landing; the "rocks," which are 10 to 20 centimeters across, are presumably made of ices. Keep in mind that you are looking at the surface of a world more than a *billion* kilometers away.

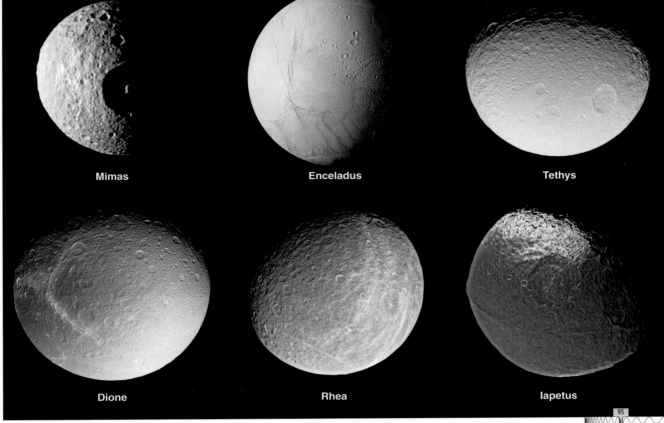

Figure 11.27 Portraits taken by the *Cassini* spacecraft of Saturn's six medium-size moons (not to scale). All but Mimas show evidence of past volcanism and/or tectonics, and the bluish grooves in the enhanced color image of Enceladus suggests current geological activity.

Mimas

Enceladus

Tethys

Dione

Rhea

Iapetus

the Death Star in the *Star Wars* movies. (The crater's official name is Herschel.) The impact that created this crater probably came close to breaking Mimas apart.

Saturn's other five medium-size moons also have heavily cratered surfaces, confirming that they lack global geological activity today. However, we find abundant evidence of volcanism and/or tectonics that must have occurred more recently. For example, the light regions visible in the photos may be places where icy lava once flowed. Enceladus even has strange grooves near its south pole, similar to some seen on Ganymede. These grooves are measurably warmer than the surrounding terrain, and they appear to be venting huge clouds of water vapor and ice crystals. Although tidal heating and radioactive heating are both present, neither is considered sufficient to explain this level of geological activity on Enceladus. Iapetus is just as bizarre, with some regions distinctly bright and others distinctly dark. The dark regions appear to be covered by a thin veneer of dark material, but we don't know what it is or how it got there. *Cassini* also imaged an astonishing ridge that spans more than a quarter of Iapetus' circumference, curiously aligned along the equator. No one knows its origin, but some kind of tectonic activity is the most likely explanation.

These moons are all considerably smaller than the Galilean moons of Jupiter; Enceladus, for example, is barely 500 kilometers across—small enough to fit inside the borders of Colorado. How, then, could they ever have had such active geology? Tidal heating sometimes plays a role, but the answer lies primarily with the icy compositions of these moons—a composition that makes them fundamentally different from terrestrial worlds despite the outward similarities of their solid surfaces. We'll investigate this issue in more depth after we complete our tour of the major jovian moons.

The Medium-Size Moons of Uranus Uranus does not have any moons that fall into our large category, but it has five medium-size moons: in order of orbital distance, their names are Miranda, Ariel, Umbriel, Titania, and Oberon (see Figure 11.14). Like other jovian moons, these moons are made largely of ice. Because of Uranus's great distance from the Sun, this ice includes a great deal of ammonia and methane as well as water.

Our only close look at these moons came during the *Voyager* flyby in 1986, but we learned enough to leave us with many unanswered questions. For example, Ariel and Umbriel are virtual twins in size, yet Ariel shows evidence of volcanism and tectonics, while the heavily cratered surface of Umbriel suggests a lack of geological activity. Titania and Oberon also are twins in size, but Titania appears to have had much more geological activity than Oberon. No one knows why these two pairs of similar-size moons should vary so greatly in geological activity.

Miranda, the smallest of Uranus's medium-size moons, is the most surprising (Figure 11.28). We might expect Miranda to be a cratered iceball like Saturn's similar-size moon Mimas. Instead, *Voyager* images of Miranda show tremendous tectonic features and relatively few craters.

cantaloupe terrain

frost deposits?

wind streaks

200 km

Triton's southern hemisphere as seen by *Voyager 2*.

50 km

This close-up shows lava-filled impact basins similar to the lunar maria, but the lava was water or slush rather than molten rock.

Figure 11.28 The surface of Miranda shows astonishing tectonic activity despite its small size. The cliff walls seen in the inset are higher than those of the Grand Canyon on Earth.

Figure 11.29 Neptune's moon Triton shows evidence of a surprising level of past geological activity.

Why should Miranda be so much more geologically active than Mimas? Our best guess is that Miranda had an episode of tidal heating billions of years ago, perhaps during a temporary orbital resonance with another moon of Uranus. Mimas apparently never had such an episode of tidal heating and thus lacks similar features.

Neptune's Triton: A Captured Moon The last stop on our tour is Neptune, so distant that only two of its moons (Triton and Nereid) were known to exist before the *Voyager 2* spacecraft visited it. Nereid is Neptune's only medium-size moon, and Triton is its only large moon. Triton holds the distinction of being the coldest world in the solar system. It is even colder than Pluto because it reflects more of the weak sunlight that reaches the outskirts of the solar system.

Triton may appear to be a typical moon, but it is not. It orbits Neptune "backward" (opposite to Neptune's rotation) and at a high inclination to Neptune's equator. These are telltale signs of a moon that was captured rather than having formed in the disk of gas around its planet. However, rather than being small and potato-shaped like most captured moons, Triton is large, spherical, and icy. In fact, it is larger than Pluto. As such, Triton presents many challenges to our understanding of satellite captures. Nonetheless, Triton almost certainly once orbited the Sun rather than Neptune. Could other, similar objects still be orbiting the Sun? As we'll discuss in the next chapter, Pluto and other large objects of the Kuiper belt are strikingly similar to Triton in character, although they orbit the Sun rather than a planet.

Triton's geology is just as surprising as its origin. Although it has numerous impact craters, its surface has clearly been reshaped by geological activity (Figure 11.29). Some regions show evidence of past volcanism, perhaps an icy equivalent to the volcanism that shaped the lunar maria. Other regions show wrinkly ridges (nicknamed

"cantaloupe terrain") that appear tectonic in nature; apparently, blobs of ice of different density have risen and fallen, contorting the crust. Triton even has a very thin atmosphere that has left some wind streaks on its surface.

What could have generated the heat needed to drive Triton's geological activity? The best guess is tidal heating. Triton probably had a very elliptical orbit and a more rapid rotation when Neptune first captured it. Tidal forces would have circularized its orbit, slowed it to synchronous rotation, and perhaps heated its interior enough to cause its geological activity. This idea would also explain why the geological activity has subsided with time.

• Why are small icy moons more geologically active than small rocky planets?

If you think about the role of planetary size in the geology of the terrestrial worlds, you might expect all of the jovian moons to be cold and geologically dead. After all, only two of the moons are even as large as Mercury, and Mercury has been geologically dead for more than 3 billion years. However, our tour of the moons has turned up numerous cases of spectacular past or present geological activity. In a few cases, we've been able to explain at least part of this activity as a result of heat generated by tidal heating. But we've found evidence of past activity even on moons that lack tidal heating. How is this possible?

We can trace the answer to a crucial difference between the jovian moons and the terrestrial worlds: composition. The jovian moons, by virtue of their formation in the cold outer solar system, contain large proportions of ices. Because ices can melt or deform at much lower temperatures than rock, icy worlds can experience geological activity at lower temperatures than rocky worlds. Indeed, most of the volcanism that has occurred in the outer solar system (except on Io) probably did not produce any hot lava at all. Instead, it produced icy lava that was essentially liquid water, perhaps mixed with methane and ammonia.

Terrestrial Planet Geology

- Internal heat, primarily from radioactive decay, can cause volcanic and tectonic activity.
- Only large planets retain enough internal heat to stay geologically active today.
- Example: Mars (photo above) probably retains some internal heat. If it had been smaller, like Mercury, it would be geologically "dead" today. If it had been larger, like Earth, it would probably have much more active and ongoing tectonics and volcanism.

Jovian Moon Geology

- Tidal heating can cause tremendous geological activity on moons with elliptical orbits around massive planets.
- Even without tidal heating, icy materials can melt and deform at lower temperatures than rock, increasing the likelihood of geological activity.
- Together, these effects explain why icy moons are much more likely to have ongoing geological activity than rocky terrestrial worlds of the same size.
- Example: Ganymede (photo above) shows evidence of recent geological activity, even though it is similar in size to the geologically dead terrestrial planet Mercury.

Figure 11.30 Jovian moons can be much more geologically active than terrestrial worlds of similar size due to their icy compositions and a heating source (tidal heating) that is not important on the terrestrial worlds.

Thus, the major lesson of our study of jovian moons is that "ice geology" is possible at much lower temperatures than "rock geology." This fact, combined with the occasional extra heating source of tidal heating, explains how the jovian moons have had such interesting geological histories despite their small sizes. In essence, the same physical properties that allowed hydrogen compounds to condense as ices in the outer solar system also allowed the worlds made from these ices to stay geologically active for long periods of time. Figure 11.30 summarizes the differences between the geology of jovian moons and that of the terrestrial worlds.

11.3 Jovian Planet Rings

We have completed our comparative study of the jovian planets and their major moons, but we have one more topic left to cover: their amazing rings. Saturn's rings have dazzled and puzzled astronomers since Galileo first saw them through his small telescope and suggested that they resembled "ears" on Saturn.

For a long time, Saturn's rings were thought to be unique in the solar system. We now know that all four jovian planets have rings. Let's begin by exploring the rings of Saturn, which we can then use as the standard for comparison when we examine the other jovian ring systems.

• What are Saturn's rings like?

You can see Saturn's rings through a backyard telescope, but learning their nature requires higher resolution (Figure 11.31). Earth-based views make the rings appear to be continuous, concentric sheets of material separated by a large gap (called the *Cassini division*). Spacecraft images reveal these "sheets" to be made of many individual rings, each separated from the next by a narrow gap. But even these appearances are somewhat deceiving. If we could wander into Saturn's rings, we'd find that they are made of countless icy particles ranging in size from dust grains to large boulders. All are far too small to be photographed even from spacecraft passing nearby.

Ring Particle Characteristics Spectroscopy reveals that Saturn's ring particles are made of relatively reflective water ice. The rings look bright where they contain enough particles to intercept sunlight and scatter it back toward us. We see gaps in places where there are few particles to reflect sunlight.

Each individual ring particle orbits Saturn independently in accord with Kepler's laws, so the rings are much like myriad tiny moons. The individual ring particles are so close together that they collide frequently. In the densest parts of the rings, each particle collides with another every few hours. However, the collisions are fairly gentle: Despite the high orbital speeds of the ring particles, nearby ring

a This Earth-based telescopic view of Saturn makes the rings look like large, concentric sheets. The dark gap within the rings is called the *Cassini division*.

b This image of Saturn's rings from the *Cassini* spacecraft reveals many individual rings separated by narrow gaps.

c Artist's conception of particles in a ring system. All the particles are moving slowly relative to one another and occasionally collide.

Figure 11.31 Interactive Photo Looming in on Saturn's rings.

particles are moving at nearly the same speed and touch only gently when they collide.

THINK ABOUT IT

Which ring particles travel faster: those closer to Saturn or those farther away? Explain why. (*Hint:* Review Kepler's third law.)

The frequent collisions explain why Saturn's rings are perhaps the thinnest known astronomical structure. They span more than 270,000 kilometers in diameter but are only a few tens of *meters* thick. Indeed, the rings are so thin that they disappear from view when we see Saturn edge-on, as we do around the equinoxes of its 29.5-year orbit of the Sun.

To understand how collisions keep the rings thin, imagine what would happen to a ring particle on an orbit slightly inclined to the central ring plane. The particle would col-

lide with other particles every time its orbit intersected the ring plane, and its orbital tilt would be reduced with every collision. Before long, these collisions would force the particle to conform to the orbital pattern of the other particles. Thus, any particle that moves away from the narrow ring plane is soon brought back within it. A similar idea explains why ring particles have almost perfectly circular orbits: Any ring particle with an elliptical orbit would quickly suffer enough collisions to force it into an orbit matching its neighbors.

Rings and Gaps Close-up photographs show an astonishing number of rings, gaps, ripples, and other features—the total number of features may be as high as 100,000. Scientists are still struggling to explain all the features, but some general ideas are now clear.

A gap moon (white dot) creates ripples as its gravity nudges particles that orbit faster than the moon (inside the gap) or slower (outside).

The 20-km-wide moon Pan clears this gap in the rings.

2000 km

a Some small moons create *gaps* within the rings.

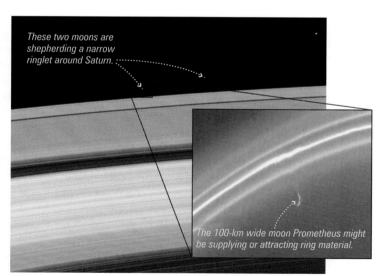

These two moons are shepherding a narrow ringlet around Saturn.

The 100-km wide moon Prometheus might be supplying or attracting ring material.

b Some moons force particles between them into a very narrow ring, as is the case with the two shepherd moons shown here. The inset shows a close-up of one of them.

Figure 11.32 Small moons within the rings have important effects on ring structure (*Cassini* photos).

Rings and gaps are caused by particles bunching up at some orbital distances and being forced out at others. This bunching happens when gravity nudges the orbits of ring particles in some particular way. One source of nudging comes from small moons located within the gaps in the rings themselves, which are sometimes called *gap moons*. The gravity of a gap moon can effectively keep the gap clear of smaller ring particles while creating ripples in the ring edges (Figure 11.32a). The ripples appear to move in opposite directions on the two sides of the gap, because ring particles on the inner side orbit Saturn slightly faster than the gap moon while those on the outer side orbit slightly slower than the gap moon. In some cases, two nearby gap moons can force particles between them into a very narrow ring (Figure 11.32b). The gap moons are often called *shepherd moons* in those cases, because they act like a shepherd forcing particles into line.

Ring particles also may be nudged by the gravity from larger, more distant moons. For example, a ring particle orbiting about 120,000 kilometers from Saturn's center will circle the planet in exactly half the time it takes the moon Mimas to orbit. Every time Mimas returns to a certain location, the ring particle will also be at its original location and therefore will experience the same gravitational nudge from Mimas. The periodic nudges reinforce one another and clear a gap in the rings—in this case, the large gap visible from Earth (the Cassini division). This type of reinforcement due to repeated gravitational tugs is another example of an *orbital resonance,* much like the orbital resonances that make Io's orbit elliptical (see Figure 11.19). Resonances can even create vast numbers of beautiful ripples and waves that start at the locations of resonances and spread out through the rings (Figure 11.33). Other orbital resonances, caused by moons both within the rings and farther out from Saturn, probably explain most of the intricate structures visible in ring photos.

• How do other jovian ring systems compare to Saturn's?

The rings of Jupiter, Uranus, and Neptune are so much fainter than Saturn's that it took almost four centuries longer to discover them. Ring particles in these three systems are far less numerous, generally smaller, and much darker. Despite these differences, a family portrait of the jovian ring systems shows many similarities (Figure 11.34). All rings lie in their planet's equatorial plane. Particle orbits are fairly circular, with small orbital tilts relative to the equator. Individual rings and gaps are probably shaped by gap moons, shepherd moons, and orbital resonances.

Uranus's rings were first discovered in 1977 during observations of a *stellar occultation*—a star passing behind Uranus as seen from Earth. During the occultation, the star "blinked" on and off nine times before it disappeared behind Uranus and nine more times as it emerged. Scientists concluded that these nine "blinks" were caused by nine thin rings encircling Uranus. Similar observations of stars passing behind Neptune yielded more confounding results:

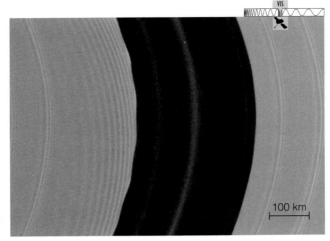

Figure 11.33 This *Cassini* image of a gap in Saturn's rings shows how the resonance with the moon Mimas creates waves that shape the rings over vast distances.

Rings appeared to be present at some times but not at others. Could Neptune's rings be incomplete or transient?

The *Voyager* spacecraft provided some answers. *Voyager* cameras first discovered thin rings around Jupiter in 1979. After next providing incredible images of Saturn's rings, *Voyager 2* photographed the rings of Uranus as it flew past in 1986. In 1989, *Voyager 2* passed by Neptune and found that it does, in fact, have partial rings—at least when seen from Earth. The space between the ring segments is filled with dust not detectable from Earth. In addition to the dust within the individual rings of Neptune, *Voyager 2* detected vast dust sheets between the widely separated rings of both Uranus and Neptune.

The differences among the ring systems present us with unsolved mysteries. The larger size, higher reflectivity, and much greater number of particles in Saturn's rings compel us to wonder whether different processes might be at work there. The reasons for the slight tilt and eccentricity of Uranus's thin rings are not yet understood, nor do we know why Neptune's rings contain dusty regions that make them appear as partial rings when viewed from Earth. Scientists hope that we may someday learn the answers with new missions to visit Uranus and Neptune, but none are yet under development. Meanwhile, advances in ground-based observing, including the new technology of adaptive optics [Section 6.3], are providing views that in some cases rival those that we obtained from the *Voyager* spacecraft.

• Why do the jovian planets have rings?

Where do the rings come from? Did they form together with the planets out of the solar nebula, or are they a more recent phenomenon? An important clue is that the rings lie close to their planet in a region where tidal forces are very strong.

Within two to three planetary radii (of any planet), the tidal forces tugging apart an object become comparable to the gravitational forces holding it together. (This region is

Jupiter

Saturn

Uranus

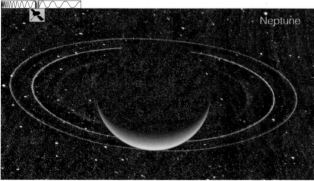

Neptune

Figure 11.34 Four ring systems (not to scale). The rings differ in appearance and in the composition and sizes of the ring particles. These photos were taken with visible light except for Uranus, which was photographed in infrared light by the Hubble Space Telescope. The Saturn photo is from *Cassini* and the Jupiter and Neptune photos are from *Voyager*.

often called the *Roche tidal zone.*) Only relatively small objects held together by nongravitational forces—such as the electromagnetic forces that hold solid rock, spacecraft, and human beings together—can avoid being ripped apart by

the strong tidal forces in this region. Because a large moon would be ripped apart by tidal forces in the region of the rings, scientists once suspected that Saturn's rings formed after a large moon came too close to the planet. However, moons don't simply "wander" away from their orbits, and a large moon could have been deflected toward Saturn only by a close gravitational encounter with an unusually large asteroid or comet that happened to pass nearby. Such encounters should be extremely rare. Back in the days when Saturn was the only planet known to have rings, it might have seemed possible that such a rare encounter was responsible for the rings. The discovery of rings around the other jovian planets all but ruled out this possibility, because it seems inconceivable that such an unlikely event could have occurred in all four cases.

If the rings did not originate with the destruction of a large moon, where did all the ring particles come from? Another idea that once seemed reasonable was that the ring particles might be leftover chunks of rock and ice that condensed in the disks of gas that orbited each jovian planet when it was young. This would explain why all four jovian planets have rings, because tidal forces near each planet would have prevented these chunks from accreting into a full-fledged moon. However, we now know that the ring particles cannot be leftovers from the births of the planets, because particles of the size we find in the rings today cannot survive for billions of years. Ring particles are continually being ground down in size, primarily by the impacts of the countless sand-size particles that orbit the Sun—the same types of particles that become meteors in Earth's atmosphere and cause micrometeorite impacts on the Moon [Section 9.3]. Millions of years of such tiny impacts would long ago have ground the existing ring particles to dust. The dust particles could not have survived either, because over time they are slightly slowed in their orbits by the pressure of sunlight, and this slowing eventually causes them to spiral into their planet. Ring particles may also be lost in other ways. For example, the thermosphere of Uranus extends up into the ring region, exerting atmospheric drag that also causes ring particles to spiral slowly into the planet. Between the effects of micrometeorite impacts and other processes, *none* of the abundant small particles that now occupy the jovian rings can have been there since the solar system formed more than 4 billion years ago.

We are therefore left with only one reasonable possibility for the origin of the rings: New particles are continually supplied to the rings to replace those that are destroyed. These new particles must come from a source that lies in each planet's equatorial plane. The most likely source is numerous small moons—moons the size of gap moons (see Figure 11.32)—that formed in the disks of material orbiting the young jovian planets. Like the ring particles themselves, tiny impacts are gradually grinding away these small moons, but they are large enough to still exist despite 4.5 billion years of such sandblasting.

The small moons contribute ring particles in two ways. First, each tiny impact releases particles from a small moon's surface, and these released particles become new, dust-size

Figure 11.35 This illustration summarizes the origin of rings around the jovian planets.

Labels in figure:
- jovian planet
- Tidal forces near the planet prevent small moonlets from accreting into larger moons.
- Moonlets are occasionally disrupted by impacts.
- Ongoing small impacts blast off dust and debris to form the rings.

ring particles. Ongoing impacts ensure that some ring particles are present at all times. Second, occasional larger impacts can shatter a small moon completely, creating a supply of boulder-size ring particles. The frequent tiny impacts then slowly grind these boulders into smaller ring particles. In summary, the dust- to boulder-size particles in rings all ultimately come from the gradual dismantling of small moons that formed during the birth of the solar system (Figure 11.35).

The collisions that shatter small moons and generate a major source of ring particles must occur only occasionally and at essentially random times, which means that the number of particles in any particular ring system must change dramatically over millions and billions of years. Rings may be broad and bright when they are full of particles and almost invisible when particles are few. Thus, the brilliant spectacle of Saturn's rings may be a special treat of our epoch, one that may not have been seen a billion years ago and that may not last long on the time scale of our solar system.

THE BIG PICTURE

Putting Chapter 11 into Context

In this chapter, we saw that the jovian planets really are a different kind of planet and, indeed, a different kind of planetary system. The jovian planets dwarf the terrestrial planets. Even some of their moons are as large as terrestrial worlds. As you continue your study of the solar system, keep in mind the following "big picture" ideas:

- The jovian planets may lack solid surfaces on which geology can occur, but they are interesting and dynamic worlds with rapid winds, huge storms, strong magnetic fields, and interiors in which common materials behave in unfamiliar ways.

- Despite their relatively small sizes and frigid temperatures, many jovian moons are geologically active by virtue of their icy compositions and tidal heating. Ironically, the cold temperatures in the solar nebula led to their icy compositions and hence their geological activity.

- Ring systems probably owe their existence to small moons formed in the disks of gas that produced the jovian planets billions of years ago. The rings we see today are composed of particles liberated from those moons surprisingly recently.

- Understanding the jovian planet systems forced us to modify many of our earlier ideas about the solar system, in particular by adding the concepts of ice geology, tidal heating, and orbital resonances. Each new set of circumstances that we discover offers further opportunities to learn how our universe works.

SUMMARY OF KEY CONCEPTS

11.1 A Different Kind of Planet

- **Are jovian planets all alike?** Jupiter and Saturn are made almost entirely of hydrogen and helium, while Uranus and Neptune are made mostly of hydrogen compounds mixed with metal and rock. These differences arose because all four planets started from ice-rich planetesimals of about the same size, but captured different amounts of hydrogen and helium gas from the solar nebula.

- **What are jovian planets like on the inside?**
The jovian planets have layered interiors with very high

internal temperatures and pressures. All have a core about 10 times as massive as Earth, consisting of hydrogen compounds, metals, and rock. They differ mainly in their surrounding layers of hydrogen and helium, which can take on unusual forms under the extreme internal conditions of the planets.

- **What is the weather like on jovian planets?**
The jovian planets all have multiple cloud layers that help

determine the colors of the planets, fast winds, and large storms. Some storms, such as the Great Red Spot, can apparently rage for centuries or longer.

- **Do jovian planets have magnetospheres like Earth's?** Jupiter has a magnetic field some 20,000 times stronger than Earth's, which leads to an enormous magnetosphere. Many of the particles in Jupiter's magnetosphere come from volcanic eruptions on Io. The other jovian planets also have magnetic fields and magnetospheres, but they are weaker and smaller than Jupiter's.

11.2 A Wealth of Worlds: Satellites of Ice and Rock

- **What kinds of moons orbit the jovian planets?** We can categorize the more than 100 known moons as small, medium-size, or large. Most of the medium-size and large moons probably formed with their planet in the disks of gas that surrounded the jovian planets when they were young. Smaller moons are often captured asteroids or comets.

- **Why are Jupiter's Galilean moons so geologically active?** Io is the most volcanically

active object in the solar system, thanks to an interior kept hot by tidal heating—which occurs because Io's close orbit is made elliptical by orbital resonances with other moons. Europa (and possibly Ganymede) may have a deep, liquid water ocean under its icy crust, also thanks to tidal heating. Callisto is the least geologically active, since it has no orbital resonance or tidal heating.

- **What is special about Titan and other major moons of the outer solar system?** Many medium-size and large moons show a surprisingly high

level of past or present volcanism or tectonics. Titan even has a thick atmosphere and ongoing erosion caused by liquid methane. Triton, which apparently was captured by Neptune, also has had relatively recent geological activity.

- **Why are small icy moons more geologically active than small rocky planets?** Ices deform and melt at much lower temperatures than rock, allowing icy volcanism and tectonics at surprisingly low temperatures. In addition, some jovian moons have a heat source—tidal heating—that is not important for the terrestrial worlds.

11.3 Jovian Planet Rings

- **What are Saturn's rings like?** Saturn's rings

are made up of countless individual particles, each orbiting Saturn independently like a tiny moon. The rings lie in Saturn's equatorial plane, and they are extremely thin. Moons within and beyond the rings create many ringlets and gaps in part through orbital resonances.

- **How do other jovian ring systems compare to Saturn's?** The other jovian planets have ring systems that are much fainter in photographs. Their ring particles are generally smaller, darker, and less numerous than Saturn's ring particles.

- **Why do the jovian planets have rings?** Ring

particles probably come from the dismantling of small moons formed in the disks of gas that surrounded the jovian planets billions of years ago. Small ring particles come from countless tiny impacts on the surfaces of these moons, while larger ones come from impacts that shatter the moons.

For instructor-assigned homework go to MasteringASTRONOMY www.masteringastronomy.com

Review Questions
Short-Answer Questions Based on the Reading

1. Briefly describe how the differences in composition among the jovian planets can be traced to their formation in the solar nebula.
2. Why is Jupiter so much denser than Saturn? Could a planet be smaller in size than Jupiter but greater in mass? Explain.
3. Briefly describe the interior structure of Jupiter and why it is layered in this way. How do the interiors of the other jovian planets compare to Jupiter?
4. Why does Jupiter have so much internal heat? What generates internal heat on other jovian planets?
5. Briefly describe Jupiter's atmospheric structure and cloud layers. How do the structures and clouds differ on the other jovian planets?

6. How do the cloud layers help explain Jupiter's colors? Why are Saturn's colors more subdued? Why are Uranus and Neptune blue?

7. Briefly describe Jupiter's weather patterns and contrast them with those on the other jovian planets. What is the *Great Red Spot*?

8. Why does Jupiter have such a strong magnetic field? Describe a few features of Jupiter's magnetosphere, and compare it to the magnetospheres of the other jovian planets.

9. Briefly describe how we categorize jovian moons by size. What is the origin of most of the medium-size and large moons? What is the origin of many of the small moons?

10. What are the key features of Jupiter's four Galilean moons? Explain the role of tidal heating and orbital resonances in shaping these features.

11. Describe the atmosphere of Titan. What did the *Cassini/Huygens* mission learn about Titan's surface?

12. What evidence suggests past geological activity on the medium-size moons of Saturn and Uranus? Describe a few of the mysteries that these moons pose.

13. Briefly describe Triton and explain why we think it is a captured moon.

14. Briefly explain why icy moons can have active geology at much smaller sizes than rocky worlds.

15. What are rings? What are they made of, and how do they differ among the four jovian planets? Briefly describe the effects of gap moons and orbital resonances on ring systems.

16. Explain why we think that ring particles must be replenished over time. Should the jovian planet rings always look the same?

Test Your Understanding

Surprising Discoveries?

Suppose someone claimed to make the discoveries described below. (These are not real discoveries.) Decide whether each discovery should be considered reasonable or surprising. More than one right answer may be possible, so explain your answer clearly.

17. Saturn's core is pockmarked with impact craters and dotted with volcanoes erupting basaltic lava.

18. Neptune's deep blue color is not due to methane, as previously thought, but instead is due to its surface being covered with an ocean of liquid water.

19. A jovian planet in another star system has a moon as big as Mars.

20. An extrasolar planet is made primarily of hydrogen and helium. It has approximately the same mass as Jupiter but is the same size as Neptune.

21. A new small moon orbits Jupiter outside the orbits of other known moons. It is smaller than Jupiter's other moons but has several large, active volcanoes.

22. A new moon orbits Neptune in the planet's equatorial plane and in the same direction that Neptune rotates, but it is made almost entirely of metals such as iron and nickel.

23. An icy, medium-size moon orbits a jovian planet in a star system that is only a few hundred million years old. The moon shows evidence of active tectonics.

24. A jovian planet is discovered in a star system that is much older than our solar system. The planet has no moons at all, but has a system of rings as spectacular as the rings of Saturn.

25. Radar measurements of Titan indicate that most of the moon is heavily cratered.

26. The *Cassini* mission finds 20 more moons of Saturn.

Quick Quiz

Choose the best answer to each of the following. Explain your reasoning with one or more complete sentences.

27. Which lists the jovian planets in order of increasing distance from the Sun? (a) Jupiter, Saturn, Uranus, Pluto (b) Saturn, Jupiter, Uranus, Neptune (c) Jupiter, Saturn, Uranus, Neptune

28. Why does Neptune appear blue and Jupiter red? (a) Neptune is hotter, which gives bluer thermal emission. (b) Methane in Neptune's atmosphere absorbs red light. (c) Neptune's air molecules scatter blue light, much as Earth's atmosphere does.

29. Why is Jupiter denser than Saturn? (a) It has a larger proportion of rock and metal. (b) It is more compressed by the Sun's gravity. (c) Its higher mass and gravity compress its interior.

30. Some jovian planets give off more energy than they receive because (a) fusion is taking place in their cores. (b) tidal heating is occurring. (c) they are still slowly contracting after formation.

31. The main ingredients of most satellites of the jovian planets are (a) rock and metal. (b) hydrogen compound ices. (c) hydrogen and helium.

32. Why is Io more volcanically active than our moon? (a) Io is much larger. (b) Io has a higher concentration of radioactive elements. (c) Io has a different internal heat source.

33. What is unusual about Triton? (a) It orbits its planet backward. (b) It does not keep the same face toward its planet. (c) It is the only moon with its own rings.

34. Which moon shows evidence of rainfall and erosion by some liquid substance? (a) Europa (b) Titan (c) Ganymede

35. Saturn's many moons affect the rings through (a) tidal forces. (b) orbital resonances. (c) magnetic field interactions.

36. Saturn's rings (a) have looked basically the same since they formed along with Saturn. (b) were created long ago when tidal forces tore apart a large moon. (c) are continually supplied by impacts into small moons.

Investigate Further

In-Depth Questions to Increase Your Understanding

Short-Answer/Essay Questions

37. *The Importance of Rotation.* Suppose the material that formed Jupiter came together without any rotation so that no "jovian nebula" formed and the planet today wasn't spinning. How else would the jovian system be different? Think of as many effects as you can, and explain each in a sentence.

38. *The Great Red Spot.* Based on the infrared and visible images in Figure 11.9, is Jupiter's Great Red Spot warmer or cooler than nearby clouds? Is it higher or lower in altitude than the nearby clouds? Explain.

39. *Comparing Jovian Planets.* You can do comparative planetology armed only with telescopes and an understanding of gravity.
   a. The small moon Amalthea orbits Jupiter at about the same distance in kilometers at which Mimas orbits Saturn, yet Mimas takes almost twice as long to orbit. From this observation, what can you conclude about how Jupiter and Saturn differ? Explain.
   b. Jupiter and Saturn are not very different in radius. When you combine this information with your answer to part (a), what can you conclude? Explain.

40. *Minor Ingredients Matter.* Suppose the jovian planet atmospheres were composed only of hydrogen and helium, with no hydrogen compounds at all. How would the atmospheres be different in terms of clouds, color, and weather? Explain.

41. *Hot Jupiters.* Many of the newly discovered planets orbiting other stars are more massive than Jupiter but orbit much closer to their stars. Assuming that they would be Jupiter-like if they orbited at a greater distance from their stars, how would you expect these new planets to differ from the jovian planets of our solar system? How would you expect their moons to differ? Explain.

42. *The New View of Titan.* What planet or moon in the solar system does Titan most resemble, in your opinion? Summarize the similarities and differences in a few sentences.

43. *Galilean Moon Formation.* Look up the densities of Jupiter's four Galilean moons in Appendix E, and notice that they follow a trend with distance from Jupiter. Based on what you've learned about condensation in the solar nebula, can you suggest a reason for this trend among the Galilean moons? Next, compare the densities of the moons with the planetary densities in Table E.1. Based on the comparison, do you think it was as hot toward the center of the nebula surrounding Jupiter as it was at the center of the solar nebula? Explain.

44. *Super Jupiter.* Suppose that the solar wind had not cleared the solar nebula until much later in our solar system's history, so that Jupiter accumulated 40 times as much mass as it actually has today.
   a. Make a sketch showing how the interior layers of this "Super Jupiter" would compare to the interior layers of the real Jupiter shown in Figure 11.4.
   b. How would you expect the cloud layers and colors to differ on Super Jupiter? Explain.
   c. How would you expect the wind speeds to be different on Super Jupiter? Explain.
   d. How would you expect the magnetic field to be different on Super Jupiter? Explain.
   e. Do you think Super Jupiter would have more or less excess heat than the real Jupiter? Explain.

45. *Observing Project: Jupiter's Moons.* Using binoculars or a small telescope, view the moons of Jupiter. Make a sketch of what you see, or take a photograph. Repeat your observations several times (nightly, if possible) over a period of a couple of weeks. Can you determine which moon is which? Can you measure the moons' orbital periods? Can you determine their approximate distances from Jupiter? Explain.

46. *Observing Project: Saturn's Rings.* Using binoculars or a small telescope, view the rings of Saturn. Make a sketch of what you see, or take a photograph. What season is it in Saturn's northern hemisphere? How far do the rings extend above Saturn's atmosphere? Can you identify any gaps in the rings? Describe any other features you notice.

Quantitative Problems

Be sure to show all calculations clearly and state your final answers in complete sentences.

47. *Disappearing Moon.* Io loses about a ton (1,000 kilograms) of sulfur dioxide per second to Jupiter's magnetosphere.
   a. At this rate, what fraction of its mass would Io lose in 4.5 billion years?

   b. Suppose sulfur dioxide currently makes up 1% of Io's mass. When will Io run out of this gas at the current loss rate?

48. *Ring Particle Collisions.* Each ring particle in the densest part of Saturn's rings collides with another about every 5 hours. If a ring particle survived for the age of the solar system, how many collisions would it undergo?

49. *Prometheus and Pandora.* These two moons orbit Saturn at 139,350 and 141,700 kilometers, respectively.
   a. Using Newton's version of Kepler's third law, find their two orbital periods. Find the percent difference in their distances and in their orbital periods.
   b. Consider the two in a race around Saturn: In one Prometheus orbit, how far behind is Pandora (in units of time)? In how many Prometheus orbits will Pandora have fallen behind by one of its own orbital periods? Convert this number of periods back into units of time. This is how often the satellites pass by each other.

50. *Orbital Resonances.* Using the data in Appendix E, identify the orbital resonance relationship between Titan and Hyperion. (*Hint:* If the orbital period of one were 1.5 times the other, we would say that they are in a 3:2 resonance.) Which medium-size moon is in a 2:1 resonance with Enceladus?

51. *Titanic Titan.* What is the ratio of Titan's mass to all the other satellites of Saturn whose mass is listed in Appendix E? Calculate the strength of gravity on Titan compared to that on Mimas. Comment on how this affects the possibility of atmospheres on each.

52. *Titan's Evolving Atmosphere.* Titan's exosphere lies nearly 1,400 kilometers above its surface. What is the escape velocity from this altitude? What is the thermal speed of a hydrogen atom at the exospheric temperature of about 200 K? Use these answers (and the method of Mathematical Insight 10.2) to comment on whether thermal escape of hydrogen is likely to be important for Titan.

53. *Saturn's Thin Rings.* Saturn's ring system is over 270,000 km wide, and only a few tens of meters thick; let's assume 50 meters thick for this problem. Assuming the rings could be shrunk down so that their diameter is the width of a dollar bill (6.6 cm), how thick would the rings be? Compare your answer to the actual thickness of a dollar bill (0.01 cm).

54. *Characterizing a newly discovered planet.* In April 2005, scientists claimed to have captured light from an extrasolar planet orbiting a brown dwarf in the constellation Hydra some 230 light-years from Earth. The scientists believe the planet is about five times the mass of Jupiter.
   a. Assume the planet has the same density as Jupiter. What is the diameter of the planet?
   b. Using Mathematical Insight 2.1, calculate the angular size of the planet as viewed from Earth, in arcseconds.

Discussion Questions

55. *Jovian Planet Mission.* We can study terrestrial planets up close by landing on them, but jovian planets have no surfaces to land on. Suppose that you are in charge of planning a long-term mission to "float" in the atmosphere of a jovian planet. Describe the technology you would use and how you would ensure survival for any people assigned to this mission.

56. *Pick a Moon.* Suppose you could choose any one moon to visit in the solar system. Which one would you pick, and why? What dangers would you face in your visit to this moon? What kinds of scientific instruments would you want to bring along for studies?

Interactive Tutorials

Tutorial Review of Key Concepts

Use the following interactive **Tutorial** at **www.mastering astronomy.com** to review key concepts from this chapter.

Formation of the Solar System Tutorial
Lesson 1 Comparative Planetology

Supplementary Tutorial Exercises

Use the interactive **Tutorial Lessons** to explore the following questions.

Formation of the Solar System Tutorial, Lesson 1

1. Contrast the general features of the jovian planets with those of the terrestrial planets.
2. Compare the properties of each of the four jovian planets. What properties do they all share? In what ways are they different?

Exploring the Sky and Solar System

Of the many activities available on the ***Voyager: SkyGazer*** CD-ROM accompanying your book, use the following files to observe key phenomena covered in this chapter.

Go to the **File: Basics** folder for the following demonstrations.
1. Tracking Jupiter and Io
2. Saturn

Go to the **File: Demo** folder for the following demonstrations.
1. Backside of Jupiter
2. Locked on Dione
3. Three Moons on Jupiter

Web Projects

1. *The Galileo Mission to Jupiter.* Learn more about the *Galileo* mission and how it ended, and about ongoing scientific study of the *Galileo* data. Write a short summary of your findings.
2. *News from Cassini.* Find the latest news about the *Cassini* mission to Saturn. What is the current mission status? Write a short report about the mission's status and recent findings.
3. *Oceans of Europa.* The possibility of subsurface oceans on Europa holds great scientific interest. Investigate plans for further study of Europa, either from Earth or with future spacecraft. Write a short summary of the plans and how they might help us learn whether an ocean really exists and, if so, what it might contain.

12

Remnants of Rock and Ice

Asteroids, Comets, and the Kuiper Belt

> As we look out into the Universe and identify the many accidents of physics and astronomy that have worked to our benefit, it almost seems as if the Universe must in some sense have known that we were coming.
>
> **Freeman Dyson**

Asteroids and comets might at first seem insignificant in comparison to the planets and moons we've discussed so far, but there is strength in numbers. The trillions of small bodies orbiting our Sun are far more important than their small size might suggest.

The appearance of comets has more than once altered the course of human history when our ancestors acted upon superstitions related to comet sightings. More profoundly, asteroids or comets falling to Earth have scarred our planet with impact craters and have altered the course of biological evolution. Asteroids and comets are also important scientifically: As remnants from the birth of our solar system, they have taught us a great deal about how our solar system formed.

In this chapter, we will explore asteroids, comets, and the rocks that fall to Earth as meteorites. Along the way, we'll see that the "ninth planet," Pluto, is actually just one of many similar objects orbiting the Sun beyond Neptune—and at least one of those objects is larger than Pluto. Finally, we will explore the dramatic effects of the occasional collisions between small bodies and large planets.

12.1 Asteroids and Meteorites

The objects we've studied so far—terrestrial planets, jovian planets, and large moons—have changed dramatically since their formation. Most planets have been transformed in almost every way: Their interiors have differentiated, their surfaces have been reshaped by geological processes, and their atmospheres have changed as they have gained and lost gas. In contrast, many small bodies remain much as they were when they first formed, some 4.5 billion years ago. Comets, asteroids, and meteorites carry the history of our solar system encoded in their compositions, locations, and numbers.

Using these small bodies to understand planetary formation is a bit like picking through the trash in a carpenter's shop to see how furniture is made. By studying the scraps, sawdust, paint chips, and glue, we can develop some understanding of how the carpenter builds furniture. Asteroids, comets, and meteorites are the "scraps" left over from the formation of our solar system, so their study can give us an understanding of how the planets and larger moons came to exist.

We'll begin our study of small bodies by focusing on asteroids and meteorites in this section. The word *asteroid* means "starlike," but asteroids are starlike only in their appearance through a telescope; as we discussed in Chapter 8,

asteroids are planetesimals (or fragments of planetesimals) left over from the birth of our solar system. Meteorites are pieces of rock that have fallen to the ground from space. Thus, asteroids and meteorites are closely related: Most meteorites are pieces of asteroids that orbited the Sun for billions of years before falling to Earth.

• What are asteroids like?

Asteroids are virtually undetectable by the naked eye and went unnoticed for almost two centuries after the invention of the telescope. The first asteroids were discovered only about 200 years ago, and it took 50 years to discover the first 10. Today, advanced telescopes can discover far more than that in a single night, and more than 150,000 asteroids have been cataloged. Asteroids can be recognized in telescopic images because over a short time period they move noticeably in relation to the stars (Figure 12.1).

Newly discovered asteroids first get a provisional name based on the discovery month and year; for example, the first asteroid discovered in January 2007 would be called Asteroid 2007 AA. An asteroid's orbit can be calculated from the law of gravity [Section 4.4], even after observation of only a fraction of a complete orbit, but only about 10% of the asteroids identified in images have been tracked long enough for their orbits to be calculated. In those cases, the discoverer may choose a name for the asteroid, subject to approval by the International Astronomical Union. The earliest discovered asteroids bear the names of mythological figures. More recent discoveries often carry names of scientists, cartoon heroes, pets, or rock stars.

Asteroid Sizes, Numbers, and Shapes   Asteroids come in a wide range of sizes. The largest, Ceres, is just under 1,000 kilometers in diameter, which is less than $\frac{1}{3}$ the diameter of our Moon. About a dozen others are large enough that we would call them medium-size moons if they orbited a planet. Smaller asteroids are far more numerous. There are probably more than a million asteroids with diameters

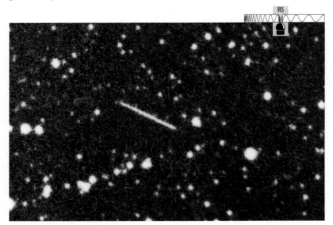

Figure 12.1   Interactive Photo   Because asteroids orbit the Sun, they move through our sky relative to the stars. In this long-exposure photograph, stars show up as white dots, while the motion of an asteroid relative to the stars makes it show up as a short streak.

greater than 1 kilometer, and many more even smaller in size. Despite their large numbers, asteroids don't add up to much in total mass. If we could put all the asteroids together and allow gravity to compress them into a sphere, they'd make an object less than 2,000 kilometers in diameter—much smaller than our Moon.

Asteroid shapes depend on the strength of their gravity. If an asteroid is relatively large, over time gravity can mold its solid rock into a spherical or near-spherical shape. Only Ceres is large enough for gravity to have compressed it into a roughly spherical shape, while the next two largest asteroids are somewhat oblong (Pallas, 540 kilometers long, and Vesta, 510 kilometers long). For smaller asteroids, gravity is too weak to reshape the rocky material. Much like potatoes, these asteroids can have almost any shape. Indeed, some objects that appear to be single asteroids are probably two or more distinct objects held in contact by a weak gravitational attraction, while other small asteroids are little more than weakly bound piles of rubble.

The handful of asteroids that have been photographed up close by spacecraft confirm these ideas (Figure 12.2). Notice that these asteroids look more like potatoes than spheres. The numerous craters tell us that asteroids, like planets and moons, have been battered by impacts. The impact that formed Mathilde's huge, central crater (the shadowed region in Figure 12.2c) would have disintegrated a solid object so Mathilde must be a loosely bound "rubble pile" that was able to absorb the shock of the impact without completely falling apart.

Given that we've photographed so few asteroids close-up and that most asteroids are far too small for even powerful telescopes to resolve, you may wonder how we can make general statements about their sizes and shapes. It's not easy, but careful analysis of asteroid light can often teach us much about them. Let's first consider measurements of size.

Asteroids shine with reflected sunlight, so an asteroid's brightness in our sky depends on its size, distance, and reflectivity. For example, if two asteroids at the same distance have the same reflectivity, the one that appears brighter must be larger in size. We can determine an asteroid's distance from its position in its orbit. Determining reflectivity

is more difficult, but we can do it by comparing the asteroid's brightness in visible light to its brightness in infrared light. The visible light is reflected sunlight, while the infrared light is thermal radiation emitted by the asteroid itself [Section 5.4]. The infrared brightness therefore depends on the asteroid's temperature, which in turn depends on how much sunlight it absorbs. Thus, the comparison between visible and infrared brightness essentially tells us the proportions of incoming sunlight that an asteroid reflects and absorbs. Once we know these proportions, the asteroid's brightness and distance tell us its size. This technique has been used to make good size estimates for more than a thousand asteroids.

THINK ABOUT IT

Suppose you discover two asteroids that are equally bright, but infrared observations tell you that Asteroid 1 is more reflective than Asteroid 2. Which one is farther away? Explain.

Determining asteroid shapes without close-up photographs is difficult but not impossible. In some cases, we can determine shapes by monitoring brightness variations as an asteroid rotates: A non-spherical asteroid with a uniformly bright surface will reflect more light when it presents its larger side toward the Sun and our telescopes. In other cases, astronomers have determined shapes by bouncing radar signals off asteroids that have passed close to Earth.

Asteroid Compositions Analysis of light can also tell us about asteroid compositions. Recall that spectra of distant objects contain spectral lines that are essentially "fingerprints" left by their chemical constituents [Section 5.4]. Thousands of asteroids have been analyzed through spectroscopy. The results are consistent with what we expect from our theory of solar system formation: Asteroids are made mostly of metal and rock, because they condensed within the frost line in the solar nebula. Those near the outskirts of the asteroid belt contain larger proportions of dark, carbon-rich material, because this material was able to condense at the

a Gaspra, photographed by the *Galileo* spacecraft.

b Ida, photographed by the *Galileo* spacecraft. The small dot to the right is Dactyl, a tiny moon orbiting Ida.

c Mathilde, photographed by the Near-Earth Asteroid Rendezvous (NEAR) spacecraft on its way to Eros.

d Eros, photographed by the NEAR spacecraft, which orbited Eros for a year before ending its mission with a soft landing on the asteroid's surface. The inset photo was taken by NEAR just before it landed.

Figure 12.2 Close-up views of asteroids studied by spacecraft.

relatively cool temperatures found in this region of the solar nebula but not in the regions closer to the Sun; some even contain small amounts of water, telling us that they formed close to the frost line. A few asteroids appear to be made mostly of metals, such as iron, suggesting that they may be fragments of the iron cores of shattered worlds.

Asteroid Masses and Densities Determining an asteroid's mass and density is much more difficult than measuring its size, shape, or composition. The only direct way to measure a distant object's mass is to observe its gravitational effect on another object. We've been able to observe the gravitational effects of asteroids in only a few cases: those in which a spacecraft has passed by close enough to have its trajectory altered by the asteroid's gravity and those in which asteroids have even smaller asteroids as tiny, orbiting "moons." For asteroids with moons, such as the asteroid Ida with its tiny moon Dactyl (see Figure 12.2b) and the asteroid Eugenia in Figure 12.3, we can determine the mass of the central asteroid by applying Newton's version of Kepler's third law to the orbital characteristics of the moon [Section 4.5]. Astronomers have found a few dozen asteroid masses by this method.

Once we measure an asteroid's mass and size, we can calculate its density. The density can offer valuable insights into the asteroid's origin and makeup. For example, Mathilde—which we argued earlier must be a "rubble pile" in order to have survived the impact that made its large central crater—has a density of only 1.5 g/cm³. This density is so low that Mathilde cannot be a solid chunk of rock, which lends further support to the idea that it is a pile of rubble

held together by weak gravity. In contrast, the asteroid Eros (see Figure 12.2d) has a density of 2.4 g/cm³, which is close to the value expected for solid rock. Eros also has a subtle pattern of surface ridges and troughs that hint at a set of parallel faults and fractures in its interior.

Asteroid Orbits The vast majority of the asteroids that have been cataloged are located in the asteroid belt between the orbits of Mars and Jupiter (Figure 12.4). All asteroids orbit the Sun in the same direction as the planets, but their orbits tend to be more elliptical and more highly inclined to the ecliptic plane (often up to 20°–30°) than those of planets.

Science fiction movies often show the asteroid belt as a crowded and hazardous place, but the asteroid belt is so large that its hundreds of thousands of asteroids are quite far apart on average. The average distance between asteroids in the asteroid belt is millions of kilometers—the equivalent of grains of sand separated by kilometers.

Not all asteroids are located in the asteroid belt. Two sets of asteroids, called *Trojan asteroids* (as in the Greek story of the Trojan War), share Jupiter's 12-year orbit around the Sun. As shown in Figure 12.4, one clump of Trojan asteroids always stays 60° ahead of Jupiter in its orbit and the other clump always stays 60° behind. The number of Trojan asteroids could be as large as the number in the asteroid belt, but the greater distance to the Trojan asteroids (and their darker surfaces) makes them more difficult to study or even count from Earth.

A relatively small number of asteroids have orbits that pass through the inner solar system—including the *near-Earth asteroids* that pass near Earth's orbit. These asteroids

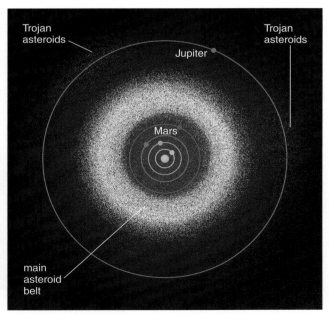

Figure 12.3 This set of images, made by a telescope using adaptive optics [Section 6.3], shows a small moon observed in five positions in its orbit around the asteroid Eugenia. The moon completes one orbit around the asteroid every 4.7 days. The asteroid itself was blocked out during the observations and is shown as the gray oval in the image.

Figure 12.4 Calculated positions of 152,942 asteroids for midnight, January 1, 2004. The asteroids themselves are much smaller than the dots on this scale. The asteroids that share Jupiter's orbit, found 60° ahead of and behind Jupiter, are called *Trojan asteroids*.

Dodge Those Asteroids!

Science fiction movies often show brave spacecraft pilots navigating through crowded fields of asteroids, dodging this way and that as they heroically pass through with only a few bumps and bruises. It's great drama, but not very realistic. The asteroid belt looks crowded when we draw it on paper as in Figure 12.4, but in reality it is an enormous region of space. Despite their large numbers, asteroids are thousands to millions of kilometers apart on average—so far apart that it would take incredibly bad luck to crash into one by accident. Indeed, spacecraft must be carefully guided to fly close enough to an asteroid to take a decent photograph. Future space travelers will have plenty of dangers to worry about, but dodging asteroids is not likely to be one of them.

are probably "impacts waiting to happen." Many asteroids with similar orbits must have hit our planet in the past. In Section 12.4, we will discuss both past impacts and the future dangers these asteroids may pose.

• Why is there an asteroid belt?

The asteroid belt between Mars and Jupiter gets its name because it is where we find the majority of asteroids. But why are asteroids concentrated in this region, and why didn't a full-fledged planet form here? The answers lie with *orbital resonances* much like those that explain the elliptical orbits of Jupiter's moons Io, Europa, and Ganymede [Section 11.2] and the gaps and other structures in Saturn's rings [Section 11.3].

Remember that an orbital resonance occurs whenever two objects periodically line up with each other. Because gravity tugs at the objects in the same direction at each alignment, the effects can build up over time, just as many small pushes can make a child swing quite high on a playground swing. Objects will periodically line up—and hence have an orbital resonance—whenever one object's orbital period is a simple ratio of another object's period, such as $\frac{1}{2}$, $\frac{1}{4}$, or $\frac{2}{5}$.

In the asteroid belt, orbital resonances occur between asteroids and Jupiter, the most massive planet by far. These resonances tend to clear gaps in the asteroid belt, much as resonances clear gaps in Saturn's rings. For example, any asteroid in an orbit that takes 6 years to circle the Sun— half of Jupiter's 12-year orbital period—would receive the same gravitational nudge from Jupiter every 12 years and thus would soon be pushed out of this orbit. The same is true for asteroids with orbital periods of 4 years ($\frac{1}{3}$ of Jupiter's period) and 3 years ($\frac{1}{4}$ of Jupiter's period). Music offers an elegant analogy: When a vocalist sings into an open piano, the strings of notes "in resonance" with the voice—not just the note being sung but also notes with a half or a quarter of the note's frequency—are "nudged" and begin to vibrate.

We can see the results of these orbital resonances on a graph of the number of asteroids with various orbital periods (Figure 12.5). For example, notice the lack of asteroids with periods exactly $\frac{1}{2}$, $\frac{1}{3}$, or $\frac{1}{4}$ of Jupiter's—periods for which the gravitational tugs from Jupiter have cleared gaps in the asteroid belt. (The gaps are often called *Kirkwood gaps,* after their discoverer.)

The orbital resonances with Jupiter likely also explain why no planet ever formed between Mars and Jupiter. When the solar system was forming, this region of the solar nebula probably contained enough rocky material to form another planet as large as Earth or Mars. However, resonances with the young planet Jupiter disrupted the orbits of this region's planetesimals, preventing them from accreting into a full-fledged planet. Over the ensuing 4.5 billion years, ongoing orbital disruptions gradually kicked pieces of this "unformed planet" out of the asteroid belt altogether. Once booted from the asteroid belt, these objects either crashed into a planet or moon or were flung out of the solar system. The asteroid belt thereby lost most of its original mass, which explains why the total mass of all its asteroids is now less than that of any terrestrial planet.

THINK ABOUT IT

Why are the gaps due to orbital resonances so easy to see in Figure 12.5 but so hard to see in Figure 12.4? (*Hint:* The top axis in Figure 12.5 is the *average* orbital distance.)

The asteroid belt is still undergoing slow change today. Jupiter's gravity continues to nudge asteroid orbits, sending asteroids on collision courses with each other and occasionally the planets. A major collision occurs somewhere in the asteroid belt every 100,000 years or so. Thus, over long

Figure 12.5 This graph shows the number of asteroids with various orbital periods, which correspond to different average distances from the Sun (labeled along the top). Notice the gaps created by orbital resonances with Jupiter.

periods of time, larger asteroids continue to be broken into smaller ones, with each collision also creating numerous dust-size particles. The asteroid belt has been grinding itself down for more than 4 billion years and will continue to do so for as long as the solar system exists.

Most orbital resonances result in gaps, but some happen to gather asteroids. Such is the case for the Trojan asteroids. Their orbits remain stable in the zones 60° ahead of and behind Jupiter, because any asteroid that wanders away from one of theses zones is nudged back into the zone by Jupiter's gravity. That explains why we see a grouping of asteroids, rather than a gap, at the $\frac{1}{1}$ resonance (where orbital periods are the same as Jupiter's) in Figure 12.5.

• Where do meteorites come from?

Because asteroids are leftovers from the birth of our solar system, studying them should teach us a lot about how Earth and the other planets formed. We can study asteroids with telescopes and spacecraft, but it would be far better if we could study an asteroid sample in a laboratory. No spacecraft has yet returned an asteroid sample, although if all goes well, the Japanese mission Hayabusa will return an asteroid sample to Earth in 2007. However, we already have pieces of asteroids nonetheless—the rocks called **meteorites** that fall from the sky.

The Difference Between Meteors and Meteorites Before we discuss what we learn from meteorites, let's be clear about what they are. In everyday language, people often use the terms *meteors* and *meteorites* interchangeably. Technically, however, they are not the same. A **meteor**, literally "a thing in the air," is only a flash of light caused by a particle entering our atmosphere at high speed, not the particle itself. Meteors are also sometimes called *shooting stars* or *falling stars*, because some people once thought they really were stars falling from the sky.

The vast majority of the particles that make meteors are no larger than peas and burn up completely before reaching the ground. Only in rare cases is a meteor caused by a chunk of rock large enough to survive the plunge through our atmosphere and leave a *meteorite* on the ground. Thus, a meteorite is an actual rock that has crashed to Earth. (The word *meteorite* means "associated with meteors.") Rocks from space that are large enough to become meteorites make unusually bright meteors, called *fireballs*, as they fall through our atmosphere. Observers find a few meteorites each year by following the trajectories of fireballs. Meteorites found in this way can be especially valuable to science, because we can trace their trajectories back to learn exactly where they came from and because they can be collected before they have much time to be contaminated by terrestrial material.

Meteorite Falls People didn't always accept that rocks fall from the sky. Stories of such events arose occasionally in human history, and sometimes even affected it. For example, stories of "fallen stars" influenced the philosophy of the

Figure 12.6 On March 26, 2003, a meteorite crashed through the roof of this home in Chicago. The main photo shows the hole in the ceiling; the inset shows the meteorite at rest on the floor. No one was hurt.

ancient Greek scientist Anaxagoras (see Figure 3.13), who concluded that planets and stars are flaming rocks in the sky. Anaxagoras's assumption that meteorites fell from the heavens made him the first person in history known to believe that the heavens and Earth are made of the same materials, even though his guess about the nature of planets and stars was not quite correct. Many later scientists regarded stories of fallen stars more skeptically. Upon hearing of a meteorite fall in Connecticut, Thomas Jefferson (who took part in science as well as politics) reportedly said, "It is easier to believe that Yankee professors would lie than that stones would fall from heaven."

Today we know that rocks really do sometimes fall from the heavens (Figure 12.6). Meteorites are often blasted apart in their fiery descent through our atmosphere, scattering fragments over an area several kilometers across. A direct hit on the head by a meteorite would be fatal, but we have no reliable accounts of human deaths from meteorites. Of course, most meteorites fall into the ocean, which covers three-fourths of Earth's surface.

Unless you actually see a meteorite fall, it can be difficult to distinguish meteorites from terrestrial rocks. Fortunately, a few clues can help. Meteorites are usually covered with a dark, pitted crust resulting from their fiery passage through the atmosphere (Figure 12.7). Some have an unusually high

Figure 12.7 This large meteorite, called the Ahnighito Meteorite, is located at the American Museum of Natural History in New York. Its dark, pitted surface comes from its fiery passage through Earth's atmosphere.

a Primitive Meteorites.

Stony primitive meteorite: Made of rocky material embedded with shiny metal flakes (arrow).

Carbon-rich primitive meteorite: Also rocky but with dark carbon compounds and small whitish spheres (arrow).

b Processed Meteorites.

Metal-rich processed meteorite: Made of iron and other metals that came from a shattered asteroid's core.

Rocky processed meteorite: Resembles volcanic rocks found on Earth.

Figure 12.8 There are two basic types of meteorites: primitive and processed. They are shown slightly larger than actual size. (The meteorites have flat faces because they have been sliced with rock saws.)

metal content, enough to attract a magnet hanging on a string. The ultimate judge of extraterrestrial origin is laboratory analysis: Meteorites often contain elements such as iridium that are very rare in Earth rocks, and even common elements in meteorites tend to have different ratios among their isotopes [Section 5.3] than rocks from Earth. If you suspect that you have found a meteorite, many museums will analyze a small chip free of charge.

Scientists have found and cataloged more than 20,000 meteorites, some found by accident and others by organized meteorite searches. Antarctica is one of the best places to hunt for meteorites, not because more fall there but because the icy surface makes them easier to find. Few terrestrial rocks end up on the ice, so a rock found on the ice has a good chance of being a meteorite. Moreover, the slow movement of Antarctic glaciers tends to carry meteorites along, concentrating them in small areas where the ice floes run into mountains. The majority of meteorites now in museums and laboratories were found in Antarctica.

Types of Meteorites   The precise origin of meteorites was long a mystery, but in recent decades we've been able to determine where in our solar system they come from. The most direct evidence comes from the relatively few meteorites whose trajectories have been observed or filmed as they fell to the ground. In every case so far, these meteorites clearly originated in the asteroid belt.

Detailed analysis of thousands of meteorites shows that they come in two basic types, each of which can be further classified into two subtypes:

●	**Primitive meteorites** (Figure 12.8a) are "primitive" in the sense of being the first (or "primary") type of meteorite to have formed. Radiometric dating [Section 8.5] shows them to be about 4.6 billion years old, making them remnants from the birth of our solar system, essentially unchanged since they first accreted in the solar nebula. Primitive meteorites come in two subtypes:

	■	*Stony primitive meteorites* are composed of rocky minerals with a small but noticeable fraction of pure metallic flakes mixed in.

	■	*Carbon-rich primitive meteorites* are generally similar to the stony primitive meteorites, but they also contain substantial amounts of carbon compounds.

Some carbon-rich meteorites even contain a small amount of water.

●	**Processed meteorites** (Figure 12.8b) apparently once were part of a larger object that "processed" the original material of the solar nebula into another form. Radiometric dating shows that processed meteorites are generally younger than the primitive meteorites by a few hundred million years. The processed meteorites can also be divided into two subtypes:

	■	*Metal-rich processed meteorites* are made mostly of high-density iron and nickel mixed with smaller amounts of other metals. Thus, they resemble the terrestrial planet cores in composition.

	■	*Rocky processed meteorites* have lower densities and are made of rock with compositions resembling that of terrestrial mantles and crusts. A few have compositions remarkably close to that of the basalts [Section 9.2] that erupt from terrestrial volcanoes.

The Origin of Primitive Meteorites   The vast majority of all meteorites are primitive, and their structures confirm what their radiometric ages tell us: They are pieces of rock that accreted in the solar nebula and orbited the Sun for billions of years before finally falling to Earth. The individual flakes may represent the tiny particles that first condensed from the gas of the solar nebula. The small, roundish features visible in Figure 12.8a may be solidified droplets splashed out in the accretion process.*

Why are some primitive meteorites stony in composition while others are carbon-rich? The answer lies in where they formed in the solar nebula. Throughout the asteroid belt, temperatures were too warm for ices to condense, so all asteroids are made of rock and metal. However, beyond about 3 AU from the Sun, temperatures were low enough for carbon compounds to condense. Thus, the carbon-rich primitive meteorites must have accreted in what are now the outer regions of the asteroid belt, while the stony primitive meteorites formed in the inner regions. Laboratory studies of primitive meteorites confirm these origins. The

---

*The roundish features are called *chondrules,* so primitive meteorites are technically known as *chondrites.* Processed meteorites are called *achondrites,* because they lack the chondrules (*a-* means "not").

composition of carbon-rich meteorites matches the composition of dark, carbon-rich asteroids in the outer part of the asteroid belt. The composition of stony meteorites matches the composition of asteroids in the inner part of the asteroid belt.

Curiously, most of the meteorites collected on Earth are of the stony variety even though most asteroids are of the carbon-rich variety. Evidently, orbital resonances are more effective at pitching material our way from the inner part of the asteroid belt than from more distant regions of the belt.

The Origin of Processed Meteorites The processed meteorites tell a more complex story. Their compositions look similar to the cores, mantles, or crusts of the terrestrial worlds. Thus, they must be fragments of larger asteroids that underwent *differentiation* [Section 9.1] in which their interiors melted so that metals sank to the center and rocks rose to the surface. This idea explains why we find two types of processed meteorites.

The rocky processed meteorites are often so close in composition to volcanic rocks on Earth that they must have been made by lava flows. We conclude that these meteorites were essentially chipped off the surfaces of asteroids that once had active volcanism, probably by collisions with smaller asteroids. Perhaps as many as a dozen large asteroids were geologically active shortly after the formation of the solar system. The old ages of processed meteorites tell us that this active period was short-lived—the interiors of these asteroids must have cooled quickly.

Metal-rich processed meteorites are less common. They have core-like compositions, suggesting that they are fragments of larger asteroids that shattered in collisions. These shattered worlds not only lost a chance to grow into planets but sent many fragments onto collision courses with other asteroids and other planets, including Earth. Thus, processed meteorites essentially present us with an opportunity to study a "dissected planet." They also represent a form of direct proof that large worlds really do undergo differentiation, confirming what we infer from seismic studies of Earth.

A forthcoming NASA mission called *Dawn* should tell us more about the connection between meteorites and asteroids. If all goes well, *Dawn* will visit several asteroids for close-up study, beginning in about 2010. Both Ceres and Vesta are included in *Dawn*'s planned itinerary. Vesta may be particularly revealing, because scientists suspect that we already have pieces of it in our laboratories: Spectroscopy tells us that the rocky processed meteorite shown in Figure 12.8b matches Vesta in composition, suggesting it was chipped from Vesta's surface.

Meteorites from the Moon and Mars In a few cases, the compositions of processed meteorites do not appear to match any known asteroids. Instead they appear to match either the Moon or Mars.

Careful analysis of these meteorite compositions makes us very confident that we really do have a few meteorites that were once part of the Moon or Mars. Moderately large impacts can blast surface material from terrestrial worlds into interplanetary space. Once they are blasted into space, the rocks orbit the Sun until they come crashing down on another world. Calculations show that it is not surprising that we should have found a few meteorites chipped off the Moon and Mars in this way.

Study of these *lunar meteorites* and *Martian meteorites* is providing new insights into conditions on the Moon and Mars. In at least one case, a Martian meteorite may offer clues about whether life ever existed on Mars [Section 24.2].

12.2 Comets

Asteroids are one of the two major categories of small bodies in the solar system. The other is comets, to which we now turn our attention. Asteroids and comets have much in common. They are both leftover planetesimals from the birth of our solar system. They both come in a wide range of sizes. The primary difference between them is their composition, which reflects where they formed. Asteroids are rocky because they formed in the inner solar system where metal and rock condensed. Comets are ice-rich because they formed beyond the frost line, where abundant hydrogen compounds condensed into ice [Section 8.3]. Note that we refer to any icy leftover planetesimal orbiting the Sun as a comet, regardless of its size, whether it has a tail, or where it resides or comes from.

• What are comets like?

For most of human history, comets were familiar only from their occasional presence in the night sky. Every few years, a comet becomes visible to the naked eye, appearing as a fuzzy ball with a long tail. Indeed, the word *comet* comes from the Greek word for "hair," because comets were recognized by the long, hairlike tails they can display in our sky (Figure 12.9).

In comet photographs, the tails make it look as if the comets are racing across the sky, but they are not. If you watch a comet for minutes or hours, you'll see it staying nearly stationary relative to the stars around it in the sky. Over many days it will rise and set just like the stars, while gradually moving relative to the constellations. You may be able to see it night after night for several weeks or more, until it finally fades from view.

Today, we know that the vast majority of comets do not have tails and never venture anywhere close to Earth. Instead, they remain perpetually in the far outer reaches of our solar system, orbiting the Sun far beyond the orbit of Neptune. The comets that appear in the night sky are the rare ones that have had their orbits changed by the gravitational influences of planets, other comets, or stars passing by in the distance. Their new orbits may carry them much closer to the Sun, eventually bringing them into the inner solar system, where they may grow the tails that allow us to see them in the sky. Most of these comets will not return

a Comet Hyakutake.

b Comet Hale-Bopp, photographed at Mono Lake, CA.

Figure 12.9 Interactive Photo Brilliant comets can appear at almost any time, as demonstrated by the back-to-back appearances of comet Hyakutake in 1996 and comet Hale-Bopp in 1997.

to the inner solar system for thousands of years, if ever. A few happen to pass near enough to a planet to have their orbits changed further, and some end up on elliptical orbits that periodically bring them close to the Sun.

Comets in Human History The occasional presence of a bright comet in the night sky was hard to miss before the advent of electric lights. In some cultures, these rare intrusions into the unchanging heavens foretold bad or good luck. The ancient Chinese, for example, believed that the appearance of a comet led to a time of major and tumultuous change.

Few ancient cultures made any attempt to explain comets in astronomical terms. In fact, comets were generally thought to be within Earth's atmosphere until 1577, when Tycho Brahe [Section 3.3] used observations made from different locations on Earth to prove that a comet lay far beyond the Moon. A century later, Newton correctly deduced that comets orbit the Sun. Then, in a book published in 1705, English scientist Edmund Halley (1656–1742) used Newton's law of gravitation to calculate the orbit of a comet that had been seen in 1682. He showed that it was the same comet that had been observed on numerous prior occasions, which meant that it orbited the Sun every 76 years. Halley predicted that the comet would return in 1758. Although he had died by that time, the comet returned as he predicted and it was then named in his honor. Thus, Halley did not really "discover" his comet but rather gained fame for being the first to correctly calculate a comet's orbit and predict its return. Halley's comet last passed through the inner solar system in 1986, and it will return in 2061.

Today, both professional and amateur astronomers routinely scan the skies with telescopes and binoculars in hopes of discovering a new comet. Several comets are discovered every year. The first observers (up to three) to report a comet discovery to the International Astronomical Union have the comet named after them. Thus, comets now bring a form of good luck to their discoverers. More generally, comets are good luck for astronomers seeking to understand the solar system. These leftover planetesimals teach us about the history of the outer solar system just as asteroids teach us about the history of the inner solar system.

The surprise champion of comet discovery is the *SOHO* spacecraft, an orbiting solar observatory. *SOHO* has detected more than 1,000 "Sun-grazing" comets, most on their last pass by the Sun (Figure 12.10).

Comet Composition We have learned about the composition of comets by studying their spectra. The results are just what we expect for objects that formed in the cold outer solar system: Comets are basically chunks of ice mixed with rocky dust and some more complex chemicals, and hence they are often described as "dirty snowballs."

Spectra of comets confirm a distant origin because they show the presence of compounds that could have condensed only in the cold outer regions of the solar nebula. For example, comet spectra invariably show emission features from hydrogen compounds, including water. They also show emission from carbon dioxide and carbon monoxide, gases that condensed only in the very coldest and most distant regions of the solar nebula. Comet spectra also show many more complex molecules, including some organic molecules, leading some scientists to speculate that much of the organic material that made life possible on Earth was brought here by comets.

This comet's tail is at least two solar diameters long—more than 3 million km.

Figure 12.10 Comet SOHO-6's final blaze of glory. The *SOHO* spacecraft observed this "Sun-grazing" comet a few hours before it passed just 50,000 kilometers above the Sun's surface. The comet did not survive its passage, due to the intense solar heating and tidal forces. In this image, the large, orange disk blocks out the Sun; the Sun's size is indicated by the white circle.

b (Above) The nucleus of comet Wild 2, photographed by the *Stardust* spacecraft along with a ground-based image of the comet. Although not visible in the *Stardust* photograph, the spacecraft found violent jets of gas and dust shooting outward from the comet's nucleus. The irregular surface probably shows effects from a combination of impacts and uneven sublimation rates in different regions.

a These photos were taken by the Deep Impact spacecraft as its 370-kg impactor crashed into Comet Tempel 1. The entire sequence, which starts just before impact at the upper left, unfolded over just 67 seconds.

Figure 12.11 Comet nuclei.

Telescopic spectroscopy tells us about the material comets shed into space, but is this material representative of their internal compositions? NASA's *Deep Impact* mission was designed to answer this and other questions about comets. The main spacecraft carried a 370-kg "impactor" that it released as it approached Comet Tempel 1. Instruments on the spacecraft then observed the results as the impactor slammed into the comet at a speed of 37,000 kilometers per hour on July 4, 2005 (Figure 12.11a). The data were only beginning to be analyzed as this book went to press, but the impactor apparently penetrated many meters into a relatively soft and porous surface before it vaporized due to the energy of the collision. The size of the ejected plume of debris exceeded predictions, giving scientists confidence that careful study will teach us a lot about what comets are made of.

We'll learn even more about comet composition when we can study comet material in the laboratory—which should be happening by the time you read this book. In early 2004, NASA's *Stardust* spacecraft passed near comet Wild 2 (pronounced "Vilt 2"), using a piece of foamed glass called "aerogel" to capture dust particles that the comet had shed into space (Figure 12.11b). *Stardust* is scheduled to return to Earth on January 15, 2006.

THINK ABOUT IT

Find the status of the *Stardust* mission. Did it return to Earth successfully? If so, what have we learned from the it?

The Flashy Lives of Comets   Comets grow tails only as they enter the inner solar system, where they are heated by the warmth of the Sun. To see what happens, let's follow the comet path shown in Figure 12.12a, and the comet anatomy in Figure 12.12b.

Far from the Sun, the comet is completely frozen—the "dirty snowball." It begins to shed gas and dust as it approaches the Sun, but it still has its central chunk of ice, which we call the **nucleus** of the comet. Our best views of

comet nuclei have come during spacecraft flybys: the first was a flyby of Halley's comet by the European Space Agency's *Giotto* spacecraft in 1986, followed more recently by the Deep Impact and Stardust missions. Despite their icy compositions, comet nuclei are darker than charcoal, reflecting less than 5% of the light that falls on them. Apparently, it doesn't take much rocky soil or carbon-rich material to darken a comet. Density measurements show that comet nuclei are not packed very solidly. For example, the density of Halley's nucleus is considerably less than the density of water (1 g/cm^3), suggesting that the nucleus is part ice and part empty space.

The scale bars in Figures 12.11 and 12.12b show that comet nuclei are typically less than about 20 km across. How can such small objects put on such spectacular shows? As a comet accelerates toward the Sun, its surface temperature increases, and ices begin to sublimate into gas that easily escapes the comet's weak gravity. Some of the escaping gas drags away dust particles from the nucleus, and the gas and dust create a huge, dusty atmosphere called a **coma**. The coma is far larger than the nucleus it surrounds. Spacecraft observations show that the release of gas from the comet nucleus can be quite violent, with jets of gas and dust shooting out at speeds of hundreds of meters per second from the interior of the nucleus into space.

The coma grows as the comet continues into the inner solar system, and tails form as extensions of the coma. Comet tails can be hundreds of millions of kilometers in length. Because they look much like exhaust trails behind rockets, many people mistakenly guess that tails extend behind comets as they travel through their orbits. In fact, comet tails generally point away from the Sun, regardless of the direction in which the comet is traveling.

Comets actually have two visible tails, one made of ionized gas, or *plasma*, and the other made of dust. The **plasma tail** consists of gas escaping from the coma. Ultraviolet light from the Sun ionizes the gas, and the solar wind then carries this gas straight outward from the Sun

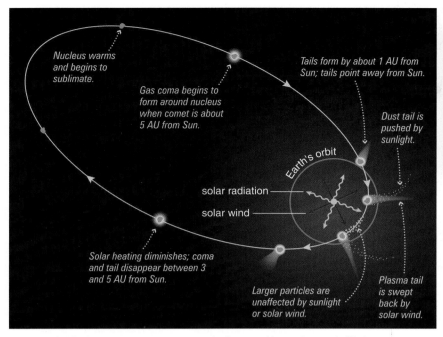

b (Above) Anatomy of a comet. The larger image shows a ground based photo of comet Hale-Bopp. The inset shows the nucleus of Halley's comet photographed by the *Giotto* spacecraft.

a (Left) This diagram shows the changes that occur when a comet's orbit takes it on a passage into the inner solar system. (Not to scale.)

Figure 12.12 A comet grows a coma and tail around its nucleus only if it happens to come close to the Sun. Most comets never do this, instead remaining perpetually frozen in the far outer solar system.

at speeds of hundreds of kilometers per second. Thus, the plasma tail extends almost directly away from the Sun at all times. The **dust tail** is made of dust-size particles escaping from the coma. These particles are not affected by the solar wind and instead are pushed away from the Sun by the much weaker pressure of sunlight itself (*radiation pressure*). Thus, while the dust tail points generally away from the Sun, it has a slight curve back in the direction the comet came from.

After the comet loops around the Sun and begins to head back outward, sublimation declines, the coma dissipates, and the tails disappear. Nothing happens until the comet again comes sunward—in a century, a millennium, a million years, or perhaps never.

Comets that repeatedly visit the inner solar system, like Halley's comet, cannot last long on the time scale of our solar system. A comet probably loses about 0.1% of its ice on every pass around the Sun, so it could not make more than a few hundred passages before losing most of its original ice. Changes in composition may end the comet's "life" even faster. In a close pass by the Sun, a comet may shed a layer of material a meter thick. Dust that is too heavy to escape accumulates on the surface. This thick, dusty deposit helps make comets dark and may eventually block the escape of interior gas, preventing the comet from growing a coma or tails on future passes by the Sun. No one is certain what happens to a comet after its ices can no longer sublime into gas and escape. Either the remaining dust disguises the dead comet as an asteroid, or the comet comes "unglued" without the binding effects of ices. In the latter case, the comet may simply disintegrate along its orbit.

Comets striking the Earth do not strew meteorites across the surface as asteroids do. The icy materials of a small comet vaporize on entry, and the rocky material in comets takes the form of dust or pebbles too small to survive the passage through Earth's atmosphere.

Comet Tails and Meteor Showers   In addition to the plasma and dust in their two visible tails, the gas escaping from comets also carries away sand- to pebble-size pieces of rocky material. These particles are too big to be affected by either the solar wind or sunlight, so they drift away slowly and spread along the orbital path. These particles essentially form a third, invisible tail that follows the comet around its orbit. They are also the particles responsible for most meteors and meteor showers.

We see a meteor lighting up the sky when one of these small particles (or a similar-size particle from an asteroid) burns up in our atmosphere. The sand- to pebble-size particles are much too small to be seen themselves, but they enter the atmosphere at such high speed—up to about 250,000 km/hr (70 km/s)—that they make the surrounding air glow with heat. It is this glow that we see as the brief but brilliant flash of a meteor. The small particles are vaporized by the heat and thus never reach the ground. An estimated 25 million particles of comet dust enter the atmosphere worldwide every day, burning up as meteors and adding hundreds of tons of comet dust to Earth daily.

Even more comet dust enters the atmosphere during a **meteor shower**. Some comet dust is sprinkled throughout the inner solar system, but the "third tails" of ejected particles make the dust much more concentrated along the orbits of comets. Thus, meteor showers occur when Earth crosses through a comet's orbit. Meteor showers recur at about the same time each year because the orbiting Earth passes through a particular comet's orbit at the same time each year. For example, the meteor shower known as the *Perseids* occurs about August 12 each year—the time when

Table 12.1 Major Annual Meteor Showers

Shower Name	Approximate Date	Associated Comet
Quadrantids	January 3	?
Lyrids	April 22	Thatcher
Eta Aquarids	May 5	Halley
Delta Aquarids	July 28	?
Perseids	August 12	Swift-Tuttle
Orionids	October 22	Halley
Taurids	November 3	Encke
Leonids	November 17	Tempel-Tuttle
Geminids	December 14	Phaeton
Ursids	December 23	Tuttle

Earth passes through the orbit of a comet called Swift-Tuttle. Table 12.1 lists major annual meteor showers and their parent comet, if known.

THINK ABOUT IT

The associated "comet" for the Geminid meteor shower is an object called Phaeton, which is classified as an asteroid because we've never seen a coma or tail associated with it. How is it possible that Phaeton looks like an asteroid today but once shed the particles that create the annual Geminid meteor shower?

You can typically see a few meteors each hour on any clear night, but you may see far more during one of the annual meteor showers. During a "good" shower at a dark site, you may see dozens of meteors per hour. Remember that you are watching the effects of comet dust entering our planet's atmosphere from space. The meteors generally appear to radiate from a particular direction in the sky, for essentially the same reason that snow or heavy rain seems to come from a particular direction in front of a moving car (Figure 12.13). Because more meteors hit Earth from the front than from behind (just as more snow hits the front windshield of a moving car), meteor showers are best observed in the predawn sky, which is the time of day when part of your sky faces in the direction of Earth's motion.

• Where do comets come from?

We've already stated that comets come from the outer solar system, but we can be much more specific. By analyzing the orbits of comets that pass close to the Sun, scientists learned that there are two major "reservoirs" of comets in the outer solar system.

Most comets that visit the inner solar system follow orbits that seem almost random. They do not orbit the Sun in the same direction as the planets, and their elliptical orbits can be pointed in any direction. Moreover, their orbits show that they come from far beyond the orbits of the planets—sometimes nearly a quarter of the distance to the nearest star. These comets must come plunging sunward from a vast, spherical region of space that scientists call the **Oort cloud** (after astronomer Jan Oort; rhymes with "court"). Be sure to note that the Oort cloud is not a cloud of gas, but rather a collection of many individual comets. Based on the number of Oort cloud comets that enter the inner solar system each year, we conclude that the Oort cloud must contain about a trillion (10^{12}) comets orbiting the Sun at distances up to about 50,000 AU.

A smaller number of the comets that visit the inner solar system have a pattern to their orbits. They travel around the Sun in the same plane and direction as the planets on elliptical orbits that carry them no more than about twice

Figure 12.13 The geometry of meteor showers.

Snowflakes and meteors appear to radiate from a single direction based on our motion relative to them.

a Meteors appear to radiate from a particular point in the sky for the same reason that we see snow or heavy rain come from a single point in front of a moving car.

b This digital composite photo, taken in Australia during the 2001 Leonid meteor shower, shows meteors as streaks of light. The large rock is Uluru, also known as Ayers Rock.

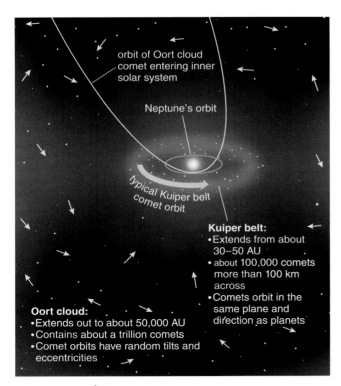

orbit of Oort cloud
comet entering inner
solar system

Neptune's orbit

typical Kuiper belt
comet orbit

Kuiper belt:
• Extends from about
  30–50 AU
• about 100,000 comets
  more than 100 km
  across
• Comets orbit in the
  same plane and
  direction as planets

Oort cloud:
• Extends out to about 50,000 AU
• Contains about a trillion comets
• Comet orbits have random tilts and
  eccentricities

Figure 12.14 The comets we occasionally see in the inner solar system come from two major reservoirs of comets in the outer solar system: the Kuiper belt and the Oort cloud.

as far from the Sun as Neptune. These comets must come from a ring of comets that orbit the Sun beyond the orbit of Neptune. This ring is usually called the **Kuiper belt** (after astronomer Gerald Kuiper; rhymes with "piper"). Figure 12.14 contrasts the general features of the Kuiper belt and the Oort cloud.

The Origin of Comets   How did comets end up in these far-flung regions of the solar system? The only answer that makes scientific sense comes from thinking about what happened to the icy, leftover planetesimals that roamed the region in which the jovian planets formed.

The leftover planetesimals that cruised the spaces between Jupiter, Saturn, Uranus, and Neptune were doomed to suffer either a collision or a close gravitational encounter with one of the young jovian planets. Recall that when a small object passes near a large planet, the planet is hardly affected but the small object may be flung off at high speed [Section 4.5]. Thus, the planetesimals that escaped being swallowed up by the jovian planets tended to be flung off in all directions. Some may have been cast away at such high speeds that they completely escaped the solar system and now drift through interstellar space. The rest ended up on orbits with very large average distances from the Sun, becoming the comets of the Oort cloud. The random directions in which they were flung explains why the Oort cloud is roughly spherical in shape. Oort cloud comets are so far from the Sun that they can be nudged by the gravity of nearby stars, preventing some of them from ever returning near the planets and sending others plummeting toward the Sun.

Gravitational encounters were much less likely for icy planetesimals that roamed in relative safety beyond the orbits of the jovian planets. These planetesimals tended to keep the orbits on which they formed in the spinning disk of the solar nebula—orbits that go in the same direction as planetary orbits and that are concentrated relatively near the ecliptic plane. Many of these planetesimals still have these orbits today, which is why they populate the ring-shaped Kuiper belt. Most remain perpetually beyond the orbit of Neptune, but gravitational resonances with the jovian planets occasionally nudge an orbit enough to send a Kuiper belt comet plunging inward toward the Sun. The Kuiper belt extends to only about twice Neptune's orbital distance, because at greater distances the density of the solar nebula was apparently too low to allow the accretion of planetesimals.

To summarize, the comets of the Kuiper belt seem to have originated farther from the Sun than the comets of the Oort cloud, even though the Oort cloud comets are now much more distant. The Oort cloud consists of leftover planetesimals that were flung outward after forming between the jovian planets, while the Kuiper belt consists of leftover planetesimals that formed and still remain in the outskirts of the planetary realm, roughly between the orbit of Neptune and twice Neptune's distance from the Sun.

Formation of the Solar System Tutorial, Lesson 3

12.3 Pluto: Lone Dog No More

You first began to learn about the solar system some time back in elementary school. You probably heard then about asteroids and comets, and you certainly would have been taught about the "nine planets." At this point in the text, we've discussed in detail the four terrestrial planets, the four jovian planets, asteroids, and comets. That leaves only one object left to discuss from your elementary school list: the "ninth planet," Pluto.

Pluto seemed a misfit among the planets almost since the day it was discovered in 1930 (see Special Topic, p. 331). We knew from the start that its 248-year orbit is more elliptical and more inclined to the ecliptic plane than that of any other planet (Figure 12.15). Note that, despite the fact that Pluto sometimes comes closer to the Sun than Neptune, there is no danger of collision: Neptune orbits the Sun precisely three times for every two Pluto orbits, and this stable orbital resonance means that Neptune is always a safe distance away whenever Pluto approaches its orbit.

As we learned more about Pluto, it proved to be a misfit in size and composition as well: It is far smaller than even the terrestrial planets, and its ice-rich composition fits neither the terrestrial nor jovian categories. These seeming oddities long made Pluto seem like a lone dog of the outer solar system. But it is not: We now know that Pluto is part of a huge pack of similar objects—essentially large comets—that orbit in its region of the solar system. In fact, Pluto is not even the largest member of this pack.

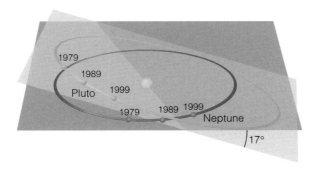

Figure 12.15 Interactive Figure Pluto's orbit is significantly elliptical and tilted relative to the ecliptic. It comes closer to the Sun than Neptune for 20 years in each 248-year orbit, as was the case between 1979 and 1999. There's no danger of a collision, however, thanks to an orbital resonance in which Neptune completes three orbits for every two of Pluto's orbits.

• How big can a comet be?

We've known since the 1950s that many of the comets that visit the inner solar system come from the reservoir that we call the Kuiper belt. We've also known that Pluto orbits the Sun near the middle of this reservoir. Over the past couple decades, scientists gradually came to realize that Pluto's location is no mere coincidence, but rather evidence of an astonishing fact: Some of the objects that populate the Kuiper belt are far larger than any of the comets that we ever see in the inner solar system. This fact forces us to reconsider just exactly what we mean by a "comet."

All the comets we've observed in the inner solar system are fairly small, with nuclei no larger than about 20 kilometers across. This shouldn't be surprising based on our understanding of comet origins, because only relatively small objects are likely to have their orbits changed enough to send them from the Kuiper belt or Oort cloud into the inner solar system. But it should make us wonder: Given that the comets we see are just rare and relatively small visitors from distant reaches of the solar system, could these outer realms hold much larger objects as well?

According to our understanding of solar system formation, the icy planetesimals of the Kuiper belt should have been able to continue their accretion as long as other icy particles were nearby. In principle, one of these planetesimals could have grown large enough to become the seed of a fifth jovian planet, beyond Neptune, but that did not occur—probably because the density of material was too low at this great distance from the Sun. Nevertheless, it's reasonable to think that many of them grew to hundreds or even thousands of kilometers in diameter.

You might think it would be easy to observe icy objects of such sizes, but it requires powerful telescope technology. Trying to see a 1000-kilometer iceball at Pluto's distance from Earth is equivalent to looking for a snowball the size of your fist that is 600 kilometers away (and in dim light, too). Aside from Pluto, the first large objects in the Kuiper belt were discovered in the early 1990s. The discoveries have come at a rapid pace ever since.

As astronomers surveyed more and more of the sky in search of large, icy objects, the record size seemed to rise with each passing year. In 2002, the record was set by an object named *Quaoar* (pronounced *kwa-o-whar*), which is more than half the diameter of Pluto. Two objects announced in 2004—2004DW and Sedna—are between about two-thirds and three-quarters of Pluto's size. It seemed only a matter of time until scientists would find an object larger than Pluto, and the time came on July 29, 2005, when Caltech astronomer Michael Brown announced the discovery of "Planet X" (Figure 12.16), which he code-named Xena (of television fame) while the world waits for the International Astronomical Union to give it an official name.

Planet X currently lies about twice as far as Pluto from the Sun. Its 560-year orbit is inclined to the ecliptic plane more than twice as much as Pluto's (which is shown in Figure 12.15), and it is so eccentric that it must sometimes come closer to the Sun than Pluto. Based on its brightness and distance, Planet X would be nearly the size of Pluto even if it were 100% reflective. Since no real object is perfectly reflective, it must be even larger in size to account for its brightness. Thus, Pluto has fallen to second place in the rankings of large objects in the Kuiper belt, and it may yet fall further—as of 2005, only about half the sky has been carefully searched for large Kuiper belt objects.

What should we call all these big iceballs? The terminology is a topic of much debate, but the science behind it seems clear. By 2005, more than 900 icy objects had been directly observed in the Kuiper belt (Figure 12.17), allowing scientists to infer that the region contains at least 100,000 objects more than 100 kilometers across. Clearly, the objects that reside in the Kuiper belt come in a wide range of sizes. Some are so small that their orbits can be easily perturbed, sending them plunging toward the Sun so that we see them as comets, while others are as big or bigger than Pluto, which has been considered a "planet" since its discovery in 1930. Regardless of size, however, the members of the Kuiper belt probably all share the same basic composition of ice and rock, making them essentially comets of

Figure 12.16 Sizes to scale of five of the largest objects in the outer solar system, along with the Moon for comparison.

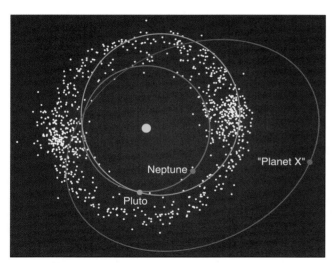

Figure 12.17 Approximate current positions of known members of the Kuiper Belt, with the orbits of Neptune, Pluto, and Planet X shown for reference. Remember that many Kuiper belt comets, including Pluto and Planet X, have large orbital inclinations (tilts relative to the ecliptic plane) that are not shown in this diagram.

different sizes. That is why we often refer to all of them as *comets* of the Kuiper belt. However, some astronomers object to calling such large objects "comets," especially since they never venture into the inner solar system to grow comas or tails. As a result, you may also hear these objects referred to as Kuiper Belt Objects (KBOs), Trans-Neptunian Objects (TNOs), or a variety of other names.

• What are the large objects of the Kuiper belt like?

Although Pluto, Planet X, and other similar objects are "large" compared to most comets of the Kuiper belt, they are still quite small compared to any of the terrestrial or jovian planets. These small sizes, combined with their great distances, makes them very difficult to study. Even our best photographs show little more than fuzzy blobs; see, for example, the photo of Pluto and its moon Charon in Figure 7.12. More commonly, photographs show only a dot of light. As a result, we generally know little about these objects besides their orbits and the fact that their reflectivities and spectra suggest that they have ice-rich compositions. To understand these objects a little better, let's first consider Pluto and then turn our attention to the rest.

Pluto   We've had more time to study Pluto than any other object of the Kuiper belt, but the most important reason we've been able to learn a lot about Pluto is its moon, Charon. Remember, for example, that we can determine an object's mass only if we can observe its gravitational effect on some other object. As a result, we didn't even know Pluto's mass well until Charon's discovery in 1978, after which Charon's orbital characteristics allowed us to find the combined Pluto–Charon mass with Newton's version of Kepler's third law [Section 4.4]. (Planet X also has a moon, which will allow us to determine its mass in the same way.)

More precise data came from some good luck in timing: From 1985 to 1990, Pluto and Charon happened to be aligned in a way that made them eclipse each other every few days as seen from Earth—something that won't happen again for more than 100 years. Detailed analysis of brightness variations during these eclipses allowed the calculation of accurate sizes, masses, and densities for both Pluto and Charon. These measurements confirmed that Pluto and Charon are both made of ice mixed with rock—just like comets. The eclipse data even allowed astronomers to construct rough maps of Pluto's surface markings, which have not yet been explained (Figure 12.18).

Comparing Pluto and Charon leads to some interesting surprises. Charon's diameter is more than half Pluto's, its mass is about one-eighth of Pluto's, and it orbits only 20,000 kilometers from Pluto. For comparison, our Moon has a mass $\frac{1}{80}$ of Earth's and orbits 400,000 kilometers away. Thus, Charon is even larger relative to its parent planet than our Moon is to Earth. Moreover, Pluto is slightly higher in density than Charon. Together, these facts have led astronomers to guess that Charon was created by a giant impact similar to the one thought to have formed our Moon [Section 8.4]. A large comet crashing into Pluto may have blasted away its low-density outer layers, which then formed a ring around Pluto and eventually reaccreted to make Charon. The impact may also explain why Pluto rotates with its rotation axis tipped almost on its side (see Figure 7.1).

Pluto is very cold, with an average temperature of only 40 K, as we would expect at its great distance from the Sun. Nevertheless, Pluto currently has a thin atmosphere of nitrogen and other gases formed by sublimation of surface ices. The atmosphere should gradually refreeze onto the surface as Pluto's 248-year orbit carries it farther from the Sun, although recent results have puzzled astronomers by showing the opposite effect.

Despite the cold, the view from Pluto would be stunning. Charon would dominate the sky, appearing almost 10 times larger in angular size than our Moon appears from Earth. Pluto and Charon's mutual tidal pulls long ago made them rotate synchronously with each other [Section 4.5]. Thus, Charon is visible from only one side of Pluto and always shows the same face to Pluto. Moreover, the synchronous rotation means that Pluto's "day" is the same length as Charon's "month" (orbital period) of 6.4 Earth days. Viewed from Pluto's surface, Charon neither rises nor sets but instead hangs motionless as it cycles through its phases every 6.4 days. The Sun would appear more than a thousand

Figure 12.18 The surface of Pluto in approximate true color, as derived with computer processing from brightness measurements made during mutual eclipses between Pluto and Charon. The many squares in the image represent the smallest regions for which we could accurately estimate colors.

times fainter than it appears here on Earth and would be no larger in angular size than Jupiter appears in our skies.

Other Kuiper Belt Objects   We don't know much about other large comets of the Kuiper belt, but we expect them to be similar in nature to Pluto and Charon. They all probably formed in the same region of the solar system, so we expect them to have similar compositions. Stronger evidence of similarities comes from careful study of orbits. Like Pluto, many Kuiper belt comets have stable orbital resonances with Neptune. In fact, hundreds of Kuiper belt comets have the same orbital period and average distance from the Sun as Pluto itself (and are nicknamed "Plutinos"). Several Kuiper belt comets besides Pluto are also known to have moons, and in those cases we can use Newton's version of Kepler's third law [Section 4.4] to calculate their masses and then infer their densities. The results confirm the idea that these objects have comet-like compositions of ice and rock.

We should learn much more about the objects of the Kuiper belt in about a decade, when the *New Horizons* spacecraft should flyby Pluto and perhaps a couple other Kuiper belt comets. Meanwhile, if you think back to what we learned in our tour of jovian moons in Chapter 11, you'll realize that we probably already have close-up photos of at least one *former* member of the Kuiper belt: Neptune's moon Triton (see Figure 11.29).

Recall that Triton's "backward" orbit indicates that it must be a captured object [Section 11.2]. Its location suggests that it was most likely captured from the Kuiper belt, since many Kuiper belt comets have orbits that cross Neptune's orbit. The fact that Triton appears to be quite similar in nature to Pluto lends further support to the idea that it once orbited the Sun within the Kuiper belt. Moreover, Triton is about 15% larger than Pluto (in diameter), which means it rivals Planet X in size. Triton's case seems to offer proof that other objects as big or bigger than Planet X were able to form in the Kuiper belt. Some may still await discovery. Others may have been destroyed in giant impacts of the type thought to have created Pluto's moon Charon. Those that did not become trapped in stable resonances may have fallen inward to be accreted by the jovian planets or been thrown outward and ejected from the solar system. All in all, it seems quite likely that the Kuiper belt, like the asteroid belt, once held more large objects than it does today.

• Is "Planet X" a planet?

The discovery of "Planet X" has forced astronomers to face the issue of whether it or Pluto should count as planets. In ancient times, a "planet" was an object that moved relative to the stars in the night sky, which meant the Sun and Moon were considered planets but Earth was not. After the Copernican revolution, the term "planet" was used for any object that orbited the Sun, putting Earth on the list and removing the Sun and Moon. This definition successfully accommodated the additions of Uranus and Neptune after their discoveries, but it was challenged by discoveries of asteroids. At first, some scientists considered asteroids to be

planets, but as their numbers grew and their small sizes were recognized, they became "minor planets," or asteroids as we know them today.

Pluto was quickly given planetary status after its discovery in 1930, mainly because early estimates of its size were quite large. These overly generous estimates came about in part because scientists significantly underestimated its reflectivity, but they were also influenced by the circumstances of its discovery. Astronomers thought that Pluto's mass could help account for reported discrepancies in Neptune's orbit—reports that later turned out to be erroneous (see Special Topic, p. 331). Indeed, similar errors in measurement led some astronomers (most notably, Percival Lowell) to suggest the existence of yet another planet, which led to a search for a tenth large planet that became popularly known as "Planet X." Note that aside from its temporary name, the "Planet X" discovered in 2005 bears no resemblance to the Planet X sought almost a century ago.

We now know that no jovian-size ninth or tenth planets exist in our solar system—if they did, by now we would have detected their gravitational influences. That, plus the fact that we now know Pluto to be just one of many similar objects, has led many people to suggest that Pluto should be demoted in much the same way that some asteroids were demoted from an initial status as planets. Because Planet X is so similar in size to Pluto, demoting Pluto would also mean that Planet X is not a planet. Our solar system would have just 8 "major planets," followed by many "minor planets" that would include both large asteroids and large comets.

However, because popular culture has embraced Pluto as a planet for 75 years, other people suggest that it should stay a planet—in which case Planet X would become the tenth planet. Under this proposal, Pluto's size would represent the minimum for a planet, and additional objects would make the planet list only if they prove to be larger than Pluto.

A third idea comes from people who object to what seem like arbitrary size cutoffs, instead arguing that the minimum size of a planet should be based on physical characteristics. Several physical ideas have been proposed, but the most popular suggests that an object should be called a planet if it is large enough for gravity to make it round. By this definition, there may be dozens of Kuiper belt comets that will make the list, along with one or two of the largest asteroids.

Perhaps the most important idea to remember is that this is a debate about words, not about science. Scientifically, there is no doubt that Pluto and Planet X are members of the Kuiper belt, making them quite different in character from either terrestrial or jovian planets. This basic fact will not change, regardless of whether we continue to call them planets.

THINK ABOUT IT

Do *you* think Planet X should be called a planet? Should Pluto or any other Kuiper belt comets be called planets? How about "planet-sized" moons, such as Triton or the Galilean moons of Jupiter? Consider several possible definitions of planet, and discuss the pros and cons of each one. With your favorite definition, how many planets are there in our solar system?

This "string of comets" was a single comet nucleus before it passed close enough to Jupiter to be ripped apart by tidal forces.

30,000 km

This crater chain was probably made long ago by a similar string of comets.

50 km

Figure 12.19 Tidal forces from Jupiter can break a single comet nucleus into a chain of smaller nuclei.

12.4 Cosmic Collisions: Small Bodies Versus the Planets

The hordes of small bodies orbiting the Sun are slowly shrinking in number through collisions with the planets and ejection from the solar system. Many more must have roamed the solar system in the days of the heavy bombardment, when most impact craters formed [Section 8.4]. Plenty of small bodies still remain, however, and cosmic collisions still occur on occasion. These collisions have had important effects on Earth and other planets.

• Have we ever witnessed a major impact?

Modern scientists have never witnessed a major impact on a solid world, but in 1994 we were privileged to witness one on Jupiter. The event was likely a rare one: Astronomers estimate that a major impact occurs on Jupiter about once every 1,000 years. The dramatic event was the impact of a comet named *Shoemaker-Levy 9*, or *SL9* for short.

Rather than having a single nucleus, SL9 consisted of a string of comet nuclei lined up in a row (Figure 12.19a). Apparently, tidal forces from Jupiter ripped apart a single comet nucleus during a previous close pass near the planet. Crater chains on Jupiter's moons are evidence that similar breakups of comets near Jupiter have occurred in the past. For example, Figure 12.19b shows a chain of craters on Callisto, presumably formed when a string of comet nuclei crashed into its icy surface.

Comet SL9 was discovered more than a year before its impact on Jupiter, and orbital calculations had told astronomers precisely when the impact would occur. Thus, when the impacts finally began, they were observed with nearly every major telescope in existence, as well as by spacecraft that were in a position to get a view. The images were astonishing (Figure 12.20). Each of the individual nuclei in comet SL9 crashed into Jupiter with an energy equivalent to that of a million hydrogen bombs. Comet nuclei barely a kilometer across left scars larger than the entire Earth. The scars lasted for months before Jupiter's strong winds finally made them fade from view.

The SL9 impacts allowed us to study both the impact process and material splashed out from deep inside Jupiter. They also provided two important sociological lessons. First, "Comet Crash Week" proved to be one of the best examples of international collaboration in history. With the aid of the Internet, scientists quickly and effectively shared data from observatories around the world. Second, extensive media coverage helped the event capture the public imagination, leading to awareness that impacts are not merely relegated to ancient geological history. If such violent impacts can happen on other planets in our lifetime, could they also happen on Earth?

• Did an impact kill the dinosaurs?

There's little doubt that major impacts have occurred on Earth in the past. Meteor Crater in Arizona (see the photo that opens this chapter, p. 359) formed about 50,000 years ago when a metallic asteroid roughly 50 meters across

a This infrared photo shows the brilliant glow of a rising fireball from one of the impacts. Jupiter is the round disk, with the impact occuring near the lower left of the disk.

b Each of the black dots in this Hubble Space Telescope photo is a scar from the impact of one of the SL9 nuclei.

Figure 12.20 The impacts of comet Shoemaker-Levy 9 on Jupiter allowed astronomers their first direct view of a cosmic collision.

c This painting shows how the impacts might have looked from the surface of Io. The impacts occurred on Jupiter's night side.

crashed to Earth with the explosive power of a 20-megaton hydrogen bomb. Although the crater is only a bit more than 1 kilometer across, the blast and ejecta probably battered an area covering hundreds of square kilometers. Meteor Crater is relatively small and recent. Despite the fact that erosion and other geological processes have erased most of Earth's impact craters, geologists have identified more than 100 impact craters on our planet. So before we consider whether an impact might occur in our lifetimes, it's worth examining the potential consequences if it did. Clearly, an impact could cause widespread physical damage. But a growing body of evidence, accumulated over the past three decades, suggests that an impact can do much more—in some cases, large impacts may have altered the entire course of evolution.

In 1978, while analyzing geological samples collected in Italy, a scientific team led by Luis and Walter Alvarez (father and son) made a startling discovery. They found that a thin layer of dark sediments deposited about 65 million years ago—about the time the dinosaurs went extinct—was unusually rich in the element iridium. Iridium is a metal that is common in meteorites but rare on Earth's surface (because it sunk to Earth's core when our planet underwent differentiation). Subsequent studies found the same iridium-rich layer in 65-million-year-old sediments around the world (Figure 12.21).* The Alvarez team suggested a stunning hypothesis: The extinction of the dinosaurs was caused by the impact of an asteroid or comet.

In fact, the death of the dinosaurs was only a small part of the biological devastation that seems to have occurred 65 million years ago. The fossil record suggests that up to 99% of all living organisms died around that time and that up to 75% of all existing *species* were driven to extinction. This makes the event a clear example of a **mass extinction**—the rapid extinction of a large fraction of all living species. Could it really have been caused by an impact?

Evidence for the Impact There's still some scientific controversy about whether the impact was the sole cause of the mass extinction or just one of many causes, but there's little doubt that a major impact coincided with the death of the dinosaurs. Key evidence comes from further analysis of the sediment layer. Besides being unusually rich in iridium, this layer contains four other features: (1) unusually high abundances of several other metals, including osmium, gold, and platinum; (2) grains of "shocked quartz," quartz crystals with a distinctive structure that indicates they experienced the high-pressure conditions of an impact; (3) spherical rock "droplets" of a type known to form when drops of molten rock cool and solidify in the air; and (4) soot (at some sites) that appears to have been produced by widespread forest fires.

All these features point to an impact. The metal abundances look much like what we commonly find in meteorites

rather than what we find elsewhere on Earth's surface. Shocked quartz is also found at other impact sites, such as Meteor Crater in Arizona. The rock "droplets" presumably were made from molten rock splashed into the air by the force and heat of the impact. The soot probably came from vast forest fires ignited by impact debris. Some debris would have been blasted so high that it rose above the atmosphere, spreading worldwide before falling back to Earth. On their downward plunge, friction would have heated the debris particles until they became a hot, glowing rain of rock.

In addition to the evidence within the sediments, scientists have identified a large impact crater that appears to match the age of the sediment layer. The crater, about 200 kilometers across, is located on the coast of Mexico's Yucatán Peninsula, about half on land and half underwater (Figure 12.22). Its size indicates that it was created by the impact of an asteroid or a comet measuring about 10 kilometers across. (It is named the *Chicxulub crater,* after a nearby fishing village.)

The Mass Extinction If the impact was indeed the cause of the mass extinction, here's how it probably happened: On that fateful day some 65 million years ago, the asteroid or comet slammed into Mexico with the force of a hundred million hydrogen bombs (Figure 12.23). It apparently hit at a slight angle, sending a shower of red-hot debris across the continent of North America. A huge tsunami sloshed more than 1,000 kilometers inland. Much of North American life may have been wiped out almost immediately. Not long after, the hot debris raining around the rest of the world ignited fires that killed many other living organisms.

The longer-term effects were even more severe. Dust and smoke remained in the atmosphere for weeks or months, blocking sunlight and causing temperatures to fall as if Earth were experiencing a global and extremely harsh winter. The reduced sunlight would have stopped photosynthesis for up to a year, killing large numbers of species throughout the food chain. This period of cold may have been followed by a period of unusual warmth. Some evidence suggests that the impact site was rich in carbonate rocks, so the impact may have released large amounts of carbon dioxide into the atmosphere. The added carbon dioxide would have strengthened the greenhouse effect, and the months of global winter may have been followed by decades or longer of global summer.

Figure 12.21 Around the world, sedimentary rock layers dating to 65 million years ago share the evidence of the impact of a comet or asteroid. Fossils of dinosaurs and many other species appear only in rocks below the iridium-rich layer.

A layer rich in iridium and soot tells us a huge impact occurred at this point in geological (and biological) history.

---

*The layer marks what geologists call the *K-T boundary,* because it separates sediments deposited in the Cretaceous and Tertiary periods (the "K" comes from the German word for Cretaceous, *Kreide*). The mass extinction that occurred 65 million years ago therefore is often called the *K-T event.*

Figure 12.22 This computer-generated image, based on measurements of small local variations in the strength of gravity, shows an impact crater (dashed circle) in the northwest corner of Mexico's Yucatán Peninsula. The crater measures about 200 kilometers across, and about half of it lies underwater off the coast. (The solid white lines show the coastline and borders of Mexican states.)

Figure 12.23 🎨 This painting shows an asteroid or comet moments before its impact on Earth, some 65 million years ago. The impact probably caused the extinction of the dinosaurs, and if it hadn't occurred the dinosaurs might still rule Earth today.

The impact probably also caused chemical reactions in the atmosphere that produced large quantities of harmful compounds, such as nitrous oxides. These compounds dissolved in the oceans, where they probably were responsible for killing vast numbers of marine organisms. Acid rain may have been another by-product, killing vegetation and acidifying lakes around the world.

Perhaps the most astonishing fact is not that 75% of all species died but that 25% survived. Among the survivors were a few small mammals. These mammals may have survived in part because they lived in underground burrows and managed to store enough food to outlast the global winter that immediately followed the impact.

The evolutionary impact of the extinctions was profound. For 180 million years, dinosaurs had diversified into a great many species large and small, while mammals (which had arisen at almost the same time as the dinosaurs) had generally remained small and rodentlike. With the dinosaurs gone, mammals became the new kings of the planet. Over the next 65 million years, the small mammals rapidly evolved into an assortment of much larger mammals—ultimately including us.

Controversies and Other Mass Extinctions There seems little doubt that a major impact coincided with the mass extinction of 65 million years ago, but does it tell the entire story? The jury is still out. The fossil record can be difficult to read when we are trying to understand events that happened over just a few years rather than a few million years. Some scientists suspect that the dinosaurs were already in decline and the impact was only the last straw. Others suggest that major volcanic eruptions also may have played a role.

Was the dinosaur extinction a unique event? Measuring precise extinction rates becomes more difficult as we look to older fossils, but there appear to have been at least four other mass extinctions during the past 500 million years. None of the other mass extinctions is as closely tied to an impact as the dinosaur extinction, but impacts almost certainly played a role. Sediments from the times of other mass extinctions reveal evidence similar to that found in the iridium-rich layer tied to the death of the dinosaurs. In some cases, impact craters have been found that date to about the right times. Much more research is needed, but impacts appear to have played a major role in shaping the history of life on Earth.

• Is the impact threat a real danger or media hype?

On June 14, 2002, the 100-meter-wide asteroid 2002MN passed within 120,000 kilometers of Earth (less than a third of the Earth-Moon distance). There was no advance warning; in fact, the asteroid wasn't discovered until after it had passed by. Although an asteroid of this size would not cause global devastation, it might kill millions of people in the unlikely event that it struck a large city. How concerned should we be about the possibility of objects like this striking Earth—or, worse yet, even larger objects such as those that may have caused past mass extinctions? We can analyze the threat by examining both the frequency of past impacts and the number of objects in space that pose a potential future threat.

No large impacts have left craters in modern times, but we know of at least one smaller impact that devastated a fairly large area. In 1908, a tremendous explosion occurred over Tunguska, Siberia (Figure 12.24). Entire forests were flattened and set on fire, and air blasts knocked over people, tents, and furniture up to 200 kilometers away. Seismic disturbances were recorded at distances of up to 1,000 kilometers, and atmospheric pressure fluctuations were detected almost 4,000 kilometers away. The explosion, now estimated to have released energy equivalent to that of

Figure 12.24 This photo shows forests burned and flattened by the 1908 impact over Tunguska, Siberia.

several atomic bombs, is thought to have been caused by a small asteroid no more than about 40 meters across. Atmospheric friction caused it to explode completely before it hit the ground, so it left no impact crater. If the asteroid had exploded over a major city instead of Siberia, it would have been the worst natural disaster in human history.

Objects of similar size to that of the Tunguska event probably strike our planet every couple hundred years or so. They have gone unnoticed, presumably because they have always hit remote areas, with most striking the oceans. Nevertheless, the death toll would be enormous if such an object struck a densely populated area.

Another way to gauge the threat is to look at asteroids that might strike Earth. Astronomers have identified more than 600 asteroids larger than 1 kilometer in diameter with orbits that pass near Earth's orbit. Orbital calculations show that none of these known objects will collide with Earth in the foreseeable future. However, the vast majority of near-Earth asteroids probably have not yet been detected. Astronomers estimate that hundreds more undiscovered near-Earth asteroids are larger than 1 kilometer across and that another 50,000 are larger than 100 meters across. NASA has set a goal of detecting 90% of all large, near-Earth asteroids by 2008.

We do not have a similar way of studying the threat from comets, because they reside so far from the Sun. By the time we saw a comet plunging toward us from the outer solar system, we'd have at best a few years to prepare for the impact.

While it seems a virtual certainty that many more large impacts will batter Earth, the chance of a major impact happening in our lifetimes is quite small. Nevertheless, the chance is not zero, which is one reason why NASA is working to predict potential asteroid threats. Figure 12.25 shows how often, on average, we expect objects of different sizes to hit Earth. Objects a few meters across probably enter Earth's atmosphere every week or two, each liberating energy equivalent to that of an atomic bomb as it burns up. Larger impacts are obviously more devastating but thankfully more rare. Hundred-meter impacts that forge craters similar to Meteor Crater in Arizona probably strike only

about every 10,000 years or so. Impacts large enough to cause mass extinctions come tens of millions of years apart.

If we were to find an asteroid or a comet on a collision course with Earth, could we do anything about it? Many people have proposed schemes to save Earth by using nuclear weapons or other means to demolish or divert an incoming asteroid, but no one knows whether current technology is really up to the task. We can only hope that the threat doesn't become a reality before we're ready.

THINK ABOUT IT

Study Figure 12.25. Based on the frequency of impacts large enough to cause serious damage, do you think we should be spending time and resources to counter the impact threat? Or should we focus resources on other threats first? Defend your opinion.

While near-Earth asteroids may pose real dangers, they may also offer opportunities. These asteroids bring valuable resources tantalizingly close to Earth. Iron-rich asteroids are particularly enticing, because they probably contain many precious metals that have mostly sunk to the core on Earth. In the future, it may prove technically feasible and financially profitable to mine metals from asteroids and bring these resources to Earth. It may also be possible to gather fuel and water from asteroids for use in missions to the outer solar system.

• How do the jovian planets affect impact rates and life on Earth?

Ancient people imagined that the mere movement of planets relative to the visible stars in our sky could somehow have an astrological influence on our lives. Although we no longer give credence to this ancient superstition, we now

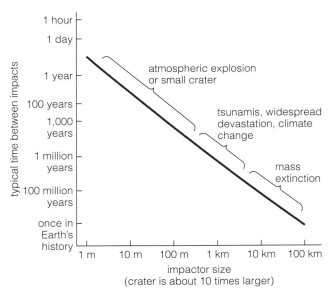

Figure 12.25 This graph shows that larger objects (asteroids or comets) hit the Earth less frequently than smaller ones. The labels describe the effects of impacts of different sizes.

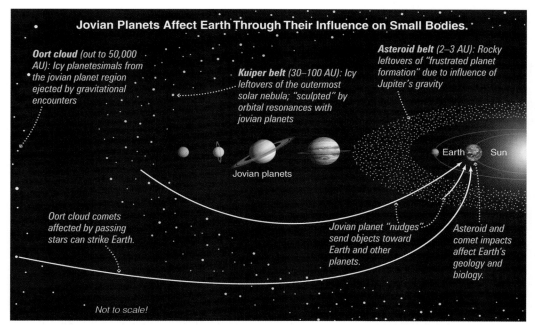

Figure 12.26 The connections between the jovian planets, small bodies, and Earth. The gravity of the jovian planets helped shape both the asteroid belt and the Kuiper belt, and the Oort cloud consists of comets ejected from the jovian planet region by gravitational encounters with these large planets. Ongoing gravitational influences sometimes send asteroids or comets heading toward Earth.

know that planets can have a real effect on life on Earth. By catapulting asteroids and comets in Earth's direction, other planets have caused cosmic collisions that helped shape Earth's destiny.

Jupiter and the other jovian planets have had the greatest effects because of their influence on the small bodies of the solar system. As we saw earlier in this chapter, Jupiter disturbed the orbits of rocky planetesimals outside Mars's orbit, preventing a planet from forming and instead creating the asteroid belt. The jovian planets ejected icy planetesimals to create the distant Oort cloud of comets, and orbital resonances with Neptune shape the orbits of many comets in the Kuiper belt. Ultimately, every asteroid or comet that has crashed to Earth since the end of the heavy bombardment was in some sense sent our way by the influence of Jupiter or one of the other jovian planets. Figure 12.26 summarizes the ways the jovian planets have controlled the motions of asteroids and comets.

Thus, because impacts of asteroids and comets have played such an important role in the history of our planet, we find a deep connection between the jovian planets and the survival of life on Earth. If Jupiter did not exist, the threat from asteroids might be much smaller, since the objects that make up the asteroid belt might instead have become part of a planet. On the other hand, the threat from comets might be much greater: Jupiter probably ejected more comets to the Oort cloud than any other jovian planet, and without Jupiter, those comets might have remained dangerously close to Earth. Of course, even if Jupiter has protected us from impacts, it's not clear whether that has been good or bad for life overall. The dinosaurs appear to have suffered from an impact, but the same impact may have paved the way for our existence. Thus, while some scientists argue

that more impacts would have damaged life on Earth, others argue that more impacts might have sped up evolution.

The role of Jupiter has led some scientists to wonder whether we could exist if our solar system had been laid out differently. Could it be that civilizations can arise only in solar systems that happen to have a Jupiter-like planet in a Jupiter-like orbit? No one yet knows the answer to this question. What we do know is that Jupiter has had profound effects on life on Earth and will continue to have effects in the future.

THE BIG PICTURE

Putting Chapter 12 into Context

In this chapter we concluded the study of our own solar system by focusing on its smallest objects, and we found that these objects can have big consequences. Keep in mind the following "big picture" ideas:

- Asteroids, comets, and meteorites may be small compared to planets, but they provide much of the evidence that has helped us understand how the solar system formed.

- The small bodies are subject to the gravitational whims of the largest. The jovian planets shaped the asteroid belt, the Kuiper belt, and the Oort cloud, and they continue to nudge objects onto collision courses with the planets.

- Collisions not only bring meteorites and leave impact craters but can profoundly affect life on Earth. An impact probably was responsible for wiping out the dinosaurs, and future impacts pose a threat that we cannot ignore.

12.1 Asteroids and Meteorites

- **What are asteroids like?** Asteroids are rocky leftovers from the era of planetary formation. Most are small, and despite their enormous numbers their total mass is less than that of any terrestrial planet.

- **Why is there an asteroid belt?** Orbital resonances with Jupiter disrupted the orbits of planetesimals in the asteroid belt, preventing them from accreting into a planet. Many were ejected but some remained and make up the asteroid belt today. Most asteroids in other regions of the inner solar system accreted into one of the planets.

- **Where do meteorites come from?** Primitive **meteorites** are remnants of the solar nebula, containing intermixed rock and metal flakes. Some are rich in carbon compounds. **Processed meteorites** are fragments of larger asteroids and therefore can be metallic like a planet's core or rocky like its mantle or crust.

12.2 Comets

- **What are comets like?** Comets are icy leftovers from the era of planet formation, and most orbit far from the Sun. If a comet approaches the Sun, its **nucleus** heats up and its ice sublimates into gas. The escaping gases carry along some dust, forming a **coma** and two tails: a **plasma tail** of ionized gas and a **dust tail**. Larger particles can also escape, becoming the particles that cause meteors and **meteor showers** on Earth.

- **Where do comets come from?** Comets come from two reservoirs: the **Kuiper belt** and the **Oort cloud**. The Kuiper belt comets still reside in the region beyond Neptune in which they formed. The Oort cloud comets formed between the jovian planets, and were kicked out to great distance by gravitational encounters with the planets.

12.3 Pluto: Lone Dog No More

- **How big can a comet be?** In the Kuiper belt, icy planetesimals were able to grow to hundreds or thousands of kilometers in size. The recently discovered "Planet X" is the largest known of these objects and Pluto is the second largest.

- **What are the large objects of the Kuiper belt like?** Like smaller comets, these objects are ice-rich in composition. They orbit the Sun roughly between the orbit of Neptune and twice that distance from the Sun. Their orbits tend to be more elliptical and more inclined to the ecliptic plane than those of the terrestrial and jovian planets. Many share orbital resonances with Neptune; Pluto, for example, completes 2 orbits for every 3 Neptune orbits. A few have moons, including Pluto.

- **Is Planet X a planet?** Whether Planet X, Pluto, and other similar objects should be called "planets" is a matter of opinion. However, their properties clearly identify them as members of the Kuiper belt.

12.4 Cosmic Collisions: Small Bodies Versus the Planets

- **Have we ever witnessed a major impact?** In 1994, we observed the impacts of comet Shoemaker-Levy 9 on Jupiter. The comet had fragmented into a string of individual nuclei, so there was a string of impacts that left Jupiter's atmosphere scarred for months.

- **Did an impact kill the dinosaurs?** It may not have been the sole cause, but a major impact clearly coincided with the **mass extinction** in which the dinosaurs died out, about 65 million years ago. Sediments from the time contain iridium and other clear evidence of an impact, and an impact crater of the right age lies near the coast of Mexico.

- **Is the impact threat a real danger or media hype?** Impacts certainly pose a threat, though the probability of a major impact in our lifetimes is fairly low. Impacts like the Tunguska event may occur every couple hundred years, and would be catastrophic if they occurred over cities.

- **How do the jovian planets affect impact rates and life on Earth?** Impacts are always linked in at least some way to the gravitational influences of Jupiter and the other jovian planets. These influences have shaped the asteroid belt, the Kuiper belt, and the Oort cloud, and continue to determine when an object is flung our way.

Review Questions

Short-Answer Questions Based on the Reading

1. Briefly describe the general characteristics of asteroids—including sizes, masses, densities, and compositions—and how we measure them.
2. How does the largest asteroid compare in size to the planets? How does the total mass of all asteroids compare to the mass of a terrestrial world?
3. Where is the *asteroid belt* located, and why? Briefly explain how orbital resonances with Jupiter have affected the asteroid belt.
4. What is the difference between a *meteor* and a *meteorite*? How can we distinguish a meteorite from a terrestrial rock?
5. Distinguish between *primitive meteorites* and *processed meteorites* in terms of both composition and origin.
6. What does a comet look like when it is far from the Sun? How does its appearance change when it is near the Sun? What happens to comets that make many passes near the Sun?
7. What produces the *coma* and tails of a comet? What is the *nucleus*? Why do tails point away from the Sun?
8. Explain how meteor showers are linked to comets. Why do meteor showers recur at about the same time each year?
9. Describe the *Kuiper belt* and *Oort cloud* in terms of their locations and the orbits of comets within them. How did comets come to exist in these two regions?
10. Briefly describe Pluto and Charon. Why won't Pluto collide with Neptune? How do we think Charon formed?
11. What is Planet X? How is it related to Pluto? What evidence suggests that these objects are really just large comets of the Kuiper Belt?
12. Briefly describe the impact of comet Shoemaker-Levy 9 on Jupiter.
13. Briefly describe the evidence suggesting that an impact caused the mass extinction that killed off the dinosaurs. How did the impact lead to the mass extinction?
14. How often should we expect impacts of various sizes on Earth? How serious a threat do we face from these impacts?

Test Your Understanding

Surprising Discoveries?

Suppose we found a solar system with the property described. (These are *not* real discoveries.) In light of what you've learned about the formation of our own solar system, decide whether the discovery should be considered reasonable or surprising. Explain.

15. A small asteroid that orbits within the asteroid belt has an active volcano.
16. Scientists discover a meteorite that, based on radiometric dating, is 7.9 billion years old.
17. An object that resembles a comet in size and composition is discovered orbiting in the inner solar system.
18. Studies of a large object in the Kuiper belt reveal that it is made almost entirely of rocky (as opposed to icy) material.
19. Astronomers discover a previously unknown comet that will be brightly visible in our night sky about 2 years from now.
20. A mission to Pluto finds that it has lakes of liquid water on its surface.

21. Geologists discover a crater from a 5-kilometer object that impacted Earth more than 100 million years ago.
22. Archaeologists learn that the fall of ancient Rome was caused in large part by an asteroid impact in Asia.
23. Astronomers discover three objects with the same average distance from the Sun (and the same orbital period) as Pluto.
24. Astronomers discover an asteroid with an orbit suggesting that it will collide with Earth in the year 2064.

Quick Quiz

Choose the best answer to each of the following. Explain your reasoning with one or more complete sentences.

25. The asteroid belt lies between the orbits of (a) Earth and Mars. (b) Mars and Jupiter. (c) Jupiter and Saturn.
26. Jupiter nudges the asteroids through the influence of (a) tidal forces. (b) orbital resonances. (c) magnetic fields.
27. Can an asteroid be pure metal? (a) No, all asteroids contain rock. (b) Yes, it must have formed where only metal could condense in the solar nebula. (c) Yes, it must have been the core of a shattered asteroid.
28. Did a large terrestrial planet ever form in the region of the asteroid belt? (a) No, because there was never enough mass there. (b) No, because Jupiter prevented one from accreting. (c) Yes, but it was shattered by a giant impact.
29. What does Pluto most resemble? (a) a terrestrial planet (b) a jovian planet (c) a comet
30. How big an object causes a typical shooting star? (a) a grain of sand or a small pebble (b) a boulder (c) an object the size of a car
31. Which have the most elliptical and tilted orbits? (a) asteroids (b) Kuiper belt comets (c) Oort cloud comets
32. Which are thought to have formed farthest from the Sun? (a) asteroids (b) Kuiper belt comets (c) Oort cloud comets
33. About how often does a 1-kilometer object strike the Earth? (a) every year (b) every million years (c) every billion years
34. What would happen if a 1-kilometer object struck the Earth? (a) It would break up in the atmosphere without causing widespread damage. (b) It would cause widespread devastation and climate change. (c) It would cause a mass extinction.

Investigate Further

In-Depth Questions to Increase Your Understanding

Short-Answer/Essay Questions

35. *The Role of Jupiter.* Suppose that Jupiter had never existed. Describe at least three ways in which our solar system would be different, and clearly explain why.
36. *Life Story of an Iron Atom.* Imagine that you are an iron atom in a processed meteorite made mostly of iron. Tell the story of how you got to Earth, beginning from the time you were part of the gas in the solar nebula 4.6 billion years ago. Include as much detail as possible. Your story should be scientifically accurate but also creative and interesting.
37. *Asteroid Discovery.* You have discovered two new asteroids and named them Albert and Isaac. Both lie at the same distance from Earth and have the same brightness when you look at them through your telescope, but Albert is twice as bright as Isaac at infrared wavelengths. What can you deduce

about the relative reflectivities and sizes of the two asteroids? Which would make a better target for a mission to mine metal? Which would make a better target for a mission to obtain a sample of a carbon-rich planetesimal? Explain.

38. *Asteroids vs. Comets.* Contrast the compositions and locations of comets and asteroids, and explain in your own words why they have turned out differently.

39. *Comet Tails.* Describe in your own words why comets have tails. Why do most comets have two distinct visible tails, and why do the tails go in different directions? Why is the third, invisible tail of small pebbles of interest to us on Earth?

40. *Oort Cloud vs. Kuiper Belt.* Explain in your own words how and why there are two different reservoirs of comets. Be sure to discuss where the two groups of comets formed and what kinds of orbits they travel on.

41. *Project: Tracking a Meteor Shower.* Armed with an expendable star chart and a dim (preferably red) flashlight, set yourself up comfortably to watch one of the meteor showers listed in Table 12.1. Each time you see a meteor, record its path on your star chart. Record at least a few dozen meteors, and try to determine the *radiant* of the shower—that is, the point in the sky from which the meteors appear to radiate. Does the meteor shower live up to its name?

42. *Project: Dirty Snowballs.* If there is snow where you live or study, make a dirty snowball. (The ice chunks that form behind tires work well.) How much dirt does it take to darken snow? Find out by allowing your dirty snowball to melt in a container and measuring the approximate proportions of water and dirt afterward.

Quantitative Problems

Be sure to show all calculations clearly and state your final answers in complete sentences.

43. *Adding Up Asteroids.* It's estimated that there are a million asteroids 1 kilometer across or larger. If a million asteroids 1 kilometer across were all combined into one object, how big would it be? How many 1-kilometer asteroids would it take to make an object as large as the Earth? (*Hint:* You can assume they're spherical. The equation for the volume of a sphere is $\left(\frac{4}{3}\right)\pi r^3$, where r is the radius—not the diameter.)

44. *Impact Energies.* A relatively small impact crater 20 kilometers in diameter could be made by a comet 2 kilometers in diameter traveling at 30 km/s (30,000 m/s).
    a. Assume that the comet has a total mass of 4.2×10^{12} kilograms. What is its total kinetic energy? (*Hint:* The kinetic energy is equal to $\frac{1}{2} \times m \times v^2$, where m is the comet's mass and v is its speed. If you use mass in kilograms and velocity in m/s, the answer for kinetic energy will have units of joules.)
    b. Convert your answer from part (a) to an equivalent in megatons of TNT, the unit used for nuclear bombs. Comment on the degree of devastation the impact of such a comet could cause if it struck a populated region on Earth. (*Hint:* One megaton of TNT releases 4.2×10^{15} joules of energy.)

45. *The "Near Miss" of Toutatis.* The 5-kilometer asteroid Toutatis passed a mere 1.5 million kilometers from Earth in 2004. Suppose Toutatis were destined to pass *somewhere* within 3 million kilometers of Earth. Calculate the probability that this "somewhere" would have meant that it slammed into Earth. Based on your result, do you think it is fair to call the 1992 passage a "near miss"? Explain. (*Hint:* You can calculate the probability by considering an imaginary dartboard of

radius 3 million kilometers in which the bull's-eye has Earth's radius, 6,378 kilometers.)

46. *Room to Roam.* It's estimated that there are a trillion comets in the Oort cloud, which extends out to about 50,000 AU. What is the total volume of the Oort cloud, in cubic AU? How much space does each comet have in cubic AU, on average? Take the cube root of the average volume per comet, to find their typical spacing in AU. (*Hints:* For the purpose of this calculation, you can assume the Oort cloud fills the whole sphere out to 50,000 AU. The equation for the volume of a sphere is $\left(\frac{4}{3}\right)\pi r^3$, where r is the radius.)

47. *Pluto and Charon Eclipses.* From 1985 to 1990, the orbit of Pluto's moon Charon caused it to pass in front of and behind Pluto during its 6-day orbit. We can use some simple examples to see how these eclipses allowed astronomers to learn about the two objects.
    a. Assume that Pluto and Charon each reflect the same fraction of the sunlight that hits them, and that their disks reflect the light uniformly. In this case, the total amount of light reflected is proportional to the areas of their disks ($A = \pi r^2$). Using the values from Appendix E, find the ratio of light reflected by Charon to that reflected by Pluto. Draw a rough "light curve" that shows how the total light from the whole system (Pluto plus Charon) varies over the 6-day orbital period, with Charon first passing in front of Pluto for a few hours, and then behind. Label these events on the plot.
    b. Now assume that Charon is much less reflective than Pluto. On the plot for part (a), using a different color show qualitatively how the light curve would be different. (*Hint:* Which has a bigger effect: a dark object passing in front of or behind a bright object?)

48. *Comet Temperatures.* Find the "no greenhouse" temperatures for a comet at distances from the Sun of 50,000 AU (in the Oort cloud), 3 AU, and 1 AU (see Mathematical Insight 10.1). Assume that the comet reflects 3% of the incoming sunlight. At which location will the temperature be high enough for water ice to sublime (about 150 K)? How do your results explain comet anatomy? Explain.

49. *Comet Dust Accumulation.* A few hundred tons of comet dust gets added to Earth daily from the millions of meteors that enter our atmosphere. Estimate the time it would take for Earth to get 0.1% heavier at this rate. Is this mass accumulation significant for Earth as a planet? Explain.

Discussion Questions

50. *Rise of the Mammals.* Suppose the impact 65 million years ago had not occurred. How do you think our planet would be different? For example, do you think that mammals still would eventually have come to dominate Earth? Would we be here? Defend your opinions.

51. *Pluto: Planet or Not?* Make your own definition of what qualifies as a planet. Do you consider Pluto a planet? Defend your opinion. If your answer is "yes," determine how many objects in our solar system meet that definition. If your answer is "no," do you believe schoolchildren should be taught that there are only eight planets?

52. *Reducing the Impact Threat.* Based on your own opinion of how the impact threat compares to other threats to our society, how much money and resources do you think should be used to alleviate it? What types of programs would you support? (Examples include programs to search for objects that could strike Earth, to develop defenses against impacts, or to build actual defenses.) Defend your opinions.

Interactive Tutorials

Tutorial Review of Key Concepts
Use the following interactive **Tutorial** at **www.mastering astronomy.com** to review key concepts from this chapter.

Formation of the Solar System Tutorial
Lesson 3 Formation of Planets

Supplementary Tutorial Exercises
Use the interactive **Tutorial Lesson** to explore the following questions.

Formation of the Solar System Tutorial, Lesson 3
1. Where were icy planetesimals able to form in the solar nebula?
2. Pluto is in the outer solar system. Why didn't it capture hydrogen and helium gas like the jovian planets?
3. Suppose the solar wind had not cleared the solar nebula until much later in the history of the solar system. Do you think Pluto or Planet X would have been different? Explain.

Exploring the Sky and Solar System

Of the many activities available on the **Voyager: SkyGazer CD-ROM** accompanying your book, use the following files to observe key phenomena covered in this chapter.

Go to the **File: Basics** folder for the following demonstrations.
1. Orbit of Hale-Bopp
2. Pluto's Orbit

Go to the **File: Demo** folder for the following demonstrations:
1. Hale-Bopp Path
2. Hyakutake at Perihelion
3. Hyakutake Nears Earth
4. Pluto's Orbit

Movies

Check out the following narrated and animated short documentaries available on **www.masteringastronomy.com** for a helpful review of key ideas covered in this chapter.

*History of the
Solar System Movie*

*Orbits in the
Solar System Movie*

Web Projects

1. *Recent Asteroid and Comet Missions.* Chose either the *NEAR* mission to the asteroid Eros or the *Stardust* mission to return comet material to Earth. Learn about the mission goals and status. What did it or will it accomplish? Write a one- to two-page summary of your findings.
2. *Future Asteroid and Comet Missions.* Learn about another proposed space mission to study asteroids or comets. For the mission you choose, write a short report about the mission plans, goals, and prospects for success.
3. *The New Horizons Mission to Pluto.* The *New Horizons* mission to visit Pluto is scheduled for launch in 2006. Find out the current status of the mission. What are its science goals? Summarize your findings in a few paragraphs.
4. *Impact Hazards.* Many groups are searching for near-Earth asteroids that might impact the Earth. They use something called the *Torino Scale* to evaluate the possible danger posed by an asteroid based on how well we know its orbit. What is this scale? What object has reached the highest level on this scale? What were the estimated chances of impact, and when?

Other Planetary Systems

The New Science of Distant Worlds

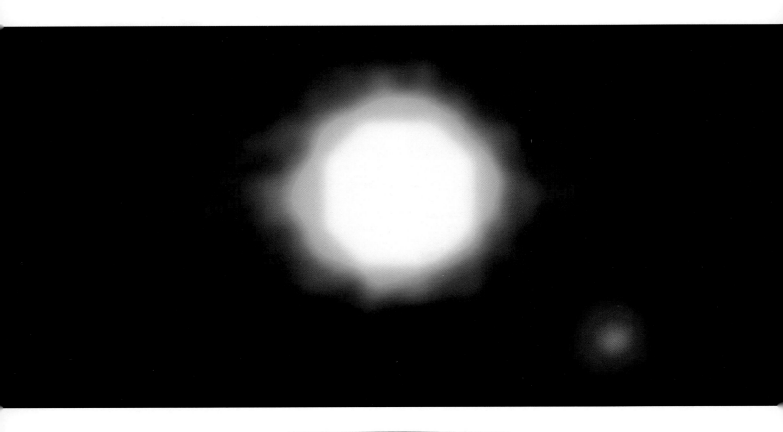

LEARNING GOALS

13.1 Detecting Extrasolar Planets
- Why is it so difficult to detect planets around other stars?
- How do we detect planets around other stars?

13.2 The Nature of Extrasolar Planets
- What have we learned about extrasolar planets?
- How do extrasolar planets compare with those in our solar system?

13.3 The Formation of Other Solar Systems
- Can we explain the surprising orbits of many extrasolar planets?
- Do we need to modify our theory of solar system formation?

13.4 Finding New Worlds
- How will we search for Earth-like planets?

How vast those Orbs must be, and how inconsiderable this Earth, the Theatre upon which all our mighty Designs, all our Navigations, and all our Wars are transacted, is when compared to them. A very fit consideration, and matter of Reflection, for those Kings and Princes who sacrifice the Lives of so many People, only to flatter their Ambition in being Masters of some pitiful corner of this small Spot."

—Christiaan Huygens, c. 1690

A little more than a decade ago, all of planetary science was based on the study of just the planets in our solar system, along with their moons and the many small bodies that orbit the Sun. Then, beginning in 1995, a dramatic change occurred. Within just 10 years, more than a hundred planets were discovered orbiting other stars. Today, the number of known planets continues to rise rapidly and we are beginning to learn about the characteristics of these distant worlds.

The discovery of planets around other stars represents a triumph of modern technology. It also has profound philosophical implications. Knowing that planets are common increases the chance that we might someday find life elsewhere, perhaps even intelligent life. Moreover, by giving us many more worlds to compare to our own, it vastly enhances our ability to learn how planets work and hence may help us better understand our home planet, Earth.

The study of other planetary systems also allows us to test in new settings the theory of our solar system's birth. Recall that the nebular theory, which we discussed in Chapter 8, successfully explains the major characteristics of our solar system. If this theory is correct, it should be able to explain the observed properties of other planetary systems as well. In this chapter, we'll focus all our attention on the exciting new science of other planetary systems. We'll begin by exploring how distant planets have been detected, then discuss the nature of these planets and how they have affected our understanding of planetary science, and finally consider future plans for detecting and studying distant planets, including missions that may soon make it possible to detect worlds similar to our own Earth.

 Detecting Extrasolar Planets Tutorial, Lessons 1–3

13.1 Detecting Extrasolar Planets

The very idea of planets around other stars, or **extrasolar planets** for short, would have shattered the worldviews of many people throughout history. After all, cultures of the western world long regarded Earth as the center of the universe, and nearly all ancient cultures imagined the heavens to be a realm distinct from Earth.

The Copernican revolution, in which we learned that Earth is a planet orbiting the Sun, opened up the possibility that planets might also orbit other stars. Still, until quite recently, no one ever succeeded in finding a planet orbiting another star. Our goals in this section are to understand why the detection of extrasolar planets presents such an extraordinary technological challenge and how astronomers have begun to meet that challenge.

Before we begin, it's worth noting that these discoveries have further complicated the question of precisely how we define a planet [Section 12.3]. Recall that the word *planet* comes from the Greek for "wandering star" and originally referred to the seven naked-eye objects that appear to wander among the constellations in our sky: the Sun, the Moon, Mercury, Venus, Mars, Jupiter, and Saturn. At the end of the Copernican revolution, the word *planet* came to mean a world orbiting the Sun. The discoveries of Uranus in 1781 and Neptune in 1846 thereby gave us a solar system with eight known planets. Meanwhile, scientists learned that comets also orbit the Sun and began to discover other small bodies—asteroids—orbiting the Sun as well. The asteroids were often called "minor planets" to distinguish them from the eight larger planets. Then came Pluto, and more recently Planet X, which beg the question of whether these large Kuiper belt objects should count as planets.

In much the same way that Pluto and Planet X raise the question of a minimum size for a planet, extrasolar planets raise the question of a maximum size. As we will see shortly, many of the known extrasolar planets are considerably more massive than Jupiter. If you recall that Jupiter has a composition very similar to that of the Sun and emits more total energy than it receives from the Sun, you can see why an even larger object might begin to beg the question of when it stops being a "very large planet" and instead should be considered a "very tiny star" (more technically called a *brown dwarf* [Section 16.3]). Moreover, our new understanding of planetary systems tells us that objects the size of planets may sometimes be ejected from their birthplaces into interstellar space; should they still count as planets once they no longer orbit a star?

Astronomers have not yet agreed on the best way to answer these questions, and as a result there is no official definition of the term *planet*. For now, we'll assume that any object less than about a dozen times the size of Jupiter can count as a planet, as long as we find it orbiting a star.

• Why is it so difficult to detect planets around other stars?

We've known for centuries that other stars are distant suns (see Special Topic box, p. 386), making it seem natural to assume that they are orbited by their own planetary systems. The nebular theory of solar system formation, well established by the middle of the twentieth century, made extrasolar planets seem even more likely. As we discussed in Chapter 8, the nebular theory explains our planetary system as a natural consequence of processes that accompanied the birth of our Sun. If the theory is correct, planets should be common throughout the universe. But are they? Prior to 1995, we lacked the evidence needed to draw a conclusion. Even today, the list of known extrasolar planets leaves many important questions unanswered, including how commonly planets form and whether Earth-like planets exist elsewhere.

What makes it so difficult to detect extrasolar planets? You already know the answer, if you think back to the scale of the solar system that we discussed in Chapter 1. Recall that on a 1-to-10-billion scale, the Sun is the size of a grapefruit, Earth is a pinhead orbiting 15 meters away, and Jupiter is a marble orbiting 80 meters away. On the same scale, the distance to the nearest stars besides the Sun is equivalent to the distance across the United States. Thus, seeing an Earth-like planet orbiting the nearest star besides the Sun would be like looking from San Francisco for a pinhead orbiting just 15 meters from a grapefruit in Washington, D.C. Seeing a Jupiter-like planet would be like trying to see a marble about 80 meters from the grapefruit-size star.

The scale alone would make the task quite challenging, but it is further complicated by the fact that a Sun-like star would be a *billion times* as bright as the light reflected from any of its planets. Because even the best telescopes blur the light from stars at least a little, the glare of scattered starlight would overwhelm the small blips of planetary light.

If you think about these challenges, you might easily come to the conclusion that detection would be near impossible. Indeed, as recently as the early 1990s, many astronomers thought that we were still decades away from meeting the challenge. But human ingenuity proved greater than the pessimists had guessed. Thanks to technological advances and clever planet-hunting strategies, we have begun to discover planets orbiting other stars. Although it is too soon to know for sure, it seems ever more likely that billions of planetary systems inhabit our Milky Way Galaxy.

• How do we detect planets around other stars?

The discovery of extrasolar planets has opened a new era in planetary science. The era is generally considered to have

---

SPECIAL TOPIC How Did We Learn That Other Stars Are Suns?

We often take it for granted that stars are other suns—meaning objects that give off enough energy to supply light and heat to orbiting planets—but this fact is not at all obvious from looking at the night sky. After all, the feeble light of stars hardly seems comparable to the majestic light of the Sun. Most ancient observers guessed that stars were much more mundane; typical guesses suggested that they were holes in the celestial sphere or flaming rocks in the sky.

The only way to realize that stars are suns is to know that they are incredibly far away; then, a simple calculation will show that they are actually as bright as or brighter than the Sun [Section 15.1]. The first person to make reasonably accurate estimates of the distances to stars was Christiaan Huygens (1629–1695). By *assuming* that other stars are indeed suns, as many earlier astronomers had guessed, Huygens successfully estimated stellar distances. The late Carl Sagan eloquently described the technique:

Huygens drilled small holes in a brass plate, held the plate up to the Sun and asked himself which hole seemed as bright as he remembered the bright star Sirius to have been the night before. The hole was effectively $\frac{1}{28,000}$ the apparent size of the Sun. So Sirius, he reasoned, must be 28,000 times farther from us than the Sun, or about half a light-year away. It is hard to remember just how bright a star is many hours after you look at it, but Huygens remembered very well. If he had known that Sirius was intrinsically brighter than the Sun, he would have come up with [a much better estimate of] the right answer: Sirius is [8.6] light-years away.*

Huygens could not actually prove that stars are suns, since his method was based on the assumption that they are. However, his results explained a fact known since ancient times: Stellar parallax

Christiaan Huygens
(1629–1695)

is undetectable to the naked eye [Section 2.4]. Recall that the lack of detectable parallax led many Greeks to conclude that Earth must be stationary at the center of the universe, but this lack also has an alternate explanation: Stars are incredibly far away. Even with his original estimate that Sirius was only half a light-year away, Huygens knew that its parallax would have been far too small to observe by naked eye or with the telescopes available at the time. Thus, Huygens "closed the loop" on the ancient mystery of the nature of stars, showing that their appearance and lack of parallax made perfect sense if they were very distant suns. This new knowledge apparently made a great impression on Huygens; as you can see from his quotation on top of p. 385, Huygens became one of the first people to express the idea that the vast size of the cosmos ought to affect our perceptions of ourselves.

---

*From *Cosmos*, by Carl Sagan (Random House, 1980). Sagan demonstrates the technique in the *Cosmos* video series, Episode 7.

begun in 1995, with the discovery of a planet orbiting a star called 51 Pegasi.* The discovery was made by Swiss astronomers Michel Mayor and Didier Queloz, and soon confirmed by a team led by Geoffrey W. Marcy and R. Paul Butler of San Francisco State University. More than 100 other extrasolar planets have been discovered since that time, many of them by these same teams of astronomers.

There are two basic ways in which astronomers can identify extrasolar planets:

1. Directly: Pictures or spectra of the planets themselves constitute direct evidence of their existence.
2. Indirectly: Precise measurements of stellar properties (such as position, brightness, or spectra) may indirectly reveal the effects of orbiting planets.

Direct detection is preferable in principle, because it can tell us far more about the planet's properties. However, current telescopes are not quite up to the challenge of direct detection, at least for planets around ordinary stars. As a result, nearly all extrasolar planets discovered to date have been found by indirect techniques. Let's now explore detection techniques in a little more detail.

Gravitational Tugs To date, nearly all extrasolar planets have been detected by observing the gravitational tugs they exert on the stars they orbit. This type of detection is indirect because we discover the planets by observing the stars they orbit without actually seeing the planets themselves.

Although we usually think of a star as remaining still while planets orbit around it, that is only approximately correct. In reality, all the objects in a star system, including the star itself, orbit the system's *center of mass* [Section 4.4]. To understand how this fact allows us to discover extrasolar planets, imagine the viewpoint of extraterrestrial astronomers observing our solar system from afar.

Let's start by considering only the influence of Jupiter, which exerts a much stronger gravitational tug on the Sun than the rest of the planets combined. The center of mass between the Sun and Jupiter lies just outside the Sun's visible surface (Figure 13.1). Thus, what we usually think of as Jupiter's 12-year orbit around the Sun is really a 12-year orbit around the center of mass; we generally don't notice this fact because the center of mass is so close to the Sun itself. In addition, because the Sun and Jupiter are always on opposite sides of the center of mass (otherwise it wouldn't be a "center"), the Sun must orbit this point with the same 12-year period as Jupiter. The Sun's orbit traces out only a very small circle (or ellipse) with each 12-year period, be-

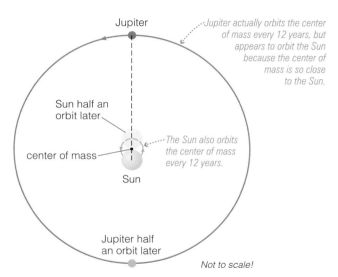

Figure 13.1 **Interactive Figure** This diagram shows how both the Sun and Jupiter actually orbit around their mutual center of mass, which lies very close to the Sun. The diagram is not to scale; the sizes of the Sun and its orbit are exaggerated about 100 times compared to the size shown for Jupiter's orbit.

cause the Sun's average orbital distance is barely larger than its own radius. Nevertheless, with sufficiently precise measurements, extraterrestrial astronomers could detect this orbital movement of the Sun. They could thereby deduce the existence of Jupiter, even without having observed Jupiter itself. They could even determine Jupiter's mass from the Sun's orbital characteristics as it goes around the center of mass. Remember that for any particular separation between the Sun and a planet, a more massive planet pulls the center of mass farther from the Sun's center, thereby giving the Sun a larger orbit and a faster orbital speed around the center of mass.

Now let's add in the effects of Saturn, which exerts the second greatest gravitational tug on the Sun. Saturn takes 29.5 years to orbit the Sun, so by itself it would cause the Sun to orbit their mutual center of mass every 29.5 years. However, because Saturn's influence is secondary to that of Jupiter, this 29.5-year period appears as an added complication on top of the Sun's 12-year orbit around its center of mass with Jupiter. In other words, every 12 years the Sun would return to *nearly* the same orbital position around its center of mass with Jupiter, but the precise point of return would make its own even tinier orbital circle (or ellipse) with Saturn's 29.5-year period. By measuring this motion carefully from afar, an extraterrestrial astronomer could deduce the existence and masses of both Jupiter and Saturn after a few decades of observing.

The other planets also exert gravitational tugs on the Sun, which create further complications to the Sun's orbital motion around the solar system's center of mass (Figure 13.2). These extra effects become increasingly difficult to measure in practice, but in principle they would allow an extraterrestrial astronomer to discover all the planets in our solar system. If we turn this idea around, you'll realize that it

---

*The first actual detection of extrasolar planet-size objects occurred in 1992, when precise timing measurements revealed the existence of three objects with Earth-like masses orbiting a type of "dead" star known as a pulsar [Section 18.2]. Because pulsars are created when stars die in supernova explosions, these "planets" must be either the charred remains of pre-existing planets or, more likely, objects that somehow formed from supernova debris. Either way, they are not planets in the same sense as those that form during star birth; in this chapter, we will focus only on planets around ordinary stars like our Sun.

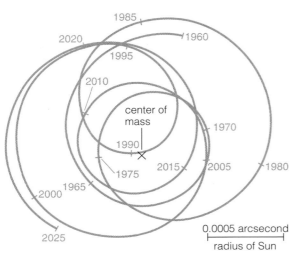

Figure 13.2 This diagram shows the orbital path of the Sun around the center of mass of our solar system as it would appear from a distance of 30 light-years away; the graph shows the motion between 1960 and 2025. Notice that the entire range of motion during this period is only about 0.0015 arcsecond, which is almost 100 times smaller than the angular resolution of the Hubble Space Telescope. Nevertheless, if alien astronomers could measure this motion, they could learn of the existence of planets in our solar system.

means we can search for planets in other star systems by carefully watching for the tiny orbital motion of a star around the center of mass of its star system.

There are two techniques for observing the small orbital motion of a star caused by the gravitational tugs of planets. The first is straightforward: We make very precise measurements of a star's position to see if it is, in fact, moving about in a way that tells us it is being tugged on by planets. This method is usually called the **astrometric technique**, because it requires very precise measurements of stellar positions in the sky (*astrometric* means "measurement of the stars"). The second scheme, known as the **Doppler technique**, detects orbital motion by searching for changing Doppler shifts in a star's spectrum. The star's orbital motion will show up with alternating blueshifts as the star moves toward us in its orbit and redshifts as it moves away. Each technique has advantages and limitations.

The Astrometric Technique The astrometric technique has been used for many decades to identify binary star systems, since two orbiting stars will move periodically around their center of mass. The technique works especially well for binary systems in which the two stars are not too close together, because the stellar motions tend to be larger in those cases. In the case of planet searches, however, the expected stellar motion is much more difficult to detect.

SPECIAL TOPIC   The Names of Extrasolar Planets

The planets in our solar system have familiar names rooted in mythology. Unfortunately, there's not yet a well-accepted scheme for naming extrasolar planets. As a result, we generally refer to extrasolar planets by the star they orbit, such as "the planet orbiting the star named . . . " Worse still, the stars themselves often have confusing or even multiple names, reflecting naming schemes used in star catalogs made by different people at different times in history.

A few hundred of the brightest stars in the sky carry proper names from ancient times. Many of these names are Arabic—such as Betelgeuse, Algol, and Aldebaran—because of the work of the Arabic scholars of the Middle Ages [Section 3.2]. In the early 1600s, German astronomer Johann Bayer developed a system that gave names to many more stars. In Bayer's system, each star gets a name based on its constellation and a Greek letter indicating its ranking in brightness within that constellation. For example, the brightest star in the constellation Andromeda is called Alpha Andromedae, the second brightest is Beta Andromedae, and so on. Of course, because there are only 24 letters in the Greek alphabet, this system works for only the 24 brightest stars in each constellation. About a century later, English astronomer John Flamsteed published a more extensive star catalog in which he used numbers once the Greek letters were exhausted. For example, 51 Pegasi gets its name from Flamsteed's catalog. (Flamsteed's numbers are based on position within a constellation rather than brightness.)

As more powerful telescopes made it possible to discover more and fainter stars, astronomers developed many new star catalogs. The names we use today usually come from one of these catalogs. For example, the star HD209458 gets its name from the Henry Draper (HD) catalog and has the number 209458 in that catalog. You may also see stars with numbers preceded by other catalog names, including Gliese, Ross, and Wolf; these catalogs are also named for the astronomers who compiled them. Moreover, because the same star is often listed in several of these catalogs, a single star can have several different names. Some of the newest planets orbit stars so faint they have not been previously cataloged. They then carry the name of the observing program that discovered them, such as TrES-1 for the first discovery of Trans-Atlantic Exoplanet Survey, or OGLE-TR-132b for the 132nd object scrutinized by the Optical Gravitational Lensing Experiment.

Objects orbiting other stars usually carry the star name plus a letter denoting their order of discovery around that star. The host star itself is designated "a." Thus, for example, HD209458b is the first object discovered to be orbiting star number 209458 in the Henry Draper catalog; Upsilon Andromedae d is the third object discovered to be orbiting the twentieth brightest star (because upsilon is the twentieth letter in the Greek alphabet) in the constellation Andromeda. Unfortunately, this scheme doesn't tell you whether the orbiting object is thought to be a planet or another star. Here's hoping that astronomers soon devise a better naming system for these wonderful new worlds!

As an example, from a distance of 10 light-years, a Jupiter-size planet orbiting 5 AU from a Sun-like star would cause its star to dance slowly over a side-to-side angular distance of only about 0.003 arcsecond—approximately the width of a hair seen from a distance of 5 kilometers. Remarkably, with careful telescope calibration astronomers can now measure displacements this small, and instruments currently under development will be five to 10 times better yet. However, two other complications add to the difficulty of the astrometric technique.

First, the farther away the star, the smaller its side-to-side movement will appear. For example, while Jupiter causes the Sun to move by about 0.003 arcsecond as seen from 10 light-years away, the observed motion is only half as large when seen from 20 light-years away and one-tenth as large when seen from 100 light-years away. Thus, the astrometric technique works best for massive planets around relatively nearby stars.

The second complication arises from the time required to detect a star's motion. It is obviously easier to detect larger movements than smaller ones, and a planet with a larger orbit has a larger effect on its star. To understand why, consider what would happen if Jupiter moved farther from the Sun. Because the center of mass of the solar system is very nearly at the balance point between the Sun and Jupiter, moving Jupiter outward would also cause the center of mass to move farther from the Sun. With the center of mass located farther from the Sun, the Sun's orbit around the center of mass—and hence its side-to-side motion as seen from a distance—would be larger. Thus, the astrometric technique works best not only for massive planets, but also in particular for massive planets that are far from their stars. However, Kepler's third law tells us that a more distant

planet takes longer to complete its orbit, which means the stellar motion it causes also takes longer. For example, while Jupiter causes the Sun to move around the center of mass with a 12-year period, Neptune's effects on the Sun show up with the 165-year period of Neptune's orbit. It would probably take a century or more of patient observation to be confident that stellar motion was occurring in a 165-year cycle.

As a result of these complications, the astrometric technique has been used to detect only one extrasolar planet through early 2005—and the planet had already been discovered by the Doppler technique. Nevertheless, as we'll discuss in Section 13.4, we expect the astrometric technique to be used extensively in the future.

The Doppler Technique   The Doppler technique has been used for the vast majority of planet discoveries to date. An orbiting planet causes its star to move alternately slightly toward and away from us, which makes the star's spectral lines shift alternately toward the blue and toward the red (Figure 13.3a). The 1995 discovery of a planet orbiting 51 Pegasi came when this star was found to be moving with a rhythmic wobble corresponding to an orbital speed of 57 meters per second (Figure 13.3b). The four-day period of the star's motion tells us the orbital period of its planet. Thus, we know that the planet lies so close to the star that its "year" lasts only four of our days, and its surface temperature is probably over 1,000 K (Figure 13.4).

Current techniques can measure a star's velocity to within about 1 meter per second—walking speed—which corresponds to a Doppler wavelength shift of only one part in 300 million. Thus, we can find planets that exert a considerably smaller gravitational tug on their stars than the

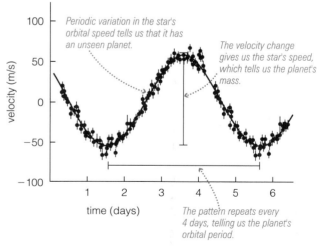

a Doppler shifts allow us to detect the slight motion of a star caused by an orbiting planet.

b A periodic Doppler shift in the spectrum of the star 51 Pegasi shows the presence of a large planet with an orbital period of about 4 days. Dots are actual data points; bars through dots represent measurement uncertainty.

Figure 13.3   Interactive Figure   The Doppler technique for discovering extrasolar planets.

Figure 13.4 🎨 Artist's conception of the planet orbiting 51 Pegasi. As we'll discuss later in the chapter, the planet's mass is probably similar to that of Jupiter and its close orbit (about one-eighth of Mercury's orbital distance from the Sun) should give it a surface temperature above 1,000 K, making it an example of what we call a "hot Jupiter."

planet orbiting 51 Pegasi. Moreover, by carefully analyzing Doppler shift data, we can not only learn of the existence of a planet but also determine its average distance from its star, the shape of its orbit, and even its mass. After all, it is the mass of the planet that causes the star to move around the system's center of mass, so for a given orbital distance, a more massive planet will cause faster stellar motion.

We derive an extrasolar planet's orbital distance using Newton's version of Kepler's third law [Section 4.4]. Recall that for a small object like a planet orbiting a much more massive object like a star, this law expresses a relationship between the star's mass, the planet's orbital period, and the

planet's average distance (semimajor axis). We generally know the masses of the stars around which we have found extrasolar planets (through methods we'll discuss in Chapter 15) and the Doppler data tell us the orbital period, so we can calculate orbital distance.

We can determine orbital shape from the shape of the Doppler data curve. A planet with a perfectly circular orbit travels at a constant speed around its star, so its data curve would be perfectly symmetric. Any asymmetry in the Doppler curve tells us that the planet is moving with varying speed and therefore must have a more eccentric ("stretched out") elliptical orbit. Figure 13.5 shows four examples of Doppler data for extrasolar planets and what we learn in each case.

THINK ABOUT IT

Study the four velocity data curves in Figure 13.5. How would each be different if the planet were: (a) closer to its star? (b) more massive? Would your answers be different if the orbit were eccentric? Do the answers depend on the orientation of the elliptical orbit?

In some cases, existing Doppler data are good enough to tell us whether the star has more than one planet. Remember that if two or more planets exert a noticeable gravitational tug on their star, the Doppler data will show the combined effect of these tugs. In 1999, such analysis was used to infer the existence of three planets around the star Upsilon Andromedae, making this the first bona fide, multiple-planet solar system known beyond our own. By early 2005, more than a dozen other multiple-planet systems had been identified.

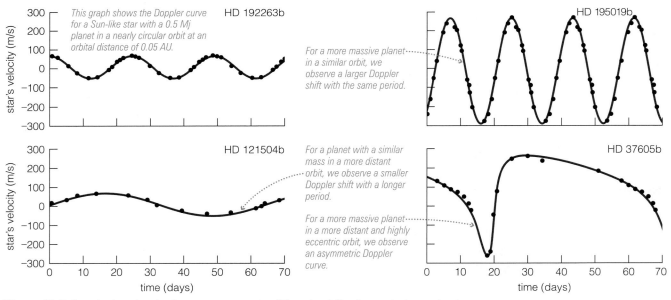

This graph shows the Doppler curve for a Sun-like star with a 0.5 Mj planet in a nearly circular orbit at an orbital distance of 0.05 AU.

HD 192263b

For a more massive planet in a similar orbit, we observe a larger Doppler shift with the same period.

HD 195019b

HD 121504b

For a planet with a similar mass in a more distant orbit, we observe a smaller Doppler shift with a longer period.

For a more massive planet in a more distant and highly eccentric orbit, we observe an asymmetric Doppler curve.

HD 37605b

Figure 13.5 Sample data showing how measurements of Doppler shifts allow us to learn about extrasolar planets for different types of orbits. The curves are based on actual planets, whose data are listed in Appendix E.

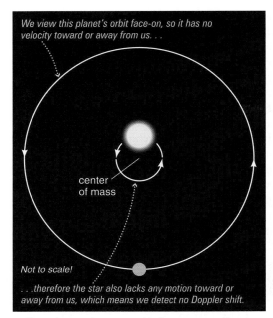

We view this planet's orbit face-on, so it has no velocity toward or away from us. . .

center of mass

Not to scale!

. . .therefore the star also lacks any motion toward or away from us, which means we detect no Doppler shift.

a If we view a planetary orbit face-on, we will not detect any Doppler shift at all.

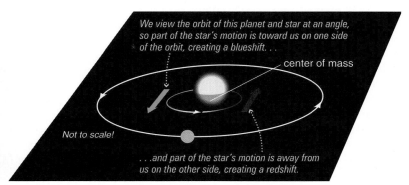

We view the orbit of this planet and star at an angle, so part of the star's motion is toward us on one side of the orbit, creating a blueshift. . .

center of mass

Not to scale!

. . .and part of the star's motion is away from us on the other side, creating a redshift.

b We can detect a Doppler shift only if the planet and star have some part of their orbital velocities directed toward or away from us. The more the orbit is tilted toward edge-on, the greater the shift we'll observe.

Figure 13.6 The amount of Doppler shift we observe in a star's spectrum depends on the orientation of the planetary orbit that causes the shift.

The Doppler technique also tells us about planetary masses, though with an important caveat. Remember that Doppler shifts reveal only the part of a star's motion directed toward or away from us (see Figure 5.23). Thus, for example, a planet whose orbit we view face-on does not cause any Doppler shift in the spectrum of its star, making it impossible to detect the planet with the Doppler technique, let alone to determine its mass (Figure 13.6a). We can observe Doppler shifts in the star's spectrum only if it has a planet orbiting at some angle other than face-on (Figure 13.6b), and the Doppler shift will tell us the star's full orbital velocity only if we are viewing it precisely edge-on. As a result, a planetary mass that we infer from its star's Doppler shift will be accurate only for a planet in an edge-on orbit. In all other cases, the Doppler shift underestimates the true orbital speed of the star and therefore also causes us to underestimate the true mass of the planet. The more the orbit is tilted away from edge-on, the greater the degree of the underestimate. Thus, planetary masses inferred from the Doppler technique alone are actually *minimum* possible masses (or "lower limits") for these planets.

We can determine a precise mass only if we somehow know that we are viewing an edge-on orbit or if we have some way to measure the precise orbital angle. For example, if we also know the planet's side-to-side motion from the astrometric technique, we can combine this information with the toward and away motions from the Doppler technique to determine the orbital angle. Unfortunately, we rarely have such knowledge, which means that nearly all planetary masses found through the Doppler technique are really minimum masses rather than actual masses. Nevertheless, statistical arguments show that in two out of three cases, the planet's true mass will be no more than double this minimum mass, so the minimum mass is still a relatively good estimate.* Of course, this also means that a third of the planets discovered with the Doppler technique have masses that are more than double the minimum mass. About 6% of the time, the planet's actual mass will be more than 10 times the minimum mass, which may raise the question of whether it is truly a planet or a very small star.

The Doppler technique is very powerful, but it has limits. In particular, it is best suited to identifying massive planets that orbit relatively close to their star. This limitation arises because gravity weakens with distance, so a planet of a given size pulls harder on its star—making the star move faster—if it is closer. Moreover, because a close-in planet has a much shorter orbital period than a more distant planet, it takes a lot less time to observe the periodic Doppler shifts caused by close-in planets. For example, while it takes only a few weeks of observation to detect a planet with a 4-day period like the one orbiting 51 Pegasi,

---

*You can understand the statistical argument if you have studied trigonometry: The actual mass is the minimum mass divided by the *cosine* of the orbital inclination (where edge-on is an angle of 0° and face-on is an angle of 90°). Thus, the actual mass is more than double the minimum mass only for cosines smaller than $\frac{1}{2}$, which means for angles between 60° and 90°. By random chance, only about $\frac{1}{3}$ of all stars will have such large orbital angles. Similarly, the actual mass is more than 10 times the minimum mass for cosines smaller than 0.1, which means angles between about 84.3° and 90°; these angles occur randomly about 6% of the time.

it would take 12 years to observe just a single orbital cycle of a planet in an orbit like that of Jupiter around our Sun. The Doppler technique also presents a practical difficulty: Because current technology allows it to be applied to just one star at a time and requires a relatively large telescope, only a relatively small number of stars can be studied with this technique.

The limitations of the Doppler technique explain what may at first seem like surprising facts: Most of the extrasolar planets discovered to date orbit relatively close to their stars, and we have yet to discover any planets at all with Earth-like masses. Both these facts may simply be *selection effects* of the Doppler technique; that is, the technique tends to find (or "select") massive planets in close orbits much more easily than any other type of planet. Planets with masses similar to Earth would have such weak gravitational effects on their stars that we could not use the Doppler technique to find them with current technol-ogy, while planets orbiting far from their stars have such long orbital periods that it might take decades of observations to detect them. Thus, the current lack of evidence for Earth-mass planets or jovian-mass planets in distant orbits does not necessarily mean that such planets are rare.

Transits and Eclipses A third indirect way of detecting distant planets does not require the observation of gravitational tugs at all. Instead, it relies on searching for slight changes in a star's brightness caused by a planet passing in front of it or behind it.

If we were to examine a large sample of stars with planets, a small number of them—typically one in several hundred—will by chance be aligned in such a way that one or more of its planets will pass directly between us and the star once each orbit. The result is a **transit**, in which the planet appears to move across the face of the star. We occasionally witness this effect in our own solar system when Mercury

---

MATHEMATICAL INSIGHT 13.1 **Finding Orbital Distances for Extrasolar Planets**

The Doppler technique directly tells us a planet's orbital period. We can then use this period to determine the planet's orbital distance. If the planet were orbiting a star of exactly the same mass as the Sun, we could find the distance by applying Kepler's third law in its simplest form: $p^2 = a^3$. In fact, many of the planets discovered to date *do* orbit Sun-like stars, so this law gives a good first estimate of orbital distance. But for more precise work, we must use Newton's version of Kepler's third law. As discussed in Mathematical Insight 4.3, this law reads:

$$p^2 = \frac{4\pi^2}{G(M_1 + M_2)} a^3$$

In the case of a planet orbiting a star, p is the planet's orbital period, a is its average orbital distance (semimajor axis), and M_1 and M_2 are the masses of the star and planet, respectively. (G is the gravitational constant; $G = 6.67 \times 10^{-11}$ m^3/(kg $\times$ s^2).) Because a star is so much more massive than a planet, the sum $M_{star} + M_{planet}$ is pretty much just M_{star}; that is, we can neglect the mass of the planet compared to the star. With this approximation, we can rearrange the equation to find the orbital distance a:

$$a \approx \sqrt[3]{\frac{GM_{star}}{4\pi^2} p_{planet}^2}$$

We will discuss how we determine stellar masses in Chapter 15; for now, we will assume the stellar masses are known so that we can calculate orbital distances of the planets.

Example 1: Doppler measurements show that the planet orbiting 51 Pegasi has an orbital period of 4.23 days, and the star's mass is 1.06 times that of our Sun. How far does the planet orbit from its star?

Solution:

 Step 1 Understand: We are given both the planet's orbital period and the star's mass, so we can use Newton's version of Kepler's third law to find the planet's orbital distance. However, to make the units consistent, we need to convert the given stellar mass to kilograms and the given orbital period to seconds; we look up the fact that the Sun's mass is about 2×10^{30} kilograms.

$$M_{star} = 1.06 \times M_{Sun} = 1.06 \times (2 \times 10^{30} \text{ kg}) = 2.12 \times 10^{30} \text{ kg}$$

$$p = 4.23 \text{ day} \times \frac{24 \text{ hr}}{1 \text{ day}} \times \frac{3600 \text{ s}}{1 \text{ hr}} = 3.65 \times 10^5 \text{ s}$$

Step 2 Solve: We now use these values of the period and mass in the above equation to find the orbital distance a:

$$a \approx \sqrt[3]{\frac{GM_{star}}{4\pi^2} p_{planet}^2}$$

$$= \sqrt[3]{\frac{6.67 \times 10^{-11} \frac{\text{m}^3}{\text{kg} \times \text{s}^2} \times 2.12 \times 10^{30} \text{ kg}}{4 \times \pi^2} (3.65 \times 10^5 \text{ s})^2}$$

$$= 7.81 \times 10^9 \text{ m}$$

Step 3 Explain: We've found that the planet orbits its star at a distance of 7.8 billion meters. It's much easier to interpret this number if we state it as 7.8 million kilometers or, better yet, convert it to astronomical units. If you recall or look up the fact that 1 AU is about 150 million kilometers or 1.50×10^{11} meters, then we can convert the orbital distance from meters to AU as follows:

$$a = 7.81 \times 10^9 \text{ m} \times \frac{1 \text{ AU}}{1.50 \times 10^{11} \text{ m}} = 0.052 \text{ AU}$$

We now see that the planet's orbital distance is only 0.052 AU—small even compared to our closest planet, Mercury, which orbits at 0.39 AU. In fact, comparing the planet's 7.8-million-kilometer distance to the size of the star itself (presumably close to the 700,000-kilometer radius of our Sun), we estimate that the planet orbits its star at a distance only a little more than 10 times the star's radius.

---

or Venus crosses in front of the Sun (see Figure S1.5). Of course, other star systems are so far away that we cannot actually see a planetary dot set against the face of the star as we can for Mercury or Venus set against the face of the Sun. Nevertheless, a transiting planet will block a little of its star's light, allowing us not only to detect the planet's existence but also to calculate the planet's size in comparison to that of its star (see Mathematical Insight 13.3). In practice, we usually know stellar sizes (through methods we'll discuss in Chapter 15), which means we can determine planetary sizes by observing transits.

Detecting planets through transits requires many repeated observations, because most stars exhibit intrinsic variations in brightness. To be confident that a dip in bright- ness is caused by an orbiting planet rather than variability in the star itself, we need to see the same dimming repeated multiple times. If the repeated dimming occurs with a regular period, then it is very likely telling us the orbital period of a transiting planet. We can then calculate the planet's orbital distance and mass.

What kind of planet is most likely to cause a transit across its star that we could observe from Earth: (a) a large planet close to its star? (b) a large planet far from its star? (c) a small planet close to its star? or (d) a small planet far from its star? Explain.

MATHEMATICAL INSIGHT 13.2 Finding Masses of Extrasolar Planets

We can find the mass of an extrasolar planet by using the law of conservation of momentum [Section 4.3]. Consider a star with a single planet, each orbiting around their common center of mass (see Figure 13.1) with the same orbital period. The system as a whole has no momentum relative to this center of mass (because it stays in a fixed place between the star and planet), so the planet's momentum must be equal to the star's momentum but in the opposite direction. Thus, if you remember that momentum is mass times velocity, then we can equate the amounts of momentum held by the star and planet:

$$M_{star} v_{star} = M_{planet} v_{planet}$$

where M stands for the mass of the star or planet, and v stands for the velocity relative to the center of mass. We can use this equation to find the planet's mass if we know the star's mass, the star's orbital velocity, and the planet's orbital velocity; we simply solve the equation for M_{planet}:

$$M_{planet} = \frac{M_{star} v_{star}}{v_{planet}}$$

The Doppler technique gives us a direct measurement of the star's velocity toward or away from us and, as discussed earlier, we generally know the star's mass. We can calculate the planet's orbital velocity from its orbital period and orbital distance; we learn the former directly from the Doppler technique and calculate the latter with the method in Mathematical Insight 13.1. Each time the planet completes an orbit, it must travel a distance of $2\pi a$, where a is the planet's average orbital distance. (Notice that we are using the formula for circumference of a circle, even though the orbits are generally elliptical.) The time it takes to complete the orbit is the orbital period p. Thus, the planet's average orbital velocity must be:

$$v_{planet} = \frac{2\pi a_{planet}}{p_{planet}}$$

You should confirm that substituting this expression for the planet's velocity into the equation above to find the planet's mass gives us the following:

$$M_{planet} = \frac{M_{star} v_{star} p_{planet}}{2\pi a_{planet}}$$

Remember that this formula with velocity data from the Doppler technique will give us the *minimum* mass of the planet unless we somehow know that we are viewing the orbital motion edge-on.

Example 1: Estimate the mass of the planet orbiting 51 Pegasi.

Solution:

Step 1 Understand: From Mathematical Insight 13.1, we already know the planet's orbital period ($p = 3.65 \times 10^5$ s), the planet's orbital distance ($a = 7.82 \times 10^9$ m), and the star's mass ($M_{star} = 2.12 \times 10^{30}$ kg). As stated earlier in the text, the star's velocity averages 57 meters per second (see Figure 13.3b). Thus, we can use the formula found above to calculate the planet's mass.

Step 2 Solve: We plug in all the values found earlier into the mass formula:

$$M_{planet} = \frac{M_{star} v_{star} p_{planet}}{2\pi a_{planet}}$$

$$= \frac{(2.12 \times 10^{30} \text{ kg}) \times \left(57 \frac{m}{s}\right) \times (3.65 \times 10^5 \text{ s})}{2\pi \times (7.82 \times 10^9 \text{ m})}$$

$$\approx 9 \times 10^{26} \text{ kg}$$

Step 3 Explain: We have found that the minimum mass of the planet is about 9×10^{26} kilograms. This answer will be more meaningful if we convert it to Jupiter masses. From Appendix E, Jupiter's mass is 1.9×10^{27} kilograms, so the planet's minimum mass is:

$$M_{planet} = 9 \times 10^{26} \text{ kg} \times \frac{1 M_{Jupiter}}{1.9 \times 10^{27} \text{ kg}} = 0.47 M_{Jupiter}$$

The planet orbiting 51 Pegasi has a mass of at least 0.47 Jupiter mass, which is just under half Jupiter's mass. Remembering that most planets will have actual masses within a factor of about 2 of the minimum mass, we see that the planet probably has a mass quite similar to that of Jupiter.

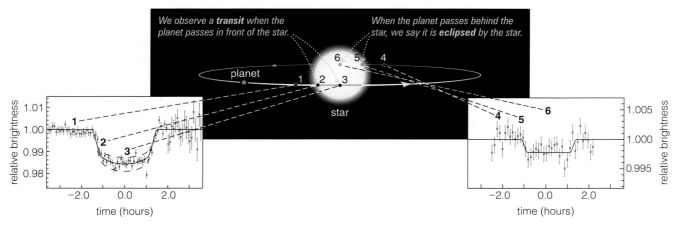

We observe a **transit** when the planet passes in front of the star.

When the planet passes behind the star, we say it is **eclipsed** by the star.

Figure 13.7 Interactive Figure ✦ The artist's conception shows the planet orbiting the star HD209458. The data shows how the star's brightness changes during transits and eclipses, which each occur once with every 3½-day orbit. During a transit, the star's brightness drops for about 2 hours by 1.7%, which tells us how the planet's radius compares to the radius of its star. During an eclipse, the infrared signal drops by 0.25%, which tells us about the planet's thermal emission.

The first success of the transit method came during follow-up studies of a planet that had already been discovered with the Doppler technique. The planet, which orbits a star called HD209458, was already known to complete an orbit every 3½ days. Thus, when astronomers observed the star to undergo dips in brightness every 3½ days (and at just the times that the Doppler measurements said the planet would be moving across our line of sight), they realized they were observing repeated transits by the planet (Figure 13.7). This discovery greatly advanced our understanding of the planet itself. First, because the transits revealed that we view the planet's orbit edge-on, it told us that the mass derived from the Doppler technique was the planet's true mass and not just a minimum mass. Second, the amount of the dip in the star's brightness allowed us to determine the planet's radius and hence its volume. We were thereby able to learn the planet's density (by dividing its mass by its volume). Because the radius and density match the values expected for a jovian planet, we had acquired the first strong evidence that a planet in another star system really does resemble the jovian planets of our own solar system.

Transits can also tell us about the composition of a planet's upper atmosphere or exosphere (Figure 13.8). To see how, consider what happens if a planet like Jupiter passes in front of its star. Although Jupiter is composed of gas, from a distance it looks like a solid disk extending out as far as its cloud tops. During the transit, this "solid" disk blocks all the starlight coming from directly behind it. Now, suppose the Jupiter-like planet also has a low-density upper atmosphere extending above its cloud tops. The gas in this upper atmosphere would absorb starlight at specific wavelengths that depend on its composition. For example, if the planet's upper atmosphere contained sodium gas, the star's spectrum would show stronger sodium absorption lines

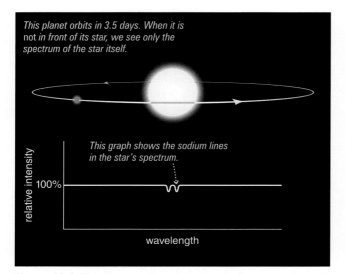

This planet orbits in 3.5 days. When it is not in front of its star, we see only the spectrum of the star itself.

This graph shows the sodium lines in the star's spectrum.

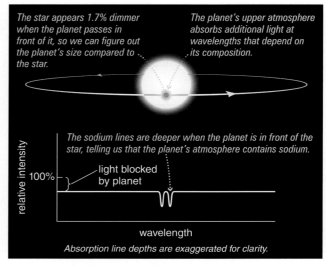

The star appears 1.7% dimmer when the planet passes in front of it, so we can figure out the planet's size compared to the star.

The planet's upper atmosphere absorbs additional light at wavelengths that depend on its composition.

The sodium lines are deeper when the planet is in front of the star, telling us that the planet's atmosphere contains sodium.

light blocked by planet

Absorption line depths are exaggerated for clarity.

Figure 13.8 This diagram shows how transit observations can give us information about the composition of an extrasolar planet's extended upper atmosphere or exosphere.

during the transit than at other times. We'd thereby learn that the planet contains sodium in its upper atmosphere.

Planets that pass in front of their stars in transits can also pass behind their stars, in which case the star blocks the light from the planet. Such an event is called an **eclipse** (see Figure 13.7), because the star blots out the light from the planet in much the same way that the Moon can blot out the light of our Sun during a total solar eclipse [Section 2.3]. Observing an eclipse is much like observing a transit; in both cases, we measure the total light from the star and planet, searching for a small dip in brightness. Because the star is so much brighter than the planet, the dip in brightness is smaller during an eclipse than a transit. To maximize the effect of the eclipse, astronomers therefore observe at infrared wavelengths, because planets emit most of their own radiation in the form of infrared thermal emission while stars emit more at visible wavelengths. Even so, the total drop in light was only 0.25% when the Spitzer Space Telescope observed an eclipse of the planet orbiting the star HD209458. Nonetheless, astronomers were able to determine the total amount of thermal emission from the planet, confirming its expected temperature of over 1,100 K. In fact, much as for planets in our own solar system, the infrared brightness can be used to determine how much starlight the planet reflects or absorbs (see Mathematical Insight 10.1), and the crude infrared spectrum can be used to identify gases in the planet's atmosphere (see Figure 5.14). The first successful observations of this type were announced in 2005, and it is likely that this technique will be improved and applied to more planets by the time you are reading this book.

The method of observing extrasolar planets through transits and eclipses has some unusual strengths and weaknesses. The most obvious weakness is that it can work only for the small fraction of planetary systems whose orbits are oriented edge-on to the Earth. A second weakness is that the method is biased in favor of planets with short orbital periods—and hence with orbits close to their stars—both because these planets transit more frequently and because we must observe repeated transits before we can be confident of a discovery. Counterbalancing these weaknesses is a very important strength: With sufficiently precise measurements of stellar brightness, the transit method ought to be able to detect planets far smaller than is currently possible with the astrometric or Doppler techniques. Indeed, as we'll discuss in Section 13.4, NASA is currently preparing a mission (called *Kepler*) specifically intended to search for transits by Earth-size planets.

Perhaps the most interesting aspect of the transit method is that it can work with both large and small telescopes. With a large telescope, searches for transits can successfully monitor stars at much greater distances than is possible with the astrometric or Doppler techniques. At the opposite extreme, a telescope as small as 4 inches in diameter has been used to discover an extrasolar planet, and it's relatively easy to confirm for yourself some of the transits already detected. What was once considered impossible can now be assigned as homework (see Problem 50 at the end of the chapter).

MATHEMATICAL INSIGHT **13.3** **Finding Sizes of Extrasolar Planets**

While finding the masses of extrasolar planets was a breakthrough suggesting that we were detecting jovian planets, it wasn't until astronomers measured extrasolar planet sizes that we gained confidence in their jovian nature. Transits provide the needed size information in a simple geometric fashion.

Like other extrasolar planet measurements, the technique relies on a good understanding of the host star. In particular, we need to know the physical size of the star's disk even though we cannot resolve it with our telescopes. If the star's light can be considered uniform over its disk, then the fractional light drop that occurs during a transit will be:

$$\frac{\text{fraction of}}{\text{light blocked}} \approx \frac{\text{area of planet's disk}}{\text{area of star's disk}} = \frac{\pi r_{\text{planet}}^2}{\pi r_{\text{star}}^2} = \frac{r_{\text{planet}}^2}{r_{\text{star}}^2}$$

Note that we used the formula for the area of a circle (πr^2) because from a distance the planet and star should look like circular disks. The equation should also make sense if you look at Figure 13.7a.

We can now rearrange the equation to find the planet's radius; you should confirm that solving for r_{planet} gives us the following formula:

$$r_{\text{planet}} \approx r_{\text{star}} \times \sqrt{\text{fraction of light blocked}}$$

Example 1: What is the radius of the planet orbiting the star HD209458? Needed data: The star's radius is 1.15 times that of our Sun, or about 800,000 kilometers, and the planet blocks 1.7% of the starlight during transit.

Solution:

Step 1 Understand: As shown in Figure 13.7b, transits of this planet cause an approximately 1.7% dip in its star's brightness. Thus, the fraction of the star's light that is blocked is 1.7% = 0.017. We now have all the information we need to use the above equation to find the planet's radius.

Step 2 Solve: We plug the numbers into the equation above to find:

$$r_{\text{planet}} \approx r_{\text{star}} \times \sqrt{\text{fraction of light blocked}}$$
$$= 800,000 \text{ km} \times \sqrt{0.017}$$
$$\approx 100,000 \text{ km}$$

Step 3 Explain: The planet's radius is close to 100,000 kilometers. From Appendix E, Jupiter's radius is about 71,500 kilometers. Thus, the planet's radius is about $\frac{100,000}{71,500} \approx 1.4$ times as large as Jupiter's. In other words, the planet is about 40% larger than Jupiter in radius (or diameter).

Direct Detection The indirect planet-hunting techniques we have discussed so far have started a revolution in planetary science by demonstrating that our solar system is just one of many planetary systems. However, these indirect techniques tell us relatively little about the planets themselves, aside from their orbital properties and sometimes their masses and radii. To learn more about their nature, we need to be able to observe the planets themselves so that we can look for details in images of their surfaces and atmospheres, or study their spectra to learn about their composition and properties.

Unfortunately, it is still not possible to obtain images or spectra of planets around Sun-like stars, primarily because of the incredible glare of the stars themselves. However, in at least one case as of early 2005, astronomers have observed a possible planet around a much fainter star. The "star" is actually a type of object known as a *brown dwarf* [Section 16.3]; brown dwarfs have masses between those of very large planets and those of very small stars and they emit very little visible light. The "candidate planet" has been photographed in the infrared by both the European Southern Observatory's Very Large Telescope (see the photo that opens this chapter) and the Hubble Space Telescope (Figure 13.9). Notice that the infrared emission from the brown dwarf itself was subtracted from the Hubble photo, allowing us to see the candidate planet more clearly. The infrared data tell us that the object is relatively cool in temperature, making it seem much more like a bona fide planet than a brown dwarf or cool star. The Very Large Telescope used adaptive optics to "unblur" effects of Earth's atmosphere [Section 6.3], thereby obtaining picture quality comparable to that from Hubble. Moreover, the Very Large Telescope was able to obtain an infrared spectrum of the candidate planet. The spectrum showed evidence of water molecules, which is consistent with what we would expect for a jovian planet. (Remember

that water is one of the most common molecules in jovian planet atmospheres, and forms one of their cloud layers [Section 11.1].)

Other Planet-Hunting Strategies The astonishing success of recent efforts to find extrasolar planets has led astronomers to think of many other possible ways of enhancing the search. One example is a project known as the Optical Gravitational Lensing Experiment (OGLE), a large survey of thousands of distant stars. Although it was not originally designed with planet detection in mind, OGLE has already detected several planets by observing transits. It has also succeeded in detecting at least one by the method in its name—*gravitational lensing*, an effect predicted by Einstein's general theory of relativity that occurs when one object's gravity bends or brightens the light of a more distant object [Sections S3.4, 22.2]. Another strategy looks not for planets themselves or their influence on stars, but rather for their effects on the vast disks of dust that surround many stars. If a planet is present within a dust disk, it can exert small gravitational tugs on the dust particles that will show up as gaps, waves, or ripples in the disk. Such is the case for the star Beta Pictoris, which is surrounded by a dust disk with ripples indicating that it also has one or more planets. Other astronomers are searching for the thermal emission from the impacts of accreting planetesimals, and even for the kinds of emission known to come from the magnetospheres of our jovian planets. As we learn more about extrasolar planets, new search methods are sure to arise.

 Detecting Extrasolar Planets Tutorial, Lessons 1–3

13.2 The Nature of Extrasolar Planets

The mere existence of planets around other stars changes our perception of our place in the universe, because it shows that our planetary system is not unique. In addition, these planets provide us with two new ways to expand our understanding of planetary science.

First, the nature of the individual planets themselves gives us an opportunity to learn more about the range of possible planets. For example, we know that the planets of our own system come in two basic types: terrestrial and jovian. Learning about the nature of extrasolar planets may tell us whether these are the only two categories of planet or whether there are others that we do not see in our own solar system.

Second, studying the arrangements of other planetary systems can tell us whether the layout of our solar system is common or rare, and thereby shed light on whether the nebular theory really does explain the origin of our solar system as neatly as we have presumed. For example, in our solar system the terrestrial planets are all located close to the Sun and the jovian planets are much farther away—an observation that the nebular theory successfully explains.

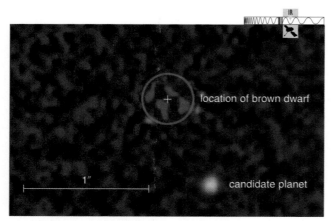

Figure 13.9 This infrared image from the Hubble Space Telescope shows a "candidate planet" in orbit around the brown dwarf known as 2M1207. The light from the brown dwarf (located within the dotted circle) was subtracted out. The red, green, and blue colors represent different wavelengths of infrared light. The candidate planet is brightest at the longest wavelengths, indicating that it is relatively cool in temperature—just as we would expect for a jovian planet.

Can the nebular theory explain the layouts of other solar systems equally well?

In this section, we'll focus our attention on the nature of the extrasolar planets themselves, discussing what we have learned about their orbits, masses, sizes, and compositions. We'll then be prepared to turn our attention in the next section to the questions of how other planetary systems compare to our own and how any differences may have arisen.

• What have we learned about extrasolar planets?

By mid-2005, we already knew of more than 160 extrasolar planets. This number is large enough that we can begin to search for patterns, trends, and groupings that might give us insight into how these planets compare to the planets of our own solar system and how they formed. Before we go into detail, let's summarize the planetary properties that we can learn with current detection techniques:

- *Orbital period:* All three indirect techniques that we discussed (astrometric, Doppler, and transits) tell us the orbital period of detected planets.
- *Orbital distance:* Once we know orbital period and the mass of the star, we can calculate orbital distance by using Newton's version of Kepler's third law.
- *Orbital shape:* We need data spanning an entire orbit to determine orbital shape—that is, whether the orbit is nearly circular or a more eccentric ellipse. Both the astrometric and Doppler techniques can provide the needed data. Transits give us data only for the small part of the orbit during which the planet passes in front of its star and hence do not tell us orbital shape.
- *Mass:* We can determine an extrasolar planet's mass from its orbital period, the mass of its star, and the speed at which it makes its star orbit their mutual center of mass. In principle, we can learn a star's full orbital speed with the astrometric technique, while the Doppler technique tells us only the part of the speed directed toward or away from us. Thus, the Doppler technique tells us a minimum mass for the planet, and gives us a precise mass only if we also know the angle of the orbit to our line of sight. We cannot learn mass from transits alone.
- *Size (radius):* We can learn a planet's size only by observing transits. The dip in the star's brightness during a transit tells us the fraction of its light blocked by the planet, which allows us to calculate the planet's radius.
- *Density:* We can calculate a planet's average density from its size and mass. Because we get size only from transits, we can determine density only for planets that produce transits and for which we also have mass data from the astrometric or Doppler techniques.
- *Composition:* We learn composition from spectra. Transits can provide limited information about the composition of a planet's upper atmosphere if the star shows absorption during the transit that is not present at other times. Eclipses can also provide limited spectral

information. More detailed information about composition requires spectra from direct detections.

We are now ready to see what we've learned about the extrasolar planets discovered to date.

Orbits Much as Johannes Kepler first appreciated the true layout of our own solar system [Section 3.3], we can now step back and see the layout of many other solar systems. Figure 13.10a shows the orbits of the first 146 known extrasolar planets all superimposed on each other, as if they were all orbiting a single star; the dots indicate the minimum masses of the planets found through the Doppler technique.

Despite the crowding of the orbits when viewed this way, at least two important facts should jump out at you. First, notice that only a handful of these planets have orbits that take them beyond about 5 AU, which is Jupiter's distance from our Sun. Most of them orbit very close to their host star. Second, notice that many of the orbits are clearly elliptical, rather than nearly circular like the orbits of planets in our own solar system. These facts are even easier to see if we display the same information on a graph (Figure 13.10b). Look first at the green dots representing planets in our own solar system; notice that they are located at the distances you should expect and all but Mercury have very small eccentricity, meaning nearly circular orbits. Now look at the red dots representing extrasolar planets. Quite a few of these planets orbit their stars more closely than Mercury orbits the Sun, and none are located as far from their stars as the jovian planets of our solar system. Many also have large orbital eccentricities, telling us that their elliptical orbits have very stretched-out shapes. As we'll see shortly, both facts give us important clues about the nature of these extrasolar planets.

THINK ABOUT IT

Should we be surprised that we haven't found many planets orbiting as far from their stars as Saturn, Uranus, and Neptune orbit the Sun? Why or why not?

At least 15 stars have so far been found to contain two or more planets, and one system has four known planets (Figure 13.11). This is not surprising, since our own solar system and our understanding of planet formation suggest that any star with planets is likely to have multiple planets. We will probably find many more multiple-planet systems as observations improve. However, one fact about these other planetary systems is surprising: Already, we've found at least seven systems that seem to have planets in orbital resonances [Section 11.2] with each other. For example, three systems have one planet that orbits in exactly half the time as another planet. In our own solar system we've seen the tremendous importance of orbital resonances in sculpting planetary rings, stirring up the asteroid belt, and even affecting the orbits of Jupiter's moons. Orbital resonances may also have profound influences for extrasolar planets, as we'll discuss in Section 13.4.

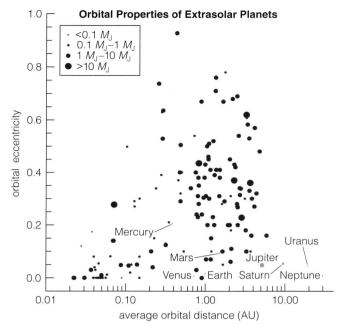

a This diagram shows all the orbits superimposed on each other, as if all the planets orbited a single star. The sizes of the dots indicate the minimum masses of the planets; the dots are located at the aphelion (farthest) point for each orbit.

b The data from (a) are shown here as a graph. Each red dot represents an extrasolar planet, and the green dots represent the planets of our own solar system. Planets that orbit closer to their stars are farther to the left on this diagram, and the planets with more circular orbits are lower down on the diagram. Notice that many extrasolar planets orbit quite close to their stars and most have larger orbital eccentricities than the planets of our solar system.

Figure 13.10 Orbital properties of the first 146 known extrasolar planets.

Masses Look again at Figure 13.10. The sizes of the dots indicate the approximate minimum masses of these planets; they are minimums because nearly all have been found with the Doppler technique. The mass data are easy to see if we display them as a bar chart (Figure 13.12). We see clearly that the planets we've detected in other star systems are generally quite massive. Most are more massive than Jupiter, and even the least massive ones are almost as massive as Uranus and Neptune. The least massive extrasolar planet found through mid-2005 has a mass about 7.5 times that of Earth. None are anywhere near as low in mass as Earth, which has a mass of about 0.003 Jupiter mass. Thus, if we go by mass alone, it seems that all the known extrasolar planets must be jovian rather than terrestrial in nature.

Should we be surprised by the lack of planets with masses like the terrestrial worlds? Not yet. Remember that nearly all these planets have been found indirectly by looking for the gravitational tugs they exert on their stars. Massive planets exert much greater gravitational tugs and are therefore much easier to detect. Thus, as we discussed earlier, the abundance of very massive planets among these early discoveries is probably a "selection effect" arising because current planet-finding techniques detect massive planets far more easily than lower mass planets.

Sizes and Densities The masses of the known extrasolar planets suggest they are jovian in nature, but mass alone cannot

rule out the possibility of "supersize" terrestrial planets—that is, very massive planets made of metal or rock. To check whether the planets are jovian in nature, we also need to know their sizes, from which we can calculate their densities. If their sizes and densities are consistent with those of the jovian planets in our solar system, then we'd have good reason to think that they really are jovian in nature.

Unfortunately, as we've already discussed, the vast majority of known extrasolar planets have been detected by the Doppler technique, which gives us reasonable mass estimates but no information about size. However, in the relatively few cases for which we also have size data from transits, the sizes and densities do indeed seem consistent with what we expect for jovian planets.

In fact, the main surprise in the size and density data to date is that most extrasolar planets actually seem to be larger and less dense than the jovian planets of our own solar system. For example, the first planet for which we obtained size data (the planet orbiting the star HD209458) is apparently lower in mass than Jupiter (0.63 Jupiter mass) but larger than Jupiter in radius (1.43 Jupiter radii). Its density is 0.3 gram per cubic centimeter, or about one-fourth of Jupiter's density. Could that mean this planet is made of different, lower-density ingredients than Jupiter? That hardly seems possible, since Jupiter is mostly made of hydrogen and helium gas—the lowest-density planetary building materials. Instead, there's probably a much simpler explanation. This

Figure 13.11 This diagram shows the orbital distances and approximate masses of the planets in the first 15 multiple-planet systems discovered. The four highlighted systems are the ones with the best data to show planets in orbital resonances. For example, the "5:2" for 47 Uma indicates that the inner of the two indicated planets completes exactly five orbits while the outer planet completes two orbits. When reading orbital periods from the graph, be sure to notice that the axis is exponential, so that each tick mark represents a period 10 times longer than the previous one.

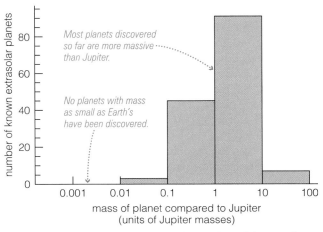

Figure 13.12 This bar chart shows the number of planets of different masses among the first 146 extrasolar planets discovered. The masses are approximate and are based on minimum masses found with the Doppler technique. Notice that the axis uses an exponential scale so that the wide range of masses can all fit on the graph.

planet orbits very close to its star: Its average orbital distance is only 0.047 AU, which is only 12% of Mercury's distance from the Sun. Thus, we expect this planet to be very hot, both because it receives so much heat from its star and because the warm environment slows the escape of the internal heat that jovian planets acquire during formation [Section 11.1]. Calculations show that its temperature is probably more than 1,000 K at its cloud tops, and it should be much warmer than Jupiter throughout its interior. This high temperature puffs up its hydrogen and helium gas, explaining why it is larger in size and lower in density than Jupiter. We now have size and density information for at least eight extrasolar planets (see Appendix E.4), confirming that most close-in worlds are larger and lower in density than Jupiter. One exception is a planet discovered in mid-2005 around the star HD149026. This planet appears to have a Saturn-like mass but a Neptune-like density of about 1.5 g/cm³ (about twice the density of Saturn). Astronomers deduce that it must contain a proportion of dense materials—rock, metal, and hydrogen compounds—more like that of Neptune than Saturn, despite its Saturn-like mass.

Compositions We have even less direct information about the compositions of extrasolar planets. Nevertheless, the available data also support the idea that these planets are jovian in nature.

Our first data about an extrasolar planet's composition came from the same transiting planet for which we first measured size. During the transits of HD209458, the Hubble Space Telescope detected additional absorption by hydrogen and sodium that is not present at other times (see Figure 13.8). From the amount of extra absorption, astron-

omers concluded that the planet has an extended upper atmosphere containing abundant hydrogen—just as we should expect for a jovian planet—and a trace of sodium. The next case of composition information came from the "candidate planet" shown in Figure 13.9 and the photo that opens this chapter. As we discussed earlier, a spectrum of this planet revealed the presence of atmospheric water, consistent with what we find in the jovian planets of our own solar system.

• How do extrasolar planets compare with those in our solar system?

Thanks to our detailed understanding of our own jovian planets (Chapter 11), we can take the limited information at hand and make some educated guesses about the nature and behavior of extrasolar planets. In particular, do planets in other star systems fit the same terrestrial and jovian categories as the planets in our solar system? The answer is a tentative "yes," with the recognition that technological limitations mean we have not yet identified even a single terrestrial planet in another star system.

For the known extrasolar planets, we have no reason to doubt that they are similar in nature to the jovian planets of our solar system. Although the only evidence for this claim comes from planetary mass in the vast majority of cases, the few cases for which we also have data on size, density, or composition lend strong support. After all, these few cases are essentially a random sample of the full set of known extrasolar planets, so the fact that these cases all appear to be jovian in nature makes it likely that most or all of the known extrasolar planets are jovian.

We have found one surprising characteristic among these extrasolar planets: All the jovian planets of our solar system orbit far from the Sun on nearly circular orbits, while many of the extrasolar planets orbit quite close to their stars or have highly elliptical orbits. We'll discuss how these planets might have gotten into these surprising orbits in the next section, but it does not change the fact that they are probably jovian planets made mostly of hydrogen and helium gas. Because their close orbits make them much higher in temperature than Jupiter, these planets have been nicknamed "hot Jupiters" (see Figure 13.4).

Assuming we are correct about the nature of the hot Jupiters, we can use models to predict a lot about their outward appearance and characteristics, even though we've never actually photographed one. We simply need to ask how Jupiter would be different if it had somehow migrated into an orbit very close to the Sun. The models show that, if Jupiter orbited just 0.05 AU from the Sun (1% of its actual distance), the extra heat would make it about 50% larger in radius—consistent with what we found for the transiting planet around HD209458.

What would the hot Jupiter look like? It would probably have clouds much like the real Jupiter but of a different type. Recall that Jupiter has multiple cloud layers made from droplets or ice flakes of compounds such as ammonia, methane, and water. The temperature on the hot Jupiter would be far too high for these gases to condense. However, other ingredients may be able to form clouds at high altitudes, including some materials that we usually think of as solids. The models suggest that a hot Jupiter with a temperature above about 1,000 K would have clouds of "rock dust" containing common minerals (especially the mineral enstatite, or $MgSiO_3$).

The hot Jupiter would probably also have a striped appearance much like that of the real Jupiter. In fact, we'd probably be able to see the stripes even on the night side, because the planet's high temperatures would make it glow at visible wavelengths. The stripes should be present because the atmospheric circulation of a hot Jupiter should still be driven by the same basic principles that apply to terrestrial and jovian planets in our solar system [Sections 10.2, 11.1]. Because the hot Jupiter is so close to its star, intense starlight will warm one side of the planet. If the planet rotates fast enough to generate a strong Coriolis effect, it should create patterns of planet-circling winds much like those on Jupiter. Can we expect a hot Jupiter to rotate rapidly? In fact, we probably already know the rotation rates of all the hot Jupiters. They are so close to their stars that strong tidal forces have almost certainly caused them to rotate synchronously with their orbits, just as the Moon rotates synchronously with its orbit around Earth [Section 4.5]. That is, their rotation periods are the same as their orbital periods, and they always show the same face toward their stars. Because the orbital periods of many of these hot Jupiters are just a few days, we expect their rotation periods also to be a few days. Although these periods mean slower rotation than that of jovian planets in our solar system, they should still be fast enough to create strong Coriolis effects and planet-circling winds on such massive planets. Figure 13.13 summarizes the similarities and differences expected between the real Jupiter and "hot Jupiters."

Jupiter

Composed primarily of hydrogen and helium
5 AU from the Sun
Orbit takes 12 Earth years
Cloud top temperatures ≈ 130 K
Clouds of various hydrogen compounds
Radius = 1 Jupiter radius
Mass = 1 Jupiter mass
Average density = 1.33 g/cm^3
Moons, rings, magnetosphere

"Hot Jupiters" orbiting other stars

Composed primarily of hydrogen and helium
As close as 0.03 AU to their stars
Orbit as short as 1.2 Earth days
Cloud top temperatures up to 1300 K
Clouds of "rock dust"
Radius up to 1.3 Jupiter radii
Mass from 0.2 to 2 Jupiter masses
Average density as low as 0.3 g/cm^3
Moons, rings, magnetospheres: unknown

Figure 13.13 A summary of the expected similarities and differences between the real Jupiter and extrasolar "hot Jupiters" orbiting Sun-like stars.

13.3 The Formation of Other Solar Systems

The discovery of extrasolar planets presents us with an opportunity to test our theory of solar system formation. Can our existing theory explain other planetary systems, or will we have to go back to the drawing board? The discoveries have already presented at least one significant challenge: The surprising orbits of the "hot Jupiters" and many other extrasolar planets.

Recall that, according to the nebular theory, planets were born in a spinning disk of material surrounding our Sun [Section 8.3]. The spinning disk was made mostly of hydrogen and helium, but a small fraction of the material consisted of hydrogen compounds (water, ammonia, methane) and an even smaller fraction of metal and rock. Terrestrial planets formed in the warm inner regions of the disk, where only metal and rock could condense into solid form. Jovian planets formed in the cold outer regions of the disk, where temperatures were low enough for hydrogen compounds to condense into ices. Because hydrogen compounds were several times as abundant as rock and metal, the icy planetesimals in the outer solar system grew to much larger sizes than the rocky planetesimals of the inner solar system. They grew large enough so that their gravity began to draw in hydrogen and helium gas from the surrounding nebula, which explains how they ultimately became jovian planets made mostly of hydrogen, helium, and hydrogen compounds. We would expect that planets should form more easily in nebulas with a higher proportion of rock and ice, and in fact more planets have been found around stars richer in the elements that make these ingredients. The nebular theory explains the nearly circular orbits of the planets in our solar system as a direct consequence of the orderly, circular motion of the spinning disk in which the planets formed.

Thus, the nebular theory clearly predicts that jovian planets should form only in the cold outer regions of star systems and that they should be born with nearly circular orbits. Known extrasolar planets, many with close-in or highly elliptical orbits, present a direct challenge to these ideas.

• Can we explain the surprising orbits of many extrasolar planets?

The nature of science demands that we question the validity of a theory whenever it is challenged by any observation or experiment [Section 3.4]. If the theory cannot explain the new observations, then it must be revised or discarded. The surprising orbits of many of the extrasolar planets have indeed caused scientists to reexamine the nebular theory of solar system formation.

The questioning began almost immediately upon the discovery of the first extrasolar planets. The close-in orbits of these massive planets made scientists wonder whether something might be fundamentally wrong with the nebular theory. For example, could it be possible that jovian planets *can* form close to their stars? Astronomers addressed this question by studying many possible models of planet formation and re-examining in detail the entire basis of the nebular theory. Several years of such re-examination did not turn up any good reasons to discard the basic theory. While it's still possible that a major flaw has gone undetected, it seems more likely that the basic outline of nebular theory is correct. Scientists therefore suspect that the extrasolar jovian planets were indeed born with circular orbits far from their stars, and that those that now have close-in or eccentric orbits underwent some sort of "planetary migration" or suffered gravitational interactions with other massive objects.

Planetary Migration   If the hot Jupiters formed in the outer regions of their star systems and then migrated inward, how did these planetary migrations occur? You might think that drag within the solar nebula could cause planets to migrate, much as atmospheric drag can cause satellites in low-Earth orbit to lose orbital energy and eventually plunge into the atmosphere; however, calculations show this drag effect to be negligible. A more likely scenario is that waves propagating through a gaseous disk lead to migration (Figure 13.14). A planet's gravity can nudge particles in the disk and thereby create waves that propagate through the disk, causing material to bunch up as the waves pass by. The "bunched up" matter in the wave peaks, in turn, exerts an extra gravitational nudge back on the planet. This gravitational nudge can cause the planet to migrate inward.

Computer models confirm that waves in the nebula can cause young planets to spiral slowly toward their star. In our own solar system, this migration is not thought to have played a significant role because the solar wind cleared out the gas before it could have much effect. But planets may form earlier in at least some other solar systems, allowing time for jovian planets to migrate substantially inward. In

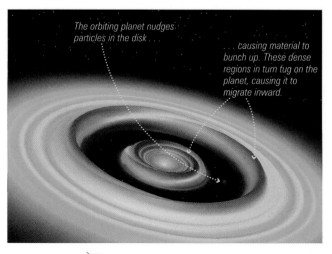

The orbiting planet nudges particles in the disk . . .

. . . causing material to bunch up. These dense regions in turn tug on the planet, causing it to migrate inward.

Figure 13.14 This figure shows a simulation of waves created by a planet embedded in a dusty disk of material surrounding its star.

some cases, the planets may form so early that they end up spiraling into their stars. Indeed, astronomers have observed some stars that have an unusual assortment of elements in their outer layers, suggesting that they may have swallowed planets

The migration hypothesis seems to work well in computer models, but as this book went to press it faces a new challenge: the discovery of a hot Jupiter in a triple-star system. The radiation from the adjacent stars should have cleared the gas from the disk quite early, in which case it is difficult to see how migration could have occurred. It's too soon to tell if this discovery requires rethinking the migration hypothesis overall, or if this is just an unusual case. One thing is certain: We are learning far more about solar system formation from the wealth of new worlds than we could from our solar system alone.

Encounters and Resonances A possible explanation for eccentric orbits is close gravitational encounters [Section 4.5] between young jovian planets in the outer regions of a disk. Close gravitational encounters between two massive planets can send one planet out of the star system entirely while the other is flung inward into a highly elliptical orbit. Alternatively, a jovian planet could migrate inward as a result of multiple close encounters with much smaller planetesimals. The jovian planets in our solar system probably migrated by this latter mechanism. As comets were ejected into the Oort cloud [Section 12.2], the jovian planets must have migrated inward, as a consequence of the law of conservation of energy.

Gravitational interactions can also affect orbits through resonances. Recall that Jupiter's moons Io, Europa, and Ganymede share orbital resonances that cause their orbits to be more elliptical than they would be otherwise (see Figure 11.20b). Models show that similar resonances between jovian planets could also cause their orbits to become more eccentric, possibly explaining the eccentricities found even for some extrasolar planets orbiting at Jupiter-like distances. In some cases, the resonance could even lead to planetary migration or to a planet's being ejected from its system altogether. In other cases, planetary migration might force two planets into a resonance that they did not have originally. The fact that we've already discovered orbital resonances in several multiple-planet systems (see Figure 13.11) lends support to the idea that resonances could play an important role in other solar systems.

• Do we need to modify our theory of solar system formation?

We began this section by asking whether the nebular theory of solar system formation can still hold up in light of our discoveries of planets around other stars. As we've seen, it probably can—provided that we allow for planets undergoing orbital changes after their births.

The bottom line is that discoveries of extrasolar planets have shown us that the nebular theory was incomplete. It explained the formation of planets and the simple layout of a solar system such as ours. However, it needs new features—such as planetary migration and gravitational encounters or resonances—to explain the differing layouts of other solar systems. A much wider range of solar system arrangements now seems possible than we had guessed before the discovery of extrasolar planets.

Look back at the discussion of the nature of science in Chapter 3, especially the definition of a scientific theory. Does our theory of solar system formation qualify as a scientific theory, even though we have recently learned that it needs modification? Does this mean the theory was "wrong" before the modifications were made? Explain.

Given the fact that the majority of planetary systems discovered to date have planets that seem to have undergone migration or other orbital changes, and that we have not yet found any other planetary system with a layout like our own, it's natural to wonder whether our own solar system is the unusual one. Could it be that the neat layout of our solar system is the result of some sort of extreme cosmic luck, and that most or all other planetary systems undergo far greater changes after their planets are born? First impressions from the current data might seem to support this idea, but a more careful look shows that we cannot yet draw such a conclusion.

Among the thousands of Sun-like stars that astronomers have so far examined in search of extrasolar planets, only about 1 in 10 show evidence of planets around them. While this could mean that planetary systems are relatively rare, a more likely hypothesis is that planets are present but more difficult to detect in the other 9 in 10 systems. Remembering that current technology makes it easier to find massive planets in close-in orbits, this latter hypothesis would suggest that in most solar systems, jovian planets orbit far from their stars—just as in our own solar system. Thus, current data are fully consistent with the hypothesis that most solar systems are more like our own than they are like the ones discovered to date. In essence, our early discoveries of extrasolar planets may be much like a glimpse of animals in a rain forest: The jungle appears full of brightly colored parrots and frogs, but far more animals fail to catch our eye. As time goes on and technology improves, we may begin to find more systems like our own.

If solar systems like our own are indeed common, then the real challenge is to explain precisely how and under what conditions a system may end up like the "unusual" ones discovered to date. This challenge has led planetary scientists to look more closely at the question of exactly how the planets in our own solar system would have interacted with one another when the solar system was young. These studies are only in preliminary stages at present, but they are already causing some scientists to wonder whether migration and gravitational interactions were more important in our solar system than previously thought. It may be that all planetary systems experience these processes to some extent, but that in most cases they stop before jovian planets end up with close-in or eccentric orbits.

All this discussion might seem fairly detailed, but in fact it has profound implications for the way we view our place in the universe. If solar systems like ours are indeed common, then it seems reasonable to imagine that Earth-like planets—and perhaps life and civilizations—might also be common. But if our solar system is a rarity or even unique, then Earth might be the lone inhabited planet in our galaxy or even the universe. We'll discuss this important issue in more depth in Chapter 24, when we turn our attention to life in the universe. In the end, however, the only way to resolve the questions is to gather more data. Thus, we turn now to the future of planetary science, and the question of how we will gather the data needed to find out whether planets like ours are rare or common.

13.4 Finding New Worlds

We have entered a new era in planetary science, one in which our understanding of planetary processes can be based on far more planets than just those of our own solar system. Although our current knowledge of extrasolar planets and their planetary systems is still quite limited, ingenious new observing techniques, dedicated observatories, and ambitious space telescopes should broaden our understanding dramatically in the coming years and decades.

In this section we'll focus on the more dramatic improvements that space missions and new ground-based methods will provide. These techniques will not only permit the discovery of Earth-like planets (if they exist), but will also give us the ability to map and study planets around other stars in far greater detail than we can now.

• How will we search for Earth-like planets?

There's probably no bigger question in planetary science than whether Earth-like planets exist around other stars. NASA's Spitzer Space Telescope has already seen the infrared glow from dust created by the collisions of rocky planetesimals in other accreting solar systems, and we have no reason to doubt that terrestrial planets have formed around other stars. Thus, we have good reason to think that Earth-like planets should be out there, but are they? When you consider the fact that searching for an Earth-size planet around even the nearest stars is like searching for a pinhead from thousands of miles away, you might be tempted to think it would be an impossible task. Remarkably, however, missions currently in preparation should be capable of doing just that. It is likely that by the end of this decade we will have surveyed tens of thousands of star systems and will have learned the definitive answer to the question of whether Earth-size planets are rare or common. Let's exam-

ine just a few of the upcoming missions that should help us answer age-old questions about our place in the universe.

Transit Missions: Kepler *and* COROT   As we discussed in Section 13.1, we should in principle be able to detect planets the size of Earth or even smaller by searching for transits—the slight dips in stellar brightness that would occur when one of these planets passes in front of its star. In practice, the search for transits by Earth-size planets poses three major technological challenges. First, the dips in brightness caused by Earth-size planets will be very small and therefore require extraordinarily precise measurement. For example, viewed from afar, a transit of our own Sun by Earth would dim the Sun's light by only about 0.008%—not quite one part in 10 thousand. Second, because stars can vary in brightness for reasons besides transits, we can be confident that we've detected a planet only if the transit repeats with a regular period. For planets with sizes and orbital periods like those of the terrestrial worlds in our solar system, this means searching for transits that last no more than a few hours and recur anywhere from every couple of months to every couple of years. Clearly, we are likely to miss the transits unless we continuously monitor stars both day and night year-round. The third challenge comes from the fact that only a tiny fraction of planetary systems will by chance be oriented in such a way that their planets pass in front of their star from our vantage point on Earth. For example, only about 1 in 200 star systems should by chance have an orientation that would allow us to see a transit by an Earth-size planet in an Earth-like orbit. We therefore must monitor thousands of stars in order to have a reasonable expectation of just a few successes.

All three challenges should be met by a NASA mission called *Kepler*, tentatively scheduled for 2008 (Figure 13.15). *Kepler* will be launched with escape velocity

Figure 13.15 Artist's conception of NASA's *Kepler* mission to find Earth-like planets around other stars.

into an orbit around the Sun, so that Earth will not get in the way of its observations. It will then be pointed so that its telescope will essentially stare continuously at the same portion of the sky (in the constellation Cygnus) for the next 4 years. Its field of view is wide enough so that it will be able to monitor about 100,000 stars, measuring their brightnesses about every 15 minutes. Its cameras are sensitive enough so that it should be able to detect transits of Earth-size planets around Sun-like stars, and transits of planets as small as Mercury around somewhat dimmer stars. Because we don't know how common Earth-size planets are, it's difficult to predict how many *Kepler* should find. However, if we assume that our solar system is typical with its two Earth-size planets (Venus and Earth), calculations show that *Kepler* should detect about 50 such planets during its 4 years of observations. Of course, it should be even more successful at detecting larger planets that block more of their star's light, and should therefore greatly add to our current collection of known extrasolar planets.

While NASA is working on *Kepler*, the European Space Agency is building its own mission capable of searching for transits. The European mission, called *COROT*, is actually scheduled for launch ahead of *Kepler*, in mid-2006. It is not sufficiently sensitive to detect planets as small as Earth, but it should be able to detect planets just a few times larger.

THINK ABOUT IT

Find the current status of the *Kepler* and *COROT* missions. Has either been launched yet? If so, has it found any planets yet? If not, when is launch expected?

Astrometric Missions: GAIA *and* SIM Recall from Section 13.1 that the astrometric technique can in principle be used to find the slight side-to-side motions of stars in our sky caused by the gravitational tugs of orbiting planets. To date, this technique has not yielded any planetary discoveries, primarily because current telescopes cannot measure stellar positions with sufficient accuracy. For example, if you look back at Figure 13.2, you'll see that from 30 light-years away, we'd need position measurements accurate to less than a milliarcsecond (a thousandth of an arcsecond) to notice the effects of Jupiter or Saturn on the Sun's position.

The European Space Agency's *GAIA* mission, slated for launch in 2010, has the ambitious goal of performing astrometric observations of a billion stars in our galaxy with an accuracy of 10 microarcseconds. (A microarcsecond is one-millionth of an arcsecond.) *GAIA* will include a telescope dedicated to obtaining spectra for Doppler measurements. By simultaneously making position and Doppler measurements, *GAIA* will allow us to precisely determine a star's motion in all three dimensions, thereby allowing us to compute very accurate masses for orbiting planets. Furthermore, *GAIA* is capable of planetary transit detections. All told, it may discover thousands of new worlds. Part of *GAIA*'s power comes from its 180 CCD detectors [Section 6.1], with a total area equivalent to a square 0.8 meter on a side.

The only way to achieve higher astrometric precision is through interferometry [Section 6.4], in which two or more telescopes work together to obtain the angular resolution of a much larger telescope. Astronomers are making rapid progress in adapting interferometry to ground-based infrared and visible-light telescopes. For example, the twin Keck telescopes on Mauna Kea may soon be capable of detecting at least some planets through the astrometric technique. A NASA mission tentatively slated for launch in 2011 should do even better: the *Space Interferometry Mission (SIM)*. As currently planned, *SIM* will be capable of measuring stellar positions to a precision of just 1 microarcsecond—10 times better than *GAIA*'s precision and good enough to detect stellar motion caused by Earth-size planets around the nearest few dozen stars and motion due to Jupiter-size planets orbiting stars as far as 3,000 light-years away.

SIM will also serve as a test bed for a new technology, called *nulling interferometry*, designed to cancel (or "null") the light from a star so that we can more easily see its orbiting planets. This technology should in principle make possible direct detection of extrasolar planets around relatively nearby stars. *SIM* itself may be able to make crude images of a few Jupiter-size planets, and follow-on missions may give us our first pictures and spectra of terrestrial worlds around other stars.

Direct Detection: TPF *and* Darwin Astronomers are optimistic that the missions described above and the ever-expanding searches conducted with ground-based telescopes will discover thousands of new extrasolar planets and answer the question of whether Earth-*size* planets are rare. But these admirable goals fall short of answering the more deep-seated question of whether other planets are Earth-*like*. To answer this question, we need actual images or spectra of distant terrestrial worlds, so that we can learn whether they are geologically dead like Mercury and the Moon, frozen like Mars or overheated like Venus, or "just right" like Earth.

To accomplish the needed direct detection, NASA is in early planning stages for two missions in the *Terrestrial Planet Finder (TPF)* project. One mission, designated *TPF-C* (with the "C" standing for coronographic), would use a 5-meter telescope equipped with specialized baffles and masks to block out starlight so that we can see surrounding planets. This mission could be launched as early as 2014. The second and more ambitious mission, designated *TPF-I* ("I" for interferometry), envisions five spacecraft flying in tightly controlled formation as an interferometer (Figure 13.16), using nulling interferometry to cancel out light from the star so that we can see planets. Scientists hope that *TPF-I* will launch by 2020. The European Space Agency is contemplating a similar mission called *Darwin* on about the same time scale, so NASA and the European Space Agency may combine efforts in this grand endeavor. Either way, the *TPF-I* or *Darwin* missions would have the ability to obtain infrared spectra of Earth-sized planets around relatively nearby stars, thereby allowing us to search for spectral signatures of life-sustaining atmospheres, and possibly, of life itself.

Figure 13.16 This painting shows one possible configuration for NASA's planned mission called *TPF-I*, with five telescopes flying in formation so that they can obtain crude images and spectra of extrasolar planets.

How do you think the discovery of other Earth-like planets would change our view of our place in the universe? Would it matter whether we knew that the planets had life? Defend your opinions.

THE BIG PICTURE

Putting Chapter 13 into Perspective

In this chapter, we have explored one of the newest areas of astronomy—the study of solar systems beyond our own. As you continue your studies, keep in mind the following important ideas:

- With the discoveries of more than 100 extrasolar planets already in hand, we now know that planetary systems are common in the universe, although we do not yet know whether most are similar to or different from our own.

- The discovery of other planetary systems has inaugurated a new era in planetary science, one in which we have far more individual worlds to study and in which we can put our theory of solar system formation to the test.

- Because nearly all detections to date have been made with indirect techniques, we do not yet know much about the planets we have discovered. However, mass estimates combined with limited information about size and composition suggest that we have so far discovered planets similar in nature to the jovian planets of our solar system.

- It is too soon to know if Earth-like planets are rare or common, but missions currently being planned should help us answer this fundamental question within the next couple of decades.

SUMMARY OF KEY CONCEPTS

13.1 Detecting Extrasolar Planets

- **Why is it so difficult to detect planets around other stars?** The great distances to stars would by itself make it very difficult to detect **extrasolar planets**, and the problem is even more difficult because typical stars are a billion times brighter than the light reflected from any of their planets.

- **How do we detect planets around other stars?** Nearly all known extrasolar planets have been

discovered indirectly. We can look for a planet's gravitational effect on its star through the **astrometric technique**, which looks for small shifts in stellar position, or the **Doppler technique**, which looks for the back-and-forth motion of stars revealed by Doppler shifts. We can also search for **transits** and **eclipses** in which a system becomes slightly dimmer as a planet passes in front of or behind its star. The Doppler technique has yielded the vast majority of extrasolar planet discoveries to date.

13.2 The Nature of Extrasolar Planets

- **What have we learned about extrasolar planets?**

The known extrasolar planets are all much more massive than Earth. Many of them orbit surprisingly close to their stars or have large orbital eccentricities. We have limited information about sizes and compositions, but these data are consistent with the idea that the planets are jovian in nature.

- **How do extrasolar planets compare with those in our solar system?**

The known planets are probably jovian in nature. The planets that orbit close to their stars are called "hot Jupiters" because they must have very high temperatures that puff them up in size and give them a lower density than Jupiter.

13.3 The Formation of Other Solar Systems

- **Can we explain the surprising orbits of many extrasolar planets?** The planets with close-in or eccentric orbits probably were born on orbits similar to those of the jovian planets in our solar system. Their orbits then changed through some sort of planetary migration due to waves in the gaseous disk in which they formed, or through gravitational encounters or resonances with other massive planets.

- **Do we need to modify our theory of solar system formation?** Our basic theory of solar system formation seems to be sound, but we have had to modify it to allow for orbital change of the type thought to have occurred with the "hot Jupiters" or planets on very eccentric orbits.

13.4 Finding New Worlds

- **How will we search for Earth-like planets?** The first systematic attempt will be made by searching for transits with the *Kepler* mission. Later missions, such as the *Terrestrial Planet Finder*, may be able to detect such planets directly.

EXERCISES AND PROBLEMS

For instructor-assigned homework go to MasteringASTRONOMY www.masteringastronomy.com

Review Questions

Short-Answer Questions Based on the Reading

1. Why are extrasolar planets hard to detect directly?
2. What are the three major methods used to detect extrasolar planets indirectly?
3. Explain why a planet can cause its star to move slightly in the sky.
4. How does the astrometric technique work? Why hasn't it been very successful in discovering planets to date?
5. How does the Doppler technique work? Explain how it can tell us a planet's orbital period, orbital distance, and orbital eccentricity.
6. Why does the Doppler technique generally allow us to determine only minimum planetary masses rather than actual planetary masses? Should we expect these minimum masses to be close to the actual masses? Explain.
7. How does the transit technique work? Could we use this method to find planets around all stars that have them? Why or why not?
8. Have any planets been detected directly? Explain.
9. Briefly summarize all the planetary properties we can in principle measure with current techniques, and state which techniques allow us to measure each of these properties.
10. How do the orbits of known extrasolar planets differ from those of jovian planets in our solar system? Why are these orbits surprising?
11. What data suggest that the known extrasolar planets are similar in nature to the jovian planets in our solar system?
12. What do we mean by a "hot Jupiter"? How should we expect a hot Jupiter to compare to the real planet Jupiter?
13. Many extrasolar jovian planets orbit surprisingly close to their stars. How might they have ended up in these orbits?
14. Many extrasolar jovian planets have high orbital eccentricities. How might they have ended up with such eccentric orbits?
15. Based on current data, does it seem likely that our solar system has a particularly unusual layout? Explain.
16. Briefly describe how the *Kepler* mission will search for evidence of Earth-size planets.
17. How will the *SIM* and *GAIA* missions aid our search for extrasolar planets?
18. What technologies offer the hope of determining whether Earth-like planets exist around other stars? What missions may give us this capability?

Test Your Understanding

Does It Make Sense?

Decide whether the statement makes sense (or is clearly true) or does not make sense (or is clearly false). Explain your reasoning. (For an example, see Chapter 1, "Does It Make Sense?")

19. An extraterrestrial astronomer surveying our solar system with the Doppler technique could discover the existence of Jupiter with just a few days of observation.
20. The fact that we have not yet discovered an Earth-mass planet tells us that such planets must be very rare.
21. Within the next few years, astronomers expect to confirm all the planet detections made with the Doppler technique by observing transits of these same planets.
22. Although "hot Jupiters" are unlikely places to find life, they could be orbited by moons that would have pleasant, Earth-like temperatures.
23. Before the discovery of planetary migration, scientists were unable to explain how Saturn could have gotten into its current orbit.
24. It's the year 2007: Astronomers have successfully photographed an Earth-size planet, showing that it has oceans and continents.
25. It's the year 2025: Astronomers have just announced that they have obtained a spectrum showing the presence of oxygen in the atmosphere of an Earth-size planet.
26. It's the year 2040: Scientists announce that our first spacecraft to reach an extrasolar planet is now orbiting a planet around a star located near the center of the Milky Way Galaxy.

27. An extrasolar planet is discovered with a year that lasts only 3 days.

28. Later this year, scientists use the Doppler technique to identify a planet whose mass is equal to Earth's mass.

29. Astronomers announce that all the Doppler technique discoveries of extrasolar planets made to date are actually more massive brown dwarfs, and we had thought they were less massive only because we didn't realize that they have nearly face-on orbits.

30. The number of extrasolar planets increases from around 150 in 2005 to more than 1,000 by the year 2010.

Quick Quiz

Choose the best answer to each of the following. Explain your reasoning with one or more complete sentences.

31. What method has detected the most extrasolar planets so far? (a) the transit method (b) Hubble images (c) the Doppler method

32. The extrasolar planets discovered so far most resemble (a) terrestrial planets. (b) jovian planets. (c) large icy worlds.

33. How many extrasolar planets have been detected? (a) between 10 and 100 (b) between a hundred and a thousand (c) more than a thousand

34. Which one of the following can the transit method tell us about a planet? (a) its mass (b) its size (c) the eccentricity of its orbit

35. Which method could detect a planet in an orbit that is face-on to the Earth? (a) Doppler method (b) transit method (c) astrometry

36. How is the planet 51 Pegasi b different from Jupiter? (a) much closer to its star (b) much longer year (c) much more massive

37. Most known extrasolar planets are more massive than Jupiter because (a) we do not expect smaller planets to exist. (b) current detection methods are more sensitive to larger planets. (c) the Doppler method usually overestimates planet masses.

38. Which detection method can be used on a backyard telescope with a CCD system? (a) Doppler method (b) transit method (c) astrometry

39. What's the best explanation for the location of "hot Jupiters"? (a) They formed closer to their stars than Jupiter did. (b) They formed farther out than Jupiter did and then migrated inward. (c) The stars' stronger gravity pulled them closer.

40. Earth-sized planets orbiting normal stars (a) have already been discovered. (b) should be discovered in the next few years by ground-based telescopes. (c) should be discovered by 2010 by a space telescope.

Investigate Further

In-Depth Questions to Increase Your Understanding

Short-Answer/Essay Questions

41. *Why So Soon?* The detection of extrasolar planets came much sooner than astronomers expected. Was this a result of planets being different than expected, or of technology improving faster? Explain.

42. *Why Not Hubble?* Of the more than 160 extrasolar planets discovered, only one possible planet has ever been imaged by the Hubble Space Telescope. What limits Hubble's ability to image planets around other stars, and why was it possible in the case of Figure 13.9?

43. *Explaining the Doppler Method.* Explain how the Doppler method works in terms an elementary school child would understand. It may help to use an analogy to explain the difficulty of direct detection and for the general phenomenon of the Doppler shift.

44. *Comparing Methods.* What are the advantages and disadvantages of the Doppler and transit methods? What kinds of planets are easiest to detect in each case? Are there certain planets that each method cannot detect, even if the planets are very large? Explain. What advantages come if a planet can be detected by both methods?

45. *No Hot Jupiters Here.* How do we think "hot Jupiters" formed? Why didn't one form in our solar system?

46. *Resonances.* How have resonances been important in affecting extrasolar planet orbits? How are these effects similar to the effects of resonances in our solar system, and how are they different?

47. *Low-Density Planets.* Only one planet in our solar system has a density less than 1 gram per cubic centimeter, but many extrasolar planets do. Explain why in a few sentences. (Hint: consider the densities of the jovian planets in our solar system, given in Figure 11.1.)

48. *A Year on HD209458b.* Imagine you're visiting the planet that orbits the star HD209458, hovering in the upper atmosphere in a suitable spacecraft. What would it be like? What would you see, and how would it look different compared to floating in Jupiter's atmosphere? Consider factors like local conditions, clouds, how the Sun would appear, and orbital motion.

49. *Lots of Big Planets.* Many of the planets discovered so far are more massive than our most massive planet. Does this mean our solar system is unusual? If so, how or why? If not, why not?

50. *Detect an Extrasolar Planet for Yourself.* Most colleges and many amateur astronomers have the equipment necessary to detect known extrasolar planets using the transit method. All that's required is a telescope 10 or more inches in diameter, a CCD camera system, and a computer system for data analysis. The basic method is to take exposures of a few minutes' duration over a period of several hours around the times of predicted transit, and to compare the brightness of the star being transited relative to other stars in the same CCD frame (Figure 13.7). For complete instructions, see MasteringAstronomy.com.

Quantitative Problems

Be sure to show all calculations clearly and state your final answers in complete sentences.

51. *Lost in the Glare.* How hard would it be for an alien astronomer to detect the light from planets in our solar system compared to light from the Sun itself?

   a. Calculate the fraction of the total emitted sunlight that is reflected by Earth. (*Hint:* Imagine a sphere around the Sun the size of the planet's orbit (area $= 4\pi a^2$). What fraction of that area does the disk of a planet (area $= \pi r^2_{\text{planet}}$) take up? Earth's reflectivity is 29%.

   b. Would detecting Jupiter be easier or harder than detecting Earth? Comment on whether you think Jupiter's larger size or greater distance has a stronger effect on its detectability. You may neglect any difference in reflectivity between Earth and Jupiter.

52. *Transit of TrES-1.* The planet orbiting this star has been detected by both the transit and Doppler methods, so we can calculate its density and get an idea of what kind of planet it is.

a. Using the method of Mathematical Insight 13.3, calculate the radius of the transiting planet. The planetary transits block 2% of the star's light. The star TrEs-1 has a radius of about 85% of our Sun's radius.

b. The mass of the planet is approximately 0.75 times the mass of Jupiter, and Jupiter's mass is about 1.9×10^{27} kilograms. Calculate the average density of the planet. Give your answer in grams per cubic centimeter. Compare this density to the average densities of Saturn (0.7 g/cm^3) and Earth (5.5 g/cm^3). Is the planet terrestrial or jovian in nature? (*Hint:* To find the volume of the planet, use the formula for the volume of a sphere: $V = \left(\frac{4}{3}\right)\pi r^3$. Be careful with unit conversions.)

53. *Planet Around 51 Pegasi.* The star 51 Pegasi has about the same mass as our Sun. A planet discovered around it has an orbital period of 4.23 days. The mass of the planet is estimated to be 0.6 times the mass of Jupiter. Use Kepler's third law to find the planet's average distance (semimajor axis) from its star. (*Hint:* Because the mass of 51 Pegasi is about the same as the mass of our Sun, you can use Kepler's third law in its original form, $p^2 = a^3$ [Section 3.3]. Be sure to convert the period into years before using this equation.)

54. *One Born Every Minute?* It's possible to make a rough estimate of how often planetary systems form by making some basic assumptions. For example, if you assume that the stars we see have been born at random times over the last 10 billion years, then the rate of star formation is simply the number of stars we see divided by 10 billion years. The fraction of planets with detected extrasolar planets is at least 5%, so this factor can be multiplied in to find the approximate rate of formation of planetary systems.

a. Using these assumptions, how often does a planetary system form in our galaxy? (Our galaxy contains at least 100 billion stars.)

b. How often does a planetary system form somewhere in the observable universe, which contains at least 100 billion galaxies?

c. Write a few sentences describing your reaction to your results. Do you think the calculation is realistic? Are the rates larger or smaller than you expected?

55. *Identical Planets?* The Doppler method allows us to find a planet's mass if we know the star's maximum velocity, the planet's period, and the star's mass (Mathematical Insight 13.2).

a. Imagine two planets orbiting a 1 solar mass star with orbits edge-on to the Earth. The peak Doppler shift for each is 50 m/s, but one has a period of 3 days and the other has a period of 300 days. Calculate the two masses and say which, if either, is larger.

b. Imagine another planet in an edge-on orbit around a star with a mass half that of our Sun. How does this planet compare to those in part (a)? Show your work.

56. *Finding the Smallest Planet.* One of the goals of extrasolar planet searches is to find ever-smaller planets that might resemble those in our solar system. What is the least massive planet that can be detected at 0.1 AU from a 1-solar-mass star if the smallest velocity measurement is 1 m/s? What about around a 0.5-solar-mass star? Repeat the calculations using a distance of 10 AU.

57. *Finding Orbit Sizes.* The Doppler method (Mathematical Insight 13.1) allows us to find a planet's semimajor axis using just the orbital period and the star's mass.

a. Imagine that a new planet is discovered around a 2-solar-mass star that has a period of 5 days. What is its semimajor axis?

b. Another planet is discovered around a 0.5-solar-mass star with a period of 100 days. Find the semimajor axis.

58. *Search Strategies.* The kinds of planets you can detect depend on your observing strategy.

a. Imagine that your search can find planets with orbital periods between 4 and 40 days. If you are searching for planets around stars with the same mass as our Sun, over what range of semimajor axes can you detect planets?

b. Repeat the calculation above for stars half and twice the Sun's mass. What combination (massive vs. low-mass star, short- vs. long-period planet) will detect the planet with the largest semimajor axis?

59. Are *They Searching Too?* Which planet in our solar system causes the largest Doppler shift of our Sun that alien astronomers might observe? Consider the cases of Mercury, Earth, and Jupiter. How do those Doppler shifts compare to our current technological limit of about 1 m/s?

60. *Habitable Planet Around 51 Pegasi?* The star 51 Pegasi is approximately as bright as our Sun and has a planet that orbits at a distance of only 0.052 AU.

a. Suppose the planet reflects 15% of the incoming sunlight. Using Mathematical Insight 10.1, calculate its "no greenhouse" average temperature. How does this temperature compare to that of Earth?

b. Repeat part (a), but assume that the planet is covered in bright clouds that reflect 80% of the incoming sunlight.

c. Based on your answers to parts (a) and (b), do you think it is likely that the conditions on this planet are conducive to life? Explain.

61. *Tricky Transits.* The transit method requires the planet's orbit to be nearly edge-on as seen from the Earth. You can use simple geometry to find out what fraction of planetary systems are likely to have a transit. We'll make the simplifying assumptions that the stars are all the size of our Sun, and that the transit can be detected as long as the middle of the planet passes in front of any part of the star. Because there's no organization to the orbits of planets around other stars, their tilts relative to edge-on are equally likely at all angles from 0 to 90°.

a. Take the case of "hot Jupiters," which orbit at about 0.05 AU. What's the maximum tilt away from edge-on at which such a planet causes a transit? What fraction of possible tilt angles can cause a transit? If you observe 10,000 stars of which 2% have "hot Jupiters," how many planets are you likely to discover? (*Hint:* draw a diagram of the system with the orbit edge-on, with Earth to the side.)

b. Repeat the calculations above for Earth-size planets orbiting at 1 AU. Assume 10% of stars have such planets.

c. Repeat for the case of Jupiter-size planets at 5 AU, assuming 10% of stars have Jupiters. About how long would you have to wait to see three transits for each planet?

Discussion Questions

62. *So What?* What is the significance of the discovery of extrasolar planets, if any? Justify your answer in the context of this book's discussion of the history of astronomy.

63. *Is It Worth It?* The cost of the *Kepler* mission is several hundred million dollars. The cost of the *Terrestrial Planet Finder* mission is likely to be several billion dollars. Are these expenses worth it, compared to the results expected? Defend your opinion.

64. *What If?* Consider the possible outcomes of the missions described in Section 13.4. What results would change our perspective on our solar system? On the possibility of life in the universe?

For self-study activities go to www.masteringastronomy.com

Interactive Tutorials

Tutorial Review of Key Concepts
Use the interactive **Tutorial** at **www.mastering astronomy.com** to review key concepts from this chapter.

Detecting Extrasolar Planets Tutorial
Lesson 1 Taking a Picture of a Planet
Lesson 2 Stars' Wobbles and Properties of Planets
Lesson 3 Planetary Transits

Supplementary Tutorial Exercises
Use the interactive **Tutorial Lessons** to explore the following questions.

Detecting Extrasolar Planets Tutorial, Lesson 2
1. Using the tool provided, explain how weekly measurements allow us to determine the orbital period of the extrasolar planet.
2. Use the tool to vary the mass of the planet. How does its mass affect the Doppler shifts in its star's light?
3. Use the tool to vary the orbital radius of the planet. How does the orbital radius affect the Doppler shifts in its star's light?

Detecting Extrasolar Planets Tutorial, Lesson 3
1. Under what conditions can we view a planetary transit of another star?
2. How does the change in brightness during a transit depend on the planet's properties?

Exploring the Sky and Solar System

Use the *Voyager: SkyGazer* CD-ROM accompanying your book to locate the brightest stars known to harbor planets. (These will be the stars in Appendix E.4 named with Greek letters preceding the constellation name.) Are they visible from where you live? At what time of the year and night? Print out a star chart and see if you can find the star in the sky. Bring along the information from the appendix to help you picture the planetary system in the sky.

Movies

Check out the following narrated and animated short documentary available on **www.masteringastronomy.com** for a helpful review of key ideas covered in this chapter.

History of the Solar System Movie

Web Projects

1. *New Planets.* Find the latest information on extrasolar planet discoveries. Create a personal "planet journal," complete with illustrations as needed, with a page for each of at least three recent discoveries of new planets. On each journal page, note the technique that was used to find the planet, give any information we have about the nature of the planet, and discuss how the planet does or does not fit in with our current understanding of planetary systems.

2. *Direct Detections.* In this chapter we saw only one example of a direct detection of a possible extrasolar planet. Search for new information on this and any other direct detections now known. Has the detection discussed in this chapter been confirmed (or disavowed) as a planet? Have we made any other direct detections, and if so, how? Summarize your findings with a short written report, including images of the directly detected planets.

3. *Extrasolar Planet Mission.* Visit the Web site for one of the future space missions discussed in this chapter and learn more about the mission design, capabilities, and goals. Write a short report on your findings.

Space and Time

LEARNING GOALS

S2.1 Einstein's Revolution
- **What are the major ideas of special relativity?**
- **What is "relative" about relativity?**

S2.2 Relative Motion
- **How did Einstein think about motion?**
- **What's surprising about the absoluteness of the speed of light?**
- **Why can't we reach the speed of light?**

S2.3 The Reality of Space and Time
- **How does relativity affect our view of time and space?**
- **Do the effects predicted by relativity really occur?**

S2.4 Toward a New Common Sense
- **How can we make sense of relativity?**
- **How does special relativity offer us a ticket to the stars?**

Henceforth space by itself, and time by itself, are doomed to fade away into mere shadows, and only a kind of union of the two will preserve an independent reality.

Hermann Minkowski, 1908

The universe consists of matter and energy moving through *space* with the passage of *time*. Up to this point in the book, we have discussed the concepts of space and time as though they are absolute and distinct—just as they appear in everyday life. But what if this appearance is deceiving?

About a century ago, Albert Einstein discovered that space and time are not what they appear to be. Instead, space and time are intertwined in a remarkable manner described by Einstein's *theory of relativity*. Because space and time are such fundamental concepts, understanding relativity is important to understanding the universe.

The theory of relativity is *not* difficult to understand, despite popular myths to the contrary. It does, however, require us to think in new ways. That is our task in this chapter and the next. By the time we are finished, you will see that Einstein brought about a revolution in human thinking with many important ramifications for understanding our place in the universe.

S2.1 Einstein's Revolution

Albert Einstein was born on March 14, 1879, in Ulm, Germany. As a toddler, he was so slow in learning to speak that relatives feared he might have brain damage. Even when he began to show promise, it was in unconventional ways. Young Albert dropped out of high school at the suggestion of a teacher who told him that he would "never amount to anything." Nevertheless, he was admitted to college in Switzerland, in large part because of his proficiency in mathematics.

Einstein hoped to teach after his college graduation, but his Jewish heritage and lack of Swiss citizenship deterred prospective employers. He therefore took a job with the Swiss patent office, starting in 1901. He devoted his free time to the study of unsolved questions in physics.

From the standpoint of history, Einstein burst onto the scene in 1905, when he earned his Ph.D. and published five papers in the *German Yearbook of Physics*. Three of the papers solved three of the greatest mysteries in physics at the time. The first paper dealt with something called the "photoelectric effect," and essentially presented the first concrete evidence of the wave-particle duality of light that we discussed in Chapter 5. The second paper explained why suspended particles in water jiggle even after the water has been still for a long time (an effect known as "Brownian motion"). Einstein's analysis led to the first direct measurements of molecular motion and molecular and atomic sizes. The third paper introduced the special theory of relativity, which forever changed our understanding of space

and time. To mark the one hundredth anniversary of the publication of these papers, the United Nations designated the year 2005 as "World Year of Physics."

Einstein returned to Germany in 1913, where he soon published his general theory of relativity. After some of its central predictions were observationally confirmed in 1919 [Section S3.4], Einstein's fame began to extend beyond the world of science. He received the Nobel Prize for Physics in 1921 and was by this time a household name.

Einstein was a visiting professor at the California Institute of Technology when Hitler came to power in Germany. Recognizing Hitler's evil, Einstein decided to remain in the United States. He played an important role in convincing President Franklin Roosevelt to start the Manhattan Project to build an atomic bomb. Nevertheless, Einstein was a committed pacifist who supported the project only because he feared that Hitler would otherwise develop an atomic bomb first. He expressed great dismay when the bomb was used against Japan, and he spent much of the rest of his life arguing for a worldwide agreement to ban the further manufacture and use of nuclear weapons. He also worked to promote human rights and equality for men and women of all races and nationalities. Einstein died in Princeton, New Jersey, on April 18, 1955. More than 50 years later, he remains the most famous scientist of modern times.

• What are the major ideas of special relativity?

We often talk about the *theory of relativity* as though it were a single theory, and in some sense it is. However, Einstein actually developed his theory in two parts that he published separately. The **special theory of relativity**, published in 1905, shows that space and time are intertwined but does not deal with the effects of gravity. The **general theory of relativity**, published in 1915, offers a surprising new view of gravity—a view that we will use to help us understand topics such as the expansion and fate of the universe and the strange objects known as *black holes*. In essence, the special theory is "special" because it deals only with the special case in which we ignore the role of gravity, while the general theory is "general" because it applies with or without gravity. We will focus on the special theory of relativity in this chapter and on the general theory in Chapter S3.

Einstein developed his theories in large part with the aid of "thought experiments"—experiments that could be carried out in principle but that would be very difficult in practice. Thought experiments allow us to think through the consequences of simple ideas or statements. Let's start with a thought experiment designed to show you why the special theory of relativity is so surprising.

Imagine that, with the aid of a long tape measure, you carefully measure the distance you walk from home to work and find it to be 5.0 kilometers. You wouldn't expect any argument about this distance. For example, if a friend drives her car along the same route and measures the distance with her car's odometer, she ought to get the same measurement of 5.0 kilometers—as long as her odometer is working properly. Likewise, you would expect agreement about how much time it takes you to walk this distance. Suppose your friend continues driving and you call her on a cellular phone just as you leave your house at 8:00 A.M. and again just as you arrive at work at 8:45 A.M. You'd certainly be surprised if she argued that your walk took an amount of time other than 45 minutes.

Distances and times appear absolute and distinct in our daily lives. We expect everyone to agree on the distance between two points, such as the locations of home and work. We also expect agreement about the time between two events, such as leaving home and arriving at work. Thus, it came as a huge surprise to everyone when Einstein showed that these expectations are not strictly correct: With extremely precise measurements, the distance you measure between home and work will be *different* from the distance measured by a friend in a car, and you and your friend will also disagree about the time it takes you to walk to work. At ordinary speeds, the differences will be so small as to be unnoticeable. But if your friend could drive at a speed close to the speed of light, the differences would be substantial.

Disagreements about distances and times are only the beginning of the astonishing ideas contained in the special theory of relativity. In the rest of this chapter, we will see how this theory leads to each of the following ideas:

- No information can travel faster than the speed of light (in a vacuum), and no material object can even reach the speed of light.

- If you observe anyone or anything moving by you at a speed close to the speed of light, you will conclude that time runs more slowly for that person or moving object. That is, a person moving by you ages more slowly than you, a clock moving by you ticks more slowly than your clock, a computer moving by you runs more slowly than your similar computer, and so on.

- If you observe two events to occur simultaneously, such as flashes of light in two different places at the same time, a person moving by you at a speed close to the speed of light may not agree that the two events were simultaneous.

- If you carefully measure the size of something moving by you at a speed close to the speed of light, you will find that its length (in the direction of its motion) is shorter than it would be if the object were not moving.

- If you could measure the mass of something moving by you at a speed close to the speed of light, you would find its mass to be greater than the mass it would have if it were stationary. As we will see, Einstein's famous equation, $E = mc^2$, follows from this fact.

Although these ideas of special relativity may sound like science fiction or fantasy, a vast body of observational and experimental evidence supports their reality. They also follow logically from a few simple premises. If you keep an open mind and think deeply as you read this chapter, you'll soon understand the basic ideas of relativity.

• What is "relative" about relativity?

You've probably heard people claim that Einstein taught that "everything is relative," but this claim is simply not true. The theory of relativity takes its name from the idea of *relativity of motion*. You can see why motion must be relative with another simple thought experiment.

Imagine a supersonic airplane that flies at a speed of 1,670 km/hr from Nairobi, Kenya, to Quito, Ecuador. How fast is the plane going? At first, this question sounds trivial—we have just said that the plane is going 1,670 km/hr.

But wait. Nairobi and Quito are both nearly on Earth's equator, and the equatorial speed of Earth's rotation is the same 1,670 km/hr speed at which the plane is flying (see Figure 1.12). Moreover, the east-to-west flight from Nairobi to Quito is opposite the direction of Earth's rotation (Figure S2.1). Thus, if you lived on the Moon, the plane would appear to stay put *while Earth rotated beneath it*. When the flight began, you would see the plane lift off the ground in Nairobi. The plane would then remain stationary while Earth's rotation carried Nairobi away from it and Quito toward it. When Quito finally reached the plane's position, the plane would drop back down to the ground.

We have two alternative viewpoints about the plane's flight. People on Earth say that the plane is traveling westward across the surface of the Earth. Observers in space say that the plane is stationary while Earth rotates eastward beneath it. Both viewpoints are equally valid.

In fact, there are many other equally valid viewpoints about the plane's flight. Observers looking at the solar system as a whole would see the plane moving at a speed of more than 100,000 km/hr—Earth's speed in its orbit around the Sun. Observers living in another galaxy would see the plane moving at about 800,000 km/hr with the rotation of the Milky Way Galaxy. The only thing all these observers would agree on is that the plane is traveling at 1,670 km/hr *relative to* the surface of the Earth.

The airplane example shows that questions like, "Who is really moving?" and "How fast are you going?" have no

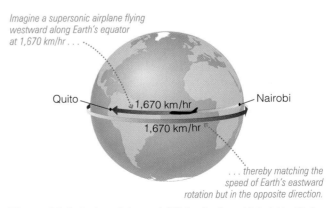

Imagine a supersonic airplane flying westward along Earth's equator at 1,670 km/hr . . .

Quito — 1,670 km/hr — Nairobi

1,670 km/hr

. . . thereby matching the speed of Earth's eastward rotation but in the opposite direction.

Figure S2.1 A plane flying at 1,670 km/hr from Nairobi to Quito (westward) travels precisely opposite Earth's eastward rotation. Thus, viewed from afar, the plane remains stationary while Earth rotates underneath it.

absolute answers. Einstein's *theory of relativity* tells us that measurements of motion, as well as measurements of time and space, make sense only when we describe whom or what they are being measured relative to.

THINK ABOUT IT

Suppose you are running on a treadmill and the readout says you are going 8 miles per hour. What is the 8 miles per hour measured relative to? How fast are you going relative to the ground? How fast would an observer on the Moon see you going? Describe a few other possible viewpoints on your speed.

The Absolutes of Relativity   The theory of relativity tells us that motion is always relative, but it does *not* say that *everything* is relative. In fact, the theory claims that two things in the universe are absolute:

1. The laws of nature are the same for everyone.
2. The speed of light is the same for everyone.

As we'll see shortly, all the astounding ideas of relativity follow directly from these two seemingly innocuous statements.

Making Sense of Relativity   One reason relativity has a reputation for being difficult to grasp, despite its underlying simplicity, is that most of its ideas and consequences are not obvious in everyday life. They become obvious only when we deal with speeds close to the speed of light or, in the case of general relativity, with gravitational fields far stronger than that of Earth. Because we don't commonly experience such extreme conditions, we have no *common* sense about them. Thus, to say that relativity violates common sense is not really accurate. The theory of relativity is perfectly consistent with everything we have come to expect in daily life.

Making sense of relativity requires only that you learn to view your everyday experiences from a new, broader perspective. Fortunately, you learned to broaden your perspective in a similar way before. At a very young age, you learned "common sense " meanings for *up* and *down: Up* is above your head, *down* is toward your feet, and things tend to fall down. One day, however, you learned that Earth is round. When you looked at a globe with the Northern Hemisphere on the top, you were immediately confronted with a **paradox**—a situation that *seems* to violate common sense or to contradict itself. Your common sense told you that Australians should fall off the Earth (Figure S2.2a), but they don't. To resolve this paradox, you were forced to accept that your "common sense" about *up* and *down* was incorrect. You therefore revised your common sense to accept that *up* and *down* are determined relative to the center of the Earth (Figure S2.2b).

THINK ABOUT IT

Why do you suppose that most maps show the Northern Hemisphere on the top and the Southern Hemisphere on the bottom? If you hung your map upside down and rewrote the words so they read right side up, would the map be equally valid?

As a teenager, Einstein wondered what the world would look like if he could travel at or beyond the speed of light. He inevitably encountered paradoxes when he thought about this question. Ultimately, he resolved the paradoxes only when he recognized that our common sense ideas about space and time must change if we are to extend them to the realm of very high speeds or very strong gravitational fields. Just as we all once learned a new common sense about up and down, we now must learn a new common sense about space and time.

S2.2 Relative Motion

We have already discussed the basic ideas of relativity, but we have not yet explained how Einstein arrived at these ideas. In this section, we will begin to explore the theory of relativity by following a series of thought experiments through to their logical conclusions.

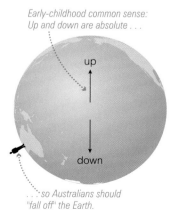

Early-childhood common sense: Up and down are absolute . . .

up

down

. . . so Australians should "fall off" the Earth.

a

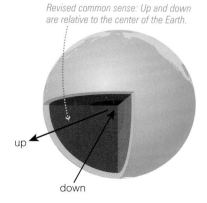

Revised common sense: Up and down are relative to the center of the Earth.

up

down

b

Figure S2.2 Learning that Earth is round helps children revise their "common sense " understanding of *up* and *down*.

• How did Einstein think about motion?

We will use slightly different thought experiments than the ones Einstein himself used, though the ideas will be essentially the same. Just as Einstein did, we will base all our thought experiments on the assumption that the two absolutes of relativity are true.

The first of the two absolutes, that the laws of nature are the same for everyone, is probably not surprising. If you're on an airplane with the shades drawn during a very smooth flight, you won't feel any sensation of motion. Thus, you should expect to get the same results from any experiments you perform on the airplane that someone else would get performing the same experiments on the ground. These equivalent results demonstrate that the laws of nature are the same in the moving airplane and on the ground. In the language of relativity, the ground and the airplane represent different **reference frames** (or *frames of reference*); we say that two people share the same reference frame if they are *not* moving relative to each other. The idea that the laws of nature are the same for everyone means that they do not depend on your reference frame.

The second absolute of relativity, that the speed of light is the same for everyone, is far more surprising. In general, we expect people in different reference frames to give different answers for the speed of the same moving object. For example, suppose you roll a ball down the aisle of an airplane. The ball rolls slowly in your airplane reference frame. However, the ball moves much faster in the reference frame of a person on the ground, because the ball has the airplane's speed in addition to the speed at which it rolls down the aisle (Figure S2.3a). Now, suppose instead that you turn on a flashlight and measure the speed of the emitted light. The idea that the speed of light is the same for everyone means that a person on the ground will measure exactly the same speed for the light beam, even though the light was emitted inside the moving airplane (Figure S2.3b). In other words, people in different reference frames can disagree about the speeds of material objects, but everyone always agrees about the speed of light, regardless of where the light comes from.

How do we know that everyone always measures the same speed of light? Einstein reached this conclusion be-cause it was the only way he could resolve the paradoxes he encountered when he thought about traveling at speeds close to the speed of light. But observations and experiments are the ultimate judge of any scientific theory. Many observations and experiments, some of which we'll study later in the chapter, have verified that the speed of light really is an absolute in nature. Thus, *the absoluteness of the speed of light is an experimentally verified fact.*

Experimental evidence also supports all the other ideas of relativity that we listed in the beginning of this chapter, which is why the theory of relativity is so well accepted. We'll discuss some of this experimental evidence later in the chapter. First, however, we'll see why these ideas follow logically from the two absolutes of relativity.

• What's surprising about the absoluteness of the speed of light?

The idea that everyone always measures the same speed of light may not sound earth-shattering, but Einstein soon realized that it leads to far-ranging consequences—consequences that force us to let go of many of our intuitive beliefs about how the universe works. We will investigate the implications of the absoluteness of the speed of light by constructing a series of thought experiments about relative motion viewed from different reference frames. Because special relativity does not deal with the effects of gravity, the thought experiments are easiest to visualize by imagining that we are in spaceships in deep space, far from any planets or stars and drifting along without engine power. Everything in and around these spaceships is weightless and floats freely, so we call the reference frames of these spaceships **free-float frames** (or *inertial reference frames*).

Thought Experiments at Ordinary Speeds   As we will see shortly, the consequences of relativity become clear when we deal with spaceships or other objects moving at very high speeds; that is, speeds that are a significant fraction of the speed of light. We generally do not notice anything strange at the more ordinary speeds we experience in everyday life. Nevertheless, before we turn to high-speed thought experiments, it's useful to start with some thought experiments involving more familiar speeds, so that you can become

If you roll a ball down the aisle at 10 km/hr . . .

900 km/hr

10 km/hr

. . . observers on the ground see the ball moving at 910 km/hr: the ball's rolling speed of 10 km/hr **plus** the airplane's speed of 900 km/hr.

But if you turn on a flashlight in the airplane . . .

900 km/hr

c

. . . both you and observers on the ground agree that light travels at the speed of light, c, regardless of the airplane's speed.

Figure S2.3 Unlike the speed of material objects, the speed of light is the same in all reference frames.

Figure S2.4 Thought Experiment 1.

Figure S2.5 Thought Experiment 2.

familiar with the basic method we will use at both low and high speeds.

Thought Experiment 1 (Figure S2.4) Imagine that you are floating freely in a spaceship. Because you feel no sensation of motion, you perceive yourself to be at rest, or traveling at zero speed. As you look out your window, you see your friend Jackie in her own spaceship, moving away at a constant speed of 90 km/hr. How does the situation appear to Jackie?

We can answer the question by logically analyzing the experimental situation. Jackie has no reason to think she is moving. Like you, she is in a free-float frame, floating freely in a spaceship with the engines off. Jackie will therefore say that *she* is at rest and that *you* are moving away from her at 90 km/hr.

Both points of view—yours and Jackie's—are equally valid. You both would find the same results for any experiments performed in your own spaceship, and your research would lead you to exactly the same laws of nature. You could argue endlessly about who is really moving, but your argument would be pointless because all motion is relative.

Thought Experiment 2 (Figure S2.5) We begin with the same situation as in Thought Experiment 1, but this time you put on your space suit and strap yourself to the outside of your spaceship. You happen to have a baseball, which you throw in Jackie's direction at a speed of 100 km/hr. How fast is the ball moving relative to Jackie?

From your point of view, Jackie and the ball are both going in the same direction. Jackie is going 90 km/hr and the ball is going 100 km/hr, so the ball is going 10 km/hr faster than she is. The ball will therefore overtake and pass her.

From Jackie's point of view, *she* is stationary and *you* are moving away from her at 90 km/hr. Thus, she would say the ball is moving even before you throw it, since it would be going with you at your speed of 90 km/hr. Once you release the ball, she therefore sees it moving toward her at 10 km/hr—the ball's speed of 100 km/hr relative to

you *minus* the 90 km/hr at which you are moving relative to her. In other words, although you and Jackie disagree about the actual speed of the ball, you both agree that the ball will pass her at a *relative* speed of 10 km/hr.

Thought Experiment 3 (Figure S2.6) This time you throw a baseball in Jackie's direction at 90 km/hr. From your point of view, the ball is traveling at exactly the same

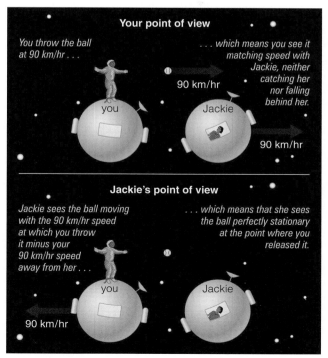

Figure S2.6 Thought Experiment 3.

speed as Jackie. Therefore, you'll see the ball forever chasing her through space, neither catching up with nor falling behind her.

From Jackie's point of view, the 90 km/hr at which you threw the ball exactly matches your 90 km/hr speed away from her. Therefore, the ball is *stationary* in her reference frame. Think about this for a moment: Before you throw the baseball, Jackie sees it moving away from her at 90 km/hr because it is in your hand. At the moment you release the baseball, it suddenly becomes stationary in Jackie's reference frame, floating in space at a fixed distance from her spaceship. Many hours later, after you have traveled far away, Jackie will still see the ball floating in the same place. If she wishes, she can put on her space suit and go out to retrieve it, or she can just leave it there. From her point of view, it's not going anywhere, and neither is she.

THINK ABOUT IT

In Thought Experiment 3, suppose instead that you throw the ball in Jackie's direction at 80 km/hr. In that case, what will Jackie see the ball doing? Next, suppose that Jackie is moving *toward* you rather than *away* from you. What would she see the ball doing in that case?

Thought Experiments at High Speeds   The absoluteness of the speed of light did not come into play in our first three thought experiments because the speeds of the objects were small compared to the speed of light. For example, 100 km/hr is less than *1 ten-millionth* of the speed of light. Now let's raise the speeds much higher and explore the strange consequences of the absoluteness of the speed of light.

Thought Experiment 4 (Figure S2.7*) Imagine that Jackie is moving away from you at 90% of the speed of light, or

---

*For simplicity, this and most other figures in this chapter ignore effects of length contraction, which would make the moving ships shorter in their direction of motion; only Figure S2.13 shows the length contraction effect.

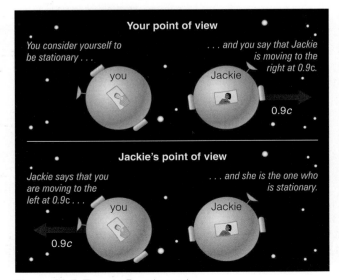

Figure S2.7 Thought Experiment 4.

0.9*c*. (Recall that *c* is the symbol for the speed of light, which is about 300,000 km/s.) How does the situation appear to Jackie?

Other than the much higher speed, this situation is just like that in Thought Experiment 1. Jackie perceives *herself* to be at rest and sees *you* moving away from her at 0.9*c*.

Thought Experiment 5 (Figure S2.8) Now, instead of throwing a baseball, you climb out of your spaceship and point a flashlight in Jackie's direction. How fast is the beam of light moving relative to Jackie?

From your point of view, Jackie and the flashlight beam are both going in the same direction. Jackie is going at 90% of the speed of light, or 0.9*c*, and the light beam is going at the full speed of light, or *c*. Thus, you see the light beam going 0.1*c* faster than Jackie.† Nothing should be surprising so far.

---

†You can't actually *see* a light beam moving forward. When we say that you "see" the beam moving at the speed of light, we really mean that you would find it to be moving at this speed if you made a careful measurement with instruments at rest in your reference frame.

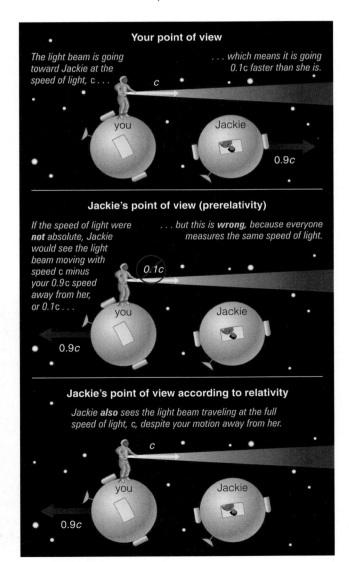

Figure S2.8 Thought Experiment 5.

From Jackie's point of view, *she* is stationary and *you* are moving away from her at 0.9c. Following the "old common sense" used in our earlier thought experiments, we would expect Jackie to see the light moving toward her at 0.1c— the light's speed of *c* minus your speed of 0.9c. *But this answer is wrong!* Relativity tells us that the speed of light is always the same for everyone. Therefore, Jackie must see the beam of light coming toward her at *c, not* at 0.1c.

In this case, you and Jackie no longer agree about her speed *relative* to the speed of the light beam: She'll see the light beam pass by her at the speed of light, *c,* but you'll see it going only 0.1c faster than she is going. By our old common sense, this result sounds preposterous. However, we found it by using simple logic, starting with the assumption that the speed of light is the same for everyone. As long as this assumption is true—and, remember, the absoluteness of the speed of light is an experimentally verified fact—our conclusions follow logically.

THINK ABOUT IT

Suppose Jackie is moving away from you at a speed of 0.99999c, which is short of the speed of light by only 0.00001c, or 3 km/s. At what relative rate would you see a light beam catch up with her? How fast will Jackie see the light going as it passes?

• Why can't we reach the speed of light?

You might be wondering what Jackie would see if she were moving away from you at the speed of light or faster— which is essentially the same question Einstein asked when

he wondered how the world would look if he could travel at or beyond the speed of light. However, once we accept the absoluteness of the speed of light, it follows that neither Jackie nor you nor any other material object can ever reach the speed of light, let alone exceed it.

Thought Experiment 6 You have just built the most incredible rocket imaginable, and you are taking it on a test ride. Soon you are going faster than anyone had ever imagined possible—and then you put the rocket into second gear! You keep going faster and faster and faster. Here is the key question: Are you ever traveling faster than the speed of light?

Before we answer this question, the fact that all motion is relative forces us to answer another question: In what reference frame is your speed being measured? Let's begin with *your* reference frame. Imagine that you turn on your rocket's headlights. Because the speed of light is the same for everyone, you must see the headlight beams traveling at the speed of light—which means they are racing away from your rocket at a speed you'll measure to be 300,000 km/s. The fact that you'll see your headlight beams racing away is true no matter how long you have been firing your rocket engines. In other words, you cannot possibly keep up with your own headlight beams.

The fact that you cannot outrace your headlight beams, combined with the fact that the speed of light is the same for everyone, means that no observer can ever see you reach or exceed the speed of light. Observers in different reference frames will measure your speed differently, but *all* observers will agree on two key points: (1) Your headlight beams are moving out ahead of you, and (2) these light beams are traveling at *c* = 300,000 km/s. Clearly, if you are being outraced

---

SPECIAL TOPIC What If Light Can't Catch You?

If you're like most students learning about relativity for the first time, you're probably already looking for loopholes in the logic of our thought experiments. For example, confronted with Thought Experiment 6, you might be tempted to ask, "What happens if you're traveling away from some planet faster than light, so the light from the planet can't catch you?" While it's surely true that light couldn't catch you if you were going faster than the speed of light, it also makes the question moot: If you can't see light from the planet, there is no way for you to know that the planet is there, so you couldn't actually make any measurement that would show your speed to be greater than the speed of light.

In fact, what relativity really tells us about the speed of light is that it is a limit on the speed at which *information* can be transmitted. There are numerous circumstances in the universe in which an object may *seem* to be exceeding the speed of light. However, these circumstances do not provide any means of sending information or objects at speeds faster than the speed of light.

As an example, imagine that you held a laser light and swept it across the sky between two stars that are separated by an angular distance of 90° and are each 10 light-years away from Earth. About 10 years from now, you'd see your laser light first make a dot (an extremely dim one!) on the surface of one star and a few seconds

later make a dot on the surface of the second star. Thus, your laser dot would seem to have traveled the many light-years from the first star to the second star in just a few seconds—which means at a speed far in excess of the speed of light. However, while the laser dot can carry information from you to each of the individual stars, it clearly cannot be transmitting any information from one star to the other. (After all, you held the laser, not someone at either star.) As a result, there's no violation of relativity.

Other examples of things that *seem* to move faster than the speed of light arise frequently in the world of quantum mechanics (see Chapter S4). According to quantum principles, measuring a particle in one place can (in certain specific circumstances) affect a particle in another place *instantaneously*—even if the particle is many light-years away. In fact, this process has been observed in laboratories over short distances. This instantaneous effect of one particle on another may at first seem to violate relativity, but it does not. The built-in randomness of quantum mechanics prevents this technique from being used to transmit useful information. Moreover, if we wish to confirm that the second particle really was affected, we will have to receive a signal carrying information about the particle—and that signal can travel no faster than the speed of light.

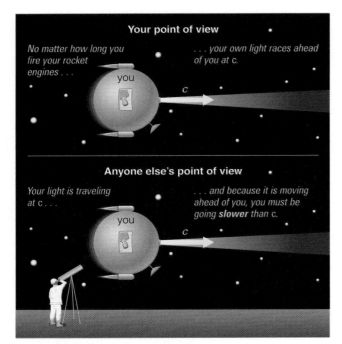

Figure S2.9 Thought Experiment 6.

by your headlights and if the headlights are traveling at the speed of light, you must be traveling *slower* than the speed of light (Figure S2.9). It does not matter who is measuring your speed. It can be you, someone on Earth, or anyone else in any other reference frame. No one can ever observe you to be traveling as fast as a light beam.

In case you are still not convinced, let's turn the situation around. Imagine that, as you race by some planet, a person on the planet turns on a light beam. Because the speed of light is absolute, you will see the light beam race past you at $c = 300,000$ km/s. The person on the planet will also see the light traveling at $c = 300,000$ km/s and will see the light outrace you. Again, everyone will agree that you are traveling slower than the speed of light.

The same argument applies to any moving object, and it is true with or without headlights. All light travels at the speed of light, including the light that reflects off an object (allowing us to see it) and the infrared light that even cool objects emit as thermal radiation [Section 5.4]. As long as the speed of light is absolute, no material object can ever keep up with the light it emits or reflects, which means no material object can reach or exceed the speed of light. Building a spaceship to travel at the speed of light is not a mere technological challenge—it simply cannot be done.

S2.3 The Reality of Space and Time

In the first section of this chapter, we listed five key ideas of special relativity. Each of these ideas is actually a prediction that arises from the two basic assumptions of the theory—that the laws of nature are the same for everyone and that everyone always measures the same speed of light.

So far, we have shown only how relativity predicts that no material object can reach or exceed the speed of light. Let's continue down the path we've started, so that we can see why our old conceptions of time and space must be revised.

• How does relativity affect our view of time and space?

The special theory of relativity tells us that measurements of time, length, and mass can all be different in different reference frames. It also tells us that different observers can disagree about whether two events are simultaneous and about the speeds of different objects. Strange as these ideas may sound, we can continue to use simple thought experiments to show why they follow directly from the premises of relativity. We'll begin by considering how relativity affects our view of time, then move on to other effects.

Time Differs in Different Reference Frames In preparation for our next thought experiment, imagine that you are on a moving train tossing a ball straight up so that it bounces straight back down from the ceiling. How do the path and speed of the ball appear to an observer along the tracks outside the train?

As shown in Figure S2.10, the outside observer sees the ball going forward with the train at the same time that it is going up and down, so the ball's path always slants forward.

Reference frame inside train

Inside the train, the ball goes straight up and down.

Reference frame outside train

Outside the train, the ball appears to be going faster: It has the same up-and-down speed, plus the forward speed of the train.

The faster the train is moving, the faster the ball appears to be going.

Figure S2.10 A ball tossed straight up and down *inside* a moving train appears to an outside observer to follow a slanted path. Thus, the outside observer sees the ball traveling at a faster total speed than does the person inside the train.

Because the outside observer sees the ball moving with this forward speed *in addition to* its up-and-down speed, while you see the ball going only up and down, he would measure a *faster* overall speed for the ball than you would. If the train is moving slowly, the outside observer sees the ball's path slant forward only slightly and would say that the ball's overall speed is only slightly faster than you would report. If the train is moving rapidly, the ball's path leans much farther forward, and the ball appears to be going considerably faster to the outside observer than to you.

Thought Experiment 7 (Figure S2.11) Inside her spaceship, Jackie has a laser on her floor that is pointed up to a mirror on her ceiling. She momentarily flashes the laser light and uses a very accurate clock to time how long it takes the light to travel from the floor to the ceiling and back. As Jackie zips by you at a speed close to the speed of light, you observe her experiment and, using your own very accurate clock, you also time the laser light's trip from Jackie's floor to her ceiling and back.

Because Jackie, the laser, and the mirror are all moving from your point of view, the light's path looks slanted as it goes from floor to ceiling and back, just as the ball tossed up and down in the train followed a slanted path to an outside observer. Thus, from your point of view, the light travels a *longer* path in going from the floor to the ceiling than it does from Jackie's point of view, just as the ball in the train took a longer path from the outside observer's point of view.

By our old common sense, this would be no big deal. You and Jackie would agree on how long it takes the light to go from the floor to the ceiling and back, just as you and an outside observer would agree on how long it takes you

to toss a ball up and down on a train. You would explain the light's longer path by saying that the light is moving faster relative to you than it is moving relative to Jackie, just as the outside observer sees the tossed ball moving faster because of the forward motion of the train. However, because we are dealing with *light,* you and Jackie must both measure its speed to be the *same*—the speed of light, or 300,000 km/s—even though you see its path slanted forward with the movement of the spaceship. And this fact has an astonishing implication.

Remember that, at any given speed, including the speed of light, traveling a longer distance must take a longer time. Because both you and Jackie see the light traveling at the same speed, but you see it traveling a longer distance, *your clock must record more time* than Jackie's as the light travels from floor to ceiling and back. And because you can see Jackie's clock as well as your own, you must see her clock running at a slower rate than yours in order for it to end up reading less elapsed time after the light completes its round trip. In other words, you will see Jackie's clock running more slowly than your own. Moreover, it doesn't matter what kind of "clock" you and Jackie use to measure the time for the light trip. *Anything* that can measure time will be going slower in Jackie's reference frame than in yours, including mechanical clocks, electrical clocks, heartbeats, and biochemical reactions. Our astonishing conclusion: From your point of view, *time itself* is running slower for Jackie.

How much slower is time running for Jackie? It depends on her speed relative to you. If she is moving slowly compared to the speed of light, you will scarcely be able to detect the slant of the light path, and your clock and Jackie's clock will tick at nearly the same rate. The faster she is moving, the more slanted the light path will appear to you and the greater the difference between the rate of her clock and that of yours. Generalizing, we reach the following conclusion:

> **From your point of view, time runs more slowly in the reference frame of anyone moving relative to you. The faster the other reference frame is moving, the more slowly time passes within it.**

This effect is called **time dilation** because it tells us that time is *dilated,* or expanded, in a moving reference frame.

The Relativity of Simultaneity Our old common sense also tells us that everyone must agree on whether two events happen at the same time or one event happens before another. For example, if you see two apples—one red and one green—fall from two different trees and hit the ground at the same time, you expect everyone else to agree that they landed at the same time (assuming you've accounted for any difference in the light travel times from the two trees). If you saw the green apple land before the red apple, you would be very surprised if someone else said that the red apple hit the ground first. Well, prepare yourself to be surprised, because our next thought experiment will show that observers in different reference frames will not necessarily

Figure S2.11 Thought Experiment 7.

agree about the order or simultaneity of events that occur in different places.

Before we go on, note that observers in different reference frames *must* agree about the order of events that occur in the *same* place. For example, suppose you grab a cookie and eat it. In your reference frame, both events (picking up the cookie and eating it) occur in the same place. Thus, it could not possibly be the case that someone else would see you eat the cookie before you pick it up.

Thought Experiment 8 Jackie has a brand-new, extra long spaceship, and she is coming toward you at a speed of 90% of the speed of light, or 0.9c (Figure S2.12a). She is in the center of her spaceship, which is totally dark except for a flashing green light at its front end and a flashing red light at its back end. Suppose that you see the green and red lights flash at *exactly the same time,* with the flashes occurring at the instant that Jackie happens to pass you. During the very short time that the light flashes are traveling toward you, Jackie's forward motion carries her toward the point where you saw the green flash occur. Thus, the green flash will reach her before the red flash, which means you'll see her illuminated first by green light and then by red light. So far, nothing should be surprising: You see the two flashes

simultaneously because you are stationary, while Jackie sees the green flash before the red flash because of her forward motion. But let's think about what happens from Jackie's point of view, in which she is the one who is stationary and you are the one who is moving.

Remember that motion cannot affect the reality of events that occur in the same place at the same time. Thus, no one can argue with the fact that you saw the green and red flashes at the same time, and no one can disagree with the fact that the green light reached Jackie before the red light. However, this agreement about events has an important implication: Because Jackie considers herself to be stationary in the center of her spaceship, the only way that the green light can reach her before the red light is if the green flash occurred first. In other words, she must see the green light flash first, and the red light flash a tiny bit later. Note that this also allows her to explain why you see the two flashes simultaneously: According to Jackie, your forward motion means the green light has to travel a little farther to reach you than the red light, so you end up seeing the two lights at the same time even though the green flash occurred first (Figure S2.12b).

Let's summarize the situation: You both agree on the indisputable reality of what each of you saw—that is, you

---

MATHEMATICAL INSIGHT **S2.1** **The Time Dilation Formula**

Look at the different paths that you and Jackie see the light take from the floor to the ceiling in Figure S2.11. We can form a right triangle from these paths, which we can use to find an exact formula for time dilation. Recall that *distance = speed × time,* and note how the sides of the triangle are constructed in Figure 1.

- *You* see the light take a slanted path from Jackie's floor to her ceiling because of her motion relative to you. Let's use *t* to represent the time that *you* measure as the light travels this path. Because the light travels at the speed of light, *c,* the length of this path is $c \times t$.
- The distance that *you* see Jackie travel while the light goes from her floor to her ceiling is $v \times t$, where *v* is her speed relative to you.
- We know that Jackie measures time differently than you, so we'll use *t'* (read "t-prime") to represent the time *she* measures as the light travels from her floor to her ceiling. The distance she sees the light travel therefore is $c \times t'$.

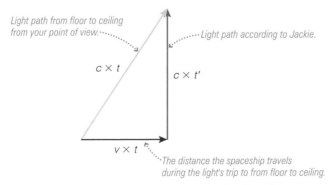

Light path from floor to ceiling from your point of view.········· ·······Light path according to Jackie.

$c \times t$

$c \times t'$

$v \times t$

········The distance the spaceship travels during the light's trip to from floor to ceiling.

Figure 1 The setup for finding the time dilation formula.

Because we have a right triangle, we can solve for the time on Jackie's clock (*t'*) in terms of the time on your clock (*t*) by using the Pythagorean theorem and a bit of algebra. (Remember that the Pythagorean theorem states that $a^2 + b^2 = c^2$ for a right triangle with side lengths *a, b,* and *c,* where *c* is the hypotenuse.)

Start with the Pythagorean theorem:

$$(ct')^2 + (vt)^2 = (ct)^2$$

Expand the squares:

$$c^2 t'^2 + v^2 t^2 = c^2 t^2$$

Subtract $v^2 t^2$ from both sides:

$$c^2 t'^2 = c^2 t^2 - v^2 t^2$$

Divide both sides by $c^2 t^2$:

$$\frac{t'^2}{t^2} = \frac{c^2 - v^2}{c^2}$$

Simplify:

$$\frac{t'^2}{t^2} = \frac{c^2}{c^2} - \frac{v^2}{c^2} = 1 - \left(\frac{v}{c}\right)^2$$

Take the square root of both sides:

$$\frac{t'}{t} = \sqrt{1 - \left(\frac{v}{c}\right)^2} \quad \text{or} \quad t' = t\sqrt{1 - \left(\frac{v}{c}\right)^2}$$

The final result, called the *time dilation formula,* tells us the *ratio* of time in a moving reference frame to time in a reference frame at rest. This ratio is graphed against speed in Figure 2. Note that $t'/t \approx 1$ at speeds that are small compared to the speed of light, meaning that clocks in both reference frames tick at about the same rate. As

Your point of view

At the instant Jackie passes by, red and green flashes occur simultaneously . . .

Jackie   0.9c

. . . and Jackie's forward motion means the green flash reaches her before the red flash.

you

a

Jackie's point of view

The green flash at the front of her spaceship occurs **before** the red flash in back . . .

Jackie

0.9c

you

. . . while your forward motion causes you to see both flashes at the same time.

b

Figure S2.12 Thought Experiment 8.

saw the two flashes at the same time, while Jackie saw the green flash before the red flash. However, you say this reality occurs because the flashes really did occur at the same time and Jackie was moving toward the green flash, while Jackie says it occurs because the green flash really occurred first while you were moving away from it. Who is right? All motion is relative, so both of you are equally correct. In other words, the flashes that are simultaneous in your reference frame occur at different times in Jackie's reference frame. That is what we mean when we say that the order or simultaneity of events depends on your frame of reference.

Effects of Motion on Length and Mass   The fact that time is different in different reference frames also implies that lengths (or distances) and masses are also affected by motion, although the reasons are a bit subtler. The following

v approaches *c*, the amount of time passing in the moving reference frame gets smaller and smaller compared to the time passing in the reference frame at rest.

Example: Suppose Jackie is moving past you at a speed of 0.9*c*. While 1 hour passes for you, how much time passes for Jackie?

Solution:

Step 1 Understand: A good way to start problems in relativity is to think about the answer we expect. The question asks about how time differs for someone moving past you, so we'll need the time dilation formula. But should your time be *t* or *t′* in the formula? Let's think about which one we expect to be longer. Because Jackie is moving past you, you should see her time running slowly. And if her time is running slowly, *less* than 1 hour should pass for Jackie while 1 hour passes for you. Notice that the square root in the time dilation formula always has a value less than 1 (since *v* is always less than *c*), which means that *t′* in the formula always ends up less than *t*. Thus, we want *t′* to represent Jackie's time and *t* to represent yours.

Step 2 Solve: We now use the time dilation formula with your time *t* = 1 hour. Jackie's speed of *v* = 0.9*c* means that *v*/*c* = 0.9. Substituting these values into the formula, we find:

$$t' = t\sqrt{1 - \left(\frac{v}{c}\right)^2}$$
$$= (1\text{ hr})\sqrt{1 - (0.9)^2}$$
$$= (1\text{ hr})\sqrt{1 - 0.81}$$
$$= (1\text{ hr})\sqrt{0.19} \approx 0.44\text{ hr}$$

Step 3 Explain: We have found that while 1 hour passes for you, only 0.44 hour, or about 26 minutes, passes for Jackie. In other words, you will see her time running less than half as fast as yours. Note that you can also find this answer from the graph in Figure 2. At a speed of *v* = 0.9*c*, the dashed lines on the graph show that the ratio of Jackie's time to your time is 0.44. Note also that Jackie's time came out shorter than yours, just as we expected when we started the problem. Although this doesn't prove that we did the calculation correctly, it at least shows that we went in the right general direction. To be confident of the actual answer, it's a good idea to recheck the calculation.

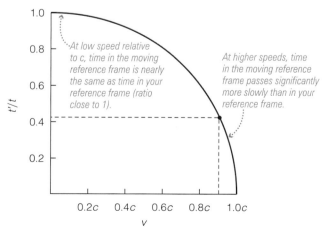

At low speed relative to c, time in the moving reference frame is nearly the same as time in your reference frame (ratio close to 1).

At higher speeds, time in the moving reference frame passes significantly more slowly than in your reference frame.

Figure 2 The time dilation factor graphed against speed.

Figure S2.13 People and objects moving relative to you are contracted in their direction of motion. The diagram shows only your point of view, but Jackie's view would be equivalent: She would see herself at "normal" size and you contracted in your direction of motion.

two thought experiments use the idea of time dilation to help us understand the effects on length and mass.

Thought Experiment 9 Jackie is back in her original spaceship, coming toward you at high speed. As usual, both you and she agree on your *relative* speed. You disagree only about who is stationary and who is moving. Now imagine that Jackie tries to measure the length of your spaceship as she passes by you. She can do this by measuring the time it takes her to pass from one end of your spaceship to the other; recall that *length = speed × time*.

In your reference frame, you say that Jackie is moving by you and therefore you see Jackie's time running slowly. Because you agree on her relative speed but her clocks record less time than yours as she passes from one end of your spaceship to the other, she must measure the length of your spaceship to be *shorter* than you measure it to be. Now, remember that your viewpoint and Jackie's viewpoint are equally valid. The fact that the laws of nature are the same for everyone means that if she measures your spaceship to be shorter than it is at rest in your reference frame, you must also measure hers to be shorter than it would be at rest, an effect called **length contraction**. Figure S2.13 shows that lengths are affected only in the direction of motion. Her spaceship is shorter from your point of view, but its height and depth are unaffected. Generalizing, we reach the following conclusion:

From your point of view, the lengths of objects moving by you (or the distances between objects moving by you) are shorter in their direction of motion than they would be if the objects were at rest. The faster the objects are moving, the shorter the lengths.

Thought Experiment 10 To see how motion affects mass, imagine that Jackie has an identical twin sister with an identical spaceship, and suppose that her sister is at rest in your reference frame while Jackie is moving by you at high speed. At the instant Jackie passes by, you give both Jackie and her sister identical pushes (Figure S2.14). That is, you push them both with the same force for the same amount of time. By our old common sense, we'd therefore expect both of them to accelerate by an identical amount, such as gaining 1 km/s of speed relative to you. But we must instead apply our new common sense, which tells us something different.

Remember that, because Jackie is moving relative to you and her sister, you'll see Jackie's time running more slowly than yours and her sister's. Thus, Jackie's clock shows that she feels the force of your push for a *shorter* time than does her sister. For example, if you and Jackie's sister measure the duration of the push to be 1 microsecond, Jackie's clock will show the duration to be less than 1 microsecond. Moreover, because Jackie feels the force of your push for a shorter time, the push must have a *smaller* effect on Jackie's velocity than it has on her sister's velocity. In other words, you'll find that your push has less effect on Jackie than on her sister, despite the fact that you gave them identical pushes. According to Newton's laws of motion [Section 4.2], the same push can have a smaller effect on Jackie's velocity only if her mass is *greater* than her sister's mass. This effect is sometimes called **mass increase:***

From your point of view, objects moving by you have greater mass than they have at rest. The faster an object is moving, the greater the increase in its mass.

---

*Newton's second law can be stated either as "force = mass × acceleration" or as "force = rate of change in momentum." (Recall that momentum = mass × velocity.) For more advanced work, most physicists prefer the second form, and they therefore prefer to think of the effect in relativity as *momentum increase* rather than mass increase. The two viewpoints are equivalent for the level of discussion in this book.

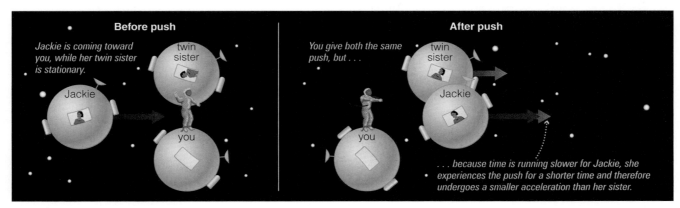

Figure S2.14 Thought Experiment 10.

Mass increase provides another way of understanding why no material object can reach the speed of light. The faster an object is moving relative to you, the greater the mass you'll find it to have. Thus, at higher and higher speeds, the same force will have less and less effect on an object's velocity. As the object's speed approaches the speed of light, you will find its mass to be heading toward infinity. No force can accelerate an infinite mass, so the object can never gain that last little bit of speed needed to push it to the speed of light.

Velocity Addition We have just one more important effect to discuss: As the next thought experiment shows, you and Jackie will disagree about the speed of a material object moving relative to both of you. The only speed you will agree on is the speed of light.

Thought Experiment 11 (Figure S2.15) Jackie is moving toward you at $0.9c$. Your friend Bob jumps into his spaceship and starts heading in Jackie's direction at $0.8c$ (from your point of view). How fast will Jackie see him approaching her?

By our old common sense, Jackie should see Bob coming toward her at $0.9c + 0.8c = 1.7c$. We know this answer is wrong, however, because $1.7c$ is faster than the speed of light. Jackie must see Bob coming toward her at a speed less than c, but he will be coming faster than the speed of $0.9c$ at which she sees you coming. Thus, she'll conclude that Bob's speed is somewhere between $0.9c$ and c. (In this case, it turns out that she'll see him coming at $0.988c$; see Mathematical Insight S2.2.)

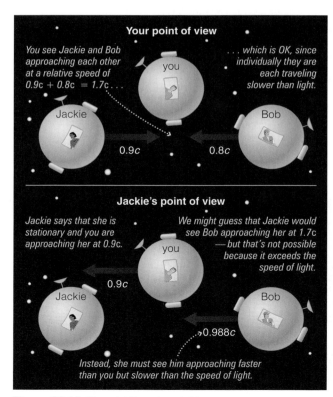

Figure S2.15 Thought Experiment 11.

• Do the effects predicted by relativity really occur?

The clear logic of our thought experiments has shown that all the major predictions of special relativity follow directly from the absoluteness of the speed of light and from the fact that the laws of nature are the same for everyone. However, remember that logic alone is not good enough in science; our conclusions remain tentative until they pass observational or experimental tests. Does relativity meet the test?

The Absoluteness of the Speed of Light The first thing we might wish to test is the surprising premise of relativity: the absoluteness of the speed of light. In principle, we can test this premise by measuring the speed of light coming from many different objects and going in many different directions and verifying that the speed is always the same.

The speed of light was first measured in the late 1600s, but experimental evidence for the *absoluteness* of the speed of light did not come until 1887, when A. A. Michelson and E. W. Morley performed their now-famous *Michelson-Morley experiment*. This experiment showed that the speed of light is not affected by Earth's motion around the Sun. Countless subsequent experiments have verified and extended the results of the Michelson-Morley experiment. The speed of light coming from opposite sides of the rotating Sun and from the orbiting stars in binary systems has also been measured, with the same result: The speed of light is always the same.

Experimental Tests of Special Relativity Although *we* cannot yet travel at speeds at which the effects of relativity should be obvious, tiny subatomic particles can reach such speeds, thereby allowing us to test the precise predictions of the formulas of special relativity.

Let's first consider one way of testing time dilation. In machines called *particle accelerators*, physicists accelerate subatomic particles to speeds near the speed of light and study what happens when the particles collide. The colliding particles have a great deal of kinetic energy, and the collisions convert some of this kinetic energy into mass-energy that emerges as a shower of newly produced particles. Many of these particles have very short lifetimes, at the end of which they decay (change) into other particles. For example, a particle called the π^+ ("pi plus") meson has a lifetime of about 18 nanoseconds (billionths of a second) when produced at rest. But π^+ mesons produced at speeds close to the speed of light in particle accelerators last much longer than 18 nanoseconds—and by precisely the amount predicted by the time dilation formula.

The same experiments allow the mass increase formula to be checked as well. The amount of energy released when high-speed particles collide depends on the particle masses and speeds. Just as relativity predicts, these masses are greater at high speed than they are at low speed—again by just the amount predicted by Einstein's formulas.

Particle accelerators even offer experimental evidence that nothing can reach the speed of light. It is relatively easy to get particles traveling at 99% of the speed of light in particle accelerators. However, no matter how much more energy is put into the accelerators, the particle speeds get only fractionally closer to the speed of light. Some particles have been accelerated to speeds within 0.00001% of the speed of light, but none have ever reached the speed of light.

Although the effects of relativity are obvious only at very high speeds, modern techniques of measuring time are so precise that effects can be measured even at ordinary speeds. For example, a 1975 experiment compared the amount of time that passed on an airplane flying in circles to the time that passed on the ground. Over 15 hours, the airborne clocks lost a bit under 6 nanoseconds to the ground clocks, matching the result expected from relativity. More recent experiments using the Space Shuttle have confirmed time dilation as well. All in all, special relativity is one of the best-tested theories in physics, and it has passed every experimental test to date with flying colors.

A Great Conspiracy? Perhaps you're thinking, "I still don't believe it." After all, how can you know that the scientists who report the experimental evidence are telling the truth? Perhaps physicists are making up the whole thing as part

of a great conspiracy designed to confuse everyone else so they can take over the world!

What you need is evidence that you can see for yourself. So . . . How about nuclear energy? Einstein's famous formula $E = mc^2$, which explains the energy released in nuclear reactions, is a direct consequence of the special theory of relativity—and one that you can derive for yourself with a bit of algebra (see Mathematical Insight S2.3). Every time you see film of an atomic bomb, or use electrical power from a nuclear power plant, or feel the energy of sunlight that the Sun generated through nuclear fusion, you are really experiencing direct experimental evidence of relativity.

Another test you can do yourself is to look through a telescope at a binary star system. If the speed of light were *not* absolute, the speed at which light from each star comes toward Earth would depend on its velocity toward us in the binary orbit (Figure S2.16). Imagine, for example, that one star is currently moving directly away from us in its orbit. If we added and subtracted speeds according to our old common sense, light from the star at this point would approach us at speed $c - v$. Some time later, when the same star is moving toward us in its orbit, its light would approach us at speed $c + v$. This light would therefore tend to catch

---

MATHEMATICAL INSIGHT S2.2 Formulas of Special Relativity

We found a formula for time dilation in Mathematical Insight S2.1. Although we will not go through the derivations, it is possible to find similar formulas for length contraction and mass increase. In summary, the three formulas are:

$$\text{time (moving frame)} = \text{time (rest frame)} \times \sqrt{1 - \left(\frac{v}{c}\right)^2}$$

$$\text{length (moving frame)} = (\text{rest length}) \times \sqrt{1 - \left(\frac{v}{c}\right)^2}$$

$$\text{moving mass} = \frac{(\text{rest mass})}{\sqrt{1 - \left(\frac{v}{c}\right)^2}}$$

There's also a simple formula for velocity addition. Suppose you see Jackie moving at speed v_1 and Jackie sees a second object moving relative to her at speed v_2. By our old common sense, you would see the second object moving at speed $v_1 + v_2$. However, the speed you actually see is:

$$\text{speed of second object} = \frac{v_1 + v_2}{1 + \left(\frac{v_1}{c} \times \frac{v_2}{c}\right)}$$

Example 1: Suppose Jackie is moving by you at $0.99c$. Because her spaceship is the same model as yours, you know that it is 100 meters long when it is at rest. How long is her spaceship as it moves by you?

Solution:

Step 1 Understand: As with all relativity problems, we start by thinking about the answer we expect. Because we are asked about the length of her spaceship when it is moving, we need the length

contraction formula. Because you see Jackie moving, you should see her length contracted, which means the answer should be a length *shorter* than the spaceship's rest length of 100 meters. Notice that the square root in the length contraction formula is the same one that is in the time dilation formula, which always has a value less than 1 (since v is always less than c). Thus, by using 100 meters as the rest length in the length contraction formula, we should come out with a moving length that is shorter than 100 meters.

Step 2 Solve: We set the rest length $= 100$ m. Jackie's speed of $0.99c$ means that $v/c = 0.99$. Substituting these values into the length contraction formula, we find:

$$\text{length (moving frame)} = (\text{rest length}) \times \sqrt{1 - \left(\frac{v}{c}\right)^2}$$
$$= (100 \text{ m}) \times \sqrt{1 - (0.99)^2} = 14 \text{ m}$$

Step 3 Explain: When Jackie is moving by you at 99% of the speed of light, you would measure her spaceship to be only 14 meters long, instead of its rest length of 100 meters. Notice that the moving length is shorter than the rest length, just as we expect.

Example 2: A fly has a mass of 1 gram at rest. It is an unusual fly, however, in that it can travel at $0.9999c$. What is the mass of the fly at that speed?

Solution:

Step 1 Understand: This time we are asked about mass, so we need the mass increase formula. In this case, we expect the mass of a moving object to be greater than its rest mass. Notice that the square root in the mass increase formula is the same one

---

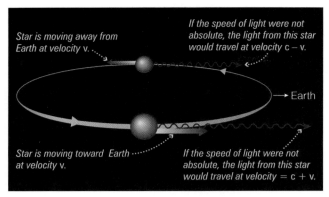

Star is moving away from Earth at velocity v.

If the speed of light were not absolute, the light from this star would travel at velocity c − v.

Earth

Star is moving toward Earth at velocity v.

If the speed of light were not absolute, the light from this star would travel at velocity = c + v.

Figure S2.16 If the speed of light were *not* absolute, the speed at which light from each star in a binary system comes toward Earth would depend on its velocity toward us in the binary orbit.

up with the light that the star emitted from the other side of its orbit. If the orbital speed and distance were just right, we might see the same star on both sides of the orbit at once! More generally, because the light from each star would come toward us at a different speed from each point in its orbit, we would see each star in multiple positions in its

orbit simultaneously, making each star appear as a short line of light rather than as a point—*if* the speed of light were not absolute. Thus, the fact that we always see distinct stars in binary systems demonstrates that the speed of light *is* absolute.*

Finally, you can explore the paradoxes that would occur if the speed of light were not absolute. For example, imagine that two cars, both traveling at about 100 km/hr, collide at an intersection (Figure S2.17). You witness the collision from far down one street. If the speed of light were *not* absolute, the light from the car that was coming toward you would have a speed of c + 100 km/hr, while the light from the other car would approach you only at a speed of c. You therefore would see the car coming toward you reach the intersection slightly *before* the other car and thus would see events unfold differently than the passengers in the car or eyewitnesses in other locations. On Earth, this difference

---

*This conclusion would not follow if light waves were carried by a medium in the same way that sound waves are carried by air. Scientists in the 1800s believed that such a medium, which they called the *ether*, permeated all of space. The Michelson-Morley experiment ruled out the existence of such a medium, so we are left with the conclusion that *c* is absolute.

---

that is in the time dilation and length contraction formulas, which means it is always less than 1. Thus, when we divide the rest mass by this square root that is less than 1, we'll end up with an answer that is greater than 1—giving us a moving mass that is larger than the rest mass, just as we expect.

Step 2 Solve: We set the fly's rest mass = 1 g, and its speed of 0.9999c means that $v/c = 0.9999$. The mass increase formula therefore gives us:

$$\text{moving mass} = \frac{(\text{rest mass})}{\sqrt{1 - \left(\dfrac{v}{c}\right)^2}}$$

$$= \frac{1\text{g}}{\sqrt{1 - (0.9999)^2}} = 70.7\text{g}$$

Step 3 Explain: At a speed of 0.9999c, or 99.99% of the speed of light, the mass of a fly that is 1 gram at rest becomes 70.7 grams, or more than 70 times its rest mass. Notice that the moving mass is greater than the rest mass, just as we expect.

Example 3: Jackie is moving toward you at 0.9c. Your friend Bob jumps into his spaceship and, from your point of view, starts heading in Jackie's direction at 0.8c (as shown in Figure S2.15). How fast will Jackie see him approaching her?

Solution:

Step 1 Understand: This is a velocity addition problem. According to Jackie, your speed is $v_1 = 0.9c$. Bob's speed *relative to you* is $v_2 = 0.8c$ in the same direction. By our old common sense, we would simply add the two velocities and find that Bob's speed

relative to Jackie is 0.9c + 0.8c = 1.7c. However, we know this cannot be right because it exceeds the speed of light. Instead, we must use the velocity addition formula given above. What should we expect for the answer? We know that since you see Jackie moving toward you at 0.9c, she must also see you moving toward her at 0.9c. We also know that Bob is moving away from you and toward Jackie, which means she must see him coming toward her even faster than she sees you coming toward her. Thus, the answer should be a speed greater than your speed of 0.9c, but less than the speed of light, c.

Step 2 Solve: We substitute the given values $v_1 = 0.9c$ and $v_2 = 0.8c$ into the velocity addition formula:

$$\begin{aligned}\text{Bob's speed} \atop \text{(relative to Jackie)} &= \frac{v_1 + v_2}{1 + \left(\dfrac{v_1}{c} \times \dfrac{v_2}{c}\right)} \\ &= \frac{0.9c + 0.8c}{1 + (0.9 \times 0.8)} \\ &= \frac{1.7c}{1.72} = 0.988c\end{aligned}$$

Step 3 Explain: Jackie sees Bob moving toward her at 0.988c, or almost 99% of the speed of light. Notice that, as we expect, Bob's speed relative to Jackie is faster than yours but slower than the speed of light.

If c were not absolute, the light from car A would be coming toward you faster than the light from car B . . .

. . . so you'd see car A reach the collision point before car B.

Result: You could not explain why the cars collide, since from your point of view they don't reach the collision point at the same time.

Figure S2.17 If the speed of light were *not* absolute, the light from a car coming toward you would approach you faster than the light from a car going across your line of sight.

would be scarcely noticeable, because 100 km/hr is only about *one-millionth* the speed of light. However, the difference would increase if you could watch from very far away. For example, suppose you had a super telescope and watched such a collision on a planet located 1 million light-years from Earth: Because the light from the first car (the one coming toward you) is coming at a speed one-millionth of the speed of light faster than the light from the second car, over a distance of 1 million light-years it would end up reaching you a full *year* ahead of the light from the second car. In other words, you would see the first car reach the intersection a *year* before the second car. This is a serious paradox: From the viewpoint of the passengers in the cars,

they have collided, yet you saw one car reach the collision point long before the other car had even started its journey! The only way to avoid the paradox is to assume that we should not have added the car's speed to the speed of light; that is, we must assume that the speed of light is absolute.

THINK ABOUT IT

The preceding paradox presents us with a choice. If the speed of light is *not* absolute, different people can witness the same events in very different ways. If the speed of light *is* absolute, measurements of time and space are relative. Einstein preferred the latter choice. Do you? Explain.

Of course, like any scientific theory, the theory of relativity can never be *proved* beyond all doubt. However, it is supported by a tremendous body of evidence, some of which you can see for yourself. This evidence is real and cannot be made to disappear. If anyone ever comes up with an alternative theory, the new theory will still have to explain the many experimental results that seem to support relativity so well.

S2.4 Toward a New Common Sense

We've used thought experiments to show that our old common sense doesn't work, and we've discussed how actual experiments verify the ideas of our thought experiments. But we haven't yet figured out what new common sense

---

SPECIAL TOPIC Measuring the Speed of Light

It is difficult to measure the speed of light, because light travels so fast. If you stand a short distance from a mirror and turn on a light, the reflection seems to appear instantaneously. Such observations led Aristotle (384–322 B.C.) to conclude that light travels at infinite speed, a view that many scientists still held as recently as the late 1600s.

One way to make the measurement easier is to place a mirror at increasingly greater distances. If the speed of light truly were infinite, the reflection would always appear instantaneously. However, if it takes time for the light to travel to and from the mirror, you should eventually find a delay between the time you turn on the light and the time you see the reflection. Galileo tried a version of this experiment using the distance between two tall hills, but he was unable to detect any delay (instead of using a mirror, he stationed an assistant on the distant hill to signal back when he saw the light). He concluded that the speed of light, if not infinite, was too fast to be measured between hills on Earth with the technology of his day.

Galileo's idea was extended to space in 1675 by the Danish astronomer Olaus Roemer. Using the four largest moons of Jupiter as his "mirrors," Roemer successfully measured a delay in the time it took for light from the Sun to be reflected and thereby made the first measurement of the speed of light. His technique worked because, by that time, the orbital periods of Jupiter's large moons were well known. Thus, it was possible to predict the precise mo-

ments at which Jupiter would eclipse each of its moons. Roemer discovered that the eclipses occurred progressively earlier than expected during those times of the year when Earth was moving toward Jupiter in its orbit and progressively later when Earth was moving away from Jupiter. He realized that the eclipses actually were occurring at the predicted times, but that the light was taking longer to reach us when Earth was farther from Jupiter. His observations proved that the speed of light is finite and allowed him to estimate its precise value. Using estimates of the Sun-Earth and Sun-Jupiter distances that were available in 1675, Roemer calculated the speed of light to be 227,000 km/s. Redoing his calculations using the presently known values of these distances yields 300,000 km/s.

As technology advanced, it became possible to measure light travel time between mirrors at much closer distances. In 1849 and 1850, the French physicists Fizeau and Foucault (also famous for the Foucault pendulum) performed a series of experiments using rotating mirrors to measure the speed of light much more precisely. Modern devices for measuring the speed of light take advantage of the wave properties of light, particularly the fact that light waves can interfere with one another. Such devices, called *interferometers*, were refined by A. A. Michelson in the early 1880s and used in the Michelson-Morley experiment. Details of how these experiments work can be found in many physics texts.

should replace the old. Fortunately, while it may take you many years to become completely comfortable with the ideas of relativity, it's relatively easy to see what you need to do to make yourself comfortable.

• How can we make sense of relativity?

Perhaps surprisingly, another thought experiment that may at first make everything seem even more bizarre will help us understand what is really going on, and thereby lead us toward a new common sense.

Thought Experiment 12 Suppose Jackie is moving by you at a speed close to the speed of light. From our earlier thought experiments, we know that you'll measure her time as running slow, her length as having contracted, and her mass as having increased. But what would *she* say?

From Jackie's point of view, she's not going anywhere—you are moving by *her* at high speed. Because the laws of nature are the same for everyone, she must reach exactly

the same conclusions from her point of view that you reach from your point of view. That is, she'll say that *your* time is running slowly, *your* length is contracted, and *your* mass has increased!

Now we have what seems to be a serious argument on our hands. Imagine that you are looking into Jackie's spaceship with a super telescope. You can clearly see that her time is running slowly because everything she does is in slow motion. You send her a radio message saying, "Hi, Jackie! Why are you doing everything in slow motion?" Because the radio message travels at the absolute speed of light, Jackie has no trouble receiving your message,* and she responds with her own radio message back to you.

As you listen to her response, you'll hear it in slow motion—"Hheeeellllloooo tthhheeerrr"—thus verifying that

---

*Of course, because of the Doppler effect [Section 5.5], she'll find that the radio waves are blueshifted from the frequency at which you send them as you come toward her, and redshifted as you move away from her. You'll find the same for her radio broadcasts. But as long as you each tune your radios to the correct Doppler-shifted frequencies, you'll have no problem hearing the broadcast message.

MATHEMATICAL INSIGHT **S2.3** **Deriving $E = mc^2$**

We can derive the formula $E = mc^2$ from the mass increase formula, which we previously wrote as:

$$\text{moving mass} = \frac{(\text{rest mass})}{\sqrt{1 - \left(\frac{v}{c}\right)^2}}$$

If we call the moving mass m and the rest mass m_0, and remember that dividing by a square root is the same as raising something to the $-\frac{1}{2}$ power, we can rewrite the mass increase formula as follows:

$$m = m_0\left(1 - \frac{v^2}{c^2}\right)^{-\frac{1}{2}}$$

To continue, we need the following mathematical approximation: If x is small compared to 1, then

$$(1 + x)^{-\frac{1}{2}} \approx 1 - \frac{1}{2}x \quad \text{(for x small compared to 1)}$$

You can verify this approximation by using your calculator to check that it holds for a few small values of x, such as $x = 0.05$ or $x = 0.001$.

Now, look again at the second equation above (the mass increase formula). The term in parentheses on the right side has the form $(1 + x)^{-\frac{1}{2}}$, as long as we identify x as $-v^2/c^2$. Thus, we can use the above approximation to rewrite this term by setting $x = -v^2/c^2$. The condition that x must be small compared to 1 now means that $-v^2/c^2$ must be small. That is, the speed v must be small compared to the speed of light, c. We start with the approximation:

$$(1 + x)^{-\frac{1}{2}} \approx 1 - \frac{1}{2}x$$

Next, we substitute $x = -v^2/c^2$:

$$\left(1 + -\frac{v^2}{c^2}\right)^{-\frac{1}{2}} \approx 1 - \frac{1}{2}\left(-\frac{v^2}{c^2}\right)$$

Finally, we simplify the result:

$$\left(1 - \frac{v^2}{c^2}\right)^{-\frac{1}{2}} \approx 1 + \frac{1}{2}\frac{v^2}{c^2}$$

Notice that the term on the left is the term that appears next to m_0 on the right side of the mass increase formula. Thus, we can replace this term with its approximation, so that the mass increase formula now reads:

$$m \approx m_0\left(1 - \frac{v^2}{c^2}\right)^{-\frac{1}{2}}$$

Next, we expand the right side by multiplying through by m_0; the formula becomes:

$$m \approx m_0 + \frac{1}{2}\frac{m_0 v^2}{c^2}$$

Finally, we multiply both sides by c^2, so the formula becomes:

$$mc^2 \approx m_0 c^2 + \frac{1}{2}m_0 v^2$$

You may recognize the last term on the right as the *kinetic energy* [Section 4.3] of an object with mass m_0. Because the other two terms also have units of mass multiplied by speed squared, they also must represent some kind of energy. Einstein recognized that the term on the left represents the *total* energy of a moving object. He then noticed that, even if the speed is *zero* ($v = 0$) so that there is *no* kinetic energy, the equation states that the total energy is *not* zero but instead is $m_0 c^2$. In other words, when an object is not moving at all, it still contains energy by virtue of its mass, and the amount of energy is equal to its rest mass times the speed of light squared. Thus, $E = mc^2$ is a direct consequence of Einstein's theory of relativity.

her time is running slowly. However, if you record the entire message and use your computer to speed up her voice so that it sounds normal, you'll hear Jackie say, "I'm not moving in slow motion, *you are!*"

You can argue back and forth all you want, but it will get neither of you anywhere. Then you come up with a brilliant idea. You hook up a video camera to your telescope and record a movie showing that Jackie's clock is moving more slowly than yours. You put your movie into a very fast rocket and shoot it off toward her. You figure that when the movie arrives and she watches it, she'll have visual proof that you are right and that she really is moving in slow motion.

Unfortunately, just before you can declare victory in the argument, you learn that Jackie had the same brilliant idea: A rocket arrives with a movie that she made. You put it into your movie player and watch what appears to be clear proof that Jackie is right—her tape shows that *you* are in slow motion! Apparently, Jackie really is seeing your time run slowly, just as you are seeing her time run slowly.

How can this be? Think back to our earlier discussion of up and down, and imagine that an American child and an Australian child are talking on the telephone. The Australian says, "Isn't the Moon beautiful up in the sky right now?" The American replies, "What are you talking about? The Moon isn't up right now!" According to childhood common sense about up and down, the two children appear to be contradicting each other and could argue endlessly. However, once they realize that up and down are measured *relative* to the center of the Earth, they realize that the argument stems only from incorrect definitions of *up* and *down* (Figure S2.18). They are both talking about the same Moon in the same place, and they argued only because they were using childhood common sense about up and down rather than the correct definitions of *up* and *down*.

In much the same way, the argument between you and Jackie arises because you are using the old common sense in which we think of space and time as absolutes and expect the speed of light to be relative. The theory of relativity tells us that we have it backward. The speed of light is the absolute, and time and space are relative.

By the new common sense, the fact that you and Jackie disagree about whose time is running slowly is no more surprising than the fact that the two children disagree about whether the Moon is up or down. The disagreement is meaningless because it relies on inadequate definitions of time and space. What counts are *results,* and every experiment that you perform will agree with every experiment that Jackie performs. You are both experiencing the same laws of nature, albeit in ways different from those our old common sense would have suggested.

• How does special relativity offer us a ticket to the stars?

One consequence of accepting the special theory of relativity is the fact that we'll never be able to travel to distant stars at a speed faster than the speed of light. However, while this fact might at first make distant stars seem forever out of reach, time dilation and length contraction actually offer a "ticket to the stars"—if we can ever build spaceships capable of traveling at speeds close to the speed of light.

Suppose you take a trip to the star Vega, about 25 light-years away, in a spaceship that travels at a speed very close to the speed of light—say, at $0.999c$. From the point of view of people on Earth, the trip to Vega must take you just over 25 years, since you are traveling at nearly the speed of light for a distance of 25 light-years. The return trip will take another 25 years. Thus, if you leave in the year 2025, you will arrive at Vega in the year 2050 and return to Earth in 2075.

However, from your point of view on the spaceship, you can legitimately claim that you and your spaceship never go anywhere; it is Earth and Vega that make the trip. That is, from your point of view, you remain stationary while Earth rushes away from you and Vega rushes toward you at $0.999c$. You'll therefore find the distance from Earth to Vega contracted from its rest length of 25 light-years; with the length contraction formula, the contracted distance turns out to be just over 1 light-year. Thus, because Vega is coming toward you at $0.999c$ and has only 1 light-year to travel to reach you from your point of view, you'll be at Vega in only about 1 year. Your return trip to Earth will also take about 1 year, so the round-trip time is only about 2 years from your point of view (Figure S2.19). If you leave at age 40, you'll return as a 42-year-old.

Although it sounds like a contradiction by our old common sense, both points of view are correct.* If you leave in 2025 at age 40, you'll return to Earth at age 42—but in the year 2075. That is, while you will have aged only 2 years,

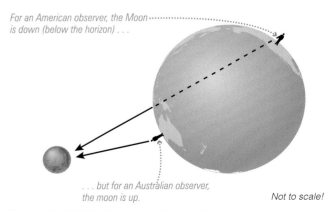

For an American observer, the Moon is down (below the horizon) . . .

. . . but for an Australian observer, the moon is up.

Not to scale!

Figure S2.18 The Moon is up for the Australian observer but down for the American observer. (Not to scale.)

---

*Given that you see the Earth-Vega distance contracted, shouldn't you also claim that it is *Earth's* time running slowly, rather than yours? This question underlies the so-called *twin paradox,* in which you make the trip while your twin sister stays home on Earth (see Chapter S3). The resolution comes from the fact that, because you must turn around at Vega, you effectively change reference frames relative to Earth at least once during your trip. A careful analysis of the changing reference frames is beyond the scope of this book, but it turns out that you do indeed measure less total time than your stay-at-home twin.

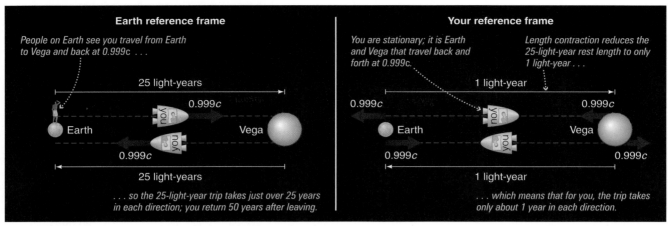

Figure S2.19 At high speed, a traveler to a distant star will age less than people back home on Earth.

all your surviving friends and family will be 50 years older than when you left.

You could make even longer trips within your lifetime with a sufficiently fast spaceship. For example, the Andromeda Galaxy is about 2.5 million light-years away, so the round trip to any star in the Andromeda Galaxy would take at least 5 million years from the point of view of observers on Earth. However, if you could travel at a speed within 50 parts in 1 trillion of the speed of light (that is, $c - 5 \times 10^{-11}c$), the trip would take only about 50 years from your point of view. You could leave Earth at age 30 and return at age 80—but you would return to an Earth on which your friends, your family, and everything you knew had been gone for 5 million years.

Thus, in terms of time, relativity offers only a one-way ticket to the stars. You can go a long distance and return to the *place* that you left, but you cannot return to the epoch from which you left.

THE BIG PICTURE

Putting Chapter S2 into Context

In this chapter, we have studied Einstein's special theory of relativity, learning that space and time are intertwined in remarkable ways that are quite different from what we might expect from everyday experience. As you look back on our new viewpoint about space and time, keep in mind the following "big picture" ideas:

- The ideas of relativity all derive from two simple facts: The laws of nature are the same for everyone, and the speed of light is absolute. The thought experiments in this chapter simply derived the consequences of these facts.

- Thought experiments are useful, but the ultimate judge of any theory is observation and experiment. The theory of relativity has been extensively tested and verified.

- Although the ideas of relativity may sound strange at first, you can understand them simply and logically if you allow yourself to develop a "new common sense" that incorporates them.

- Space and time are properties of the universe itself, and the new understanding of them gained through the theory of relativity will enable you to better appreciate how the universe works.

SUMMARY OF KEY CONCEPTS

S2.1 Einstein's Revolution

- **What are the major ideas of special relativity?** Special relativity tells us that different observers can measure time, distance, and mass differently, even though everyone always agrees on the speed of light. It also tells us that no material object can reach or exceed the speed of light (in a vacuum).

- **What is "relative" about relativity?** The theory of relativity is based on the idea that all *motion* is relative. That is, there is no correct answer to the question of who or what is really moving in the universe, so motion can be described only for one object relative to another.

S2.2 Relative Motion

- **How did Einstein think about motion?** All the predictions of the special theory of relativity are based on two assumptions: (1) The laws of nature are the same for everyone; and (2) the speed of light is the same for everyone.

- **What's surprising about the absoluteness of the speed of light?** In our everyday lives, we expect

velocities to add simply; for example, an observer on the ground should see a ball in an airplane traveling at the airplane's speed plus the ball's speed. However, this is not true for light, which everyone always measures to travel at the same speed.

- **Why can't we reach the speed of light?** Light

always travels at the same speed, so your own light (the light that you emit or reflect) is always moving ahead of you at the speed of light. All other observers will also see your light moving at the speed of light—and because it is moving ahead of you, the observers will always conclude that you are moving slower than the speed of light.

S2.3 The Reality of Space and Time

- **How does relativity affect our view of time and space?** If you observe an object moving by you

at high speed, you'll find that its time is running more slowly than yours, its length is shorter than its length when at rest, and its mass is greater

than its mass when at rest. Moreover, observers in different reference frames may disagree about whether two events are simultaneous and about the speeds of different objects.

- **Do the effects predicted by relativity really occur?** Experiments with light confirm that its speed is always the same. Experiments with subatomic particles in particle accelerators confirm the predictions of time dilation and mass increase at speeds close to the speed of light, and time dilation has been verified at relatively low speeds in aircraft and spacecraft. Nuclear power plants and nuclear bombs release energy in accordance with the formula $E = mc^2$, which is also a prediction of special relativity.

S2.4 Toward a New Common Sense

- **How can we make sense of relativity?** A person

moving by you at high speed will see exactly the same effects on you as you see on her; for example, she'll see your time running slowly while you see her time running slowly. Although this might sound contradictory, it simply tells us that time and space must be relative in much the same way that up and down are relative on Earth.

- **How does special relativity offer us a ticket to the stars?** Although the theory tells us that journeys to the stars will always take many years from the point of view of Earth, it also tells us that the time for the passengers will be much shorter if they travel at speeds close enough to the speed of light. Thus, the passengers may be able to make very distant journeys within their lifetimes, even though their friends back on Earth will not be there to greet them when they return.

EXERCISES AND PROBLEMS

For instructor-assigned homework go to Mastering**ASTRONOMY**   www.masteringastronomy.com

Review Questions

Short-Answer Questions Based on the Reading

1. What is the *theory of relativity*? How does *special relativity* differ from *general relativity*?
2. List five major predictions of the special theory of relativity.
3. What is a *paradox*? How can a paradox lead us to a deeper understanding of an issue?
4. According to the theory of relativity, what are the two absolutes in the universe? Which one is more surprising, and why?
5. What do we mean by a *frame of reference*? What is a *free-float frame*?
6. Suppose you see a friend moving by you at some constant speed. Explain why your friend can equally well say that she is stationary and you are moving by her.

7. Construct your own variation on Thought Experiment 6 to prove that no material object can be observed to travel at or above the speed of light.
8. What is *time dilation*? Explain how and why your measurements of time will differ from those of someone moving by you.
9. Explain why observers in different reference frames will not necessarily agree about the order of two events that occur in different places.
10. What is *length contraction*? How will your measurements of the size of a spaceship differ if the spaceship is moving by you compared to when it is at rest in your reference frame?
11. What is *mass increase*? How does the mass of an object moving by you compare to its rest mass?

12. Construct your own variation on Thought Experiment 9 to show why velocities must add differently than we would expect according to our "old common sense."

13. Briefly describe several experimental tests that support special relativity.

14. Why do tests of $E = mc^2$ also test special relativity? Describe several tests of $E = mc^2$ that you can see for yourself.

15. If you watch a friend moving by you, you'll say that her time is running slowly, her length is contracted, and her mass is greater than her rest mass. How will she perceive her own time, length, and mass? Why? How will she perceive *your* time, length, and mass?

16. Suppose that you could take a trip to a distant star at a speed very close to the speed of light. Explain how relativity makes it possible for you to make this trip in a reasonably short amount of time. What will you find when you return home to Earth?

Test Your Understanding

Does It Make Sense? ❓

Decide whether the statement makes sense (or is clearly true) or does not make sense (or is clearly false). Explain your reasoning. (For an example, see Chapter 1, "Does It Make Sense?")

17. Einstein proved that everything is relative.

18. An object moving by you at very high speed will appear to have a higher density than it has at rest. (*Hint:* Think about the effects on both length and mass.)

19. Suppose you and a friend are standing at opposite sides of a room, and you each pop a peanut into your mouth at precisely the same instant. According to the theory of relativity, it is possible for a person moving past you at high speed to observe that you ate your peanut before your friend ate hers.

20. Suppose you and a friend are standing at opposite sides of a room, and you each pop a peanut into your mouth at precisely the same instant. According to the theory of relativity, it is possible for a person moving past you at high speed to observe that you ate cashews rather than peanuts.

21. Relativity is "only a theory," and we really have no way to know whether any of its predictions would really occur at speeds close to the speed of light.

22. The detonation of a nuclear bomb is a test of the special theory of relativity.

23. If you could travel away from Earth at a speed close to the speed of light, you would find yourself feeling uncomfortably heavy because of your increased mass.

24. We do not have rockets that can reach the speed of light today, but someday we will be able to build more powerful rockets that will allow us to travel much faster than the speed of light.

25. If you see someone's time running slowly in a different reference frame, that person must see your time running fast.

26. If you had a sufficiently fast spaceship, you could leave today, make a round trip to a star 500 light-years away, and return home to Earth in the year 2050.

Quick Quiz

Choose the best answer to each of the following. Explain your reasoning with one or more complete sentences.

27. Which of the following is *not* relative in the special theory of relativity? (a) motion (b) time (c) the speed of light

28. Which of the following must be true of a person who shares the same *reference frame* as you? (a) The person must be sitting right next to you. (b) The person must be the same size as you. (c) The person must not be moving relative to you.

29. Which of the following best describes why your rocket could never reach the speed of light? (a) The absoluteness of the speed of light means you could never keep up with the light coming from you and your rocket. (b) Our technology has not advanced enough to make faster-than-light travel possible. (c) A rocket that could reach the speed of light would have to be bigger than the entire Earth.

30. Carla is traveling past you at a speed close to the speed of light. According to *you*, while 1 minute passes for you, how much time passes for Carla? (a) 1 minute (b) less than 1 minute (c) more than 1 minute

31. Carla is traveling past you at a speed close to the speed of light. According to *her*, while 1 minute passes for her, how much time passes for you? (a) 1 minute (b) less than 1 minute (c) more than 1 minute

32. A subatomic particle that normally decays in 1 microsecond is created in a particle accelerator and is traveling at close to the speed of light. If you measure its lifetime in that case, you'll find that it is (a) less than 1 microsecond. (b) more than 1 microsecond. (c) exactly 0.95 microsecond.

33. What does the famous formula $E = mc^2$ have to do with special relativity? (a) Nothing; it comes from a different theory. (b) It is one of the two starting assumptions of special relativity. (c) It is a direct consequence of the theory, and hence a way of testing the theory's validity.

34. The idea that everyone always measures the same speed of light is one of the two key starting points for special relativity. What provides the strongest evidence that it is true? (a) It is an experimentally verified fact. (b) Thought experiments show that we would encounter paradoxes if it were not true. (c) Einstein said so, and he was very smart.

35. If you observe people moving by you at very high speed, you will say that their time runs slowly, their lengths are contracted in the direction of motion, and their masses are increased from their rest masses. What will they say about you? (a) Your time runs slowly, your length is contracted in the direction of motion, and your mass is increased from your rest mass. (b) Your time runs fast, your length is expanded in the direction of motion, and your mass is decreased from your rest mass. (c) Your time runs fast, but your length and mass are unaffected.

36. Suppose you had a spaceship so fast that you could make a round-trip journey of 1 million light-years (in Earth's reference frame) in just 50 years of ship time. If you left in the year 2010, you would return to Earth (a) in the year 2060. (b) in the year 2110. (c) a million years from now.

Investigate Further

In-Depth Questions to Increase Your Understanding

Short-Answer/Essay Questions

37. *Stationary Bike.* Suppose you are riding on a stationary bike and the speedometer says you are going 30 km/hr. What does this number mean? What does it tell you about the idea of relative motion?

38. *Walking Trip.* Suppose you go for a walk down the street. In your own reference frame, you are going quite slowly,

probably just a few kilometers per hour. About how fast would you appear to be going according to an astronaut on the Moon? Explain.

39. *Relative Motion Practice I.* In all the following, assume that you and your friends are in free-float reference frames.
    a. Bob is coming toward you at a speed of 75 km/hr. You throw a baseball in his direction at 75 km/hr. What does he see the ball doing?
    b. Marie is traveling away from you at a speed of 120 km/hr. She throws a baseball that, according to her, is going 100 km/hr in your direction. What do you see the ball doing?
    c. José is traveling away from you at 99% of the speed of light when he turns on a flashlight and points it in your direction. How fast will the beam of light be going when it reaches you?

40. *Relative Motion Practice II.* In all the following, assume that you and your friends are in free-float reference frames.
    a. Carol is going away from you at 75 km/hr, and Sam is going away from you in the opposite direction at 90 km/hr. According to Carol, how fast is Sam going?
    b. Consider again the situation in part (a). Suppose you throw a baseball in Sam's direction at a speed of 120 km/hr. What does Sam see the ball doing? What does Carol see the ball doing?
    c. Cameron is traveling toward you at 99.9999% of the speed of light when he turns on a flashlight and points it in your direction. How fast will the beam of light be going when it reaches you?

41. *Moving Spaceship.* Suppose you are watching a spaceship go past you at a speed close to the speed of light.
    a. How do clocks on the spaceship run, compared to your own clocks?
    b. If you could measure the length, width, and height of the spaceship as it passed by, how would these measurements compare to your measurements of the spaceship's size if it were stationary?
    c. If you could measure the mass of the spaceship, how would it compare to its rest mass?
    d. How would a passenger on the spaceship view your time, size, and mass?

42. *Relativity of Simultaneity.* Consider the situation in Thought Experiment 8, about the green and red flashes of light at opposite ends of Jackie's spaceship. Suppose your friend Bob is traveling in a spaceship in the opposite direction from Jackie. Further imagine that he is also precisely aligned with you and Jackie at the instant the two flashes of light occur (in your reference frame).
    a. According to Bob, is Jackie illuminated first by the green flash or the red flash? Explain.
    b. According to you, which flash illuminates Bob first? Why?
    c. According to Bob, which flash occurred first? Explain. How does Bob's view of the order of the flashes compare to your view and to Jackie's view?

Quantitative Problems

Be sure to show all calculations clearly and state your final answers in complete sentences.

43. *Time Dilation 1.* A clever student, after learning about the theory of relativity, decides to apply his knowledge in order to prolong his life. He decides to spend the rest of his life

in a car, traveling around the freeways at 55 miles per hour (89 km/hr). Suppose he drives for a period of time during which 70 years pass in his house. How much time will pass in the car? (*Hint:* If you are unable to find a difference, be sure to explain why.)

44. *Time Dilation 2.* An even more clever student, upon realizing the folly of the student in the previous problem, decides on a better approach for prolonging her life. She decides to spend time cruising around the local solar neighborhood at a speed of 0.95c (95% of the speed of light). How much time will pass on her spacecraft during a period in which 70 years pass on Earth?

45. *Length Contraction 1.* Your friend Marta flies past you at 75% of the speed of light, traveling in a spaceship that would measure 50 meters from end to end if she were at rest in your reference frame. If you measure the length of her spaceship as she goes by, how long will it be from end to end?

46. *Length Contraction 2.* The star Sirius is located 8.6 light-years from Earth (in our Earth-based reference frame). Suppose you travel from Earth to Sirius at a speed of 92% of the speed of light. During your trip, how long would you measure the distance from Earth to Sirius to be?

47. *Mass Increase 1.* Suppose a spaceship has a rest mass of 500,000 tons. If you could measure its mass when it is traveling at half the speed of light, what would it be?

48. *Mass Increase 2.* A fly has a mass of 1 gram at rest. How fast would it have to be traveling to have the mass of a large SUV, which is about 3,000 kilograms?

49. *Time Dilation with Subatomic Particles.* Recall that a π^+ meson produced at rest has a lifetime of 18 nanoseconds $(1.8 \times 10^{-8} \text{ s})$. Thus, in its own reference frame, a π^+ meson will always "think" it is at rest and therefore will decay after 18 nanoseconds. Suppose a π^+ meson is produced in a particle accelerator at a speed of 0.998c. How long will scientists see the particle last before it decays? Briefly explain how an experiment like this helps verify the special theory of relativity.

50. *Time Dilation in the Space Shuttle.* The Space Shuttle orbits Earth at a speed of about 30,000 km/hr. While 1 hour passes on Earth, how much time passes on the Shuttle? For this problem, assume that both the Shuttle and Earth are free-float frames, although in reality they are not. (*Hint:* Don't forget to convert the Shuttle's speed to km/s before determining its fraction of the speed of light.)

51. *Travel to the Stars.* Suppose you stay home on Earth while your twin sister takes a trip to a distant star and back in a spaceship that travels at 99% of the speed of light. If both of you are 25 years old when she leaves and you are 45 years old when she returns, how old is your sister when she gets back?

52. *The Betelgeuse Red Stars.* Like the fans in Boston, the fans of interstellar baseball at Betelgeuse (in the constellation Orion) waited a long time for a championship team. In fact, the Betelgeuse Red Stars recently won the Universe Series for the first time in nearly 200,000 years. (They did, however, come very close to winning 75,000 years ago. Their defeat was sealed only when a routine ground ball went through the legs of a player named Zargon Buckner.) In hopes of winning more championships before their star explodes as a supernova—an event expected within the next 100,000 years—the Red Stars management has decided to break some league rules (hopefully without getting caught) by recruiting play-

ers from Earth. The team convinced Johan Santana to accept a lucrative offer, though in an interview with the *Intergalactic Press*, Santana said it was the travel opportunity that lured him to Betelgeuse, rather than the money or extended life span. Santana was given a ticket to travel to Betelgeuse on an express spaceship at 95% of the speed of light. During the trip, he found that, with the replacement body parts provided by the Red Stars management, his fastball was considerably improved: He was now able to throw a pitch at 80% of the speed of light. Assuming that he throws a pitch in the same direction the spacecraft is traveling, use the formula for velocity addition to calculate how fast *we* would see the ball moving if we could watch it from Earth.

53. *Racing a Light Beam, Part 1.* A long time ago, in a galaxy far, far away, there was a civilization whose inhabitants hosted an Olympic competition every four of their years. Unfortunately, the competition had become tarnished by the use of illegal substances designed to aid performance, and many great athletes were disqualified and stripped of their medals. One famed sprinter, named Jo, finally decided that it was pointless to continue racing humans. Instead, he held a press conference to announce that he would race a beam of light! Sponsors lined up, crowds gathered, and the event was sold to a pay-per-view audience. Everything was set. The starting gun was fired. Jo raced out of the starting blocks and shattered the old world record, running 100 meters in 8.7 seconds. The light beam, represented by a flashlight turned on at precisely the right moment, of course emerged from its "blocks" at the speed of light. How long did it take the light beam to cover the 100-meter distance? What can you say about the outcome of this race?

54. *Racing a Light Beam, Part 2.* Following his humiliation in the first race against the light beam (Problem 53), Jo went into hiding for the next 2 years. By that time, most people had forgotten about both him and the money they had wasted on the pay-per-view event. However, Jo was secretly in training during this time. He worked out hard and tested new performance-enhancing substances. One day, he emerged from hiding and called another press conference. "I'm ready for a rematch," he announced. Sponsors were few this time and spectators scarce in the huge Olympic stadium where Jo and the flashlight lined up at the starting line. But those who

were there will never forget what they saw, although it all happened very quickly. Jo blasted out of the starting blocks at 99.9% of the speed of light. The light beam, emitted from the flashlight, took off at the speed of light. The light beam won again—but barely! After the race, TV commentators searched for Jo, but he seemed to be hiding again. Finally, they found him in a corner of the locker room, sulking under a towel. "What's wrong? You did great!" said the commentators. Jo looked back sadly, saying, "Two years of training and experiments, for nothing!" Let's investigate what happened.

a. As seen by spectators in the grandstand, how much faster than Jo is the light beam?

b. As seen by Jo, how much faster is the light beam than he is? Explain your answer clearly.

c. Using your results from parts (a) and (b), explain why Jo can say that he was beaten just as badly as before, while the spectators can think he gave the light beam a good race.

d. Although Jo was disappointed by his performance against the light beam, he did experience one pleasant surprise: The 100-meter course seemed short to him. In Jo's reference frame during the race, how long was the 100-meter course?

Discussion Questions

55. *Common Sense.* Discuss the meaning of the term *common sense.* How do we develop common sense? Can you think of other examples, besides the example of the meanings of *up* and *down,* in which you've had to change your common sense? Do you think that the theory of relativity contradicts common sense? Why or why not?

56. *Photon Philosophy.* Extend the ideas of time dilation and length contraction to think about how the universe would look if you were a photon traveling at the speed of light. Do you think there's any point to thinking about how a photon "perceives" the universe? If so, discuss any resulting philosophical implications. If not, explain why not.

57. *Ticket to the Stars.* Suppose that we someday acquire the technology to travel among the stars at speeds near the speed of light. Imagine that many people make journeys to many places. Discuss some of the complications that would arise from people aging at different rates depending on their travels.

MEDIA EXPLORATIONS

For self-study activities go to **Mastering**ASTRONOMY www.masteringastronomy.com

Web Projects

1. *Relativity Simulations.* Explore some of the simulations of the effects of special relativity available on the Web. Write a short report on what you learn.

2. *Einstein's Life.* Learn more about Einstein's life and work and how he has influenced the modern world. Write a short essay describing some aspect of his life or work.

3. *The Michelson-Morley Experiment.* Find details about the famous Michelson-Morley experiment. Write a one-page description of the experiment and its results, including a diagram of the experimental setup.

Spacetime and Gravity

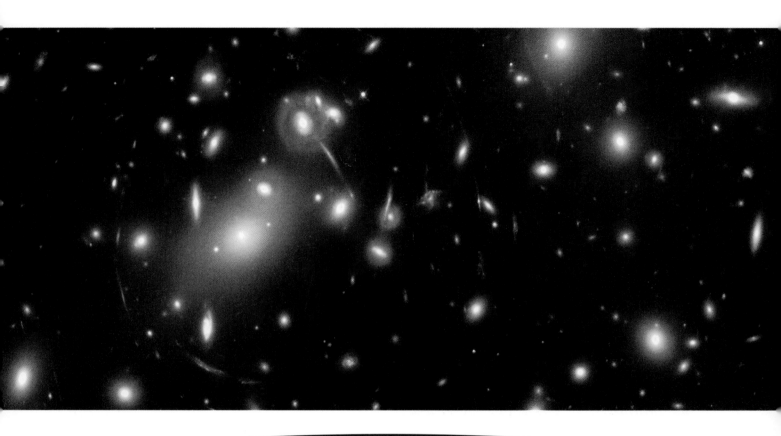

LEARNING GOALS

S3.1 Einstein's Second Revolution
- What are the major ideas of general relativity?
- Is all motion relative?

S3.2 Understanding Spacetime
- What is spacetime?
- What is curved spacetime?

S3.3 A New View of Gravity
- What is gravity?
- What is a black hole?
- How does gravity affect time?

S3.4 Testing General Relativity
- How do we test the predictions of the general theory of relativity?
- What are gravitational waves?

S3.5 Hyperspace, Wormholes, and Warp Drive
- Where does science end and science fiction begin?

The eternal mystery of the world is its comprehensibility. The fact that it is comprehensible is a miracle.

Albert Einstein

W̶hat is gravity? Newton considered gravity to be a mysterious force that somehow reaches across vast distances of space to hold the Moon in orbit around Earth and the planets in orbit around the Sun. His law of gravity explained the actions and consequences of this mysterious force but said nothing about how the force is transmitted through space.

Einstein removed the mystery of how gravity acts at a distance. As he extended his theory of relativity, Einstein found that he could explain gravity in terms of the structure of space and time. In his view, the orbits of the Moon and the planets are as natural as motion in a straight line.

As we investigate Einstein's revolutionary view of gravity, we will see that the consequences of his discoveries abound in astronomy and explain phenomena ranging from the peculiar orbit of Mercury to black holes. We will also see how space and time merge into a four-dimensional *spacetime* that determines the overall structure of the universe.

S3.1 Einstein's Second Revolution

Imagine that you and everyone around you believe the Earth to be flat. As a wealthy patron of the sciences, you decide to sponsor an expedition to the far reaches of the world. You select two fearless explorers and give them careful instructions. Each is to journey along a perfectly straight path, but they are to travel in opposite directions. You provide each with a caravan for land-based travel and boats for water crossings, and you tell each to turn back only after discovering "something extraordinary."

Some time later, the two explorers return. You ask, "Did you discover something extraordinary?" To your surprise, they answer in unison, "Yes, but we both discovered the same thing: We ran into each other, despite having traveled in opposite directions along perfectly straight paths."

Although this outcome would be extraordinarily surprising if you truly believed the Earth to be flat, we are not really surprised because we know that Earth is round (Figure S3.1a). In a sense, the explorers followed the *straightest possible paths,* but these "straight" lines follow the curved surface of the Earth.

Now let's consider a somewhat more modern scenario. You are floating freely in a spaceship somewhere out in space. Hoping to learn more about space in your vicinity, you launch two small probes along straight paths in opposite directions. Each probe is equipped with a camera that transmits pictures back to your spaceship. Imagine that, to your astonishment, the probes one day transmit pictures of each other! That is, although you launched them in op-

a Travelers going in opposite directions along paths that are as straight as possible will meet as they go around the Earth, a fact that we attribute to the curvature of Earth's surface.

b Two space probes launched in opposite directions in Earth orbit will meet as they orbit Earth, a fact that we usuallly attribute to the mysterious force of gravity.

Figure S3.1 Travelers on Earth's surface and orbiting objects follow similar-shaped paths, but we usually explain these paths in very different ways.

posite directions and neither has ever fired its engines, the probes have somehow met. This might at first sound surprising, but in fact this situation arises quite naturally with orbiting objects. If you launch two probes in opposite directions from a space station, they will meet as they orbit Earth (Figure S3.1b).

Since the time of Newton, we've generally explained the curved paths of the two probes as an effect caused by the force of gravity. However, by analogy with the explorers journeying in opposite directions on Earth, might we instead conclude that the probes meet because *space* is somehow curved? The idea that space could be curved certainly sounds strange at first. While it's easy to visualize a surface curving *through* space, our minds cannot visualize three-dimensional space itself as being curved. The idea that space can be curved lies at the heart of Einstein's second revolution—a revolutionary view of gravity contained in his *general theory of relativity,* published in 1915.

• What are the major ideas of general relativity?

Recall that the special theory of relativity applies only to situations in which we are not concerned with the effects of gravity. From this theory, we already know that space and time are inextricably linked. More specifically, special relativity tells us that the three dimensions of space and the one dimension of time together form an inseparable, *four-*dimensional combination called **spacetime**. When Einstein extended the theory of relativity to the general case that includes gravity, he discovered that matter shapes the "fabric" of spacetime in a manner analogous to the way heavy weights distort a taut rubber sheet or trampoline (Figure S3.2). Of course, we cannot place weights "upon" spacetime because all matter exists *within* spacetime, and we cannot visualize distortions of spacetime. However, using a rubber sheet as an analogy, we can begin to appreciate the principles of general relativity.

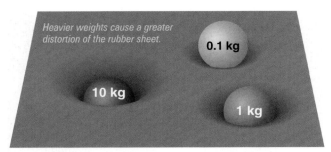

Heavier weights cause a greater distortion of the rubber sheet.

0.1 kg

10 kg

1 kg

Figure S3.2 A rubber sheet analogy to spacetime: Matter distorts the "fabric" of four-dimensional spacetime in a manner analogous to the way heavy weights distort a taut, two-dimensional rubber sheet. The greater the mass, the greater the distortion of spacetime.

It is difficult to overstate the significance of general relativity to our understanding of the universe. For example, the following ideas all come directly from Einstein's general theory of relativity:

- Gravity arises from distortions of spacetime. It is *not* a mysterious force that acts at a distance. The presence of mass causes the distortions, and the resulting distortions determine how other objects move through spacetime.
- Time runs slowly in gravitational fields. The stronger the gravity, the more slowly time runs.
- *Black holes* can exist in spacetime, and falling into a black hole means leaving the observable universe.
- The universe has no boundaries and no center, yet it might still have a finite volume.
- Large masses that undergo rapid changes in motion or structure emit *gravitational waves* that travel at the speed of light.

• Is all motion relative?

Special relativity tells us that there is no single, absolute answer to the question, "Who is moving?" when two people pass each other at a constant velocity in free-float frames. Each individual can claim to be at rest, and each claim is equally valid. However, the situation seems quite different if one of the reference frames is accelerating rather than traveling at constant velocity. Understanding this apparent difference is a good way to begin our study of the general theory of relativity.

A Thought Experiment with Acceleration   Imagine that you and your friend Jackie from Chapter S2 are both floating freely in space when you decide to fire your rocket engines. Before you fire the engines, you and Jackie are both in free-float reference frames, which means you are both floating weightlessly in your spaceships. Thus, as we saw in Chapter S2, you cannot determine who is really moving no matter what your relative velocity, because you will both experience the laws of nature in exactly the same way. Now, suppose your rocket engines start to fire with enough thrust to give you an acceleration of 1*g*, or 9.8 m/s²—which is the acceleration of gravity on Earth [Section 4.1]. Jackie sees

you accelerating away, with your speed growing ever faster, so she sends you a radio message saying, "Good-bye, have a nice trip!"

━━━━ **THINK ABOUT IT** ━━━━

Suppose you start from rest in Jackie's reference frame and she sees you accelerate at 1*g* (= 9.8 m/s²). Approximately how fast will Jackie see you going after 1 second? After 10 seconds? After a minute? (*Hint:* See Section 4.1 to review the meaning of acceleration.)

If all motion is relative, you should be free to claim that you are still stationary and that it is Jackie who is receding into the distance at ever-faster speeds. You might therefore wish to reply: "Thanks, but I'm not going anywhere. You're the one accelerating into the distance." However, this situation has a new element that was not present when we dealt with constant velocities: Because your rocket engines are firing, you feel a force inside your spaceship (Figure S3.3). In fact, because the engines are giving you an acceleration of 1*g*, you'll feel a force holding you to the floor of your spaceship that is the same as the force you feel when you stand on Earth's surface. In other words, you'll no longer be weightless but instead will be able to walk on the spaceship floor with your normal Earth weight. Thus, Jackie may respond back: "Oh, yeah? If you're not going anywhere, why are you stuck to the floor of your spaceship, and why do you have your engines turned on? Furthermore, if I'm accelerating, why am I weightless?"

You must admit that Jackie is asking very good questions. It certainly *looks* as if you really are the one who is accelerating, in which case you *cannot* legitimately claim to be stationary. In other words, it seems that motion is no longer relative when we introduce acceleration. This idea did not sit well with Einstein, because he believed that *all* motion should be relative, regardless of whether the motion was at constant velocity or included an acceleration. Einstein

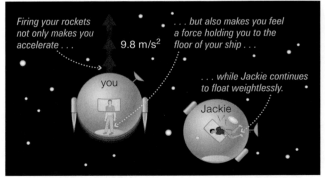

Firing your rockets not only makes you accelerate . . .

9.8 m/s²

. . . but also makes you feel a force holding you to the floor of your ship . . .

. . . while Jackie continues to float weightlessly.

you

Jackie

Figure S3.3 Out in deep space with no engines firing, you and Jackie would both float freely in your spaceships. But as soon as you start firing your engines, you feel a force that presses you against the floor of your spaceship. Jackie sees you accelerate away. You might wish to claim that *she* is accelerating away from you, but how, then, do you explain the force that you feel and the fact that Jackie remains weightless?

therefore needed some way to explain the force you feel due to firing your rocket engines without necessarily assuming that you are accelerating through space. (Einstein did not really do his thought experiments with spaceships, but the basic ideas of his thought experiments were the same.)

The Equivalence Principle In 1907, Einstein hit upon what he later called "the happiest thought of my life." His revelation consisted of the idea that, whenever you feel weight (as opposed to weightlessness), you can equally well attribute it to effects of either acceleration or gravity. This idea is called the **equivalence principle**. Stated more precisely, it says:

> **The effects of gravity are exactly equivalent to the effects of acceleration.***

To clarify the meaning of the equivalence principle, imagine that you are sitting inside with doors closed and window shades pulled down when your room is magically removed from Earth and sent hurtling through space with an acceleration of 1*g* (Figure S3.4). According to the equivalence principle, you would have no way of knowing that you've left Earth. Any experiment you performed, such as dropping balls of different weights, would yield the same results you'd get on Earth.

Now let's go back to Jackie's questions. She claims that *you* must be accelerating because you feel weight as you stand in your spaceship. However, the equivalence principle tells us that, with equal validity, you can claim to feel weight because of gravity. From this perspective, space is filled with a gravitational field pointing "downward" toward the floor of your spaceship. You are stationary only because your rocket engine *prevents* you from falling (thereby keep-

---

*Technically, this equivalence holds only within small regions of space. Over larger regions, the gravity of a massive object varies in ways that would not occur due to acceleration; for example, such variation gives rise to *tidal forces* that do not arise from acceleration.

The Equivalence Principle

You cannot tell the difference between being in a closed room on Earth . . .

. . . and being in a closed room accelerating through space at 1g.

9.8 m/s²

Figure S3.4 The equivalence principle states that the effects of gravity are exactly equivalent to the effects of acceleration. Thus, according to the equivalence principle, you cannot tell the difference between being in a closed room on Earth and being in a closed room accelerating through space at 1g.

You feel weight when you are hovering in a gravitational field . . .

you

. . . while Jackie is weightless because she is in free-fall through the gravitational field.

Jackie

9.8 m/s²

Figure S3.5 According to the equivalence principle, you can claim to be stationary in a gravitational field, using your engines to prevent you from falling. You feel weight due to gravity, while Jackie is weightless because she is in free-fall.

ing you stationary), and you feel weight just as you would if you were hovering in a helicopter on Earth. And how do you explain Jackie's weightlessness? Easy: Because she is not using her engines, she and her spaceship are falling through the gravitational field that fills space around you—and anyone in free-fall feels weightless [Section 4.1]. In essence, you can claim that the situation is much as it would be if you were hovering over a cliff while Jackie had fallen over the edge (Figure S3.5). Thus, you can respond: "Sorry, Jackie, but I still say that you have it backward. I'm using my engines to prevent my spaceship from falling, and I feel weight because of *gravity*. You're weightless because you're in free-fall. I hope you won't be hurt by hitting whatever lies at the bottom of this gravitational field!"

Because the equivalence principle says that acceleration affects the laws of physics in exactly the same way as gravity, it allows us to treat all motion as relative. Thus, it is the starting point for general relativity. Just as the consequences of special relativity follow directly from the ideas that the laws of nature are the same for everyone and that everyone always measures the same speed of light, the astounding predictions of general relativity follow from the equivalence principle. Moreover, just as special relativity led us to recognize some underlying truths about nature—such as that space and time are different for observers in different reference frames—general relativity also leads us to a new and deeper understanding of the universe.

S3.2 Understanding Spacetime

It's easy to *say* that you can equally well attribute your weight to effects of gravity or of acceleration, but the two effects tend to *look* very different. A person standing on the surface of the Earth appears to be motionless, while an astronaut accelerating through space continually gains speed. How can gravity and acceleration produce such similar effects when they look so different? According to the general

Given that the similarities in the effects of gravity and of acceleration were well known to scientists as far back as the time of Newton, you may be wondering why the equivalence principle is so surprising. The answer is that the similarities were generally attributed to coincidence—although a very puzzling coincidence. It was as if other scientists imagined that nature was showing them two boxes, one labeled "effects of gravity" and the other labeled "effects of acceleration." They shook, weighed, and kicked the boxes but could never find any obvious differences between them. They concluded: "What a strange coincidence! The boxes seem the same from the outside even though they contain different things." Einstein's revelation was, in essence, to look at the boxes and say that it is not a coincidence at all. The boxes appear the same from the outside because they contain the same thing.

In many ways, Einstein's assertion of the equivalence principle represented a leap of faith, although it was a faith he would willingly test through scientific experiment. He proposed the equivalence principle because he thought the universe would make more sense if it were true, not because of any compelling observational or experimental evidence for it at the time. This leap of faith sent him on a path far ahead of his scientific colleagues.

From a historical viewpoint, special relativity was a "theory waiting to happen" because it was needed to explain two significant problems left over from the nineteenth century: the perplexing constancy of the speed of light, demonstrated in the Michelson-Morley experiment [Section S2.3], and some seeming peculiarities of the laws of electromagnetism. Indeed, several other scientists were very close to discovering the ideas of special relativity when Einstein published the theory in 1905, and *someone* was bound to come up with special relativity around that time.

General relativity, in contrast, was a *tour de force* by Einstein. He recognized that unsolved problems remained after completing the theory of special relativity, and he alone took the leap of faith required to accept the equivalence principle. Without Einstein, general relativity probably would have remained undiscovered for at least a couple of decades beyond 1915, the year he completed and published the theory.

---

theory of relativity, the answer is that they look different only because we're not seeing the whole picture. Instead of looking just at the three dimensions of space, we must learn to "look" at the *four* dimensions of spacetime.

• What is spacetime?

The first step in understanding spacetime is understanding what we mean when we say, for example, that something is two dimensional or three dimensional. The concept of **dimension** describes the number of independent directions in which movement is possible (Figure S3.6). A **point** has zero dimensions. If you were a geometric prisoner confined to a point, you'd have no place to go. Sweeping a point back and forth along one direction generates a **line**. The line is one dimensional because only one direction of motion is possible (going backward is considered the same as going forward by a negative distance). Sweeping a line back and forth generates a two-dimensional **plane**. The two directions of possible motion are, say, lengthwise and widthwise. Any other direction is just a combination of these two. If we sweep a plane up and down, it fills **three-dimensional space**, with the three independent directions of length, width, and depth.

We live in three-dimensional space and thus cannot visualize any direction that is distinct from length, width, and depth (and combinations thereof). However, just because we cannot *see* "other" directions doesn't mean they don't exist. Thus, we can imagine sweeping space back and forth in some "other" direction to generate a **four-dimensional space**.

Although we have no hope of visualizing a four-dimensional space, we can describe it mathematically. In algebra, we do one-dimensional problems with the single variable x, two-dimensional problems with the variables x and y, and three-dimensional problems with the variables x, y, and z. A four-dimensional problem simply requires adding a fourth variable, as in x, y, z, and w. We could continue to five dimensions, six dimensions, and many more. Any space with more than three dimensions is called a **hyperspace**, which means "beyond space."

Spacetime Spacetime is a four-dimensional space in which the four directions of possible motion are length, width,

Figure S3.6 An object's number of dimensions is the number of independent directions in which movement is possible within the object. It is zero for a point, one for a line, two for a plane, and three for space.

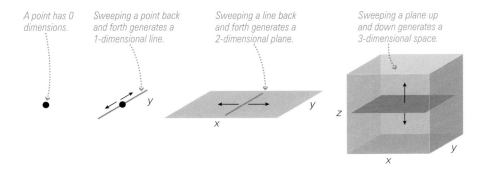

A point has 0 dimensions.

Sweeping a point back and forth generates a 1-dimensional line.

Sweeping a line back and forth generates a 2-dimensional plane.

Sweeping a plane up and down generates a 3-dimensional space.

depth, and *time*. Note that time is not "the" fourth dimension. It is simply one of the four. (However, time differs in an important way from the other three dimensions. See Mathematical Insight S3.1.)

We cannot picture all four dimensions of spacetime at once, but we can imagine what things would look like if we could. In addition to the three spatial dimensions of spacetime that we ordinarily see, every object would be stretched out through time. Objects that we see as three-dimensional in our ordinary lives would appear as four-dimensional objects in spacetime. If we could see in four dimensions, we could look through time just as easily as we look to our left or right. If we looked at a person, we could see every event in that person's life. If we wondered what really happened during some historical event, we'd simply look to find the answer.

THINK ABOUT IT

Try to imagine how you would look in four dimensions. How would your body, stretched through time, appear? Imagine that you bumped into someone on the bus yesterday. What would this event look like in spacetime?

This spacetime view of objects provides a new way of understanding why different observers can disagree about measurements of time and distance. Because we can't visualize four dimensions, we'll use a three-dimensional analogy. Suppose you give the same book to many different people and ask each person to measure the book's dimensions. Everyone will get the same results, agreeing on the three-dimensional structure of the book (Figure S3.7a). Now, suppose instead that you show each person only a two-dimensional picture of the book rather than the book itself. The pictures may look very different, even though they show the same book in all cases (Figure S3.7b). If the people believed that the two-dimensional pictures reflected reality, they might argue endlessly about what the book really looks like.

In our ordinary lives, we perceive only three dimensions, and we assume that this perception reflects reality. But spacetime is actually four dimensional. Just as different people can see different two-dimensional pictures of the same three-dimensional book, different observers can see different three-dimensional "pictures" of the same spacetime reality. These different "pictures" are the differing perceptions of time and space of observers in different reference frames. Thus, different observers will get different results when they measure time, length, or mass, even though they are all actually looking at the same spacetime reality. In the words of a famous textbook on relativity:

> *Space is different for different observers.*
> *Time is different for different observers.*
> *Spacetime is the same for everyone.* *

---

*From E. F. Taylor and J. A. Wheeler, *Spacetime Physics*, 2d ed., Freeman, 1992.

a A book has an unambiguous three-dimensional shape.

b Two-dimensional pictures of the book can look very different.

Figure S3.7 Two-dimensional views of a three-dimensional object (like a book) can appear different, even though the object has only a single real shape. In a similar way, observers in different reference frames may measure space and time differently (because they perceive only three dimensions at once) even though they are all observing the same four-dimensional spacetime reality.

Spacetime Diagrams Suppose you drive your car along a straight road from home to work as shown in Figure S3.8a. At 8:00 A.M., you leave your house and accelerate to 60 km/hr. You maintain this speed until you come to a red light, where you decelerate to a stop. After the light turns green, you accelerate again to 60 km/hr, which you maintain until you slow to a stop when you reach work at 8:10. What does your trip look like in spacetime?

If we could see all four dimensions of spacetime, we'd see all three dimensions of your car and your trip stretched out through the 10 minutes of time taken for your trip. We can't visualize all four dimensions at once, but in this case we have a special situation: Your trip progressed along only one dimension of space because you took a straight road. Therefore, we can represent your trip in spacetime by drawing a graph showing your path through one dimension of space on the horizontal axis and your path through time on the vertical axis (Figure S3.8b). This type of graph is called a **spacetime diagram**.

The car's path through four-dimensional spacetime is called its **worldline**. Any particular point along a worldline represents a particular **event**. That is, an event is a specific place and time. For example, the lowest point on the worldline in Figure S3.8b represents the event of your leaving your house: The place is 0 kilometers from home, and the time is 8:00 A.M. You can see three very important properties of any worldline in Figure S3.8b:

8:10   Car comes to a stop at work.
8:09:30   Car begins to decelerate.

        Car maintains 60 km/hr.

8:06   Car reaches 60 km/hr.

8:05   Car begins to accelerate from rest.
        Car at rest.
8:04   Car comes to stop at stop sign.
8:03:30   Car begins to decelerate.

        Car maintains 60 km/hr.

8:01   Car reaches 60 km/hr.

8:00   Car accelerates away from home.

a This diagram shows the events that occur during a 10-minute car trip from home to work on a straight road.

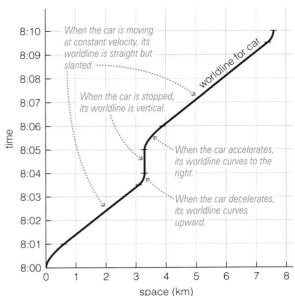

b We make a spacetime diagram for the trip by putting space (in this case, the car's distance from home) on the horizontal axis and time on the vertical axis.

Figure S3.8 A spacetime diagram allows us to represent one dimension of space and the dimension of time on a single graph.

1. The worldline of an object at rest is vertical (that is, parallel to the *time* axis). The object is going nowhere in space, but it still moves through time.
2. The worldline of an object moving at constant velocity is straight but slanted. The more slanted the worldline, the faster the object is moving.
3. The worldline of an accelerating object is curved. If the object's speed is increasing, its worldline curves toward the horizontal. If its speed is decreasing, its worldline gradually becomes more vertical.

In Figure S3.8b, we use units of minutes for time and kilometers for distance. In relativity, it is usually easier to work with spacetime diagrams in which we use units related to the speed of light, such as *seconds* for time and *light-*

seconds for distance. In that case, light follows 45° lines on a spacetime diagram because it travels 1 light-second of distance with each second of time. For example, suppose you are sitting still in your chair, so that your worldline is vertical (Figure S3.9a). If at some particular time you flash a laser beam pointed to your right, the worldline of the light goes diagonally to the right. If you flash the laser to the left a few seconds later, its worldline goes diagonally to the left. Worldlines for several other objects are shown in Figure S3.9b.

THINK ABOUT IT

Explain why, in Figure S3.9, the worldlines of all the objects we see in our everyday lives would be nearly vertical.

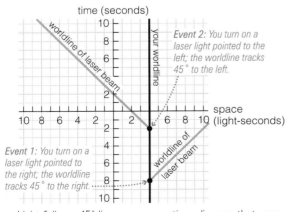

a Light follows 45° lines on a spacetime diagram that uses units of seconds for time and light-seconds for space.

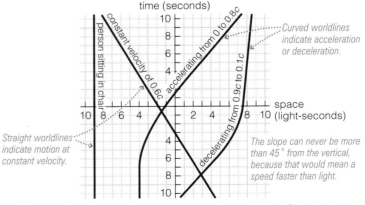

b This spacetime diagram shows several sample worldlines. Objects at rest have vertical worldlines, objects moving at constant velocity have straight but slanted worldlines, and accelerating objects have curved worldlines.

Figure S3.9 Spacetime diagrams marked with units of seconds for time and light-seconds for space.

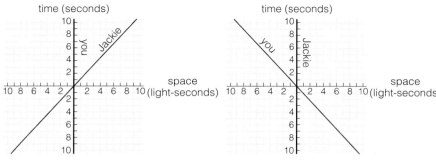

a The spacetime diagram from your point of view.

b The spacetime diagram from Jackie's point of view.

Figure S3.10 Spacetime diagrams for the situation in which Jackie is moving by you at 0.9c.

We can use spacetime diagrams to clarify the relativity of time and space. Suppose you see Jackie moving past you in a spaceship at 0.9c. Figure S3.10a shows the spacetime diagram from your point of view: You are at rest and therefore have a vertical worldline, while Jackie is moving and has a slanted worldline. Of course, Jackie claims that *you* are moving by her and therefore would draw the spacetime diagram shown in Figure S3.10b, in which her worldline is vertical and yours is slanted. Special relativity tells us

that you would see Jackie's time running slowly, while she sees *your* time running slowly (along with effects on length and mass). We know there is no contradiction here, but simply a problem with our old common sense about space and time.

From a four-dimensional perspective, the problem is that the large angle between your worldline and Jackie's means that neither of you is looking at the other "straight-on" in spacetime. Thus, you and Jackie are both looking

---

MATHEMATICAL INSIGHT S3.1 Spacetime Geometry

Spacetime geometry is easy in principle, because it simply requires adding a fourth dimension (time) to our usual three dimensions of space. However, spacetime has some surprising properties, because time enters the equations of spacetime geometry differently than do the three dimensions of space.

To gain insight into the nature of spacetime geometry, consider two points in a plane separated by amounts $x = 3$ along the horizontal axis and $y = 4$ along the vertical axis (see the left side of the figure below). You may recall from geometry that the *distance* between the two points is given by $\sqrt{x^2 + y^2}$. Now consider the same two points viewed from a coordinate system that happens to be rotated so that both points lie along the *x*-axis, as in the right side of the figure. In this coordinate system, the horizontal separa-

tion is $x = 5$ and the vertical separation is $y = 0$. However, the distance $\sqrt{x^2 + y^2}$ between the points is still the same.

This fact should not be surprising: *Distance* is a real, physical quantity, while the *x* and *y* separations are artifacts of a chosen coordinate system. The same idea holds if we add a *z*-axis, perpendicular to both *x* and *y* (you can represent the *z*-axis with a pencil that sticks straight up out of the page), to make a three-dimensional coordinate system. Different stationary observers using different coordinate systems can disagree about the *x*, *y*, and *z* separations, but they will always agree on the distance $\sqrt{x^2 + y^2 + z^2}$.

We can think of spacetime as having a fourth axis, which we will call the *t*-axis, for time. We might expect that, just as different observers always agree on the three-dimensional distance between two points, they will also agree on some kind of four-dimensional "distance" that has the formula $\sqrt{x^2 + y^2 + z^2 + t^2}$. However, it turns out that different observers will instead agree on the value of the quantity $\sqrt{x^2 + y^2 + z^2 - t^2}$. This value is called the *interval*. (Technically, the interval formula should use *ct* rather than *t* so that all the terms have units of length.) That is, different observers can disagree about the values of *x*, *y*, *z*, and *t* separating two events, but all will agree on the interval between the two events.

The *minus sign* that goes with the time dimension in the interval formula is what makes the geometry of spacetime surprisingly complex. For example, the three-dimensional distance between two points can be zero only if the two points are in the same place, but the interval between two events can be zero even if they are in different places in spacetime, as long as $x^2 + y^2 + z^2 = t^2$. For example, the interval is zero between any two events connected by a light path on a spacetime diagram. If you study general relativity further, you will see many more examples of how this strange geometry comes into play.

The x- and y-coordinates of two points can be different in different coordinate systems . . .

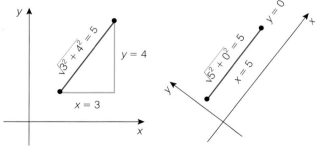

. . . but the distance between them (red line) is the same either way.

The distance between two points in a plane is the same regardless of how we set up a coordinate system.

at the same four-dimensional reality but from different three-dimensional perspectives. It's not surprising that, like two people looking at each other cross-eyed, if you see Jackie's time running slowly, she sees the same thing when she looks at you.

• **What is curved spacetime?**

So far, we've been viewing spacetime diagrams drawn on the flat pages of this book. However, as we discussed in the beginning of this chapter, spacetime can be curved. What do we mean by curved spacetime?

It's easy to visualize the curvature of a *two*-dimensional surface, such as the surface of a bent sheet of paper or the surface of the Earth. These surfaces are still two-dimensional despite their curvature because they still allow only two independent directions of travel. For example, we usually identify the two independent directions of travel on Earth's surface as north-south (changes in latitude) and east-west (changes in longitude). Unfortunately, we have no hope of visualizing the curvature of three-dimensional space, let alone of four-dimensional spacetime. After all, we can visualize the curvature of two-dimensional surfaces because they curve *through* the third dimension of space. By analogy, we'd therefore need extra dimensions to visualize the curvature of space or spacetime, but we cannot see any dimensions beyond those of three-dimensional space. Nevertheless, we can determine whether space or spacetime is curved by identifying the rules of geometry that apply.

Three Basic Types of Geometry Because we cannot visualize curved space or spacetime, we'll instead use two-dimensional surfaces as an analogy. Consider Earth's curved, two-dimensional surface. Because Earth's surface is curved everywhere, there really is no such thing as a "straight" line on the Earth. For example, if you take a piece of string and lay it across a globe, it will inevitably curve around the surface. Thus, we can immediately see a fundamental difference between the rules of geometry on a curved surface like that of Earth and the more familiar rules of geometry on a flat plane. Recall that the shortest

distance between two points in a flat plane is always a straight line. Clearly, this rule cannot be true on Earth's surface, because there is no such thing as a straight line!

What rule replaces this shortest-distance rule on Earth's surface? If you experiment by measuring pieces of string stretched in different ways between two points on a globe, you'll find that the shortest and *straightest possible* path between two points on Earth's surface is a piece of a **great circle**—a circle whose center is at the center of Earth (Figure S3.11a). For example, the equator is a great circle, and any "line" of longitude is part of a great circle. Note that circles of latitude (besides the equator) are *not* great circles because their centers are *not* at the center of Earth. Thus, if you are seeking the shortest and straightest route between two cities, you must follow a *great-circle route*. For example, Philadelphia and Beijing are both at about 40°N latitude, but the shortest route between them does *not* follow the circle of 40°N latitude. Instead, it follows a great-circle route that extends far to the north (Figure S3.11b).

Find a globe and locate New Orleans and Katmandu (Nepal). Explain why the shortest route between these two cities goes almost directly over the North Pole. Why do you think airplanes try to follow great-circle routes as closely as possible?

Other familiar rules of geometry in a flat plane also are different on Earth's curved surface. Figure S3.12a shows the straight-line rule and several other geometrical rules on a flat surface, and Figure S3.12b shows how these rules differ for a spherical surface like that of Earth; in the latter case, we must draw "lines" as portions of great circles, because those are the shortest and straightest possible paths. Notice, for example, that lines that are anywhere parallel on a flat plane stay parallel forever, while lines that start out parallel on a sphere eventually converge (just as lines of longitude start out parallel at Earth's equator but all converge at the North and South Poles). Similarly, the sum of the angles in a triangle is always 180° in a flat plane but is greater than 180° on the surface of a sphere, and the circumference of

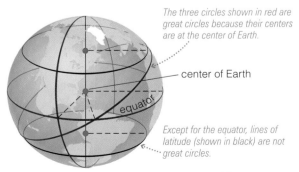

a A great circle is any circle on the surface of Earth that has its center at the center of Earth.

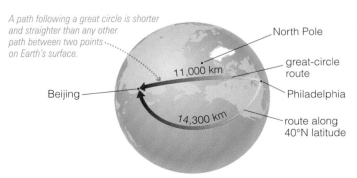

b The shortest and straightest possible path between two points on Earth is always a piece of a great circle.

Figure S3.11 The straightest possible path between any two points on a sphere must be a segment of a great circle.

a Rules of flat geometry.

b Rules of spherical geometry.

c Rules of saddle-shaped geometry.

Figure S3.12 These diagrams contrast three basic types of geometry.

a circle is $2\pi r$ in a flat plane but is *less* than $2\pi r$ on a spherical surface. Generalizing these ideas to more than two dimensions, we say that space, or spacetime, has a **flat geometry** if the rules of geometry for a flat plane hold. (Flat geometry is also known as *Euclidean geometry,* after the Greek mathematician Euclid [c. 325–270 B.C.].) For example, if the circumference of a circle in space really *is* $2\pi r$, then space has a flat geometry. However, if the circumference of a circle in space turns out to be less than $2\pi r$, we say that space has a **spherical geometry** because the rules are those that hold on the surface of a sphere.

Flat and spherical geometries are two of three general types of geometry. The third general type of geometry is called **saddle-shaped geometry** (also called *hyperbolic geometry*) because its rules are most easily visualized on a two-dimensional surface shaped like a saddle (Figure S3.12c). In this case, lines that start out parallel eventually diverge, the sum of the angles in a triangle is *less* than 180°, and the circumference of a circle is *greater* than $2\pi r$.

The Geometry of the Universe The actual geometry of spacetime turns out to be a mixture of all three general types. Earth's surface again provides a good analogy. When we view only small portions of Earth's surface, some regions appear flat while others are curved with hills and valleys. However, when we expand our view to the entire Earth, it's clear that the overall geometry of Earth's surface is like

the surface of a sphere. In a similar way, but in two more dimensions, the four-dimensional spacetime of our universe obeys different geometrical rules in different regions but presumably has some overall shape. This overall shape must be one of the three general types of geometry shown in Figure S3.12: flat, spherical, or saddle shaped. This fact explains how it is possible for the universe to have no center and no edges, an idea we first discussed in Chapter 1.

In geometry, a plane is infinite in extent, which means it has no center or edges. An idealized, saddle-shaped (hyperbolic) surface is also infinite in extent. Thus, if the universe has either a flat or a saddle-shaped geometry overall, then spacetime is infinite and the universe has no center and no edges.

In contrast, if the overall geometry of the universe is spherical, then spacetime is finite, much like the surface of Earth. However, it would still have no center or edges. Just as you can sail or fly around Earth's surface endlessly, you could fly through the universe forever and never encounter an edge. And just as the *surface* of Earth has no center— New York is no more "central" than Beijing or any other place on Earth's surface—there would be no center to the universe. (Of course, the three-dimensional Earth *does* have a center, but this center is not part of the two-dimensional surface of Earth and therefore plays no role in our analogy.)

"Straight" Lines in Curved Spacetime The rules of geometry give us a way to determine the geometry of any localized region of spacetime. In particular, we can learn the geometry of spacetime by observing the paths of objects that are following the *straightest possible* paths between two points in spacetime. For example, if the straightest possible path is truly straight, we know that spacetime is flat in that region. If the straightest possible path is curved, then the shape of the curve must be telling us the shape of spacetime. However, given that we can visualize neither the time part of spacetime nor the curvature of spacetime, how can we know whether an object is traveling on the straightest possible path through spacetime?

Einstein used the equivalence principle to provide the answer. According to the equivalence principle, we can attribute a feeling of weight either to experiencing a force generated by acceleration or to being in a gravitational field. Similarly, any time we feel weightless, we may attribute it either to being in free-fall or to traveling at constant velocity far from any gravitational fields. Because traveling at constant velocity means traveling in a straight line, Einstein reasoned that objects experiencing weightlessness for *any* reason must be traveling in a "straight" line—that is, along a line that is the straightest possible path between two points in spacetime. In other words:

> **If you are floating freely, then your worldline is following the *straightest possible path* through spacetime. If you feel weight, then you are not on the straightest possible path.**

This fact provides us with a remarkable way to examine the geometry of spacetime. Recall that any orbit is a free-fall trajectory [Section 4.1]. The Space Station is always free-falling toward Earth, but its forward velocity always moves it ahead just enough to "miss" hitting the ground. Earth is constantly free-falling toward the Sun, but our planet's orbital speed keeps us going around and around instead of ever hitting the Sun. According to the equivalence principle, all orbits must therefore represent paths of objects that are following the straightest possible path through spacetime. Thus, the shapes and speeds of orbits reveal the geometry of spacetime, which leads us to an entirely new view of gravity.

THINK ABOUT IT

Suppose you are standing on a scale in your bathroom. Is your worldline following the straightest possible path through spacetime? Explain.

S3.3 A New View of Gravity

Newton's law of gravity claims that every mass exerts a gravitational attraction on every other mass, no matter how far away they are from each other. However, on close examination, this idea of "action at a distance" is rather mysterious. For example, how does Earth feel the Sun's attraction and know to orbit it? Newton himself was troubled by this idea. A few years after publishing his law of gravity in 1687, Newton wrote:

That one body may act upon another at a distance through a vacuum, . . . and force may be conveyed from one to another, is to me so great an absurdity, that I believe no man, who has . . . a competent faculty in thinking, can ever fall into it.*

---

*Letter from Newton, 1692–1693, as quoted in J. A. Wheeler, *A Journey into Gravity and Spacetime,* Scientific American Library, 1990, p. 2.

Nevertheless, for more than 230 years after Newton published his gravitational law, no one found any better way to explain gravity's mysterious "action at a distance." Einstein changed all that, when he realized that the equivalence principle allowed him to explain the action of gravity without requiring any long-distance force.

• What is gravity?

Einstein's general theory of relativity removes the idea of "action at a distance" by stating that Earth feels *no* force tugging on it in its orbit, and therefore follows the straightest possible path through spacetime. Thus, the fact that Earth goes around the Sun tells us that spacetime itself is curved near the Sun. In other words:

What we perceive as gravity arises from the curvature of spacetime.

Rubber Sheet Analogy We cannot actually picture the curvature of spacetime, but a two-dimensional analogy can help us understand the idea. We represent spacetime in the analogy with a stretched rubber sheet. To make the analogy work, we have to ignore any effects of friction on the rubber sheet, because there is no friction in space.

Figure S3.13a shows a flat rubber sheet representing spacetime in a region where it has a flat geometry. Notice that the radial distances between each of the circles shown on the sheet are the same, and all the circles have circumferences that follow the flat geometry formula of $2\pi r$. If you rolled a marble across this frictionless sheet, it would roll in a straight line at constant speed. This fact essentially illustrates Newton's first law of motion, in which objects move at constant velocity when they are not affected by gravity or any other forces.

Figure S3.13b shows what happens to spacetime around the Sun. We represent the Sun with a heavy mass on the rubber sheet, which causes the sheet to curve and form

In flat regions of spacetime, freely moving objects move in straight lines.

The mass of the Sun causes spacetime to curve . . .

. . . so freely moving objects (such as planets and comets) follow the straightest possible paths allowed by the curvature of spacetime.

Mars

comet

Earth

Sun

Circles that were evenly spaced in flat spacetime become more widely spaced near the central mass.

a On a flat rubber sheet, evenly spaced circles all have circumference $2\pi r$.

b The Sun curves spacetime much like a heavy weight curves a rubber sheet.

Figure S3.13 According to general relativity, planets orbit the Sun for much the same reason that you can make a marble go around in a salad bowl: Each planet is going as straight as it can, but the curvature of spacetime causes its path through space to go round and round.

a bowl-like depression. The circles that were evenly spaced on the flat sheet now become more widely separated (with circumferences increasingly less than $2\pi r$) near the bottom of the bowl, showing that gravity becomes stronger and the curvature of spacetime becomes greater as we approach the Sun's surface. (Notice that the curvature does *not* continue to increase with depth inside the Sun, because the strength of gravity actually weakens near the Sun's center.) If you rolled marbles on this rubber sheet, they could not go in straight lines because the sheet itself is curved. Instead, the marbles would follow the *straightest possible* paths given the curvature of the sheet. A particular marble's path would depend on the speed and direction with which you rolled it. You'd find that marbles rolled relatively slowly and close to the center would follow circular or elliptical "orbits" around the center of the bowl, while marbles rolled from farther away or at higher speeds could loop around the center on unbound parabolic or hyperbolic paths.

By analogy, general relativity tells us that, depending on their speed and direction, planets or other objects moving freely in space can follow circular, elliptical, or unbound parabolic or hyperbolic orbits—the same orbital shapes that Newton's universal law of gravitation allows [Section 4.4]. However, the explanation for these orbits is now quite different from that in Newton's view of gravity. Rather than orbiting because of a mysterious force exerted on them by the distant Sun, the planets orbit because they follow the *straightest possible paths* allowed by the shape of spacetime around them. The central mass of the Sun is not grabbing them, communicating with them, or doing anything else to influence their motion. Instead, it is simply dictating the shape of spacetime around it. In other words:

A mass like the Sun causes spacetime to curve, and the curvature of spacetime determines the paths of freely moving masses like the planets.

Weightlessness in Space This idea gives us a new way to explain the weightlessness of astronauts in space. Just as the Sun curves spacetime into a "bowl shape" (but in four dimensions) that makes the straightest possible paths of the planets go round and round, Earth also curves spacetime in a way that makes orbiting spacecraft go round and round. In other words, spacecraft orbit Earth because, as long as their engines are off and they are unaffected by atmospheric drag, circular or elliptical orbits are the straightest possible paths they can follow through spacetime in Earth's vicinity. Thus, instead of having to invoke the idea of free-fall caused by a gravitational attraction to Earth, we can explain the weightlessness of astronauts in the Space Station simply by recognizing that they are following the straightest possible paths through spacetime.

The same idea holds true for any other orbital trajectory. For example, if we launched a human mission to Mars, we would need to give the spaceship escape velocity from Earth. In the rubber sheet analogy, this means launching it with enough speed so that it can escape from the bowl-shaped region around Earth, like a marble shot fast enough to roll

out of the bowl and onto the flatter region far away from it. Except when their rockets are firing, the astronauts would still be weightless throughout the trip because they would be following the straightest possible path. Firing the engines, either to accelerate away from Earth or to decelerate near Mars, would make the spaceship deviate from the straightest possible path, so the astronauts would feel weight during those portions of their journey.

Limitations of the Analogy The rubber sheet analogy is useful for understanding how mass affects spacetime, but it also has limitations because it is a two-dimensional representation of a four-dimensional reality. In particular, the analogy has three important limitations that you should keep in mind whenever you use it:

- The rubber sheet is supposed to represent the universe, but it makes no sense to think of placing a mass like the Sun "upon" the universe. Instead, we should think of the masses as being *within* the rubber sheet.
- The rubber sheet allows us to picture only two of the three dimensions of space. For example, it allows us to show that different planets orbit at different distances from the Sun and that some have more highly elliptical orbits than others, but it does not allow us to show the fact that the planets do not all orbit the Sun in precisely the same plane.
- The rubber sheet analogy does not show the *time* part of spacetime at all. Bound orbits on the sheet or in space appear to return to the same point with each circuit of the Sun. However, objects cannot return to the same point in spacetime, because they always move forward through time. For example, with each orbit of the Sun, Earth returns to the same place in space (relative to the Sun) but to a time that is a year later (Figure S3.14).

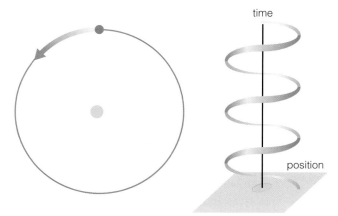

a If we ignore time, Earth appears to return to the same point with each orbit of the Sun.

b If we include a time axis, we see that Earth never returns to the same point in spacetime because it always moves forward in time.

Figure S3.14 Earth's path through spacetime.

• What is a black hole?

Greater curvature of spacetime means stronger gravity, and the rubber sheet analogy suggests two basic ways to increase the strength of gravity. First, a larger mass causes greater curvature at any particular distance away from it. For example, the Sun curves spacetime more than any planet, and Earth curves spacetime more than the Moon. Note that this idea is consistent with Newton's law of gravity, in which increasing the mass of an object increases the gravitational attraction at all distances.

The second way to increase the curvature of spacetime around an object is to leave its mass alone but increase its density by making it smaller in size. For example, suppose we could compress the Sun into a type of "dead" star called a *white dwarf* [Section 18.1]. Because its total mass would still be the same, there would be no effect on the curvature of spacetime far from the Sun. However, spacetime would be much more curved near the compressed Sun's surface, reflecting the fact that gravity is much stronger on the surface of a compressed white dwarf than on the Sun. Again, the idea that the surface gravity on an object of a particular mass grows stronger as the object shrinks in radius is consistent with Newton's law of gravity.

Now, imagine that we could continue to compress the Sun to smaller and smaller sizes. Far from the Sun, this compression would have no effect at all, because the same total mass would still be causing the curvature of spacetime. Near the Sun's surface, however, spacetime would become increasingly curved as we shrank the Sun in size. In fact, if we shrank the Sun enough, we could eventually curve spacetime so much that it would become a bottomless pit—a hole in the observable universe. This is what we call a **black hole** (Figure S3.15). Note that Newton's view of gravity does not really have any analog to a black hole, because it does not envision the possibility of holes in the universe.

Thus, a black hole is a place where spacetime is so curved that nothing that falls into it can ever escape. The boundary that marks the "point of no return" is called the **event horizon**, because events that occur within this boundary can have no influence on our observable universe. The idea

of black holes is so bizarre that for decades after Einstein published his general theory of relativity, most scientists did not think they could really exist. However, we now have very strong evidence suggesting that black holes are in fact quite common. We'll discuss the nature of black holes in more detail in Chapter 18, and will discuss evidence for their existence in both Chapters 18 and 21.

• How does gravity affect time?

Given that gravity arises from the curvature of spacetime, you should not be surprised to learn that gravity affects time as well as space. We can learn about the effects of gravity on time by considering the effects of accelerated motion and then invoking the equivalence principle.

Imagine that you and Jackie are floating weightlessly at opposite ends of a spaceship. You both have watches that flash brightly each second, which you synchronized beforehand. Because you are both floating freely with no relative motion between you, you are both in the same reference frame. Therefore, you will see each other's watches flashing at the same rate.

Now suppose you fire the spaceship engines so that the spaceship begins to accelerate, with you at the front and Jackie at the back. When the ship begins accelerating, you and Jackie will no longer be weightless. The acceleration introduces an even more important change into the situation, which we can understand by imagining the view of someone floating weightlessly outside the spaceship. Remember that observers moving at different relative speeds are in different reference frames. When the spaceship is accelerating, its speed is constantly increasing relative to the outside observer, which means that both you and Jackie are constantly changing reference frames. Moreover, the flashes from your watches take a bit of time to travel the length of the spaceship. Thus, by the time a particular flash from Jackie's watch reaches you (or a flash from your watch reaches Jackie), both your reference frames are different from what they were at the time the flash was emitted.

Because you are in the *front* of the accelerating spaceship, your changing reference frames are always carrying

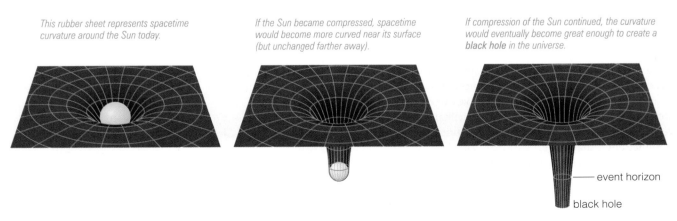

This rubber sheet represents spacetime curvature around the Sun today.

If the Sun became compressed, spacetime would become more curved near its surface (but unchanged farther away).

If compression of the Sun continued, the curvature would eventually become great enough to create a **black hole** in the universe.

event horizon

black hole

Figure S3.15 Interactive Figure ▶ According to general relativity, a black hole is like a bottomless pit in spacetime. Once an object crosses the event horizon, it has left our observable universe.

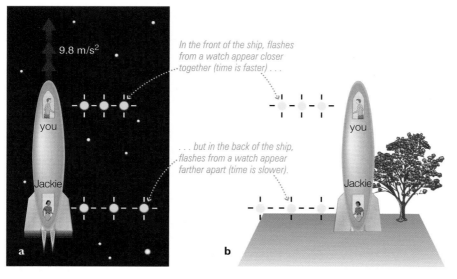

In the front of the ship, flashes from a watch appear closer together (time is faster) . . .

. . . but in the back of the ship, flashes from a watch appear farther apart (time is slower).

a In an *accelerating* spaceship (but *not* in one at constant velocity), time must run faster at the front end and more slowly at the back end. The yellow dots represent the flashes from the watches, and the spacing between the dots represents the time between the flashes.

b By the equivalence principle, time must also run more slowly at lower altitudes in a gravitational field.

Figure S3.16 Gravity causes time to run more slowly at lower altitudes than at higher altitudes, an effect called *gravitational time dilation*. (Note that the effect occurs even in a uniform gravitational field; that is, it does not depend on the additional fact that gravity tends to weaken at higher altitudes.)

you *away* from the point at which each of Jackie's flashes is emitted. Thus, the light from each of her flashes will take a little *longer* to reach you than it would if the ship were not accelerating. As a result, instead of seeing Jackie's flashes 1 second apart, you'll see them coming a little *more* than 1 second apart. That is, you'll see Jackie's watch flashing more slowly than yours (Figure S3.16a). You will therefore conclude that time is running more slowly at the back end of the spaceship.

From Jackie's point of view at the *back* of the accelerating spaceship, her changing reference frames are always carrying her *toward* the point at which each of your flashes is emitted. Thus, the light from each of your flashes will take a little *less* time to reach her than it would if the ship were not accelerating, so she'll see them coming a little *less* than 1 second apart. She will see your watch flashing *faster* than hers and conclude that time is running *fast* at the front end of the spaceship. Note that you and Jackie agree: Time is running more slowly at the back end of the spaceship and faster at the front end. The greater the acceleration of the spaceship, the greater the difference in the rate at which time passes at the two ends of the spaceship.

Now we apply the equivalence principle, which tells us that we should get the same results for a spaceship at rest in a gravitational field as we do for a spaceship accelerating through space. Thus, if the spaceship were at rest on a planet, time would also have to be running more slowly at the bottom of the spaceship than at the top (Figure S3.16b). That is, time must run more slowly at lower altitudes than at higher altitudes in a gravitational field. This effect is known as **gravitational time dilation**.

The stronger the gravity—and hence the greater the curvature of spacetime—the larger the factor by which time runs slowly. On an object with relatively weak gravity, like Earth, the slowing of time is barely detectable compared to the rate at which time passes in deep space. However, time

runs noticeably more slowly on the surface of the Sun than on Earth, and more slowly on the surface of a white dwarf star than on the Sun. Perhaps you've already guessed that the extreme case is a black hole: To anyone watching from a distance, time comes to a stop at the event horizon. If you could observe clocks placed at varying distances from the black hole, you'd see that clocks nearer the event horizon run more slowly and clocks *at* the event horizon would show time to be frozen.

THINK ABOUT IT

Where would you age more slowly, on Earth or on the Moon? Would you expect the difference to be significant? Explain.

S3.4 Testing General Relativity

Starting from the principle of equivalence, we've used logic and analogies to develop the ideas of general relativity. However, as always, we should not accept these logical conclusions unless they withstand observational and experimental tests.

• How do we test the predictions of the general theory of relativity?

Like the predictions of special relativity, those of general relativity have faced many tests and have passed with flying colors. Let's examine some of the most important tests of general relativity.

Mercury's Peculiar Orbit The first observational test passed by the theory of general relativity concerned the orbit of the planet Mercury. Newton's law of gravity predicts that Mercury's orbit should precess slowly around the Sun because of the gravitational influences of other planets (Figure S3.17). Careful observations of Mercury's orbit during the 1800s showed that it does indeed precess, but careful calculations made with Newton's law of gravity could not completely account for the observed precession. Although the discrepancy was small, further observations verified that it was real.

Einstein was aware of this discrepancy and, from the time he first thought of the equivalence principle in 1907, he hoped he would be able to explain it. When he finally succeeded in November 1915, he was so excited that he was unable to work for the next three days. He later called the moment of this success the high point of his scientific life.

In essence, Einstein showed that the discrepancy arose because Newton's law of gravity assumes that time is absolute and space is flat. In reality, time runs more slowly and space is more curved on the part of Mercury's orbit that is nearer the Sun. The equations of general relativity take this distortion of spacetime into account, providing a predicted orbit for Mercury that precisely matches its observed orbit.

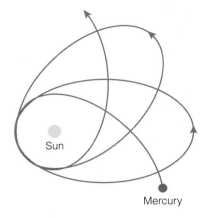

Note: The amount of precession with each orbit is highly exaggerated in this picture.

Figure S3.17 Mercury's orbit slowly precesses around the Sun.

THINK ABOUT IT

Suppose the perihelion of Mercury's orbit were even closer to the Sun than it actually is. Would you expect the discrepancy between the actual orbit and the orbit predicted by Newton's laws to be greater than or less than it actually is? Explain.

Gravitational Lensing We can also test Einstein's claim that space is curved by observing the trajectories of light rays moving through the universe. Because light always travels at the same speed, which means it never accelerates or decelerates, light must always follow the straightest possible path. If space itself is curved, the\n light paths will appear curved as well.

Suppose we could carefully measure the angular separation between two stars during the daytime just when the light from one of the stars passes near the Sun. The curvature of space near the Sun should cause the light beam passing closer to the Sun to curve more than the light beam

---

SPECIAL TOPIC The Twin Paradox

Imagine two twins, one of whom stays on Earth while the other takes a high-speed trip to a distant star and back. In Chapter S2, we said that the twin who takes the trip will age less than the twin who stays home on Earth. Shouldn't the traveling twin be allowed to claim that she stayed stationary while Earth made a trip away from her and back? And in that case, shouldn't the twin on Earth be the one who ages less? This question underlies the so-called *twin paradox,* which can be analyzed in several different ways. We will take an approach that offers some insights into the nature of spacetime.

Suppose you and Jackie are floating weightlessly next to each other with synchronized watches. While you remain weightless, Jackie uses her engines to accelerate a short distance away from you, decelerate to a stop a bit farther away, and then turn around and return. From your point of view, Jackie's motion means that you'll see her watch ticking more slowly than yours. Thus, upon her return, you expect to find that less time has passed for her than for you. But how does Jackie view the situation?

The two of you can argue endlessly about who is really moving, but one fact is obvious to both of you: During the trip, you remained weightless while Jackie felt *weight* holding her to the floor of her spaceship. Jackie can account for her weight in either of two ways. First, she can agree with you that she was the one who accelerated. However, because we know that time runs more slowly in an accelerating spaceship, she'll therefore agree that her watch ran more slowly than yours. Alternatively, she can claim that she felt weight because her engines counteracted a gravitational field in which she was stationary while you fell freely, but we also know that time runs slowly in gravitational fields. Therefore, she'll still agree that her watch ran more slowly than yours. Thus, no matter how you or Jackie looks at it, the result is the same: Less time passes for Jackie.

The left side of Figure 1 shows a spacetime diagram for this experiment. You and Jackie both moved between the same two events

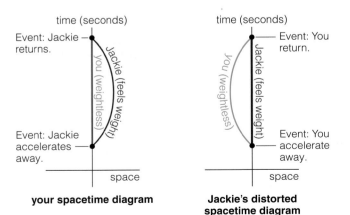

your spacetime diagram

Jackie's distorted spacetime diagram

Figure 1 A person floating weightlessly must be following the straightest possible path through spacetime (left). Because this is not the case in Jackie's diagram (right), her diagram must be distorted.

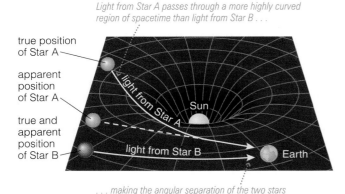

Light from Star A passes through a more highly curved region of spacetime than light from Star B . . .

true position of Star A

apparent position of Star A

true and apparent position of Star B

. . . making the angular separation of the two stars appear smaller than their true angular separation.

Figure S3.18 When we see starlight that passes near the Sun during a total eclipse, the curvature of spacetime causes a shift in the star's apparent position.

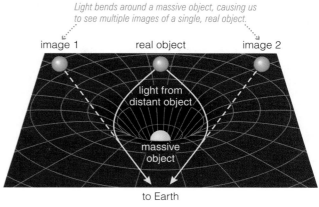

Light bends around a massive object, causing us to see multiple images of a single, real object.

image 1   real object   image 2

light from distant object

massive object

to Earth

Figure S3.19 Gravitational lensing can create distorted or multiple images of a distant object whose light passes by a massive object on its way to Earth.

from the other star (Figure S3.18). Therefore, the angular separation of the two stars should appear smaller than their true angular separation (which we would know from nighttime measurements). This effect was first observed in 1919, when astronomers traveled far and wide to measure stellar positions near the Sun during a total eclipse. The results agreed with Einstein's predictions, and the media attention drawn by the eclipse expeditions brought Einstein worldwide fame.

Even more dramatic effects occur when a distant star or galaxy, as seen from Earth, lies directly behind another ob-

ject with a strong gravitational field (Figure S3.19). The mass of the intervening object curves spacetime in its vicinity, altering the trajectories of light beams passing nearby. Different light paths can curve so much that they end up converging at Earth, grossly distorting the appearance of the star or galaxy. Depending on the precise four-dimensional geometry of spacetime between us and the observed star or galaxy, the image we see may be magnified or distorted into arcs, rings, or multiple images of the same object (Figure S3.20). This type of distortion is called **gravitational lensing**, analogous to the lensing of light when it is bent by

in spacetime: the start and end points of Jackie's trip. However, your path between the two events is shorter than Jackie's. Because we have already concluded that less time passes for Jackie, we are led to a remarkable insight about the passage of time:

> **Between any two events in spacetime, more time passes on the shorter (and hence straighter) path.**

The maximum amount of time you can record between two events in spacetime occurs if you follow the straightest possible path—that is, the path on which you are weightless.

The subtlety arises if Jackie chooses to claim that she is at rest and attributes her weight to gravity. In that case, she might be tempted to draw the spacetime diagram on the right in Figure 1, on which she appears to have the shorter path through spacetime. The rule that more time passes on shorter paths would then seem to imply that *your* watch should have recorded less time than Jackie's, contradicting our earlier claims. The contradiction is an illusion. If Jackie wishes to assert that she felt gravity, she must also claim that the gravity she felt implies that spacetime is curved in her vicinity. Therefore, she should not have drawn a spacetime diagram on a flat piece of paper.

Jackie's problem is analogous to that of a pilot who plans a trip from Philadelphia to Beijing on a flat map of Earth (Figure 2). On the flat map, it appears that the pilot has plotted the straightest possible path. However, this appearance is an illusion: The shortest and straightest path really is the great-circle route shown in Figure S3.11b, which

appears curved on the flat map of Earth. A flat map of Earth distorts reality because the actual geometry of Earth's surface is spherical.

Just as the distortions in a map of the world do not change the actual distances between cities, the way we choose to draw a spacetime diagram does not alter the reality of spacetime. The solution to the twin paradox is that the two twins do not share identical situations. The twin who turns around at the distant star must have a more strongly curved worldline than the stay-at-home twin. Thus, more time must pass for the stay-at-home twin, and the traveler does indeed age less during the journey.

11,000 km

14,300 km

Beijing   Philadelphia

Figure 2 This flat map shows the same two paths on Earth shown in Figure S3.11b. However, the distortion involved in making the map flat means that what looks like a straight line is not really as straight or as short as possible.

a In this case of gravitational lensing, called Einstein's Cross, the gravity of a foreground galaxy (center) bends light from a single bright background object so that it reaches us along four different paths—creating four distinct images of a single object.

b When one galaxy lies directly behind another, the foreground galaxy can bend light on all sides so that the light converges on Earth, forming an Einstein Ring like that pictured here.

Figure S3.20 Examples of gravitational lensing.

a glass lens. We'll see more examples of gravitational lensing in Chapter 22.

Gravitational Time Dilation We can test the prediction of gravitational time dilation by comparing clocks located in places with different gravitational field strengths. Even in Earth's weak gravity, experiments demonstrate that clocks at low altitude tick more slowly than identical clocks at higher altitude. Although the effect would add up to only a few billionths of a second over a human lifetime, the differences agree precisely with the predictions of general relativity. In fact, the global positioning system (GPS) takes these effects into account; if it didn't, it would be far less accurate in locating positions on Earth.

Surprisingly, it's even easier to compare the passage of time on Earth with the passage of time on the surface of the Sun and other stars. Because stellar gases emit and absorb *spectral lines* with particular frequencies [Section 5.4], they serve as natural atomic clocks. Suppose that, in a laboratory on Earth, we find that a particular type of gas emits a spectral line with a frequency of 500 trillion cycles per second. If this same gas is present on the Sun, it will also emit a spectral line with a frequency of 500 trillion cycles per second.

However, general relativity claims that time should be running slightly more slowly on the Sun than on Earth. That is, 1 second on the Sun lasts *longer* than 1 second on Earth, or, equivalently, a second on Earth is shorter than a second on the Sun. Thus, during 1 second on Earth, we will see *fewer* than 500 trillion cycles from the gas on the Sun. Because lower frequency means longer or redder wavelengths, the spectral lines from the Sun ought to be *redshifted*. This redshift has nothing to do with the Doppler shifts that we see from moving objects [Section 5.5]. Instead, it is a **gravitational redshift**, caused by the fact that time runs slowly in gravitational fields. Gravitational redshifts have been measured for spectral lines from the Sun and from many other stars. The results agree with the predictions of general relativity, confirming that time slows down in stronger gravitational fields.

Measuring Earth's Effect on Spacetime Earth has a relatively weak gravitational field, which means that it causes a fairly small amount of spacetime curvature. Nevertheless, Earth's effects on spacetime should be measurable in principle. In the 1960s, a group of physicists at Stanford University began to contemplate ways of measuring the curvature of spacetime around Earth. Such measurements require extraordinary precision, and in 2004 a satellite designed to make them was finally launched into space. Known by the name "Gravity Probe B," the satellite uses precision gyroscopes—which consist of the most perfectly round objects ever made—to look for subtle effects of spacetime curvature and the effect of Earth's rotation on spacetime. Results from Gravity Probe B are expected some time during 2006, and scientists anxiously await them to see whether they match the predictions of Einstein's general theory of relativity.

THINK ABOUT IT

Go to the Gravity Probe B Web site to find its current status. Are the results yet in? If so, do they agree with the predictions of general relativity?

• What are gravitational waves?

According to general relativity, a sudden change in the curvature of space in one place should propagate outward through space like ripples on a pond. For example, the effect of a star suddenly imploding or exploding should be rather like the effect of dropping a rock into a pond, and two massive stars orbiting each other closely and rapidly should generate ripples of curvature in space rather like those of a blade turning in water. Einstein called these ripples **gravitational waves**. Similar in character to light waves but far weaker, gravitational waves are predicted to have no mass and to travel at the speed of light. But do they actually exist?

The distortions of space carried by gravitational waves should compress and expand objects they pass by. In principle, we could detect gravitational waves by looking for such waves of compression and expansion, but these effects are expected to be extremely weak. No one has yet succeeded in detecting them. However, a new observatory dedicated to the search for gravitational waves has recently begun operations. Called the Laser Interferometer Gravitational Wave Observatory, or LIGO, it consists of two large detectors—one in Louisiana and one in Washington State—that search in tandem for telltale signs of gravitational waves. Several other nations, including Germany, Italy, and Japan, have recently built or are working on similar gravitational wave detectors.

Despite the lack of direct detection, scientists are quite confident that gravitational waves exist because of a special set of observations carried out over the past 30 years. In 1974, astronomers Russell Hulse and Joseph Taylor discovered an unusual binary star system in which both stars are highly compressed *neutron stars* [Section 18.2]. The small sizes of these objects allow them to orbit each other extremely

Figure S3.21 The decrease in the orbital period of the Hulse-Taylor binary star system matches what we expect if the system is emitting gravitational waves.

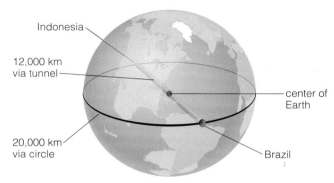

Figure S3.22 If you could take a shortcut *through* the Earth, the trip from Brazil to Indonesia would be shorter than is possible on the *surface* of the Earth.

closely and rapidly. General relativity predicts that this system should be emitting a substantial amount of energy in gravitational waves. If the system is losing energy to these waves, the orbits of the two stars should steadily decay. Observations show that the rate at which the orbital period is decreasing matches the prediction of general relativity, a strong suggestion that the system really is losing energy by emitting gravitational waves (Figure S3.21). Indeed, in 1993 Hulse and Taylor received the Nobel Prize for their discovery, indicating that the scientific community believes their work all but settled the case for gravitational waves. In 2003, astronomers announced the discovery of another neutron binary system with orbits decaying as expected due to emission of gravitational waves. The neutron stars in this system are currently orbiting each other every 2.4 hours, and the energy they are losing to gravitational waves will cause them to collide with one another "just" 85 million years from now.

S3.5 Hyperspace, Wormholes, and Warp Drive

If you're a fan of *Star Trek, Star Wars,* or other science fiction, you've seen spaceships bounding about the galaxy with seemingly little regard for Einstein's prohibition on traveling faster than the speed of light. In fact, these stories do not necessarily have to violate the precepts of relativity as long as they exploit potential "loopholes" in the known laws of nature. General relativity just might provide the necessary loopholes.

• Where does science end and science fiction begin?

Let's begin with an analogy. Suppose you want to take a trip between Brazil and Indonesia, which happen to lie diametrically opposite on Earth's surface (Figure S3.22).

Ordinarily, we are restricted to traveling along Earth's surface by car, boat, or plane, and the most direct trip would cover a distance of about 20,000 kilometers. However, suppose you could somehow drill a hole through the center of the Earth and fly through the hole from Brazil to Indonesia. In that case, the trip would be only about 12,000 kilometers. You could thereby fly between Brazil and Indonesia in much less time than we would expect if we thought you could travel only along the surface.

Now consider a trip from Earth to the star Vega, about 25 light-years away. From the point of view of someone who stays home on Earth, this trip must take at least 25 years in each direction. However, suppose space happens to be curved in such a way that Earth and Vega are much closer together as viewed from a multidimensional *hyperspace,* just as Brazil and Indonesia are closer together if we can go through the Earth than if we must stay on its surface. Further, suppose there is a tunnel through hyperspace, often called a **wormhole,** through which we can travel (Figure S3.23). If the tunnel is short—say, just a few kilometers in length—then a spaceship would need to travel only a few kilometers through the wormhole to go from Earth to Vega. The trip might then take only a few minutes in each direction! Relativity is not violated because the

The distance through our universe between Earth and Vega is 25 light-years . . .

. . . but the distance would be much shorter if we could travel through a wormhole.

Figure S3.23 The curved sheet represents our universe, in which a trip from Earth to Vega covers a distance of 25 light-years. This trip could be much shorter if a wormhole existed that created a shortcut through hyperspace.

spaceship has not exceeded the speed of light. It has simply taken a shortcut through hyperspace.

If no wormhole is available, perhaps we might discover a way to "jump" through hyperspace and return to the universe anywhere we please. Such hyperspace jumps are the fictional devices used for space travel in the *Star Wars* movies. Alternatively, we might discover a way to warp spacetime to our own specifications, thereby allowing us to make widely separated points in space momentarily touch in hyperspace. This fictional device is the basic premise behind *warp drive* in the *Star Trek* series.

Do wormholes really exist and, if so, could we really travel through them? Is it possible that we might someday discover a way to jump into hyperspace or create a warp drive? Our current understanding of physics is insufficient to answer these questions definitively. For the time being, the known laws of physics do not prohibit any of these exotic forms of travel. These loopholes are therefore ideal for science fiction writers, because they might allow rapid travel among distant parts of the universe without violating the established laws of relativity.

However, many scientists believe we will eventually find that these exotic forms of travel are *not* possible. Their primary objection is that wormholes seem to make time travel possible. If you could jump through hyperspace to another place in our universe, couldn't you also jump back to another *time*? If you used a trip through hyperspace to travel into the past, could you prevent your parents from ever meeting?

The paradoxes we encounter when we think about time travel are severe and seem to have no resolution. Most scientists therefore believe that time travel will prove to be impossible, even though we don't yet know of any laws of physics that prohibit it. In the words of physicist Stephen Hawking, time travel should be prohibited "to keep the world safe for historians."

If time travel is not possible, it is much more difficult to see how shortcuts through hyperspace could be allowed. Nevertheless, neither time travel nor travel through hyperspace can yet be ruled out in the same way that we can rule out the possibility of exceeding the speed of light. Until we learn otherwise, the world remains safe for science fiction writers who choose their fictional space travel techniques with care, avoiding any conflicts with relativity and other known laws of nature.

S3.6 The Last Word

We now know that space and time are intertwined in ways that would have been difficult to imagine before Einstein's work. For the last word in our study of relativity, we turn to Einstein himself. Here is what he said about a month before his death on April 18, 1955:

"Death signifies nothing. . . . the distinction between past, present, and future is only a stubbornly persistent illusion."*

THE BIG PICTURE

Putting Chapter S3 into Context

Just as the Copernican revolution overthrew the ancient belief in an Earth-centered universe, Einstein's revolution overthrew the common belief that space and time are distinct and absolute. We have explored Einstein's revolution in some detail in the past two chapters. Keep in mind the following "big picture" ideas:

- We live in four-dimensional spacetime. Disagreements among different observers about measurements of time and space occur because the different observers are looking at a single four-dimensional reality from different three-dimensional perspectives.

- Gravity arises from curvature of spacetime. Once we recognize this fact, the orbits of planets, moons, and all other objects are perfectly natural consequences of the curvature, rather than results of a mysterious "force" acting over great distances.

- Although the predictions of relativity may seem quite bizarre, they have been well verified by many observations and experiments.

- Some questions remain well beyond our current understanding. In particular, we do not yet know whether travel through hyperspace might be possible, allowing some of the imaginative ideas of science fiction to become reality.

---

*This quotation was found with the aid of Alice Calaprice, author of *The Quotable Einstein*, Princeton University Press, 1996.

SUMMARY OF KEY CONCEPTS

S3.1 Einstein's Second Revolution

- **What are the major ideas of general relativity?** General relativity tells us that gravity arises from curvature of spacetime, and that the curvature arises from the presence of masses. This idea leads us to a view of gravity in which time runs more slowly in gravitational fields, black holes can exist in spacetime, and the universe

has no center or edges. It also predicts the existence of gravitational waves propagating through space.

- **Is all motion relative?** Special relativity shows that motion at constant velocity is always relative, but the relative nature of motion is less evident when gravity and acceleration enter the picture. However, the **equivalence principle** allows us to continue treating all motion as relative because it tells us that the effects of gravity are exactly equivalent to the effects of acceleration. The physical effects of being in an accelerating reference frame are identical to those on someone who is stationary in a gravitational field.

S3.2 Understanding Spacetime

- **What is spacetime?** **Spacetime** is the four-dimensional combination of space and time that forms the "fabric" of our universe.

- **What is curved spacetime?**

Spacetime can be curved much as a sheet of paper can be curved, but in more dimensions. We can recognize spacetime curvature from the rules of geometry. The three possible geometries are a **flat geometry**, in which the ordinary laws of flat (Euclidean) geometry apply; a **spherical geometry**, in which lines that start out parallel tend to converge; and a **saddle-shaped geometry**, in which lines that start out parallel tend to diverge. Spacetime may have different geometries in different places.

S3.3 A New View of Gravity

- **What is gravity?**

Gravity arises from curvature of spacetime. Mass causes spacetime to curve, and the curvature of spacetime determines the paths of freely moving masses.

- **What is a black hole?**

A **black hole** is a place where spacetime is curved so much that it essentially forms a bottomless pit, making it a true hole in spacetime.

- **How does gravity affect time?** Time runs more slowly in places where gravity is stronger, an effect called **gravitational time dilation**.

S3.4 Testing General Relativity

- **How do we test the predictions of the general theory of relativity?**

Observations of the precession of Mercury's orbit match the precession predicted by Einstein's theory. Observations of stars during eclipses and photos of **gravitational lensing** provide spectacular confirmation of the idea that light can travel curved paths through space. **Gravitational redshifts** observed in the light of objects with strong gravity confirm the slowing of time predicted by general relativity, a prediction that has also been confirmed with clocks at different altitudes on Earth.

- **What are gravitational waves?** General relativity predicts that accelerating masses produce **gravitational waves** that travel at the speed of light. Observations of binary neutron stars provide solid indirect evidence that gravitational waves really exist.

S3.5 Hyperspace, Wormholes, and Warp Drive

- **Where does science end and science fiction begin?** No known physical laws prevent hyperspace, **wormholes**, or warp drive from offering "loopholes" that could allow us to get from one place to another in less time than we could by traveling through ordinary space. However, if any one of them proves to be real, then cause and effect might not be absolute, a proposition troubling to many scientists.

EXERCISES AND PROBLEMS

For instructor-assigned homework go to **Mastering**ASTRONOMY www.masteringastronomy.com

Review Questions
Short-Answer Questions Based on the Reading

1. Explain what we mean by the *straightest possible path* on the Earth's surface.
2. What do we mean by *spacetime*?
3. List five major ideas that come directly from the general theory of relativity.
4. What is the *equivalence principle*? Give an example that clarifies its meaning.
5. What do we mean by *dimension*? Describe a *point*, a *line*, a *plane*, a *three-dimensional space*, and a *four-dimensional space*. What does *hyperspace* mean?
6. Explain the meaning of the statement, *"Space is different for different observers. Time is different for different observers. Spacetime is the same for everyone."*

7. What is a *spacetime diagram*? What is an *event*? What is a *worldline*?

8. Consider a spacetime diagram that shows worldlines for three objects: one worldline is vertical, one is straight but slanted, and one is curved. In the reference frame of the person who drew the diagram, how is each object moving?

9. How do rules of geometry differ depending on whether the geometry is *flat*, *spherical*, or *saddle shaped*?

10. Explain how the idea of spacetime geometry means that the universe has no center and no edges.

11. How can you tell whether you are following the straightest possible path through spacetime?

12. According to general relativity, what is gravity? With this view, why does the Earth orbit the Sun?

13. Describe the idea behind the "rubber sheet" analogy to space-time, and the limitations of the analogy. In the analogy, why does the curvature of spacetime near a star depend on both the star's mass and its size?

14. What is a *black hole*? What do we mean by the event horizon of a black hole?

15. What is *gravitational time dilation*? What determines how much time is slowed in a gravitational field?

16. Briefly describe several observational tests that support general relativity.

17. What is *gravitational lensing*? According to general relativity, why does it occur?

18. Why do we see a *gravitational redshift* when we look at spectral lines from the Sun?

19. What are *gravitational waves*? What evidence supports their existence?

20. What is the current evidence regarding the possibility of travel through hyperspace, wormholes, or warp drive? Explain how a *wormhole* might allow us to travel between two places that are 10 light-years apart in less than 10 years.

Test Your Understanding

Does It Make Sense?

Decide whether the statement makes sense (or is clearly true) or does not make sense (or is clearly false). Explain your reasoning. (For an example, see Chapter 1, "Does It Make Sense?")

21. The equivalence principle tells us that experiments performed on a spaceship accelerating through space at $1g$ will give the same results as experiments performed on Earth.

22. The equivalence principle tells us that there's no difference at all between a planet and a human-made spaceship.

23. A person moving by you at high speed will measure time and space differently than you, but you will both agree that there is just a single reality in spacetime.

24. With a sufficiently powerful telescope, we could search for black holes by looking for funnel-shaped objects in space.

25. Time runs slightly more slowly on the surface of the Sun than it does here on Earth.

26. Telescopes sometimes see multiple images of a single object, just as we should expect from the general theory of relativity.

27. When I walk in circles, I am causing curvature of spacetime.

28. Although special relativity deals only with relativity of motion, general relativity tells us that *everything* is relative.

29. The shortest distance between two points is always a straight line.

30. According to the general theory of relativity, it is impossible to travel through hyperspace or to use anything like *Star Trek*'s "warp drive."

Quick Quiz

Choose the best answer to each of the following. Explain your reasoning with one or more complete sentences.

31. Spacetime is (a) another name for gravity. (b) the combination of time and the three dimensions of space. (c) a curved rubber sheet.

32. The equivalence principle tells us that effects of these two things are indistinguishable: (a) space and time. (b) gravity and acceleration. (c) gravity and curvature of spacetime.

33. The surface of Earth has _____ dimensions: (a) one (b) two (c) three

34. If two lines begin parallel but later diverge, the geometry is (a) flat. (b) spherical. (c) saddle shaped.

35. You know that you are following the straightest possible path through spacetime if (a) you are standing still. (b) you are moving directly from one place to another. (c) you are weightless.

36. According to general relativity, Earth goes around the Sun rather than flying straight off into space because (a) the force of gravity acts like an invisible bond holding Earth and the Sun together. (b) Earth is going as straight as possible, but the shape of spacetime makes this path go round and round. (c) in its own reference frame, Earth can consider itself to be stationary.

37. Of the following objects, on which one would time on its surface run most slowly? (a) the Sun (b) an object with the same mass as the Sun but twice as large in radius (c) an object with the same mass as the Sun but only half as large in radius

38. Gravitational lensing occurs because (a) gravity causes light to slow down. (b) gravity curves space and light always follows the straightest possible path through space. (c) the effects of gravity are indistinguishable from the effects of glass lenses.

39. Why do we think that gravitational waves really exist? (a) We have observed them with large telescopes in space. (b) We have observed orbiting objects that are losing precisely the amount of energy we expect them to be losing to gravitational waves. (c) They are predicted to exist by the general theory of relativity.

40. If wormholes are real, which of the following best describes what one is? (a) a place where it is possible to travel faster than light (b) a shortcut between distant parts of the universe (c) a black hole with a wormlike shape

Investigate Further

In-Depth Questions to Increase Your Understanding

Short-Answer/Essay Questions

41. *Einstein's Quotation.* Read and think about the quotation from Einstein in Section S3.6. Based on what you have learned about relativity, what do you think he meant? Write a short essay explaining your answer.

42. *Northward Journey.* Suppose two people start out traveling due north from the same latitude, one near Denver and one near San Francisco. Are their paths parallel when they start? Will their paths stay parallel, or will they meet? If they will meet, where will they meet? Summarize your answers with a one- or two-paragraph description of their journey, and explain what the results tell us about the geometry of Earth.

43. *Alternative Geometries.* For each of the three basic types of geometry—*flat*, *spherical*, and *saddle shaped*—find an everyday object that obeys the rules of this geometry. Describe the object and explain how you know which geometry it has.

44. *Funnels in Space?* Many people have seen rubber sheet diagrams that show black holes looking like funnels, as in Figure S3.15, and therefore they assume that black holes really are funnel shaped. Imagine that you are talking to a person who believes this misconception to be true. Write a one- or two paragraph explanation in words the person would understand that explains how the misconception arises and what a black hole would really look like if we could see one. (For example, what shape would it be?)

45. *Black Hole Sun.* Suppose the Sun were magically replaced by a black hole of precisely the same mass.
    a. How would the curvature of spacetime change in the region where the Sun used to be located?
    b. How would the curvature of spacetime change in the region of Earth's orbit?
    c. What effect would the change have on Earth's orbit? Explain.

46. *Galileo and the Equivalence Principle.* Galileo demonstrated that all objects near Earth's surface should fall with the same acceleration, regardless of their mass. According to general relativity, why shouldn't the mass of a falling object affect its rate of fall? Explain in one or two paragraphs.

47. *Long Trips at Constant Acceleration.* On a realistic trip to the stars, we could not suddenly jump to a speed near the speed of light without being killed by the forces associated with the sudden acceleration. Thus, a more realistic trip would have us accelerate at a comfortable rate, such as 1*g*, until we are halfway to our destination and then decelerate at the same rate until we reach our destination. Explain why we would be comfortable with this acceleration. By our own reckoning, would we notice anything unusual about lengths, masses, or the passage of time on our spaceship? Why or why not?

48. *Movie Science Fiction.* Choose a popular science fiction movie that depicts interstellar travel and study it closely as you watch it. What aspects of the movie are consistent with relativity and other laws of physics? What aspects of the movie violate the laws of relativity or other laws of physics? Write a two- to three-page summary of your findings.

49. *Research: The Eötvös Experiment.* Galileo's result that all objects fall to Earth with the same acceleration (neglecting air resistance) is very important to general relativity—if it were not true, general relativity would be in serious trouble. Describe the experiments of Baron Roland von Eötvös, who tested Galileo's conclusions in the late 1800s. How did the results of these experiments influence Einstein as he worked on general relativity? Write a one- to two-page summary of your findings.

50. *Research: Wormholes.* Some scientists have thought seriously about wormholes and their consequences. Find and read a popular article or book about wormholes. Write a short summary of your reading, and discuss your opinion of the implications of wormholes raised in it.

Quantitative Problems

Be sure to show all calculations clearly and state your final answers in complete sentences.

51. *Worldlines at Low Speed.* Make a spacetime diagram and draw worldlines for each of the following situations. Explain your drawings.
    a. A person sitting still in a chair.
    b. A person driving by at a constant velocity of 50 km/hr.
    c. A person driving by at a constant velocity of 100 km/hr.
    d. A person accelerating from a stop sign to a speed of 50 km/hr.
    e. A person decelerating from 50 km/hr to a stop.

52. *Worldlines at High Speed.* Make a spacetime diagram on which the time axis is marked in seconds and the space axis is marked in light-seconds. Assume you are floating weightlessly and therefore consider yourself at rest.
    a. Draw your own worldline.
    b. Draw a worldline for Sebastian, who is moving to your right at 0.5*c*.
    c. Draw a worldline for Michaela, who is moving to your left at 0.7*c*.
    d. Draw a worldline for Angela, who is traveling away from you at a speed of 100,000 km/s.

53. *Highly Sloped Worldline.* Make a spacetime diagram on which the time axis is marked in seconds and the space axis is marked in light-seconds. Draw a worldline with a slope of 30° (from the horizontal). At what speed would an object have to be traveling to have this worldline? Can any object have this worldline? Explain.

54. *Triangle on Earth.* Draw a simple sketch of Earth (a plain sphere will do), with a triangle on it that connects the following three points: (1) the equator at longitude 0°; (2) the equator at longitude 90°W; (3) the North Pole. What is the sum of the angles in this triangle? Explain how you know the angles, and what the sum tells you about the geometry of Earth's surface.

55. *Long Trips at Constant Acceleration: Earth Time.* Suppose you stay on Earth and watch a spaceship leave on a long trip at a constant acceleration of 1*g*.
    a. At an acceleration of 1*g*, approximately how long will it take before you see the spaceship traveling away from Earth at *half* the speed of light? Explain. (Use $g = 9.8$ m/s^2.)
    b. Describe how you will see its speed change as it continues to accelerate. Will it keep gaining speed at a rate of 9.8 m/s each second? Why or why not?
    c. Suppose the ship travels to a star that is 500 light-years away. According to you back on Earth, *approximately* how long will this trip take? Explain.

56. *Long Trips at Constant Acceleration: Spaceship Time.* Consider again the spaceship on a long trip with a constant acceleration of 1*g* (as in Problem 55). Although the derivation is beyond the scope of this book, it is possible to show that, as long as the ship is gone from Earth for many years, the amount of time that passes on the spaceship during the trip is approximately:

$$t_{ship} = \frac{2c}{g} \ln\left(\frac{g \times D}{c^2}\right)$$

In this formula, *D* is the distance to the destination and ln stands for the natural logarithm. (Your calculator probably has a key for taking natural logarithms [usually labeled "ln"], so you can use this formula even if you are not familiar with them.) If *D* is in meters, $g = 9.8$ m/s^2, and $c = 3 \times 10^8$ m/s, the answer will be in units of *seconds*. Use the formula to determine how much time will pass on the ship during its trip to a star that is 500 light-years away. Compare to the amount of time that passes on Earth. (*Hint:* Be sure you convert the distance from light-years to meters and convert your answer from seconds to years.)

57. *Trip to the Center of the Galaxy.* Use the same scenario as in Problem 56, but this time suppose the ship travels to the center of the Milky Way Galaxy, about 28,000 light-years

away. How much time will pass on the ship? Compare this to the amount of time that passes on Earth.

58. *Trip to Another Galaxy.* The Andromeda Galaxy is about 2.5 million light-years away. Suppose you had a spaceship that could constantly accelerate at 1*g*. Could you go to the Andromeda Galaxy and come back within your lifetime? Explain. What would you find when you returned to Earth? (*Hint:* You'll need the formula from Problem 56.)

59. *Gravitational Time Dilation on Earth.* If a gravitational field is relatively weak, such as that of a planet or an ordinary star, the following formula tells us the fractional amount of gravitational time dilation at a distance *r* from the center of an object of mass M_{object}:

$$\frac{1}{c^2} \times \frac{GM_{\text{object}}}{r}$$

(Recall that $G = 6.67 \times 10^{-11}\ \text{m}^3/(\text{kg} \times \text{s}^2)$ and $c = 3 \times 10^8$ m/s.) For example, while 1 hour passes in deep space far from the object, the amount of time that passes at a distance *r* is 1 hour multiplied by the factor above. Note that the formula does *not* apply to strong gravitational fields, like those near black holes. Use the formula to calculate the amount of time that passes on Earth's surface while 1 hour passes in deep space. Based on your answer, should we expect effects of general relativity to be obvious on Earth? Explain.

60. *Gravitational Time Dilation on the Sun.* Use the formula given in Problem 59 to calculate the percentage by which time runs slower on the surface of the Sun than in deep space. Based on your answer, approximately how much of a gravitational redshift should you expect for a spectral line with a rest wavelength of 121.6 nm?

Discussion Questions

61. *Relativity and Fate.* In principle, if we could see all four dimensions of spacetime, we could see future events as well as past events. In his novel *Slaughterhouse Five*, writer Kurt Vonnegut used this idea to argue that our futures are predetermined and that there is no such thing as free will. Do you agree with this argument? Why or why not?

62. *Philosophical Implications of Relativity.* According to our description of spacetime, you exist in spacetime as a "solid" object stretching through time. In that sense, you cannot erase anything you've ever said or done from spacetime. If we could see in four dimensions, we would be able to see everything you've ever said or done. Do you think these ideas have any important philosophical implications? Discuss.

63. *Wormholes and Causality.* Suppose it turns out that travel through wormholes *is* possible and that it is possible to travel into the past. Discuss some of the paradoxes that would occur. In light of these paradoxes, do you believe that travel through wormholes will turn out to be prohibited? Why or why not?

MEDIA EXPLORATIONS

For self-study activities go to MasteringASTRONOMY www.masteringastronomy.com

Interactive Tutorials

Tutorial Review of Key Concepts

Use the following interactive **Tutorial** at **www.mastering astronomy.com** to review key concepts from this chapter.

Black Holes Tutorial
Lesson 1 What Are Black Holes?
Lesson 2 The Search for Black Holes

Supplementary Tutorial Exercises

Use the interactive **Tutorial Lessons** to explore the following questions.

Black Holes Tutorial, Lesson 1

1. Define a black hole in terms of escape velocity. How does this definition compare to the definition in terms of curvature of spacetime?
2. What is the Schwarzschild radius? How is it related to the event horizon?

3. How does the Schwarzschild radius depend on the mass of the object that forms the black hole?
4. Are there limits to the mass of a black hole (on either the low or high ends)? Explain.

Black Holes Tutorial, Lesson 2

1. What evidence makes us think that black holes really exist?
2. How do we think black holes may form?
3. What do we mean by supermassive black holes? Where do we think they exist?

Web Projects

1. *Person of the Century.* Read the article "A Brief History of Relativity" by Stephen Hawking (on the *Time* magazine Web site) to learn why *Time* chose Einstein as their "Person of the Century" for the twentieth century. Write a short essay explaining the reasons and whether you agree with the choice. Defend your opinion.
2. *Gravity Probe B.* Learn more about Gravity Probe B. Write a short report on one impressive aspect of the mission, such as the construction of its gyroscopes, how it tests general relativity, or some other precision technology that it uses.
3. *Gravitational Wave Detectors.* Learn more about the Laser Interferometer Gravitational Wave Observatory (LIGO) or other current or planned gravitational wave observatories. Write a short report about how the observatory seeks to detect gravitational waves and its prospects for success.

Building Blocks of the Universe

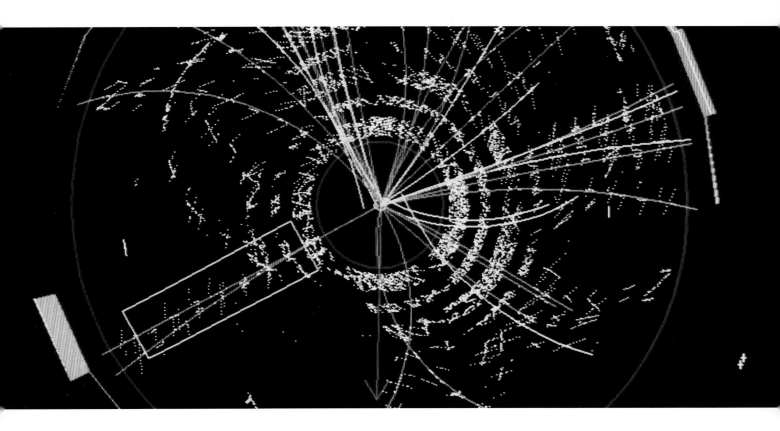

LEARNING GOALS

S4.1 The Quantum Revolution
- How has the quantum revolution changed our world?

S4.2 Fundamental Particles and Forces
- What are the basic properties of subatomic particles?
- What are the fundamental building blocks of matter?
- What are the fundamental forces in nature?

S4.3 Uncertainty and Exclusion in the Quantum Realm
- What is the uncertainty principle?
- What is the exclusion principle?

S4.4 Key Quantum Effects in Astronomy
- How do the quantum laws affect special types of stars?
- How is "quantum tunneling" crucial to life on Earth?
- How empty is empty space?
- Do black holes last forever?

There is a theory which states that if ever anyone discovers exactly what the Universe is for and why it is here, it will instantly disappear and be replaced by something even more bizarre and inexplicable.

There is another which states that this has already happened.

Douglas Adams, from *The Restaurant at the End of the Universe*

The microscopic realm of atoms and nuclei seems far removed from the vast realm of planets, stars, and galaxies. Nevertheless, much of what we know about the cosmos today would have remained mysterious without a thorough understanding of these tiny particles. They are the building blocks from which everything is made, and the behavior of very large objects frequently depends on the laws that govern their tiniest pieces.

We've already seen that matter and energy behave in some strange ways when we break them down into small units. For example, we know that electrons in atoms can have only specific energies and that photons act sometimes like particles and sometimes like waves.

In this chapter, we will examine the laws of nature that underlie the structure of matter. We will look more deeply into the building blocks of nature, investigating current knowledge of the fundamental particles and forces that make up the universe. We will see that the laws of the microscopic world play a crucial role in diverse processes such as nuclear fusion in the Sun and the collapse of a star into a black hole.

S4.1 The Quantum Revolution

Around the same time that Einstein was discovering the principles of relativity, he and others were also investigating the behavior of matter and energy. Their discoveries in this area were no less astonishing. In 1905, the same year he published his special theory of relativity, Einstein showed that light behaves like particles (photons) in addition to behaving like waves [Section 5.2]. In 1911, British physicist Ernest Rutherford (1871–1937) discovered that atoms consist mostly of empty space, raising the question of how matter can ever feel solid. In 1913, Danish physicist Niels Bohr (1885–1962) suggested that electrons in atoms can have only particular energies—that is, electron energies are *quantized* [Section 5.3]. Thus, the realm of the very small is often called the *quantum realm,* and the science of the quantum realm is called **quantum mechanics**.

• How has the quantum revolution changed our world?

Other scientists soon built upon the work of Einstein, Rutherford, and Bohr. By the mid-1920s, our ideas about the structure and nature of atoms and subatomic particles were undergoing a total revolution. The repercussions of this *quantum revolution* continue to reverberate today. They have forced us to reexamine our "common sense" about the nature of matter and energy. They have also driven a technological revolution, because the laws of quantum mechanics make modern electronics possible. Most important, at least from an astronomical point of view, the combination of new ideas and new technology is enabling us to look ever deeper into the heart of matter and energy—the ultimate building blocks of the universe.

This chapter discusses key ideas of the quantum revolution that are important to the study of astronomy. The following are among these surprising ideas:

- Protons, neutrons, and electrons are not the ultimate building blocks of matter. According to present understanding, the building blocks of ordinary matter are *quarks* and *leptons,* which, in turn, belong to a category called *fermions.* Photons belong to an entirely distinct category of particles called *bosons.*
- *Antimatter* is real and is readily produced in the laboratory. When a particle and its antiparticle meet, the result is mutual annihilation and the release of energy.
- Just four forces govern all interactions between particles: gravity, electromagnetism, and the strong and weak nuclear forces. In fact, these four forces are themselves manifestations of a smaller number of truly fundamental forces. Perhaps a single unified force rules all of nature.
- Our everyday common sense tells us that particles and waves are different, but the quantum laws show that *all* tiny particles exhibit the same *wave-particle duality* that Einstein demonstrated for photons.
- The quantum laws have important astronomical consequences. For example, a quantum effect called *degeneracy pressure* can prevent the core of a dying star from collapsing, *quantum tunneling* helps make nuclear fusion possible in the Sun, and phantom-like *virtual particles* may be important to the ultimate fate of black holes and of the universe itself.

No matter how bizarre the quantum laws may seem, they lead to concrete predictions that can be tested experimentally and observationally. Some of these tests require sophisticated technological equipment found only in advanced physics laboratories. Others require billion-dollar particle accelerators. But some are performed every day, right before your eyes. Every time you see a ray of sunlight, you are seeing the product of nuclear reactions made possible by the quantum laws. Every time you turn on a computer, a calculator, a television, or any other high-tech electronic device, the laws of quantum mechanics are being put to work for your benefit. Indeed, almost every aspect of our "information society"—from radios and televisions to cell phones, computers, and the Internet—has been made possible by the science of quantum mechanics.

S4.2 Fundamental Particles and Forces

More than 2,400 years ago, Democritus proposed that matter is made from building blocks that he called *atoms* [Section 5.3]. He believed that atoms were **fundamental particles**, meaning that they were the most basic units of matter, impossible to divide any further. By this definition, the particles we now call atoms are not truly fundamental. By the 1930s, we had learned that atoms themselves are made of protons, neutrons, and electrons. Following this realization, scientists briefly hoped that these three particles were the true fundamental building blocks of the universe. However, under more extreme conditions, matter starts to display greater variety, and strange new particles begin to appear.

Physicists can generate many unusual particles with the aid of *particle accelerators* (sometimes called *atom smashers*), such as Fermilab near Chicago (Figure S4.1), the Stanford Linear Accelerator in California, the Relativistic Heavy Ion Collider at Brookhaven Labs in New York, and CERN on the border between Switzerland and France. The large magnets inside particle accelerators accelerate familiar particles such as electrons or protons to very high speeds—often extremely close to the speed of light. When these particles collide with one another or with a stationary target, they release a substantial amount of energy within a very small space. Some of this energy spontaneously turns into mass, producing a shower of particles. (Recall that $E = mc^2$ tells us not only that mass can turn into energy, but also that energy can turn into mass.) Scientists recognize particles of different types by their differing behavior. Whenever scientists observe a particle that behaves in previously unseen ways, they catalog it as a new type of particle and give it a name.

Many dozens of different particles had been discovered by the early 1960s, and scientists began to wonder whether

Figure S4.1 Aerial photograph of the Fermilab particle accelerator in Illinois.

any of them were truly fundamental. At about that time, physicist Murray Gell-Mann proposed a scheme in which all these particles could be built from just a few fundamental components. Gell-Mann's scheme has since blossomed into what physicists call the **standard model** for the structure of matter. The standard model has proved very successful and has even predicted the existence of new particles later discovered in particle accelerators. In this section, we briefly describe the fundamental particles according to the standard model, along with their special relationships with the forces of nature.

• What are the basic properties of subatomic particles?

Each particular type of subatomic particle, such as an electron, a proton, or a neutron, has its own peculiar behavior. Just a few basic properties determine this behavior; the most important of these properties are *mass, charge,* and *spin.* Mass is already quite familiar to you, and the effects of charge, such as static electricity or lightning, are also part of your everyday experience. Spin, on the other hand, is a property evident only in the quantum realm of the very small.

The word *spin* indicates that this property is related to angular momentum. Recall that a spinning ice-skater or a spinning baseball has rotational angular momentum [Section 4.3]. By analogy, we say that a subatomic particle, such as an electron, has **spin angular momentum**—or **spin**, for short—as it "spins" on its axis. However, because an electron is not a particle in the same sense as a baseball, it does not really spin like a spinning ball. *Spin* is simply a term used to describe the angular momentum that inherently belongs to the electron, even when it has no circular or orbital motion. Just as all electrons have exactly the same mass and electric charge, all electrons have exactly the same amount of spin. Similar considerations hold for all other subatomic particles: Every particle of a particular type has a particular amount of mass, charge, and spin.

Spin is a particularly important property for subatomic particles. All particles fall into one of two broad classes based on their spin: the **fermions**, named for Enrico Fermi (1901–1954), and the **bosons**, named for Satyendra Bose (1894–1974).* The most familiar fermions are electrons, neutrons, and protons. The most familiar bosons are photons. As we'll discuss momentarily, fermions are further subdivided into two categories: *quarks* and *leptons.* Figure S4.2 summarizes the basic classification scheme for fundamental particles.

One important aspect of spin is its orientation: For fermions, spin can be oriented in two ways, usually called *spin up*

---

*Physicists measure the angular momentum of subatomic particles in units of Planck's constant divided by 2π. In these special units, *fermions* have half-integer spins $\left(\frac{1}{2}, \frac{3}{2}, \frac{5}{2}, \ldots\right)$, and *bosons* have integer spins $(0, 1, 2, \ldots)$.

Fundamental Particle Classification

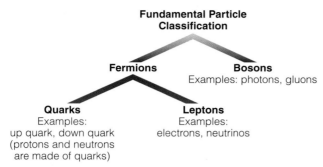

Fermions

Bosons
Examples: photons, gluons

Quarks
Examples:
up quark, down quark
(protons and neutrons
are made of quarks)

Leptons
Examples:
electrons, neutrinos

Figure S4.2 Classification of fundamental particles in the standard model.

A proton is composed of 2 up quarks (u) and 1 down quark (d).

Total charge:
+ 2/3 + 2/3 − 1/3 = +1

A neutron is composed of 1 up quark (u) and 2 down quarks (d).

Total charge:
+ 2/3 − 1/3 − 1/3 = 0

Figure S4.4 Protons and neutrons are made from quarks.

and *spin down*. These two orientations, which correspond to the two opposite senses of rotation (clockwise and counterclockwise), are often represented by arrows (Figure S4.3). Keep in mind that denoting a particle's spin by a small dot with an arrow is a representation of convenience and that subatomic particles are *not* tiny spinning balls.

• What are the fundamental building blocks of matter?

While electrons appear to be truly fundamental, the standard model holds that protons and neutrons are made from even smaller particles, called **quarks**. Protons and neutrons each contain two different types of quarks: the **up quark**, which has an electric charge of $+\frac{2}{3}$, and the **down quark**, which has an electric charge of $-\frac{1}{3}$. (Charge values are given relative to the -1 charge of electrons.) Two up quarks and one down quark form a proton, giving it an overall charge of $+\frac{2}{3} + \frac{2}{3} - \frac{1}{3} = +1$. One up quark and two down quarks form a neutron, making it neutral: $+\frac{2}{3} - \frac{1}{3} - \frac{1}{3} = 0$ (Figure S4.4).

To sum up so far, the fundamental building blocks of atoms are the up quark, the down quark, and the electron. But what about the hundreds of other particles discovered by scientists? The standard model organizes all these particles into a relatively simple hierarchy. First, as we discussed earlier, particles are classified as either fermions or bosons, depending on their *spin*. Quarks, electrons, protons, and neutrons are all fermions. The fermions in turn fall into two groups: those made from quarks and those not made from quarks. Fermions *not* made from quarks, such as electrons,

are called **leptons**.* (More technically, quarks interact with one another via the *strong force* while leptons do not.)

In most high school science classes, students are told that the fundamental building blocks of atoms are protons, neutrons, and electrons. However, based on the preceding discussion, explain why our current understanding holds that the fundamental building blocks of atoms are *quarks* and *leptons*.

Particles Made from Quarks Protons and neutrons are the only particles made from quarks that we find in the matter we encounter in everyday life. However, scientists have identified many other particles made from quarks in experiments in particle accelerators. These other particles are generally short-lived, decaying a fraction of a second after they are produced in collisions. Nevertheless, scientists can observe their behavior during their short lives, usually with detectors placed in or near the particle accelerators. The detectors are designed to record the tracks that the particles leave as they pass through, decay, or undergo other collisions. The photo that opens this chapter shows an example of such particle tracks. Careful analysis of such tracks allows scientists to determine a particle's quark composition.

Experiments in particle accelerators have turned up several important facts that are now incorporated into the standard model. For example, while scientists have found many particles composed of either two quarks or three quarks, and recent experiments suggest the existence of particles with five quarks, no single quark has ever been detected in isolation. Thus, when we speak of scientists detecting a particular type of quark, we really mean that they found a particle that contains this type of quark. In addition, the experiments show that the two quark types that make up protons and neutrons—the up quark and the down quark—are not enough to explain the diversity of particle behavior that we have observed. Instead, the standard model tells us that quarks come in six different types (or "flavors"). The other four types of quarks have the rather exotic names *strange, charmed, top,* and *bottom*. The

Figure S4.3 The two possible states of an electron's, neutron's, or proton's spin are spin up and spin down, represented here by arrows.

spin up   spin down

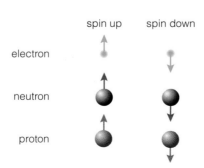

electron

neutron

proton

---

*The technical name for particles made from quarks is *hadrons*. Moreover, the hadrons are subdivided into those made from two quarks, which are called *mesons*, and those made from three quarks, which are called *baryons*.

experimental detection of the top quark in 1995—more than 20 years after it was first predicted to exist—was a particularly impressive success of the standard model. (All the other quarks were detected in earlier experiments.)

The Leptons and the Importance of Fundamental Particles The electron is the only type of lepton that we find in the matter of everyday life, but the standard model also predicts the existence of six lepton types to match up with the six quark types. All the predicted lepton types have been identified experimentally. The six types are known as the electron and the *electron neutrino*, the *muon* and the *mu neutrino*, and the *tauon* and the *tau neutrino*. Table S4.1 lists the six types of quarks and leptons.

At this point, you may be wondering what all these bizarrely named particles have to do with astronomy. In part, the answer is simply that these particles appear to be the fundamental building blocks of everything, from atoms to people, planets, stars, and galaxies. In addition, in Chapter 23 we'll see that the events that unfolded during the first fraction of a second after the Big Bang depended critically on the various types of fundamental particles. We'll also see that *neutrinos*, in particular, are important in several astronomical processes, including nuclear fusion, the explosion of stars, and the fate of the universe.

Incidentally, neutrinos get their name, which means "little neutral ones," from the fact that they are electrically neutral and extremely lightweight—far less massive than electrons. No one has yet succeeded in measuring the precise masses of neutrinos, but they turn out to be extremely common—neutrinos outnumber protons, neutrons, and electrons combined by a factor of roughly a billion. As a result, their total mass could be significant, though it is unlikely to be more than a tiny fraction of the total mass of the universe.

Antimatter Anyone who has watched *Star Trek* has heard of **antimatter**. Although *Star Trek* is science fiction, antimatter is not. It really exists. In fact, every quark and every lepton has a corresponding *antiquark* and *antilepton*. The antiparticle is an exact opposite of its corresponding ordinary particle. For example, an *antielectron* (also called a *positron*) is identical to an ordinary electron except that it has a positive charge instead of a negative charge. Note that

a An electron and positron (antielectron) collide, resulting in annihilation in which all their energy emerges as a pair of photons.

b The energy of two photons combines to create an electron and positron.

Figure S4.5 Every particle of matter has a corresponding antiparticle. When a particle meets its antiparticle, the result is mutual annihilation. Pair production creates a particle and an antiparticle from energy.

while particles and antiparticles are opposites, they are not opposite in mass; as far as we know, there is no such thing as negative mass, so the mass of an antiparticle is precisely the same as the mass of its corresponding particle.

When a particle and its corresponding antiparticle meet, the result is mutual **annihilation** (Figure S4.5a). The combined mass of the particle and antiparticle turns completely into energy in accord with $E = mc^2$. Because the matter in our universe is predominantly of the ordinary type, antimatter generally does not last very long. Whenever an antiparticle is produced, it quickly meets an ordinary particle, and the two annihilate each other to make energy.

This process also works in reverse. When conditions are right, pure energy can turn into a particle-antiparticle pair. For example, whenever an electron "pops" into existence, an antielectron also pops into existence with it (Figure S4.5b). This process of **pair production** conserves charge (and satisfies other conservation laws as well). It happens routinely in particle accelerators here on Earth and on a much grander scale in outer space. In fact, during the first few moments after the Big Bang, the universe's energy fields were so intense that particle-antiparticle pairs popped rapidly in and out of existence at virtually every point in space.

When we include antiparticles, the total number of types of quarks and leptons really is twice that shown in Table S4.1. Thus, there are really 12 types of quarks: the six quarks listed and their six corresponding antiquarks (for example, the up quark and the anti-up quark). Similarly, there are 12 types of leptons: the six listed and their six corresponding antileptons. The net total of 24 different fundamental particles seems quite complex.* As we'll see shortly, this complexity leads many scientists to believe that simplifying principles of particle physics still await discovery.

---

*Moreover, each of the six quarks and six antiquarks is believed to come in three distinct varieties, called *colors*. (The term *color* is not meant to be literal. Rather, it describes a property of quarks that we cannot visualize.) For example, the up quark comes in three colors, often called red, green, and blue.

Table S4.1 Fundamental Fermions

The Quarks	The Leptons
Up	Electron
Down	Electron neutrino
Strange	Muon
Charmed	Mu neutrino
Top	Tauon
Bottom	Tau neutrino

• What are the fundamental forces in nature?

Without forces, the universe would be infinitely boring, a uniform sea of fundamental particles drifting aimlessly about. Forces supply the means through which particles interact and exchange momentum, attracting or repelling one another depending on their properties. For example, particles with mass interact with one another via the force of *gravity,* and particles with charge interact with one another via the *electromagnetic* force.

The Four Forces   According to present understanding, only four fundamental forces operate under ordinary conditions in the universe today. **Gravity** and **electromagnetism** are the most familiar of the four. The other two known forces are called the **strong force** and the **weak force**.

The strong and weak forces act only on extremely short distance scales—so short that these forces can be felt only *within* atomic nuclei. You can see why the strong force must exist by remembering that the nuclei of all elements except hydrogen (which has just a single proton) contain more than one proton. Protons are positively charged, so the electromagnetic force pushes them apart. If the electromagnetic force were unopposed, atomic nuclei would fly apart. The strong force, so named because it is strong enough to overcome electromagnetic repulsion, is what holds nuclei together. The weak force, also very important in nuclear reactions, is a bit more subtle. All particles made from quarks respond to the strong force, but neutrinos, for example, feel only the weak force.

According to the standard model, each force is transmitted by an **exchange particle** that transfers momentum between two interacting particles. For example, photons are the exchange particle for the electromagnetic force. To understand the idea, think about how we see a star. Motions of electrons in the star create photons. The photons then cross light-years of space to our eyes, where they generate an electromagnetic disturbance that we see as starlight. Thus, the photons have carried the electromagnetic force from the particles in the star to particles in our eyes.

In a similar way, exchange particles called *gravitons* are thought to carry the gravitational force through the uni-

verse.* *Gluons,* which get their name because they act like glue to bind nuclei together, carry the strong force. All of the exchange particles are bosons, so the particles that carry the weak force are generally called *weak bosons.*

Relative Strength of the Four Forces   Table S4.2 lists the four forces and their exchange particles. It also has two columns that describe the relative strengths of the four forces. One column shows the relative strengths of the forces within an atomic nucleus, and the next shows relative strengths outside nuclei. The values in these columns represent the relative attraction or repulsion that would be felt between two particles affected by the forces. For example, consider the relative strengths indicated for the electromagnetic force and gravity. The numbers in the table mean that if we took two protons and put them a certain distance apart, the electromagnetic force of repulsion between them would be 10^{43} times as strong as the gravitational force of attraction between them. That is why two protons will repel due to their electric charge rather than attracting due to their gravity.

The relative strengths explain why the different forces tend to be important in different ways. For example, notice that within the nucleus, the strong force is about 100 times as strong as the electromagnetic force. That is why the strong force can hold nuclei together despite the electromagnetic repulsion between protons. The zero relative strength for the strong and weak forces outside the nucleus expresses the fact that these forces vanish beyond very short distances, leaving only gravity and electromagnetism to govern the interactions we notice in our daily lives.

In fact, if you remember that a typical atom (with its electrons) is 100,000 times as large as an atomic nucleus [Section 5.3], you'll realize that the strong and weak forces cannot even affect interactions between atoms. Gravity is almost equally ineffective between atoms, because it is so incredibly weak compared to electromagnetism. Thus, electromagnetism is the only force that affects the inter-

---

*General relativity tells us that gravitational waves transmit changes in spacetime curvature through the universe, so the correspondence between *gravitons* and gravitational waves is analogous to the correspondence between photons and electromagnetic waves. Gravitons have not yet been detected.

Table S4.2 The Four Forces

Force	Relative Strength Within Nucleus*	Relative Strength Beyond Nucleus	Exchange Particles	Major Role
Strong	100	0	Gluons	Holding nuclei together
Electromagnetic	1	1	Photons	Chemistry and biology
Weak	10^{-5}	0	Weak bosons	Nuclear reactions
Gravity	10^{-43}	10^{-43}	Gravitons	Large-scale structure

*The relative values for the strong and weak forces are rough approximations.

actions of atoms—which means it is the one and only force that can group electrons and nuclei into atoms, atoms into molecules, and molecules into living cells. Remarkable as it may seem, the force of electromagnetism governs every aspect of chemistry and biology, and in essence is responsible for the existence of life itself.

Given the extreme weakness of gravity in comparison to the other forces, you might wonder why it is important at all. The answer is that for very massive objects, gravity generally wins a game of elimination. As we've already seen, the strong and weak forces cannot operate on large distance scales. The electromagnetic force loses importance for massive objects because it effectively cancels itself out. Like gravity, the electromagnetic force can act on large distance scales; in fact, its strength declines with distance in the same way that the gravitational force declines with distance (an inverse square law). However, the electromagnetic force differs from gravity in a very important way: It can be either attractive or repulsive depending on the charges of the particles, while gravity always attracts. Large objects always contain virtually equal numbers of protons and electrons, because if they didn't, the strength of the electromagnetic force would quickly either drive out the excess charge or draw in oppositely charged particles to balance it. The charge balance means that large objects are electrically

neutral overall, and therefore unresponsive to the electromagnetic force. Thus, gravity is the only force left to act between massive enough objects, and it grows only stronger as an object's mass grows larger. That is why gravity dominates the structure of the universe on large scales, despite its weakness among individual particles.

The Quest for Simplicity   The standard model has four forces that mediate interactions between particles built from six types of quarks and six types of leptons, plus their corresponding antiparticles. It explains many experimental and observational results quite successfully. However, many scientists think that this model is still too complicated. They seek an even more basic theory of matter that reduces the number of forces and fundamental particles.

Theoretical work in the 1970s, verified experimentally in the 1980s, showed that the electromagnetic and weak forces are really just two different aspects of a single force, called the *electroweak force*. Many scientists hope that future discoveries will show three or even all four forces to be different aspects of a single, unified force governing *all* interactions in nature. As we will see in Chapter 23, these *unified theories* might be necessary to understanding the goings-on during the first fraction of a second after the Big Bang.

SPECIAL TOPIC What Is String Theory?

During the past 400 years, the history of physics has been marked by discoveries that reveal nature to be ever simpler. For example, Newton showed that we need only one set of physical laws for both Earth and the heavens, Einstein showed that gravity could be viewed more simply than as a mysterious "action at a distance," and the quantum laws show that forces can be understood through the interactions of exchange particles. Perhaps as a result of this history, most physicists today suspect that nature is even simpler than we now understand it to be in the standard model of fundamental particles and forces discussed in this chapter. Many physicists have devoted their lives to the search for a simpler theory of nature. Even Einstein spent much of the latter part of his life searching for such a "theory of everything."

Today, much of our attempt to discover a simpler model of nature goes by the name of *string theory*. String theory is not really a single theory, but rather a family of theories that share a common idea—that all the particles and forces of the standard model arise from tiny structures called *strings*. The term *string* comes from an analogy to little pieces of string that can be shaped into loops or that can vibrate like strings on a violin. However, the strings of string theory are hypothesized to exist in 10 or more dimensions. Aside from the three ordinary dimensions of space and the one dimension of time, the remaining six or more dimensions are thought to be "folded up" in a way that does not allow them to be detected on macroscopic scales. Some variations on string theory include multidimensional analogies to membranes, called "branes," and an idea called *M-theory* unites several different versions of string theory.

Despite all the work that has gone into string theory (and M-theory) in the past couple of decades, scientists do not yet know whether the basic tenets of the theory are correct. The problem is that, at least so far, most of the predictions of string theory would be observable only at energies far in excess of what current particle accelerators can achieve. In other words, we haven't really been able to put string theory to the test through new observations or experiments. Thus, by our usual definitions in science, string theory should really be called "string hypothesis," because it does not meet the high standard of verification required for a scientific theory. (String theory does make some predictions that are consistent with current knowledge of the universe, giving theorists at least some hope that it is on the right track.)

Indeed, in its current form, string theory is really more of a mathematical theory than a scientific one. That is, scientists and mathematicians are developing new theorems as they work out the mathematics of multidimensional strings and branes. Their hope is that as they learn more about the mathematics, they will eventually be able to use it to make testable predictions. Meanwhile, the mathematical results give tantalizing hints of a new view of nature, one that could potentially be as revolutionary as any past revolution in physics.

If you would like to learn more about string theory, you can read about it in numerous popular articles and books, such as *The Elegant Universe* by Brian Greene. It is a fascinating topic, regardless of whether strings ultimately prove to be at the heart of nature, or a dead-end idea that scientists will eventually abandon in their ongoing quest to discover nature's underlying simplicity.

S4.3 Uncertainty and Exclusion in the Quantum Realm

So far, we've used the word *particles* as if we were talking about tiny balls of matter. However, subatomic particles do not behave like tiny baseballs. Rather, in much the same way that photons can act as both particles and waves, particles of matter can also act like waves. In other words, for reasons that we will discuss shortly, all subatomic particles—whether particles of matter like protons and electrons or exchange particles like photons—exhibit what we call **wave-particle duality**.

The dual wave-particle nature of subatomic particles explains why even professional physicists have difficulty forming a mental "picture" of atoms; after all, the idea that a baseball could sometimes spread out like a water wave is counterintuitive. Nevertheless, the science of quantum mechanics allows us to predict with great accuracy the properties of matter, which is why we have been able to use it to build computers and other modern electronic devices. Two fundamental laws lie at the heart of quantum mechanics. They are known as the *uncertainty principle* and the *exclusion principle,* and we'll discuss each of them in turn.

• What is the uncertainty principle?

The **uncertainty principle** was first described by Werner Heisenberg (1901–1976) in 1927. Here is one way of stating it:

> **The Uncertainty Principle: The more we know about where a particle is located, the less we can know about its momentum, and conversely, the more we know about its momentum, the less we can know about its location.**

We can illuminate the meaning of the uncertainty principle by considering how we might measure the trajectories of a baseball and an electron. In the case of a baseball, we could photograph it with a blinking strobe light. The resulting photograph would show us both where the ball was and where it was going at each moment in time (Figure S4.6). In scientific terms, knowing the path of the baseball means that we are measuring both its *location* and its *velocity* at each instant, or, equivalently, its location and its *momentum*. (Recall that momentum is mass times velocity.)

Now imagine trying to observe an electron in the same way. We will detect the electron only if it manages to scatter some of the photons streaming by it. However, while the photons are tiny particles of light compared to a baseball, they are quite large compared to the minuscule electron. The precision with which we can pinpoint the electron's location depends on the wavelengths of the photons. If we use visible light with a wavelength of 500 nanometers, we can measure the electron's location to within only 500 nanometers—about *5,000 times* the size of a typical atom. That is, if we see a flash from a row of 5,000 atoms, we do not

Figure S4.6 A photograph of a ball taken with a blinking strobe light allows us to determine both where the ball was and where it was going at each moment in time.

even know which atom contains the electron that caused the flash!

To locate the electron more precisely, we must use shorter-wavelength light, such as ultraviolet light or X rays. Now we encounter our next problem. To determine the electron's path, we must observe the flashes from its interactions with one photon after another. Yet each photon's energy delivers a "kick" that disturbs the electron and thereby changes the momentum we are trying to measure. The shorter the wavelength of the photon—which means the higher its energy—the more it alters the electron's momentum.

It is almost as if nature is playing a perverse trick on us. Locating the electron precisely requires hitting it with a short-wavelength photon, but the high energy of this photon prevents us from determining the electron's momentum. Conversely, measuring the electron's momentum requires hitting it with a low-energy photon that will not disturb it much. But because low-energy light has long wavelengths, we'll no longer have a very good idea of where the electron is located.

THINK ABOUT IT

Colloquially, we often express the uncertainty principle by saying that we can't know both where a particle is and where it is going. How does this statement relate to the more precise statement that we can't know both the particle's location and its momentum? In what ways is the colloquial statement accurate, and in what ways is it an oversimplification?

The uncertainty principle applies to *all* particles, not just to electrons. In fact, it applies even to large "particles" such as baseballs, but it is unnoticeable at this level. Consider what happens when we look at a baseball with visible light with a 500-nanometer wavelength. Just as with the electron, we can locate any part of the baseball to within only 500 nanometers. However, an uncertainty of 500 nanometers (about 0.00002 inch) is negligible in comparison to the size of the baseball. Moreover, the energy of visible light is so small compared to the energy of the baseball (including its mass-energy) that it has no noticeable effect on the baseball's momentum. That is why Newton's laws work perfectly well when we deal with the motions of baseballs, cars, planets, or other objects in the macroscopic world. (The prefix *macro* comes from the Greek word for "large," and the term *macroscopic* is used to contrast the world of objects large enough to be visible to the naked eye with the microscopic world, which is visible only through microscopes.) Newton's laws fail us only in the microscopic quantum realm, where we must deal with the implications of the uncertainty principle.

Wave-Particle Duality   Our thought experiment about measuring an electron's position and momentum with photons might at first seem to suggest that the electron somehow "hides" its precise path from us. However, the uncertainty principle runs deeper than this—it implies that the electron *does not even have* a precise path. From this point of view, the concepts of location and momentum do not exist independently for electrons in the way they appear to exist for objects in everyday life. Instead of imagining the electron to be following some complex but hidden path, we need to think of the electron as being "smeared out" over some volume of space [Section 5.3]. This "smearing out" of electrons and other particles holds the essence of the idea of wave-particle duality.

If we choose to regard the electron as a particle, we are imagining that we can locate it precisely by hitting it with a short-wavelength photon. In that case, we can measure the electron's precise location at each instant in time but can never predict where it will be at the next instant. In other words, we can see the electron in one place at one moment and in another place at another moment, but we'll have no idea how it passed through the regions in between. If we made many such measurements, we could thereby make a map of an atom in which we use lots of dots to represent electron locations at different times, but we could not draw the electron's actual path between the dots. In fact, a mathematical description of quantum mechanics allows us to calculate the *probability* that we'll find the electron in any particular place at any particular time. For example, Figure S4.7 shows the probability patterns for the electron in several energy levels of hydrogen. The brighter regions in the figure show where the electron has a higher probability of being found at any particular instant.

Alternatively, we can choose to regard the electron as a wave. We can measure the momentum of a wave, such as that of a ripple moving across a pond, but we cannot say that a wave has a single, precise location (Figure S4.8). Instead, the wave is spread out over some region of the water. In the same way, regarding an electron as a wave means that we'd see it spread out over some region of space. More precisely, the electron would be a three-dimensional wave surrounding an atom. Thus, the wave viewpoint also leads us to conclude that the electron is "smeared out" over some volume of space.

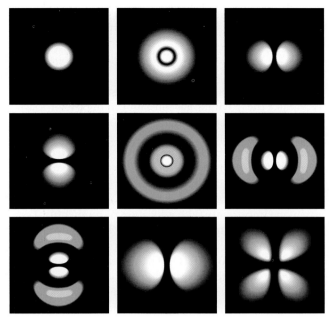

Figure S4.7 If we regard an electron as a particle, we can map out its location within the atom at different times. These diagrams show calculated probabilities for finding the electron at various places in a hydrogen atom; brighter regions correspond to a higher probability of finding the electron in that region. The nucleus, which consists of a single proton, is in the center in each case. The nine patterns shown here represent the probability patterns for an electron in nine different energy levels.

Figure S4.8 A wave has a well-defined momentum (represented by the arrow) but not a single, precise location. Thus, if we regard an electron as a wave, we can imagine it as a three-dimensional wave occupying some volume around an atom, but we cannot pinpoint its location.

The fact that electrons exhibit both particle and wave properties demonstrates that our common sense from the macroscopic world does not translate well to the quantum world. It also explains what we mean when we say that electrons, like photons of light, exhibit wave-particle duality. In fact, all subatomic "particles" exhibit wave-particle duality. We call them *particles* only for convenience. Like a photon of light, each particle has a wavelength.* When the wavelength of the particle is small, we can locate the particle fairly precisely, but then the particle's momentum is highly uncertain. When the wavelength of the particle is large, its momentum becomes well defined, but its location grows fuzzy.

---

*Electrons usually have very small wavelengths and therefore can be used to locate other particles with high precision. This is the principle behind *electron microscopes*, in which short-wavelength electrons are used to study microscopic objects. Whereas the resolution of visible-light microscopes is limited to the roughly 500-nanometer wavelength of visible light, electron microscopes can achieve resolutions of less than 0.1 nanometer.

Quantifying the Uncertainty Principle We can quantify the uncertainty principle with a simple mathematical statement:

$$\text{uncertainty in location} \times \text{uncertainty in momentum} \approx \text{Planck's constant}$$

Planck's constant is a fundamental constant in nature, rather like the gravitational constant (*G*) in Newton's law of gravity and the speed of light. Its numerical value is 6.626×10^{-34} Joule $\times$ s. (We encountered Planck's constant earlier in the book in the formula for the energy of a photon [Section 5.2].)

This formula quantifies what we have already learned: Because the product of the uncertainties is roughly constant, when one uncertainty (either location or momentum) goes down, the other must go up. For example, if we determine the location of an electron with a particular amount of uncertainty, such as to within 500 nanometers, we can use the formula to calculate the amount of uncertainty in the electron's momentum. The numerical value of Planck's constant is quite small, which explains why uncertainties are scarcely noticeable for macroscopic objects.

A second way of writing the uncertainty principle is mathematically equivalent but leads to additional insights. Instead of expressing the uncertainty principle in terms of location and momentum, this alternative version expresses it in terms of the amount of *energy* that a particle has and *when* it has this energy. This version reads:

$$\text{uncertainty in energy} \times \text{uncertainty in time} \approx \text{Planck's constant}$$

An amusing way to gain further appreciation for the uncertainty principle in both its forms is to imagine a game of "quantum baseball." Suppose a quantum pitcher is pitching an electron that you try to hit with your quantum bat. You'll find it extremely difficult! With the first version of the uncertainty principle, the problem is that you'll never know both where the electron is and where it is headed next. You might see it right in front of you, but because you don't know in which direction it's going, you can't know whether to swing straight, up, down, or sideways. With the second version of the uncertainty principle, your problem is that you might know the electron's energy, which tells you how hard you need to swing, but you'll never know *when* to swing. Either way, your chances of hitting the electron are completely random, governed by probabilities that can be calculated with the equations of quantum mechanics.

THINK ABOUT IT

Explain why real baseball players don't have the same problems that would arise with quantum baseball. (*Hint:* How do the quantum uncertainties compare to the size of a real baseball and a real bat?)

• What is the exclusion principle?

The second fundamental law of quantum mechanics, called the **exclusion principle**, was first described by Wolfgang Pauli (1900–1958) in 1925. In its simplest sense, the exclusion principle says that two particles cannot be in the same

---

MATHEMATICAL INSIGHT **S4.1** **Electron Waves in Atoms**

In the text, we've said that an electron in an atom is "smeared out" over some volume of space. In fact, the physics is much more precise than this vague statement implies. If we choose to view the electron as a wave, an electron in an atom can be regarded as a *standing wave*.

You are probably familiar with standing waves on a string that is anchored in place at its two ends, such as a violin string. Such waves are called standing waves because each point on the string vibrates up and down but the wave does not appear to move along the length of the string. Moreover, because the string is anchored at both ends, only wave patterns with an integer or half-integer $(0, \frac{1}{2}, 1, \frac{3}{2}, 2, \dots)$ number of wavelengths along the string are possible. Other patterns, such as having three-fourths of a wavelength along the string, are not possible without breaking one end of the string away from its anchor point (see the figure).

An electron viewed as a standing wave is anchored by the electromagnetic force holding it in the atom. Like the case of waves on a string, only particular wave patterns are possible. However, these patterns are more complex than waves on a string because they are three-dimensional wave patterns. The allowed wave patterns for an electron in an atom can be calculated with the famous *Schrödinger equation*, developed by Erwin Schrödinger in 1926. These allowed wave patterns correspond directly to the allowed energies of the elec-

tron in the atom. Thus, the Schrödinger equation enables scientists to predict the allowed energies in different atoms. The fact that the Schrödinger equation successfully predicts the energy levels that are measured in the laboratory (by analysis of spectral lines) is one of the great triumphs of quantum mechanics.

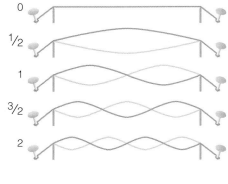

For a string anchored at both ends, waves can have any integer or half-integer number of wavelengths . . .

0
1/2
1
3/2
2

. . . but no other wave patterns can exist without breaking one end away from its anchor.

3/4

place at the same time. A more complete understanding of the exclusion principle requires investigating the properties of particles a little more deeply.

The Quantum State of a Particle Scientists use the term *state* to describe the current conditions of an object. For example, if you are relaxing in a chair, a scientist might say that you are in a "state of rest." A more precise description of your state in the chair might be something like, "Your current state is a velocity of zero (at rest), a heart rate of 65 beats per minute, a breathing rate of 12 breaths per minute, a body temperature of 37°C, a metabolic rate of 200 Calories per hour," and so on.

Suppose you have the following information about the current state of a friend: velocity of 5 km/hr, heart rate of 160 beats per minute, metabolic rate of 1,200 Calories per hour. Which of the following is your friend most likely doing: walking slowly as she reads a book, driving in her car, riding a bicycle down a hill, or swimming at a hard pace? How do you know?

Completely describing a person's state is quite complicated. Fortunately, describing the state of a subatomic particle such as an electron, proton, or neutron is much easier. In general, a complete description of the state of a subatomic particle—called its **quantum state**—specifies its location, momentum, orbital angular momentum, and spin to the extent the uncertainty principle allows. Like the energy of an electron in an atom, each property that describes a particle's

quantum state is *quantized*, meaning that it can take on only particular values and not other values in between.

Statement and Meaning of the Exclusion Principle Earlier, we described the exclusion principle in simple terms by saying that two particles cannot be in the same place at the same time. Now it is time to state the exclusion principle more precisely. First, the exclusion principle applies only to particles whose spin qualifies them as fermions, not bosons; thus, for example, it applies to protons, neutrons, and electrons because they are all fermions, but it does not apply to photons because they are bosons. With that in mind, here is a more precise statement of the exclusion principle:

The Exclusion Principle: Two fermions of the same type cannot occupy the same quantum state at the same time.

The exclusion principle has many important implications. One of the most important is in chemistry, where it dictates how electrons occupy the various energy levels in atoms. For example, an electron occupying the lowest energy level in an atom necessarily has a particular amount of orbital angular momentum and a restricted range of locations. The electron's energy level fully determines its quantum state, except for its spin. Because electrons have only two possible spin states, up or down, only two electrons can occupy the lowest energy level. If you tried to put a third electron into the lowest energy level, it would have the same spin—and hence the same quantum state—as one of the two electrons already there. The exclusion principle won't

---

SPECIAL TOPIC Does God Play Dice?

Suppose we knew all the laws of nature and, at some particular moment in time, what every single particle in the universe was doing. Could we then predict the future of the universe for all time?

Until the twentieth century, nearly every philosopher would have answered "yes." The idea was so pervasive that many philosophers concluded that God was like a watchmaker: God simply started up the universe, and the future was forever after determined. This idea that everything in the universe is predictable from its initial state is called *determinism,* and a universe that runs predictably, like a watch, is called a *deterministic universe.*

The discovery of the uncertainty principle shattered the idea of a deterministic universe because it tells us that, at best, we can make statements only about the *probability* of the precise future location of a subatomic particle. Because everything is made of subatomic particles, the uncertainty principle implies a degree of randomness built into the universe.

The idea that nature is governed by probability rather than certainty unsettled many people, including Einstein. Although he was well aware that the theories of quantum physics had survived many experimental tests, Einstein maintained a belief that the theories were incomplete. He believed that scientists would one day discover a deeper level of nature at which uncertainty would be removed.

To summarize his philosophical objections to uncertainty, Einstein said, "God does not play dice."

Einstein did more than simply object on philosophical grounds. He also proposed a number of thought experiments in which he showed that the uncertainty principle implies paradoxical results. Claiming that such paradoxes made no sense, he argued that the uncertainty principle must not be correct. In the years since Einstein's death in 1955, advances in technology have made it possible to actually perform some of Einstein's quantum thought experiments. The results have proved to be just as paradoxical as Einstein claimed they would be, even as they have confirmed the uncertainty principle.

What can we make of an idea, such as the uncertainty principle, that seems to violate common sense at the same time that it survives every experimental test? Under the tenets of science, experiment is the ultimate judge of theory, and we must accept the results despite philosophical objections. In a sense, Einstein's objection that "God does not play dice" reflected his beliefs about how the universe *should* behave. Niels Bohr argued instead that nature need not fit our preconceptions, and he famously replied to Einstein by saying: "Stop telling God what to do."

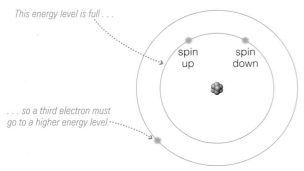

This energy level is full . . .

spin up   spin down

. . . so a third electron must go to a higher energy level.

Figure S4.9 The exclusion principle tells us that only two electrons, one with spin up and the other with spin down, can share a single energy level in an atom.

Lower temperature means lower thermal pressure.

Higher temperature means higher thermal pressure, making the balloon expand.

Figure S4.10 Thermal pressure is the familiar type of pressure that keeps a balloon inflated. Molecules move faster at higher temperatures, thereby increasing the pressure. (The dots in this diagram represent molecules, and the lengths of the arrows represent their speeds.)

allow that, however, so the third electron must go into a higher energy level (Figure S4.9). If you take a course in chemistry, you'll learn how a similar analysis of higher energy levels explains the chemical properties of all the elements, including their arrangement in the periodic table of the elements (see Appendix D).

The uncertainty principle and the exclusion principle together determine the sizes of atoms and of everything made of atoms, including your own body. The uncertainty principle ensures that electrons cannot be packed into infinitesimally tiny spaces. If you tried to confine an electron in too small a space in an atom, the uncertainty principle means that its momentum would become so large that the electromagnetic attraction of the nucleus could no longer retain it. The exclusion principle ensures that each electron has to have its own space. Thus, despite the fact that atoms are almost entirely empty, the two fundamental laws of quantum mechanics explain the solidity of matter. In the words of physicist Richard Feynman (1918–1988), "It is the fact that electrons cannot all get on top of each other that makes tables and everything else solid."

The exclusion and uncertainty principles also govern the behavior of tightly grouped protons, neutrons, and all other kinds of fermions. Just as these principles give atoms their size, they also determine the size of nuclei, because they limit how closely protons and neutrons can pack together. As we will see shortly, these quantum effects can even influence the lives of stars.

Before we turn to astronomical implications of the quantum laws, it's worth noting that there are also important implications to the fact that photons and other bosons do *not* obey the exclusion principle. For example, laser beams are so intense because many photons *can* be in the same state at the same time. In addition, it is possible under special conditions for two or more fermions to act together like a single boson. Such conditions lead to some amazing behavior, including *superconductivity,* in which electricity flows without any resistance, and *superfluidity,* in which extremely cold liquids can climb up the walls of a container. Many popular books on quantum laws explain these phenomena in more detail.

S4.4 Key Quantum Effects in Astronomy

The uncertainty principle and the exclusion principle have clear consequences in the subatomic realm. Amazingly, this microscopic behavior also produces important effects on much larger scales. In fact, we cannot fully understand how stars are born, shine brightly throughout their lives, and die unless we understand the implications of the quantum laws. In this section, we investigate four quantum effects of great importance in astronomy: degeneracy pressure, quantum tunneling, virtual particles, and the evaporation of black holes.

• How do the quantum laws affect special types of stars?

Under ordinary conditions in gases, pressure and temperature are closely related. For example, suppose we inflate a balloon, filling it with air molecules. The individual molecules zip around inside the balloon, continually bouncing off its walls (Figure S4.10). The force of these molecules striking the walls of the balloon creates **thermal pressure**, which keeps the balloon inflated. If we cool the balloon by, say, putting it into a freezer, the molecules slow down. Slowing the molecules reduces the force with which they strike the walls, reducing the thermal pressure and shrinking the balloon. Heating the balloon speeds up the molecules, raising the thermal pressure and causing the balloon to expand. Thus, thermal pressure gets its name because it depends on temperature.

Thermal pressure is the dominant type of pressure at the low to moderate densities that we experience in everyday life. However, quantum effects produce an entirely different type of pressure under conditions of extremely high density, one that does not depend on temperature at all.

Consider what happens when we compress a *plasma,* a mixture of positively charged ions and free electrons [Section 5.3]. At first, the energy we expend in crushing the plasma makes the electrons and protons move faster and faster, increasing the pressure and the temperature. Now suppose we let the plasma cool off for a while. As the plasma cools, its pressure drops, enabling us to compress it further. Continuing this process of cooling and compression, we can in principle squeeze the plasma down to a very dense state. However, we cannot continue this process indefinitely. According to the exclusion principle, no two electrons in the plasma can have exactly the same position, momentum, and spin. That is, just as in an atom, all the electrons can't get on top of one another at once. Thus, the compression must stop at some point, no matter how cold the plasma. This resistance to compression that stems from the exclusion principle is called **degeneracy pressure**.

The following analogy might help you visualize how degeneracy pressure works. Imagine a small number of people moving from chair to chair in an auditorium filled with folding chairs. Each person can move freely about and sit in any empty chair. The chairs represent available quantum states, the people represent electrons darting from place to place, and their motions represent thermal pressure as the electrons move from one quantum state to the next. The exclusion principle corresponds to the rule that two people cannot sit in the same chair at the same time. As long as the chairs greatly outnumber the people, two people will rarely fight over the same chair (Figure S4.11a).

Suppose ushers begin removing chairs from the front of the auditorium (compression), gradually forcing people to move to the back. Soon everyone has to crowd toward the back of the auditorium, where the number of remaining chairs is just slightly larger than the number of people. Now when a person moves to a particular chair, there's a good chance that it's already occupied. Ultimately, when the number of people equals the number of chairs, people can still move from place to place, but only if they swap seats with somebody else (Figure S4.11b). The ushers can't take away any more chairs, and the compression must stop. In other words, all the available states are filled.

The uncertainty principle also influences degeneracy pressure, though in a way that does not perfectly fit this analogy. In a highly compressed plasma, the available space for each electron is very small, which in essence means that each electron's position is precisely defined. According to the uncertainty principle, its momentum must then be extremely uncertain. For the many electrons in the plasma, the great uncertainty in momentum means they must be moving very fast on average. This requirement that highly compressed electrons have to move quickly holds even if the object is very cold.* This quantum-mechanical trade-off between position and momentum is at the root of degeneracy pressure. If you want to compress lots of electrons into a tiny space, you need to exert an enormous force to rein in their momentum.

Degeneracy pressure caused by the crowding of electrons, or **electron degeneracy pressure**, affects the lives of stars in several different ways. In some cases, it can prevent a collapsing cloud of gas from becoming a star in the first place, creating what is called a *brown dwarf* [Section 16.3]. In stars like the Sun, it determines how they begin burning helium near the ends of their lives [Section 17.2]. When stars die, most leave behind an extremely dense stellar corpse called a *white dwarf,* which is also supported by electron degeneracy pressure [Section 18.1].

Not all stars meet this fate, because electron degeneracy pressure cannot grow infinitely strong. Under extreme compression, the average speed of the electrons begins to approach the speed of light. Nothing can go faster than the speed of light, so we eventually reach a limit to how much degeneracy pressure the electrons can exert. Once a dying star reaches that limit, electron degeneracy pressure cannot prevent it from shrinking further. The star then collapses until it becomes a ball of neutrons, called a *neutron star* [Section 18.2]. Neutron stars support themselves through **neutron degeneracy pressure**, which is just like electron degeneracy pressure except that it is caused by neutrons and occurs at much higher densities. Neutron degeneracy pressure comes into play only at much higher densities, because neutrons have much greater mass than electrons.

---

*In this sense, an object is *cold* if there's no way to get heat from it. Consider a plasma in which all the available momentum states are filled up to a certain level. To extract heat from the plasma, you'd have to slow down some of its particles. That means moving them to lower-momentum states, which are already taken. The exclusion principle thus prevents any energy from escaping, so the plasma is cold even though the electrons may be moving at high speeds.

Figure S4.11 The auditorium analogy to degeneracy pressure. Chairs represent available quantum states, and people who must keep moving from chair to chair represent electrons.

a When there are many more available quantum states (chairs) than electrons (people), an electron is unlikely to try to enter the same state as another electron. The only pressure comes from the temperature-related motion of the electrons, which is the thermal pressure.

b When the number of electrons (people) approaches the number of available quantum states (chairs), finding an available state requires that the electrons move faster than they would otherwise. This extra motion creates degeneracy pressure.

A neutron near the speed of light possesses over 1,800 times more momentum than an electron at the same speed. Thus, the positions of neutrons can be over 1,800 times more precise, enabling them to occupy a much smaller volume of space.

Neutron degeneracy pressure cannot grow infinitely strong either. It begins to fail when the speed of the neutrons approaches the speed of light. At that point, gravity can make an object shrink further, and according to our present understanding, nothing can stop the collapse of an object once its gravity overcomes neutron degeneracy pressure. Such an object collapses indefinitely, becoming the mysterious type of structure in spacetime that we call a *black hole* [Section S3.3].

• How is "quantum tunneling" crucial to life on Earth?

The next quantum effect we'll investigate arises from the uncertainty principle and has important implications not only in astronomy, but also to our very existence and to modern technology. Let's start with an analogy.

Imagine that, as the unfortunate result of a case of mistaken identity, you're sitting on a bench in a locked jail cell (Figure S4.12a). Another bench is on the other side of the cell wall. If you could magically transport yourself from the bench on the inside to the bench on the outside, you'd be free.

Alas, no such magic ever occurs for humans. But what if you were an electron? In that case, the uncertainty principle would prevent us from predicting your precise location. At best, we could state only the probability of your being in various locations. While the probability that you remain in your cell might be very high, there is always some small probability that you will be found outside your cell, despite the existence of the wall. Thus, with a bit of luck, you might suddenly find yourself free, thanks to the uncertainty principle (Figure S4.12b). Although electrons don't get put into jails, an analogous process in which electrons or other subatomic particles "magically" go through

A person is confined to the bench inside the jail cell, even though it would take no more energy to sit on the outside bench . . .

. . . but an electron can move to the other side of the wall and become free through the process of quantum tunneling.

Figure S4.12 A jail cell analogy to explain quantum tunneling.

wall-like barriers really does happen. We call it **quantum tunneling**.

We can gain a deeper understanding of quantum tunneling by thinking in terms of the *energy* needed to cross a barrier. If you are sitting in a jail cell, the barrier is the wall of the cell. The reason you cannot escape the cell is that you don't have enough energy to crash through the wall. Just as the cell wall keeps you imprisoned, a barrier of electromagnetic repulsion can imprison an electron that does not have enough energy to crash through it. However, remember that we can write the uncertainty principle in the following form:

$$\genfrac{}{}{0pt}{}{\text{uncertainty}}{\text{in energy}} \times \genfrac{}{}{0pt}{}{\text{uncertainty}}{\text{in time}} \approx \text{Planck's constant}$$

Because of the uncertainty inherent in energy, there is always *some* chance that the electron will have more energy than we think at a particular moment, allowing it to cross the barrier that lies in its way. Thus, from this point of view, quantum tunneling comes about because of uncertainty in energy rather than uncertainty in location.

Both points of view on quantum tunneling are equivalent, but the latter viewpoint illustrates a rather remarkable "loophole" in the law of conservation of energy. To cross the barrier, the particle must briefly gain some excess energy. Thanks to the uncertainty principle, this "stolen" energy need not come from anywhere as long as the particle returns it within a time period *shorter* than the uncertainty in time. In that case, we cannot be certain that any energy was ever missing! It's like stealing a dollar and putting it right back before anyone notices, so that no harm is done— except that the particle uses the stolen energy to cross the barrier before returning it.

This tale of phantom quantum energy thefts may at first sound utterly ridiculous, but the process of quantum tunneling is readily observed and extremely important. In fact, the "microchips" used in all modern computers and many other modern electronic devices work because of quantum tunneling by electrons. We can control the rate of quantum tunneling, and hence the electric current, by adjusting the "height" of the energy barrier. The higher the energy barrier, the less likely it is that particles will tunnel through it and the lower the electric current. This control over the electric current is critical in modern electronics.

Even more amazingly, our universe would look much different were it not for quantum tunneling. The nuclear fusion reactions that power stars occur when atomic nuclei smash together so hard that they stick. However, nuclei tend to repel each other, because they are positively charged (they contain only positive protons and neutral neutrons) and like charges repel. This repulsion creates an electromagnetic barrier that prevents nuclear fusion under most conditions. Even at the high temperatures found deep in the cores of stars like our Sun, atomic nuclei don't have enough energy to crash through the electromagnetic barrier. Instead, they rely on quantum tunneling to "sneak" through the barrier into the region where the strong force dominates. In other words, quantum tunneling is what makes nuclear fusion possible in stars like our Sun—which means that life on Earth could not exist without it. (Quantum tunneling is not as crucial to fusion in stars that have core temperatures much higher than that of the Sun.)

• How empty is empty space?

The same "loophole" in the law of conservation of energy that allows particles to tunnel through otherwise impenetrable barriers also means that space can never be truly empty. According to the uncertainty principle, particles can "pop" into existence from nowhere—their mass made from stolen energy—as long as they "pop" back out of existence before anyone can verify that they ever existed. A somewhat fanciful analogy will help us understand this concept.

Imagine that the law of conservation of energy is enforced by a "great cosmic accountant." (In reality, of course, the law is enforced naturally.) Further, imagine that the cosmic accountant keeps a storehouse of energy in a large bank vault and ensures that any time something borrows some energy, it returns the energy in a precisely equal amount. A particle that pops into existence is like a bank robber who steals some energy from the vault. If the cosmic accountant catches the particle holding its stolen energy, someone will have to pay for the theft. However, the particle won't be caught as long as it pops back out of existence quickly enough, returning the stolen energy before the cosmic accountant notices that anything was ever missing. The length of time the particle has in which to pop back out of existence and return the stolen energy is a time short enough that the uncertainty principle prevents anyone from knowing that energy is missing.

Particles that pop in and out of existence before anyone can possibly detect them are called **virtual particles**. In fact, modern theories of the universe propose that empty space—what we call a *vacuum*—actually "bubbles" with virtual particles that pop rapidly in and out of existence.

THINK ABOUT IT

Imagine that you write a check for $100, but you have no money in your checking account. Your check is not necessarily doomed to bounce—as long as you deposit the needed $100 before the check clears. Explain how this $100 of "virtual money" is similar in concept to virtual particles.

You might wonder why a vacuum would seem so empty if it's actually teeming with virtual particles. The reason is that there's usually no way to extract the stolen energy from those virtual particles before they pop back out of existence. Consequently, the real particles that we can detect with our instruments cannot gain any energy by interacting with the virtual particles of the vacuum, making these vacuum particles essentially undetectable. However, the fact that we cannot detect these particles doesn't mean that they are unimportant; theory predicts that virtual particles should exert

some measurable effects on real particles, and such effects have indeed been observed in specially designed experiments. For example, theory predicts that placing certain metal surfaces very close together should change the number of virtual photons in the vacuum between them, which should cause a change in the force acting between the metal surfaces. This change in force has been measured and agrees with the theoretical predictions.

Moreover, the combined energy of the teeming sea of virtual particles in the vacuum, known as the **vacuum energy** (or *zero point energy*), may have a major impact on the universe as a whole. When we apply Einstein's general theory of relativity to the quantum idea of virtual particles, we find that the energy associated with the virtual particles in a vacuum can oppose the gravitational force associated with mass. In essence, the vacuum energy can act as a repulsive force, driving the objects in space farther apart even while gravity tries to pull them together. In principle, if the vacuum energy is great enough, it could exert so much repulsion that the expansion of the universe would accelerate with time. As we will discuss in Chapter 22, in recent years astronomers have found evidence suggesting that the universal expansion is indeed accelerating, leading some theorists to suggest that the acceleration may be a result of the repulsive force associated with virtual particles. However, calculations based on current theories of particle physics predict that if the virtual particles are responsible for acceleration, the acceleration would be far greater (by many orders of magnitude) than the acceleration we observe. Scientists are actively working to improve the theories, in order to understand why their predictions disagree with reality. For now, the question of whether vacuum energy can explain the acceleration of the expansion remains unanswered.

• Do black holes last forever?

Virtual particles are also thought to limit the lifetimes of black holes. We can understand why by extending our cosmic accountant analogy to the case of a virtual electron popping into existence near a black hole. Because electric charge must be conserved (there's no loophole for this conservation law), the virtual electron cannot pop into existence alone. Instead, it must be accompanied by a virtual positron (an antielectron) so that the total electric charge of the two particles is zero. Like all virtual particles, the virtual electron-positron pair essentially comes into existence with stolen energy, so to remain undetected they must return the stolen energy before our imaginary cosmic accountant catches them. The pair returns the energy by quickly annihilating each other (Figure S4.13a).

However, suppose a virtual electron-positron pair pops into existence very close to, but outside, the event horizon of a black hole. The electron and positron are supposed to annihilate each other quickly, but a terrible problem comes up: One of the particles crosses the event horizon during its brief, virtual existence (Figure S4.13b). From the

Most of Space

Virtual pairs of electrons and positrons continually appear and annihilate each other.

Space Near a Black Hole

Near a black hole, one particle may cross the event horizon . . .

. . . leaving the other particle as a real particle.

a Pairs of virtual electrons and positrons can pop into existence, as long as they annihilate each other before they can be detected.

b Near the event horizon of a black hole, this same process leads to the creation of real particles, not just virtual ones. The energy to make these particles comes at the expense of the black hole, which loses as much mass as the new particles gain.

Figure S4.13 Virtual particles may become real particles near a black hole.

perspective of our cosmic accountant, this virtual particle was never accounted for in the first place, so there's no problem with the fact that it will never be seen again. But the other particle is suddenly caught like a deer in headlights. Without its virtual mate, it has no way to annihilate itself and is caught red-handed by the cosmic accountant. Because the particle is now holding real rather than virtual energy, the cosmic accountant demands that someone or something pay for the stolen energy. The particle itself cannot pay, because it now has the energy by virtue of its existence and has no way to give it back because its virtual partner is gone. The cosmic accountant therefore makes the black hole pay for the energy, and the black hole pays by giving up a little bit of its gravitational potential energy.

If we strip away the fanciful imagery of a cosmic accountant, the end result is the creation of *real* particles, not virtual ones, just outside the event horizon of a black hole. Nothing is escaping from inside the black hole. Rather, these real particles are created from the gravitational potential energy of the black hole. Around the black hole, these *real* electrons and positrons annihilate each other, producing real photons that are radiated into space. To an outside observer, the black hole would therefore appear to be radiating, even though nothing ever escapes from inside it. This effect was first predicted by Stephen Hawking in the 1970s and is therefore called **Hawking radiation**. As we've just seen, the ultimate source of Hawking radiation is the gravitational potential energy, and hence the mass, of the black

hole. Thus, the continual emission of Hawking radiation causes the black hole to shrink slowly in mass, or *evaporate*, over very long periods of time.

The idea that black holes can evaporate remains untested, but if it is true, it may have profound implications for both the origin and the fate of the universe. Some scientists speculate that black holes of all sizes might have been created during the Big Bang. If so, some of the smaller ones should be evaporating today, emitting bursts of gamma rays during their final moments of existence. The fact that we have not yet detected any such evaporation sets limits on the number and size of small black holes that might have formed in the Big Bang.

At the other end of time, if the universe continues to expand forever, black holes may be the last large masses left after all the stars have died. In that case, the slow evaporation caused by Hawking radiation will mean that even black holes cannot last forever, and the universe eventually will contain nothing but a fog of photons and subatomic particles, separated from one another by incredible distances as the universe continues to grow in size.

THE BIG PICTURE

Putting Chapter S4 into Context

In this chapter, we have studied the quantum revolution and its astronomical consequences. As you look back, keep in mind the following "big picture" ideas:

- The quantum revolution can be considered the third great revolution in our understanding of the universe. The first was the Copernican revolution, which demolished the ancient belief in an Earth-centered universe. The second was Einstein's discovery of relativity, which radically revised our ideas about space, time, and gravity. The quantum revolution has changed our ideas about the fundamental nature of matter and energy.

- Strange as the laws of quantum mechanics may seem, they can be readily tested and confirmed through observation and experiment. The quantum laws, like relativity, now stand on very solid ground. So far, they have passed every experimental test devised for them.

- The tiny quantum realm may seem remote from the large scales we are accustomed to in astronomy, but it is exceedingly important. The laws of quantum mechanics are necessary for understanding many astronomical processes, including nuclear fusion in the Sun, the degeneracy pressure that supports stellar corpses, and the possible evaporation of black holes.

S4.1 The Quantum Revolution

- **How has the quantum revolution changed our world?** Quantum mechanics has revolutionized our understanding of particles and forces and made possible the development of modern electronic devices such as computers.

S4.2 Fundamental Particles and Forces

- **What are the basic properties of subatomic particles?** The three most important properties of a particle are its *mass, charge,* and *spin.* All particles fall into one of two classes by spin: **fermions** and **bosons**. Fermions include all the particles that make up atoms, including electrons, neutrons, and protons. Bosons include photons and other particles that transmit forces.

- **What are the fundamental building blocks of matter?** **Quarks** and **leptons**, both of which are fermions, are thought to be the fundamental building blocks of matter. There are six known types of each. Two of the six known types of quarks make up protons and neutrons, while electrons are one of the six known types of leptons. Each type of particle of matter also has a corresponding particle of **antimatter**.

- **What are the fundamental forces in nature?** The four fundamental forces are **gravity**, **electromagnetism**, the **weak force**, and the **strong force**.

S4.3 Uncertainty and Exclusion in the Quantum Realm

- **What is the uncertainty principle?** The **uncertainty principle** tells us that we cannot simultaneously know the precise value of an object's position and momentum—or, equivalently, its energy and the precise time during which it has this energy.

- **What is the exclusion principle?** The **exclusion principle** tells us that two fermions of the same type cannot occupy the same quantum state at the same time. This principle does not apply to bosons.

S4.4 Key Quantum Effects in Astronomy

- **How do the quantum laws affect special types of stars?** **Degeneracy pressure** is a type of pressure that can occur even in the absence of heat. It arises from the combination of the exclusion principle and the uncertainty principle. It is the dominant form of pressure in the astronomical objects known as brown dwarfs, white dwarfs, and neutron stars.

- **How is "quantum tunneling" crucial to life on Earth?** **Quantum tunneling** is a process in which subatomic particles can "tunnel" from one place to another even when they don't actually have enough energy to overcome an energy barrier between the two places. Without this process, fusion in the Sun would not be possible and life on Earth could not exist. It is also important to many other astronomical processes and to modern technology.

- **How empty is empty space?** According to the uncertainty principle, a vacuum cannot be completely empty but must instead be filled with unobservable **virtual particles** that are constantly popping in and out of existence. While the virtual particles generally do not affect the particles that we can observe, the **vacuum energy** associated with them may oppose gravity, and in principle could cause the expansion of the universe to accelerate.

- **Do black holes last forever?** No. According to current theory, isolated black holes can gradually evaporate through quantum tunneling, emitting **Hawking radiation** in the process. Although the theoretical basis of this idea seems solid, it has not yet been observed in nature.

EXERCISES AND PROBLEMS

For instructor-assigned homework go to **Mastering ASTRONOMY** www.masteringastronomy.com

Review Questions

Short-Answer Questions Based on the Reading

1. What do we mean by the *quantum* realm? What is *quantum mechanics?*
2. List five major ideas that come from the laws of quantum mechanics.
3. What do we mean by *fundamental particles?* How do we investigate fundamental particles in *particle accelerators?* What is the *standard model?*
4. What is *spin?* What are the two basic categories of particles based on spin?
5. What are *quarks* and *leptons?* Explain why we say that quarks and leptons are subsets of the fermions.

6. List the six quarks and six leptons in the standard model. Describe the quark composition of a proton and of a neutron.

7. What are *neutrinos*? Why might they be important to the overall mass of the universe?

8. What is *antimatter*? How does a positron differ from an electron? What happens when a particle and its antiparticle meet?

9. How is antimatter produced, and why is it always produced along with matter in *pair production*?

10. List the four basic forces in nature, and the *exchange particles* for each.

11. Describe the relative strengths of the four forces. Why does gravity dominate on large scales, even though it is by far the weakest of the four forces?

12. Why do many scientists believe that the standard model will eventually be replaced by an even simpler model of nature?

13. What is the *uncertainty principle*? How is it related to the idea of wave-particle duality?

14. Describe two ways of quantifying the uncertainty principle, and give an example showing the meaning of each.

15. What do we mean by the *quantum state* of a particle?

16. What is the *exclusion principle*? What types of particles obey it? Briefly explain how the exclusion principle determines how electrons fill energy levels in atoms.

17. What is *degeneracy pressure*? How does it differ from *thermal pressure*? How is it important in astronomy?

18. Explain why the uncertainty principle implies that the particles in a highly compressed plasma must move at very high speeds. How does this fact explain why there is a limit to the strength of degeneracy pressure? Why is this limit higher for *neutron degeneracy pressure* than for *electron degeneracy pressure*?

19. What is *quantum tunneling*? How is it important to modern electronics? How is it important to nuclear fusion in the Sun?

20. What do we mean by *virtual particles*? How might they help explain the observed acceleration of the expansion of the universe? How might they lead to gradual evaporation of black holes through *Hawking radiation*?

Test Your Understanding

Does It Make Sense?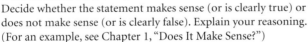

Decide whether the statement makes sense (or is clearly true) or does not make sense (or is clearly false). Explain your reasoning. (For an example, see Chapter 1, "Does It Make Sense?")

21. Although there are six known types of quarks, ordinary atoms contain only two of these types.

22. If you put a quark and a lepton close together, they'll annihilate each other.

23. There's no such thing as antimatter, except in science fiction.

24. Some particle accelerators have been known to build up a huge electrical charge because of the electrons produced inside them.

25. According to the uncertainty principle, we can never be certain whether one theory is really better than another.

26. The exclusion principle describes the cases in which the uncertainty principle is excluded from being true.

27. No known astronomical objects exhibit any type of degeneracy pressure.

28. Although we speak of four fundamental forces—gravity, electromagnetic, strong, and weak—it is likely that these forces are different manifestations of a smaller number of truly fundamental forces.

29. Imagine that, somewhere in deep space, you meet a person made entirely of antimatter. Shaking that person's hand would be very dangerous.

30. Someday, we may detect radiation coming from an evaporating black hole.

Quick Quiz

Choose the best answer to each of the following. Explain your reasoning with one or more complete sentences.

31. The fundamental particles of matter are (a) atoms and molecules. (b) quarks and leptons. (c) electrons, protons, and neutrons.

32. When an electron is produced from energy in a particle accelerator, the following also always happens: (a) An anti-electron (positron) is produced. (b) A quark is produced. (c) The electron soon disintegrates into a bunch of quarks.

33. Within an atomic nucleus, the strongest of the four forces is (a) gravity. (b) electromagnetism. (c) the strong force.

34. Across a distance of 1 millimeter, the strongest force acting between two protons is (a) gravity. (b) electromagnetism. (c) the strong force.

35. If we measure a subatomic particle's position very precisely, then (a) its momentum will be highly uncertain. (b) its spin will be highly uncertain. (c) we would be violating the uncertainty principle.

36. Aside from photons, which of the following also behave sometimes as waves and sometimes as particles? (a) fermions (b) bosons (c) all other subatomic particles

37. Which of the following is not allowed by the exclusion principle? (a) knowing both the precise position and precise momentum of an electron (b) having two electrons in the same quantum state (c) having an electron with a spin that is neither up nor down

38. The strength of degeneracy pressure in an object (such as a white dwarf) depends on (a) its temperature. (b) its density. (c) both its temperature and density.

39. In which one of the following objects does degeneracy pressure play the most important role? (a) a neutron star (b) the Sun (c) a star 10 times as massive as the Sun

40. According to the laws of quantum mechanics, empty space (a vacuum) is (a) truly empty. (b) bubbling with virtual particles. (c) filled with tiny black holes.

Investigate Further
In-Depth Questions to Increase Your Understanding

Short-Answer/Essay Questions

41. *The Strong Force.* The strong force is the force that holds the protons and neutrons in the nucleus together. Based on the fact that atomic nuclei can be stable, briefly explain how you can conclude that the strong force must be even stronger than the electromagnetic force, at least over very short distances.

42. *Chemistry and Biology.* All chemical and biological reactions involve the creation and breaking of chemical bonds, which are bonds between the electrons of one atom and the electrons of others. Given this fact, explain why the electromagnetic force governs all chemical and biological reactions. Also explain why the strong force, the weak force, and gravity play no role in these reactions.

43. *Gravity.* In one or two paragraphs, explain both (i) what we mean when we say that gravity is the weakest of the four forces; and (ii) why it nevertheless dominates the universe on large scales.

44. *Quantum Tunneling and Life.* In one or two paragraphs, explain the role of quantum tunneling in creating the elements from which we are made. (*Hint:* Recall that we are *star stuff* in the sense that the elements of our bodies were produced by nuclear fusion inside stars.)

45. *Antimatter Engines.* In the *Star Trek* series, starships are powered by matter-antimatter annihilation. Explain why we should expect matter-antimatter annihilation to be the most efficient possible source of power. What practical problems would we face in developing matter-antimatter engines?

46. *Exchange of Information.* Suppose an electron is moving up and down at some place located far in the distance. How could we learn that the electron is moving? How long after the movement would we know that it had occurred? Does this idea have any practical importance in astronomy? Explain. (*Hint:* Remember that the photon is the exchange particle for the electromagnetic force.)

47. *Nonquantum Baseball.* In your own words, explain why a baseball player does not have to take into account the uncertainty principle when up for bat.

48. *The Electron.* Based on what you have learned in this chapter, how would you explain an electron to a friend? Be sure that you explain why physicists don't think of electrons as tiny, negatively charged, spinning balls. Write your explanation in one or two paragraphs.

Quantitative Problems

Be sure to show all calculations clearly and state your final answers in complete sentences.

49. *Gravity and the EM Force.* In this problem, we compare the strength of gravity to the strength of the electromagnetic (EM) force for two interacting electrons. Because both electrons are negatively charged, they will want to *repel* each other according to the EM force. Because electrons have mass, they will want to *attract* each other according to gravity. Let's see which effect will dominate. You will need the following information for this problem:

● The force law for gravitation is:

$$F_g = G\frac{M_1 M_2}{d^2}\left(G = 6.67 \times 10^{-11}\frac{N \times m^2}{kg^2}\right)$$

where M_1 and M_2 are the masses of the two objects, d is the distance between them, and G is the gravitational constant. ("N" is the abbreviation for *newton,* the metric unit of force.)

● The force law for electromagnetism is:

$$F_{EM} = k\frac{q_1 q_2}{d^2}\left(k = 9.0 \times 10^9\frac{N \times m^2}{Coul^2}\right)$$

where q_1 and q_2 are the *charges* of the two objects (in *Coulombs,* the standard unit of charge), d is the distance between them, and k is a constant. ("Coul" is an abbreviation for *Coulomb.*)

● The *mass* of an electron is 9.10×10^{-31} kg.

● The *charge* of an electron is -1.6×10^{-19} Coul.

a. Calculate the gravitational force, in newtons, that attracts the two electrons if a distance of 10^{-10} meter (about the diameter of an atom) separates them.

b. Calculate the electromagnetic force, in newtons, that repels the two electrons at the same distance.

c. How many times stronger is the electromagnetic repulsion than the gravitational attraction for the two electrons?

50. *Large-Scale Gravity.* Suppose the Sun and Earth each had an excess charge equivalent to the charge of just one electron. Using the data from Problem 49, compare the gravitational force between the Earth and Sun to the electromagnetic force between them in that case. What does your answer tell you about why gravity dominates on large scales? (*Hint:* See Appendix E for the data you'll need about the Earth and Sun.)

51. *Evaporation of Black Holes.* The time it takes for a black hole to evaporate through the process of Hawking radiation can be calculated using the following formula, in which M is the mass of the black hole in kilograms and t is the lifetime of the black hole in seconds:

$$t = 10{,}240\pi^2\frac{G^2 M^3}{hc^4}\left(h = 6.63 \times 10^{-34}\frac{kg \times m^2}{s};\right.$$
$$\left. G = 6.67 \times 10^{-11}\frac{m^3}{kg \times s^2}\right)$$

Without doing any calculations, explain how this formula implies that lower-mass black holes have much shorter lifetimes than more massive ones and that the evaporation process accelerates as a black hole loses mass.

52. *Solar Mass Black Holes.* Use the formula from Problem 51 to calculate the lifetime of a black hole with the mass of the Sun ($M_{Sun} = 2.0 \times 10^{30}$ kg). How does your answer compare to the current age of the universe?

53. *Long-Lived Black Holes.* Some scientists speculate that in the far distant future, the universe will consist only of gigantic black holes and scattered subatomic particles. The largest black holes that conceivably might form would have a mass of about a trillion (10^{12}) Suns. Using the formula from Problem 51, calculate the lifetime of such a giant black hole. How does your answer compare to the current age of the universe? (*Hint:* Your calculator may not be able to handle the large numbers involved in this problem, in which case you will need to rearrange the numbers so that you can calculate the powers of 10 without your calculator.)

54. *Mini-Black Holes.* Some scientists speculate that black holes of many different masses might have been formed during the early moments of the Big Bang. Some of these black holes might be mini-black holes, much smaller in mass than those that can be formed by the crush of gravity in today's universe. Use the formula from Problem 51 to calculate the lifetime of a mini-black hole with the mass of Earth (about 6×10^{24} kg). Compare this to the current age of the universe.

55. *Black Holes Evaporating Today.* Suppose that mini-black holes formed in the Big Bang. Starting with the formula from Problem 51, calculate the mass of a black hole formed in the Big Bang that would be completing its evaporation today. Compare your answer to the mass of Earth. Assume that the universe is 14 billion years old.

56. *Your Quantum Uncertainty.* Suppose you are running at a speed of about 10 km/hr, but there is an uncertainty of 0.5 km/hr in your precise speed. Given your mass, you can calculate your momentum and the uncertainty of that momentum. What is the corresponding quantum limit to the measurement of your position? Is this significant? Why or why not? (*Hint:* You'll need the first form of the uncertainty principle given on p. 466; you can use the value of Planck's constant, h, from Problem 51.)

57. *An Electron's Quantum Uncertainty.* You are conducting an experiment in which you can measure the location of individual electron collisions to within 10^{-10} meter. What is the

theoretical limit to which you can simultaneously measure the momentum of those collisions? What is the uncertainty in the electron's speed? (The electron has a rest mass of 9.1×10^{-31} kg.)

Discussion Questions

58. *Common Sense vs. Experiment.* Even the most highly trained physicists find the results of quantum mechanics to be strange and counter to their everyday common sense, yet the predictions of quantum mechanics have passed every experimental test yet posed for them. Discuss whether it is important to reconcile our common sense with experimental results and, if so, how we can do so.

59. *The Meaning of the Uncertainty Principle.* When they first hear about it, many people assume that the uncertainty principle means that we cannot *measure* the position and momentum of a particle precisely. According to current understanding, it really tells us that the particle *does not have* a precise position and momentum in the sense that we would expect from everyday life. How do these two viewpoints differ? Discuss the different philosophical consequences of these two viewpoints.

MEDIA EXPLORATIONS

For self-study activities go to MasteringASTRONOMY www.masteringastronomy.com

Web Projects

1. *Fermilab and CERN.* Visit the Web site for either the Fermilab or CERN particle accelerator. Take a virtual tour, and learn about the work at the laboratory. Write a short report about what you learn.

2. *The Higgs Particle.* Learn about how the so-called Higgs particle fits into the standard model and how physicists are searching for it. Write a short summary of your findings, including your opinion as to when we are likely to acquire evidence in support of (or against) the existence of the Higgs particle.

3. *Quantum Computing.* Learn how computer scientists hope to harness quantum effects to build computers much more powerful than any existing today. Briefly summarize the ideas, and write an essay stating your opinion concerning the benefits and drawbacks of developing this technology.

4. *Beyond the Standard Model.* Research some of the ideas that physicists are considering as possible improvements upon the standard model. Choose one such idea and write a short essay describing the idea, its potential effect on physics if it is correct, and how the idea may be tested.

14

Our Star

LEARNING GOALS

14.1 A Closer Look at the Sun
- Why was the Sun's energy source a major mystery?
- Why does the Sun shine?
- What is the Sun's structure?

14.2 The Cosmic Crucible
- How does nuclear fusion occur in the Sun?

- How does the energy from fusion get out of the Sun?
- How do we know what is happening inside the Sun?

14.3 The Sun-Earth Connection
- What causes solar activity?
- How does solar activity affect humans?
- How does solar activity vary with time?

I say Live, Live, because of the Sun, The dream, the excitable gift.

—Anne Sexton (1928–1974)

Astronomy today encompasses the study of the entire universe, but the root of the word *astronomy* comes from the Greek word for "star." Although we have learned a lot about the universe up to this point in the book, only now do we turn our attention to the study of the stars, the namesakes of astronomy.

When we think of stars, we usually think of the beautiful points of light visible on a clear night. But the nearest and most easily studied star is visible only in the daytime—our Sun. Of course, the Sun is important to us in many more ways than as an object for astronomical study. The Sun is the source of virtually all light, heat, and energy reaching Earth, and life on Earth's surface could not survive without it.

In this chapter, we will study the Sun in some depth. We will learn how the Sun generates the energy that supports life on Earth. Equally important, we will study our Sun as a star so that in subsequent chapters we can more easily understand stars throughout the universe.

 The Sun Tutorial, Lesson I

14.1 A Closer Look at the Sun

Most of us take it for granted today that the Sun is a star—a giant ball of hot gas that generates light and shines it brightly in all directions. However, scientists discovered this fact only quite recently in human history. In this section, we'll consider the age-old question of what makes the Sun shine, and then take an imaginary plunge into the Sun that will get us better acquainted with its general features.

• Why was the Sun's energy source a major mystery?

The Sun's energy is absolutely vital to human existence. Ancient peoples certainly recognized that fact. Some worshipped the Sun as a god. Others created mythologies to explain its daily rise and set. But no one who lived before the twentieth century knew how the Sun provides us with light and heat.

Most ancient thinkers assumed the Sun to be some type of fire, perhaps a lump of burning coal or wood. The Greek philosopher Anaxagoras (c. 500–428 B.C.) imagined the Sun to be a very hot, glowing rock about the size of the Greek peninsula of Peloponnesus (comparable in size to Massachusetts). It was a reasonable suggestion for the times, since science had not yet advanced to the point where the idea could be tested, and it made him one of the first people in history to believe that objects in the heavens are composed of the same types of materials that we find on Earth.

The question of how the Sun shines became more pressing once scientists learned the Sun's true size and distance from Earth. Even the best ancient estimates of the Sun's distance had put the Sun much closer to Earth than it really is, so ancient people did not recognize the Sun's truly enormous energy output. But the Sun's size and distance were reasonably well known by the mid-1800s [see Special Topic, p. 218] making it clear that the Sun had to be generating an astonishing amount of energy to heat and illuminate the Earth from 150 million kilometers away.

What could possibly make the Sun shine so brightly? Familiar energy sources on Earth offered only two possibilities: Either the Sun was a cooling ember that had once been much hotter, or it generated energy from some type of chemical burning similar to the burning of coal or wood. Simple calculations showed that a cooling or chemically burning Sun could shine for a few thousand years, which squared well with biblically based estimates of Earth's age that were popular at the time. However, both these ideas suffered from fatal flaws. If the Sun were a cooling ember, it would have been much hotter just a few hundred years earlier, making it too hot for civilization to have existed. Chemical burning was ruled out when calculations showed that it could not possibly account for the Sun's huge energy output.

In the late 1800s, astronomers came up with what seemed at first like a much better idea: They suggested that the Sun generates energy by slowly contracting in size, a process called **gravitational contraction** (or *Kelvin-Helmholtz contraction,* after the scientists who proposed the mechanism). Recall that a shrinking gas cloud heats up because some of the gravitational potential energy of gas particles far from the cloud center is converted into thermal energy as the gas moves inward (see Figure 4.14). A gradually shrinking Sun would always have some gas moving inward, converting gravitational potential energy into thermal energy. This thermal energy would keep the inside of the Sun hot. Because of its large mass, the Sun would need to contract only very slightly each year to maintain its temperature—so slightly that the contraction would have been unnoticeable to nineteenth-century astronomers. Moreover, calculations showed that gravitational contraction could have kept the Sun shining steadily for up to about 25 million years, which many astronomers thought was plenty long enough. However, geologists disagreed: Their studies of rocks and fossils had already led them to conclude that Earth was at least hundreds of millions and possibly billions of years old, which meant that gravitational contraction could not be the mechanism by which the Sun generates its energy.

Geologists and astronomers debated Earth's age for many years, but by the beginning of the twentieth century the geological evidence had convinced most astronomers that Earth really is much older than 25 million years. Thus, astronomers were left with a major mystery: They knew the Sun had been producing energy at a great rate for at least hundreds of millions of years, but they knew of no mechanism that could explain this energy output.

Figure 14.1 An acrobat stack is in gravitational equilibrium: The lowest person supports the most weight and feels the greatest pressure, and the overlying weight and underlying pressure decrease for those higher up.

• Why does the Sun shine?

The Sun's true energy generation mechanism did not become clear until Einstein published his special theory of relativity in 1905. Einstein's theory included his famous discovery of $E = mc^2$ [Sections 4.3, S2.3]. This equation shows that mass itself contains an enormous amount of potential energy. Calculations immediately showed that the Sun's mass contained more than enough energy to account for billions of years of sunshine, if only there were some way for the Sun to convert the energy of mass into thermal energy. It took a few decades for scientists to work out the details, but by the end of the 1930s we had learned that the Sun converts mass into energy through *nuclear fusion* of hydrogen into helium [Section 1.1].

Thus, the basic answer to the question of why the Sun shines is that it generates energy through nuclear fusion. This cannot be the entire story, however, because fusion requires extremely high temperatures and densities (for reasons we will discuss in the next section). Thus, we still must answer two deeper questions. First, given that hot gases tend to expand, what keeps the Sun stable in size even while it generates so much light and heat? Second, how did the Sun get hot enough for fusion in the first place? Let's look at each of these questions, so that we can finally answer the age-old question of why the Sun shines.

The Stable Sun The high temperatures and densities that sustain nuclear fusion are found deep in the Sun's core. Thus, for the Sun to shine steadily, it must have a way of keeping the core hot and dense. It maintains these internal conditions through a natural balance between two competing forces: gravity pulling inward and pressure pushing outward. This balance is called **gravitational equilibrium** (or *hydrostatic equilibrium*).

A stack of acrobats provides a simple example of gravitational equilibrium (Figure 14.1). The bottom person supports the weight of everybody above him, so his arms must push upward with enough pressure to support all this weight. At each higher level, the overlying weight is less, so it's a little easier for each additional person to hold up the rest of the stack.

Gravitational equilibrium works much the same in the Sun, except the outward push against gravity comes from internal gas pressure rather than an acrobat's arms. The Sun's internal pressure precisely balances gravity at every point within it, thereby keeping the Sun stable in size (Figure 14.2). Because the weight of overlying layers is greater as we look deeper into the Sun, the pressure must increase with depth. Deep in the Sun's core, the pressure makes the gas hot and dense enough to sustain nuclear fusion. The energy released by fusion, in turn, heats the gas and maintains the pressure that keeps the Sun in balance against the inward pull of gravity.

THINK ABOUT IT

Earth's atmosphere is also in gravitational equilibrium, with the pressure in lower layers supporting the weight of upper layers. Use this idea to explain why the air gets thinner at higher altitudes.

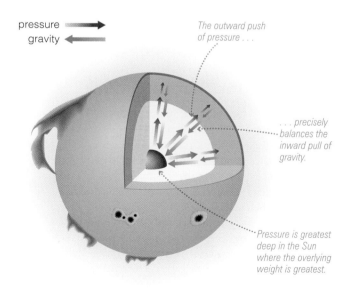

pressure
gravity

The outward push of pressure . . .

. . . precisely balances the inward pull of gravity.

Pressure is greatest deep in the Sun where the overlying weight is greatest.

Figure 14.2 Gravitational equilibrium in the Sun: At each point inside, the pressure pushing outward balances the weight of the overlying layers.

How Fusion Started The steadiness of our Sun's gravitational equilibrium is fairly easy to understand if we think only about how it works at present: Gravity pushes inward while the energy released by fusion maintains the pressure that pushes outward. But how did the Sun become hot enough for fusion to begin in the first place?

The answer invokes the mechanism of gravitational contraction that astronomers of the late 1800s mistakenly thought might be responsible for the Sun's heat today. Recall that our Sun was born about 4.6 billion years ago from a collapsing cloud of interstellar gas [Section 8.2]. The contraction of the cloud released gravitational potential energy. Much of this energy was radiated away from the cloud's surface as thermal radiation, but the rest of it remained inside, causing the interior temperature and pressure to rise. When the central temperature and density finally grew high enough to sustain nuclear fusion, energy generation in the Sun's interior came into balance with the energy lost from the surface in the form of radiation. This energy balance stabilized the size of the Sun, bringing it into a state of gravitational equilibrium that has persisted to this day.

Calculations show that the Sun was born with enough hydrogen in its core to shine steadily and maintain its gravitational equilibrium for about 10 billion years. Thus, the Sun is only about halfway through its 10-billion-year "lifetime." About 5 billion years from now, when the Sun finally exhausts its nuclear fuel, the internal pressure will drop and gravitational contraction of the core will begin once again. As we will see in later chapters, some of the most important and spectacular processes in astronomy arise from the changes that occur as the crush of gravity begins to overcome a star's internal sources of pressure.

In summary, the answer to the question, "Why does the Sun shine?" is that about 4.6 billion years ago, *gravitational contraction* made the Sun hot enough to sustain nuclear fusion in its core. Ever since, energy liberated by fusion has maintained the Sun's *gravitational equilibrium* and kept the Sun shining steadily, supplying the light and heat that sustain life on Earth.

• What is the Sun's structure?

We've already stated that the Sun is a giant ball of hot gas. To be more precise, the Sun is a ball of *plasma*—a gas in which many of the atoms are ionized because of the high temperature [Section 5.3]. A plasma behaves much like an ordinary gas, but its many positively charged ions and freely moving electrons mean it can create and respond to magnetic fields.

The differing temperatures and densities of the plasma at different depths give the Sun the layered structure shown in Figure 14.3. To make sense of what you see in the figure, let's imagine that you have a spaceship that can somehow withstand the immense heat and pressure of the Sun, and take an imaginary journey from Earth to the center of the Sun.

The Sun from Afar As you begin your voyage from Earth, the Sun appears as a whitish ball of glowing gas. Just as

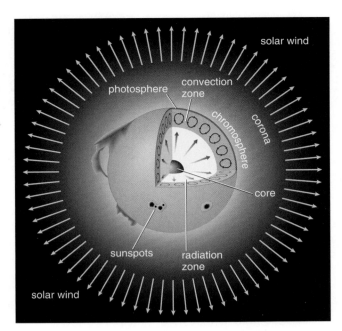

Figure 14.3 Interactive Figure. The basic structure of the Sun.

astronomers have done in real life, you can use simple observations to determine basic properties of the Sun. Spectroscopy [Section 5.4] tells you that the Sun is made almost entirely of hydrogen and helium. From the Sun's angular size and distance, you can determine that its radius is nearly 700,000 kilometers, or more than 100 times the radius of Earth. Even **sunspots**, which appear as dark splotches on the Sun's surface, can be larger in size than Earth (Figure 14.4).

You can calculate the Sun's mass from its gravitational effects on the planets by using Newton's version of Kepler's third law [Section 4.4]. You'll find that the Sun's mass is 2×10^{30} kilograms—about 300,000 times the mass of Earth and more than 1,000 times the mass of all the planets in our solar system put together. You can determine the Sun's rotation rate by observing the motion of sunspots or by measuring Doppler shifts on opposite sides of the Sun.

Figure 14.4 This photo of the visible surface of the Sun shows several dark sunspots, each large enough to swallow our entire planet.

You'll find that, unlike a spinning ball, the entire Sun does *not* rotate at the same rate. Instead, the solar equator completes one rotation in about 25 days, and the rotation period increases with latitude to about 30 days near the solar poles.

THINK ABOUT IT

As stated above, we can use Newton's version of Kepler's third law to determine the Sun's mass. What two properties of Earth do we need to know in order to apply this law?

The Sun releases an enormous amount of radiative energy into space, which you can measure through the window of your spacecraft. Recall that in science we measure energy in units called joules [Section 4.3] and define *power* as the *rate* at which energy is used or released [Section 5.1]. The standard unit of power is the *watt*, defined as 1 joule of energy per second; that is, 1 watt = 1 joule/s. For example, a 100-watt light bulb requires 100 joules of energy for every second it is left turned on. The Sun's total power output, or **luminosity**, is an incredible 3.8×10^{26} watts. If we could somehow capture and store just 1 second's worth of the Sun's luminosity, it would be enough to meet current human energy demands for roughly the next 500,000 years. Table 14.1 summarizes the basic properties of the Sun.

Of course, only a tiny fraction of the Sun's total energy output reaches Earth, with the rest dispersing in all directions into space. Most of this energy is radiated in the form of visible light, but after you've left the protective blanket of Earth's atmosphere, you encounter significant amounts of other types of solar radiation, including dangerous ultraviolet and X rays. Your spaceship will require substantial shielding to protect you from the serious radiation burns these high-energy forms of light cause.

Entering the Sun's Atmosphere Even at great distances from the Sun, you and your spacecraft can feel slight effects from the **solar wind**—a stream of charged particles continually blown outward in all directions from the Sun. The solar wind helps shape the magnetospheres of planets [Sections 10.1, 11.1] and blows back the material that forms the plasma tails of comets [Section 12.2]. The solar wind also cleared away the gas of the solar nebula at the end of the era of planet formation, some 4.5 billion years ago [Section 8.4].

As you approach the Sun more closely, you'll begin to encounter the low-density gas that represents what we usually think of as the Sun's atmosphere. The outermost layer of this atmosphere, called the **corona**, extends several million kilometers above the visible surface of the Sun. The temperature of the corona is astonishingly high—about 1 million K—explaining why this region emits most of the Sun's X rays. However, the corona's density is so low that your spaceship feels relatively little heat despite the million-degree temperature [Section 4.3].

Nearer the surface, the temperature suddenly drops to about 10,000 K in the **chromosphere**, the middle layer of the solar atmosphere and the region that radiates most of the Sun's ultraviolet light. Then you plunge through the

Table 14.1 Basic Properties of the Sun

Radius (R_{Sun})	696,000 km (about 109 times the radius of Earth)
Mass (M_{Sun})	2×10^{30} kg (about 300,000 times the mass of Earth)
Luminosity (L_{Sun})	3.8×10^{26} watts
Composition (by percentage of mass)	70% hydrogen, 28% helium, 2% heavier elements
Rotation rate	25 days (equator) to 30 days (poles)
Surface temperature	5,800 K (average); 4,000 K (sunspots)
Core temperature	15 million K

lowest layer of the atmosphere, or **photosphere**, which is the visible surface of the Sun. Although the photosphere looks like a well-defined surface from Earth, it consists of gas far less dense than Earth's atmosphere. The temperature of the photosphere averages just under 6,000 K, and its surface seethes and churns like a pot of boiling water. The photosphere is also where you'll find sunspots; the wild swings of your compass needle will tell you that sunspots are actually regions with very strong magnetic fields.

Plunging into the Interior Up to this point in your journey, you may have seen Earth and the stars when you looked

COMMON MISCONCEPTIONS

Is the Sun on Fire?

We are accustomed to saying that the Sun is "burning," a description that conjures up images of a giant bonfire in the sky. However, the Sun does not burn in the same sense as a fire burns on Earth. Fires generate light through chemical changes that consume oxygen and produce a flame. The glow of the Sun has more in common with the glowing embers left over after the flames have burned out. Much like hot embers, the Sun's surface shines because it is hot enough to emit thermal radiation that includes visible light [Section 5.4].

Hot embers stop glowing as they cool, because cooler objects emit only infrared light and not visible light. The Sun keeps shining with visible light because energy rising from the Sun's core keeps its surface hot. The energy rising from the core is generated by nuclear fusion, so we sometimes say that it is the result of "nuclear burning"—a term that suggests nuclear changes in much the same way that "chemical burning" suggests chemical changes. Thus, while it is reasonable to say that the Sun undergoes nuclear burning in its core, the Sun's surface shines primarily with thermal radiation and not through any kind of burning fire.

back. But as you slip beneath the photosphere, blazing light engulfs you. You are inside the Sun, and incredible turbulence tosses your spacecraft about. If you can hold steady long enough to see what is going on around you, you'll notice spouts of hot gas rising upward, surrounded by cooler gas cascading down from above. You are in the **convection zone**, where energy generated in the solar core travels upward, transported by the rising of hot gas and falling of cool gas called *convection* [Section 9.1]. With some quick thinking, you may realize that the photosphere above you is the top of the convection zone and that convection is the cause of the Sun's seething, churning appearance.

About a third of the way down to the center, the turbulence of the convection zone gives way to the calmer plasma of the **radiation zone**, where energy moves outward primarily in the form of photons of light. The temperature rises to almost 10 million K, and your spacecraft is bathed in X rays trillions of times more intense than the visible light at the solar surface.

No real spacecraft could survive, but your imaginary one keeps plunging straight down to the solar **core**. There you finally find the source of the Sun's energy: nuclear fusion transforming hydrogen into helium. At the Sun's center, the temperature is about 15 million K, the density is more than 100 times that of water, and the pressure is 200 billion times that on the surface of Earth. The energy produced in the core today will take a few hundred thousand years to reach the surface.

With your journey complete, it's time to turn around and head back home. We'll continue this chapter by studying fusion in the solar core and then tracing the flow of the energy generated by fusion as it moves outward through the Sun.

The Sun Tutorial, Lessons 2–3

14.2 The Cosmic Crucible

The prospect of turning common metals like lead into gold enthralled those who pursued the medieval practice of alchemy. Sometimes they tried primitive scientific approaches, such as melting various ores together in a vessel called a *crucible*. Other times they tried magic. However, their get-rich-quick schemes never managed to work. Today we know that there is no easy way to turn other elements into gold, but it *is* possible to transmute one element or isotope into another.

If a nucleus gains or loses protons, its atomic number changes and it becomes a different element. If it gains or loses neutrons, its atomic mass changes and it becomes a different isotope [Section 5.3]. The process of splitting a nucleus into two smaller nuclei is called **nuclear fission**. The process of combining nuclei to make a nucleus with a greater number of protons or neutrons is called **nuclear fusion** (Figure 14.5). Human-built nuclear power plants rely on nuclear fission of uranium or plutonium. The nuclear power plant at the center of the Sun relies on nuclear fusion, turning hydrogen into helium.

Figure 14.5 Nuclear fission splits a nucleus into smaller nuclei (not usually of equal size), while nuclear fusion combines smaller nuclei into a larger nucleus.

• How does nuclear fusion occur in the Sun?

Fusion occurs within the Sun because the 15 million K plasma in the solar core is like a "soup" of hot gas, with bare, positively charged atomic nuclei (and negatively charged electrons) whizzing about at extremely high speeds. At any one time, some of these nuclei are on high-speed collision courses with each other. In most cases, electromagnetic forces deflect the nuclei, preventing actual collisions, because positive charges repel one another. If nuclei collide with sufficient energy, however, they can stick together to form a heavier nucleus (Figure 14.6).

Sticking positively charged nuclei together is not easy. The **strong force**, which binds protons and neutrons together in atomic nuclei, is the only force in nature that can overcome the electromagnetic repulsion between two positively charged nuclei [Section S4.2]. In contrast to gravitational and electromagnetic forces, which drop off gradually as the distances between particles increase (by an inverse square law [Section 4.4]), the strong force is more like glue or Velcro: It overpowers the electromagnetic force over very small distances but is insignificant when the distances between particles exceed the typical sizes of atomic nuclei. The key to nuclear fusion, therefore, is to push the positively charged nuclei close enough together for the strong force to outmuscle electromagnetic repulsion.

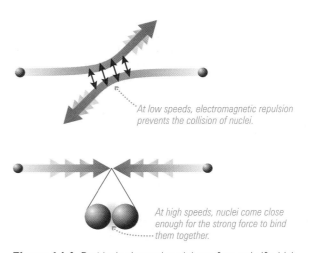

At low speeds, electromagnetic repulsion prevents the collision of nuclei.

At high speeds, nuclei come close enough for the strong force to bind them together.

Figure 14.6 Positively charged nuclei can fuse only if a high-speed collision brings them close enough for the strong force to come into play.

The high pressures and temperatures in the solar core are just right for fusion of hydrogen nuclei into helium nuclei. The high temperature is important because the nuclei must collide at very high speeds if they ⟨ enough together to fuse. (Quantum tunı portant to this process [Section S4.4].) T.

MATHEMATICAL INSIGHT 14.1 Mass-Energy Conversion in Hydroge

The fusion of hydrogen into helium releases energy because the four hydrogen nuclei (protons) that go into the overall reaction have a slightly greater mass than the helium nucleus that comes out. Laboratory measurements show that a single proton has a mass of 1.6726×10^{-27} kg, so four protons have four times as much mass, or a total of 6.690×10^{-27} kg. Measurements show that a single helium-4 nucleus has a slightly lower mass, 6.643×10^{-27} kg. Thus, when four protons fuse to make one helium-4 nucleus, the amount of mass that "disappears" and becomes energy is:

$$6.690 \times 10^{-27} \text{ kg} - 6.643 \times 10^{-27} \text{ kg} = 0.047 \times 10^{-27} \text{ kg}$$

If we divide this mass loss by the original mass of the four protons, we find the fractional loss of mass that occurs in hydrogen fusion:

$$\frac{\text{fractional mass loss}}{\text{in hydrogen fusion}} = \frac{0.047 \times 10^{-27} \text{ kg}}{6.69 \times 10^{-27} \text{ kg}} = 0.007$$

That is, a fraction 0.007, or 0.7%, of the original hydrogen mass is converted into energy during the hydrogen fusion process. For example, when 1 kilogram of hydrogen fuses, the resulting helium weighs only 993 grams, while 7 grams of mass turns into energy. We can use these facts about mass loss during fusion with Einstein's equation $E = mc^2$ to learn about what happens in the Sun.

Example 1: How much hydrogen is converted to helium each second in the Sun?

Solution:

Step 1 Understand: The first step in understanding this question is to realize that we can figure out how much hydrogen is being converted to helium in the Sun only if we know the Sun's total energy output. Fortunately, we already know this (see Table 14.1): The Sun's luminosity is 3.8×10^{26} watts, which means it produces 3.8×10^{26} joules of energy each second. Knowing the amount of energy produced each second means we can use Einstein's equation $E = mc^2$ to calculate the amount of mass that must be turned into energy each second. Then we can figure out how much hydrogen must fuse in order to turn this amount of mass into energy, using the fact that 0.7% of the hydrogen mass becomes energy during the fusion process.

Step 2 Solve: As described above, we can use Einstein's equation to calculate the amount of mass that must be converted to energy each second in the Sun. We start by solving Einstein's equation for the mass, m:

$$E = mc^2 \quad \Rightarrow \quad m = \frac{E}{c^2}$$

Now we plug in 3.8×10^{26} joules for the energy (E) that the Sun produces each second and recall that the speed of light is $c = 3 \times 10^8$ m/s; to make sure we can work with the units properly, we use the fact that 1 joule is equivalent to 1 kg $\times$ m²/s²:

$$m = \frac{E}{c^2} = \frac{3.8 \times 10^{26} \frac{\text{kg} \times \text{m}^2}{\text{s}^2}}{\left(3 \times 10^8 \frac{\text{m}}{\text{s}}\right)^2} = 4.2 \times 10^9 \text{ kg}$$

We've found the amount of mass converted to energy in the Sun each second. We know that this mass is only 0.7% ($= 0.007$) of the mass of the hydrogen that goes into the fusion process. Thus, the total mass of the hydrogen that undergoes fusion each second must be:

$$\text{mass of hydrogen fused} = \frac{4.2 \times 10^9 \text{ kg}}{0.007} = 6.0 \times 10^{11} \text{ kg}$$

Step 3 Explain: We have found that the Sun fuses 600 billion kilograms of hydrogen each second. We also found that about 4.2 billion kilograms of this mass is converted to energy. Thus, of the 600 billion kilograms of hydrogen the Sun fuses each second, about 4 billion kilograms turns into energy and the remaining 596 billion kilograms becomes helium. The numbers may be easier to understand if we convert them to metric tons, using the fact that 1 metric ton = 1,000 kilograms. In that case, we say that the Sun fuses about 600 million metric tons of hydrogen each second, producing about 596 million metric tons of helium and turning 4 million metric tons into energy.

Example 2: How many times does the fusion reaction occur each second in the Sun?

Solution:

Step 1 Understand: There are several ways we could approach this problem, but perhaps the simplest is just to divide the total amount of mass lost each second in the Sun by the mass loss that occurs each time four hydrogen nuclei are fused into one helium nucleus. The result will tell us how many times the fusion reaction must occur each second.

Step 2 Solve: From Example 1, we know that the Sun converts a total of 4.2×10^9 kg of mass into energy each second, and from the discussion above we know that each individual fusion reaction converts 0.047×10^{-27} kg of mass into energy. We divide these masses to find the number of fusion reactions that occur each second in the Sun:

$$\begin{aligned} \text{number of} \\ \text{fusion reactions} \\ \text{(per second)} \end{aligned} = \frac{\text{total mass lost through fusion (per second)}}{\text{mass lost in each fusion reaction}}$$

$$= \frac{4.2 \times 10^9 \text{ kg}}{0.047 \times 10^{-27} \text{ kg}}$$

$$= 8.9 \times 10^{37}$$

Step 3 Explain: Notice that our result, 8.9×10^{37}, is just a little less than 10^{38}. Thus, we have found that the overall hydrogen fusion reaction, which fuses four hydrogen nuclei into one helium nucleus, occurs nearly 10^{38} times per second in the Sun.

are, the harder the collisions, making fusion reac-more likely at higher temperatures. The high pressure the overlying layers is necessary because without it the hot plasma of the solar core would simply explode into space, shutting off the nuclear reactions.

THINK ABOUT IT

The Sun generates energy by fusing hydrogen into helium, but as we'll see in later chapters, some stars fuse helium and even heavier elements. Do temperatures need to be higher or lower for the fusion of heavier elements than for the fusion of hydrogen? Why? (*Hint:* How does the amount of charge of a nucleus affect the difficulty of fusing it to another nucleus?)

The Proton-Proton Chain   Let's investigate the fusion process in the Sun in a little more detail. Recall that hydrogen nuclei are simply individual protons, while the most common form of helium consists of two protons and two neutrons. Thus, the overall hydrogen fusion reaction transforms four individual protons into a helium nucleus containing two protons and two neutrons:

$$4 \; {}^{1}\text{H} \qquad\qquad 1 \; {}^{4}\text{He}$$

This overall reaction actually proceeds through several steps involving just two nuclei at a time. The sequence of steps that occurs in the Sun is called the **proton-proton chain**, because it begins with collisions between individual protons (hydrogen nuclei). Figure 14.7 illustrates the steps in the proton-proton chain:

1. *Step 1.* Two protons fuse to form a nucleus consisting of one proton and one neutron, which is the isotope

of hydrogen known as *deuterium*. Note that this step converts a proton into a neutron, reducing the total nuclear charge from +2 for the two fusing protons to +1 for the resulting deuterium nucleus. The lost positive charge is carried off by a *positron* (antielectron), the antimatter version of an electron with a positive rather than a negative charge [Section S4.2]. A **neutrino**—a subatomic particle with a very tiny mass—is also produced in this step.* (The positron won't last long, because it soon meets up with an ordinary electron, resulting in the creation of two gamma-ray photons through matter-antimatter annihilation.) This step must occur twice in the overall reaction, since it requires a total of four protons.

2. *Step 2.* A fair number of deuterium nuclei are always present along with the protons and other nuclei in the solar core, since Step 1 occurs so frequently in the Sun (about 10^{38} times per second). Step 2 occurs when one of these deuterium nuclei collides and fuses with a proton. The result is a nucleus of helium-3, a rare form of helium with two protons and one neutron, along with the production of a gamma-ray photon. This step also occurs twice in the overall reaction.

3. *Step 3.* The third and final step of the proton-proton chain requires the addition of another neutron to the helium-3, thereby making normal helium-4. This final step can proceed in several different ways, but the most common is through a collision of two helium-3

---

*Producing a neutrino is necessary because of a law called *conservation of lepton number:* The number of leptons (e.g., electrons or neutrinos [Chapter S4]) must be the same before and after the reaction. The lepton number is zero before the reaction because there are no leptons. Among the reaction products, the positron (antielectron) has lepton number -1 because it is antimatter, and the neutrino has lepton number $+1$. Thus, the total lepton number remains zero.

Hydrogen Fusion by the Proton-Proton Chain

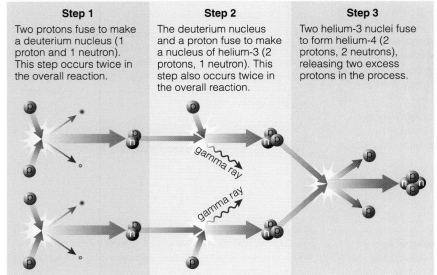

Step 1
Two protons fuse to make a deuterium nucleus (1 proton and 1 neutron). This step occurs twice in the overall reaction.

Step 2
The deuterium nucleus and a proton fuse to make a nucleus of helium-3 (2 protons, 1 neutron). This step also occurs twice in the overall reaction.

Step 3
Two helium-3 nuclei fuse to form helium-4 (2 protons, 2 neutrons), releasing two excess protons in the process.

Overall reaction
gamma ray

gamma ray

Key:
- neutron
- proton
- gamma ray
- neutrino
- positron

Figure 14.7   Interactive Figure   In the Sun, four hydrogen nuclei (protons) fuse into one helium-4 nucleus by way of the proton-proton chain. Gamma rays and subatomic particles known as neutrinos and positrons carry off the energy released in the reaction.

nuclei. Each of these helium-3 nuclei resulted from a prior, separate occurrence of Step 2 somewhere in the solar core. The final result is a normal helium-4 nucleus and two protons.

Notice that a total of six protons enter the reaction during Steps 1 and 2, with two coming back out in Step 3. Thus, the overall reaction is just as described above, with four protons combining to make one helium nucleus. The gamma rays and subatomic particles (neutrinos and positrons) carry off the energy released in the reaction.

Fusion of hydrogen into helium generates energy because a helium nucleus has a mass slightly less (by about 0.7%) than the combined mass of four hydrogen nuclei (see Mathematical Insight 14.1). Thus, when four hydrogen nuclei fuse into a helium nucleus, a little bit of mass disappears. The disappearing mass becomes energy in accord with Einstein's formula $E = mc^2$. About 98% of the energy emerges as kinetic energy of the resulting helium nuclei and radiative energy of the gamma rays. As we will see, this energy slowly percolates to the solar surface, eventually emerging as the sunlight that bathes Earth. Neutrinos carry off the other 2% of the energy. Overall, fusion in the Sun converts about 600 million tons of hydrogen into 596 million tons of helium every second, which means that 4 million tons of matter is turned into energy each second. Although this sounds like a lot, it is a minuscule fraction of the Sun's total mass and thus does not affect the overall mass of the Sun in any measurable way.

The Solar Thermostat Nuclear fusion is the source of all the energy the Sun releases into space. If the fusion rate varied, so would the Sun's energy output, and large variations in

the Sun's luminosity would almost surely be lethal to life on Earth. Fortunately, the Sun fuses hydrogen at a steady rate, thanks to a natural feedback process that acts as a thermostat for the Sun's interior. We have encountered a natural thermostat once before, when we discussed how Earth's carbon dioxide cycle keeps our planet's surface temperature relatively steady [Section 10.6]. The solar thermostat is even more efficient.

Solar energy production remains steady because the rate of nuclear fusion is very sensitive to temperature. A slight increase in the Sun's core temperature would mean a much higher fusion rate, and a slight decrease in temperature would mean a much lower fusion rate. Either kind of change in the fusion rate would alter the Sun's gravitational equilibrium—the balance between the pull of gravity and the push of internal pressure—in a way that would quickly restore the original temperature and fusion rate. To see how, let's examine what would happen if there were a small change in the core temperature (Figure 14.8).

Suppose the Sun's core temperature rose very slightly. Protons in the core would then move faster and collide more often. The rate of nuclear fusion would therefore soar, generating lots of extra energy. Because energy moves so slowly through the Sun, this extra energy would be bottled up in the core, causing an increase in the core pressure. The push of this pressure would temporarily exceed the pull of gravity, causing the core to expand and cool. This cooling, in turn, would cause the fusion rate to drop back down until the core was restored to its original size and temperature, thus returning the fusion rate back to its original value.

A slight drop in the Sun's core temperature would trigger an opposite chain of events. The reduced core tempera-

Figure 14.8 Interactive Figure The solar thermostat. Gravitational equilibrium regulates the Sun's core temperature. Everything is in balance if the amount of energy leaving the core equals the amount of energy produced by fusion. A rise in core temperature triggers a chain of events that causes the core to expand, lowering its temperature to its original value. A decrease in core temperature triggers the opposite chain of events, also restoring the original core temperature.

ture would lead to a decrease in the rate of nuclear fusion, causing a drop in pressure and contraction of the core. As the core shrank, its temperature would rise until the fusion rate returned to normal and restored the core to its original size and temperature.

Notice that the solar thermostat depends on *two* kinds of balance. The first is gravitational equilibrium—pressure must balance gravity at each point in the Sun. Second, the flow of energy through the Sun must remain balanced: The rate of energy production in the core must be the same as the rate at which energy flows outward through each layer of the Sun. If either of these forms of balance is somehow disrupted, natural feedback processes will automatically restore it, keeping the Sun's central temperature and

---

MATHEMATICAL INSIGHT 14.2 **Pressure in the Sun: The Ideal Gas Law**

The pressure that resists gravity inside the Sun comes from the thermal motions of gas particles. Recall that particles in a hot gas move more quickly and collide with more force than those in a cooler gas [Section 5.3], which means they also exert more pressure on any surface that tries to contain them. (Remember also that pressure is defined as the force per unit area exerted on any surface.) However, pressure also depends on *how many* particles are colliding with each unit area of that surface each second, which means that pressure depends on the *number density* of gas particles—the number of gas particles contained in each cubic centimeter. Mathematically, we express this relationship between temperature, density, and pressure with the **ideal gas law**, which can be written as:

$$P = nkT$$

where P represents gas pressure, n represents the number density of particles (in particles per cubic centimeter), T represents gas temperature (in Kelvin), and $k = 1.38 \times 10^{-23}$ joules/K, which is known as Boltzmann's constant. This law applies to all gases consisting of simple, freely flying particles, like those in the Sun. In the units given here, the ideal gas law gives pressure in units of joules per cubic centimeter (J/cm^3); these units of energy per unit volume are equivalent to units of force per unit area.

Example 1: The Sun's core contains about 10^{26} particles per cubic centimeter at a temperature of 15 million K. How does the gas pressure in the core of the Sun compare to the pressure of Earth's atmosphere at sea level, where there are about 2.4×10^{19} particles per cubic centimeter at a temperature of roughly 300 K?

Solution:

Step 1 Understand: In order to determine pressure, we need to know both the temperature and the number density of particles in a gas. We have that information for both the Sun's core and the Earth's atmosphere, so we have all the information we need. Because we are asked to compare the pressures, we don't need to calculate them individually; instead, we need only find their ratio by dividing the larger (the Sun's core pressure) by the smaller (Earth's atmospheric pressure).

Step 2 Solve: Dividing the Sun's core pressure by Earth's atmospheric pressure, we find that their ratio is:

$$\frac{P_{core}}{P_{atmos}} = \frac{n_{core}kT_{core}}{n_{atmos}kT_{atmos}} = \frac{10^{26} \text{ cm}^{-3} \times 1.5 \times 10^7 \text{ K}}{2.4 \times 10^{19} \text{ cm}^{-3} \times 300 \text{ K}}$$

$$= 2 \times 10^{11}$$

Step 3 Explain: We have found that the Sun's core pressure is about 200 billion (2×10^{11}) times as great as atmospheric pressure on Earth. This extreme pressure explains why the Sun can

steadily fuse hydrogen into helium without blowing itself apart. The gravitational forces responsible for compressing the core to this extreme pressure are strong enough to keep that plasma in place while it releases fusion energy.

Example 2: How would the Sun's core pressure change if the Sun fused all its core hydrogen into helium without shrinking and without its core temperature changing? Use your answer to explain what should really happen as fusion occurs in the real Sun, in which the core *can* shrink and heat up. To simplify the problem, assume that the Sun's core begins with pure hydrogen that is fully ionized, so that there are two particles for every hydrogen nucleus (the proton plus one electron that must also be present for charge balance in the core), and ends as pure helium with three particles for each helium nucleus (the nucleus plus two electrons to balance the two protons in helium).

Solution:

Step 1 Understand: The core pressure will change if either the temperature or the number density of particles changes. We are told that the temperature does not change in this hypothetical example. We are also told that the core does not shrink, meaning that its volume remains constant, so a change in number density occurs only because there are fewer particles after fusion than before it. Thus, we simply need to determine how fusion changes the number of particles in the core.

Step 2 Solve: In this simplified example, the core initially contained two particles for every hydrogen nucleus and finished with three particles for every helium nucleus. However, we also know that it takes four hydrogen nuclei to make one helium nucleus. Thus, for each fusion reaction, there are eight particles before the reaction (four electrons and four protons) and three particles after it (one helium nucleus and two electrons)—meaning that there are $\frac{3}{8}$ as many particles after fusion as before it. Because the temperature and volume don't change in this example, the core pressure is also $\frac{3}{8}$ of its original value after all the hydrogen fuses into helium.

Step 3 Explain: In this simplified example, the core pressure at the end of the fusion process would have decreased to $\frac{3}{8}$ of its initial value. In reality, a decrease in core pressure would tip the balance of gravitational equilibrium in favor of gravity. The core must therefore shrink as fusion progresses in the Sun, in response to the gradually decreasing number of particles. This shrinkage decreases the volume of the core, allowing the number density of particles to stay approximately constant. The gradual shrinkage also produces a slight but gradual rise in core temperature, and therefore in the fusion rate, which is why the Sun's luminosity gradually increases with time during its hydrogen-burning life.

---

fusion rate steady and the Sun's overall energy output steady as well.

The Gradually Brightening Sun While the processes involved in gravitational equilibrium prevent erratic changes in the fusion rate, they also ensure that the fusion rate gradually rises over billions of years. This rise in the fusion rate explains a fact we first encountered when we discussed climate change factors in Chapter 10: the gradual brightening of the Sun with time.

Remember that each fusion reaction converts *four* hydrogen nuclei into *one* helium nucleus. Thus, the total number of *independent particles* in the solar core gradually falls with time. This gradual reduction in the number of particles causes the solar core to shrink. The slow shrinkage, in turn, gradually increases the core temperature and fusion rate, keeping the core pressure high enough to counteract the stronger compression of gravity. Theoretical models indicate that the Sun's core temperature should have increased enough to raise its fusion rate and the solar luminosity by about 30% since the Sun was born 4.6 billion years ago.

• How does the energy from fusion get out of the Sun?

The solar thermostat balances the Sun's fusion rate so that the amount of nuclear energy generated in the core equals the amount of energy radiated from the surface as sunlight. However, the journey of solar energy from the core to the photosphere takes hundreds of thousands of years.

Most of the energy released by fusion starts its journey out of the solar core in the form of photons. Although photons travel at the speed of light, the path they take through the Sun's interior zigzags so much that it takes them a very long time to make any outward progress. Deep in the solar interior, the plasma is so dense that a photon can travel only a fraction of a millimeter in any one direction before it interacts with an electron. Each time a photon "collides" with an electron, the photon gets deflected into a new and random direction. Thus, the photon bounces around the dense interior in a haphazard way (sometimes called a *random walk*) and only very gradually works its way outward from the Sun's center (Figure 14.9). The technical term for this slow, outward migration of photons is **radiative diffusion**; to *diffuse* means to "spread out" and *radiative* refers to the photons of light or radiation.

THINK ABOUT IT

Radiative diffusion is just one type of diffusion. Another is the diffusion of dye through a glass of water. If you place a concentrated spot of dye at one point in the water, each individual dye molecule begins a random walk as it bounces among the water molecules. The result is that the dye gradually spreads through the entire glass. Can you think of any other examples of diffusion in the world around you?

Energy released by fusion moves outward through the Sun's radiation zone (see Figure 14.3) primarily by way of

Figure 14.9 A photon in the solar interior bounces randomly among electrons, slowly working its way outward.

these randomly bouncing photons. At the top of the radiation zone, where the temperature has dropped to about 2 million K, the solar plasma absorbs photons more readily (rather than just bouncing them around). This absorption creates the conditions needed for convection [Section 9.1] and hence marks the beginning of the Sun's convection zone.

In the convection zone, convection occurs because hot gas is less dense than cool gas. Like a hot-air balloon, a hot bubble of solar plasma rises upward through the cooler plasma above it. Meanwhile, cooler plasma from above slides around the rising bubble and sinks to lower layers, where it is heated. The rising of hot plasma and sinking of cool plasma form a cycle that transports energy outward from the base of the convection zone to the solar surface, or photosphere (Figure 14.10a). There, the density of the gas becomes so low that photons can escape to space, which is why we see the photosphere as the "surface" of the Sun.

The convecting gas gives the photosphere the mottled appearance that we see in close-up photographs (Figure 14.10b): We see bright blobs where hot gas is welling up from inside the Sun and darker borders around those blobs where the cooler gas is sinking.* If we watched a movie of the photosphere, we'd see its surface bubbling rather like a pot of boiling water. Just as bubbles in a pot of boiling water burst on the surface and are replaced by new bubbles, each hot blob lasts only a few minutes before being replaced by others bubbling upward. The average temperature of the gas in the photosphere is about 5,800 K, but the convection means that the precise temperature varies significantly from place to place.

To summarize, energy produced by fusion in the Sun's core works its way slowly through the radiation zone by random bounces of photons, then gets carried upward by convection in the Sun's convection zone. The photosphere lies at the top of the convection zone and marks the place where the density of gas has become low enough that photons can escape to space. Thus, the energy produced hundreds of thousands of years earlier in the solar core finally emerges from the Sun as thermal radiation [Section 5.4] coming from the 5,800 K gas of the photosphere. Once in space, the photons travel straight away at the speed of light, bathing the planets in sunlight.

---

*The blobs are formally called *granules*, and the photosphere's mottled appearance is sometimes referred to as *solar granulation*.

Bright spots appear on Sun's surface where hot gas is rising . . .

. . . then the gas sinks after it has cooled off.

a This diagram shows convection beneath the Sun's surface: hot gas (white arrows) rises while cooler gas (orange/black arrows) descends around it.

Hot gas is rising here . . .

. . . and cooler gas is sinking here.

b This photograph shows the mottled appearance of the Sun's photosphere. The bright spots, each about 1,000 kilometers across, correspond to the rising plumes of hot gas in the diagram in part (a).

Figure 14.10 The Sun's photosphere churns with rising hot gas and falling cool gas as a result of underlying convection.

• How do we know what is happening inside the Sun?

We cannot see inside the Sun, so you may wonder how we can claim to know so much about what goes on underneath its surface and in its core. In fact, we can study the Sun's interior in three different ways: through mathematical models of the Sun, observations of solar vibrations, and observations of solar neutrinos.

Mathematical Models The primary way we learn about the interior of the Sun (and other stars) is by creating *mathematical models* that use the laws of physics to predict internal conditions. A basic model uses the Sun's observed composition and mass as inputs to equations that describe gravitational equilibrium, the solar thermostat, and the rate at which solar energy moves from the core to the photosphere. With the aid of a computer, we can use the model to calculate the Sun's temperature, pressure, and density at any depth. We can then predict the rate of nuclear fusion in the solar core by combining these calculations with knowledge about nuclear fusion gathered in laboratories here on Earth.

If a model is a good description of the Sun's interior, it should correctly "predict" the radius, surface temperature, luminosity, age, and many other observable properties of the Sun. Current models do indeed predict these properties quite accurately, giving us confidence that we really do understand what is going on inside the Sun.

Solar Vibrations A second way to learn about the inside of the Sun is to observe vibrations of the Sun's surface that are somewhat similar to the vibrations that earthquakes cause on Earth. These vibrations result from movement of gas within the Sun, which generates waves of pressure that travel through the Sun like sound waves moving through air. We can observe these vibrations on the Sun's surface by looking for Doppler shifts [Section 5.5]. Light from portions of the surface that are rising toward us is slightly blueshifted, while light from portions that are falling away from us is slightly redshifted. The vibrations are relatively small but measurable (Figure 14.11).

In principle, we can deduce a great deal about the solar interior by carefully analyzing these vibrations. (By analogy to seismology on Earth, this type of study of the Sun is

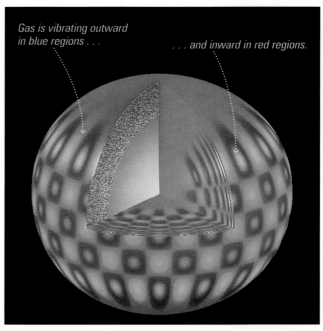

Gas is vibrating outward in blue regions . . .

. . . and inward in red regions.

Figure 14.11 This schematic diagram shows one of the Sun's many possible vibration patterns, which we can measure from Doppler shifts in spectra of the Sun's surface.

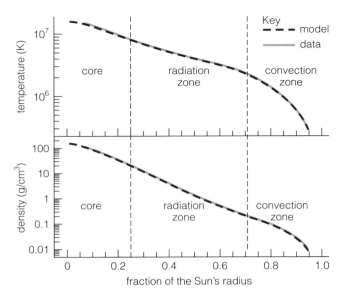

Figure 14.12 This graph shows the close agreement between predictions about the Sun's interior structure based on mathematical models and actual data obtained from observations of the Sun's surface vibrations. The agreement gives us confidence that the models are a good representation of the Sun's interior.

called *helioseismology—helios* means "sun.") Results to date confirm that our mathematical models of the solar interior are on the right track (Figure 14.12). At the same time, they provide data that help to improve the models further.

Solar Neutrinos   A third way to study the Sun's interior is to observe subatomic particles produced by fusion reactions in the core. Don't panic, but as you read this sentence, about a thousand trillion solar neutrinos will zip through your body. Traveling at nearly the speed of light, they reach us just minutes after being produced in fusion reactions in the Sun's core (see Figure 14.7). Fortunately, they won't do any damage, because neutrinos rarely interact with anything. Neutrinos created by fusion in the solar core fly quickly through the Sun as if passing through empty space. In fact, while an inch of lead will stop an X ray, stopping an average neutrino would require a slab of lead more than a light-year thick! Clearly, counting neutrinos is dauntingly difficult, because virtually all of them stream right through any detector built to capture them.

Nevertheless, neutrinos *do* occasionally interact with matter, and it is possible to capture a few solar neutrinos with a large enough detector. To distinguish neutrino captures from reactions caused by other particles, scientists usually place neutrino detectors deep underground in mines. The overlying rock blocks most other particles, but the neutrinos have no difficulty passing through.

The first major solar neutrino detector, built in the 1960s, was located 1,500 meters underground in the Homestake gold mine in South Dakota (Figure 14.13). The detector for this "Homestake experiment" consisted of a 400,000-liter vat of chlorine-containing dry-cleaning fluid. It turns out that, on very rare occasions, a chlorine nucleus can capture a neutrino and change into a nucleus of radioactive argon.

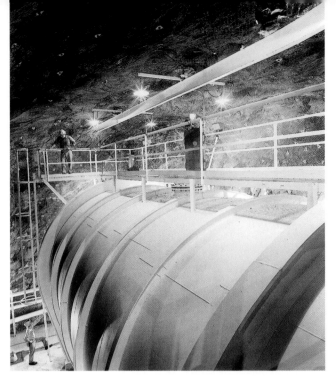

Figure 14.13 This tank of dry-cleaning fluid (visible underneath the catwalk), located deep within South Dakota's Homestake mine, was a solar neutrino detector. The chlorine nuclei in the cleaning fluid turned into argon nuclei when they captured neutrinos from the Sun.

By looking for radioactive argon in the tank of cleaning fluid, experimenters could count the number of neutrinos captured in the detector.

From the many trillions of solar neutrinos that passed through the tank of cleaning fluid each second, experimenters expected to capture an average of just one neutrino per day. This predicted capture rate was based on measured properties of chlorine nuclei and models of nuclear fusion in the Sun. However, over a period of more than two decades, neutrinos were captured only about once every three days on average. That is, the Homestake experiment detected only about one-third of the predicted number of neutrinos. This disagreement between model predictions and actual observations came to be called the **solar neutrino problem**.

For more than three decades, the solar neutrino problem was one of the great mysteries in astronomy. Other experiments found the same shortfall as the Homestake experiment. These results allowed only two possible conclusions: Either something was wrong with our understanding of fusion in the Sun, or some of the Sun's neutrinos were somehow hiding themselves from us.

Today, scientists are fairly convinced that the neutrinos were hiding. Here's how neutrinos are thought to hide: Scientists have known for decades that neutrinos come in three types, called *electron neutrinos, muon neutrinos*, and *tau neutrinos* [Section S4.2] Fusion reactions in the Sun produce only electron neutrinos, and until recently most solar neutrino detectors could detect only electron neutrinos. Thus, if some of the electron neutrinos produced by fusion change into neutrinos of the other two types on their way to Earth, then our detectors would count fewer electron neutrinos than we would have expected.

Recent experiments support the idea that neutrinos can change type on their way to Earth, probably during the portion of their trip from the Sun's core to its surface. Most significantly, a new detector in Canada called the Sudbury Neutrino Observatory is capable of detecting all three neutrino types (Figure 14.14). Early results from this observatory suggest that the total number of neutrinos of all types is indeed what we expect from our models of nuclear fusion in the Sun. If these results hold up as more data are collected, it will mean not only that the solar neutrino problem has been solved, but also that neutrino counts verify that nuclear fusion occurs in the Sun in just the way our models predict.

14.3 The Sun-Earth Connection

Energy liberated by nuclear fusion in the Sun's core eventually reaches the solar surface. There, before it is ultimately released into space as sunlight, the energy stimulates a wide variety of phenomena that we can observe from Earth. Sunspots are only the most obvious of these phenomena. Because sunspots and other features of the Sun's surface change with time, they constitute what we call *solar weather* or **solar activity**. The "storms" associated with solar weather are not just of academic interest. Sometimes they are so violent that they affect our day-to-day life on Earth. In this section, we'll explore solar activity and its far-reaching effects.

• What causes solar activity?

Most of the Sun's surface churns constantly with rising and falling gas and thus looks like the close-up photo shown in Figure 14.10b. However, larger features sometimes appear, including sunspots, the huge explosions known as solar

Figure 14.14 This photograph shows the main tank of the Sudbury Neutrino Observatory in Canada, located at the bottom of a mine shaft more than 2 kilometers underground. The large sphere, 12 meters in diameter, contains 1,000 tons of ultrapure "heavy water." (Heavy water is water in which the two hydrogen atoms are replaced by deuterium, making each molecule heavier than a molecule of ordinary water.) Neutrinos of all three types can cause reactions in the heavy water, and detectors surrounding the tank can record these reactions when they occur.

a This close-up view of the Sun's surface shows two large sunspots and several smaller ones. Each of the big sunspots is roughly as large as Earth.

Figure 14.15 Sunspots are regions of strong magnetic fields.

Outside a sunspot we see a single spectral line . . .

. . . but the strong magnetic field inside a sunspot splits that line into three lines.

b Very strong magnetic fields split the absorption lines in spectra of sunspot regions. The dark vertical bands are absorption lines in a spectrum of the Sun. Notice that these lines split where they cross the dark horizontal bands corresponding to sunspots.

flares, and gigantic loops of hot gas that extend high into the Sun's corona. All these features are created by magnetic fields, which form and change easily in the convecting plasma in the outer layers of the Sun.

Sunspots and Magnetic Fields Sunspots are the most striking features of the solar surface (Figure 14.15a). If you could look directly at a sunspot (but you should *never* stare at the Sun), you would find it blindingly bright. Sunspots appear dark in photographs only because they are less bright than the surrounding photosphere. They are less bright because they are cooler: The temperature of the plasma in sunspots is about 4,000 K, significantly cooler than the 5,800 K plasma that surrounds them.

If you think about this for a moment, you may wonder how sunspots can be so much cooler than their surroundings. Gas can usually flow easily, so you might expect the hotter gas from outside a sunspot to mix with the cooler gas within it, quickly warming the sunspot. The fact that sunspots stay relatively cool means that something must prevent hot plasma from entering them, and that "something" turns out to be magnetic fields.

Detailed observations of the Sun's spectral lines reveal sunspots to be regions with strong magnetic fields. These magnetic fields can alter the energy levels in atoms and ions and therefore can alter the spectral lines they produce. More specifically, magnetic fields cause some spectral lines to split into two or more closely spaced lines (Figure 14.15b). Wherever we see this effect (called the *Zeeman effect*), we know that magnetic fields must be present. Scientists can map magnetic fields on the Sun by looking for the splitting of spectral lines in light from different parts of the solar surface.

To understand how sunspots stay cooler than their surroundings, we must investigate the nature of magnetic fields

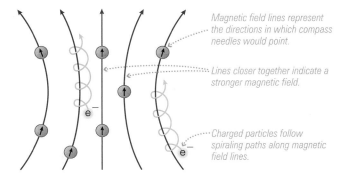

Figure 14.16 We draw magnetic field lines (red) to represent invisible magnetic fields.

in a little more depth. Magnetic fields are invisible, but we can represent them by drawing **magnetic field lines** (Figure 14.16). These lines represent the directions in which compass needles would point if we placed them within the magnetic field. The lines are closer together where the field is stronger, and farther apart where the field is weaker. Because these imaginary field lines are so much easier to visualize than the magnetic field itself, we usually discuss magnetic fields by talking about how the field lines would appear. Charged particles, such as the ions and electrons in the solar plasma, cannot easily move perpendicular to the field lines but instead follow spiraling paths along them.

Solar magnetic field lines act somewhat like elastic bands, being twisted into contortions and knots by turbulent motions in the solar atmosphere. Sunspots occur where the most taut and tightly wound magnetic fields poke nearly straight out from the solar interior (Figure 14.17a). These tight magnetic field lines suppress convection within the sunspot and prevent surrounding plasma from entering the sunspot. With hot plasma unable to enter the region, the sunspot plasma becomes cooler than that of the rest of

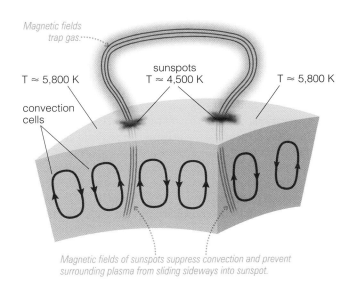

a Pairs of sunspots are connected by tightly wound magnetic field lines.

b This X-ray photo (from NASA's TRACE mission) shows hot gas trapped within looped magnetic field lines.

Figure 14.17 Strong magnetic fields keep sunspots cooler than the surrounding photosphere, while magnetic loops can arch from the sunspots to great heights above the Sun's surface.

the photosphere. Individual sunspots typically last up to a few weeks, dissolving when their magnetic fields weaken and allow hotter plasma to flow in.

Sunspots tend to occur in pairs, connected by a loop of magnetic field lines that can arc high above the Sun's surface (Figure 14.17b). Gas in the Sun's chromosphere and corona becomes trapped in these loops, making giant **solar prominences**. Some prominences rise to heights of more than 100,000 kilometers above the Sun's surface (Figure 14.18). Individual prominences can last for days or even weeks, disappearing only when the magnetic fields finally weaken and release the trapped gas.

Solar Storms The magnetic fields winding through sunspots and prominences sometimes undergo dramatic and sudden change, producing short-lived but intense storms on the Sun. The most dramatic of these storms are **solar flares**, which send bursts of X rays and fast-moving charged particles shooting into space (Figure 14.19).

Flares generally occur in the vicinity of sunspots, which is why we think they are created by changes in magnetic fields. The leading model for solar flares suggests that they occur when the magnetic field lines become so twisted and knotted that they can no longer bear the tension. They are thought to suddenly snap and reorganize themselves into a less twisted configuration. The energy released in the process heats the nearby plasma to 100 million K over the next few minutes or hours, generating X rays and accelerating some of the charged particles to nearly the speed of light.

Heating of the Chromosphere and Corona As we've seen, many of the most dramatic weather patterns and storms on the

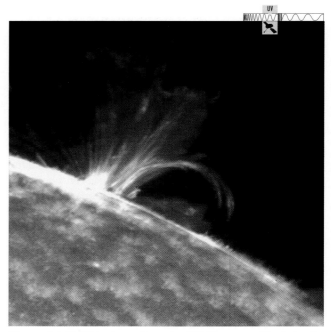

Figure 14.19 This photo (from the *TRACE* spacecraft) of ultraviolet light emitted by hydrogen atoms shows a solar flare erupting from the Sun's surface.

Sun show up when we look at the very hot gas of the Sun's chromosphere and corona. But why is that gas so hot in the first place? This question has perplexed scientists for decades.

Remember that temperatures gradually decline as we move outward from the Sun's core to the top of its photosphere. We might expect the decline to continue in the Sun's atmosphere, but instead it reverses, making the chromosphere and corona much hotter than the Sun's surface. Some aspects of this atmospheric heating remain a mystery today, but we have at least a general explanation: The Sun's strong magnetic fields carry energy upward from the churning solar surface to the chromosphere and corona.

More specifically, the rising and falling of gas in the convection zone probably shakes tightly wound magnetic field lines beneath the solar surface. The magnetic field lines carry this energy upward to the solar atmosphere, where they deposit this energy as heat. Thus, the same magnetic fields that keep sunspots cool make the overlying plasma of the chromosphere and corona hot.

Observations confirm the connection between magnetic fields and the structure of the chromosphere and corona. The density of gas in the chromosphere and corona is so low that we cannot see this gas with our eyes except during a total eclipse, when we can see the faint visible light scattered by electrons in the corona (see Figure 2.29). However, we can observe the chromosphere and corona at any time with ultraviolet and X-ray telescopes in space: The roughly 10,000 K plasma of the chromosphere emits strongly in the ultraviolet, and the million K plasma of the corona is the source of virtually all X rays coming from the Sun. Figure 14.20 shows an X-ray image of the Sun. The X-ray emission is brightest in regions where hot gas is being trapped and heated in loops of the magnetic field. In fact, the bright

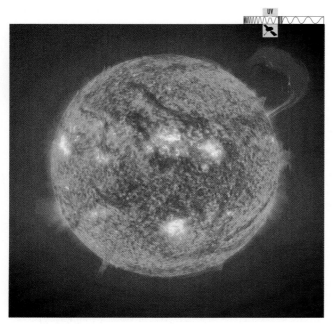

Figure 14.18 A gigantic solar prominence erupts from the solar surface at the upper right of this ultraviolet light photo (from the *SOHO* mission). The height of the prominence is over 20 times the diameter of Earth.

Figure 14.20 Interactive Photo An X-ray image of the Sun reveals the million-degree gas of the corona. Brighter regions of this image (yellow) correspond to regions of stronger X-ray emission. The darker regions (such as near the north pole at the top of this photo) are the coronal holes from which the solar wind escapes. (From the *Yohkoh* space observatory.)

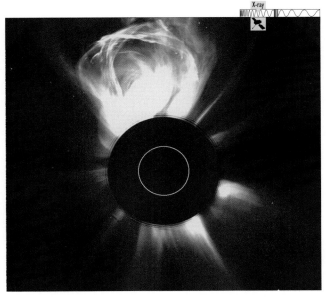

Figure 14.21 This X-ray image from the *SOHO* spacecraft shows a coronal mass ejection (the bright arc of gas headed almost straight upward as oriented here) during the solar storms of 2003. The central red disk blocks the Sun itself, and the white circle represents the size of the Sun in this picture.

spots in the corona tend to be directly above sunspots in the photosphere, confirming that they are created by the same magnetic fields.

Notice that some regions of the corona, called **coronal holes**, barely show up in X-ray images. More detailed analyses show that the magnetic field lines in coronal holes project out into space like broken rubber bands, allowing particles spiraling along them to escape the Sun altogether. These particles streaming outward from the corona are the source of the *solar wind*, which blows through the solar system at an average speed of about 500 kilometers per second and has important effects on planetary surfaces, atmospheres, and magnetospheres. Well beyond the planets, the pressure of interstellar gas must eventually halt the solar wind. The *Pioneer* and *Voyager* spacecraft that visited the outer planets in the 1970s and 1980s [Section 7.3] are still traveling outward from our solar system and may soon encounter this "boundary" (called the *heliopause*) marking the outer limit of the realm of the Sun.

The solar wind also gives us something tangible to study. In the same way that meteorites provide us with samples of asteroids we've never visited, solar wind particles captured by satellites provide us with a sample of material from the Sun. Analysis of these solar particles has reassuringly verified that the Sun is made mostly of hydrogen, just as we conclude from studying the Sun's spectrum.

• How does solar activity affect humans?

Flares and other solar storms sometimes eject large numbers of highly energetic charged particles from the Sun's corona. These particles travel outward from the Sun in huge bubbles that we call **coronal mass ejections** (Figure 14.21). The bubbles have strong magnetic fields and can reach

Earth in a couple of days if they happen to be aimed in our direction. Once a coronal mass ejection reaches Earth, it can create a *geomagnetic storm* in Earth's magnetosphere. On the positive side, these storms can lead to unusually strong auroras (see Figure 10.11) that can be visible throughout much of the United States. On the negative side, they can hamper radio communications, disrupt electrical power delivery, and damage the electronic components in orbiting satellites.

During a particularly powerful magnetic storm on the Sun in March 1989, the U.S. Air Force temporarily lost track of more than 2,000 satellites, and powerful currents induced in the ground circuits of the Quebec hydroelectric system caused it to collapse for more than 8 hours. The combined cost of the loss of power in the United States and Canada exceeded $100 million. In January 1997, AT&T lost contact with a $200 million communications satellite, probably because of damage caused by particles coming from another powerful solar storm. More recently, in the fall of 2003, a series of extremely powerful solar flares once again threatened Earth's communications and electrical systems, but the systems escaped major damage, due in part to our improved preparedness for solar storms.

Satellites in low Earth orbit are particularly vulnerable during periods of strong solar activity, when the increase in solar X rays and energetic particles heats Earth's upper atmosphere, causing it to expand. The density of the gas surrounding low-flying satellites therefore rises, exerting drag that saps their energy and angular momentum. If this drag proceeds unchecked, the satellites ultimately plummet back to Earth. Thus, satellites in low orbits, including the Hubble Space Telescope and the Space Station, require occasional boosts to prevent them from falling out of the sky.

• How does solar activity vary with time?

Solar weather is just as unpredictable as weather on Earth. Individual sunspots can appear or disappear at almost any time, and we have no way to know that a solar storm is coming until we observe it through our telescopes. However, long-term observations have revealed overall patterns in solar activity that make sunspots and solar storms more common at some times than at others.

The Sunspot Cycle The most notable pattern in solar activity is the **sunspot cycle**—a cycle in which the average number of sunspots on the Sun gradually rises and falls (Figure 14.22). At the time of *solar maximum,* when sunspots are most numerous, we may see dozens of sunspots on the Sun at one time. In contrast, we may see few if any sunspots at the time of *solar minimum.* The frequency of prominences, flares, and coronal mass ejections also follows the sunspot cycle, with these events being most common at solar maximum and least common at solar minimum.

As you can see in Figure 14.22a, the sunspot cycle varies from one period to the next. Some maximums have much greater numbers of sunspots than others. The average length of time between maximums is 11 years, but we have observed it to be as short as 7 years and as long as 15 years. The locations of sunspots on the Sun also vary with the sunspot cycle (Figure 14.22b). As a cycle begins at solar minimum, sunspots form primarily at midlatitudes (30° to 40°) on the Sun. The sunspots tend to form at lower latitudes as the cycle progresses, appearing very close to the solar equator as the next solar minimum approaches. Then the sunspots of the next cycle begin to form near midlatitudes again.

A less obvious feature of the sunspot cycle is that something peculiar happens to the Sun's magnetic field at each solar minimum: The Sun's entire magnetic field flip-flops, turning magnetic north into magnetic south and vice versa. We know this because the magnetic field lines connecting pairs of sunspots (see Figure 14.17) on the same side of the solar equator all tend to point in the same direction throughout an 11-year cycle. For example, all compass needles might point from the easternmost sunspot to the westernmost sunspot in each sunspot pair north of the solar equator. However, as the cycle ends at solar minimum, the magnetic field reverses: In the subsequent solar cycle, the field lines connecting pairs of sunspots point in the opposite direction. Thus, the Sun's *complete* magnetic cycle (sometimes called the *solar cycle*) really averages 22 years, since it takes two 11-year sunspot cycles before the magnetic field is back the way it started.

Longer-Term Changes in the Sunspot Cycle Figure 14.22 shows how the sunspot cycle has varied in length and intensity during the past century, the time during which we've had the most complete records of sunspot number. However, astronomers have observed the Sun telescopically for nearly 400 years, and these longer-term observations suggest that the sunspot cycle can change even more dramatically (Figure 14.23). For example, astronomers observed virtually no sunspots between the years 1645 and 1715, a period sometimes called the *Maunder minimum* (after E. W. Maunder, who identified it in historical sunspot records).

Is it possible that the Maunder minimum is part of a longer-term cycle of solar activity, one lasting much longer than 11 or 22 years? Some scientists have hypothesized that such cycles might exist, but little evidence has been found

Figure 14.22 Sunspot cycle during the past century.

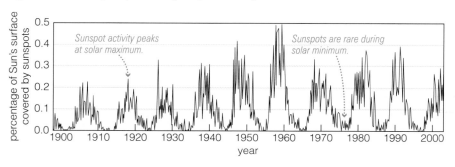

a This graph shows how the number of sunspots on the Sun changes with time. The vertical axis shows the percentage of the Sun's surface covered by sunspots. The cycle has a period of approximately 11 years.

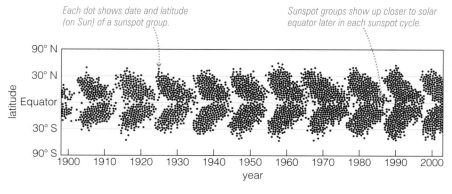

b This graph shows how the latitudes at which sunspot groups appear tend to shift during a single sunspot cycle.

Figure 14.23 This graph reconstructs the sunspot cycle over the past 400 years, based on available data from telescopic observations.

to back up these claims. Of course, the search for long-term variations is difficult, because we have very limited observational data predating the invention of the telescope. Chinese astronomers kept records of naked-eye observations of sunspots going back almost 2,000 years, but these records are sparse and naked-eye observations may not be very reliable.

The search for long-term cycles therefore relies on less direct evidence. For example, we can make educated guesses about past solar activity from descriptions of solar eclipses recorded around the world: When the Sun is more active, the corona tends to have longer and brighter "streamers" visible to the naked eye. We can also gauge past solar activity by studying the amount of carbon-14 in tree rings. High-energy *cosmic rays* [Section 19.2] coming from beyond our own solar system produce radioactive carbon-14 in Earth's atmosphere. During periods of high solar activity, the solar wind tends to grow stronger, shielding Earth from some of these cosmic rays. Thus, production of carbon-14 drops when the Sun is more active. Because trees steadily incorporate atmospheric carbon into their rings (through their respiration of carbon dioxide), we can estimate the level of solar activity during each year of a tree's life by measuring the level of carbon-14 in the corresponding ring. These data have not yet turned up any clear evidence of longer-term cycles of solar activity, but the search goes on.

The Cause of the Sunspot Cycle   The precise reasons for the sunspot cycle are not fully understood, but solar physicists believe they understand the general nature of the processes involved. The leading model ties the sunspot cycle to a combination of convection and the way in which the Sun's rotation rate varies with latitude. Convection is thought to dredge up weak magnetic fields generated in the solar interior, amplifying them as they rise. The Sun's rotation— faster at its equator than near its poles—then stretches and shapes these fields.

THINK ABOUT IT

Suppose you take a photograph of the Sun and notice two sunspots: one near the equator and one directly north of it at higher latitude. If you look again at the Sun in a few days, would you still find one sunspot directly north of the other? Why or why not? Explain how you could use this type of observation to learn how the Sun rotates.

Imagine what happens to magnetic field lines that start out running along the Sun's surface from south to north

(Figure 14.24). At the equator, the lines circle the Sun every 25 days, but at higher latitudes they lag behind. As a result, the lines gradually get wound more and more tightly around the Sun. This process, operating at all times over the entire Sun, produces the contorted field lines that generate sunspots and other solar activity.

The detailed behavior of these magnetic fields is quite complex, so scientists attempt to study it with sophisticated computer models. Using these models, scientists have successfully replicated many features of the sunspot cycle, including changes in the number and latitude of sunspots and the magnetic field reversals that occur about every 11 years. However, much still remains mysterious, including why the period of the sunspot cycle varies and why solar activity is different from one cycle to the next.

Over extremely long time periods—hundreds of millions to billions of years—these theoretical models predict a gradual lessening of solar activity. Recall that, according to our theory of solar system formation, the Sun must have rotated much faster when it was young [Section 8.3], and a faster rotation rate should have meant much more activity. Observations of other stars that are similar to the Sun but rotate faster confirm that these stars are much more active. We find evidence for many more "starspots" on these stars than on the Sun, and their relatively bright ultraviolet and X-ray emissions suggest that they have brighter chromospheres and coronas—just as we would expect if they are more active than the Sun.

The Sunspot Cycle and Earth's Climate   Despite the changes that occur during the sunspot cycle, the Sun's total output of energy barely changes at all—the largest measured changes have been less than 0.1% of the Sun's average luminosity. However, the ultraviolet and X-ray output of the Sun, which comes from the magnetically heated gas of the chromosphere and corona, can vary much more significantly. Could any of these changes affect the weather or climate on Earth?

Some data suggest connections between solar activity and Earth's climate. For example, the period from 1645 to 1715, when solar activity seems to have virtually ceased (see Figure 14.23), was a time of exceptionally low temperatures in Europe and North America known as the *Little Ice Age*. However, no one really knows whether the low solar activity caused these low temperatures, or whether it was just a coincidence. Similarly, some researchers have claimed that certain weather phenomena, such as drought cycles or frequencies of storms, are correlated with the 11- or 22-year

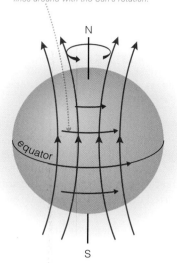

Charged particles tend to push the field lines around with the Sun's rotation.

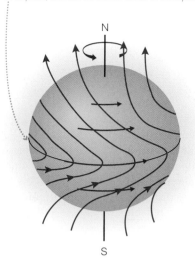

Because the Sun rotates faster near its equator than at its poles, the field lines bend ahead at the equator.

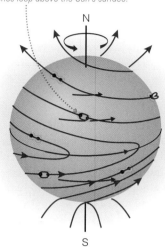

The field lines become more and more twisted with time, and sunspots form when the twisted lines loop above the Sun's surface.

Figure 14.24 The Sun rotates more quickly at its equator than it does near its poles. Because gas circles the Sun faster at the equator, it drags the Sun's north-south magnetic field lines into a more twisted configuration. The magnetic field lines linking pairs of sunspots, depicted here as green and black blobs, trace out the directions of these stretched and distorted field lines.

cycle of solar activity. A few scientists have even claimed that changes in the Sun during the last century have had a greater impact on global warming [Section 10.6] than emissions of greenhouse gases. Again, the data supporting these claims are weak, though we cannot entirely rule them out. The bottom line is that we do not yet know the extent to which solar activity affects weather on Earth. Given the importance of the question, it is certain to remain a hot topic of scientific research.

THE BIG PICTURE

Putting Chapter 14 into Context

In this chapter, we have examined our Sun, the nearest star. When you look back at this chapter, make sure you understand these "big picture" ideas:

- The ancient riddle of why the Sun shines is now solved. The Sun shines with energy generated by fusion of hydrogen into helium in the Sun's core. After a journey through the solar interior lasting several hundred thou-

sand years and an 8-minute journey through space, a small fraction of this energy reaches Earth and supplies sunlight and heat.

- Gravitational equilibrium, the balance between pressure and gravity, determines the Sun's interior structure and helps create a natural thermostat that keeps the fusion rate steady in the Sun. If the Sun were not so steady, life on Earth might not be possible.

- The Sun's atmosphere displays its own version of weather and climate, governed by solar magnetic fields. Some solar weather, such as coronal mass ejections, clearly affects Earth's magnetosphere. Other claimed connections between solar activity and Earth's climate may or may not be real.

- The Sun is important not only as our source of light and heat, but also because it is the only star near enough for us to study in great detail. In the coming chapters, we will use what we've learned about the Sun to help us understand other stars.

14.1 A Closer Look at the Sun

- **Why was the Sun's energy source a major mystery?** Before we learned the Sun's distance from Earth, no one realized just how much energy the Sun releases into space. Once we learned the Sun's energy output, scientists were at a loss to explain it until Einstein showed that mass itself contains a tremendous amount of energy.

- **Why does the Sun shine?** The Sun shines because **gravitational equilibrium** keeps its core hot and dense enough to release energy through nuclear fusion. The core originally became hot through the release of energy by **gravitational contraction**, as gravity made the Sun's birth cloud contract.

- **What is the Sun's structure?** The Sun's interior layers, from the inside out, are the **core**, the **radiation zone**, and the **convection zone**. Atop the convection zone lies the **photosphere**, the layer from which photons can freely escape into space. Above the photosphere, which is essentially the surface of the Sun, are the **chromosphere** and the **corona**.

14.2 The Cosmic Crucible

- **How does nuclear fusion occur in the Sun?** The core's extreme temperature and density are just right for fusion of hydrogen into helium, which occurs via the **proton-proton chain**. Because the fusion rate is so sensitive to temperature, gravitational equilibrium acts as a thermostat that keeps the fusion rate steady.

- **How does the energy from fusion get out of the Sun?** Energy moves through the deepest layers of the Sun—the core and the radiation zone—through **radiative diffusion**, in which photons bounce randomly among gas particles. After energy emerges from the radiation zone, convection carries it the rest of the way to the photosphere, where it is radiated into space as sunlight. Energy produced in the core takes hundreds of thousands of years to reach the photosphere.

- **How do we know what is happening inside the Sun?** We can construct theoretical models of the solar interior using known laws of physics and then check the models against observations of the Sun's size, surface temperature, and energy output. We also use studies of solar vibrations and solar **neutrinos**.

14.3 The Sun-Earth Connection

- **What causes solar activity?** Convection combined with the rotation pattern of the Sun—faster at the equator than at the poles—causes **solar activity** because these gas motions stretch and twist the Sun's magnetic field. These contortions of the magnetic field are responsible for phenomena such as **sunspots**, **flares**, **prominences**, and **coronal mass ejections**, and for heating the gas in the chromosphere and corona.

- **How does solar activity affect humans?** Bursts of charged particles ejected from the Sun during periods of high solar activity can hamper radio communications, disrupt electrical power generation, and damage orbiting satellites.

- **How does solar activity vary with time?** The **sunspot cycle**, or the variation in the number of sunspots on the Sun's surface, has an average period of 11 years. The magnetic field flip-flops every 11 years or so, resulting in a 22-year magnetic cycle. Sunspots first appear at midlatitudes at solar minimum, then become increasingly more common near the Sun's equator as the next minimum approaches. The number of sunspots can vary dramatically from one cycle to the next, and sometimes sunspots seem to be absent altogether.

Review Questions

Short-Answer Questions Based on the Reading

1. Briefly describe how *gravitational contraction* generates energy. When was it important in the Sun's history? Explain.

2. What two forces are balanced in *gravitational equilibrium?* Describe how gravitational equilibrium makes the Sun hot and dense in its core.

3. State the Sun's luminosity, mass, radius, and average surface temperature, and put the numbers into perspective that makes them meaningful.

4. Briefly describe the distinguishing features of each of the layers of the Sun shown in Figure 14.3.

5. What is the difference between nuclear *fission* and nuclear *fusion?* Which one is used in nuclear power plants? Which one does the Sun use?

6. Why does nuclear fusion require high temperatures and pressures?

7. What is the overall nuclear fusion reaction in the Sun? Briefly describe the proton-proton chain.

8. Does the Sun's fusion rate remain steady or vary wildly? Describe the feedback process that regulates the fusion rate.

9. Why has the Sun gradually brightened with time?

10. Why does the energy produced by fusion in the solar core take so long to reach the solar surface? Describe the processes by which energy generated by fusion makes its way to the Sun's surface.

11. Explain how mathematical models allow us to predict conditions inside the Sun. How can we be confident that the models are on the right track?

12. What are *neutrinos?* What was the *solar neutrino problem,* and why do we think it has now been solved?

13. What do we mean by *solar activity?* Describe some of the features of solar activity, including *sunspots, solar prominences, solar flares,* and *coronal mass ejections.*

14. Describe the appearance and temperature of the Sun's photosphere. Why does the surface look mottled? How are sunspots different from the surrounding photosphere?

15. How do magnetic fields keep sunspots cooler than the surrounding plasma? Explain.

16. Why is the chromosphere best viewed with ultraviolet telescopes? Why is the corona best viewed with X-ray telescopes? Briefly explain how we think the chromosphere and corona are heated.

17. What is the sunspot cycle? Describe how the changes that occur during the cycle, including magnetic field reversals at each solar minimum. Are there any longer-term changes that occur in solar activity?

18. Describe the leading model for explaining the sunspot cycle. Does the sunspot cycle influence Earth's climate? Explain.

Test Your Understanding

Does It Make Sense?

Decide whether the statement makes sense (or is clearly true) or does not make sense (or is clearly false). Explain your reasoning. (For an example, see Chapter 1, "Does It Make Sense?")

19. Before Einstein, gravitational contraction appeared to be a perfectly plausible mechanism for solar energy generation.

20. A sudden temperature rise in the Sun's core is nothing to worry about, because conditions in the core will soon return to normal.

21. If fusion in the solar core ceased today, worldwide panic would break out tomorrow as the Sun began to grow dimmer.

22. Astronomers have recently photographed magnetic fields churning deep beneath the solar photosphere.

23. Neutrinos probably can't harm me, but just to be safe I think I'll wear a lead vest.

24. There haven't been many sunspots this year, but there ought to be many more in about 5 years.

25. News of a major solar flare today caused concern among professionals in the fields of communications and electrical power generation.

26. By observing solar neutrinos, we can learn about nuclear fusion deep in the Sun's core.

27. If the Sun's magnetic field somehow disappeared, there would be no more sunspots on the Sun.

28. Scientists are currently building an infrared telescope designed to observe fusion reactions in the Sun's core.

Quick Quiz

Choose the best answer to each of the following. Explain your reasoning with one or more complete sentences.

29. Which of these groups of particles has the greatest mass? (a) a helium nucleus with two protons and two neutrons (b) four electrons (c) four individual protons

30. Which of these layers of the Sun is coolest? (a) photosphere (b) chromosphere (c) corona

31. Which of these layers of the Sun is coolest? (a) core (b) radiation zone (c) photosphere

32. Scientists estimate the central temperature of the Sun using (a) probes that measure changes in Earth's atmosphere. (b) mathematical models of the Sun. (c) laboratories that create miniature versions of the Sun.

33. Why do sunspots appear darker than their surroundings? (a) They are cooler than their surroundings. (b) They block some of the sunlight from the photosphere. (c) They do not emit any light.

34. At the center of the Sun, fusion converts hydrogen into (a) plasma. (b) radiation and elements like carbon and nitrogen. (c) helium, energy, and neutrinos.

35. Solar energy leaves the core of the Sun in the form of (a) photons. (b) rising hot gas. (c) sound waves.

36. How does the number of neutrinos passing through your body at night compare with the number passing through your body during the day? (a) about the same (b) much smaller (c) much larger

37. What is the most common kind of particle in the solar wind, in which the overall charge is neutral? (a) proton (b) electron (c) helium nucleus

38. Which of these things poses the greatest hazard to communications satellites? (a) photons from the Sun (b) solar magnetic fields (c) protons from the Sun

Investigate Further

In-Depth Questions to Increase Your Understanding

Short-Answer/Essay Questions

39. *The End of Fusion I.* Describe what would happen in the Sun if fusion reactions abruptly shut off.

40. *The End of Fusion II.* If fusion reactions were to suddenly shut off in the Sun, how would we be able to tell?

41. *A Really Strong Force.* How would the interior temperature of the Sun be different if the strong force that binds nuclei together were 10 times stronger?

42. *Measuring the Sun's Rotation.* Suppose you observe a sunspot from Earth, taking a picture of the Sun each day. Using these photos, you count the number of days for the sunspot to return to the same position it had in the first picture. Now suppose a friend of yours standing on Pluto does the same thing. Would you and your friend measure the same number of days for the sunspot to return to where it started? Explain your answer.

43. *Covered with Sunspots.* Describe what the Sun would look like from Earth if the entire photosphere were the same temperature as a sunspot.

44. *Inside the Sun.* Describe how scientists determine what the interior of the Sun is like. Why have we not yet sent a probe into the Sun to measure what is happening there?

45. *Solar Energy Output.* The Sun's output of visible light remains very steady, staying well within 1% of its current value for the past century. However, the Sun's maximum X-ray output can be as much as 10 times greater than its minimum output. Explain why the changes in X-ray output can be so much more pronounced than those in the output of visible light.

46. *An Angry Sun.* A *Time* magazine cover once suggested that an "angry Sun" was becoming more active as human activity changed Earth's climate through global warming. It's certainly possible for the Sun to become more active at the same time that humans are affecting Earth, but is it possible that the Sun could be responding to human activity? Can humans affect the Sun in any significant way? Explain.

Quantitative Problems

Be sure to show all calculations clearly and state your final answers in complete sentences.

47. *Chemical Burning and the Sun.* Estimate how long the Sun would last if it were merely a huge fire that was releasing chemical energy. Assume that the Sun begins with roughly 10^8 joules per kilogram, a chemical energy content typical of atomic matter.

48. *The Lifetime of the Sun.* The total mass of the Sun is about 2×10^{30} kg, of which about 75% was hydrogen when the Sun formed. However, only about 13% of this hydrogen ever becomes available for fusion in the core. The rest remains in layers of the Sun where the temperature is too low for fusion.
   a. Based on the given information, calculate the total mass of hydrogen available for fusion over the lifetime of the Sun.
   b. Combine your results from part (a) and the fact that the Sun fuses about 600 billion kilograms of hydrogen each second to calculate how long the Sun's initial supply of hydrogen can last. Give your answer in both seconds and years.

   c. Given that our solar system is now about 4.6 billion years old, when will we need to worry about the Sun running out of hydrogen for fusion?

49. *Solar Power Collectors.* This problem leads you through the calculation and discussion of how much solar power can be collected by solar cells on Earth.
   a. Imagine a giant sphere surrounding the Sun with a radius of 1 AU. What is the surface area of this sphere, in square meters? (*Hint:* The formula for the surface area of a sphere is $4\pi r^2$.)
   b. Because this imaginary giant sphere surrounds the Sun, the Sun's entire luminosity of 3.8×10^{26} watts must pass through it. Calculate the power passing through each square meter of this imaginary sphere in *watts per square meter.* Explain why this number represents the maximum power per square meter that a solar collector in Earth orbit can collect.
   c. List several reasons why the average power per square meter collected by a solar collector on the ground will always be less than what you found in part (b).
   d. Suppose you want to put a solar collector on your roof. If you want to optimize the amount of power you can collect, how should you orient the collector? (*Hint:* The optimum orientation depends on both your latitude and the time of year and day.)

50. *Solar Power for the United States.* The total annual U.S. energy consumption is about 2×10^{20} joules.
   a. What is the average power requirement for the United States, in watts? (*Hint:* 1 watt = 1 joule/s.)
   b. With current technologies and solar collectors on the ground, the best we can hope is that solar cells will generate an average (day and night) power of about 200 watts/m². (You might compare this to the maximum power per square meter you found in Problem 49b.) What total area would we need to cover with solar cells to supply all the power needed for the United States? Give your answer in both square meters and square kilometers.
   c. The total surface area of the United States is about 2×10^7 km². What fraction of the U.S. area would have to be covered by solar collectors to generate all of the U.S. power needs? In one page or less, describe potential environmental impacts of covering so much area with solar collectors. Also discuss whether you think these environmental impacts would be greater or less than the impacts of using current energy sources such as coal, oil, nuclear power, and hydroelectric power.

51. *The Color of the Sun.* The Sun's average surface temperature is about 5,800 K. Use Wien's law (see Mathematical Insight 6.2) to calculate the wavelength of peak thermal emission from the Sun. What color does this wavelength correspond to in the visible-light spectrum? Why do you think the Sun appears white or yellow to our eyes?

52. *The Color of a Sunspot.* The typical temperature of a sunspot is about 4,000 K. Use Wien's law (see Mathematical Insight 6.2) to calculate the wavelength of peak thermal emission from a sunspot. What color does this wavelength correspond to in the visible-light spectrum? How does this color compare with that of the Sun?

53. *Solar Mass Loss.* Estimate how much mass the Sun loses through fusion reactions during its 10-billion-year life. You can simplify the problem by assuming the Sun's energy

output remains constant. Compare the amount of mass lost with Earth's mass.

54. *Pressure of the Photosphere.* The gas pressure of the photosphere changes substantially from its upper levels to its lower levels. Near the top of the photosphere the temperature is about 4,500 K and there are about 1.6×10^{16} gas particles per cubic centimeter. In the middle the temperature is about 5,800 K and there are about 1.0×10^{17} gas particles per cubic centimeter. At the bottom of the photosphere the temperature is about 7,000 K and there are about 1.5×10^{17} gas particles per cubic centimeter. Compare the pressures of each of these layers and explain the reason for the trend in pressure that you find. How do these gas pressures compare with Earth's atmospheric pressure at sea level?

55. *Tire Pressure.* Air pressure at sea level produces a force of about 15 pounds per square inch. The recommended air pressure in your car tires is about 30 pounds per square inch. How does the density of gas particles inside your tires compare with the density of gas particles in the air outside your tires? What happens to gas pressure in the tire if it springs a leak and loses gas particles? How does your tire respond to this loss of gas particles? How is the tire's response like the response of the Sun's core to the slowly declining number of independent particles within it? How is the tire's response different?

56. *Your Energy Content.* The power needed to operate your body is about 100 watts. Suppose your body could run on fusion power and could convert 0.7% of its mass into energy. How much energy would be available through fusion? For how long could your body then operate on fusion power?

Discussion Questions

57. *The Role of the Sun.* Briefly discuss how the Sun affects us here on Earth. Be sure to consider not only factors such as its light and warmth, but also how the study of the Sun has led us to new understandings in science and to technological developments. Overall, how important has solar research been to our lives?

58. *The Sun and Global Warming.* One of the most pressing environmental issues on Earth is the extent to which human emissions of greenhouse gases are warming our planet. Some people claim that part or all of the observed warming over the past century may be due to changes in the Sun, rather than to anything humans have done. Discuss how a better understanding of the Sun might help us comprehend the threat posed by greenhouse gas emissions. Why is it so difficult to develop a clear understanding of how the Sun affects Earth's climate?

Interactive Tutorials

Tutorial Review of Key Concepts
Use the following interactive **Tutorials** at **www.mastering astronomy.com** to review key concepts from this chapter.

The Sun Tutorial
Lesson 1 Structure and Gravitational Equilibrium
Lesson 2 Fusion of Hydrogen into Helium
Lesson 3 Why Does Fusion Only Occur at High Temperatures?

Web Projects

1. *Current Solar Weather.* Daily information about solar activity is available at NASA's Web site spaceweather.com. Where are we in the sunspot cycle right now? When is the next solar maximum or minimum expected? Have there been any major solar storms in the past few months? If so, did they have any significant effects on Earth? Summarize your findings in a one- to two-page report.

2. *Solar Observatories in Space.* Visit NASA's Web site for the Sun-Earth connection and explore some of the current and planned space missions designed to observe the Sun. Choose one mission to study in greater depth, and write a one- to two-page report on the mission status and goals and what it has taught or will teach us about the Sun.

3. *Sudbury Neutrino Observatory.* Visit the Web site for the Sudbury Neutrino Observatory (SNO) and learn how it has helped to solve the solar neutrino problem. Write a one- to two-page report describing the observatory, any recent results, and what we can expect from it in the future.

4. *Nuclear Power.* There are two basic ways to generate energy from atomic nuclei: through nuclear fission (splitting nuclei) and through nuclear fusion (combining nuclei). All current nuclear reactors are based on fission, but fusion would have many advantages if we could develop the technology. Research some of the advantages of fusion and some of the obstacles to developing fusion power. Do you think fusion power will be a reality in your lifetime? Explain.

Surveying the Stars

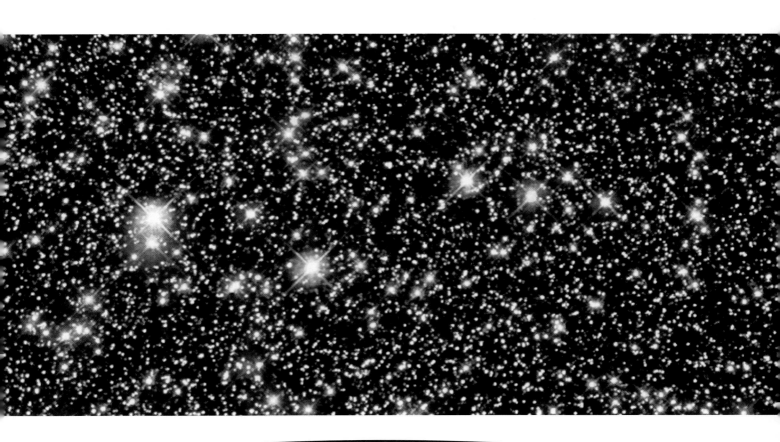

"All men have the stars," he answered, *"but they are not the same things for different people. For some, who are travelers, the stars are guides. For others they are no more than little lights in the sky. For others, who are scholars, they are problems. For my businessman they were wealth. But all these stars are silent. You—you alone—will have the stars as no one else has them."*

—Antoine de Saint-Exupéry, from *The Little Prince*

On a clear, dark night, a few thousand stars are visible to the naked eye. Many more become visible through binoculars, and a powerful telescope reveals so many stars that we could never hope to count them. Like each individual person, each individual star is unique. And yet, like the human family, all stars share much in common.

Today, we know that stars are born from clouds of interstellar gas, shine brilliantly by nuclear fusion for millions to billions of years, and then die, sometimes in dramatic ways. In this chapter, we'll discuss how we study and categorize stars and how we have come to realize that stars, like people, change over their lifetimes.

15.1 Properties of Stars

Imagine that an alien spaceship flies by Earth on a simple but short mission: The visitors have just 1 minute to learn everything they can about the human race. In 60 seconds, they will see next to nothing of any individual person's life. Instead, they will obtain a collective "snapshot" of humanity showing people from all stages of life engaged in their daily activities. From this snapshot alone, they must piece together their entire understanding of human beings and their lives, from birth to death.

We face a similar problem when we look at the stars. Compared with stellar lifetimes of millions or billions of years, the few hundred years humans have spent studying stars with telescopes is rather like the aliens' 1-minute glimpse of humanity. We see only a brief moment in any star's life, and our collective snapshot of the heavens consists of such frozen moments for billions of stars. From this snapshot, we try to reconstruct the life cycles of stars.

Thanks to the efforts of hundreds of astronomers studying this snapshot of the heavens, stars are no longer mysterious points of light in the sky. We now know that all stars share much in common with the Sun. They all form in great clouds of gas and dust, and each one begins its life with roughly the same chemical composition as the Sun: About three-quarters of a star's mass at birth is hydrogen, and about one-quarter is helium, with no more than 2% consisting of elements heavier than helium. Nevertheless, stars are not all the same; they differ in such properties as size, age, brightness, and temperature. We'll devote most of this and the next two chapters to understanding how and why stars differ. First, however, let's explore how we measure three of the most fundamental properties of stars: luminosity, surface temperature, and mass.

• How do we measure stellar luminosities?

If you go out on any clear night, you'll immediately see that stars differ in brightness. Some stars are so bright that we can use them to identify constellations [Section 2.1]. Others are so dim that our naked eyes cannot see them at all. However, these differences in brightness do not by themselves tell us anything about how much light these stars are really generating, because the brightness of a star depends on its distance as well as on how much light it actually emits. For example, the stars Procyon and Betelgeuse, which make up two of the three corners of the winter triangle (see Figure 2.2), appear about equally bright in our sky. Betelgeuse actually emits about 5,000 times as much light as Procyon. However, its brightness is similar because it is much farther away.

Because two similar-looking stars can be generating very different amounts of light, we need to distinguish clearly between a star's brightness in our sky and the actual amount of light that it emits into space (Figure 15.1):

- When we talk about the way stars look in our sky, we are talking about **apparent brightness**—the brightness of a star as it appears to our eyes. More specifically, we define the apparent brightness of any star in our sky as the amount of power (energy per second) reaching us *per unit area*. (A more technical term for apparent brightness is *flux*.)
- When we talk about how bright stars are in an absolute sense, regardless of their distance, we are talking about **luminosity**—the total amount of power that a star emits into space.

We can understand the difference between apparent brightness and luminosity by thinking about a 100-watt light bulb. The bulb always puts out the same amount of light, so its luminosity doesn't vary. However, its apparent brightness depends on your distance from the bulb: It will look quite bright if you stand very close to it, but quite dim if you are far away.

The Inverse Square Law for Light We can measure the apparent brightness of a star in our sky, but we cannot directly measure luminosity. However, we can calculate a star's luminosity, because apparent brightness follows an *inverse square law* with distance, much like the inverse square law that describes the force of gravity [Section 4.4]. For example, if we viewed the Sun from twice Earth's distance, it would appear dimmer by a factor of $2^2 = 4$. If we viewed it from 10 times Earth's distance, it would appear $10^2 = 100$ times dimmer.

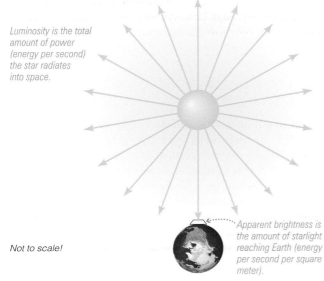

Luminosity is the total amount of power (energy per second) the star radiates into space.

Not to scale!

Apparent brightness is the amount of starlight reaching Earth (energy per second per square meter).

Figure 15.1 Luminosity is a measure of power, and apparent brightness is a measure of power per unit area.

The same amount of starlight passes through each sphere.

The surface area of a sphere depends on the square of its radius (distance from the star) . . .

. . . so the amount of light passing through each unit of area depends on the inverse square of distance from the star.

Figure 15.2 The inverse square law for light: The apparent brightness of a star declines with the square of its distance.

THINK ABOUT IT

Suppose Star A is four times as luminous as Star B. How will their apparent brightnesses compare if they are both the same distance from Earth? How will their apparent brightnesses compare if Star A is twice as far from Earth as Star B? Explain.

Measuring Apparent Brightness We can measure a star's apparent brightness by using a detector, such as a CCD, that records how much energy strikes its light-sensitive surface each second. For example, such a detector would record an apparent brightness of 2.7×10^{-8} watt per square meter from Alpha Centauri A (the brightest of the three stars in the Alpha Centauri system). The only difficulties we face in measuring apparent brightness are making sure the detector is properly calibrated and, for ground-based telescopes, taking into account the absorption of light by Earth's atmosphere.

No detector can record light of all wavelengths, so we necessarily measure apparent brightness in only some small range of the complete spectrum. For example, the human eye is sensitive to visible light but does not respond to ultraviolet or infrared photons. Thus, when we perceive a star's brightness, our eyes are measuring the apparent brightness only in the visible region of the spectrum.

When we measure the apparent brightness in visible light, we can calculate only the star's *visible-light luminosity*. Similarly, when we observe a star with a spaceborne X-ray telescope, we measure only the apparent brightness in X rays and can calculate only the star's *X-ray luminosity*. We will use the terms **total luminosity*** and **total apparent brightness** to describe the luminosity and apparent brightness we would measure *if* we could detect photons across the entire electromagnetic spectrum.

One important complication often arises when we try to calculate a star's luminosity from its apparent brightness: The inverse square law for light works perfectly only if the

Figure 15.2 shows why apparent brightness follows an inverse square law. The same total amount of light must pass through each imaginary sphere surrounding the star. If we focus on the light passing through the small square on the sphere located at 1 AU, we see that the same amount of light must pass through *four* squares of the same size on the sphere located at 2 AU. Thus, each square on the sphere at 2 AU receives only $\frac{1}{2^2} = \frac{1}{4}$ as much light as the square on the sphere at 1 AU. Similarly, the same amount of light passes through *nine* squares of the same size on the sphere located at 3 AU. Thus, each of these squares receives only $\frac{1}{3^2} = \frac{1}{9}$ as much light as the square on the sphere at 1 AU. Generalizing, the amount of light received per unit area decreases with increasing distance by the square of the distance—an inverse square law.

This inverse square law leads to a very simple and important formula relating the apparent brightness, luminosity, and distance of any light source. We will call it the **inverse square law for light**:

$$\text{apparent brightness} = \frac{\text{luminosity}}{4\pi \times \text{distance}^2} \rightarrow \text{surface area of sphere}$$

Because the standard units of luminosity are watts [Section 14.1], the units of apparent brightness are *watts per square meter*. (The 4π in the formula for the inverse square law for light comes from the fact that the surface area of a sphere is given by $4\pi \times (\text{radius})^2$.)

In principle, we can always determine a star's apparent brightness by carefully measuring the amount of light we receive from the star per square meter. Thus, we can use the inverse square law to calculate a star's luminosity if we can first measure its distance, or to calculate a star's distance if we somehow know its luminosity.

---

*Astronomers sometimes refer to the total luminosity as the *bolometric* luminosity.

starlight follows an uninterrupted path to Earth. In reality, the light of most stars passes through at least some clouds containing interstellar dust on its way to Earth, and this dust can absorb or scatter some of the star's light [Section 16.1]. Today, thanks largely to our modern scheme of stellar classification, we can usually measure the effect of interstellar dust and account for it when we apply the inverse square law for light. A century ago, before astronomers knew of the existence of interstellar dust, astronomers often underestimated stellar distances because they did not realize that the dust was making stars appear less bright than they really are.

Measuring Cosmic Distances Tutorial, Lesson 2

Measuring Distance Through Stellar Parallax If we want to calculate a star's luminosity from its apparent brightness, we must also know its distance. The most direct way to measure a star's distance is with *stellar parallax,* the small annual shifts in a star's apparent position caused by Earth's motion around the Sun [Section 2.4]. Recall that you can observe parallax of your finger by holding it at arm's length and looking at it with first one eye closed and then the other eye closed. Astronomers measure stellar parallax by comparing observations of a nearby star made six months apart (Figure 15.3). The nearby star appears to shift against the background of more distant stars because we are observing it from two opposite points of Earth's orbit. The star's *parallax angle*—the angle labeled *p* in Figure 15.3—is defined as *half* the star's annual back-and-forth shift. Notice

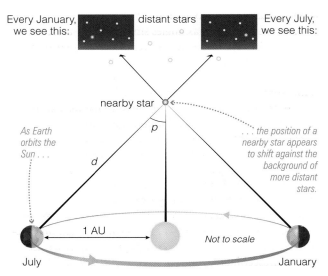

Figure 15.3 Interactive Figure Parallax makes the apparent position of a nearby star shift back and forth with respect to distant stars over the course of each year. The angle *p*, called the *parallax angle*, represents half the total parallax shift each year. If we measure *p* in arcseconds, the distance *d* to the star in parsecs is $1/p$. The angle in this figure is greatly exaggerated: All stars have parallax angles of less than 1 arcsecond.

that this angle would be smaller if the star were farther away, so we conclude that more distant stars have *smaller* parallax angles.

Stars are so far away that they have very small parallax angles, which explains why the ancient Greeks were never

---

MATHEMATICAL INSIGHT 15.1 **The Inverse Square Law for Light**

We can derive the inverse square law for light by extending the idea illustrated in Figure 15.2. Suppose we are located a distance *d* from a star with luminosity *L*. The apparent brightness of the star is the power per unit area that we receive at our distance *d*. We can find this apparent brightness by imagining that we are part of a giant sphere with radius *d*, similar to any one of the three spheres in Figure 15.2. The surface area of this giant sphere is $4\pi \times d^2$, and the star's entire luminosity *L* must pass through this surface area. (The surface area of any sphere is $4\pi \times \text{radius}^2$.) We find the apparent brightness at distance *d* by dividing the total luminosity by the surface area of the sphere, giving us our formula for the inverse square law for light:

$$\text{apparent brightness} = \frac{\text{star's luminosity}}{\text{surface area of imaginary sphere}}$$

$$= \frac{L}{4\pi \times d^2}$$

Example: What is the Sun's apparent brightness as seen from Earth?

Solution:

Step 1 Understand: The Sun's apparent brightness at Earth is the power per unit area that we receive in the form of sunlight. We can find this power per unit area with the inverse square law for light from the Sun's luminosity and Earth's distance from the Sun; for unit consistency, we'll want the Earth-Sun distance in meters.

Step 2 Solve: The Sun's luminosity is $L_{\text{Sun}} = 3.8 \times 10^{26}$ watts, and Earth's distance from the Sun is $d = 1.5 \times 10^{11}$ meters. Thus, the Sun's apparent brightness is:

$$\frac{L}{4\pi \times d^2} = \frac{3.8 \times 10^{26} \text{ watts}}{4\pi \times (1.5 \times 10^{11} \text{ m})^2}$$

$$= 1.3 \times 10^3 \text{ watts/m}^2$$

Step 3 Explain: The Sun's apparent brightness is about 1,300 watts per square meter at Earth's distance. This is the maximum power per unit area that could be collected by a detector on Earth that directly faces the Sun, such as a solar power (or *photovoltaic*) cell. In reality, solar collectors usually collect less power because Earth's atmosphere absorbs some sunlight, particularly when it is cloudy.

able to measure parallax with their naked eyes. Even the nearest stars have parallax angles smaller than 1 arcsecond—well below the approximately 1 arcminute angular resolution of the naked eye [Section 6.2]. For increasingly distant stars, the parallax angles quickly become too small to measure even with our highest-resolution telescopes. Current technology allows us to measure parallax accurately only for stars within a few hundred light-years—not much farther than what we call our *local solar neighborhood* in the vast, 100,000-light-year-diameter Milky Way Galaxy.

By definition, the distance to an object with a parallax angle of 1 arcsecond is 1 **parsec (pc)**. (The word *parsec* comes from combining the words *parallax* and *arcsecond*.) Because all stars have parallax angles smaller than one arcsecond, they are all farther than 1 parsec away. If we use units of arcseconds for the parallax angle, p, a simple formula allows us to calculate distances in parsecs:

$$d \text{ (in parsecs)} = \frac{1}{p \text{ (in arcseconds)}}$$

For example, the distance to a star with a parallax angle of $\frac{1}{2}$ arcsecond is 2 parsecs, the distance to a star with a paral-lax angle of $\frac{1}{10}$ arcsecond is 10 parsecs, and the distance to a star with a parallax angle of $\frac{1}{100}$ arcsecond is 100 parsecs.

Astronomers often state distances in parsecs, kiloparsecs (1,000 parsecs), or megaparsecs (1 million parsecs). However, with a bit of geometry, it's possible to show that 1 parsec is equivalent to 3.26 light-years (see Mathematical Insight 15.2). Thus, we can modify the above formula slightly to give distances in light-years:

$$d \text{ (in light-years)} = 3.26 \times \frac{1}{p \text{ (in arcseconds)}}$$

In this book, we'll generally state distances in light-years rather than parsecs.

Parallax was the first reliable technique astronomers developed for measuring distances to stars, and it remains the only technique that directly tells us the distances of stars, without any assumptions about the nature of stars. If we know a star's distance from parallax, we can calculate its luminosity with the inverse square law for light. We now have parallax measurements for thousands of stars, which is a large enough number that astronomers have been able to draw some general conclusions about stars. As we'll see later, these lessons have taught astronomers how

---

MATHEMATICAL INSIGHT **15.2** **The Parallax Formula**

Here is one of several ways to derive the formula relating a star's distance and parallax angle. Figure 15.3 shows that the parallax angle p is part of a right triangle, the side opposite p is the Earth-Sun distance of 1 AU, and the hypotenuse is the distance d to the object. You may recall from trigonometry that the *sine* of an angle in a right triangle is the length of its opposite side divided by the length of the hypotenuse. In this case, we find:

$$\sin p = \frac{\text{length of opposite side}}{\text{length of hypotenuse}} = \frac{1 \text{ AU}}{d}$$

If we solve for d, the formula becomes:

$$d = \frac{1 \text{ AU}}{\sin p}$$

By definition, 1 parsec is the distance to an object with a parallax angle of 1 arcsecond (1″), or $\frac{1}{3,600}$ degree (because $1° = 60'$ and $1' = 60''$). Substituting these numbers into the parallax formula and using a calculator to find that $\sin 1'' = 4.84814 \times 10^{-6}$, we get:

$$1 \text{ pc} = \frac{1 \text{ AU}}{\sin 1''} = \frac{1 \text{ AU}}{4.84814 \times 10^{-6}} = 206,265 \text{ AU}$$

That is, 1 parsec = 206,265 AU, which is equivalent to 3.09×10^{13} km or 3.26 light-years. (Recall that 1 AU = 149.6 million km.)

We need one more fact from geometry to derive the parallax formula given in the text. As long as the parallax angle, p, is small, $\sin p$ is proportional to p. For example, $\sin 2''$ is twice as large as $\sin 1''$, and $\sin \frac{1}{2}''$ is half as large as $\sin 1''$. (You can verify these examples with your calculator.) Thus, if we use $\frac{1}{2}''$ instead of $1''$ for the parallax angle in the formula above, we get a distance of 2 pc instead of 1 pc. Similarly, if we use a parallax angle of $\frac{1}{10}''$, we get a distance of 10 pc. Generalizing, we get the simple parallax formula given in the text:

$$d \text{ (in parsecs)} = \frac{1}{p \text{ (in arcseconds)}}$$

Example 1: Sirius, the brightest star in our night sky, has a measured parallax angle of 0.379″. How far away is Sirius in parsecs? In light-years?

Solution:

Step 1 Understand: We are given the parallax angle for Sirius in arcseconds, so we can use the parallax formula to find its distance. Because the parallax angle is between 0.1″ and 1″, we expect the answer to be a distance between 1 and 10 parsecs.

Step 2 Solve: Substituting the parallax angle of 0.379″ into the formula, we find that the distance to Sirius in parsecs is:

$$d \text{ (in pc)} = \frac{1}{0.379} = 2.64 \text{ pc}$$

Because 1 pc = 3.26 light-years, this distance is equivalent to:

$$2.64 \text{ pc} \times 3.26 \frac{\text{light-years}}{\text{pc}} = 8.60 \text{ light-years}$$

Step 3 Explain: From its measured parallax angle, we have found that the distance to Sirius is 2.64 parsecs, or 8.60 light-years.

to estimate luminosities for many more stars, even without first knowing their distances. Thus, astronomers today often use the inverse square law for light to calculate distances to objects for which we can reliably estimate luminosities, as well as for calculating luminosities of objects for which we have measured distances.

Parallax measurements have given us detailed knowledge of what our local solar neighborhood is like. For example, we know of more than 300 stars within about 33 light-years (10 parsecs) of the Sun. About half are binary star systems consisting of two orbiting stars or multiple star systems containing three or more stars. Most are tiny, dim red stars—so dim that we cannot see them with the naked eye, despite the fact that they are relatively close. A few nearby stars, such as Sirius (8.6 light-years), Altair (17 light-years), Vega (25 light-years), and Fomalhaut (25 light-years), are white in color and bright in our sky, but most of the brightest stars in the sky lie farther away. Because so many nearby stars appear dim while many more distant stars appear bright, stellar luminosities must span a wide range.

The Luminosity Range of Stars Now that we have discussed how we determine stellar luminosities, it's time to take a quick look at the results; these results have been drawn both from stars for which we have parallax measurements and from those for which we determine distance in other ways. We usually state stellar luminosities in comparison to the Sun's luminosity, which we write as L_{Sun} for short. For example, Proxima Centauri, the nearest of the three stars in the Alpha Centauri system and hence the nearest star besides our Sun, is only about 0.0006 times as luminous as the Sun, or $0.0006L_{Sun}$. Betelgeuse, the bright left-shoulder star of Orion, has a luminosity of $38,000L_{Sun}$, meaning that it is 38,000 times as luminous as the Sun. Overall, studies of the luminosities of many stars have taught us two particularly important lessons:

- Stars have a wide range of luminosities, with our Sun somewhere in the middle. The dimmest stars have luminosities $\frac{1}{10,000}$ times that of the Sun ($10^{-4}L_{Sun}$), while the brightest stars are about 1 million times as luminous as the Sun ($10^{6}L_{Sun}$).

MATHEMATICAL INSIGHT 15.3 The Modern Magnitude Scale

The modern magnitude system is defined so that each difference of five magnitudes corresponds to a factor of exactly 100 in brightness. Thus, a single magnitude corresponds to a factor of $(100)^{1/5} \approx 2.512$ in brightness. Given this fact, the following formula allows us to calculate the ratio of the apparent brightnesses of two stars from their apparent magnitudes:

$$\frac{\text{apparent brightness of Star 1}}{\text{apparent brightness of Star 2}} = (100^{1/5})^{m_2 - m_1}$$

where m_1 and m_2 are the apparent magnitudes of Stars 1 and 2, respectively.

If we replace the apparent magnitudes with absolute magnitudes (designated M instead of m), the same formula allows us to calculate the ratio of stellar luminosities:

$$\frac{\text{luminosity of Star 1}}{\text{luminosity of Star 2}} = (100^{1/5})^{M_2 - M_1}$$

Example 1: On a clear night, stars dimmer than magnitude 5 are quite difficult to see. However, sensitive instruments on large, modern telescopes can detect objects as faint as magnitude 30. How much more sensitive are such telescopes than the human eye?

Solution:

Step 1 Understand: We imagine that our eye sees "Star 1" with magnitude 5 and the telescope detects "Star 2" with magnitude 30. Because every five steps up in magnitude corresponds to a drop of a factor of 100 in apparent brightness, the apparent brightness of Star 2 must be far smaller than that of Star 1. In order to determine that difference in apparent brightness, we can use the first formula above.

Step 2 Solve: Substituting $m_1 = 5$ and $m_2 = 30$ into the formula, we find:

$$\frac{\text{apparent brightness of Star 1}}{\text{apparent brightness of Star 2}} = (100^{1/5})^{m_2 - m_1} = (100^{1/5})^{30-5}$$

$$= (100^{1/5})^{25} = 100^5 = 10^{10}$$

Step 3 Explain: The magnitude 5 star is 10^{10}, or 10 billion, times brighter than the magnitude 30 star, so the telescope is 10 billion times more sensitive than the human eye.

Example 2: The Sun has an absolute magnitude of about 4.8. Polaris, the North Star, has an absolute magnitude of -3.6. How much more luminous is Polaris than the Sun?

Solution:

Step 1 Understand: Luminosity and absolute magnitude are two different ways of describing a star's total power output, and the second formula above gives the relationship between them. We can therefore use that formula with Polaris as Star 1 and the Sun as Star 2 to compare the luminosities of the two stars.

Step 2 Solve: Substituting $M_1 = -3.6$ for Polaris and $M_2 = 4.8$ for the Sun into the formula, we find:

$$\frac{\text{luminosity of Polaris}}{\text{luminosity of Sun}} = (100^{1/5})^{M_2 - M_1} = (100^{1/5})^{4.8-(-3.6)}$$

$$= (100^{1/5})^{8.4} = 100^{1.7} \approx 2,500$$

Step 3 Explain: From their absolute magnitudes, we have found that Polaris is about 2,500 times as luminous as the Sun.

• Dim stars are far more common than bright stars. For example, even though our Sun is roughly in the middle of the overall range of stellar luminosities, it is brighter than the vast majority of stars in our galaxy.

The Magnitude System   The methods we've discussed for describing apparent brightness and luminosity work perfectly well, but many amateur and professional astronomers still describe these quantities in another way: They use the ancient **magnitude system** devised by the Greek astronomer Hipparchus (c. 190–120 B.C.).

The magnitude system originally classified stars according to how bright they look to human eyes, which were the only instruments available to measure brightness in ancient times. The brightest stars received the designation "first magnitude," the next brightest "second magnitude," and so on. The faintest visible stars were magnitude 6. We call these descriptions **apparent magnitudes** because they compare how bright different stars *appear* in the sky. Thus, apparent magnitudes are directly related to apparent brightness, except the scale runs backward: A larger apparent magnitude means a dimmer apparent brightness. For example, a star of magnitude 4 is dimmer in the sky than a star of magnitude 1. Star charts (such as those in Appendix J) often use dots of different sizes to represent the apparent magnitudes of stars. Because larger dots are meant to represent brighter stars, the dots are larger for the smaller magnitude numbers.

In modern times, the magnitude system has been extended and more precisely defined. In particular, each difference of five magnitudes is defined to represent a factor of exactly 100 in brightness. For example, a magnitude 1 star is 100 times brighter than a magnitude 6 star, and a magnitude 3 star is 100 times brighter than a magnitude 8 star. As a result of this precise definition, stars can have fractional apparent magnitudes and a few bright stars have apparent magnitudes *less than* 1—which means *brighter* than magnitude 1. For example, the brightest star in the night sky, Sirius, has an apparent magnitude of −1.46. Appendix F gives apparent magnitudes and actual luminosities for both the nearest stars and the brightest stars visible in the sky.

The modern magnitude system also defines **absolute magnitudes** as a way of describing stellar luminosities. A star's absolute magnitude is the apparent magnitude it would have *if* it were at a distance of 10 parsecs (32.6 light-years) from Earth. For example, the Sun's absolute magnitude is about 4.8, meaning that the Sun would have an apparent magnitude of 4.8 *if* it were 10 parsecs away from us—bright enough to be visible but not conspicuous on a dark night. Although many articles and books still quote apparent and absolute magnitudes, comparisons between stars are much easier when we think about how apparent brightness depends on luminosity according to the inverse square law. Thus, we'll stick to the inverse square law in this book.

• How do we measure stellar temperatures?

A second fundamental property of a star is its surface temperature. You might wonder why we emphasize *surface* temperature rather than interior temperature. The answer is that only surface temperature is directly measurable; interior temperatures are inferred from mathematical models of stellar interiors [Section 14.2]. Thus, whenever you hear astronomers speak of the "temperature" of a star, you can be pretty sure they mean surface temperature unless they state otherwise.

Measuring a star's surface temperature is somewhat easier than measuring its luminosity, because the star's distance doesn't affect the measurement. Instead, we determine surface temperature from either the star's color or its spectrum. Let's briefly investigate how each technique works.

Color and Temperature   Take a careful look at Figure 15.4. Notice that stars come in almost every color of the rainbow. Simply looking at the colors tells us something about the surface temperatures of the stars. For example, a red star is cooler than a yellow star, which in turn is cooler than a white or blue star.

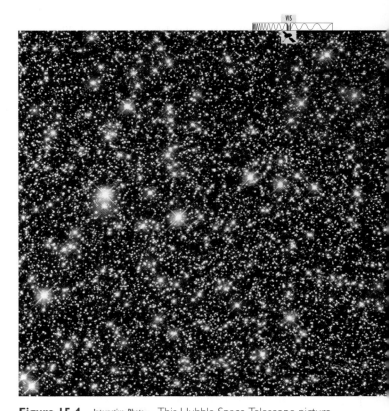

Figure 15.4 Interactive Photo This Hubble Space Telescope picture shows a wide variety of stars that differ in color and brightness. Most of the stars in this picture are at roughly the same distance, about 2,000 light-years from the center of our galaxy. Clouds of gas and dust obscure our view of visible light from most of our galaxy's central regions, but a gap in the clouds allows us to see the stars in this photo.

Photos of Stars

Photographs of stars, star clusters, and galaxies convey a great deal of information, but they also contain a few artifacts that are not real. For example, different stars seem to have different sizes in photographs such as Figure 15.4, but stars are so far away that they should all appear as mere points of light. The sizes are an artifact of how our instruments record light. Bright stars tend to be overexposed in photographs, making them appear larger in size than dimmer stars. Overexposure also explains why the centers of globular clusters and galaxies usually look like big blobs in photographs: The central regions of these objects contain many more stars than the outskirts, and the combined light of so many stars tends to get overexposed to make a big blob.

Spikes around bright stars in photographs, often making the pattern of a cross with a star at the center, are another such artifact. You can see these spikes around many of the brightest stars in Figure 15.4. These spikes are not real but rather are created by the interaction of starlight with the supports holding the secondary mirror in the telescope [Section 6.2]. The spikes generally occur only with point sources of light like stars, and not with larger objects like galaxies. When you look at a photograph showing many galaxies (for example, Figure 20.1), you can tell which objects are stars by looking for the spikes.

Table 15.1 The Spectral Sequence

Spectral Type	Example(s)	Temperature Range
O	Stars of Orion's Belt	>30,000 K
B	Rigel	30,000 K–10,000 K
A	Sirius	10,000 K–7,500 K
F	Polaris	7,500 K–6,000 K
G	Sun, Alpha Centauri A	6,000 K–5,000 K
K	Arcturus	5,000 K–3,500 K
M	Betelgeuse, Proxima Centauri	<3,500 K

Stars come in different colors because they emit thermal radiation [Section 5.4]. Recall that an ideal thermal radiation spectrum depends only on the (surface) temperature of the object that emits it (see Figure 5.19). For example, the Sun's 5,800 K surface temperature causes it to emit most strongly in the middle of the visible portion of the spectrum, which is why the Sun looks yellow or white in color. A cooler star, such as Betelgeuse (surface temperature 3,400 K), looks red because it emits much more red light than blue light. A hotter star, such as Sirius (surface temperature 9,400 K), emits a little more blue light than red light and therefore has a slightly blue color to it.

Astronomers can measure surface temperature fairly precisely by comparing a star's apparent brightness in two different colors of light. For example, by comparing the amount of blue light and red light coming from Sirius, astronomers can measure how much more blue light it emits than red light. Because thermal radiation spectra have a very distinctive shape (again, see Figure 5.19), this difference in blue and red light output allows astronomers to calculate a surface temperature.

Spectral Type and Temperature A star's spectral lines provide a second way to measure its surface temperature. Moreover, because interstellar dust can affect the apparent colors of stars, temperatures determined from spectral lines are

generally more accurate than temperatures determined from colors alone. Stars displaying spectral lines of highly ionized elements must be fairly hot, because it takes a high temperature to ionize atoms. Stars displaying spectral lines of molecules must be relatively cool, because molecules break apart into individual atoms at relatively cool temperatures. Thus, the types of spectral lines present in a star's spectrum provide a direct measure of the star's surface temperature.

Astronomers classify stars according to surface temperature by assigning a **spectral type** determined from the spectral lines present in a star's spectrum. The hottest stars, with the bluest colors, are called spectral type O, followed in order of declining surface temperature by spectral types B, A, F, G, K, and M.* The time-honored mnemonic for remembering this sequence, OBAFGKM, is "Oh Be A Fine Girl/Guy, Kiss Me!" Table 15.1 summarizes the characteristics of each spectral type.

Each spectral type is subdivided into numbered subcategories (such as B0, B1, . . . , B9). The larger the number, the cooler the star. For example, the Sun is designated spec-

---

*The sequence of spectral types has recently been extended beyond type M to include spectral types L and T, representing starlike objects even cooler than stars of spectral type M. However, as we will see in Chapter 16, most of these objects are not true, hydrogen-burning stars.

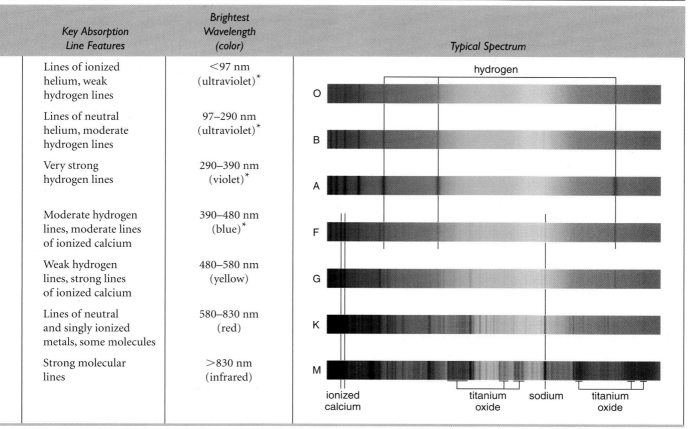

Key Absorption Line Features	Brightest Wavelength (color)	Typical Spectrum
Lines of ionized helium, weak hydrogen lines	<97 nm (ultraviolet)*	O
Lines of neutral helium, moderate hydrogen lines	97–290 nm (ultraviolet)*	B
Very strong hydrogen lines	290–390 nm (violet)*	A
Moderate hydrogen lines, moderate lines of ionized calcium	390–480 nm (blue)*	F
Weak hydrogen lines, strong lines of ionized calcium	480–580 nm (yellow)	G
Lines of neutral and singly ionized metals, some molecules	580–830 nm (red)	K
Strong molecular lines	>830 nm (infrared)	M

*All stars above 6,000 K look more or less white to the human eye because they emit plenty of radiation at all visible wavelengths.

tral type G2, which means it is slightly hotter than a G3 star but cooler than a G1 star.

The range of surface temperatures for stars is much narrower than the range of luminosities. The coolest stars, of spectral type M, have surface temperatures as low as 3,000 K. The hottest stars, of spectral type O, have surface temperatures that can exceed 40,000 K. Cool, red stars are much more common than hot, blue stars.

THINK ABOUT IT

Invent your own mnemonic for the OBAFGKM sequence. To help get you thinking, here are two examples: (1) Only Bungling Astronomers Forget Generally Known Mnemonics; and (2) Only Business Acts For Good, Karl Marx.

History of the Spectral Sequence You may wonder why the spectral types follow the peculiar order of OBAFGKM. The answer lies in the history of stellar spectroscopy.

Astronomical research never paid well, and many astronomers of the 1800s were able to do research only because of family wealth. One such astronomer was Henry Draper (1837–1882), an early pioneer of stellar spectroscopy. After Draper died in 1882, his widow made a series of large donations to Harvard College Observatory for the purpose of building upon his work. The observatory director, Edward

Pickering (1846–1919), used the gifts to improve the facilities and to hire numerous assistants, whom he called "computers." Pickering added money of his own, as did other wealthy donors.

Most of Pickering's hired computers were women who had studied physics or astronomy at women's colleges such as Wellesley and Radcliffe. Women had few opportunities to advance in science at the time. Harvard, for example, did not allow women to enroll as students and would not hire them as faculty. Pickering's project of studying and classifying stellar spectra provided plenty of work and opportunity for his computers, and many of the Harvard Observatory women ended up among the most prominent astronomers of the late 1800s and early 1900s.

One of the first computers was Williamina Fleming (1857–1911). Following Pickering's suggestion, Fleming classified stellar spectra according to the strength of their hydrogen lines: type A for the strongest hydrogen lines, type B for slightly weaker hydrogen lines, and so on, to type O, for stars with the weakest hydrogen lines. Pickering published Fleming's classifications of more than 10,000 stars in 1890.

As more stellar spectra were obtained and the spectra were studied in greater detail, it became clear that the classification scheme based solely on hydrogen lines was inadequate. Ultimately, the task of finding a better classification scheme fell to Annie Jump Cannon (1863–1941), who joined

Figure 15.5 Women astronomers pose with Edward Pickering at Harvard College Observatory in 1913. Annie Jump Cannon is fifth from the left in the back row.

Pickering's team in 1896 (Figure 15.5). Building on the work of Fleming and another of Pickering's computers, Antonia Maury (1866–1952), Cannon soon realized that the spectral classes fell into a natural order—but not the alphabetical order determined by hydrogen lines alone. Moreover, she found that some of the original classes overlapped others and could be eliminated. Cannon discovered that the natural sequence consisted of just a few of Pickering's original classes in the order OBAFGKM and also added the subdivisions by number.

Cannon became so adept that she could properly classify a stellar spectrum with little more than a momentary glance. In the course of her career, she personally classified over 400,000 stars. She became the first woman ever awarded an honorary degree by Oxford University, and in 1929 the League of Women Voters named her one of the 12 greatest living American women.

The astronomical community adopted Cannon's system of stellar classification in 1910. However, no one at that time knew *why* spectra followed the OBAFGKM sequence. Many astronomers guessed, incorrectly, that the different sets of spectral lines reflected different compositions for the

Cecilia Payne-Gaposchkin

stars. The correct answer—that all stars are made primarily of hydrogen and helium and that a star's surface temperature determines the strength of its spectral lines—was discovered by Cecilia Payne-Gaposchkin (1900–1979), another woman working at Harvard Observatory.

Relying on insights from what was then the newly developing science of quantum mechanics, Payne-Gaposchkin showed that the differences in spectral lines from star to star merely reflected changes in the ionization level of the emitting atoms. For example, O stars have weak hydrogen lines because, at their high surface temperatures, nearly all their hydrogen is ionized. Without an electron to "jump" between energy levels, ionized hydrogen can neither emit nor absorb its usual specific wavelengths of light. At the other end of the spectral sequence, M stars are cool enough for some particularly stable molecules to form, explaining their strong molecular absorption lines. Payne-Gaposchkin described her work and her conclusions in a dissertation published in 1925. A later review of twentieth-century astronomy called her work "undoubtedly the most brilliant Ph.D. thesis ever written in astronomy."

• How do we measure stellar masses?

Mass is generally more difficult to measure than surface temperature or luminosity. The most dependable method for "weighing" a star relies on Newton's version of Kepler's third law [Section 4.4]. Recall that this law can be applied only when we can observe one object orbiting another, and it requires that we measure both the orbital period and the average orbital distance of the orbiting object. For stars, these requirements generally mean that we can apply the law to measure masses only in **binary star systems**—systems in which two stars continually orbit one another. Before we consider how we determine the orbital periods and distances needed to use Newton's version of Kepler's third law, let's look briefly at the different types of binary star systems that we can observe.

Types of Binary Star Systems Surveys show that about half of all stars orbit a companion star of some kind and thus are members of binary star systems. These star systems fall into three classes:

● A **visual binary** is a pair of stars that we can see distinctly (with a telescope) as the stars orbit each other. Some-

| 1900 | 1910 | 1920 | 1930 | 1940 | 1950 | 1960 | 1970 |

Figure 15.6 Each frame represents the relative positions of Sirius A and Sirius B at 10-year intervals from 1900 to 1970. The back-and-forth "wobble" of Sirius A allowed astronomers to infer the existence of Sirius B even before the two stars could be resolved in telescopic photos. The average orbital separation of the binary system is about 20 AU.

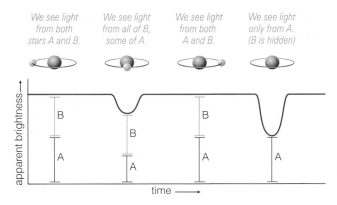

We see light from both stars A and B.

We see light from all of B, some of A.

We see light from both A and B.

We see light only from A. (B is hidden)

apparent brightness →

B

A

B

B

A

B

A

A

time →

Figure 15.7 Interactive Figure The apparent brightness of an eclipsing binary system drops when either star eclipses the other.

On one side of its orbit, star B is approaching us . . .

. . . so its spectrum is blueshifted.

B

A

to Earth →

B

On the other side of its orbit, star B is receding from us . . .

. . . so its spectrum is redshifted.

Figure 15.8 Interactive Figure The spectral lines of a star in a binary system are alternately blueshifted as it comes toward us in its orbit and redshifted as it moves away from us.

Mizar is a visual binary . . .

Alcor

Mizar B

Mizar

Mizar A

. . . and spectroscopy shows that each of the visual "stars" is itself binary.

Figure 15.9 Mizar looks like one star to the naked eye but is actually a system of four stars. Through a telescope, Mizar appears to be a visual binary made up of two stars, Mizar A and Mizar B, that gradually change positions, indicating that they orbit each other every few thousand years. Moreover, each of these two "stars" is itself a spectroscopic binary, making a total of four stars. (The star Alcor appears to be very close to Mizar to the naked eye but does *not* orbit it. The ring around Mizar A is an artifact of the photographic process, not a real feature.)

times we observe a star slowly shifting position in the sky as if it were a member of a visual binary, but its companion is too dim to be seen. For example, slow shifts in the position of Sirius, the brightest star in the sky, revealed it to be a binary star long before its companion was discovered (Figure 15.6).

● An **eclipsing binary** is a pair of stars that orbit in the plane of our line of sight (Figure 15.7). When neither star is eclipsed, we see the combined light of both stars. When one star eclipses the other, the apparent brightness of the system drops because some of the light is blocked from our view. A *light curve*, or graph of apparent brightness against time, reveals the pattern of the eclipses. The most famous example of an eclipsing binary is Algol, the "demon star" in the constellation Perseus (*algol* is Arabic for "the ghoul"). Algol's brightness drops to only a third of its usual level for a few hours about every three days as the brighter of its two stars is eclipsed by its dimmer companion.

● If a binary system is neither visual nor eclipsing, we may be able to detect its binary nature by observing Doppler shifts in its spectral lines [Section 5.5]. Such systems are called **spectroscopic binary** systems. If one star is orbiting another, it periodically moves toward us and away from us in its orbit. Its spectral lines show blueshifts and redshifts as a result of this motion (Figure 15.8). Sometimes we see two sets of lines shifting back and forth—one set from each of the two stars in the system (a *double-lined* spectroscopic binary). Other times we see a set of shifting lines from only one star because its companion is too dim to be detected (a *single-lined* spectroscopic binary).

Some star systems combine two or more of these binary types. For example, telescopic observations reveal Mizar (the second star in the handle of the Big Dipper) to be a visual binary. Spectroscopy then shows that each of the two stars in the visual binary is itself a spectroscopic binary (Figure 15.9).

Measuring Masses in Binary Systems Even for a binary system, we can apply Newton's version of Kepler's third law only if

we can measure both the orbital period and the separation of the two stars. Measuring orbital period is fairly easy. In a visual binary, we simply observe how long each orbit takes. In an eclipsing binary, we measure the time between eclipses. In a spectroscopic binary, we measure the time it takes the spectral lines to shift back and forth.

Determining the average separation of the stars in a binary system is usually much more difficult. Except in the rare cases in which we can measure the separation directly, we can calculate the separation only if we know the actual orbital speeds of the stars from their Doppler shifts. Unfortunately, a Doppler shift tells us only the portion of a star's velocity that is directly toward us or away from us (see Figure 5.23). Because orbiting stars generally do not move directly along our line of sight, their actual velocities can be significantly greater than those we measure through the Doppler effect.

The exceptions are eclipsing binary stars. Because these stars orbit in the plane of our line of sight, their Doppler shifts can tell us their true orbital velocities.* Eclipsing binaries are therefore particularly important to the study of stellar masses. As an added bonus, eclipsing binaries allow us to measure stellar radii directly. Because we know how fast the stars are moving across our line of sight as one

eclipses the other, we can determine their radii by timing each eclipse.

Through careful observations of eclipsing binaries and other binary star systems, astronomers have established the masses of many different kinds of stars. The overall range extends from as little as 0.08 times the mass of the Sun ($0.08 M_{Sun}$) to about 150 times the mass of the Sun ($150 M_{Sun}$). We'll discuss the reasons for that mass range in Chapter 16.

---

*In other binaries, we can calculate an actual orbital velocity from the velocity obtained by the Doppler effect if we also know the system's orbital inclination. Astronomers have developed techniques for determining orbital inclination in a relatively small number of cases.

MATHEMATICAL INSIGHT 15.4 Measuring Stellar Masses

As described in the text, we rely on Newton's version of Kepler's third law (see Mathematical Insight 4.3) to measure the masses of stars in binary systems. It's generally easy to measure the orbital period p for binary star systems. We can rarely measure the semimajor axis a directly, but we can calculate it in cases in which we can measure the orbital velocity of one star relative to the other. If we assume that the first star traces a circle of radius a around its companion, the circumference of its orbit is $2\pi a$. Because the star makes one circuit of this circumference in one orbital period p, its velocity relative to its companion is:

$$v = \frac{\text{distance traveled in one orbit}}{\text{period of one orbit}} = \frac{2\pi a}{p}$$

Solving for a, we find:

$$a = \frac{pv}{2\pi}$$

Once we know both p and a, Newton's version of Kepler's third law allows us to calculate the *sum* of the masses of the two stars ($M_1 + M_2$). We can then calculate the individual masses of the two stars by taking advantage of the fact that the relative velocities of the two stars around their common center of mass are inversely proportional to their relative masses.

Example: The spectral lines of two stars in an eclipsing binary system shift back and forth with a period of 2 years ($p = 6.3 \times 10^7$ seconds). The lines of one star (Star 1) shift twice as far as the lines of the other (Star 2). The amount of Doppler shift indicates an orbital speed of $v = 100,000$ m/s for Star 1 relative to Star 2. What are the masses of the two stars? Assume that each of the two stars traces a circular orbit around their center of mass.

Solution:

Step 1 Understand: Newton's version of Kepler's third law relates the masses in an orbiting system to the period p and semimajor axis a of the system. This law was given in Mathematical Insight 4.3:

$$p^2 = \frac{4\pi^2}{G(M_1 + M_2)} a^3$$

Here, we will need to solve this law for the masses. You should confirm that a little algebra allows us to rearrange the equation into the following form:

$$(M_1 + M_2) = \frac{4\pi^2}{G} \times \frac{a^3}{p^2}$$

To use this law, we need to know the star system's orbital period p and semimajor axis a. We are given p, but not a. However, as described above, we can find a from the orbital velocity v for Star 1. Thus, we have all the information we need to determine the sum of the masses, and we can determine the relative masses of the stars in the system by comparing their Doppler shifts. The overall solution therefore breaks down into three steps: (1) Determine the semimajor axis a. (2) Plug the semimajor axis a and the orbital period p into Newton's version of Kepler's third law to determine the sum of the masses $M_1 + M_2$. (3) Determine how much of that total mass belongs to each star, based on the fact that the mass of each star is inversely proportional to the amount of its Doppler shift. Because the lines of Star 1 shift twice as far as those of Star 2, we know that Star 1 moves twice as fast as Star 2, and hence that Star 1 is half as massive as Star 2.

Step 2 Solve: First, we find the semimajor axis a of the system from the given orbital velocity v:

$$a = \frac{pv}{2\pi} = \frac{(6.3 \times 10^7 \text{ s}) \times (100{,}000 \text{ m/s})}{2\pi}$$
$$= 1.0 \times 10^{12} \text{ m}$$

Second, we use this value, the value of the gravitational constant G (see Appendix A), and the given orbital period $p = 6.3 \times 10^7$ s to find the sum of the masses with Newton's version of Kepler's third law.

$$(M_1 + M_2) = \frac{4\pi^2}{\left(6.67 \times 10^{-11} \dfrac{\text{m}^3}{\text{kg} \times \text{s}^2}\right)} \times \frac{(1.0 \times 10^{12} \text{ m})^3}{(6.3 \times 10^7 \text{ s})^2}$$
$$= 1.5 \times 10^{32} \text{ kg}$$

Third, we know from the Doppler shifts of the stars that Star 2 is twice as massive as Star 1. Using this fact and their combined mass of 1.5×10^{32} kg, we conclude that the mass of Star 2 is 1.0×10^{32} kg and the mass of Star 1 is 0.5×10^{32} kg.

Step 3 Explain: Newton's version of Kepler's third law, combined with the relative Doppler shifts of the stars in the binary system, tell us that the mass of Star 1 is 0.5×10^{32} kg and the mass of Star 2 is 1.0×10^{32} kg. Dividing these masses by 2×10^{30} kg/M_{Sun} converts them to units of solar masses, showing that Star 1 has a mass of $50 M_{Sun}$ and Star 2 has a mass of $25 M_{Sun}$.

15.2 Patterns Among Stars

We have seen that stars vary widely in their luminosities, surface temperatures, and masses. But are these characteristics randomly distributed among stars, or can we find patterns that might tell us something about stellar lives? The key that finally unlocked the secrets of stars was the development of an appropriate classification system.

Before the twentieth century, people classified stars primarily by their brightness and location in our sky. The names of the brightest stars within each constellation still bear Greek letters designating their order of brightness. For example, the brightest star in the constellation Centaurus is Alpha Centauri, the second brightest is Beta Centauri, the third brightest is Gamma Centauri, and so on. However, a star's brightness and membership in a constellation tell us little about its true nature. We have already seen that a star that appears bright can be either extremely luminous or unusually nearby, and two stars that appear right next to each other in our sky might not be true neighbors if they lie at significantly different distances from Earth (see Figure 2.3).

The great breakthrough in stellar classification came in the first decade of the twentieth century through the work of Danish astronomer Ejnar Hertzsprung and American astronomer Henry Norris Russell. Building upon the work of Annie Jump Cannon and others, Hertzsprung and Russell independently decided to make graphs of stellar properties by plotting stellar luminosities on one axis and spectral types on the other. These graphs revealed previously unsuspected patterns among the properties of stars, and ultimately helped unlock the secrets of stellar life cycles.

• What is a Hertzsprung-Russell diagram?

Graphs showing how luminosity depends on spectral type are now called **Hertzsprung-Russell (H-R) diagrams**. These diagrams quickly became one of the most important tools in astronomical research, and they remain central to the study of stars.

Basics of the H-R Diagram   Figure 15.10 displays an example of an H-R diagram. All you need to know to plot a star on an H-R diagram is its luminosity and its spectral type.

- The horizontal axis represents stellar surface temperature, which, as we've discussed, corresponds to spectral type. Temperature *decreases* from left to right because Hertzsprung and Russell based their diagrams on the spectral sequence OBAFGKM.
- The vertical axis represents stellar luminosity, in units of the Sun's luminosity (L_{Sun}). Stellar luminosities span a wide range, so we keep the graph compact by making each tick mark represent a luminosity 10 times as large as the prior tick mark.

Each location on the diagram represents a unique combination of spectral type and luminosity. For example, the dot representing the Sun in Figure 15.10 corresponds to the Sun's spectral type, G2, and its luminosity, $1 L_{Sun}$. Because luminosity increases upward on the diagram and surface temperature increases leftward, stars near the upper left are hot and luminous. Similarly, stars near the upper right are cool and luminous, stars near the lower right are cool and dim, and stars near the lower left are hot and dim.

> **THINK ABOUT IT**
>
> Explain how the colors of the stars in Figure 15.10 help indicate stellar surface temperature. Do these colors tell us anything about interior temperatures? Why or why not?

The H-R diagram also provides direct information about stellar radii, because a star's luminosity depends on both its surface temperature and its surface area or radius (see Mathematical Insight 15.5). If two stars have the same surface temperature, one can be more luminous than the other only if it is larger in size. Stellar radii therefore must increase as we go from the high-temperature, low-luminosity corner on the lower left of the H-R diagram to the low-temperature, high-luminosity corner on the upper right. Notice the diagonal lines that represent different stellar radii in Figure 15.10.

Patterns in the H-R Diagram   Stars do not fall randomly throughout an H-R diagram like Figure 15.10 but instead cluster into four major groups:

- Most stars fall somewhere along the **main sequence**, the prominent streak running from the upper left to the lower right on the H-R diagram. Notice that our Sun is one of these *main-sequence stars*.
- The stars in the upper right are called **supergiants** because they are very large in addition to being very bright.
- Just below the supergiants are the **giants**, which are somewhat smaller in radius and lower in luminosity (but still much larger and brighter than main-sequence stars of the same spectral type).
- The stars near the lower left are small in radius and appear white in color because of their high temperatures. We call these stars **white dwarfs**.

Luminosity Classes   In addition to the four major groups we've just listed, stars sometimes fall into "in-between" categories. For more precise work, astronomers therefore assign each star to a **luminosity class**, designated with a Roman numeral from I through V. The luminosity class describes the region of the H-R diagram in which the star falls; thus, despite the name, a star's luminosity class is more closely related to its size than to its luminosity. The basic luminosity classes are I for supergiants, III for giants, and V for main-sequence stars. Luminosity classes II and IV are intermediate to the others; for example, luminosity class IV represents stars with radii larger than those of main-sequence

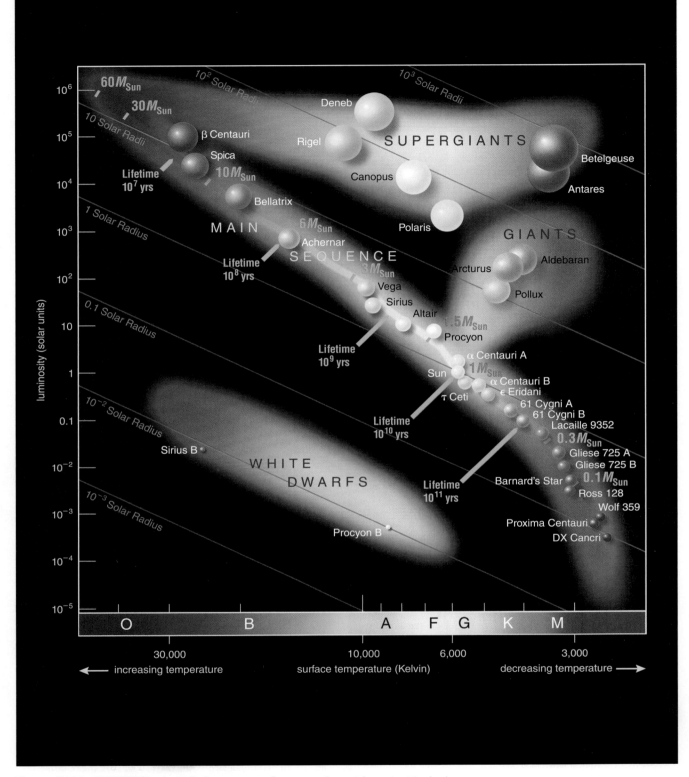

Figure 15.10 Interactive Figure An H-R diagram, one of astronomy's most important tools, shows a star's surface temperature or spectral type along the horizontal axis and its luminosity along the vertical axis. Several of the brightest stars in the sky are plotted here, along with a few of those closest to Earth. They are not drawn to scale—the diagonal lines, labeled in solar radii, indicate how large they are compared to the Sun. The lifetime and mass labels apply only to main-sequence stars. (Star positions on this diagram are based on data from the *Hipparcos* satellite.)

stars but not quite large enough to qualify them as giants. Table 15.2 summarizes the luminosity classes. White dwarfs fall outside this classification system and instead are often assigned the luminosity class "wd."

Complete Stellar Classification We have now described two different ways of categorizing stars:

- A star's *spectral type*, designated by one of the letters OBAFGKM, tells us its surface temperature and color. O stars are the hottest and bluest, while M stars are the coolest and reddest.
- A star's *luminosity class*, designated by a Roman numeral, is based on its luminosity but also tells us about the star's radius. Luminosity class I stars have the largest radii, with radii decreasing to luminosity class V.

We use both spectral type and luminosity class to fully classify a star. For example, the complete classification of our Sun is G2 V. The G2 spectral type means it is yellow-white in color, and the luminosity class V means it is a hydrogen-burning, main-sequence star. Betelgeuse is M2 I, making it a red supergiant. Proxima Centauri is M5 V— similar in color and surface temperature to Betelgeuse, but far dimmer because of its much smaller size.

Table 15.2 Stellar Luminosity Classes

Class	Description
I	Supergiants
II	Bright giants
III	Giants
IV	Subgiants
V	Main-sequence stars

THINK ABOUT IT

By studying Figure 15.10, determine the approximate spectral type, luminosity class, and radius of the following stars: Bellatrix, Vega, Antares, Pollux, and Proxima Centauri.

• What is the significance of the main sequence?

Most of the stars we observe, including our Sun, have properties that place them on the main sequence of the H-R diagram. Main-sequence stars fall along a particular line on the H-R diagram because they all share a particular type of

---

MATHEMATICAL INSIGHT 15.5 Calculating Stellar Radii

Almost all stars are too distant for us to measure their radii directly. However, we can calculate a star's radius from its luminosity with the aid of the thermal radiation laws. As given in Mathematical Insight 5.2, the amount of thermal radiation emitted by a star of surface temperature T is

$$\text{emitted power (per square meter of surface)} = \sigma T^4$$

where the constant $\sigma = 5.7 \times 10^{28} \text{ watt/(m}^2 \times \text{Kelvin}^4)$.

The luminosity L of a star is its power per unit area multiplied by its total surface area. If the star has radius r, its surface area is given by the formula $4\pi r^2$. Thus:

$$L = 4\pi r^2 \times \sigma T^4$$

With a bit of algebra, we can solve this formula for the star's radius r:

$$r = \sqrt{\frac{L}{4\pi\sigma T^4}}$$

Example: The red supergiant star Betelgeuse, which is the star marking the upper left shoulder of the constellation Orion, has a luminosity of $38{,}000 L_{\text{Sun}}$ and a surface temperature of about 3,400 K. What is its radius?

Solution:

Step 1 Understand: We are given both the luminosity L and surface temperature T, so we have the basic information we need to find the radius of Betelgeuse with the formula given above. However, we must make the units of the luminosity consistent

with the units of the constant σ by converting the luminosity of Betelgeuse into watts.

Step 2 Solve: First, we convert the given luminosity into watts. Remembering that $L_{\text{Sun}} = 3.8 \times 10^{26}$ watts, we find:

$$\begin{aligned} L_{\text{Bet}} &= 38{,}000 \times L_{\text{Sun}} \\ &= 38{,}000 \times 3.8 \times 10^{26} \text{ watts} \\ &= 1.4 \times 10^{31} \text{ watts} \end{aligned}$$

Now we can use the formula derived above to calculate the radius of Betelgeuse:

$$\begin{aligned} r &= \sqrt{\frac{L}{4\pi\sigma T^4}} \\ &= \sqrt{\frac{1.4 \times 10^{31} \text{ watts}}{4\pi \times \left(5.7 \times 10^{-8} \frac{\text{watt}}{\text{m}^2 \times \text{K}^4}\right) \times (3{,}400 \text{ K})^4}} \\ &= \sqrt{\frac{1.4 \times 10^{31} \text{ watts}}{9.6 \times 10^7 \frac{\text{watts}}{\text{m}^2}}} = 3.8 \times 10^{11} \text{ m} \end{aligned}$$

Step 3 Explain: The radius of Betelgeuse is about 380 billion meters or, equivalently, 380 million kilometers. We can make this number more meaningful by comparing it to the Earth-Sun distance of about 150 million kilometers; notice that the radius of Betelgeuse is more than twice the Earth-Sun distance. Thus, the orbits of all the inner planets of our solar system could fit easily inside Betelgeuse.

relationship between luminosity and surface temperature. The significance of this relationship became clear when astronomers started measuring the masses of main-sequence stars in binary star systems. These measurements showed that a star's position along the main sequence is very closely related to its mass: All stars along the main sequence are fusing hydrogen into helium in their cores, just like the Sun, but they differ in surface temperature and luminosity because the rate of hydrogen fusion depends strongly on a star's mass.

Masses Along the Main Sequence   If you look along the main sequence in Figure 15.10, you'll notice purple labels indicating stellar masses and green labels indicating stellar lifetimes. To make them easier to see, Figure 15.11 repeats the same data but shows only the main sequence rather than the entire H-R diagram.

Let's focus first on mass: Notice that *stellar masses decrease downward along the main sequence.* At the upper end of the main sequence, the hot, luminous O stars can have masses as high as 100 times that of the Sun ($100M_{Sun}$). On the lower end, cool, dim M stars may have as little as 0.08 times the mass of the Sun ($0.08M_{Sun}$). Many more stars fall on the lower end of the main sequence than on the upper end, which tells us that low-mass stars are much more common than high-mass stars.

The orderly arrangement of stellar masses along the main sequence tells us that *mass* is the most important attribute of a hydrogen-burning star. As we've discussed, luminosity depends directly on mass because the weight of a star's outer layers determines the nuclear fusion rate in its core. More weight means the star must sustain a higher nuclear

fusion rate in order to maintain gravitational equilibrium. The nuclear fusion rate, and hence the luminosity, is very sensitive to mass. For example, a $10M_{Sun}$ star on the main sequence is about 10,000 times as luminous as the Sun.

The relationship between mass and surface temperature is a little subtler. In general, a very luminous star must either be very large or have a very high surface temperature, or some combination of both. The most massive main-sequence stars are many thousands of times more luminous than the Sun but only about 10 times the size of the Sun in radius. Thus, their surfaces must be significantly hotter than the Sun's surface to account for their high luminosities. Main-sequence stars more massive than the Sun therefore have higher surface temperatures than the Sun, and those less massive than the Sun have lower surface temperatures. That is why the main sequence slices diagonally from the upper left to the lower right on the H-R diagram.

The fact that mass, surface temperature, and luminosity are all related means that we can estimate a main-sequence star's mass just by knowing its spectral type. For example, any hydrogen-burning, main-sequence star that has the same spectral type as the Sun (G2) must have about the same mass and luminosity as the Sun. Similarly, any main-sequence star of spectral type B1 must have about the same mass and luminosity as Spica (see Figure 15.10). Note that only main-sequence stars follow this simple relationship between mass, temperature, and luminosity; it does not hold for giants or supergiants.

Lifetimes Along the Main Sequence   A star is born with a limited supply of core hydrogen and therefore can remain as a hydrogen-fusing, main-sequence star for only a limited time—the star's **main-sequence lifetime**. Because stars spend the vast majority of their lives as main-sequence stars, we sometimes refer to the main-sequence lifetime as simply the "lifetime." Like masses, stellar lifetimes vary in an orderly way as we move up the main sequence: Massive stars near the upper end of the main sequence have *shorter* lives than less massive stars near the lower end (see Figure 15.11).

Why do more massive stars have shorter lives? A star's lifetime depends on both its mass and its luminosity. Its mass determines how much hydrogen fuel the star initially contains in its core. Its luminosity determines how rapidly the star uses up its fuel. Massive stars start their lives with a larger supply of hydrogen, but they fuse this hydrogen into helium so rapidly that they end up with shorter lives. For example, a 10-solar-mass star ($10M_{Sun}$) is born with 10 times as much hydrogen as the Sun. However, its luminosity of $10,000L_{Sun}$ means that it burns through this hydrogen at a rate 10,000 times as fast as the rate in the Sun. Thus, because a 10-solar-mass star has only 10 times as much hydrogen and burns through it 10,000 times faster, its lifetime is only $\frac{1}{1,000}$ as long as the Sun's lifetime. Since the Sun's core hydrogen-burning lifetime is about 10 billion years [Section 14.1], a 10-solar-mass star must have a lifetime of only about 10 million years. Its actual lifetime is

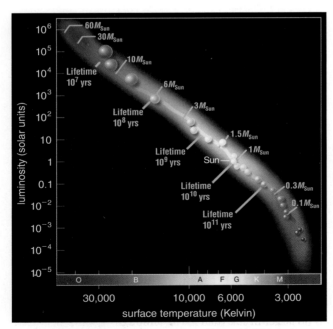

Figure 15.11 The main sequence from Figure 15.10 is isolated here so that you can more easily see how masses and lifetimes vary along it. Notice that more massive hydrogen-burning stars are brighter and hotter but have shorter lifetimes. (Stellar masses are given in units of solar masses: $1M_{Sun} = 2 \times 10^{30}$ kg.)

a little longer than this, because it can use more of its core hydrogen for fusion than can the Sun.

Cosmically speaking, the several-million-year lifetimes of very massive stars are a remarkably short time. That is one reason why massive stars are so rare: Most of the massive stars that have ever been born are long since dead. A second reason is that higher-mass stars are born in smaller numbers to begin with [Section 16.3]. Indeed, the fact that massive stars exist at all at the present time tells us that stars must form continuously in our galaxy. The massive, bright O stars in our galaxy today formed only recently and will die long before they have a chance to complete even one orbit around the center of the galaxy.

On the other end of the scale, a 0.3-solar-mass main-sequence star emits a luminosity just 0.01 times that of the Sun and consequently lives roughly $\frac{0.3}{0.01} = 30$ times as long as the Sun, or some 300 billion years. In a universe that is now about 14 billion years old, even the most ancient of these small, dim, red stars of spectral type M still survive and will continue to shine faintly for hundreds of billions of years to come.

Mass: A Star's Most Fundamental Property Astronomers began classifying stars by their spectral type and luminosity class before they understood why stars vary in these properties. Today, we know that the most fundamental property of any star is its *mass*. As we have discussed, a star's mass determines both its surface temperature and luminosity throughout the main-sequence portion of its life, and these properties in turn explain why higher-mass stars have shorter lifetimes. Figure 15.12 compares four main-sequence stars, showing how they differ because of their different masses.

THINK ABOUT IT

Which star shown in Figure 15.10 has the longest lifetime? Explain.

• What are giants, supergiants, and white dwarfs?

Main-sequence stars are fusing hydrogen into helium in their cores, but what about the other classes of stars on the H-R diagram? These other classes all represent stars that have exhausted the supply of hydrogen in their central cores, so that they can no longer generate energy in the same way as our Sun.

Giants and Supergiants Even a quick glance at an H-R diagram shows that giants and supergiants do not follow the relationship between surface temperature and luminosity that we observe for hydrogen-burning, main-sequence stars. As we'll discuss in Chapter 17, we now know that giants and supergiants are stars nearing the ends of their lives. They have already exhausted the supply of hydrogen fuel in their central cores, and are in essence facing an energy crisis as they try to stave off the inevitable crushing force of gravity. This crisis causes these stars to release fu-

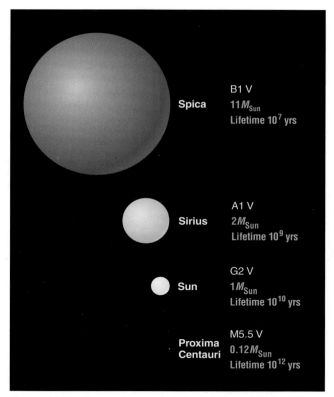

Figure 15.12 Four main-sequence stars shown to scale. The mass of a main-sequence star determines its fundamental properties of luminosity, surface temperature, radius, and lifetime. More massive main-sequence stars are hotter and brighter than less massive ones, but have shorter lifetimes.

sion energy at a furious rate, while the need to radiate away this huge amount of energy causes them to expand to enormous size. That is why giants and supergiants are so large and luminous.

Because giants and supergiants are so bright, we can see them even if they are not especially close to us. Many of the brightest stars in our sky are giants or supergiants, often identifiable by their reddish colors. Overall, however, giants and supergiants are considerably rarer than main-sequence stars. In our snapshot of the heavens, we catch most stars in the act of hydrogen burning and relatively few in a later stage of life.

White Dwarfs Giants and supergiants eventually run out of fuel entirely. A giant with a mass similar to that of our Sun ultimately ejects its outer layers, leaving behind a "dead" core in which all nuclear fusion has ceased. White dwarfs are these remaining embers of former giants. They are hot because they are essentially exposed stellar cores, but they are dim because they lack an energy source and radiate only their leftover heat into space. A typical white dwarf is no larger in size than Earth, although it may have a mass as great as that of our Sun. Clearly, white dwarfs must be made of matter compressed to an extremely high density, unlike anything found on Earth. We'll discuss the nature of white dwarfs and other stellar corpses in Chapter 18.

Figure 15.13 A typical light curve for a pulsating variable star. This particular star is a Cepheid variable star with a pulsation period of about 50 days.

• Why do the properties of some stars vary?

Not all stars shine steadily like our Sun. Any star that varies significantly in brightness with time is called a *variable star*. Certain types of variable stars have the peculiar problem of not being able to achieve the proper balance between the power welling up from the core and the power being radiated from the surface. Sometimes the upper layers of such a star are too opaque, so energy and pressure build up beneath the photosphere and the star expands in size. However, this expansion puffs the outer layers outward, making them too transparent. So much energy then escapes that the underlying pressure drops, and the star contracts again.

In a futile quest for a steady equilibrium, the atmosphere of such a **pulsating variable star** alternately expands and contracts, causing the star to rise and fall in luminosity. Figure 15.13 shows a typical light curve for a pulsating variable

star, with the star's brightness graphed against time. Any pulsating variable star has its own particular period between peaks and valleys in luminosity, which we can discover easily from its light curve. These periods can range from as short as several hours to as long as several years.

Most pulsating variable stars inhabit a strip (called the *instability strip*) on the H-R diagram that lies between the main sequence and the red giants (Figure 15.14). A special category of very luminous pulsating variable stars lies in the upper portion of this strip. They are known as *Cepheid variable stars*, and they are significant both because they are so bright and because their pulsation periods turn out to be closely related to their luminosities. As a result, Cepheids have played a key role in helping us establish the distances to many galaxies beyond the Milky Way, thereby revealing the overall scale of the cosmos. We will discuss the cosmic distance scale further in Chapter 20.

Stellar Evolution Tutorial, Lessons 1, 4

15.3 Star Clusters

All stars are born from giant clouds of gas. Because a single interstellar cloud can contain enough material to form many stars, stars almost inevitably form in groups [Section 16.1]. In our snapshot of the heavens, many stars still congregate in the groups in which they formed.

These groups are known as *star clusters*, and they are extremely useful to astronomers for two key reasons:

1. All the stars in a cluster lie at about the same distance from Earth.
2. All the stars in a cluster formed at about the same time (within a few million years of one another).

Astronomers can therefore use star clusters as laboratories for comparing the properties of stars that all have similar ages, and we shall see that these features of star clusters enable us to use them as cosmic clocks.

• What are the two types of star clusters?

Star clusters come in two basic types: modest-size **open clusters** and densely packed **globular clusters**. The two types differ not only in how densely they are packed with stars, but also in their locations and ages. Recall that most of the stars, gas, and dust in the Milky Way Galaxy, including our Sun, lie in the relatively flat *galactic disk*; the region above and below the disk is called the *halo* of the galaxy (see Figure 1.14). Open clusters of stars are always found in the disk of the galaxy, while most globular clusters are found in the halo. Moreover, as we'll discuss shortly, open clusters tend to be young in age while the globular clusters belonging to our galaxy are always quite old.

Open clusters can contain up to several thousand stars and typically are about 30 light-years across. The most famous open cluster is the *Pleiades*, a prominent clump of stars in the constellation Taurus (Figure 15.15). The Pleiades

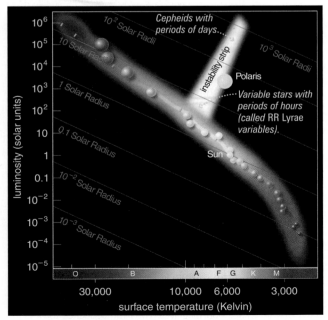

Figure 15.14 An H-R diagram with the instability strip highlighted. Notice that Polaris, the North Star, is a Cepheid variable star.

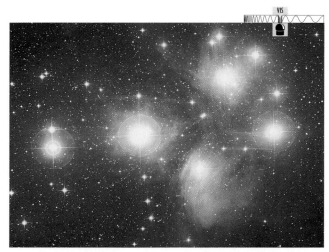

Figure 15.15 A photo of the Pleiades, a nearby open cluster of stars. The most prominent stars in this open cluster are of spectral type B, indicating that the Pleiades are no more than 100 million years old, relatively young for a star cluster. The region shown is about 11 light-years across.

are often called the *Seven Sisters*, although only six of the cluster's several thousand stars are easily visible to the naked eye. Other cultures have other names for this beautiful group of stars. In Japan it is called *Subaru*, which is why the logo for Subaru automobiles is a diagram of the Pleiades.

In contrast to the few thousand stars in open clusters, globular clusters can contain more than a million stars. These clusters are called "globular" because their stars are concentrated in the shape of a ball. A globular cluster is typically between about 60 and 150 light-years in diameter. Its central region can have 10,000 stars packed into a space just a few light-years across (Figure 15.16). The view from a planet in a globular cluster would be marvelous, with thousands of stars lying closer than Alpha Centauri is to the Sun.

Because a globular cluster's stars nestle so closely together, they engage in an intricate and complex dance choreographed by gravity. Some stars zoom from the cluster's

core to its outskirts and back again at speeds approaching escape velocity from the cluster, while others orbit the dense core more closely. When two stars pass especially close to each other, the gravitational pull between them deflects their trajectories, altering their speeds and sending them careening off in new directions. Occasionally, a close encounter boosts one star's velocity enough to eject it from the cluster. Through such ejections, globular clusters gradually lose stars and grow more compact.

• How do we measure the age of a star cluster?

We can use clusters as clocks, because we can determine their ages by plotting their stars in an H-R diagram. To understand how the process works, look at Figure 15.17, which shows an H-R diagram for the Pleiades. Most of the stars in the Pleiades fall along the standard main sequence, with one important exception: The Pleiades' stars trail away to the right of the main sequence at the upper end. That is, the hot, short-lived stars of spectral type O are missing from the main sequence. Apparently, the Pleiades are old enough for its main-sequence O stars to have already ended their hydrogen-burning lives. At the same time, the cluster is young enough for some of its stars of spectral type B to still survive as hydrogen-burning stars on the main sequence.

The precise point on the H-R diagram at which the Pleiades' main sequence diverges from the standard main sequence is called its **main-sequence turnoff** point. In this cluster, it occurs around spectral type B6. The main-sequence lifetime of a B6 star is roughly 100 million years,

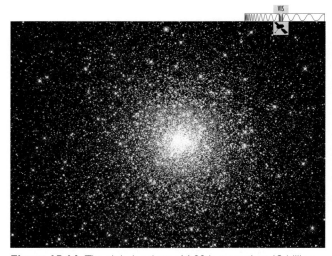

Figure 15.16 The globular cluster M 80 is more than 12 billion years old. The prominent reddish stars in this Hubble Space Telescope photo are red giant stars nearing the ends of their lives. The central region pictured here is about 15 light-years across.

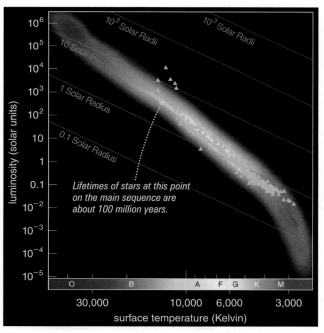

Figure 15.17 An H-R diagram for the stars of the Pleiades. Triangles represent individual stars. The Pleiades cluster is missing its upper main-sequence stars, indicating that these stars have already ended their hydrogen-burning lives. The main-sequence turnoff point at about spectral type B6 tells us that the Pleiades are about 100 million years old.

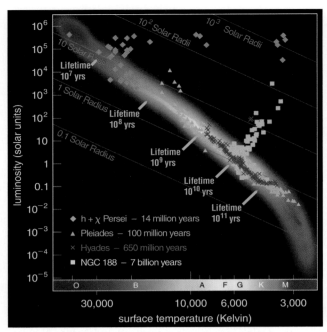

Figure 15.18 This H-R diagram shows stars from four clusters. Their differing main-sequence turnoff points indicate very different ages.

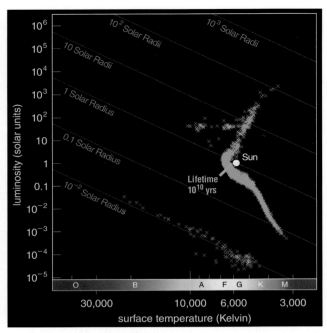

Figure 15.19 This H-R diagram shows stars from the globular cluster M4. The main-sequence turnoff point is in the vicinity of stars like our Sun, indicating an age for this cluster of around 10 billion years. A more technical analysis of this cluster places its age at around 13 billion years.

so this must be the age of the Pleiades. Any star in the Pleiades that was born with a main-sequence spectral type hotter than B6 had a lifetime shorter than 100 million years and hence is no longer found on the main sequence. Stars with lifetimes longer than 100 million years are still fusing hydrogen in their cores and hence remain as main-sequence stars. Over the next few billion years, the B stars in the Pleiades will die out, followed by the A stars and the F stars. Thus, if we could make an H-R diagram for the Pleiades every few million years, we would find that the main sequence gradually grows shorter.

Comparing the H-R diagrams of other open clusters makes this effect more apparent (Figure 15.18). In each case, we determine the cluster's age from the lifetimes of the stars at its main-sequence turnoff point. More precisely, we can determine the age of any star cluster if we can make an H-R diagram showing the spectral types and luminosities of its stars: *The age of the cluster is equal to the lifetimes of stars at its main-sequence turnoff point.* Stars in a particular cluster that once resided above the turnoff point on the main sequence have already exhausted their core supply of hydrogen, while stars below the turnoff point remain on the main sequence.

The technique of identifying main-sequence turnoff points is our most powerful tool for evaluating the ages of star clusters. We've learned that most open clusters are relatively young, with very few older than about 5 billion years. In contrast, the stars at the main-sequence turnoff points in globular clusters are usually less massive than our Sun (Figure 15.19). Because stars like our Sun have a lifetime of about 10 billion years and these stars have already died in globular clusters, we conclude that globular-cluster stars are older than 10 billion years.

More precise studies of the turnoff points in globular clusters, coupled with theoretical calculations of stellar lifetimes, place the ages of these clusters at about 13 billion years, making them the oldest known objects in the galaxy. In fact, globular clusters place a lower limit on the possible age of the universe: If stars in globular clusters are 13 billion years old, then the universe must be at least that old. Recent observations suggesting that the universe is about 14 billion years old therefore fit well with the ages of these stars and tell us that the first stars began to form by the time the universe was a billion years old.

Suppose a star cluster is precisely 10 billion years old. Where would you expect to find its main-sequence turnoff point? Would you expect this cluster to have any main-sequence stars of spectral type A? Would you expect it to have main-sequence stars of spectral type K? Explain. (*Hint:* What is the lifetime of our Sun?)

THE BIG PICTURE

Putting Chapter 15 into Context

We have classified the diverse families of stars visible in the night sky. Much of what we know about stars, galaxies, and the universe itself is based on the fundamental properties of stars introduced in this chapter. Make sure you understand the following "big picture" ideas:

- All stars are made primarily of hydrogen and helium, at least at the time they form. The differences between stars come about primarily because of differences in mass and stage of life.

- Stars spend most of their lives as main-sequence stars that fuse hydrogen into helium in their cores. The most massive stars, which are also the hottest and most luminous, live only a few million years. The least massive stars, which are coolest and dimmest, will survive until the universe is many times its present age.

- The key to recognizing these patterns among stars was the H-R diagram, which shows stellar surface temperatures on the horizontal axis and luminosities on the vertical axis. The H-R diagram is one of the most important tools of modern astronomy, and we'll make much more use of it in the rest of the book.

- Much of what we know about the universe comes from studies of star clusters. For example, we can measure a star cluster's age by plotting its stars in an H-R diagram and determining the hydrogen-burning lifetime of the brightest and most massive stars still on the main sequence.

SUMMARY OF KEY CONCEPTS

15.1 Properties of Stars

- **How do we measure stellar luminosities?**
The **apparent brightness** of a star in our sky depends on both its **luminosity**—the total amount of light it emits into space—and its distance from Earth, as expressed by the **inverse square law for light**. Thus, we can calculate a star's luminosity from its apparent brightness and distance; we can measure the latter through stellar parallax.

- **How do we measure stellar temperatures?**
We measure a star's surface temperature from its color or spectrum, and we classify spectra according to the sequence of **spectral types** OBAFGKM, which runs from hottest to coolest. Cool, red stars of spectral type M are much more common than hot, blue stars of spectral type O.

- **How do we measure stellar masses?** We can measure the masses of stars in **binary star systems** using Newton's version of Kepler's third law if we can measure the orbital period and separation of the two stars.

15.2 Patterns Among Stars

- **What is a Hertzsprung-Russell diagram?**
An **H-R diagram** plots stars according to their surface temperatures (or spectral types) and luminosities. Stars spend most of their lives fusing hydrogen into helium in their cores, and stars in this stage of life are found in the H-R diagram in a narrow band known as the **main sequence**. **Giants** and **supergiants** are to the upper right of the main sequence and **white dwarfs** are to the lower left.

- **What is the significance of the main sequence?** Stars on the main sequence are all fusing hydrogen into helium in their cores, and a star's position along the main sequence depends on its mass. High-mass stars are at the upper left end of the main sequence, and the masses of stars become progressively smaller as we move toward the lower right end. Lifetimes vary in the opposite way, because higher-mass stars live shorter lives.

- **What are giants, supergiants, and white dwarfs?** Giants and supergiants are stars that have exhausted their central core supplies of hydrogen for fusion and are undergoing other forms of fusion at a prodigious rate as they near the ends of their lives. White dwarfs are the exposed cores of stars that have already died, meaning they have no further means of generating energy through fusion.

- **Why do the properties of some stars vary?**
Some stars fail to achieve a proper balance between the amount of fusion energy welling up from their cores and the amount of radiative energy emanating from their surfaces. The surfaces of these stars therefore pulsate in and out, periodically rising and falling in luminosity.

15.3 Star Clusters

- **What are the two types of star clusters?**
Open clusters contain up to several thousand stars and   are found in the disk of the galaxy. **Globular clusters** contain hundreds of thousands of stars, all closely packed together. They are found mainly in the halo of the galaxy.

- **How do we measure the age of a star cluster?**
Because all of a cluster's stars were born at the same time, we can measure a cluster's age by finding the **main-sequence turnoff** point on an H-R diagram of its stars. The cluster's age is equal to the hydrogen-burning lifetime of the hottest, most luminous stars that remain on the main sequence. We've learned that open clusters are much younger than globular clusters, which can be as old as about 13 billion years.

Review Questions

Short-Answer Questions Based on the Reading

1. How is a star's *apparent brightness* related to its *luminosity*? Explain by describing the *inverse square law for light*.

2. Briefly explain how we use *stellar parallax* to determine a star's distance. Once we know a star's distance, how can we determine its luminosity?

3. What do we mean by a star's apparent and absolute *magnitudes*? How are they related to apparent brightness and luminosity?

4. What do we mean by a star's spectral type? How is a star's spectral type related to its surface temperature and color? Which stars are hottest and coolest in the spectral sequence OBAFGKM?

5. How was the spectral sequence discovered? Describe the roles of a few of the most prominent scientists in the early history of stellar astronomy.

6. What are the three basic types of *binary star systems?* Why are *eclipsing binaries* so important to measuring masses of stars?

7. Draw a sketch of a basic Hertzsprung-Russell (H-R) diagram. Label the main sequence, giants, supergiants, and white dwarfs. Where on this diagram do we find stars that are cool and dim? Cool and luminous? Hot and dim? Hot and luminous?

8. What do we mean by a star's *luminosity class?* What does the luminosity class tell us about the star? Briefly explain how we classify stars by spectral type and luminosity class.

9. What is the defining characteristic of a *main-sequence* star? Briefly explain why massive main-sequence stars are more luminous and have hotter surfaces than less massive main-sequence stars.

10. Which stars have longer lifetimes: massive stars or less massive stars? Explain why.

11. Why is a star's birth mass its most fundamental property?

12. How do *giants* and *supergiants* differ from main-sequence stars? What are *white dwarfs?*

13. How does the luminosity of a pulsating variable star change with time?

14. Describe in general terms how open clusters and globular clusters differ in their numbers of stars, ages, and locations in the galaxy.

15. Explain why H-R diagrams look different for star clusters of different ages. How does the location of the main-sequence turnoff point tell us the age of the star cluster?

Test Your Understanding

Does It Make Sense?

Decide whether the statement makes sense (or is clearly true) or does not make sense (or is clearly false). Explain your reasoning. (For an example, see Chapter 1, "Does It Make Sense?")

16. Two stars that look very different must be made of different kinds of elements.

17. Two stars that have the same apparent brightness in the sky must also have the same luminosity.

18. Sirius looks brighter than Alpha Centauri, but we know that Alpha Centauri is closer because its apparent position in the sky shifts by a larger amount as Earth orbits the Sun.

19. Stars that look red-hot have hotter surfaces than stars that look blue.

20. Some of the stars on the main sequence of the H-R diagram are not converting hydrogen into helium.

21. The smallest, hottest stars are plotted in the lower left-hand portion of the H-R diagram.

22. Stars that begin their lives with the most mass live longer than less massive stars because they have so much more hydrogen fuel.

23. Star clusters with lots of bright, blue stars of spectral type O and B are generally younger than clusters that don't have any such stars.

24. All giants, supergiants, and white dwarfs were once main-sequence stars.

25. Most of the stars in the sky are more massive than the Sun.

Quick Quiz

Choose the best answer to each of the following. Explain your reasoning with one or more complete sentences.

26. If the star Alpha Centauri were moved to a distance 10 times farther than it is now, its parallax angle would (a) get larger. (b) get smaller. (c) stay the same.

27. What do we need to measure in order to determine a star's luminosity? (a) apparent brightness and mass (b) apparent brightness and temperature (c) apparent brightness and distance

28. What two pieces of information would you need in order to measure the masses of stars in an eclipsing binary system? (a) the time between eclipses and the average distance between the stars (b) the period of the binary system and its distance from the Sun (c) the velocities of the stars and the Doppler shifts of their absorption lines

29. Which of these stars has the coolest surface temperature? (a) an A star (b) an F star (c) a K star

30. Which of these stars is the most massive? (a) a main-sequence A star (b) a main-sequence G star (c) a main-sequence M star

31. Which of these stars has the longest lifetime? (a) a main-sequence A star (b) a main-sequence G star (c) a main-sequence M star

32. Which of these stars has the largest radius? (a) a supergiant A star (b) a giant K star (c) a supergiant M star

33. Which of these stars has the greatest surface temperature? (a) a $30M_{Sun}$ main-sequence star (b) a supergiant A star (c) a Cepheid variable star

34. Which of these star clusters is youngest? (a) a cluster whose brightest main-sequence stars are white (b) a cluster whose brightest stars are red (c) a cluster containing stars of all colors

35. Which of these star clusters is oldest? (a) a cluster whose brightest main-sequence stars are white (b) a cluster whose brightest main-sequence stars are yellow (c) a cluster containing stars of all colors

Investigate Further
In-Depth Questions to Increase Your Understanding

Short-Answer/Essay Questions

36. *Stellar Data.* Consider the following data table for several bright stars. M_v is absolute magnitude, and m_v is apparent magnitude. (*Hint:* Remember that the magnitude scale runs backward, so that brighter stars have smaller (or more negative) magnitudes.)

Star	M_v	m_v	Spectral Type	Luminosity Class
Aldebaran	−0.2	+0.9	K5	III
Alpha Centauri A	+4.4	0.0	G2	V
Antares	−4.5	+0.9	M1	I
Canopus	−3.1	−0.7	F0	II
Fomalhaut	+2.0	+1.2	A3	V
Regulus	−0.6	+1.4	B7	V
Sirius	+1.4	−1.4	A1	V
Spica	−3.6	+0.9	B1	V

Answer each of the following questions, and include a brief explanation with each answer.
  a. Which star appears brightest in our sky?
  b. Which star appears faintest in our sky?
  c. Which star has the greatest luminosity?
  d. Which star has the least luminosity?
  e. Which star has the highest surface temperature?
  f. Which star has the lowest surface temperature?
  g. Which star is most similar to the Sun?
  h. Which star is a red supergiant?
  i. Which star has the largest radius?
  j. Which stars have finished burning hydrogen in their cores?
  k. Among the main-sequence stars listed, which one is the most massive?
  l. Among the main-sequence stars listed, which one has the longest lifetime?

37. *Data Tables.* Study the spectral types listed in Appendix F for the 20 brightest stars and for the stars within 12 light-years. Why do you think the two lists are so different? Explain.

38. *Interpreting the H-R Diagram.* Using the information in Figure 15.10, describe how Proxima Centauri differs from Sirius.

39. *Parallax from Jupiter.* Suppose you could travel to Jupiter and observe changes in positions of nearby stars during one orbit of Jupiter around the Sun. Describe how those changes would be different from what we measure from Earth. How would your ability to measure the distances to stars be different from the vantage point of Jupiter?

40. *An Expanding Star.* Describe what would happen to the surface temperature of a star if its radius doubled in size with no change in luminosity.

41. *Colors of Eclipsing Binaries.* Figure 15.7 shows an eclipsing binary system consisting of a small blue star and a larger red star. Explain why the decrease in apparent brightness of the combined system is greater when the blue star is eclipsed than when the red star is eclipsed.

42. *Visual and Spectroscopic Binaries.* Suppose you are observing two binary star systems at the same distance from Earth. Both are spectroscopic binaries consisting of similar types of stars, but only one of these binary systems is a visual binary. Which of these star systems would you expect to have the greater Doppler shifts in its spectra? Explain your reasoning.

43. *Life of a Star Cluster.* Imagine you could watch a star cluster from the time of its birth to an age of 13 billion years. Describe in one or two paragraphs what you would see happening during that time.

Quantitative Problems
Be sure to show all calculations clearly and state your final answers in complete sentences.

44. *The Inverse Square Law for Light.* Earth is about 150 million kilometers from the Sun, and the apparent brightness of the Sun in our sky is about 1,300 watts/m^2. Using these two facts and the inverse square law for light, determine the apparent brightness that we would measure for the Sun *if* we were located at the following positions.
  a. Half Earth's distance from the Sun.
  b. Twice Earth's distance from the Sun.
  c. Five times Earth's distance from the Sun.

45. *The Luminosity of Alpha Centauri A.* Alpha Centauri A lies at a distance of 4.4 light-years and has an apparent brightness in our night sky of 2.7×10^{-8} watt/m^2. Recall that 1 light-year = 9.5×10^{12} km = 9.5×10^{15} m.
  a. Use the inverse square law for light to calculate the luminosity of Alpha Centauri A.
  b. Suppose you have a light bulb that emits 100 watts of visible light. (*Note:* This is *not* the case for a standard 100-watt light bulb, in which most of the 100 watts goes to heat and only about 10–15 watts is emitted as visible light.) How far away would you have to put the light bulb for it to have the same apparent brightness as Alpha Centauri A in our sky? (*Hint:* Use 100 watts as L in the inverse square law for light, and use the apparent brightness given above for Alpha Centauri A. Then solve for the distance.)

46. *More Practice with the Inverse Square Law for Light.* Use the inverse square law for light to answer each of the following questions.
  a. Suppose a star has the same luminosity as our Sun (3.8×10^{26} watts) but is located at a distance of 10 light-years. What is its apparent brightness?
  b. Suppose a star has the same apparent brightness as Alpha Centauri A (2.7×10^{-8} watt/m^2) but is located at a distance of 200 light-years. What is its luminosity?
  c. Suppose a star has a luminosity of 8×10^{26} watts and an apparent brightness of 3.5×10^{-12} watt/m^2. How far away is it? Give your answer in both kilometers and light-years.
  d. Suppose a star has a luminosity of 5×10^{29} watts and an apparent brightness of 9×10^{-15} watt/m^2. How far away is it? Give your answer in both kilometers and light-years.

47. *Parallax and Distance.* Use the parallax formula to calculate the distance to each of the following stars. Give your answers in both parsecs and light-years.
  a. Alpha Centauri: parallax angle of 0.7420.
  b. Procyon: parallax angle of 0.2860.

48. *The Magnitude System.* Use the definitions of the magnitude system to answer each of the following questions.
   a. Which is brighter in our sky, a star with apparent magnitude 2 or a star with apparent magnitude 7? By how much?
   b. Which has a greater luminosity, a star with absolute magnitude −4 or a star with absolute magnitude +6? By how much?

49. *Measuring Stellar Mass.* The spectral lines of two stars in a particular eclipsing binary system shift back and forth with a period of 6 months. The lines of both stars shift by equal amounts, and the amount of the Doppler shift indicates that each star has an orbital speed of 80,000 m/s. What are the masses of the two stars? Assume that each of the two stars traces a circular orbit around their center of mass. (*Hint:* See Mathematical Insight 15.4.)

50. *Calculating Stellar Radii.* Sirius A has a luminosity of $26L_{Sun}$ and a surface temperature of about 9,400 K. What is its radius? (*Hint:* See Mathematical Insight 15.5.)

51. *Lifetime as a Red Giant.* The H-R diagram in Figure 15.19 shows a star cluster with a large number of red giants in it.
   a. What is the approximate mass of the most massive stars left on the main sequence of this star cluster?
   b. What is the luminosity of the most luminous stars in the cluster?
   c. Compute the ratio between the luminosity from part (b) to the mass from part (a). How does that ratio compare with the Sun's ratio of luminosity to mass?
   d. Estimate the maximum amount of time these very luminous stars can last as red giants from your answer to part (c).

Discussion Question

52. *Classification.* About a century ago Annie Jump Cannon and other astronomers at Harvard University made many important contributions to astronomy, particularly in the area of systematic stellar classification. Why do you think rapid advances in our understanding of stars followed so quickly on the heels of their efforts? Can you think of other areas in science where huge advances in understanding followed directly from improved systems of classification?

MEDIA EXPLORATIONS

For self-study activities go to www.masteringastronomy.com

Interactive Tutorials

Tutorial Review of Key Concepts
Use the interactive **Tutorials** at **www.mastering astronomy.com** to review key concepts from this chapter.

Measuring Cosmic Distances Tutorial
Lesson 2 Stellar Parallax

Hertzsprung-Russell Diagram Tutorial
Lesson 1 The Hertzsprung-Russell (H-R) Diagram
Lesson 2 Determining Stellar Radii
Lesson 3 The Main Sequence

Stellar Evolution Tutorial
Lesson 1 Main-Sequence Lifetimes
Lesson 4 Cluster Dating

Supplementary Tutorial Exercises
Use the interactive **Tutorial Lessons** to explore the following questions.

Measuring Cosmic Distances Tutorial, Lesson 2
1. Explain how we measure distances with stellar parallax. Give an example.
2. Explain why we cannot use parallax to measure the distance to *all* stars.

Hertzsprung-Russell Diagram Tutorial, Lessons 1–3
1. If one star appears brighter than another, can you be sure that it is more luminous? Why or why not?
2. Answer each part of this question with either high, low, left, or right. On an H-R diagram, where will a star be if it is hot? Cool? Bright? Dim?
3. Why is there a relationship between stellar radii and locations on the H-R diagram?
4. Why is there a relationship between luminosity and mass for main-sequence stars on the H-R diagram?

Stellar Evolution Tutorial, Lesson 4

1. In the animation at the beginning of Lesson 4, list the order in which you saw the three differently colored stars in the cluster disappear, and explain why they disappeared in this order.
2. In the second animation in Lesson 4, in what order did you see stars on the main sequence disappear? Explain the reason for this.
3. How does the age of a dim star cluster of mostly small stars compare to a bright cluster with some giants in it? Explain your answer.

 Exploring the Sky and Solar System

Of the many activities available on the *Voyager: SkyGazer* **CD-ROM** accompanying your book, use the following files to observe key phenomena covered in this chapter.

Go to the **File: Basics** folder for the following demonstrations.

1. Large Stars
2. More Stars
3. Star Color and Size

Go to the **File: Demo** folder for the following demonstrations.

1. Circling the Hyades
2. Flying Around Pleiades
3. The Tail of Scorpius

Web Projects

1. *Women in Astronomy.* Until fairly recently, men greatly outnumbered women in professional astronomy. Nevertheless, many women made crucial discoveries in astronomy throughout history—including discovering the spectral sequence for stars. Do some research about the life and discoveries of a female astronomer from any time period and write a two- to three-page scientific biography.
2. *The Hipparcos Mission.* The European Space Agency's *Hipparcos* mission, which operated from 1989 to 1993, made precise parallax measurements for more than 40,000 stars. Learn about how *Hipparcos* allowed astronomers to measure smaller parallax angles than they could from the ground and how *Hipparcos* discoveries have affected our knowledge of the universe. Write a one- to two-page report on your findings.
3. *The GAIA Mission.* The European Space Agency's *GAIA* mission, slated for launch in 2010, should make even better parallax measurements. Research the *GAIA* mission and what scientists hope to learn from it. Write a one- to two-page report on your findings.

16

Star Birth

I can hear the sizzle of newborn stars, and know anything of meaning, of the fierce magic emerging here. I am witness to flexible eternity, the evolving past, and know I will live forever, as dust or breath in the face of stars, in the shifting pattern of winds.

—Joy Harjo, *Secrets from the Center of the World*

The stars in the night sky shine so steadily that it's easy to think of them as fixtures of the heavens. However, their seeming permanence is an illusion, one that occurs only because a human lifetime is so short compared with the lifetime of a star. The stars we observe today are in many different stages of life. Some are just starting to fuse hydrogen into helium in their cores, while others are running out of fuel and have expanded to become giants or supergiants. Our goal in this chapter and the next is to understand how the lives of stars unfold over millions or billions of years.

In this chapter, we will focus on star birth. We will see how and why a cloud of interstellar gas can begin to collapse to make a new star, we will examine the processes that can make star birth surprisingly violent before the onset of nuclear fusion, and we will discuss why stars can have masses only within a particular range. Ultimately, we will see that the collapse of a gas cloud to make a star marks the beginning of a "battle" between pressure and gravity that will last the rest of the star's life: Gravity inexorably tries to squeeze a star's gas into an ever more compact form, while gas pressure resists the crush of gravity. So let's begin at the beginning, in the murky depths of interstellar space.

16.1 Stellar Nurseries

Today, we know a great deal about the life cycles of stars. We know that all stars are born in clouds of interstellar gas, shine with energy produced by nuclear fusion in their cores, and die when they finally exhaust all sources of fuel for fusion. However, we cannot actually observe individual stars going through all these life stages, because they unfold over millions or billions of years. Instead, as we discussed in Chapter 15, astronomers have pieced together the life stories of stars by studying and comparing the many stars we can observe in the sky.

Our understanding of star birth comes from studying young stars and stellar birthplaces. Recall that we can determine the ages of star clusters by looking for their main-sequence turnoff points on H-R diagrams [Section 15.3]. In the vicinity of the youngest star clusters we always find dark clouds of gas and dust, indicating that such clouds are the birthplaces of stars. Theoretical models of star formation confirm this idea, telling us how the physical processes that occur in gas clouds ultimately give birth to stars. However, stars do not form within just any interstellar cloud. Instead, they form only in clouds that are particularly cold and dense. In this section, we'll explore the types of clouds that can give birth to stars, and how gravity overcomes gas pressure to make stars from the material in these clouds.

• Where do stars form?

Star formation is a common process. Counting the number of young stars nearby tells us that new stars form in our galaxy at a rate of about two or three per year, if our part of the galaxy is typical. We are seeing the birthing grounds of those stars with our own eyes whenever we look into the Milky Way on a dark night [Section 2.1]. The black patches within the Milky Way are interstellar gas clouds that appear dark because they block our view of the stars and any glowing gas behind them. These clouds are the birthplaces of stars, although the dusty gas they contain makes it difficult to see the newborn stars inside the clouds, even with a telescope (Figure 16.1). Nevertheless, we have learned a great

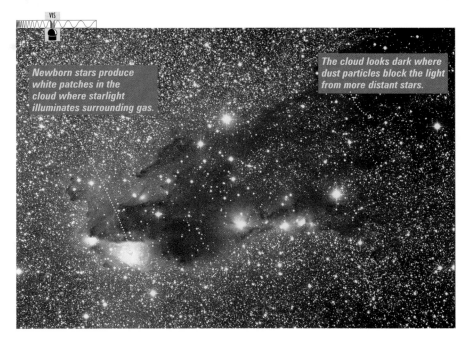

VIS

Newborn stars produce white patches in the cloud where starlight illuminates surrounding gas.

The cloud looks dark where dust particles block the light from more distant stars.

Figure 16.1 A star-forming cloud in the constellation Scorpius. The region pictured here is about 50 light-years across.

deal about these dark clouds and the way in which they give birth to stars.

The Interstellar Medium We often think of space as being "empty," and it is indeed a superb vacuum by earthly standards. However, space is not completely empty. We find at least some gas and dust everywhere we look. We refer to the gas and dust that fill the spaces between stars within a galaxy as the **interstellar medium**.

The gas between the stars is composed mostly of hydrogen and helium, a fact that we can trace back to the origin of the universe. Recall that, as we discussed briefly in Chapter 1, hydrogen and helium were the only chemical elements produced in the Big Bang. The very first stars must have been born from clouds made only of hydrogen and helium gas. Since that time, stars have transformed a small fraction of the original hydrogen and helium into heavier elements, as we will discuss in detail in Chapter 17.

We use spectroscopy to measure the abundances of the new elements that stars have added to the interstellar medium. The most straightforward technique is to observe the spectrum of a star lying behind a cloud of gas and measure the absorption lines the cloud leaves in the star's spectrum [Section 5.4]. The wavelengths of the lines tell us which chemical elements are present in the cloud, and comparing the amounts of light absorbed by different elements allows us to determine the relative amount of each element in the cloud (Figure 16.2). Measurements like this show that the interstellar medium in our region of the Milky Way consists of 70% hydrogen, 28% helium, and 2% heavier elements, by mass.

Virtually all the gas between stars in the Milky Way Galaxy has this same chemical composition, but the interstellar medium differs from place to place in its temperature and density. In some places, interstellar material is extremely hot and low in density. In other places, the interstellar medium contains clouds that are cold and relatively dense. Most of the interstellar medium has temperatures and densities in between these extremes. We'll discuss the reasons

for these temperature and density differences in Chapter 19, where we'll look more closely at the way in which our galaxy functions as a cosmic recycling plant. For the moment, we'll focus our attention only on the clouds that give birth to stars.

Star-Forming Clouds Stars are born in the coldest and highest-density types of interstellar clouds. These clouds are usually called **molecular clouds** because they are cold enough and dense enough to allow atoms to combine together into molecules. The temperature of a molecular cloud is typically only 10–30 K. (Recall that 0 K is absolute zero, and temperatures on Earth are around 300 K; see Figure 4.12.) The density is typically about 300 molecules occupying each cubic centimeter—quite high by interstellar standards, though almost a million trillion times less dense than the air at sea level on Earth. However, that is just an average density. Molecular clouds are in fact very lumpy, with high-density regions that are hundreds of times denser than average.

Molecular hydrogen (H_2) is by far the most abundant molecule in these clouds, because hydrogen and helium are the most abundant elements and helium atoms do not combine together into molecules. However, the relatively rare atoms of elements heavier than helium combine to make small amounts of a wide variety of other molecules. We generally identify these molecules through the emission lines that they produce in the radio portion of the spectrum. More than 120 molecules have been identified in molecular clouds by their radio emission lines. Carbon monoxide (CO) is the most abundant of these molecules, and others include water (H_2O), ammonia (NH_3), and ethyl alcohol (C_2H_5OH).

Molecular hydrogen itself is difficult to detect because the temperatures in molecular clouds are usually too cold for the gas to produce H_2 emission lines. Only in regions of molecular clouds that are much hotter than normal can we see strong emission from molecular hydrogen. Most of what we know about molecular clouds therefore comes from observing spectral lines of molecules that make up only a tiny fraction of a cloud's mass, especially carbon monoxide (Figure 16.3).

Interstellar Dust Not all of the material in a molecular cloud is gaseous. About half the atoms of elements heavier than helium are found in tiny, solid grains of **interstellar dust**. Keeping in mind that atoms heavier than helium make up about 2% of the mass of the interstellar medium, we conclude that interstellar dust grains constitute about 1% of a molecular cloud's mass. Individual dust grains are microscopic in size, usually less than 1 micrometer across, which is smaller than a single cell of bacteria. This small size makes them more like the solid microscopic particles found in smoke than like the grains of sand you would find at the beach (Figure 16.4). Some are made mostly of carbon, while others are combinations of elements like silicon, oxygen, and iron.

Figure 16.2 This spectrum of a star has absorption lines that allow us to measure the proportions of different elements in an interstellar cloud lying between Earth and the star. (See Figure 5.14 to review how spectral lines form.)

Figure 16.3 This image shows radio emission from carbon monoxide (CO) molecules between the stars in our galaxy. The bright regions are giant molecular clouds containing large numbers of CO molecules. The angular size of this patch of sky is about 20 degrees by 10 degrees, roughly the same angular size as the Big Dipper.

Despite the tiny size of these grains, there are so many of them that they profoundly affect how light travels through a molecular cloud. We can see the two most important effects of dust on light in Figure 16.5a. First, dust grains scatter or absorb virtually all the visible light that enters a molecular cloud, preventing us from seeing stars that lie behind it. That is why regions of the sky with molecular clouds look so black. Second, stars seen near the edges of a

Figure 16.4 This photograph shows a microscopic dust grain from interplanetary space that is suspected to predate the formation of the solar system. Interstellar dust grains are thought to be much like this one. It is slightly less than 1 micrometer (1 μm) across, putting it on the large side for an interstellar dust grain.

molecular cloud appear redder than similar stars outside the cloud, a phenomenon known as **interstellar reddening**.

This phenomenon occurs for much the same reason that our Sun appears redder when viewed through smoke or smog: Dust grains block shorter-wavelength (bluer) photons of visible light more easily than longer-wavelength (redder) photons. Because more red light than blue light gets through the edges of the cloud, where the stars are only partially obscured, these stars appear redder than their true colors.

We can distinguish interstellar reddening from a Doppler shift because reddening doesn't change the wavelengths of a star's spectral lines. As a result, we can determine the amount of reddening by comparing a star's observed color to the color expected for a star of its spectral type [Section 15.1]. The amount of reddening tells us how much dusty gas lies between Earth and a distant star, and many such measurements have allowed us to map out the distribution of interstellar dust in our region of the galaxy.

The fact that longer-wavelength light passes more easily through dusty gas is even more helpful when we try to observe stars directly behind a molecular cloud and not just at its edges. Most molecular clouds are thick enough to block all forms of red light from stars directly behind them, but longer wavelength infrared light can often pass right through (Figure 16.5b). These clouds appear more and more transparent as we observe light of progressively longer wavelengths. Infrared observations therefore allow us to see directly through molecular clouds to stars lying at much greater distances. More importantly, infrared observations

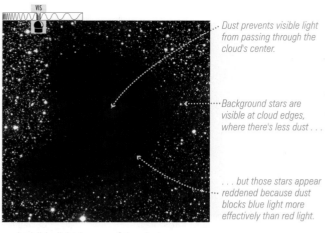

Dust prevents visible light from passing through the cloud's center.

Background stars are visible at cloud edges, where there's less dust . . .

. . . but those stars appear reddened because dust blocks blue light more effectively than red light.

Dust is less effective at blocking infrared light, so we can see stars lying behind the cloud in this infrared photo.

a A visible-light image of the dark molecular cloud Barnard 68.

b An infrared image of Barnard 68 showing the stars that lie behind the cloud.

Figure 16.5 We can see the effects of interstellar dust on starlight by comparing visible and infrared photographs of a molecular cloud. The cloud shown here is about half a light-year in size.

reveal new stars embedded within the clouds themselves, caught in the very act of birth (Figure 16.6).

Much of the radiation produced by new stars within a molecular cloud does not escape the cloud directly, because dust grains within the cloud absorb the visible light and even some of the infrared light from these stars. The radiative energy of the absorbed light does not disappear but instead goes into heating the dust grains. Additional radiative energy coming from newly formed stars just outside the cloud can heat the grains even further. The dust grains therefore emit thermal radiation in the infrared and microwave bands of the electromagnetic spectrum. Consequently, clouds that appear dark in visible-light photos often glow when observed in the longest-wavelength infrared light (Figure 16.7).

• Why do stars form?

Stars form in molecular clouds for a simple reason: Gravity causes the cloud to contract, and the contraction contin-

ues until the central object becomes hot enough to sustain nuclear fusion in its core, at which point it has become a star. However, the role of gravity is only one part of the story.

For a more complete understanding of star formation, we must also pay attention to how gas pressure resists gravity. Recall that our Sun remains stable in size because the outward push of gas pressure balances the inward pull of gravity, a balance that we call *gravitational equilibrium* [Section 14.1]. The gas pressure is far weaker in the low-density gas of interstellar space, but so is the pull of gravity. In most places within our galaxy, gravity is not strong enough to overcome the internal pressure of interstellar gas, which is why star formation does not occur everywhere. Molecular clouds are the exception—they are the only places in which gravity can win the battle against pressure and start the formation of stars.

Gravity Versus Pressure Gravity can create stars only if it can overcome the outward push of the pressure within a gas

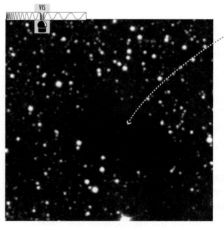

This visible light photo shows a cloud that seems to have no stars at all.

An infrared photo of this same cloud shows a newborn star in the cloud along with several stars lying behind the cloud.

a

b

Figure 16.6 A newborn star buried inside a dark molecular cloud known as L1014. The colors correspond to different wavelengths of infrared light. The region in the picture is about 1 light-year across.

a This infrared image from the Spitzer Space Telescope shows a star-forming cloud in the constellation Cepheus. The pink color represents infrared light from heated dust grains in the cloud's cold, dense gas. The inset photo shows the same region in visible light. The region in the picture is about 15 light-years across.

b This picture shows infrared radiation from star-forming regions in the constellation Orion. The colors correspond to the temperature of the dust grains producing the infrared light. Star formation is most intense in the hot, bright regions, including the Orion Nebula. The region pictured here is about 80 light-years across at the distance of the Orion Nebula.

Figure 16.7 Interactive Photo ↖ Star-forming molecular clouds are colder than stars and therefore emit infrared rather than visible light.

cloud, which depends on both the density and temperature of the cloud. To see why, consider gas particles inside a balloon. The outward pressure the gas exerts on the balloon is just the collective force that the colliding gas particles apply to each unit of area on the balloon's surface [Section 5.3]. The pressure goes up if you increase the density of the gas in the balloon (by blowing more air into it), because the higher density means more particles crashing into each little area of the balloon's surface. The pressure also goes up if you heat the balloon, because higher temperature means the gas particles move faster and therefore hit the surface both more frequently and with greater force. (These ideas are quantified in the *ideal gas law*; see Mathematical Insight 14.2.)

This type of pressure, which depends on both density and temperature, affects not only balloons and interstellar clouds but virtually all the gases we've encountered so far in this book, including the atmospheres of planets and the plasma throughout the interior of the Sun. However, it is not the only type of pressure that can occur in gas; as we discussed in Chapter S4 and will review later in this chapter, the laws of quantum mechanics introduce a very different type of pressure, called *degeneracy pressure*, that does not depend on temperature. To make it easier to distinguish the two types of pressure, we will hereafter refer to the temperature-dependent pressure in ordinary gas clouds as **thermal pressure**.

Thermal pressure can resist gravity in most interstellar gas clouds, because their gas densities are comparatively low, keeping gravity quite weak. Only in molecular clouds, where gas densities are tens to thousands of times greater than average, is gravity strong enough to overcome thermal pressure (see Mathematical Insight 16.1). Despite the high density of particles, thermal pressure in a molecular cloud isn't much different than in other parts of the interstellar medium because of the low temperature. However, with more particles and therefore more mass packed into each cubic centimeter of volume, the force of gravity in a low-temperature molecular cloud is stronger than elsewhere in interstellar space, giving gravity the upper hand.

As we will discuss further in Chapter 19, observations suggest that gravity can form stars more easily if some other force begins to compress a molecular cloud. For example, a collision between two molecular clouds can compress both clouds, driving up the gas density within each of them. The increase in density strengthens the force of gravity within the clouds, tipping the balance between gravity and pressure in favor of gravity, and thereby triggering star formation. Similarly, debris from an exploding star can potentially trigger star formation when it collides with and compresses a molecular cloud.

Preventing a Pressure Buildup Regions of a molecular cloud in which the gravitational attraction is stronger than the thermal pressure are forced to contract. We have already seen the consequences of *gravitational contraction* in our study of the Sun [Section 14.1]. As gravity makes a gas cloud shrink, it converts some of the cloud's gravitational potential energy into thermal energy. If the cloud cannot get rid of that thermal energy as quickly as it is released, then thermal energy builds up inside the cloud. If all this thermal energy remained within the cloud, it would raise the cloud's temperature, and thus its thermal pressure, eventually bringing the process of star formation to a halt.

Molecular clouds avoid this fate because they quickly rid themselves of any thermal energy that builds up. Collisions between gas molecules in the cloud transform the thermal energy into photons by exciting the rotational and vibrational energy levels of those molecules, which then produce emission lines in the infrared and radio portions of the spectrum [Section 5.4]. As long as the photons produced by these colliding molecules can escape the cloud, the cloud's temperature can remain low. With no significant buildup of thermal energy, gravity will continue to dominate over thermal pressure, and all parts of the cloud that remain cool can contract to form stars.

Clustered Star Formation Observations show that most stars are born in clusters containing thousands of stars. Part of the reason that stars are born in clusters stems from the difficulty that gravity has in overcoming thermal pressure.

Greater mass means stronger gravity, so gravity can more easily overpower thermal pressure in a high-mass molecular cloud than in a low-mass molecular cloud. Precise calculations show that at the temperatures and densities of typical molecular clouds, gravity can begin to overcome the thermal pressure only in clouds with a minimum mass of around 100 times the mass of the Sun (see Mathematical Insight 16.1). However, star-forming clouds often hold much more mass than this minimum. A typical cloud contains thousands of times the mass of the Sun. Something besides thermal pressure must explain why these clouds grow so large before they begin to form stars.

We are still uncertain of all the processes that enable massive clouds to resist the crush of gravity but at least two factors probably play important roles. First, Doppler shifts of emission lines from gas clumps within molecular clouds show that these clumps often move at substantial speeds

---

MATHEMATICAL INSIGHT 16.1 Gravity Versus Pressure

We can determine whether a cloud is able to form stars by comparing the force due to thermal pressure inside the cloud to the compressing force of gravity: If gravity is stronger, the cloud can contract, but it cannot contract if the force due to pressure is stronger. To simplify the problem of comparing the forces, let's consider a spherical cloud of mass M and radius r; the volume of the cloud is therefore $\frac{4}{3}\pi r^3$.

The thermal pressure P inside the cloud depends on both the temperature T of the cloud and its number density n, which is equal to the number of gas particles contained within each cubic centimeter (see Mathematical Insight 14.2):

$$P = nkT$$

If each gas particle has a mass m, then the total number of particles within the cloud is M/m. Dividing this number by the spherical cloud's volume gives us its number density:

$$n \approx \frac{M}{m} \div \left(\frac{4}{3}\pi r^3\right) \approx \frac{3M}{4\pi mr^3}$$

Because thermal pressure is the amount of force that colliding gas particles apply to each unit of area inside the cloud, we need to multiply by an area to get the total amount of force due to pressure. In the case of a spherical cloud, we can imagine dividing the cloud into two halves that meet over a circular surface with an area πr^2. If both halves have the same density and temperature, then the total force that each half exerts on the other is the thermal pressure (nkT) multiplied by the area (πr^2). The first step below shows this equation, and the following steps then show what happens when we substitute the expression we found above for the number density n:

$$\begin{aligned} F\,(\text{due to pressure}) &= \text{pressure} \times \text{area} \\ &\approx nkT \times \pi r^2 \\ &\approx \frac{3M}{4\pi mr^3}\,kT \times \pi r^2 \\ &\approx \frac{3MkT}{4mr} \end{aligned}$$

This is the expression we need for the force arising from thermal pressure. Now we need a similar expression for the force due to gravity. Again, we can think of a spherical cloud cut in half, so that each half exerts a gravitational force on the other half. The mass of each half is $M/2$ and the centers of the two halves are separated by a distance approximately equal to r. We can therefore find the approximate gravitational force between them with Newton's universal law of gravity. Recall that this law says that the force due to gravity is GM_1M_2/r^2 [Section 4.4]; in this case, we set both M_1 and M_2 equal to half the cloud's mass, or $M/2$:

$$\begin{aligned} F\,(\text{due to gravity}) &\approx G\frac{\left(\frac{M}{2}\right)\left(\frac{M}{2}\right)}{r^2} \\ &\approx G\frac{M^2}{4r^2} \end{aligned}$$

These expressions for the forces of gravity and thermal pressure illustrate how the battle between these forces depends on a cloud's mass, temperature, and density. An increase in temperature raises the force due to thermal pressure without changing the force of gravity. An increase in mass raises both the total pressure and the gravitational force, but the force of gravity goes up more because it depends on mass squared. An increase in density corresponds to a decrease in radius, also causing a rise in both the pressure and the gravitational force. However, in a cloud of fixed temperature, the strength of gravity rises more than the strength of pressure because gravity depends on $1/r^2$ while the force due to pressure depends only on $1/r$.

In fact, we could find the precise balance point at which thermal pressure and gravity are equally matched by equating the two expressions we've found for the force of gravity and the force due to thermal pressure. However, we made a few approximations along the way, and our assumption of a perfectly spherical cloud with a uniform pressure is not realistic. Although we will not go through the full derivation here, the same general ideas we've used above lead to an expression for the balance point between thermal pressure and gravity in a more realistic molecular cloud. Expressed in convenient units, the resulting formula is:

and that the overall gas motion is turbulent, much like the water in the wake of a rapidly moving boat. In other words, different parts of the cloud are moving in different directions at speeds that would tear the cloud apart if not for the attractive force of gravity. Stars can form in such a cloud only if gravity is strong enough to overcome the turbulent gas motion, and overcoming that motion can require considerably more than the 100 solar masses of material needed to overcome thermal pressure alone.

Magnetic fields may provide a second source of resistance to gravity. We can tell that molecular clouds have magnetic fields by how they interact with light. Waves of starlight usually travel through space with their electric and magnetic fields vibrating in random directions. However, starlight that reaches Earth after passing through a molecular cloud often has its vibrating electric and magnetic fields aligned in particular directions, telling us that the dust grains

through which the starlight has passed must be aligned in the cloud like iron filings near a bar magnet. (This type of alignment of electric and magnetic fields in light is called *polarization*; see the Special Topic box on page 150.) Magnetic fields threading the clouds are thought to be what causes these dust grains to align, and they can also affect the motions of charged particles within the cloud.

Recall that magnetic fields bend the paths of charged particles so that these particles tend to move on spiral paths *along* magnetic field lines and have difficulty moving perpendicular to these field lines (see Figure 14.16). A magnetic field threading through a molecular cloud therefore tends to prevent charged particles inside the cloud from moving in certain directions. The overall proportion of charged particles in a molecular cloud is relatively small, but these particles exert friction on other (neutral) particles that are moving perpendicular to the field lines. The

$$M_{\text{balance}} = 18 M_{\text{Sun}} \sqrt{\frac{T^3}{n}}$$

where the gas temperature T is in Kelvin and the number density n is stated in units of particles per cubic centimeter. This formula tells us the mass of a cloud in which gravity and pressure are evenly matched. Gravity is stronger than thermal pressure in a cloud with a mass greater than the mass M_{balance}, and thermal pressure is stronger than gravity in a cloud with mass less than this mass. (The M_{balance} is often called the *Jeans mass*, because the formula is based on ideas first worked out by astronomer Sir James Jeans.)

Example 1: A typical star-forming cloud has a temperature of 30 K and an average particle density of 300 particles per cubic centimeter. What is the minimum mass that such a cloud must have in order to form stars? What does this result tell you about the nature of star formation?

Solution:

Step 1 Understand: In order to form stars, a cloud must have a great enough mass for gravity to overpower thermal pressure. Because we know the cloud's temperature and density, we can determine the mass at which gravity and pressure are balanced in a cloud with those characteristics, so we have all the information we need.

Step 2 Solve: We calculate the mass at which gravity and pressure balance from the formula for M_{balance} with $T = 30$ K and $n = 300$ particles per cubic centimeter:

$$M_{\text{balance}} = 18 M_{\text{Sun}} \sqrt{\frac{30^3}{300}} = 18 M_{\text{Sun}} \sqrt{90} \approx 171 M_{\text{Sun}}$$

Step 3 Explain: A star-forming cloud with a typical temperature and density can form stars only if its mass is greater than about $171 M_{\text{Sun}}$. This result explains why stars tend to form in clusters. The mass of a typical star is much less than $171 M_{\text{Sun}}$, so a typical star-forming cloud contains enough mass to form many stars.

In fact, most star-forming clouds are much more massive than this, and produce clusters with thousands of stars.

Example 2: If a star-forming cloud is massive enough for gravity to overpower pressure, the cloud will begin to contract, increasing the number density of particles within it. Suppose the cloud's gas is able to radiate away the thermal energy released within it as the gas grows denser, keeping its temperature at 30 K. What is the minimum mass needed for star formation when the gas density reaches 300,000 particles per cubic centimeter? What does this result tell you about the nature of star formation?

Solution:

Step 1 Understand: If a cloud's temperature remains constant while its gas density increases, the minimum mass needed for star formation decreases. In this example, the cloud's temperature is 30 K, and we are looking for the minimum mass needed for star formation at this temperature and a density of 300,000 particles per cubic centimeter. We can use the equation above for the mass at which pressure and gravity are balanced, and we expect the mass to be lower than the mass we found in Example 1.

Step 2 Solve: We use the formula for M_{balance}, this time with $T = 30$ K and $n = 300,000$ particles per cubic centimeter:

$$M_{\text{balance}} = 18 M_{\text{Sun}} \sqrt{\frac{30^3}{300,000}} = 18 M_{\text{Sun}} \sqrt{0.09} \approx 5.4 M_{\text{Sun}}$$

Step 3 Explain: When the gas density in a star-forming cloud reaches 300,000 particles per cubic centimeter, the minimum mass for star formation is only $5.4 M_{\text{Sun}}$, as long as its temperature stays at 30 K. Because molecular clouds are generally lumpy, each lump with a mass greater than about $6 M_{\text{Sun}}$ can go on to form a star. That is why a single star-forming molecular cloud can fragment into many stars. Once some smaller lumps have formed within it, gravity can overcome pressure in each of those lumps.

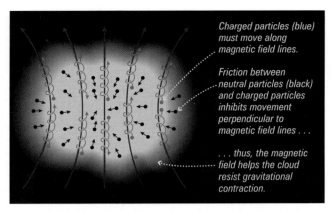

Charged particles (blue) must move along magnetic field lines.

Friction between neutral particles (black) and charged particles inhibits movement perpendicular to magnetic field lines . . .

. . . thus, the magnetic field helps the cloud resist gravitational contraction.

Figure 16.8 Magnetic field lines threading a molecular cloud can hinder its collapse by preventing particles from traveling perpendicular to the field lines.

friction is great enough that a magnetic field can effectively inhibit the movement of all the gas in a cloud (Figure 16.8). Depending on the strength of the magnetic field, this process can slow or even halt the gravitational collapse of a molecular cloud.

Fragmentation of the Molecular Cloud Gravity in a molecular cloud with a large enough mass is strong enough to surmount all these obstacles, which is why the cloud continues to collapse. But why does a large molecular cloud form many individual stars instead of a single extremely massive star? The answer goes back to the battle between gravity and pressure.

Recall that the force of gravity follows an inverse square law [Section 4.4], so the strength of gravity increases as the cloud shrinks in size. As long as the photons emitted by

molecules in the cloud can carry away the gravitational potential energy released by contraction (keeping the interior temperature and thermal pressure low), we have a situation in which gravity is strengthening more than thermal pressure. The contraction of the cloud therefore gives gravity a growing advantage in the battle against pressure, which in turn means that a smaller total amount of mass is necessary for gravity to win the battle.

Because molecular clouds are turbulent and lumpy, there are plenty of small, dense clumps within a contracting cloud that are soon able to shrink on their own. The accelerating nature of this process makes it inevitable that a large molecular cloud will split into numerous individual cloud fragments (sometimes called *molecular cloud cores*). Each of these fragments will go on to become a star system containing either a single star or two or more stars that orbit each other (Figure 16.9).

Isolated Star Formation A molecular cloud does not necessarily need to be very massive to form a star, as long as it is unusually dense and cold. Gravity can overcome pressure in clouds with a few tens of thousands of molecules per cubic centimeter and a temperature of 10 K, even if the cloud contains only a few solar masses of material (Figure 16.10). Small, isolated molecular clouds with these properties are indeed observed, and they tend to form just one or a few stars at a time. However, it is not yet clear how these small, dense clouds become compressed to such high densities in isolated regions of space.

The First Generation of Stars Spectroscopic observations of our Sun and many other nearby stars show that their outer layers have about the same composition that we find in

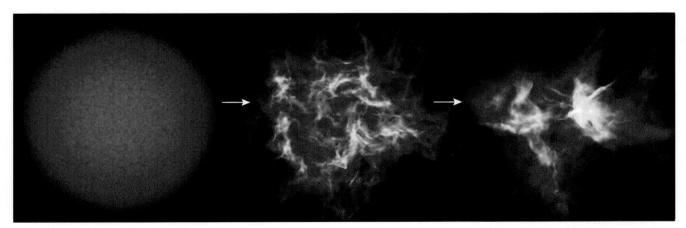

a The simulation begins with a turbulent gas cloud 1.2 light-years across and containing 50M_{Sun} of gas.

b Random motions in the cloud cause it to become lumpy, with some regions denser than others. If gravity can overcome thermal pressure in these dense regions they can collapse to form even denser lumps of matter.

c The large cloud therefore fragments into many smaller lumps of matter, and each lump can go on to form one or more new stars.

Figure 16.9 Computer simulation of a fragmenting molecular cloud. Gravity attracts matter to the densest regions of a molecular cloud. If gravity can overcome thermal pressure in these dense regions, they collapse to form even denser knots of gaseous matter. Each of these knots can form one or more new stars. Notice the similarity between the structures in this simulation and those in Figure 16.3.

interstellar clouds today: about 70% hydrogen and 28% helium, with the remaining 2% consisting of elements heavier than helium, which astronomers sometimes call *heavy elements*.* These measurements tell us that the composition of gas clouds in our galaxy, and of the stars they produce, has not changed much during the last 5 billion years or so. However, if we go far enough back in time, the composition of interstellar gas must have been different, because virtually all elements heavier than helium have been produced through the lives and deaths of stars [Section 1.1]. The abundances of these elements in the oldest known stars confirm this idea: Stars in the oldest globular clusters, with ages greater than 12 billion years, contain less than 0.1% of their mass in the form of elements heavier than helium, just as we should expect for stars that were born before most of the heavy elements had been made [Section 19.3]. Extending this idea even further back in time, the very first generation of stars must have been born before any heavy elements had been produced, and thus they must have formed in clouds of gas containing only hydrogen and helium from the Big Bang.

We will explore in Chapter 19 what these important observations tell us about the formation and evolution of our galaxy. Here we will concentrate on their implications for the earliest stars. These first-generation stars cannot have been born in quite the same way that stars are born today, because their birth clouds could not have contained molecules like carbon monoxide to radiate away thermal energy. Given the importance of emission lines from carbon monoxide to the process of molecular cloud contrac-

---
*Astronomers sometimes refer to elements heavier than helium as *metals*— a very different use of the term than that used in daily life, where *metal* refers to substances like copper, silver, and gold that shine when polished and often are good electrical conductors.

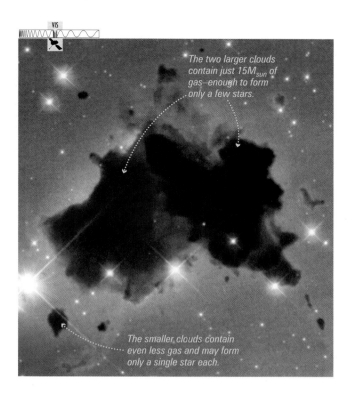

The two larger clouds contain just $15M_{Sun}$ of gas—enough to form only a few stars.

The smaller clouds contain even less gas and may form only a single star each.

Figure 16.10 These relatively small molecular clouds, silhouetted against a background of glowing red gas, each contain only enough mass to form just a few stars. The larger clouds are about 1.4 light-years across.

tion that we have discussed, how did this first generation of stars form?

We have not yet found any stars that were born without elements other than hydrogen and helium, but we can use supercomputer simulations to model how they might have formed (Figure 16.11). These models show that emission lines from hydrogen molecules in the birth clouds of these

600 lt-yr

a The orange blob is one of the first gas clouds that gravity forms in this simulation of the early universe.

60 lt-yr

b Hydrogen molecules near the center of the blob are enabling the gas there to cool to 100–300 K.

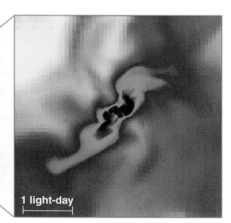

1 light-day

c At the very center of the cloud, about $200M_{Sun}$ of gas is collecting into the first protostar. The cloud cannot fragment further because there are no complex molecules to make it cooler than 100 K.

Figure 16.11 Frames from a supercomputer simulation modeling the formation of one of the first stars in the universe. Color indicates gas temperature, with blue and green representing cooler gas (100–300 K) while orange and red represent warmer gas (500–3,000 K).

stars could have helped keep them cool, but not as cool as carbon monoxide and other molecules keep today's molecular clouds. More specifically, the first generation of stars must have been born in clouds that never cooled below a temperature of about 100 K—considerably warmer than the 10–30 K temperatures of molecular clouds today—because temperatures exceeding 100 K are necessary to excite the emission lines of hydrogen molecules.

The relatively high temperatures of these first-generation molecular clouds would have made it more difficult for gravity to overcome pressure, requiring stars to form in cloud fragments of unusually large masses. The models suggest that only stars with masses of at least 30 times that of our Sun could have been born. Such massive stars have very short lifetimes [Section 15.3], which would explain why we can't find any examples of first-generation stars in today's universe: They would all have died off long ago. The large masses also help explain why the proportion of heavy elements rose quickly in the young universe. As we'll discuss in the next chapter, massive stars produce most of the heavy elements and release those elements into space when they die. The massive first-generation stars would therefore have seeded the interstellar medium with substantial quantities of heavy elements that could be incorporated into all subsequent generations of stars.

 Formation of the Solar System Tutorial, Lesson 2

16.2 Stages of Star Birth

We have seen that gravity can overcome pressure in a molecular cloud, forcing a large cloud to contract and fragment into many smaller clouds that eventually become stars. But how exactly does a contracting cloud become a full-fledged star that shines with energy released by nuclear fusion? Once again, the answer lies in the ongoing battle between the inward pull of gravity and the outward push of thermal pressure. As long as the interior of the cloud can continue to radiate away its thermal energy, the cloud remains cool and the pressure remains too weak to slow the crush of gravity. However, as the cloud becomes smaller and denser, it becomes more difficult for radiation to escape from the interior. The thermal energy becomes trapped, raising the interior temperature enough that thermal pressure begins to slow the cloud's collapse. In this section, we'll investigate what happens from the time a cloud fragment first becomes isolated from other fragments of the molecular cloud to the time its internal temperature grows so high that nuclear fusion can begin.

• What slows the contraction of a star-forming cloud?

As we discussed in the previous section, a cloud fragment will continue to collapse as long as it remains cold, and it can remain cold as long as photons emitted by molecules within the cloud can carry away the energy released by gravitational contraction. In essence, gravitational contraction converts gravitational potential energy into thermal energy within the cloud, but photons carry that thermal energy away before it can build up inside the cloud. It's easy for photons emitted by molecules to escape from the cloud while it is still fairly low in density and spread out, because the photons are unlikely to run into any other molecules after they are emitted. This situation begins to change, however, as continued contraction packs the cloud's molecules closer together.

Trapping of Thermal Energy Cloud contraction makes it increasingly difficult for the emission line photons to escape, because the growing density makes it more and more likely that a photon on its way out will run into a molecule of the same type that emitted it. That molecule can then absorb the photon, leaving the molecule in an excited state. In a dense enough cloud, an excited molecule can change the absorbed photon's energy back into thermal energy by colliding with another molecule. The more molecules that stand in the photon's way, the greater the likelihood that its energy will be deposited back into the cloud as thermal energy, raising the cloud's internal temperature and pressure. A similar effect happens with dust grains. For a while, dust grains can help keep the cloud cool by emitting infrared light, but as contraction packs the dust grains closer together, even infrared photons have trouble escaping the cloud.

The central region of the cloud fragment eventually grows dense enough to trap almost all the radiation inside it. When that happens, the inner regions of the contracting cloud can no longer radiate away their heat. The central temperature and pressure begin to rise dramatically, and the rising pressure pushes back against the crush of gravity, slowing the contraction. This change marks the first stage of star formation. The dense center of the cloud fragment is now a **protostar**—a clump of gas that will become a new star. Seen from the outside, protostars look starlike, with surface temperatures and luminosities similar to those of true stars. However, a protostar is not yet a true star because its core is not yet hot enough for nuclear fusion.

Growth of a Protostar by Gas Infall A protostar's mass grows with time because the contraction of a molecular cloud fragment proceeds in an "inside-out" fashion. Gravity is strongest at the center of the cloud fragment, where the gas density is greatest, so it is this central region that first becomes the protostar. The gas in the outer part of the cloud fragment feels a weaker gravitational pull, so it initially remains behind as the protostar forms.

In essence, the gas layers lying above the innermost regions of the cloud fragment suddenly behave as if they were standing on a trapdoor that has just opened. The gas beneath these layers has contracted into a compact protostar, leaving them with little pressure support from below. As a result, the gas in these upper layers begins to rain down onto the protostar, gradually increasing its mass.

This rain of mass onto the protostar continues until the gas surrounding the protostar is gone, either because the gas runs out or because something blows the remaining gas away. The star itself can blow the surrounding gases away through the combined effects of its radiation and stellar wind, much as the Sun's radiation and solar wind is thought to have cleared out excess gas during the formation of our solar system [Section 8.3]. Observations show that young stars tend to have strong stellar winds, lending support to the idea that gas can be blown out of the system. Alternatively, a protostar may lose some of its surrounding gas because it is blown away by the winds of other nearby young stars (Figure 16.12).

• How does a cloud's rotation affect star birth?

So far, our description of star birth may make it sound like the gravitational equivalent of squishing a piece of paper into a tight ball. However, observations of young stars show that their births can be quite violent. For example, many young stars appear to be shooting jets of gas into deep space. These phenomena are consequences of the cloud's rotation, and in particular the law of *conservation of angular momentum* [Section 4.3].

Protostellar Disks Random motions of gas particles inevitably give a gas cloud some small, overall rotation, although it may be imperceptibly slow. However, as the cloud contracts in size, the law of conservation of angular momentum ensures that the rotation will become much faster [Section 4.3]. Like an ice-skater pulling in her arms, a shrinking cloud must rotate increasingly fast in order to keep its total angular momentum the same. This rapid rotation prevents gas from raining directly down onto a protostar. Instead, as shown in Figure 16.13, it settles into a **protostellar disk**

similar to the spinning disk of gas from which the planets of the solar system formed [Section 8.2].

Observations of newly forming stars show that they do indeed have disks around them (see Figure 8.4). Sometimes these disks coalesce into planetary systems like our solar system. We do not yet know how often this occurs, but our rapidly growing list of known extrasolar planets suggests that it must be quite common [Section 13.2]. In the context of star formation, the most important feature of the disk around a protostar is that it helps to transfer angular momentum away from the infalling gas, enabling the protostar to grow more massive.

Gas in a protostellar disk can gradually spiral inward toward the protostar because of friction. Individual gas particles in the disk obey Kepler's laws [Section 3.3], just like anything else that orbits a massive body. Gas in the inner parts of the disk moves faster than gas in the outer parts. Because of these differences in orbital speed, gas in any particular part of the disk "rubs" against slower-moving gas

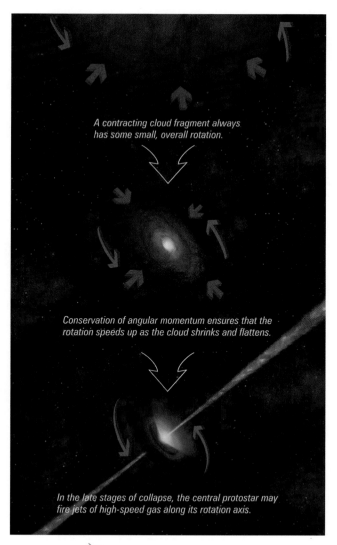

A contracting cloud fragment always has some small, overall rotation.

Conservation of angular momentum ensures that the rotation speeds up as the cloud shrinks and flattens.

In the late stages of collapse, the central protostar may fire jets of high-speed gas along its rotation axis.

Figure 16.13 Artist's conception of star birth.

Gas driven by young stars outside the image area is flowing from right to left.

This newly forming star blocks the flowing gas like a rock in a stream.

As a result, the outflowing gas forms a structure around the star similar to the wake of a boat.

Figure 16.12 The curved structure around this forming star shows that the bright young stars outside the lower right side of the frame are blowing gas past it, possibly stripping away gas that has not yet fallen onto it.

Hidden Protostar

Flow of jet

1,000 AU

a This photograph shows a jet of material being shot far into interstellar space by a protostar. The structure near the far right is formed as the jet material rams into surrounding interstellar gas.

Protostellar Disk

Flow of jet

1,000 AU

b This photograph shows a close-up view of jets (red) and a disk of gas (green) around a protostar. We are seeing the disk nearly edge-on. The top and bottom surfaces of the disk are glowing, but we cannot see the darker middle layers of the disk.

Figure 16.14 These photos show jets of gas shot from protostars into interstellar space.

just outside it. The "rubbing" generates friction and heat in the same way that rubbing your palms together makes them warm. This friction slowly causes the orbits of individual gas particles to decay, making the orbits smaller and smaller until the gas particles fall onto the surface of the protostar. Because the process in which material falls onto another body is called *accretion* [Section 8.3], a disk in which friction causes material to spiral inward is often called an *accretion disk*. Astronomers sometimes refer to protostellar disks as accretion disks, because the material within them can gradually accrete onto the protostar, thereby increasing the protostar's mass.

The protostellar disk probably also plays a large role in eventually slowing the rotation of the protostar itself. The protostar's rapid rotation generates a strong magnetic field. As the magnetic field lines sweep through the protostellar disk, they transfer some of the angular momentum to outlying material, slowing the protostar's rotation (see Figure 8.9). The strong magnetic field may also help to generate a strong **protostellar wind**—an outward flow of particles similar to the *solar wind* [Section 14.1]. The wind can carry additional angular momentum from the protostar to interstellar space, accounting for observations showing that older stars rotate more slowly than young ones.

Protostellar Jets Observations show that many young protostars fire high-speed streams of gas, or **jets**, into interstellar space (Figure 16.14). We generally see two jets shooting outward in opposite directions along the protostar's rotation axis. Sometimes the jets are lined with glowing blobs of gas, which are presumably clumps of matter swept up as the jets plow into the surrounding interstellar material.[*]

No one knows exactly how protostars generate these jets, but the fact that they are aligned with the disk's rotation axis indicates that angular momentum plays a large role. The most promising models for explaining the jets rely on magnetic fields to link the angular momentum of the protostar's disk to the outflowing gas in the jet. Magnetic field lines passing through the protostellar disk get twisted into a ropelike configuration by the disk's rotation, and this twisted field may help channel jets of charged particles along the rotation axis.

Together, winds and jets can help clear away the cocoon of gas that surrounds a forming star, revealing the protostar within. As they carry away some of the protostar's angular momentum, shooting it out into interstellar space,

---

[*]These clumps of matter are often called *Herbig-Haro objects* after the two astronomers who discovered them; their nature was unknown at the time, which is why they were given the generic name "objects."

they also pump significant amounts of kinetic energy into the surrounding molecular cloud. The disruptive effects of jets are therefore likely to cause some of the turbulent motions we observe in molecular clouds.

Single Star or Binary? Angular momentum is also part of the reason why so many stars exist in binary star systems rather than as single stars like our Sun. As a molecular cloud contracts, breaks up into fragments, and forms protostars, some of those protostars end up quite close to one another. Gravity can then pull two neighboring protostars closer together, but they don't crash into each other. Instead, they go into orbit around each other, because each pair of protostars has a certain amount of angular momentum. Pairs with large amounts of angular momentum have large orbits, and those with smaller amounts orbit closer together.

In some cases, gravitational interactions between a pair of protostars and other protostars and gas clumps in their vicinity can remove angular momentum from the binary system. If that happens, then their orbit gets smaller, and the two stars can end up quite close to each other. The resulting pair is called a **close binary** system. These star systems typically have orbital separations less than 0.1 AU and orbital periods of only a few days.

• How does nuclear fusion begin in a newborn star?

Once a protostar has accreted a significant amount of mass, its interior grows quite hot. Ultimately, the temperature at the protostar's center becomes high enough for nuclear fusion. However, a protostar with a mass similar to the Sun's must wait millions of years for fusion to begin, because the process of gravitational contraction slows once the star starts trapping its thermal energy inside.

From Protostar to the Main Sequence The central temperature of a protostar may be only a million degrees or so when its wind and jets blow away the surrounding gas. The protostar looks starlike, but its energy comes only from gravitational contraction, not from fusion.* To ignite fusion, the protostar needs to contract further to boost the central temperature.

Paradoxically, radiation of thermal energy from the surface of a protostar is what enables its central temperature to rise, because only half the thermal energy released by gravitational contraction is radiated away. The other half remains in the protostar's interior, raising its temperature. If the protostar did not lose thermal energy from its surface, it would not contract, and its central temperature would remain fixed. Early in this period of contraction, convection carries the protostar's thermal energy to the surface, as

in the convection zone of the Sun [Section 14.2]. However, as the interior of the protostar heats up, photons can flow more easily, and energy flows outward primarily by radiative diffusion, the same process that carries energy outward in the Sun's radiation zone.

A protostar becomes a true star when its core temperature exceeds 10 million K, making it hot enough for hydrogen fusion to operate efficiently by the *proton-proton chain* (see Figure 14.7). The ignition of fusion halts the protostar's gravitational contraction and marks what we consider the birth of a star. The new star's interior structure stabilizes because the energy produced in the center matches the amount radiated from its surface. The star is now a hydrogen-burning, *main-sequence star* [Section 15.2].

The length of time from the formation of a protostar to the birth of a main-sequence star depends on the star's mass. Massive stars do everything faster. The contraction of a high-mass protostar into a main-sequence star of spectral type O or B may take only a million years or less. A star like our Sun takes about 30 million years to go from the beginning of the protostellar stage to becoming a main-sequence star. A very low-mass star of spectral type M may spend more than a hundred million years as a protostar. Thus, the most massive stars in a young star cluster may live and die before the smallest stars even begin to fuse hydrogen in their cores.

The Surface of a Protostar The surface temperature of a protostar remains surprisingly constant while its core temperature gradually rises to 10 million K. That is because the surface temperature depends primarily on the ease with which energy flows through a protostar's outer layers. When a protostar first begins to contract, its surface gradually heats up as the protostar shrinks. However, when the surface temperature reaches about 3,000 K, photons suddenly have a harder time escaping from the protostar's outer layers. The reason is that the temperature has risen to the point at which hydrogen atoms begin to become ionized, releasing some of their electrons. A few of these electrons stick to other hydrogen atoms in the gas, making negatively charged ions known as H^-. Because H^- ions interact strongly with visible light, they tend to trap photons within gas that is hotter than 3,000 K. The convecting gas below a protostar's surface therefore rises until it reaches a layer where the temperature is 3,000 K. At that point, the H^- ions become rare and photons can escape to space. The visible surface of the protostar is defined by this 3,000 K layer of gas, so the surface keeps this same temperature throughout much of the protostar's contraction.

Birth Stages on a Life Track We can summarize the transitions that occur during star birth with a special type of H-R diagram. A standard H-R diagram shows luminosities and surface temperatures for many different stars [Section 15.2]. This special H-R diagram shows part of a **life track** (also called an *evolutionary track*) for a single star in relation to

---

*Protostars that have reached this stage are often called *pre-main-sequence stars*, but they are not yet true stars because their energy isn't coming from fusion.

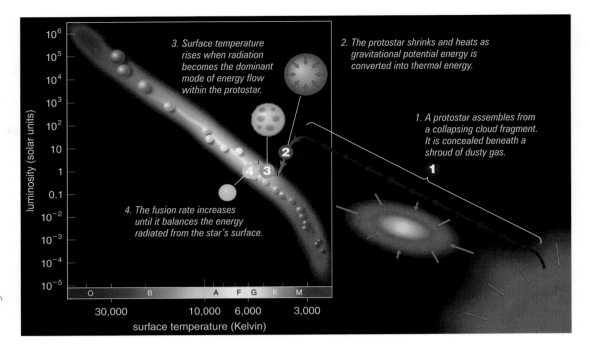

Figure 16.15

Interactive Figure The life track of a $1M_{Sun}$ star from protostar to main-sequence star.

Figure labels (within the image):

3. Surface temperature rises when radiation becomes the dominant mode of energy flow within the protostar.

2. The protostar shrinks and heats as gravitational potential energy is converted into thermal energy.

1. A protostar assembles from a collapsing cloud fragment. It is concealed beneath a shroud of dusty gas.

4. The fusion rate increases until it balances the energy radiated from the star's surface.

the standard main sequence. Each point along a star's life track represents its surface temperature and luminosity at some point during its life.

Figure 16.15 shows a life track leading to the birth of a $1M_{Sun}$ star like our Sun. This prebirth period includes four distinct stages:

Stage 1—Assembly of a Protostar. When gravity first assembles a protostar from a collapsing cloud fragment, the protostar is concealed within a shroud of dusty molecular gas. The protostellar wind and jets, along with radiation from the protostar, then disrupt this shroud, revealing the protostar itself. At this time, energy moves to the surface of the protostar primarily through convection. At the end of this stage, the photosphere's temperature is about 3,000 K, placing it on the right side of the H-R diagram, and its surface is many times larger than that of the Sun, producing a luminosity between $10L_{Sun}$ and $100L_{Sun}$.

Stage 2—Convective Contraction. The protostar's surface temperature remains near 3,000 K as long as convection remains the dominant mechanism for transporting energy to the surface. During this stage of a protostar's development, gravitational contraction leads to a decrease in its luminosity, because its radius becomes smaller while its surface temperature stays nearly constant. Consequently, the protostar's life track drops almost straight downward on the H-R diagram.

Stage 3—Radiative Contraction. When energy transport within the protostar switches from convection to radiative diffusion, the surface temperature begins to rise. This rise in temperature brings a slight rise in lumi-

nosity, even though the protostar continues to contract. The life track therefore bends toward higher surface temperature and slightly higher luminosity. During this stage, hydrogen nuclei begin to fuse into helium nuclei, but the energy released is small compared with the amount of energy radiated away.

Stage 4—Self-Sustaining Fusion. The core temperature and rate of fusion continue to increase gradually for a few tens of millions of years. Finally, the rate of fusion becomes high enough to balance the rate at which radiative energy escapes from the surface. At this point, fusion becomes self-sustaining. The star settles into its hydrogen-burning, main-sequence life.

Protostars of different masses go through similar stages as they approach the main sequence, but progress through those stages at different rates. Figure 16.16 illustrates life tracks for several protostars of different masses.

THINK ABOUT IT

Explain in your own words what we mean by a *life track* for a star. Why do we say that Figures 16.15 and 16.16 show only *pre-main-sequence* life tracks? In general terms, predict the appearance on Figure 16.16 of pre-main-sequence life tracks for a $25M_{Sun}$ star and a $0.1M_{Sun}$ star.

16.3 Masses of Newborn Stars

When we examined the properties of stars in Chapter 15, we saw that the masses of stars span a wide range. Stars at the low end of the range have masses less than $0.1M_{Sun}$, while stars at the high end can be as massive as $60M_{Sun}$ or

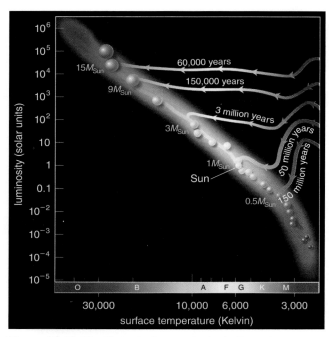

Figure 16.16 The life tracks from protostar to the main sequence for stars of several different masses.

maybe even $150M_{Sun}$. But why aren't there stars with masses outside this range? The answer once again comes down to the battle between gravity and pressure.

• What is the smallest mass a newborn star can have?

Contracting clouds with too low a mass never become stars because their central temperatures never climb above the 10 million K threshold needed for efficient nuclear fusion. Instead, a type of pressure called **degeneracy pressure** halts gravitational contraction before hydrogen burning can begin. Like thermal pressure, degeneracy pressure pushes outward against the force of gravity, but it differs from thermal pressure in one crucial way: Degeneracy pressure depends only on density and not on temperature. As we'll see, this feature of degeneracy pressure prevents objects with mass below about $0.08M_{Sun}$ from becoming true stars. Note that $0.08M_{Sun}$ is about 80 times the mass of Jupiter.

The Origin of Degeneracy Pressure We discussed the quantum mechanical origins of degeneracy pressure in some detail in Chapter S4. If you did not cover that chapter, the basic idea is that the laws of quantum mechanics prevent subatomic particles from getting too close together. These quantum laws are the same ones that give rise to distinct energy levels in atoms [Section 5.3]. In much the same way that electrons in atoms face restrictions that allow them to occupy only particular energy levels, there are restrictions on how close together electrons can be in a gas. Under most circumstances, these restrictions have little effect on the motions or locations of the electrons, and hence little effect

on the pressure. However, in a protostar with a mass below $0.08M_{Sun}$, the electrons become packed closely enough for these restrictions to matter. We can see how this fact leads to degeneracy pressure with a simple analogy.

Imagine an auditorium in which the laws of quantum mechanics dictate the spacing between chairs and people represent electrons (see Figure S4.11). As though playing a game of musical chairs, the people are always moving from seat to seat, just as electrons must remain constantly in motion. Most objects are like auditoriums with many more available chairs than people, so that the people (electrons) can easily find chairs as they move about. However, the cores of protostars with masses below $0.08M_{Sun}$ are like much smaller auditoriums with so few chairs that the people (electrons) fill nearly all of them. Because there are virtually no open seats, the people (electrons) can't all squeeze into a smaller section of the auditorium.

This resistance to squeezing is the origin of degeneracy pressure. If the people were really like electrons, quantum laws would also require them to move faster and faster to find open seats as they were squeezed into a smaller section. However, their speeds would have nothing to do with temperature. Degeneracy pressure and the particle motion that goes with it arise *only* because of the restrictions on where the particles can go, which is why temperature does not affect them.

Brown Dwarfs Because degeneracy pressure halts the contraction of a protostar with less than $0.08M_{Sun}$ before the release of fusion energy can balance the energy radiated from the protostar's surface, the result is a "failed star" that slowly radiates away its internal thermal energy, gradually cooling with time. Such objects, called **brown dwarfs**, occupy a fuzzy gap between what we call a planet and what we call a star. Because degeneracy pressure does *not* rise and fall with temperature, the gradual cooling of a brown dwarf's interior does not weaken its degeneracy pressure. In the constant battle of any "star" to resist the crush of gravity, brown dwarfs are winners, albeit dim ones. Their degeneracy pressure will not diminish with time, so gravity will never gain the upper hand.

Despite their name, brown dwarfs radiate primarily in the infrared and actually look deep red or magenta in color (Figure 16.17). They are far dimmer than normal stars and therefore are extremely difficult to detect, even if they are quite nearby. The first brown dwarf was discovered in 1995— it is an object (called Gliese 229B) with an estimated mass of $0.05M_{Sun}$ in orbit around a much brighter star (Gliese 229A). Many more brown dwarfs have since been discovered (Figure 16.18).

With the discovery of so many brown dwarfs, astronomers have begun to consider differences between them and how they might be classified. Recall that we classify ordinary stars using the spectral sequence OBAFGKM, in which spectral type O represents stars with the hottest surface temperatures and spectral type M represents stars with the

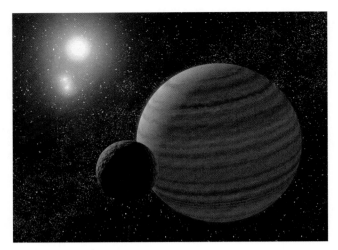

Figure 16.17 🎨 Artist's conception of a brown dwarf in a system with multiple stars. The brown dwarf is the banded object toward the right of the picture. Its reddish appearance approximates the color a brown dwarf would have if you could see one with your own eyes. The artist has added bands to the surface because we expect brown dwarfs to be more like giant jovian planets than stars. A small, rocky planet in the foreground orbits the brown dwarf. The system's hydrogen-burning stars can be seen in the background at the upper left.

lowest surface temperatures. Astronomers have recently extended the spectral sequence to include starlike objects even cooler than M stars by adding two new spectral types—called L and T. Objects with surface temperatures between 2,200 K and about 1,400 K are classified as "L dwarfs,"

and those cooler than that are "T dwarfs." The boundary between the two classes corresponds to the temperature at which methane absorption lines begin to appear in the spectra of these objects. All T dwarfs are too cool to be stars and must therefore be brown dwarfs. The L dwarfs seem to include both young brown dwarfs and some true hydrogen-burning stars at the extreme lower end of the stellar mass range. The only sure way to determine whether a T dwarf is a brown dwarf or a hydrogen-burning star is to measure its mass.

• What is the greatest mass a newborn star can have?

The maximum mass a star can have is not as well defined as the minimum mass. Observations of newly formed star clusters show stars with masses up to about $100 M_{\text{Sun}}$ but none as massive as $200 M_{\text{Sun}}$. Careful analysis of these observations suggests that the maximum mass is around $150 M_{\text{Sun}}$ (Figure 16.19).

The reason that stars have a maximum mass is yet another form of pressure, called **radiation pressure**, which gets its name because it is caused by light. Photons of light exert a slight amount of pressure when they strike matter. We don't ordinarily notice this pressure, because the force from any light source on Earth is quite small compared with the force of gravity. In space, however, the pressure of sunlight can in principle be used to accelerate spacecraft that unfurl large *solar sails* [Section 24.5]. Now, recall that the

Figure 16.18 An infrared image showing brown dwarfs (circled) in the constellation Orion. They are easier to spot in star-forming regions like this one than elsewhere in our galaxy, because young brown dwarfs still have much of the thermal energy left by the process of gravitational contraction. They therefore emit measurable amounts of infrared light.

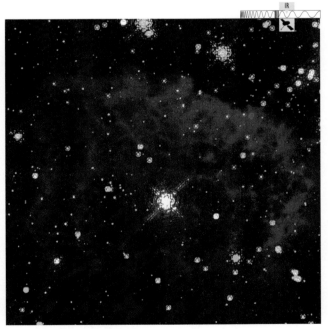

Figure 16.19 An infrared image of the Pistol Star, one of the most massive stars known (about $150 M_{\text{Sun}}$). This star lies near the center of our galaxy, behind dense interstellar clouds, which is why it is hidden from view in visible-light photos. The glowing gas that surrounds the star is being driven outward by the star's powerful radiation pressure. The image shows a region about 5 light-years across.

most massive stars are millions of times as luminous as the Sun. In a very massive star, nuclear fusion generates so much energy that the radiation bouncing around inside the star exerts a pressure stronger than thermal pressure. Theoretical models indicate that stars with masses above about $150M_{Sun}$ generate energy so furiously that gravity cannot resist the force of radiation pressure. Such stars effectively blow away any extra mass by driving their outer layers into space.

In fact, radiation pressure is so powerful in the highest-mass stars that it's unclear how they formed in the first place. Some theoretical models predict that radiation pressure should stop the accretion of matter onto a protostar when its mass exceeds about $10M_{Sun}$. Protostars that accumulate matter through an accretion disk (rather than direct infall) can grow somewhat more massive, because the inner parts of the disk shield the rest of the disk from the star's flow of photons, but models of this process still cannot account for the existence of stars with masses of $100M_{Sun}$ or more. Perhaps the most massive stars form through the collisions and mergers of several smaller protostars. Such collisions can occur in the centers of particularly dense, young star clusters. Once an extremely massive star forms, it rapidly loses mass as radiation pressure drives its outer layers into space. Models of such stars indicate that they cannot last more than a few hundred thousand years before blowing themselves apart.

• What are the typical masses of newborn stars?

We have seen that the masses of stars range from 0.08 solar mass to about 150 solar masses, but what are the more typical stellar masses? Astronomers address that question in two ways. First, they count how many stars of each mass are present in a young star cluster to determine the relative proportions of stars of each spectral type. However, this technique usually underestimates the number of low-mass stars, because these stars are not bright enough to be seen in any but the nearest of star clusters. Thus, astronomers use a second technique to get a handle on the overall number of low-mass stars: They count the number of low-mass stars in the immediate vicinity of the Sun, and use this count to estimate how many such stars have been made during the entire history of our galaxy.

The results show that in a newly formed star cluster, stars with low masses must greatly outnumber stars with high masses (Figure 16.20). For every star with a mass between 10 and 150 solar masses, there are typically 10 stars with masses between 2 and 10 solar masses, 50 stars with masses between 0.5 and 2 solar masses, and a few hundred stars with masses below 0.5 solar mass. We do not yet fully understand the processes that determine how many stars form in each mass range, but they probably have to do with the turbulent gas motions in the birth cloud and the subsequent pattern of fragmentation as the cloud contracts to form stars.

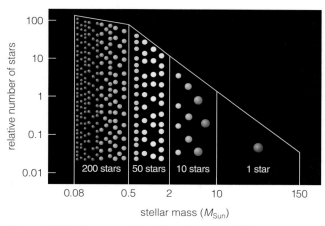

Figure 16.20 Demographics of newborn stars. This schematic diagram shows how many stars of each mass are produced for every star greater than $10M_{Sun}$ in an episode of star formation. Very massive stars ($10M_{Sun}$ to $150M_{Sun}$) are relatively rare; lower-mass stars ($0.08M_{Sun}$ to $2M_{Sun}$) are much more common.

Notice that, although the Sun lies toward the middle of the overall range of stellar masses, most stars in a new star cluster are less massive than the Sun. High-mass stars are rare, and low-mass stars are extremely common. With the passage of time, the balance tilts even more in favor of the low-mass stars as the high-mass stars die away. We'll see in the next chapter why the more massive stars die younger, as we investigate the life stories of stars.

Putting Chapter 16 into Context

In this chapter we saw how a star's life begins. As you look back over the chapter, remember these "big picture" ideas:

- In essence, star birth occurs when gravity squeezes a cloud of gas to the point that it forms a star with a core hot enough to sustain nuclear fusion. The key to understanding this process—and all other aspects of a star's life—lies in analyzing the ever-present "battle" between the inward pull of gravity and the outward push of pressure.

- Stars form in molecular clouds because these clouds have the relatively high densities and low temperatures needed to give gravity the upper hand in its battle against pressure. A single molecular cloud typically fragments into many pieces as it contracts, leaving behind a star cluster with many more low-mass stars than high-mass stars.

- Stars are born with a wide range of masses, but lower-mass stars are born in far greater numbers than higher-mass stars. In fact, the process of star formation also gives birth to many objects—the brown dwarfs—that are too small to shine with energy generated by nuclear fusion.

16.1 Stellar Nurseries

- **Where do stars form?** Stars form in cold, relatively dense **molecular clouds**. Solid grains of **interstellar dust** prevent visible light from passing through these clouds, but we can use infrared observations to see what's going on inside them.

- **Why do stars form?** A star can form in a molecular cloud only when gravity is strong enough to overpower the outward push of **thermal pressure**. Stars tend to form in clusters because gravity can more easily overcome pressure in more massive molecular clouds. A large cloud fragments into many smaller clumps of gas as it contracts, because the advantage of gravity over pressure increases as a clump of gas shrinks in size.

16.2 Stages of Star Birth

- **What slows the contraction of a star-forming cloud?** A contracting cloud begins its transformation into a star when its core starts trapping the thermal energy released by gravitational contraction. As pressure begins to push back harder, the contraction slows down and the central part of the cloud becomes a **protostar**. Meanwhile, matter from the surrounding cloud rains down upon the protostar, increasing its mass.

- **How does a cloud's rotation affect star birth?** Conservation of angular momentum ensures that a young protostar spins rapidly, and much of the material falling inward toward a protostar ends up in a spinning **protostellar disk**. Friction in the disk can transfer angular momentum away from the inner parts of the disk, allowing gas to accrete more easily onto the protostar. Some protostars drive powerful **jets** outward along the disk's rotation axis.

Along with strong **protostellar winds**, these jets can disrupt gas in the surrounding molecular cloud.

- **How does nuclear fusion begin in a newborn star?** Nuclear fusion becomes self-sustaining when a protostar's core temperature rises above 10 million degrees. Paradoxically, in order to reach this temperature, the protostar must keep radiating some of its thermal energy, so that it can continue contracting. During the late phases of star formation, the protostar's luminosity declines and its surface temperature increases, and we can represent these changes with a **life track** on the H-R diagram.

16.3 Masses of Newborn Stars

- **What is the smallest mass a newborn star can have?** A quantum mechanical effect known as **degeneracy pressure** halts the contraction of a protostar with a mass less than 0.08 solar masses before its temperature grows high enough to sustain fusion. Thus, $0.08 M_{Sun}$ is the minimum mass for a star. Starlike objects with masses below this limit are called **brown dwarfs**.

- **What is the greatest mass a newborn star can have?** The flood of photons coming from an extremely massive star exerts **radiation pressure** that can drive a star's outer layers into interstellar space. This form of pressure blows apart stars with masses greater than about $150 M_{Sun}$.

- **What are the typical masses of newborn stars?** Low-mass stars are far more numerous than high-mass stars. For every star with a mass above $10 M_{Sun}$ in a newborn star cluster, there are typically 10 stars with masses between 2 and $10 M_{Sun}$, 50 stars with masses between 0.5 and $2 M_{Sun}$, and a few hundred stars with masses below $0.5 M_{Sun}$.

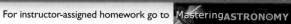
Review Questions

Short-Answer Questions Based on the Reading

1. What is the *interstellar medium?* What is its chemical composition and how do we measure it?
2. What is a *molecular cloud?* How does a molecular cloud's temperature and density compare with the rest of the interstellar medium?
3. What is *interstellar dust?* How does it interact with visible light? What are the consequences for our view of the heavens and how is that view different in infrared light?
4. What features of molecular clouds make the conditions favorable for star formation?
5. What happens to the thermal energy released into molecular clouds as gravity makes them contract? Why doesn't it build up and stop star formation?
6. Why do stars tend to form in clusters? Describe the process by which a single cloud gives birth to an entire cluster of stars.
7. Under what conditions can a small molecular cloud give birth to a single star?
8. Why do we think the very first stars were much more massive than the Sun?
9. What happens to a contracting cloud when its thermal energy can no longer escape the cloud's interior in the form of photons? How does the trapped thermal energy affect the process of star formation?
10. What is a *protostar?* How does it form? Why does its mass increase with time?
11. What is a *protostellar disk?* Describe how these disks enable additional matter to accrete onto the protostar.
12. What are *jets?* Why do we think they are related to the protostar's rotation? How do they affect the cloud surrounding the protostar?
13. Why does radiation of thermal energy from the surface of a protostar enable its central temperature to rise? Describe the final stages a protostar goes through before fusion begins in its core. How are these stages represented on a *life track?*
14. What is *degeneracy pressure,* and how does it differ from *thermal pressure?* Explain why degeneracy pressure can support a stellar core against gravity even when the core becomes very cold.
15. What is the minimum mass for a star, and why can't objects with lower masses be true stars? What is a *brown dwarf?*
16. What is the maximum mass of a star? What kind of pressure limits how massive a star can be?
17. How do the numbers of low-mass stars compare with those of higher-mass stars in new star clusters?

Test Your Understanding

Does It Make Sense?

Decide whether the statement makes sense (or is clearly true) or does not make sense (or is clearly false). Explain your reasoning. (For an example, see Chapter 1, "Does It Make Sense?")

18. If you want to get a more accurate count of the number of stars in our galaxy, use an infrared telescope to observe them instead of a visible-light telescope.

19. A molecular cloud needs to trap all the energy released by gravitational contraction in order for its center to become hot enough for fusion.
20. Low-mass stars form more easily in clouds that are unusually cold and dense.
21. The current mass of any star is the same as the mass it had when it first became a protostar.
22. The rotation of a protostar always speeds up with time because it is surrounded by a spinning disk.
23. Some of the stars in a star cluster live their entire lives and then die off before many of the cluster's stars initiate fusion.
24. Protostars are generally best observed in ultraviolet light because their surfaces have to get very hot before fusion can begin.
25. Degeneracy pressure exists only in objects that are very cold.
26. If Jupiter were 10 times more massive, we would consider it a brown dwarf, and if it were 100 times more massive, we would consider it a star.
27. Most of the stars that formed from the same cloud as the Sun have already died off.

Quick Quiz

Choose the best answer to each of the following. Explain your reasoning with one or more complete sentences.

28. Which of these colors of light passes most easily through interstellar clouds? (a) yellow light (b) green light (c) blue light
29. Molecular clouds stay cool because their molecules emit photons. Which of these molecules produces the largest number of photons from a molecular cloud? (a) molecular hydrogen (H_2) (b) carbon monoxide (CO) (c) water (H_2O)
30. What happens to a cloud's thermal pressure if its temperature falls while its density rises? (a) Thermal pressure goes up. (b) Thermal pressure goes down. (c) More information is needed to determine what thermal pressure does.
31. What happens within a contracting cloud in which gravity is stronger than pressure if its temperature remains the same as it contracts? (a) It breaks into smaller fragments. (b) Thermal pressure starts to push back more effectively against gravity. (c) It is trapping all the energy released by gravitational contraction.
32. Why are the very first stars thought to have been much more massive than the Sun? (a) The clouds that made them were much more massive than today's star-forming clouds. (b) The temperatures of the clouds that made them were higher because they consisted entirely of hydrogen and helium. (c) Star-forming clouds were much denser early in time.
33. What slows down the contraction of a star-forming cloud when it makes a protostar? (a) production of fusion energy (b) magnetic fields (c) trapping of thermal energy inside the protostar
34. What effects are jets and magnetic fields thought to have on a protostar? (a) They carry away angular momentum, helping the protostar grow in mass. (b) They carry away thermal energy, helping the protostar shrink. (c) They transfer gravitational energy to the core, causing fusion to start.

35. Which kind of pressure prevents stars of extremely large mass from forming? (a) thermal pressure (b) radiation pressure (c) degeneracy pressure

36. Which kinds of stars are most common in a newly formed star cluster? (a) O stars (b) G stars (c) M stars

Investigate Further
In-Depth Questions to Increase Your Understanding

Short-Answer/Essay Questions

37. *Interstellar Dust.* Describe how our view of the night sky would be different if there were no dust grains in the interstellar medium.

38. *Pressure vs. Gravity.* Suppose pressure and gravity are perfectly balanced within a certain molecular cloud. Describe what would happen to that balance if the temperature suddenly dropped. What would happen if the temperature suddenly rose? What would happen if the density suddenly increased without a change in temperature? What would happen if the cloud gained a little bit of mass?

39. *Protostars.* Describe the life history of a protostar from its beginnings as part of a molecular cloud to the moment hydrogen fusion begins. Give as many details as possible. How would that life story be different if the protostar formed in a cloud without any angular momentum?

40. *Approach to the Main Sequence.* As a protostar approaches the main sequence, its radius shrinks because it is converting gravitational potential energy into thermal energy to replace the thermal energy it is radiating away. Figure 16.15 shows the stages that a star like the Sun goes through before it becomes a main-sequence star. How do you think the time the star spends as a $100L_{Sun}$ protostar compares with the time it spends as a $10L_{Sun}$ protostar? Explain your reasoning.

41. *Understanding Life Tracks.* Compare the life tracks of pre-main-sequence stars in Figure 16.16. Which star's luminosity changes the least as it approaches the main sequence? Which star's luminosity changes the most? Which star's surface temperature changes the most? Which star's surface temperature changes the least?

42. *Degeneracy Pressure.* Describe how Jupiter would be different if there were no such thing as degeneracy pressure.

43. *Brown Dwarfs.* How are brown dwarfs like jovian planets? In what ways are brown dwarfs like stars?

44. *A Newborn Star Cluster.* Sketch the development of a newborn star cluster on a series of H-R diagrams. Assume that all the stars are initially protostars with a surface temperature that places them on the extreme right-hand side of the H-R diagram in Figure 16.16. What would the H-R diagram of this star cluster look like after 100,000 years? One million years? Ten million years? One hundred million years?

Quantitative Problems

Be sure to show all calculations clearly and state your final answers in complete sentences.

45. *Water in Molecular Clouds.* About one-millionth of the mass of a molecular cloud is in the form of water molecules. The mass density of such a cloud is roughly 10^{-21} g/cm^{-3}. Estimate the volume of a piece of molecular cloud that has the same amount of water as your body. How does this volume compare with the volume of the entire Earth?

46. *Interstellar Dust Grains.* Dark interstellar gas clouds contain so many dust grains that starlight cannot pass through, even though the dust grains are tiny and the spaces between them

are quite large by earthly standards. A typical dust grain has a radius of about 10^{-7} meter and a mass of about 10^{-14} gram.
   a. Estimate how many dust particles there are in a cloud containing $1,000M_{Sun}$ of dusty gas, if 1% of the cloud's mass is in the form of dust grains.
   b. Estimate the total surface area these grains would cover if you could put them side by side. You can assume that the grains are approximately spherical so that each grain covers an area πr^2, where r is the grain's radius. State your answer in square light-years.
   c. Estimate the total surface area the cloud covers, assuming that its matter density is like that of a typical molecular cloud, about 10^{-21} g/cm^{-3}. (*Hint:* First calculate the cloud's volume from its mass and density, then determine the cloud's radius using the formula for the radius of a sphere $R = (3 \times \text{volume}/4\pi)^{1/3}$.) State your answer in square light-years.
   d. Based on your answers to parts (b) and (c), what do you think the chances are that a photon passing through the cloud will hit a dust grain?

47. *A Large Star-Forming Cloud.* Typical molecular clouds have a temperature of 30 K and a particle density of 300 per cubic centimeter. Estimate how large the radius of a molecular cloud with this temperature and density needs to be for the cloud to harbor star formation. Give your answer in both meters and light-years. (For the purposes of this estimate, you can assume the cloud is perfectly spherical with a volume $4\pi r^3/3$ and that it is made entirely of hydrogen molecules of mass 3.3×10^{-24} g.)

48. *An Isolated Star-Forming Cloud.* Isolated molecular clouds can have a temperature as low as 10 K and a particle density as great as 100,000. What is the minimum mass that a cloud with these properties needs in order to form a star? Estimate the size of such a cloud using the same assumptions as in Problem 47.

49. *Masses of the First Stars.* Models of the first star-forming clouds indicate that they had a temperature of roughly 200 K and a particle density of roughly 300,000 particles per cubic centimeter at the time they started trapping their internal thermal energy. Estimate the mass at which thermal pressure balances gravity for these values of pressure and temperature. How does that mass compare with the Sun's mass? What is the estimated lifetime of a star with that mass?

50. *Internal Temperature of the Sun.* The Sun is essentially a gas cloud in which the forces of pressure and gravity balance each other. We can therefore use the equation in Mathematical Insight 16.1 to estimate the interior temperature of the Sun from its mass and particle density.
   a. What is the average number density of particles within the Sun, given that the average mass per particle is about 10^{-24} gram? (*Hint:* The volume of a sphere of radius r is equal to $4\pi r^3/3$.)
   b. What is the approximate temperature necessary for gas pressure to balance gravity within the Sun, given the average particle density from part (a)?
   c. How does your estimate compare with the internal core temperature of the Sun?

51. *Internal Temperature of a Brown Dwarf.* The maximum temperature inside a brown dwarf is the temperature at which thermal pressure would balance gravity. In reality, degeneracy pressure halts the contraction of a brown dwarf before it can reach that maximum value. You can use the formula in Mathematical Insight 16.1 to estimate the maximum interior

temperature of a brown dwarf with a mass of $0.05M_{Sun}$ and a radius of $0.1R_{Sun}$.

a. What is the average number density of particles inside the brown dwarf, given that the average mass per particle is about 10^{-24} gram? (*Hint:* The volume of a sphere of radius r is equal to $4\pi r^3/3$.)

b. What is the approximate temperature necessary for gas pressure to balance gravity within the brown dwarf, given the average particle density from part (a)?

c. How does that temperature compare with the 1×10^7 K temperature needed to sustain hydrogen fusion?

d. Explain how conditions inside the brown dwarf would change if you raised its mass to $0.1M_{Sun}$.

52. *Angular Momentum of a Close Binary.* Some close binary star systems have orbital periods as short as several days. Here we will estimate the angular momentum of such a system with two $1M_{Sun}$ stars in a circular orbit having an orbital period of 10 days.

a. Determine the average orbital separation of this system using Kepler's version of Newton's third law.

b. Determine the average speed of each star with respect to the center of mass of the system.

c. The center of mass of this system lies at the midpoint between the two stars because their masses are equal. The distance of each star from the center of mass is therefore equal to half the orbital separation. What is the angular momentum of each star with respect to the center of mass?

d. Suppose the Sun had the same angular momentum as these two stars combined. Estimate the speed at which the Sun's surface would be moving around the Sun's rotation axis.

e. How does speed from part (d) compare with the escape velocity from the Sun's surface? What would happen if the Sun were spinning this fast?

53. *Mass of a Brown Dwarf.* Suppose you observe a binary system containing a main-sequence star and a brown dwarf. The orbital period of the system is 1 year, and the average separation of the system is 1 AU. You then measure the Doppler shifts of the spectral lines from the main-sequence star and the brown dwarf, finding that the orbital speed of the brown dwarf in the system is 20 times greater than that of the main-sequence star. How massive is the brown dwarf?

54. *Light from a Newborn Star Cluster.* Suppose a new star cluster is born with one O star, 10 A stars, 100 G stars, and 1,000 M stars. Which stellar type dominates the light output from the cluster? What would the color of this star cluster appear to be if you observed it from a distance so large that you could not make out the individual stars?

Discussion Questions

55. *Life in a Molecular Cloud?* As far as we know, molecular clouds are the only place other than planets that contain the kinds of complex molecules needed to support life. They contain not only water molecules but also more complicated molecules based on chains of carbon atoms, known as *organic molecules*. Do you think it is possible for life to exist in a molecular cloud? What would life have to be like to survive there?

56. *A Star Is "Born."* Our discussion of star formation in this chapter talks about star "birth," even though stars are not really living things like humans, plants, or animals. In what sense is star birth like the birth of a living being? How is it different? Do you think it is appropriate to use the word *birth* in connection with star formation?

MEDIA EXPLORATIONS

For self-study activities go to **Mastering**ASTRONOMY www.masteringastronomy.com

Interactive Tutorials

Tutorial Review of Key Concepts
Use the interactive **Tutorial** at **www.mastering astronomy.com** to review key concepts from this chapter.

Supplementary Tutorial Exercises
Use the interactive **Tutorial Lessons** to explore the following questions.

Formation of the Solar System, Lesson 2
1. Most stars are thought to form in the same way that our Sun formed. Study the animation of solar system formation. Why does the star-forming cloud spin faster as it collapses?
2. Why does a newly-forming star develop a disk around it?
3. What does the disk around a protostar look like? Draw a series of pictures illustrating how it looks from different points of view.

Movies
Check out the following narrated and animated short documentary available on **www.masteringastronomy.com** for a helpful review of key ideas covered in this chapter.

Lives of Stars Movie

Web Projects

1. *Star Birth and the Spitzer Telescope.* The Spitzer Space Telescope is one of astronomers' best tools for learning about star birth. Go to the Web site and look for the latest findings about star formation. Summarize what scientists are learning in a one- to two-page report.

2. *Molecules in Space.* Over a hundred different kinds of molecules have been identified in molecular clouds, and astronomers continue to search for new ones. Use the Internet to compile a list of molecules that have been found in space, and give examples of substances on Earth that contain these molecules.

17

Star Stuff

> *We are, therefore, made out of star stuff ... we feed upon sunbeams, we are kept warm by the radiation of the Sun, and we are made out of the same materials that constitute the stars.*
>
> **—Harlow Shapley, in *The Universe of Stars*, 1929**

We inhale oxygen with every breath. Iron-bearing hemoglobin in our blood carries this oxygen through our bodies. Chains of carbon and nitrogen form the backbone of the proteins, fats, and carbohydrates in our cells. Calcium strengthens our bones, while sodium and potassium ions moderate communications of the nervous system. What does all this biology have to do with astronomy? The profound answer, recognized only in the second half of the twentieth century, is that life is based on elements created by stars.

We've already discussed in general terms how the elements in our bodies came to exist. Hydrogen and helium were produced in the Big Bang, and stars later created heavier elements that stellar explosions scattered into space. There, in the spaces between the stars, these elements mixed with interstellar gas and became incorporated into subsequent generations of stars.

In this chapter, we will discuss the origins of the elements in greater detail by delving into the lives of stars. As you read, keep in mind that no matter how far removed the stars may seem from our everyday lives, they actually are connected to us in the most intimate way possible: Without the lives and deaths of stars, none of us would be here. We are truly made from "star stuff."

17.1 Lives in the Balance

The story of a star's life is in many ways the story of an extended battle between two opposing forces: gravity and pressure. We studied the early stages of this battle in the previous chapter. We examined the conditions under which gravity can overcome pressure in interstellar gas, causing fragments of a molecular cloud to contract into protostars, and we saw that gravity's advantage over pressure continues until fusion begins in a star's core. Once hydrogen fusion begins, the energy it generates balances the energy the star radiates into space. With energy balanced, the star's internal pressure stabilizes and halts the crush of gravity. The star is then in a state of balance much like our Sun, with thermal pressure balancing gravity and fusion energy from the core balancing the flow of radiative energy from the star's surface [Section 14.2].

Stars can remain in this state of balance for millions to billions of years, but they will eventually exhaust the hydrogen in their cores. When that happens, fusion ceases in the core, and gravity regains the upper hand over pressure. The battle between pressure and gravity then grows increasingly more dramatic, with a final outcome that depends on the star's mass at birth.

• How does a star's mass affect nuclear fusion?

We discussed the significance of stellar birth masses in Chapter 15. Main-sequence stars with large masses have much greater luminosities than ones with small masses, which means that their cores must release fusion energy at much greater rates. The reason that more massive stars have greater fusion rates goes back to the idea of *gravitational equilibrium*, the balance between pressure and gravity that we studied in Chapter 14.

All main-sequence stars are in gravitational equilibrium, with the outward push of gas pressure balancing the inward pull of gravity. Because the force of gravity is greater within more massive stars, the pressure inside these stars must also be greater. To provide this greater pressure, the cores of stars more massive than the Sun must be somewhat hotter and denser than the core of the Sun. If you recall that the rate of fusion is very sensitive to temperature [Section 14.2], you'll understand why stars more massive than the Sun are so luminous and short-lived: The great strength of gravity within these massive stars leads to higher core temperatures and hence much higher fusion rates than in the Sun, and as a result they burn through their hydrogen gas much more rapidly than the Sun. Conversely, stars with masses smaller than the Sun's mass have weaker gravity, lower core temperature, and hence much lower fusion rates than the Sun. That is why these low-mass stars are much less luminous and have much longer main-sequence lifetimes than the Sun.

A star's mass also determines what happens when the star finally exhausts its core supply of hydrogen. Once the hydrogen is gone, fusion shuts down and the central core can no longer support itself against the crush of gravity. The core contracts, and the star's mass determines whether it eventually becomes hot enough to fuse helium or heavier elements. Even at the end of its life, when a star can no longer generate energy through fusion of any kind, its final fate depends on the mass it had at birth. As we will see, relatively low-mass stars like our Sun will end up as white dwarfs, while high-mass stars die violently and leave behind either a neutron star or black hole.

In order to organize our discussion of stellar lives and the role of mass, it's useful to divide stars into three basic groups:

- **Low-mass stars** are stars born with less than about two times the mass of our Sun, or less than 2 *solar masses* ($2M_{\mathrm{Sun}}$) of material.
- **Intermediate-mass stars** have birth weights between about 2 and 8 solar masses.
- **High-mass stars** are those stars born with masses greater than about 8 solar masses.

Our primary focus in this chapter will be on the dramatic differences between the lives of low- and high-mass stars. Because the life stages of intermediate-mass stars are quite similar to the corresponding stages of high-mass stars until the very ends of their lives, we include them in our discussion of high-mass stars.

As we discuss the life stories of stars in detail, you might wonder how we can claim to know what happens inside distant stars over time periods of millions and billions of years. After all, we can't even see what's happening inside any star right now. As always in science, astronomers' confidence in the life stories of stars comes from comparisons of theoretical models with detailed observations. On the theoretical side, we use mathematical models based on the known laws of physics to predict the interior structures and life cycles of stars. On the observational side, we study stars in star clusters (Figure 17.1). Recall that we can determine the ages of star clusters by finding their main-sequence turnoff points on H-R diagrams [Section 15.3]. By comparing clusters of different ages, we learn what stars of different masses are like at these ages. Occasionally, we even catch a star in its death throes. Our theoretical predictions of the life cycles of stars now agree quite well with the observations and confirm the idea that nuclear fusion in stars has produced essentially all the elements heavier than helium. In the remainder of this chapter, we will examine in detail our modern understanding of the life stories of stars and how they manufacture the variety of elements—the *star stuff*—that make our lives possible.

 Stellar Evolution Tutorial, Lesson 2

17.2 Life as a Low-Mass Star

In the grand hierarchy of stars, our Sun ranks as rather mediocre. We should be thankful for this mediocrity. If the Sun had been a high-mass star, it would have lasted only a few million years, dying before life could have arisen on Earth. Instead, the Sun has shone steadily for nearly 5 billion years, providing the light and heat that have allowed life to thrive on our planet. Other low-mass stars have similarly long lives. In this section, we investigate the lives and deaths of low-mass stars like our Sun.

• What are the life stages of a low-mass star?

Our Sun is currently in the middle of its roughly 10-billion-year life as a hydrogen-burning, main-sequence star. We therefore expect our Sun to continue to shine steadily for billions of years to come. Eventually, however, the Sun will exhaust its hydrogen and undergo a series of dramatic changes leading up to its death. Let's begin our study of a low-mass star's life stages with a look at what will happen to the Sun.

Main-Sequence Life: Slow and Steady As we discussed in Chapter 14, the Sun slowly and steadily fuses hydrogen into helium in its core via the *proton-proton chain* [Section 14.2]. The Sun shines steadily because of the self-regulating processes that we called the *solar thermostat*, in which gravitational equilibrium and the balance between the core energy

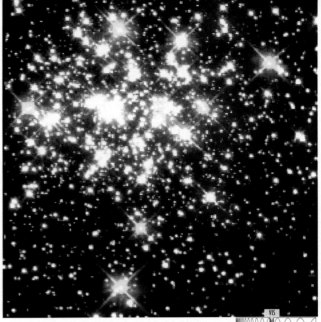

Figure 17.1 This star cluster contains both blue main-sequence stars and red supergiants. Its main-sequence turnoff point on an H-R diagram (not shown) indicates an age of 50 million years, meaning that stars with a birth mass of about $7M_{Sun}$ are just now running out of core hydrogen. The red supergiants in this cluster must therefore have been born with slightly higher masses, in the neighborhood of $8M_{Sun}$.

production rate and the rate at which energy escapes into space work together to keep the Sun's fusion rate and overall luminosity quite steady.

Other low-mass stars generate energy and shine steadily in much the same way as our Sun throughout their main-sequence lives. Models of those stars indicate that their interior structure is generally like the Sun's, with some minor differences in the way energy travels from the core to the surface. As in the Sun, the energy released by nuclear fusion in other low-mass stars takes hundreds of thousands of years to travel from the core to the surface, where it finally escapes into space as the star's luminosity. The energy moves outward from the core through a combination of *radiative diffusion* and *convection* [Section 14.2]. Radiative diffusion transports energy through the random bounces of photons from one electron to another, and convection transports energy through the rising of hot plasma and the sinking of cool plasma.

The depth of a star's convection zone depends on its internal temperature and hence on its mass (Figure 17.2). Deep inside a star like the Sun, the high temperatures allow radiative diffusion to carry energy outward at the same rate as fusion produces it in the core. Convection occurs only in the Sun's outer layers, where the cooler temperatures make it more difficult for photons to transport energy [Section 14.2]. In the Sun, the transition from radiative diffusion to convection occurs about 70% of the way from the center to the surface. Stars less massive than the Sun have cooler interiors and hence deeper convection zones. In very-low-mass stars, the convection zone extends all the way down to the core. Higher-mass stars have hotter interiors and hence shallower convection zones, and the highest-mass stars have no convection zone at all near their

high-mass star

$1M_{Sun}$ star

very low-mass star

Figure 17.2 Among main-sequence stars, convection zones extend deeper in lower-mass stars. High-mass stars have convective cores but no convection zones near their surfaces.

surfaces. However, high-mass stars can have convective cores, because they produce energy so furiously that radiative diffusion cannot transport it out of the core quickly enough. Convection therefore transports energy out of the cores of these high-mass stars, but radiative diffusion takes over throughout the rest of their interiors.

Convection plays a major role in determining whether a star has activity similar to that of the sunspot cycle on our Sun [Section 14.3]. Recall that the Sun's activity arises from the twisting and stretching of its magnetic fields by convection and rotation. The most dramatically active stars are very-low-mass stars (spectral type M) that happen to have fast rotation rates in addition to their deep convection zones. The churning interiors of these stars are in a constant state of turmoil, twisting and knotting their magnetic field lines. When these field lines suddenly snap and reconfigure themselves, releasing energy from the magnetic field, the result can be a spectacular flare. For a few minutes or hours, the flare can produce more radiation in X rays than the total amount of light coming from the star in infrared and visible light. Life on a planet near one of these **flare stars** might be quite difficult.

Aside from flares and other kinds of surface activity, the long lives of low-mass stars remain relatively uneventful as long as hydrogen fusion continues in the core. During that time, the luminosities of low-mass stars gradually rise for reasons we discussed in Chapter 14. Just as in the Sun, fusion in the core of a low-mass star reduces the number of independent particles in the core: Each fusion reaction converts four independent protons into just one independent helium nucleus. As the number of particles drops, the core must shrink and heat up in order to keep pressure in balance with gravity. This slight but continual rise in core temperature slowly raises the fusion rate and therefore the luminosity of the star as it ages. Much more dramatic change occurs only when nuclear fusion finally exhausts the star's central supply of hydrogen.

Red Giant Stage Hydrogen fusion supplies the thermal energy that maintains a star's thermal pressure and holds gravity at bay during its life as a main-sequence star, but that period of time as a well-balanced star is only temporary. When a star's core hydrogen is finally depleted, nu-

clear fusion ceases in the star's core. With no fusion to supply thermal energy and maintain the pressure, the star is once again out of balance, much like it was as a protostar. The core can no longer resist the crush of gravity and it begins to shrink much more rapidly than it did during the star's main-sequence life. Let's now look at the dramatic changes that will eventually happen in our own Sun, which will serve as an example of the changes that occur in all low-mass stars at the ends of their lives.

About 5 billion years from now, the Sun's core will run out of hydrogen and will begin to shrink. Surprisingly, the Sun's outer layers will expand outward while its core is shrinking. On an H-R diagram, its life track will move almost horizontally to the right as the Sun grows in size to become a **subgiant** (Figure 17.3). Then, as the expansion of the outer layers continues, the Sun's luminosity will begin to increase substantially, and its life track will turn upward on the H-R diagram. Over a period of about a billion years, the Sun will slowly grow in size and luminosity to become a **red giant**. At the peak of its red giant phase, the Sun will be more than 100 times larger in radius and more than 1,000 times brighter in luminosity than it is today.

But why will the Sun expand in overall size when its core is shrinking? We can find the answer by considering how its interior structure will change. After the core exhausts its hydrogen, it will be made almost entirely of helium, because helium is the "ash" left behind by hydrogen fusion. However, the gas surrounding the core will still contain plenty of fresh hydrogen that has never previously undergone fusion. Because gravity shrinks both the *inert* (non-burning) helium core and the surrounding *shell* of hydrogen, the hydrogen shell soon becomes hot enough to sustain hydrogen fusion (Figure 17.4). In fact, the shell will become

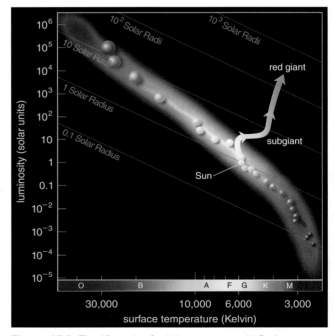

Figure 17.3 The life track of a $1M_{Sun}$ star on an H-R diagram from the end of its main-sequence life until it becomes a red giant.

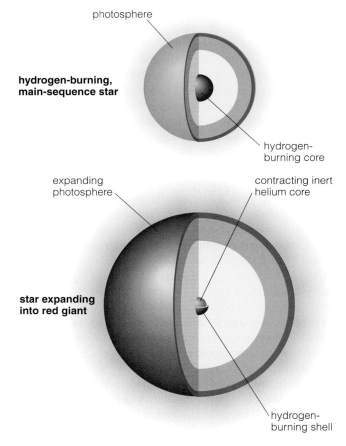

photosphere

hydrogen-burning, main-sequence star

hydrogen-burning core

expanding photosphere

contracting inert helium core

star expanding into red giant

hydrogen-burning shell

Figure 17.4 Interactive Figure. After a star ends its main-sequence life, its inert helium core contracts while hydrogen shell burning begins. The high rate of fusion in the hydrogen shell forces the star's upper layers to expand outward.

so hot that this **hydrogen shell burning** will proceed at a higher rate than core hydrogen fusion does today. That is why the Sun will become more luminous. It will also grow in size because energy transport within its interior cannot keep pace with this larger energy-generation rate. The new thermal energy will effectively be trapped within the Sun, so thermal pressure will build up and push the surface outward. However, even though the red giant Sun will be large on the outside, most of its mass will remain buried deep in its shrunken stellar core.

The situation will grow more extreme as long as the helium core remains inert. Today, the self-correcting feedback process of the solar thermostat regulates the Sun's fusion rate: A rise in the fusion rate causes the core to inflate and cool until the fusion rate drops back down. In contrast, thermal energy generated in the hydrogen-burning shell of a red giant cannot do anything to inflate the inert core that lies underneath. Instead, newly produced helium keeps adding to the mass of the helium core, amplifying its gravitational pull and shrinking it further. The hydrogen-burning shell shrinks along with the core, growing hotter and denser. The fusion rate in the shell consequently rises, feeding even more helium ash to the core. The star is caught in a vicious circle with a broken thermostat.

The core and shell will continue to shrink in size, and the ever-growing fusion rate will continue to push the Sun's upper layers outward, until the temperature of the inert helium core reaches about 100 million K. At that point, it will be hot enough for helium nuclei to begin to fuse together, and the Sun will enter the next stage of its life. Meanwhile, the Sun's increasing radius will weaken the pull of gravity at its surface, allowing large amounts of mass to escape in a *stellar wind*. Observation of winds from red giants shows that they carry away much more matter than the solar wind carries away from the Sun today, but at much slower speeds.

Other low-mass stars expand into red giants in the same way as the Sun, though the time it takes depends on their precise masses: Like all phases of stellar lives, the process occurs faster for more massive stars and more slowly for less massive stars. Stars more massive than the Sun reach the red giant phase in less than 10 billion years, and many lower-mass stars have such long main-sequence lifetimes that none of them can yet have reached the red giant phase in a 14-billion-year-old universe. In some very-low-mass stars, the inert helium core may never become hot enough to fuse helium. In that case, the core collapse will instead be halted by degeneracy pressure, leaving the star's corpse as a *helium white dwarf*.

THINK ABOUT IT

Before you read on, briefly summarize why a star grows larger and brighter after it exhausts its core hydrogen. When does the growth of a red giant finally halt, and why? How would a star's red giant stage be different if the temperature required for helium fusion were around 200 million K, rather than 100 million K? Why?

Helium Burning Recall that fusion occurs only when two nuclei come close enough together for the attractive *strong force* to overcome electromagnetic repulsion [Section 14.2]. Helium nuclei have two protons (and two neutrons) and hence a greater positive charge than the single proton of a hydrogen nucleus. The greater charge means that helium nuclei repel one another more strongly than hydrogen nuclei. **Helium fusion** therefore occurs only when nuclei slam into one another at much higher speeds than those needed for hydrogen fusion, which means that helium fusion requires much higher temperatures than hydrogen fusion.

The helium fusion process (often called the *triple-alpha reaction* because helium nuclei are sometimes called *alpha particles*) converts three helium nuclei into one carbon nucleus:

3 ^{4}He   1 ^{12}C   energy

Energy is released because the carbon-12 nucleus has a slightly lower mass than the three helium-4 nuclei, and the lost mass becomes energy in accord with $E = mc^2$.

The ignition of helium burning in a low-mass star like the Sun has one subtlety. Theoretical models show that the thermal pressure in the inert helium core is too low to counteract gravity. Instead, according to the models, the pressure fighting against gravity is *degeneracy pressure*—the same strange type of pressure that supports brown dwarfs [Section 16.3]. Because degeneracy pressure does *not* increase with temperature, the onset of helium fusion heats the core rapidly without causing it to inflate. The rising temperature causes the helium fusion rate to soar upward in what is called a **helium flash**.

The helium flash releases an enormous amount of energy into the core. In a matter of seconds, the temperature rises so much that thermal pressure again becomes dominant, and degeneracy pressure is no longer important. In fact, the thermal pressure becomes strong enough to push back against gravity, and the core actually begins to expand. This core expansion pushes the hydrogen-burning shell outward, lowering its temperature and its fusion rate. The result is that, even though the star now has core helium fusion *and* hydrogen shell burning taking place simultaneously (Figure 17.5), the total energy production falls from its peak during the red giant phase. The reduced total energy output of the star reduces its luminosity, allowing its outer layers to contract from their peak size during the red giant phase. As the outer layers contract, the star's surface temperature also increases somewhat. Thus, after the Sun spends about a billion years expanding into a luminous red giant, its size and luminosity will decline as it becomes a

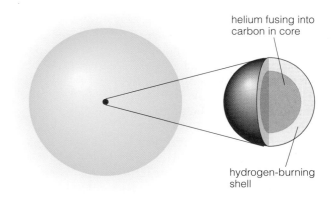

Figure 17.5 Core structure of a helium-burning star. Helium fusion causes the core and hydrogen-burning shell to expand and slightly cool, thereby reducing the overall energy-generation rate in comparison to the rate during the red giant stage. The outer layers shrink back, so a helium-burning star is smaller than a red giant of the same mass.

helium-burning star. With fusion once again operating in the core, the star regains the same sort of balance it had as a main-sequence star, except now it is helium fusion that keeps the central temperature steady.

Because the helium-burning star is now smaller and hotter than it was as a red giant, its life track on the H-R diagram drops downward and to the left (Figure 17.6a). The helium cores of all low-mass stars fuse helium into carbon at about the same rate, so these stars all have about

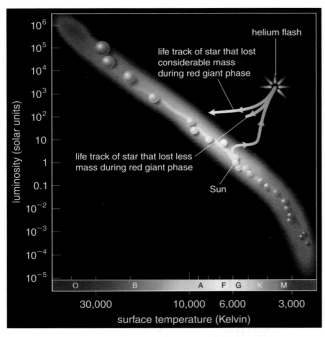

a Helium fusion begins with the helium flash, after which the star's surface shrinks and heats, making the star's life track move downward and to the left on the H-R diagram.

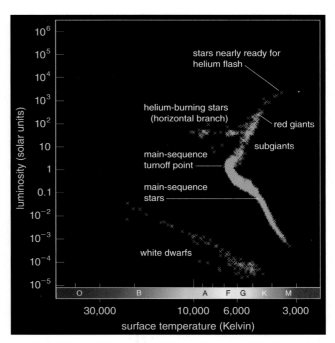

b An H-R diagram of a globular cluster shows low-mass stars in several different life stages.

Figure 17.6 After a low-mass star exhausts its core hydrogen, hydrogen-shell burning causes it to grow into a red giant. Once its helium core becomes hot enough for fusion, it temporarily settles down as a helium-burning star.

the same luminosity. However, the outer layers of these stars can have different masses depending on how much mass they have expelled through their stellar winds. Stars that expelled more mass end up with smaller radii and higher surface temperatures and hence are farther to the left on the H-R diagram.

We can see the results for many low-mass stars in the H-R diagram of a globular cluster (Figure 17.6b). Stars along the lower right of the H-R diagram, below the main-sequence turnoff point, are still in the hydrogen-burning, main-sequence phase that lasts most of a star's life. Just to the right of the main-sequence turnoff point we see subgiants—stars that have just begun their expansion into red giants as their cores have shut down and hydrogen shell burning has begun. The longer a star undergoes hydrogen shell burning, the larger and more luminous it becomes, which is why the diagram shows a continuous line of stars extending up to the most luminous red giants. These are the red giants on the verge of helium flash. The stars that have already undergone the helium flash and become helium-burning stars appear slightly below and to the left of the red giants, because they are somewhat smaller, hotter, and less luminous than they were at the moment of helium flash. Because these helium-burning stars all have about the same luminosity but can differ in surface temperature, they trace out a horizontal line on the H-R diagram known as the **horizontal branch**.

• How does a low-mass star die?

It is only a matter of time until a helium-burning star fuses all its core helium into carbon. In the Sun, the core helium will run out after about a hundred million years of burning. When the core helium is exhausted, fusion will again cease, and the star will once again go out of balance. The core, now made of the carbon "ash" from helium fusion, will begin to shrink under the crush of gravity. In a low-mass star like the Sun, the carbon core will never become hot enough to sustain carbon fusion. After having shined brightly for more than 10 billion years, the Sun will at last be on its deathbed.

Last Gasps The exhaustion of core helium will cause the Sun to expand once again, just as it did in becoming a red giant. This time, the trigger for the expansion will be helium fusion in a shell around the inert carbon core. Meanwhile, the hydrogen shell will still burn atop the helium layer. Thus, the Sun will have become a *double-shell burning giant*. Both shells will contract along with the inert core, driving their temperatures and fusion rates so high that the Sun will expand to an even greater size and luminosity than it had in its previous red giant phase. Theoretical models show that helium burning inside such a star never reaches equilibrium but instead proceeds in a series of **thermal pulses** during which the fusion rate spikes upward every few thousand years.

The furious burning in the helium and hydrogen shells cannot last long—maybe a few million years or less. The

Sun's only hope of extending its life will then lie with its carbon core, but this is a false hope for a low-mass star like the Sun. Carbon fusion is possible only at temperatures above about 600 million K. Before the core of a low-mass star ever reaches such a lofty temperature, degeneracy pressure halts its gravitational collapse.

The huge size of a dying star gives it only a very weak grip on its outer layers. As the star's luminosity and radius rise, increasing amounts of matter flow outward with the star's stellar wind. Meanwhile, during each thermal pulse, strong convection dredges up carbon from the core, enriching the surface of the star with carbon. Red giants whose photospheres become especially carbon-rich in this way are called **carbon stars**.

Carbon stars have cool, low-speed stellar winds, and the temperature of the gas in these winds drops with distance from the stellar surface. At the point at which the temperature has dropped to 1,000–2,000 K, some of the gas atoms in these slow-moving winds begin to stick together in microscopic clusters, forming small, solid particles of dust. These dust particles continue their slow drift with the stellar wind into interstellar space, where they become the interstellar dust grains that we discussed in Chapter 16. The process of particulate formation is similar to the formation of smoke particles in a fire. Thus, in a sense, carbon stars are the most voluminous polluters in the universe. However, this "carbon smog" is essential to life: Most of the carbon in your body (and in all life on Earth) was manufactured in carbon stars and blown into space by their stellar winds.

THINK ABOUT IT

Suppose the universe contained only low-mass stars. Would elements heavier than carbon exist? Why or why not?

Planetary Nebula The Sun's final end will be beautiful to those who witness it, as long as they are plenty far away. Through winds and other processes, the Sun will eject its outer layers into space, creating a huge shell of gas expanding away from the inert carbon core. The exposed core will still be very hot and will therefore emit intense ultraviolet radiation. This radiation will ionize the gas in the expanding shell, making it glow brightly as a **planetary nebula**. We have photographed many examples of planetary nebulae around other low-mass stars that have recently died in this very same way (Figure 17.7). Note that, despite their name, planetary nebulae have nothing to do with planets. The name comes from the fact that nearby planetary nebulae look much like planets through small telescopes, appearing as simple disks.

The glow of the planetary nebula will fade as the exposed core cools and the ejected gas disperses into space. The nebula will disappear within a million years, leaving the Sun's cooling carbon core behind as a *white dwarf*. Recall from Chapter 15 that white dwarfs are small in radius and often quite hot. We can now understand why: They are small in radius because they are the exposed cores of dead stars,

a Ring Nebula   **b** Eskimo Nebula   **c** Spirograph Nebula   **d** Hourglass Nebula

Figure 17.7 Hubble Space Telescope photos of planetary nebulae, which form when low-mass stars in their final death throes cast off their outer layers of gas. The central white dots are the remaining hot cores of the stars that ejected the gas. These hot cores ionize and energize the shells of gas that surround them. As the gas of the nebula disperses into space, the hot core remains as a white dwarf.

supported against the crush of gravity by degeneracy pressure. They are often hot because some of them were only recently in the center of a star and have not yet had time to cool much.

In the ongoing battle between gravity and a star's internal pressure, white dwarfs are in a sort of stalemate. As long as no mass is added to the white dwarf from some other source (such as a companion star in a binary system), neither the strength of gravity nor the strength of the degeneracy pressure that holds gravity at bay will ever change. Thus, a white dwarf is little more than a decaying corpse that will cool for the indefinite future, eventually disappearing from view as it becomes too cold to emit any more visible light.

Figure 17.8 summarizes the life stages of a $1M_{Sun}$ star on an H-R diagram, starting from the time it reaches the main sequence and continuing until it ejects a planetary nebula and leaves a white dwarf behind. We have already examined the life track to the point at which the star becomes a helium-burning star (see Figures 17.3 and 17.6). On this new diagram, we can see what happens after core helium burning ends. The life track again turns upward as the star enters its second red giant phase, this time with energy generated by fusion in shells of both helium and hydrogen. As the star ejects the planetary nebula, the dashed curve indicates that we are shifting from plotting the surface temperature of a red giant to plotting the surface temperature of the exposed stellar core left behind. The curve becomes solid again near the lower left, indicating that this core is a hot white dwarf. From that point, the curve continues downward and to the right as the remaining ember cools and fades.

The Fate of the Earth The death of the Sun will obviously have consequences for Earth, and some of these consequences will begin even before the Sun enters the final stages of its life. The Sun will gradually brighten during its remaining

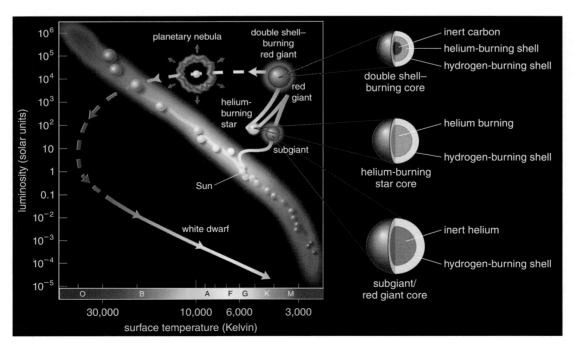

Figure 17.8 The life track of a $1M_{Sun}$ star from the time it first becomes a hydrogen-burning, main-sequence star to the time it dies as a white dwarf. Core structure is shown at key stages.

time as a main-sequence star, just as it has been brightening since its birth more than 4 billion years ago [Section 14.2]. The Sun's past brightening has not threatened the long-term survival of life on Earth, because Earth's climate self-regulates by adjusting the strength of the greenhouse effect (through the carbon dioxide cycle [Section 10.6]). However, this climate regulation will eventually break down as the Sun warms.

We still do not understand climate regulation well enough to be certain when the warming Sun will begin to overheat Earth. Some climate models predict that the oceans will begin to evaporate about a billion years from now, while other models suggest that our planet's climate may remain stable much longer. All models agree that, by about 3–4 billion years from now, the Sun will have brightened enough to doom Earth to a runaway greenhouse effect like that on Venus [Section 10.5]. The oceans will boil away, presumably spelling the end for any living organisms that have not stored water in well-protected enclosures.

Temperatures on Earth will rise even more dramatically when the Sun finally exhausts its core supply of hydrogen, somewhere around the year A.D. 5,000,000,000, and conditions will become even worse as the Sun grows into a red giant over the next several hundred million years. Just before helium flash, the Sun will be more than 1,000 times as luminous as it is today, and this huge luminosity will heat Earth's surface to more than 1,000 K (Figure 17.9a). Clearly, any surviving humans will need to have found a new home. Saturn's moon Titan [Section 11.2] might not be a bad choice. Its surface temperature will have risen from well below freezing today to about the present temperature of Earth.

The Sun will shrink and cool somewhat after helium flash turns it into a helium-burning star, providing a temporary lull in the incineration of Earth. However, this respite will last only 100 million years or so, and then Earth will suffer one final disaster.

After exhausting its core helium, the Sun will expand again during its last million years. Its luminosity will soar to thousands of times what it is today, and its radius will grow to nearly the present radius of Earth's orbit—so large that solar prominences might lap at Earth's surface (Figure 17.9b). Finally, the Sun will eject its outer layers as a planetary nebula that will engulf Jupiter and Saturn and drift on past Pluto into interstellar space. If Earth is not destroyed, its charred surface will be cold and dark in the faint, fading light of the white dwarf that the Sun will become.

 Stellar Evolution Tutorial, Lesson 3

17.3 Life as a High-Mass Star

Human life would be impossible without both low-mass stars and high-mass stars. The long lives of low-mass stars allow evolution to proceed for billions of years, but only high-mass stars produce the full array of elements on which life depends. Low-mass stars can't produce these elements because their cores never get hot enough to fuse elements heavier than helium. The greater electric charges of heavier nuclei create greater repulsive forces that prevent fusion from happening except at extremely high temperatures. Only the cores of high-mass stars reach high enough temperatures, and that happens only near the very end of a high-mass star's life, when the immense weight of its overlying layers bears down on a core that has already exhausted its hydrogen fuel.

The early stages of a high-mass star's life are similar to the early stages of the Sun's life, except they proceed much more rapidly. However, in the final stages of life, the highest-mass stars proceed to fuse increasingly heavier elements until they have exhausted all possible fusion sources. When fusion finally stops for good, gravity causes the core to implode suddenly. As we will soon see, the implosion of the core causes the star to self-destruct in the titanic explosion

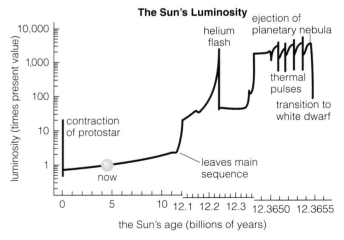

a Changes in the Sun's luminosity over time.

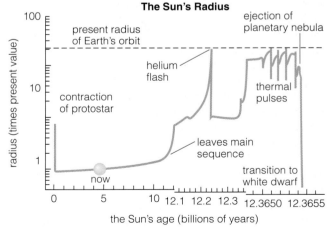

b Changes in the Sun's radius over time.

Figure 17.9 Evolution of the Sun. These graphs show results from a theoretical model of how the Sun's luminosity (left) and radius (right) change throughout its life. (This model gives a main-sequence lifetime of about 11 billion years, slightly greater than the more commonly quoted 10-billion-year lifetime.)

The Sun's demise in about 5 billion years might at first seem worrisome, but 5 billion years is a very long time. It is longer than Earth has yet existed, and human time scales pale by comparison. A single human lifetime, if we take it to be about 100 years, is only 2×10^{-8}, or two-hundred-millionths , of 5 billion years. Because 2×10^{-8} of a human lifetime is about 1 minute, we can say that a human lifetime compared to the life expectancy of the Sun is roughly the same as 60 heartbeats compared to a human lifetime.

What about human creations? The Egyptian pyramids have often been described as "eternal," but they are slowly eroding due to wind, rain, air pollution, and the impact of tourists. All traces of them will have vanished within a few hundred thousand years. While a few hundred thousand years may seem like a long time, the Sun's remaining lifetime is more than 1,000 times longer.

On a more somber note, we can gain perspective on 5 billion years by considering evolutionary time scales. During the past century, our species has acquired sufficient technology and power to destroy human life totally, if we so choose. However, even if we make

that unfortunate choice, some species (including many insects) are likely to survive.

Would another intelligent species ever emerge on Earth? We have no way to know [Section 24.4], but we can look to the past for guidance. Many species of dinosaurs were biologically quite advanced, if not truly intelligent, when they were suddenly wiped out about 65 million years ago. Some small rodentlike mammals survived, and here we are 65 million years later. We therefore might guess that another intelligent species could evolve some 65 million years after a human extinction. If these beings also destroyed themselves, another species could evolve 65 million years after that, and so on.

Even at 65 million years per shot, Earth would have *nearly 80* more chances for an intelligent species to evolve in 5 billion years (5 billion/65 million = 77). Perhaps one of those species will not destroy itself, and future generations might move on to other star systems by the time the Sun finally dies. Perhaps this species will be our own.

---

we call a *supernova*. The fast-paced life and cataclysmic death of a high-mass star is surely among the great dramas of the universe.

• What are the life stages of a high-mass star?

Like all other stars, a high-mass star forms out of a cloud fragment that gravity forces to contract into a protostar. Hydrogen burning begins when the gravitational potential energy released by the contracting protostar makes the core hot enough for fusion. However, hydrogen fusion inside a high-mass star proceeds through a different set of steps than hydrogen fusion in a low-mass star, which is part of the reason why high-mass stars live such brief but brilliant lives.

Hydrogen Fusion in a High-Mass Star   Recall that a low-mass star like our Sun fuses hydrogen into helium through the *proton-proton chain* (see Figure 14.7). You already know that the rate of fusion via the proton-proton chain increases substantially at higher temperatures. The even hotter core temperatures of high-mass stars enable protons to slam into carbon, oxygen, or nitrogen nuclei as well as into other protons. Although carbon, nitrogen, and oxygen make up less than 2% of the material from which stars form in interstellar space, this 2% is more than enough to be useful in a stellar core. The carbon, nitrogen, and oxygen act as catalysts for hydrogen fusion, making it proceed at a far higher rate than would be possible by the proton-proton chain alone. (A *catalyst* is something that aids the progress of a reaction without being consumed in the reaction.) This faster chain of hydrogen fusion reactions is called the **CNO cycle**, with the letters *CNO* standing for carbon, nitrogen, and oxygen, respectively. Figure 17.10 shows the six steps of the CNO cycle.

Notice that the overall reaction of the CNO cycle is the same as that of the proton-proton chain: Four hydrogen nuclei fuse into one helium-4 nucleus. The amount of energy generated in each reaction cycle therefore is also the same—it is equal to the difference in mass between the four hydrogen nuclei and the one helium nucleus multiplied

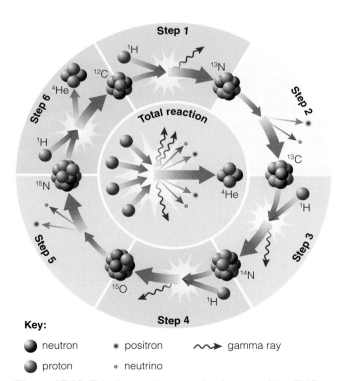

Key:

● neutron   ● positron   ∿→ gamma ray
● proton   ∘ neutrino

Figure 17.10 This diagram illustrates the six steps of the CNO cycle by which massive stars fuse hydrogen into helium. Note that the overall result is the same as that of the proton-proton chain: Four hydrogen nuclei fuse to make one helium nucleus. The carbon, nitrogen, and oxygen nuclei are catalysts that help the cycle proceed but are neither consumed nor created in the overall cycle.

by c^2. However, the CNO cycle allows hydrogen fusion to proceed at a far higher rate than would be possible by the proton-proton chain alone. That is why the luminosities of high-mass stars are so much higher than those of low-mass stars and why their lives are so much shorter.

Did the very first high-mass stars in the history of the universe produce energy through the CNO cycle? Explain.

The enormous fusion rates in high-mass stars generate remarkable amounts of power. Many more photons stream from the photospheres of high-mass stars than from the Sun, and many more photons are bouncing around inside. These photons exert a significant amount of *radiation pressure* in high-mass stars. Recall that radiation pressure ultimately blows apart the highest-mass stars, which is why there is an upper limit to stellar masses [Section 16.3]. Near the photosphere of a very-high-mass star, the radiation pressure can drive strong, fast-moving stellar winds. The wind from such a star can expel as much as 10^{-5} solar mass of gas per year at speeds greater than 1,000 km/s. This wind would cross the United States in about 5 seconds and would send a mass equivalent to that of our Sun hurtling into space in only 100,000 years. Such a wind cannot last long because it would blow away all the mass of even a very massive star in just a few million years.

Becoming a Supergiant With hydrogen fusion proceeding at a fast rate via the CNO cycle, high-mass stars soon begin to run low on core hydrogen fuel. For example, a 25-solar-mass star can last only a few million years as a hydrogen-burning, main-sequence star. As its core hydrogen runs out, a high-mass star responds much like a low-mass star, but much faster. It develops a hydrogen-burning shell, and its outer layers begin to expand outward, ultimately turning it into a *supergiant* star. At the same time, the core contracts, and this gravitational contraction releases energy that raises the core temperature until it becomes hot enough to fuse helium into carbon. However, there is no helium flash in stars of more than 2 solar masses, because their core temperatures are high enough that thermal pressure remains strong, preventing degeneracy pressure from being a factor. Helium burning therefore ignites gradually, just as hydrogen burning did at the beginning of the star's life.

A high-mass star fuses helium into carbon so rapidly that it is left with an inert carbon core after no more than a few hundred thousand years. Once again, the absence of fusion leaves the core without an energy source to fight off the crush of gravity. The inert carbon core shrinks, the crush of gravity intensifies, and the core pressure, temperature, and density all rise. Meanwhile, a helium-burning shell forms between the inert core and the hydrogen-burning shell. The star's outer layers swell further.

Up to this point, the life stories of intermediate-mass stars ($2-8 M_{Sun}$) and high-mass stars ($>8 M_{Sun}$) are very similar, except that all stages proceed more rapidly in higher-mass stars. However, degeneracy pressure prevents the cores of intermediate-mass stars from reaching the temperatures required to burn carbon or oxygen into anything much heavier. These stars eventually blow away their upper layers and finish their lives as white dwarfs. The rest of a high-mass star's life, on the other hand, is unlike anything that a low- or intermediate-mass star ever experiences.

• How do high-mass stars make the elements necessary for life?

A low-mass star can't make elements heavier than carbon because degeneracy pressure halts the contraction of its inert carbon core before it can get hot enough for fusion. A high-mass star has no such problem. The crush of gravity in a high-mass star keeps its carbon core so hot that degeneracy pressure never comes into play. After helium fusion stops, the gravitational contraction of the carbon core continues until it reaches the 600 million K required to fuse carbon into heavier elements.

Carbon fusion provides the core with a new source of energy that restores gravitational equilibrium, but only temporarily. In the highest-mass stars, carbon burning may last only a few hundred years. When the core carbon has been depleted, the core again begins to collapse, shrinking and heating once more until it can fuse a still-heavier element. The star is engaged in the final phases of a desperate battle against the ever-strengthening crush of gravity. The star will ultimately lose the battle, but it will be a victory for life in the universe: In the process of its struggle against gravity, the star will produce the heavy elements of which Earth-like planets and life are made.

Advanced Nuclear Burning The nuclear reactions in a high-mass star's final stages of life become quite complex, and many different reactions may take place simultaneously. The simplest sequence of fusion stages occurs through successive **helium-capture reactions**—reactions in which a helium nucleus fuses into some other nucleus (Figure 17.11a). Helium capture can fuse carbon into oxygen, oxygen into neon, neon into magnesium, and so on.*

At high enough temperatures, a star's core plasma can fuse heavy nuclei to one another. For example, fusing carbon to oxygen creates silicon, fusing two oxygen nuclei creates sulfur, and fusing two silicon nuclei generates iron (Figure 17.11b). Some of these heavy-element reactions release free neutrons, which may fuse with heavy nuclei to make still rarer elements. The star is forging the variety of elements that, in our solar system at least, became the stuff of life.

Each time the core depletes the elements it is fusing, it shrinks and heats until it becomes hot enough for other fusion reactions. Meanwhile, a new type of shell burning

---

*These reactions can still proceed even after the star has used up its initial supply of core helium because there are other reactions that release helium nuclei. For example, when two carbon nuclei fuse together, the reaction can produce a neon nucleus and a helium nucleus.

a Helium-capture reactions.

b Other reactions. (Note: Fusion of two silicon nuclei actually first produces nickel-56, but this decays rapidly to cobalt-56 and then to iron-56).

Figure 17.11 A few of the many nuclear reactions that occur in the final stages of a high-mass star's life.

ignites between the core and the overlying shells of fusion. Near the end, the star's central region resembles the inside of an onion, with layer upon layer of shells burning different elements (Figure 17.12). During the star's final few days, iron begins to pile up in the silicon-burning core.

Despite the dramatic events taking place in its interior, the high-mass star's outer appearance changes only slowly. As each stage of core fusion ceases, the surrounding shell burning intensifies and further inflates the star's outer layers. Each time the core flares up, the outer layers contract somewhat but the star's overall luminosity remains about

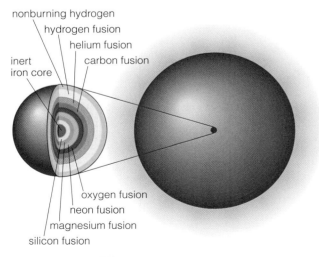

Figure 17.12 Interactive Figure The multiple layers of nuclear burning in the core of a high-mass star during the final days of its life.

the same. The result is that the star's life track zigzags across the top of the H-R diagram (Figure 17.13). In the most massive stars of all, the core changes happen so quickly that the outer layers don't have time to respond, and the star progresses steadily toward becoming a red supergiant.

One of these massive, red supergiant stars happens to be relatively nearby: Betelgeuse, the upper left shoulder of Orion. Its radius is over 500 solar radii, or more than twice the distance from the Sun to Earth. We have no way of knowing what stage of nuclear burning is now taking place in Betelgeuse's core. Betelgeuse may have a few thousand years of nuclear burning still ahead, or we may be seeing it as iron piles up in its core. If the latter is the case, then sometime in the next few days we will witness one of the most dramatic events that ever occurs in the universe.

Iron: Bad News for the Stellar Core As a high-mass star develops an inert core of iron, the core continues shrinking and heating while iron continues to pile up from nuclear burning in the surrounding shells. If iron were like the other elements in prior stages of nuclear burning, this core contraction would stop when iron fusion ignited. However, iron is unique among the elements in a very important way: It is the one element from which it is *not* possible to generate any kind of nuclear energy.

To understand why iron is unique, remember that only two basic processes can release nuclear energy: *fusion* of light elements into heavier ones, and *fission* of very heavy elements into not-so-heavy ones (see Figure 14.5). Recall that hydrogen fusion converts four protons (hydrogen

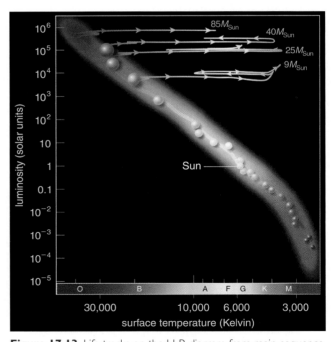

Figure 17.13 Life tracks on the H-R diagram from main-sequence star to red supergiant for selected high-mass stars. Labels on the tracks give the star's mass at the beginning of its main-sequence life. Because of the strong wind from such a star, its mass can be considerably smaller when it leaves the main sequence. (Based on models from A. Maeder and G. Meynet.)

nuclei) into a helium nucleus that consists of two protons and two neutrons. Thus, the total number of *nuclear particles* (protons and neutrons combined) does not change. However, this fusion reaction generates energy (in accord with $E = mc^2$) because the *mass* of the helium nucleus is less than the combined mass of the four hydrogen nuclei that fused to create it—despite the fact that the *number* of nuclear particles is unchanged.

In other words, fusing hydrogen into helium generates energy because helium has a lower *mass per nuclear particle* than hydrogen. Similarly, fusing three helium-4 nuclei into one carbon-12 nucleus generates energy because carbon has a lower mass per nuclear particle than helium, which means that some mass disappears and becomes energy in this fusion reaction. This decrease in mass per nuclear particle from hydrogen to helium to carbon is part of a general trend shown in Figure 17.14.

The mass per nuclear particle tends to decrease as we go from light elements to iron, which means that fusion of light nuclei into heavier nuclei generates energy. This trend reverses beyond iron: The mass per nuclear particle tends to *increase* as we look to still heavier elements. As a result, elements heavier than iron can generate nuclear energy only through fission into lighter elements. For example, uranium has a greater mass per nuclear particle than lead, so uranium fission (which ultimately leaves lead as a by-product) must convert some mass into energy.

Iron has the lowest mass per nuclear particle of all nuclei and therefore cannot release energy by either fusion or fission. Thus, once the matter in a stellar core turns to iron, it can generate no further energy. The iron core's only hope of resisting the crush of gravity lies with degeneracy pressure, but iron keeps piling up until even degeneracy pressure

cannot support the core. What ensues is the ultimate nuclear waste catastrophe: The star explodes as a supernova, scattering all the newly made elements into interstellar space.

THINK ABOUT IT

How would the universe be different if hydrogen, rather than iron, had the lowest mass per nuclear particle? Why?

Evidence for the Origin of Elements Before we look at how a supernova happens, let's consider the evidence that indicates we actually understand the origin of the elements. We cannot see inside stars, so we cannot directly observe elements being created in the ways we've discussed. However, the signature of nuclear reactions in massive stars is written in the patterns of elemental abundances across the universe.

For example, if massive stars really produce heavy elements (that is, elements heavier than hydrogen and helium) and scatter these elements into space when they die, the total amount of these heavy elements in interstellar gas should gradually increase with time (because additional massive stars have died). We should expect stars born recently to contain a greater proportion of heavy elements than stars born in the distant past, because these stars formed from interstellar gas that contained more heavy elements.

Stellar spectra confirm this prediction: Older stars do indeed contain smaller amounts of heavy elements than younger stars. For very old stars in globular clusters, elements besides hydrogen and helium typically make up as little as 0.1% of the total mass. In contrast, young stars that formed in the recent past contain about 2–3% of their mass in the form of heavy elements.

We gain even more confidence in our model of elemental creation when we compare the abundances of various elements in the cosmos. For example, because helium-capture reactions add two protons (and two neutrons) at a time, we expect nuclei with even numbers of protons to outnumber those with odd numbers of protons that fall between them. Sure enough, even-numbered nuclei such as carbon, oxygen, and neon are relatively abundant (Figure 17.15). Similarly, because elements heavier than iron are made primarily by rare fusion reactions shortly before and during a supernova, we expect these elements to be extremely rare.* Again, observations verify this prediction made by our model of nuclear creation.

• How does a high-mass star die?

Let's return now to our high-mass star, with iron piling up in its core. As we've discussed, it has no hope of generating any energy by fusion of this iron. After shining brilliantly for a few million years, the star will not live to see another day.

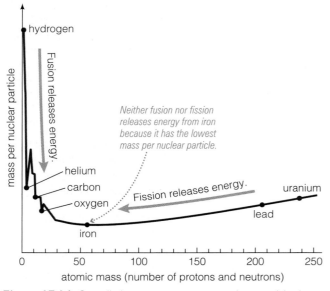

Figure 17.14 Overall, the average mass per nuclear particle declines from hydrogen to iron and then increases. Selected nuclei are labeled to provide reference points. (This graph shows the most general trends only. A more detailed graph would show numerous up-and-down bumps superimposed on the general trends. The vertical scale is arbitrary, but shows the general idea.)

---

*Nuclei heavier than iron grow by capturing neutrons, which then decay into protons, releasing an electron and an antineutrino with each decay. Because this process requires energy, it happens most rapidly in supernova explosions, but it can also happen slowly to heavy nuclei in the helium-burning zones of giant and supergiant stars.

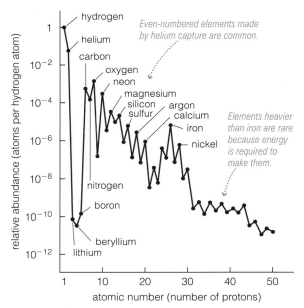

Figure 17.15 This graph shows the observed relative abundances of elements in the galaxy in comparison to the abundance of hydrogen (given as 1 in this comparison). For example, the abundance of nitrogen is about 10^{-4}, which means that there are about $10^{-4} = 0.0001$ times as many nitrogen atoms in the galaxy as hydrogen atoms.

In the graph: Even-numbered elements made by helium capture are common. Elements heavier than iron are rare because energy is required to make them.

The Supernova Explosion The degeneracy pressure that briefly supports the inert iron core arises because the laws of quantum mechanics prohibit electrons from getting too close together [Section S4.4]. Once gravity pushes the electrons past the quantum mechanical limit, however, they can no longer exist freely. In an instant, the electrons disappear by combining with protons to form neutrons, releasing neutrinos in the process (Figure 17.16). The electron degeneracy pressure suddenly vanishes, and gravity has free rein.

In a fraction of a second, an iron core with a mass comparable to that of our Sun and a size larger than that of Earth collapses into a ball of neutrons just a few kilometers across. The collapse halts only because the neutrons have a degeneracy pressure of their own. The entire core then resembles a giant atomic nucleus. If you recall that ordinary atoms are made almost entirely of empty space [Section 5.3] and that almost all their mass is in their nuclei, you'll realize that a giant atomic nucleus must have an astoundingly high density.

The gravitational collapse of the core releases an enormous amount of energy—more than a hundred times what the Sun will radiate over its entire 10-billion-year lifetime. Where does this energy go? It drives the outer layers off into space in the titanic explosion that we call a **supernova**. The ball of neutrons left behind is called a **neutron star**. In some cases, the remaining mass may be so large that gravity also overcomes neutron degeneracy pressure, and the core continues to collapse until it becomes a *black hole* [Sections S3.3, 18.3].

Theoretical models of supernovae successfully reproduce the observed energy outputs of real supernovae, but the precise mechanism of the explosion is not yet clear.

Figure 17.16 During the final, catastrophic collapse of a high-mass stellar core, electrons and protons combine to form neutrons, accompanied by the release of neutrinos.

Two general processes could contribute to the explosion. In the first process, neutron degeneracy pressure halts the gravitational collapse, causing the core to rebound slightly and ram into overlying material that is still falling inward. Until recently, most astronomers thought that this core-bounce process ejected the star's outer layers. Current models of supernovae, however, suggest that the more important process involves the neutrinos formed when electrons and protons combine to make neutrons. Although these ghostly particles rarely interact with anything [Section 14.2], so many are produced when the core implodes that they drive a shock wave that propels the star's upper layers into space.

The shock wave sends the star's former surface zooming outward at a speed of 10,000 kilometers per second—fast enough to travel the distance from the Sun to Earth in only about 4 hours. The heat of the explosion makes the gas shine with dazzling brilliance. For about a week, a supernova blazes as powerfully as 10 billion Suns, rivaling the luminosity of a moderate-size galaxy. The ejected gases slowly cool and fade in brightness over the next several months, but they continue to expand outward until they eventually mix with other gases in interstellar space. The scattered debris from the supernova carries with it the variety of elements produced in the star's nuclear furnace, as well as additional elements created when some of the neutrons produced during the core collapse slam into other nuclei. Millions or billions of years later, this debris may be incorporated into a new generation of stars. We are truly "star stuff," because we and our planet were built from the debris of stars that exploded long ago.

Historical Supernovae Observations The study of supernovae owes a great debt to astronomers of many different epochs and cultures. Careful scrutiny of the night skies allowed ancient people to identify several supernovae whose remains can still be seen. The most famous example is the Crab Nebula in the constellation Taurus. The Crab Nebula is a **supernova remnant**—an expanding cloud of debris from a supernova explosion (Figure 17.17).

A spinning neutron star lies at the center of the Crab Nebula, providing evidence that supernovae really do create neutron stars. Photographs taken years apart show that the nebula is growing larger at a rate of several thousand kilometers per second. Calculating backward from its present size, we can trace the nebula's birth to somewhere near A.D. 1100. Thanks to observations made by ancient astronomers, we can be even more precise.

Figure 17.17 Interactive Photo The Crab Nebula is the remnant of the supernova observed in A.D. 1054. This photograph was taken with the Very Large Telescope at the European Southern Observatory in Chile.

The official history of the Sung Dynasty in China contains a record of a remarkable celestial event:

In the first year of the period Chih-ho, the fifth moon, the day chi-ch'ou, a guest star appeared approximately several [degrees] southeast of Thien-kuan. After more than a year it gradually became invisible.

This description of the sudden appearance and gradual dimming of a "guest star" matches what we expect for a supernova, and the location "southeast of Thien-kuan" corresponds to the Crab Nebula's location in Taurus. Moreover, the Chinese date described in the excerpt corresponds to July 4, 1054, telling us precisely when the Crab supernova became visible from Earth. Descriptions of this particular supernova also appear in Japanese astronomical writings and in an Arabic medical textbook. Some people have claimed that it is even recorded in Native American paintings in the southwestern United States, but these claims now seem doubtful. Curiously, European records do not mention this supernova even though it would have been clearly visible.

Other historical records of supernovae have allowed us to age-date other supernova remnants, which in turn allows us to determine the kinds of supernovae that produced the remnants and to assess how frequently stars explode in our region of the Milky Way Galaxy. At least four supernovae have been observed during the past thousand years, appearing as brilliant new stars for a few months in the years 1006, 1054, 1572, and 1604. The supernova of 1006, the brightest of these four, could be seen during the daytime and cast shadows at night.

Supernovae may even have influenced human history. The Chinese were meticulous in recording their observations because they believed that celestial events foretold the future, and they may have acted in accord with such fortune-telling. The 1572 supernova was witnessed by Tycho

Brahe and helped convince him and others that the heavens were not as perfect and unchanging as Aristotle had imagined [Section 3.3]. Kepler saw the 1604 supernova at a time when he was struggling to make planetary orbits fit perfect circles. Perhaps this "imperfection" of the heavens helped push him to test elliptical orbits instead.

THINK ABOUT IT

When Betelgeuse explodes as a supernova, it will be more than 10 times as bright as the full moon in our sky. If our ancestors had seen Betelgeuse explode a few hundred or a few thousand years ago, do you think it could have had any effect on human history? How do you think our modern society would react if we saw Betelgeuse explode tomorrow?

Modern Supernova Observations No supernova has been seen in our own galaxy since 1604, but today astronomers routinely discover supernovae in other galaxies. The nearest of these extragalactic supernovae, and the only one near enough to be visible to the naked eye, burst into view in 1987. Because it was the first supernova detected that year, it was given the name **Supernova 1987A**. Supernova 1987A was the explosion of a star in the *Large Magellanic Cloud*, a small galaxy that orbits the Milky Way and is visible only from southern latitudes. The Large Magellanic Cloud is about 150,000 light-years away, so the star really exploded some 150,000 years ago.

As the nearest supernova witnessed in four centuries, Supernova 1987A provided a unique opportunity to study a supernova and its debris in detail. Astronomers from all over the planet traveled to the Southern Hemisphere to observe it, and several orbiting spacecraft added observations in many different wavelengths of light.

Older photographs of the Large Magellanic Cloud allowed astronomers to determine precisely which star had exploded (Figure 17.18). It turned out to be a blue star, not the red supergiant expected when core fusion has ceased. The most likely explanation is that the star's outer layers were unusually thin and warm near the end of its life, changing its appearance from that of a red supergiant to a blue one. The surprising color of the pre-explosion star demonstrates that we still have much to learn about supernovae. Reassuringly, most other theoretical predictions of stellar life cycles were well matched by observations of Supernova 1987A.

One of the most remarkable findings from Supernova 1987A was a burst of neutrinos, recorded by neutrino detectors in Japan and Ohio. The neutrino data confirmed that the explosion released most of its energy in the form of neutrinos, suggesting that we are correct in believing that the stellar core undergoes sudden collapse to a ball of neutrons. The capture of neutrinos from Supernova 1987A also spurred scientific interest in building more purposeful "neutrino telescopes." Many are now operating or are in development, and scientists hope that these neutrino telescopes will soon provide us with a new way of studying events in the distant universe.

Before. The arrow points to the star observed to explode in 1987.

After. The supernova actually appeared as a bright point of light. It appears larger than a point in this photograph only because of overexposure.

Figure 17.18 Before-and-after photos of the location of Supernova 1987A.

In the meantime, the drama of Supernova 1987A is not yet over. After the discovery of the supernova, astronomers spotted three glowing rings of gas around it (Figure 17.19a). The rings glow because they were flash-heated by ultraviolet light from the supernova, but it's not clear where the gaseous rings came from in the first place. They may be made of gas from the wind of the star that exploded, or they might be made of gas left over from the star's formation. Either way, astronomers quickly predicted that debris flying outward from the supernova would ram into the innermost ring within a couple of decades, making the ring glow even more brightly. Recent observations in visible light, X rays, and radio waves show that the collision has already begun (Figure 17.19b). As the collision continues to unfold over the next decade, new observations may shed light on both the origin of the mysterious rings and on the debris that came from the exploding star.

17.4 The Roles of Mass and Mass Exchange

Throughout this chapter we have focused on the key role of birth mass in determining a star's destiny. However, we have so far treated stars as if they live in isolation, even though nearly half the stars we see in the sky are actually binary star systems. In this final section, we will first summarize what we have learned about the lives of stars when they are not part of close binary star systems. We will then examine how a star's life story can change if it happens to orbit another star closely enough so that mass can sometimes flow from one star to the other.

• How does a star's mass determine its life story?

A star's birth mass determines its life cycle because it governs how nuclear fusion progresses in the core. Fusion proceeds relatively slowly in low-mass stars and does not make elements much heavier than carbon. These stars therefore live long lives and die in planetary nebulae, leaving behind white dwarfs composed mostly of carbon. Fusion reactions proceed somewhat faster in the hotter cores of intermediate-mass stars. However, these stars never manage to make iron and also die in planetary nebulae, leaving white dwarfs that can contain elements heavier than carbon. Fusion proceeds most quickly in the very hot cores of high-mass stars, eventually leading to iron production. When too much iron accumulates, the high-mass star explodes as a supernova, leaving behind a neutron star or black hole.

Figure 17.20 summarizes the stages in the life cycles of stars by focusing on two illustrative cases: a high-mass star with a mass 25 times that of our Sun, and a

Continued on p. 566

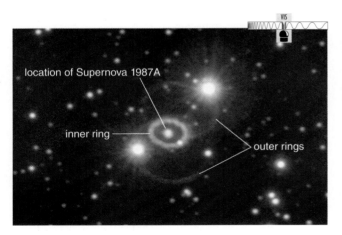

a The gaseous rings around Supernova 1987A glow because they were heated by ultraviolet light from the supernova explosion. In three dimensions, the rings outline an hourglass shape with the outer rings at the top and bottom of the hourglass and the inner ring (0.7 light-years in radius) at the waist of the hourglass.

Figure 17.19 Rings around Supernova 1987A.

b This montage shows how the inner ring around Supernova 1987A has gradually brightened in different wavelengths. It is expected to brighten even more as additional supernova debris rams into it over the next few years. The visible light images were taken by Hubble Space Telescope, the X-ray images are from the Chandra Observatory, and the radio wave images are from the Australia Telescope Compact Array.

Protostars: A star system forms when a cloud of interstellar gas collapses under gravity. The central protostar is surrounded by a protostellar disk in which planets may eventually form.

Life of a 25M_{Sun} Star.
Main-sequence lifetime: 5 million years
Duration of later stages: <1 million years

Blue main-sequence star: Star is fueled by hydrogen fusion in its core. In high-mass stars, hydrogen fusion proceeds by the series of reactions known as the CNO cycle.

Red supergiant: After core hydrogen is exhausted, the core shrinks and heats. Hydrogen shell burning begins around the inert helium core, causing the star to expand into a red supergiant.

Helium-burning supergiant: Helium fusion begins when enough helium has collected in the core. The core then expands, slowing the fusion rate and allowing the star's outer layers to shrink somewhat. Hydrogen shell burning continues at a reduced rate.

Multiple shell–burning supergiant: After core helium is exhausted, the core shrinks until carbon fusion begins, while helium and hydrogen continue to burn in shells surrounding the core. Late in its life, the star fuses heavier elements like carbon and oxygen in shells while iron collects in the inert core.

Neutron star or black hole: The core collapse that initiates the supernova forms a ball of neutrons, which may remain behind as a neutron star or collapse further to make a black hole.

Supernova: Iron cannot provide fusion energy, so it accumulates in the core until degeneracy pressure can no longer support it. Then the core collapses, leading to the catastrophic explosion of the star.

Life of a 1 M_{Sun} Star.
Main-sequence lifetime: 10 billion years
Duration of later stages: 1 billion years

Figure 17.20 Summary of stellar lives. The life stages of a high-mass star (on the left) and a low-mass star (on the right) are depicted in clockwise sequences beginning with the protostellar stage in the upper left corner. (Stars not drawn to scale.)

Yellow main-sequence star: Star is fueled by hydrogen fusion in its core, which converts four hydrogen nuclei into one helium nucleus. In low-mass stars, hydrogen fusion proceeds by the series of reactions known as the proton-proton chain.

Red giant star: After core hydrogen is exhausted, the core shrinks and heats. Hydrogen shell burning begins around the inert helium core, causing the star to expand into a red giant.

Helium-burning star: Helium fusion, in which three helium nuclei fuse to form a single carbon nucleus, begins when enough helium has collected in the core. The core then expands, slowing the fusion rate and allowing the star's outer layers to shrink somewhat. Hydrogen shell burning continues at a reduced rate.

Double shell–burning red giant: After core helium is exhausted, the core again shrinks and heats. Helium shell burning begins around the inert carbon core and the star enters its second red giant phase. Hydrogen shell burning continues.

Planetary nebula: The dying star expels its outer layers in a planetary nebula, leaving behind the exposed inert core.

White dwarf: The remaining white dwarf is made primarily of carbon and oxygen because the core never grew hot enough to fuse these elements into anything heavier.

low-mass star with the same mass as our Sun. Remember that all types of stars grow larger and redder when they exhaust their core hydrogen, and during these late stages they make the elements necessary for human existence. Much of the carbon in our bodies was made in low- and intermediate-mass stars and then blown into space by their stellar winds and planetary nebulae. Most of the heavier elements that our bodies rely upon were made in higher-mass stars and expelled in supernova explosions.

• How are the lives of stars with close companions different?

For the most part, stars in binary systems proceed from birth to death as if they were isolated and alone. However, exceptions can occur in close binary star systems. Algol, the "demon star" in the constellation Perseus, provides a good example of such exceptions. Algol appears as a single star to our eyes and telescopes, but it is actually an eclipsing binary star system [Section 15.1] consisting of two stars that orbit each other closely: a $3.7M_{Sun}$ main-sequence star and a $0.8M_{Sun}$ subgiant.

A moment's thought reveals that something quite strange is going on. The stars of a binary system are born at the same time and therefore must both be the same age. We know that more massive stars live shorter lives, and therefore the more massive star must exhaust its core hydrogen and become a subgiant before the less massive star does. How, then, can Algol's less massive star be a subgiant while the more massive star is still burning hydrogen as a main-sequence star?

This so-called *Algol paradox* reveals some of the complications in ordinary stellar life cycles that can arise in close binary systems. The two stars in a close binary are near enough to exert significant tidal forces on each other [Section 4.5]. The gravity of each star attracts the near side of the other star more strongly than it attracts the far side. The stars therefore stretch into football-like shapes rather than remain spherical. In addition, the stars become *tidally locked* so that they always show the same face to each other, much as the Moon always shows the same face to Earth.

During the time that both stars are main-sequence stars, the tidal forces have little effect on their lives. However, the situation can change when one of the stars begins to expand into a red giant. The more massive star has a shorter main-sequence lifetime and therefore exhausts its core hydrogen first. The massive star swells in size to becomes a red giant, allowing the gravity of the small companion to pull gas away from its huge, tidally distorted outer layers. The result is **mass exchange**: As gas spills over from the red giant to the companion, the companion gains mass at the giant's expense.

The solution to the Algol paradox should now be clear (Figure 17.21). The $0.8M_{Sun}$ subgiant *used to be* much more

Algol shortly after its birth. The higher-mass star (left) evolved more quickly than its lower-mass companion (right).

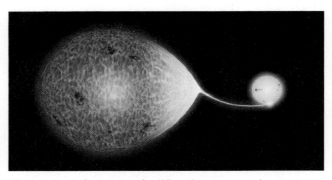

Algol at onset of mass transfer. When the more massive star expanded into a red giant, it began losing some of its mass to its normal, hydrogen-burning companion.

Algol today. As a result of the mass transfer, the red giant has shrunk to a subgiant, and the normal star on the right is now the more massive of the two stars.

Figure 17.21 Artist's conception of the development of the Algol close binary system.

massive. As the more massive star, it was the first to begin expanding into a red giant. As it expanded, however, so much of its matter spilled over onto its companion that it is now the less massive star.

The future may hold even more interesting events for Algol. The $3.7M_{Sun}$ star is still gaining mass from its subgiant companion. Thus, its life cycle is actually accelerating as its increasing gravity raises its core hydrogen fusion rate. Millions of years from now, it will exhaust its hydrogen and begin to expand into a red giant itself. At that point, it can begin to transfer mass *back* to its companion. Even more amazing things can happen in other mass-exchange systems, particularly when one of the stars is a white dwarf or a neutron star. But that is a topic for the next chapter.

Putting Chapter 17 into Context

In this chapter, we answered the question of the origin of elements that we first discussed in Chapter 1. As you look back over this chapter, remember these "big picture" ideas:

● Virtually all elements besides hydrogen and helium were forged in the nuclear furnaces of stars and released into space from these stars. Thus, we and our planet are made of stuff produced in stars that lived and died long ago.

● Low-mass stars like our Sun live long lives and die with the ejection of a planetary nebula, leaving behind a white dwarf.

● High-mass stars live fast and die young, exploding dramatically as supernovae and leaving behind neutron stars or black holes.

● Close binary stars can exchange mass, altering the usual course of stellar evolution.

SUMMARY OF KEY CONCEPTS

17.1 Lives in the Balance

• **How does a star's mass affect nuclear fusion?** Stars of greater mass have hotter core temperatures, causing fusion to proceed more quickly and enabling fusion of heavier elements to take place. A star's *mass* at birth therefore determines almost every aspect of its life and death. To understand the general characteristics of stellar lives, we divide stars into three groups by mass: **low-mass stars** with masses less than $2M_{Sun}$; **intermediate-mass stars** with masses between 2 and $8M_{Sun}$; and **high-mass stars** with masses above $8M_{Sun}$.

17.2 Life as a Low-Mass Star

• **What are the life stages of a low-mass star?** A low-mass star spends most of its life generating energy

by fusing hydrogen in its core via the proton-proton chain. When core hydrogen is exhausted, the core begins to shrink while the star as a whole expands to become a **red giant**, with **hydrogen shell burning** around an inert helium core. When the core becomes hot enough, a **helium flash** initiates **helium fusion** in the core, which fuses helium into carbon. This phase lasts until core helium is exhausted. Low-mass stars never become hot enough for carbon fusion, so at this point their lives must come to an end.

• **How does a low-mass star die?** The core again shrinks after core helium

burning ceases. Helium shell burning begins around the inert carbon core and beneath the hydrogen-burning shell. The outer layers expand again, making the star into a double-shell burning red giant. The star's energy generation never reaches equilibrium during this time; instead, the star experiences a series of **thermal pulses** and ultimately expels

its outer layers into space as a **planetary nebula**. The remaining "dead" stellar core is a **white dwarf**.

17.3 Life as a High-Mass Star

• **What are the life stages of a high-mass star?** A high-mass star lives a much shorter life than a low-mass

star, fusing hydrogen into helium via the **CNO cycle**. After exhausting its core hydrogen, a high-mass star begins hydrogen shell burning and then goes through a series of stages burning successively heavier elements. The furious rate of this fusion makes the star swell in size to become a supergiant.

• **How do high-mass stars make the elements necessary for life?** In its final stages of life, a high-

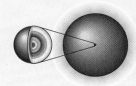

mass star's core becomes hot enough to fuse carbon and other heavy elements. The variety of different fusion reactions produces a wide range of elements—including all the elements necessary for life—that are then released into space when the star dies.

• **How does a high-mass star die?** A high-mass

star dies in the cataclysmic explosion of a **supernova**, scattering newly produced elements into space and leaving a neutron star or black hole behind. The supernova occurs after fusion begins to pile up iron in the high-mass star's core. Because iron fusion cannot release energy, the core cannot hold off the crush of gravity for long. In the instant that gravity overcomes degeneracy pressure, the core collapses and the star explodes. The expelled gas may be visible for a few thousand years as a **supernova remnant**.

17.4 The Roles of Mass and Mass Exchange

- **How does a star's mass determine its life story?** A star's mass determines how it lives its life. Low-mass stars never get hot enough to fuse carbon or heavier elements in their cores, and they end their lives by expelling their outer layers and leaving a white dwarf behind. High-mass stars live short but brilliant lives, ultimately dying in supernova explosions.

- **How are the lives of stars with close companions different?** When one star in a close binary system begins to swell in size at the end of its hydrogen-burning life, it can begin to transfer mass to its companion. This **mass exchange** can change the remaining life histories of both stars.

EXERCISES AND PROBLEMS

For instructor-assigned homework go to MasteringASTRONOMY www.masteringastronomy.com

Review Questions

Short-Answer Questions Based on the Reading

1. Why is mass so important to a star's life? How and why do we divide stars into groups by mass?
2. What do all low-mass stars share in common? Why do they differ in their levels of surface activity? What are *flare stars*?
3. When a star exhausts its core hydrogen fuel, the core contracts but the star as a whole expands. Why?
4. What is the basic reaction of helium fusion, and why does it require much higher temperatures than hydrogen fusion? Briefly explain why helium fusion in the Sun will begin with a *helium flash*.
5. Why does the H-R diagram of a globular cluster show a *horizontal branch*? What are the characteristics of the stars on the horizontal branch?
6. What happens to a low-mass star after it exhausts its core helium? Why can't it fuse carbon into heavier elements?
7. What are *carbon stars*, and how are they important to life on Earth?
8. What is a *planetary nebula*? What happens to the core of a star after a planetary nebula occurs?
9. What will happen to Earth as the Sun changes in the future?
10. Summarize the stages of life that we see on the Sun's life track in Figure 17.8. Be sure to explain both the changes that occur in the Sun's core with each stage and the changes that are observable from outside the Sun.
11. In broad terms, explain how the life of a high-mass star differs from that of a low-mass star. How do intermediate-mass stars fit into this picture?
12. Describe some of the nuclear reactions that can occur in high-mass stars after they exhaust their core helium. Why does this continued nuclear burning occur in high-mass stars but not in low-mass stars?
13. Why can't iron be fused to release energy?
14. Summarize some of the observational evidence supporting our ideas about how heavy elements form in massive stars.
15. What event initiates a supernova? Explain what happens during the explosion, and why a neutron star or black hole is left behind. What observational evidence supports our understanding of supernovae?
16. What is the Algol paradox and its resolution? Why can the lives of close binary stars differ from those of single stars?

Test Your Understanding

Does It Make Sense?

Decide whether the statement makes sense (or is clearly true) or does not make sense (or is clearly false). Explain your reasoning. (For an example, see Chapter 1, "Does It Make Sense?")

17. The iron in my blood came from a star that blew up more than 4 billion years ago.
18. Humanity will eventually have to find another planet to live on, because one day the Sun will blow up as a supernova.
19. I sure am glad hydrogen has a higher mass per nuclear particle than many other elements. If it had the lowest mass per nuclear particle, none of us would be here.
20. I just discovered a $3.5M_{Sun}$ main-sequence star orbiting a $2.5M_{Sun}$ red giant. I'll bet that red giant was more massive than $3M_{Sun}$ when it was a main-sequence star.
21. If the Sun had been born as a high-mass star some 4.6 billion years ago rather than as a low-mass star, the planet Jupiter would probably have Earth-like conditions today, while Earth would be hot like Venus.
22. If you could look inside the Sun today, you'd find that its core contains a much higher proportion of helium and a lower proportion of hydrogen than it did when the Sun was born.
23. Most of the supernova explosions that occur in a star cluster happen during its first 100 million years.
24. Globular clusters generally contain lots of white dwarfs.
25. After hydrogen fusion stops in a low-mass star, its core cools off until the star becomes a red giant.
26. The gold in my new ring came from a supernova explosion.

Quick Quiz

Choose the best answer to each of the following. Explain your reasoning with one or more complete sentences.

27. Which of these stars has the hottest core? (a) a white main-sequence star (b) an orange main-sequence star (c) a red main-sequence star
28. Which of these stars has the hottest core? (a) a blue main-sequence star (b) a red supergiant (c) a red main-sequence star
29. Which of these stars does not have fusion occurring in its core? (a) a red giant (b) a red main-sequence star (c) a blue main-sequence star

30. What happens to a low-mass star after helium flash? (a) Its luminosity goes up. (b) Its luminosity goes down. (c) Its luminosity stays the same.

31. What would stars be like if hydrogen had the smallest mass per nuclear particle? (a) Stars would be brighter. (b) All stars would be red giants. (c) Nuclear fusion would not occur in stars of any mass.

32. What would stars be like if carbon had the smallest mass per nuclear particle? (a) Supernovae would be more common. (b) Supernovae would never occur. (c) High-mass stars would be hotter.

33. What would you be most likely to find if you returned to the solar system in 10 billion years? (a) a neutron star (b) a white dwarf (c) a black hole

34. Which of these stars has the shortest life expectancy? (a) an isolated $1M_{Sun}$ star (b) a $1M_{Sun}$ star in a close binary with a $0.8M_{Sun}$ star (c) a $1M_{Sun}$ star in a close binary with a $2M_{Sun}$ star

35. What happens to the core of a high-mass star after it runs out of hydrogen? (a) It shrinks and heats up. (b) It shrinks and cools down. (c) Helium fusion begins right away.

36. Which of these elements had to be made in a supernova explosion? (a) calcium (b) uranium (c) oxygen

Investigate Further
In-Depth Questions to Increase Your Understanding

Short-Answer/Essay Questions

Homes to Civilization? We do not yet know how many stars have Earth-like planets, nor do we know the likelihood that such planets might harbor advanced civilizations like our own. However, some stars can probably be ruled out as candidates for advanced civilizations. For example, given that it took a few billion years for humans to evolve on Earth, it seems unlikely that advanced life would have had time to evolve around a star that is only a few million years old. For each of the following stars, decide whether you think it is possible that it could harbor an advanced civilization. Explain your reasoning in one or two paragraphs.

37. A $10M_{Sun}$ main-sequence star.
38. A $1.5M_{Sun}$ main-sequence star.
39. A $1.5M_{Sun}$ red giant.
40. A $1M_{Sun}$ helium-burning star.
41. A red supergiant.
42. A flare star.
43. A carbon star.
44. *Rare Elements.* Lithium, beryllium, and boron are elements with atomic numbers 3, 4, and 5, respectively. Despite their being three of the five simplest elements, Figure 17.15 shows that they are rare compared to many heavier elements. Suggest a reason for their rarity. (*Hint:* Consider the process by which helium fuses into carbon.)
45. *Future Skies.* As a red giant, the Sun will have an angular size in Earth's sky of about 30°. What will sunset and sunrise be like? About how long will they take? Do you think the color of the sky will be different from what it is today? Explain.
46. *Research: Historical Supernovae.* Historical accounts exist for supernovae in the years 1006, 1054, 1572, and 1604. Choose one of these supernovae and learn more about historical records of the event. Did the supernova influence human history in any way? Write a two- to three-page summary of your research findings.

Quantitative Problems
Be sure to show all calculations clearly and state your final answers in complete sentences.

47. *Density of a Red Giant.* Near the end of its life, the Sun's radius will extend nearly to the distance of Earth's orbit. Estimate the volume of the Sun at that time using the formula for the volume of a sphere $(4\pi r^3/3)$. Using that result, estimate the average matter density of the Sun at that time. How does that density compare with the density of water (1 g/cm^3)? How does it compare with the density of Earth's atmosphere at sea level (about 10^{-3} g/cm^3)?

48. *Escape Velocity from a Red Giant.* What is the escape velocity from a red giant with a mass of $1M_{Sun}$ and a radius of $100R_{Sun}$? How does that velocity compare with the escape velocity from the Sun? Describe how your results help account for the fact that red giants have strong stellar winds.

49. *Roasting the Earth.* During its final days as a red giant, the Sun will reach a peak luminosity of about $3,000L_{Sun}$. Earth will therefore absorb about 3,000 times as much solar energy as it does now, and it will need to radiate 3,000 times as much thermal energy to keep its surface temperature in balance. Estimate the temperature Earth's surface will need to attain in order to radiate that much thermal energy. You will need to use the formula for emitted power per unit area in Mathematical Insight 15.5.

50. *Supernova Betelgeuse.* The distance of the red supergiant Betelgeuse is approximately 427 light-years. If it were to explode as a supernova, it would be one of the brightest stars in the sky. Right now, the brightest star other than the Sun is Sirius, with a luminosity of $26L_{Sun}$ and a distance of 26 light-years. How much brighter than Sirius would the Betelgeuse supernova be in our sky if it reached a maximum luminosity of $10^{10}L_{Sun}$?

51. *Construction of Elements.* Using the periodic table in Appendix D, determine which elements are made by the following nuclear fusion reactions. (You can assume the total number of protons in the reaction remains constant.)
    a. Fusion of a carbon nucleus with another carbon nucleus
    b. Fusion of a carbon nucleus with a neon nucleus
    c. Fusion of an iron nucleus with a helium nucleus

52. *Expansion of the Crab Nebula.* Observations of the Crab Nebula taken over several decades show that gas blobs that are now 100 arcseconds from the center of the nebula are moving away from the center by about 0.11 arcsecond per year. Use that information to estimate the year in which the explosion ought to have been observed. How does that year compare with the year in which the supernova that produced the nebula was actually observed?

53. *Algol's Orbital Separation.* The Algol binary system consists of a $3.8M_{Sun}$ star and a $0.8M_{Sun}$ star with an orbital period of 2.87 days. Use Newton's version of Kepler's third law to calculate the orbital separation of the system. How does that separation compare with the typical size of a red giant star?

54. *The Speed of Supernova Debris.* Compute the speed of the debris that was seen hitting the inner ring around Supernova 1987A in the year 2001. Assume that the radius of the inner ring is 0.7 light-year. How does the speed you find compare with the speed of light?

Discussion Questions

55. *Connections to the Stars.* In ancient times, many people believed that our lives were somehow influenced by the patterns of the stars in the sky. Modern science has not found

any evidence to support this belief, but instead has found that we have a connection to the stars on a much deeper level: We are "star stuff." Discuss in some detail our real connections to the stars as established by modern astronomy. Do you think these connections have any philosophical im-

plications in terms of how we view our lives and our civilization? Explain.

56. *Humanity in* A.D. *5,000,000,000.* Do you think it is likely that humanity will survive until the Sun begins to expand into a red giant 5 billion years from now? Why or why not?

 Interactive Tutorials

Tutorial Review of Key Concepts

Use the interactive **Tutorial** at **www.mastering astronomy.com** to review key concepts from this chapter.

Stellar Evolution Tutorial
Lesson 1 Main-Sequence Lifetimes
Lesson 2 Evolution of a Low-Mass Star
Lesson 3 Late Stages of a High-Mass Star

Supplementary Tutorial Exercises

Use the interactive **Tutorial Lessons** to explore the following questions.

Stellar Evolution Tutorial, Lesson 1
1. Use the tool for calculating stellar lifetimes to estimate the lifetimes of 10 stars of different masses. Record the mass and lifetime for each of your 10 stars.
2. Make a graph of your results from question 1, plotting mass on the *x*-axis and lifetime on the *y*-axis.
3. Based on your graph from question 2, briefly describe in words how lifetime depends on mass for a main-sequence star.

Stellar Evolution Tutorial, Lessons 2, 3
1. Study the animations for both the low-mass (Lesson 2) and high-mass (Lesson 3) stars. How are the lives of low-mass and high-mass stars similar? How are they different?
2. How do low-mass stars "move" on the H-R diagram as they go through their various stages of life?
3. How do high-mass stars "move" on the H-R diagram as they go through their various stages of life?

 Exploring the Sky and Solar System

Of the many activities available on the *Voyager: SkyGazer* CD-ROM accompanying your book, use the following files to observe key phenomena covered in this chapter.

Go to the **File: Demo** folder for the following demonstrations.
1. Crab from Finland

 Movies

Check out the following narrated and animated short documentaries available on **www.masteringastronomy.com** for a helpful review of key ideas covered in this chapter.

Lives of Stars Movie   *Double Stars Movie*

Web Projects

1. *Fireworks in Supernova 1987A.* The light show from Supernova 1987A is not yet over. In particular, between now and about 2010, the expanding cloud of gas from the supernova is expected to continue ramming into the material surrounding the star, and the heat generated by the impact is expected to create a new light show. Learn more about how Supernova 1987A is changing and what we might expect to see from it in the future. Summarize your findings in a one- to two-page report.

2. *Picturing Star Birth and Death.* Photographs of stellar birthplaces (i.e., molecular clouds) and death places (e.g., planetary nebulae and supernova remnants) can be strikingly beautiful, but only a few such photographs are included in this chapter. Search the Web for additional photographs of this type. Look not only for photos taken in visible light, but also for photographs made from observations in other wavelengths of light. Put each photograph you find into a personal online journal, along with a one-paragraph description of what the photograph shows. Try to compile a journal of at least 20 such photographs.

The Bizarre Stellar Graveyard

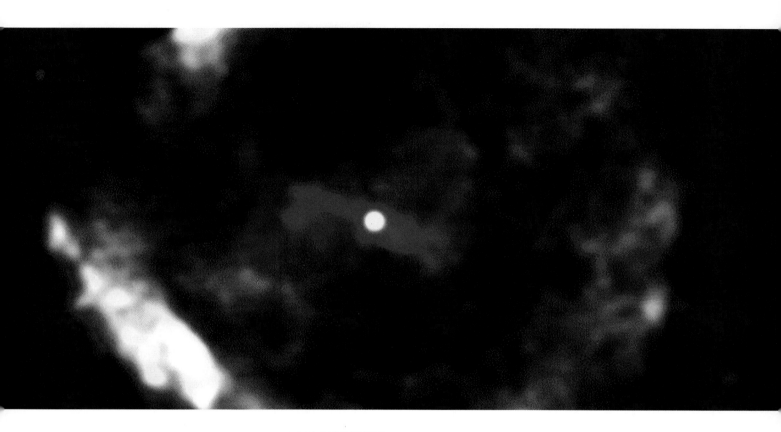

LEARNING GOALS

18.1 White Dwarfs
- **What is a white dwarf?**
- **What can happen to a white dwarf in a close binary system?**

18.2 Neutron Stars
- **What is a neutron star?**
- **How were neutron stars discovered?**
- **What can happen to a neutron star in a close binary system?**

18.3 Black Holes: Gravity's Ultimate Victory
- **What is a black hole?**
- **What would it be like to visit a black hole?**
- **Do black holes really exist?**

18.4 The Mystery of Gamma-Ray Bursts
- **Where do gamma-ray bursts come from?**
- **What causes gamma-ray bursts?**

> *Now, my suspicion is that the universe is not only queerer than we suppose, but queerer than we can suppose.*
>
> **—J. B. S. Haldane, *Possible Worlds*, 1927**

Sirius A, the brightest star in the night sky, appears relatively dim in this X-ray photo . . .

. . . because its hot white dwarf companion Sirius B is brighter in ultraviolet and X-ray light.

Figure 18.1 The binary star system Sirius as seen by the Chandra X-Ray Observatory. Sirius B, the white dwarf in the binary, has a much hotter surface than its companion, Sirius A, and is therefore much brighter in ultraviolet and X-ray light. (The spikes emanating from Sirius B are not real. They are artifacts created by the telescope's optics.)

Welcome to the afterworld of stars, the fascinating domain of white dwarfs, neutron stars, and black holes. To scientists, these dead stars are ideal laboratories for testing the most extreme predictions of general relativity and quantum theory. To most other people, the eccentric behavior of stellar corpses demonstrates that the universe is stranger than they ever imagined.

Dead stars behave in unusual and unexpected ways that challenge our minds and stretch the boundaries of what we believe is possible. Stars that have finished nuclear burning have only one hope of staving off the crushing power of gravity: the quantum mechanical effect of degeneracy pressure. However, even this form of pressure cannot save the most massive stellar cores. In this chapter, we will study the bizarre properties and occasional catastrophes of the stellar corpses known as white dwarfs, neutron stars, and black holes. Prepare to be amazed by the eerie inhabitants of the stellar graveyard!

Stellar Evolution Tutorial, Lesson 1

18.1 White Dwarfs

In the previous two chapters, we've seen that an ongoing "battle" between the inward crush of gravity and the outward push of pressure governs a star's life from the time it first begins to form in an interstellar cloud to the time it finally exhausts its nuclear fuel. This battle continues when a star dies, and its outcome determines the ultimate fate of the star's remains.

Stellar corpses come in three basic types: white dwarfs, neutron stars, and black holes. Recall that low-mass stars like the Sun leave behind white dwarfs when they die. Higher-mass stars die in the titanic explosions known as supernovae, leaving behind a neutron star or black hole. Because low-mass stars are so much more numerous than high-mass stars, white dwarfs are by far the most common type of stellar corpse. We'll therefore begin our study of the stellar graveyard with white dwarfs, and later turn our attention to the even more exotic properties of neutron stars and black holes.

• What is a white dwarf?

A **white dwarf** is essentially the exposed core of a star that has died and shed its outer layers in a planetary nebula [Section 17.2]. It is quite hot when it first forms, because it was recently the core of a star, but slowly cools with time. White dwarfs are stellar in mass but small in radius [Section 15.2], which is why they are generally dim compared to an average star like the Sun. However, the hottest white dwarfs can

shine quite brightly in high-energy light such as ultraviolet and X rays (Figure 18.1).

A white dwarf's combination of a starlike mass with a small radius makes gravity very strong near its surface. If gravity were unopposed, it would crush the white dwarf to an even smaller size, so some sort of pressure must be pushing back equally hard to keep the white dwarf stable. Because there is no fusion to maintain heat and pressure inside a white dwarf, the pressure that opposes gravity must come from some other source. As we discussed in Chapters S3 and 17, the source is *degeneracy pressure*—a type of pressure that arises when subatomic particles are packed as closely as the laws of quantum mechanics allow. More specifically, the degeneracy pressure in white dwarfs arises from closely packed electrons, so we call it **electron degeneracy pressure**. Thus, a white dwarf exists in a state of balance because the outward push of electron degeneracy pressure matches the inward crush of its gravity.

White Dwarf Composition, Density, and Size Because a white dwarf is the core left over after a star has ceased nuclear burning, its composition reflects the products of the star's final nuclear-burning stage. The white dwarf left behind by a $1M_{\text{Sun}}$ star like our Sun will be made mostly of carbon, since stars like the Sun fuse helium into carbon in their final stage of life. The cores of very-low-mass stars never become hot enough to fuse helium and thus end up as helium white dwarfs. Some intermediate-mass stars progress to carbon burning but do not create any iron (and hence do not explode as supernovae). These stars leave behind white dwarfs containing large amounts of oxygen or even heavier elements.

Despite these ordinary-sounding compositions, a scoop of matter from a white dwarf would be unlike anything ever seen on Earth. A typical white dwarf has the mass of the Sun ($1M_{Sun}$) compressed into an object the size of Earth. If you recall that Earth is smaller than a typical sunspot, you can probably imagine that packing the entire mass of the Sun into the volume of Earth is no small feat. The density of such a white dwarf is so high that a teaspoon of its material would weigh as much as a small truck if you could bring that material to Earth.

More massive white dwarfs are actually smaller in size than less massive ones. For example, a $1.3M_{Sun}$ white dwarf is half the diameter of a $1.0M_{Sun}$ white dwarf (Figure 18.2). The more massive white dwarf is smaller because its greater gravity can compress its matter to a much greater density. According to the laws of quantum mechanics, the electrons in a white dwarf must respond to this compression by moving faster, which makes the degeneracy pressure strong enough to resist the increased force of gravity. The most massive white dwarfs are therefore the smallest.

The idea that more massive white dwarfs are smaller in size also explains why red giants become more luminous as they age [Section 17.2]. Degeneracy pressure supports the inert helium core of a low-mass red giant, so this core is essentially a white dwarf buried inside a star. As the hydrogen-burning shell above it deposits more helium ash onto the degenerate core, the mass of the core continually increases. Just as more massive white dwarfs must be smaller in size, adding mass to the degenerate core must cause it to contract. The surrounding shell of hydrogen fusion shrinks along with it, and this shrinking makes the shell hotter and increases its fusion rate. Thus, as long as the core remains inert and fusion occurs in a shell overlying it, the luminosity of the red giant must steadily increase.

The White Dwarf Limit The fact that electron speeds are higher in more massive white dwarfs leads to a fundamental limit on the maximum mass of a white dwarf. Theoretical calculations show that the electron speeds would reach the speed of light in a white dwarf with a mass of about 1.4 times the mass of the Sun ($1.4M_{Sun}$). Because neither electrons nor anything else can travel faster than the speed of light, no white dwarf can have a mass greater than this $1.4M_{Sun}$ **white dwarf limit**. (The white dwarf limit is often called the *Chandrasekhar limit*, after its discoverer.)

Strong observational evidence supports this theoretical limit on the mass of a white dwarf. Many known white dwarfs are members of binary systems, and hence we can measure their masses [Section 15.1]. In every observed case, the white dwarfs have masses below $1.4M_{Sun}$, just as we expect from theory.

• What can happen to a white dwarf in a close binary system?

Left to itself, a white dwarf will never again shine as brightly as the star it once was. With no source of fuel for fusion, it will simply cool with time into a cold, black dwarf. Its size will never change, because its electron degeneracy pressure will forever keep it stable against the crush of gravity. However, the situation can be quite different for a white dwarf in a close binary system.

Accretion Disks A white dwarf in a close binary system can gradually gain mass if its companion is a main-sequence or giant star (Figure 18.3). When a clump of mass first spills over from the companion to the white dwarf, it has some small orbital velocity. The law of conservation of angular momentum dictates that the clump must orbit faster and faster as it falls toward the white dwarf's surface. The in-falling matter therefore forms a whirlpool-like disk around the white dwarf. Because the process in which material falls onto another body is called *accretion* [Sections 8.3, 16.2], this rapidly rotating disk is called an **accretion disk**.

Accretion disks form around white dwarfs for much the same reason that infalling gas creates *protostellar disks* around protostars. (Indeed, a protostellar disk is a type of accretion disk [Section 16.2].) In both cases, gas particles in the disk move on orbits that obey Kepler's laws, so that gas in the inner parts of the disk orbits faster than gas in the outer parts. The differences in gas particle speeds lead to friction that removes orbital energy from the inner parts of the accretion disk, and the loss of energy means that gas gradually spirals inward and eventually settles onto the protostar or white dwarf. The major differences between protostellar disks and the accretion disks around white dwarfs are size and orbital speed. The small size and high density of a white dwarf make its surface gravity far stronger than that of a protostar, which means the gas in the accretion disk around a white dwarf moves at much higher orbital speeds than the gas in a protostellar disk. Because of the higher orbital speeds, more orbital energy can be turned into heat, making the disk around a white dwarf much hotter than the one around a protostar.

Accretion can provide a "dead" white dwarf with a new energy source as long as its companion keeps feeding matter into the accretion disk. Theory predicts that the heat generated by friction should make the accretion disk hot enough to radiate optical and ultraviolet light, and sometimes even X rays. Thus, although accretion disks around white dwarfs are far too small to be seen directly, we should be able to detect their intense ultraviolet or X-ray radiation.

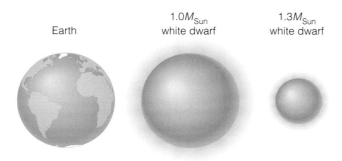

Earth $1.0M_{Sun}$ $1.3M_{Sun}$
               white dwarf   white dwarf

Figure 18.2 Contrary to what you might expect, more massive white dwarfs are actually *smaller* (and thus denser) than less massive white dwarfs. Earth is shown for scale.

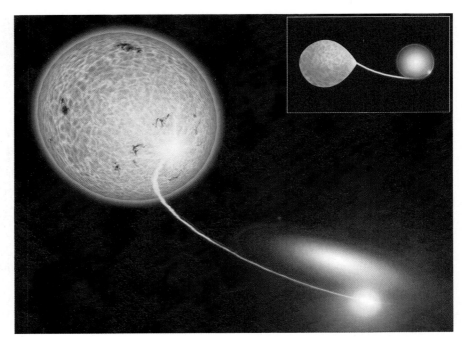

Figure 18.3 This artist's conception shows how mass spilling from a companion star (left) toward a white dwarf (right) forms an accretion disk around the white dwarf. The white dwarf itself is in the center of the accretion disk—too small to be seen on this scale. Matter streaming onto the disk creates a hot spot at the point of impact. The inset shows how the system looks from above rather than from the side.

Searches for this radiation have turned up strong evidence for accretion disks around many white dwarfs. In some cases, the brightness of these systems is highly variable—sudden increases in brightness by a factor of 10 or more may persist for a few days and then fade away, only to repeat a few weeks or months later. Such brightening events (sometimes called *dwarf novae*) probably arise when instabilities in the accretion disk cause some of the matter to fall suddenly onto the white dwarf's surface, with an accompanying release of gravitational potential energy.

Novae Accreting white dwarfs occasionally flare up even more dramatically. Remember that a white dwarf is "dead" because it has no hydrogen left to fuse. This situation changes for an accreting white dwarf. The gas spilling onto an accreting white dwarf comes from the upper layers of its companion star and thus is composed mostly of hydrogen. The white dwarf's strong gravity compresses this hydrogen gas into a thin surface layer that gradually builds in mass as more and more material rains down from the accretion disk. Both the pressure and temperature rise as the shell grows, and hydrogen fusion ignites when the temperature at the bottom of the layer exceeds 10 million K.

With the onset of fusion in the hydrogen shell, the white dwarf suddenly blazes back to life. This thermonuclear flash causes the binary system to shine for a few glorious weeks as a **nova** (Figure 18.4a). A nova can shine as brightly as 100,000 Suns. It generates heat and pressure, ejecting most of the material that has accreted onto the white dwarf. This material expands outward, creating a *nova remnant* that may remain visible for many years after the nova explosion (Figure 18.4b).

Accretion resumes after a nova explosion subsides, so the entire process can repeat itself. The time between successive novae in a particular system depends on the rate at which hydrogen accretes on the white dwarf surface and on how highly compressed this hydrogen becomes. The compression of hydrogen is greatest for the most massive white dwarfs, which have the strongest surface gravities. In some cases, novae have been observed to repeat after just a few decades. More commonly, accreting white dwarfs may have 10,000 years between nova outbursts.

Notice that, according to our modern definitions, a nova and a supernova are quite different events: A nova is a relatively minor detonation of hydrogen fusion on the surface of a white dwarf in a close binary, while a supernova is the total explosion of a star. However, the word *nova* simply means "new." Historically, a nova referred to any star that appeared to the naked eye where none was visible before. Because supernovae generate far more power than novae—the light of 10 billion Suns in a supernova versus that of 100,000 Suns in a nova—a very distant supernova can appear as bright in our sky as a nova that is relatively close. Thus, people could not distinguish between novae and supernovae until modern astronomical methods enabled us to measure their distances.

White Dwarf Supernovae Each time a nova occurs, the white dwarf ejects some of its mass. Each time a nova subsides, the white dwarf begins to accrete matter again. Theoretical models cannot yet tell us whether the net result should be a gradual increase or decrease in the white dwarf's mass. Nevertheless, observations show that in at least some cases, accreting white dwarfs in binary systems continue to gain mass as time passes. If such a white dwarf gains enough mass, it can one day approach the $1.4 M_{Sun}$ white dwarf limit. This day is the white dwarf's last.

Remember that most white dwarfs are made largely of carbon. As the white dwarf's mass approaches $1.4 M_{Sun}$, its interior temperature rises high enough for carbon fusion to ignite. When carbon fusion begins, it ignites almost instantly throughout the star. This "carbon bomb" detonation

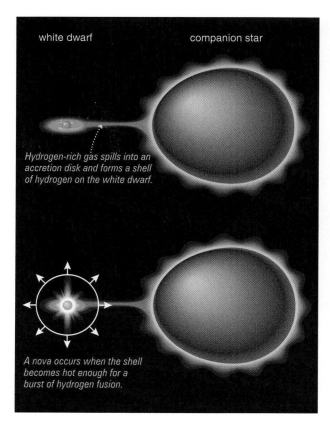

Hydrogen-rich gas spills into an accretion disk and forms a shell of hydrogen on the white dwarf.

A nova occurs when the shell becomes hot enough for a burst of hydrogen fusion.

a Diagram of the nova process.

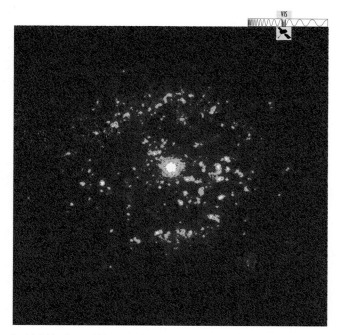

b Hubble Space Telescope image showing blobs of gas ejected from the nova T Pyxidis. The bright spot at the center of the blobs is the binary star system that generated the nova

Figure 18.4 A nova occurs when hydrogen fusion ignites on the surface of a white dwarf in a binary star system.

is similar to the helium flash that occurs in low-mass red giants [Section 17.2], but it releases far more energy because all the carbon undergoes fusion at once. The white dwarf explodes completely in what we will call a **white dwarf supernova**.

THINK ABOUT IT

According to our understanding of novae and white dwarf supernovae, can either of these events ever occur with a white dwarf that is *not* a member of a binary star system? Explain.

The "carbon bomb" detonation that creates a white dwarf supernova is quite different from the iron catastrophe that leads to a supernova ending the life of a high-mass star [Section 17.3]. To distinguish the two types of supernova, we'll refer to the latter as a **massive star supernova**. Both types of supernova shine brilliantly, with peak luminosities about 10 billion times that of the Sun ($10^{10}L_{Sun}$). Astronomers can distinguish between the two types by studying their light.* Spectra of white dwarf supernovae lack hydrogen lines, because white dwarfs contain very little hydrogen. In contrast, massive stars usually have plenty of hydro-

gen in their outer layers when they explode, so hydrogen lines are prominent in the spectra of most massive star supernovae. In addition, the luminosities of white dwarf supernovae fade steadily with time, while massive star supernovae fade in two distinct stages. Figure 18.5 contrasts typical light curves for white dwarf and massive star supernovae.

In addition to being dramatic, white dwarf supernovae are important to astronomers because they provide one of the primary means by which we measure large distances in the universe. Because white dwarf supernovae

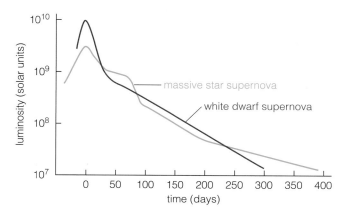

Figure 18.5 Interactive Figure The curves on this graph show how the luminosities of the two types of supernovae fade with time. The white dwarf supernova fades steadily after it peaks, while the massive star supernova fades in two distinct stages. (Notice that the luminosity scale is given in powers of 10.)

---

*Observationally, astronomers classify supernovae as *Type II* if their spectra show hydrogen lines, and *Type I* otherwise. All Type II supernovae are assumed to be massive star supernovae. However, a Type I supernova can be either a white dwarf supernova or a massive star supernova in which the star blew away all its hydrogen before exploding. Type I supernovae appear in three classes whose light curves differ, called *Type Ia*, *Type Ib*, and *Type Ic*. Only Type Ia supernovae are thought to be white dwarf supernovae.

always occur in white dwarfs that have just reached the $1.4 M_{Sun}$ limit, their light curves all look amazingly similar, and their maximum luminosities are nearly identical; the same is not true of massive star supernovae, because they occur in stars of many different masses. The fact that all white dwarf supernovae are nearly identical is extremely useful. Once we know the true luminosity of one white dwarf supernova, we essentially know the luminosities of them all. Thus, whenever we discover a white dwarf supernova in a distant galaxy, we can determine the galaxy's distance by using the inverse square law for light [Section 15.1]. In Chapter 20, we'll look more deeply at the role of white dwarf supernovae in helping us establish the cosmic distance scale.

 Stellar Evolution Tutorial, Lesson 3

18.2 Neutron Stars

White dwarfs with densities of 5 tons per teaspoon may seem incredible, but neutron stars are stranger still. The possibility that neutron stars might exist was first proposed in the 1930s, but many astronomers thought it preposterous that nature could truly make anything so bizarre. Nevertheless, a vast amount of evidence now makes it quite clear that neutron stars really exist. In this section, we'll examine their properties, their discovery, and the strange things that can happen to them in close binary star systems.

• What is a neutron star?

A **neutron star** is the ball of neutrons created by the collapse of the iron core in a massive star supernova (Figure 18.6). Typically just 10 kilometers in radius yet more massive than the Sun, neutron stars are essentially giant atomic nuclei made almost entirely of neutrons and held together by gravity. Like white dwarfs, neutron stars resist the crush of gravity with the strange degeneracy pressure that arises when particles are packed as closely as nature allows. In the case of neutron stars, however, it is neutrons rather than electrons that are closely packed, so we say that **neutron degeneracy pressure** supports them against the crush of gravity.

Neutron Star Gravity The force of gravity at the surface of a neutron star is truly awe-inspiring. Escape velocity is about half the speed of light. If you foolishly chose to visit a neutron star's surface, your body would be squashed immediately into a microscopically thin puddle of subatomic particles.

Things would be only slightly less troubling if a bit of neutron star could somehow come to visit you. A paper clip with the density of neutron star material would outweigh Mount Everest. If such a paper clip magically appeared in your hand, you could not prevent it from falling. Down it would plunge, passing through the Earth like a rock falling through air. It would gain speed until it reached the center of the Earth, and its momentum would carry it onward until

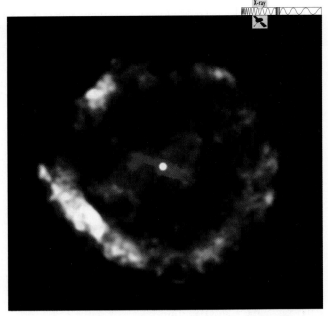

Figure 18.6 This X-ray image from the Chandra X-Ray Observatory shows the supernova remnant G11.2-03, the remains of a supernova observed by Chinese astronomers in A.D. 386. The white dot at the center represents X rays from the neutron star left behind by the supernova. The different colors correspond to different X-ray wavelength bands. The region pictured is about 23 light-years across.

it slowed to a stop on the other side of our planet. Then it would fall back down again. If it came in from space, each plunge of the neutron star material would drill a different hole through the rotating Earth. In the words of astronomer Carl Sagan, the inside of the Earth would "look briefly like Swiss cheese" (until the melted rock flowed to fill in the holes) by the time friction finally brought the piece of neutron star to rest at the center of the Earth.

In the unfortunate event that an *entire* neutron star came to visit you, it would not fall at all. Because it would be only about 10 kilometers in radius, the neutron star would probably fit in your hometown. Remember, however, that it would be 300,000 times as massive as Earth. As a result, the neutron star's immense surface gravity would quickly destroy your hometown and the rest of civilization. By the time the dust settled, the former Earth would be a shell no thicker than your thumb on the surface of the neutron star.

Neutron Star Structure Theoretical models of neutron stars predict that the state of the matter within a neutron star depends on its depth. Despite the fact that neutron stars are made primarily of neutrons, matter at the surface can be composed largely of electrons and positively charged atomic nuclei because the pressure there is not much different from the pressure inside a white dwarf. Those surface layers constitute the neutron star's *crust*. However, deeper inside a neutron star, the immense pressure of the overlying layers has caused individual atomic nuclei to disintegrate and has forced almost all the electrons to combine with protons. The inner layers of a neutron star are therefore made almost entirely of neutrons.

The properties of neutron-rich matter under the extreme pressures that prevail inside neutron stars are still somewhat uncertain. Our knowledge of those properties is incomplete because it is so difficult to re-create those conditions in laboratories on Earth. Observations of neutron stars are therefore particularly useful in helping us learn how matter behaves at extremely high density and pressure. The theoretical models suggest that the matter inside a neutron star has a property known as *superfluidity*, meaning that it flows around inside the neutron star without experiencing any friction whatsoever. Some of it may also be *superconducting*, meaning that electricity can pass through it without experiencing any resistance.

• How were neutron stars discovered?

The first observational evidence for neutron stars came in 1967, when a 24-year-old graduate student named Jocelyn Bell discovered a strange source of radio waves. Bell had helped her adviser, Anthony Hewish, build a radio telescope ideal for discovering fluctuating sources of radio waves. She was busily trying to interpret the flood of data pouring out of this instrument in October 1967 when she noticed a peculiar signal. After ruling out other possibilities, she concluded that *pulses* of radio waves were arriving from somewhere near the direction of the constellation Cygnus at precise 1.337301-second intervals (Figure 18.7).

Pulsars The pulses coming from Cygnus were very surprising because no known astronomical object pulsated so regularly. In fact, the pulsations came at such precise intervals that they were nearly as reliable for measuring time as the most precise human-made clocks. For a while, the mysterious source of the radio waves was dubbed "LGM" for Little Green Men—only half-jokingly. Today we refer to such rapidly pulsing radio sources as **pulsars**.

The mystery of pulsars was soon solved. By the end of 1968, astronomers had found two smoking guns: Pulsars sat at the centers of both the Crab Nebula and the Vela Nebula, the gaseous remnants of supernovae (Figure 18.8). The pulsars are neutron stars left behind by supernova explosions.

The pulsations arise because the neutron star is spinning rapidly as a result of the conservation of angular momentum [Section 4.3]: As an iron core collapses into a neutron star, its rotation rate must increase as it shrinks in size. The collapse also bunches the magnetic field lines running

Figure 18.7 About 20 seconds of data from the first pulsar discovered by Jocelyn Bell in 1967.

through the core far more tightly, greatly amplifying the strength of the magnetic field. As a result, the neutron star's magnetic field is a trillion times as strong as Earth's shortly after it forms in the supernova event. This intense magnetic field directs beams of radiation out along the magnetic poles, although we do not yet know exactly how. If a neutron star's magnetic poles are not aligned with its rotation axis, the beams of radiation sweep round and round (Figure 18.9). Like lighthouses, the neutron stars actually emit a fairly steady beam of light, but we see a pulse of light each time the beam sweeps past Earth.

Pulsars are not quite perfect clocks. The continual twirling of a pulsar's magnetic field generates electromagnetic radiation that carries away energy and angular momentum, causing the neutron star's rotation rate to slow gradually. The pulsar in the Crab Nebula, for example, currently spins about 30 times per second. Two thousand years from now, it will spin less than half as fast. Eventually, a pulsar's spin slows so much and its magnetic field becomes so weak that we can no longer detect it. In addition, some spinning neutron stars may be oriented so that their beams do not sweep past our location. Thus, we have the following rule: All pulsars are neutron stars, but not all neutron stars are pulsars.

THINK ABOUT IT

Suppose we do *not* see pulses from a particular neutron star and hence do not call it a pulsar. Is it possible that a civilization living in some other star system would see this neutron star as a pulsar? Explain.

We know that pulsars must be neutron stars because no other massive object could spin so fast. A white dwarf, for example, can spin no faster than about once per second. An increase in spin would tear it apart because its surface would be rotating faster than the escape velocity. Pulsars

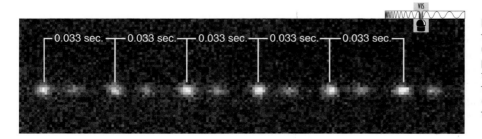

Figure 18.8 This time-lapse image of the pulsar at the center of the Crab Nebula, a supernova remnant, shows its main pulse recurring every 0.033 second. The fainter pulses are thought to come from the pulsar's other lighthouse-like beam. (Photo from the Very Large Telescope of the European Southern Observatory.)

A pulsar is a rotating neutron star that beams radiation along its magnetic axis.

Figure 18.9 Radiation from a neutron star can appear to pulse like a lighthouse if the neutron star is rotating. Each time one of the pulsar's beams sweeps across Earth, we see a pulse of radiation.

If the magnetic axis is not aligned with the rotation axis, the pulsar's beams sweep through space like lighthouse beams.

have been discovered that rotate as fast as 625 times per second. Only an object as extremely small and dense as a neutron star could spin so fast without breaking apart.

Pulsar Timing and Pulsar Planets Our ability to time the intervals between a pulsar's pulses with great precision has led to a number of interesting discoveries. For example, as we discussed in Chapter S3, observations of a known binary system in which both objects are neutron stars have been used to test Einstein's general theory of relativity. According to Einstein's theory, the orbits of the two neutron stars should slowly decay, because they lose energy by radiating *gravitational waves.* Careful tracking of the orbital period of that system has verified that the orbits decay exactly as the theory predicts (see Figure S3.21), giving scientists confidence that gravitational waves really exist, even though they have not yet been detected directly.

What made those observations possible was that one of the two neutron stars in the system is a pulsar. As it moves toward us in its orbit, its pulsation period appears to speed up because each successive pulse comes from closer to Earth. As it moves away, the pulses appear to slow down. Thus, because we can measure the arrival times of the pulses very precisely, we also obtain very accurate measurements of the pulsar's orbit.

Another discovery made through pulsar timing concerns a very unusual set of planets. As we discussed in Chapter 13, we now know of more than 150 planets orbiting other stars, most orbiting stars that are similar to our own Sun. However, the very first planets discovered outside our solar system actually orbit a pulsar. In 1992, the pulsar known as PSR B1257+12 was found to have a pulsation rate that periodically becomes very slightly shorter or longer. A detailed analysis of the changes to the pulsation rate suggests that they can be explained by gravitational tugs of planet-like bodies on the pulsar: Apparently, at least three planets

orbit PSR B1257+12, with orbital periods of about 25, 67 and 98 days, respectively.

This discovery came as an immense surprise to astronomers, because these planets are orbiting the remains of a star that exploded. Why weren't the planets destroyed during their star's red supergiant stage? Why didn't they fly off into space when the supernova ejected most of the star's mass? Why didn't the supernova explosion kick the pulsar away from its planets? The answer is that the planets probably formed *after* the explosion. It's likely that the pulsar once had a stellar companion that came too close. Tidal forces would have ripped the companion star apart, with the debris forming a protoplanetary disk that then coalesced into planets.

• What can happen to a neutron star in a close binary system?

Like their white dwarf brethren, neutron stars in close binary systems can brilliantly burst back to life. As is the case with white dwarfs in close binaries, gas overflowing from a companion star can create a hot, swirling accretion disk around the neutron star. However, in the neutron star's mighty gravitational field, infalling matter releases an amazing amount of gravitational potential energy. Dropping a brick onto a neutron star would liberate as much energy as an atomic bomb.

X-Ray Binaries The huge amount of energy released by infalling matter from the companion star makes a neutron star's accretion disk much hotter and much more luminous than the accretion disk around a white dwarf. The high temperatures in the inner regions of the accretion disk cause it to radiate powerfully in X rays. Some close binaries with neutron stars emit 100,000 times as much energy in X rays as our Sun emits in all wavelengths of light combined. Due to this intense X-ray emission, close binaries that contain accreting neutron stars are often called **X-ray binaries**.

The emission from most X-ray binaries pulsates rapidly as the neutron star spins, much like the radio pulsations of the neutron stars we see as pulsars. However, the pulsation rates of X-ray binaries tend to accelerate with time, rather than slowing like radio pulsars. Apparently, matter accreting onto the neutron star adds angular momentum, speeding up the neutron star's rotation (Figure 18.10). Some of these neutron stars rotate so fast that they pulsate every few thousandths of a second (and hence are sometimes called *millisecond pulsars*).

One of the strangest X-ray binaries contains a pulsar known as the *Black Widow Pulsar* because it appears to be devouring its companion. The companion star must once have had a reasonable mass, but almost all of it has spilled over onto the Black Widow Pulsar. The result is that the pulsar now spins every 1.6 milliseconds, while the companion orbits it every 9 hours. The companion also eclipses the pulsar for about an hour of each orbit. Such a long eclipse means that the companion must still have an extended atmosphere despite the fact that its mass is a mere $0.02 M_{Sun}$. Evidently, the pulsar's energy output gradually causes matter to evaporate from the surface of its companion. In a few million years, the companion star will have vanished entirely, leaving behind a solitary millisecond pulsar.

X-ray Bursts   Like accreting white dwarfs that occasionally erupt into novae, accreting neutron stars sporadically erupt with a pronounced spike in luminosity (Figure 18.11). Because these eruptions release their energy primarily in the form of X rays, we call them **X-ray bursts** and the systems that produce them are known as **X-ray bursters**. Much like novae, X-ray bursts result from the sudden ignition of nu-

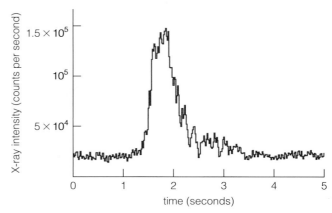

Figure 18.11 Light curve of an X-ray burst. In this particular burst the X-ray luminosity of the neutron star spiked to over six times its usual brightness in a matter of seconds.

clear fusion. However, while novae occur when hydrogen fusion ignites on the surface of a white dwarf in a close binary system, X-ray bursts arise from the ignition of helium fusion on the neutron star in a close binary system.

Hydrogen-rich material from the companion star continually spirals through the accretion disk to the neutron star's surface, coating the surface with a thin layer of fresh hydrogen gas. Pressures at the bottom of this hydrogen layer are high enough to maintain steady fusion, which produces a layer of helium beneath the hydrogen. The steady fusion keeps the hydrogen layer from becoming more than about 1 meter thick, while the helium "ash" from the fusion gradually adds to the mass of the underlying helium layer.

The temperature of the helium layer slowly climbs as helium accumulates within it. When the temperature reaches about 100 million K, the layer suddenly ignites with helium fusion. The helium fuses rapidly to make carbon and heavier elements, generating a burst of energy that escapes the neutron star in the form of X rays. An X-ray burst converts both the helium in its helium layer and the overlying hydrogen into heavier elements. However, fresh hydrogen from the companion star soon begins to replenish the hydrogen layer, and the entire process is soon ready to repeat. Typical X-ray bursters flare every few hours to every few days. Each burst lasts only a few seconds, but during those seconds the system radiates 100,000 times as much power as the Sun, all in X rays. Within a minute after a burst, the X-ray burster cools back down and resumes accreting.

 Black Holes Tutorial, Lessons 1, 2

18.3 Black Holes: Gravity's Ultimate Victory

The story of stellar corpses would be strange enough if it ended with white dwarfs and neutron stars, but it does not. Sometimes, the gravity in a stellar corpse becomes so strong that nothing can prevent the corpse from collapsing under its own weight. The stellar corpse collapses without end,

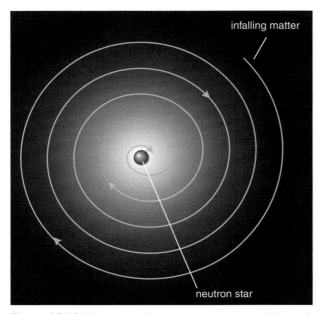

Figure 18.10 Matter accreting onto a neutron star adds angular momentum, increasing the neutron star's rate of spin.

infalling matter

neutron star

crushing itself out of existence and forming perhaps the most bizarre type of object in the universe: a *black hole*.

• What is a black hole?

The idea of a black hole was first suggested in the late 1700s by British philosopher John Mitchell and French physicist Pierre Laplace. It was already known from Newton's laws that the escape velocity from any object depends only on its mass and size. Making an object of a particular mass more compact raises its escape velocity [Section 4.5]. Mitchell and Laplace speculated about objects so compact that their escape velocity exceeds the speed of light. Because they worked in a time before it was known that light always travels at the same speed, they assumed that light emitted from such an object would behave like a rock thrown upward, eventually slowing to a stop and falling back down.

Einstein's general theory of relativity eventually showed that black holes are considerably stranger than Mitchell and Laplace thought, but their basic idea was correct. It is indeed possible for an object's gravity to be so strong that not even light can escape. Physicist John Wheeler named such objects "black holes" in 1967, long after they were first proposed. The "black" in the name comes from the fact that no light can escape. The "hole" part of the name captures the idea that a black hole is like a *hole* in the observable universe in the following sense: If you enter a black hole, you leave the region of the universe that we can observe with a telescope and can never return.

The Event Horizon   The boundary between the inside of a black hole and the universe outside is called the **event horizon**. The event horizon essentially marks the point of no return for objects entering a black hole: It is the boundary around a black hole at which the escape velocity equals the speed of light. The boundary tends to be spherical because the velocity needed to escape a black hole's gravity depends on the distance to its center, which is the same for every point on the event horizon. Nothing that passes within this boundary can ever escape. Thus, the event horizon gets its name because we have no hope of learning about any events that occur within it.

The structure of space and time near the event horizon is extraordinarily strange. Space and time are not distinct, as we usually think of them, but instead are bound up together as four-dimensional **spacetime** [Section S3.2]. Moreover, Einstein's general theory of relativity tells us that what we perceive as gravity arises from *curvature of spacetime*. The concept "curvature of spacetime" is challenging to think about because we cannot even visualize four dimensions at once, let alone visualize their curvature. However, as discussed in more detail in Chapter S3, we can understand the basic idea with a two-dimensional analogy.

Let's consider the geometry of a two-dimensional slice through spacetime that passes directly through the middle of a spherical object, dividing it into two equal halves. Fig-

a A two-dimensional representation of "flat" spacetime. The radial distance is the same between each of the circles shown.

b Gravity arises from curvature of spacetime, represented here by a mass pushing down on the rubber sheet. Notice how the circles become more widely separated near the mass, showing that the curvature is greater as we approach the mass on the sheet.

event horizon

c The curvature of spacetime becomes greater and greater as we approach a black hole, and a black hole itself is a bottomless pit in spacetime.

Figure 18.12   Interactive Figure   We can use two-dimensional rubber sheets to show an analogy to curvature in four-dimensional spacetime.

ure 18.12 represents the curvature of that two-dimensional slice by depicting it as a rubber sheet. In this analogy, the sheet is flat in a region far from any mass, meaning that gravity is weak there (Figure 18.12a). Near a massive object with strong gravity, the sheet becomes curved (Figure 18.12b)— the stronger the gravity, the more curved it gets. Because gravity gets stronger and stronger as we get closer to the black hole, the sheet curves more and more, illustrating that space and time are becoming extremely distorted near the event horizon (Figure 18.12c). A black hole is therefore like a bottomless pit in spacetime. Keep in mind, however, that a black hole is really spherical, not funnel-shaped. The illustration is only a two-dimensional analogy representing the degree to which spacetime is distorted in the vicinity of the event horizon.

The Size of a Black Hole We usually think of the "size" of a black hole as the radius of its event horizon. As the rubber sheet analogy in Figure 18.12c shows, it is possible to draw a series of concentric circles around a black hole. However, it is not really possible to measure a *radius* for these circles, because their centers lie within the event horizon and hence are not part of our observable universe. We therefore define the radius of a circle around a black hole as the radius it *would* have if geometry were flat (Euclidean), as it is in Figure 18.12a. (That is, we simply define the radius to be the circle's circumference divided by 2π.) The radius of the event horizon is known as the **Schwarzschild radius**, named for Karl Schwarzschild (1873–1916). Schwarzschild computed his famous radius from Einstein's general theory of relativity, and he did so only a month after Einstein published the theory. Moreover, Schwarzschild did this work while serving in the German Army on the Russian front during World War I. Sadly, he died less than a year later of an illness contracted during the war.

The Schwarzschild radius of a black hole depends only on its mass. A black hole with the mass of the Sun has a Schwarzschild radius of about 3 kilometers—only a little smaller than the radius of a neutron star of the same mass. More massive black holes have larger Schwarzschild radii.

For example, a black hole with 10 times the mass of the Sun has a Schwarzschild radius of about 30 kilometers.

Properties of a Black Hole A collapsing stellar core becomes a black hole at the moment it shrinks to a size smaller than its Schwarzschild radius. At that moment, the core disappears within its own event horizon. The black hole still contains all the mass and exerts the full amount of gravity associated with that mass, but its outward appearance tells us nothing about what fell in. Thus, any information about the object that collapsed to make the black hole is lost. For example, if we found an isolated black hole floating through space, we'd have no way to know whether it was made from the collapse of the iron core of a massive star during a supernova, from the merger of two neutron stars, or from some completely different process. The black hole similarly destroys nearly all information about matter that falls into it at later times. If you tossed 1 kilogram of lead into a black hole, it would have precisely the same effect as tossing 1 kilogram of precious diamonds into it—in both cases, the black hole's mass would increase by 1 kilogram, and there would be no other evidence left to tell us what went in.

In fact, theoretical models tell us that mass is one of only three measurable properties of a black hole. The second

---

MATHEMATICAL INSIGHT 18.1   **The Schwarzschild Radius**

The Schwarzschild radius (R_S) of a black hole is given by a simple formula:

$$\text{Schwarzschild radius} = R_S = \frac{2GM}{c^2}$$

where M is the black hole's mass, $G = 6.67 \times 10^{-11}$ m^3/(kg $\times$ s^2) is the gravitational constant, and $c = 3 \times 10^8$ m/s is the speed of light. With a bit of calculation, this formula can also be written as:

$$\text{Schwarzschild radius} = R_S = 3.0 \times \frac{M}{M_{Sun}} \text{ km}$$

Example 1: What is the Schwarzschild radius of a black hole with a mass of $10M_{Sun}$?

Solution:

Step 1 Understand: We are given the black hole's mass, which is the only information we need to compute its Schwarzschild radius. Because we are given the mass in units of solar masses, the second version of the formula above will be easier to use.

Step 2 Solve: We set $M = 10M_{Sun}$ and use the formula to find the Schwarzschild radius:

$$R_S = 3 \times \frac{10M_{Sun}}{M_{Sun}} = 30 \text{ km}$$

Step 3 Explain: We have found that the Schwarzschild radius of a $10M_{Sun}$ black hole is about 30 kilometers. Notice that

this is quite small—no larger than the radius of a large city on Earth.

Example 2: As far as we know, black holes in the present-day universe form only when an object's mass exceeds the roughly $3M_{Sun}$ neutron star limit. However, Stephen Hawking and others have speculated that much less massive *mini–black holes* might have formed during the Big Bang. Suppose a mini–black hole has the mass of Earth (about 6×10^{24} kg). What is its Schwarzschild radius?

Solution:

Step 1 Understand: Again, we are given the black hole's mass, which is the only information we need to compute its Schwarzschild radius. But in this case, we are given the mass in kilograms, so it is easier to use the first version of the formula above.

Step 2 Solve: We set $M = 6 \times 10^{24}$ kg and use the formula with the given values of G and c:

$$R_S = 2 \times \frac{\left(6.67 \times 10^{-11} \frac{\text{m}^3}{\text{kg} \times \text{s}^2}\right) \times (6 \times 10^{24} \text{ kg})}{\left(3 \times 10^8 \frac{\text{m}}{\text{s}}\right)^2}$$

$$\approx 0.009 \text{ m}$$

Step 3 Explain: The mini–black hole would have a Schwarzschild radius of only 9 millimeters, making it small enough to fit on the tip of your finger. But don't try to hold it—it would weigh as much as the entire Earth!

measurable property is electric charge, but this is relatively unimportant: If a black hole had any positive or negative charge, it would quickly attract oppositely charged particles from the interstellar medium, thereby making it electrically neutral. The third measurable property of a black hole is its angular momentum. Conservation of angular momentum dictates that black holes should be rotating rapidly when they form, because stellar cores normally rotate. Much like an ice-skater pulling in her arms, a rotating stellar core should rotate faster and faster as it collapses to make a black hole.

Under the right circumstances, the rotation of the black hole can affect its surroundings. Einstein's general theory of relativity predicts that a massive rotating object actually drags neighboring regions of spacetime around in circles. The effect (known as *frame-dragging*) is usually very small, although it has been successfully measured even around Earth. However, near the event horizon of a black hole, the dragging of spacetime should be quite strong, making it very difficult for objects to fall directly into the black hole. Instead, the dragging of spacetime would tend to accelerate the infalling object in the direction of the black hole's rotation, thereby altering its orbit. This effect can actually change the shape of a black hole's event horizon because matter and energy moving along with the black hole's rotation can resist falling into the black hole more easily than matter and energy moving in the opposite direction.

Singularity and the Limits to Knowledge What lies inside a black hole? We cannot answer this question observationally, even in principle, because no information can ever emerge from within the event horizon. Nevertheless, we can use our understanding of the laws of physics to predict what must occur inside a black hole. Because nothing can stop the crush of gravity in a black hole, we expect that all the matter that forms a black hole must ultimately be crushed to an infinitely tiny and dense point in the black hole's center. We call this point a **singularity**. (In a rapidly rotating black hole, the singularity may stretch into the shape of a ring around the center, rather than being a precise point.)

Unfortunately, this idea of a singularity pushes up against the limits to scientific knowledge today. The problem is that two very successful theories make very different predictions about the nature of a singularity. Einstein's theory of general relativity, which seems to explain successfully how gravity works throughout the universe, predicts that spacetime should grow infinitely curved as it enters the pointlike singularity. Quantum physics, which successfully explains the nature of atoms and the spectra of light, predicts that spacetime should fluctuate chaotically near the singularity. These are clearly different claims, and no theory that can reconcile them has yet been found. We will not fully understand how a singularity behaves until scientists develop a new theory that encompasses both general relativity and quantum mechanics. Some promising candidates for such a theory are currently being pursued, but the mathematical challenges are extremely daunting. In the meantime, the uncertainty in our current knowledge is a gold mine for science fiction writers, who speculate about using objects like black holes for exotic forms of travel through spacetime [Section S3.5].

• What would it be like to visit a black hole?

Imagine that you are a pioneer of the future, making the first visit to a black hole. You've selected a black hole with a mass of $10M_{Sun}$ and a Schwarzschild radius of 30 kilometers. As your spaceship approaches the black hole, you fire its engines to put the ship on a circular orbit a few thousand kilometers above the event horizon. This orbit will be perfectly stable—there is no need to worry about getting "sucked in."

Your first task is to test Einstein's general theory of relativity. This theory predicts that time should run more slowly as the force of gravity grows stronger. It also predicts that light coming out of a strong gravitational field should show a redshift, called a *gravitational redshift,* that is due to gravity rather than to the Doppler effect. You test these predictions with the aid of two identical clocks whose numerals glow with blue light. You keep one clock aboard the ship and push the other one, with a small rocket attached, directly toward the black hole (Figure 18.13). The small rocket automatically fires its engines just enough so that the clock falls only gradually toward the event horizon. Sure enough, the clock on the rocket ticks more slowly as it heads toward the black hole, and its light becomes increasingly redshifted. When the clock reaches a distance of about 10 kilometers above the event horizon, you see it ticking only half as fast as the clock on your spaceship, and its numerals are red instead of blue.

The rocket has to expend fuel rapidly to keep the clock hovering in the strong gravitational field, and it soon runs out of fuel. The clock plunges toward the black hole. From your safe vantage point inside the spaceship, you see the clock ticking more and more slowly as it falls. However, you soon need a radio telescope to "see" it, as the light from the clock face shifts from the red part of the visible spectrum, through the infrared, and on into the radio. Finally,

Figure 18.13 Interactive Figure Time runs more slowly on the clock nearer to the black hole, and a gravitational redshift makes its glowing blue numerals appear red from your orbiting spaceship.

its light is so far redshifted that no conceivable telescope could detect it. Just as the clock vanishes from view, you see that the time on its face has frozen to a stop.

Curiosity overwhelms the better judgment of one of your colleagues. He hurriedly climbs into a space suit, grabs the other clock, resets it, and jumps out of the air lock on a trajectory aimed straight for the black hole. Down he falls, clock in hand. He watches the clock, but because he and the clock are traveling together, its time seems to run normally and its numerals stay blue. From his point of view, time seems to neither speed up nor slow down. In fact, he'd say that *you* were the one with the strange time, as he would see your time running increasingly fast and your light becoming increasingly blueshifted. When his clock reads, say, 00:30, he and the clock pass through the event horizon. There is no barrier, no wall, no hard surface. The event horizon is a mathematical boundary, not a physical one. From his point of view, the clock keeps ticking. He is inside the event horizon, the first human being ever to vanish into a black hole.

Back on the spaceship, you watch in horror as your overly curious friend plunges to his death. Yet, from your point of view, he will *never* cross the event horizon. You'll see time come to a stop for him and his clock just as he vanishes from view due to the huge gravitational redshift of light. When you return home, you can play a video for the judges at your trial, proving that your friend is still outside the black hole. Strange as it may seem, all this is true according to Einstein's theory. From your point of view, your friend takes *forever* to cross the event horizon (even though he vanishes from view due to his ever-increasing redshift).

From his point of view, it is but a moment's plunge before he passes into oblivion.

The truly sad part of this story is that your friend did not live to experience the crossing of the event horizon. The force of gravity he felt as he approached the black hole grew so quickly that it actually pulled much harder on his feet than on his head, simultaneously stretching him lengthwise and squeezing him from side to side (Figure 18.14). In essence, your friend was stretched in the same way the oceans are stretched by the tides, except that the *tidal force* near the black hole is trillions of times stronger than the tidal force of the Moon on Earth [Section 4.5]. No human could survive it.

If he had thought ahead, your friend might have waited to make his jump until you visited a much larger black hole, like one of the *supermassive black holes* thought to reside in the centers of many galaxies [Section 21.4]. A 1 billion M_{Sun} black hole has a Schwarzschild radius of 3 billion kilometers—about the distance from our Sun to Uranus. Although the gravitational forces at the event horizon of all black holes are equally great, the larger size of a supermassive black hole makes its tidal forces much weaker and hence nonlethal. Your friend could safely plunge through the event horizon.

Again, from your point of view, the crossing would take forever, and you would see time come to a stop for him just as he vanished from sight because of the gravitational redshift. Again, he would experience time running normally and would see time in the outside universe running increasingly fast as he approached the event horizon. Unfortunately, anything he saw would do him little good as he plunged to oblivion inside the black hole.

Black Holes Don't Suck

What would happen if our Sun suddenly became a black hole? For some reason, it has become part of our popular culture for most people to believe that Earth and the other planets would inevitably be "sucked in" by the black hole. That is not true. Although the sudden disappearance of light and heat from the Sun would be bad news for life, Earth's orbit would not change.

Newton's law of gravity tells us that the allowed orbits in a gravitational field are ellipses, hyperbolas, and parabolas [Section 4.4]. Note that "sucking" is not on the list! A spaceship would get into trouble only if it came so close to a black hole—within about three times its Schwarzschild radius—that the force of gravity would deviate significantly from what Newton's law predicts. Otherwise, a spaceship passing near a black hole would simply swing around it on an ordinary orbit (ellipse, parabola, or hyperbola). In fact, because most black holes are so small—typical Schwarzschild radii are far smaller than any star or planet—a black hole is actually one of the most difficult things in the universe to fall into by accident.

squeezed →   ← squeezed

stretched

Figure 18.14 Tidal forces would be lethal near a black hole formed by the collapse of a star. The black hole would pull more strongly on the astronaut's feet than on his head, stretching him lengthwise and squeezing him from side to side.

• Do black holes really exist?

As was the case for neutron stars, at first most astronomers who contemplated the idea of black holes thought them too strange to be true. Today, however, our understanding of physics gives us reason to think that black holes ought to be fairly common, and observational evidence strongly suggests that black holes really exist.

The Formation of a Black Hole   The idea that black holes ought to exist comes from considering how they might form. Recall that white dwarfs cannot exceed $1.4M_{Sun}$, because gravity overcomes electron degeneracy pressure above that mass. Calculations show that the mass of a neutron star has a similar limit that lies somewhere between about 2 and 3 solar masses. Above this mass, neutron degeneracy pressure cannot hold off the crush of gravity in a collapsing stellar core.

A supernova occurs when the electron degeneracy pressure supporting the iron core of a massive star succumbs to gravity, causing the core to collapse catastrophically into a ball of neutrons [Section 17.3]. That is why most supernovae leave neutron stars behind. However, theoretical models show that the most massive stars might not succeed in blowing away all their upper layers. If enough matter falls back onto the neutron core, its mass may rise above the neutron star limit.

As soon as the core exceeds the neutron star limit, gravity overcomes the neutron degeneracy pressure and the core collapses once again. This time, no known force can keep the core from collapsing into oblivion as a black hole. Moreover, another effect of Einstein's theory of relativity makes it highly unlikely that any other as-yet-unknown force could intervene and prevent the collapse.

Recall that Einstein's theory tells us that energy is equivalent to mass ($E = mc^2$) [Section 4.3], implying that energy, like mass, must also exert some gravitational attraction. The gravity of pure energy usually is negligible, but in a stellar core collapsing beyond the neutron star limit energy itself becomes a powerful source of gravity. Once a star collapses beyond the point where it could have been a neutron star, the energy associated with the enhanced internal temperature and pressure acts like additional mass, making the crushing power of gravity even stronger. The more the star collapses, the stronger gravity gets. To the best of our understanding, *nothing* can halt the crush of gravity at this point. The core collapses without end, forming a black hole. Gravity has achieved its ultimate triumph.

Observational Evidence for Black Holes   The fact that black holes emit no light might make it seem as if they should be impossible to detect. However, a black hole's gravity can influence its surroundings in a way that reveals its presence. Astronomers have discovered many objects that show the telltale signs of an unseen gravitational influence that has a large enough mass to suggest that it is a black hole.

Strong observational evidence for black holes formed by supernovae comes from studies of X-ray binaries. Recall that the accretion disks around neutron stars in close binary systems can emit strong X-ray radiation, making an X-ray binary. The accretion disk forms because the neutron star's strong gravity pulls mass from the companion star. Because a black hole has even stronger gravity than a neutron star, a black hole in a close binary system should also be surrounded by a hot, X-ray-emitting accretion disk. Thus, some X-ray binaries may contain black holes rather than neutron stars. The trick to learning which type of corpse resides in an X-ray binary depends on measuring the object's mass.

One of the most promising black hole candidates is in an X-ray binary called Cygnus X-1 (Figure 18.15). This system contains an extremely luminous star with an estimated mass of $18M_{Sun}$. Based on Doppler shifts of its spectral lines, astronomers have concluded that this star orbits a compact, unseen companion with a mass of about $10M_{Sun}$. Although there is some uncertainty in these mass estimates, the mass of the invisible accreting object clearly exceeds the

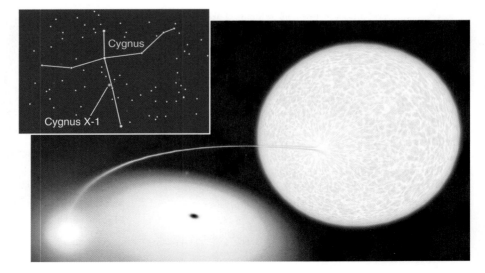

Figure 18.15   Artist's conception of the Cygnus X-1 system, which gets its name because it is the brightest X-ray source in the constellation Cygnus. The X rays come from the high-temperature gas in the accretion disk surrounding the black hole. The inset shows the location of Cygnus X-1 in the sky.

$3M_{Sun}$ neutron star limit. Thus, it is too massive to be a neutron star, so by current knowledge it cannot be anything other than a black hole.

THINK ABOUT IT

Recall that some X-ray binaries that contain neutron stars emit frequent X-ray bursts and are called X-ray bursters. Could an X-ray binary that contains a black hole exhibit the same type of X-ray bursts? Why or why not? (*Hint:* Where do the X-ray bursts occur in an X-ray binary with a neutron star?)

A few dozen other X-ray binaries offer similarly strong evidence for black holes formed from the collapse of massive stellar cores. In addition, an even greater body of evidence suggests the existence of far more massive black holes residing at the centers of many galaxies. These so-called *supermassive black holes* often have masses millions or billions of times that of our Sun, so they must have formed in a different way than the stellar-mass black holes in X-ray binaries. We will discuss the evidence for supermassive black holes in Chapter 21, and we will also see how they are thought to power some of the most luminous objects in the universe.

Of course, confirming that black holes are real with 100% certainty is very difficult. However, our current theories successfully explain neutron stars, and the general theory of relativity that leads to the idea of black holes is also on solid ground. Unless something is dramatically wrong in our current theories about the mass limit of neutron stars or some other, unknown type of compact object can have a huge mass, black holes must be real.

SPECIAL TOPIC Too Strange to Be True?

Theoretical calculations predicted the existence of neutron stars and black holes long before their observational discovery, but many astronomers considered these theoretical results too strange to be true.

The story begins with Subrahmanyan Chandrasekhar, an astrophysicist from India. Chandrasekhar was only 19 when he completed the calculations showing that there is a white dwarf limit of $1.4M_{Sun}$, and he boldly predicted that a more massive white dwarf would collapse under the force of gravity. He did this work in 1931 while traveling by ship to England,

Subrahmanyan Chandrasekhar

where he hoped to impress the eminent British astrophysicist Sir Arthur Stanley Eddington. However, Eddington ridiculed Chandrasekhar for believing that white dwarfs could collapse. Neutrons had not yet been discovered, little was known about fusion, and no one had any idea what supernovae were. The idea of gravity achieving an ultimate victory seemed nonsensical to Eddington, who speculated that some type of force must prevent gravity from crushing any object.

Sir Arthur Stanley Eddington

A few more radical thinkers took collapsing stars more seriously. A Russian physicist, Lev Davidovich Landau, independently computed the white dwarf limit in 1932. Neutrons were discovered just a few months later, and Landau speculated that stellar corpses above the white dwarf limit might collapse until neutron degeneracy pressure halted the crush of gravity. While most astronomers found the idea of neutron stars to be unacceptably weird, two European scientists who had emigrated to California, Fritz Zwicky and Walter Baade, were not so skep-

Lev Davidovich Landau

tical. Without knowing of Landau's ideas, they also concluded that neutron stars were possible. In 1934, they suggested that a supernova might result when a stellar core collapses and forms a neutron star—an extraordinarily insightful guess. By 1938, physicist Robert Oppenheimer, working at Berkeley, was contemplating whether neutron stars had a limiting mass of their own. He and his coworkers concluded that the answer was yes and that neutron degeneracy pressure could not resist the crush of gravity when the mass rose above a few solar masses. Because no known force could keep such a star from collapsing indefinitely, Oppenheimer speculated that gravity would achieve ultimate victory, crushing the star into a black hole.

Robert Oppenheimer

Astronomers gradually came to accept Chandrasekhar's $1.4M_{Sun}$ white dwarf limit, because observations found no white dwarfs more massive than this. However, most astronomers held to a belief that high-mass stars would inevitably shed enough mass late in life to prevent the formation of a more massive collapsed object. Jocelyn Bell's 1967 discovery of pulsars shattered this belief. Within a few months, Thomas Gold of Cornell University correctly suggested that the pulsars were spinning neutron stars. These discoveries forced astronomers to admit that nature was far stranger than they had expected. The verification that neutron stars really did exist made the prospect of the still-stranger black holes much less difficult to accept.

Jocelyn Bell

Chandrasekhar, who had long since moved to the University of Chicago, was awarded a Nobel Prize in 1984 for his lifelong contributions to astronomy. Landau won a Nobel Prize in 1962 for his work on condensed states of matter. Oppenheimer went on to lead the *Manhattan Project* that developed the atomic bomb in 1945. Eddington died in 1944, still convinced that white dwarf stars could not collapse.

18.4 The Mystery of Gamma-Ray Bursts

In the early 1960s, the United States began launching a series of top-secret satellites designed to look for gamma rays emitted by nuclear bomb tests. The satellites soon began detecting occasional bursts of gamma rays, typically lasting a few seconds (Figure 18.16). It took several years for military scientists to become convinced that these **gamma-ray bursts** were coming from space, not from some sinister human activity. They publicized the discovery in 1973. At first, astronomers assumed that these gamma-ray bursts were just a variation on the X-ray bursts that can occur on neutron stars in close binary systems. However, the truth turns out to be much more spectacular. According to recent observations, at least some gamma-ray bursts seem to come from the formation of black holes at very great distances from Earth, although the process that produces the gamma rays still remains mysterious.

• Where do gamma-ray bursts come from?

When gamma-ray bursts were first discovered, astronomers had very little information about where they come from. The problem was that gamma-ray photons carry so much energy that they are very difficult to focus. Instead of reflecting off a telescope's mirror, they tend to penetrate right through it.

Our ignorance about the origins of gamma-ray bursts began to lift with the 1991 launch of NASA's *Compton Gamma Ray Observatory*, or *Compton* for short, which operated in Earth orbit for almost a decade. Compton carried an array of eight detectors designed expressly to study gamma-ray bursts. By comparing the data recorded by all eight detectors, scientists could determine the direction of a gamma-ray burst within about 1°. The results were stunning. Compton detected a gamma-ray burst about once a day on average and soon had found enough bursts to show their distribution on a map of the sky. To almost everyone's

surprise, the bursts were *not* concentrated in the disk of the Milky Way Galaxy, where the X-ray binaries tend to be. That finding ruled out the possibility that gamma-ray bursts come from the same types of systems as X-ray bursts. Instead, the gamma-ray bursts seemed to come randomly from all directions in space.

The even distribution of gamma-ray bursts across the sky suggested that they must come from far outside our own galaxy. If the bursts had been coming from objects distributed spherically about the Milky Way Galaxy, we would have seen a concentration of them in the direction of the galactic center. (Remember that we are located more than halfway out from the center of our galaxy.)

Direct evidence that gamma-ray bursts do indeed come from far outside our galaxy arrived in 1997, thanks to a breakthrough in our observational capabilities: For the first time, it became possible to respond almost immediately to the detection of a gamma-ray burst with observations in other wavelength bands. When a space-based observatory detected a gamma-ray burst, its operators quickly trained X-ray telescopes on the region of sky from which the burst came and relayed its location to astronomers at visible-light telescopes on the ground. In this way, astronomers were able to observe X rays and visible light from the afterglow of the burst. The higher resolution possible in these other wavelengths allowed astronomers to pinpoint the location of the burst, confirming that it came from a distant galaxy. Since then, numerous other bursts have been traced to explosions in galaxies billions of light-years away (Figure 18.17). At least some gamma-ray bursts, but maybe not all of them, must therefore be extremely powerful explosions that occur at extremely large distances from Earth.

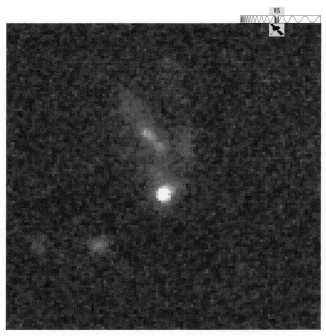

Figure 18.17 The bright dot near the center of this image is the visible-light afterglow of a gamma-ray burst, as seen by the Hubble Space Telescope. The elongated blob extending above the dot is the distant galaxy in which the burst occurred.

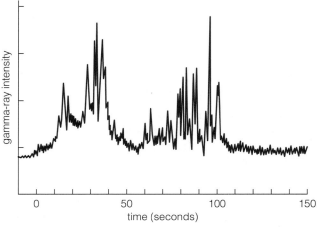

Figure 18.16 The intensity of a gamma-ray burst can fluctuate dramatically over time periods of just a few seconds.

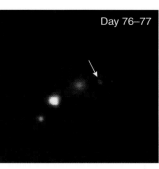

Figure 18.18 The bright object marked "Source of Gamma Ray Burst" in the first panel of this sequence is the visible-light afterglow of a gamma-ray burst in the outskirts of a distant galaxy (marked "Host Galaxy"). Continued monitoring of the light from this object once the initial afterglow faded showed that it had the light curve of a massive star supernova.

• What causes gamma-ray bursts?

Learning that gamma-ray bursts generally come from far outside our galaxy only deepened the mystery surrounding them. The afterglows of some gamma-ray bursts can be seen with binoculars, even though they are coming from galaxies billions of light-years away—making them by far the most powerful bursts of energy that ever occur in the universe. If these bursts shine their light equally in all directions, like a light bulb, then the total luminosity of a burst can briefly exceed the combined luminosity of a million galaxies like our Milky Way! Because such a high luminosity is very difficult to explain, some scientists speculate that gamma-ray bursts channel their energy into narrow searchlight beams, like pulsars, so that only some are visible from Earth. But even in this case, a burst's luminosity would still temporarily surpass that of thousands of galaxies like our Milky Way.

What could cause such massive outbursts of energy? At least some gamma-ray bursts appear to come from extremely powerful supernova explosions. An ordinary supernova that forms a neutron star does not release enough energy to explain the luminosity of the brightest gamma-ray bursts. However, a supernova that forms a black hole crushes even more matter into an even smaller radius, releasing many times more gravitational potential energy than one that forms a neutron star. This kind of event (sometimes called a *hypernova*) might be powerful enough to explain the most extreme gamma-ray bursts.

The primary evidence linking gamma-ray bursts to exploding stars comes from observing them in other wavelengths of light, including visible light and X rays. Several recently launched gamma-ray telescopes—especially NASA's *Swift* satellite launched in 2004—are capable of rapidly pinpointing the location of a gamma-ray burst in the sky, thereby allowing other types of telescopes to be pointed at the gamma-ray source almost as soon as the burst is detected. Observations like these have shown that at least some gamma-ray bursts coincide in both time and location with powerful supernova explosions (Figure 18.18). In a number of other cases, the gamma-ray bursts are observed to come from distant galaxies that are actively forming new stars. Those observations also support the idea that gamma-ray bursts are produced in the explosions of extremely massive stars because such stars are very short-lived and therefore should be found only in places where stars are actively forming.

The identification of a few gamma-ray bursts with powerful supernovae did not fully solve the mystery of their origin, because bursts come in at least two distinct types. Bursts of the second type last only a few seconds, making them more difficult to catch in action with telescopes for other wavelengths. Nevertheless, in mid-2005 *Swift* and a satellite called HETE each captured a short burst both in gamma rays and X rays; the latter burst was pinpointed well enough for follow-up observations with the Chandra X-ray Observatory and the Hubble Space Telescope. Results clearly show that the short bursts do *not* come from supernovae. What could they be coming from? The leading hypothesis is a collision in a binary system containing either two neutron stars or a neutron star and a black hole. Einsten's general theory of relativity predicts that such binary systems must lose substantial energy to *gravitational waves* [Section S3.4], gradually causing the two objects to spiral in toward each other until they collide. Observations of such systems confirm their orbital decay (see Figure S3.21). Thus, short gamma-ray bursts may be the signatures of catastrophic collisions between two orbiting members of the bizarre stellar graveyard.

THE BIG PICTURE

Putting Chapter 18 into Context

We have now seen the mind-bending consequences of stellar death. As you think about the bizarre objects described in this chapter, try to keep in mind these "big picture" ideas:

- Despite the strange nature of stellar corpses, clear evidence exists for white dwarfs and neutron stars, and the case for black holes is very strong.

- White dwarfs, neutron stars, and black holes can all have close stellar companions from which they accrete matter. These binary systems produce some of the most spectacular events in the universe, including novae, white dwarf supernovae, and X-ray bursters.

- Black holes are holes in the observable universe that strongly warp space and time around them. The nature of black hole singularities remains beyond the frontier of current scientific understanding.

18.1 White Dwarfs

- **What is a white dwarf?** A white dwarf is the core left over from a low-mass star, supported against the crush of gravity by **electron degeneracy pressure**. A white dwarf typically has the mass of the Sun compressed into a size no larger than Earth. No white dwarf can have a mass greater than $1.4M_{Sun}$.

- **What can happen to a white dwarf in a close binary system?** A white dwarf in a close binary

system can acquire hydrogen from its companion through an **accretion disk** that swirls toward the white dwarf's surface. As hydrogen builds up on the white dwarf's surface, it may ignite with nuclear fusion to make a **nova** that, for a few weeks, may shine as brightly as 100,000 Suns. In extreme cases, accretion may continue until the white dwarf's mass exceeds the **white dwarf limit** of $1.4M_{Sun}$, at which point it will explode as a **white dwarf supernova**.

18.2 Neutron Stars

- **What is a neutron star?** A neutron star is the ball of neutrons created by the collapse of the iron core in a massive star supernova. It resembles a giant atomic nucleus 10 kilometers across but more massive than the Sun.

- **How were neutron stars discovered?** Neutron

stars spin rapidly when they are born, and their strong magnetic fields can direct beams of radiation that sweep through space as the neutron star spins. We see such neutron stars as **pulsars**, and these pulsars provided the first direct evidence for the existence of neutron stars.

- **What can happen to a neutron star in a close binary system?** Neutron stars in close binary systems can accrete hydrogen from their companions, forming dense, hot accretion disks. The hot gas emits strongly in X rays, so we see these systems as **X-ray binaries**. In some of these systems, frequent bursts of helium fusion ignite on the neutron star's surface, emitting **X-ray bursts**.

18.3 Black Holes: Gravity's Ultimate Victory

- **What is a black hole?** A black hole is a place where

gravity has crushed matter into oblivion, creating a hole in the universe from which nothing can ever escape, not even light. The **event horizon** marks the boundary between our observable universe and the inside of the black hole; the size of the black hole is characterized by its **Schwarzschild radius**. A black hole has only three measurable properties: its mass, electric charge, and angular momentum.

- **What would it be like to visit a black hole?** You could orbit a black hole just like any other object of the same mass. However, you'd see strange effects for an object falling toward the black hole: Time would seem to run slowly for the object, and its light would be increasingly redshifted as it approached the black hole. The object would never quite reach the event horizon, but it would soon disappear from view as its light became so redshifted that no instrument could detect it.

- **Do black holes really exist?** No known force can

stop the collapse of a stellar corpse with a mass above the neutron star limit of 2 to 3 solar masses, and theoretical studies of supernovae suggest that such objects should sometimes form. Observational evidence supports this idea: Some X-ray binaries include compact objects far too massive to be neutron stars, making it likely that they are black holes.

18.4 The Mystery of Gamma-Ray Bursts

- **Where do gamma-ray bursts come from?** **Gamma-ray bursts** generally come from distant galaxies

that can be billions of light-years from Earth. Because their sources are both bright and very distant, they must be the most powerful bursts of energy we observe anywhere in the universe.

- **What causes gamma-ray bursts?** No one knows exactly how gamma-ray bursts occur, but the energy required to make one suggests that gamma-ray bursts occur when certain kinds of black holes are formed. At least some gamma-ray bursts appear to come from unusually powerful supernova explosions that may create black holes.

Review Questions

Short-Answer Questions Based on the Reading

1. What is degeneracy pressure, and how is it important to the existence of white dwarfs and neutron stars? What is the difference between electron degeneracy pressure and neutron degeneracy pressure?

2. Describe the mass, size, and density of a typical white dwarf. How does the size of a white dwarf depend on its mass?

3. What happens to the electron speeds in a more massive white dwarf, and how does this behavior lead to a limit on the mass of a white dwarf? What is the *white dwarf limit*?

4. What are *accretion disks,* and why do we find them only in close binary systems? Explain how the accretion disk provides a white dwarf with a new source of energy that we can detect from Earth.

5. What is a *nova*? Describe the process that creates a nova, and what a nova looks like.

6. What causes a *white dwarf supernova*? Observationally, how do we distinguish white dwarf and massive star supernovae?

7. Describe the mass, size, and density of a typical neutron star. What would happen if a neutron star came to your hometown?

8. How do we know that *pulsars* are neutron stars? Are all neutron stars also pulsars? Explain.

9. Explain how the presence of a neutron star can make a close binary star system appear to us as an *X-ray binary*. Why do some of these systems appear to us as *X-ray bursters*?

10. What do we mean when we say that a black hole really is a hole in *spacetime*? What is the *event horizon* of a black hole, and how is it related to the *Schwarzschild radius*? What are the three measurable properties of a black hole?

11. What do we mean by the *singularity* of a black hole? How do we know that our current theories are inadequate to explain what happens at the singularity?

12. Suppose you are falling into a black hole. How will you perceive the passage of your own time? How will you perceive the passage of time in the universe around you? Briefly explain why your trip is likely to be lethal.

13. Why do we think that supernovae should sometimes form black holes? What observational evidence supports the existence of black holes?

14. How do we know that gamma-ray bursts do *not* come from the same sources as X-ray bursts? Summarize current ideas about the cause of gamma-ray bursts.

Test Your Understanding

Does It Make Sense?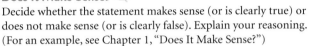

Decide whether the statement makes sense (or is clearly true) or does not make sense (or is clearly false). Explain your reasoning. (For an example, see Chapter 1, "Does It Make Sense?")

15. Most white dwarf stars have masses close to that of our Sun, but a few white dwarf stars are up to three times as massive as the Sun.

16. The radii of white dwarf stars in close binary systems gradually increase as they accrete matter.

17. If you want to find a pulsar, you might want to look near the remnant of a supernova described by ancient Chinese astronomers.

18. If a black hole 10 times as massive as our Sun were lurking just beyond Pluto's orbit, we'd have no way of knowing it was there.

19. If the Sun suddenly became a $1 M_{Sun}$ black hole, the orbits of the nine planets would not change at all.

20. We can detect black holes with X-ray telescopes because matter falling into a black hole emits X rays after it smashes into the event horizon.

21. If two black holes merge together, the resulting black hole is even smaller than the original ones because of its stronger gravity.

22. If gamma-ray bursts really channel their energy into narrow beams, then the total number of gamma-ray bursts that occur is probably far greater than the number we detect.

23. If the Sun suddenly became a $1 M_{Sun}$ black hole, Earth's tides would become much greater.

24. The pulsation period of a pulsar appears to speed up if the pulsar is moving toward us.

Quick Quiz

Choose the best answer to each of the following. Explain your reasoning with one or more complete sentences.

25. Which of these objects has the smallest radius? (a) a $1.2 M_{Sun}$ white dwarf (b) a $0.6 M_{Sun}$ white dwarf (c) Jupiter

26. Which of these objects has the largest radius? (a) a $1.2 M_{Sun}$ white dwarf (b) a $1.5 M_{Sun}$ neutron star (c) a $3.0 M_{Sun}$ black hole

27. Which of these things has the smallest radius? (a) a $1.2 M_{Sun}$ white dwarf (b) the event horizon of a $3.0 M_{Sun}$ black hole (c) the event horizon of a $10 M_{Sun}$ black hole

28. What would happen if the Sun suddenly became a black hole without changing its mass? (a) The black hole would quickly suck in the Earth. (b) Earth would gradually spiral into the black hole. (c) Earth's orbit would not change.

29. Which of these isolated neutron stars must have had a binary companion? (a) a pulsar inside a supernova remnant that pulses 30 times per second (b) an isolated pulsar that pulses 600 times per second (c) a neutron star that does not pulse at all

30. What would happen to a neutron star with an accretion disk orbiting in a direction opposite to the neutron star's spin? (a) Its spin would speed up. (b) Its spin would slow down. (c) Its spin would stay the same.

31. Which of these binary systems is most likely to contain a black hole? (a) an X-ray binary containing an O star and another object of equal mass (b) a binary with an X-ray burster (c) an X-ray binary containing a G star and another object of equal mass

32. How would a flashing red light appear as it fell into a black hole? (a) It would appear to flash more quickly. (b) Its flashes would appear bluer. (c) Its flashes would shift to the infrared part of the spectrum.

33. Which of these black holes exerts the weakest tidal forces on something near its event horizon? (a) a $10 M_{Sun}$ black hole (b) a $100 M_{Sun}$ black hole (c) a $10^6 M_{Sun}$ black hole

34. Where do gamma-ray bursts tend to come from? (a) neutron stars in our galaxy (b) black holes in our galaxy (c) extremely distant galaxies

Investigate Further
In-Depth Questions to Increase Your Understanding

Short-Answer/Essay Questions

Life Stories of Stars. Write a one- to two-page life story for the following scenarios in Problems 35 through 38. Each story should be detailed and scientifically correct but also creative. That is, it should be entertaining while at the same time prove that you understand stellar evolution. Be sure to state whether "you" are a member of a binary system.

35. You are a white dwarf of $0.8M_{Sun}$.
36. You are a neutron star of $1.5M_{Sun}$.
37. You are a black hole of $10M_{Sun}$.
38. You are a white dwarf in a close binary system that is accreting matter from its companion star.
39. *Census of Stellar Corpses.* Which kind of object do you think is most common in our galaxy: white dwarfs, neutron stars, or black holes? Explain your reasoning.
40. *Fate of an X-Ray Binary.* The X-ray bursts that happen on the surface of an accreting neutron star are not powerful enough to accelerate the exploding material to escape velocity. Predict what will happen in an X-ray binary system in which the companion star eventually feeds over 3 solar masses of matter into the neutron star's accretion disk.
41. *Why Black Holes Are Safe.* Explain why the principle of conservation of angular momentum makes it very difficult to fall into a black hole.
42. *Surviving the Plunge.* The tidal forces near a black hole with a mass similar to a star would tear a person apart before that person could fall through the event horizon. Black hole researchers have recently pointed out that a fanciful "black hole life preserver" could help counteract those tidal forces. The life preserver would need to have a mass similar to that of an asteroid and would need to be shaped like a flattened hoop placed around the person's waist. In what direction would the gravitational force from the hoop pull on the person's head? In what direction would it pull on the person's feet? Based on your answers, explain in general terms how the gravitational forces from the "life preserver" would help to counteract the black hole's tidal forces.

Quantitative Problems

Be sure to show all calculations clearly and state your final answers in complete sentences.

43. *A Black Hole I?* You've just discovered a new X-ray binary, which we will call *Hyp-X1* ("Hyp" for hypothetical). The system Hyp-X1 contains a bright, B2 main-sequence star orbiting an unseen companion. The separation of the stars is estimated to be 20 million kilometers, and the orbital period of the visible star is 4 days.
    a. Use Newton's version of Kepler's third law to calculate the sum of the masses of the two stars in the system. (*Hint:* See Mathematical Insight 15.4.) Give your answer in both kilograms and solar masses ($M_{Sun} = 2.0 \times 10^{30}$ kg).
    b. Determine the mass of the unseen companion. Is it a neutron star or a black hole? Explain. (*Hint:* A main-sequence star with spectral type B2 has a mass of about $10M_{Sun}$.)
44. *A Black Hole II?* You've just discovered another new X-ray binary, which we will call *Hyp-X2* ("Hyp" for hypothetical). The system Hyp-X2 contains a bright, G2 main-sequence star orbiting an unseen companion. The separation of the

stars is estimated to be 12 million kilometers, and the orbital period of the visible star is 5 days.
    a. Use Newton's version of Kepler's third law to calculate the sum of the masses of the two stars in the system. (*Hint:* See Mathematical Insight 15.4.) Give your answer in both kilograms and solar masses ($M_{Sun} = 2.0 \times 10^{30}$ kg).
    b. Determine the mass of the unseen companion. Is it a neutron star or a black hole? Explain. (*Hint:* A main-sequence star with spectral type G2 has a mass of about $1M_{Sun}$.)
45. *Neutron Star Density.* A typical neutron star has a mass of about $1.5M_{Sun}$ and a radius of 10 kilometers.
    a. Calculate the average density of a neutron star, in *kilograms per cubic centimeter.*
    b. Compare the mass of 1 cm^3 of neutron star material to the mass of Mount Everest ($\approx 5 \times 10^{10}$ kg).
46. *Schwarzschild Radii.* Calculate the Schwarzschild radius (in kilometers) for each of the following.
    a. A $10^8 M_{Sun}$ black hole in the center of a quasar.
    b. A $5M_{Sun}$ black hole that formed in the supernova of a massive star.
    c. A mini–black hole with the mass of the Moon.
    d. A mini–black hole formed when a superadvanced civilization decides to punish you (unfairly) by squeezing you until you become so small that you disappear inside your own event horizon.
47. *The Crab Pulsar Winds Down.* Theoretical models of the slowing of pulsars predict that the age of a pulsar is approximately equal to $p/2r$, where p is the pulsar's current period and r is the rate at which the period is slowing with time. Observations of the pulsar in the Crab nebula show that it pulses 30 times a second, so that $p = 0.0333$ second, but the time interval between pulses is growing longer by 4.2×10^{-13} second with each passing second, so that $r = 4.2 \times 10^{-13}$ second per second. Using that information, estimate the age of the Crab pulsar. How does your estimate compare with the true age of the pulsar, which was born in the supernova observed in A.D. 1054?
48. *A Water Black Hole.* A clump of matter does not need to be extraordinarily dense in order to have an escape velocity greater than the speed of light, as long as its mass is large enough. You can use the formula for the Schwarzschild radius R_s to calculate the volume $\frac{4}{3}\pi R_s^3$ inside the event horizon of a black hole of mass M. What does the mass of a black hole need to be in order for its mass divided by its volume to be equal to the density of water (1 g/cm^3)?
49. *Energy of a Supernova.* In a massive star supernova explosion, a stellar core collapses down to form a neutron star roughly 10 kilometers in radius. The gravitational potential energy released in such a collapse is approximately equal to GM^2/r, where M is the mass of the neutron star, r is its radius, and $G = 6.67 \times 10^{-11}$ m^3/kg $\times$ s^2 is the gravitational constant. Using this formula, estimate the amount of gravitational potential energy released in a massive star supernova explosion. How does it compare with the amount of energy released by the Sun during its entire main-sequence lifetime?
50. *Challenge Problem: A Neutron Star Comes to Town.* Suppose a neutron star were suddenly to appear in your hometown. How thick a layer would Earth form as it wraps around the neutron star's surface? To make the problem easier, you may assume that the layer formed by Earth has the same average density as the neutron star. (*Hint:* Consider the mass of Earth to be distributed in a spherical *shell* over the surface

of the neutron star and then calculate the thickness of such a shell with the same mass as Earth. The volume of a spherical shell is approximately its surface area times its thickness: $V_{shell} = 4\pi r^2 \times$ thickness. Because the shell will be thin, you can assume that its radius is the radius of the neutron star.)

Discussion Questions

51. *Too Strange to Be True?* Despite strong theoretical arguments for the existence of neutron stars and black holes, many scientists rejected the possibility that such objects could really exist until they were confronted with very strong observational evidence. Some people claim that this type of scientific skepticism demonstrates an unwillingness on the part of scientists to give up their deeply held scientific beliefs. Others claim that this type of skepticism is necessary for scientific advancement. What do you think? Defend your opinion.

52. *Black Holes in Popular Culture.* Phrases such as "it disappeared into a black hole" are now common in popular culture. Give a few examples in which the term *black hole* is used in popular culture but is not meant to be taken literally. In what ways are these uses correct in their analogies to real black holes? In what ways are they incorrect? Why do you think such an esoteric scientific idea as that of a black hole has so captured the public imagination?

MEDIA EXPLORATIONS

For self-study activities go to Mastering ASTRONOMY   www.masteringastronomy.com

 Interactive Tutorials

Tutorial Review of Key Concepts

Use the interactive **Tutorials** at **www.mastering astronomy.com** to review key concepts from this chapter.

Stellar Evolution Tutorial

Lesson 1 Main-Sequence Lifetimes
Lesson 2 Evolution of a Low-Mass Star
Lesson 3 Late Stages of a High-Mass Star

Black Holes Tutorial

Lesson 1 What Are Black Holes?
Lesson 2 The Search for Black Holes

Supplementary Tutorial Exercises

Use the interactive **Tutorial Lessons** to explore the following questions.

Stellar Evolution Tutorial, Lessons 2, 3

1. What kind of "corpse" does a low-mass star leave behind? Why?
2. What kind of "corpse" does a high-mass star leave behind? Why?

Black Holes Tutorial, Lesson 1

1. What would have to happen to Earth to turn it into a black hole? Could this really happen?
2. Use the tool in the simulation to determine the Schwarzschild radii for 10 black holes of different masses. Record the mass and Schwarzschild radius for each black hole.
3. Make a graph of your results from question 2, plotting mass on the *x*-axis and Schwarzschild radius on the *y*-axis.
4. Based on your graph from question 3, briefly describe in words how the Schwarzschild radius depends on the mass of a black hole.

Black Holes Tutorial, Lesson 2

1. Briefly explain how scientists can detect a black hole.
2. Why does the search for black holes generally begin with a search for X-ray sources in the sky?

Web Projects

1. *Gamma-Ray Bursts.* Go to the Web site for a mission studying gamma-ray bursts (such as *HETE*, *INTEGRAL*, or *Swift*) and find the latest information about gamma-ray bursts. Write a one- to two-page essay on recent discoveries and how they may shed light on the mystery of gamma-ray bursts.
2. *Black Holes.* Andrew Hamilton, a professor at the University of Colorado, maintains a Web site with a great deal of information about black holes and what it would be like to visit one. Visit his site and investigate some aspect of black holes that you find to be of particular interest. Write a short report on what you learn.

Our Galaxy

The infinitude of creation is great enough to make a world, or a Milky Way of worlds, look in comparison with it what a flower or an insect does in comparison with the Earth.

—Immanuel Kant

I n previous chapters, we saw how stars forge new elements and expel them into space. We also studied how interstellar gas clouds enriched with these stellar by-products form new stars and planetary systems. These processes do not occur in isolation. Instead, they are part of a dynamic system that acts throughout our Milky Way Galaxy.

You are probably familiar with the idea that all living species on Earth interact with one another and with the land, water, and air to form a large, interconnected ecosystem. In a similar way, but on a much larger scale, our galaxy is a nearly self-contained system that cycles matter from stars into interstellar space and back into stars again. The birth of our solar system and the evolution of life on Earth would not have been possible without this "galactic ecosystem."

In this chapter, we will study our Milky Way Galaxy. We will investigate the galactic processes that maintain an ongoing cycle of stellar life and death, examine the structure and motion of the galaxy, probe its history, and explore the mysteries of the galactic center. Through it all, we will see that we are not only "star stuff" but also "galaxy stuff"—the product of eons of complex recycling and reprocessing of matter and energy in the Milky Way Galaxy.

19.1 The Milky Way Revealed

On a dark night, you can see a faint band of light slicing across the sky through several constellations, including Sagittarius, Cygnus, Perseus, and Orion. This band of light looked like a flowing ribbon of milk to the ancient Greeks, and we now call it the *Milky Way* (see Figure 2.1). In the early 1600s, Galileo used his telescope to prove that the light of the Milky Way comes from myriad individual stars. Together these stars make up the kind of stellar system we call a *galaxy*, echoing the Greek word for "milk," *galactos*.

The true size and shape of our Milky Way Galaxy are hard to guess from the way it looks in our night sky. Because we live inside the galaxy, trying to determine its structure is somewhat like trying to draw a picture of your house without ever leaving your bedroom. You can make a reasonable guess about what your house looks like by looking out your bedroom window at other houses, but you can't get all the specifics that way. Making the task even more difficult is the fact that much of our galaxy's visible light is hidden from our view. Only recently have we had the technology to observe our galaxy in other wavelengths of light. Nevertheless, by carefully observing our galaxy and comparing it to others that we see from the outside, we now have a good

understanding of the processes that shape our galaxy. In this section, we'll begin our exploration of the galaxy by investigating its basic structure and orbital motion.

• What does our galaxy look like?

Our Milky Way Galaxy holds over 100 billion stars and is just one among tens of billions of galaxies in the observable universe [Section 1.1]. It is a vast **spiral galaxy**, so named because of the spectacular **spiral arms** illustrated in Figure 19.1a. If we viewed the galaxy edge-on, as shown in Figure 19.1b, we'd see that the spiral arms are part of a fairly flat **disk** of stars surrounding a bright central **bulge**. The entire disk is surrounded by a dimmer, rounder **halo**. Most of the galaxy's bright stars reside in its disk. The most prominent stars in the halo are found in about 200 **globular clusters** of stars [Section 15.3].

The Milky Way is a relatively large galaxy. The galactic disk is about 100,000 light-years in diameter, but only about 1,000 light-years thick. Our Sun is located in the disk about 28,000 light-years from the galactic center—a little more than halfway out from the center to the edge of the disk. Within our Local Group of galaxies, only the Andromeda Galaxy is comparable in size to the Milky Way (see Figure 1.1).

It took us a long time to learn these facts about the Milky Way's true size and shape. Recall that the galactic disk is

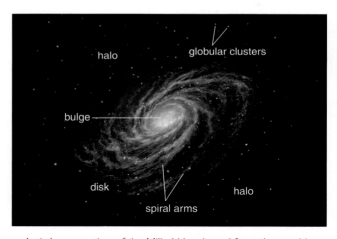

a Artist's conception of the Milky Way viewed from the outside.

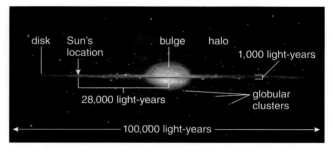

b Edge-on schematic view of the Milky Way.

Figure 19.1   Interactive Figure   The Milky Way Galaxy.

filled with interstellar gas and dust—known collectively as the *interstellar medium* [Section 16.1]—that obscures our view when we try to peer directly through it. The dusty, smoglike nature of the interstellar medium hides most of our galaxy from us when we try to observe it in visible light, and as a result it long fooled astronomers into believing that we lived near our galaxy's center (see Special Topic below).

Astronomer Harlow Shapley finally proved otherwise in the 1920s, when he demonstrated that the Milky Way's globular clusters are centered on a point tens of thousands of light-years from our Sun. He concluded that this point, not our Sun, must be the center of the galaxy. Thus, it has been less than a century since we established that our Sun holds no special, central place in the galaxy or the universe.

The Milky Way's strong gravity influences smaller galaxies in its vicinity. For instance, two small galaxies known as the *Large Magellanic Cloud* and the *Small Magellanic Cloud* orbit the Milky Way at distances of some 150,000 and 200,000 light-years, respectively. Both Magellanic Clouds are visible to the naked eye from the Southern Hemisphere.

Two other small galaxies lie even closer to the Milky Way than the Magellanic Clouds. These two galaxies are each in the process of colliding with the Milky Way's disk, and our own galaxy's tidal forces will ultimately rip them apart. Both were discovered only recently, because their visible light is obscured from view by the gas and dust in the Milky Way's disk. One galaxy, known as the Sagittarius

Dwarf, was discovered in 1994. The other, called the Canis Major Dwarf, was discovered in 2003.

Keep in mind that while these four nearby galaxies are quite small as galaxies go, they are still vast objects, each containing perhaps 1 billion to a few billion stars, making them 1,000 or more times the size of typical globular clusters. These galaxies are small only when we compare them to the enormous size of our own Milky Way Galaxy, which is hundreds of times as large as its smaller companions.

• How do stars orbit in our galaxy?

Now that we have discussed the basic structure of our galaxy, let's turn our attention to the orbital motions of the stars within it. A spiral galaxy like ours may look as though it should rotate like a giant pinwheel, but the galaxy is not a solid structure. Instead, each individual star follows its own orbital path around the center of the galaxy.

Nearly all stars in the Milky Way orbit the galaxy's center following one of two basic orbital patterns. Stars in the disk orbit in roughly circular paths that all go in the same direction in nearly the same plane. In contrast, stars in the bulge and halo soar high above and below the disk on randomly oriented orbits.

Orbits of Disk Stars   If you could stand outside the Milky Way and watch it for a few billion years, the disk would

---

SPECIAL TOPIC **How Did We Learn the Structure of the Milky Way?**

For most of human history, we knew the Milky Way only as an indistinct river of light in the sky. In 1610, Galileo used his telescope to discover that the Milky Way is made up of innumerable stars, but no one yet knew the size or shape of our galaxy.

In the late 1700s, British astronomers William and Caroline Herschel (brother and sister) tried to determine the shape of the Milky Way more accurately by counting how many stars lie in each direction. Their approach suggested that the Milky Way's width is five times its thickness. More than a century later, in the early 1900s, Dutch astronomer Jacobus Kapteyn and his colleagues used a more sophisticated star-counting method to gauge the size and shape of the Milky Way. Their results seemed to confirm the general picture found by the Herschels and suggested that the Sun lies near the center of the galaxy.

Kapteyn's results made astronomers with a sense of history slightly nervous. Only four centuries earlier, before Copernicus challenged the Ptolemaic system, astronomers had believed Earth was the center of the universe. Kapteyn's placement of the Sun near the Milky Way's center seemed to be giving Earth a central place again. Kapteyn knew that obscuring material could deceive us by hiding the rest of the galaxy like some kind of interstellar fog, but he found no evidence for such a fog.

While Kapteyn was counting stars, American astronomer Harlow Shapley was studying globular clusters. He found that these clusters appeared to be centered on a location tens of thousands of light-years from the Sun. (His original estimate was that the center of the galaxy was 45,000 light-years from the Sun, which was later

refined to the current 28,000 light-years.) Shapley concluded that this point marked the true center of our galaxy and that Kapteyn must be wrong.

Today we know that Shapley was right. The Milky Way's interstellar medium is the "fog" that misled Kapteyn. Robert Trumpler, working at California's Lick Observatory in the 1920s, established the existence of this dusty gas by studying open clusters of stars. By assuming that all open clusters had about the same diameter, he estimated their distances from their apparent sizes in the sky, much as you might estimate the distances of cars at night from the apparent separation of their headlights. He found that stars in distant clusters appear dimmer than expected based on their estimated distance, just as a car's headlights might appear in foggy weather. Trumpler concluded that light-absorbing material fills the spaces between the stars, partially obscuring the distant clusters and making them appear fainter than they would appear otherwise. Thus, we learned that interstellar material had been deceiving earlier astronomers, and that the stars visible in the night sky occupy a minuscule portion of the observable universe.

Subsequent studies established the spiral nature of our Milky Way Galaxy. Understanding the effects of interstellar dust allowed astronomers to take it into account in their observations, helping us learn the size and shape of our galaxy. Then, beginning in about the 1950s, careful studies of the motions of stars and gas clouds (made with radio observations of the 21-centimeter line from atomic hydrogen gas) gradually uncovered the detailed structure of the galactic disk, showing us the locations and motions of spiral arms.

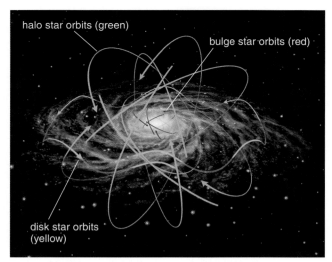

Figure 19.2 Characteristic orbits of disk stars (yellow), bulge stars (red), and halo stars (green) around the galactic center. (The yellow path exaggerates the up-and-down motion of the disk star orbits.)

resemble a huge merry-go-round. Like horses on a merry-go-round, individual stars bob up and down through the disk as they orbit. The general orbit of a star around the galaxy arises from its gravitational attraction toward the galactic center, while the bobbing arises from the localized pull of gravity within the disk itself (Figure 19.2). A star that is "too far" above the disk is pulled back into the disk by gravity. Because the density of interstellar gas is too low to slow the star, it flies through the disk until it is "too far" *below* the disk on the other side. Gravity then pulls it back in the other direction. This ongoing process produces the bobbing of the stars.

The up-and-down motions of the disk stars give the disk its thickness of about 1,000 light-years—a great distance by human standards, but only about 1% of the disk's 100,000-light-year diameter. In the vicinity of our Sun, each star's

orbit takes more than 200 million years, and each up-and-down "bob" takes a few tens of millions of years.

The galaxy's rotation is unlike that of a merry-go-round in one important respect: On a merry-go-round, horses near the edge move much faster than those near the center. But in our galaxy's disk, the orbital velocities of stars near the edge and those near the center are about the same. Stars closer to the center therefore complete each orbit in less time than stars farther out.

Orbits of Halo and Bulge Stars The orbits of stars in the halo and bulge are much less organized than the orbits of stars in the disk (see Figure 19.2). Individual bulge and halo stars travel around the galactic center on more or less elliptical paths, but the orientations of these paths are relatively random. Neighboring halo stars can circle the galactic center in opposite directions. They swoop from high above the disk to far below it and back again, plunging through the disk at velocities so great that the disk's gravity hardly alters their trajectories. Several fast-moving halo stars are currently passing through the disk not too far from our own solar system.

These swooping orbits explain why the bulge and halo are much puffier than the disk. Halo and bulge stars soar to heights above the disk far greater than the up-and-down bobbing of the disk stars. The great differences between orbits in the disk and orbits in the bulge and halo indicate that disk stars and halo stars have very different origins. As we'll discuss in Section 19.3, these orbital differences provide an important clue to how our galaxy formed.

THINK ABOUT IT

Is there much danger that a halo star swooping through the disk of the galaxy will someday hit the Sun or Earth? Why or why not? (*Hint:* Think about the typical distances between stars, as illustrated by use of the 1-to-10-billion scale in Chapter 1.)

SPECIAL TOPIC How Do We Determine Stellar Orbits?

Astronomers learned how stars orbit the Milky Way by measuring the motions of many different stars relative to the Sun. Although these measurements are easy to make in principle, they can be difficult in practice.

Determining a star's precise motion relative to the Sun requires knowing its true velocity through space. However, our primary means of measuring speeds in the universe—the Doppler effect—can tell us only a star's *radial velocity*, the component of its velocity directed toward or away from us (see Figure 5.23). If we want to know the star's true velocity, we must also measure its *tangential velocity*, the component of its velocity directed across our line of sight.

Tangential velocity is difficult to measure because of the vast distances to stars. Over tens of thousands of years, the tangential velocities of stars cause their apparent positions in our sky to change, gradually changing the shapes of the constellations. These changes are far too small for human eyes to notice. However, we can mea-

sure tangential velocities for many stars by comparing telescopic photographs taken years or decades apart. For example, if photographs taken 10 years apart show that a star has moved across our sky by an angle of 1 arcsecond, we know the star is moving at an angular rate of 0.1 arcsecond per year. We can convert this angular rate of motion (often called the star's *proper motion*) to a tangential velocity if we also know the star's distance. For a given angular rate, the tangential velocity is greater for more distant stars.

Because the earliest telescopic photographs date only to the late nineteenth century, we can measure tangential velocities only for objects that have moved noticeably since that time. In general, this limits us to stars within a few hundred light-years of Earth. Thus, we can determine precise stellar orbits only for relatively nearby stars. For more distant stars (and galaxies), we can usually measure only radial velocities. This is one reason why we have only limited knowledge of large-scale motion in the universe.

Stellar Orbits and the Mass of the Galaxy The Sun's orbital path is fairly typical of disk stars located near our 28,000-light-year distance from the galactic center. By measuring our Sun's speed relative to many globular clusters in our galaxy's halo, we've determined that the Sun and its neighbors orbit the center of the Milky Way at a speed of about 220 kilometers per second (about 800,000 km/hr [Section 1.3]). Even at this speed, it takes the Sun about 230 million years to complete one orbit around the galactic center. Early dinosaurs were just emerging on Earth when our Sun last visited this side of the galaxy.

MATHEMATICAL INSIGHT 19.1 Using Stellar Orbits to Measure Galactic Mass

As discussed in the text, Newton's version of Kepler's third law allows us to use the orbital characteristics of the Sun or any other star or gas cloud to calculate the mass of the galaxy *within* the object's orbit. Because the mass of the galaxy is much greater than the mass of any individual star or gas cloud that orbits within it, we start with Newton's version of Kepler's third law in the form that is valid when one object is much more massive than the other (see Mathematical Insight 4.3);

$$p^2 = \frac{4\pi^2}{G \times M} \times a^3$$

where M is the mass of the massive object; p and a are the orbital period and semimajor axis of a smaller orbiting object, respectively; and the gravitational constant is $G = 6.67 \times 10^{-11} \frac{m^3}{kg \times s^2}$. Since we are interested in finding the galaxy's mass, we now solve the above equation for M; you should confirm that the equation becomes:

$$M = \frac{4\pi^2 \times a^3}{G \times p^2}$$

In principle, we could use this equation as is. However, when we study the motions of stars and gas clouds in the galaxy, we usually measure their orbital velocities rather than their orbital periods. Thus, it is easier to use this equation if we convert into a form that uses velocity (v) rather than period (p). To do so, recall the expression we found in Mathematical Insight 15.4 for the orbital speed of an object with a circular orbit:

$$v = \frac{\text{distance traveled in one orbit}}{\text{orbital period}} = \frac{2\pi a}{p}$$

Solving for p, we find:

$$p = \frac{2\pi a}{v}$$

We now substitute this expression for p into our equation for M above:

starting equation: $M = \dfrac{4\pi^2 \times a^3}{G \times p^2}$

substitute $p = \dfrac{2\pi a}{v}$: $M = \dfrac{4\pi^2 \times a^3}{G \times \left(\dfrac{2\pi a}{v}\right)^2}$

expand square and cancel like terms: $M = \dfrac{4\pi^2 \times a^{\cancel{3}}}{G \times \dfrac{4\pi^2 a^2}{v^2}}$

simplify: $M = \dfrac{a \times v^2}{G}$

This is the formula we need, but we can make it easier to remember with two further changes. First, because we are dealing with stars or gas clouds on circular orbits around the galactic center, we can replace a with the radius r of the object's orbit. Second, because this formula tells us only the mass of the galaxy *within* the distance r from the center, we will refer to the mass as M_r. We write our final equation as:

$$M_r = \frac{r \times v^2}{G}$$

We will call this equation the *orbital velocity law*, because it allows us to calculate the galaxy's mass from the orbital velocity (and orbital distance) of a star or gas cloud.

Example: Calculate the mass of the Milky Way Galaxy within the Sun's orbit.

Solution:

Step 1 Understand: The orbital velocity law allows us to use the Sun's orbital velocity and distance to calculate the mass of the galaxy within the Sun's orbit. From the discussion in the chapter, we know that the Sun orbits the center of the Milky Way Galaxy with velocity $v = 220$ km/s $= 2.2 \times 10^5$ m/s; we put it in meters per second so that the units will agree with the units given for G. The Sun's orbital distance is 28,000 light-years, which we must also convert to meters for unit consistency. Using the fact that a light-year is about 9.46×10^{12} kilometers (see Mathematical Insight 1.1), which is equivalent to 9.46×10^{15} meters, the Sun's orbital radius is $r = 2.6 \times 10^{20}$ m. We now have all the information we need to find the mass M_r within the Sun's orbit.

Step 2 Solve: We now calculate the mass M_r by substituting the above values for v and r into the orbital velocity law:

$$
\begin{aligned}
M_r &= \frac{r \times v^2}{G} \\
&= \frac{(2.6 \times 10^{20}\, \text{m}) \times (2.2 \times 10^5\, \tfrac{m}{s})^2}{6.67 \times 10^{-11}\, \frac{m^3}{kg \times s^2}} \\
&= 1.9 \times 10^{41}\, \text{kg}
\end{aligned}
$$

Step 3 Explain: We have found that the mass of the Milky Way Galaxy within the Sun's orbit is about 2×10^{41} kg. This number will be easier to interpret if we convert it to solar masses. Dividing the mass we've found by the Sun's mass of about 2×10^{30} kg, we find that the mass of the galaxy within the Sun's orbit is about 10^{11}, or 100 billion, solar masses.

The orbital motion of the Sun and other stars gives us a way to determine the mass of the galaxy. Recall that Newton's law of gravity determines how quickly objects orbit one another. This fact, embodied in Newton's version of Kepler's third law [Section 4.4], allows us to determine the mass of a relatively large object when we know the period and average distance of a much smaller object in orbit around it.

For example, we can use the Sun's orbital velocity and its distance from the galactic center to determine the mass of our galaxy lying *within* the Sun's orbit (see Mathematical Insight 19.1). To understand why the Sun's orbital motion allows us to calculate only the mass within the Sun's orbit, rather than the total mass of the galaxy, we need to consider the difference between the gravitational effects of mass within the Sun's orbit and of mass beyond its orbit. Every part of the galaxy exerts gravitational forces on the Sun as it orbits, but the net force from matter outside the Sun's orbit is relatively small because the pulls from opposite sides of the galaxy virtually cancel one another. In contrast, the net gravitational forces from mass within the Sun's orbit all pull the Sun in the same direction—toward the galactic center. Thus, the Sun's orbital velocity responds almost exclusively to the gravitational pull of matter inside its orbit. By using the Sun's 28,000-light-year distance and 220-km/s orbital velocity in Newton's version of Kepler's third law, we find that the total amount of mass within the Sun's orbit is about 2×10^{41} kilograms, or about 100 billion times the mass of the Sun.

Similar calculations based on the orbits of more distant stars in the Milky Way have revealed one of the greatest mysteries in astronomy—one that we first encountered in Chapter 1 (see Figure 1.15). Photographs of spiral galaxies make it appear that most of their mass is concentrated near their centers. However, orbital motions tell us that just the opposite is true. If most of the mass were concentrated near the galaxy's center, the orbital speeds of more distant stars would be slower, just as the orbital speeds of the planets decline with distance from the Sun because the Sun's gravity is weaker at greater distances [Section 4.4]. Instead, we find that orbital speeds remain about the same out to great distances from the galactic center, telling us that most of the galaxy's mass resides far from the center and is distributed throughout the halo. Because we see very few stars and virtually no gas or dust in the halo, we conclude that most of the galaxy's mass must not give off any light that we can detect, and hence we refer to it as *dark matter*. We will discuss the evidence for dark matter in more detail in Chapter 22, along with ideas about its nature and its implications for the history and fate of the universe.

19.2 Galactic Recycling

The Milky Way Galaxy is home to our solar system, but its importance to our existence runs much deeper. The birth of the Sun and the planets of our solar system could not have occurred without the galactic recycling that takes place within the disk of the galaxy and its interstellar medium.

We have already discussed the general idea of galactic recycling in Chapters 16 and 17. Stars are born when gravity causes the collapse of molecular clouds. Stars shine for millions or billions of years with energy produced by nuclear fusion, dying only when they've exhausted their fuel for fusion. In death, stars return much of their material back to space, either through a planetary nebula in the case of low-mass stars or through a supernova in the case of high-mass stars.

In addition to making new generations of stars possible, galactic recycling gradually changes the chemical composition of the interstellar medium, although different regions of the galaxy change in composition at different rates. Recall that the early universe contained only the chemical elements hydrogen and helium; all heavier elements have been produced by stars. The newly created elements mix with other interstellar gas and thereby become incorporated into new generations of stars, and that is how our solar system came to have the elements from which our planet was made. Today, thanks to more than 10 billion years of galactic recycling, elements heavier than helium constitute about 2% of the galaxy's gaseous content by mass. The remaining 98% still consists of hydrogen (about 70%) and helium (about 28%).

This general picture of the galactic recycling process may sound simple enough, but it leaves some important questions unanswered. In particular, the supernova explosions that scatter most heavy elements into space send debris flying out at speeds of several thousand kilometers per second—much faster than the escape velocity from the galaxy. So how have the chemical riches produced by stars managed to remain in our galaxy? The answer turns out to depend on interactions between the matter expelled by supernovae and the interstellar medium that fills the galactic disk. In this section, we'll examine these interactions in more detail, so that we will understand why our existence owes as much to the functioning of our galaxy as it does to the manufacturing of heavy elements by stars.

• How is gas recycled in our galaxy?

The galactic recycling process proceeds in several stages, making up what we will call the **star-gas-star cycle** that is summarized in Figure 19.3. From Chapters 16 and 17, you are already familiar with the basic ideas in the lower three frames shown in the figure: Stars are born in molecular clouds, they shine and produce heavier elements through nuclear fusion, and they return much of their content back into the interstellar medium through supernovae and stellar winds. Let's complete the loop, starting with the gas ejected by dying stars and finishing back at star birth.

Star–Gas–Star Cycle

hot bubbles
Fig. 19.6

atomic-hydrogen clouds
Fig. 19.12a

molecular clouds
Fig. 19.10

returning gas: supernovae and stellar winds
Fig. 19.5

star formation
Fig. 19.11

stellar lives: nuclear fusion/ heavy-element formation
Fig. 19.13

Figure 19.3 A pictorial representation of the star-gas-star cycle. These photos appear individually later in the chapter. Their figure numbers are indicated.

Gas from Dying Stars As we've discussed, all stars return much of their original mass to interstellar space in two basic ways: through stellar winds that blow throughout their lives, and through "death events" of planetary nebulae (for low-mass stars) or supernovae (for high-mass stars). Low-mass stars generally have weak stellar winds while they are on the main sequence. Their winds grow stronger and carry more material into space when they become red giants. By the time a low-mass star like the Sun ends its life with the ejection of a planetary nebula [Section 17.2], it has returned almost half its original mass to the interstellar medium (Figure 19.4).

High-mass stars lose mass much more dynamically and explosively, returning almost all their original mass to interstellar space. The powerful winds from supergiants and massive O and B stars recycle large amounts of matter, and at the ends of their lives, these stars explode as supernovae. The high-speed gas ejected into space by supernovae or powerful stellar winds from hot stars sweeps up surrounding interstellar material, excavating a **bubble** of hot, ionized gas (gas in which atoms are missing some of their electrons [Section 5.3]) around the exploding star (Figure 19.5). These

Figure 19.4 A dying low-mass star, like this one photographed by the Hubble Space Telescope, returns gas to the interstellar medium in a planetary nebula. This particular planetary nebula is known as the Retina Nebula. It is about 1 light-year across in the longer direction.

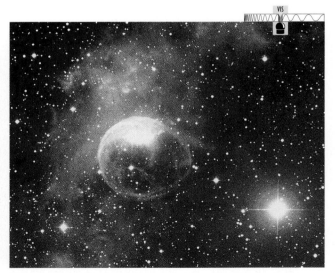

Figure 19.5 This photo shows a bubble of hot, ionized gas blown by a wind from the hot star near its center. Although it looks much like a soap bubble, it is actually an expanding shell of hot gas about 10 light-years in diameter. It glows where gas piles up as the bubble sweeps outward through the interstellar medium.

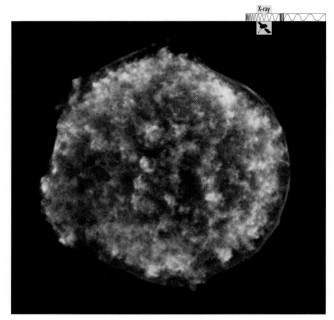

Figure 19.6 This image from the Chandra X-Ray Observatory shows X-ray emission from hot gas in a young supernova remnant—the remnant from the supernova observed by Tycho Brahe in 1572. The most energetic X rays (blue) come from 20-million-degree gas just behind the expanding shock wave. Less energetic X rays (green and red) come from the 10-million-degree debris ejected by the exploded star. The remnant is about 20 light-years across.

hot, tenuous bubbles are quite common in the disk of the galaxy, but they are not always easy to detect. While some emit strongly in visible light and others are hot enough to emit profuse amounts of X rays, many bubbles are evident only through radio emission from the shells of gas that surround them.

The bubbles created by supernovae can have even more dramatic effects on the interstellar medium than those created by fast-moving stellar winds. Supernovae generate *shock waves*—waves of pressure that move faster than the speed

of sound. A shock wave sweeps up surrounding gas as it travels, creating a "wall" of fast-moving gas on its leading edge. When we observe a *supernova remnant* [Section 17.3], we are seeing the aftermath of its shock wave, which compresses, heats, and ionizes all the interstellar gas it encounters.

Figure 19.6 shows a young supernova remnant whose shocked gas is hot enough to emit X rays. In contrast, the older supernova remnant shown in Figure 19.7 is cooler because its shock wave has swept up more material, thereby distributing its energy more widely. Eventually, the shocked gas will radiate away most of its original energy, and the expanding wall of gas will slow to subsonic speeds and merge with the surrounding interstellar medium.

Hot bubbles are fairly easy to spot with X-ray telescopes when they lie at some distance from us, because of their distinctive shapes. However, the nearest bubble is not quite as obvious because we're living inside it. Observations of X rays from the interstellar medium reveal hot, X-ray-emitting gas in all directions around the Sun and the nearby stars. Beyond this hot gas, at distances ranging up to a few hundred light-years, lies a region of much cooler gas. The existence of this *Local Bubble* means that one or more supernovae must have detonated within the Sun's neighborhood during the past several million years.

In addition to their role as the movers and shakers of the interstellar medium, shock waves from supernovae can act as subatomic particle accelerators. Some of the electrons in supernova remnants accelerate to nearly the speed of light as they interact with the shock wave. These fast electrons emit radio waves as they spiral around magnetic field lines threading the supernova remnant (Figure 19.8). (This radio emission is sometimes called *synchrotron radiation*.)

Supernovae may also generate the **cosmic rays** that permeate the interstellar medium and bombard Earth. Cosmic rays are made of electrons, protons, and atomic nuclei that zip through interstellar space at close to the speed of light. Some cosmic rays penetrate Earth's atmosphere and reach Earth's surface. On average, about one cosmic-ray particle

a This photograph shows the entire supernova remnant glowing in visible light. The angular size of this remnant in our sky is six times that of the Moon, and it is about 130 light-years across.

b The close-up view, from Hubble Space Telescope, displays the fine filamentary structure in a small piece of the remnant. The spectacular colors come from emission lines of the atoms and ions indicated in the spectrum in part (c).

c A visible-light spectrum from the Cygnus Loop shows the strong emission lines that account for the distinct colors in the Hubble Space Telescope photo.

Figure 19.7 Emission of visible light from an older supernova remnant, the Cygnus Loop.

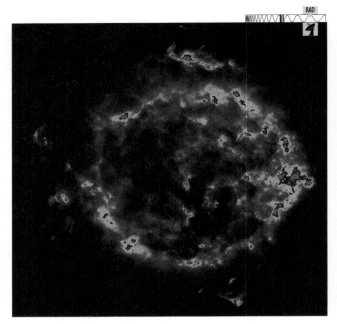

a Radio emission caused by electrons spiraling around magnetic field lines in the young supernova remnant Cassiopeia A. This remnant is about 10 light-years in diameter.

Figure 19.8 Supernova remnant Cassiopeia A.

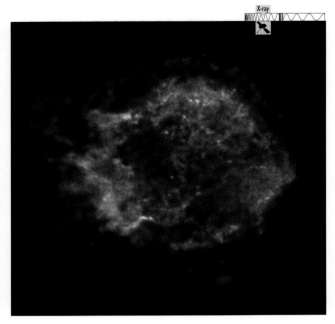

b This photograph shows X ray emission from the hot gas of Cassiopeia A, as seen by the Chandra X-Ray Observatory. Red indicates the lowest-energy X rays and blue indicates the highest-energy X rays.

strikes your body each second. Cosmic rays can cause genetic mutations in living organisms, so it is possible that mutations induced by cosmic rays have played an important part in the evolution of life on Earth. It is also possible that too much exposure to cosmic rays could lead to cancer. The cosmic-ray bombardment rate is 100 times greater at the high altitudes at which jet planes fly, and even more cosmic rays funnel along magnetic field lines to Earth's magnetic poles. That is why some airlines restrict how often their flight crews cross the Earth's polar regions.

COMMON MISCONCEPTIONS

The Sound of Space

In many science fiction movies, a thunderous sound accompanies the demolition of a spaceship. If the moviemakers wanted to be more realistic, they would silence the explosion. On Earth, we perceive sound when sound waves—which are waves of alternately rising and falling pressure—cause trillions of gas atoms to push our eardrums back and forth. Although sound waves can and do travel through interstellar gas, the extremely low density of this gas means that only a handful of atoms per second would collide with something the size of a human eardrum. As a result, it would be impossible for a human ear (or a similar-size microphone) to register any sound. Despite the presence of sound waves and shock waves in space, the sound of space is silence.

Superbubbles and Fountains The bubble created by a single supernova can grow to a diameter of about a hundred light-years before it slows and merges with surrounding interstellar gas. But in some regions of the Milky Way, we see cavities of hot gas that are more than a thousand light-years across. These huge cavities presumably arise when many individual bubbles combine to form a much larger **superbubble**. Recall that stars tend to form in clusters in which all the stars are about the same age. As a result, the hottest, most massive stars in a cluster can end their lives and explode within a few hundred thousand years of one another. The shock waves from the individual supernovae soon overlap, combining their energy into one very powerful shock wave that we see as the superbubble. Subsequent supernovae from the cluster explode inside the superbubble, adding even more energy to the shock wave.

In many places in our galactic disk, we see what appear to be elongated superbubbles extending from young clusters of stars to distances of 3,000 light-years or more above the disk. These probably are places where superbubbles have grown so large that they cannot be contained within the Milky Way's disk. Once the superbubble breaks out of the disk, where nearly all the Milky Way's gas resides, nothing remains to slow its expansion except gravity. The result is a **blowout** that is in some ways similar to a volcanic eruption, but on a galactic scale: Hot gas erupts from the disk, spreading out as it shoots upward into the galactic halo.

Theoretical modeling of blowouts suggests that they continually cycle gas between the Milky Way's disk and the halo (Figure 19.9). According to this model, sometimes called

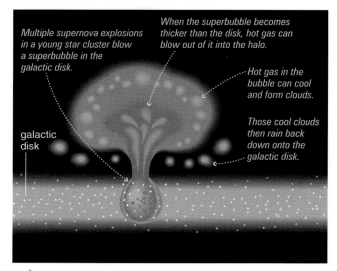

Multiple supernova explosions in a young star cluster blow a superbubble in the galactic disk.

When the superbubble becomes thicker than the disk, hot gas can blow out of it into the halo.

Hot gas in the bubble can cool and form clouds.

Those cool clouds then rain back down onto the galactic disk.

galactic disk

a 🎨 The galactic fountain model.

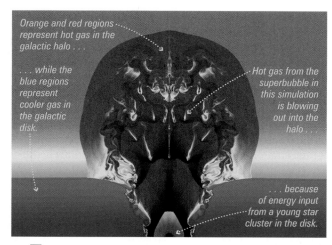

Orange and red regions represent hot gas in the galactic halo . . .

. . . while the blue regions represent cooler gas in the galactic disk.

Hot gas from the superbubble in this simulation is blowing out into the halo . . .

. . . because of energy input from a young star cluster in the disk.

b 🖥️ This image from a supercomputer simulation shows a superbubble of hot gas (red) blowing out of the cooler gas (blue) in the galactic disk. Such blowouts can be quite turbulent, as is evident from the complex structure in this image.

Figure 19.9 According to the galactic fountain model, superbubbles formed by the shock waves of many supernovae may blow gas out of the galactic disk, which then cools into gas clouds that rain back down on the disk.

the **galactic fountain**, the gravity of the galactic disk slows and eventually halts the rise of the gas from a blowout. Near the top of its trajectory, the ejected gas starts to cool and form clouds. Gravity then causes these clouds to rain back down into the disk, where their contents mix with the gas throughout a large region of the galaxy. If the galactic fountain model is correct, then superbubbles and blowouts play an important role in the galaxy-wide recycling system that has made our existence possible.

The galactic fountain model is plausible but difficult to verify. Observationally, we do indeed see some hot gas high above the galaxy's disk. We also see cooler clouds that appear to be raining down from the halo. However, we can see this "rain" only directly above and below us, making it difficult to demonstrate beyond a doubt that galactic fountains circulate the products of supernovae throughout the Milky Way.

Cooling and Cloud Formation The hot, ionized gas in bubbles, superbubbles, and fountains is dynamic and widespread, but it is a relatively small fraction of the gas in the Milky Way. Most of the gas is much cooler—cool enough so that hydrogen atoms remain neutral rather than being ionized. We therefore refer to this gas as **atomic hydrogen gas**, although the hydrogen is mixed with neutral atoms of helium and heavy elements in the usual proportions for the galaxy (70% hydrogen, 28% helium, and 2% heavy elements by mass). After the gas that makes bubbles or superbubbles cools, it becomes part of this widespread atomic hydrogen gas in the galaxy.

We can map the distribution of atomic hydrogen gas in the Milky Way with radio observations. Atomic hydrogen emits a spectral line with a wavelength of 21 centimeters, which lies in the radio portion of the electromagnetic spec-

trum (see Figure 5.7).* We see the radio emission from this **21-centimeter line** coming from all directions, telling us that atomic hydrogen gas is distributed throughout the galactic disk. Based on the overall strength of the 21-centimeter emission, the total amount of atomic hydrogen gas in our galaxy must be about 5 billion solar masses, which is a few percent of the galaxy's total mass.

Atomic hydrogen gas tends to be found in two distinct forms: large, tenuous clouds of warm (10,000 K) atomic hydrogen and smaller, denser clouds of cool (100 K) atomic hydrogen. If you were to take an interstellar voyage across the Milky Way, you would spend the majority of your time cruising through regions of warm atomic hydrogen interspersed with bubbles of hot, ionized gas. Every thousand light-years or so, you would encounter a cooler, denser cloud of atomic hydrogen.

In the warm regions, you would detect about one atom per cubic centimeter and a weak magnetic field (weaker than Earth's magnetic field by a factor of about 100,000). In the cooler clouds, you would find a density of about 100 atoms per cubic centimeter and a stronger magnetic field. As we discussed in Chapter 16, these magnetic fields may provide support to clouds that gravity would otherwise cause to collapse and form stars.

---

*If you recall that transitions between electron energy levels in hydrogen usually produce infrared, visible, or ultraviolet spectral lines (see Figure 5.15), you may wonder how the 21-centimeter line forms. The line does not come from a transition of the type we discussed earlier, but instead comes from a low-energy transition involving the electron's *spin*. Recall that electrons always have either "spin up" or "spin down" (see Figure S4.3). The electron has slightly more energy when its spin is aligned with the spin of the proton in the nucleus than when the proton and electron have opposite spins. When the electron changes its spin from the higher-energy to the lower-energy state—the so-called *spin-flip transition*— a photon is emitted with wavelength 21 centimeters.

Matter remains in the warm atomic hydrogen stage of the star-gas-star cycle for millions of years. Gravity slowly draws blobs of this gas together into tighter clumps that radiate energy more efficiently as they grow denser. The blobs therefore cool and contract, forming the smaller clouds of cool atomic hydrogen. This process of shrinkage and coagulation takes a much longer time than the other steps in the journey from star death to star birth. The slowness of the transition from warm gas to cool clouds accounts for the large amount of gaseous matter in the atomic hydrogen stage of the star-gas-star cycle. During this long period of cooling, disturbances produced by supernova explosions occurring in the vicinity of the cooling gas keep it stirred up, which is probably why cold, star-forming clouds are often observed to be turbulent [Section 16.1].

Clouds of atomic hydrogen also contain a small but important amount of interstellar dust. Recall that interstellar *dust grains* are tiny, solid flecks of carbon and silicon minerals that resemble particles of smoke and form in the winds of red giant stars [Section 16.1]. Once formed, dust grains remain in the interstellar medium until they are heated and destroyed by a passing shock wave or incorporated into a protostar. Dust grains make up only about 1% of the mass of atomic hydrogen clouds, but they are responsible for the absorption of visible light that prevents us from seeing through the disk of the galaxy.

From Atomic to Molecular Clouds   As the temperature drops further in the center of a cool cloud of atomic hydrogen, hydrogen atoms combine into molecules, making a *molecular cloud* [Section 16.1]. Recall that molecular clouds are the coldest, densest collections of gas in the interstellar medium, and they are the birthplaces of stars. They often congregate into *giant molecular clouds* that contain up to a million solar masses of gas (Figure 19.10). The total mass of molecular clouds in the Milky Way is somewhat uncertain, but it is probably about the same as the total mass of atomic hydrogen gas—about 5 billion solar masses. Throughout much of this molecular gas, temperatures hover only a few degrees above absolute zero.

Molecular clouds are heavy and dense compared to the rest of the interstellar gas and therefore tend to settle toward the central layers of the Milky Way's disk. This tendency creates a phenomenon you can see with your own eyes: the dark lanes running through the luminous band of light in our sky that we call the Milky Way (see Figure 2.1).

Completing the Cycle   As we discussed in Chapter 16, a large molecular cloud fragments and gives birth to a cluster of stars. Once a few stars form in a newborn cluster, their radiation begins to erode the surrounding gas in the molecular cloud. Ultraviolet photons from high-mass stars heat and ionize the gas, and winds and radiation pressure push the ionized gas away. This kind of feedback pre-

Figure 19.10 This image shows the complex structure of a molecular cloud in the constellation Orion. The picture was made by measuring Doppler shifts of emission lines from carbon monoxide molecules, and the colors indicate gas motions: Bluer parts are moving toward us and redder parts are moving away from us (relative to the cloud as a whole). This enormous cloud is about 1,600 light-years distant and several hundred light-years across.

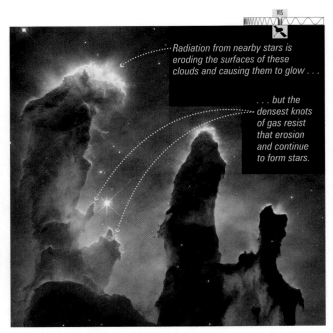

RAD

VIS

Radiation from nearby stars is eroding the surfaces of these clouds and causing them to glow . . .

. . . but the densest knots of gas resist that erosion and continue to form stars.

Figure 19.11 A portion of the Eagle Nebula, as seen by the Hubble Space Telescope. The dark columns of gas are molecular clouds, and stars are currently forming in the densest parts of these clouds. Arrows indicate two of the locations where dense knots of gas are giving birth to stars. The region pictured here is about 5 light-years across.

vents much of the gas in a molecular cloud from turning into stars.

The process of molecular cloud erosion is vividly illustrated in the Eagle Nebula, a complex of clouds where new stars are forming (Figure 19.11). The dark, lumpy columns are molecular clouds. To the upper right (outside the picture), newly formed massive stars glow with ultraviolet radiation. This radiation sears the surface of the molecular clouds, destroying molecules and stripping electrons from atoms. As a result, matter "evaporates" from the molecular clouds and joins the hotter ionized gas encircling them. Only the densest knots of gas resist evaporation. Stars are forming in some of these dense knots, which remain compact while the rest of the cloud erodes. These star-forming knots are the tips of the dark protrusions on the columns of molecular gas in the figure.

We have arrived back where we started in the star-gas-star cycle. The most massive stars now forming in the Eagle Nebula will explode within a few million years, filling the region with bubbles of hot gas and newly formed heavy elements. The expanding bubbles will slow and cool as their gas merges with the widespread atomic hydrogen gas in the galaxy. Further in the future, this gas will cool further and coalesce into molecular clouds, forming new stars, new planets, and maybe even new civilizations.

Despite the recycling of matter from one generation of stars to the next, the star-gas-star cycle cannot go on forever. With each new generation of stars, some of the galaxy's gas becomes permanently locked away in brown dwarfs that never return material to space and in stellar corpses left behind when stars die (white dwarfs, neutron stars, and black holes). The interstellar medium therefore is slowly running out of gas, and the rate of star formation will gradually taper off over the next 50 billion years or so. Eventually, star formation will cease.

Putting It All Together: The Distribution of Gas in the Milky Way
Different regions of the galaxy are in different stages of the star-gas-star cycle. Because the cycle proceeds over such a long period of time compared to a human lifetime, each

stage appears to us as a snapshot. We therefore see the interstellar medium in a wide variety of manifestations, ranging from the tenuous million-degree gas of bubbles to the cold, dense gas of molecular clouds. Table 19.1 summarizes the different states in which we see interstellar gas in the galactic disk.

We can see how these different states of gas are arranged in our galaxy by observing the galaxy in many different wavelengths of light. Figure 19.12 shows seven views of the disk of the Milky Way Galaxy. Each view represents a panorama in a particular wavelength band, made by photographing the Milky Way's disk in every direction from Earth.

- Figure 19.12a shows variations in the intensity of radio emission from the 21-centimeter line of atomic hydrogen. Thus, it maps the distribution of atomic hydrogen gas, demonstrating that this gas fills much of the galactic disk.
- Figure 19.12b shows variations in the intensity of radio emission lines from carbon monoxide (CO) and therefore maps the distribution of molecular clouds. These cold, dense clouds are concentrated in a narrow layer near the midplane of the galactic disk.
- Figure 19.12c shows variations in the intensity of long-wavelength infrared emission from interstellar dust grains. Notice that the regions of strongest emission from dust correspond to the locations of the molecular clouds in Figure 19.12b.
- Figure 19.12d shows shorter-wavelength infrared light from stars at wavelengths that penetrate clouds of gas and dust. Thus, this image shows how our galaxy would look if there were no dust blocking our view. The galactic bulge is clearly evident at the center.
- Figure 19.12e shows the galactic disk in visible light, just as it appears in the night sky. (Of course, only part of the Milky Way is above the horizon at any one time.) Because visible light cannot penetrate interstellar dust, the dark blotches correspond closely to the bright patches of molecular radio emission and infrared dust emission in Figure 19.12b and c.

Table 19.1 Typical States of Gas in the Interstellar Medium

State of Gas	Primary Constituent	Approximate Temperature	Approximate Density ($atoms\ per\ cm^3$)	Description
Hot bubbles	Ionized hydrogen	1,000,000 K	0.01	Pockets of gas heated by supernova shock waves
Warm atomic gas	Atomic hydrogen	10,000 K	1	Fills much of galactic disk
Cool atomic clouds	Atomic hydrogen	100 K	100	Intermediate stage of star-gas-star cycle
Molecular clouds	Molecular hydrogen	30 K	300	Regions of star formation
Molecular cloud cores	Molecular hydrogen	60 K	10,000	Star-forming clouds

a 21-cm radio emission from atomic hydrogen gas.

b Radio emission from carbon monoxide reveals molecular clouds.

c Infrared (60–100 μm) emission from interstellar dust.

d Infrared (1–4 μm) emission from stars that penetrates most interstellar material.

e Visible light emitted by stars is scattered and absorbed by dust.

f X-ray emission from hot gas bubbles (diffuse blobs) and X-ray binaries (pointlike sources).

g Gamma-ray emission from collisions of cosmic rays with atomic nuclei in interstellar clouds.

Figure 19.12 Interactive Photo. Panoramic views of the Milky Way in different bands of the spectrum. The center of the galaxy, which lies in the direction of the constellation Sagittarius, is in the center of each strip. The rest of each strip shows all other directions in the Milky Way disk as seen from Earth. (Imagine attaching the left and right ends of each strip to form a circular band that corresponds to the 360° band of the Milky Way in our sky.)

● Figure 19.12f shows X-ray light from the galactic disk. The pointlike blotches in this view are mostly X-ray binaries [Section 18.2]. The rest of the X-ray emission comes primarily from hot gas bubbles. Because hot gas tends to rise into the halo, it is less concentrated toward the midplane than the atomic and molecular gas.

● Figure 19.12g shows gamma-ray emission from the Milky Way. Most of the gamma-ray emission is produced by collisions between cosmic-ray particles and atomic nuclei in interstellar clouds. Such collisions happen most frequently where gas densities are highest, so the gamma-ray emission corresponds closely to the locations of molecular and atomic gas.

THINK ABOUT IT

Carefully compare and contrast the views of the Milky Way's disk in Figure 19.12. Why do regions that appear dark in some views appear bright in others? What general patterns do you notice?

• Where do stars tend to form in our galaxy?

The star-gas-star cycle has operated continuously since the Milky Way's birth, yet new stars are not spread evenly across the galaxy. Some regions seem much more fertile than others. Galactic environments rich in molecular clouds tend to spawn new stars easily, while gas-poor environments do not. However, molecular clouds are dark and hard to see. Certain other signatures of star formation are much more obvious. A quick tour of some star-forming galactic environments will help you spot where the action is.

Hallmarks of Star-Forming Regions Wherever we see hot, massive stars, we know that we have spotted a region of active star formation. Because these stars live fast and die young, they never get a chance to move very far from their birthmates. Thus, they signal the presence of star clusters in which many of their lower-mass companions are still forming.

Regions of active star formation can be extraordinarily picturesque. Near hot stars we often find colorful, wispy blobs of glowing gas known as **ionization nebulae** (sometimes called *emission nebulae* or *H II regions*). These nebulae glow because electrons in their atoms are raised to high energy levels or ionized when they absorb ultraviolet photons from the hot stars, so they emit light as the electrons fall back to lower energy levels [Section 5.4]. The Orion Nebula, about 1,500 light-years away in the "sword" of the constellation Orion, is among the most famous. Few

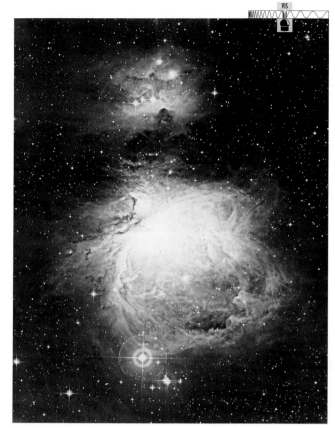

Figure 19.13 A photo of the Orion Nebula, an ionization nebula energized by ultraviolet photons from hot stars.

Figure 19.14 The spectrum of an ionization nebula in blue and green light. The prominent emission lines reveal the atoms and ions that emit most of the light. Through careful study of these lines, we can determine the nebula's chemical composition.

Figure 19.15 The blue tints in this nebula in the constellation Scorpius are produced by reflected light.

astronomical objects can match its spectacular beauty (Figure 19.13; also see Figure 1.3).

Most of the striking colors in an ionization nebula come from particular spectral lines produced by particular atomic transitions. For example, the transition in which an electron falls from energy level 3 to energy level 2 in a hydrogen atom generates a red photon with a wavelength of 656 nanometers (see Figure 5.15). Ionization nebulae appear predominantly red in photographs because of all the red photons released by this particular transition. (Jumps from level 2 to level 1 are even more common, but they produce ultraviolet photons that can be studied only with ultraviolet telescopes in space.) Transitions in other elements produce other spectral lines of different colors (Figure 19.14).

The blue and black tints in some star-forming regions have a different origin. Starlight reflected from dust grains produces the blue colors, because interstellar dust grains scatter blue light much more readily than red light (Figure 19.15). These so-called *reflection nebulae* are always bluer in color than the stars supplying the light; the effect is similar to the scattering of sunlight in our atmosphere that makes the sky blue [Section 10.1]. The black regions of nebulae are dark, dusty gas clouds that block our view of the stars beyond them. Figure 19.16 shows a multicolored nebula characteristic of a hot-star neighborhood.

THINK ABOUT IT

In Figure 19.16, identify the red ionized regions, the blue reflecting regions, and the dark obscuring regions. Briefly explain the origin of the colors in each region.

Spiral Arms Taking a broader view of our galaxy, we can see that the spiral arms must be full of newly forming stars because they bear all the hallmarks of star formation. They are home to both molecular clouds and numerous clusters of

Figure 19.16 A photo of the Horsehead Nebula and its surroundings. (The region pictured is about 150 light-years across.)

Dark patches on inner edge of spiral arm show where gas clouds are packing together . . .

. . . and compression of these clouds triggers star formation in the arm.

Blue specks are young stars that formed in the spiral arm.

Red patches are ionization nebulae around the hottest, youngest stars.

Flow of gas and stars through spiral arm

Figure 19.17 Interactive Photo Spiral arms in the Galaxy M51. This photo from the Hubble Space Telescope shows M51's two magnificent spiral arms along with a smaller galaxy that is currently interacting with one of those arms. Notice that the spiral arms are much bluer in color than the central bulge. Because massive, blue stars live only for a few million years, the relative blueness of the spiral arms tells us that stars must be forming more actively within them than elsewhere in the galaxy. (The large image shows a region roughly 90,000 light-years wide.)

young, bright, blue stars surrounded by ionization nebulae. Detailed images of other spiral galaxies show these characteristics more clearly (Figure 19.17). Hot, blue stars and ionization nebulae trace out the arms while the stars between the arms are generally redder and older. We also see enhanced

amounts of molecular and atomic gas in the spiral arms, and streaks of interstellar dust often obscure the inner sides of the arms themselves. Thus, spiral arms contain both young stars and the material necessary to make new stars.

At first glance, spiral arms look as if they ought to move with the stars, like the fins of a giant pinwheel in space. However, we know that spiral arms cannot be fixed groupings of stars that rotate along with the galaxy. The reason is that stars near the center of the galaxy complete an orbit in much less time than stars far from the center. If the spiral arms simply moved along with the stars, the central parts of the arms would complete several orbits around the galaxy as the outer parts orbited just once. This difference in orbital periods would eventually wind up the spiral arms into a tight coil. Because we generally don't see such tightly wound spiral arms in galaxies, we conclude that spiral arms are more like swirling ripples in a whirlpool than like the fins of a giant pinwheel.

In fact, we now think that spiral arms are enormous waves of star formation that propagate through the gaseous disk of a spiral galaxy like the Milky Way. Theoretical models suggest that disturbances called **spiral density waves** are responsible for the spiral arms. According to these models, spiral arms are places in a galaxy's disk where stars and gas

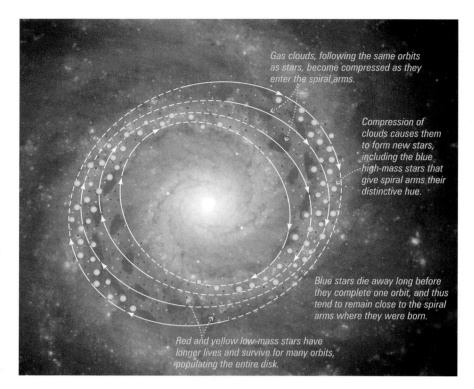

Gas clouds, following the same orbits as stars, become compressed as they enter the spiral arms.

Compression of clouds causes them to form new stars, including the blue high-mass stars that give spiral arms their distinctive hue.

Blue stars die away long before they complete one orbit, and thus tend to remain close to the spiral arms where they were born.

Red and yellow low-mass stars have longer lives and survive for many orbits, populating the entire disk.

Figure 19.18 Schematic illustration of star formation in a spiral density wave. Stars and gas clouds pass through spiral arms as they orbit the galaxy. Each arm is a self-sustaining pattern because gravity slows stars down as they pass through an arm, causing the stars to be more densely packed in the arms.

clouds are more densely packed. Pushing the stars closer together has little effect on the stars themselves—they are still much too widely separated to collide with each other. However, the large gas clouds do collide, and packing the clouds closer together enhances the force of gravity within them, triggering the formation of many new stars [Section 16.1]. As new star clusters form in the arms, supernova explosions from the most massive stars in those clouds can compress the surrounding clouds even more, triggering even more star formation.

To visualize how spiral density waves propagate through a galaxy's disk, consider how traffic backs up behind a slow-moving tractor on a rural highway. Cars approaching the tractor slow down and bunch together. After cars pass the tractor, they speed up and spread out again. Thus, a pack of cars is always bunched up behind the tractor, even though the cars themselves are constantly flowing past it.

In a spiral density wave, gravity plays the role of the tractor, while stars and gas clouds play the role of the cars. The stars and gas clouds of a galaxy's disk are constantly flowing through its spiral arms, but the extra density of matter in the spiral arm alters that flow. The extra matter exerts a gravitational force that pulls stars and gas clouds into the arm and hinders their escape as they move out the other side (Figure 19.18). This gravitational pull is not strong enough to trap the stars and gas clouds. However, like the tractor, it temporarily slows them down, and this temporary slowdown produces a long-lasting pattern.

We call this kind of propagating disturbance a *wave* because, like a wave in water, it moves through matter without carrying that matter along with it. Also, just as with water waves, some sort of disturbance is needed to generate a density wave within a galaxy's disk. Gravitational tugs from

other galaxies passing nearby are one possible source of disturbance. The whirlpool-like rotation of the disk then stretches the wave initiated by such a tug into a spiral shape. Once a gravitational tug sets a spiral density wave in motion, the wave will continue to move through the galaxy's disk, perhaps for billions of years.

To sum up, spiral arms are sites of prolific star formation. Stars are created more readily in spiral arms because gravity bunches interstellar gas clouds more tightly in these arms than elsewhere in a galaxy's disk. Collisions between these gas clouds compress the gas inside, increasing the strength of gravity in the cloud and thereby triggering star formation. The underlying spiral density pattern that initiates this star formation does not move with the stars but instead propagates through the disk like a wave. Massive blue stars that form as gas clouds pass through a spiral arm die out quickly. These luminous hot stars therefore are found close to the spiral arms in which they formed, making the arms look bluer than the rest of the galaxy. Yellow and red stars live long enough to pass through many spiral arms and therefore are distributed more evenly throughout the galactic disk.

19.3 The History of the Milky Way

Now that we have discussed the basic properties of the Milky Way, including the star-gas-star cycle of the disk, we are ready to turn our attention to the history of our galaxy. All the galaxy's properties provide clues to its history. However, some of the most important clues come from a detailed comparison of disk stars with halo stars. We'll begin with this comparison and then discuss a basic model of galaxy formation that explains many of the differences between these two groups of stars.

• What clues to our galaxy's history do halo stars hold?

We have already seen how the disorderly orbits of halo stars differ from the generally circular orbits of disk stars. Two other differences between halo stars and disk stars give us further clues to their origins. First, we don't find any young stars in the halo, while in the disk we see stars of many different ages. Second, the spectra of halo stars show that they contain fewer heavy elements than do disk stars. Because of these striking differences, astronomers divide the Milky Way's stars into two distinct populations.

1. The **disk population** (sometimes called *Population I*) contains both young stars and old stars, all of which have heavy-element proportions of about 2%, like our Sun.
2. The **spheroidal population** (or *Population II*) consists of stars in the halo and bulge, both of which are roughly spherical in shape. Stars in this population are always old and therefore low in mass, and those in the halo sometimes have heavy-element proportions as low as 0.02%—meaning that heavy elements are about 100 times rarer in these stars than in the Sun.

We can understand why disk and halo stars differ by looking at how the Milky Way's gas is distributed. The halo does not contain the cold, dense molecular clouds required for star formation. In fact, the halo contains very little gas at all, and that small amount of gas is generally quite hot. Because star-forming molecular clouds are found only in the disk, new stars can be born only in the disk and not in the halo.

The relative lack of heavy elements in halo stars indicates that they must have formed early in the galaxy's history—before many supernovae had exploded and added heavy elements to star-forming clouds. Thus, we conclude that the halo has lacked the gas needed for star formation for a very long time. Apparently, all the Milky Way's cool gas settled into the disk long ago. The only stars that still survive in the halo are long-lived, low-mass stars. Any more massive stars that were once born in the halo died long ago.

THINK ABOUT IT

How does the halo of our galaxy resemble the distant future fate of the galactic disk? Explain.

• How did our galaxy form?

Any model for our galaxy's formation must account for the differences between disk stars and halo stars. The most basic model proposes that our galaxy began as a giant **protogalactic cloud** containing all the hydrogen and helium gas that the galaxy eventually turned into stars. Gravity would have caused such a cloud to contract and fragment, just as in present-day star-forming clouds [Section 16.1].

According to this model, the stars of the spheroidal population (bulge and halo) formed first. Early on, the gravity associated with our protogalactic cloud drew in matter from all directions, creating a cloud that was blobby in shape and had little or no measurable rotation. The orbits of stars forming within such a cloud could have had any orientation, which accounts for the randomly oriented orbits of stars in the spheroidal population.

Later, the remaining gas settled into a flattened, spinning disk as it contracted under the force of gravity because of conservation of angular momentum (Figure 19.19). This process was much like the process that leads to the formation of spinning disks of gas around young stars [Sections 8.2, 16.2], but on a much larger scale. Collisions among gas particles tended to average out their random motions, leading them to acquire orbits in the same direction and in the same plane. Stars that form within this spinning disk are born on orbits moving at the same speed and in the same direction as their neighbors and thus become members of the disk population of stars.

We can test this basic model by studying the "fossil record" written within the stars of the Milky Way. The clues found to date support the basic picture but suggest that the full story of galaxy formation may be somewhat more complex.

All available evidence confirms that the stars in the Milky Way's halo are indeed old. The main-sequence turnoff points in H-R diagrams of globular clusters show that their stars were born at least 12 billion years ago [Section 15.3]. Individual halo stars (those not in globular clusters) and some of the bulge stars appear similarly old. Furthermore, the proportions of heavy elements in halo stars are much lower than in the Sun, indicating that they formed before many generations of supernovae had a chance to enrich the Milky Way's interstellar medium.

However, careful study of heavy-element proportions suggests that our galaxy formed from a few different gas clouds. If the Milky Way had formed from a single protogalactic cloud, it would have steadily accumulated heavy elements during its inward collapse as stars formed and exploded within it. In that case, the outermost stars in the halo would be the oldest and the most deficient in heavy elements. Stars belonging to different globular clusters in the Milky Way's halo do indeed differ in age and heavy-element content, but these variations do not seem to depend on the stars' distance from the galactic center.

The easiest way to account for the variations is to suppose that the Milky Way's earliest stars formed in relatively small protogalactic clouds, each with a few globular clusters, and that these clouds later collided and combined to create the full protogalactic cloud that became the Milky Way (Figure 19.20). In fact, similar processes may still be happening at some level. As we discussed earlier, the Sagittarius and Canis Major Dwarf galaxies are currently crashing through the Milky Way's disk and are being torn apart in the process. A billion years from now, their stars will

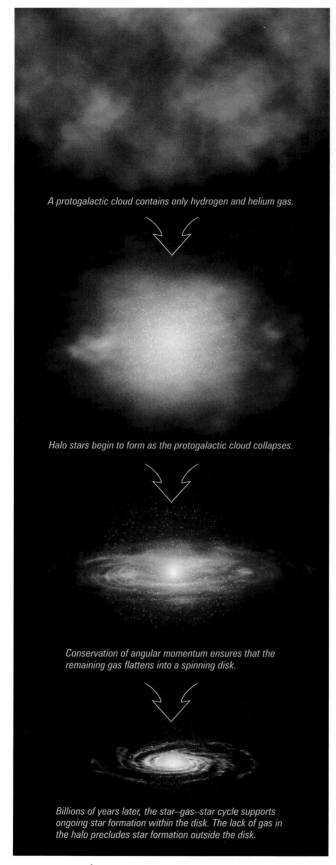

A protogalactic cloud contains only hydrogen and helium gas.

Halo stars begin to form as the protogalactic cloud collapses.

Conservation of angular momentum ensures that the remaining gas flattens into a spinning disk.

Billions of years later, the star–gas–star cycle supports ongoing star formation within the disk. The lack of gas in the halo precludes star formation outside the disk.

Figure 19.19 Interactive Figure This four-picture sequence illustrates a simple schematic model of galaxy formation, showing how a spiral galaxy might develop from a protogalactic cloud of hydrogen and helium gas.

Figure 19.20 This painting shows a model of how the Milky Way's halo may have formed. The characteristics of stars in the Milky Way's halo suggest that several smaller gas clouds, already bearing some stars and globular clusters, may have merged to form the Milky Way's protogalactic cloud. These stars and star clusters remained in the halo while the gas settled into the Milky Way's disk.

be indistinguishable from halo stars because they will all be circling the Milky Way on orbits that carry them high above the disk.

THINK ABOUT IT

If the preceding scenario is true, then the Milky Way suffered several collisions early in its history. Explain why we should not be surprised that galaxy collisions (or collisions between protogalactic clouds) were rather common in the distant past. (*Hint:* How did the average separations of galaxies in the past compare to their average separations today?)

Once the Milky Way's full protogalactic cloud was in place, its collapse and heavy-element enrichment should have continued in a more orderly fashion than previously. Support for this scenario comes from a layer of stars intermediate between the disk and the halo. The heavy-element content of stars and globular clusters in this intermediate layer does indeed depend on their distances from the galactic center. These stars are nearly as old as halo stars but formed just before the spinning protogalactic cloud finished flattening into a disk. Their proportions of heavy elements suggest that, at the time the Milky Way's disk formed, it contained about 10% as much material in the form of heavy elements as it does today. However, that proportion of heavy elements must have differed from place to place in the galaxy. Observations of old stars in the bulge of our galaxy show that some of them have a chemical composition similar to that of our Sun, even though their ages exceed 10 billion years. Apparently, many of the heavy elements produced during the initial collapse of our galaxy ended up near the center, thereby becoming incorporated into the bulge stars that subsequently formed.

After the formation of the disk, generations of stars and the star-gas-star cycle increased the abundance of heavy

a This infrared image shows stars and gas clouds within 1,000 light-years of the center of the Milky Way.

b This radio image shows vast threads of emission tracing magnetic field lines near the galactic center.

c This radio image zooms in on gas swirling around the suspected black hole known as Sgr A* (located at the position of the white dot).

d This infrared image shows stars within about 1 light-year of Sgr A*. The two arrows point to the precise location of Sgr A*.

Figure 19.21 Zooming into the galactic center at infrared and radio wavelengths.

elements in the disk more gradually than in the bulge. Because the star-gas-star cycle has been operating continuously in the Milky Way's disk ever since the disk formed, the ages of disk stars range from newly born to 10 billion or more years old. Furthermore, new stars will continue to be born in the disk as long as enough gas remains within it.

 Black Holes Tutorial, Lessons 1–2

19.4 The Mysterious Galactic Center

The center of the Milky Way Galaxy lies in the direction of the constellation Sagittarius. This region of the sky does not look particularly special to our unaided eyes. However, if we could remove the interstellar dust that obscures our view, the galaxy's central bulge would be one of the night sky's most spectacular sights. Moreover, a gigantic black hole with more than 3 million times the mass of our Sun may lurk deep within the bulge, at the very center of our galaxy.

• What lies in the center of our galaxy?

Although the Milky Way's clouds of gas and dust prevent us from seeing visible light from the center of the galaxy, we can peer into the heart of our galaxy with radio, infrared, and X-ray telescopes. Figure 19.21 shows a series of infrared and radio views that look ever deeper into the galaxy's center. Within about 1,000 light-years of the center, we find swirling clouds of gas and a cluster of several million stars. Bright radio emission traces out the magnetic fields that thread this turbulent region. In the very center we find a source of radio emission named Sagittarius A* (pronounced "Sagittarius A-star"), or Sgr A* for short, that is quite unlike any other radio source in our galaxy. Several hundred stars crowd the region within about 1 light-year of Sgr A*.

The motions of stars and gas near Sgr A* indicate that it contains a few million solar masses within a very small region of space. Observations show that there are not nearly enough stars in this region to account for so much mass, even though stars here are much more crowded together than in our region of the galaxy. As a result, astronomers have long suspected that Sgr A* contains a very massive black hole. These suspicions received a big boost in 2002 when astronomers monitoring infrared light from the

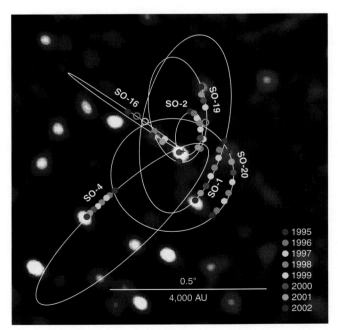

Figure 19.22 This diagram shows observed stellar positions (colored dots indicate year of observation) and calculated orbits for several stars near the very center of the galaxy. By applying Newton's version of Kepler's third law to these orbits, we infer that the central object has a mass 3 to 4 million times that of our Sun, packed into a space so small that it is almost certainly a black hole. (The 4,000 AU shown on the scale bar is equivalent to about 23 light-days.)

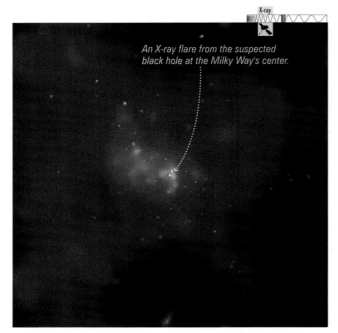

An X-ray flare from the suspected black hole at the Milky Way's center.

Figure 19.23 This X-ray image from the Chandra X-Ray Observatory shows the central 60 light-years of our galaxy. The circled white dot in the middle of the image is an X-ray flare from the suspected black hole at the Milky Way's center.

galactic center observed stars swooping within a few light-hours of this massive object in their orbits (Figure 19.22). By applying Newton's version of Kepler's third law to the orbits of these stars, they concluded that the central object must have a mass of 3 to 4 million solar masses, all packed into a region of space just a little larger than our solar system. An object that massive within such a small space is almost certainly a black hole. However, it must have a different origin than the black holes that form in supernova explosions for its mass to be so large. We will see in Chapter 21 that the formation of such a large black hole seems linked to galaxy formation itself, although the exact formation mechanism remains mysterious.

The behavior of the suspected black hole at our galaxy's center is rather puzzling. Most other suspected black holes are thought to accumulate matter through accretion disks that radiate brightly in X rays. These include black holes in binary star systems like Cygnus X-1 [Section 18.3] and some giant black holes at the centers of other galaxies that we will discuss in Chapter 21. If the black hole at the center of our galaxy had an accretion disk like these others, its X-ray light would easily penetrate the dusty gas of our gal-

axy and it would appear fairly bright to our X-ray telescopes. Yet the X-ray emission from Sgr A* has usually been relatively faint.

Recent observations of Sgr A* made with the Chandra X-Ray Observatory are helping us understand this surprising behavior. In October 2000, an enormous X-ray flare lasting 3 hours was observed coming from the location of the suspected black hole (Figure 19.23). This sudden change in X-ray brightness seems to have come from energy released by a comet-size lump of matter that tidal forces tore apart just before it disappeared into the black hole. If we continue to observe similar X-ray flares from Sgr A*, then the explanation for its generally low X-ray brightness may be that matter falls into it in big chunks instead of in the smooth, swirling flow of an accretion disk. Until we better understand Sgr A*, it is sure to remain a favorite target for X-ray telescopes, infrared telescopes, and radio telescopes alike.

THE BIG PICTURE

Putting Chapter 19 into Context

In this chapter, we have explored the structure, motion, and history of our galaxy, along with the recycling of gas that has made our existence possible. When you review this chapter, pay attention to these "big picture" ideas:

- The inability of visible light to penetrate deeply through interstellar gas and dust concealed the true nature of our galaxy until recent times. Modern astronomical instruments reveal the Milky Way Galaxy to be a dynamic system of stars and gas that continually gives birth to new stars and planetary systems.

- Stellar winds and explosions make interstellar space a violent place. Hot gas tears through the atomic hydrogen gas that fills much of the galactic disk, leaving expanding bubbles and fast-moving clouds in its wake. All this violence might seem quite dangerous, but it performs the great service of mixing new heavy elements into the gas of the Milky Way.

- Although the elements from which we are made were forged in stars, we could not exist if stars were not organized into galaxies. The Milky Way Galaxy acts as a giant recycling plant, converting gas expelled from past generations of stars into later generations, sometimes leading to the formation of planets like our own.

19.1 The Milky Way Revealed

- **What does our galaxy look like?** The Milky Way Galaxy consists of a thin **disk** about 100,000 light-years in diameter with a central bulge and a spherical region called the **halo** that surrounds the entire disk. The disk contains the gas and dust of the interstellar medium, while the halo contains very little gas.

- **How do stars orbit in our galaxy?** Stars in the disk all orbit the galactic center in about the same plane and in the same direction. Halo and bulge stars also orbit the center of the galaxy, but their orbits are randomly inclined to the disk of the galaxy. Orbital motions of stars allow us to determine the distribution of mass in our galaxy.

19.2 Galactic Recycling

- **How is gas recycled in our galaxy?** Stars are born from the gravitational collapse of gas clumps in molecular clouds. Massive stars explode as supernovae when they die, creating hot **bubbles** in the interstellar medium that contain the new elements made by these stars. Eventually, this gas cools and mixes into the surrounding interstellar medium, turning into atomic hydrogen and then cooling further, producing molecular clouds. These molecular clouds then form stars, completing the **star-gas-star cycle**.

- **Where do stars tend to form in our galaxy?** Active star-forming regions, marked by the presence of hot, massive stars and ionization nebulae, are particularly common in **spiral arms**. The spiral arms represent regions where a **spiral density wave** has caused gas clouds to crash into each other, thereby compressing them and making star formation more likely.

19.3 The History of the Milky Way

- **What clues to our galaxy's history do halo stars hold?** The stars of the bulge and halo, together known as the **spheroidal population** of stars, are old low-mass stars with a much smaller proportion of heavy elements than stars in the **disk population**. Thus, halo stars must have formed early in the galaxy's history, before the gas settled into a disk.

- **How did our galaxy form?** The galaxy probably began as a huge blob of gas called a **protogalactic cloud**. Gravity caused the cloud to shrink in size, and conservation of angular momentum caused the gas to form the spinning disk of our galaxy. Stars in the halo formed before the gas finished collapsing into the disk.

19.4 The Mysterious Galactic Center

- **What lies in the center of our galaxy?** Motions of stars near the center of our galaxy suggest that it contains a black hole about 3 to 4 million times as massive as the Sun. The black hole appears to be powering a bright source of radio emission known as Sgr A*.

For instructor-assigned homework go to MasteringASTRONOMY www.masteringastronomy.com

Review Questions

Short-Answer Questions Based on the Reading

1. Draw simple sketches of our galaxy as it would appear face-on and edge-on. Identify the *disk, bulge, halo,* and *spiral arms,* and indicate the galaxy's approximate dimensions.
2. What are the Large and Small Magellanic Clouds, and the Sagittarius and Canis Major Dwarfs?
3. Describe the basic characteristics of stellar orbits in the bulge, disk, and halo of our galaxy.
4. How can we use orbital properties to learn about the mass of the galaxy? What have we learned?
5. Summarize the stages of the star-gas-star cycle illustrated in Figure 19.3.
6. What creates a *bubble* of hot, ionized gas? What happens to the gas in the bubble over time?

7. Why do star clusters make *superbubbles*? What happens to those bubbles when they grow thicker than the galactic disk?

8. How does a *galactic fountain* help circulate new elements within the Milky Way?

9. What are *cosmic rays,* and where are they thought to come from?

10. What do we mean by *atomic hydrogen gas?* How common is it, and how do we map its distribution in the galaxy?

11. Briefly summarize the different types of gas present in the disk of the galaxy, and describe how they appear when we view the galaxy in different wavelengths of light.

12. What are ionization nebulae, and why are they found near hot, massive stars?

13. How do we know that spiral arms do not rotate like giant pinwheels? What makes spiral arms bright?

14. What triggers star formation within a spiral arm? How do we think spiral arms are maintained?

15. Briefly describe the characteristics that distinguish the galaxy's *disk population* of stars from its *spheroidal population* of stars.

16. How do the different ages of disk stars and halo stars support the idea that our galaxy formed from the gravitational collapse of a *protogalactic cloud?*

17. Why do we think that the Milky Way's full protogalactic cloud may have been formed from the merger of several smaller protogalactic clouds?

18. What is Sgr A*? What evidence suggests that it contains a massive black hole?

Test Your Understanding

Does It Make Sense?

Decide whether the statement makes sense (or is clearly true) or does not make sense (or is clearly false). Explain your reasoning. (For an example, see Chapter 1, "Does It Make Sense?")

19. We did not understand the true size and shape of our galaxy until NASA launched satellites into the galactic halo, enabling us to see what the Milky Way looks like from the outside.

20. Planets like Earth probably didn't form around the very first stars because there were so few heavy elements back then.

21. If I could see infrared light, the galactic center would look much more impressive.

22. Many spectacular ionization nebulae are seen throughout the Milky Way's halo.

23. The carbon in my diamond ring was once part of an interstellar dust grain.

24. The Sun's velocity around the Milky Way tells us that most of our galaxy's dark matter lies in the galactic disk near the center of the galaxy.

25. We know that a black hole lies at our galaxy's center because numerous stars near it have vanished over the past several years, telling us that they've been sucked in.

26. If we could watch a time-lapse movie of a spiral galaxy over millions of years, we'd see many stars being born and dying within the spiral arms.

27. The star-gas-star cycle will keep the Milky Way looking just as bright in 100 billion years as it looks now.

28. Halo stars orbit the center of our galaxy much faster than the disk stars.

Quick Quiz

Choose the best answer to each of the following. Explain your reasoning with one or more complete sentences.

29. Where are most of the Milky Way's globular clusters found? (a) in the disk (b) in the bulge (c) in the halo

30. Why do disk stars bob up and down as they orbit the galaxy? (a) because the gravitational pull of other disk stars always pulls them toward the disk (b) because of friction with the interstellar medium (c) because the halo stars keep knocking them back into the disk

31. How do we determine the Milky Way's mass outside the Sun's orbit? (a) from the Sun's orbital velocity and its distance from the center of our galaxy (b) from the orbits of halo stars near the Sun (c) from the orbits of stars and gas clouds orbiting the galactic center at greater distances than the Sun

32. Which part of the galaxy has gas with the hottest average temperature? (a) the disk (b) the halo (c) the bulge

33. What is the typical hydrogen content of stars that are forming right now in the vicinity of the Sun? (a) 100% hydrogen (b) 75% hydrogen (c) 70% hydrogen

34. Which of these forms of radiation passes most easily through the disk of the Milky Way? (a) red light (b) blue light (c) infrared light

35. Where would you least expect to find an ionization nebula? (a) in the halo (b) in the disk (c) in a spiral arm

36. Where would be the most likely place to find an ionization nebula? (a) in the halo (b) in the bulge (c) in a spiral arm

37. Which kind of star is most likely to be part of the spheroidal population? (a) an O star (b) an A star (c) an M star

38. We measure the mass of the black hole at the galactic center from (a) the orbits of stars in the galactic center. (b) the orbits of gas clouds in the galactic center. (c) the amount of radiation coming from the galactic center.

Investigate Further

In-Depth Questions to Increase Your Understanding

Short-Answer/Essay Questions

39. *Unenriched Stars.* Suppose you discover a star made purely of hydrogen and helium. How old do you think it would be? Explain your reasoning.

40. *Enrichment of Star Clusters.* The gravitational pull of an isolated globular cluster is rather weak—a single supernova explosion can blow all the interstellar gas out of a globular cluster. How might this fact be related to observations indicating that stars ceased to form in globular clusters long ago? How might it be related to the fact that globular clusters are deficient in elements heavier than hydrogen and helium? Summarize your answers in one or two paragraphs.

41. *High-Velocity Star.* The average speed of stars relative to the Sun in the solar neighborhood is about 20 km/s (i.e., the speed at which we see stars moving toward or away from the Sun—*not* their orbital speed around the galaxy). Suppose you discover a star in the solar neighborhood that is moving relative to the Sun at a much higher speed, say, 200 km/s. What kind of orbit does this star probably have around the Milky Way? In what part of the galaxy does it spend most of its time? Explain.

42. *Research: Discovering the Milky Way.* Humans have been looking at the Milky Way since long before recorded history, but only in the past century did we verify the true shape of the galaxy and our location within it. Learn more about how conceptions of the Milky Way developed through history. What names did different cultures give the band of light they saw? What stories did they tell about it? How have ideas about the galaxy changed in the past few centuries? Try to locate

diagrams that illustrate these changes. Write a two- to three-page summary of your findings.

43. *Future of the Milky Way.* Describe in one or two paragraphs how the Milky Way would look from the outside if you could watch it for the next 100 billion years. How would its appearance change over that time period?

44. *Orbits at the Galactic Center.* Using the information in Figure 19.22, identify which two stars reach the highest orbital speeds. (Each star is labeled with a name consisting of "SO-" followed by a number.) Explain how the orbits of those two stars illustrate Kepler's first two laws.

45. *A Nonspinning Galaxy.* How would the development of the Milky Way Galaxy have been different if its original protogalactic cloud had no angular momentum? Describe how you think our galaxy would look today and explain your reasoning.

46. *Gas Distribution in the Milky Way.* Make a sketch of the gas distribution in the plane of the Milky Way based on the photographs in Figure 19.12. In your sketch, map out where the molecular clouds are and draw the layer of atomic hydrogen gas that surrounds them. Then add the locations of the most prominent bubbles of hot gas. Explain why each of those components of the Milky Way's interstellar medium is found in the location where you have drawn it.

Quantitative Problems

Be sure to show all calculations clearly and state your final answers in complete sentences.

47. *Mass of the Milky Way's Halo.* The Large Magellanic Cloud is a small galaxy that orbits the Milky Way. It is currently orbiting the Milky Way at a distance of roughly 160,000 light-years from the galactic center at a velocity of about 300 km/s. Use these values in the orbital velocity law to get an estimate of the Milky Way's mass within 160,000 light-years from the center. (The value you obtain is a fairly rough estimate because the orbit of the Large Magellanic Cloud is not circular.)

48. *Mass of the Central Black Hole.* Suppose you observe a star orbiting the galactic center at a speed of 1,000 km/s in a circular orbit with a radius of 20 light-days. What would your estimate be for the mass of the object that the star is orbiting?

49. *Mass of a Globular Cluster.* Stars in a typical globular cluster orbit at speeds of about 10 km/s, and stars at the outskirts of a globular cluster are about 50 light-years from the center. Use this information to estimate the mass of a typical globular cluster.

50. *Mass of Saturn.* The innermost rings of Saturn orbit in a circle with a radius of 67,000 kilometers at a speed of 23.8 km/s. Use the orbital velocity law to compute the mass contained within the orbit of those rings. Compare your answer with the mass of Saturn listed in Appendix E.

51. *Pressure vs. Gravity in Hot Ionized Gas.* Use the formula in Mathematical Insight 19.1 to estimate the mass of a gas cloud in which pressure balances gravity for a gas temperature of 10^6 K and a number density of 0.01 particle per cubic centimeter. Use your answer to explain why hot gas must cool down before it can collect into star-forming clouds.

52. *Pressure vs. Gravity in Warm Atomic Gas.* Use the formula in Mathematical Insight 19.1 to estimate the mass of a gas cloud in which pressure balances gravity for a gas temperature of 10^4 K and a number density of 1 particle per cubic centimeter. Compare your answer to the mass of a giant molecular cloud. Use your answer to explain why additional cooling is necessary in order to create such clouds.

53. *Pressure vs. Gravity in Cool Atomic Gas.* Use the formula in Mathematical Insight 19.1 to estimate the mass of a gas cloud in which pressure balances gravity for a gas temperature of 10^2 K and a number density of 100 particles per cubic centimeter. Use your answer to explain why star-forming clouds typically contain enough gas to form several thousand stars.

54. *The Speed of Supernova Debris.* The kinetic energy $E_{kinetic}$ of an amount of mass M traveling at a velocity v is given by the formula $E_{kinetic} = Mv^2/2$. The total kinetic energy of the matter ejected from a supernova explosion is about 10^{44} joules. Determine the typical speed at which that matter is ejected from a supernova with a mass of $10M_{Sun}$. Compare that speed with the Sun's orbital speed around our galaxy. Based on your comparison, do you think the galaxy's gravity would be strong enough to retain the supernova debris if there were no interstellar medium to slow it down? Explain.

Discussion Questions

55. *Galactic Ecosystem.* We have likened the star-gas-star cycle in our Milky Way to the ecosystem that sustains life on Earth. Here on our planet, water molecules cycle from the sea to the sky to the ground and back to the sea. Our bodies convert atmospheric oxygen molecules into carbon dioxide, and plants convert the carbon dioxide back into oxygen molecules. How are the cycles of matter on Earth similar to the cycles of matter in the galaxy? How do they differ? Do you think the term *ecosystem* is appropriate in discussions of the galaxy?

56. *Galaxy Stuff.* In the chapters on stars, we learned why we are "star stuff." Based on what you've learned in this chapter, explain why we are also "galaxy stuff." Does the fact that the entire galaxy was involved in bringing forth life on Earth change your perspective on Earth or on life in any way? If so, how? If not, why not?

Interactive Tutorials

Tutorial Review of Key Concepts

Use the interactive **Tutorials** at **www.mastering astronomy.com** to review key concepts from this chapter.

Detecting Dark Matter in a Spiral Galaxy Tutorial

Lesson 1 Introduction to Rotation Curves
Lesson 2 Determining the Mass Distribution

Black Holes Tutorial

Lesson 1 What Are Black Holes?
Lesson 2 The Search for Black Holes

Supplementary Tutorial Exercises

Use the interactive **Tutorial Lessons** to explore the following questions.

Detecting Dark Matter in a Spiral Galaxy Tutorial, Lesson 1

1. What is a rotation curve? Use the tool in the simulation to determine how the rotation curve for a star system changes as the mass of the central star changes. Describe the change in words.
2. Compare the rotation curve of a galaxy with the rotation curve of the solar system.

Detecting Dark Matter in a Spiral Galaxy Tutorial, Lesson 2

1. How does orbital speed depend on the mass and distance of the central object?
2. What can we infer about the distribution of mass in our galaxy by observing the orbital speeds of stars as they get farther from the center of our galaxy? Explain your answer.

Black Holes Tutorial, Lesson 2

1. Explain how we determine the mass of the central black hole in the Milky Way Galaxy.
2. Do other galaxies have giant black holes in their centers? Explain.

 Exploring the Sky and Solar System

Of the many activities available on the **Voyager: SkyGazer** CD-ROM accompanying your book, use the following files to observe key phenomena covered in this chapter.

Go to the **File: Basics** folder for the following demonstrations.

1. Milky Way
2. Wide Field Milky Way
3. Winter Milky Way
4. Lagoon Nebulae

Go to the **Explore** menu for the following demonstration.

1. Solar Neighborhood

 Movies

Check out the following narrated and animated short documentary available on **www.masteringastronomy.com** for a helpful review of key ideas covered in this chapter.

The Milky Way Galaxy Movie

Web Projects

1. *Images of the Star-Gas-Star Cycle.* Explore the Web to find pictures of nebulae and other forms of interstellar gas in different stages of the star-gas-star cycle. Assemble the pictures into a sequence that tells the story of interstellar recycling, with a one-paragraph explanation accompanying each image.
2. *The Galactic Center.* Search the Web for recent images of the galactic center, along with information about whether the center hides a massive black hole. Present a two- to three-page report, with pictures, giving an update on current knowledge about the center of the Milky Way Galaxy.

Galaxies

and the Foundation of Modern Cosmology

Thus the explorations of space end on a note of uncertainty. . . . Eventually, we reach the dim boundary— the utmost limits of our telescopes. There we measure shadows, and we search among ghostly errors of measurement for landmarks that are scarcely more substantial.

—Edwin Hubble

Far beyond the Milky Way, we see billions of other galaxies scattered throughout space. Some look similar to our own galaxy, while others look quite different. The sight of all these galaxies inspires us not only to wonder how they came to be, but also to ask fundamental questions about our universe: How old is the universe? How big is it? How is the universe changing with time? Such questions might have seemed ridiculously speculative a century ago. Today, we believe we know the answers with respectable accuracy.

Edwin Hubble, the man for whom the Hubble Space Telescope is named, provided the key discovery when he proved conclusively that galaxies exist beyond the Milky Way. The distances he measured to those galaxies revealed an astonishing fact: The more distant a galaxy is, the faster it moves away from us. Hubble's discovery dealt a mortal blow to the traditional belief in a static, eternal, and unchanging universe. The motions of the galaxies away from

one another imply instead that the entire universe is expanding and that its age is finite.

In this chapter, we will get acquainted with these islands in space, the galaxies of stars. We will discuss how we measure their distances, and how doing so has helped us learn about the age and size of our universe. Along the way, we will see how Hubble's work led to the profound revelation that we live in an expanding universe, and how this discovery heralded the birth of modern cosmology.

20.1 Islands of Stars

Figure 20.1 shows an amazing image of a tiny patch of the sky, called the *Hubble Deep Field*, taken by the Hubble Space Telescope. The telescope pointed in a single direction in the sky and collected all the light it could for 10 days in 1995. If you held a grain of sand at arm's length, the angular size of the grain would match the angular size of the entire patch of sky in this picture. Almost every blob of light in the large image is a galaxy—an island of stars bound together by gravity. Like our own Milky Way, each galaxy is a dynamic system that has cycled hydrogen gas through stars for billions of years, producing new elements for future generations of stars.

The sheer number of galaxies in this tiny patch of sky is truly astounding. Counting the number of galaxies it contains and multiplying by the number of such patches it

Figure 20.1 The Hubble Deep Field, an image composed of 10 days of exposures taken by the Hubble Space Telescope. Some of the galaxies pictured are located three-quarters of the way across the observable universe. The zoom-out sequence shows the location of the field within the Big Dipper, recognizable in the lower left frame.

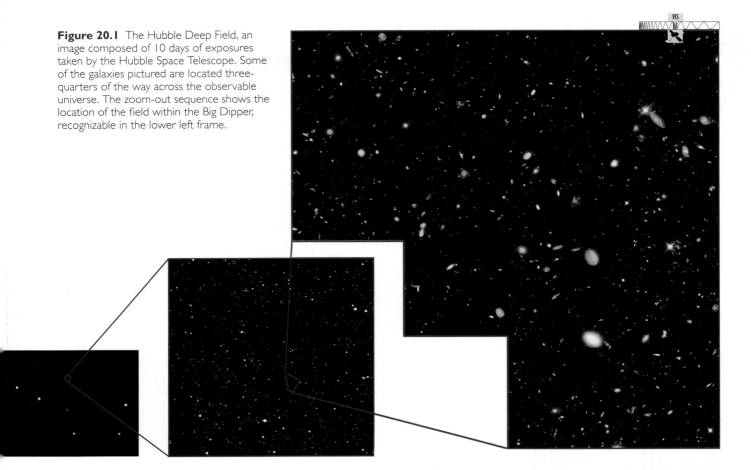

would take to cover the entire sky, astronomers found that the observable universe contains more than 80 billion galaxies. The Hubble Space Telescope has since taken an even more sensitive photo of a different patch of the sky, shown as the opener to Chapter 1 on page 1. This photo, called the *Hubble Ultra Deep Field*, was made with an 11-day exposure using a newer camera installed by Shuttle astronauts in 2002, and it shows that the observable universe contains well over 100 billion galaxies. Our quest in this chapter and the next is to understand how all these galaxies formed, why their properties differ, and what galaxies have to tell us about the nature of the universe itself.

• How are the lives of galaxies connected with the history of the universe?

Even without knowing the distances to the individual galaxies in Figure 20.1, we can see that galaxies come in many sizes, colors, and shapes. Some look large, some small. Some are reddish, some whitish. Some appear round, and some appear flat. We would like to understand why galaxies differ in these ways, but it is not easy to learn their histories. Just as with stars, our observations capture only the briefest instant in any galaxy's life, leaving us to piece together the life story of a typical galaxy from pictures of galaxies at various life stages.

However, the nature of the universe complicates the task because it forces us to deal with an important issue that does not arise when we study the lives of stars: While we see stars of many different ages all around us, the same is not true for galaxies. Most of the galaxies located relatively nearby appear to be similar in age to our own Milky Way Galaxy, meaning that they formed more than 10 billion years ago, when the universe was only a few billion years old. To study young galaxies, we must therefore look far back into the universe's past—which means looking to great distances. Remember that the farther away we look in space, the further back we look in time. Thus, when we observe galaxies that are very far away, we see them as they were when they and the universe were young (see Figure 1.5). In fact, almost all of the galaxies in Figure 20.1 are billions of light-years away.

The relationship between a galaxy's age, its distance, and the age of the universe makes it impossible to consider the histories of galaxies without at the same time considering the evolution of the universe as a whole. The study of galaxies is therefore intimately connected with what we call **cosmology**—the study of the overall structure and evolution of the universe. We will begin our quest to understand the lives of galaxies with an attempt to categorize the galaxies we see nearby, whose details are easier to observe than the ones in Figure 20.1. Then we will study how astronomers measure the distances to galaxies. That will lead us to the story of Edwin Hubble and how he used such measurements to discover the expansion of the universe, a discovery that is the cornerstone of modern cosmology. With that foundation in place, we'll then be prepared to study the lives of galaxies in the next chapter.

• What are the three major types of galaxies?

Astronomers classify galaxies into three major categories:

- **Spiral galaxies**, such as our own Milky Way, look like flat white disks with yellowish bulges at their centers. The disks are filled with cool gas and dust, interspersed with hotter ionized gas, and they usually display beautiful spiral arms.
- **Elliptical galaxies** are redder, more rounded, and often longer in one direction than in the other two, like a football. Compared with spiral galaxies, elliptical galaxies contain very little cool gas or dust, though they often contain very hot, ionized gas.
- **Irregular galaxies** appear neither disklike nor rounded.

The differing colors of galaxies arise from the different kinds of stars that populate them: Spiral and irregular galaxies look white because they contain stars of all different colors and ages, while elliptical galaxies look redder because old, reddish stars produce most of their light. Galaxies also come in a wide range of sizes, from *dwarf galaxies* containing as few as 100 million (10^8) stars to *giant galaxies* with more than 1 trillion (10^{12}) stars. Let's take a closer look at galaxies of each type.

THINK ABOUT IT

Take a moment and try to classify the larger galaxies in Figure 20.1. How many appear spiral? Elliptical? Irregular? Do the colors of galaxies seem related to their shapes?

Spiral Galaxies   Like the Milky Way, other spiral galaxies also have a thin *disk* extending outward from a central *bulge* (Figure 20.2). The bulge itself merges smoothly into a *halo* that can extend to a radius of more than 100,000 light-years. However, the halo is difficult to see in photographs because its stars are generally dim and spread over a large volume of space.

Recall that the Milky Way is made up of two distinct populations of stars [Section 19.3]. The *disk population* includes stars of all ages and masses that orbit in the disk of the galaxy. The *spheroidal population* consists of halo and bulge stars, with the halo stars generally being old and low in mass. We find the same two populations in other spiral galaxies, and we use them to define two primary components of galaxies:

- The **disk component** is the flat disk in which stars follow orderly, nearly circular orbits around the galactic center. The disk component always contains an *interstellar medium* of gas and dust, but the amounts and proportions of the interstellar medium in molecular, atomic, and ionized forms differ from one spiral galaxy to the next. Spiral galaxies with large bulges generally have less interstellar gas and dust than those with small bulges.
- The bulge and halo together make up the **spheroidal component**, which gets its name from its rounded shape. Stars in the spheroidal component have orbits with many

a NGC 6744, a spiral galaxy thought to be very similar to our Milky Way.

b NGC 4414, a spiral galaxy whose disk is somewhat tilted to our line of sight.

c NGC 891, a spiral galaxy seen nearly edge-on. Note the central streak of dust associated with the disk.

Figure 20.2 This three-photo sequence shows spiral galaxies from three different perspectives ranging from nearly face-on to nearly edge-on. In each case, the region pictured is about 100,000 light-years across. ("NGC" stands for the *New General Catalog*, a listing of more than 7,000 objects published in 1888.)

different inclinations, and the spheroidal component generally contains little cool gas or dust. Figure 20.3 shows a spiral galaxy with an unusually large bulge that illustrates the general shape of the spheroidal component. Although no clear boundary divides the pieces of the spheroidal component, astronomers usually consider stars within 10,000 light-years of the center to be members of the bulge and those outside this radius to be members of the halo.

All spiral galaxies have both a disk and a spheroidal component, but there are some variations on the general theme. Some spiral galaxies appear to have a straight bar of stars cutting across the center, with spiral arms curling away from the ends of the bar. Such galaxies are known as *barred spiral galaxies* (Figure 20.4). Astronomers suspect that the Milky Way itself is a barred spiral galaxy, because our galaxy's bulge appears to be somewhat elongated.

Other galaxies have disk and spheroidal components like spiral galaxies but appear to lack spiral arms (Figure 20.5). These so-called *lenticular galaxies* (*lenticular* means "lens-shaped") are sometimes considered an intermediate class between spirals and ellipticals, because they tend to have less cool gas than normal spirals but more than ellipticals.

Among large galaxies in the universe, most (75–85%) are spiral or lenticular. Galaxies with obvious disks are much rarer among small galaxies.

Elliptical Galaxies The major difference between elliptical and spiral galaxies is that ellipticals lack a significant disk component. An elliptical galaxy has only a spheroidal component and looks much like the bulge and halo of a spiral galaxy. (In fact, elliptical galaxies are sometimes called *spheroidal galaxies*.)

Elliptical galaxies come in a particularly wide range of sizes (Figure 20.6). Although most large galaxies in the universe are spiral, some of the most massive galaxies in the universe are *giant elliptical galaxies*. Nevertheless, the vast

Figure 20.3 NGC 4594 (the Sombrero Galaxy) is a spiral galaxy with a large bulge and a dusty disk that we see almost edge-on. A much larger but nearly invisible halo surrounds the entire galaxy. The bulge and halo together make up the spheroidal component of the galaxy. This image shows a region of the galaxy about 82,000 light-years across.

Figure 20.4 NGC 1300, a barred spiral galaxy about 110,000 light-years in diameter.

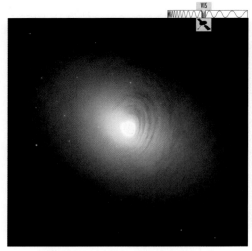

Figure 20.5 The central part of NGC 2787, a lenticular galaxy. A few streaks of dusty gas can be seen in this galaxy's disk, but it does not contain any noticeable spiral arms. The region pictured is about 4,400 light-years across.

majority of elliptical galaxies are small, and these small elliptical galaxies are the most common type of galaxy in the universe. Particularly small ellipticals with fewer than about a billion stars, known as *dwarf elliptical galaxies*, are often found near larger spiral galaxies. For example, at least 10 dwarf elliptical galaxies belong to the Local Group.

Elliptical galaxies usually contain very little dust or cool gas, although some have relatively small and cold gaseous disks rotating at their centers. However, large elliptical galaxies can contain substantial amounts of very hot gas. This low-density, X-ray-emitting gas is much like the gas in the hot bubbles created by supernovae and powerful stellar winds in the Milky Way [Section 19.2].

The lack of cool gas in elliptical galaxies means that, like the halo of the Milky Way, they generally have little or no ongoing star formation. Thus, elliptical galaxies tend to look red or yellow in color because they do not have any of the hot, young, blue stars found in the disks of spiral galaxies.

Irregular Galaxies Some of the galaxies we see nearby fall into neither of the two major categories. This irregular class of galaxies is a miscellaneous class, encompassing small galaxies such as the Magellanic Clouds and larger "peculiar" galaxies that appear to be in disarray (Figure 20.7). These blobby star systems are usually white and dusty, like the disks of spirals. Their colors tell us that they contain young, massive stars.

Among nearby galaxies, only a small percentage of galaxies as large as the Milky Way are irregular. Telescopic observations probing farther into the universe show that distant galaxies are more likely to be irregular in shape than nearby galaxies. Because the light of more distant galaxies has taken longer to reach us, these observations tell us that irregular galaxies were more common when the universe was younger.

Hubble's Galaxy Classes Edwin Hubble invented a system for classifying galaxies that remains widely used. It assigns the letter *E*, followed by a number, to each elliptical galaxy. The larger the number, the flatter the galaxy. An E0 galaxy is round, and an E7 galaxy is highly elongated. A spiral galaxy is assigned an uppercase *S*, or *SB* if it has a bar, and a lowercase *a*, *b*, or *c* that indicates the size of the bulge and the dustiness of the disk. The bulge size decreases from *a* to *c*, while the amount of dusty gas increases. For example, an Sa galaxy has a large bulge and a modest amount of dusty

a M 87, a giant elliptical galaxy in the Virgo Cluster, is one of the largest galaxies in the universe. The region shown is over 120,000 light-years across.

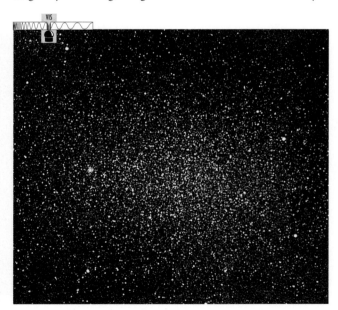

b Leo I is one of at least 10 dwarf elliptical galaxies in the Local Group. It is only about 2,500 light-years across.

Figure 20.6 Elliptical galaxies come in a wide range of sizes.

a The Large Magellanic Cloud, a small companion to the Milky Way. It is about 30,000 light-years across.

b The small Magellanic Cloud, a smaller companion to the Milky Way. It is about 18,000 light-years across.

c NGC 1313, an irregular galaxy with scattered patches of star formation. The region pictured is about 50,000 light-years across.

Figure 20.7 Irregular galaxies.

gas, and an SBc galaxy has a bar, a small bulge, and lots of dusty gas. Lenticular galaxies are designated S0, and irregular galaxies are designated Irr. Figure 20.8 shows Hubble's scheme.

Astronomers had once hoped that the classification of galaxies might yield deep insights, just as the classification of stars did in the early twentieth century. The Hubble classification scheme itself was suspected for a time to be an evolutionary sequence in which galaxies flattened and spread out as they aged. Unfortunately for astronomers, galaxies turn out to be far more complex than stars, and classification schemes like this one have not led to easy answers about their nature.

• How are galaxies grouped together?

Although some galaxies travel solo through the universe, many are gravitationally bound together with neighboring galaxies. Spiral galaxies are often found in loose collections of up to a few dozen galaxies, called groups. Our Local Group is one example (see Figure 1.1). Figure 20.9 shows another galaxy group.

Elliptical galaxies appear to be more "social" than spiral galaxies. They are particularly common in clusters of galaxies, which can contain hundreds and sometimes thousands of galaxies extending over more than 10 million light-years

Figure 20.8 This "tuning fork" diagram illustrates Hubble's galaxy classes.

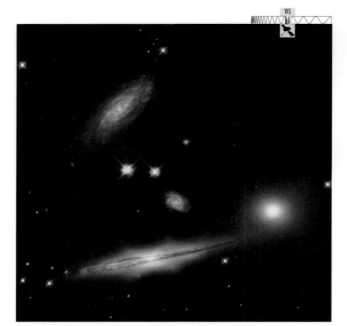

Figure 20.9 Hickson Compact Group 87, a small group of galaxies consisting of a large edge-on spiral galaxy at the bottom of this photo, two smaller spiral galaxies at the center and upper left, and an elliptical galaxy to the right. The whole group is about 170,000 light-years in diameter. (The other objects in this photograph are foreground stars in our own galaxy.)

(Figure 20.10). Elliptical galaxies make up about half the large galaxies in the central regions of clusters, while they represent only a small minority (about 15%) of the large galaxies found outside clusters.

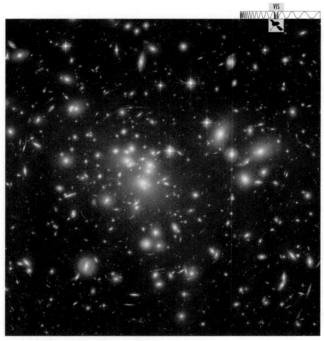

Figure 20.10 Central part of the galaxy cluster Abell 1689. Almost every object in this photograph is a galaxy belonging to the cluster. Yellowish elliptical galaxies outnumber the whiter spiral galaxies. The region pictured is about 2 million light-years across. (A few stars from our own galaxy appear in the foreground, looking like white dots with four spikes in the form of a cross.)

20.2 Measuring Galactic Distances

To learn more about galaxies than just their shape, color, and type, we need to know how far away they are. Measuring the distances to galaxies is one of the most challenging tasks we face when trying to understand galaxies and the universe as a whole, but the payoff is enormous. These measurements not only tell us galactic distances, but, as we will see in the next section, they also reveal the size and age of the observable universe.

• How do we measure the distances to galaxies?

Our determinations of astronomical distances depend on a chain of methods in which each step allows us to measure greater distances in the universe. We have already discussed the measurement of distances to nearby stars using parallax [Section 15.1]. Because parallax is the apparent shift in a star's position as Earth orbits the Sun, measuring distances by parallax requires knowing the precise Sun-Earth distance, or astronomical unit (AU). Astronomers measure the AU with a technique called **radar ranging**, in which radio waves are transmitted from Earth and bounced off Venus (see box, "How did we learn the scale of the solar system?," p. 218). Because radio waves travel at the speed of light, the round-trip travel time for the radar signals tells us Venus's distance from Earth. We can then use Kepler's laws and a little geometry to calculate the length of an AU. Thus, using radar ranging to measure the AU represents the first link in the distance chain, and parallax measurements of distances to nearby stars represent the second link. We will now follow the rest of this chain, link by link, to the outermost reaches of the observable universe.

Standard Candles   Once we have measured distances to nearby stars through parallax, we can begin to measure distances to other stars in the same way that we might estimate the distance to a street lamp at night. If the street lamp does not look very bright, then it's probably far away. If it looks very bright, then it's probably quite close.

We can determine the lamp's distance more accurately if we can measure its apparent brightness. For example, suppose we see a distant street lamp and know that all street lamps of its type put out 1,000 watts of light. If we then measure the street lamp's apparent brightness, we can calculate its distance by using the inverse square law for light [Section 15.1].

An object such as a street lamp, for which we are likely to know the true luminosity, represents what astronomers call a **standard candle**. The term *standard candle* is meant to suggest a light source of a known, standard luminosity. Unlike light bulbs, astronomical objects do not come marked with wattage. An astronomical object can serve as a standard candle only if we have some way of knowing its true luminosity without first measuring its apparent brightness and distance. Fortunately, many astronomical objects meet this requirement. For example, any star that is a twin of our Sun—that is, a main-sequence star with spectral type G2—should have about the same luminosity as the Sun. If we measure the apparent brightness of a Sun-like star, we can assume it has the same luminosity as the Sun (3.8×10^{26} watts) and then use the inverse square law for light to estimate its distance.

Beyond the few hundred light-years for which we can measure distances by parallax, we use standard candles for most cosmic distance measurements. These distance measurements always have some uncertainty, because no astronomical object is a perfect standard candle. The challenge of measuring astronomical distances comes down to the challenge of finding the objects that make the best standard candles. The more confidently we know an object's true luminosity, the more certain we are of its distance.

Main-Sequence Fitting We can use Sun-like stars as standard candles because we know that they are similar to the Sun and because we can measure the Sun's luminosity quite easily. However, Sun-like stars are relatively dim, and we cannot detect them at great distances. To measure distances beyond 1,000 light-years or so, we need brighter standard candles.

Brighter main-sequence stars represent an obvious first choice, because all main-sequence stars of a particular spectral type have about the same luminosity [Section 15.2]. However, before we can use any main-sequence star as a standard candle, we must first have some way of knowing its true luminosity. Thus, we must follow two steps to use bright main-sequence stars as standard candles:

1. We identify a star cluster that is close enough for us to determine its distance by parallax and plot its H-R diagram. Because we know the distances to the cluster's stars, we can use the inverse square law for light to establish their luminosities from their apparent brightnesses.
2. We can look at stars in other clusters that are too far away for parallax measurements and measure their apparent brightnesses. If we assume that main-sequence stars in other clusters have the same luminosities as their counterparts in the nearby cluster, we can calculate their distances from the inverse square law for light.

Twentieth-century astronomers laid the groundwork for this technique by calibrating the luminosities on a "standard" H-R diagram. This calibration relied largely on a single, nearby star cluster—the Hyades Cluster in the constellation Taurus, whose distance is now known from its parallax. We can find the distances to other star clusters by

---

MATHEMATICAL INSIGHT **20.1** **Standard Candles**

The inverse square law for light tells us that the apparent brightness of a light source decreases with the square of its distance away from us. As we saw in Mathematical Insight 15.1, we can write this law in the following form:

$$\text{apparent brightness} = \frac{\text{luminosity}}{4\pi \times (\text{distance})^2}$$

We can usually measure an object's apparent brightness, which means we generally use the inverse square law for light either to calculate luminosity when we know distance or to calculate distance when we know luminosity. For nearby objects, we can often measure distance through parallax and thus use the formula to determine luminosity. For objects that are too far away for parallax measurements, we can use the formula to estimate their distances *if* we have a good estimate of their luminosities. This idea is the essence of all standard candle techniques: In cases where we think we know an object's luminosity, we can find its distance simply by rearranging the formula for the inverse square law for light. You should confirm that with a little algebra, the formula becomes:

$$\text{distance} = \sqrt{\frac{\text{luminosity}}{4\pi \times (\text{apparent brightness})}}$$

Example: You observe a star with the same spectral type and luminosity class as the Sun, measuring its apparent brightness to be 1.0×10^{-12} watt/m^2. Can you use it as a standard candle? How far away is it?

Solution:

Step 1 Understand: We can use the star as a standard candle because its spectral classification tells us that it must have about the same luminosity as our Sun. From Table 14.1, we know that the Sun's luminosity is $L = 3.8 \times 10^{26}$ watts, so we assume the same luminosity for the star. We are given the star's apparent brightness, so we have all the information we need to calculate its distance with the inverse square law for light.

Step 2 Solve: We now plug the assumed luminosity and measured apparent brightness into the inverse square law in the form found above and solve for distance:

$$\text{distance} = \sqrt{\frac{3.8 \times 10^{26} \text{ watts}}{4\pi \times \left(1.0 \times 10^{-12} \frac{\text{watt}}{\text{m}^2}\right)}} \approx 5.5 \times 10^{18} \text{ m}$$

Step 3 Explain: The distance to the star is about 5.5×10^{18} meters. This number will be easier to interpret if we convert it to light-years. From Mathematical Insight 1.1, a light-year is about 9.5×10^{13} kilometers, which is 9.5×10^{16} meters. Dividing the distance in meters by 9.5×10^{16}, we find that it is equivalent to about 580 light-years. Thus, by assuming that the star has the same luminosity as our Sun, we have found that the star is about 580 light-years away. Of course, if its actual luminosity is different from the Sun's, the distance we found will also be in error.

comparing the apparent brightnesses of their main-sequence stars with those in the Hyades Cluster and assuming that all main-sequence stars of the same color have the same luminosity (Figure 20.11). This technique of determining distances by comparing main sequences in different star clusters is called **main-sequence fitting**.

The Hyades Cluster does not contain stars of every spectral type. Building a complete, standard H-R diagram therefore required that astronomers use main-sequence fitting to find the distances to many nearby star clusters until every spectral type was represented. Today, with the standard H-R diagram well established (see Figure 15.10), we can use main-sequence fitting to measure distances to any star cluster near enough for us to identify individual main-sequence stars.

In principle, we need to measure the apparent brightness of only a single star in a cluster to determine the cluster's distance. Once we have determined the star's spectral type, we can use its expected true luminosity and the inverse square law for light to calculate its distance. This method is often used for measuring distances to isolated stars that are not members of clusters. In practice, however, main-sequence fitting gives more precise results. The advantage of main-sequence fitting over using a single star in a cluster is that it reduces the uncertainty in the distance calculation. By comparing entire main sequences, we are essentially comparing many different stars at once, and thus we achieve much more precise distance measurements.

Cepheid Variables   Main-sequence fitting works well for measuring distances to star clusters throughout the Milky Way, but not for measuring distances to other galaxies.

Most main-sequence stars are too faint to be seen in other galaxies, even with our largest telescopes. Instead, we need very bright stars to serve as standard candles for distance measurements beyond the Milky Way.

The most useful bright stars for measuring the distances to galaxies are called **Cepheid variable stars**, or **Cepheids** for short. These stars vary in brightness in our sky, alternately becoming dimmer and brighter with periods ranging between a few days and a few months. Each Cepheid has its own particular time period between peaks in luminosity; we saw one example earlier in Figure 15.13.

In 1912, Henrietta Leavitt discovered that the periods of Cepheids are very closely related to their luminosities: The longer the period, the more luminous the star (Figure 20.12). We say that Cepheids obey a **period-luminosity relation** that allows us to determine (within about 10%) a Cepheid's luminosity simply by measuring the time period over which its brightness varies. For example, Figure 20.12 shows that a Cepheid variable whose brightness peaks every 30 days is effectively screaming out, "Hey, everybody, my luminosity is 10,000 times that of the Sun!" Thus, once we measure a Cepheid's period, we know its luminosity and we can use the inverse square law for light to determine its distance.

Leavitt discovered the period-luminosity relation with careful observations of Cepheids in a nearby galaxy (the Large Magellanic Cloud), but she did not know why Cepheids vary in this special way. We now know that Cepheids are very bright examples of *pulsating variable stars* [Section 15.2]: They vary in luminosity because they actually pulsate in and out in size, growing brighter as they grow larger and then dimming as they shrink back in size. The period-luminosity relation holds because larger (and hence more luminous) Cepheids take longer to pulsate in and out in size.

Cepheids have been used for almost a century to measure distances to nearby galaxies. As we'll discuss shortly,

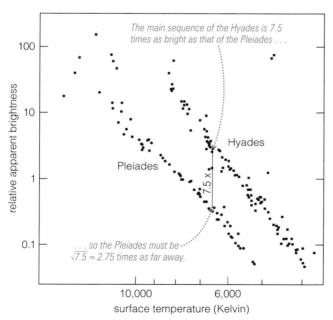

Figure 20.11 To use the technique of main-sequence fitting, we compare the apparent brightness of the main sequence of a cluster of unknown distance (the Pleiades, in this case) to that of a cluster whose distance we already know (such as the Hyades, whose distance is known from parallax).

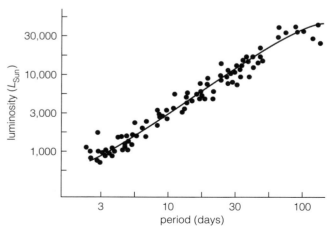

Figure 20.12 Cepheid period-luminosity relation. The data show that all Cepheids of a particular period have very nearly the same luminosity. Thus, by measuring a Cepheid's period, we can determine its luminosity and hence its distance. (Cepheids actually come in two types with two different period-luminosity relations. The relation here is for Cepheids with heavy-element content similar to that of our Sun, or "Type I Cepheids.")

they played a critical role in Edwin Hubble's discoveries. More recently, one of the main missions of the Hubble Space Telescope was to measure accurate distances to galaxies up to 100 million light-years away by studying Cepheids within them. This distance may sound very large, but it is still quite small compared with the distances of galaxies in Figure 20.1. In order to go further, we use the distances determined with these Cepheids to measure the luminosities of even brighter standard candles.

Distant Standard Candles Astronomers have discovered several techniques for estimating distances beyond those for which we can observe Cepheids. In recent years, the most valuable of these techniques has proved to be the use of white dwarf supernovae as standard candles.

Recall that white dwarf supernovae are thought to be exploding white dwarf stars that have reached the 1.4-solar-mass white dwarf limit [Section 18.1]. These supernovae should all have nearly the same luminosity, because they all come from stars of the same mass that explode in the same way. Although white dwarf supernovae are quite rare in any individual galaxy, several have been detected during the past century in galaxies within about 50 million light-years of the Milky Way (Figure 20.13). Astronomers kept careful records of those events, so that today we can determine the true luminosities of these supernovae by using Cepheids to measure the distances to the galaxies in which they occurred. These measurements confirm that the luminosities of all white dwarf supernovae are about the same.

Because white dwarf supernovae are so bright—about 10 billion solar luminosities at their peak—we can detect them even when they occur in galaxies billions of light-years away (Figure 20.14). Thus, white dwarf supernovae are standard candles that allow us to measure the distances of galaxies in the far reaches of the observable universe, completing the distance chain. However, the number of galaxies whose distances we can measure with this technique is relatively small because white dwarf supernovae occur only once every few hundred years in a typical galaxy.

Distant galaxies before supernova explosions

The same galaxies after supernova explosions

Figure 20.14 White dwarf supernovae in galaxies approximately 10 billion light-years away. White arrows in the lower images indicate the supernovae, and the upper images show what these galaxies looked like without supernovae.

Other techniques can be used to estimate the distances of galaxies in which we have not observed a white dwarf supernova. One useful method relies on a close relationship that has been observed between the total luminosities of spiral galaxies and the rotation speeds of their disks: The faster a spiral galaxy's rotation speed, the more luminous it is (Figure 20.15). This relationship, called the **Tully-Fisher relation** (after its discoverers), holds because both luminosity and rotation speed depend on the galaxy's mass. A galaxy's luminosity depends on the number of stars it contains, which is related to the total amount of matter within

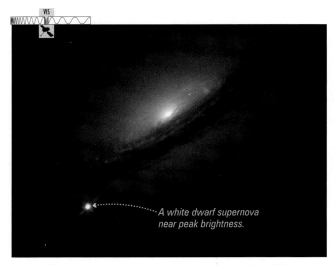

A white dwarf supernova near peak brightness.

Figure 20.13 A white dwarf supernova observed in 1994 in the galaxy NGC 4256.

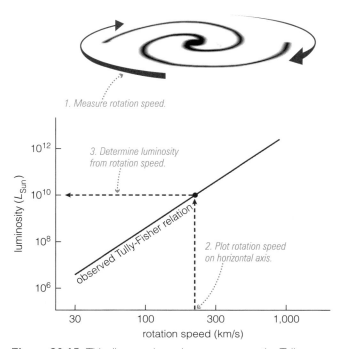

1. Measure rotation speed.

3. Determine luminosity from rotation speed.

2. Plot rotation speed on horizontal axis.

observed Tully-Fisher relation

luminosity (L_{Sun})

rotation speed (km/s)

Figure 20.15 This diagram shows how we can use the Tully-Fisher relation to estimate the distance of a spiral galaxy. (The precise relationship differs among different subclasses of spiral galaxies.)

it, and the total amount of matter determines a galaxy's rotation speed [Section 22.1].

We can measure the rotation speed of a spiral galaxy's disk by, for example, comparing the Doppler shifts of the portion of the disk rotating toward us and the portion rotating away from us. Once we have measured the rotation speed of a spiral galaxy, the Tully-Fisher relation tells us its true luminosity, which makes the galaxy itself a standard candle that we can use to determine its distance.

Distance Chain Summary Figure 20.16 summarizes the chain of measurements that allows us to determine ever-greater distances. With each link in the distance chain, however, uncertainties become somewhat greater. Thus, although we know the Earth-Sun distance at the beginning of the chain extremely accurately, distances to the farthest reaches of the observable universe remain uncertain by about 10%. Let's follow the chain shown in the figure, proceeding from left to right:

- *Radar ranging:* We measure the Earth-Sun distance by bouncing radio waves off planets and using some geometry.
- *Parallax:* We measure the distances to nearby stars by observing how their positions change, relative to the background stars, as Earth orbits the Sun. These distances thus rely on our knowledge of the Earth-Sun distance, determined with radar ranging to planets.
- *Main-sequence fitting:* We know the distance to the Hyades star cluster in our Milky Way Galaxy through parallax. Comparing the apparent brightnesses of its main-sequence stars to those of stars in other clusters gives us the distances to these other star clusters in our galaxy.
- *Cepheid variables:* By studying Cepheids in star clusters with distances measured by main-sequence fitting, we learn the precise period-luminosity relation for Cepheids.

When we find a Cepheid in a more distant star cluster or galaxy, we can determine its luminosity by measuring the period between its peaks in brightness and then use this luminosity to determine the distance.

- *Distant standards:* By measuring distances to relatively nearby galaxies with Cepheids, we learn the true luminosities of white dwarf supernovae and other distant standard candles, enabling us to measure great distances throughout the universe.

Using these distant standards, astronomers have been able to calibrate yet another tool, known as Hubble's law, that can be used to measure distances to galaxies throughout the universe. However, the significance of Hubble's law to our understanding of galaxies and the history of the universe itself goes well beyond just distance measurement. To fully appreciate its importance, we will devote the rest of this chapter to Edwin Hubble, his revolutionary discoveries, and their role in the birth of modern cosmology.

 Hubble's Law Tutorial, Lessons 1–3

20.3 Hubble's Law

The ability to measure distances to galaxies underlies much of our modern understanding of the size and age of the universe. Yet, less than a century ago, we did not even know the distances to the nearest galaxies. A revolution in our understanding was about to happen, and the discoveries of Edwin Hubble are what set that revolution in motion. Let's now explore how Hubble discovered the famous law that bears his name, and how this law has helped us answer fundamental questions about the universe in which we live.

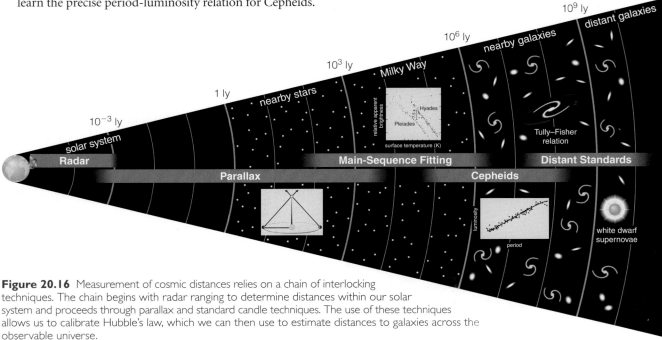

Figure 20.16 Measurement of cosmic distances relies on a chain of interlocking techniques. The chain begins with radar ranging to determine distances within our solar system and proceeds through parallax and standard candle techniques. The use of these techniques allows us to calibrate Hubble's law, which we can then use to estimate distances to galaxies across the observable universe.

How did Hubble prove that galaxies lie far beyond the Milky Way?

The only galaxies easily visible to the naked eye are the Andromeda Galaxy and the Large and Small Magellanic Clouds, but they appear as little more than fuzzy, glowing blobs. With so little to go on, naked-eye observers had no way of even guessing the true nature of these fuzzy objects.

The Spiral Nebulae Our understanding of galaxies took a step forward with the advent of telescopic observations. By the mid-1700s, telescopes revealed that the Andromeda Galaxy was really a much larger spiral-shaped structure and that many similar spiral-shaped structures littered the sky. You might at first think it would have been obvious that these objects were other galaxies, but remember that no one yet knew what a galaxy was, let alone that many galaxies have spiral shapes. Moreover, the telescopes available at the time still could not make out the individual stars in these objects. Thus, many astronomers assumed that the spirals were merely spinning gas clouds located within the Milky Way, and they therefore called them "spiral nebulae."

A few creative thinkers correctly guessed the true nature of the spiral nebulae. For example, here is a 1755 quotation from German philosopher Immanuel Kant, as translated by Edwin Hubble in his book *The Realm of the Nebulae:*

> We see that scattered through space out to infinite distances, there exist similar systems of stars [that is, galaxies], and that creation, in the whole extent of its infinite grandeur, is everywhere organized into systems whose members are in relation with one another. . . . A vast field lies open to discoveries and observation alone will give the key.

Following Kant, some later writers referred to galaxies as "island universes," because they appeared to be vast, self-contained communities of stars, isolated from one another by chasms of empty space. However, even though these views ultimately proved correct, guessing correctly is never good enough in science. Science demands hard evidence, and little of relevance to the nature of the spiral nebulae was forthcoming for 150 years after Kant's good guesswork.

The Shapley-Curtis Debate During the first two decades of the twentieth century, advances in astronomical technology and data collection began to arm astronomers with more facts about the spiral nebulae, and for a while these facts fueled proponents on both sides of the debate about their nature. The issue became so contentious that in 1920 the National Academy of Sciences sponsored what has since become known in astronomical circles as The Great Debate, or the Shapley-Curtis Debate.

The Great Debate took place in Washington, D.C, on April 26, 1920; the official topic was "The Scale of the Universe." Harlow Shapley, then a rising young astronomer who would the following year become director of the Har-

vard Observatory, took the view that the spiral nebulae were gas clouds internal to the Milky Way, and therefore that the universe was composed only of our own, single galaxy. Heber Curtis, of the Lick Observatory, argued in favor of Kant's island universes, which implied that the universe is immense and possibly infinite.

Both Shapley and Curtis knew that the key issue was the distance to the spiral nebulae, but neither was able to directly measure a distance from the data at hand. The debate therefore rested on indirect evidence. Shapley argued that the spiral nebulae must be relatively nearby because an apparent nova outburst seen in the Andromeda Galaxy in 1885 had an apparent brightness similar to nova outbursts in our own galaxy. He also pointed out observations of another spiral nebula called M 101 that seemed to show its spiral pattern rotating with a period of 100,000 years; if M 101 really were a separate galaxy, this rotation period would have given a rotation speed faster than the speed of light, in violation of Einstein's special theory of relativity published more than a decade earlier. Curtis countered that all the other nova outbursts observed in the Andromeda Galaxy were much dimmer than the one seen in 1885 and that the data on M 101 did not support the claimed rotation speed.

Neither Shapley nor Curtis scored a convincing victory at the time of the debate, and many years passed before astronomers learned that two key pieces of data in Shapley's argument had in fact been misconstrued: Andromeda's "nova" of 1885 was actually a supernova and the claimed rotation of M 101 was a misinterpretation of the data. Nevertheless, a decisive resolution to The Great Debate came only a few years after the debate itself.

Hubble and the Andromeda Galaxy Edwin Hubble was finishing graduate school in 1917 when he was invited to join the staff at the Mount Wilson Observatory in Pasadena, California. It was a prestigious invitation for someone so young, because work was just finishing on Mount Wilson's jewel, a 100-inch telescope that would be the world's largest for the next 30 years (Figure 20.17). But in April 1917, the

Figure 20.17 The 100-inch telescope on Mount Wilson, outside Los Angeles.

Figure 20.18 Edwin Hubble at the Mount Wilson Observatory.

United States entered the war against Germany, and instead of accepting the invitation, Hubble volunteered for service. He telegraphed the observatory, saying, "Regret cannot accept your invitation. Am off to the war." He served in France, rising to the rank of major before returning to the United States, where he finally joined the staff at Mount Wilson in 1919.

Once he settled in at Mount Wilson, Hubble turned his attention to the study of the spiral nebulae (Figure 20.18). With the 100-inch telescope, he could see what looked like individual stars in the Andromeda Galaxy, which strongly suggested that Kant's idea was right. But his breakthrough came when he discovered that some of these stars were dimming and brightening with a regular period, marking them as Cepheids. By comparing photographs of the galaxy taken on different nights, Hubble measured the periods of the Cepheids. He then used Henrietta Leavitt's period-luminosity relation to estimate the luminosities of the Cepheids in Andromeda, and from there he could estimate the distance to the Andromeda Galaxy from the inverse square law for light.

With hindsight, we now know that Hubble underestimated the true distance by about half, because no one yet knew that Cepheids actually come in two different types that obey two different period-luminosity relations.* Leavitt's period-luminosity relation actually applied to the dimmer type, but the Cepheids that Hubble was observing were of the brighter type. Thus, when he applied Leavitt's

---

*The two types of Cepheids, creatively called Type I and Type II, differ in their heavy-element content. Type I Cepheids have Sun-like heavy-element content and are found in the disk of the galaxy. Type II Cepheids have much lower heavy-element content and are found in the halo of the galaxy. For any particular period of variability, a Type I Cepheid is about four times as luminous as a Type II Cepheid.

relation, he underestimated their actual luminosities and therefore underestimated their distances. Nevertheless, Hubble still succeeded in showing that the Cepheids in Andromeda lie far beyond the outer reaches of stars in the Milky Way. The debate was over: Hubble had proved once and for all that the Andromeda "nebula" was in fact a separate galaxy.

This single scientific discovery dramatically changed our view of the universe. Rather than thinking the Milky Way was the entire universe, we suddenly knew that it is just one among many galaxies in an enormous universe. The stage was set for an even greater discovery.

• What is Hubble's law?

Hubble's determination of the distance to the Andromeda Galaxy ensured him a permanent place in the history of astronomy, but he didn't stop there. Hubble proceeded to estimate the distances to many more galaxies. Within just a few years, he made one of the most astonishing discoveries in the history of science: that the universe is expanding.

Distance and Redshift Astronomers had known since the 1910s that the spectra of most spiral galaxies tend to be *redshifted* (Figure 20.19). Recall that redshifts occur when the object emitting the radiation is moving away from us [Section 5.5]. Because Hubble had not yet proved that spiral galaxies are in fact separate from the Milky Way, no one understood the true significance of their motions.

Following his discovery of Cepheids in Andromeda, Hubble and his coworkers spent the next few years busily measuring the redshifts of galaxies and estimating their distances. Because even Cepheids were too dim to be seen in most of these galaxies, Hubble needed brighter standard candles for his distance estimates. One of his favorite techniques was to use the brightest object he could see in each galaxy as a standard candle, because he assumed these objects to be very bright stars that would always have about the same luminosity.

In 1929, Hubble announced his conclusion: The more distant a galaxy, the greater its redshift and hence the faster

Figure 20.19 Redshifted galaxy spectrum.

it moves away from us. As we discussed in Chapter 1 (see Figure 1.16), this discovery implies that the entire universe is expanding. Thus, Hubble had discovered the expansion of the universe.

Hubble's original assertion was based on an amazingly small sample of galaxies (Figure 20.20). Even more incredibly, he had grossly underestimated the luminosities of his standard candles. The "brightest stars" he had been using as standard candles were really entire *clusters* of bright stars. Fortunately, Hubble was both bold and lucky. Subsequent studies of much larger samples of galaxies showed that they are indeed receding from us, but they are even farther away than Hubble thought.

A Formula for Hubble's Law We express the idea that more distant galaxies move away from us faster with a very simple formula, now known as **Hubble's law**:

$$v = H_0 \times d$$

where v stands for a galaxy's velocity away from us (sometimes called a *recession velocity*), d stands for its distance, and H_0 (pronounced "H-naught") is a number called **Hubble's constant**. We usually write Hubble's law in this

Figure 20.20 Hubble's original velocity-distance diagram. Although Hubble underestimated the true galactic distances, he still discovered the general trend (represented by the straight lines) in which more distant galaxies move away from us at higher speeds. He determined the galaxy speeds by measuring redshifts in the galaxy spectra. (Units on the vertical axis are km/s.)

form to express the idea that galaxy speeds depend on their distances. However, astronomers more often use the law in reverse—measuring a galaxy's speed from its redshift and then using Hubble's law to estimate its distance.

MATHEMATICAL INSIGHT 20.2 Redshift

The *redshift* of an object, usually designated by the letter z, is defined as the fractional difference between the observed wavelength ($\lambda_{observed}$) of a line in the object's spectrum and the wavelength the line would have if the object were standing still (λ_{rest}). (The Greek letter λ is usually used to represent wavelength.) Thus, we can calculate redshift with the following equation:

$$\text{redshift}(z) = \frac{\lambda_{observed} - \lambda_{rest}}{\lambda_{rest}}$$

Note that a redshift affects an entire spectrum in the same way, so the value of z is the same for every line in a moving object's spectrum.

For a relatively nearby galaxy, with a redshift z that is much less than 1, we can find its velocity away from us with the following simple formula:

$$v = c \times z$$

where $c = 3.0 \times 10^5$ km/s is the speed of light. (We can use a more complex formula when the redshift is *not* much less than 1, but as we will discuss later in this chapter, the idea of speed becomes hard to define for such distant galaxies.)

Example: Two emission lines from hydrogen in the visible spectrum have rest wavelengths of 656.3 nanometers and 486.1 nanometers. Suppose you see this pair of hydrogen emission lines at 662.9 nanometers and 491.0 nanometers in the spectrum of a distant galaxy. What is the redshift of the galaxy? How fast is it moving away from us?

Solution:

Step 1 Understand: We are given the observed wavelengths and rest wavelengths for two different spectral lines in the galaxy's

spectrum. We can therefore calculate the redshift, z, for each line—and we expect it to be the same for both lines since they both are in the same object's spectrum. (Astronomers generally try to measure the redshifts of at least two lines in any particular spectrum to make sure that they aren't mistaking a misidentified line for a shifted line.) Once we have the redshift, we can use the formula $v = c \times z$ to calculate the galaxy's speed.

Step 2 Solve: Let's first calculate the redshift for each line. For the first line, we have $\lambda_{rest} = 656.3$ nm and $\lambda_{observed} = 662.9$ nm, which gives a redshift of:

$$z = \frac{662.9 \text{ nm} - 656.3 \text{ nm}}{656.3 \text{ nm}} = 0.010$$

For the second line, $\lambda_{rest} = 486.1$ nm and $\lambda_{observed} = 491.0$ nm, which gives a redshift of:

$$z = \frac{491.0 \text{ nm} - 486.1 \text{ nm}}{486.1 \text{ nm}} = 0.010$$

As we expect, both lines give the same value for the redshift. Because this redshift is much less than 1, we can use the formula $v = c \times z$ to calculate the galaxy's speed:

$$v = (3.0 \times 10^5 \text{ km/s}) \times 0.010 = 3{,}000 \text{ km/s}$$

Step 3 Explain: We have found that the galaxy has a redshift of $z = 0.010$, which means it is moving away from us at a speed of 3,000 km/s. Notice that this is about 1% of the speed of light—slow compared to light, but far faster than any spacecraft we have ever built.

Because Hubble's law in principle applies to all distant galaxies for which we can measure a redshift, it is the most useful technique for determining distances to galaxies that are very far away. Nevertheless, we encounter two important practical difficulties when we try to use Hubble's law to measure galactic distances:

1. Galaxies do not obey Hubble's law perfectly. Hubble's law would give an exact distance only for a galaxy whose speed is determined solely by the expansion of the universe. In reality, nearly all galaxies experience gravitational tugs from other galaxies, and these tugs alter their speeds from the "ideal" values predicted by Hubble's law.
2. Even when galaxies closely obey Hubble's law, the distances we find with it are only as accurate as our best measurement of Hubble's constant.

The first problem is most serious for nearby galaxies. Within the Local Group, for example, Hubble's law does not work at all: The galaxies in the Local Group are gravitationally bound together with the Milky Way and therefore are *not* moving away from us in accord with Hubble's law. However, Hubble's law works fairly well for more distant galaxies. The recession speeds of galaxies at large distances are so great that any motions caused by the gravitational tugs of neighboring galaxies are tiny in comparison.

The second problem means that, even for distant galaxies, we can know only *relative* distances until we pin down the true value of H_0. For example, Hubble's law tells us that a galaxy moving away from us at 20,000 km/s is twice as far away as one moving at 10,000 km/s, but we can determine the actual distances of the two galaxies only if we know H_0. Until recently, estimates of the value of H_0 were quite uncertain.

MATHEMATICAL INSIGHT **20.3** **Understanding Hubble's Law**

Hubble's law is one of the most important tools of modern astronomy. The units we use for Hubble's constant depend on the units we use to measure distances and speeds. Virtually all astronomers quote the speeds of distant galaxies in units of kilometers per second. Distances may be quoted either in millions of light-years or in megaparsecs. In this book, we will use millions of light-years, so the units of Hubble's constant are *kilometers per second per million light-years* (km/s/Mly). However, you may also hear astronomers quote Hubble's constant in units of *kilometers per second per megaparsec* (km/s/Mpc). (Because 1 parsec = 3.26 light-years, you can convert from km/s/Mly to km/s/Mpc simply by multiplying by 3.26.)

What do these strange-sounding units mean? They tell us that if we measure the distance to a distant galaxy in millions of light-years, then Hubble's law will give us the galaxy's velocity in kilometers per second. The idea should become clear with some simple examples. Since the current best estimate of Hubble's constant puts its value between 20 and 24 kilometers per second per million light-years, let's use the middle value of 22 kilometers per second per million light-years:

● For a galaxy located 10 million light-years away, Hubble's law tells us to expect it to be moving away from us at a speed of:

$$v = H_0 \times d = 22 \frac{\text{km/s}}{\text{Mly}} \times 10 \text{ Mly} = 220 \text{ km/s}$$

● For a galaxy located 11 million light-years away, Hubble's law tells us to expect it to be moving away from us at a speed of:

$$v = H_0 \times d = 22 \frac{\text{km/s}}{\text{Mly}} \times 11 \text{ Mly} = 242 \text{ km/s}$$

Notice that this is 22 km/s faster than the galaxy located at a distance of 10 million light-years—just as we should expect since Hubble's constant tells us that each additional million light-years of distance means an additional 22 km/s of speed.

● For a galaxy located 700 million light-years away, Hubble's law tells us to expect it to be moving away from us at a speed of:

$$v = H_0 \times d = 22 \frac{\text{km/s}}{\text{Mly}} \times 700 \text{ Mly} = 15,400 \text{ km/s}$$

Notice that this speed is quite high, representing more than 5% of the speed of light. In reality, astronomers rarely use Hubble's law to calculate velocities, because they can determine a galaxy's velocity directly by measuring its redshift. Instead, astronomers use Hubble's law to estimate distances to far-off galaxies whose distances cannot be measured with other techniques in the distance chain.

Example: Estimate the distance to a galaxy whose redshift indicates that it is moving away from us at a speed of 22,000 km/s. Assume that the value of Hubble's constant is $H_0 = 22$ km/s/Mly.

Solution:

Step 1 Understand: In this case we are given the galaxy's speed, found from its redshift, and are asked to find its distance. Thus, we must solve Hubble's law for the distance d. We can do so by dividing both sides of the law in its standard form ($v = H_0 \times d$) by H_0; the result is:

$$d = \frac{v}{H_0}$$

In this form, we can find the galaxy's distance simply by plugging in its recession speed and Hubble's constant.

Step 2 Solve: Putting in the given values, we find:

$$d = \frac{v}{H_0} = \frac{22,000 \text{ km/s}}{22 \frac{\text{km/s}}{\text{Mly}}} = 1,000 \text{ Mly}$$

Step 3 Explain: We have found that the galaxy's distance is 1,000 million light-years, which is the same as 1 billion light-years. Remember that this distance estimate is accurate only if Hubble's constant really is 22 km/s/Mly. If the true value of Hubble's constant is larger, then the galaxy is closer than the answer we've found here (because Hubble's constant is in the denominator of the distance equation above). If the true value of Hubble's constant is smaller, then the galaxy is more distant than we've found here.

One of the main missions of the Hubble Space Telescope has been to obtain an accurate value of H_0. Astronomers used the telescope to discover Cepheid variables in galaxies out to about 60 million light-years and then used those distances to determine the luminosities of distant standard candles such as white dwarf supernovae. Plotting galactic distances measured with those distant standard candles against the velocities indicated by their redshifts has pinned down the value of H_0 to somewhere between 20 and 24 *kilometers per second per million light-years* (Figure 20.21). In other words, for every million light-years of distance away from us, a galaxy's speed away from us is between 20 and 24 kilometers per second. For example, with this range of values for Hubble's constant, Hubble's law predicts that a galaxy located 100 million light-years away would be moving away from us at a speed between 2,000 and 2,400 kilometers per second.

• How do distance measurements tell us the age of the universe?

Hubble's law is a remarkably powerful tool for understanding the universe. Not only does it tell us that the universe is expanding and give us a way to measure galactic distances, but it also helps us determine the age and size of the observable universe. To see how, we must consider the expansion of the universe in a little more detail.

Universal Expansion We first discussed the expansion of the universe in Chapter 1. Galaxies all across the universe are moving away from one another, and this fact implies that galaxies must have been closer together in the past. Tracing this convergence back in time, we reason that all the matter in the observable universe started very close together and that the entire universe came into being at a single moment.

It may be tempting to think of the expanding universe as a ball of galaxies expanding into a void, but this impression is mistaken. To the best of our knowledge, the universe is not expanding *into* anything. As far as we can tell, there is no edge to the distribution of galaxies in the universe. On very large scales, the distribution of galaxies appears to be relatively smooth, meaning that the overall appearance of the universe around you would look more or less the same no matter where you are located. The idea that the matter in the universe is uniformly distributed, without a center or an edge, is often called the **Cosmological Principle**. Although we cannot prove it to be true, it is completely consistent with all our observations of the universe [Section 22.3].

So how can the universe be expanding if it's not expanding *into* anything? Back in Chapter 1, we likened the expanding universe to a raisin cake baking [Section 1.3], but a cake has a center and edges that grow into empty space as it bakes. A better analogy is something that can expand but that has no center and no edges. The surface of a balloon can fit the bill, as can an infinite surface such as a sheet of rubber that extends to infinity in all directions.

Because it's hard to visualize infinity, let's use the surface of a balloon as our analogy to the expanding universe (Figure 20.22). Note that this analogy uses the balloon's two-dimensional *surface* to represent all three dimensions of space. Thus, the surface of the balloon represents the entire universe, and the spaces inside and outside the balloon have no meaning in this analogy. Aside from the reduced number of dimensions, the analogy works well because the balloon's spherical surface has no center and no edges, just as no city is the center of Earth's surface and no edges exist where you could walk or sail off the Earth. Now, remember that although the universe as a whole is expanding, individual galaxies and galaxy clusters do not expand, because gravity holds them together. We can therefore represent

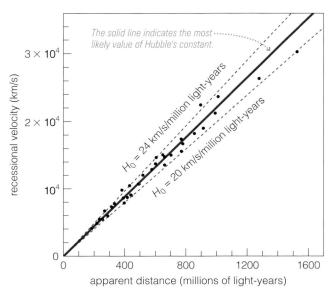

Figure 20.21 Interactive Figure White dwarf supernovae can be used as standard candles to establish Hubble's law out to very large distances. The points on this figure show the apparent distances of white dwarf supernovae and the recession velocities of the galaxies in which they exploded. The fact that these points all fall close to a straight line demonstrates that these supernovae are good standard candles.

The solid line indicates the most likely value of Hubble's constant.

$H_0 = 24$ km/s/million light-years
$H_0 = 20$ km/s/million light-years

recessional velocity (km/s)
3×10^4
2×10^4
10^4
0

apparent distance (millions of light-years)
0   400   800   1200   1600

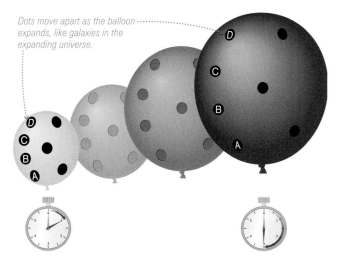

Dots move apart as the balloon expands, like galaxies in the expanding universe.

Figure 20.22 Interactive Figure As the balloon expands, the dots move apart in the same way that galaxies move apart in our expanding universe.

galaxies (or clusters of galaxies) with plastic dots attached to the balloon, and we can make our model universe expand by inflating the balloon. The dots move apart as the surface of the balloon expands, but they themselves do not grow in size.

The Age of the Universe   We can now see how Hubble's law leads us to an age estimate for the universe. Imagine that some miniature scientists are living on dot B in Figure 20.22. Suppose that, 3 seconds after the balloon begins to expand, they measure the following:

Dot A is 3 centimeters away and moving at 1 cm/s.

Dot C is 3 centimeters away and moving at 1 cm/s.

Dot D is 6 centimeters away and moving at 2 cm/s.

They could summarize these observations as follows: *Every dot is moving away from our home with a speed that is 1 cm/s for each 3 centimeters of distance.* Because the expansion of the balloon is uniform, scientists living on any other dot would come to the same conclusion. Each scientist living on the balloon would determine that the following formula relates the distances and velocities of other dots on the balloon:

$$v = \left(\tfrac{1}{3} \text{s}\right) \times d$$

where v and d are the velocity and distance of any dot, respectively.

Confirm that this formula gives the correct values for the speeds of dots C and D, as seen from dot B, 3 seconds after the balloon begins expanding. How fast would a dot located 9 centimeters from dot B move, according to the scientists on dot B?

If the miniature scientists think of their balloon as a bubble, they might call the number relating distance to velocity—the term $\tfrac{1}{3}$ s in the preceding formula—the "bubble constant." An especially insightful miniature scientist might flip over the "bubble constant" and find that it is exactly equal to the time since the balloon started expanding. That is, the "bubble constant" $\tfrac{1}{3}$ s tells them that the balloon has been expanding for 3 seconds. Perhaps you see where we are heading.

Just as the inverse of the "bubble constant" tells the miniature scientists that their balloon has been expanding for 3 seconds, the inverse of the Hubble constant, or $\tfrac{1}{H_0}$, tells us something about how long our universe has been expanding. The "bubble constant" for the balloon depends on when it is measured, but it is always equal to 1 divided

---

MATHEMATICAL INSIGHT 20.4   Age from Hubble's Constant

The reciprocal of Hubble's constant, or $\tfrac{1}{H_0}$, is related to the age of the universe: It tells us what the age of the universe would be *if* the expansion rate has remained constant since the beginning. We can calculate this age from the measured value of H_0. Let's do so using the assumption that $H_0 = 22$ km/s/Mly.

The easiest way to find the reciprocal of H_0 is to first rewrite it in more convenient units. First, because a million is 10^6, we can write a million light-years as 10^6 light-years. Next, because a light-year is the speed of light ($c = 300{,}000$ km/s) times 1 year, we can rewrite our value of Hubble's constant as:

$$H_0 = 22 \,\frac{\text{km/s}}{10^6 \text{ ly}} = 22 \,\frac{\text{km/s}}{10^6 \times (3 \times 10^5 \text{ km/s}) \times 1 \text{ yr}}$$

Taking the reciprocal, we find:

$$\frac{1}{H_0} = \frac{10^6 \times (3 \times 10^5 \text{ km/s}) \times 1 \text{ yr}}{22 \text{ km/s}}$$
$$= \frac{3 \times 10^{11} \text{ yr} \times \text{km/s}}{22 \text{ km/s}} = 1.36 \times 10^{10} \text{ yr}$$

The reciprocal of Hubble's constant is 1.36×10^{10} yr, or 13.6 billion years. In other words, if the expansion rate has never changed, a value of Hubble's constant of $H_0 = 22$ km/s/Mly implies that the universe is 13.6 billion years old.

What if the expansion rate *has* changed? If the rate of expansion has slowed with time (due to gravity, for example), then the expansion rate would have been faster in the past. In that case, the universe would have reached its current size faster than we would guess by assuming a constant expansion rate, which means the universe is *younger* than the age we get from $\tfrac{1}{H_0}$. Conversely, if the rate of expan-

sion has increased with time (due to "dark energy," for example, see Chapter 22), then the universe is *older* than the age we get from $\tfrac{1}{H_0}$.

Example: How would our estimate for the age of the universe be different if the measured value of H_0 were 44 km/s/Mly rather than 22 km/s/Mly?

Solution:

Step 1 Understand: The question asks how the age of the universe would be different if Hubble's constant were twice as large as the value we assumed in our calculations above. Because our estimate for the age of the universe depends on the *reciprocal* of Hubble's constant, a larger value for the constant means a smaller estimated age.

Step 2 Solve: In this example, Hubble's constant is taken to be two times as large as the current value, and the reciprocal of 2 is $\tfrac{1}{2}$. Thus, our estimate for the universe's age would be $\tfrac{1}{2}$ of the estimate we found earlier. Half of 13.6 billion years is 6.8 billion years.

Step 3 Explain: We have found that if Hubble's constant were twice as large as our best current measurement of it, the age of the universe would be only about half what we find from current measurements. We can understand this result by looking at what Hubble's constant says about the relationship between a galaxy's speed and its distance. The value of H_0 is essentially the speed of a galaxy divided by its distance. Doubling Hubble's constant would mean that the speeds of galaxies at a given distance are twice as fast. It would therefore have taken them only half as long to reach their current distances.

---

by the time since the balloon started expanding. Similarly, the Hubble constant actually changes with time but stays roughly equal to 1 divided by the age of the universe. We call it a constant because it is the same at all locations in the universe, and because its value does not change noticeably on the time scale of human civilization.

Current estimates based on the value of Hubble's constant put the age of the universe between about 12 and 15 billion years. To derive a more precise value for the universe's age, we need to know whether the expansion has been speeding up or slowing down over time, a question we will examine more closely in Chapter 22. If the gravitational pull of each galaxy on every other galaxy has significantly slowed the expansion rate, then the universe's age is somewhat less than $\frac{1}{H_0}$. If some mysterious force has accelerated the expansion rate, then the universe's age is somewhat more than $\frac{1}{H_0}$. The best available evidence as of 2006 suggests that the universe is about 14 billion years old.

• How does the universe's expansion affect our distance measurements?

The expansion of the universe leads to a complication in discussing galaxy distances that we have ignored up to this point. To understand the complication, imagine observing a supernova in a distant galaxy. Suppose the supernova occurred 400 million years ago but we are only just now seeing it. The supernova's light must have traveled 400 million light-years to reach Earth, but this simple statement about how far the light has traveled does not translate easily into a distance for the galaxy in which the supernova occurred. The problem is that the universe is expanding, making the distance between Earth and the supernova greater today than it was at the time of the supernova event. Thus, if we simply say that the galaxy is "400 million light-years away," it's not clear whether we mean its distance now, its distance at the time of the supernova, or something in between.

Lookback Time Because distances between galaxies are always changing, it is easier to speak about faraway objects in terms of how much time their light takes to reach us— 400 million years in the case of the supernova. We call this the **lookback time** to the supernova. In other words, a distant object's lookback time is the difference between the current age of the universe and the age of the universe when the light left the object. Unlike a statement about distance, the meaning of a lookback time is unambiguous: If the lookback time is 400 million years, it means the light really traveled through space for a period of 400 million years to reach us.

One way to visualize the relationship between distance, expansion, and lookback time is with a *spacetime diagram* [Section S3.3] for the supernova and our observation of it. The horizontal axis in Figure 20.23 shows distance (through space) from the Milky Way, and the vertical axis shows time. The Milky Way's path on the diagram is vertical because the Milky Way is staying in the same place as time

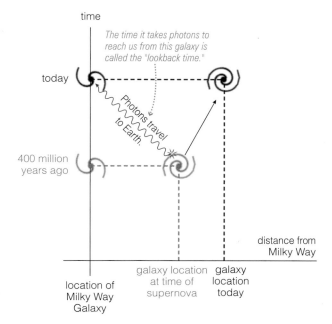

Figure 20.23 Interactive Figure A spacetime diagram for a supernova in a distant galaxy. The lookback time to this supernova is 400 million years because that is how long its light took to reach us.

passes, at least from our perspective. In contrast, the distance of the galaxy in which the supernova exploded is increasing with time, so its path through the diagram is tilted diagonally away from the Milky Way. Photons of light from the supernova travel away from it in all directions, but the diagram shows only those photons that are arriving at the Milky Way today. Their path is tilted diagonally toward the Milky Way because their distance decreases as they approach us. Notice that the distance to the other galaxy is changing during the entire time that the photons are traveling, but the vertical axis of the diagram shows clearly that the entire trip took 400 million years. That is why we say the galaxy has a lookback time of 400 million years.

THINK ABOUT IT

Explain why there is no question about meaning when we talk about distances to objects within the Local Group, but distances become difficult to define when we talk about objects much farther away.

Cosmological Redshift An object's lookback time is directly related to its redshift. Recall that the redshifts of galaxies tell us how quickly they are moving away from us. In the context of an expanding universe, redshifts have an additional, more fundamental interpretation. Let's return again to the universe on the balloon. Suppose you draw wavy lines on the balloon's surface to represent light waves. As the balloon inflates, these wavy lines stretch out, and their wavelengths increase (Figure 20.24). This stretching closely resembles what happens to photons in an expanding universe. The expansion of the universe stretches out all the photons within it, shifting them to longer, redder wavelengths. We call this effect a **cosmological redshift**.

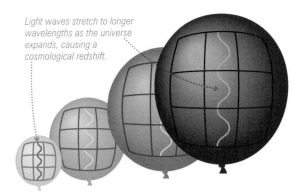

Light waves stretch to longer wavelengths as the universe expands, causing a cosmological redshift.

Figure 20.24 Interactive Figure. As the universe expands, photon wavelengths stretch like the wavy lines on this expanding balloon.

In a sense, we have a choice when we interpret the redshift of a distant galaxy: We can think of the redshift as being caused either by the Doppler effect as the galaxy moves away from us or by a photon-stretching, cosmological redshift. However, as we look to very distant galaxies, the ambiguity in the meaning of *distance* also makes it difficult to specify precisely what we mean by a galaxy's *speed*. Thus, it becomes preferable to interpret the redshift as being due to photon stretching in an expanding universe.

From this perspective, it is better to think of space itself as expanding, carrying the galaxies along for the ride, than to think of the galaxies as projectiles flying through a static universe. The cosmological redshift of a galaxy thus tells us how much space has expanded during the time since light from the galaxy left on its journey to us.

The Horizon of the Universe When we began our discussion of the expanding universe, we stressed that the universe as a whole does not seem to have an edge. Yet the universe does have a horizon, a place beyond which we cannot see. The cosmological horizon that marks the limits of the observable universe is a boundary in time, not in space. It exists because we cannot see back to a time before the universe began. For example, if the universe is 14 billion

---

MATHEMATICAL INSIGHT **20.5** **Cosmological Redshift and the Stretching of Light**

The cosmological interpretation of redshift in terms of the stretching of light waves offers us a simple way to measure how the average distances between galaxies change with time. In Mathematical Insight 20.2, we learned that the difference between the observed wavelength of a spectral line from a galaxy ($\lambda_{observed}$) and the wavelength that line would have if the galaxy were standing still (λ_{rest}) can be expressed in terms of the redshift $z = (\lambda_{observed} - \lambda_{rest})/\lambda_{rest}$. Adding the number 1 to both sides of that expression and simplifying the right-hand side leads to the equation

$$1 + z = \frac{\lambda_{observed}}{\lambda_{rest}}$$

Because the redshift of a very distant galaxy is cosmological in nature, the ratio $\lambda_{observed}/\lambda_{rest}$ for such a galaxy tells us how much the light waves from this galaxy have been stretched since the time the light was emitted. Because this stretching comes from the expansion of the universe itself, it is also proportional to the average distance between galaxies. Thus, if we can measure the redshift z of a distant galaxy, we can then determine the factor $1 + z$ by which the distances between galaxies have changed during the time it took for light to reach us from that galaxy.

Example: Measurements based on several lines in the spectrum of a distant galaxy indicate that its redshift is $z = 1.5$. By what factor have the average distances between galaxies increased since the time the light we see today left that distant galaxy?

Solution:

Step 1 Understand: As discussed above, the relationship between redshift and the expansion of the universe is based on $1 + z$. In this case, we are given the redshift, z, for the distant galaxy, so we have all the information we need.

Step 2 Solve: We simply use the given value of $z = 1.5$ for the galaxy. With this value, we find that: $1 + z = 1 + 1.5 = 2.5$.

Step 3 Explain: Our result of $1 + z = 2.5$ tells us that the average distances between galaxies today are 2.5 times as large as they were when light left the distant galaxy for which we measure a redshift of $z = 1.5$.

years old, then no object can have a lookback time greater than 14 billion years (see Figure 1.5). Thus, the age of the universe fundamentally limits the size of our observable universe.

Our quest to study galaxies and measure their distances has brought us to the very limits of the observable universe. The galaxies in Figure 20.1 at the beginning of this chapter extend from relatively nearby almost all the way to the cosmological horizon. So what have these galaxies been doing during the 14-billion-year history of the universe? That is the topic we will turn to in our next chapter.

SUMMARY OF KEY CONCEPTS

20.1 Islands of Stars

• **How are the lives of galaxies connected with the history of the universe?** Because nearby galaxies tend to be similar in age to the Milky Way, we know that most galaxies formed when the universe was much younger and have aged along with the universe itself. Thus, the lives of galaxies are intimately connected with the evolution of the universe, making the study of galaxies a part of **cosmology**—the study of the overall structure and evolution of the universe.

• **What are the three major types of galaxies?**

(1) **Spiral galaxies** have prominent disks and spiral arms.

(2) **Elliptical galaxies** are rounder and redder than spiral galaxies and contain less cool gas and dust.

(3) **Irregular galaxies** are neither disklike nor rounded in appearance.

• **How are galaxies grouped together?** Spiral galaxies tend to collect in groups that contain up to several dozen galaxies. Elliptical galaxies are more common in clusters of galaxies, which contain hundreds to thousands of galaxies, all bound together by gravity.

20.2 Measuring Galactic Distances

• **How do we measure the distances to galaxies?** Our measurements of galaxy distances depend on a chain of methods. The chain begins with radar ranging in our own solar system and parallax measurements of distances to nearby stars; then it relies on **standard candles** to measure greater distances.

20.3 Hubble's Law

- **How did Hubble prove that that galaxies lie far beyond the Milky Way?** Using the largest telescope in the world at the time, Hubble identified individual **Cepheid variable stars** in the Andromeda Galaxy. He was then able to use the **period-luminosity relation** for Cepheids to determine their luminosities and hence their distances, proving that Andromeda is much too far away to be part of the Milky Way.

- **What is Hubble's law?** Hubble's law tells us that

more distant galaxies are moving away faster: $v = H_0 \times d$, where H_0 is **Hubble's constant**. It allows us to determine a galaxy's distance from the speed at which it is moving away from us, which we can measure from its Doppler shift.

- **How do distance measurements tell us the age of the universe?** Combining distance measurements with velocity measurements tells us Hubble's constant, and the inverse of Hubble's constant tells us how long it would have taken the universe to reach its present size *if* the expansion rate had never changed. Based on Hubble's constant and estimates of how it has changed with time, we now estimate the age of the universe at about 14 billion years.

- **How does the universe's expansion affect our distance measurements?** Distances between

galaxies are always changing because of the expansion of the universe. It is therefore best to express the distance to a faraway galaxy in terms of its **lookback time**—the time it has taken for the galaxy's light to reach us. The expansion of the universe during that time stretches the light coming from the galaxy, leading to a **cosmological redshift** directly related to the galaxy's lookback time.

EXERCISES AND PROBLEMS

For instructor-assigned homework go to MasteringASTRONOMY   www.masteringastronomy.com

Review Questions

Short-Answer Questions Based on the Reading

1. Why do we need to understand the evolution of the universe in order to understand the lives of galaxies?
2. What are the three major types of galaxies, and how do they look different from one another?
3. Describe the differences between normal spiral galaxies, barred spiral galaxies, and lenticular galaxies.
4. Distinguish between the *disk component* and the *spheroidal component* of a spiral galaxy. Which component includes cool gas and active star formation?
5. What is the major difference between spiral and elliptical galaxies? Answer in terms of the presence or absence of disk and spheroidal components. How does this difference explain the lack of hot, young stars in elliptical galaxies?
6. How are galaxy types different in clusters of galaxies than among smaller groups and isolated galaxies?
7. What do we mean by a *standard candle*? Explain how we can use standard candles to measure distances.
8. Summarize each of the major links in the distance chain. Why are *Cepheid variable stars* so important? Why are white dwarf supernovae so useful, even though they are quite rare?
9. Explain how Hubble used Cepheid variable stars to prove that the Andromeda Galaxy lies beyond the bounds of the Milky Way.
10. What is *Hubble's law*? What is *Hubble's constant*? Explain what we mean when we say that Hubble's constant is between 20 and 24 *kilometers per second per million light-years*.

11. What is the Cosmological Principle, and how is it important to our understanding of the universe?
12. How is the expansion of the surface of an inflating balloon similar to the expansion of the universe? Use the balloon analogy to explain why Hubble's constant is related to the age of the universe.
13. What do we mean by the *lookback time* to a distant galaxy? Briefly explain why lookback times are less ambiguous than distances when discussing objects very far away.
14. What is the *cosmological horizon*, and what determines how far away it lies?
15. What do we mean by a *cosmological redshift*? How does our interpretation of a distant galaxy's redshift differ if we think of it as a cosmological redshift rather than as a Doppler shift?

Test Your Understanding

Does It Make Sense?

Decide whether the statement makes sense (or is clearly true) or does not make sense (or is clearly false). Explain your reasoning. (For an example, see Chapter 1, "Does It Make Sense?")

16. If you want to find a lot of elliptical galaxies, you'll have better luck looking in clusters of galaxies than elsewhere in the universe.
17. Cepheid variables make good standard candles because they all have exactly the same luminosity.
18. If the standard candles you are using are less luminous than you think they are, then the distances you determine from them will be too small.

19. Galaxy A is moving away from me twice as fast as galaxy B. That probably means it's twice as far away.

20. After measuring the galaxy's redshift, I used Hubble's law to estimate its distance.

21. The center of the universe is more crowded with galaxies than any other place in the universe.

22. The lookback time to the Andromeda Galaxy is about 2.5 million light-years.

23. I'd love to live in one of the galaxies near our cosmological horizon, because I want to see the black void into which the universe is expanding.

24. If someone in a galaxy with a lookback time of 4.6 billion years had a superpowerful telescope, they could see our solar system in the process of its formation.

25. We can't see galaxies beyond the cosmological horizon because they are moving away from us faster than the speed of light.

Quick Quiz

Choose the best answer to each of the following. Explain your reasoning with one or more complete sentences.

26. Which of these galaxies is most likely to be oldest? (a) a galaxy in the Local Group (b) a galaxy observed at a distance of 5 billion light-years (c) a galaxy observed at a distance of 10 billion light-years

27. Which of these galaxies would you most likely find at the center of a large cluster of galaxies? (a) a large spiral galaxy (b) a large elliptical galaxy (c) a small irregular galaxy

28. In which of these galaxies would you be least likely to find an ionization nebula? (a) a large spiral galaxy (b) a large elliptical galaxy (c) a small irregular galaxy

29. About how many galaxies are there in a typical cluster of galaxies? (a) about 10 (b) a few dozen (c) a few hundred

30. If all the stars on the main sequence of a star cluster are typically only one-hundredth as bright as their main-sequence counterparts in the Hyades Cluster, then that cluster's distance is (a) 100 times as far as the Hyades's distance. (b) 30 times as far as the Hyades's distance. (c) 10 times as far as the Hyades's distance.

31. Which kind of object is the best standard candle for measuring distances to extremely distant galaxies? (a) a white dwarf (b) a Cepheid variable star (c) a white dwarf supernova

32. When the ultraviolet light from hot stars in very distant galaxies finally reaches us, it arrives at Earth in the form of (a) X rays. (b) slightly more energetic ultraviolet light. (c) visible light.

33. Why do virtually all the galaxies in the universe appear to be moving away from our own? (a) because we are located near where the Big Bang happened (b) because we are located near the center of the universe (c) Observers in all galaxies observe a similar phenomenon because of the universe's expansion.

34. If you observed the redshifts of galaxies at a given distance to be twice as large as they are now, then you would determine a value for Hubble's constant that is (a) twice as large as its current value. (b) equal to its current value. (c) half its current value.

35. What would your estimate be for the age of the universe if you measured Hubble's constant to be 11 kilometers per second per million light-years? (a) 7 billion years (b) 14 billion years (c) 28 billion years

Investigate Further
In-Depth Questions to Increase Your Understanding

Short-Answer/Essay Questions

36. *The Hubble Deep Fields.* Both the Hubble Deep Field in Figure 20.1 and the Hubble Ultra Deep Field shown at the beginning of Chapter 1 were chosen for observation in large part because they are completely ordinary parts of the sky. Why do you think astronomers would want to devote so much precious telescope time to observing totally ordinary regions of the sky in such great detail? Explain your reasoning.

37. *Supernovae in Other Galaxies.* In which type of galaxy would you be most likely to observe a massive star supernova, in a giant elliptical galaxy or in a large spiral galaxy? Explain your reasoning.

38. *Hubble's Galaxy Types.* How would you classify the following galaxies using the system illustrated in Figure 20.8? Justify your answers.
    a. Galaxy NGC 4594 (Figure 20.3)
    b. Galaxy NGC 6744 (Figure 20.2a)
    c. Galaxy NGC 4414 (Figure 20.2b)
    d. Galaxy NGC 1300 (Figure 20.4)
    e. Galaxy M 87 (Figure 20.6a)

39. *Distance Measurements.* The techniques astronomers use to measure distances are not so very different from the ones you use every day. Describe how standard candle measurements are similar to the way you estimate the distance to an oncoming car at night.

40. *Cepheids as Standard Candles.* Suppose you are observing Cepheids in a nearby galaxy. You observe one Cepheid with a period of 8 days between peaks in brightness, and another with a period of 35 days. Estimate the luminosity of each star. Explain how you arrived at your estimate. (*Hint:* See Figure 20.12.)

41. *Galaxies at Great Distances.* The most distant galaxies that astronomers have observed are much easier to see in infrared light than in visible light. Explain why that is the case.

42. *Universe on a Balloon.* In what ways is the surface of a balloon a good analogy for the universe? In what ways is this analogy limited? Explain why a miniature scientist living in a polka dot on the balloon would observe all other dots to be moving away, with more distant dots moving away faster.

43. *Light Paths in a Spacetime Diagram.* The path of the galaxy in Figure 20.23 has a steeper slope than the path of the photons emitted by the supernova in that galaxy. Explain why the path of the photons has a shallower slope than the path of the galaxy.

Quantitative Problems

Be sure to show all calculations clearly and state your final answers in complete sentences.

44. *Counting Galaxies.* Estimate how many galaxies are pictured in Figure 20.1. Explain the method you used to arrive at this estimate. This picture shows about $\frac{1}{30,000,000}$ of the sky, so multiply your estimate by 30,000,000 to obtain an estimate of how many galaxies like these fill the entire sky.

45. *Distances to Star Clusters.* The distance of the Hyades Cluster is known from parallax to be 151 light-years. Estimate the distance to the Pleiades Cluster using the information in Figure 20.11.

46. *Cepheids in M 100.* Scientists using the Hubble Space Telescope have observed Cepheids in the galaxy M 100. Here are the actual data for three Cepheids in M 100:
    - Cepheid 1: luminosity = 3.9×10^{30} watts, brightness 9.3×10^{-19} watt/m^2
    - Cepheid 2: luminosity = 1.2×10^{30} watts, brightness 3.8×10^{-19} watt/m^2
    - Cepheid 3: luminosity = 2.5×10^{30} watts, brightness 8.7×10^{-19} watt/m^2

    Compute the distance to M 100 with data from each of the three Cepheids. Do all three distance computations agree? Based on your results, estimate the uncertainty in the distance you have found.

47. *Supernovae as Standard Candles.* The peak luminosity of a white dwarf supernova is around $10^{10}L_{Sun}$, and it remains above 10^8L_{Sun} for about 150 days. In comparison, the luminosity of a bright Cepheid variable star is about $10,000L_{Sun}$. The Hubble Space Telescope is sensitive enough to make accurate measurements of apparent brightness for Cepheid variables at distances up to about 100 million light-years. Estimate the distance of a fading white dwarf supernova of luminosity 10^8L_{Sun} whose apparent brightness is comparable to that of a bright Cepheid variable star 100 million light-years from Earth. How does the distance of that supernova compare with the size of the observable universe?

48. *Redshift and Hubble's Law.* Imagine that you have obtained spectra for several galaxies and have measured the observed wavelength of a hydrogen emission line that has a rest wavelength of 656.3 nanometers. Here are your results:
    - Galaxy 1: Observed wavelength of hydrogen line is 659.6 nanometers.
    - Galaxy 2: Observed wavelength of hydrogen line is 664.7 nanometers.
    - Galaxy 3: Observed wavelength of hydrogen line is 679.2 nanometers.
    a. Calculate the redshift, z, for each of these galaxies.
    b. From their redshifts, calculate the speed at which each of the galaxies is moving away from us. Give your answers in both km/s and as a fraction of the speed of light.
    c. Estimate the distance to each galaxy from Hubble's law. Assume that $H_0 = 22$ km/s/Mly.

49. *Estimating the Universe's Age.* What would be your estimate of the age of the universe if you measured a value for Hubble's constant of $H_0 = 33$ km/s/Mly? You can assume that the expansion rate has remained unchanged during the history of the universe.

50. *Hubble's First Attempt.* Edwin Hubble's first attempt to measure the universe's expansion rate was flawed because the standard candles he was using were not properly calibrated. Look at Figure 20.20. Estimate the value of H_0 corresponding to the solid line in the figure. Remember that 10^6 parsecs = 3.26 million light-years. What is the approximate age of the universe indicated by that erroneous value of H_0?

51. *Extremely Distant Galaxies.* The most distant galaxies observed to date have a redshift of approximately $z = 7$. How does the wavelength of light we observe from those galaxies compare with its original wavelength when it was emitted?

52. *Stretching of the Universe.* The most distant white dwarf supernova observed as of 2006 had a redshift $z = 1.7$. How does the average distance between galaxies now compare with the average distance between galaxies at the time that supernova exploded?

53. *Lookback Time and the Ages of Galaxies.* Suppose you observe a distant galaxy with a lookback time of 10 billion years. What was the maximum possible age for that galaxy when the light we are now observing from it began its journey to Earth? Explain your reasoning. (*Hint:* Assume the galaxy was born less than a billion years after the Big Bang.)

Discussion Questions

54. *Cosmology and Philosophy.* One hundred years ago, many scientists believed that the universe was infinite and eternal, with no beginning and no end. When Einstein first developed his general theory of relativity, he found that it predicted that the universe should be either expanding or contracting. He believed so strongly in an eternal and unchanging universe that he modified the theory, a modification he was said to have called his "greatest blunder." Why do you think Einstein and others assumed that the universe had no beginning? Do you think that a universe with a definite beginning in time, some 14 billion or so years ago, has any important philosophical implications? Explain.

55. *Hubble's Revolution.* Just a century ago, astronomers were still not sure whether the Milky Way was the only large collection of stars in the universe. Then Edwin Hubble measured the distance of the Andromeda Galaxy, proving once and for all that our Milky Way was just one among many galaxies in the universe. How was this change in our "cosmic perspective" similar to the Copernican revolution? How was it different?

Interactive Tutorials

Tutorial Review of Key Concepts
Use the interactive Tutorials at **www.mastering astronomy.com** to review key concepts from this chapter.

Measuring Cosmic Distances Tutorial
Lesson 1 Radar
Lesson 2 Stellar Parallax
Lesson 3 Standard Candles: Main-Sequence Stars and Cepheid Variables
Lesson 4 Standard Candles: White Dwarf Supernovae and Spiral Galaxies

Hubble's Law Tutorial
Lesson 1 Hubble's Law
Lesson 2 The Expansion of the Universe
Lesson 3 The Age of the Universe

Supplementary Tutorial Exercises
Use the interactive Tutorial Lessons to explore the following questions.

Measuring Cosmic Distances Tutorial, Lessons 3–4
1. What makes it possible for a streetlight to appear much brighter than a star?
2. How can we measure the distance to a faraway galaxy in which we observe a white dwarf supernova?
3. How can we measure the distance to a faraway spiral galaxy in which we do *not* observe a supernova?

Hubble's Law Tutorial, Lesson 1
1. How do we determine the motion of distant galaxies toward or away from us?
2. How can we use Hubble's law to determine the distance to a faraway galaxy?

Exploring the Sky and Solar System

Of the many activities available on the *Voyager: SkyGazer* CD-ROM accompanying your book, use the following file to observe key phenomena covered in this chapter.

Go to the File: Basics folder for the following demonstration.
1. Galaxies in Coma

Movies

Check out the following narrated and animated short documentary available on **www.masteringastronomy.com** for a helpful review of key ideas covered in this chapter.

From the Big Bang to the Galaxies Movie

Web Projects

1. *Galaxy Gallery.* Many fine images of galaxies are available on the Web. Collect several images of each major type and build a galaxy gallery of your own. Supply a descriptive paragraph about each galaxy.
2. *Greatest Lookback Time.* Look for recent discoveries of objects with the largest lookback times (or redshifts). What is the current record for the most distant known object? What kind of object is it?

21

Galaxy Evolution

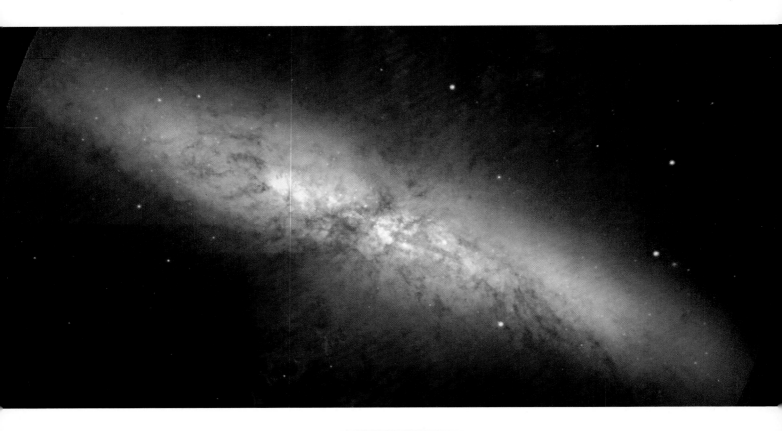

Reality provides us with facts so romantic that imagination itself could add nothing to them.

—Jules Verne

The spectacle of galaxies strewn like beautiful islands across the universe invites us to ponder their origins. Some galaxies have majestic spiral arms, some are elliptical, and some are irregular. What processes sculpted them into their current shapes?

If we look closely, we see even more spectacular sights. A few nearby galaxies are engaged in titanic collisions with other galaxies. Many of these colliding galaxies are full of star-forming clouds in which new stars are born and massive stars explode 100 times more frequently than in the Milky Way. Other galaxies appear to harbor enormous black holes surrounded by gigantic accretion disks that generate extraordinary luminosities. Narrow streams of matter jet from a few of these galaxies into intergalactic space at nearly the speed of light.

The origins of galaxies and the incredible phenomena they sometimes display puzzled astronomers for much of the twentieth century, but today we are assembling a rudimentary understanding of how galaxies evolve. In this chapter, we will sift through the fascinating clues that hint at how galaxies formed and developed, pausing now and again to admire the fantastic spectacles that we have uncovered in this search for our origins.

21.1 Looking Back Through Time

In Chapter 20, we learned how the distances to galaxies are measured and how such measurements revealed the expansion of our universe. We also saw how the expansion rate indicates an age for the universe of about 14 billion years. With that understanding we are now ready to turn our attention back to the study of how galaxies form and develop in our expanding universe—a subject known as **galaxy evolution**.

As we study how galaxies evolve, keep in mind that we would not be here if not for the galaxy-wide recycling processes that gradually transform the Milky Way's primordial gases into stars and planets [Section 19.2]. The same basic star-gas-star cycle that operates in our own galaxy governs the development of all galaxies, even though the outward appearances of galaxies can greatly differ. The differences we observe among galaxies turn out to provide many clues to how galaxies form and develop. Thus, if we want to understand our cosmic origins, we must learn both about how galaxies first formed and how they can end up looking so different from one another.

• How do we observe the life histories of galaxies?

The first set of clues we will consider comes from deep images of the universe, such as the Hubble Deep Field and the Hubble Ultra Deep Field (see Figure 20.1 and p. 1 of the book). Remember that we can use powerful telescopes like time machines to observe the life histories of galaxies—the farther we look into the universe, the further we can see back in time [Section 1.1]. When we look outward as far as we can, the most distant galaxies we observe have a look-back time of about 13 billion years, meaning that we are seeing them as they were when the universe was only about a billion years old. Because we can see starlight from those galaxies, they must already have had some stars in place about 13 billion years ago, about the same time that the oldest stars in our own galaxy formed. It therefore seems safe to assume that many or most galaxies began to form at about this time, in which case most galaxies would be roughly the same age today.

Applying this linkage between a galaxy's distance and its age more generally, we can assemble entire "family albums" of galaxies. Simply by photographing galaxies at different distances, we can observe galaxies in different stages of development. Pictures of the farthest galaxies show galaxies in their childhood, and pictures of the nearest show mature galaxies as they are today. Figure 21.1 shows partial family albums for elliptical, spiral, and irregular galaxies. Each individual photograph shows a single galaxy at a single stage in its life, and measuring the redshift of each galaxy allows us to place it on a time line of the universe. Grouping these photographs by galaxy type then allows us to see how galaxies of a particular type have changed through time.

In the preceding paragraph, we used the term *today* in a very broad sense. For example, if we look at a relatively nearby galaxy—one located, say, 20 million light-years away—we see it as it was 20 million years ago. In what sense is this "today"? (*Hint:* How does 20 million years compare to the age of the universe? To the lifetime of a massive star?)

The photographs in Figure 21.1, taken by the Hubble Space Telescope, show galaxies extending back to a time when the universe was just 1 or 2 billion years old. However, we suspect that the first stars and galaxies formed even earlier than this. Observing these first stars and galaxies is a challenge not even Hubble can meet. Detecting such far-away galaxies will require larger telescopes, and the extreme redshifts we expect for such galaxies mean the telescopes will need to be particularly sensitive to infrared light. Within a decade or so, NASA hopes to launch a much larger infrared-sensitive successor to the Hubble Space Telescope (called the James Webb Space Telescope), but for now we do not have much direct information about galaxy birth.

• How did galaxies form?

We have little hope of seeing galaxies at a time before their first stars formed, because even if gas was already collecting into galaxies, there was no starlight to illuminate this gas. Thus, if we hope to "look" back to the time when the first

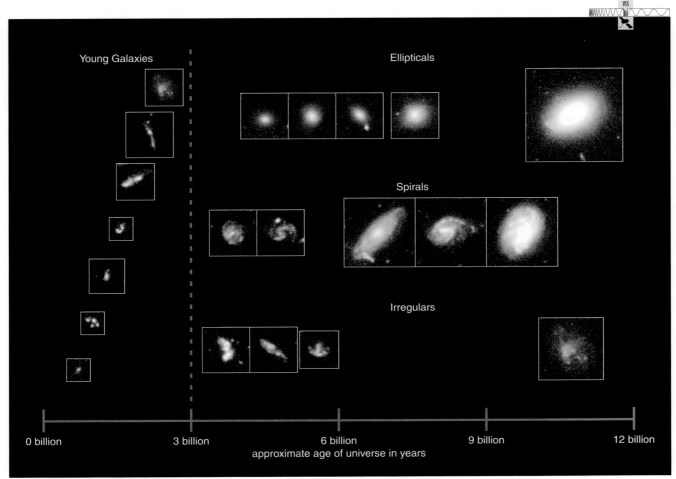

Figure 21.1 Family albums for elliptical, spiral, and irregular galaxies of different ages, plus some very young galaxies shown on the far left. These photos are all zoomed-in images of galaxies shown in Figure 20.1. We see more distant galaxies as they were when they were younger, indicated by the approximate age of the universe along the horizontal axis. The younger galaxies appear smaller because they are more distant. (The time scale runs to 12 billion years, rather than to the current age of the universe of 14 billion years, because the nearest galaxies in Figure 21.1 are about 2 billion light-years away.)

galaxies began to form, we must use theoretical modeling. The most successful models for galaxy formation assume the following:

- Hydrogen and helium gas filled all of space more or less uniformly when the universe was very young—say, in the first million years after its birth.
- The distribution of matter in the universe was not perfectly uniform—certain regions of the universe started out ever so slightly denser than others.

Beginning from these assumptions, which a mounting body of evidence supports [Section 23.2], we can model galaxy formation using well-established laws of physics to trace how the denser regions in the early universe grew into galaxies (Figure 21.2). The models show that the regions of enhanced density originally expanded along with the rest of the universe. However, the slightly greater pull of gravity in these regions gradually slowed their expansion. Within about a billion years, the expansion of these denser regions halted and reversed, and the material within them began to contract into *protogalactic clouds* like the cloud of matter

that eventually formed our Milky Way [Section 19.3]. Just as we suspect for the Milky Way, many galaxies may have formed from the merger of multiple protogalactic clouds (see Figure 19.20).

According to the models, the protogalactic clouds that eventually formed spiral galaxies initially cooled as they contracted, radiating away their thermal energy, and the first generation of stars grew from the densest, coldest clumps of gas. These first-generation stars were probably quite massive [Section 16.1], living and dying within just a few million years—a short time compared to the time required for the collapse of a protogalactic cloud into a mature galaxy. The supernovae of these and later generations of massive stars seeded the galaxy with its first smattering of heavy elements, generating shock waves that heated the surrounding interstellar gas. This heating slowed the collapse of young galaxies and the rate at which stars formed within them, allowing time for the rest of the gas to settle into a rotating disk.

This picture explains many of the basic features of galaxies. In particular, it explains why other spiral galaxies

642   part VI   •   Galaxies and Beyond

Early in time, the gas in this cubic region of the universe is almost uniformly distributed.

Gravity draws gas into the denser regions of space as time passes.

time

Protogalactic clouds form in the densest regions and go on to become galaxies.

Figure 21.2 🖥 Interactive Figure. A computer simulation of the formation of protogalactic clouds. The simulated region of space is about 500 million light-years wide and will go on to form numerous galaxies.

have the same basic structure as the Milky Way: a *disk population* of stars that orbit the galactic center in a fairly flat plane, and a *spheroidal population* of stars with more randomly oriented orbits (see Figure 19.2). The spheroidal population consists of stars that were born before the galaxy's rotation became organized, which is why they have orbits oriented in many different ways around the galactic center. The disk population consists of stars born after the gas in the protogalactic cloud settled into a rotating disk, which is why they all have similar orbits around the center of the galaxy (see Figure 19.19).

However, this basic picture leaves at least two major questions unanswered. First, our models assume that galaxies formed in regions of slightly enhanced density, but they do not tell us where these density enhancements came from. The nature and origin of density enhancements in the early universe is one of the major puzzles in astronomy, and we'll revisit it in Chapter 23. Second, our basic picture explains the origin of spiral galaxies quite well, but it does not tell us why some galaxies are elliptical and others irregular. That is the question to which we turn our attention next.

21.2 The Lives of Galaxies

The distinct differences between spiral, elliptical, and irregular galaxies clearly tell us that their life stories must be quite different. We would like to tell the life story of each different type of galaxy from beginning to end as completely as we told the life stories of stars, but too many aspects of galaxy evolution remain mysterious. Because so many important questions remain unanswered, the study of galaxy evolution is one of the most active research areas in astronomy. As a result, we now know the general outline of a gal-

axy's life story, and we have some promising ideas about why galaxies differ from one another. In this section we'll examine some of those ideas and show how they fit into the overall picture of galaxy evolution, even though the picture itself is not yet complete.

• Why do galaxies differ?

Our best models for galaxy formation suggest that all galaxies began their lives in the same basic way, with gravity pulling matter into a patch of the universe that was slightly denser than its surroundings. That patch then contracted into a protogalactic cloud and began to form stars. From that point onward, all galaxies obeyed the same laws of physics but somehow ended up looking different. Our models suggest that those differences might have arisen in two distinct ways: First, galaxies may have ended up looking different because they began with slightly different conditions in their protogalactic clouds; alternatively, galaxies may have begun their lives quite similarly but changed at later times due to interactions with other galaxies. It now seems likely that both factors have played a role in explaining the differences we see among galaxies. To help differentiate the role of conditions in the protogalactic cloud from the role of subsequent interactions, we will explore a single key question: Why do spiral galaxies have gas-rich disks, while elliptical galaxies do not?

Conditions in the Protogalactic Cloud Two plausible explanations for the differences between spiral galaxies and elliptical galaxies trace a galaxy's type back to the protogalactic cloud from which it formed:

- *Protogalactic spin* (Figure 21.3a). A galaxy's type might be determined by the spin of the protogalactic cloud from which it formed. If the original cloud had a significant amount of angular momentum, it would have rotated quickly as it collapsed. The galaxy it produced would therefore have tended to form a disk, and the resulting galaxy would be a spiral galaxy. If the protogalactic cloud had little or no angular momentum, its gas might not have formed a disk at all, and the resulting galaxy would be elliptical.
- *Protogalactic density* (Figure 21.3b). A galaxy's type might be determined by the density of the protogalactic cloud from which it formed. A protogalactic cloud with relatively high gas density would have radiated energy more effectively and cooled more quickly, thereby allowing more rapid star formation. If the star formation proceeded fast enough, all the gas could have been turned into stars before any of it had time to settle into a disk. The resulting galaxy would therefore lack a disk, making it an elliptical galaxy. In contrast, a lower-density cloud would have formed stars more slowly, leaving plenty of gas to form the disk of a spiral galaxy.

Evidence for the latter scenario comes from a few giant elliptical galaxies at very great distances. These galaxies look very red even after we have accounted for their large

A protogalactic cloud that starts with little or no angular momentum . . .

. . . may end up as an elliptical galaxy.

A protogalactic cloud with more angular momentum spins faster as it contracts . . .

. . . ending up as a spiral galaxy.

a The angular momentum of a protogalactic cloud may determine whether it ends up spiral or elliptical.

A dense protogalactic cloud may form its stars before it settles into a disk . . .

. . . and ends up as an elliptical galaxy.

Gas in a less dense protogalactic cloud forms stars more slowly and settles into a disk . . .

. . . ending up as a spiral galaxy.

b The density of a protogalactic cloud may determine whether it ends up spiral or elliptical.

Figure 21.3 🎨 Interactive Figure These diagrams show two ways in which a galaxy's birth conditions may have determined whether it ended up spiral or elliptical.

redshifts (Figure 21.4). They apparently have no blue or white stars at all, indicating that new stars no longer form within these galaxies—even though we are seeing them as they were when the universe was only a few billion years old. This finding suggests that all the stars in these elliptical galaxies formed almost simultaneously and very early in the history of the universe, which is consistent with the idea that all the stars formed before a disk could develop.

Galactic Collisions The two preceding scenarios, in which the formation of a gas-rich disk depends on the angular momentum or density of the protogalactic cloud, probably describe important parts of the overall story. However, they ignore one key fact: Galaxies rarely evolve in perfect isolation.

Think back to our scale-model solar system in Chapter 1. Using a scale on which the Sun was the size of a grapefruit, the nearest star was like another grapefruit a few thousand kilometers away. Because the average distances between stars are so huge compared to the sizes of stars, collisions between stars are extremely rare. However, if we rescale the universe so that our *galaxy* is the size of a grapefruit, the Andromeda Galaxy is like another grapefruit only about 3 meters away, and a few smaller galaxies lie considerably closer. Thus, the average distances between galaxies are not much larger than the sizes of galaxies, and collisions between

galaxies are inevitable. Even the Milky Way Galaxy is not immune. About 80,000 light-years away from us, directly behind the galactic bulge, a small elliptical galaxy (the

Figure 21.4 The light we observe from the distant elliptical galaxy called LBDS53W091 (arrow) left that galaxy when the universe was about 4 billion years old. The redness of its light at this age indicates that all its stars formed during the first billion years or so after the birth of the universe.

This collision between two spiral galaxies stripped out long tidal tails of stars . . .

. . . and triggered a burst of star formation, producing many young blue star clusters.

a A pair of colliding spiral galaxies known as the Antennae (NGC 4038/4039). The image from the ground (left) reveals their vast tidal tails, and the close-up from the Hubble Space Telescope shows the fine details of the burst of star formation at the center of the collision.

Figure 21.5 Galaxy collisions.

Streamers of stars could be old tidal tails from a previous collision.

b Galaxy NGC 6240. The disturbed appearance of this galaxy suggests that it has undergone a collision much like the one now ongoing in the Antennae, but we are seeing it at a later stage when the two galaxies have almost merged into one.

Sagittarius Dwarf [Section 19.1]) is currently crashing through the Milky Way's disk.

Collisions between galaxies are spectacular events that unfold over hundreds of millions of years (Figure 21.5). In our short lifetimes, we can at best see a snapshot of a collision in progress, distorting the shapes of the colliding galaxies. Galactic collisions must have been even more frequent in the distant past, when the universe was smaller and galaxies were even closer together. Observations confirm that distorted-looking galaxies—probably galaxy collisions in progress—were more common in the early universe than they are today (Figure 21.6).

We can learn much more about galactic collisions with the aid of computer simulations that allow us to "watch" collisions that in nature take hundreds of millions of years to unfold. These computer models show that a collision between two spiral galaxies can create an elliptical galaxy (Figure 21.7). Tremendous tidal forces between the colliding

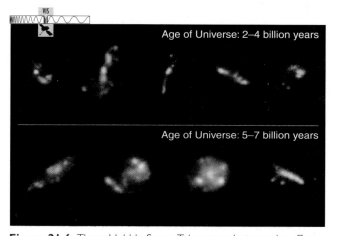

Age of Universe: 2–4 billion years

Age of Universe: 5–7 billion years

Figure 21.6 These Hubble Space Telescope photographs offer a zoomed-in view of some of the young galaxies shown in Figure 21.1. Notice that these galaxies do not look like either the spiral or elliptical galaxies that are common in the present-day universe, and instead appear like galaxies undergoing collisions. We therefore infer that galaxy collisions were much more common in the early universe than they are today.

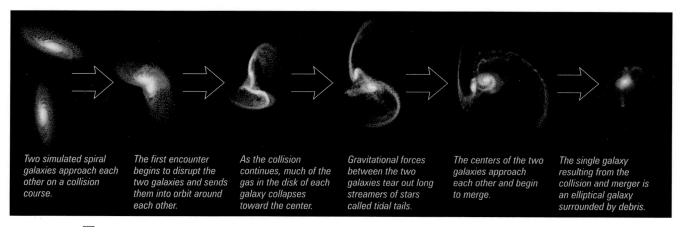

Two simulated spiral galaxies approach each other on a collision course.

The first encounter begins to disrupt the two galaxies and sends them into orbit around each other.

As the collision continues, much of the gas in the disk of each galaxy collapses toward the center.

Gravitational forces between the two galaxies tear out long streamers of stars called tidal tails.

The centers of the two galaxies approach each other and begin to merge.

The single galaxy resulting from the collision and merger is an elliptical galaxy surrounded by debris.

Figure 21.7 Several stages in a supercomputer simulation of a collision between two spiral galaxies that results in an elliptical galaxy. At least some of the elliptical galaxies in the present-day universe formed in this way. The whole sequence spans about 1.5 billion years.

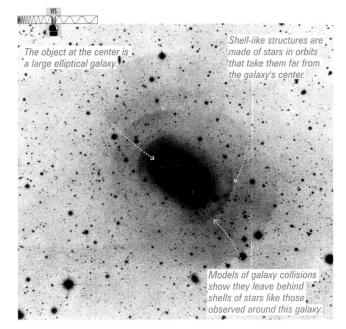

The object at the center is a large elliptical galaxy.

Shell-like structures are made of stars in orbits that take them far from the galaxy's center.

Models of galaxy collisions show they leave behind shells of stars like those observed around this galaxy.

a The central region of elliptical galaxy NGC 3923 is surrounded by several distinct shells of stars. These stars probably formed after gas was stripped out of the galaxy during a past collision. The image is a negative in which starlight is black and space is white because it makes it easier to see faint details.

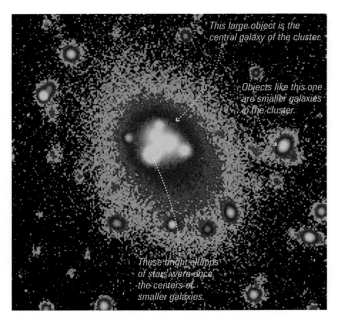

This large object is the central galaxy of the cluster.

Objects like this one are smaller galaxies in the cluster.

These bright clumps of stars were once the centers of smaller galaxies.

b This image shows the central dominant galaxy of the cluster Abell 3827, which has apparently grown by consuming smaller galaxies that have collided with it. Notice the multiple clumps of stars that probably once were the centers of individual galaxies. The colors in this image are not real but instead represent levels of brightness, because it makes the brightest clumps (yellow) easier to see.

Figure 21.8 Evidence for past collisions in elliptical galaxies.

galaxies tear apart the two disks, randomizing the orbits of their stars. Meanwhile, a large fraction of their gas sinks to the center of the collision and rapidly forms new stars. Supernovae and stellar winds eventually blow away the rest of the gas. When the cataclysm finally settles down, the merger of the two spirals has produced a single elliptical galaxy. Little gas is left for a disk, and the orbits of the stars have random orientations.

Galaxies in Clusters Observations of galaxies in clusters support the idea that at least some elliptical galaxies result from collisions and subsequent mergers. Elliptical galaxies dominate the galaxy populations at the cores of dense clusters of galaxies, where collisions should be most frequent. This fact may mean that any spirals once present became ellipticals through collisions.

Stronger evidence comes from structural details of elliptical galaxies, which often attest to a violent past. Some elliptical galaxies have stars and gas clouds with orbits suggesting that they are leftover pieces of galaxies that merged in a past collision. One such example is the elliptical galaxy NGC 3923 shown in Figure 21.8a. Although it is not a member of a cluster, this galaxy's unusual structure is probably the result of a collision. The shells surrounding this galaxy are made of stars on orbits that are hard to explain if the entire galaxy came from a single protogalactic cloud. The stars that we see in the shells probably plunge back and forth through the central part of the galaxy, swinging from one side to the other like a pendulum. The shells represent the extreme ends of these plunging orbits, where the stars spend most of their time.

The most decisive evidence that collisions affect the evolution of elliptical galaxies comes from observations of the **central dominant galaxies** found at the centers of many dense clusters. Central dominant galaxies are gigantic elliptical galaxies that apparently grew to a huge size by consuming other galaxies through collisions. They frequently contain several tightly bound clumps of stars, each of which was probably the center of an individual galaxy that was swallowed by the giant (Figure 21.8b). This process of *galactic cannibalism* can create central dominant galaxies more than 10 times as massive as the Milky Way, making them the largest galaxies in the universe.

Cluster observations also suggest yet another mechanism by which spiral galaxies might become ellipticals. The central regions of dense galaxy clusters tend to be filled with very hot gas [Section 22.2]. When a spiral galaxy cruises through the center of such a cluster, the hot gas exerts drag forces that slow the galaxy's gas but not its stars. Thus, the galaxy's stars continue to move freely along their way, but the gas is left behind. If the disk has not yet formed many stars, then the galaxy will evolve to look more like an elliptical galaxy as its massive stars die away. Its disk will fade, while its bulge and halo will remain prominent. If the disk has already formed a large number of stars when its gas is stripped, the remaining galaxy will look like a spiral galaxy without its gaseous disk, which makes it what we call a *lenticular galaxy* (see Figure 20.5).

An Incomplete Answer We have discussed several mechanisms that can determine a galaxy's type. It is possible that its type is determined at birth, through the conditions of the galaxy's protogalactic cloud. If the protogalactic cloud began with either unusually slow rotation or unusually high density, the result may be an elliptical galaxy; otherwise, it will be a spiral galaxy. However, interactions with other galaxies can change a galaxy's type later in life. Both observations and models suggest that collisions can turn spiral galaxies into elliptical galaxies. Spiral galaxies may also have their cool gaseous disks stripped out as they pass through the hot gas in the central regions of galaxy clusters.

Birth conditions and subsequent interactions probably both play important roles in galaxy evolution, but we are not yet certain which is more influential. Nevertheless, when we consider both kinds of mechanisms together, they do seem to account for the basic differences between galaxy types. The formation scenarios explain why the vast majority of galaxies are either spiral or elliptical in shape. The interaction scenarios explain why ellipticals are more common in clusters while spirals are more common outside clusters, and gas stripping may explain why we see significant numbers of lenticular galaxies. Even the relatively small fraction of galaxies that are irregular may be explained by these ideas: At least some irregulars probably are galaxies undergoing some sort of disruptive interaction.

What are starbursts?

Our discussion of galaxies' lives has so far made it seem like all spiral galaxies form their stars steadily and slowly and that all the stars in elliptical galaxies formed in the distant past. However, observations show that a small percentage of the galaxies in the present-day universe are currently forming stars at a prodigious rate. We call these galaxies **starburst galaxies**, and they represent a stage of evolution that many galaxies may have gone through at least once during their lives.

A starburst must be a temporary stage in a galaxy's life because the rates of star formation in starburst galaxies are unsustainable. For example, the Milky Way Galaxy produces an average of about one new star per year. At this rate, the Milky Way won't exhaust the interstellar gas in its disk until long after the Sun has died. In contrast, some starburst galaxies form new stars at rates exceeding 100 stars per year, more than 100 times the star formation rate in the Milky Way. At their current rates of star formation, those starburst galaxies would consume *all* their interstellar gas in just a few hundred million years. After its starburst is over, a starburst galaxy presumably returns to a spiral, elliptical, or irregular state. Let's look a little closer at this spectacular but temporary stage of galaxy evolution.

Observations of Starburst Galaxies   The impressive rates of star formation in starburst galaxies were generally recognized only about two decades ago. These voracious consumers of interstellar gas look peculiar at visible wavelengths because they are filled with star-forming molecular clouds, but these clouds conceal much of the action. Dust grains in the molecular clouds absorb most of the visible and ultraviolet radiation streaming from the many hot, young stars of a starburst galaxy.

Because we cannot see the active star formation in starburst galaxies with visible light, astronomers didn't recognize their nature until they began to study them in long-wavelength infrared light—light that can be observed only with telescopes in space. These galaxies emit strongly in the infrared because of their interstellar dust. The visible and ultraviolet radiation from the galaxy's many hot, young stars heats its dust grains to higher temperatures than we find for dust in the Milky Way, and the dust ultimately reemits all this absorbed energy as infrared light. Figure 21.9 shows an infrared image of an especially luminous starburst galaxy, along with its spectrum from infrared through visible wavelengths. Note that the visible output of this galaxy is about 10 billion solar luminosities ($10^{10} L_{Sun}$), not very different from the total luminosity of the Milky Way. However, its infrared output is a trillion times that of our Sun ($10^{12} L_{Sun}$), making it 100 times brighter in infrared light than in visible light.

Galactic Winds   A star formation rate 100 times that of the Milky Way also means that supernovae will occur at 100 times the Milky Way's rate. Just as in the Milky Way, each supernova in a starburst galaxy generates a shock wave that creates a *bubble* of hot gas [Section 19.2]. The shock waves from several nearby supernovae quickly overlap and blend into a much larger *superbubble*. In the Milky Way, the story usually ends here, but in a starburst galaxy the drama is just beginning. Supernovae continue to explode inside the superbubble, adding to its thermal and kinetic energy. When the superbubble starts to break through the disrupted gaseous disk, it expands even faster. Hot gas erupts into intergalactic space, creating a **galactic wind**.

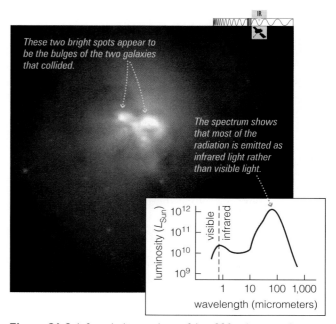

These two bright spots appear to be the bulges of the two galaxies that collided.

The spectrum shows that most of the radiation is emitted as infrared light rather than visible light.

Figure 21.9 Infrared observations of Arp 220, a large starburst galaxy. The two central bulges suggest that the starburst has been caused by a collision between two galaxies that are now merging together. The inset spectrum shows that the galaxy emits most of its energy as infrared light rather than visible light. (The region shown in the photo is about 10,000 light-years across.)

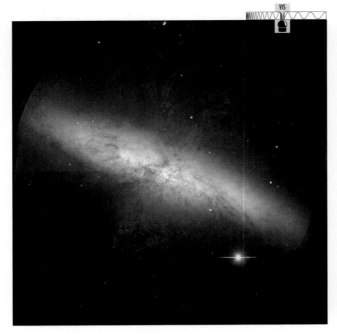

a This visible-light photograph (from the Subaru telescope) shows violently disturbed gas (red) blowing out both above and below the disk.

Figure 21.10 Visible-light and X-ray views of a starburst galaxy called M 82, showing its galactic wind. Both images show the same region, which is about 16,000 light-years across.

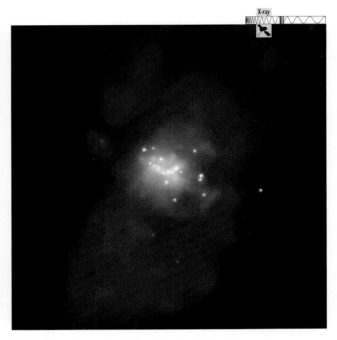

b This X-ray image from the Chandra X-Ray Observatory shows the same region as the visible-light photograph in (a). The reddish region represents X-ray emission from very hot gas blowing out of the disk. The bright dots in the galactic disk probably represent X-ray emission from accretion disks around black holes or neutron stars produced by recent supernovae.

Galactic winds consist of low-density but extremely hot gas, typically with temperatures of 10–100 million Kelvin (Figure 21.10). They do not emit much visible light, but they do generate X rays. X-ray telescopes in orbit have detected pockets of X-ray emission surrounding the disks of some starburst galaxies, presumably coming from the outflowing galactic wind. Sometimes we also see the glowing remnants of a punctured superbubble extending out into space.

Supernova-driven galactic winds have an even more dramatic impact when they occur in small starburst galaxies (Figure 21.11). The winds can blow out of small galaxies on all sides, driving away much of their gas. As a result, star formation in these galaxies may shut down for billions of years. Some of the small elliptical galaxies in the Local Group apparently have burst like this at least twice during their lifetimes. In one example, about half the stars in the small galaxy were born more than about 12 billion years ago, and the rest only 5–8 billion years ago. Presumably, each of these bursts of star formation created enough supernovae to eject nearly all the gas that remained in the galaxy. Star formation was then put on hold for several billion years until enough gas could reaccumulate within the galaxy for a new starburst to ignite.

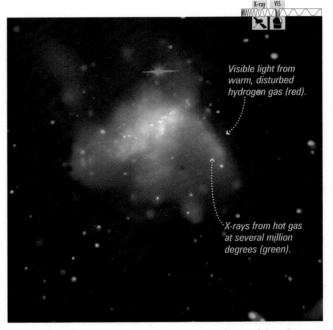

Visible light from warm, disturbed hydrogen gas (red).

X-rays from hot gas at several million degrees (green).

Figure 21.11 This X-ray/visible composite photo of dwarf starburst galaxy NGC 1569 shows hot gas blowing out in several different directions. X rays from escaping gas at a temperature of several million degrees are shown in green, and visible light from disturbed hydrogen gas is shown in red.

THINK ABOUT IT

Dwarf galaxies that have undergone bursts of star formation tend to have fewer heavy elements than large galaxies. Why do you think that is? (*Hint:* What happens to the heavy elements produced by the burst of star formation?)

Causes of Starbursts Many of the most luminous starburst galaxies appear to be violently disturbed, suggesting that a collision between galaxies triggered the starburst. For example, the colliding galaxy pair shown in Figure 21.5a is currently undergoing a starburst. The picture clearly shows many young, blue star clusters strewn amidst the dark, disturbed gas clouds. Thus, starbursts help explain why elliptical galaxies lack young stars and cool gas. The starburst uses up most of the cool gas; the galactic wind blows away what remains; and all the hot, massive stars die out within just a few hundred million years after the starburst ends. By the time the collision ends and the merger into an elliptical galaxy is complete, there is simply no cool gas left to support ongoing star formation.

The causes of smaller-scale starbursts are not yet clear. At least some small irregular galaxies look irregular because they are currently undergoing collisions and starbursts, but not all irregular galaxies are colliding. For example, the Large Magellanic Cloud (which orbits the Milky Way) is an irregular galaxy that is also undergoing a period of rapid star formation. The starburst leading to this galaxy's irregular appearance might have been triggered not by a collision but rather by a close encounter with the Milky Way. No matter what their exact causes turn out to be, starbursts clearly represent an important piece in the overall puzzle of galaxy evolution.

 Black Holes Tutorial, Lessons 1, 2

21.3 Quasars and Other Active Galactic Nuclei

Starbursts may be spectacular, but some galaxies display even more incredible phenomena: extreme amounts of radiation and sometimes powerful jets of material emanating from deep in their centers (Figure 21.12). These unusually bright galactic centers are usually called **active galactic nuclei**. The very brightest active galactic nuclei are known as **quasars**, and they are fantastically luminous. The most powerful quasars produce more light than 1,000 galaxies the size of the Milky Way.

Like starbursts, quasars are yet another temporary stage in the process of galaxy evolution. Because we find quasars primarily at great distances, we know that these blazingly luminous objects were most common billions of years ago, when galaxies were in their youth. Within a few hundred million light-years of the Milky Way, we find no quasars and relatively few nearby galaxies with any type of active galactic nucleus. Apparently, the objects that shine as quasars in young galaxies must become dormant as the galaxies age. Many nearby galaxies that now look quite normal must therefore have centers that once shone brilliantly as quasars. We do not yet know exactly how quasars tie in with the overall story of galaxy evolution, but mounting evidence suggests that the development of quasars is intimately connected with the growth of galaxies.

• What are quasars?

What could possibly be the source of the incredible luminosities of quasars, and why did quasars fade away? A growing body of evidence points to a single answer: The energy output of quasars comes from gigantic accretion disks surrounding **supermassive black holes**—black holes with masses millions to billions of times that of our Sun. Before we study how these incredible powerhouses work, let's investigate the evidence that points to their existence.

The Discovery of Quasars   In the early 1960s, a young professor at the California Institute of Technology named Maarten Schmidt was busy identifying cosmic sources of radio-wave emission. Radio astronomers would tell him the coordinates of newly discovered radio sources, and he would try to match them with objects seen through visible-light telescopes. Usually the radio sources turned out to be normal-looking galaxies, but one day Schmidt discovered a major mystery: A radio source called 3C 273 looked like a blue star through a telescope but had strong emission lines at wavelengths that did not appear to correspond to any known chemical element. (The designation 3C 273 stands for 3rd Cambridge Radio Catalogue, object 273.)

After months of puzzlement, Schmidt suddenly realized that the emission lines were not coming from an unfamiliar element at all. Instead, they were emission lines of hydrogen that were hugely redshifted from their normal wavelengths (Figure 21.13). Schmidt calculated that the expansion of the universe was carrying 3C 273 away from us at 17% of the speed of light.

Schmidt computed the distance to 3C 273 using Hubble's law. Then he used the distance and apparent brightness of 3C 273 to estimate its luminosity from the inverse square law for light [Section 15.1]. What he found was astonishing: 3C 273 has a luminosity of about 10^{39} watts, or well

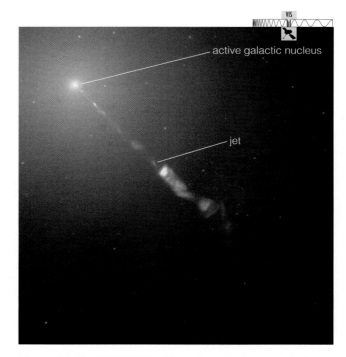

Figure 21.12 Interactive Photo   The active galactic nucleus in the elliptical galaxy M 87. The bright yellow spot is the active nucleus, and the blue streak is a jet of particles shooting outward from the nucleus at nearly the speed of light.

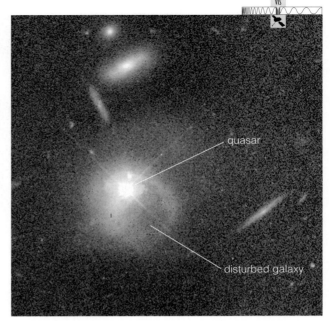

Figure 21.13 The top photograph shows the spectrum of the quasar 3C 273, and the bottom shows a comparison spectrum with spectral lines at their rest wavelengths. The lines labeled Hβ, Hγ, and Hδ are hydrogen emission lines. Note their significant redshift in the quasar spectrum relative to the comparison spectrum.

Figure 21.14 Photographs like this one help confirm that quasars are extremely far away. The quasar appears to be located in the middle of a distant galaxy. Moreover, the quasar, the rest of the galaxy that surrounds it, and the other galaxies in the photo all have the same redshift, proving that the quasar is truly distant and is an extremely bright active galactic nucleus.

over a trillion (10^{12}) times that of our Sun—making it hundreds of times more luminous than the entire Milky Way Galaxy. Discoveries of similar but even more distant objects soon followed. Because the first few of these objects were strong sources of radio emission that looked like stars through visible-light telescopes, they were named "quasi-stellar radio sources," or *quasars* for short. Later, astronomers learned that most quasars are not such powerful radio emitters, but the name has stuck.

For many years, a debate raged among astronomers over whether Hubble's law could really be used to determine quasar distances. Some argued that quasars might have high redshifts for reasons other than distance and therefore might be much nearer to us than Hubble's law would suggest. The vast majority of astronomers now consider this debate settled. Improved images show that quasars are indeed the centers of extremely distant galaxies and often are members of very distant galaxy clusters (Figure 21.14).

Most quasars lie more than halfway to the cosmological horizon. The lines in typical quasar spectra are shifted to more than three times their rest wavelengths, which tells us that the light from these quasars emerged when the universe was less than a third of its present age. The light we see from the farthest known quasars began its journey when the universe was less than 1 billion years old.

The extraordinary power output of quasars explains why they were at first so mysterious. They generally emit so much light that they swamp the light of the galaxies that contain

them, making the surrounding galaxies hard to detect. That is why quasars look "quasi-stellar." Moreover, quasars emit their energy across an unusually wide swath of the electromagnetic spectrum. Quasars radiate approximately equal amounts of power from infrared wavelengths all the way through to gamma rays (Figure 21.15). They also produce strong emission lines. By comparison, most stars and galaxies emit primarily visible light. The wide spread of photon energies coming from quasars implies that they contain matter with a wide range of temperatures. These facts give us some hints about the power source of quasars, but we need more data to pin down the source.

Evidence from Nearby Active Galactic Nuclei Quasars are difficult to study in detail because they are so far away. Luckily, some quasarlike objects are much closer to home. About 1% of present-day galaxies—that is, galaxies we see nearby—have active galactic nuclei that look very much like quasars,

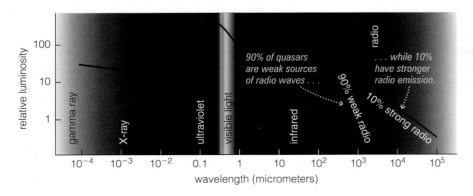

Figure 21.15 This schematic quasar spectrum represents the average emission spectrum of many quasars. (The dashed portion of the spectrum represents wavelengths for which we lack good data.)

RAD

Figure 21.16 This image, made with the Very Large Array in New Mexico, shows radio-wave emission from the radio galaxy Cygnus A. Brighter regions represent stronger radio emission. Notice that the strongest emission comes from two radio lobes that lie far beyond the bounds of the galaxy that we see with visible light (inset photo), but the lobes are clearly connected to the central active galactic nucleus by two long jets of particles.

except that they are less powerful.* These objects have spectra that look much the same as quasar spectra, with strong emission lines and energy radiated from infrared to gamma rays, suggesting that the same type of phenomenon is producing their spectra. However, because these active galactic nuclei are less luminous than quasars, it is easier to see the galaxies that surround them and hence easier to learn about their nature.

The light-emitting regions of active galactic nuclei are so small that even the sharpest images do not resolve them. Our best visible-light images show only that active galactic nuclei must be smaller than 100 light-years across. Radio-wave images made with the aid of interferometry [Section 6.4] show that these nuclei are even smaller: less than 3 light-years across. Rapid changes in the luminosities of some active galactic nuclei point to an even smaller size.

To understand how variations in luminosity give us clues about an object's size, imagine that you are a master of the universe and you want to signal one of your fellow masters a billion light-years away. An active galactic nucleus would make an excellent signal beacon, because it is so bright. However, suppose the smallest nucleus you can find is 1 light-year across. Each time you flash it on, the photons from the front end of the source reach your fellow master a full year before the photons from the back end. Thus, if you flash it on and off more than once a year, your signal will be smeared out. Similarly, if you find a source that is 1 light-day across, you can transmit signals that flash on and off no more than once a day. If you want to send signals just a few hours apart, you need a source no more than a few light-hours across.

Occasionally, the luminosity of an active galactic nucleus doubles in a matter of hours. The fact that we see a clear signal indicates that the source must be less than a few light-hours across. In other words, *the incredible luminosities of active galactic nuclei and quasars are apparently being generated in a volume of space not much bigger than our solar system.*

Radio Galaxies and Jets Another clue to the nature of quasars came in the early 1950s, a decade before the discovery of

*The galaxies that contain these active galactic nuclei are often called *Seyfert galaxies* after astronomer Carl Seyfert, who in 1943 grouped galaxies with active galactic nuclei into a special class.

quasars. Radio astronomers noticed that certain galaxies, now called **radio galaxies**, emit unusually strong radio waves. Upon closer inspection, they learned that much of the radio emission comes not from the galaxies themselves but rather from pairs of huge *radio lobes,* one on either side of the galaxy. Today, radio telescopes resolve the structure of radio galaxies in vivid detail (Figure 21.16).

At the center of a radio galaxy we see emission from an active galactic nucleus less than a few light-years across. We often see two gigantic jets of plasma shooting out of the active galactic nucleus in opposite directions; both jets are easy to see in Figure 21.16, but some radio galaxies are oriented so that we can see only the jet tilted in our direction. Using time-lapse radio images taken several years apart, we can track the motions of various plasma blobs in the jets. Some of these blobs move at close to the speed of light. The lobes lie at the ends of the jets, which are sometimes as far as a million light-years from the galactic center.

Putting all this information together gives us a basic picture of what occurs in a radio galaxy: The active galactic nucleus is the power source, and it drives two jets of particles that stream outward in opposite directions at nearly the speed of light. These jets shoot out far beyond the bounds of the stars in the radio galaxy, but they eventually ram into surrounding intergalactic gas. The places where the jets ram into the gas show up as the hot spots within radio lobes. The particles are then deflected from these hot spots to make the larger radio lobes, much like the spray of water from a fire hose is deflected when it hits a wall.

The relative prominence of the active galactic nucleus, jets, and lobes can vary greatly from one radio galaxy to another, largely because of differences in the luminosity of the nucleus and the densities of particles in the jets and the surrounding intergalactic gas. The shapes of the jets and lobes can also vary, especially if the galaxy is moving relative to surrounding intergalactic gas. Figure 21.17 shows a small gallery of typical radio galaxies.

As a result of these observations, we now suspect that quasars and radio galaxies are the same types of objects viewed in slightly different ways. We have discovered that many quasars have jets and radio lobes like those seen in radio galaxies (Figure 21.18). Moreover, the active galactic nuclei of many radio galaxies seem to be concealed beneath

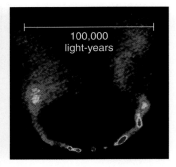

a Radio galaxy 3C 353.

b Radio galaxy 3C 31.

c Radio galaxy NGC 1265. The lobes are swept back because the galaxy is moving relative to the surrounding intergalactic gas.

Figure 21.17 Radio galaxy gallery. Note the double-lobe structure in all cases, but with very different shapes and sizes.

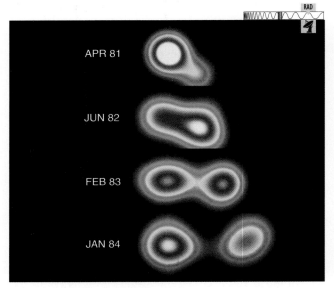

Figure 21.18 These radio images, taken over a period of several years, show a blob of plasma moving at almost the speed of light in a jet extending from the quasar 3C 345; the quasar is at the far left in each image.

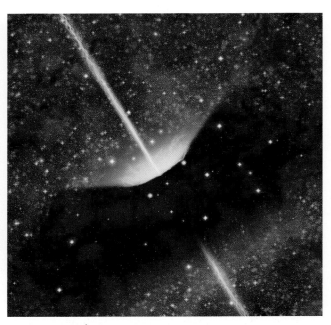

Figure 21.19 Artist's conception of the central few hundred light-years of a radio galaxy. The active galactic nucleus, obscured by a ring of dusty molecular clouds, lies at the point from which the jets emerge. If viewed along a direction closer to the jet axis, the active nucleus would not be obscured and it would therefore look more like a quasar.

donut-shaped rings of dark molecular clouds (Figure 21.19). Such structures may look like quasars when they are oriented so that we can see the active galactic nucleus at the center and look like the nuclei of radio galaxies when the ring of dusty gas dims our view of the central object.*

• What is the power source for quasars and other active galactic nuclei?

Astronomers have worked hard to envision physical processes that might explain how radio galaxies, quasars, and other active galactic nuclei release so much energy within such small central volumes. Only one explanation seems to fit: The energy comes from matter falling into a supermassive black hole.

---

*A subset of active galactic nuclei called *BL Lac objects* (after the prototype object in the constellation Lacerta) probably represents the centers of radio galaxies whose jets happen to point directly at us.

The idea that the huge energy outputs of active galactic nuclei can be traced to supermassive black holes is much like the idea we used earlier to explain the emission from X-ray binary star systems [Section 18.2]. The gravitational potential energy of matter falling toward the black hole is converted into kinetic energy, and collisions between infalling particles convert the kinetic energy into thermal energy. The resulting heat causes this matter to emit the intense radiation we observe. As in X-ray binaries, we expect that the infalling matter swirls through an accretion disk before it disappears beneath the event horizon of the black hole (Figure 21.20). However, in order to produce the enormous luminosity of a quasar, an amount of matter greater than that of the Sun needs to be passing through the accretion disk and falling into the black hole each year.

Figure 21.20 Artist's conception of an accretion disk surrounding a supermassive black hole. This picture represents only the very center of an object like that shown in Figure 21.19.

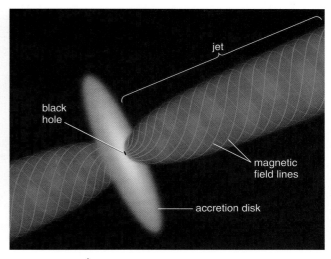

Figure 21.21 This schematic drawing illustrates one hypothesis for how supermassive black holes create jets. The hypothesis relies on the magnetic field lines thought to thread the accretion disk surrounding a black hole. As the accretion disk spins, it twists the magnetic field lines. Charged particles at the accretion disk's surface can then fly outward along the twisted magnetic field lines.

If the supermassive black hole model is correct, it should be able to explain the major observed features of quasars and other active galactic nuclei. In particular, it should explain their extreme luminosities, the fact that they emit radiation over a broad range of wavelengths from radio waves to X rays, and the presence of their powerful jets.

Explaining the extreme luminosities of quasars was the main motivation for the supermassive black hole model. Matter falling into a black hole can generate awesome amounts of energy. During its fall to the event horizon of a black hole, as much as 10–40% of the mass-energy ($E = mc^2$) of a chunk of matter can be converted into thermal energy and ultimately to radiation. (The precise value for a particular black hole depends on its rotation rate: Faster rotation allows more energy to be released.) Thus, accretion by black holes can produce light far more efficiently than nuclear fusion, which converts less than 1% of mass-energy into photons. Remember that the light is coming not from the black hole itself but rather from the hot gas in the accretion disk that surrounds it.

The environment surrounding a supermassive black hole explains why active galactic nuclei emit light across such a broad wavelength range. Hot gas in and above the accretion disk produces enormous amounts of ultraviolet and X-ray photons. This radiation ionizes surrounding interstellar gas, creating ionization nebulae that emit visible light. (The emission lines produced by these nebulae are the same ones Maarten Schmidt used to measure the first quasar redshifts.) Dust grains in the molecular clouds that encircle the active galactic nucleus (see Figure 21.19) absorb high-energy light and reemit it as infrared light. Finally, the fast-moving electrons that we sometimes see jetting from these nuclei at nearly the speed of light can produce the radio emission from active galactic nuclei.

The powerful jets emerging from active galactic nuclei are more difficult to explain, but certainly there is plenty of energy available for flinging material outward at nearly the speed of light. One plausible model for jet production relies on the twisted magnetic fields thought to accompany accretion disks (Figure 21.21). As an accretion disk spins, it pulls the magnetic field lines that thread it around in circles. Charged particles fly outward along the field lines like beads on a twirling string, forming a jet that shoots out into space.

On the whole, the supermassive black hole model seems to explain the major observed features of quasars and other active galactic nuclei. However, several important mysteries remain unsolved. For example, we do not yet know why quasars eventually run out of gas to accrete and stop shining so brightly. We also do not know how those black holes formed in the first place. Nevertheless, the hypothesis that gigantic black holes are responsible for quasars, nearby active galactic nuclei, and radio galaxies is so far withstanding the tests of time and thousands of observations.

• Do supermassive black holes really exist?

The idea that such monster black holes really do exist has been hard for some astronomers to swallow. Proving that we have found a black hole is tricky. Black holes themselves do not emit any light, so we need to infer their existence from the ways in which they alter their surroundings. In the vicinity of a black hole, matter should be orbiting at high speed around something invisible. We have already examined the evidence for a black hole at the center of our Milky Way [Section 19.4], but what about other galaxies?

Hunting for Supermassive Black Holes Detailed observations of matter orbiting at the centers of nearby galaxies suggest that supermassive black holes are quite common. In fact, it is possible that *all* galaxies contain supermassive black holes at their centers. One prominent example is the relatively nearby galaxy M 87, which features a bright, active galactic nucleus and a jet that emits both radio and visible light (see Figure 21.12). Thus, it was already a prime black

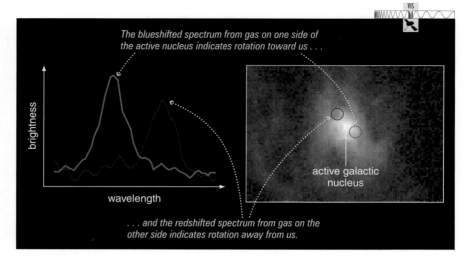

The blueshifted spectrum from gas on one side of the active nucleus indicates rotation toward us . . .

active galactic nucleus

. . . and the redshifted spectrum from gas on the other side indicates rotation away from us.

brightness

wavelength

Figure 21.22 This Hubble Space Telescope photo shows gas near the center of the galaxy M 87, and the graph shows Doppler shifts of spectra from gas 60 light-years from the center on opposite sides (the circled regions in the photo). The Doppler shifts tell us that gas is orbiting around the galactic center, and precise measurements tell us that orbital speed is about 800 km/s. From the orbital speed and Newton's laws, we find that the central object must have a mass 2–3 billion times that of the Sun.

hole suspect when astronomers pointed the Hubble Space Telescope at its core in 1994 (Figure 21.22). The spectra they gathered showed blueshifted emission lines on one side of the nucleus and redshifted emission lines on the other. This pattern of Doppler shifts is the characteristic signature of orbiting gas: On one side of the orbit the gas is coming toward us and hence is blueshifted, while on the other side it is moving away from us and is redshifted. The magnitude

---

MATHEMATICAL INSIGHT 21.1   Feeding a Black Hole

As discussed in the text, 10–40% of the mass-energy of matter falling into a black hole can be radiated away as energy. Suppose 10% of the mass-energy is radiated away. Then the amount of energy radiated by mass m falling into a black hole is $E = 1/10 \; mc^2$. We can use this fact to determine how much mass is accreting around the black hole in an active galactic nucleus, because we can determine the amount of energy radiated into space from the object's luminosity and distance. We just need to solve the energy equation for the mass m; you should confirm that the result is:

$$\text{accreted mass} = m = 10 \times \frac{E}{c^2}$$

Remember that this result is based on the assumption that 10% of the accreted mass becomes energy that is radiated away. If the percentage is larger, then a *smaller* amount of mass is needed to produce the same energy. For example, if 20% of the mass becomes radiated energy, then only half as much mass is needed compared with what we'd calculate under the assumption that 10% of the mass-energy is radiated away.

Example: The most powerful quasars have luminosities of about 10^{40} watts. Assuming that 10% of the mass-energy of consumed mass is radiated away, how many solar masses of material must the central black hole consume each year? How would the answer differ if the quasar radiates away 25% of the mass-energy of the consumed mass rather than 10%?

Solution:

Step 1 Understand: We are given the quasar luminosity in watts, which is a unit of power, but we will need energy units to determine the amount of mass accreted with the formula above. Recall that 1 watt = 1 joule/s; thus, the black hole must accrete enough mass to radiate $E = 10^{40}$ joules each second. To make the units work out correctly, we also need to remember that 1 joule = 1 kg $\times$ m^2/s^2. To determine how much mass

the black hole accretes in a year, we'll need to multiply the amount per second by the number of seconds in 1 year, and to convert to solar masses we need the fact (from Chapter 14) that $1 \; M_{Sun} = 2.0 \times 10^{30}$ kg. By using these numbers in the above formula, we'll find the amount of mass the black hole accretes under the assumption that 10% of the mass-energy is radiated away. If 25% were radiated away, then the amount of mass needed to explain the radiation would be only $10/25 = 0.4$ times as much.

Step 2 Solve: To find the amount of mass radiated each second, we use the above formula with the energy $E = 10^{40}$ kg $\times$ m^2/s^2 and the speed of light $c = 3 \times 10^8$ m/s:

$$m = 10 \times \frac{E}{c^2} = 10 \times \frac{10^{40} \frac{kg \; \times \; m^2}{s^2}}{\left(3 \times 10^8 \frac{m}{s}\right)^2} = 1.1 \times 10^{24} \; kg$$

To find the amount of mass accreted each year, we multiply this result by the number of seconds in 1 year:

$$1.1 \times 10^{24} \frac{kg}{s} \times 60 \frac{s}{min} \times 60 \frac{min}{hr} \times 24 \frac{hr}{day} \times 365 \frac{day}{yr}$$
$$= 3.5 \times 10^{31} \frac{kg}{yr}$$

Finally, we convert this result to solar masses:

$$\frac{3.5 \times 10^{31} \frac{kg}{yr}}{2.0 \times 10^{30} \frac{kg}{M_{Sun}}} \approx 17 \; M_{Sun}/yr$$

Step 3 Explain: We have found that, in order to account for the luminosity of the quasar, its central supermassive black hole must accrete the equivalent of about 17 Suns per year if it converts 10% of the accreted mass into energy that it radiates away. If it radiates away 25% of the mass-energy, then it needs to accrete $0.4 \times 17 \approx 7$ solar masses of material each year.

---

of these Doppler shifts shows that the gas, located up to 60 light-years from the center, is orbiting something invisible at a speed of hundreds of kilometers per second. This high-speed orbital motion indicates that the central object has a mass some 2–3 billion times that of our Sun.

Observations of NGC 4258, another galaxy with a visible jet, have delivered even more persuasive evidence. A ring of molecular clouds orbits the nucleus of this galaxy in a circle less than 1 light-year in radius. We can pinpoint these clouds because they amplify the microwave emission lines of water molecules, generating beams of microwaves very similar to laser beams. (The word *laser* stands for "*l*ight *a*mplification by *s*timulated *e*mission of *r*adiation." These clouds contain water *masers*. The word *maser* stands for "*m*icrowave *a*mplification by *s*timulated *e*mission of *r*adiation.") The Doppler shifts of these emission lines allow us to determine the orbits of the clouds very precisely. Their orbital motion tells us that the clouds are circling a single, invisible object with a mass of 36 million solar masses.

As the search for black holes in active galactic nuclei progresses, we are continuing to find examples like these. In each case, a supermassive black hole seems to be the only explanation for the enormous orbital speeds. We may never be 100% certain that these objects are indeed giant black holes. The best we can do is rule out all other possibilities. However, a supermassive black hole is the only thing we know of that could be so massive while remaining unseen.

Black Holes and Galaxy Formation Evidence for supermassive black holes is also found in galaxies whose centers are not currently active, and the masses of those black holes follow

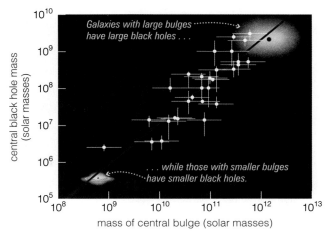

Figure 21.23 The relationship between bulge mass and the mass of a supermassive black hole.

a very interesting pattern: The mass of the black hole at the center of a galaxy appears to be closely related to the properties of the galaxy's spheroidal component. Detailed studies of the orbital speeds of stars and gas clouds in the centers of nearby galaxies show that the mass of the central black hole is typically about $\frac{1}{500}$ of the mass of the galaxy's bulge (Figure 21.23). Because this relationship holds for galaxies with a wide range of properties, from small spiral galaxies with a bulge mass of less than 10^8 M_{Sun} to giant elliptical galaxies whose spheroidal component exceeds 10^{11} M_{Sun}, we conclude that the growth of a central black hole must be closely linked with the process of galaxy formation.

MATHEMATICAL INSIGHT **21.2** **Weighing Supermassive Black Holes**

We weigh supermassive black holes the same way we weigh almost everything else in the universe: by measuring the velocity v and orbital radius r of the matter circling the central black hole. Given these measurements, we can apply the orbital velocity law from Mathematical Insight 19.1 to find the mass M_r within a distance r of the galactic center:

$$M_r = \frac{r \times v^2}{G}$$

Example: Doppler shifts show that ionized gas in the nucleus of the active galaxy M 87 orbits at a speed of about 800 km/s at a radius of 60 light-years, which is approximately 5.6×10^{17} meters. Calculate the amount of mass that lies within 60 light-years of the galactic center.

Solution:

Step 1 Understand: We are given both the orbital radius (r) and the orbital velocity (v) of gas clouds orbiting the center of the galaxy, so we have all the information we need to apply the orbital velocity law and determine the mass that lies within the distance r of the galactic center. For the units to work out properly, we will want the orbital radius in units of meters and the

orbital velocity in units of meters per second. We are already given the former; the velocity of 800 km/s is equivalent to 800,000 m/s, or 8.0×10^5 m/s.

Step 2 Solve: Substituting the given values into the orbital velocity law, we find:

$$M_r = \frac{r \times v^2}{G}$$
$$= \frac{(5.6 \times 10^{17} \text{ m}) \times \left(8.0 \times 10^5 \frac{m}{s}\right)^2}{6.67 \times 10^{-11} \frac{m^3}{kg \times s^2}}$$
$$= 5.4 \times 10^{39} \text{ kg}$$

Step 3 Explain: The mass that resided interior to the gas clouds is about 5.3×10^{39} kg. This mass will be easier to interpret if we convert it to solar masses:

$$M_r = 5.4 \times 10^{39} \text{ kg} \times \left(\frac{1 \ M_{Sun}}{2.0 \times 10^{30} \text{ kg}}\right) = 2.7 \times 10^9 \ M_{Sun}$$

Thus, the mass of the central region of the galaxy is equivalent to that of about 2.7 billion Suns. Presumably, nearly all this mass is in the central black hole.

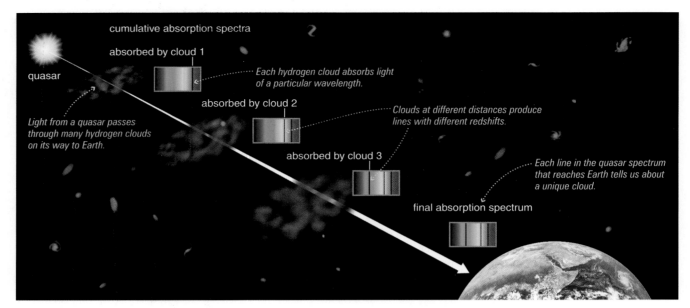

cumulative absorption spectra

absorbed by cloud 1

quasar

Each hydrogen cloud absorbs light of a particular wavelength.

Light from a quasar passes through many hydrogen clouds on its way to Earth.

absorbed by cloud 2

Clouds at different distances produce lines with different redshifts.

absorbed by cloud 3

Each line in the quasar spectrum that reaches Earth tells us about a unique cloud.

final absorption spectrum

Figure 21.24 Interactive Figure This schematic illustration shows how interstellar hydrogen clouds leave their mark on the spectra of quasars. Each line in the quasar spectrum that reaches Earth tells us about a unique hydrogen cloud along the line of sight. As we study progressively redder lines, we learn about clouds in progressively earlier stages of the universe's development.

Astronomers have long suspected that galaxy evolution goes hand in hand with the formation of supermassive black holes because quasars were so much more common early in time, when galaxies were growing rapidly. Unfortunately, we do not yet know how that process works. Some scientists have suggested that the black holes formed first out of gas at the center of a protogalactic cloud and that their energy output regulated the growth of the galaxy around them. Other scientists have suggested that clusters of neutron stars resulting from extremely dense starbursts at the centers of young galaxies might have somehow coalesced to form an enormous black hole, but these speculations are still unverified. The origins of supermassive black holes and their connection to galaxy evolution remain mysterious.

• How do quasars let us study gas between the galaxies?

The most mysterious part of galaxy evolution is the part we've not yet observed: the formation and development of protogalactic clouds. In recent years, the study of quasars has begun to shed light on even this very early stage of galaxy evolution. The most distant quasars inhabit the outskirts of our observable universe. Photons from some of these quasars began journeying to Earth when the universe was less than 1 or 2 billion years old. Along the way, these photons have passed through numerous intergalactic hydrogen clouds. The vast majority of these clouds are too diffuse and wispy ever to become galaxies, but a few have as much hydrogen as the disk of the Milky Way. The thickest of these clouds may well be protogalactic clouds in the process of becoming galaxies.

Quasar spectra therefore contain valuable information about the properties of hydrogen clouds in the early universe. Because atoms tend to absorb light at very specific wavelengths, every time a light beam from a quasar passes through an intergalactic or protogalactic cloud, some of the atoms in the cloud absorb photons from the beam, cre-

ating an absorption line (Figure 21.24). Studies of these absorption lines in quasar spectra can tell us what happened in protogalactic clouds during the epoch of galaxy formation and thus provide clues about how galaxies evolve.

We are only beginning to learn how to read the clues that hydrogen absorption lines have etched into the spectra of quasars, but the evidence gathered so far supports our general picture of spiral galaxy evolution. The most prominent hydrogen absorption lines, thought to be associated with newly forming galaxies, are typically produced by the most distant clouds, indicating that the youngest galaxies are made mostly of gas. In fact, they contain about the same amount of mass in the form of hydrogen gas as older galaxies contain in the form of stars. The hydrogen lines from nearer clouds, which arise in more mature galaxies, are not nearly as strong. Presumably, a greater fraction of the gas in these galaxies has already collected into stars.

Absorption lines from elements other than hydrogen also support this picture. The lines from these heavy elements are more prominent in mature galaxies than in the youngest galaxies, implying that mature galaxies have experienced more supernovae, which have added heavy elements to their interstellar gas. This overall pattern of gradual heavy-element enrichment accompanied by the gradual diminishing of interstellar hydrogen agrees well with what we know about the Milky Way. All across the universe, stars in the gaseous disks of galaxies appear to have been forming steadily for over 10 billion years.

With every new study of a quasar spectrum, we learn more about the galaxies and intergalactic clouds that have left their mark on the light we observe from quasars. Perhaps someday soon these observations will help us fit together all the pieces of the galaxy evolution puzzle, and we at last will understand the whole glorious history of galaxy evolution in our universe. Until that time, we will keep peering deep into space and back into time, searching for the clues that will unlock the mysteries of cosmic evolution.

Putting Chapter 21 into Context

We have not yet solved the whole puzzle of galaxy evolution, but in this chapter we have described some of its crucial pieces. As you look back, keep sight of these "big picture" ideas:

- We can study galaxy evolution by looking back through time: At great distances we see galaxies as they were when the universe was young, and nearby we see mature galaxies as they exist today. These observations, along with theoretical modeling, are rapidly helping us understand the lives of galaxies.

- Galaxies probably all began as protogalactic clouds, but they do not always evolve peacefully. Some galaxies suffer gargantuan collisions with neighbors, often with dramatic results.

- The tremendous energy outputs of quasars and other active galactic nuclei, including those of radio galaxies, are probably powered by gas accreting onto supermassive black holes. The centers of many present-day galaxies must still contain the supermassive black holes that once enabled them to shine as quasars.

21.1 Looking Back Through Time

- **How do we observe the life histories of galaxies?** Today's telescopes enable us to observe galaxies of many different ages because they are powerful enough to detect light from objects with lookback times almost as large as the age of the universe. We can therefore assemble "family albums" of galaxies at different distances and lookback times.

- **How did galaxies form?** The most successful models of galaxy formation assume that galaxies formed as gravity pulled together regions of the universe that were ever so slightly denser than their surroundings. Gas collected in protogalactic clouds, and stars began to form as the gas cooled.

21.2 The Lives of Galaxies

- **Why do galaxies differ?** Differences between present-day galaxies probably can arise both from conditions in their protogalactic clouds and from collisions with other galaxies. Slowly rotating or high-density protogalactic clouds may form elliptical rather than spiral galaxies. Ellipticals may also form through the collision and merger of two spiral galaxies.

- **What are starbursts?** A **starburst galaxy** is a galaxy that is forming new stars at a very high rate—sometimes more than 100 times the star formation rate of the Milky Way. This high rate of star formation can lead to a supernova-driven **galactic wind.** Many starbursts apparently result from collisions between galaxies. Some starbursts may occur as a result of close encounters with other galaxies rather than from direct collisions.

21.3 Quasars and Other Active Galactic Nuclei

- **What are quasars?** Some galaxies have unusually bright centers known as **active galactic nuclei.** A **quasar** is a particularly bright active galactic nucleus. Quasars are generally found at very great distances, telling us that they were much more common early in the history of the universe.

- **What is the power source for quasars and other active galactic nuclei?** **Supermassive black** **holes** are thought to be the power sources for active galactic nuclei. As matter falls into a supermassive black hole through an accretion disk, its gravitational potential energy is transformed with enormous efficiency into thermal energy and then into light.

- **Do supermassive black holes really exist?** Observations of orbiting stars and gas clouds in the nuclei of galaxies suggest that *all* galaxies may harbor supermassive black holes at their centers. The masses we measure for those black holes are closely related to the properties of the galaxy around them, suggesting that the growth of these black holes is closely tied to the process of galaxy evolution.

- **How do quasars let us study gas between the galaxies?** Each cloud of gas through which the quasar's light passes on its long journey to Earth produces a hydrogen absorption line in the quasar spectrum. Study of these absorption lines in quasar spectra allows us to study matter—including protogalactic clouds—that we cannot otherwise detect.

Review Questions

Short-Answer Questions Based on the Reading

1. What do we mean by *galaxy evolution*? How do telescopic observations allow us to study galaxy evolution? How do theoretical models help us study galaxy formation?

2. Briefly describe the starting assumptions for galaxy formation, and how models suggest that these starting conditions can lead to the formation of a spiral galaxy.

3. Describe two ways in which conditions in a protogalactic cloud might lead to the birth of an elliptical rather than a spiral galaxy.

4. Why happens when two galaxies collide? How might elliptical galaxies form from collisions of spiral galaxies? What evidence supports the idea that galaxy collisions sometimes occur?

5. Briefly explain why we expect collisions between galaxies to be relatively common, while collisions between stars are extremely rare. Why should galaxy collisions have been more common in the past than they are today?

6. What is a *starburst galaxy*? How do observations of starbursts help us understand why the stars in elliptical galaxies are so old?

7. Briefly explain why starburst galaxies often appear ordinary when they are observed in visible light but extraordinary when they are observed in infrared light.

8. What is a galactic wind? What causes it? How is it similar to a superbubble in the Milky Way, and how is it different?

9. Briefly describe the discovery of *quasars*. What evidence convinced astronomers that the high redshifts of quasars really do imply great distances? Why can we learn more about quasars by studying nearby *active galaxies* and *radio galaxies*?

10. Briefly explain how we can use variations in luminosity to set limits on the size of an object's emitting region. For example, if an object doubles its luminosity in 1 hour, how big can it be?

11. Summarize the supermassive black hole model for the energy output of quasars and other active galactic nuclei. What evidence suggests that such black holes really exist?

12. What is a radio galaxy? Describe jets and radio lobes. Why do we think that the ultimate energy sources of radio galaxies lie in quasarlike galactic nuclei?

Test Your Understanding

Does It Make Sense?

Decide whether the statement makes sense (or is clearly true) or does not make sense (or is clearly false). Explain your reasoning. (For an example, see Chapter 1, "Does It Make Sense?")

13. Galaxies that are more than 10 billion years old are too far away for us to see even with our most powerful telescopes.

14. Several protogalactic clouds have recently been discovered in the neighborhood of the Milky Way.

15. Elliptical galaxies are more likely to form in denser regions of space.

16. If the Andromeda Galaxy someday collides and merges with the Milky Way, the resulting galaxy may be elliptical.

17. NGC 9645 is a starburst galaxy that has been forming stars at the same furious pace for some 10 billion years.

18. The energy from supernova explosions can drive a large proportion of the interstellar gas out of a small galaxy.

19. Astronomers proved that quasar 3C 473 contains a supermassive black hole when they discovered that its center is completely dark.

20. The black hole at the center of our own galaxy may once have powered an active galactic nucleus.

21. Radio galaxies emit only radio waves and no visible light.

22. Analyses of quasar light can tell us about intergalactic clouds that might otherwise remain invisible.

Quick Quiz

Choose the best answer to each of the following. Explain your reasoning with one or more complete sentences.

23. When we observe a distant galaxy whose photons have traveled for 10 billion years before reaching Earth, we are seeing that galaxy as it was when the universe was (a) 10 billion years old. (b) 7 billion years old. (c) 4 billion years old.

24. Which of these items is a key assumption in our most successful models for galaxy formation? (a) The distribution of matter was perfectly uniform early in time. (b) Some regions of the universe were slightly denser than others. (c) Galaxies formed around supermassive black holes.

25. A collision between two large spiral galaxies is likely to produce (a) a large elliptical galaxy. (b) a large spiral galaxy. (c) one large spiral galaxy and one large elliptical galaxy.

26. A collision and merger of two large elliptical galaxies will eventually produce (a) a large elliptical galaxy. (b) a large spiral galaxy. (c) a large irregular galaxy.

27. Starburst galaxies are especially bright in (a) visible light. (b) ultraviolet light. (c) infrared light.

28. The rate at which supernovae explode in a starburst galaxy that is forming stars 10 times faster than the Milky Way is (a) about the same as in the Milky Way. (b) about 10 times higher than in the Milky Way. (c) about 100 times higher than in the Milky Way.

29. The luminosity of a quasar is generated in a region the size of (a) the Milky Way. (b) a star cluster. (c) the solar system.

30. The primary source of a quasar's energy is (a) chemical energy. (b) nuclear energy. (c) gravitational potential energy.

31. Supermassive black holes found at the centers of galaxies are related to the properties of those galaxies in the following way (a) the mass of the black hole is related to the mass of the galaxy's bulge. (b) the luminosity of the active nucleus is related to the mass of the galaxy's disk. (c) the luminosity of the active nucleus is related to the luminosity of the galaxy's bulge.

32. Which of the following quasars would you expect to have the largest number of hydrogen absorption lines in its spectrum? (a) a quasar with a redshift $z = 1.0$ (b) a quasar with a redshift $z = 3.0$ (c) a quasar with a redshift $z = 6.0$

Investigate Further

In-Depth Questions to Increase Your Understanding

Short-Answer/Essay Questions

33. *Life Story of a Spiral.* Imagine that you are a spiral galaxy. Describe your life history from birth to the present day. Your story should be detailed and scientifically consistent, but also

creative. That is, it should be entertaining while at the same time incorporating current scientific ideas about the formation of spiral galaxies.

34. *Life Story of an Elliptical.* Imagine that you are an elliptical galaxy. Describe your life history from birth to the present. There are several possible scenarios for the formation of elliptical galaxies, so choose one and stick to it. Be creative while also incorporating scientific ideas that demonstrate your understanding.

35. *The Color of an Elliptical Galaxy.* Explain how the color of an old elliptical galaxy has changed during the last 10 billion years. How might you be able to use the galaxy's color to determine when it formed?

36. *Very Early Collisions.* Suppose two protogalactic clouds collide while they are still made primarily of gas, before forming many stars. Could the resulting system end up as a spiral galaxy? Explain your reasoning.

37. *A Small-Scale Starburst.* The Large Magellanic Cloud, a small companion galaxy to the Milky Way, is currently undergoing a small-scale starburst. Inspect the picture of the Large Magellanic Cloud in Figure 20.7a for evidence of widespread star formation. Describe the evidence you find.

38. *Quasar Redshifts.* Some astronomers did not immediately accept that the large redshifts of quasars means that they must be at extremely large distances. What properties would an object need in order to have a large redshift but still be no further than a billion light-years from the Milky Way? How might you be able to test the hypothesis that redshifts of quasars are not cosmological in origin?

39. *Orbits Around Supermassive Black Holes.* The data in Figure 21.22 show the Doppler shifts of emission lines from gas at a distance of 60 light-years from the center of the galaxy M 87. Suppose you observed emission lines from gas 30 light-years from the center. How would you expect the Doppler shifts of those lines to be different, assuming that the gas really is orbiting a supermassive black hole? What about gas at 120 light-years from the center?

40. *Absorption Lines in Quasar Spectra.* Based on your understanding of galaxy evolution, where in the spectrum of a distant quasar would you expect to find the largest number of hydrogen absorption lines? Would they have redshifts near that of the quasar itself or would you tend to find more hydrogen absorption lines with smaller redshifts? Would you expect to see hydrogen absorption lines with redshifts greater than that of the quasar? Explain your reasoning.

Quantitative Problems

Be sure to show all calculations clearly and state your final answers in complete sentences.

41. *Distances Between Galaxies.* If you were to divide the present-day universe up into cubes whose sides are 10 million light-years long, each cube would contain, on average, about one galaxy similar in size to the Milky Way. Now suppose you travel back in time, to an era when the average distance between galaxies is one-quarter of its current value, corresponding to a cosmological redshift $z = 3$. How many galaxies similar in size to the Milky Way would you expect to find, on average, in cubes of that same size? In order to simplify the problem, assume that the total number of galaxies of each type has not changed between then and now. Based on your answer, would you expect collisions to be much more frequent at that time or only moderately more frequent?

42. *A Nearby Starburst.* The galaxy M 82, shown in Figure 21.10, is one of the nearest starburst galaxies. Even though it is considerably smaller than the Milky Way, it is currently forming stars more quickly, converting approximately 10 M_{Sun} of gas into stars each year. The total amount of gas in its interstellar medium is currently about 10^9 M_{Sun}. About how much longer can the starburst in M 82 last? Describe how the appearance of M 82 will change with time, based on your answer.

43. *Your Last Hurrah.* Suppose you fell into an accretion disk that swept you into a supermassive black hole. On your way down, the disk radiates 10% of your mass-energy, $E = mc^2$.
   a. What is your mass in kilograms? (Recall that 1 kg = 2.2 lb on Earth.) Calculate how much radiative energy will be produced by the accretion disk as a result of your fall into the black hole.
   b. Calculate approximately how long a 100-watt light bulb would have to burn to radiate this same amount of energy.

44. *The Black Hole in NGC 4258.* The molecular clouds circling the center of the active galaxy NGC 4258 orbit at a speed of about 1,000 km/s, with an orbital radius of 0.49 light-year = 4.8×10^{15} meters. Use the orbital velocity law (see Mathematical Insight 21.2) to calculate the mass of the central black hole. Give your answer both in kilograms and in solar masses (1 $M_{Sun} = 2.0 \times 10^{30}$ kg).

45. *The Black Hole in M 31.* Measurements of star motions at the center of the Andromeda Galaxy, also known as M 31, show that stars about 3 light-years from the center are orbiting at a speed of 400 km/s. These stars are suspected to be orbiting a supermassive black hole. Use the orbital velocity law from Mathematical Insight 21.2 to estimate the mass of this black hole.

46. *Building a Supermassive Black Hole.* The black hole in the galaxy M 87 has a mass of about 3 billion solar masses. Let us assume that most of that mass flowed into the black hole through an accretion disk that radiated 10% of the mass-energy passing through it. In that case, what would be the total amount of energy radiated by the accretion disk during the history of the black hole? What would be the average luminosity of the accretion disk, if it continuously radiated that energy over a period of 10 billion years? How does that average luminosity compare with the luminosity of the Milky Way?

47. *Stars and Clouds in Colliding Galaxies.* When two spiral galaxies collide, the stars generally do not run into each other, but the gas clouds generally do collide, triggering a burst of new star formation.
   a. Estimate the probability that our Sun would collide with another star in the Andromeda Galaxy, if a collision between the Milky Way and the Andromeda Galaxy were happening at the present time. To simplify the problem, assume that each galaxy has 100 billion stars exactly like the Sun spread evenly over a circular disk with a radius of 100,000 light-years. (*Hint:* First calculate the total area of 100 billion circles with the radius of the Sun and then compare that total area to the area of the galactic disk.)
   b. Estimate the probability that a gas cloud in our galaxy would collide with another gas cloud in the Andromeda Galaxy. To simplify the problem, assume that each galaxy contains 100,000 clouds of warm hydrogen gas that each has a radius of 300 light-years, and that these clouds are spread evenly over a circular disk with a radius of 100,000 light-years. (*Hint:* Use the same method as in part (a).)
   c. Compare the probabilities you found in parts (a) and (b) and use them to explain what you see in Figure 21.5.

48. *The Case for Supermassive Black Holes.* The evidence for supermassive black holes at the center of galaxies is strong. However, it is very difficult to prove absolutely that they exist because the black holes themselves emit no light. We can infer their existence only from their powerful gravitational influences on surrounding matter. How compelling do you find the evidence? Do you think astronomers have proved the case for black holes beyond a reasonable doubt? Defend your opinion.

49. *Life in Colliding Galaxies.* Suppose the Milky Way were currently undergoing a collision with another large spiral galaxy. Do you think this collision would affect life on Earth? Why or why not? How different would the night sky look if our galaxy were in the midst of such a collision?

MEDIA EXPLORATIONS

For self-study activities go to www.masteringastronomy.com

 Interactive Tutorials

Tutorial Review of Key Concepts
Use the interactive Tutorials at **www.mastering astronomy.com** to review key concepts from this chapter.

Black Holes Tutorial
Lesson 1 What Are Black Holes?
Lesson 2 The Search for Black Holes

Supplementary Tutorial Exercises
Use the interactive Tutorial Lessons to explore the following questions.

Black Holes Tutorial, Lessons 1, 2
1. What is the Schwarzschild radius for a black hole of 1 million solar masses? For black holes of 10 million and 100 million solar masses?
2. Describe how scientists determine whether the center of a galaxy contains a supermassive black hole.

 Exploring the Sky and Solar System

Of the many activities available on the *Voyager: SkyGazer* CD-ROM accompanying your book, use the following file to observe key phenomena covered in this chapter.

Go to the File: Basics folder for the following demonstration.
1. Galaxies in Coma

 Movies

Check out the following narrated and animated short documentary available on **www.masteringastronomy.com** for a helpful review of key ideas covered in this chapter.

From the Big Bang to the Galaxies Movie

Web Projects

1. *Future Observatories.* The subject of galaxy evolution is a very active area of research. Look for information on current and future observatories involved in investigating galaxy evolution (such as the James Webb Space Telescope). How big are the planned telescopes? What wavelengths will they look at? When will they be built? Write a short summary of a proposed mission.
2. *Greatest Redshift.* As of summer 2005, the most distant quasar known has a redshift $z = 6.4$, meaning that the wavelengths of its light are $1 + 6.4 = 7.4$ longer than normal. Find the current record holder for the largest redshift. Write a one-page report describing the object and its discovery.
3. *The Quasar Controversy.* For many years, some astronomers argued that quasars are not really as distant as Hubble's law indicates. Research the history of the discovery of quasars and the debates that followed. What evidence led some astronomers to think quasars might be nearer than Hubble's law suggests? Why did most astronomers eventually conclude that quasars really are far away? Write a one- to two-page report summarizing your findings.

Dark Matter, Dark Energy, and the Fate of the Universe

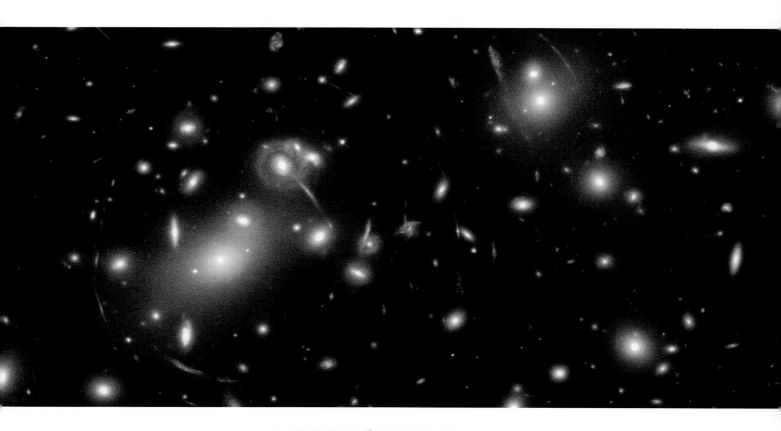

It is difficult beyond description to conceive that space can have no end; but it is more difficult to conceive an end. It is difficult beyond the power of man to conceive an eternal duration of what we call time; but it is more impossible to conceive a time when there shall be no time.

Thomas Paine, *The Age of Reason* (1796)

In prior chapters, we explored current understanding of the evolution of galaxies in an expanding universe. However, we left at least two crucial questions unaddressed: First, what is the source of the gravity that causes galaxies to form and holds them together? Second, what will happen to the expansion of the universe in the future? Both questions would be interesting in their own right, but they've taken on even greater importance because our attempts to answer them have led us to two of the greatest mysteries in science.

If we are interpreting the data correctly, then the dominant source of gravity in the universe is an unidentified type of mass, known as *dark matter*, that is completely invisible to our eyes and telescopes. The gravity of dark matter must therefore be the "glue" that holds galaxies like our own together. Moreover, the very fate of the universe seems to hinge on the total amount of dark matter and on the existence of an even more mysterious force or energy—often called *dark energy*—that may counteract the effects of gravity on large scales.

In this chapter, we will explore the evidence for dark matter and dark energy and discuss why their natures remain so mysterious. We'll also investigate how dark matter determines the current structure of the universe and the implications of dark energy for the fate of the universe, if dark matter and dark energy indeed prove to be real. Whatever the ultimate answers turn out to be, the quest to resolve these mysteries offers an outstanding opportunity to observe how science progresses. Keep an eye on the science section of your daily newspaper to see how the story unfolds.

22.1 Unseen Influences in the Cosmos

What is the universe made of? Ask an astronomer this seemingly simple question, and you might see a professional scientist blush with embarrassment. Based on all the available evidence today, the answer to this simple question is "We do not know."

It might seem incredible that we still do not know the composition of most of the universe, but you might also wonder why we should be so clueless. After all, astronomers can measure the chemical composition of distant stars and galaxies from their spectra, so we know that stars and gas clouds are made almost entirely of hydrogen and helium,

with small amounts of heavier elements mixed in. But notice the key words "chemical composition." When we say these words, we are talking about the composition of material built from atoms of elements such as hydrogen, helium, carbon, and iron. While it is true that all familiar objects—including people, planets, and stars—are built from atoms, the same may not be true of the universe as a whole.

In fact, we now have good reason to think that the vast majority of the matter in the universe is *not* composed of atoms. Instead, the universe may consist largely of a mysterious form of mass known as *dark matter* and a mysterious form of energy known as *dark energy*. As we'll discuss in Chapter 23, some recent observations tell us the precise percentages of the universe that consist of dark energy, dark matter, and ordinary atoms, if we're interpreting these observations properly. Nevertheless, the actual nature of dark matter and dark energy remains completely unknown.

• What do we mean by dark matter and dark energy?

It's easy to talk about dark matter and dark energy, but what do these terms really mean? They are nothing more than names given to unseen influences in the cosmos. In both cases we have been led to think that there is something out there, even though we cannot identify it, because we have observed phenomena that otherwise do not make sense.

We might naively think that the major source of gravity that holds galaxies together is the same gas and dust that make up their stars. However, decades of observations suggest otherwise: By carefully observing gravitational effects on matter that we can see, such as stars or glowing clouds of gas, we've learned that there must be far more matter than meets the eye. Because this matter appears to give off little or no light, we call it **dark matter**. Thus, dark matter is simply a name we give to whatever unseen influence is causing the observed gravitational effects. We've already discussed dark matter briefly in Chapters 1 and 19, noting that studies of the Milky Way's rotation suggest that most of our galaxy's mass is distributed throughout its halo while most of the galaxy's light comes from stars and gas clouds in the thin galactic disk (see Figure 1.15).

We infer the existence of the second unseen influence from careful studies of the expansion of the universe. From the time that Hubble first discovered the expansion, it was generally assumed that gravity must slow the expansion with time. In just the past few years, however, mounting evidence has suggested that the expansion of the universe is actually accelerating. If so, some mysterious type of force or energy must be able to counteract the effects of gravity on very large scales.

Dark energy is the most common name given to whatever it is that may be causing the expansion to accelerate, but it is not the only name; you may occasionally hear the same unseen influence attributed to *quintessence* or to a *cosmological constant*. The term *dark energy* has become popular because it echoes the term *dark matter*, but there's nothing unusually "dark" about it—after all, we don't expect to see light from the mere presence of a force or energy field. Thus, despite the similarity in their names, dark matter and dark energy are thought to exist for completely different reasons.

Astronomers are generally more comfortable with the notion that dark matter exists than with the notion that dark energy exists. Scientific confidence in dark matter has been building for decades and is now at the point where dark matter seems almost indispensable to explaining the current structure of the universe. That is why we will devote most of this chapter to a discussion of dark matter and its presumed role as the dominant source of gravity in our universe—there is simply a lot more we can say about it. We will therefore save discussion of dark energy for the end of the chapter, where we will present the evidence that it also exists and that its nature will determine the fate of the universe.

Before we continue, it's important to think about dark matter and dark energy in the context of science. Upon first hearing of these ideas, you might be tempted to think that astronomers have "gone medieval," arguing about unseen influences in the same way that scholars in medieval times supposedly argued about the number of angels that could dance on the head of a pin. However, strange as the ideas of dark matter and dark energy may seem, they have emerged from careful scientific study conducted in accordance with the hallmarks of science that we discussed in Chapter 3 (see Figure 3.26). Dark matter and dark energy were each proposed to exist because they seemed the simplest ways to explain observed motions in the universe. They've each gained credibility because models of the universe that assume their existence make testable predictions and, at least so far, further observations have borne out some of those predictions. Thus, even if we someday conclude that we were wrong to infer the existence of dark matter or dark energy, we will still need alternative explanations for the observations made to date. One way or the other, what we learn as we explore the mysteries of these unseen influences will forever change our view of the universe.

 Detecting Dark Matter in a Spiral Galaxy Tutorial, Lessons 1–3

22.2 Evidence for Dark Matter

We are now ready to begin investigating dark matter in greater detail. In this section, we'll examine the evidence for the existence of dark matter and what the evidence indicates about the nature of dark matter.

• What is the evidence for dark matter in galaxies?

Let's begin our discussion of the evidence for dark matter by examining the case for dark matter in our own Milky Way Galaxy. We'll then proceed to other galaxies and clusters of galaxies.

Distribution of Mass in the Milky Way In Chapter 19 we saw how the Sun's motion around the galaxy reveals the total amount of mass within its orbit. Similarly, we can use the orbital motion of any other star to measure the mass of the Milky Way within that star's orbit. In principle, we could determine the complete distribution of mass in the Milky Way by doing the same thing with the orbits of stars at every different distance from the galactic center.

In practice, interstellar dust obscures our view of disk stars more than a few thousand light-years away from us, making it very difficult to measure stellar velocities. However, radio waves penetrate this dust, and clouds of atomic hydrogen gas emit a spectral line at the radio wavelength of 21 centimeters [Section 19.2]. Measuring the Doppler shift of this 21-centimeter line tells us a cloud's velocity toward or away from us. With the help of a little geometry, we can then determine the cloud's orbital velocity.

A diagram called a **rotation curve**, which plots the *rotational velocity* of stars or gas clouds against their distance from the center of the galaxy, summarizes the results of these orbital velocity measurements. As a simple example of the concept, let's construct a rotation curve for a merry-go-round. Every object on a merry-go-round goes around the center with the same rotational period, but objects farther from the center move in larger circles. Thus, objects farther from the center move at faster speeds, and the rotation curve for a merry-go-round is a straight line that rises steadily outward (Figure 22.1a).

In contrast, the rotation curve for our solar system drops off with distance from the Sun, because inner planets orbit at faster speeds than outer planets (Figure 22.1b). This drop-off in speed with distance occurs because virtually all the mass of the solar system is concentrated in the Sun. The gravitational force holding a planet in its orbit decreases with distance from the Sun, and a smaller force means a lower orbital speed. The rotation curve of any astronomical system whose mass is concentrated toward the center must drop similarly.

Figure 22.1c shows the rotation curve for the Milky Way Galaxy. Each individual dot represents the distance from the galactic center and the orbital speed of a particular star or gas cloud. The curve running through the dots represents a "best fit" to the data. Notice that the orbital velocities remain approximately constant beyond the inner few thousand light-years, making the rotation curve look flat. This behavior contrasts sharply with the steeply declining rotation curve of the solar system. Thus, unlike the case for the solar system, most of the mass of the Milky Way must *not* be concentrated at its center. Instead, the orbits of progressively more distant gas clouds must encircle more and more mass. The Sun's orbit encompasses about 100 billion solar masses, but a circle twice as large surrounds twice as much mass, and a larger circle surrounds even more mass. Because of a A rotation curve for a merry-go-round.

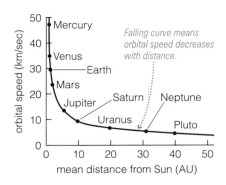

b The rotation curve for the planets in our solar system.

c The rotation curve for the Milky Way Galaxy. Dots represent stars or gas clouds whose rotational speeds have been measured.

Figure 22.1 Interactive Figure Rotation curves show how the orbital speed of a system depends on distance from its center. The solar system's rotation curve declines with radius because its mass is concentrated at the center. The Milky Way's rotation curve is flat, indicating that the Milky Way's mass extends well beyond the Sun's orbit.

the difficulty of finding clouds to measure on the outskirts of the galaxy, we have not yet found the "edge" of this mass distribution.

The flatness of the Milky Way's rotation curve therefore implies that most of our galaxy's mass lies well beyond our Sun, tens of thousands of light-years from the galactic center. A more detailed analysis suggests that most of this mass is located in the spherical halo that surrounds the disk of our galaxy, and that the total amount of this mass might be *10 times* the total mass of all the stars in the disk. Because we have detected very little radiation coming from this enormous amount of mass, it qualifies as dark matter. Thus, if we are interpreting the evidence correctly, the luminous part of the Milky Way's disk must be rather like the tip of an iceberg, marking only the center of a much larger clump of mass (Figure 22.2).

MATHEMATICAL INSIGHT 22.1 Mass-To-Light Ratio

We define an object's mass-to-light ratio as its total mass in units of *solar masses* divided by its total *visible* luminosity in units of *solar luminosities*. Thus, by definition, the mass-to-light ratio of the Sun is:

$$\frac{M}{L} \text{ for Sun} = \frac{1\,M_{Sun}}{1\,L_{Sun}} = 1\,\frac{M_{Sun}}{L_{Sun}}$$

We read this answer with its units as "1 solar mass per solar luminosity." Note that we use M/L as an abbreviation for the mass-to-light ratio in equations, and we always write the mass and luminosity in solar units.

The following examples should clarify the idea of the mass-to-light ratio and show what it can tell us about the existence of dark matter.

Example 1: What is the mass-to-light ratio of a 1 M_{Sun} red giant with luminosity 100 L_{Sun}? What does the answer tell us about how mass-to-light ratio depends on stellar type?

Solution:

Step 1 Understand: Finding a mass-to-light ratio simply requires knowing an object's total mass in solar masses and its total luminosity in solar luminosities. We are told that the object is a star of 1 solar mass with a luminosity 100 times that of the Sun, so we have all the information we need.

Step 2 Solve: We divide the star's mass by its luminosity, both in solar units:

$$\frac{M}{L} = \frac{1\,M_{Sun}}{100\,L_{Sun}} = 0.01\,\frac{M_{Sun}}{L_{Sun}}$$

Step 3 Explain: The red giant has a mass-to-light ratio of 0.01 solar mass per solar luminosity. The ratio is *less* than 1 because a red giant puts out *more* light per unit mass than the Sun. Massive stars also put out more light per unit mass than the Sun, because main-sequence luminosities rise much faster than main-sequence masses [Section 15.2]. Thus, in general, stars *more luminous* than the Sun have mass-to-light ratios *less* than 1 solar mass per solar luminosity. Conversely, stars that are *less luminous* than the Sun generally have mass-to-light ratios *greater* than 1 solar mass per solar luminosity.

Example 2: The Sun's orbit around the Milky Way tells us that the part of the galaxy within the Sun's orbit contains about 90 billion (9×10^{10}) solar masses of material. Observations tell us that the total luminosity of stars within that same region is about 15 billion (1.5×10^{10}) solar luminosities. What is the mass-to-light ratio of the matter in our galaxy within the Sun's orbit? What does this imply?

Suppose we made a rotation curve for the moons orbiting Jupiter. Would it rise, increase, or stay flat with increasing distance of Jupiter?

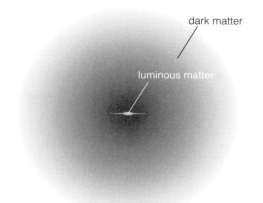

Dark Matter in Other Spiral Galaxies   Other galaxies also seem to contain vast quantities of dark matter. We can determine the amount of dark matter in a galaxy by comparing the galaxy's mass to its luminosity. (More formally, astronomers calculate the galaxy's *mass-to-light ratio*; see Mathematical Insight 22.1.) The procedure is fairly simple in principle. First, we use the galaxy's luminosity to estimate the amount of mass that the galaxy contains in the form of stars. Next, we determine the galaxy's total mass by applying the law of gravity to observations of the orbital velocities of stars and gas clouds. If this total mass is larger than the mass that we can attribute to stars, then we infer that the excess mass must be dark matter.

Measuring a galaxy's luminosity is relatively easy, as long as we can determine its distance with one of the techniques discussed in Chapter 20. We simply point a telescope at the galaxy in question, measure its apparent brightness, and calculate its luminosity from its distance and the inverse square law of light [Section 15.1]. Measuring the galaxy's total mass requires measuring orbital speeds of stars or gas clouds as far from the galaxy's center as possible. Atomic hydrogen gas clouds can be found in spiral galaxies at greater distances from the center than

Figure 22.2 The dark matter associated with a spiral galaxy like the Milky Way occupies a much larger volume than the galaxy's luminous matter. The radius of this dark-matter halo may be 10 times as large as the galaxy's halo of stars.

stars, so most of our data come from radio observations of the 21-centimeter line from these clouds. We use Doppler shifts of the 21-centimeter line to determine how fast a cloud is moving toward us or away from us (Figure 22.3). We then combine observations for clouds at varying orbital distances to construct the galaxy's rotation curve.

The rotation curves of most spiral galaxies turn out to be remarkably flat as far out as we can see (Figure 22.4). Just as in the Milky Way, these flat rotation curves imply that a great deal of matter lies far out in the halos of these other

---

Solution:

Step 1 Understand: We are asked to find the mass-to-light ratio for the region of our galaxy within the Sun's orbit. We are given both the mass and luminosity of this region in solar units, so we have all the information we need.

Step 2 Solve: We divide the region's mass by its luminosity, both in solar units:

$$\frac{M}{L} = \frac{9 \times 10^{10}\,M_{Sun}}{1.5 \times 10^{10}\,L_{Sun}} = 6\,\frac{M_{Sun}}{L_{Sun}}$$

Step 3 Explain: The mass-to-light ratio of the matter within the Sun's orbit is about six solar masses per solar luminosity. Because this mass-to-light ratio is *greater* than the Sun's ratio of 1 solar mass per solar luminosity, it tells us that most matter in this region is *dimmer* per unit mass than our Sun. This is not surprising, because we know that most stars are smaller and dimmer than our Sun.

Example 3: The rotation curve of a nearby spiral galaxy shows that it contains 5×10^{11} solar masses within a radius of 150,000 light-years of its center. Its apparent brightness and distance tell us that its total luminosity is 1.5×10^{10} solar luminosities. What is its mass-to-light ratio? What does this imply?

Solution:

Step 1 Understand: In this case, we are given the galaxy's full mass (out to the large distance of 150,000 light-years from its center), determined from its rotation curve and its total luminosity. Thus, we again have all the information we need to calculate the mass-to-light ratio.

Step 2 Solve: We divide the galaxy's mass by its luminosity, both in solar units:

$$\frac{M}{L} = \frac{5 \times 10^{11}\,M_{Sun}}{1.5 \times 10^{10}\,L_{Sun}} = 33\,\frac{M_{Sun}}{L_{Sun}}$$

Step 3 Explain: The mass-to-light ratio of this galaxy is about 33 solar masses per solar luminosity, which is more than five times as large as the mass-to-light ratio for the Milky Way Galaxy within the Sun's orbit. This mass-to-light ratio tells us that on average, the mass in this galaxy is much less luminous than the mass found in the inner regions of the Milky Way. We were able to explain a mass-to-light ratio of six solar masses per solar luminosity as a simple consequence of the fact that most stars are dimmer than the Sun. However, this far greater ratio of 33 solar masses per solar luminosity is not so easily explained by stars; instead, we conclude that the large mass-to-light ratio is telling us that the galaxy must contain a lot of mass that emits little or no light; that is, it contains dark matter.

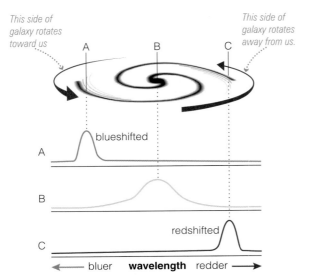

A   B   C

A   blueshifted

B

C   redshifted

← bluer   **wavelength**   redder →

Figure 22.3 Measuring the rotation of a spiral galaxy with the 21-centimeter line of atomic hydrogen. Blueshifted lines on the left side of the disk show how fast that side is rotating toward us. Red-shifted lines on the right side show how fast that side is rotating away from us.

spiral galaxies. A detailed analysis tells us that these other spiral galaxies also have at least 10 times as much mass in dark matter as they do in stars. In other words, typical spiral galaxies are made of 90% or more dark matter, and 10% or less matter in stars.

Dark Matter in Elliptical Galaxies   We must use a different technique to weigh elliptical galaxies, because most of them contain very little atomic hydrogen gas and hence do not produce detectable 21-centimeter radiation. We generally weigh the inner parts of elliptical galaxies by observing the motions of the stars themselves.

The motions of stars in an elliptical galaxy are disorganized, so we cannot assemble their velocities into a sensible rotation curve. Nevertheless, the velocity of each individual star still depends on the mass inside the star's orbit. At any particular distance from an elliptical galaxy's center, some stars are moving toward us and some are moving away

Figure 22.4 Actual rotation curves of four spiral galaxies. They are all nearly flat over a wide range of distances from the center, indicating that dark matter is common in spiral galaxies.

from us. As a result, every star has a slightly different Doppler shift. When we look at spectral lines from the galaxy as a whole, we see the combined effect of all these Doppler shifts. Together, they change the spectral line from a nice narrow line at a particular wavelength to a *broadened* line spanning a range of wavelengths (Figure 22.5). The greater the broadening of the spectral line, the faster the stars must be moving.

When we compare spectral lines from different regions of an elliptical galaxy, we find that the speeds of the stars remain fairly constant as we look farther from the galaxy's center. Thus, just as in spirals, most of the matter in elliptical galaxies must lie beyond the distance where the light trails off and hence must be dark matter. The evidence for dark matter becomes even more convincing in cases in which we can measure the speeds of globular star clusters orbiting at large distances from the center of an elliptical galaxy. These measurements suggest that elliptical galaxies, like spirals, contain far more matter than we can see in the form of stars.

• What is the evidence for dark matter in clusters of galaxies?

The evidence we have discussed so far indicates that stars make up only about 10% of a galaxy's mass—the remaining mass consists of dark matter. Observations of galaxy clusters suggest that the total proportion of dark matter is even greater. The mass of dark matter in clusters appears to be as much as 50 times the mass in stars.

The evidence for dark matter in clusters comes from three different ways of measuring cluster masses: measuring the speeds of galaxies orbiting the center of the cluster, studying the X-ray emission from hot gas between the cluster galaxies, and observing how the clusters bend light as

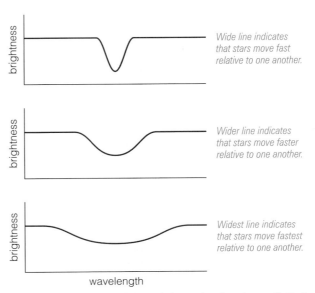

Figure 22.5 The broadening of absorption lines in an elliptical galaxy's spectrum tells us how fast its stars move relative to one another.

gravitational lenses. Let's investigate each of these techniques more closely.

Orbits of Galaxies in Clusters   The problem of dark matter in astronomy is not particularly new. In the 1930s, astronomer Fritz Zwicky was already arguing that clusters of galaxies held enormous amounts of this mysterious stuff (Figure 22.6). Few of his colleagues paid attention, but later observations supported Zwicky's claims.

Zwicky was one of the first astronomers to think of galaxy clusters as huge swarms of galaxies bound together by gravity. It seemed natural to him that galaxies clumped closely together in space should all be orbiting one another, just like the stars in a star cluster. He therefore assumed that he could measure cluster masses by observing galaxy motions and applying Newton's laws of motion and gravitation.

Armed with a spectrograph, Zwicky measured the redshifts of the galaxies in a particular cluster and used these redshifts to calculate the speeds at which the individual galaxies are moving away from us. He determined the recession speed of the cluster as a whole—that is, the speed at which the expansion of the universe carries it away from us—by averaging the speeds of its individual galaxies.

Once he knew the recession speed for the cluster, Zwicky could subtract this speed from each individual galaxy's speed to determine the speeds of galaxies relative to the cluster center. Of course, this method told him only the average *radial* component (the speed toward or away from us) of the actual galaxy velocities [Section 5.5], but by averaging enough individual galaxies, Zwicky could get a good average orbital velocity for the cluster galaxies as a whole. Once he knew the average orbital velocity of the galaxies, he could use Newton's universal law of gravitation to estimate the cluster's mass (see Mathematical Insight 22.2). Finally, he compared the cluster's mass to its luminosity.

To his surprise, Zwicky found that clusters of galaxies have much greater masses than their luminosities would suggest. That is, when he estimated the total mass of stars necessary to account for the overall luminosity of a cluster, he found that it was far less than the mass he measured

---

MATHEMATICAL INSIGHT **22.2** **Cluster Masses from Galaxy Orbits**

The first method described in the text for measuring the masses of galaxy clusters relies on the *orbital velocity law* introduced in Mathematical Insight 19.1:

$$M_r = \frac{r \times v^2}{G}$$

where M_r is the total mass contained *within* a distance r of a galaxy's center, v is the average velocity of objects orbiting the center of the galaxy, and G is the gravitational constant ($G = 6.67 \times 10^{-11}$ m^3/(kg $\times$ s^2)). Recall that this law is derived from Newton's version of Kepler's third law, which in turn is derived from Newton's universal law of gravitation. Thus, the orbital velocity law is simply a convenient way of using the law of gravitation to calculate a galaxy's mass from the speeds of orbiting objects.

This law also applies to galaxy clusters (and other objects) if we simply consider r as the distance from the center of the cluster (or other object) rather than from the center of an individual galaxy. However, it is exact only for galaxies on circular orbits.

Example: A galaxy cluster has a radius of 6.2 million light-years, and Doppler shift measurements show that its galaxies orbit the center of the cluster with an average speed of approximately 1,350 km/s. Find the cluster's mass.

Solution:

Step 1 Understand: We can find the cluster's mass from the orbital velocity law above because we are given both the cluster radius (r in the formula) and the average orbital speed of the galaxies (v in the formula). However, we cannot directly use the given values because their units are inconsistent: The radius is given in light-years but the speed is given in kilometers per second. Moreover, we are given the gravitational constant G in units that use meters. Thus, before we can solve the problem, we'll need to convert the radius into meters and the speed into meters per second.

Step 2 Solve: We first need to perform the unit conversions on the given values; then we can plug the results into the orbital velocity law. You should confirm for yourself that the radius (r) of 6.2 million light-years is equivalent to 5.9×10^{22} meters and that the speed (v) of 1,350 km/s is equivalent to 1.35×10^6 m/s. Using these values, we find:

$$\begin{aligned} M_r &= \frac{r \times v^2}{G} \\ &= \frac{(5.9 \times 10^{22} \text{ m}) \times (1.35 \times 10^6 \text{ m/s})^2}{6.67 \times 10^{-11} \frac{\text{m}^3}{\text{kg} \times \text{s}^2}} \\ &= 1.6 \times 10^{45} \text{ kg} \end{aligned}$$

Step 3 Explain: We have found that the mass of the cluster is 6.7×10^{44} kilograms. However, it's difficult to interpret such a large mass expressed in kilograms, so we need to convert it to a more appropriate unit. In this case, where we are dealing with the mass of a galaxy, the most appropriate unit is solar masses. Looking back at Chapter 14 or in Appendix E, you'll find that the Sun's mass is 2.0×10^{30} kilograms. Thus, the cluster mass in units of solar masses is:

$$M_r = 1.6 \times 10^{45} \text{ kg} \times \frac{1 \text{ M}_{\text{Sun}}}{2 \times 10^{30} \text{ kg}} \approx 8.0 \times 10^{14} \text{ M}_{\text{Sun}}$$

The cluster mass is about 8.0×10^{14}, or 800 trillion, solar masses. If you recall that the total mass of the Milky Way (including dark matter) is somewhere near 1 trillion solar masses, you'll see that the cluster contains the equivalent mass of about 800 galaxies as large as the Milky Way.

by studying galaxy speeds. He concluded that most of the matter within these clusters must not be in the form of stars and instead must be almost entirely dark. Many astronomers disregarded Zwicky's result, believing that he must have done something wrong to arrive at such a strange answer. Today, far more sophisticated measurements of galaxy orbits in clusters confirm Zwicky's original finding.

THINK ABOUT IT

What would happen to a cluster of galaxies if you instantly removed all the dark matter without changing the velocities of the galaxies?

Hot Gas in Clusters A second method for measuring the mass of a cluster of galaxies relies on observing X rays from the hot gas that fills the space between the galaxies in the cluster (Figure 22.7). This gas (sometimes called the *intracluster medium*) is so hot that it emits primarily X rays and therefore went undetected until the 1960s, when X-ray telescopes were finally launched above Earth's atmosphere. The temperature of this gas is tens of millions of degrees in many clusters and can exceed 100 million degrees in the largest clusters. This hot gas can also contain a great deal of mass. Large clusters have up to seven times as much mass in the form of X-ray-emitting gas than they do in the form of stars.

The hot gas can tell us about dark matter because its temperature depends on the total mass of the cluster. The gas in most clusters is nearly in a state of *gravitational equilibrium*—that is, the outward gas pressure balances gravity's inward pull [Section 14.1]. In this state of balance, the average kinetic energies of the gas particles are determined primarily by the strength of gravity and hence by the amount of mass within the cluster. Because the temperature of a gas

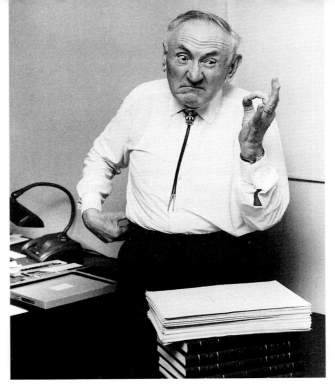

Figure 22.6 Fritz Zwicky, discoverer of dark matter in clusters of galaxies. Zwicky had an eccentric personality, but some of his ideas that seemed strange in the 1930s proved correct many decades later.

reflects the average kinetic energies of its particles, the gas temperatures we measure with X-ray telescopes tell us the average speeds of the X-ray-emitting particles (see Mathematical Insight 22.3). We can then use these particle speeds to determine the cluster's total mass.

The results obtained with this method agree well with the results found by studying the orbital motions of the cluster's galaxies. Even after we account for the mass of the

---

SPECIAL TOPIC Pioneers of Science

Scientists always take a risk when they publish what they think are groundbreaking results. If their results turn out to be in error, their reputations may suffer. When it came to dark matter, the pioneers in its discovery risked their entire careers. A case in point is Fritz Zwicky, and his proclamations in the 1930s about dark matter in clusters of galaxies. Most of his colleagues considered him an eccentric who leapt to premature conclusions.

Another pioneer in the discovery of dark matter was Vera Rubin, an astronomer at the Carnegie Institution. Working in the 1960s, she became the first woman to observe under her own name at California's Palomar Observatory, then the largest telescope in the world. (Another woman, Margaret Burbidge, was permitted to observe at Palomar earlier but was required to apply for time under the name of her husband, also an astronomer.) Rubin first saw the gravitational signature of dark matter in spectra that she recorded of stars in the Andromeda Galaxy. She noticed that stars in the outskirts of Andromeda moved at surprisingly high speeds, suggesting a stronger gravitational attraction than the mass of the galaxy's stars alone could explain. In other words, she found that the rotation curve for

Andromeda is relatively flat to great distances from the center, just as we now know is also the case for the Milky Way.

Working with a colleague, Kent Ford, Rubin constructed rotation curves for the hydrogen gas in many other spiral galaxies (by studying Doppler shifts in the spectra of hydrogen gas) and discovered that flat rotation curves are common. Although Rubin and Ford did not immediately recognize the significance of the results, they were soon arguing that the universe must contain substantial quantities of dark matter.

For a while, many other astronomers had trouble believing the results. Some astronomers suspected that the bright galaxies studied by Rubin and Ford were unusual for some reason. So Rubin and Ford went back to work, obtaining rotation curves for fainter galaxies. They found flat rotation curves—a signature of dark matter—even in these galaxies. By the 1980s, the evidence that Rubin, Ford, and other astronomers measuring rotation curves had compiled was so overwhelming that even the critics came around. Either the theory of gravity was wrong, or the astronomers measuring rotation curves had discovered dark matter in spiral galaxies.

Figure 22.7 A distant cluster of galaxies in both visible light and X-ray light. The visible-light photo shows the individual galaxies as whitish blobs. The blue overlay represents X-ray emission from extremely hot gas (around 100 million Kelvin) in the cluster. Evidence for dark matter comes both from the observed motions of the visible galaxies and from the temperature of the hot gas. (The photo shows a region about 8 million light-years across.)

hot gas, we find that the amount of dark matter in clusters of galaxies is up to 50 times that of the combined mass of the stars in the cluster's galaxies. In other words, the gravity of dark matter seems to be binding the galaxies of a cluster together in much the same way that gravity helps bind individual galaxies together.

Gravitational Lensing   Until very recently, astronomers relied exclusively on methods based on Newton's laws to measure galaxy and cluster masses. These laws keep telling us that the universe holds far more matter than we can see. Can we trust these laws? One way to check is to measure masses in a different way. Today, astronomers have another tool for measuring masses: *gravitational lensing*.

Gravitational lensing occurs because masses distort spacetime—the "fabric" of the universe [Section S3.3]. Massive objects can therefore act as **gravitational lenses** that bend light beams passing nearby. This prediction of Einstein's general theory of relativity was first verified in 1919 during an eclipse of the Sun [Section S3.4]. Because the light-bending angle of a gravitational lens depends on the mass of the object doing the bending, we can measure the masses of objects by observing how strongly they distort light paths.

Figure 22.8 shows a striking example of how a cluster of galaxies can act as a gravitational lens. Many of the yellow elliptical galaxies concentrated toward the center of

---

MATHEMATICAL INSIGHT 2 2 . 3   **Cluster Masses from Gas Temperature**

To find a cluster's mass from the temperature of its hot, X-ray-emitting gas, we need to know how the gas temperature is related to the speeds of individual gas particles. Although we will not present a derivation here, the following formula tells us the approximate speed at which hydrogen nuclei move around a cluster if its hot gas has a temperature T:

$$v_H = 140 \tfrac{m}{s} \times \sqrt{T}$$

where v_H is the average orbital speed of the hydrogen nuclei and T is the temperature in Kelvin. We can use this formula for the hot gas in galaxy clusters because most of the gas is composed of hydrogen. Once we find the speeds of the hydrogen nuclei, we can use them in the orbital velocity law to find the cluster mass.

Example: Consider the galaxy cluster described in the Example in Mathematical Insight 22.2, with a radius of 6.2 million light-years. Suppose the cluster is filled with hot gas that has a temperature of 9×10^7 K. Use this temperature to find the cluster's mass.

Solution:

Step 1 Understand: We are given a gas temperature and asked to find a cluster mass. From Mathematical Insight 22.2, we know that we can use the orbital velocity law to find the cluster mass if we know the cluster's radius (r) and the average velocity (v) of orbiting particles. We are given the cluster's radius of 6.2 million light-years, and we already found that this is equivalent

to 5.9×10^{22} meters. We can use the formula given above to find the average orbital speed of hydrogen nuclei, which we can use as the velocity (v) in the orbital velocity law.

Step 2 Solve: First, we find the average orbital speed of the hydrogen nuclei from the given temperature of 9×10^7 K:

$$v_H = 140 \tfrac{m}{s} \times \sqrt{T}$$
$$= 140 \tfrac{m}{s} \times \sqrt{9 \times 10^7}$$
$$= 1.3 \times 10^6 \tfrac{m}{s}$$

Now we use this value as v in the orbital velocity law, along with the cluster radius $r = 4.5 \times 10^{22}$ meters. We find that the cluster mass is:

$$M_r = \frac{r \times v^2}{G}$$
$$= \frac{(5.9 \times 10^{22} \text{ m}) \times (1.3 \times 10^6 \text{ m/s})^2}{6.67 \times 10^{-11} \frac{\text{m}^3}{\text{kg} \times \text{s}^2}}$$
$$\approx 1.5 \times 10^{45} \text{ kg}$$

Step 3 Explain: We have found that the mass of the cluster is 1.5×10^{45} kilograms. You should confirm for yourself that this is equivalent to about 750 trillion solar masses. Note that this is just slightly lower than the 800 trillion solar masses that we found in Mathematical Insight 22.2 from the galaxy speeds, showing that the two methods of estimating mass agree well in this example.

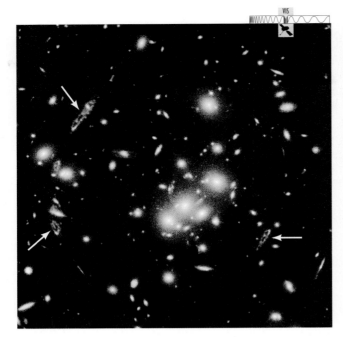

Figure 22.8 Interactive Photo ↖ This Hubble Space Telescope photo shows a galaxy cluster acting as a gravitational lens. The yellow elliptical galaxies are cluster members. The many small blue ovals (such as those indicated by the arrows) are multiple images of a single galaxy that lies almost directly behind the cluster's center. (The picture shows a region about 1.4 million light-years across.)

the picture belong to the cluster, but at least one of the galaxies pictured does not. At several positions on various sides of the central clump of yellow galaxies, you will notice multiple images of the same blue galaxy. Each one of these images, whose sizes differ, looks like a distorted oval with an off-center smudge.

The blue galaxy seen in these multiple images lies almost directly behind the center of the cluster, at a much greater distance. We see multiple images of this single galaxy because photons do not follow straight paths as they travel from the galaxy to Earth. Instead, the cluster's gravity bends the photon paths, allowing light from the galaxy to arrive at Earth from a few slightly different directions (Figure 22.9). Each alternative path produces a separate, distorted image of the blue galaxy.

Multiple images of a gravitationally lensed galaxy are rare. They occur only when a distant galaxy lies directly behind the lensing cluster. However, single, distorted images of gravitationally lensed galaxies are quite common. Figure 22.10 shows a typical example. This picture shows numerous normal-looking galaxies and several arc-shaped galaxies. The oddly curved galaxies are not members of the cluster, nor are they really curved. They are normal galaxies lying far beyond the cluster whose images have been distorted by the cluster's gravity.

Careful analyses of the distorted images created by clusters enable us to measure cluster masses without resorting to Newton's laws. Instead, Einstein's general theory of relativity tells us how massive these clusters must be to generate the observed distortions. It is reassuring that cluster masses derived in this way generally agree with those de-

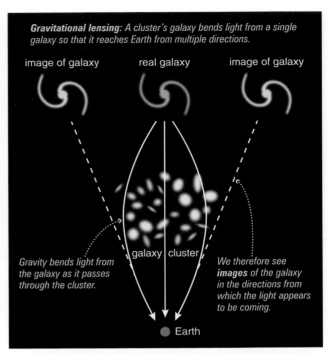

Gravitational lensing: A cluster's galaxy bends light from a single galaxy so that it reaches Earth from multiple directions.

image of galaxy   real galaxy   image of galaxy

Gravity bends light from the galaxy as it passes through the cluster.

galaxy cluster

We therefore see **images** *of the galaxy in the directions from which the light appears to be coming.*

● Earth

Result: *Through a telescope on Earth, we see multiple images of what is really a single galaxy.*

Figure 22.9 Interactive Figure ↖ A cluster's powerful gravity bends light paths from background galaxies to Earth. If light can arrive from several different directions, we see multiple images of the same galaxy.

rived from galaxy velocities and X-ray temperatures. The three different methods all indicate that clusters of galaxies hold very substantial amounts of dark matter.

• Does dark matter really exist?

Astronomers have made a strong case for the existence of dark matter, but could they be completely off base? Is it possible that dark matter is a figment of human imagination, and that there's a completely different explanation for the observations we've discussed? Addressing these questions gives us a chance to see how science progresses.

All the evidence for dark matter rests on our understanding of gravity. For individual galaxies, the case for dark matter rests primarily on applying Newton's laws of motion and gravity to observations of the orbital speeds of stars and gas clouds. We've used the same laws to make the case for dark matter in clusters, along with additional evidence based on gravitational lensing predicted by Einstein's general theory of relativity. Thus, it seems that one of the following must be true:

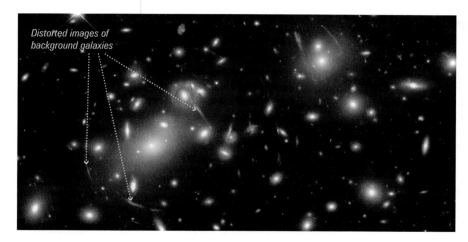

Distorted images of background galaxies

Figure 22.10 Hubble Space Telescope photo of the cluster Abell 2218. The thin, elongated galaxies around the main clump of galaxies on the left side of the picture are the images of background galaxies distorted by the cluster's gravity. By measuring these distortions, astronomers can determine the total amount of mass in the cluster. (The region pictured is about 1.4 million light-years from side to side.)

1. Dark matter really exists, and we are observing the effects of its gravitational attraction.
2. There is something wrong with our understanding of gravity that is causing us to mistakenly infer the existence of dark matter.

We cannot yet rule out the second possibility, but most astronomers consider it very unlikely. Newton's laws of motion and gravity are among the most trustworthy tools in science. We have used them time and again to measure masses of celestial objects from their orbital properties. We found the masses of Earth and the Sun by applying Newton's version of Kepler's third law to objects that orbit them [Section 4.4]. We used this same law to calculate the masses of stars in binary star systems, revealing the general relationships between the masses of stars and their outward appearances. Newton's laws have also told us the masses of things we can't see directly, such as the masses of orbiting neutron stars in X-ray binaries and of black holes in active galactic nuclei. Einstein's general theory of relativity likewise stands on solid ground, having been repeatedly tested and verified to high precision in many observations and experiments. Thus, we have good reason to trust our current understanding of gravity.

Moreover, many scientists have already made valiant efforts to come up with alternate theories of gravity that could account for the observations implying the existence of dark matter in some other way. (After all, there's a Nobel Prize waiting for anyone who can substantiate a new theory of gravity.) So far, no one has succeeded in doing so in a way that can still explain the many other observations accounted for by our current theories of gravity.

In essence, our high level of confidence in our current understanding of gravity gives us equally high confidence that dark matter really exists. Thus, while we should always keep an open mind about the possibility of future changes in our understanding, we will proceed for now under the assumption that dark matter is real. With that assumption in mind, let's turn our attention to the nature of what seems to be the most common form of matter in the universe.

THINK ABOUT IT

Should the fact that we have three different ways of measuring cluster masses give us greater confidence that we really do understand gravity and that dark matter really does exist? Why or why not?

• What might dark matter be made of?

What is all this dark stuff in galaxies and clusters of galaxies? We don't yet know. Nevertheless, we can make at least some educated guesses.

At least some of the dark matter is likely to be *ordinary,* made of protons, neutrons, and electrons. The only unusual thing about dark matter of this kind is that it doesn't emit much detectable radiation. Otherwise, it is made of the same stuff as all the "bright matter" that we can see. However, as we'll discuss shortly, it's also likely that some of the dark matter is *extraordinary,* made of particles that we have yet to discover.

A bit of terminology will be useful. Because the protons and neutrons that make up most of the mass of ordinary matter belong to a category of particles called **baryons**, ordinary matter is sometimes called **baryonic matter**. By extension, extraordinary matter is called **nonbaryonic matter**.

Ordinary Dark Matter   Matter need not be extraordinary to be dark. In astronomy, "dark" merely means "not as bright as a normal star and therefore not visible across vast distances of space." Your body is dark matter, because you would be far too dim for our telescopes to detect if you were somehow flung into the halo of our galaxy. Everything you own is dark matter. Earth and the rest of the planets are dark matter as well. In fact, the "failed stars" known as *brown dwarfs* [Section 16.3] and even faint red main-sequence stars of spectral type M [Section 15.2] are too dim for current telescopes to see in the halo and therefore qualify as dark matter. Thus, it's conceivable that trillions of faint red stars, brown dwarfs, and Jupiter-size objects left over from the Milky Way's formation still roam our galaxy's halo, providing much of its mass. These objects are sometimes called

MACHOs, for *massive compact halo objects,* although they are better thought of as dim, starlike (or planetlike) objects.

MACHOs such as brown dwarfs and dim, red stars are too faint for us to see directly, but there are other ways to search for them. One innovative technique takes advantage of gravitational lensing on a much smaller scale than the examples we studied for clusters of galaxies. If trillions of these dim stars and similar objects really roam the halo of the galaxy, every once in a while one of them should drift across our line of sight to a more distant star. When the object lies almost directly between us and the farther star, its gravity will focus more of the star's light directly toward Earth. The distant star will appear much brighter than usual for several days or weeks as the lensing object passes in front of it (Figure 22.11). We cannot see the lensing object itself, but the duration of the lensing event reveals its mass.

Gravitational lensing events such as these are rare, happening to about one star in a million each year. To detect such lensing events, we therefore must monitor huge numbers of stars. Current large-scale monitoring projects now record numerous lensing events annually. These events demonstrate that dim, starlike objects (MACHOs) do indeed populate our galaxy's halo, but not in large enough numbers to account for all the Milky Way's dark matter. Similar measurements rule out the possibility that the dark matter consists of large numbers of black holes formed by the deaths of massive stars.[*] Something else lurks unseen in the outer reaches of our galaxy.

Extraordinary Dark Matter   A more exotic possibility is that most of the dark matter in galaxies and clusters of galaxies is not made of ordinary, baryonic matter at all. Let's begin to explore this possibility by taking another look at those nonbaryonic particles we discussed in Section 14.2: neutrinos. These unusual particles are dark by their very nature, because they have no electrical charge and hence cannot emit electromagnetic radiation of any kind. Moreover, they are never bound together with charged particles in the way that neutrons are bound in atomic nuclei, so their presence cannot be revealed by associated light-emitting particles.

Particles like neutrinos interact with other forms of matter through only two of the four forces: gravity and the *weak force* [Section S4.2]. For this reason, they are said to be *weakly interacting particles.* If you recall that trillions of neutrinos from the Sun are passing through your body at this very moment without doing any damage, you'll see why the name *weakly interacting* fits well.

The dark matter in galaxies cannot be made of neutrinos, because these very-low-mass particles travel through the universe at enormous speeds and can easily escape a galaxy's gravitational pull. (However, neutrinos make up a

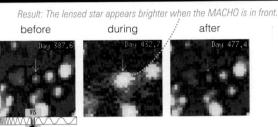

Result: The lensed star appears brighter when the MACHO is in front.

before   during   after

Figure 22.11 When a small, starlike object (MACHO) passes in front of a more distant star, gravitational lensing temporarily makes the star appear brighter. Searches for such events show that our galaxy's halo does indeed contain dim, starlike objects, but that these objects do not constitute the majority of the galaxy's dark matter.

small amount of the dark matter outside galaxies.) What if other weakly interacting particles exist that are similar to neutrinos but considerably heavier? They too would evade direct detection, but they would move more slowly so that their mutual gravity could hold together a large collection of them. Such hypothetical particles are called **WIMPs,** for *weakly interacting massive particles.* Note that WIMPs are subatomic particles, so the "massive" in their name is relative—they are "massive" only in comparison to lightweight particles like neutrinos. (They are also often called *cold dark matter* to set them apart from the faster-moving neutrinos.) WIMPs could make up most of the mass of a galaxy or cluster of galaxies, but they would be completely invisible in all wavelengths of light. Most astronomers now consider it likely that WIMPs make up the vast majority of dark matter, and hence the majority of all matter in the universe.

It might surprise you that scientists suspect the universe to be filled with particles they haven't yet discovered. However, this hypothesis would also explain why dark matter seems to be distributed throughout spiral galaxy halos rather than concentrated in flattened disks like the visible matter. Recall that galaxies are thought to have formed as gravity

---

[*]Lensing observations cannot yet rule out the presence of black holes with masses less than those of stars. Such low-mass black holes could conceivably be left over from the Big Bang, but our best theoretical models of the early universe do not predict a large enough number of them to account for the dark matter.

pulled together matter in regions of slightly enhanced density in the early universe [Section 21.1]. This matter would have consisted mostly of dark matter mixed with some ordinary (hydrogen and helium) gas. The ordinary gas could collapse to form a rotating disk because individual gas particles could lose orbital energy: Collisions among gas particles can convert some of their orbital energy into radiative energy that escapes from the galaxy in the form of photons. In contrast, WIMPs cannot produce photons and they rarely interact and exchange energy with other particles. Thus, as the gas collapsed to form a disk, WIMPs would have remained stuck on orbits far out in the galactic halo—just where most dark matter seems to be located.

By itself, the agreement between the measured distribution of dark matter in galaxies and what we'd expect from dark matter made of WIMPs doesn't prove that extraordinary dark matter exists. However, as we'll discuss in the next chapter, there are additional reasons many astronomers believe that baryons represent only a minority of the universe's mass and hence that WIMPs are the most common form of matter in the universe.

What do you think of the idea that much of the universe is made of as-yet-undiscovered particles? Can you think of other instances in the history of science in which the existence of something was predicted before it was discovered?

22.3 Structure Formation

Dark matter remains enigmatic, but every year we are learning more about its role in the universe. Because galaxies and clusters of galaxies seem to contain much more dark matter than luminous matter, dark matter's gravitational pull must be the primary force holding these structures together. Thus, we strongly suspect that the gravitational attraction of dark matter is what pulled galaxies and clusters together in the first place.

• What is the role of dark matter in galaxy formation?

Stars, galaxies, and clusters of galaxies are all *gravitationally bound systems*—their gravity is strong enough to hold them together. In most of the gravitationally bound systems we have discussed so far, gravity has completely overwhelmed the expansion of the universe. That is, while the universe as a whole is expanding, space is *not* expanding within our solar system, our galaxy, or our Local Group of galaxies.

Our best guess at how galaxies formed, outlined in Section 21.2, envisions them growing from slight density enhancements that were present in the very early universe. During the first few million years after the Big Bang, the universe expanded everywhere. Gradually, the stronger gravity in regions of enhanced density pulled in matter until

these regions stopped expanding and became protogalactic clouds, even as the universe as a whole continued (and still continues) to expand.

If dark matter is indeed the most common form of mass in galaxies, it must have provided most of the gravitational attraction responsible for creating the protogalactic clouds. The hydrogen and helium gas in the protogalactic clouds collapsed inward and gave birth to stars, while weakly interacting dark matter remained in the outskirts because of its inability to radiate away its orbital energy. According to this model, the luminous matter in each galaxy must still be nestled inside the larger cocoon of dark matter that initiated the galaxy's formation (see Figure 22.2), just as observational evidence seems to suggest.

The formation of galaxy clusters probably echoes the formation of galaxies. Early on, all the galaxies that will eventually constitute a cluster are flying apart with the expansion of the universe, but the gravity of the dark matter associated with the cluster eventually reverses the trajectories of these galaxies. The galaxies ultimately fall back inward and start orbiting each other randomly, like the stars in the halo of our galaxy.

Some clusters of galaxies apparently have not yet finished forming, because their immense gravity is still drawing in new members. For example, the large and relatively nearby Virgo Cluster of galaxies (about 60 million light-years away) appears to be drawing in the Milky Way and other galaxies of the Local Group. The evidence comes from careful study of galaxy speeds. Plugging the Virgo Cluster's distance into Hubble's law tells us the speed at which the Milky Way and the Virgo Cluster should be drifting apart due to the universal expansion [Section 20.2]. However, the measured speed is about 400 km/s slower than the speed we predict from Hubble's law alone. We conclude that this 400 km/s discrepancy (sometimes called a *peculiar velocity*) arises because the Virgo Cluster's gravity is pulling us back against the flow of universal expansion. In other words, while the Milky Way and other galaxies of our Local Group are still moving away from the Virgo Cluster with the expansion of the universe, the rate at which we are separating from the cluster is slowing with time. Eventually, the cluster's gravity may stop the separation altogether, at which point the cluster will begin pulling in the galaxies of our Local Group, ultimately making them members of the cluster.

Similar processes are taking place on the outskirts of other large clusters of galaxies, where we see many galaxies whose speeds indicate that the cluster's gravity is pulling on them. Eventually, some or perhaps all of these galaxies will fall into the cluster. Thus, many clusters are still attracting galaxies, adding to the hundreds they already contain. On even larger scales, clusters themselves seem to be tugging on one another, hinting that they might be parts of even bigger gravitationally bound systems, called **superclusters,** that are still in the early stages of formation (Figure 22.12). But some structures are even larger than superclusters.

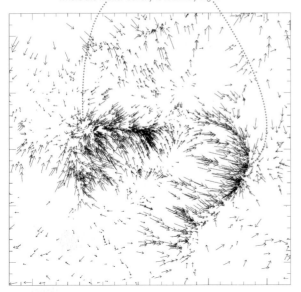

Figure 22.12 This graph represents the motions of galaxies attributable to effects of gravity. Each arrow represents the amount by which a galaxy's actual velocity (inferred from a combination of observations and modeling) differs from the velocity we'd expect it to have from Hubble's law alone. The Milky Way is at the center of the picture, which shows an area about 600 million light-years from side to side. (Only a representative sample of galaxies is shown.) Note how the galaxies tend to flow into regions where the density of galaxies is already high. These vast, high-density regions are probably superclusters in the process of formation.

THINK ABOUT IT

State whether each of the following is a gravitationally bound system, and explain why: (a) Earth; (b) a hurricane (on Earth); (c) the Orion Nebula; (d) a supernova.

• What are the largest structures in the universe?

Beyond about 300 million light-years from Earth, deviations from Hubble's law owing to gravitational tugs are insignificant compared with the universal expansion, so Hubble's law becomes our primary method for measuring galaxy distances [Section 20.2]. Using this law, astronomers can make maps of the distribution of galaxies in space. Such maps reveal **large-scale structures** much vaster than clusters of galaxies.

Mapping Large-Scale Structures Making maps of galaxy locations requires an enormous amount of data. A long-exposure photo showing galaxy positions is not enough, because it does not tell us the galaxy distances. We must also measure the redshift of each individual galaxy so that we can estimate its distance by applying Hubble's law. These measurements once required intensive labor, and up until just over

a decade ago it took years of effort to map the locations of just a few hundred galaxies. However, astronomers have recently developed technology that allows redshift measurements for hundreds of galaxies during a single night of telescopic observation. As a result, we now have redshift measurements—and hence estimated distances—for many thousands of distant galaxies.

Figure 22.13 shows the distribution of galaxies in three slices of the universe, each extending farther out in distance. Our Milky Way Galaxy is located at the vertex at the far left, and each dot represents another entire galaxy of stars. The slice at the left comes from one of the first surveys of large-scale structures performed in the 1980s. This map, which required years of effort by many astronomers, dramatically revealed the complex structure of our corner of the universe. It showed that galaxies are not scattered randomly through space but are instead arranged in huge chains and sheets that span many millions of light-years. Clusters of galaxies are located at the intersections of these chains. Between these chains and sheets of galaxies lie giant empty regions called **voids**. The other two slices show data from the more recent Sloan Digital Sky Survey. The Sloan Survey has measured redshifts for nearly a million galaxies spread across about one-fourth of the sky.

Some of the structures in these pictures are amazingly large. The so-called Sloan Great Wall, clearly visible in the center slice, extends more than 1 billion light-years from end to end. Immense structures such as these apparently have not yet collapsed into randomly orbiting, gravitationally bound systems.

The universe may still be growing structures on these very large scales. However, there seems to be a limit to the size of the largest structures. If you look closely at the rightmost slice in Figure 22.13, you'll notice that the overall distribution of galaxies appears nearly uniform on scales larger than about a billion light-years. In other words, on very large scales the universe looks much the same everywhere, in agreement with what we expect from the *Cosmological Principle* [Section 20.2].

The Origin of Large Structures Why is gravity collecting matter on such enormous scales? Just as we suspect that galaxies formed from regions of slightly enhanced density in the early universe, we suspect that these larger structures were also regions of enhanced density. Galaxies, clusters, superclusters, and the Sloan Great Wall probably all started as mildly high-density regions of different sizes. The voids in the distribution of galaxies probably started as mildly low-density regions.

If this picture of structure formation is correct, then the structures we see in today's universe mirror the original distribution of dark matter very early in time. Supercomputer models of structure formation in the universe can now simulate the growth of galaxies, clusters, and larger structures from tiny density enhancements as the universe evolves (Figure 22.14). Models of extremely large regions

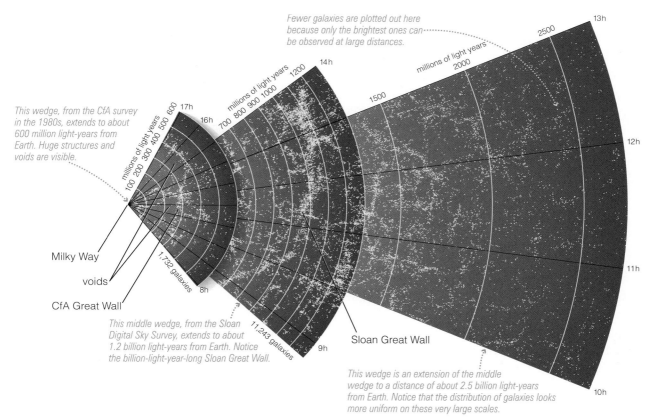

This wedge, from the CfA survey in the 1980s, extends to about 600 million light-years from Earth. Huge structures and voids are visible.

Fewer galaxies are plotted out here because only the brightest ones can be observed at large distances.

millions of light years
100 200 300 400 500 600

17h
16h

14h
millions of light years
700 800 900 1000 1200

13h
2500
millions of light years
2000
1500

Milky Way

voids

CfA Great Wall

1,732 galaxies

8h

This middle wedge, from the Sloan Digital Sky Survey, extends to about 1.2 billion light-years from Earth. Notice the billion-light-year-long Sloan Great Wall.

11,243 galaxies

9h

Sloan Great Wall

12h

11h

10h

This wedge is an extension of the middle wedge to a distance of about 2.5 billion light-years from Earth. Notice that the distribution of galaxies looks more uniform on these very large scales.

Figure 22.13 Each of these three wedges shows a "slice" of the universe extending outward from our own Milky Way Galaxy. The dots represent galaxies, shown at their measured distances from Earth. We see that galaxies are not scattered randomly but instead trace out long chains and sheets surrounded by huge voids containing very few galaxies. (The wedges are shown flat but actually are a few angular degrees in thickness; the CfA wedge at left does not actually line up with the two Sloan wedges.)

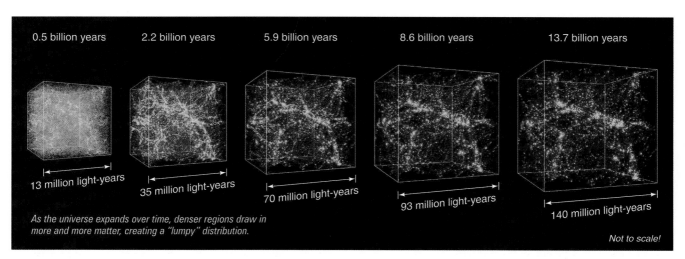

0.5 billion years   2.2 billion years   5.9 billion years   8.6 billion years   13.7 billion years

13 million light-years

35 million light-years

70 million light-years

93 million light-years

140 million light-years

As the universe expands over time, denser regions draw in more and more matter, creating a "lumpy" distribution.

Not to scale!

Figure 22.14 🖥️ Interactive Figure Frames from a supercomputer simulation of structure formation. These five boxes depict the development of a cubical region that is now 140 million light-years across. The labels above the boxes give the age of the universe, and the labels below give the size of the box as it expands with time. Notice that the distribution of matter is only slightly lumpy when the universe is young (left frame). Structures grow more pronounced with time as the densest lumps draw in more and more matter.

Figure 22.15 These images from an extremely large computer simulation illustrate the structure of dark matter in the universe. The largest image shows a region similar in size to our observable universe, and the image sequence zooms in on a massive cluster of galaxies. The images show structure as it would appear if we could see dark matter—the brightest clumps in the image represent the highest densities of dark matter in the computer simulation. Notice that the large-scale distribution of dark matter has a uniform web-like pattern.

reveal how dark matter should be distributed throughout the entire observable universe (Figure 22.15). The results of these models look remarkably similar to the slices of the universe in Figure 22.13, bolstering our confidence in this scenario. However, the models do not tell us *why* the universe started with these slight density enhancements—that is a topic for the next chapter. Nevertheless, it seems increasingly clear that these "lumps" in the early universe were the seeds of all the marvelous structures we see in the universe today.

 Fate of the Universe Tutorial, Lessons 1–3

22.4 The Fate of The Universe

Some say the world will end in fire,
Some say in ice.
From what I've tasted of desire
I hold with those who favor fire.
But if it had to perish twice,
I think I know enough of hate

To say that for destruction ice
Is also great
And would suffice.

Robert Frost

We now arrive at one of the ultimate questions in astronomy: How will the universe end? Edwin Hubble's work established that the galaxies in the universe are rapidly flying away from one another [Section 20.2], but the gravitational pull of each galaxy on every other galaxy acts to slow the expansion. The possible outcomes would appear to fall into two general categories. If gravity is strong enough, the expansion will someday halt and the universe will begin collapsing, eventually ending in a cataclysmic crunch. Alternatively, if the expansion can overcome the pull of gravity, the universe will continue to expand forever, growing ever colder as its galaxies grow ever farther apart. The fate of the universe thus seems to boil down to a simple question: Is the universe expanding fast enough to escape its own gravitational pull and keep on expanding forever?

• Will the universe continue expanding forever?

Let's begin by considering the fate of the universe as it seemed just a few years ago, before the discoveries that suggested the presence of dark energy in the universe. In the absence of dark energy, we would expect gravity to be slowing the expansion of the universe with time. In that case, the fate of the universe hinges on the overall strength of the universe's gravitational pull. The strength of this pull depends on the density of matter in the universe: The greater the density, the greater the overall strength of gravity and the higher the likelihood that gravity will someday halt the expansion.

Precise calculations show that gravity can win out over expansion if the current density of the universe exceeds a seemingly minuscule 10^{-29} gram per cubic centimeter, which is roughly equivalent to a few hydrogen atoms in a volume the size of a closet. The precise density marking the dividing line between eternal expansion and eventual collapse is called the **critical density**. (Remember that for the moment, we are assuming a universe without dark energy.)

Observations of the luminous matter in galaxies show that the mass contained in stars falls far short of the critical density. The visible parts of galaxies contribute about 0.5% of the matter density needed to halt the universe's expansion. The fate of the universe would thus seem to rest with the dark matter. Is there enough dark matter to halt the expansion of the universe?

Because stars seem to contribute about 0.5% of the matter density needed to halt the expansion, the expansion could halt only if the total mass of dark matter were at least 200 times that of the mass in stars. Our studies of individual galaxies suggest that they contain at least 10 times as much dark matter as matter in stars, and studies of clusters of galaxies raise that number further to about 50 times as much dark matter as matter in stars. However, this is still only about a quarter of the amount of dark matter needed to halt the expansion. Thus, if the proportion of dark matter in the universe at large is similar to that in clusters, the universe seems destined to expand forever. For gravity to reverse the expansion and pull the universe back together, even more dark matter would have to lie beyond the boundaries of clusters.

If large-scale structures really did contain a higher proportion of dark matter than do clusters, the influence of that extra dark matter should show up in the velocities of galaxies near those large-scale structures: Larger amounts of dark matter would be causing greater deviations from Hubble's law. As of 2005, however, studies of galaxy velocities are holding the line near the value we infer from clusters, which is about 25% of the critical density required to reverse the expansion. If that is the case, the universe seems destined to expand forever, even if there is no dark energy affecting the rate of expansion.

• Is the expansion of the universe accelerating?

In the past few years, observations of distant white dwarf supernovae have enabled us to probe the fate of the universe in an entirely new way. Because white dwarf supernovae are such good standard candles [Section 20.2], we can use them to determine whether gravity has been slowing the universe's expansion, as it must if the universe is destined to end in a cataclysmic crunch. However, the astronomers who set out to measure gravity's influence over the universe by observing supernovae discovered something quite unexpected.

Instead of slowing because of gravity, the expansion of the universe appears to be speeding up, suggesting that some mysterious repulsive force—the so-called *dark energy*—is pushing all the universe's galaxies apart. This discovery, if it holds up to further scrutiny, has far-reaching implications for both the fate of the universe and our understanding of the forces that govern its behavior on large scales. To understand the evidence for an accelerating expansion, we must become more familiar with the possible futures of the universe.

Four Expansion Patterns   Astronomers subdivide the two general possibilities for the fate of the universe—expanding forever or someday collapsing—into four broad categories. Each represents a particular pattern of change in the future expansion rate (Figure 22.16). We will call these four possible expansion patterns *recollapsing, critical, coasting,* and *accelerating.* The first three possibilities assume that gravity is the only force that affects the expansion rate of the universe, while the fourth adds a repulsive force (dark energy) that opposes gravity:

● **A recollapsing universe:** If there is no dark energy and the matter density of the universe is *larger* than the critical density, then the collective gravity of all its matter will eventually halt the universe's expansion and reverse it. The galaxies will come crashing back together, and the entire universe will end in a fiery "Big Crunch." We call this a recollapsing universe because the final state, with all matter collapsed together, would look much like the state in which the universe began in the Big Bang. (A recollapsing universe is sometimes called a *closed universe,* because the overall geometry of spacetime would close upon itself like the surface of a sphere [Section S3.2].)

● **A critical universe:** If there is no dark energy and the matter density of the universe *equals* the critical density, then the collective gravity of all its matter is exactly the amount needed to balance the expansion. The universe will never collapse but will expand more and more slowly as time progresses. We call this a critical universe because its density is the critical density. (Mathematically speaking, a critical universe stops expanding after

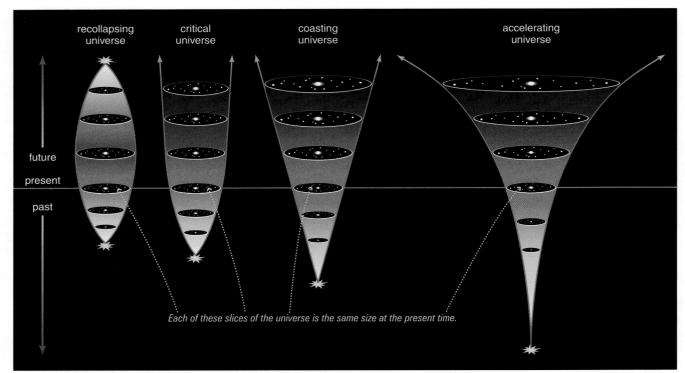

Figure 22.16 Four models for the fate of the universe. Each diagram shows how the size of a circular slice of the universe changes with time in a particular model. The slices are the same size at the present time, marked by the red line, but the models make different predictions about the sizes of the slices in the past and future. The first three cases assume that there is no dark energy, so that the fate of the universe depends only on how its actual density compares to the critical density. The last case assumes that a repulsive force—the so-called dark energy—is accelerating the expansion over time. (The diagram assumes continuous acceleration, but it is also possible that the universe initially slowed before the acceleration began.)

infinite time, and its overall geometry is "flat"—like the surface of a table but in more dimensions. Thus, a critical universe is one example of what astronomers call a *flat universe*.)

- **A coasting universe:** If there is no dark energy and the matter density of the universe is *smaller* than the critical density, then the collective gravity of all its matter cannot halt the expansion. The universe will keep expanding forever, with little change in its rate of expansion; that is, the expansion will continue to coast forever. (A coasting universe is sometimes called an *open universe*, because its overall geometry is more like the open surface of a saddle than like the closed surface of a sphere.)

- **An accelerating universe:** If dark energy exerts a repulsive force that causes the expansion of the universe to *accelerate* with time, then the expansion rate will grow with time. Galaxies will recede from one another with increasing speed, and the universe will become cold and dark more quickly than it would in a coasting universe. (Depending on the strength of gravity relative to the repulsive force, the overall geometry of an accelerating universe could be flat, open, or closed. As we'll discuss in Chapter 23, current evidence suggests a flat geometry for this case.)

THINK ABOUT IT

Do you think that one of the potential fates of the universe is preferable to the others? Why or why not?

Evidence for Acceleration Figure 22.17 illustrates how the average distance between galaxies should change with time for each possibility. The lines for the accelerating, coasting, and critical universes always continue upward as time increases, because in these cases the universe is always expanding. The steeper the slope, the faster the expansion. In the recollapsing case, the line begins on an upward slope but eventually turns around and declines as the universe contracts. All the lines pass through the same point and have the same slope at the moment labeled "now," because the current separation between galaxies and the current expansion rate in each case must agree with observations of the present-day universe.

Note that the age that we infer for the universe from its expansion rate differs in each case. A recollapsing universe requires the least amount of time to arrive at the current separation between galaxies—the example in Figure 22.17 goes from zero separation to the current separation in less than 5 billion years. The cases for which gravity is less important require more time to achieve the current separa-

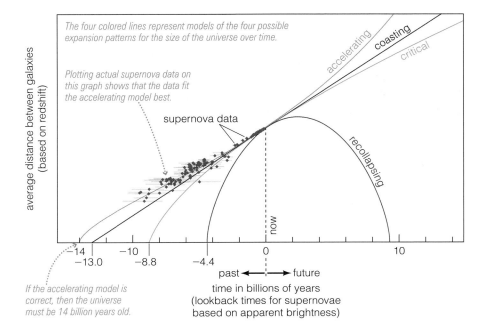

The four colored lines represent models of the four possible expansion patterns for the size of the universe over time.

Plotting actual supernova data on this graph shows that the data fit the accelerating model best.

supernova data

accelerating coasting critical

recollapsing

now

past ◄——►future

−14
−13.0
−10
−8.8
−4.4
0
10

time in billions of years
(lookback times for supernovae
based on apparent brightness)

If the accelerating model is correct, then the universe must be 14 billion years old.

average distance between galaxies (based on redshift)

Figure 22.17 Data from white dwarf supernovae are shown along with four possible models for the expansion of the universe. Each curve shows how the average distance between galaxies changes with time for a particular model. A rising curve means that the universe is expanding, and a falling curve means that the universe is contracting. Notice that the supernovae data fit the accelerating universe better than the other models.

tion between galaxies. The ages that we would infer from the examples in Figure 22.17 are 8.8 billion years for a critical universe, 13 billion years for a coasting universe, and around 14 billion years for an accelerating universe.

This relationship between the age of the universe and its expansion pattern enables us to determine the expansion pattern from observations of white dwarf supernovae. Because these supernovae are so bright and make such excellent standard candles, we can identify them and measure their distances and redshifts even when they are more than

halfway across the observable universe. The distance we measure essentially tells us the lookback time to the supernova, while its redshift tells us how much the universe has expanded since the time of the supernova explosion. Combining these two pieces of information for supernova explosions at different times in the past therefore reveals how fast the universe was expanding over that entire time period.

Observations of such distant supernovae are still very difficult, but we have some data that are plotted as dots in Figure 22.17. Although there is some scatter in these data,

---

SPECIAL TOPIC Einstein's Greatest Blunder

Shortly after Einstein completed his general theory of relativity in 1915, he found that it predicted that the universe could not be standing still: The mutual gravitational attraction of all the matter would make the universe collapse. Because Einstein thought at the time that the universe should be eternal and static, he decided to alter his equations. In essence, he inserted a "fudge factor" called the *cosmological constant* that acted as a repulsive force to counteract the attractive force of gravity.

Had he not been so convinced that the universe should be standing still, Einstein might instead have come up with the correct explanation for why the universe is not collapsing: because it is still expanding from the event of its birth. After Hubble discovered the universal expansion, Einstein supposedly called his invention of the cosmological constant "the greatest blunder" of his career.

Recently, however, astronomers have begun to take the idea of a cosmological constant more seriously. In the mid-1990s, a few observations suggested that the oldest stars were slightly older than the age of the universe derived from Hubble's constant under the assumption that gravity is the only force affecting the universe's expansion. Clearly, stars cannot be older than the universe. If these observations were being interpreted correctly, the universe had to

be older than Hubble's constant implies. If the expansion rate has accelerated, so that the universe is expanding faster today than it was in the past, then the age of the universe would be greater than that ordinarily found from Hubble's constant (see Figure 22.16). What could cause the expansion of the universe to accelerate over time? The repulsive force represented by a cosmological constant, of course.

Further study of the troubling observations has shown that the oldest stars probably are *not* older than the age of the universe derived from Hubble's constant. However, measurements of distances to high-redshift galaxies using white dwarf supernovae as standard candles now suggest that the expansion *is* accelerating. A cosmological constant arising from dark energy could account for this startling finding, but we'll need more observations before we can be sure it is correct. Interestingly, the observations we currently have indicate that dark energy has properties identical to the repulsive force that Einstein originally proposed. The amount of dark energy in each volume of space seems to remain unchanged while the universe expands, as if the vacuum of space itself were constantly rippling with energy [Section S4.5]. Einstein's greatest blunder, it seems, just won't go away.

they appear to fit the curve for an accelerating universe better than any of the other models, and they do not fit either a critical or a recollapsing universe. In other words, the observations to date seem to favor an accelerating universe.

Exactly why the expansion of the universe might be accelerating remains a deep mystery. No known force would act to push the universe's galaxies apart, and an enormous amount of energy would be required. Of course, our lack of understanding does not stop us from giving a name to the repulsive force that is causing the acceleration, and we have already discussed why it is often dubbed *dark energy*. Keep in mind, however, that we do not yet have any idea of what the dark energy might actually be. Nevertheless, if dark energy really exists, it is the most prevalent form of energy in the universe, outstripping the total mass-energy of all the matter in the universe—including the dark matter. Only continued observations will tell us whether the dark energy is real or an artifact of our still-limited data.

A Never-Ending Expansion?   Whether or not the dark energy exists, and whatever dark energy might turn out to be, it now seems likely that the universe is indeed doomed to expand forever, its galaxies receding ever more quickly into an icy, empty future. After all, our examination of the strength of gravity showed that it is too weak to stop the expansion even without dark energy, and the acceleration due to dark energy would seem only to seal this fate. Some scientists even hypothesize that the dark energy could eventually cause galaxies, stars, and planets to break apart and disperse with the never-ending expansion.

However, before we convince ourselves that we now know the fate of the universe, we should bear in mind that forever is a very long time. The universe may hold other surprises that we haven't yet discovered, surprises that might force us to rethink what might happen between now and the end of time.

> *This is the way the world ends*
> *This is the way the world ends*
> *This is the way the world ends*
> *Not with a bang but a whimper.*

From *The Hollow Men* by T. S. Eliot

Putting Chapter 22 into Context

We have found that there may be much more to the universe than meets the eye. Dark matter too dim for us to see seems to far outweigh the stars, and a mysterious dark energy may be even more prevalent. Here are some key "big picture" points to remember about this chapter:

- Dark matter and dark energy sound very similar, but they are each hypothesized to explain different observations. Dark matter is thought to exist because we detect its gravitational influence. "Dark energy" is a term given to whatever strange, repulsive force may be accelerating the expansion of the universe.

- Either dark matter exists, or we do not understand how gravity operates across galaxy-size distances. There are many reasons to be confident about our understanding of gravity, leading most astronomers to conclude that dark matter is real.

- Dark matter seems by far to be the most abundant form of mass in the universe. We still do not know what it is, but we suspect it is largely made up of some type of as-yet-undiscovered subatomic particles.

- If dark matter is indeed the dominant source of gravity in the universe, then it is the glue that binds together galaxies, clusters, superclusters, and other large-scale structures in the universe. All this structure has probably grown from regions in the early universe where the density of dark matter was slightly enhanced.

- The fate of the universe depends on whether gravity can ever halt the present expansion. The total strength of gravity seems too weak to do so even when we account for dark matter, and the evidence indicating that the expansion is accelerating only reinforces the idea that expansion will never cease.

SUMMARY OF KEY CONCEPTS

22.1 Unseen Influences in the Cosmos

- **What do we mean by dark matter and dark energy?**   **Dark matter** and **dark energy** have never been directly observed, but each has been proposed to exist because it seems the simplest way to explain a set of observed motions in the universe. "Dark matter" is the name given to the unseen mass whose gravity governs the observed motions of stars and gas clouds. "Dark energy" is the name given to whatever may be causing the expansion of the universe to accelerate.

22.2 Evidence for Dark Matter

- **What is the evidence for dark matter in galaxies?**   The orbital velocities of stars and gas clouds in galaxies do not change much with distance from the center of the galaxy. Applying Newton's laws of gravitation and motion to these orbits leads to the conclusion that the total mass of a galaxy is far larger than the mass of its stars. Because no detectable visible light is coming from this matter, we call it **dark matter**.

- **What is the evidence for dark matter in clusters of galaxies?** We have three different ways of measuring the amount of dark matter in clusters of galaxies: from galaxy orbits, from the temperature of the hot gas in clusters, and from the **gravitational lensing** predicted by Einstein. All these methods are in agreement, indicating that the total mass of a cluster is about 50 times the mass of its stars, implying huge amounts of dark matter.

- **Does dark matter really exist?** We infer that dark matter exists from its gravitational influence on the matter we can see, leaving two possibilities: Either dark matter exists, or there is something wrong with our understanding of gravity. We cannot rule out the latter possibility, but we have good reason to be confident about our current understanding of gravity and the idea that dark matter is real.

- **What might dark matter be made of?** Some of the dark matter could be ordinary or **baryonic matter** in the form of dim stars or planetlike objects, but there does not appear to be enough ordinary matter to account for all the dark matter. Most of it is probably extraordinary or **nonbaryonic matter** consisting of undiscovered particles that we call **WIMPs**.

22.3 Structure Formation

- **What is the role of dark matter in galaxy formation?** Because most of a galaxy's mass is in the form of dark matter, the gravity of that dark matter is probably what formed protogalactic clouds and then galaxies from slight density enhancements in the early universe.

- **What are the largest structures in the universe?** Galaxies appear to be distributed in gigantic chains and sheets that surround great voids. These giant structures trace their origin directly back to regions of slightly enhanced density early in time.

22.4 The Fate of the Universe

- **Will the universe continue expanding forever?** Even before we consider the possibility of a mysterious dark energy, the evidence points to eternal expansion. The **critical density** is the average matter density that the universe would need for the strength of gravity to eventually halt the expansion. The overall matter density of the universe appears to be only about 25% of the critical density.

- **Is the expansion of the universe accelerating?** Observations of distant supernovae indicate that the expansion of the universe is speeding up. No one knows the nature of the mysterious force (dark energy) that could be causing this acceleration.

EXERCISES AND PROBLEMS

For instructor-assigned homework go to **Mastering**ASTRONOMY www.masteringastronomy.com

Review Questions
Short-Answer Questions Based on the Reading

1. Define *dark matter* and *dark energy*, and clearly distinguish between them. What types of observations have led scientists to propose the existence of each of these unseen influences?

2. What is a *rotation curve*? Describe the rotation curve of the Milky Way, and explain how it indicates the presence of large amounts of dark matter.

3. How do we construct rotation curves for other spiral galaxies? What do they tell us about the galaxy masses and dark matter?

4. How do we measure the masses of elliptical galaxies? What do these masses lead us to conclude about dark matter in elliptical galaxies?

5. Briefly describe the three different ways of measuring the mass of a cluster of galaxies. Do the results from the different methods agree? What do they tell us about dark matter in galaxy clusters?

6. What is *gravitational lensing?* Why does it occur? How can we use it to estimate the masses of lensing objects?

7. Briefly explain why the conclusion that dark matter exists rests on assuming that we understand gravity correctly. Is it possible that our understanding of gravity is not correct? Explain.

8. In what sense is dark matter "dark"? Briefly explain why objects like you, planets, and even dim stars could qualify as dark matter.

9. What do we mean by *MACHOs?* How can we search for them? Briefly describe why these searches suggest that star-like (or planetlike) objects and black holes *cannot* account for all the dark matter in the halo of our galaxy.

10. Explain what we mean when we say that a neutrino is a *weakly interacting particle.* Why can't the dark matter in galaxies be made of neutrinos?

11. What do we mean by *WIMPs?* Why does it seem likely that dark matter consists of these particles, even though we do not yet know what they are?

12. Briefly explain why dark matter is thought to have played a major role in the formation of galaxies and larger structures in the universe. What evidence suggests that large structures are still forming?

13. What does the large-scale structure of the universe look like? Explain why we think this structure reflects the density patterns of the early universe.

14. What do we mean by the *critical density* of the universe? According to current evidence, how does the actual density of matter in the universe compare to the critical density?

15. Describe and compare four possible patterns for the expansion of the universe: recollapsing, critical, coasting, and accelerating. Observationally, how can we decide which of the four possible expansion models is the right one?

16. Assuming the accelerating expansion of the universe is real, what does it imply for the fate of the universe? What does current evidence suggest for the fate of the universe if the acceleration is not real? Explain.

Test Your Understanding

Does It Make Sense? ❓

Decide whether the statement makes sense (or is clearly true) or does not make sense (or is clearly false). Explain your reasoning. (For an example, see Chapter 1, "Does It Make Sense?")

17. Strange as it may sound, most of both the mass and energy in the universe may take forms that we are unable to detect directly.

18. A cluster of galaxies is held together by the mutual gravitational attraction of all the stars in the cluster's galaxies.

19. We can estimate the total mass of a cluster of galaxies by studying the distorted images of galaxies whose light passes through the cluster.

20. Clusters of galaxies are the largest structures that we have so far detected in the universe.

21. The primary evidence for an accelerating universe comes from observations of young stars in the Milky Way.

22. There is no doubt remaining among astronomers that the fate of the universe is to expand forever.

23. Dark matter is called "dark" because it blocks light from traveling between the stars.

24. If the universe has more dark matter than we think, then it is also younger than we think.

25. The distance to a white dwarf supernova with a particular redshift is larger in an accelerating universe than in a universe with no acceleration.

26. If dark matter consists of WIMPs, then we should be able to observe photons produced by collisions between these particles.

Choose the best answer to each of the following. Explain your reasoning with one or more complete sentences.

27. Dark matter is inferred to exist because: (a) we see lots of dark patches in the sky. (b) it explains how the expansion of the universe can be accelerating. (c) we can observe its gravitational influence on visible matter.

28. Dark energy has been hypothesized to exist in order to explain: (a) observations suggesting that the expansion of the universe is accelerating. (b) the high orbital speeds of stars far from the center of our galaxy. (c) explosions that seem to create giant voids between galaxies.

29. The flat part of the Milky Way Galaxy's rotation curve tells us that stars in the outskirts of the galaxy: (a) orbit the galactic center just as fast as stars closer to the center. (b) rotate rapidly on their axes. (c) travel in straight, flat lines rather than elliptical orbits.

30. Strong evidence for the existence of dark matter comes from observations of: (a) our solar system. (b) the center of the Milky Way. (c) clusters of galaxies.

31. A photograph of a cluster of galaxies shows distorted images of galaxies that lie behind it at greater distances. This is an example of what astronomers call: (a) dark energy. (b) spiral density waves. (c) a gravitational lens.

32. Based on the observational evidence, is it possible that dark matter doesn't really exist? (a) No, the evidence for it is too strong to think it could be in error. (b) Yes, but only if there is something wrong with our current understanding of how gravity should work on large scales. (c) Yes, but only if all the observations themselves are in error.

33. Based on current evidence, which of the following is considered a likely candidate for the majority of the dark matter in galaxies? (a) subatomic particles that we have not yet detected in particle physics experiments (b) swarms of relatively dim, red stars (c) supermassive black holes

34. Which region of the early universe was most likely to become a galaxy? (a) a region whose matter density was lower than average (b) a region whose matter density was higher than average (c) a region with an unusual concentration of dark energy

35. The major evidence for the idea that the expansion of the universe is accelerating comes from observations of: (a) white dwarf supernovae. (b) the orbital speeds of stars within galaxies. (c) the evolution of quasars.

36. Which of the following possible types of universe would *not* expand forever? (a) a critical universe (b) an accelerating universe (c) a recollapsing universe

Investigate Further

In-Depth Questions to Increase Your Understanding

Short-Answer/Essay Questions

37. *Dark Matter.* Overall, how convincing do you consider the case for the existence of dark matter? Write a short essay in which you describe what we mean by dark matter, describe the evidence for its existence, and discuss your opinion about the strength of the evidence.

38. *Dark Energy.* Overall, how convincing do you consider the case for the existence of dark energy? Write a short essay in which you describe what we mean by dark energy, describe the evidence for its existence, and discuss your opinion about the strength of the evidence.

39. *The Future Universe.* Based on current evidence concerning the growth of structure in the universe, briefly describe what you would expect the universe to look like on large scales about 10 billion years from now.

40. *Dark Matter and Life.* State and explain at least two reasons one might argue that dark matter is (or was) essential for life to exist on Earth.

41. *Rotation Curves.* Draw and label a rotation curve for each of the following three hypothetical situations. Make sure the radius axis has approximate distances labeled.
    a. All the mass of the galaxy is concentrated in the center of the galaxy.
    b. The Galaxy has a constant mass density inside 20,000 light-years, and zero density outside that.
    c. The galaxy has a constant mass density inside 20,000 light-years, and its enclosed mass increases proportionally to the radius outside that.

42. *Dark Energy and Supernova Brightness.* When astronomers began measuring the brightnesses and redshifts of distant white dwarf supernovae, they expected to find that expansion of the universe was slowing down. Instead they found that it was speeding up! Were the distant supernovae brighter or fainter than expected? Explain why. (*Hint:* In Figure 22.16, the position of a supernova point on the vertical axis depends on its redshift. Its position on the horizontal axis depends on its brightness—supernovae seen farther back in time are not as bright as those seen closer in time.)

43. *What Is Dark Matter?* Describe at least three possible constituents of dark matter. Explain how we would expect each to interact with light, and how we might go about detecting its existence.

44. *Alternative Gravity.* How would gravity have to be different in order to explain the rotation curves of galaxies without the need for dark matter? Would gravity need to be stronger or weaker than expected at very large distances? Explain.

Quantitative Problems

Be sure to show all calculations clearly and state your final answers in complete sentences.

45. *White Dwarf M/L.* What is the mass-to-light ratio of a 1 M_{Sun} white dwarf with luminosity 0.001 L_{Sun}?

46. *Supergiant M/L.* What is the mass-to-light ratio of a 30 M_{Sun} supergiant star with luminosity 300,000 L_{Sun}?

47. *Solar System M/L.* What is the mass-to-light ratio of the solar system?

48. *Mass from Rotation Curve.* Study the rotation curve for the spiral galaxy NGC 7541, which is shown in Figure 22.4.

    a. Use the orbital velocity law to determine the mass (in solar masses) of NGC 7541 enclosed within a radius of 30,000 light-years from its center. (*Hint:* 1 light-year = 9.461×10^{15} m.)
    b. Use the orbital velocity law to determine the mass of NGC 7541 enclosed within a radius of 60,000 light-years from its center.
    c. Based on your answers to parts (a) and (b), what can you conclude about the distribution of mass in this galaxy?

49. *Weighing a Cluster.* A cluster of galaxies has a radius of about 5.1 million light-years (4.8×10^{22} m) and an intracluster medium with a temperature of 6×10^7 K. Estimate the mass of the cluster. Give your answer in both kilograms and solar masses. Suppose the combined luminosity of all the stars in the cluster is 8×10^{12} L_{Sun}. What is the cluster's mass-to-light ratio?

50. *Cluster Mass from Hot Gas.* The gas temperature of the Coma cluster of galaxies is about 9×10^7 K. What is the mass of this cluster within 15 million light-years of the cluster center?

51. *How Many MACHOs?* Imagine a galaxy whose stars are all identical to the Sun but that has an overall mass-to-light ratio of 30.
    a. What is the ratio of dark matter to luminous matter in this galaxy?
    b. Suppose all the dark matter consists of MACHOs similar to Jupiter, each with a mass of 0.001 solar mass. How many of these MACHOs must the galaxy contain for each ordinary star? Explain.

52. *From Newton to Dark Matter.* Show that the equation $M = r \times v^2/G$ from Mathematical Insight 22.2 is equivalent to Newton's version of Kepler's third law from Mathematical Insight 4.3. Assume that one mass is much larger than the other mass and that the orbit is circular. (*Hint:* What is the mathematical relationship between period p and orbital velocity v and orbital radius r for a circular orbit?)

Discussion Questions

53. *Dark Matter or Revised Gravity.* One possible explanation for the evidence we find for dark matter is that we are currently using the wrong law of gravity to measure the masses of very large objects. If we really do misunderstand gravity, then many fundamental theories of physics, including Einstein's theory of general relativity, will need to be revised. Which explanation for our observations do you find more appealing, dark matter or revised gravity? Explain why. Why do you suppose most astronomers find dark matter more appealing?

54. *Our Fate.* Scientists, philosophers, and poets alike have speculated on the fate of the universe. How would you prefer the universe as we know it to end, in a "Big Crunch" or through eternal expansion? Explain the reasons behind your preference.

Interactive Tutorials

Tutorial Review of Key Concepts

Use the following interactive **Tutorials** at **www.mastering astronomy.com** to review key concepts from this chapter.

Detecting Dark Matter in a Spiral Galaxy Tutorial
Lesson 1 Introduction to Rotation Curves
Lesson 2 Determining the Mass Distribution
Lesson 3 Where Is the Dark Matter?

Fate of the Universe Tutorial
Lesson 1 The Role of Gravity
Lesson 2 The Role of Dark Energy
Lesson 3 Fate and History of the Universe

Supplementary Tutorial Exercises

Use the interactive **Tutorial Lessons** to explore the following questions.

Detecting Dark Matter in Spiral Galaxies Tutorial, Lesson 3
1. Explain how stellar rotation curves suggest the existence of dark matter.
2. How does the amount of dark matter change as you move away from the center of a galaxy?

Fate of the Universe Tutorial, Lesson 1
1. If gravity were the only factor affecting the fate of the universe, what would the possible fates be?
2. How would the mass density of the universe determine which fate would be in store for the universe?

Fate of the Universe Tutorial, Lesson 2
1. What does dark energy do to the expansion of the universe?
2. How would the existence of dark energy change the fate of the universe from what it would be if only gravity played a role in its fate?

Fate of the Universe Tutorial, Lesson 3
1. How would the existence of dark energy change estimates of the age of the universe?
2. What evidence suggests that dark energy really exists?

 Movies

Check out the following narrated and animated short documentary available on **www.masteringastronomy.com** for a helpful review of key ideas covered in this chapter.

*From the Big Bang to
the Galaxies Movie*

Web Projects

1. *Gravitational Lenses.* Gravitational lensing occurs in numerous astronomical situations. Compile a catalog of examples from the Web. Try to find pictures of lensed stars, quasars, and galaxies. Give a one-paragraph explanation of what is going on in each picture.
2. *Accelerating Universe.* Search for the most recent information about possible acceleration of the expansion of the universe. Write a one- to three-page report on your findings.
3. *The Nature of Dark Matter.* Find and study recent reports on new ideas about the possible nature of dark matter. Write a one- to three-page report that summarizes the latest ideas about what dark matter is made of.

23

The Beginning of Time

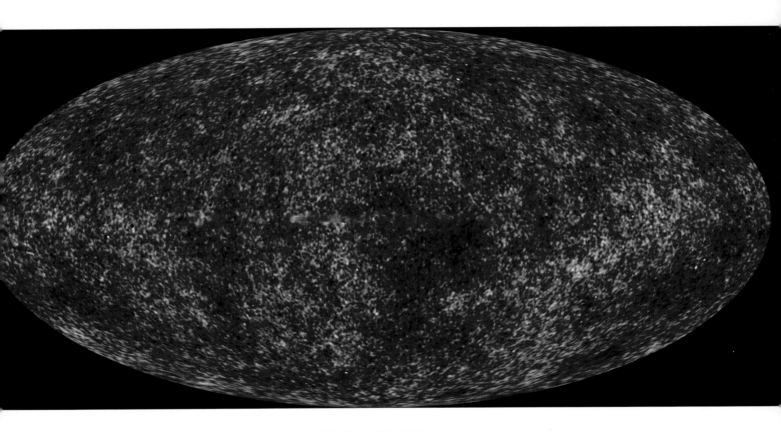

LEARNING GOALS

23.1 The Big Bang
- What were conditions like in the early universe?
- What is the history of the universe according to the Big Bang theory?

23.2 Evidence for the Big Bang
- How do we observe the radiation left over from the Big Bang?
- How do the abundances of elements support the Big Bang theory?

23.3 The Big Bang and Inflation
- What aspects of the universe were originally unexplained by the Big Bang model?
- How does inflation explain these features of the universe?
- How can we test the idea of inflation?

23.4 Observing the Big Bang for Yourself
- Why is the darkness of the night sky evidence for the Big Bang?

Somewhere, something incredible is waiting to be known.

Carl Sagan

The universe has been expanding for about 14 billion years. During that time, matter has collected into galaxies and stars have formed in those galaxies, producing heavy elements that have been recycled into later generations of stars. One of these late-coming stars formed about 4.6 billion years ago, in a remote corner of a galaxy called the Milky Way. This star was born with a host of planets that formed in a flattened disk surrounding it. One of these planets soon became covered with life that gradually evolved into ever more complex forms. Today, the most advanced life-form on this planet, human beings, can look back on this series of events and marvel at how the universe created conditions suitable for life.

To this point in the book, we have discussed how the matter produced in the early universe gradually assembled into planets, stars, and galaxies. However, we have not yet answered one big question: Where did the *matter itself* come from? To answer this ultimate question, we must go beyond the most distant galaxies and even beyond what we can see near the horizon of the universe. We must go back not only to the origins of matter and energy but to the beginning of time itself.

 Hubble's Law Tutorial, Lessons 1–3

23.1 The Big Bang

Is it really possible to study the origin of the entire universe? Not long ago, questions about the origin of everything we see were considered unfit for scientific study. That attitude began to change with Hubble's discovery that the universe is expanding. This discovery led to the insight that all things very likely sprang into being at a single moment in time, in an event that we have come to call the *Big Bang*. Today, powerful telescopes allow us to view how galaxies have changed over the past 14 billion years, and at great distances we see young galaxies still in the process of forming [Section 21.1]. These observations confirm that the universe is gradually aging, just as we should expect if the entire universe really was born some 14 billion years ago.

Unfortunately, we cannot see back to the very beginning of time. Light from the most distant galaxies shows us what the universe looked like when it was 1 or 2 billion years old. Beyond these galaxies, we have not yet found any objects shining brightly enough for us to see them. Ultimately, we face an even more fundamental problem. The universe is filled with a faint glow of radiation that appears to be the remnant heat of the Big Bang. This faint glow is light that has traveled freely through space since the universe was about 380,000 years old, which is when the universe first became transparent to light. Before that time, light could not pass freely through the universe, so there is no possibility of seeing light from earlier times. Thus, just as we must

rely on mathematical modeling to determine what the Sun is like on the inside, we must also use modeling to investigate what the universe was like during its earliest moments.

• What were conditions like in the early universe?

Scientific models of the conditions that prevailed in the early universe are based on fundamental principles of physics. The universe is cooling and becoming less dense as it expands, so it must have been hotter and denser in the past. Calculating exactly how hot and dense the universe must have been when it was more compressed is similar to calculating the temperature and density of gas in a balloon when you squeeze it, except that the conditions become much more extreme. Figure 23.1 shows just how hot the universe was during its earliest moments, according to such calculations.

The universe was so hot during the first few seconds that photons could transform themselves into matter, and vice versa, in accordance with Einstein's formula $E = mc^2$ [Section 4.3]. Reactions that create and destroy matter are now relatively rare in the universe at large, but physicists can reproduce many such reactions in their laboratories.

One such reaction is the creation or destruction of an *electron-antielectron pair* (Figure 23.2). When two photons collide with a total energy greater than twice the mass-energy of an electron (the electron's mass times c^2), they can create two brand-new particles: a negatively charged electron and its positively charged twin, the *antielectron* (also known as a *positron*). The electron is a particle of

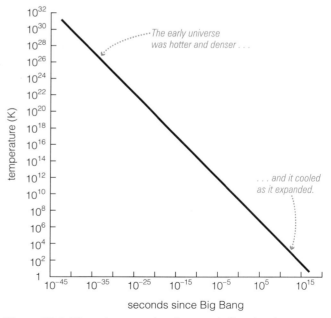

Figure 23.1 The universe cools as it expands. By using the laws of physics and the current temperature of the universe (about 3 K), we can calculate how hot the universe must have been in the past. This graph shows the results. Notice that both axis scales use powers of 10. (The graph extends to the present; 10^{10} yr $\approx 3 \times 10^{17}$ s.)

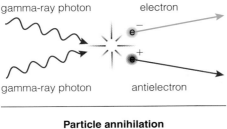

Particle creation

gamma-ray photon   electron

gamma-ray photon   antielectron

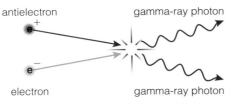

Particle annihilation

antielectron   gamma-ray photon

electron   gamma-ray photon

Figure 23.2 Electron-antielectron creation and annihilation. Reactions like these constantly converted photons to particles and vice versa in the early universe.

matter, and the antielectron is a particle of **antimatter**. The reaction that creates an electron-antielectron pair also runs in reverse. When an electron and an antielectron meet, they *annihilate* each other totally, transforming all their mass-energy back into photon energy. In order to conserve both energy and momentum, an annihilation reaction must produce two photons instead of just one.

Similar reactions can produce or destroy any particle-antiparticle pair, such as a proton and antiproton or a neutron and antineutron. The early universe therefore was filled with an extremely hot and dense blend of photons, matter, and antimatter, converting furiously back and forth. Despite all these vigorous reactions, describing conditions in the early universe is straightforward, at least in principle. We simply need to use the laws of physics to calculate the proportions of the various forms of radiation and matter at each moment in the universe's early history. The only difficulty is our incomplete understanding of the laws of physics.

To date, physicists have investigated the behavior of matter and energy at temperatures as high as those that existed in the universe just *one ten-billionth* (10^{-10}) of a second after the Big Bang, giving us confidence that we actually understand what was happening at that early moment in the history of the universe. Our understanding of physics is less certain under the more extreme conditions that prevailed even earlier, but we do have some ideas about what the universe was like when it was a mere 10^{-38} second old, and perhaps a glimmer of what it was like at the age of just 10^{-43} second. These tiny fractions of a second are so small that, for all practical purposes, we are studying the very moment of creation—the Big Bang itself.

• What is the history of the universe according to the Big Bang theory?

The **Big Bang theory**—the scientific theory of the universe's earliest moments—is based on applying known and tested

laws of physics to the idea that all we see today, from Earth to the cosmic horizon, began as an incredibly tiny, hot, and dense collection of matter and radiation. The Big Bang theory describes how expansion and cooling of this unimaginably intense mixture of particles and photons could have led to the present universe of stars and galaxies, and it explains several aspects of today's universe with impressive accuracy. We will discuss the evidence supporting the Big Bang theory later in this chapter. First, in order to help you understand the significance of the evidence, we'll examine the history of the universe according to this theory.

Figure 23.3 summarizes the story by dividing the overall history of the universe into a series of *eras*, or time periods. (Some scientists further subdivide the eras described here.) Each era is distinguished from the next by some major change in the conditions of the universe. You'll find it easiest to keep track of the various eras if you refer back to this figure as we discuss each era in detail. Notice that most of the key events in the history of the universe occurred in a very short period of time. It will take you longer to read this chapter than it took the universe to progress through the first five eras we will discuss, by which point the chemical composition of the universe had already been determined.

The Planck Era As we work our way back through time, we ultimately reach the limit of our current scientific ability to understand the physical conditions when the universe was an incomprehensibly young 10^{-43} second old. This instant in time is called the *Planck time* after physicist Max Planck, one of the founders of the science of quantum mechanics. We refer to all times prior to the Planck time as the **Planck era**; that is, the Planck era represents the first 10^{-43} second in the history of the universe.

Current theories cannot adequately describe the extreme conditions that must have existed during the Planck era. According to the laws of quantum mechanics, there must have been substantial energy fluctuations from point to point in the very early universe. Because energy and mass are equivalent, Einstein's theory of general relativity tells us that these energy fluctuations must have generated a rapidly changing gravitational field that randomly warped space and time. During the Planck era, these random energy fluctuations were so large that our current theories are inadequate to describe what might have been happening. In essence, the problem is that we do not yet have a theory that links quantum mechanics (our successful theory of the very small) and general relativity (our successful theory of the very big). Perhaps someday we will be able to merge these theories of the very small and the very big into a single "theory of everything" (see Special Topic, p. 463). Until that happens, science cannot describe the universe before the Planck time.

The GUT Era Understanding the transition that marked the beginning of the next era requires thinking in terms of the *forces* that operate in the universe. Everything that happens in the universe today is governed by four distinct

Eras and Characteristics

Era of Galaxies
Stars, galaxies and clusters of galaxies (made of atoms and plasma)

Era of Atoms
Atoms and plasma (stars begin to form)

Era of Nuclei
Plasma of hydrogen and helium nuclei plus electrons

Era of Nucleosynthesis
Protons, neutrons, electrons, neutrinos (antimatter rare)

Particle Era
Elementary particles (antimatter common)

Electroweak Era
Elementary particles

GUT Era
Elementary particles?

Planck Era
????

Timeline and Events Since Big Bang

14 billion years (present day): Humans observe the cosmos.

1 billion years: First galaxies form.

380,000 years: Atoms form; photons fly free and become microwave background.

3 minutes: Fusion ceases; normal matter is 75% hydrogen, 25% helium, by mass.

0.001 second: Matter annihilates antimatter.

10^{-10} *second:* Electromagnetic and weak forces become distinct.

10^{-38} *second:* Strong force becomes distinct, perhaps causing inflation of universe.

10^{-43} *second:* Gravity becomes distinct from other forces?

Key

neutron
proton
electron
neutrino
antiproton
antineutron
antielectrons
quarks

Figure 23.3 Interactive Figure This diagram summarizes the eras of the universe. The names of the eras and the state of matter during each era are indicated on the left, and the events and times marking the end of each era are indicated on the right.

forces: *gravity, electromagnetism,* the *strong force,* and the *weak force.* We have already encountered each of these forces individually.

Gravity is the most familiar of the four forces, providing the "glue" that holds planets, stars, and galaxies together. The electromagnetic force, which depends on the electrical charge of a particle instead of its mass, is far stronger than gravity. It is therefore the dominant force between particles in atoms and molecules, responsible for all chemical and biological reactions. However, the existence of both positive and negative electrical charges causes the electromagnetic force to lose out to gravity on large scales, even though

both forces decline with distance by an inverse square law. Most large astronomical objects (such as planets and stars) are electrically neutral overall, making the electromagnetic force unimportant on that scale. Gravity therefore becomes the dominant force for such objects, because more mass always means more gravity.

The strong and weak forces operate only over extremely short distances, making them important within atomic nuclei but not on larger scales. The strong force binds atomic nuclei together [Section 14.2]. The weak force plays a crucial role in nuclear reactions such as fission and fusion, and it is the only force besides gravity that affects weakly interacting

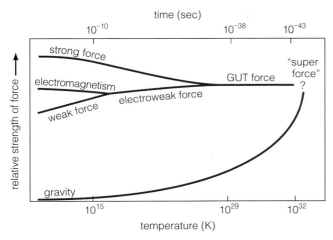

time (sec)

strong force

electromagnetism

GUT force

"super force" ?

electroweak force

weak force

relative strength of force

gravity

temperature (K)

Figure 23.4 The four forces are distinct at low temperatures but may merge at very high temperatures.

particles such as neutrinos or WIMPs (weakly interacting massive particles [Section 22.2]).

Although the four forces behave quite differently from one another, we now believe that they are actually just different aspects of a smaller number of more fundamental forces, probably only one or two (Figure 23.4). At the high temperatures that prevailed in the early universe, the four forces were not so distinct as they are today.

As an analogy, think about ice, liquid water, and water vapor. These three substances are quite different from one another in appearance and behavior, yet they are just different phases of the single substance H$_2$O. In a similar way, experiments have shown that the electromagnetic and weak forces lose their separate identities under conditions of very high temperature or energy and merge together into a single **electroweak force**. At even higher temperatures and energies, the electroweak force may merge with the strong force and ultimately with gravity. Theories that predict the merger of the electroweak and strong forces are called **grand unified theories**, or **GUTs** for short. Thus, the merger of the strong, weak, and electromagnetic forces is often called the *GUT force*. Many physicists suspect that at even higher energies, the GUT force and gravity merge into a single "super force" that governs the behavior of everything. (Among the names you may hear for theories linking all four forces are *supersymmetry, superstrings,* and *supergravity.*)

Calculations from general relativity and quantum mechanics suggest that this unified "super force" may have reigned in the universe during the Planck era. If so, the Planck time (10^{-43} second) marks the instant when gravity became distinct from the other three forces, which were still merged as the GUT force. By analogy to ice crystals forming as a liquid cools, we say that gravity "froze out" at the Planck time. The universe subsequently entered the **GUT era**, when two forces operated in the universe: gravity and the GUT force.

The GUT era is thought to have lasted but a tiny fraction of a second, coming to an end when the universe had cooled to 10^{29} K at an age of about 10^{-38} second. Grand

unified theories predict that the strong force froze out from the GUT force at this point, leaving three forces operating in the universe: gravity, the strong force, and the electroweak force. The freezing out of the strong force may have released an enormous amount of energy. This energy release may have caused the universe to undergo a sudden and dramatic expansion that we call **inflation**. In the space of a mere 10^{-36} second, pieces of the universe the size of an atomic nucleus may have grown to the size of our solar system. Inflation sounds bizarre, but as we will discuss later, it explains several important features of today's universe.

The Electroweak Era The end of the GUT era marked the beginning of the **electroweak era**, so named because the electromagnetic and weak forces were still unified in the electroweak force. Intense radiation filled all of space, as it had since the Planck era, spontaneously producing matter and antimatter particles that almost immediately annihilated each other and turned back into photons. The universe continued to expand and cool throughout the electroweak era, dropping to a temperature of 10^{15} K when it reached an age of 10^{-10} second. This temperature is still 100 million times hotter than the temperature in the core of the Sun today, but it was low enough for the electromagnetic and weak forces to freeze out from the electroweak force. After this instant (10^{-10} second), all four forces were forever distinct in the universe.

The end of the electroweak era marked an important transition not only in the physical universe, but also in human understanding of the universe. The theory that unified the weak and electromagnetic forces, developed in the 1970s, predicted the emergence of new types of particles (called the W and Z bosons, or *weak bosons*) at temperatures above the 10^{15} K temperature of the universe when it was 10^{-10} second old. In 1983, experiments performed in a huge particle accelerator near the French/Swiss border (CERN) reached energies equivalent to such high temperatures for the first time. The new particles showed up just as predicted, produced from extremely high-energy particle collisions in accord with $E = mc^2$.

Thus, we have direct experimental evidence concerning the conditions in the universe at the end of the electroweak era. We do *not* have any direct experimental evidence of conditions prior to that time. Our theories concerning the earlier parts of the electroweak era and the GUT era consequently are much more speculative than our theories describing the universe from the end of the electroweak era to the present.

The Particle Era As long as the universe was hot enough for the spontaneous creation and annihilation of particles to continue, the total number of particles was roughly in balance with the total number of photons. Once the universe became too cool for this spontaneous exchange of matter and energy to continue, photons became the dominant form of energy. We refer to the time between the end of the electroweak era and the moment when spontaneous particle production ceased as the **particle era**, to emphasize the importance of subatomic particles during this period.

During the early parts of the particle era (and earlier eras), photons turned into all sorts of exotic particles that we no longer find freely existing in the universe today, including *quarks*—the building blocks of protons and neutrons. By the end of the particle era, all quarks had combined into protons and neutrons, which shared the universe with other particles such as electrons, neutrinos, and perhaps WIMPs. The particle era came to an end when the universe reached an age of 1 millisecond (0.001 second) and the temperature had fallen to 10^{12} K. At this point, it was no longer hot enough to produce protons and antiprotons (or neutrons and antineutrons) spontaneously from pure energy.

If the universe had contained equal numbers of protons and antiprotons (or neutrons and antineutrons) at the end of the particle era, all the pairs would have annihilated each other, creating photons and leaving essentially no matter in the universe. From the obvious fact that the universe contains a significant amount of matter, we conclude that protons must have slightly outnumbered antiprotons at the end of the particle era.

We can estimate the size of the imbalance between matter and antimatter by comparing the present numbers of protons and photons in the universe. The two numbers should have been similar in the very early universe, but today photons outnumber protons by about a billion to one. This ratio indicates that for every billion antiprotons in the early universe, there must have been about a billion and one protons. Thus, for each 1 billion protons and antiprotons that annihilated each other at the end of the particle era, a single proton was left over. This seemingly slight excess of matter over antimatter makes up all the ordinary matter in the present-day universe. Some of those protons (and neutrons) left over from when the universe was 0.001 second old are the very ones that make up our bodies.

The Era of Nucleosynthesis   So far, everything we have discussed occurred within the first 0.001 second of the universe's existence—a time span shorter than the time it takes you to blink an eye. At this point, the protons and neutrons left over after the annihilation of antimatter began to fuse into heavier nuclei. However, the heat of the universe remained so high that most nuclei broke apart as fast as they formed. This dance of fusion and breakup marked the **era of nucleosynthesis**.

The era of nucleosynthesis ended when the universe was about 3 minutes old. After this time, the density in the expanding universe had dropped so much that fusion no longer occurred, even though the temperature was still about a billion Kelvin (10^9 K)—much hotter than the temperature at the center of the Sun today. When fusion ceased, about 75% of the mass of the ordinary (baryonic) matter in the universe remained as individual protons, or hydrogen nuclei. The other 25% of this mass had fused into helium nuclei, with trace amounts of deuterium (hydrogen with a neutron) and lithium (the next heaviest element after hydrogen and helium). Except for the relatively small amount of matter that stars later forged into heavier elements, the chemical composition of the universe remains the same today.

The Era of Nuclei   At the end of the era of nucleosynthesis, the universe consisted of a very hot plasma of hydrogen nuclei, helium nuclei, and free electrons. This basic picture held for about the next 380,000 years as the universe continued to expand and cool. The fully ionized nuclei moved independently of electrons during this period (rather than being bound with electrons in neutral atoms), which we call the **era of nuclei**. Throughout this era, photons bounced rapidly from one electron to the next, just as they do deep inside the Sun today [Section 14.2], never managing to travel far between collisions. Any time a nucleus managed to capture an electron to form a complete atom, one of the photons quickly ionized it.

The era of nuclei came to an end when the expanding universe was about 380,000 years old. At this point the temperature had fallen to about 3,000 K—roughly half the temperature of the Sun's surface today. Hydrogen and helium nuclei finally captured electrons for good, forming stable, neutral atoms for the first time. With electrons now bound into atoms, the universe suddenly became transparent, as if a thick fog had suddenly lifted. Photons, formerly trapped among the electrons, began to stream freely across the universe. We still see these photons today as the *cosmic microwave background*, which we will discuss shortly.

The Era of Atoms and the Era of Galaxies   We've already discussed the rest of the story in earlier chapters of the book. The end of the era of nuclei marked the beginning of the **era of atoms**, when the universe consisted of a mixture of neutral atoms and plasma (ions and electrons), along with a large number of photons. Thanks to the slight density enhancements present in the universe at this time and the gravitational attraction of dark matter, the atoms and plasma slowly assembled into protogalactic clouds. Stars formed in these clouds, transforming the gas clouds into galaxies. The first full-fledged galaxies had formed by the time the universe was about 1 billion years old, beginning what we call the **era of galaxies**.

The era of galaxies continues to this day. Generation after generation of star formation in galaxies steadily builds elements heavier than helium and incorporates them into new star systems. Some of these star systems develop planets, and on at least one of these planets, life burst into being a few billion years ago. Now here we are, thinking about it all. Describing both the follies and the achievements of the human race, Carl Sagan once said, "These are the things that hydrogen atoms do—given 15 billion years of cosmic evolution." (Sagan died in 1996, before we refined the age of the universe to 14 billion years.)

23.2 Evidence for the Big Bang

Like any scientific theory, the Big Bang theory is a model of nature designed to explain a set of observations. If it is close to the truth, it should be able to make predictions about the real universe that we can verify through more observations

Figure 23.5 Arno Penzias and Robert Wilson, discoverers of the cosmic microwave background, with the Bell Labs microwave antenna.

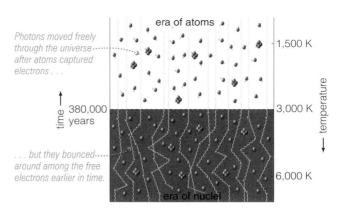

Figure 23.6 Interactive Figure Photons (yellow squiggles) frequently collided with free electrons during the era of nuclei and thus could travel freely only after electrons became bound into atoms. This transition was something like the transition from a dense fog to clear air. The photons released at the end of the era of nuclei, when the universe was about 380,000 years old, make up the cosmic microwave background. Precise measurements of these microwaves tell us what the universe was like at this moment in time.

or experiments. The Big Bang model has gained wide scientific acceptance for two key reasons:

- The Big Bang model predicts that the radiation that began to stream across the universe at the end of the era of nuclei should still be present today. Sure enough, we find that the universe is filled with what we call the **cosmic microwave background**. Its characteristics precisely match what we expect according to the Big Bang model.
- The Big Bang model predicts that some of the original hydrogen in the universe should have fused into helium during the era of nucleosynthesis. Observations of the actual helium content of the universe closely match the amount of helium predicted by the Big Bang model.

Let's take a closer look at this evidence, starting with the cosmic microwave background.

• How do we observe the radiation left over from the Big Bang?

The first major piece of evidence supporting the Big Bang theory was announced in 1965. Arno Penzias and Robert Wilson, two physicists working at Bell Laboratories in New Jersey, were calibrating a sensitive microwave antenna designed for satellite communications (Figure 23.5). (*Microwaves* fall within the radio portion of the electromagnetic spectrum; see Figure 5.7.) Much to their chagrin, they kept finding unexpected "noise" in every measurement they made with the antenna.

Fearing that they were doing something wrong, Penzias and Wilson worked frantically to discover and eliminate all possible sources of background noise. They even climbed up on their antenna to scrape off pigeon droppings, on the off-chance that these were somehow causing the noise. No matter what they did, however, the microwave noise wouldn't go away. The noise was the same no matter where they

pointed their antenna, indicating that it came from all directions in the sky and ruling out the possibility that it came from any particular astronomical object or from any place on Earth. Embarrassed by their inability to explain the noise, Penzias and Wilson prepared to "bury" their discovery about the noise at the end of a long scientific paper about their antenna.

Meanwhile, physicists at nearby Princeton University were busy calculating the expected characteristics of the radiation left over from the heat of the Big Bang. They concluded that, if the Big Bang had really occurred, this radiation should be permeating the entire universe and should be detectable with a microwave antenna. On a fateful airplane trip home from an astronomical meeting, Penzias sat next to an astronomer who told him of the Princeton calculations. The Princeton group soon met with Penzias and Wilson to compare notes. The "noise" in the Bell Labs antenna was not an embarrassment after all. Instead, it was the cosmic microwave background—and the first strong evidence that the Big Bang had really happened. Penzias and Wilson received the 1978 Nobel Prize in physics for their discovery of the cosmic microwave background.*

The cosmic microwave background consists of photons arriving at Earth directly from the end of the era of nuclei, when the universe was about 380,000 years old. Because neutral atoms finally could remain stable, they captured most of the electrons in the universe. With no more free electrons to block them, the photons from that epoch have flown unobstructed through the universe ever since (Figure 23.6). Thus, when we observe the cosmic microwave background, we essentially are seeing back to a time when the universe

---

*The dramatic story of the discovery of the cosmic microwave background is told in greater detail, along with much more scientific history, in Timothy Ferris, *The Red Limit* (New York: Quill, 1983). The possible existence of microwave radiation left over from the Big Bang was first predicted by George Gamow and his colleagues in the late 1940s, but neither Penzias and Wilson nor the Princeton group were aware of his work.

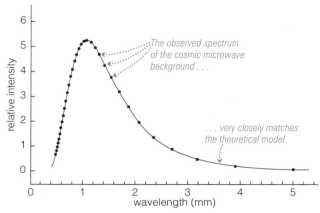

Figure 23.7 This graph shows the spectrum of the cosmic microwave background recorded by NASA's *COBE* satellite. A theoretically calculated thermal radiation spectrum (smooth curve) for a temperature of 2.73 K perfectly fits the data (dots). This excellent fit is important evidence in favor of the Big Bang theory.

The observed spectrum of the cosmic microwave background . . .

. . . very closely matches the theoretical model.

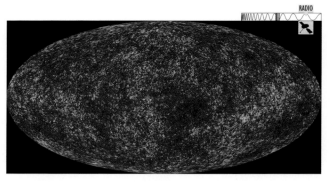

RADIO

Figure 23.8 Interactive Photo This all-sky map shows temperature differences in the cosmic microwave background measured by *WMAP*. The background temperature is about 2.73 K everywhere, but the brightest regions of this picture are about 0.0001 K hotter than the darkest regions—indicating that the early universe was very slightly lumpy at the end of the era of nuclei. We are essentially seeing what the universe was like at the surface marked "380,000 years" in Figure 23.3. Gravity later drew matter toward the centers of these lumps, forming the structures we see in the universe today.

was only 380,000 years old. In that sense, we are seeing light from the most distant regions of the observable universe—only 380,000 light-years from our cosmological horizon [Section 20.3].

Surprisingly, it does not take a particularly powerful telescope to "see" this radiation. In fact, you can pick it up with an ordinary television antenna. If you set an antenna-fed television (that is, *not* cable or satellite TV) to a channel for which there is no local station, you will see a screen full of static "snow." About 1% of this static is due to photons in the cosmic microwave background. Try it. If your friends ask why you are watching nothing, tell them that you are actually watching the most incredible sight ever seen on a television screen: the Big Bang, or at least as close to it as we'll ever get.

The cosmic microwave background came from the heat of the universe itself and therefore should have an essentially perfect thermal radiation spectrum [Section 5.4]. When this radiation first broke free 380,000 years after the Big Bang, the temperature of the universe was about 3,000 K, not too different from that of a red giant star's surface. Thus, the spectrum of the cosmic microwave background originally peaked in visible light, just like the thermal radiation from a red star, with wavelengths of a few hundred nanometers. However, the universe has expanded by a factor of about 1,000 since that time, stretching the wavelengths of these photons by the same amount [Section 20.3]. Their wavelengths have therefore shifted to about a millimeter, squarely in the microwave portion of the spectrum and corresponding to a temperature of a few degrees above absolute zero.

In the early 1990s, a NASA satellite called the *Cosmic Background Explorer (COBE)* was launched to test these ideas about the cosmic microwave background. The results were a stunning success for the Big Bang theory. As shown in Figure 23.7, the cosmic microwave background does indeed have a perfect thermal radiation spectrum, with a peak corresponding to a temperature of 2.73 K. In a very

real sense, the temperature of the night sky is a frigid 3 degrees above absolute zero.

THINK ABOUT IT

Suppose the cosmic microwave background did not really come from the heat of the universe itself but instead came from many individual stars and galaxies. Explain why, in that case, we would not expect it to have a perfect thermal radiation spectrum. How does the spectrum of the cosmic microwave background lend support to the Big Bang theory?

COBE achieved an even greater success mapping the temperature of the cosmic microwave background in all directions. It was already known that the cosmic microwave background is extraordinarily uniform throughout the universe. Conditions in the early universe must have been extremely uniform to produce such a smooth radiation field. For a time, this uniformity was considered a strike against the Big Bang theory because, as we discussed in Chapters 21 and 22, the universe must have contained some regions of enhanced density in order to explain the formation of galaxies. The *COBE* measurements restored confidence in the Big Bang theory because they showed that the cosmic microwave background is *not quite* perfectly uniform. Instead, its temperature varies very slightly from one place to another by a few parts in 100,000.*

More recently, NASA's *Wilkinson Microwave Anisotropy Probe (WMAP)* has provided even more dramatic confirmation of the small temperature variations, with a map of the cosmic microwave background released in 2003 (Figure 23.8). These variations in temperature indicate that the density of the early universe really did differ slightly from

---

*Earth's motion (such as our orbit of the Sun and the Sun's orbit around the center of the galaxy) means that we are moving relative to the cosmic microwave background radiation. Thus, we see a slight blueshift (about 0.12%) in the direction we're moving and a slight redshift in the opposite direction. We must first subtract these effects before analyzing the temperature of the background radiation.

As shown in Figure 23.7, the spectrum of the cosmic microwave background is a nearly perfect thermal radiation spectrum for an object at a temperature of 2.73 K. Thus, we infer that the current temperature of the universe itself is a little less than 3 K. We can calculate the wavelength of the photons at the peak of the spectrum using Wien's law, which we discussed in Mathematical Insight 5.2:

$$\lambda_{max} \approx \frac{2,900,000}{T(\text{Kelvin})}\text{nm} = \frac{2,900,000}{2.73}\text{nm} = 1.1 \times 10^6 \text{ nm}$$

If you remember that 10^6 nm = 1 mm, then you'll see that this value is equivalent to 1.1 millimeters. In other words, the spectrum of the cosmic microwave background currently peaks at a wavelength of about 1 millimeter (notice that this agrees with Figure 23.7). But what was the wavelength of the cosmic microwave photons in the past?

Remember that the expansion of the universe stretches the photons within it, changing their wavelengths through the effect we call the *cosmological redshift* [Section 20.3]. From Mathematical Insight 20.5, we know that the universe has grown in size by a factor of $(1 + z)$ since the time light left objects that we observe to have a redshift z. Thus, to find the peak wavelength of photons in the cosmic microwave background at any past time in terms of the redshift z, we simply need to divide the current peak wavelength by $(1 + z)$:

$$\lambda_{max}(\text{at redshift } z) \approx \frac{1.1 \text{ mm}}{1 + z}$$

Because higher values of z correspond to earlier times in the history of the universe, this formula tells us that the peak wavelength of the background radiation was *shorter* at earlier moments in time. Moreover, for a thermal radiation spectrum, a shorter peak wavelength means a *higher* temperature. Combining our result for the peak wavelength at redshift z with Wien's law, we find a simple formula for the temperature of the universe at any earlier time at which we see objects with redshift z:

$$T_{universe}(\text{at redshift } z) \approx 2.73 \text{ K} \times (1 + z)$$

Notice that, as we expect, this formula will give a higher temperature at times in the past that correspond to higher redshift z.

Example 1: What was peak wavelength of the photons in background radiation of the universe at the time they began to travel freely through the universe? Assume that the temperature of the universe at this time was 3,000 K.

Solution:

Step 1 Understand: We are given the temperature of the universe and are asked to find the peak wavelength of the thermal radiation corresponding to that temperature. We therefore need to use Wien's law relating peak wavelength to temperature.

Step 2 Solve: We plug the temperature of 3,000 K into Wien's law:

$$\lambda_{max} \approx \frac{2,900,000}{T(\text{Kelvin})}\text{nm} = \frac{2,900,000}{3,000}\text{nm} = 970 \text{ nm}$$

Step 3 Explain: We have found that the peak wavelength of the background radiation at the time it began to travel freely through the universe was about 970 nanometers, which is in the infrared portion of the electromagnetic spectrum fairly close to the wavelength of red visible light (see Figure 5.7). This fact should not be surprising, because 3,000 K is roughly the same temperature as the surface of a red giant. In other words, the entire universe at the time would have been filled with light as bright as you would see if you were just beneath the surface of a red giant star.

Example 2: How much has the expansion of the universe stretched the wavelengths of the background radiation since it began to travel freely through the universe?

Solution:

Step 1 Understand: One way to answer this question would be to use Wien's law to calculate the wavelength of the background radiation both now and when the universe was at a temperature of 3,000 K and compare the results. However, an easier way is to use the formula we found above that relates the temperature of the background radiation to the cosmological redshift z. We are given the 3,000 K temperature, so we simply need to find the stretching factor $(1 + z)$.

Step 2 Solve: We solve the formula found above for the factor $(1 + z)$ by dividing both sides of the equation by the current temperature of the universe, 2.73 K; you should confirm that the formula becomes:

$$1 + z = \frac{T_{universe}(\text{at redshift } z)}{2.73 \text{ K}}$$

In this case, we are looking for the stretching factor corresponding to the time when the universe had a temperature of 3,000 K. Plugging this value into the formula, we find:

$$1 + z = \frac{3,000 \text{ K}}{2.73 \text{ K}} \approx 1,100$$

Step 3 Explain: We have found that the expansion of the universe has stretched photons by a factor of about 1,000 since the time they first began to travel freely across the universe, when the universe was about 380,000 years old. Notice that the answer has no units because it tells us the *ratio* of the size of the universe now to the size of the universe then.

Although the Big Bang theory enjoys wide acceptance among scientists today, alternative ideas have been proposed and considered. One of the cleverest alternatives, developed in the late 1940s, was called the *steady state* universe. This hypothesis accepted the fact that the universe is expanding but rejected the idea of a Big Bang, instead postulating that the universe is infinitely old. The steady state hypothesis may seem paradoxical at first: If the universe has been expanding forever, shouldn't every galaxy be infinitely far away from every other galaxy? Proponents of the steady state universe answered by claiming that new galaxies continually form in the gaps that open up as the universe expands, thereby keeping the same average distance between galaxies at all times. In a sense, the

steady state hypothesis says that the creation of the universe is an ongoing and eternal process rather than a single event that occurred all at once with a Big Bang.

Two key discoveries caused the steady state universe to lose favor. First, the 1965 discovery of the cosmic microwave background matched a prediction of the Big Bang theory but was not adequately explained by the steady state hypothesis. Second, a steady state universe should look about the same at all times, which is inconsistent with observations showing that galaxies at great distances look younger than nearby galaxies. As a result of these predictive failures, most astronomers no longer take the steady state universe seriously.

---

place to place. The seeds of structure formation were indeed present during the era of nuclei.

The discovery of density enhancements bolstered the idea that some of the dark matter consists of WIMPs that we have not yet identified [Section 22.2] and that the gravity of this dark matter drove the formation of structure in the universe. Regions of enhanced density can grow into galaxies because the extra gravity in these regions draws matter together even while the rest of the universe expands. The greater the density enhancements, the faster matter should have collected into galaxies.

Detailed calculations show that, to explain the fact that galaxies formed within a few billion years, the density enhancements at the end of the era of nuclei must have been significantly greater than the few parts in 100,000 suggested by the temperature variations in the cosmic microwave background. Because WIMPs are weakly interacting and do not interact with photons, we do not expect them to influence the temperature of the cosmic microwave background directly. However, the gravity of the WIMPs can collect ordinary baryonic matter into clumps that *do* interact with photons. Thus, the small density enhancements detected by microwave telescopes appear to echo much larger density enhancements made up of WIMPs. Careful modeling of these temperature variations shows that they are consistent with dark-matter density enhancements large enough to account for the structure we see in the universe today.

• How do the abundances of elements support the Big Bang theory?

The discovery of the cosmic microwave background in 1965 quickly solved another long-standing astronomical problem: the origin of cosmic helium. Everywhere in the universe, about one-quarter of the mass of ordinary matter (not including dark matter) is helium. The Milky Way's helium fraction is about 28%, and no galaxy has a helium fraction lower than 25%. A small proportion of this helium comes from hydrogen fusion in stars, but most does not: Fusion of hydrogen to helium in stars could have produced only about 10% of the observed helium.

The majority of the helium in the universe must already have been present in the protogalactic clouds that preceded the formation of galaxies. In other words, the universe itself must once have been hot enough to fuse hydrogen into helium. The current microwave background temperature of 2.73 K tells us precisely how hot the universe was in the distant past and exactly how much helium it should have made. The result—25% helium—is another impressive success of the Big Bang theory.

Helium Formation in the Early Universe   In order to see why 25% of ordinary matter became helium, we need to understand what protons and neutrons were doing during the 3 minutes that marked the era of nucleosynthesis. Early in this era, when the universe's temperature was 10^{11} K, nuclear reactions could convert protons into neutrons, and vice versa. As long as the universe remained hotter than 10^{11} K, these reactions kept the numbers of protons and neutrons nearly equal. But as the universe cooled, neutron-proton conversion reactions began to favor protons.

Neutrons are slightly more massive than protons, and therefore reactions that convert protons to neutrons require energy to proceed (in accordance with $E = mc^2$). As the temperature fell below 10^{11} K, the required energy for neutron production was no longer readily available, so the rate of these reactions slowed. In contrast, reactions that convert neutrons into protons release energy and thus are unhindered by cooler temperatures. By the time the temperature of the universe fell to 10^{10} K, protons began to outnumber neutrons because the conversion reactions ran in only one direction. Neutrons changed into protons, but the protons didn't change back.

For about the next 3 minutes, the universe was still hot and dense enough for nuclear fusion to operate. Protons and neutrons constantly combined to form *deuterium*—the rare form of hydrogen nuclei that contains a neutron in addition to a proton—and deuterium nuclei fused to form helium (Figure 23.9). However, during the early part of the era of nucleosynthesis, the helium nuclei were almost immediately blasted apart by one of the many gamma rays that filled the universe.

Step 1	Step 2	Step 3	Key:
Proton and neutron fuse to form a deuterium nucleus.	Two deuterium nuclei fuse to make hydrogen-3.	Hydrogen-3 fuses with deuterium to create helium-4.	● neutron ● proton

Figure 23.9 During the 3-minute-long era of nucleosynthesis, virtually all the neutrons in the universe fused with protons to form helium-4. This figure illustrates one of several possible reaction pathways.

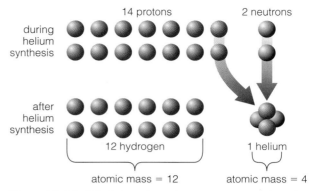

Figure 23.10 Calculations show that protons outnumbered neutrons 7 to 1, which is the same as 14 to 2, during the era of nucleosynthesis. The result was 12 hydrogen nuclei (individual protons) for each helium nucleus. Thus, the hydrogen-to-helium mass ratio is 12 to 4, which is the same as 75% to 25%. The agreement between this prediction and the observed abundance of helium is important evidence in favor of the Big Bang theory.

Fusion began to create long-lasting helium nuclei when the universe was about 1 minute old and had cooled to a temperature at which it contained few destructive gamma rays. Calculations show that the proton-to-neutron ratio at this time should have been about 7 to 1. Moreover, almost all the available neutrons should have become incorporated into nuclei of helium-4. Figure 23.10 shows that, based on the 7-to-1 ratio of protons to neutrons, the universe should have had a composition of 75% hydrogen and 25% helium by mass at the end of the era of nucleosynthesis.

Thus, the Big Bang theory makes a very concrete prediction about the chemical composition of the universe: It should be 75% hydrogen and 25% helium by mass. The fact that observations confirm this predicted ratio of hydrogen to helium is another striking success of the Big Bang theory.

THINK ABOUT IT

Briefly explain why it should not be surprising that some galaxies contain a little more than 25% helium, but it would be very surprising if some galaxies contained less. (*Hint:* Think about how the relative amounts of hydrogen and helium in the universe are affected by fusion in stars.)

Abundances of Other Light Elements Why didn't the Big Bang produce heavier elements? By the time stable helium nuclei formed, when the universe was about a minute old, the temperature and density of the rapidly expanding universe had already dropped too far for a process like carbon production (three helium nuclei fusing into carbon [Section 17.2]) to occur. Reactions between protons, deuterium nuclei, and helium were still possible, but most of these reactions led nowhere. In particular, fusing two helium-4 nuclei results in a nucleus that is unstable and falls apart in a fraction of a second, as does fusing a proton to a helium-4 nucleus.

A few reactions involving hydrogen-3 (also known as *tritium*) or helium-3 can create long-lasting nuclei. For example, fusing helium-4 and hydrogen-3 produces lithium-7. However, the contributions of these reactions to the overall composition of the universe were minor because hydrogen-3 and helium-3 were so rare. Models of element production in the early universe show that, before the cooling of the

universe shut off fusion entirely, such reactions generated only trace amounts of lithium, the next lightest element after helium. Thus, aside from hydrogen, helium, and lithium, all other elements were forged much later in the nuclear furnaces of stars. (Beryllium and boron, which are heavier than lithium but lighter than carbon, were created later when high-energy particles broke apart heavier nuclei that formed in stars.)

The Density of Ordinary Matter Calculations made with the Big Bang model allow scientists to estimate the density of ordinary (baryonic) matter in the universe from the observed amount of deuterium in the universe today. Remember that, during the era of nucleosynthesis, protons and neutrons first fused into deuterium and the deuterium nuclei then fused into helium. The fact that some deuterium nuclei still exist in the universe indicates that this process stopped before all the deuterium nuclei were used up. The amount of deuterium in the universe today therefore tells us about the density of protons and neutrons (baryons) during the era of nucleosynthesis: The higher the density, the more efficiently fusion would have proceeded. Thus, a higher density in the early universe would have left less deuterium in the universe today, and a lower density would have left more deuterium.

Observations show that about one out of every 40,000 hydrogen atoms contains a deuterium nucleus—that is, a nucleus with a neutron in addition to its proton. Calculations based on this deuterium abundance show that the density of ordinary (baryonic) matter in the universe is about 4% of the critical density (Figure 23.11). (Recall that the critical density is the density required if the expansion of the universe is to stop and reverse someday [Section 22.4].) Similar calculations based on the observed abundance of lithium and helium-3 lead to the same conclusion, adding to our confidence in the Big Bang model.

These results also lead to an astonishing prediction about the nature of dark matter. Recall that the overall density

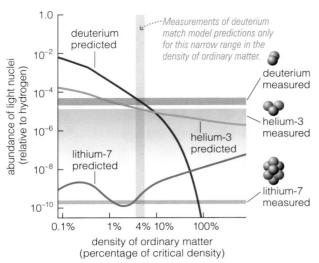

Figure 23.11 This graph shows how the measured abundances of deuterium, helium-3, and lithium-7 lead to the conclusion that the density of ordinary matter is about 4% of the critical density. The three horizontal swaths show measured abundances; the thickness of each swath represents the range of uncertainty in the measurements. (The upper edge of the blue swath indicates the upper limit on the helium-3 abundance; a lower limit has not yet been established.) The three curves represent models based on the Big Bang theory; these curves show how the abundance of each type of nucleus is expected to depend on the density of ordinary matter in the universe. Notice that the predictions (curves) match up with the measurements (horizontal swaths) only in the gray vertical strip, which represents a density of about 4% of the critical density.

of the universe appears to be close to 25% of the critical density [Section 22.4]. Because this is about six times as large as the 4% of critical density that we find for ordinary matter, we conclude that the universe contains about six times as much extraordinary (nonbaryonic) dark matter as it does of ordinary (baryonic) matter. Unless we are missing something fundamental in our understanding of all these issues, the Big Bang model predicts that extraordinary (nonbaryonic) dark matter such as WIMPs constitutes the majority of the universe's mass. That is why most astronomers think that dark matter consists mostly of WIMPs, and why many scientists are actively trying to find ways to detect WIMPs and learn about their properties.

THINK ABOUT IT

The ideas just discussed point to a rather amazing fact: Although we have yet to discover any WIMPs, we suspect they dominate the total mass of the universe. Briefly explain how this is possible, and comment on how confident we can be that weakly interacting particles make up the bulk of dark matter.

23.3 The Big Bang and Inflation

The Big Bang model relies heavily on our knowledge of particle physics, which has been tested to temperatures of 10^{15} K, corresponding to temperatures at the end of the electroweak era, when the universe was a mere 10^{-10} second old. Our knowledge of earlier times rests on a weaker

foundation because we are less certain of the physical laws at work. In fact, the best laboratory for studying the laws of physics at such high temperatures is the Big Bang itself.

Different models of how matter might behave at such high energies predict different outcomes for the universe we see today. If a particular model predicts that our universe should look different from the way it really does, then that model must be wrong. On the other hand, if a new model of particle physics explains some previously unexplained aspects of the universe, then it may be on the right track. The grand unified theories of particle physics discussed in Section 23.1 have not yet been extensively tested, but many scientists believe that these theories are on the right track for just this reason. In particular, the grand unified theories suggest that the universe at a very early age underwent the dramatic burst of expansion that we call *inflation,* and inflation seems to explain several key aspects of our universe that are otherwise left unexplained by the Big Bang theory.

• What aspects of the universe were originally unexplained by the Big Bang model?

The Big Bang theory has gained wide acceptance because of the strong evidence from the cosmic microwave background and the measured abundances of light elements in the universe. However, without the addition of more speculative physics such as that of the grand unified theories and their prediction of inflation, the Big Bang theory leaves several major aspects of our universe unexplained. Three of the most pressing questions are the following:

● *Where does structure come from?* Recall that our models of the formation of galaxies and larger structures all assume that gravity collected matter around regions of slightly enhanced density in the early universe. Thus, explaining the origin of structure requires that the Big Bang must have somehow produced these slight density enhancements. The subtle temperature differences seen in the cosmic microwave background (see Figure 23.8) tell us that regions of enhanced density did indeed exist at the end of the era of nuclei, when the universe was 380,000 years old. But we still need to explain where the density enhancements came from.

● *Why is the large-scale universe so uniform?* Although the slight temperature variations in the cosmic microwave background show that the universe is not *perfectly* uniform on large scales, the overall smoothness is nonetheless remarkable. Observations of the cosmic microwave background tell us that the density of the universe at the end of the era of nuclei varied from place to place by no more than about 0.01%, and this uniformity explains why distant reaches of the universe look so similar today. Without inflation, the Big Bang theory does not explain why distant reaches of the universe look so similar.

● *Why is the density of the universe close to the critical density?* The total density of dark matter plus dark energy in the universe [Section 22.4] appears to be remarkably close to the critical density—so close that it is difficult to consider it a coincidence. After all, there is no obvious reason why the density could not have been, say, 1,000 times the critical density or 0.0000001 times the critical density. But without inflation, the Big Bang model is unable to explain the near-critical density of the universe as anything other than luck.

• How does inflation explain these features of the universe?

Physicist Alan Guth realized in 1981 that grand unified theories could potentially answer all three questions. These theories predict that the separation of the strong force from the GUT force should have released enormous energy, causing the universe to expand dramatically, perhaps by a factor of 10^{30} in less than 10^{-36} second. This dramatic expansion is what we call *inflation*. While it sounds outrageous to talk about something that happened in the first trillion-trillion-trillionth of a second in the history of the universe, inflation may have shaped the way the universe looks today.

Structure: Giant Quantum Fluctuations   To understand how inflation explains the origin of structure, we need to recognize a special feature of energy fields. Laboratory-tested principles of quantum mechanics, especially the uncertainty principle [Section S4.3], tell us that on very small scales, the energy fields at any point in space are always fluctuating. Thus, the distribution of energy through space is very slightly irregular, even in a complete vacuum. The tiny quantum "ripples" that make up the irregularities can be characterized by a wavelength that corresponds roughly to their size. In principle, quantum ripples in the very early universe could have been the seeds for density enhancements that later grew into galaxies. However, the wavelengths of the original ripples were far too small to explain density enhancements like those we see imprinted on the cosmic microwave background.

Inflation would have dramatically increased the wavelengths of these quantum fluctuations. The fantastic growth of the universe during the period of inflation would have stretched tiny ripples from a size smaller than an atomic nucleus to the size of our solar system (Figure 23.12), making them large enough to become the density enhancements from which galaxies and larger structures later formed. Amazingly, the structure of today's universe may have started as tiny quantum fluctuations just before the period of inflation.

Uniformity: Equalizing Temperatures and Densities   We'll next consider how inflation explains the overall uniformity of the universe on large scales, but first let's look more closely at the question of why the uniformity is surprising. The idea that different parts of the universe were very similar shortly after the Big Bang may seem quite natural, but on further inspection the uniformity of the universe becomes difficult to explain.

Imagine observing the cosmic microwave background in a certain part of the sky. You are seeing microwaves that have traveled through the universe since the end of the era of nuclei, just 380,000 years after the Big Bang. Thus, you are seeing a region of the universe as it was some 14 billion years ago, when the universe was only 380,000 years old. Now imagine turning around and looking at the background radiation coming from the opposite direction. You are also seeing this region at an age of 380,000 years, and it looks virtually identical in temperature and density. The surprising part is this: The two regions are billions of light-years apart on opposite sides of our observable universe but we are seeing them as they were when they were only 380,000 years old. They can't possibly have exchanged light or any other information (Figure 23.13). A signal traveling at the speed of light from one to the other would barely have started

size of ripple before inflation = size of atomic nucleus

size of ripple after inflation = size of solar system

Figure 23.12 During inflation, ripples in spacetime would have stretched by a factor of perhaps 10^{30}. The peaks of these ripples then would have become the density enhancements that produced all the structure we see in the universe today.

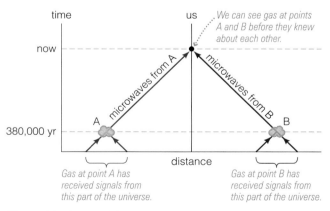

Figure 23.13 Light left the microwave-emitting regions we see on opposite sides of the universe long before they could have communicated with each other and equalized their temperatures, yet their temperatures are virtually identical. Without inflation these similar temperatures are a puzzle.

its journey. So how did they come to have the same temperature and density?

Inflation answers this question by saying that even though the two regions cannot have had any contact *since* the time of inflation, they were in contact prior to that time. Before the onset of inflation, when the universe was 10^{-38} second old, the two regions were less than 10^{-38} light-second away from one another. Thus, radiation traveling at the speed of light would have had time to bounce between the two regions, and this exchange of energy equalized their temperatures and densities. Inflation then pushed these equalized regions to much greater distances, far out of contact with one another (Figure 23.14). Like criminals getting their stories straight before being locked in separate jail cells, the two regions (and all other parts of the observable universe) came to the same temperature and density before inflation spread them far apart.

Because inflation caused different regions of the universe to separate so vastly in such a short period of time, many people wonder whether inflation violates Einstein's theories, saying that nothing can move faster than the speed of light. It does not, because nothing actually *moves* through space as a result of inflation or the ongoing expansion of the universe. Instead, the expansion of the universe is the expansion of *space itself.* Objects may be separating from one another at a speed faster than the speed of light, but no matter or radiation is able to travel between them during that time. In essence, inflation opens up a huge gap in space between objects that were once close together. The objects get very far apart, but nothing ever travels between them at a speed that exceeds the speed of light.

Density: Balancing the Universe   The third question answered by inflation asks why the matter density of the universe is

so close to the critical density. Another way to say that the universe's density is close to critical is to say that the overall geometry of the universe is remarkably "flat." To understand this idea, we must consider the overall geometry of the universe in a little more detail.

Recall that Einstein's general theory of relativity tells us that the presence of matter can curve the structure of spacetime [Section S3.3]. Although we cannot visualize this curvature in all three dimensions of space (or all four dimensions of spacetime), we can detect its presence by its effects on light. For example, observations of gravitational lensing [Section 22.2] tell us that we are seeing light that has passed through a curved region of space. Although the curvature of the universe can vary from place to place, the universe as a whole must have some overall shape. Almost any shape is possible, but all the possibilities fall into just three general categories (Figure 23.15). By analogy to objects that we can see in three dimensions, scientists refer to these three categories of shape as *flat* (or critical), *spherical* (or closed), and *saddle shaped* (or open).

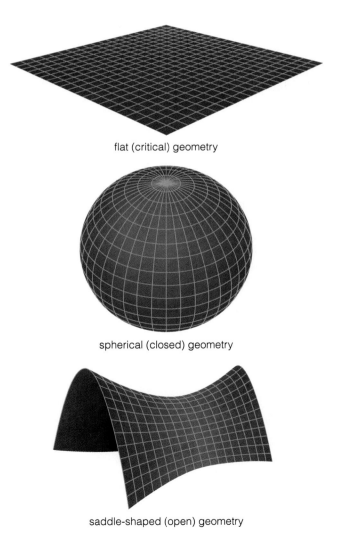

flat (critical) geometry

spherical (closed) geometry

saddle-shaped (open) geometry

Figure 23.15 Analogies to the three possible categories of overall geometry for the universe. Keep in mind that the real universe has these "shapes" in more dimensions than we can see.

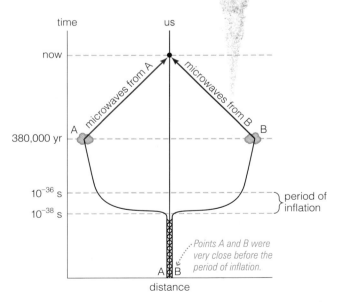

Figure 23.14   Interactive Figure   Before inflation, regions A and B were near enough to communicate and equalize their temperatures. Inflation then pushed them far apart. Today, we can see both A and B, but they are too far apart to see each other.

According to general relativity, the overall geometry would be flat if the matter density of the universe were precisely equal to the critical density, in which case the kinetic energy of expansion would precisely balance the universe's overall gravitational pull. In the absence of dark energy, any imbalance in these energies causes curvature of space-time, and deviations from precise balance grow more severe as the universe evolves. For example, if the universe had been 10% denser at the end of the era of nuclei, it would have collapsed long ago. On the other hand, if it had been 10% less dense at that time, galaxies would never have formed before expansion spread all the matter too thin. Thus, the universe had to start out remarkably balanced to be even remotely close to flat today.

Inflation can explain this precise balance by its effects on the geometry of the universe. In terms of Einstein's theory, the effect of inflation on spacetime curvature is similar to the flattening of a balloon's surface when you blow into it (Figure 23.16). The flattening of space during the period of inflation would have been so enormous that any curvature the universe might have had previously would be noticeable only on size scales much larger than the observable universe. Thus, inflation predicts that the overall geometry of the universe should appear perfectly flat, in which case the overall density of matter plus energy should be precisely equal to the critical density.

The fine balance predicted by inflation turns out to be both a success and a potential pitfall of the inflation theory. On one hand, it explains how the universe managed to have a density just right to allow the birth of galaxies. On the other hand, its prediction of a perfectly flat universe seems to disagree with observations showing that the total density of matter (including dark matter) is only about 25% of the critical density.

Dark energy might explain this shortfall in the density of dark matter. Remember that Einstein's theory of relativity tells us that mass can be transformed into energy and vice versa, which means that energy must be able to curve spacetime in the same way that mass can curve spacetime

[Section S3.3]. Thus, the dark energy associated with a large-scale repulsive force could compensate for the shortfall in the matter density, making the present-day universe flatter than it would otherwise be. Remarkably, the supernova measurements indicating that the expansion of the universe is accelerating [Section 22.4] also show that the strength of the implied repulsive force is about right to render the universe perfectly flat, just as inflation predicts. In other words, the total density of matter and dark energy together may indeed be precisely equal to the critical density.

• How can we test the idea of inflation?

We've seen that inflation answers some outstanding mysteries about the universe, but did it really happen? We cannot directly observe the universe at the very early time when inflation is thought to have occurred. Nevertheless, we can test the idea of inflation by exploring whether its predictions are consistent with our observations of the universe at later times. Scientists are only beginning to make observations that test inflation, but the findings to date are consistent with the idea that an early inflationary episode made the universe uniform and flat while planting the seeds of structure formation.

The strongest tests of inflation to date come from detailed studies of the cosmic microwave background, and in particular of the map made by the *WMAP* satellite (see Figure 23.8). Remember that this map shows tiny temperature differences corresponding to density variations in the universe at the end of the era of nuclei, when the universe was about 380,000 years old. However, according to the idea of inflation, these density enhancements were actually created much earlier, when inflation caused tiny quantum ripples to expand into seeds of structure. Thus, careful observations of the temperature variations in the microwave background can tell us about the structure of the universe during its first instant of existence.

For example, complex calculations show that the largest temperature differences in the cosmic microwave background should typically be between patches of sky separated by about 1° if the overall geometry of the universe is flat. (Similar calculations show that this angular separation would be smaller than 1° if the universe were open and curved like a saddle, and larger than 1° if the universe were closed and curved like a sphere [Section 22.4].) The strongest temperature differences are indeed observed at angular separations of 1°, indicating that the universe is geometrically flat, as predicted by inflation.

In fact, the overall pattern of temperature differences agrees with the predictions of models based on inflation, which is why these results tend to support the idea that inflation really occurred. Figure 23.17 shows an analysis of the temperature variations observed by *WMAP* in the cosmic microwave background, along with additional data from other microwave telescopes. The graph shows how the typical temperature differences between patches of sky depend on their angular size on the celestial sphere. The

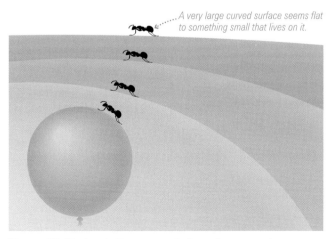

A very large curved surface seems flat to something small that lives on it.

Figure 23.16 As a balloon expands, its surface seems increasingly flat to an ant crawling along it. Inflation is thought to have made the universe seem flat in a similar way.

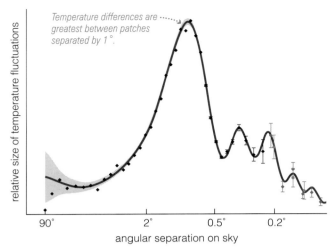

Temperature differences are greatest between patches separated by 1°.

relative size of temperature fluctuations

90° 2° 0.5° 0.2°

angular separation on sky

Figure 23.17 This graph shows how detailed analysis of temperature differences in the cosmic microwave background supports the idea of inflation. The data points indicate how the typical temperature differences between patches of sky depend on their angular size on the celestial sphere. (Black dots come from the *WMAP* data shown in Figure 23.8, and the blue and green points represent data from other telescopes.) The red curve shows the prediction of a model that relies on inflation to produce a universe whose ordinary matter density, dark-matter density, and expansion rate are all similar to their observed values. Close agreement between the data points and the model indicates that the universe we live in has many of the characteristics predicted by that model.

dots represent data from the observations, and the red curve shows the inflation-based model that best fits the observations. This model makes specific predictions not only about the data shown in the figure, but also about other characteristics of our universe such as its overall geometry, composition, and age. Thus, in a sense, these new observations of the cosmic microwave background are revealing the characteristics of the seeds from which our universe has grown. To the extent that we have been able to observe those seeds, their nature aligns reassuringly well with the universe we observe about us at the present time. According to the model shown in the figure, today's universe should have the following features:

- The overall geometry is flat, implying that the total mass-energy of the universe must be equivalent to the critical density.
- The density of ordinary (baryonic) matter is 4.4% of the critical density, in agreement with observations of deuterium in the universe [Section 23.2].
- The total matter density is 27% of the critical density. Subtracting the 4.4% for ordinary matter, we conclude that extraordinary (nonbaryonic) dark matter makes up about 23% of the critical density, in agreement with what we infer from measurements of the masses of clusters of galaxies [Section 22.4].
- The combination of a flat geometry and a matter density lower than the critical density implies the existence of a repulsive "dark energy" that currently accelerates the expansion, in agreement with observations of distant

supernovae [Section 22.4]. Because the total mass-energy of the universe is the critical density, and matter accounts for only 27% of this, dark energy must account for the remaining 73% of the mass-energy of the universe.

- The universe's age should be about 13.7 billion years at the current microwave temperature of 2.73 K, in agreement with what we infer from Hubble's constant [Section 20.3] and the ages of the oldest stars [Section 15.3].

This close correspondence between the seeds inherent in the universe at an age of 380,000 years and our observations of the present-day universe, some 14 billion years later, is persuasive evidence in favor of the Big Bang in general and inflation in particular. The bottom line is that, all things considered, inflation does a remarkable job of explaining features of our universe that are otherwise unaccounted for in the Big Bang theory. Many astronomers and physicists therefore suspect that some process akin to inflation did affect the early universe, but the details of the interaction between high-energy particle physics and the evolving universe remain unclear. If these details can be worked out successfully, we face an amazing prospect—a breakthrough in our understanding of the very smallest particles, achieved by studying the universe on the largest observable scales.

23.4 Observing the Big Bang for Yourself

You might occasionally read an article in a newspaper or a magazine questioning whether the Big Bang really happened. We will never be able to prove with absolute certainty that the Big Bang theory is correct. However, no one has come up with any other model of the universe that so successfully explains so much of what we see. As we have discussed, the Big Bang model makes at least two specific predictions that we have observationally verified: the characteristics of the cosmic microwave background and the composition of the universe. It also explains quite naturally many other features of the universe. So far, at least, we know of nothing that is inconsistent with the Big Bang model.

The Big Bang theory's very success has also made it a target for respected scientists, skeptical nonscientists, and crackpots alike. The nature of scientific work requires that we test established wisdom to make sure it is valid. A sound scientific disproof of the Big Bang theory would be a discovery of great importance. However, stories touted in the news media as disproofs of the Big Bang usually turn out to be disagreements over details rather than fundamental problems that threaten to bring down the whole theory. Yet scientists must keep refining the theory and tracking down disagreements, because once in a while a small disagreement blossoms into a full-blown scientific revolution.

You don't need to accept all you have read without question. The next time you are musing on the universe's origins, try an experiment for yourself. Go outside on a clear night, look at the sky, and ask yourself why it is dark.

According to the Big Bang theory, time and space have a beginning in the Big Bang. Do they also have an end? If we live in a recollapsing universe [Section 22.4], the answer seems to be a clear "yes." Sometime in the distant future, the universal expansion will cease and reverse, and the universe will eventually come to an end in a fiery "Big Crunch." However, it now appears unlikely that we live in a recollapsing universe, in which case the universe seems destined to expand forever.

If dark energy is real, our Local Group of galaxies will become increasingly more lonely as the acceleration of the universe carries the more distant galaxies ever faster away from us. Some scientists speculated that the repulsive force of dark energy might even strengthen with time. In that case, the galaxies of the Local Group will someday separate, and the growing repulsive force could eventually tear apart our galaxy, our solar system, and even matter itself in a catastrophic event sometimes called the "Big Rip." However, the Big Rip is quite speculative; a more plausible scenario for the end of a perpetually expanding universe proceeds as follows.

The star-gas-star cycle in galaxies cannot continue forever, because not all the material is recycled. With each generation of stars, more mass becomes locked up in planets, brown dwarfs, white dwarfs, neutron stars, and black holes. Eventually, about a trillion years from now, even the longest-lived stars will burn out, and the galaxies will fade into darkness.

At this point, the only new action in the universe will occur on the rare occasions when two objects—such as two brown dwarfs or two white dwarfs—collide within a galaxy. The vast distances separating star systems in galaxies make such collisions extremely rare. For example, the probability of our Sun (or the white dwarf that it will become) colliding with another star is so small that it would be expected to happen only once in a quadrillion (10^{15}) years. Forever is a long time, however, and even low-probability events will eventually happen many times. If a star system experiences a collision once in a quadrillion years, it will experience about 100 collisions in 100 quadrillion (10^{17}) years. By the time the universe reaches an age of 10^{20} years, star systems will have suffered an average of 100,000 collisions each, making a time-lapse history of any galaxy look like a cosmic game of pinball.

These multiple collisions will severely disrupt galaxies. As in any gravitational encounter, some objects lose energy in such collisions and some gain energy. Objects that gain enough energy will be flung into intergalactic space, where they will drift ever farther from all other objects with the expansion of the universe. Objects that lose energy will eventually fall to the galactic center and will enter the black hole that sits there, adding to its mass and ultimately creating a gigantic black hole where our galaxy used to be. The remains of the universe will consist of black holes with masses as great as a trillion solar masses that are widely separated from a few scattered planets, brown dwarfs, and stellar corpses. If Earth somehow survives, it will be a frozen chunk of rock in the darkness of the expanding universe, billions of light-years from any other solid object.

If grand unified theories are correct, Earth still cannot last forever. These theories predict that protons will eventually fall apart. The predicted lifetime of protons is extremely long: a half-life of at least 10^{33} years. However, if protons really do decay, then by the time the universe is 10^{40} years old, Earth and all other atomic matter will have disintegrated into radiation and subatomic particles such as electrons and neutrinos.

The final phase may come through a mechanism proposed by Stephen Hawking. According to Hawking's theory, even black holes cannot last forever. Instead, they slowly "evaporate," their mass-energy turning into radiation. The process is so slow that we do not expect to be able to see it from any existing black holes, but if Hawking is correct, then black holes in the distant future will disappear in brilliant bursts of radiation. The largest black holes last longest, but even trillion-solar-mass black holes will evaporate sometime after the universe reaches an age of 10^{100} years. From then on, the universe will consist only of individual photons and subatomic particles, each separated by enormous distances from the others. Nothing new will ever happen, and no events will ever occur that would allow an omniscient observer to distinguish past from future. In a sense, the universe will finally have reached the end of time.

Lest any of this sound depressing, keep in mind that 10^{100} years is an indescribably long time. As an example, imagine that you wanted to write on a piece of paper a number that consisted of a 1 followed by 10^{100} zeros (that is, the number $10^{10^{100}}$). It sounds easy, but a piece of paper large enough to hold all those zeros *would not fit in the observable universe* today. If that still does not alleviate your concerns, you may be glad to know that a few creative thinkers speculate about ways in which the universe might undergo rebirth, even after the end of time.

• Why is the darkness of the night sky evidence for the Big Bang?

If the universe were infinite, unchanging, and everywhere the same, then the entire night sky would blaze as brightly as the Sun. Johannes Kepler [Section 3.3] was one of the first people to reach this conclusion, but we now refer to the idea as **Olbers' paradox** after Heinrich Olbers, a German astronomer of the 1800s.

To understand how Olbers' paradox comes about, imagine that you are in a dense forest on a flat plain. If you look in any direction, you'll likely see a tree. If the forest is small, you might be able to see through some gaps in the trees to the open plains, but larger forests have fewer gaps (Figure 23.18). An infinite forest would have no gaps at all—a tree trunk would block your view along any line of sight.

The universe is like a forest of stars in this respect. In an unchanging universe with an infinite number of stars, we would see a star in every direction, making every point in the sky as bright as the Sun's surface. Even the presence of obscuring dust would not change this conclusion. The intense starlight would heat the dust over time until it too glowed like the Sun or evaporated away.

There are only two ways out of this dilemma. Either the universe has a finite number of stars, in which case we would not see a star in every direction, or it changes over time in some way that prevents us from seeing an infinite number of stars. For several centuries after Kepler first

a In a large forest, a tree will block your view no matter where you look. Similarly, in an unchanging universe with an infinite number of stars, we would expect to see stars in every direction, making the sky bright even at night.

b In a small forest with a smaller number of trees, you can see open spaces beyond the trees. Because the night sky is dark, the universe must similarly have spaces in which we see nothing beyond the stars, which means either that the number of stars is finite or that the universe changes in a way that prevents us from seeing an infinite number of them.

Figure 23.18 Olbers' paradox is similar to the view through a forest.

recognized the dilemma, astronomers leaned toward the first option. Kepler himself preferred to believe that the universe had a finite number of stars because he thought it had to be finite in space, with some kind of dark wall surrounding everything. Astronomers in the early twentieth century preferred to believe that the universe was infinite in space but that we lived inside a finite collection of stars. They thought of the Milky Way as an island floating in a vast black void. However, subsequent observations showed that galaxies fill all of space more or less uniformly. We are therefore left with the second option: The universe changes over time.

The Big Bang theory solves Olbers' paradox in a particularly simple way. It tells us that we can see only a finite number of stars because the universe began at a particular moment. While the universe may contain an infinite number of stars, we can see only those that lie within the observable universe, inside our cosmological horizon [Section 20.3]. There are other ways in which the universe could change over time and prevent us from seeing an infinite number of stars, so Olbers' paradox does not *prove* that the universe began with a Big Bang. However, there must be some explanation for why the sky is dark at night, and no explanation besides the Big Bang also explains so many other observed properties of the universe so well.

THE BIG PICTURE

Putting Chapter 23 into Context

Our "big picture" is now about as complete as it gets. We've discussed the universe from Earth outward, and from the beginning to the end. When you think back on this chapter, keep in mind the following ideas:

- Predicting conditions in the early universe is straightforward, as long as we know how matter and energy behave under such extreme conditions.

- Our current understanding of physics allows us to reconstruct the conditions that prevailed in the universe all the way back to the first 10^{-10} second. Our understanding is less certain back to 10^{-38} second. Beyond 10^{-43} second, we run up against the present limits of human knowledge.

- Although it may sound strange to talk about the universe during its first fraction of a second, our ideas about the Big Bang rest on a solid foundation of observational, experimental, and theoretical evidence. We cannot say with absolute certainty that the Big Bang really happened, but no other model ever proposed has so successfully explained how our universe came to be as it is.

23.1 The Big Bang

- **What were conditions like in the early universe?** The early universe was filled with radiation and elementary particles. It was so hot and dense that the energy of radiation could turn into particles of **matter** and **antimatter**, which then collided and turned back into radiation.

- **What is the history of the universe according to the Big Bang theory?** The universe has been through a series of eras, each marked by unique physical conditions. We know little about the **Planck era**, when the four forces may have all behaved as one. Gravity became distinct at the start of the **GUT era**, which may have ended with the rapid expansion called **inflation**. Electromagnetism and the weak force became distinct at the end of the **electroweak era**. Matter particles annihilated all the antimatter particles at the end of the **particle era**. Fusion of protons and neutrons into helium ceased at the end of the **era of nucleosynthesis**. Hydrogen nuclei captured all the free electrons, forming hydrogen atoms at the end of the **era of nuclei**. Galaxies began to form at the end of the **era of atoms**. The **era of galaxies** continues to this day.

23.2 Evidence for the Big Bang

- **How do we observe the radiation left over from the Big Bang?** Telescopes that can detect microwaves allow us to observe the **cosmic microwave background**—radiation left over from the Big Bang. Its spectrum matches the characteristics expected of the radiation released at the end of the era of nuclei, spectacularly confirming a key prediction of the Big Bang theory.

- **How do the abundances of elements support the Big Bang theory?** The Big Bang theory predicts the ratio of protons to neutrons during the era of nucleosynthesis, and from this ratio predicts that the chemical composition of the universe should be about 75% hydrogen and 25% helium (by mass). This prediction matches observations of the cosmic abundances, another spectacular confirmation of the Big Bang theory.

23.3 The Big Bang and Inflation

- **What aspects of the universe were originally unexplained by the Big Bang model?** (1) The origin of the density enhancements that turned into galaxies and larger structures. (2) The overall smoothness of the universe on large scales. (3) The fact that the actual density of matter is close to the critical density.

- **How does inflation explain these features of the universe?** (1) The episode of inflation stretched tiny, random quantum fluctuations to sizes large enough for them to become the density enhancements around which structures later formed. (2) The universe is smooth on large scales because, prior to inflation, everything we can observe today was close enough together for temperatures and densities to equalize. (3) Inflation caused the universe to expand so much that the observable universe appears geometrically flat, implying that its overall density of mass plus energy equals the critical density.

- **How can we test the idea of inflation?** Models of inflation make specific predictions about the temperature patterns we should observe in the cosmic microwave background. The observed patterns seen in recent observations by microwave telescopes match those predicted by inflation.

23.4 Observing the Big Bang for Yourself

- **Why is the darkness of the night sky evidence for the Big Bang?** **Olbers' paradox** tells us that if the universe were infinite, unchanging, and filled with stars, the sky everywhere would be as bright as the surface of the Sun, and it would not be dark at night. The Big Bang theory solves this paradox by telling us that the night sky is dark because the universe has a finite age, which means we can see only a finite number of stars in the sky.

Review Questions

1. What is *antimatter*? How were particle-antiparticle pairs created in the early universe? How were they destroyed?

2. Explain what we mean by the *Big Bang theory.*

3. Make a list of the major eras in the history of the universe, summarizing the important events thought to have occurred during each era.

4. Why can't our current theories describe the history of the universe during the *Planck era*?

5. What are the four forces that operate in the universe today? Why do we think there were fewer forces operating in the early universe?

6. What are *grand unified theories*? According to these speculative theories, how many forces operated during the *GUT era*? How are these forces thought to be related to the four forces that operate today?

7. What do we mean by *inflation,* and when do we think it occurred?

8. Why do we think there was a slight imbalance between matter and antimatter in the early universe? What happened to all the antimatter, and when?

9. How long did the *era of nucleosynthesis* last? Explain why this era was so important in determining the chemical composition of the universe forever after.

10. When we observe the *cosmic microwave background,* at what age are we seeing the universe? How long have the photons in the background been traveling through space? Explain.

11. Briefly describe how the cosmic microwave background was discovered. How does the existence and nature of this radiation support the Big Bang theory?

12. How does the chemical abundance of helium in the universe support the Big Bang theory? Explain.

13. How do measurements of deuterium and lithium tell us about the density of the universe, and why do they suggest that most dark matter consists of WIMPs?

14. Describe each of the three major questions left unanswered by the Big Bang theory without inflation, and explain how inflation answers each of them.

15. How can observations of the cosmic microwave background—radiation released when the universe was 380,000 years old—tell us about the universe at the much earlier time when inflation occurred? Summarize the geometry, composition, and age of the universe according to observations made to date.

16. What is *Olbers' paradox,* and how is it resolved by the Big Bang theory?

Test Your Understanding

Does It Make Sense?

Decide whether the statement makes sense (or is clearly true) or does not make sense (or is clearly false). Explain your reasoning. (For an example, see Chapter 1, "Does It Make Sense?")

17. Although the universe today appears to be made mostly of matter and not antimatter, the Big Bang theory suggests that the early universe had nearly equal amounts of matter and antimatter.

18. According to the Big Bang theory, the cosmic microwave background was created when energetic photons ionized the neutral hydrogen atoms that originally filled the universe.

19. While the existence of the cosmic microwave background is consistent with the Big Bang theory, we can also easily explain it by assuming that it comes from individual stars and galaxies.

20. According to the Big Bang theory, most of the helium in the universe was created by nuclear fusion in the cores of stars.

21. The idea of inflation suggests that the structure of the universe today may have originated as tiny quantum fluctuations.

22. The fact that the night sky is dark tells us that the universe cannot be infinite, unchanging, and everywhere the same.

23. We'll never know whether inflation actually happened because our model for inflation doesn't make any predictions we can test.

24. In the distant past, the radiation that we call the cosmic microwave background actually consisted primarily of infrared light.

25. The main reason that the night sky is dark is that stars are generally so far away.

26. The cosmic microwave background is our main source of information about what the universe was like at an age of about 3 minutes.

Quick Quiz

Choose the best answer to each of the following. Explain your reasoning with one or more complete sentences.

27. What is the current temperature of the universe? (a) absolute zero (b) a few degrees (c) a few thousand degrees

28. What is the charge of an antiproton? (a) positive (b) negative (c) neutral

29. What happens when a proton collides with an antiproton? (a) They repel each other. (b) They fuse together. (c) They convert into two photons.

30. Which of the following does *not* provide strong evidence for the Big Bang theory? (a) observations of the cosmic microwave background (b) observations of the amount of hydrogen in the universe (c) observations of the ratio of helium to hydrogen in the universe

31. If the current density of normal matter in the universe were 10 times as great as it is now, we would expect to observe (a) more deuterium. (b) less deuterium. (c) about the same amount of deuterium.

32. Which of the following does inflation help to explain? (a) the uniformity of the cosmic microwave background (b) the amount of helium in the universe (c) the temperature of the cosmic microwave background

33. Which of the following does inflation help to explain? (a) the origin of hydrogen (b) the origin of galaxies (c) the origin of atomic nuclei

34. Which of these pieces of evidence supports the idea that inflation really happened? (a) the enormous size of the observable universe (b) the large amount of dark matter that the universe contains (c) observations of the cosmic microwave background that indicate a flat geometry for the universe

35. What is the earliest time in the universe that we can directly observe? (a) a few hundred million years after the Big Bang (b) a few hundred thousand years after the Big Bang (c) a few minutes after the Big Bang

36. Which of these options is the best explanation for why the night sky is dark? (a) The universe is not infinite in space. (b) The universe has not always looked the way it looks today. (c) The distribution of matter in the universe is not uniform on very large scales.

Investigate Further
In-Depth Questions to Increase Your Understanding

Short-Answer/Essay Questions

37. *Life Story of a Proton.* Tell the life story of a proton from its formation shortly after the Big Bang to its presence in the nucleus of an oxygen atom you have just inhaled. Your story should be creative and imaginative, but it should also demonstrate your scientific understanding of as many stages in the proton's life as possible. You can draw on material from the entire book, and your story should be three to five pages long.

38. *Creative History of the Universe.* The story of creation as envisioned by the Big Bang theory is quite dramatic, but it is usually told in a fairly straightforward, scientific way. Write a more dramatic telling of the story, in the form of a short story, play, or poem. Be as creative as you wish, but be sure to remain accurate according to the science as it is understood today.

39. *Re-Creating the Big Bang.* Particle accelerators on Earth can push particles to extremely large speeds. When these particles collide, the amount of energy associated with the colliding particles is much greater than the mass-energy these particles have when at rest. As a result, these collisions can produce many other particles out of pure energy. Explain in your own words how the conditions that occur in these accelerators are similar to the conditions that prevailed shortly after the Big Bang. Also, point out some of the differences between what happens in particle accelerators and what happened in the early universe.

40. *Betting on the Big Bang Theory.* If you had $100, how much money would you wager on the proposition that we have a reasonable scientific understanding of what the universe was like when it was 1 minute old? Explain the reasoning behind your choice in terms of the scientific evidence presented in this chapter.

41. *"Observing" the Early Universe.* The only way we have of studying the era of nucleosynthesis is through the abundances of the nuclei that it left behind. Explain why we will never be able to observe that era through direct detection of the radiation emitted at that time.

42. *Element Production in the Big Bang.* Nucleosynthesis in the early universe was unable to produce more than trace amounts of elements heavier than helium. Using the information in Figure 17.14, which shows the mass per nuclear particle for many different elements, explain why producing elements like lithium (3 protons), boron (4 protons), and beryllium (5 protons) was so difficult.

43. *Darkness at Night.* Suppose you are Kepler, pondering the darkness of the night sky without any knowledge of the Big Bang or the expanding universe. Come up with an explanation for the darkness of the night sky that does not depend on the Big Bang theory. Propose an experiment that future scientists might be able to perform to test your hypothesis.

44. *Evidence for the Big Bang.* Alternatives to the Big Bang theory need to propose alternative explanations for many of the observations scientists have made of the universe. Make a list of at least seven observed features of the universe that are satisfactorily explained by the Big Bang theory when it is combined with the idea of inflation.

Quantitative Problems

Be sure to show all calculations clearly and state your final answers in complete sentences.

45. *Energy from Antimatter.* The total annual U.S. power consumption is about 2×10^{20} joules. Suppose you could supply that energy by combining pure matter with pure antimatter. Estimate the total mass of matter-antimatter fuel you would need to supply the United States with energy for 1 year. How does that mass compare with the amount of matter in your car's gas tank? (A gallon of gas has a mass of about 4 kilograms.)

46. *Gravity vs. the Electromagnetic Force.* The amount of electromagnetic force between two charged objects can be computed with an inverse square law similar to that of Newton's universal law of gravitation; for the electromagnetic force, the law is:

$$F = k \times \frac{(\text{charge of object 1}) \times (\text{charge of object 2})}{d^2}$$

In this formula, the charges must be given in units of Coulombs (abbreviated C), the distance d between the objects' centers must be in meters, and the constant $k = 9 \times 10^9 \, \text{kg} \times \text{m}^3/(\text{C}^2 \times \text{s}^2)$.

a. Compute the gravitational force between your body and Earth using Newton's universal law of gravitation (see Section 4.4 or Appendix B).

b. Now suppose all the electrons suddenly disappeared from Earth, making it positively charged, and all the protons in your body suddenly changed into neutrons, making you negatively charged. Compute the strength of the electromagnetic force between the electrons in your body and the protons in Earth. Assume that the charge per unit of mass of both you and Earth is 5×10^7 C/kg.

c. Compare the electromagnetic force from part (b) to the gravitational force from part (a). Use that result to explain why gravity is considered weaker than the electromagnetic force.

47. *Background Radiation during Galaxy Formation.* What was the peak wavelength of the background radiation at the time that light left the most distant galaxies we can currently see? Assume those galaxies have a cosmological redshift of $z = 7.0$. What is the temperature corresponding to that peak wavelength?

48. *Expansion since the Era of Nucleosynthesis.* Compare the peak wavelength of the radiation in the universe at the end of the era of nucleosynthesis to its current peak wavelength. Assume the temperature at the end of the era of nucleosynthesis was 10^9 K. How much have the wavelengths of the photons in the universe been stretched since that time?

49. *Temperature of the Universe.* What will the temperature of the cosmic microwave background be when the average distances between galaxies are twice as large as they are today?

50. *Uniformity of the Cosmic Microwave Background.* The temperature of the cosmic microwave background differs by only a few parts in 100,000 across the sky. Compare that level of uniformity to the surface of a table in the following way. Consider a table that is 1 meter in size. How big would the largest bumps on that table be if its surface were smooth to one part in 100,000? Could you see bumps of that size on the table's surface?

51. 10^{100} *Years.* In the box "How Will the Universe End?" we found that the final stage in the history of a perpetually expanding universe will come about 10^{100} years from now. Such a large number is easy to write but difficult to understand. This problem investigates some of the incredible properties of very large numbers.

    a. The current age of the universe is around 10^{10} years. How much longer is a trillion years than this current age? How much longer is 10^{15} years? 10^{20} years?

    b. Suppose protons decay with a half-life of 10^{32} years. When will the number of remaining protons be half the current amount? When will it be a quarter of its current amount? How many half-lives will have gone by when the universe reaches an age of 10^{34} years? What fraction of the original protons will remain at this time? Based on your answers, is it reasonable to conclude that *all* protons in today's universe will be gone by the time the universe is 10^{40} years old? Explain. (*Hint:* See Mathematical Insight 8.1.)

    c. Suppose you were trying to write 10^{100} zeros on a piece of paper and could write microscopically, so that each zero (including the thickness of the pencil mark) occupied a volume of 1 cubic micrometer—about the size of a bacterium. Could 10^{100} zeros of this size fit in the observable universe? Explain. (*Hint:* Calculate the volume of the observable universe in cubic micrometers by assuming it is a sphere with a radius of 14 billion light-years. The volume of a sphere is $\frac{4}{3}\pi r^3$; 1 light-year $\approx 10^{15}$ meters; 1 cubic meter $= 10^{18}$ cubic micrometers.)

52. *Daytime at "Night."* According to Olbers' paradox, the entire sky would be as bright as the surface of a typical star if the universe were infinite in space, unchanging in time, and the same everywhere. However, conditions would not need to be quite that extreme for the "nighttime" sky to be as bright as the daytime sky.

    a. Using the inverse square law for light from Mathematical Insight 15.1, determine the apparent brightness of the Sun in our sky.

    b. Using the inverse square law for light, determine the apparent brightness our Sun would have if it were at a distance of 10 billion light-years.

    c. From your answers to parts (a) and (b), estimate how many stars like the Sun would need to exist at a distance of 10 billion light-years for their total apparent brightness to equal that of our Sun.

    d. Compare your answer to part (c) with the estimate of the total number of stars in our observable universe from Mathematical Insight 1.3. Use your answer to explain why the night sky is much darker than the daytime sky. How much larger would the total number of stars need to be for "night" to be as bright as day?

Discussion Questions

53. *The Moment of Creation.* You've probably noticed that, in discussing the Big Bang theory, we never quite talk about the first instant. Even our most speculative theories at present take us back only to within 10^{-43} second of creation. Do you think it will *ever* be possible for science to consider the moment of creation itself? Will we ever be able to answer questions such as *why* the Big Bang happened? Defend your opinions.

54. *The Big Bang.* How convincing do you find the evidence for the Big Bang model of the universe's origin? What are the strengths of the theory? What does it fail to explain? Overall, do *you* think the Big Bang really happened? Defend your opinion.

Interactive Tutorials

Tutorial Review of Key Concepts

Use the following interactive **Tutorial** at **www.mastering astronomy.com** to review key concepts from this chapter.

Hubble's Law Tutorial
Lesson 1 Hubble's Law
Lesson 2 The Expansion of the Universe
Lesson 3 The Age of the Universe

Supplementary Tutorial Exercises

Use the interactive **Tutorial Lesson** to explore the following questions.

Hubble's Law Tutorial, Lesson 2

1. How does the observed rate of expansion allow us to determine distances between galaxies in the distant past?
2. Suppose we have an uncertainty of 10% in the rate of expansion. Would this cause significant changes in what we conclude about temperatures in the early universe? Why or why not?

Movies

Check out the following narrated and animated short documentary available on **www.masteringastronomy.com** for a helpful review of key ideas covered in this chapter.

From the Big Bang to the Galaxies Movie

Web Projects

1. *New Tests of the Big Bang Theory.* The *Cosmic Background Explorer (COBE)* satellite provided striking confirmation of several predictions of the Big Bang theory. Newer satellites have been designed to test the Big Bang theory further, primarily by observing the subtle variations in the cosmic microwave background with a much higher sensitivity than *COBE.* Use the Web to gather pictures and information about the *COBE* mission and its successors, such as *WMAP* or *Planck.* Write a one- to two-page report about the strength of the evidence compiled by *COBE* and *WMAP* and how much more we might learn from upcoming missions.

2. *Decay of the Proton.* One of the most startling predictions of grand unified theories is that protons will eventually decay, albeit with a half-life of more than 10^{32} years. If this is true, it may be possible to observe an occasional proton decay, despite the extraordinarily long half-life. Several experiments to look for proton decay are under way or being planned. Find out about one or more of these experiments, and write a one- to two-page summary in which you describe the experiment(s), any results to date, and what these results (or potential results) mean to the grand unified theories.

3. *New Ideas in Inflation.* The idea of inflation solves many of the puzzles associated with the standard Big Bang theory, but we are still a long way from finding evidence confirming that inflation really occurred. Find recent articles that discuss some of the latest ideas about inflation and how we might test these ideas. Write a two- to three-page summary of your findings.

Life in the Universe

LEARNING GOALS

24.1 Life on Earth
- **When did life arise on Earth?**
- **How did life arise on Earth?**
- **What are the necessities of life?**

24.2 Life in The Solar System
- **Could there be life on Mars?**
- **Could there be life on Europa or other jovian moons?**

24.3 Life Around Other Stars
- **Are habitable planets likely?**
- **Are Earth-like planets rare or common?**

24.4 The Search for Extraterrestrial Intelligence
- **How many civilizations are out there?**
- **How does SETI work?**

24.5 Interstellar Travel and Its Implications to Civilization
- **How difficult is interstellar travel?**
- **Where are the aliens?**

We, this people, on a small and lonely planet
Travelling through casual space
Past aloof stars, across the way of indifferent suns
To a destination where all signs tell us
It is possible and imperative that we learn
A brave and startling truth.

Maya Angelou, Excerpted from
A Brave and Startling Truth

We have covered a lot of ground in this book, discussing fundamental questions about the nature and origin of our planet, our star, our galaxy, and our universe. But we have not yet discussed one of the most profound questions of all: Are we alone? The universe seems to be filled with worlds beyond imagination—more than 100 billion star systems in our galaxy alone, and some 100 billion galaxies in the observable universe—yet we do not know whether any other world has ever been home to life.

In this chapter, we will discuss the possibility of life beyond Earth. We'll begin by considering the history of life on Earth, which should help us understand the prospects for finding life elsewhere. We'll then consider the possibility of finding microbial life elsewhere in our solar system or beyond and examine efforts to search for extraterrestrial intelligence (SETI). Finally, we'll discuss the astonishing implications that the search for life may hold for the future of our own civilization.

24.1 Life on Earth

It may seem that aliens are everywhere. Popular shows and movies, such as *Star Trek* and *Star Wars,* show aliens from many different worlds traveling easily among the stars. Closer to home, supermarket tabloids routinely carry headlines about the latest alien atrocities or about alien corpses hidden by the government at "Area 51."

Despite their popularity in the media, any alien visitors have been sadly negligent in leaving scientific evidence of their trespass. Decades of scientific observation and study have not turned up a single piece of undeniable evidence that aliens have been here (see the box "Are Aliens Already Here?," p. 730). Why, then, do so many scientists believe it is worth the effort to search for life beyond Earth?

For a long time, the answer was simply that it seemed natural for other worlds to be inhabited. Belief in life on other worlds was common even among ancient Greek philosophers, and many famous scientists of the past few centuries took it as a given that intelligent beings exist on other planets. For example, Kepler suggested that the Moon was inhabited, and William Herschel (co-discoverer of Uranus with his sister Caroline) spoke of life on virtually all the planets in our solar system. Most famously, in the late nineteenth century Percival Lowell claimed to see networks of canals on Mars, which he argued were the mark of an advanced civilization [Section 9.4]. Lowell's ideas became the basis for H. G. Wells's novel *The War of the Worlds,* upon which a recent movie was based.

Hopes of finding life on our planetary neighbors were diminished by the bleak images of Mars returned by spacecraft and the discovery of the runaway greenhouse effect on Venus. For a few decades, scientific interest in extraterrestrial life waned. However, that interest has been on the upswing again during the past couple of decades. Part of the renewed interest has come from the discovery of planets around other stars [Section 13.1], which confirms that there are plenty of places to look for life beyond our own solar system. But new discoveries about life on Earth have played an even more important role. In particular, three recent developments in the study of life on Earth have made it seem much more likely that we could find life elsewhere:

● We have learned that life arose quite early in Earth's history, suggesting that life might also form quickly on other worlds with the right conditions.
● Laboratory experiments have shown that the chemical constituents thought to have been common on the young Earth combine readily into complex organic molecules. We may never know exactly how life arose on Earth, but these experiments suggest that life might arise through naturally occurring chemistry—in which case the same chemistry could have given rise to life on many other worlds.
● We have discovered microscopic living organisms that can survive in conditions similar to those found on at least some other worlds in our solar system, suggesting that the basic necessities of life may be quite common in the universe.

Because these three ideas are so important to understanding the modern science of life in the universe, let's examine each of them in greater detail. Along the way, we'll also briefly discuss the nature and history of life on Earth. We will then be prepared to consider how we might actually search for life on other worlds.

• When did life arise on Earth?

Our planet was born about 4.5 billion years ago, but we would not expect life to have had much of a chance to survive during the first several hundred million years of Earth's history. This was the time of the *heavy bombardment* [Section 8.4], when large asteroids or comets should have repeatedly struck Earth. At least a few of these impacts probably had sufficient energy to vaporize the early oceans completely, killing off any life that might already have been present on Earth. The ancestors of living organisms today could not have arisen much before the end of the heavy bombardment.

Studies of craters on the Moon suggest that the last major impacts of the heavy bombardment occurred between about 4.2 and 3.9 billion years ago. Remarkably, we now have evidence suggesting that life was thriving prior to

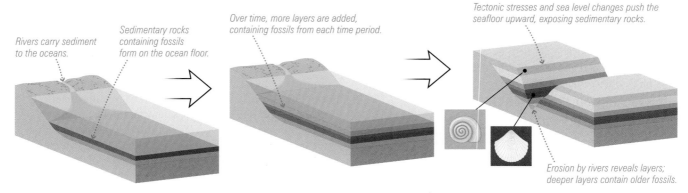

Rivers carry sediment to the oceans.

Sedimentary rocks containing fossils form on the ocean floor.

Over time, more layers are added, containing fossils from each time period.

Tectonic stresses and sea level changes push the seafloor upward, exposing sedimentary rocks.

Erosion by rivers reveals layers; deeper layers contain older fossils.

Figure 24.1 Formation of sedimentary rock. Each layer represents a particular time and place in Earth's history and is characterized by fossils of organisms that lived in that time and place.

3.85 billion years ago. If we are interpreting the evidence correctly, it means that life arose in a geological blink of an eye from the moment when conditions first allowed it. To understand this evidence and its importance, we need to discuss how scientists study the history of life on Earth.

The Geological Time Scale   We learn about the history of life on Earth through the study of **fossils**, relics of organisms that lived and died long ago. Most fossils form when dead organisms fall to the bottom of a sea (or other body of water) and are gradually buried by layers of sediment. The sediments are produced by erosion on land and carried by rivers to the sea. Over millions of years, sediments pile up on the seafloor, and the weight of the upper layers compresses underlying layers into rock. Erosion or tectonic activity can later expose the fossils (Figure 24.1). In some places, such as the Grand Canyon, the sedimentary layers record billions of years of Earth's history (Figure 24.2). Some fossils are remarkably well preserved (Figure 24.3), though the vast majority of dead organisms decay completely and leave no fossils behind.

The key to reconstructing the history of life is to determine the dates at which fossil organisms lived. The *relative* ages of fossils found in different layers are easy to determine: Deeper layers formed earlier and contain more ancient fos-

sils. Radiometric dating [Section 8.5] confirms these relative ages and gives us fairly precise absolute ages for fossils. Based on the layering of rocks and fossils, geologists divide Earth's history into a set of distinct intervals that make up what we call the **geological time scale**. Figure 24.4 shows the names of the various intervals on a timeline, along with numerous important events in Earth's history.

Study Figure 24.4. How does the length of time during which animals and plants have lived on land compare to the length of time during which life has existed? How does the length of time during which humans have existed compare to the length of time since mammals and dinosaurs first arose?

Fossil Evidence for the Early Origin of Life   You might wonder why the geological time scale shows so much more detail in the last few hundred million years than it does for earlier times. The answer is that fossils become increasingly difficult to find as we look deeper into Earth's history, for three major reasons: First, older rocks are much rarer than younger rocks, because most of Earth's surface is geologically young. Second, even when we find very old rocks, they often turn out to have been subject to transformations (caused by heat and pressure) that would have destroyed any fossil

Figure 24.2 The rock layers of the Grand Canyon record 2 billion years of Earth's history.

Figure 24.3 Dinosaur fossils.

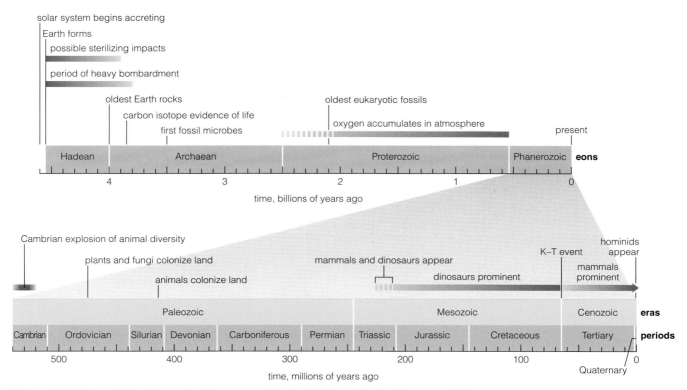

Figure 24.4 The geological time scale. Be sure to notice that the lower timeline is an expanded view of the last portion of the upper timeline. The eons, eras, and periods are defined by changes observed in the fossil record. The absolute ages come from radiometric dating. (The K–T event is the geological term for the impact linked to the mass extinction of the dinosaurs [Section 12.4].)

evidence they may have contained. Third, all life prior to a few hundred million years ago was microscopic, and microscopic fossils are much more difficult to identify than dinosaur bones.

Despite these difficulties, geologists have found a few very old rocks that suggest life was already thriving on Earth by 3.5 billion years ago, and possibly for several hundred million years before that. One strong line of evidence comes from rocks called *stromatolites*. These rocks show structure nearly identical to that found in large bacterial mats today (Figure 24.5). Radiometric dating shows that some stromatolites are 3.5 billion years old.

The fact that organisms were already advanced enough to build stromatolites by 3.5 billion years ago suggests that more primitive organisms must have lived even earlier. Some evidence supports this idea. More ancient rocks that have undergone too much change to leave fossils intact may still hold carbon that was once part of living organisms.

Carbon has two stable isotopes: carbon-12, with six protons and six neutrons in its nucleus, and carbon-13, which

a These large mats photographed at Shark Bay, Western Australia, are colonies of microbes known as "living stromatolites."

b The bands visible in this section of a modern-day mat are formed by layers of sediment adhering to different types of microbes.

c This section of a 3.5-billion-year-old stromatolite shows a structure nearly identical to that of a living mat. This structure, along with more detailed study, strongly suggests that stromatolites are the fossilized remains of microbes, including some photosynthetic ones, that lived as much as 3.5 billion years ago.

Figure 24.5 Rocks called stromatolites offer evidence of photosynthetic life as early as 3.5 billion years ago.

has one extra neutron (see Figure 5.9). Living organisms incorporate carbon-12 slightly more easily than carbon-13. As a result, the fraction of carbon-13 is always a bit lower in fossils than in rock samples that lack fossils. All life and all fossils tested to date show the same characteristic ratio of the two carbon isotopes. Dating very ancient rocks can be difficult, but some rocks that are more than 3.85 billion years old show the same ratio of carbon-12 to carbon-13, suggesting that these rocks contain remnants of life.

To summarize, fossil evidence points to life already thriving by 3.5 billion years ago, and carbon isotope evidence suggests that life was present more than 3.85 billion years ago. Thus, even under the conservative assumption that the last major impact of the heavy bombardment occurred 4.2 billion years ago, we conclude that life arose in a period of no more than about 300 million years. If the last major impact occurred 3.9 billion years ago, then life arose within a few tens of millions of years or less.

• How did life arise on Earth?

We have discussed when life arose on Earth, but how did life arise? To answer this question, we must first consider how life today differs from life at the time of its origin on Earth. Then we can ask how the first organisms may have originated.

The Theory of Evolution   The fossil record clearly shows that life has gone through great changes over time (see Figure 24.4). Thus, if we are going to understand how life arose, we must understand what causes this change, so that we can trace life back to its origin. The unifying theory through which scientists understand the history of life on Earth is the **theory of evolution**, proposed by Charles Darwin (1809–1882) in his 1859 book called *The Origin of Species.*

Evolution simply means "change with time," and strong fossil evidence for evolution had already been gathered before Darwin published his theory. Indeed, most scientists of Darwin's time had already accepted the idea that species change through time; what was missing prior to Darwin was an explanation of how and why species might undergo change. In essence, the fossil record provides strong evidence that evolution *has* occurred, while Darwin's theory of evolution explains *how* it occurs.

Darwin provided extensive evidence for the occurrence of evolution and, based on this evidence, he laid out a clear model to explain how evolution occurs. As is the case with most scientific theories, the underlying logic of his model is really very simple. As described by biologist Stephen Jay Gould (1941–2002), Darwin built his model from "two undeniable facts and an inescapable conclusion":

● *Fact 1: overproduction and struggle for survival.* Any localized population of a species has the potential to produce far more offspring than the local environment can possibly support with resources such as food and shelter. This overproduction leads to a struggle for survival among the individuals of the population.

● *Fact 2: individual variation.* Individuals in a population of any species vary in many heritable traits (that is, traits passed from parents to offspring). No two individuals are exactly alike, and some individuals possess traits that make them better able to compete for food and other vital resources.

● *The inescapable conclusion: unequal reproductive success.* In the struggle for survival, those individuals whose traits best enable them to survive and reproduce will, on average, leave the largest number of offspring that in turn survive to reproduce. Therefore, in any local environment, heritable traits that enhance survival and successful reproduction will become progressively more common in succeeding generations. It is this unequal reproductive success that Darwin called **natural selection**; that is, nature "selects" the advantageous trait to be passed on to future generations.

Of course, even the strongest logic still needs evidence to back it up if it is to gain acceptance as a scientific theory [Section 3.4], and Darwin's greatest achievement may have been in compiling a large body of extensive evidence in support of his theory. By studying the variety of species in places like the Galapagos Islands, Darwin demonstrated that natural selection had given the individuals of different species important survival advantages. Moreover, his work showed that when enough small individual variations accumulate, natural selection can even give rise to an entirely new species. Darwin's evidence was so complete and so well documented that the vast majority of biologists accepted his theory within a decade of the publication of his book.

The Mechanism of Evolution   In the nearly 150 years since Darwin first published his theory, ongoing research has only given further support to Darwin's idea. Perhaps the strongest support for the theory has come from the discovery of **DNA** (short for *deoxyribonucleic acid*), the genetic material of all life on Earth. As a result of this discovery, scientists now understand the mechanism of evolution on a molecular level.

As you have probably learned in past biology classes, a molecule of DNA consists of two long strands—somewhat like the interlocking strands of a zipper—wound together in the spiral shape known as a *double helix* (Figure 24.6). The instructions for assembling a living organism are written in the precise order of four chemical bases that make up the interlocking portions of the DNA "zipper": adenine (A), thymine (T), guanine (G), and cytosine (C). These bases pair up in a way that ensures that each of the two strands of a DNA molecule contains the same genetic information. Thus, by unwinding and allowing new strands to form alongside the original ones (with the new strands made from chemicals floating around inside a cell), a single DNA molecule can give rise to two identical copies of itself.

Living organisms reproduce by copying DNA molecules and passing them on to their descendants. Evolution occurs because the passing of genetic information from one generation to the next is not always perfect. An organism's DNA

Figure 24.6 DNA is the genetic material of all life on Earth. Left: A model of a small piece of a DNA molecule. Right: This diagram shows that a DNA molecule is made from two interlocking strands wound in the shape of a double helix. The interlocking pieces (which look like ladder steps in the figure) are composed of four chemical bases, labeled by their first letters A, T, G, and C. Note that A always connects to T, and G always connects to C. Thus, each strand contains the same basic information, so when the two strands separate and "helper molecules" bring new chemical bases to attach to each one, the result is two new molecules each identical to the original one that separated. This ability of DNA to copy itself is the key to heredity.

may be altered by occasional copying errors or by external influences, such as ultraviolet light from the Sun or exposure to toxic or radioactive chemicals. Any change in the sequence of the chemical bases in an organism's DNA is called a **mutation**. Many mutations are lethal, killing the cell in which the mutation occurs. Some, however, may improve a cell's ability to survive and reproduce, allowing the cell to pass this improvement to its offspring.

Our molecular understanding of the mechanism of natural selection has put the theory of evolution on a stronger foundation than ever. While no theory can ever be proved true beyond all doubt, the theory of evolution today is as solid as any theory in science, including the theory of gravity and the theory of atoms. Indeed, biologists today routinely witness evolution occurring before their eyes among laboratory microorganisms, or over periods of just a few decades among plants and animals subjected to some kind of environmental stress. Moreover, the theory of evolution has become the underpinning of virtually all modern biology, medicine, and agriculture. For example, agricultural scientists apply the idea of natural selection to develop pest control strategies that reduce the populations of harmful insects without harming the populations of beneficial ones; medical researchers test new drugs on animals that are genetically similar to humans, because the theory of evolution tells us that genetically similar species should have at least somewhat similar physiological responses; and biologists study the relationships between organisms by comparing their DNA.

The First Living Organisms The basic chemical nature of DNA is virtually identical among all living organisms. This fact, along with other biochemical similarities shared by all living organisms (such as the fact that all organisms build proteins from the same set of amino acids and use essentially the same genetic code), tells us that all life on Earth today can trace its origins to a common ancestor. In other words, sometime before about 3.85 billion years ago, a single organism arose and gradually evolved into every other organism that has ever lived.

What did the first living organism look like? Although we may never have fossils of the earliest organisms, in principle we can learn about early life through careful studies of the DNA of living organisms. Biologists can determine the evolutionary relationships among living species by comparing the sequences of bases in their DNA. For example, two organisms whose DNA sequences differ in five places for a particular gene are probably more distantly related than two organisms whose gene sequences differ in only one place.

Many such DNA comparisons suggest that all living organisms are related in a way depicted schematically by the "tree of life" in Figure 24.7. Although details in the structure of this tree remain uncertain, it confirms that living organisms share a common ancestry. As it is currently understood, the tree of life shows that life on Earth is divided into three major groupings (known as the *domains* Bacteria, Archaea, and Eukarya). Notice that plants and animals represent only two tiny branches of the great diversity of life on Earth.

Comparison of DNA sequences also allows at least reasonable guesses as to which organisms found on Earth today most resemble the common ancestor. The answer appears to be microscopic organisms living in the deep oceans around seafloor volcanic vents called *black smokers* (after

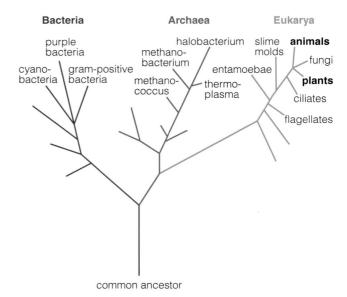

Figure 24.7 The tree of life, showing evolutionary relationships determined by comparison of DNA sequences in different organisms. Just two small branches at the upper right represent *all* plant and animal species.

the dark, mineral-rich water that flows out of them) and in hot springs in places such as Yellowstone (Figure 24.8). These organisms thrive in temperatures as high as 110°C (230°F). (The high pressures at the seafloor prevent the water from boiling despite the high temperature.) Unlike most life at Earth's surface, which depends on sunlight, these organisms get energy from chemical reactions in water heated volcanically by Earth itself.

That early organisms might have lived in such "extreme" conditions may at first seem surprising, but it makes sense

---

SPECIAL TOPIC Evolution and the Schools

You're probably familiar with the public debate about how evolution should be taught in schools. Much of the controversy comes from misunderstandings on both sides. On the antievolution side, there's a general lack of understanding of why scientists are near unanimous in accepting the theory of evolution. On the scientists' side, there's often a lack of respect for people's individual religious beliefs.

The heart of the debate surrounds the question of what counts as science and what does not. You can see why by considering a common question: Do you believe in creationism or evolution? This question may be philosophically and theologically interesting, but it is poorly phrased from a scientific standpoint because science is about evidence, not beliefs. A tremendous weight of evidence supports the idea that the universe is much older than the Earth, that Earth is some $4\frac{1}{2}$ billion years old, that life on Earth has existed for nearly 4 billion years and has undergone dramatic change during that time, and that this change has occurred through the mechanism of evolution by natural selection. Nevertheless, all the evidence in the world can never prove a scientific theory to be true beyond all doubt. To take a somewhat extreme case, you could accept the overwhelming evidence for evolution and still believe that God created Earth and the universe a mere 6,000 years ago by, for example, assuming that God put all the evidence for evolution in place. Because the evidence would look the same in this case as if evolution actually occurred, science cannot say anything about the validity of such a belief.

We can gain deeper insight by fitting the theory of evolution to the hallmarks of science illustrated in Figure 3.26. The first hallmark says that science seeks explanations for observed phenomena that rely on natural causes, and the theory of evolution does exactly that: It explains the fossil record and observed differences between species in terms of natural selection. Note that this hallmark does not in any way preclude the possibility of divine intervention; it just says that if divine intervention has occurred, we cannot study it scientifically. The second and third hallmarks concern the way evidence is gathered and interpreted: In science, we create and test models designed to explain our observations, and we modify or discard those models if they fail the tests. Again, the theory of evolution exhibits these hallmarks because it is continually subject to testing and verification, and it has indeed been modified as new evidence has come to light. For example, Darwin's original theory discussed natural selection, but the molecular basis for it wasn't added until the discovery of DNA. Similarly, Darwin thought that evolution would be a gradual process occurring in a large population of a given species, but more recent evidence suggests that evolution often occurs in small, isolated populations, and often in episodic bursts. The ongoing debate about the precise nature of evolution sometimes makes nonscientists think there is scientific controversy over the basic theory, but in reality it shows why the process of science has put the theory of evolution by natural selection on such a solid footing.

The fact that evolution is such a well-established and important scientific theory makes it an obvious part of the science curriculum for the schools, and few people disagree. However, many people have suggested teaching alternative ideas such as *creationism* or *intelligent design* alongside evolution in school curricula. The scientific response is very simple: Scientists take no position on the general school curriculum, but the *science curriculum* should not include these alternate ideas because they are not science. Again, we can see why by considering whether these ideas reflect the hallmarks of science.

Let's start with *creationism*, the idea that the world was divinely created. In fact, we can't even define this idea clearly enough to test it against the hallmarks of science. Many religions incorporate stories of creation, and these stories are often quite different from one another. For example, many Native American religious beliefs speak of creation in terms that bear little resemblance to the story in Genesis. Even among Christians who claim a literal belief in the Bible, there are differences of interpretation about creation. While some Biblical literalists argue that the creation must have occurred in just 6 days, as the first chapter of Genesis seems to say, others suggest that the term "day" in Genesis does not necessarily mean 24 hours and that Genesis is compatible with a much older Earth. Thus, while creationism is a perfectly valid religious belief, it does not qualify as a scientific theory.

The more recent idea of *intelligent design* holds that the complexity of life is too great to have arisen naturally, and therefore there must have been a "designer." Intelligent design can sound much like science, especially because most of its proponents accept both the idea of an old Earth and the idea that some living things have changed through time as indicated by the fossil record. However, the assumption of an intelligent designer does not yield testable predictions, because it posits that the designer is beyond our scientific comprehension; in essence, the idea of intelligent design means that there is no point in continuing to look for natural causes, because a supernatural process made the universe and life. This idea may or may not be true, but it isn't science. As a simple analogy, consider the collapse of a bridge. If an engineer declares that the collapse was an act of God, he could well be right—but that belief won't help him learn how to design a better bridge. It is the scientific quest for a natural understanding of life—embodied in the theory of evolution—that has led to the discovery of genetics, DNA, and virtually all modern medicine.

So where does this discussion leave us with regard to the question of teaching evolution in schools? Let's start with what our discussion does *not* tell us. It does not tell us that everyone needs to accept the theory of evolution—whether you choose to believe the theory is up to you. Nor does it tell us whether divine intervention has guided evolution—science does not say anything about the existence or role of God. What it *does* tell us is that the theory of evolution is a clear and crucial part of the study of science, and that without it we cannot fully understand modern science and its impact on our world.

a This photograph shows a black smoker—a volcanic vent on the ocean floor that spews out hot, mineral-rich water.

b This aerial photo shows a hot spring in Yellowstone National Park. The different colors are from different microbes that survive in water of different temperatures. For a sense of scale, note the walkway winding along the lower right.

Figure 24.8 DNA evidence suggests that the ancestor of all life on Earth resembled organisms that live today in hot water near volcanic vents or hot springs.

when we think about it more deeply. Sunlight can provide energy only to organisms capable of photosynthesis, which is a very complex molecular process. The chemical pathways for extracting energy from the heat of Earth itself are actually much simpler and therefore far more likely to have been used by the earliest organisms. Moreover, chemical reactions tend to proceed faster and with greater diversity in hotter water than cooler water, so the hot water near volcanic vents may have offered a natural laboratory in which chemistry "experiments" were constantly under way.

The Transition from Chemistry to Biology The theory of evolution tells us how a single organism can have evolved into the great diversity of life on Earth today. But where did the first organism come from? We may never know for sure, because it is unlikely that any fossil evidence could have recorded the transition from nonlife to life. However, over the past several decades, scientists have conducted many laboratory experiments designed to mimic the conditions that existed on the young Earth. These experiments suggest that life could have arisen through natural chemical reactions.

The first such experiments were performed in the 1950s (Figure 24.9), and they have been greatly refined and improved since that time. In essence, the experiments simply mix chemical ingredients thought to have been present on the early Earth and then "spark" the chemicals with electricity to simulate lightning or other energy sources. The chemical reactions that follow have produced all the major molecules of life, including all the amino acids and DNA bases. Many of these same molecules are found today in meteorites, suggesting that impacts may also have been an important source of organic molecules on the early Earth.

Laboratory experiments also show that mixing a warm dilute solution of organic molecules with naturally occurring sand or clay allows the molecules to assemble themselves into much more complex molecules. Strands of RNA (ribonucleic acid), an important molecule in all living organisms that looks much like a single strand of DNA, have been produced in the laboratory with lengths of up to nearly

100 bases. Because some RNA molecules are capable of self-replication, many biologists now presume that RNA was the original genetic material of life on Earth, with DNA coming later.

Other experiments have shown that microscopic enclosed membranes also form under conditions expected

Figure 24.9 This photograph shows chemist Stanley Miller with a reproduction of the experimental setup he first used in the 1950s to study pathways to the origin of life. (He worked with Harold Urey, so the experiment is called the Miller-Urey experiment.) The flasks are filled with liquids or gases thought to have been present on the early Earth, and an electrical discharge supplies energy. The result is the production of many complex organic molecules. Scientists now think the early atmosphere was different from that used in Miller's original experiment, but similar experiments with the correct gases still yield similar results.

a These microscopic spheres were made by cooling a warm water solution of amino acids. They are not alive, but they exhibit many lifelike properties.

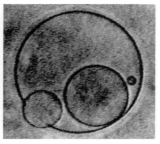

b These microscopic membranes are made from lipids that, when mixed with water, spontaneously form enclosed droplets.

Figure 24.10 These microscopic photos show enclosed membranes that self-assemble in laboratory experiments mimicking conditions on the early Earth.

on the early Earth (Figure 24.10). Self-replicating RNA molecules may have become enclosed in such membranes, making what are sometimes called "pre-cells." Those pre-cells in which RNA replicated faster and more accurately were more likely to spread, leading to a type of positive feedback that would have encouraged even faster and more accurate replication. Given millions of years of chemical reactions occurring all over the Earth, RNA might eventually have evolved into DNA, making true living organisms.

Figure 24.11 summarizes the steps by which chemistry might have become biology on Earth. We may never know whether life really arose in this way, but it certainly seems possible.

Could Life Have Migrated to Earth?   Our scenario suggests that life could have arisen naturally here on Earth. However, an alternative possibility is that life arose somewhere else first—perhaps on Venus or Mars—and then migrated to Earth on meteorites. Remember that we have collected meteorites that were blasted by impacts from the surfaces of the Moon and Mars [Section 12.1]. Calculations suggest that Venus, Earth, and Mars all should have exchanged many tons of rock, especially in the early days of the solar system when impacts were more common.

The idea that life could travel through space to land on Earth, sometimes called *panspermia*, once seemed out-

landish. After all, it's hard to imagine a more forbidding environment than that of space, with no air, no water, and constant bombardment by dangerous radiation from the Sun and stars. However, the presence of organic molecules in meteorites and comets tells us that the building blocks of life can indeed survive in space, and tests have shown that some microbes can survive in space for years.

In a sense, Earth, Venus, and Mars have been "sneezing" on each other for billions of years. Life could conceivably have originated on any of these three planets and been transported to the others. It's an intriguing thought, but it does not change our basic scenario for the origin of life—it simply moves it from one planet to another.

A Brief History of Life on Earth   We are now ready to take a quick look at the history of life on Earth, as summarized in Figure 24.4. We will focus on events that may help us understand what we are looking for when we search for life on other worlds.

Chemical reactions probably began producing organic molecules not long after Earth's formation more than 4.5 billion years ago. Some mineral evidence suggests that Earth already had oceans by 4.3 to 4.4 billion years ago, and the early oceans would have been natural laboratories for chemistry that could have led to life. However, even if life arose at such early times, it was almost certainly extinguished by major impacts of the heavy bombardment. The common ancestor of life today arose after the impacts had subsided.

Whether through chemical reactions in the oceans or through migration from Venus or Mars, living organisms quickly took hold on Earth after the end of the heavy bombardment. Starting from the first organism, evolution rapidly diversified life on Earth.

Despite the rapid pace of evolution, nearly all life remained single-celled for at least a billion years. Land remained inhospitable even longer, because of the lack of a protective ozone layer. Continents much like those of today were surrounded by oceans teeming with life, but, despite pleasant temperatures and plentiful rainfall, the land itself was probably as barren as Mars is now. Life on land probably became possible only after atmospheric oxygen accumulated to the point where an ozone layer could form.

Nearly all the oxygen in our atmosphere was originally released through photosynthesis by single-celled organisms

1. *Organic precursor molecules appear.*   2. *RNA molecules become self-replicating.*   3. *Membrane-enclosed precells arise.*   4. *True cells with RNA genome appear.*   5. *Modern cells with DNA genome evolve.*

Figure 24.11 A summary of the steps by which chemistry on the early Earth may have led to the origin of life.

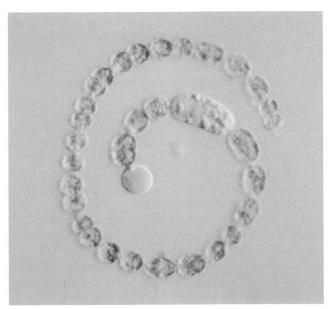

Figure 24.12 This microscopic photo shows a chain of modern cyanobacteria. The ancestors of these living organisms produced essentially all the oxygen in Earth's atmosphere.

known as cyanobacteria (Figure 24.12). These microbes may have been producing oxygen through photosynthesis as early as 3.5 billion years ago. However, the oxygen did not immediately begin to accumulate in the atmosphere. For more than a billion years, chemical reactions with surface rocks pulled oxygen back out of the atmosphere as fast as the cyanobacteria could produce it. But these tiny organisms were abundant and persistent, and eventually the surface rock was so saturated by oxygen that the rate of oxygen removal slowed down. At that point, some 2 billion years ago, oxygen began to accumulate in the atmosphere, though it may not have reached a level that we could have breathed until just a few hundred million years ago.

Today, we often think of oxygen as a necessity for life. However, oxygen was probably poisonous to most organisms living before about 2 billion years ago (and remains a poison to many microbes still living today). The rise of atmospheric oxygen therefore caused tremendous evolutionary pressure and may well have been a major factor in the eventual evolution of complex plants and animals.

There were undoubtedly many crucial changes as primitive microbes gradually evolved into multicellular organisms and early plants and animals, but the fossil record does not allow us to pinpoint the times at which all these changes occurred. However, we see a dramatic change in the fossil record dating to about 540 million years ago. Over the next 40 million years or so, animal life evolved from tiny and primitive organisms into all the basic body plans (phyla) that we find on Earth today. This remarkable flowering of animal diversity occurred in such a short time relative to the history of Earth that it is often called the *Cambrian explosion*. (*Cambrian* is the name geologists give to the period from about 540 million to 500 million years ago; see Figure 24.4.)

Early dinosaurs and mammals arose around the same time, some 225 to 250 million years ago, but dinosaurs at first proved more successful and dominated the landscape for well over 100 million years. Their sudden demise 65 million years ago, probably as a result of an asteroid or comet impact [Section 12.4], paved the way for the evolution of large mammals—including humans. The earliest humans appeared on the scene only a few million years ago, or after 99.9% of Earth's history to date had already gone by. Our few centuries of industry and technology have come after 99.99999% of Earth's history.

Despite our recent arrival (in geological terms), modern humans are arguably the most successful species ever to inhabit Earth. Humans survive and prosper in virtually every type of land environment. With proper equipment, we can survive underwater, and a few people have lived under the sea for extended periods of time in experimental habitats. We are even developing the technology to survive away from our planet in the inhospitable environment of space. Indeed, it is our very success that poses the greatest threat to our future survival, which now depends on our ability to limit the damage we do to the environment that makes our existence possible [Section 10.6].

• What are the necessities of life?

If we are going to look for other worlds that might harbor life, a first step is to understand the necessities of life. While it is always possible that life on other worlds could be quite different from life on Earth, it's easiest to begin the search for life by looking for conditions that would be comfortable for organisms from Earth. So what exactly does Earth life need to survive?

If we think about ourselves, the requirements for life seem fairly stringent: We need abundant oxygen in an atmosphere that is otherwise not poisonous, we need temperatures in a fairly narrow range of conditions, and we need abundant and varied food sources. However, the discovery of life in "extreme" environments—such as in the hot water near black smokers and hot springs—shows that many microbes (often called *extremophiles*) can survive under a much wider range of conditions.

The organisms living in hot water prove that at least some microbes can survive in much higher temperatures than we would have guessed. Other organisms live in other extremes. In the freezing cold but very dry valleys of Antarctica, the surface appears barren but microbes have been found living *inside* rocks, surviving on tiny droplets of liquid water and energy from sunlight. Other microscopic organisms have been found living up to several kilometers underground in water that fills pores within the subterranean rock. We have found life thriving in environments so acidic, alkaline, or salty that we humans would be poisoned almost instantly. We have even found microbes that can survive high doses of radiation, making it possible for them to live for many years in the radiation-filled environment of space.

If we compare the different forms of life on Earth, we find that life as a whole has only three basic requirements:

1. A source of nutrients (elements and molecules) from which to build living cells.
2. Energy to fuel the activities of life, either from sunlight, from chemical reactions, or from the heat of Earth itself.
3. Liquid water.

These requirements give us a basic road map for the search for life elsewhere. If we want to find life on another world, it makes sense to start by searching for worlds that offer these basic necessities.

Interestingly, only the third requirement (liquid water) seems to pose much of a constraint. Organic molecules are present almost everywhere—even on meteorites and comets. Many worlds are large enough to retain internal heat that could provide energy for life, and virtually all worlds have sunlight (or starlight) bathing their surfaces, although the inverse square law for light [Section 15.1] means that light provides less energy to worlds farther from their sun or star. Thus, nutrients and energy should be available to at least some degree on almost every planet and moon. In contrast, liquid water seems to be relatively rare, and the search for liquid water therefore drives the search for life in our universe.

24.2 Life in The Solar System

If we assume that life elsewhere would be at least a little bit like life on Earth, then the search for life begins with a search for **habitable worlds**—worlds that contain the basic necessities for life as we know it, including liquid water. The liquid water requirement rules out most of the worlds in our solar system. For example, Mercury and the Moon are barren and dry; Venus is too hot for liquid water; and Pluto, Planet X, other comets, and most of the small moons of the outer solar system are too cold. The jovian planets may have droplets of liquid water in some of their clouds, but the strong vertical winds on these planets make their clouds unlikely homes for life. That leaves two major possibilities besides Earth for habitability in our solar system: (1) Mars and (2) a few of the large moons orbiting jovian planets, most notably Europa [Section 11.2]. Let's discuss the current evidence about the potential habitability of these worlds.

SPECIAL TOPIC What Is Life?

You may have noticed that while we've been talking about life, we haven't actually defined the term. In fact, it's surprisingly difficult to draw a clear boundary between life and nonlife. Life can be so difficult to define that we may be tempted to fall back on the famous words of Supreme Court Justice Potter Stewart, who, in avoiding the difficulty of defining pornography, wrote: "I shall not today attempt further to define [it]. . . . But I know it when I see it." If living organisms on other worlds turn out to be much like those found on Earth, it may prove true that we'll know them when we see them. But if the organisms are fairly different from those on Earth, we'll need clearer guidelines to decide whether they are truly "living."

One way to seek distinguishing features of life is to study living organisms, looking for common characteristics. Given the difficulty of defining life, you probably won't be surprised to learn that there are exceptions to almost any "rule" we think of. Nevertheless, biologists have identified at least six key properties that most or all living organisms on Earth appear to share:

1. *Order:* Living organisms are not random collections of molecules but rather have molecules arranged in orderly patterns that form cell structures.
2. *Reproduction:* Organisms are capable of reproducing.
3. *Growth and development:* Living organisms grow and develop in patterns determined at least in part by heredity.
4. *Energy utilization:* Living organisms use energy to fuel their many activities.
5. *Response to the environment:* Life actively responds to changes in its surroundings. For example, simple organisms may alter their chemistry in the presence of a food source, while warm-blooded mammals may sweat, pant, or adjust blood flow to maintain a constant internal temperature.

6. *Evolutionary adaptation:* Life evolves through natural selection, as organisms pass on traits that make them better adapted to survival in their local environments.

These six properties are all important, but biologists today regard evolution as the most fundamental and unifying of them. Evolutionary adaptation is the only property that can explain the great diversity of life on Earth. Moreover, understanding how evolution works allows us to understand how all the other properties came to be. As a result, the simplest definition of life might be "something that can reproduce and evolve through natural selection."

For most practical purposes, this definition of life would probably suffice. However, some cases may still challenge this definition. For example, computer scientists can now write programs (that is, lines of computer code) that can reproduce themselves (that is, create additional sets of identical lines of code). By adding instructions that allow random changes to the programs, they can even make "artificial life" that evolves on a computer. Should this "artificial life," which consists of nothing but electronic signals processed by computer chips, be considered alive?

The fact that we have such difficulty distinguishing the living from the nonliving on Earth suggests that we should be very cautious about constraining our search for life elsewhere. No matter what definition of life we choose, there's always the possibility that we'll someday encounter something that challenges it. Nevertheless, the properties of reproduction and evolution seem very likely to be shared by most if not all life in the universe and therefore provide a useful starting point as we consider how to explore the possibility of extraterrestrial life.

Is it possible for a world to be habitable but not actually have life? Is it possible for a world to have life but not be habitable? Explain.

• Could there be life on Mars?

We now have sufficiently detailed images of Mars to be quite confident that no civilizations have ever existed there. The "canals" seen by Percival Lowell were some type of mirage, presumably formed by a combination of real Martian features and his own imagination. Nevertheless, we have good reason to believe that liquid water flowed on Mars in the past [Section 9.4], making it at least seem possible that life could once have found a home there. Moreover, since Mars today contains subsurface ice, it's conceivable that life might still survive near sources of volcanic heat, where pockets of liquid water might persist underground.

Missions to Mars   Mars is not only the best candidate in our solar system for life beyond Earth but is also the only place where we've begun an actual search for life. Our first attempt to search for life on Mars came with the *Viking* missions to Mars in the 1970s. Two *Viking* landers arrived on the surface of Mars in 1976. Each was equipped with a robotic arm for scooping up soil samples, which were fed into several onboard robotically controlled experiments.

Three of the *Viking* experiments were designed expressly to look for signs of life. None of these experiments could actually "see" life but rather looked for chemical changes that could be attributed to living organisms. Although all three experiments gave results that initially seemed consistent with life, further study suggested that chemical reactions could have produced the same results. Moreover, a fourth experiment, which analyzed the content of Martian soil, found no measurable level of organic molecules—the opposite of what we would expect if life were present. As a result, most scientists have concluded that no life was present at the locations where *Viking* sampled the soil.

More recent missions to Mars, including the Pathfinder lander in 1997 and the *Spirit* and *Opportunity* rovers that landed in 2004, have taken a more measured approach to the search for life (Figure 24.13). Rather than conducting direct experiments like those on *Viking*, these missions have been designed to help us better understand Martian conditions, so that we'll have a better idea of how and where to search for life with future missions.

Within a decade or so, NASA hopes to launch a mission to Mars that will bring back surface samples for study in laboratories on Earth. Later, we may send more advanced robots or humans to Mars, where they could search for fossils or living organisms in deep canyons like Valles Marineris, in ancient valley bottoms and dried-up lake beds, or in underground pockets of water near not-quite-dead volcanoes.

Figure 24.13 The *Opportunity* rover studies a rock on Mars during its mission in 2004. Such studies are helping scientists understand whether and when Mars may have been habitable.

The search will not be easy, but we should eventually learn whether life has ever existed on Mars.

Methane on Mars   In 2004, scientists working with the European Space Agency's *Mars Express* mission, which entered Mars orbit in 2003, announced an unexpected and potentially important result: Instruments on the spacecraft had detected methane gas in the Martian atmosphere. This news came as a surprise to most scientists, because methane gas cannot last more than a few centuries in the Martian atmosphere before chemical reactions transform it into other gases.* Thus, if the methane detection is real, it implies that Mars has a source of ongoing methane production somewhere on or beneath its surface.

Some scientists still doubt that the data are being properly interpreted, but a few ground-based observations made to check the result seem to confirm a methane signature in spectra of Mars. Where could the methane be coming from? We know of at least three possibilities: comet impacts, volcanic activity, or life. The first possibility is highly unlikely, since impacts are such rare events and an impact would have to have occurred quite recently for methane to remain in the atmosphere. That leaves us with volcanism or life.

A few scientists have looked at correlations between the location of methane and water vapor in the Martian atmosphere and used the results to argue that we might actually be seeing the signature of microscopic life on Mars. The vast majority of scientists consider any such conclusion to be premature at best, arguing that volcanism is a more likely source for the methane. After all, as we discussed in Chapter 9, we suspect that Mars has at least some low level of ongoing volcanic activity. But even this result could have

*The methane is oxidized to form water and carbon dioxide; the oxidation occurs because there is some free oxygen in the Martian atmosphere, made by breakup of the abundant atmospheric carbon dioxide.

implications for the existence of life: If there is enough volcanic activity to produce a detectable level of methane, it may also mean there is enough volcanic heat to make pockets of liquid water in some places beneath the Martian surface, raising the possibility that life could exist in some of this liquid water.

The Debate over Martian Meteorites   An entirely different approach to searching for life on Mars relies on studies of the couple of dozen known *Martian meteorites*—meteorites whose chemical and isotopic composition suggests they came from Mars [Section 12.1]. One of these meteorites, designated ALH84001, generated particular excitement when a group of scientists claimed that it contained evidence of past life on Mars (Figure 24.14).

ALH84001 was found on the Antarctic ice in 1984. Careful study of the meteorite shows that it landed in Antarctica about 13,000 years ago, following a 16-million-year journey through space after an impact blasted it from Mars. The rock itself dates to 4.5 billion years ago, indicating that it solidified shortly after Mars formed and therefore resided on Mars throughout the times when Mars may have been warmer and wetter.

Painstaking analysis of the meteorite reveals several lines of evidence that could indicate the past presence of life on Mars. For example, the rock contains layered carbonate minerals and complex organic molecules (polycyclic aromatic hydrocarbons, or PAHs) that are associated with life when found in Earth rock, as well as microscopic chains of magnetite crystals quite similar to chains made in Earth rocks by living bacteria. Most intriguingly, highly magnified images reveal rod-shaped structures that look much like recently discovered *nanobacteria* on Earth. The terrestrial nanobacteria are about $\frac{1}{100}$ the size of ordinary bacteria, and some biologists question whether they truly are living organisms. However, they appear to contain DNA, which suggests that they are indeed a form of life. If so, could the similar-looking structures in the Martian meteorite be fossil life from Mars?

It's possible, but each of the tantalizing hints of Martian life can also be explained in a nonbiological way. Subsequent studies have shown that chemical and geological processes can produce structures very similar to those found in the Martian meteorite. In addition, terrestrial bacteria have been found living inside the meteorite, indicating that it was contaminated by Earth life during the 13,000 years it resided in Antarctica. This contamination may explain the presence of the complex molecules found in the rock.

On balance, most scientists now doubt that the Martian meteorite shows true evidence of Martian life. Nevertheless, studies of ALH84001 and other Martian meteorites are continuing, and they may yet turn up surprises.

• Could there be life on Europa or other jovian moons?

After Mars, the next most likely candidates for life in our solar system are some of the moons of the jovian planets—especially Jupiter's moons Europa, Ganymede, and Callisto and Saturn's moon Titan [Section 11.2]. These moons are all large enough that they would be considered planets if they orbited the Sun independently.

The strongest of these candidates for harboring life is Jupiter's moon Europa, on which tidal heating probably creates a deep ocean beneath an icy crust (see Figure 11.21). The ice and rock from which Europa formed undoubtedly included the necessary chemical ingredients for life, and Europa's internal heating (primarily due to tidal heating) is strong enough to power volcanic vents on the sea bottom. Thus, it's easy to imagine places on Europa's ocean floor that look much like black smokers on Earth (see Figure 24.8a). If life on Earth first really arose near such undersea volcanic vents, Europa would seem to have everything needed for an origin of life.

The possibility of life on Europa is especially interesting because, unlike any potential life on Mars, it would not necessarily have to be microscopic. After all, the several kilometers of surface ice that hide Europa's ocean (if it exists) could also hide large creatures swimming within it. However, the potential energy sources for life on Europa are far more limited than the energy sources for life on Earth, mainly because sunlight could not fuel photosynthesis in the subsurface ocean. As a result, most scientists suspect

a The Martian meteorite ALH84001, before it was cut open for detailed study. The block on the right provides a sense of scale; it measures 1 cubic centimeter, about the size of a typical sugar cube.

b This photo shows rod-shaped structures found in a highly magnified slice of ALH84001. They measure about 100 nanometers in length and are as small as 10–20 nanometers in width.

c This photo shows terrestrial nanobacteria in a sample of volcanic rock from Sicily. They are close in size to the structures seen in ALH84001.

Figure 24.14 Controversial meteorite evidence for life on Mars.

that any life that might exist on Europa would probably be quite small and primitive.

As we discussed in Chapter 11, some evidence suggests that Jupiter's moons Ganymede and Callisto may also have subsurface oceans. However, these moons have much less internal heat and therefore would have even less energy for life than Europa. If they have life at all, it is almost certainly small and primitive. Nevertheless, Europa, Ganymede, and Callisto offer the astonishing possibility that Jupiter alone could be orbited by more worlds with life than we find in all the rest of the solar system.

Titan offers another enticing place to look for life. Its surface is far too cold for liquid water, but it may have lakes and rivers of liquid methane [Section 11.2]. Although many biologists think it unlikely, it is possible that these liquids could support life much as does water on Earth. Titan also appears to have many other organic molecules on its surface, and, at the very least, Titan offers an incredible natural laboratory of interesting organic chemistry.

Moreover, we can't completely rule out water-based life on Titan, even if it seems improbable. While liquid water does not exist on Titan's surface today, pockets of liquid water may have existed in the past following impacts of asteroids or comets. The heat of a large impact could have melted water ice and kept it liquid for as long as a few thousand years. If water-based life arose in such temporary pockets of water, it might have found a way to survive as the water froze, perhaps by migrating deep underground where Titan's internal heat might still keep some water liquid.

 Detecting Extrasolar Planets Tutorial, Lessons 1–3

24.3 Life Around Other Stars

We already know of planets orbiting more than 150 stars besides our Sun, and it's quite likely that billions of planetary systems inhabit our galaxy. These numbers might immediately make prospects for life elsewhere seem quite good, but numbers alone don't tell the whole story. In this section, we'll consider the prospects for life on worlds orbiting other stars.

Before we begin, we must distinguish between *surface* life like that on Earth and *subsurface* life like that we envision as a possibility on Mars or Europa. While large telescopes could in principle allow us to discover surface life on distant planets, no foreseeable technology will allow us to find life that is hidden deep underground in other star systems (unless the subsurface life has a noticeable effect on the planet's atmosphere). We therefore will focus on the search for life on planets with habitable surfaces—surfaces with temperatures and pressures that could allow liquid water to exist.

• Are habitable planets likely?

We have not yet discovered any extrasolar planets that seem likely to be habitable. Presumably because of the limitations of current detection techniques [Section 13.1], nearly all the extrasolar planets found to date are much more massive than Earth, suggesting that they are more like the jovian planets of our solar system than the terrestrial planets. Such planets are unlikely candidates for life, though some of them have orbits that could conceivably allow them to have moons with life.

However, the existence of other jovian planets makes it seem quite reasonable to suppose that terrestrial planets are also common. Our technology is just not yet up to the task of finding them. Thus, if we hope to find other habitable worlds, the first task is to decide whether we should expect them to exist, and if so, to determine where to look.

Constraints on Star Systems Before we consider planets themselves, it's useful to ask how many stars have any chance of having planets with life. In other words, which stars would make good "Suns," providing sufficient heat and light to the surfaces of terrestrial planets that happen to orbit them?

The first requirement for a star to have life-bearing worlds is that it be old enough that life could have arisen. More massive stars live shorter lives, and the most massive stars live no more than a few million years [Section 17.3]. Given that life on Earth did not arise for hundreds of millions of years after our solar system was born, we can rule out any star with more than a few times the mass of our Sun. However, because less massive stars are far more common than more massive stars, the lifetime constraint rules out only about 1% of all stars.

A second requirement is that the star allow planets to have stable orbits. Surveys suggest that roughly half of all stars are in binary or multiple star systems [Section 15.1], in which stable planetary orbits are less likely than around single stars. If life is not possible in such systems, then we can rule out life around half the stars in our galaxy. Of course, the other half—still some 100 billion stars or more—remain possible homes for life. Moreover, under some circumstances, stable planetary orbits are possible in multiple star systems, so we shouldn't entirely rule out life in such systems.

A third constraint on the likelihood of finding habitable planets is the size of a star's **habitable zone**—the region in which a terrestrial planet of the right size could have a surface temperature that might allow for liquid water and life. Figure 24.15 shows the approximate sizes, to scale, of the habitable zones around our Sun, around a star with about half the mass of the Sun (spectral type K), and around a star with about $\frac{1}{10}$ the mass of our Sun (spectral type M). Although habitable planets seem possible in all three cases, the smaller size of the habitable zones around the less massive stars makes it less likely that suitable planets would have formed in these regions.

All in all, it seems that the vast majority of stars are at least potentially capable of having life-bearing planets. Moreover, even very conservative assumptions suggest enormous numbers of possibilities. For example, limiting the search for habitable planets to stars very similar to our Sun (that is, spectral type G) would still mean billions of potential other Suns in the Milky Way Galaxy.

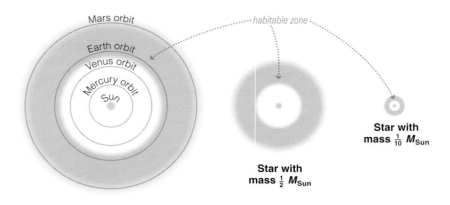

Figure 24.15 Interactive Figure The approximate habitable zones around our Sun, a star with half the mass of the Sun (spectral type K), and a star with $\frac{1}{10}$ the mass of the Sun (spectral type M), shown to scale. The habitable zone becomes increasingly smaller and closer-in for stars of lower mass and luminosity.

Solar System

Star with mass $\frac{1}{2}$ M_{Sun}

Star with mass $\frac{1}{10}$ M_{Sun}

Finding Habitable Planets Finding Earth-size planets is a daunting technological challenge [Section 13.1]. Recall that looking for an Earth-like planet around a nearby star is like standing on the East Coast of the United States and looking for a pinhead on the West Coast. Nevertheless, advancing technology should soon put Earth-like planets within our telescopic reach. The Kepler mission, scheduled for launch in 2008, will look for transits of Earth-size planets in front of their stars (see Figure 13.15). Scientists hope that Kepler will detect hundreds of Earth-size planets, and orbital properties measured by Kepler will tell us whether these planets lie within their stars' habitable zones. The Space Interferometry Mission (SIM), which NASA hopes to launch by 2011, may also be capable of detecting Earth-size planets around some nearby stars.

We will need images or spectra to determine whether the planets really are habitable or have life. Scientists are actively working on technologies that may provide such data. If all goes well, within about a decade NASA hopes to follow SIM with the *Terrestrial Planet Finder (TPF)*, an orbiting interferometer that will be capable of obtaining low-resolution spectra and crude images (a few pixels) of Earth-like planets around nearby stars [Section 13.4].

Astronomers hope to deploy even more powerful interferometers in later decades, either in space or on the Moon. Within the lifetimes of today's college students, we could conceivably have optical interferometers with dozens of telescopes spread across hundreds of kilometers. Such telescopes will be able not only to detect Earth-like planets around nearby stars but also to obtain fairly clear images and high-resolution spectra of those planets.

Signatures of Life Images from future telescopes may tell us whether the planets have continents and oceans like Earth and perhaps even allow us to monitor seasonal changes. Spectra from future telescopes should prove even more important to the search for life. Moderate-resolution infrared spectra can reveal the presence and abundance of many atmospheric gases, including carbon dioxide, ozone, methane, and water vapor (Figure 24.16). Careful analysis of atmospheric makeup might tell us whether a planet has life.

On Earth, for example, the large abundance of oxygen (21% of our atmosphere) is a direct result of photosynthetic life. Abundant oxygen in the atmosphere of a distant world might similarly indicate the presence of life, since we know of no nonbiological way to produce an oxygen abundance as high as Earth's. Other evidence might come from the ratio of oxygen to the other detected gases. Scientists engaged in the search for life in other planetary systems are working to improve our understanding of how life influences atmospheric chemistry in hopes that we will be able to recognize particular gas combinations as unmistakable signatures of life.

• Are Earth-like planets rare or common?

We will not know for certain whether habitable planets exist or how common such planets may be until we survey many star systems with telescopes capable of detecting such

Figure 24.16 The infrared spectra of Venus, Earth, and Mars, as they might be seen from afar, showing absorption features that point to the presence of carbon dioxide (CO_2), ozone (O_3), and sulfuric acid (H_2SO_4) in their atmospheres. While carbon dioxide is present in all three spectra, only our own planet has appreciable oxygen (and hence ozone)—a product of photosynthesis. If we could make similar spectral analyses of distant planets, we might detect atmospheric gases that would indicate life.

small planets. Nevertheless, the existence of more than one Earth-size planet in our own solar system and our understanding of planetary formation make it seem reasonable to imagine that many Earth-size planets exist within the habitable zones of other stars. But should we expect these planets to have Earth-like conditions in which life could arise and evolve?

Most scientists think so, but a few have raised some very interesting questions. In essence, these scientists suggest that Earth's hospitality is the result of several rare kinds of planetary luck. According to this idea, sometimes called the "rare Earth hypothesis," the specific circumstances that have allowed life on Earth to survive and evolve into complex forms (such as oak trees and people) might be so rare that ours could be the only planet in the galaxy that harbors anything but the simplest life. Let's briefly examine some of the key issues in the rare Earth hypothesis.

Galactic Constraints   Proponents of the rare Earth hypothesis suggest that Earth-like planets can form in only a relatively small region of the Milky Way Galaxy, making the number of potential homes for life far smaller than we might otherwise expect it to be. In essence, they argue that there is a fairly narrow ring at about our solar system's distance from the center of the Milky Way that makes up a *galactic habitable zone* analogous to the habitable zone around an individual star (Figure 24.17).

According to the arguments for a galactic habitable zone, outer regions of our galaxy are unlikely to have terrestrial planets because of a low abundance of elements besides hydrogen and helium. Recall that the fraction of heavy elements varies among different stars, from less than 0.1% among the old stars in globular clusters to more than 2% among young stars in the galactic disk [Section 19.3]. Even

within the galactic disk, the abundance tends to decline with distance from the center of the galaxy. Because terrestrial planets are made almost entirely of heavy elements, a lower abundance of these elements might lessen the chance that terrestrial planets could form. The inner regions of the galaxy are ruled out primarily through an argument concerning supernova rates. Supernovae are more common in the more crowded, inner regions of the galactic disk, making it more likely that a terrestrial planet would be exposed to the intense radiation from a nearby supernova. By assuming that this radiation would be detrimental to life, the proponents of a galactic habitable zone argue against finding habitable planets in the inner regions of the galaxy. Together, the constraints on finding Earth-like planets in the inner and outer regions of the galaxy leave the galactic habitable zone as a relatively narrow ring encompassing no more than about 10% of the stars in the galactic disk.

However, other scientists offer counterarguments to both sets of galactic constraints. For the heavy-element abundance, they note that Earth's mass is less than $\frac{1}{100,000}$ of the mass of the Sun. Thus, even a very small heavy-element abundance could be enough to make one or more Earth-like planets. Unless there is something about the planetary accretion process that prevents terrestrial planets from forming in systems with low heavy-element abundances, then we might find terrestrial planets around almost any star. Regarding the radiation danger from supernovae, we do not really know whether such radiation would be detrimental to life. A planet's atmosphere might protect life against the effects of the radiation. It is even possible that the radiation could be beneficial to life by increasing the rate of mutations and thereby accelerating the pace of evolution. If these counterarguments are correct, then Earth-like planets might be found throughout much or all of the galaxy.

Impact Rates and Jupiter   Another issue raised by rare Earth proponents is the impact rate on planets in other star systems. We have seen that Earth was probably subjected to numerous large impacts—some large enough to vaporize the oceans and sterilize the planet—during the heavy bombardment that went on during the first half-billion years or so after our planet's birth. In our solar system, the impact rate lessened dramatically after that. Might the impact rate remain high much longer in other planetary systems?

The most numerous small objects in our solar system are the trillion or so comets of the distant Oort cloud [Section 12.2]. Fortunately for us, nearly all these myriad objects are essentially out of reach, posing no threat to our planet. However, the reason they are out of reach can be traced directly to Jupiter: Recall that the Oort cloud's comets are thought to have formed among the jovian planets, later being "kicked out" to their current orbits by close encounters with these planets, especially Jupiter (see Figure 12.24). Thus, if Jupiter did not exist, many of the comets might have remained in regions of the solar system where they could pose a danger to Earth. In that case, the heavy bombardment might never have ended, and huge impacts would

Figure 24.17   The green ring in this diagram of the Milky Way Galaxy represents what some scientists believe is a galactic habitable zone—the only region of the galaxy in which Earth-like planets are likely to be found. However, other scientists doubt the claims that underlie the galactic habitable zone, in which case Earth-like planets could be far more widespread.

continue to this day. From this viewpoint, our existence on Earth has been possible only because of the "luck" of having Jupiter as a planetary neighbor.

The primary question in this case is just how "lucky" this situation might be. Our discoveries of extrasolar planets so far suggest that Jupiter-size planets are in fact quite common. However, we've also found that many large planets migrate inward [Section 13.3], perhaps disrupting terrestrial planet orbits along the way. Whether having a Jupiter in the right place for ejecting comets is lucky or typical remains an open question.

Climate Stability   Another issue affecting the rarity of Earth-like planets is climate stability. Earth's climate has been stable enough for liquid water to exist throughout the past 4 billion years. This climate stability has almost certainly played a major role in allowing complex life to evolve on our planet. If our planet had frozen over like Mars or overheated like Venus, we would not be here today. Advocates of the rare Earth hypothesis point to at least two pieces of "luck" related to Earth's stable climate.

The first piece of "luck" is the existence of plate tectonics. Recall that plate tectonics plays a major role in regulating Earth's climate through the carbon dioxide cycle [Section 10.6]. Plate tectonics probably was not necessary to the origin of life, but it seems to have been very important in keeping the climate stable enough for the subsequent evolution of plants and animals. But are we really "lucky" to have plate tectonics, or should this geological process be common on similar-size planets elsewhere? We do not yet know. The apparent lack of plate tectonics on Venus [Section 9.5], which is quite similar in size to Earth, might seem to argue for plate tectonics being rare. On the other hand, some scientists suspect that the lack of plate tectonics on Venus can be traced to its runaway greenhouse effect [Section 10.5], which may have baked out subsurface water and caused Venus's lithosphere to thicken too much to allow plate movements. Venus suffered its runaway greenhouse effect because it is not quite far enough from the Sun to be within the Sun's habitable zone. In that case, it's possible that any Earth-size planet within a star's habitable zone would have plate tectonics.

The second piece of "luck" in climate stability is the existence of Earth's relatively large Moon. Models of Earth's rotation and orbit show that if the Moon did not exist, gravitational tugs from other planets would cause large swings in Earth's axis tilt over periods of tens to hundreds of thousands of years. (Such swings in axis tilt are thought to occur on Mars.) Changes in axis tilt would affect the severity of the seasons, which in turn could cause deeper ice ages and more intense periods of warmth. Given that the Moon probably formed as a result of a random giant impact [Section 8.4], we might seem to be very lucky to have the Moon and the climate stability it brings.

Again, however, there are other ways to look at the issue. Changes in axis tilt might warm or cool different parts of the planet dramatically, but the changes would probably occur slowly enough for life to adapt or migrate as the cli-

mate changed. In addition, our Moon's presumed formation in a random giant impact does not necessarily mean that large moons will be rare. At least a few giant impacts should be expected in any planetary system. Indeed, Earth may not be the only object in our own solar system that ended up with a large moon because of a giant impact. Pluto's moon (Charon) may have formed in the same way [Section 12.3]. Thus, while luck was certainly involved in Earth's having a large moon, it might not be a very rare kind of luck.

The Bottom Line   The bottom line is that while the rare Earth hypothesis offers some intriguing arguments, it is too early to say whether any of them will hold up over time. For each potential argument that Earth has been lucky, we've seen counterarguments suggesting otherwise. There's no doubt that our solar system and our world have "personality"—they exhibit properties that might be found only occasionally in other star systems—but we have no clear reason to think that these special properties are essential to the existence of complex or even intelligent life. Indeed, it may be that we have missed out on some helpful phenomena that could have sped evolution on Earth. We might be less lucky than we recognize, and creatures on other worlds might regard the nature of our planet with disappointment. Until we learn much more about other planets in the universe, we cannot know whether Earth-like planets and complex life are common or rare.

24.4 The Search for Extraterrestrial Intelligence

So far, we have focused on search strategies for microbial or other nonintelligent life. However, if intelligent beings and civilizations exist elsewhere, we might be able to find them with a completely different type of search strategy. Instead of searching for hard-to-find spectroscopic signs of life, we might simply listen for signals that intelligent beings are sending into interstellar space, either in deliberate attempts to contact other civilizations or as a means of communicating among themselves. The search for signals from other civilizations is generally known as the **search for extraterrestrial intelligence**, or **SETI** for short.

• How many civilizations are out there?

SETI efforts have a chance to succeed only if other advanced civilizations are broadcasting signals that we could receive. Thus, in order to judge the chances of SETI success, we'd like to know how many civilizations are broadcasting such signals right now.

Given that we do not even know whether microbial life exists anywhere beyond Earth, it should be clear that we cannot yet know whether other civilizations exist, let alone how many there might be. Nevertheless, for the purposes of planning a search for extraterrestrial intelligence, it is useful to have an organized way of thinking about the number of civilizations that might be out there. To keep our discussion simple, let's consider only the number of potential civilizations in our own galaxy. We can always extend our estimate to the rest of the universe by simply multiplying the result we find for our galaxy by 100 billion, the approximate number of galaxies in the observable universe.

The Drake Equation   The first scientific conference on the search for extraterrestrial intelligence was held in 1961 in Green Bank, West Virginia, where a pioneering search for an alien signal had recently been conducted. During the meeting, astronomer Frank Drake wrote a simple equation designed to summarize the factors that would determine the number of civilizations we might contact (Figure 24.18).

This equation is now known as the **Drake equation**, and in principle it gives us a simple way to calculate the number of civilizations capable of interstellar communication that are currently sharing the Milky Way Galaxy with us. In a form slightly modified from the original, the Drake equation looks like this:

$$\text{Number of civilizations} = N_{HP} \times f_{life} \times f_{civ} \times f_{now}$$

This equation will make sense once you understand the meaning of each factor:

- N_{HP} is the number of habitable planets in the galaxy; that is, the number of planets that are at least capable of having life.
- f_{life} is the fraction of habitable planets that actually *have* life. For example, if $f_{life} = 1$ it would mean that all habitable planets have life, and if $f_{life} = \frac{1}{1,000,000}$ it would mean that only 1 in a million habitable planets have life. Thus, the product $N_{HP} \times f_{life}$ tells us the number of life-bearing planets in the galaxy.
- f_{civ} is the fraction of the life-bearing planets upon which a civilization capable of interstellar communication *has at some time* arisen. For example, if $f_{civ} = \frac{1}{1,000}$ it would mean that such a civilization has existed on 1 out of 1,000 planets with life, while the other 999 out of 1,000 have not had a species that learns to build radio transmitters, high-powered lasers, or other devices for interstellar conversation. When we multiply this factor by the first two factors to form the product $N_{HP} \times f_{life} \times f_{civ}$, we get the total number of planets upon which intelligent beings have evolved and developed a communicating civilization at some time in the galaxy's history.
- f_{now} is the fraction of these civilization-bearing planets that happen to have a civilization *now,* as opposed to, say, millions or billions of years in the past. This factor is important because we can hope to contact only civilizations that are broadcasting signals we could receive at present.*

---

*For the purposes of the Drake equation, we'll assume that the term f_{now} takes into account the light-travel time for signals from other stars; for example, if a star with a civilization is 10,000 light-years away, it counts in determining f_{now} if the civilization existed 10,000 years ago, because signals broadcast at that time would just now be arriving at Earth.

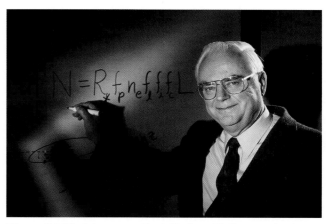

Figure 24.18 Astronomer Frank Drake, with the equation he first wrote in 1961. (With Dr. Drake's approval, we use a slightly modified form of his equation in this book.)

Because the product of the first three factors tells us the total number of civilizations that have *ever* arisen in the galaxy, multiplying by f_{now} tells us how many civilizations we could potentially make contact with today. Thus, the result of the Drake equation is the number of civilizations that we might hope to contact.

Unfortunately, we don't yet know the value of any of the factors in the Drake equation, so we cannot actually calculate the number of civilizations in our galaxy. Nevertheless, the equation is a useful way of organizing our thinking, as we can see by thinking about the potential values for each of its factors.

Try the following sample numbers in the Drake equation. Suppose there are 1,000 habitable planets in our galaxy, that 1 in 10 habitable planets has life, that 1 in 4 planets with life has at some point had an intelligent civilization, and that 1 in 5 civilizations that have ever existed is in existence now. How many civilizations would exist at present? Explain.

The Number of Life-Bearing Planets   Let's begin with the first two factors in the Drake equation, whose product ($N_{HP} \times f_{life}$) tells us the number of life-bearing planets in our galaxy. We can make a reasonably educated guess only about the first factor, the number of habitable planets (N_{HP}). Current understanding of solar system formation (see Figure 8.13) suggests that terrestrial planets ought to form fairly easily and, as we discussed earlier, there ought to be many billions of stars in our galaxy with habitable zones large enough to have Earth-like planets. Thus, unless some of the "rare Earth" ideas prove to be correct, it seems entirely reasonable to suppose that our galaxy has billions of habitable planets.

The factor f_{life} presents more difficulty. For the moment, we have no reliable way to estimate the fraction of habitable planets upon which life actually arose. The problem is that we cannot generalize when we have only one example to study—our own Earth. Still, the fact that life arose rapidly on Earth suggests that the origin of life was fairly "easy." In that case, we might expect most or all habitable planets to also have life, making the fraction f_{life} close to 1. Of course, until we have solid evidence that life arose anywhere else, such as on Mars, it is also possible that Earth might really have been very lucky and that f_{life} might be so close to zero that life has never arisen on any other planet in our galaxy.

The Question of Intelligence   Even if life-bearing planets are very common, civilizations capable of interstellar communication might not be. The fraction of life-bearing planets that at some time have such civilizations, f_{civ}, depends on at least two things: First, a planet would have to have a species evolve with sufficient intelligence to develop interstellar communication. In other words, the planet needs a species at least as smart as we are. Second, that species would have to develop a civilization with technology at least as advanced as ours.

Although we cannot really be sure, most scientists suspect that only the first requirement is difficult to meet. A fundamental assumption in nearly all science today is that we are not "special" in any particular way. We live on a fairly typical planet orbiting an ordinary star in a normal galaxy, and we assume that living creatures elsewhere—whether they prove to be rare or common—would be subjected to evolutionary pressures quite similar to those that have operated on Earth. Thus, if species with intelligence similar to ours have arisen elsewhere, we assume that they would have similar sociological drives that would eventually lead them to develop the technology necessary for interstellar communication.

If this assumption is correct, then the fraction f_{civ} depends primarily on the question of whether sufficient intelligence is rare or common among life-bearing planets. As with the question of life of any kind, the short answer to this question is we just don't know, but we can get at least some insight by considering what happened on Earth.

Look again at Figure 24.4. While life arose quite quickly on Earth, nearly all life remained microbial until just a few hundred million years ago, and it took nearly 4 billion years for us to arrive on the scene. This slow progress toward intelligence might suggest that producing a civilization is very difficult even when life is present. On the other hand, roughly half the stars in the Milky Way are older than our Sun, so if Earth's case is typical, then plenty of planets have existed long enough for intelligence to arise. Thus, the value of f_{civ} depends on whether the 4 billion years it took for humans to evolve on Earth is typical, fast, or slow. If it was unusually fast and the average time needed to develop complex species is much longer than 4 billion years, then most life-bearing planets may be covered with nothing more advanced than bacteria. If the time we took to evolve on Earth is typical or slow, then intelligent species might be quite common.

A second way to address the question of intelligence is by considering our level of intelligence in comparison to that of other animals on Earth. It's difficult to get animals to take IQ tests, but we can get a rough measure of intelligence by comparing brain mass to total body mass (a mea-

sure sometimes called the *encephalization quotient,* or EQ). Figure 24.19 shows the brain weights for a sampling of birds and mammals (including primates) plotted against their body weights. There is a clear and expected trend in that heavier animals have heavier brains. By drawing a straight line that fits these data, we can define an average value of brain mass for each body mass. Animals whose brain mass falls above the line are smarter than average, while animals whose brain mass falls below the line are less mentally agile. Keep in mind that it is the *vertical* distance above the line that tells us how much smarter a species is than the average, and that the scale goes in powers of 10 on both axes. If you look closely, you'll see that the data point for humans lies significantly farther above the line than the data point for any other species. Thus, by this measure of intelligence, we are far smarter than any other species that has ever existed on Earth.

Some people use this fact to argue that even on a planet with complex life, a species as intelligent as we are would be very rare. They say that even if there is an evolutionary drive toward intelligence in general, it takes extreme luck to reach our level of intelligence. After all, while it's evolutionarily useful to have enough intelligence to capture prey and evade other predators, it's not clear why natural selection would lead to brains big enough to build spacecraft. However, the same data can be used to reach an opposite conclusion. The scatter in the levels of intelligence among different animals tells us that some variation should be expected, and statistical analysis shows that we are not unreasonably far above the average. It might therefore be inevitable that some species would develop our level of intelligence on any planet with complex life.

Technological Lifetimes   For the sake of argument, let's assume that life and intelligence are at least reasonably likely, so

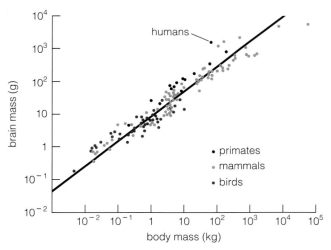

Figure 24.19 This graph shows how brain mass compares to body mass for some mammals (including primates) and birds. The straight line represents an average of the ratio of brain mass to body mass, so that animals that fall above the line are smarter than average and animals that fall below the line are less smart. Note that the scale uses powers of 10 on both axes. (Adapted from Sagan, 1977.)

that thousands or millions of planets in our galaxy have at some time given birth to a civilization. In that case, the final factor in the Drake equation, f_{now}, determines the likelihood of there being someone whom we could contact now. The value of this factor depends on the survivability of civilizations.

Consider our own example. In the roughly 12 billion years during which our galaxy has existed, we have been capable of interstellar communication via radio for only about 60 years. Thus, if we were to destroy ourselves tomorrow (saving you the unpleasantness of a final exam), then other civilizations could have received signals from us during only 60 years out of the galaxy's 12-billion-year existence, equivalent to 1 part in 200 million of the galaxy's history. If such a short technological lifetime is typical of civilizations, then f_{now} would be only $\frac{1}{200,000,000}$, and some 200 million civilization-bearing planets would need to have existed at one time or another in the Milky Way in order for us to have a good chance of finding another civilization out there now.

However, we'd expect f_{now} to be so small only if we are on the brink of self-destruction—after all, the fraction will grow larger for as long as our civilization survives. Thus, if civilizations are at all common, survivability is the key factor in whether any are out there now. If most civilizations self-destruct shortly after achieving the technology for interstellar communication, then we are almost certainly alone in the galaxy at present. But if most survive and thrive for thousands or millions of years, the Milky Way may be brimming with civilizations—most of them far more advanced than our own.

Describe a few reasons why a civilization capable of interstellar communication would also be capable of self-destruction. Overall, do you believe our civilization can survive for thousands or millions of years? Defend your opinion.

• How does SETI work?

If there are indeed other civilizations out there, then in principle we ought to be able to make contact with them. Based on our current understanding of physics, it seems likely that even very advanced civilizations would communicate much as we do—by encoding signals in radio waves or other forms of light. Most SETI researchers use large radio telescopes to search for alien radio signals (Figure 24.20). A few researchers are beginning to check other parts of the electromagnetic spectrum. For example, some scientists use visible-light telescopes to search for communications encoded as laser pulses. Of course, advanced civilizations may well have invented communication technologies that we cannot even imagine. In that case, SETI efforts will not detect them.

A good way to think about our chances of picking up an alien signal is to imagine what aliens would need to do to pick up signals from us. We have been sending relatively

Figure 24.20 The 64-meter Parkes radio telescope in New South Wales, Australia. A SETI experiment "piggybacks" on this telescope while it is engaged in other astronomical research.

high-power transmissions into space since about the 1950s in the form of television broadcasts. Thus, in principle, anyone within about 50 light-years of Earth could watch our old television shows (perhaps a frightening thought). However, in order to detect our broadcasts, they would need far larger and more sensitive radio telescopes than we have today. If their technology were at the same level as ours, they could receive a signal from us only if we deliberately broadcast an unusually high-powered transmission.

To date, humans have made only a few attempts to broadcast our existence in this way. The most powerful of these occasional transmissions was made in 1974 and lasted only 3 minutes. The powerful planetary radar transmitter on the Arecibo radio telescope (see Figure 6.24) was used to send a small pictorial message to the globular cluster M 13 (Figure 24.21). This target was chosen in part because it contains a few hundred thousand stars, seemingly offering a good chance that at least one has a civilization around it.

Figure 24.21 In 1974, a short message was broadcast to the globular cluster M 13 using the Arecibo radio telescope. The picture shown here was encoded by using two different radio frequencies, one for "on" and one for "off" (the colors shown here are arbitrary). To decode the message, the aliens would need to realize that the bits are meant to be arranged in a rectangular grid as shown, but that should not be difficult: The grid has 73 rows and 23 columns, and aliens would presumably know that these are both prime numbers. The picture represents the Arecibo radio dish, our solar system, a human stick figure, and a schematic of DNA and the eight simple molecules used in its construction.

However, M 13 is about 21,000 light-years from Earth, so it will take some 21,000 years for our signal to get there and another 21,000 years for any response to make its way back to Earth.

Several SETI projects under way or in development would be capable of detecting signals like the one we broadcast from Arecibo if they came from civilizations within a few hundred light-years. However, we could detect the signal only if we had the receiver tuned to the frequency of the broadcast—just as you can listen to your favorite radio station only by calling up the correct frequency on your radio dial. What radio frequency would aliens use? In the past, some astronomers made guesses about popular alien frequencies based on things such as frequencies emitted by common molecules in interstellar space. Today, SETI efforts generally seek to bypass this question by scanning millions of frequency bands simultaneously. Thus, if anyone nearby is deliberately broadcasting on an ongoing basis, we have a good chance of detecting the signals.

THINK ABOUT IT

SETI efforts are often controversial because of their cost and uncertain chances of success. However, SETI supporters say the cost is justified because contact with an extraterrestrial intelligence would be one of the most important discoveries in human history. Do you agree? Defend your opinion.

24.5 Interstellar Travel and Its Implications to Civilization

So far, we have discussed ways of detecting distant civilizations without ever leaving the comfort of our own planet. Could we ever actually visit other worlds in other star systems? A careful analysis of this question turns out to have profound implications for the future of our civilization. To see why, we first need to consider the prospects for achieving interstellar travel.

• How difficult is interstellar travel?

In many science fiction movies, our descendants travel among the stars as routinely as we jet about Earth in airplanes. They race around the galaxy in starships of all sizes and shapes, circumventing nature's prohibition on faster-than-light travel by entering hyperspace, wormholes, or warp drive. They witness firsthand incredible cosmic phenomena such as stars and planets in all stages of development, accretion disks around white dwarfs and neutron stars, and the distortion of spacetime near black holes. Along the way they encounter numerous alien species, most of which happen to look and act a lot like us.

Unfortunately, Einstein's theory of relativity tells us that real interstellar travel must be at speeds slower than the speed of light [Section S2.2], and even approaching that speed will require overcoming huge technological hurdles. Never-

theless, we have already sent out our first emissaries to the stars, and there's no reason to believe that we won't develop better technologies in the future.

The Challenge of Interstellar Travel   To date, we have launched four spacecraft that will leave our solar system and eventually travel among the stars: the planetary probes *Pioneer 10, Pioneer 11, Voyager 1,* and *Voyager 2.* These spacecraft are traveling about as fast as anything ever built by humans, but their speeds are still less than $\frac{1}{10,000}$ the speed of light. It would take each of them some 100,000 years just to reach the next nearest star system (Alpha Centauri), but their trajectories won't take them anywhere near it. Instead, they will simply continue their journey without passing close to any nearby stars, wandering the Milky Way for millions or even billions of years to come. Each carries a greeting from Earth, just in case someone comes across one of them someday (Figure 24.22).

If we want to make interstellar journeys within human lifetimes, we will need starships that can travel at speeds close to the speed of light. We will need entirely new types of engines to reach such high speeds. The energy requirements of interstellar spacecraft may pose an even more daunting challenge. For example, the energy needed to accelerate a single ship the size of *Star Trek*'s *Enterprise* to just half the speed of light would be more than 2,000 times the total annual energy use of the world today. Clearly, interstellar travel will require vast new sources of energy. In addition, fast-moving starships will require new types of shielding to protect crew members from instant death. As a starship travels through interstellar gas at near-light speed, ordinary atoms and ions will hit it like a deadly flood of high-energy cosmic rays.

If we succeed in building starships capable of traveling at speeds close to the speed of light, the crews will face significant social challenges. According to well-tested principles of Einstein's theory of relativity, time will run much slower on a spaceship that travels at high speed to the stars than it does here on Earth [Section S2.4]. For example, in a ship traveling at an average speed of 99.9% of the speed of light, the 50-light-year round trip to the star Vega would take the travelers aboard only about 2 years—but more than 50 years would pass on Earth while they were gone. Thus, the crew will need only 2 years' worth of provisions and will age only 2 years during the voyage, but they will return to a world quite different from the one they left. Family and friends will be older or deceased, new technologies might have made their knowledge and skills obsolete, and many political and social changes may have occurred in their absence. Clearly, the crew will face a difficult adjustment when they come home to Earth.

Starship Design   Despite all the difficulties, some scientists and engineers have already proposed designs that could in principle take us to nearby stars. In the 1960s a group of scientists proposed an approach, called *Project Orion,* that envisioned accelerating a spaceship with repeated detona-

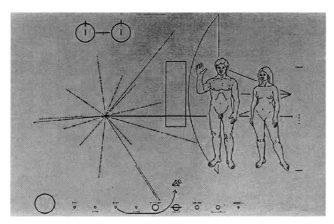

a The *Pioneer* plaque, about the size of an automobile license plate. The human figures are shown in front of a drawing of the spacecraft to give them a sense of scale. The "prickly" graph to their left shows the Sun's position relative to nearby pulsars, and Earth's location around the Sun is shown below. Binary code indicates the pulsar periods; because pulsars slow with time, the periods will allow someone reading the plaque to determine when the spacecraft was launched.

b *Voyagers* 1 and 2 carry a phonograph record—a 12-inch gold-plated copper disk containing music, greetings, and images from Earth.

Figure 24.22 Messages aboard the *Pioneer* and *Voyager* spacecraft, which are bound for the stars.

tions of relatively small hydrogen bombs. Each explosion would take place a few tens of meters behind the spaceship and would propel the ship forward as the vaporized debris hit a "pusher plate" on the back of the spacecraft (Figure 24.23). Calculations showed that a spaceship accelerated by the rapid-fire detonation of a million H-bombs could reach Alpha Centauri in just over a century. In principle, we could build an Orion spacecraft with existing technology, though it would be very expensive and would require an exception to the international treaty banning nuclear detonations in space.

A century to the nearest star isn't bad, but it still wouldn't make interstellar travel easy. Unfortunately, no available technology could go much faster. The problem is weight:

Figure 24.23 Artist's conception of the Project Orion starship, showing one of the small hydrogen-bomb detonations that would propel it. Debris from the detonation strikes the flat disk, called the pusher plate, at the back of the spaceship. The central sections (enclosed in a lattice) hold the bombs, and the front sections house the crew.

Making a rocket faster requires more fuel, but adding fuel adds weight and makes it more difficult for the rocket to accelerate. Calculations show that even in the best case, rockets carrying nuclear fuel could achieve speeds no more than a few percent of the speed of light. Nevertheless, we can envision some possible future technologies that might get around this problem.

One idea suggests powering starships with engines that generate energy through matter-antimatter annihilation. Whereas nuclear fusion converts less than 1% of the mass of atomic nuclei into energy, matter-antimatter annihilation [Section 23.1] converts *all* the annihilated mass into energy. Starships with matter-antimatter engines could probably reach speeds of 90% or more of the speed of light. At these speeds, the slowing of time predicted by relativity becomes noticeable, putting many nearby stars within a few years' journey for the crew members. However, because no natural reservoirs of antimatter exist, we would have to be able to manufacture many tons of antimatter and then store it safely for the trip—capabilities that are far beyond our present means.

An even more speculative and futuristic design, known as an *interstellar ramjet*, would collect interstellar hydrogen with a gigantic scoop, using the collected gas as fuel for its nuclear engines (Figure 24.24). By collecting fuel along the way, the ship would not need to carry the weight of fuel onboard. However, because the density of interstellar gas is so low, the scoop would need to be enormous. As astronomer Carl Sagan said, we are talking about "spaceships the size of worlds."

The bottom line is that while we face enormous obstacles to achieving interstellar travel, there's no reason to think it's impossible. If we can avoid self-destruction and if we continue to explore space, it seems quite likely that our descendants will make journeys to the stars.

Figure 24.24 Artist's conception of a spaceship powered by an interstellar ramjet. The giant scoop in the front (left) collects interstellar hydrogen for use as fusion fuel.

• Where are the aliens?

Imagine that we survive and become interstellar travelers and that we begin colonizing habitable planets around nearby stars. As the colonies grow at each new location, some of the people may decide to set out for other star systems. Even if our starships traveled at relatively low speeds—say, a few percent of the speed of light—we could have dozens of outposts around nearby stars within a few centuries. In 50,000 years, our descendants would be spread among stars within a few hundred light-years of Earth. In a few million years, we could have outposts throughout the Milky Way Galaxy. We will have become a true galactic civilization.

SPECIAL TOPIC Are Aliens Already Here?

In this chapter, we have discussed contact with intelligent aliens as a possibility, not a reality. However, public opinion polls suggest that up to half the American public believes that aliens are already visiting us. What can science say about this remarkable notion?

The bulk of the claimed evidence for alien visitation consists of sightings of UFOs—unidentified flying objects. Many thousands of UFOs are reported each year, and no one doubts that unidentified objects are being seen. The question is whether they are alien spacecraft.

Aliens have long been a staple of science fiction, but modern interest in UFOs began with a widely reported sighting in 1947. While flying a private plane near Mount Rainier in Washington State, businessman Kenneth Arnold saw nine mysterious objects streaking across the sky. He told a reporter that the objects "flew erratic, like a saucer if you skip it across the water." (In fact, he may have seen meteors skipping across the atmosphere, though no one knows for sure.) He did *not* say that the objects were saucer-shaped, but the reporter nevertheless wrote up Arnold's experience as a sighting of "flying saucers." The story was front-page news throughout America, and within a decade "flying saucers" had invaded popular culture, if not our planet.

The flying saucer reports also interested the U.S. Air Force, largely out of concern that the UFOs might represent new types of aircraft developed by the then–Soviet Union. For two decades, the Air Force hired teams of academics to study UFO reports. In the overwhelming majority of cases, these experts were able to specify a plausible identification of the UFO. The explanations included bright stars and planets, aircraft and gliders, rocket launches, balloons, birds, ball lightning, meteors, atmospheric phenomena, and the occasional hoax. For a minority of the sightings, the investigators could not deduce what was seen, but their overall conclusion was that there was no reason to believe the UFOs were either highly advanced Soviet craft or visitors from other worlds. The Air Force ultimately dropped its investigations of the UFO phenomenon.

Believers discounted the Air Force denials, claiming to have other evidence of alien visitation. So far, none of this evidence has ever withstood scientific scrutiny. Photographs and film clips are nearly always too fuzzy to clearly show alien spacecraft, except in cases that are obviously faked. UFO witnesses are frequently credible (they include seasoned pilots), but generally there are several possible explanations for what they've seen besides alien spacecraft.

Crop circles are easily made by pranksters. Stories of alien abductions are dramatic but cannot be verified. Pieces of metal that "UFO experts" say could not have been made by humans have turned out to be pieces of cars or refrigerators. Champions of alien visitation generally explain away the lack of clear evidence in one of two ways: government cover-ups or a failure of the mainstream scientific community to take the relevant phenomena seriously. Neither explanation is compelling.

It's certainly conceivable that a secretive government might *try* to put the lid on evidence of alien visits, though the motivation for doing so is unclear. The usual explanations are that the public couldn't handle the news or that the government is taking secret advantage of the alien materials to design new military hardware (via "reverse engineering"). Both explanations are silly. Half the population already believes in alien visitors and would hardly be shocked if newspapers announced that aliens were stacked up in government warehouses. As for reverse engineering extraterrestrial spacecraft, we should keep in mind how difficult it is to travel from star to star. Any society that could do so routinely would be technologically far beyond our own. Reverse engineering their spaceships is as unlikely as expecting Neanderthals to construct personal computers just because a laptop somehow landed in their cave. In addition, while a government might successfully hide evidence for a short time, does it really seem possible that evidence could remain secret for decades? After all, talk show fame and riches would await anyone who uncovered the conspiracy. Moreover, unless the aliens landed only in the United States, can we seriously believe that *every* government has cooperated in hiding the evidence?

Alleged disinterest on the part of the scientific community is an equally unimpressive claim. Scientists are constantly competing with one another to be the first with a great discovery, and clear evidence of alien visitors would certainly rank high on the all-time list. Countless researchers would work evenings and weekends, without pay, if they thought they could make such a discovery. The fact that few scientists are engaged in such study reflects not a lack of interest, but a lack of evidence worthy of study.

Of course, absence of evidence is not evidence of absence. Most scientists are open to the possibility that we might someday find evidence of alien visits, and many would welcome aliens with open arms. So far, however, we have no hard evidence to support the belief that aliens are already here.

Now, if we take the idea that *we* could develop a galactic civilization within a few million years and combine it with the reasonable (though unproved) idea that civilizations ought to be common, we are led to an astonishing conclusion: Someone else should already have created a galactic civilization. In fact, it should have been done a long time ago.

To see why, let's take some sample numbers. Suppose the overall odds of a civilization arising around a star are about the same as your odds of winning the lottery, or 1 in a million. Taking a low estimate of 100 billion stars in the Milky Way Galaxy, this means some 100,000 civilizations in our galaxy alone. Further, suppose we are a fairly typical civilization, so that civilizations generally arise when their stars are approaching 5 billion years old. Given that the galaxy is some 12 billion years old, the first of these 100,000 civilizations would have arisen at least 7 billion years ago. Others would have arisen, on average, about every 70,000 years. Under these assumptions, we would expect the youngest civilization besides ourselves to be some 70,000 years ahead of us technologically, and most would be millions or billions of years ahead of us.

Thus, we encounter a strange paradox: Plausible arguments suggest that a galactic civilization should already exist, yet we have so far found no evidence of such a civilization. This paradox is often called *Fermi's paradox*, after the Nobel Prize–winning physicist Enrico Fermi. During a 1950 conversation with other scientists about the possibility of extraterrestrial intelligence, Fermi responded to speculations by asking, "So where is everybody?"

This paradox has many possible solutions, but broadly speaking we can group them into three categories:

1. *We are alone.* There is no galactic civilization because civilizations are extremely rare—so rare that we are the first to have arisen on the galactic scene, perhaps even the first in the universe.
2. *Civilizations are common, but no one has colonized the galaxy.* There are at least three possible reasons why this might be the case. Perhaps interstellar travel is much harder or vastly more expensive than we have guessed, and civilizations are unable to venture far from their home worlds. Perhaps the desire to explore is unusual, and other societies either never leave their home star systems or stop exploring before they've colonized much of the galaxy. Most ominously, perhaps many civilizations have arisen, but they have all destroyed themselves before achieving the ability to colonize the stars.
3. *There IS a galactic civilization,* but it has not yet revealed its existence to us.

We do not know which, if any, of these explanations is the correct solution to the question "Where are the aliens?" However, each category of solution has astonishing implications for our own species.

Consider the first solution—that we are alone. If this is true, then our civilization is a remarkable achievement. It implies that through all of cosmic evolution, among countless star systems, we are the first piece of our galaxy or the universe ever to know that the rest of the universe exists. Through us, the universe has attained self-awareness. Some philosophers and many religions argue that the ultimate purpose of life is to become truly self-aware. If so, and if we are alone, then the destruction of our civilization and the loss of our scientific knowledge would represent an inglorious end to something that took the universe some 14 billion years to achieve. From this point of view, humanity becomes all the more precious, and the collapse of our civilization would be all the more tragic.

The second category of solutions has much more terrifying implications. If thousands of civilizations before us have all failed to achieve interstellar travel on a large scale, what hope do we have? Unless we somehow think differently than all other civilizations, this solution says that we will never go far in space. Because we have always explored when the opportunity arose, this solution almost inevitably leads to the conclusion that failure will come about because we destroy ourselves. We can only hope that this answer is wrong.

The third solution is perhaps the most intriguing. It says that we are newcomers on the scene of a galactic civilization that has existed for millions or billions of years before us. Perhaps this civilization is deliberately leaving us alone for the time being and will someday decide the time is right to invite us to join it.

No matter what the answer turns out to be, learning it is sure to mark a turning point in the brief history of our species. Moreover, this turning point is likely to be reached within the next few decades or centuries. We already have the ability to destroy our civilization. If we do so, then our fate is sealed. But if we survive long enough to develop technology that can take us to the stars, the possibilities seem almost limitless.

THE BIG PICTURE

Putting Chapter 24 into Context

Throughout our study of astronomy, we have taken the "big picture" view of trying to understand how we fit into the universe. Here, at last, we have returned to Earth and examined the role of our own generation in the big picture of human history. Tens of thousands of past human generations have walked this Earth. Ours is the first generation with the technology to study the far reaches of our universe, to search for life elsewhere, and to travel beyond our home planet. It is up to us to decide whether we will use this technology to advance our species or to destroy it.

Imagine for a moment the grand view, a gaze across the centuries and millennia from this moment forward. Picture our descendants living among the stars, having created or joined a great galactic civilization. They will have the privilege of experiencing ideas, worlds, and discoveries far beyond our wildest imagination. Perhaps, in their history lessons, they will learn of our generation—the generation that history placed at the turning point and that managed to steer its way past the dangers of self-destruction and onto the path to the stars.

24.1 Life on Earth

- **When did life arise on Earth?** Fossil evidence puts the origin of life at least 3.5 billion years ago, and carbon isotope evidence pushes this date to more than 3.85 billion years ago. Thus, life arose within a few hundred million years after the last major impact of the heavy bombardment, and possibly in a much shorter time.

- **How did life arise on Earth?** Genetic evidence suggests that all life on Earth evolved from a common ancestor, and this ancestor was probably similar to microbes that live today in hot water near undersea volcanic vents or hot springs. We do not know how this first organism arose, but laboratory experiments suggest that it may have been the result of natural chemical processes on the early Earth. Once life arose, it rapidly diversified and evolved through **natural selection**.

- **What are the necessities of life?** Life on Earth thrives in a wide range of environments, and in general seems to require only three things: a source of nutrients, a source of energy, and liquid water.

24.2 Life in The Solar System

- **Could there be life on Mars?** Mars once had conditions that may have been conducive to an origin of life. If life arose, it might still survive in pockets of liquid water underground.

- **Could there be life on Europa or other jovian moons?** Europa probably has a subsurface ocean of liquid water, and may have undersea volcanoes on its ocean floor. If so, it has conditions much like those in which life on Earth probably arose, making it a good candidate for life beyond Earth. Ganymede and Callisto might have oceans as well. Titan may have other liquids on its surface, though it is too cold for liquid water. Perhaps life can survive in these other liquids, or perhaps Titan has liquid water deep underground.

24.3 Life Around Other Stars

- **Are habitable planets likely?** Billions of stars have at least moderate-size **habitable zones** in which life-bearing planets might exist. We do not yet have the technology to search for habitable planets directly, but several planned missions should be able to begin the search soon.

- **Are Earth-like planets rare or common?** We don't know. Arguments can be made on both sides of the question, and at present we lack the data to determine their validity.

24.4 The Search for Extraterrestrial Intelligence

- **How many civilizations are out there?** We don't know, but the **Drake equation** gives us a way to organize our thinking about the question. The equation (in a modified form) says that the number of civilizations in the Milky Way Galaxy with whom we could potentially communicate is $N_{HP} \times f_{life} \times f_{civ} \times f_{now}$, where N_{HP} is the number of habitable planets in the galaxy, f_{life} is the fraction of habitable planets that actually have life on them, f_{civ} is the fraction of life-bearing planets upon which a civilization capable of interstellar communication has at some time arisen, and f_{now} is the fraction of all these civilizations that exist now.

- **How does SETI work?** SETI, the search for extraterrestrial intelligence, generally refers to efforts to detect signals—such as radio or laser communications— coming from civilizations on other worlds.

24.5 Interstellar Travel and Its Implications to Civilization

- **How difficult is interstellar travel?** Convenient interstellar travel remains well beyond our technological capabilities, because of the technological requirements for engines, the enormous energy needed to accelerate spacecraft to speeds near the speed of light, and the difficulties of shielding the crew from radiation. Nevertheless, people have proposed ways around all these difficulties, and it seems reasonable to think that we will someday achieve interstellar travel if we survive long enough.

- **Where are the aliens?** It seems that we should be capable of colonizing the galaxy in a few million years or less, and the galaxy was around for at least 7 billion years before Earth was even born. Thus, it seems that someone should have colonized the galaxy long ago—yet we have no evidence of other civilizations. Every possible category of explanation for this surprising fact has astonishing implications for our species and our place in the universe.

Review Questions

Short-Answer Questions Based on the Reading

1. Describe three recent developments in the study of life on Earth that make it seem much more likely that we could find life elsewhere.
2. How do we study the history of life on Earth? Describe the *geological time scale* and a few of the major events along it.
3. Summarize the evidence pointing to an early origin of life on Earth. How far back in Earth's history did life exist?
4. What is the theory of evolution, and why is it so critical to our understanding of the history of life on Earth? Briefly describe the logic that led Darwin to propose the theory of evolution.
5. What is *natural selection,* and how does it explain the evolution of life on Earth? What is *DNA*, and how do *mutations* in DNA explain the molecular basis of natural selection?
6. Give a brief overview of the history of life on Earth. What evidence points to a common ancestor for all life? How and when did oxygen accumulate in Earth's atmosphere? When did larger animals diversify on Earth?
7. Where do we think life on Earth first arose, and why?
8. Describe how laboratory experiments are helping us study the origin of life on Earth.
9. Is it possible that life migrated to Earth from elsewhere? Explain.
10. Describe the range of environments in which life thrives on Earth. What three basic requirements apply to life in all these environments?
11. What is a *habitable world?* Which worlds in our solar system seem potentially habitable, and why?
12. Briefly summarize the debate over possible fossil evidence of life in a meteorite from Mars.
13. What do we mean by a star's *habitable zone?* Do we expect many stars to be capable of having habitable planets? Explain.
14. What is the *rare Earth hypothesis?* Summarize the arguments on both sides regarding the validity of this hypothesis.
15. What is the *Drake equation?* Define each of its factors, and describe the current state of understanding about the potential values of each factor.
16. What is SETI? Describe the capabilities of current SETI efforts.
17. Summarize the factors that make interstellar travel difficult, and describe a few technologies that might someday make it possible nonetheless.
18. What is Fermi's paradox? Describe several potential solutions to the paradox, and the implications of each to the future of our civilization.

Test Your Understanding

Fantasy or Science Fiction?

Each of the following describes some futuristic scenario that, while perhaps common and entertaining, may or may not be plausible. In each case, decide whether the scenario is plausible according to our present understanding of science or whether it is unlikely to be possible. Explain your reasoning.

19. The first human explorers on Mars discover that the surface is littered with the ruins of an ancient civilization including remnants of tall buildings and temples.

20. The first human explorers on Mars drill a hole into a Martian volcano to collect a sample of soil from several meters underground. Upon analysis of the soil, they discover that it holds living microbes resembling terrestrial bacteria but with a different biochemistry.
21. In 2020, a spacecraft lands on Europa and melts its way through the ice into the Europan ocean. It finds numerous strange, living microbes, along with a few larger organisms that feed on the microbes.
22. It's the year 2075. A giant telescope on the Moon, consisting of hundreds of small telescopes linked together across a distance of 500 kilometers, has just captured a series of images of a planet around a distant star that clearly show seasonal changes in vegetation.
23. A century from now, after completing a careful study of planets around stars within 100 light-years of Earth, we've discovered that the most diverse life exists on a planet orbiting a young star that formed just 100 million years ago.
24. In 2030, a brilliant teenager working in her garage builds a coal-powered rocket that can travel at half the speed of light.
25. In the year 2750, we receive a signal from a civilization around a nearby star telling us that the *Voyager 2* spacecraft recently crash-landed on its planet.
26. Crew members of the matter-antimatter spacecraft *Star Apollo,* which left Earth in the year 2165, return to Earth in the year 2450, looking only a few years older than when they left.
27. Aliens from a distant star system invade Earth with the intent to destroy us and occupy our planet, but we successfully fight them off when their technology proves no match for ours.
28. A single great galactic civilization exists. It originated on a single planet long ago but is now made up of beings from many different planets, each of which was assimilated into the galactic culture in turn.

Quick Quiz

Choose the best answer to each of the following. Explain your reasoning with one or more complete sentences.

29. Fossil evidence suggests that life on Earth arose (a) almost immediately after Earth formed. (b) very soon after the end of the heavy bombardment. (c) about a billion years before the rise of the dinosaurs.
30. The theory of evolution is (a) a scientific theory, meaning that it is backed by extensive evidence. (b) one of several competing scientific models that all seem equally successful in explaining the nature of life on Earth. (c) essentially just a guess about how life changes through time.
31. Plants and animals are (a) the two major forms of life on Earth. (b) the only organisms that have DNA. (c) just two small branches of the diverse "tree of life" on Earth.
32. Which of the following is a reason why early living organisms on Earth could not have survived on the surface? (a) the lack of an ozone layer (b) the lack of oxygen in the atmosphere (c) the fact that these organisms were single-celled
33. According to current understanding, the key requirement for life is (a) photosynthesis. (b) liquid water. (c) an ozone layer.

34. Which of the following worlds is not considered a candidate for harboring life? (a) Europa (b) Mars (c) the Moon

35. How does the habitable zone around a star of spectral type G compare to that around a star of spectral type M? (a) It is larger. (b) It is hotter. (c) It is closer to its star.

36. In the Drake equation, suppose that the term $f_{life} = \frac{1}{2}$. What would this mean? (a) Half the stars in the Milky Way Galaxy have a planet with life. (b) Half of all life forms in the universe are intelligent. (c) Half of the habitable worlds in the galaxy actually have life, while the other half don't.

37. The amount of energy that would be needed to accelerate a large spaceship to a speed close to the speed of light is (a) about 100 times as much energy as is needed to launch the Space Shuttle. (b) more than the total amount of energy used by the entire world in a year. (c) more than the amount of energy that our Sun emits into space in a year.

38. According to current scientific understanding, the idea that the Milky Way Galaxy might be home to a civilization millions of years more advanced than ours is (a) a virtual certainty. (b) extremely unlikely. (c) one reasonable answer to Fermi's paradox.

Investigate Further
In-Depth Questions to Increase Your Understanding

Short-Answer/Essay Questions

39. *Artificial Selection.* Suppose you lived hundreds of years ago (before we knew about genetic engineering) and wanted to breed a herd of cows that provided more milk than the cows in your current herd. How would you have gone about it? How would this process of "artificial selection" be similar to natural selection? How would it be different?

40. *Statistics of One.* Much of the search for life in the universe is based on what we know from the example of life on Earth. Write a short essay discussing the pros and cons of basing our general understanding of biology on the example of Earth.

41. *Most Likely to Have Life.* Suppose you were asked to vote in a contest to name the world in our solar system (besides Earth) "most likely to have life." Which world would you cast your vote for? Explain and defend your choice in a one-page essay.

42. *Likely Suns.* Study the stellar data for nearby stars given in Appendix F, Table F.1. Which star on the list would you expect to have the largest habitable zone? Which would have the second-largest habitable zone? If we rule out multiple-star systems, which star would you expect to have the highest probability of having a habitable planet? Explain your answers.

43. *Are Earth-Like Planets Common?* Based on what you have learned in this book, form an opinion as to whether you think Earth-like planets will ultimately prove to be rare, common, or something in between. Write a one- to two-page essay explaining and defending your opinion.

44. *Solution to the Fermi Paradox.* Among the various possible solutions to the question "Where are the aliens?," which do you think is most likely? (If you have no opinion on their likelihood, which do you like best?) Write a one- to two-page essay in which you explain why you favor this solution.

45. *What's Wrong with This Picture?* Many science fiction stories have imagined the galaxy divided into a series of empires, each having arisen from a different civilization on a different world, that hold each other at bay because they are all at about the same level of military technology. Is this a realistic scenario? Explain.

46. *Aliens in the Movies.* Choose a science fiction movie (or television show) that involves an alien species. Do you think aliens like this could really exist? Write a one- to two-page critical review of the movie, focusing primarily on the question of whether the movie portrays the aliens in a scientifically reasonable way.

Quantitative Problems
Be sure to show all calculations clearly and state your final answers in complete sentences.

47. *Nearest Civilization I.* Suppose there are 10,000 civilizations existing in the Milky Way Galaxy right now. If the civilizations are randomly distributed throughout the disk of the galaxy, about how far (on average) would it be to the nearest civilization? (*Hint:* Start by finding the area of the Milky Way's disk, assuming that it is circular and 100,000 light-years in diameter. Then find the average area per civilization, and use the distance across this area to estimate the distance between civilizations.)

48. *Nearest Civilization II.* Repeat Problem 47, but this time assume that there are only 100 civilizations in the Milky Way Galaxy right now.

49. *SETI Search.* Suppose there are 10,000 civilizations broadcasting radio signals in the Milky Way Galaxy right now. On average, how many stars would we have to search before we would expect to hear a signal? Assume there are 500 billion stars in the galaxy. How does your answer change if there are only 100 civilizations instead of 10,000?

50. *SETI Signal.* Consider a civilization broadcasting a signal with a power of 10,000 watts. The Arecibo radio telescope, which is about 300 meters in diameter, could detect this signal if it is coming from as far away as 100 light-years. Suppose instead that the signal is being broadcast from the other side of the Milky Way Galaxy, about 70,000 light-years away. How large a radio telescope would we need to detect this signal? (*Hint:* You can do this problem with the inverse square law for light.)

51. *Cruise Ship Energy.* Suppose we have a spaceship about the size of a typical ocean cruise ship today, which means it has a mass of about 100 million kilograms, and we want to accelerate the ship to a speed of 10% of the speed of light.
   a. How much energy would be required? (*Hint:* You can find the answer simply by calculating the kinetic energy of the ship when it reaches its cruising speed; because 10% of the speed of light is still small compared to the speed of light, you can use the formula that tells us that kinetic energy $= \frac{1}{2} \times m \times v^2$.)
   b. How does your answer compare to total world energy use at present, which is about 5×10^{22} joules per year?
   c. The typical cost of energy today is roughly 5¢ per 1 million joules. Using this price, how much would it cost to generate the energy needed by this spaceship?

52. *Matter-Antimatter Engine.* Consider the spaceship from Problem 51. Suppose you want to generate the energy to get it to cruising speed using matter-antimatter annihilation. How much antimatter would you need to produce and take on the ship? (*Hint:* Remember that when matter and antimatter meet, they turn all their mass into energy equivalent to mc^2.)

Discussion Questions

53. *Funding the Search for Life.* Imagine that you are a member of Congress, so that your job includes deciding how

much government funding goes to research in different areas of science. How much would you allot to the search for life in the universe compared to the amount allotted to research in other areas of astronomy and planetary science? Why?

54. *Distant Dream or Near Reality -Reality?* Considering all the issues surrounding interstellar flight, when (if ever)

do you think we are likely to begin traveling among the stars? Why?

55. *The Turning Point.* Discuss the idea that our generation has acquired a greater responsibility for the future than any previous generation. Do you agree with this assessment? If so, how should we deal with this responsibility? Defend your opinions.

MEDIA EXPLORATIONS

For self-study activities go to **Mastering**ASTRONOMY   www.masteringastronomy.com

Interactive Tutorials

Tutorial Review of Key Concepts
Use the following interactive **Tutorial** at **www.mastering astronomy.com** to review key concepts from this chapter.

Detecting Extrasolar Planets Tutorial
Lesson 1 Taking a Picture of a Planet
Lesson 2 Stars' Wobbles and Properties of Planets
Lesson 3 Planetary Transits

Supplementary Tutorial Exercises
Use the interactive **Tutorial Lesson** to explore the following questions.

Detecting Extrasolar Planets Tutorial, Lessons 1–3
1. Give two reasons why visual detection of planets orbiting other stars is extremely difficult.
2. As you move away from two objects, what happens to the apparent angle between them?
3. How can we detect extrasolar planets?
4. How might transits allow us to detect Earth-size planets around other stars?

Movies

Check out the following narrated and animated short documentary available on **www.masteringastronomy.com** for a helpful review of key ideas covered in this chapter.

The Search for Extraterrestrial Life Movie

Web Projects

1. *Astrobiology News.* Go to NASA's astrobiology home page and read some of the recent news from the search for life in the universe. Choose one recent news article and write a one-to two-page summary of the research and how it relates to the question of life in the universe in general.

2. *Martian Meteorites.* Find information about the latest discoveries concerning Martian meteorites and whether the meteorites contain evidence of life. Choose one recent discovery that seems important, and write a short summary of how you think it alters the debate about the habitability of Mars or about life on Mars.

3. *The Search for Extraterrestrial Intelligence.* Go to the home page for the SETI Institute. Learn more about how SETI is funded and carried out. In one page or less, describe the SETI Institute and its work.

4. *Starship Design.* Find more details about a proposal for starship propulsion or design. How would the starship work? What new technologies would be needed, and what existing technologies could be applied? Summarize your findings in a one- to two-page report.

5. *Advanced Spacecraft Technologies.* NASA supports many efforts to incorporate new technologies into spaceships. Although few of them reach the level of being suitable for interstellar colonization, most are innovative and fascinating. Learn about one such NASA project, and write a short summary of your findings.

Appendixes

A Useful Numbers

Astronomical Distances

1 AU $\approx 1.496 \times 10^8$ km

1 light-year $\approx 9.46 \times 10^{12}$ km

1 parsec (pc) $\approx 3.09 \times 10^{13}$ km ≈ 3.26 light-years

1 kiloparsec (kpc) $= 1{,}000$ pc $\approx 3.26 \times 10^3$ light-years

1 megaparsec (Mpc) $= 10^6$ pc $\approx 3.26 \times 10^6$ light-years

Astronomical Times

1 solar day (average) $= 24^h$

1 sidereal day $\approx 23^h\,56^m\,4.09^s$

1 synodic month (average) ≈ 29.53 solar days

1 sidereal month (average) ≈ 27.32 solar days

1 tropical year ≈ 365.242 solar days

1 sidereal year ≈ 365.256 solar days

Universal Constants

Speed of light: $c = 3 \times 10^5$ km/s $= 3 \times 10^8$ m/s

Gravitational constant: $G = 6.67 \times 10^{-11} \dfrac{\text{m}^3}{\text{kg} \times \text{s}^2}$

Planck's constant: $h = 6.626 \times 10^{-34}$ joule $\times$ s

Stefan–Boltzmann constant: $\sigma = 5.7 \times 10^{-8} \dfrac{\text{watt}}{\text{m}^2 \times \text{Kelvin}^4}$

mass of a proton: $m_{\text{p}} = 1.67 \times 10^{-27}$ kg

mass of an electron: $m_{\text{e}} = 9.1 \times 10^{-31}$ kg

Useful Sun and Earth Reference Values

Mass of the Sun: $1\ M_{\text{Sun}} \approx 2 \times 10^{30}$ kg

Radius of the Sun: $1\ R_{\text{Sun}} \approx 696{,}000$ km

Luminosity of the Sun: $1\ L_{\text{Sun}} \approx 3.8 \times 10^{26}$ watts

Mass of the Earth: $1\ M_{\text{Earth}} \approx 5.97 \times 10^{24}$ kg

Radius (equatorial) of the Earth: $1\ R_{\text{Earth}} \approx 6{,}378$ km

Acceleration of gravity on Earth: $g = 9.8$ m/s^2

Escape velocity from surface of Earth: $v_{\text{escape}} = 11$ km/s $= 11{,}000$ m/s

Energy and Power Units

Basic unit of energy: 1 joule $= 1 \dfrac{\text{kg} \times \text{m}^2}{\text{s}^2}$

Basic unit of power: 1 watt $= 1$ joule/s

Electron-volt: 1 eV $= 1.60 \times 10^{-19}$ joule

B Useful Formulas

■ Universal law of gravitation for the force between objects of mass M_1 and M_2, distance d between their centers:

$$F = G\frac{M_1 M_2}{d^2}$$

■ Newton's version of Kepler's third law; p and a are period and semimajor axis, respectively, of either orbiting mass:

$$p^2 = \frac{4\pi^2}{G(M_1 + M_2)} a^3$$

■ Escape velocity at distance R from center of object of mass M:

$$v_{escape} = \sqrt{\frac{2GM}{R}}$$

■ Relationship between a photon's wavelength (λ), frequency (f), and the speed of light (c):

$$\lambda \times f = c$$

■ Energy of a photon of wavelength λ or frequency f:

$$E = hf = \frac{hc}{\lambda}$$

■ Stefan–Boltzmann law for thermal radiation at temperature T (in Kelvin):

$$\text{emitted power per unit area} = \sigma T^4$$

■ Wien's law for the peak wavelength (λ_{max}) thermal radiation at temperature T (in Kelvin):

$$\lambda_{max} = \frac{2{,}900{,}000}{T}\text{nm}$$

■ Doppler shift (radial velocity is positive if the object is moving away from us and negative if it is moving toward us):

$$\frac{\text{radial velocity}}{\text{speed of light}} = \frac{\text{shifted wavelength} - \text{rest wavelength}}{\text{rest wavelength}}$$

■ Angular separation (α) of two points with an actual separation s, viewed from a distance d (assuming d is much larger than s):

$$\alpha = \frac{s}{2\pi d} \times 360°$$

■ Inverse Square Law for Light:

$$\text{apparent brightness} = \frac{\text{luminosity}}{4\pi d^2}$$

(where d is the distance to the object)

■ Parallax formula (distance d to a star with parallax angle p in arcseconds):

$$d\text{ (in parsecs)} = \frac{1}{p\text{ (in arcseconds)}}$$

$$\text{or, } d\text{ (in light-years)} = 3.26 \times \frac{1}{p\text{ (in arcseconds)}}$$

■ The orbital velocity law, to find the mass M_r contained within the circular orbit of radius r for an object moving at speed v:

$$M_r = \frac{r \times v^2}{G}$$

C A Few Mathematical Skills

This appendix reviews the following mathematical skills: powers of 10, scientific notation, working with units, the metric system, and finding a ratio. You should refer to this appendix as needed while studying the textbook.

C.1 Powers of 10

Powers of 10 simply indicate how many times to multiply 10 by itself. For example:

$$10^2 = 10 \times 10 = 100$$

$$10^6 = 10 \times 10 \times 10 \times 10 \times 10 \times 10 = 1,000,000$$

Negative powers are the reciprocals of the corresponding positive powers. For example:

$$10^{-2} = \frac{1}{10^2} = \frac{1}{100} = 0.01$$

$$10^{-6} = \frac{1}{10^6} = \frac{1}{1,000,000} = 0.000001$$

Table C.1 lists powers of 10 from 10^{-12} to 10^{12}. Note that powers of 10 follow two basic rules:

1. A positive exponent tells how many zeros follow the 1. For example, 10^0 is a 1 followed by no zeros, and 10^8 is a 1 followed by eight zeros.

2. A negative exponent tells how many places are to the right of the decimal point, including the 1. For example, $10^{-1} = 0.1$ has one place to the right of the decimal point; $10^{-6} = 0.000001$ has six places to the right of the decimal point.

Multiplying and Dividing Powers of 10

Multiplying powers of 10 simply requires adding exponents, as the following examples show:

$$10^4 \times 10^7 = \underbrace{10,000}_{10^4} \times \underbrace{10,000,000}_{10^7} = \underbrace{100,000,000,000}_{10^{4+7} \, = \, 10^{11}} = 10^{11}$$

$$10^5 \times 10^{-3} = \underbrace{100,000}_{10^5} \times \underbrace{0.001}_{10^{-3}} = \underbrace{100}_{10^{5+(-3)} = 10^2} = 10^2$$

$$10^{-8} \times 10^{-5} = \underbrace{0.00000001}_{10^{-8}} \times \underbrace{0.00001}_{10^{-5}} = \underbrace{0.0000000000001}_{10^{-8+(-5)} \, = \, 10^{-13}} = 10^{-13}$$

Table C.1 Powers of 10

Zero and Positive Powers			Negative Powers		
Power	Value	Name	Power	Value	Name
10^0	1	One			
10^1	10	Ten	10^{-1}	0.1	Tenth
10^2	100	Hundred	10^{-2}	0.01	Hundredth
10^3	1,000	Thousand	10^{-3}	0.001	Thousandth
10^4	10,000	Ten thousand	10^{-4}	0.0001	Ten thousandth
10^5	100,000	Hundred thousand	10^{-5}	0.00001	Hundred thousandth
10^6	1,000,000	Million	10^{-6}	0.000001	Millionth
10^7	10,000,000	Ten million	10^{-7}	0.0000001	Ten millionth
10^8	100,000,000	Hundred million	10^{-8}	0.00000001	Hundred millionth
10^9	1,000,000,000	Billion	10^{-9}	0.000000001	Billionth
10^{10}	10,000,000,000	Ten billion	10^{-10}	0.0000000001	Ten billionth
10^{11}	100,000,000,000	Hundred billion	10^{-11}	0.00000000001	Hundred billionth
10^{12}	1,000,000,000,000	Trillion	10^{-12}	0.000000000001	Trillionth

Dividing powers of 10 requires subtracting exponents, as in the following examples:

$$\frac{10^5}{10^3} = \underbrace{100,000}_{10^5} \div \underbrace{1,000}_{10^3} = \underbrace{100}_{10^{5-3}\ =\ 10^2} = 10^2$$

$$\frac{10^3}{10^7} = \underbrace{1,000}_{10^3} \div \underbrace{10,000,000}_{10^7} = \underbrace{0.0001}_{10^{3-7}\ =\ 10^{-4}} = 10^{-4}$$

$$\frac{10^{-4}}{10^{-6}} = \underbrace{0.0001}_{10^{-4}} \div \underbrace{0.000001}_{10^{-6}} = \underbrace{100}_{10^{-4-(-6)}\ =\ 10^2} = 10^2$$

Powers of Powers of 10

We can use the multiplication and division rules to raise powers of 10 to other powers or to take roots. For example:

$$(10^4)^3 = 10^4 \times 10^4 \times 10^4 = 10^{4+4+4} = 10^{12}$$

Note that we can get the same end result by simply multiplying the two powers:

$$(10^4)^3 = 10^{4\times3} = 10^{12}$$

Because taking a root is the same as raising to a fractional power (e.g., the square root is the same as the $\frac{1}{2}$ power, the cube root is the same as the $\frac{1}{3}$ power, etc.), we can use the same procedure for roots, as in the following example:

$$\sqrt{10^4} = (10^4)^{1/2} = 10^{4\times(1/2)} = 10^2$$

Adding and Subtracting Powers of 10

Unlike with multiplication and division, there is no shortcut for adding or subtracting powers of 10. The values must be written in longhand notation. For example:

$$10^6 + 10^2 = 1,000,000 + 100 = 1,000,100$$

$$10^8 + 10^{-3} = 100,000,000 + 0.001 = 100,000,000.001$$

$$10^7 - 10^3 = 10,000,000 - 1,000 = 9,999,000$$

Summary

We can summarize our findings using n and m to represent any numbers:

- To *multiply* powers of 10, *add* exponents: $10^n \times 10^m = 10^{n+m}$

- To *divide* powers of 10, *subtract* exponents: $\dfrac{10^n}{10^m} = 10^{n-m}$

- To *raise* powers of 10 to other powers, multiply exponents: $(10^n)^m = 10^{n \times m}$

C.2 Scientific Notation

When we are dealing with large or small numbers, it's generally easier to write them with powers of 10. For example, it's much easier to write the number 6,000,000,000,000 as 6×10^{12}. This format, in which a number *between* 1 and 10 is multiplied by a power of 10, is called **scientific notation**.

Converting a Number to Scientific Notation

We can convert numbers written in ordinary notation to scientific notation with a simple two-step process:

1. Move the decimal point to come after the *first* nonzero digit.

2. The number of places the decimal point moves tells you the power of 10; the power is *positive* if the decimal point moves to the left and *negative* if it moves to the right.

   Examples:

   $$3,042 \xrightarrow[\text{3 places to left}]{\text{decimal needs to move}} 3.042 \times 10^3$$

   $$0.00012 \xrightarrow[\text{4 places to right}]{\text{decimal needs to move}} 1.2 \times 10^{-4}$$

   $$226 \times 10^2 \xrightarrow[\text{2 places to left}]{\text{decimal needs to move}} (2.26 \times 10^2) \times 10^2 = 2.26 \times 10^4$$

Converting a Number from Scientific Notation

We can convert numbers written in scientific notation to ordinary notation by the reverse process:

1. The power of 10 indicates how many places to move the decimal point; move it to the *right* if the power of 10 is positive and to the *left* if it is negative.

2. If moving the decimal point creates any open places, fill them with zeros.

Examples:

$$4.01 \times 10^2 \xrightarrow[\text{2 places to right}]{\text{move decimal}} 401$$

$$3.6 \times 10^6 \xrightarrow[\text{6 places to right}]{\text{move decimal}} 3{,}600{,}000$$

$$5.7 \times 10^{-3} \xrightarrow[\text{3 places to left}]{\text{move decimal}} 0.0057$$

Multiplying or Dividing Numbers in Scientific Notation

Multiplying or dividing numbers in scientific notation simply re-quires operating on the powers of 10 and the other parts of the number separately.

Examples:

$$(6 \times 10^2) \times (4 \times 10^5) = (6 \times 4) \times (10^2 \times 10^5) = 24 \times 10^7 = (2.4 \times 10^1) \times 10^7 = 2.4 \times 10^8$$

$$\frac{4.2 \times 10^{-2}}{8.4 \times 10^{-5}} = \frac{4.2}{8.4} \times \frac{10^{-2}}{10^{-5}} = 0.5 \times 10^{-2-(-5)} = 0.5 \times 10^3 = (5 \times 10^{-1}) \times 10^3 = 5 \times 10^2$$

Note that, in both these examples, we first found an answer in which the number multiplied by a power of 10 was *not* between 1 and 10. We therefore followed the procedure for converting the final answer to scientific notation.

Addition and Subtraction with Scientific Notation

In general, we must write numbers in ordinary notation before adding or subtracting.

Examples:

$$(3 \times 10^6) + (5 \times 10^2) = 3{,}000{,}000 + 500 = 3{,}000{,}500 = 3.0005 \times 10^6$$

$$(4.6 \times 10^9) - (5 \times 10^8) = 4{,}600{,}000{,}000 - 500{,}000{,}000 = 4{,}100{,}000{,}000 = 4.1 \times 10^9$$

When both numbers have the *same* power of 10, we can factor out the power of 10 first.

Examples:

$$(7 \times 10^{10}) + (4 \times 10^{10}) = (7 + 4) \times 10^{10} = 11 \times 10^{10} = 1.1 \times 10^{11}$$

$$(2.3 \times 10^{-22}) - (1.6 \times 10^{-22}) = (2.3 - 1.6) \times 10^{-22} = 0.7 \times 10^{-22} = 7.0 \times 10^{-23}$$

C.3 Working with Units

Showing the units of a problem as you solve it usually makes the work much easier and also provides a useful way of checking your work. If an answer does not come out with the units you expect, you probably did something wrong. In general, working with units is very similar to working with numbers, as the following guidelines and examples show.

Five Guidelines for Working with Units

Before you begin any problem, think ahead and identify the units you expect for the final answer. Then operate on the units along with the numbers as you solve the problem. The following five guidelines may be helpful when you are working with units:

1. Mathematically, it doesn't matter whether a unit is singular (e.g., meter) or plural (e.g., meters); we can use the same abbreviation (e.g., m) for both.

2. You cannot add or subtract numbers unless they have the *same* units. For example, 5 apples + 3 apples = 8 apples, but the expression 5 apples + 3 oranges cannot be simplified further.

3. You *can* multiply units, divide units, or raise units to powers. Look for key words that tell you what to do.

   ▪ *Per* suggests division. For example, we write a speed of 100 kilometers per hour as:

   $$100 \, \frac{km}{hr} \quad \text{or} \quad 100 \, \frac{km}{1 \, hr}$$

   ▪ *Of* suggests multiplication. For example, if you launch a 50-kg space probe at a launch cost *of* $10,000 per kilogram, the total cost is:

   $$50 \, \cancel{kg} \times \frac{\$10,000}{\cancel{kg}} = \$500,000$$

   ▪ *Square* suggests raising to the second power. For example, we write an area of 75 square meters as 75 m^2.

   ▪ *Cube* suggests raising to the third power. For example, we write a volume of 12 cubic centimeters as 12 cm^3.

4. Often the number you are given is not in the units you wish to work with. For example, you may be given that the speed of light is 300,000 km/s but need it in units of m/s for a particular problem. To convert the units, simply multiply the given number by a *conversion factor:* a fraction in which the numerator (top of the fraction) and denominator (bottom of the fraction) are equal, so that the value of the fraction is 1; the number in the denominator must have the units that you wish to change. In the case of changing the speed of light from units of km/s to m/s, you need a conversion factor for kilometers to meters. Thus, the conversion factor is:

   $$\frac{1,000 \, m}{1 \, km}$$

   Note that this conversion factor is equal to 1, since 1,000 meters and 1 kilometer are equal, and that the units to be changed (km) appear in the denominator. We can now convert the speed of light from units of km/s to m/s simply by multiplying by this conversion factor:

   $$\underbrace{300,000 \, \frac{\cancel{km}}{s}}_{\substack{\text{speed of light} \\ \text{in km/s}}} \times \underbrace{\frac{1,000 \, m}{1 \, \cancel{km}}}_{\substack{\text{conversion from} \\ \text{km to m}}} = \underbrace{3 \times 10^8 \, \frac{m}{s}}_{\substack{\text{speed of light} \\ \text{in m/s}}}$$

   Note that the units of km cancel, leaving the answer in units of m/s.

5. It's easier to work with units if you replace division with multiplication by the reciprocal. For example, suppose you want to know how many minutes are represented by 300 seconds. We can find the answer by dividing 300 seconds by 60 seconds per minute:

   $$300 \, s \div 60 \, \frac{s}{min}$$

However, it is easier to see the unit cancellations if we rewrite this expression by replacing the division with multiplication by the reciprocal (this process is easy to remember as "invert and multiply"):

$$300 \text{ s} \div 60 \frac{\text{s}}{\text{min}} = 300 \cancel{\text{s}} \times \frac{1 \text{ min}}{60 \cancel{\text{s}}} = 5 \text{ min}$$

$$\underbrace{\underbrace{\phantom{\frac{1 \text{ min}}{60}}}_{\text{invert}}}_{\text{and multiply}}$$

We now see that the units of seconds (s) cancel in the numerator of the first term and the denominator of the second term, leaving the answer in units of minutes.

More Examples of Working with Units

Example 1. How many seconds are there in 1 day?

Solution: We can answer the question by setting up a *chain* of unit conversions in which we start with 1 *day* and end up with *seconds*. We use the facts that there are 24 hours per day (24 hr/day), 60 minutes per hour (60 min/hr), and 60 seconds per minute (60 s/min):

$$1 \underbrace{\cancel{\text{day}}}_{\substack{\text{starting} \\ \text{value}}} \times \underbrace{\frac{24 \cancel{\text{hr}}}{\cancel{\text{day}}}}_{\substack{\text{conversion} \\ \text{from} \\ \text{day to hr}}} \times \underbrace{\frac{60 \cancel{\text{min}}}{\cancel{\text{hr}}}}_{\substack{\text{conversion} \\ \text{from} \\ \text{hr to min}}} \times \underbrace{\frac{60 \text{ s}}{\cancel{\text{min}}}}_{\substack{\text{conversion} \\ \text{from} \\ \text{min to s}}} = 86,400 \text{ s}$$

Note that all the units cancel except *seconds,* which is what we want for the answer. There are 86,400 seconds in 1 day.

Example 2. Convert a distance of 10^8 cm to km.

Solution: The easiest way to make this conversion is in two steps, since we know that there are 100 centimeters per meter (100 cm/m) and 1,000 meters per kilometer (1,000 m/km):

$$\underbrace{10^8 \text{ cm}}_{\substack{\text{starting} \\ \text{value}}} \times \underbrace{\frac{1 \text{ m}}{100 \text{ cm}}}_{\substack{\text{conversion} \\ \text{from} \\ \text{cm to m}}} \times \underbrace{\frac{1 \text{ km}}{1,000 \text{ m}}}_{\substack{\text{conversion} \\ \text{from} \\ \text{m to km}}} = 10^8 \cancel{\text{cm}} \times \frac{1 \cancel{\text{m}}}{10^2 \cancel{\text{cm}}} \times \frac{1 \text{ km}}{10^3 \cancel{\text{m}}} = 10^3 \text{ km}$$

Alternatively, if we recognize that the number of kilometers should be smaller than the number of centimeters (because kilometers are larger), we might decide to do this conversion by dividing as follows:

$$10^8 \text{ cm} \div \frac{100 \text{ cm}}{\text{m}} \div \frac{1,000 \text{ m}}{\text{km}}$$

In this case, before carrying out the calculation, we replace each division with multiplication by the reciprocal:

$$10^8 \text{ cm} \div \frac{100 \text{ cm}}{\text{m}} \div \frac{1,000 \text{ m}}{\text{km}} = 10^8 \text{ cm} \times \frac{1 \text{ m}}{100 \text{ cm}} \times \frac{1 \text{ km}}{1,000 \text{ m}}$$

$$= 10^8 \cancel{\text{cm}} \times \frac{1 \cancel{\text{m}}}{10^2 \cancel{\text{cm}}} \times \frac{1 \text{ km}}{10^3 \cancel{\text{m}}}$$

$$= 10^3 \text{ km}$$

Note that we again get the answer that 10^8 cm is the same as 10^3 km, or 1,000 km.

Example 3. Suppose you accelerate at 9.8 m/s² for 4 seconds, starting from rest. How fast will you be going?

Solution: The question asked "how fast?" so we expect to end up with a speed. Therefore, we multiply the acceleration by the amount of time you accelerated:

$$9.8 \, \frac{m}{s^2} \times 4 \, s = (9.8 \times 4) \, \frac{m \times \cancel{s}}{s^{\cancel{2}}} = 39.2 \, \frac{m}{s}$$

Note that the units end up as a speed, showing that you will be traveling 39.2 m/s after 4 seconds of acceleration at 9.8 m/s².

Example 4. A reservoir is 2 km long and 3 km wide. Calculate its area, in both square kilometers and square meters.

Solution: We find its area by multiplying its length and width:

$$2 \, km \times 3 \, km = 6 \, km^2$$

Next we need to convert this area of 6 km² to square meters, using the fact that there are 1,000 meters per kilometer (1,000 m/km). Note that we must square the term 1,000 m/km when converting from km² to m²:

$$6 \, km^2 \times \left(1,000 \, \frac{m}{km}\right)^2 = 6 \, km^2 \times 1,000^2 \, \frac{m^2}{km^2} = 6 \, \cancel{km^2} \times 1,000,000 \, \frac{m^2}{\cancel{km^2}}$$

$$= 6,000,000 \, m^2$$

The reservoir area is 6 km², which is the same as 6 million m².

C.4 The Metric System (SI)

The modern version of the metric system, known as *Système Internationale d'Unites* (French for "International System of Units") or **SI**, was formally established in 1960. Today, it is the primary measurement system in nearly every country in the world with the exception of the United States. Even in the United States, it is the system of choice for science and international commerce.

The basic units of length, mass, and time in the SI are:

- The **meter** for length, abbreviated m

- The **kilogram** for mass, abbreviated kg

- The **second** for time, abbreviated s

Multiples of metric units are formed by powers of 10, using a prefix to indicate the power. For example, *kilo* means 10^3 (1,000), so a kilometer is 1,000 meters; a microgram is 0.000001 gram, because *micro* means 10^{-6}, or one millionth. Some of the more common prefixes are listed in Table C.2.

Table C.2 SI (Metric) Prefixes

Small Values			Large Values		
Prefix	**Abbreviation**	**Value**	**Prefix**	**Abbreviation**	**Value**
Deci	d	10^{-1}	Deca	da	10^1
Centi	c	10^{-2}	Hecto	h	10^2
Milli	m	10^{-3}	Kilo	k	10^3
Micro	μ	10^{-6}	Mega	M	10^6
Nano	n	10^{-9}	Giga	G	10^9
Pico	p	10^{-12}	Tera	T	10^{12}

Metric Conversions

Table C.3 lists conversions between metric units and units used commonly in the United States. Note that the conversions between kilograms and pounds are valid only on Earth, because they depend on the strength of gravity.

Example 1. International athletic competitions generally use metric distances. Compare the length of a 100-meter race to that of a 100-yard race.

Solution: Table C.3 shows that 1 m = 1.094 yd, so 100 m is 109.4 yd. Note that 100 meters is almost 110 yards; a good "rule of thumb" to remember is that distances in meters are about 10% longer than the corresponding number of yards.

Example 2. How many square kilometers are in 1 square mile?

Solution: We use the square of the miles-to-kilometers conversion factor:

$$(1 \text{ mi}^2) \times \left(\frac{1.6093 \text{ km}}{1 \text{ mi}} \right)^2 = (1 \text{ mi}^2) \times \left(1.6093^2 \frac{\text{km}^2}{\text{mi}^2} \right) = 2.5898 \text{ km}^2$$

Therefore, 1 square mile is 2.5898 square kilometers.

C.5 Finding a Ratio

Suppose you want to compare two quantities, such as the average density of the Earth and the average density of Jupiter. The way we do such a comparison is by dividing, which tells us the *ratio* of the two quantities. In this case, the Earth's average density is 5.52 grams/cm^3 and Jupiter's average density is 1.33 grams/cm^3 (see Figure 8.1), so the ratio is:

$$\frac{\text{average density of Earth}}{\text{average density of Jupiter}} = \frac{5.52 \text{ g/cm}^3}{1.33 \text{ g/cm}^3} = 4.15$$

Notice how the units cancel on both the top and bottom of the fraction. We can state our result in two equivalent ways:

■ The ratio of the Earth's average density to Jupiter's average density is 4.15.

■ The Earth's average density is 4.15 times Jupiter's average density.

Sometimes, the quantities that you want to compare may each involve an equation. In such cases, you could, of course, find the ratio by first calculating each of the two quantities individually and then dividing. However, it is much easier if you first express the ratio as a fraction, putting the equation for one quantity on top and the other on the bottom. Some of the terms in the equation may then cancel out, making any calculations much easier.

Example 1. Compare the kinetic energy of a car traveling at 100 km/hr to that of a car traveling at 50 km/hr.

Table C.3 Metric Conversions

To Metric	From Metric
1 inch = 2.540 cm	1 cm = 0.3937 inch
1 foot = 0.3048 m	1 m = 3.28 feet
1 yard = 0.9144 m	1 m = 1.094 yards
1 mile = 1.6093 km	1 km = 0.6214 mile
1 pound = 0.4536 kg	1 kg = 2.205 pounds

Solution: We do the comparison by finding the ratio of the two kinetic energies, recalling that the formula for kinetic energy is $\frac{1}{2}mv^2$. Since we are not told the mass of the car, you might at first think that we don't have enough information to find the ratio. However, notice what happens when we put the equations for each kinetic energy into the ratio, calling the two speeds v_1 and v_2:

$$\frac{\text{K.E. car at } v_1}{\text{K.E. car at } v_2} = \frac{\frac{1}{2}m_{car}v_1^{\,2}}{\frac{1}{2}m_{car}v_2^{\,2}} = \frac{v_1^{\,2}}{v_2^{\,2}} = \left(\frac{v_1}{v_2}\right)^2$$

All the terms cancel except those with the two speeds, leaving us with a very simple formula for the ratio. Now we put in 100 km/hr for v_1 and 50 km/hr for v_2:

$$\frac{\text{K.E. car at 100 km/hr}}{\text{K.E. car at 50 km/hr}} = \left(\frac{100 \text{ km/hr}}{50 \text{ km/hr}}\right)^2 = 2^4 = 4$$

The ratio of the car's kinetic energies at 100 km/hr and 50 km/hr is 4. That is, the car has four times as much kinetic energy at 100 km/hr as it has at 50 km/hr.

Example 2. Compare the strength of gravity between the Earth and the Sun to the strength of gravity between the Earth and the Moon.

Solution: We do the comparison by taking the ratio of the Earth–Sun gravity to the Earth–Moon gravity. In this case, each quantity is found from the equation of Newton's law of gravity. (See Section 4.4.) Thus, the ratio is:

$$\frac{\text{Earth–Sun gravity}}{\text{Earth–Moon gravity}} = \frac{G\dfrac{M_{Earth}M_{Sun}}{(d_{Earth–Sun})^2}}{G\dfrac{M_{Earth}M_{Moon}}{(d_{Earth–Moon})^2}} = \frac{M_{Sun}}{(d_{Earth–Sun})^2} \times \frac{(d_{Earth–Moon})^2}{M_{Moon}}$$

Note how all but four of the terms cancel; the last step comes from replacing the division with multiplication by the reciprocal (the "invert and multiply" rule for division). We can simplify the work further by rearranging the terms so that we have the masses and distances together:

$$\frac{\text{Earth–Sun gravity}}{\text{Earth–Moon gravity}} = \frac{M_{Sun}}{M_{Moon}} \times \frac{(d_{Earth–Moon})^2}{(d_{Earth–Sun})^2}$$

Now it is just a matter of looking up the numbers (see Appendix E) and calculating:

$$\frac{\text{Earth–Sun gravity}}{\text{Earth–Moon gravity}} = \frac{1.99 \times 10^{30} \text{ kg}}{7.35 \times 10^{22} \text{ kg}} \times \frac{(384.4 \times 10^3 \text{ km})^2}{(149.6 \times 10^6 \text{ km})^2} = 179$$

In other words, the Earth–Sun gravity is 179 times stronger than the Earth–Moon gravity.

D The Periodic Table of the Elements

Key

12	— Atomic number
Mg	— Element's symbol
Magnesium	— Element's name
24.305	— Atomic mass*

*Atomic masses are fractions because they represent a weighted average of atomic masses of different isotopes—in proportion to the abundance of each isotope on Earth.

1																	2
H Hydrogen 1.00794																	**He** Helium 4.003
3 **Li** Lithium 6.941	4 **Be** Beryllium 9.01218											5 **B** Boron 10.81	6 **C** Carbon 12.011	7 **N** Nitrogen 14.007	8 **O** Oxygen 15.999	9 **F** Fluorine 18.988	10 **Ne** Neon 20.179
11 **Na** Sodium 22.990	12 **Mg** Magnesium 24.305											13 **Al** Aluminum 26.98	14 **Si** Silicon 28.086	15 **P** Phosphorus 30.974	16 **S** Sulfur 32.06	17 **Cl** Chlorine 35.453	18 **Ar** Argon 39.948
19 **K** Potassium 39.098	20 **Ca** Calcium 40.08	21 **Sc** Scandium 44.956	22 **Ti** Titanium 47.88	23 **V** Vanadium 50.94	24 **Cr** Chromium 51.996	25 **Mn** Manganese 54.938	26 **Fe** Iron 55.847	27 **Co** Cobalt 58.9332	28 **Ni** Nickel 58.69	29 **Cu** Copper 63.546	30 **Zn** Zinc 65.39	31 **Ga** Gallium 69.72	32 **Ge** Germanium 72.59	33 **As** Arsenic 74.922	34 **Se** Selenium 78.96	35 **Br** Bromine 79.904	36 **Fr** Krypton 83.80
37 **Rb** Rubidium 85.468	38 **Sr** Strontium 87.62	39 **Y** Yttrium 88.9059	40 **Zr** Zirconium 91.224	41 **Nb** Niobium 92.91	42 **Mo** Molybdenum 95.94	43 **Tc** Technetium (98)	44 **Ru** Ruthenium 101.07	45 **Rh** Rhodium 102.906	46 **Pd** Palladium 106.42	47 **Ag** Silver 107.868	48 **Cd** Cadmium 112.41	49 **In** Indium 114.82	50 **Sn** Tin 118.71	51 **Sb** Antimony 121.75	52 **Te** Tellurium 127.60	53 **I** Iodine 126.905	54 **Xe** Xenon 131.29
55 **Cs** Cesium 132.91	56 **Ba** Barium 137.34		72 **Hf** Hafnium 178.49	73 **Ta** Tantalum 180.95	74 **W** Tungsten 183.85	75 **Re** Rhenium 186.207	76 **Os** Osmium 190.2	77 **Ir** Iridium 192.22	78 **Pt** Platinum 195.08	79 **Au** Gold 196.967	80 **Hg** Mercury 200.59	81 **Ti** Thallium 204.383	82 **Pb** Lead 207.2	83 **Bi** Bismuth 208.98	84 **Po** Polonium (209)	85 **At** Astatine (210)	86 **Rn** Radon (222)
87 **Fr** Francium (223)	88 **Ra** Radium 226.0254		104 **Rf** Rutherfordium (261)	105 **Db** Dubnium (262)	106 **Sg** Seaborgium (263)	107 **Bh** Bohrium (262)	108 **Hs** Hassium (265)	109 **Mt** Meitnerium (266)	110 **Uun** Unumnilium (269)	111 **Uuu** Unununium (272)	112 **Uub** Ununbium (277)						

Lanthanide Series

57 **La** Lanthanum 138.906	58 **Ce** Cerium 140.12	59 **Pr** Praseodymium 140.908	60 **Nd** Neodymium 144.24	61 **Pm** Promethium (145)	62 **Sm** Samarium 150.36	63 **Eu** Europium 151.96	64 **Gd** Gadolinium 157.25	65 **Tb** Terbium 158.925	66 **Dy** Dysprosium 162.50	67 **Ho** Holmium 164.93	68 **Er** Erbium 167.26	69 **Tm** Thulium 168.934	70 **Yb** Ytterbium 173.04	71 **Lu** Lutetium 174.967

Actinide Series

89 **Ac** Actinium 227.028	90 **Th** Thorium 232.038	91 **Pa** Protactinium 231.036	92 **U** Uranium 238.029	93 **Np** Neptunium 237.048	94 **Pu** Plutonium (244)	95 **Am** Americium (243)	96 **Cm** Curium (247)	97 **Bk** Berkelium (247)	98 **Cf** Californium (251)	99 **Es** Einsteinium (252)	100 **Fm** Fermium (257)	101 **Md** Mendelevium (258)	102 **No** Nobelium (259)	103 **Lr** Lawrencium (260)

E Planetary Data

Table E.1 Physical Properties of the Sun and Planets

Name	Radius (Eq[a]) (km)	Radius (Eq) (Earth units)	Mass (kg)	Mass (Earth units)	Average Density (g/cm^3)	Surface Gravity (Earth = 1)	Escape Velocity (km/s)
Sun	695,000	109	1.99×10^{30}	333,000	1.41	27.5	—
Mercury	2,440	0.382	3.30×10^{23}	0.055	5.43	0.38	4.43
Venus	6,051	0.949	4.87×10^{24}	0.815	5.25	0.91	10.4
Earth	6,378	1.00	5.97×10^{24}	1.00	5.52	1.00	11.2
Mars	3,397	0.533	6.42×10^{23}	0.107	3.93	0.38	5.03
Jupiter	71,492	11.19	1.90×10^{27}	317.9	1.33	2.36	59.5
Saturn	60,268	9.46	5.69×10^{26}	95.18	0.70	0.92	35.5
Uranus	25,559	3.98	8.66×10^{25}	14.54	1.32	0.91	21.3
Neptune	24,764	3.81	1.03×10^{26}	17.13	1.64	1.14	23.6
Pluto	1,160	0.181	1.31×10^{22}	0.0022	2.05	0.07	1.25
Planet X[b]	1,430	0.22	—	—	—	—	—

[a]Eq = equatorial.

[b] Little is known about this newly-discovered object, formally known as 2003 UB313. Even its size could be 25% larger or smaller, depending on its actual reflectivity. Its mass is therefore unknown to a factor of two or more.

Table E.2 Orbital Properties of the Sun and Planets

Name	Distance from Sun[a] (AU)	(10⁶ km)	Orbital Period (years)	Orbital Inclination[b] (degrees)	Orbital Eccentricity	Sidereal Rotation Period (Earth days)[c]	Axis Tilt (degrees)
Sun	—	—	—	—	—	25.4	7.25
Mercury	0.387	57.9	0.2409	7.00	0.206	58.6	0.0
Venus	0.723	108.2	0.6152	3.39	0.007	−243.0	177.3
Earth	1.00	149.6	1.0	0.00	0.017	0.9973	23.45
Mars	1.524	227.9	1.881	1.85	0.093	1.026	25.2
Jupiter	5.203	778.3	11.86	1.31	0.048	0.41	3.08
Saturn	9.539	1,427	29.42	2.48	0.056	0.44	26.73
Uranus	19.19	2,870	84.01	0.77	0.046	−0.72	97.92
Neptune	30.06	4,497	164.8	1.77	0.010	0.67	29.6
Pluto	39.48	5,906	248.0	17.14	0.248	−6.39	112.5
Planet X[d]	67.89	10,156	560	44.0	0.438	—	—

[a] Semimajor axis of the orbit.

[b] With respect to the ecliptic.

[c] A negative sign indicates rotation is backward relative to other planets.

[d] The orbit of this newly-discovered object (formally known as 2003 UB313) is well known, but little is known about the object itself.

Table E.3 Satellites of the Solar System (as of 2005)[a]

Planet Satellite	Radius or Dimensions[b] (km)	Distance from Planet (10^3 km)	Orbital Period[c] (Earth days)	Mass[d] (kg)	Density[d] (g/cm^3)	Notes About the Satellite
Earth						**Earth**
Moon	1,738	384.4	27.322	7.349×10^{22}	3.34	*Moon:* Probably formed in giant impact.
Mars						**Mars**
Phobos	$13 \times 11 \times 9$	9.38	0.319	1.3×10^{16}	1.9	*Phobos, Deimos:* Probable captured asteroids.
Deimos	$8 \times 6 \times 5$	23.5	1.263	1.8×10^{15}	2.2	
Jupiter						**Jupiter**
Small inner moons (4 moons)	8 to 83	128–222	0.295–0.6745	—	—	*Metis, Adrastea, Amalthea, Thebe:* Small moonlets within and near Jupiter's ring system.
Io	1,821	421.6	1.769	8.933×10^{22}	3.57	*Io:* Most volcanically active object in the solar system.
Europa	1,565	670.9	3.551	4.797×10^{22}	2.97	*Europa:* Possible oceans under icy crust.
Ganymede	2,634	1,070.0	7.155	1.482×10^{23}	1.94	*Ganymede:* Largest satellite in solar system; unusual ice geology.
Callisto	2,403	1,883.0	16.689	1.076×10^{23}	1.86	*Callisto:* Cratered iceball.
Irregular group 1 (7 moons)	4–85	7,200–17,000	130–457	—	—	*Themisto, Leda, Himalia, Lysithea, Elara, and 44 others:* Probable captured moons with inclined orbits.
Irregular group 2 (48 moons)	1–30	15,900–29,500	−490 to −983	—	—	*Ananke, Carme, Pasiphae, Sinope, and 46 others:* Probable captured moons in inclined backward orbits.
Saturn						**Saturn**
Small inner moons (11)	3–89	134–212	0.574–1.1	—	—	*Pan, Atlas, Prometheus, Pandora, Epimetheus, Janus and 5 more:* Small moonlets within and near Saturn's ring system.
Mimas	199	185.52	0.942	3.70×10^{19}	1.17	*Mimas, Enceladus, Tethys:* Small and medium-size iceballs, many with interesting geology.
Enceladus	249	238.02	1.370	1.2×10^{20}	1.24	
Tethys	530	294.66	1.888	6.17×10^{20}	1.26	
Calypso & Telesto	8–12	294.66	1.888	—	—	*Calypso, Telesto:* Small moonlets sharing Tethys's orbit.
Dione	559	377.4	2.737	1.08×10^{21}	1.44	*Dione:* Medium-size iceball, with interesting geology.
Helene & Polydeuces	2–16	377.4	2.737	1.6×10^{16}	—	*Helene & Polydeuces:* Small moonlets sharing Dione's orbit.
Rhea	764	527.04	4.518	2.31×10^{21}	1.33	*Rhea:* Medium-size iceball, with interesting geology.
Titan	2,575	1,221.85	15.945	1.35×10^{23}	1.88	*Titan:* Dense atmosphere shrouds surface; ongoing geological activity.
Hyperion	$180 \times 140 \times 112$	1,481.1	21.277	2.8×10^{19}	—	*Hyperion:* Only satellite known not to rotate synchronously.

Moon	Radius (km)	Distance	Period (days)	Mass (kg)	Density	Notes
Iapetus	718	3,561.3	79.331	1.59×10^{21}	1.21	*Iapetus:* Bright and dark hemispheres show greatest contrast in the solar system.
Phoebe	110	12,952	−550.4	1×10^{19}	—	*Phoebe:* Very dark; material ejected from Phoebe may coat one side of Iapetus.
Irregular groups (25 moons)	2–16	11,400–23,400	450–930 and −550 to −1,320	—	—	Probable captured moons with highly inclined and/or backward orbits.
Uranus						**Uranus**
Small inner moons (13 moons)	5–81	49–98	0.4–0.9	—	—	*Cordelia, Ophelia, Bianca, Cressida, Desdemona, Juliet, Portia, Rosalind, Belinda, Puck, 1986 U10, 2003 U1, 2003 U3:* Small moonlets within and near Uranus's ring system.
Miranda	236	129.8	1.413	6.6×10^{19}	1.26	*Miranda, Ariel, Umbriel, Titania, Oberon:* Small and medium-size iceballs, with some interesting geology.
Ariel	579	191.2	2.520	1.35×10^{21}	1.65	
Umbriel	584.7	266.0	4.144	1.17×10^{21}	1.44	
Titania	788.9	435.8	8.706	3.52×10^{21}	1.59	
Oberon	761.4	582.6	13.463	3.01×10^{21}	1.50	
Irregular group (9 moons)	5–95	4,280–21,000	580–2,820	—	—	*Caliban, Sycorax, Stephano, Prospero, Setebos, Trinculo, 2001 U2, 2001 U3, 2003 U3:* Probable captured moons; several in backward orbits.
Neptune						**Neptune**
Small inner moons (5 moons)	29–86	48–74	0.30–0.55	—	—	*Naiad, Thalassa, Despina, Galatea, Larissa:* Small moonlets within and near Neptune's ring system.
Proteus	218 × 208 × 201	117.6	1.121	6×10^{19}	—	
Triton	1,352.6	354.59	−5.875	2.14×10^{22}	2.0	*Triton:* Probable captured Kuiper belt object—largest captured object in solar system.
Nereid	170	5,588.6	360.125	3.1×10^{19}	—	*Nereid:* Small, icy moon; very little known.
Irregulars (5 moons)	15–27	16,600–48,600	1,870–9,412	—	—	*2002 N1, N2, N3, N4, 2003 N1:* Possible captured moons in inclined or backward orbit.
Pluto						**Pluto**
Charon	593	19.6	6.38718	1.56×10^{21}	1.6	*Charon:* Unusually large compared to its planet; may have formed in giant impact.
S/2005 P1 and S/2005 P2	32–100	—	—	—	—	Newly discovered moons outside Charon's orbit.

[a] *Note:* Authorities differ substantially on many of the values in this table.

[b] a × b × c values for the Dimensions are the approximate lengths of the axes (center to edge) for irregular moons.

[c] Negative sign indicates backward orbit.

[d] Masses and densities are most accurate for those satellites visited by a spacecraft on a flyby. Masses for the smallest moons have not been measured but can be estimated from the radius and an assumed density.

Table E.4 Fifty Extrasolar Planets of Note (listed in order of distance from their star)

Name	Detection Methods[2]	Minimum mass (Jupiter masses)	Semimajor Axis (AU)	Period (days)	Radius (Jupiter radii)	Notes[1]
Gliese 876 d	radial velocity	0.023	0.021	1.94	—	"Hot Jupiter"; smallest known semimajor axis of 2005
OGLE-TR-56 b	radial velocity, transit	1.45	0.023	1.21	1.23	"Hot Jupiter"; first planet discovered by transit
GJ 436 b	radial velocity	0.067	0.028	2.64	—	"Hot Jupiter"; sub-Saturn mass
OGLE-TR-132 b	radial velocity, transit	1.19	0.031	1.69	1.13	"Hot Jupiter"
55 Cnc e	radial velocity	0.045	0.038	2.81	—	"Hot Jupiter"; sub-Neptune mass
TrES-1	transit, radial velocity, eclipse	0.61	0.039	3.03	1.08	"Hot Jupiter"
HD 46375 b	radial velocity	0.25	0.041	3.02	—	"Hot Jupiter"; sub-Saturn mass
OGLE-TR-10 b	transit, radial velocity	0.54	0.042	3.10	1.16	"Hot Jupiter"
HD 149026 b	radial velocity, transit	0.36	0.042	2.88	0.73	"Hot Jupiter"
HD 188753A b	radial velocity	1.14	0.045	3.35	—	"Hot Jupiter"; first planet discovered in triple-star system
HD 209458 b	radial velocity, transit, eclipse	0.69	0.045	3.52	1.32	"Hot Jupiter"; first planet and first atmosphere successfully detected by transit
HD 88133 b	radial velocity	0.22	0.047	3.41	—	"Hot Jupiter"; sub-Saturn mass
OGLE-TR-111 b	radial velocity, transit	0.53	0.047	4.02	1.00	"Hot Jupiter"
51 Peg b	radial velocity	0.47	0.052	4.23	—	"Hot Jupiter"; first exoplanet discovered around sun-like star
Ups And b	radial velocity	0.69	0.059	4.62	—	"Hot Jupiter"; in first multi-planet system discovered around sun-like star
HD 160691 d	radial velocity	0.044	0.090	9.55	—	"Hot Jupiter"; sub-Neptune mass
55 Cnc b	radial velocity	0.78	0.115	14.7	—	—
HD 190360 c	radial velocity	0.057	0.128	17.1	—	Sub-Saturn mass
Gliese 876 c	radial velocity	0.56	0.130	30.1	—	"Eccentric"[3]
HD 102117 b	radial velocity	0.14	0.149	20.7	—	Sub-Saturn mass
Gliese 876 b	radial velocity	1.94	0.208	60.9	—	—
55 Cnc c	radial velocity	0.22	0.240	43.9	—	"Eccentric"; sub-Saturn mass
HD 16141 b	radial velocity	0.23	0.350	75.6	—	Sub-Saturn mass
HD 80606 b	radial velocity	3.4	0.439	112	—	"Eccentric"
HD 82943 c	radial velocity	0.88	0.730	222	—	"Eccentric"

Ups And c	radial velocity	1.89	242	0.829	"Eccentric"; in first multi-planet system discovered
HR 810 b	radial velocity	1.94	311	0.910	—
HD 210277 b	radial velocity	1.24	436	1.097	"Eccentric"; planet mass inferred from surrounding disk
HD 114783 b	radial velocity	0.99	501	1.200	—
HD 41004A b	radial velocity	2.3	655	1.310	"Eccentric"; planet in a system with two stars and brown dwarf
HD 183263 b	radial velocity	3.7	634	1.520	"Eccentric"
HD 114386 b	radial velocity	0.99	872	1.620	"Eccentric"
HD 4208 b	radial velocity	0.80	812	1.670	—
HD 13189 b	radial velocity	14.0	472	1.850	"Eccentric"
Gamma Cephei b	radial velocity	1.59	902	2.030	First extrasolar planet discovered in close stellar binary system
47 UMa b	radial velocity	2.5	1089	2.090	"Eccentric"
HD 50554 b	radial velocity	4.9	1279	2.380	"Eccentric"
Ups And d	radial velocity	3.8	1284	2.530	"Eccentric"; in first multi-planet system discovered
HD 202206 c	radial velocity	2.4	1383	2.550	"Eccentric"
HD 12661 c	radial velocity	1.57	1445	2.560	—
HD 168443 c	radial velocity	17.1	1740	2.870	—
HD 37124 d	radial velocity	0.66	2295	3.190	—
Epsilon Eridani b	radial velocity, astrometry	0.86	2502	3.300	"Eccentric"; star surrounded by dust disk
HD 74156 c	radial velocity	6.2	2025	3.400	"Eccentric"
HD 169830 c	radial velocity	4.0	2102	3.600	"Eccentric"
HD 38529 c	radial velocity	12.7	2174	3.680	"Eccentric"
47 UMa c	radial velocity	0.76	2594	3.730	—
HD 50499 b	radial velocity	1.71	2583	3.860	—
HD 72659 b	radial velocity	3.0	3177	4.160	"Eccentric"
55 Cnc d	radial velocity	3.9	4517	5.257	"Eccentric"; largest known semimajor axis as of 2005

Notes: [1]The list includes all planets detected by two methods, most planets in multiple systems, most "Hot Jupiters," and a representative sample of other extrasolar planets. More than 100 known extrasolar planets are not listed.

[2]If two detection methods are listed, the discovery method is given first.

[3]"Eccentric" means eccentricity > 0.25

F Stellar Data

Table F.1 Stars Within 12 Light-Years

Star	Distance (ly)	Spectral Type		RA h	RA m	Dec °	Dec '	Luminosity (L/L_{Sun})
Sun	0.000016	G2	V	—	—	—	—	1.0
Proxima Centauri	4.2	M5.5	V	14	30	−62	41	0.0006
α Centauri A	4.4	G2	V	14	40	−60	50	1.6
α Centauri B	4.4	K0	V	14	40	−60	50	0.53
Barnard's Star	6.0	M4	V	17	58	+04	42	0.005
Wolf 359	7.8	M6	V	10	56	+07	01	0.0008
Lalande 21185	8.3	M2	V	11	03	+35	58	0.03
Sirius A	8.6	A1	V	06	45	−16	42	26.0
Sirius B	8.6	DA2	—	06	45	−16	42	0.002
UV Ceti	8.7	M5.5	V	01	39	−17	57	0.0009
BL Ceti	8.7	M6	V	01	39	−17	57	0.0006
Ross 154	9.7	M3.5	V	18	50	−23	50	0.004
Ross 248	10.3	M5.5	V	23	42	+44	11	0.001
ε Eridani	10.5	K2	V	03	33	−09	28	0.37
Lacaille 9352	10.7	M1.5	V	23	06	−35	51	0.05
Ross 128	10.9	M4	V	11	48	+00	49	0.003
EZ Aquarii A	11.3	M5	V	22	39	−15	18	0.0006
EZ Aquarii B	11.3	M6	V	22	39	−15	18	0.0004
EZ Aquarii C	11.3	M6.5	V	22	39	−15	18	0.0003
61 Cygni A	11.4	K5	V	21	07	+38	42	0.15
61 Cygni B	11.4	K7	V	21	07	+38	42	0.09
Procyon A	11.4	F5	IV–V	07	39	+05	14	7.4
Procyon B	11.4	DA	—	07	39	+05	14	0.0005
Gliese 725 A	11.5	M3	V	18	43	+59	38	0.02
Gliese 725 B	11.5	M3.5	V	18	43	+59	38	0.01
GX Andromedae	11.15	M1.5	V	00	18	+44	01	0.03
GQ Andromedae	11.15	M3.5	V	00	18	+44	01	0.003
DX Cancri	11.8	M6.5	V	08	30	+26	47	0.0003
ε Indi A	11.8	K5	V	22	03	−56	45	0.26
ε Indi B	11.8	T1.0	Brown Dwarf	22	04	−56	46	—
ε Indi C	11.8	T1.0	Brown Dwarf	22	04	−56	46	—
τ Ceti	11.9	G8	V	01	44	−15	57	0.59
GJ 1061	11.9	M5.5	V	03	36	−44	31	0.0009

Note: These data were provided by the RECONS project, courtesy of Dr. Todd Henry (July, 2005). The luminosities are all total (bolometric) luminosities. The DA stellar types are white dwarfs. The coordinates are for the year 2000. The bolometric luminosity of the brown dwarfs is primarily in the infrared, and has not been measured accurately yet.

Table F.2 Twenty Brightest Stars

Star	Constellation	RA h	RA m	Dec °	Dec ′	Distance (ly)	Spectral Type		Apparent Magnitude	Luminosity (L/L_{Sun})
Sirius	Canis Major	6	45	−16	42	8.6	A1	V	−1.46	26
Canopus	Carina	6	24	−52	41	313	F0	Ib-II	−0.72	13,000
α Centauri	Centaurus	14	40	−60	50	4.4	G2	V	−0.01	1.6
							K0	V	1.3	0.53
Arcturus	Boötes	14	16	+19	11	37	K2	III	−0.06	170
Vega	Lyra	18	37	+38	47	25	A0	V	0.04	60
Capella	Auriga	5	17	+46	00	42	G0	III	0.75	70
							G8	III	0.85	77
Rigel	Orion	5	15	−08	12	772	B8	Ia	0.14	70,000
Procyon	Canis Minor	7	39	+05	14	11.4	F5	IV–V	0.37	7.4
Betelgeuse	Orion	5	55	+07	24	427	M2	Iab	0.41	38,000
Achernar	Eridanus	1	38	−57	15	144	B5	V	0.51	3,600
Hadar	Centaurus	14	04	−60	22	525	B1	III	0.63	100,000
Altair	Aquila	19	51	+08	52	17	A7	IV–V	0.77	10.5
Acrux	Crux	12	27	−63	06	321	B1	IV	1.39	22,000
							B3	V	1.9	7,500
Aldebaran	Taurus	4	36	+16	30	65	K5	III	0.86	350
Spica	Virgo	13	25	−11	09	260	B1	V	0.91	23,000
Antares	Scorpio	16	29	−26	26	604	M1	Ib	0.92	38,000
Pollux	Gemini	7	45	+28	01	34	K0	III	1.16	45
Fomalhaut	Piscis Austrinus	22	58	−29	37	25	A3	V	1.19	18
Deneb	Cygnus	20	41	+45	16	2,500	A2	Ia	1.26	170,000
β Crucis	Crux	12	48	−59	40	352	B0.5	IV	1.28	37,000

Note: Three of the stars on this list, Capella, α Centauri, and Acrux, are binary systems with members of comparable brightness. They are counted as single stars because that is how they appear to the naked eye. All the luminosities given are total (bolometric) luminosities. The coordinates are for the year 2000.

G Galaxy Data

Table G.1 Galaxies of the Local Group

Galaxy Name	Distance (millions of ly)	Type[a]	RA h	RA m	Dec °	Dec '	Luminosity (millions of L_{Sun})
Milky Way	—	Sbc	—	—	—	—	15,000
WLM	3.0	Irr	00	02	−15	30	50
IC 10	2.7	dIrr	00	20	+59	18	160
Cetus	2.5	dE	00	26	−11	02	0.72
NGC 147	2.4	dE	00	33	+48	30	131
And III	2.5	dE	00	35	+36	30	1.1
NGC 185	2.0	dE	00	39	+48	20	120
NGC 205	2.7	E	00	40	+41	41	370
And VIII	2.7	dE	00	42	+40	37	240
M 32	2.6	E	00	43	+40	52	380
M 31	2.5	Sb	00	43	+41	16	21,000
And I	2.6	dE	00	46	+38	00	4.7
SMC	0.19	Irr	00	53	−72	50	230
And IX	2.9	dE	00	52	+43	12	—
Sculptor	0.26	dE	01	00	−33	42	2.2
LGS 3	2.6	dIrr	01	04	+21	53	1.3
IC 1613	2.3	Irr	01	05	+02	08	64
And V	2.9	dE	01	10	+47	38	—
And II	1.7	dE	01	16	+33	26	2.4
M 33	2.7	Sc	01	34	+30	40	2,800
Phoenix	1.5	dIrr	01	51	−44	27	0.9
Fornax	0.45	dE	02	40	−34	27	15.5
EGB0427 + 63	4.3	dIrr	04	32	+63	36	9.1
LMC	0.16	Irr	05	24	−69	45	1,300
Carina	0.33	dE	06	42	−50	58	0.4
Canis Major	0.025	dIrr	07	15	−28		—
Leo A	2.2	dIrr	09	59	+30	45	3.0
Sextans B	4.4	dIrr	10	00	+05	20	41
NGC 3109	4.1	Irr	10	03	−26	09	160
Antlia	4.0	dIrr	10	04	−27	19	1.7
Leo I	0.82	dE	10	08	+12	18	4.8
Sextans A	4.7	dIrr	10	11	−04	42	56
Sextans	0.28	dE	10	13	−01	37	0.5
Leo II	0.67	dE	11	13	+22	09	0.6
GR 8	5.2	dIrr	12	59	+14	13	3.4
Ursa Minor	0.22	dE	15	09	+67	13	0.3
Draco	2.7	dE	17	20	+57	55	0.3
Sagittarius	0.08	dE	18	55	−30	29	18
SagDIG	3.5	dIrr	19	30	−17	41	6.8
NGC 6822	1.6	Irr	19	45	−14	48	94
DDO 210	2.6	dIrr	20	47	−12	51	0.8
IC 5152	5.2	dIrr	22	03	−51	18	70
Tucana	2.9	dE	22	42	−64	25	0.5
UKS2323-326	4.3	dE	23	26	−32	23	5.2
And VII	2.6	dE	23	38	+50	35	—
Pegasus	3.1	dIrr	23	29	+14	45	12
And VI	2.8	dE	23	52	+24	36	—

[a] Types beginning with S are spiral galaxies classified according to Hubble's system (see Chapter 15). Type E galaxies are elliptical or spheroidal. Type Irr galaxies are irregular. The prefix d denotes a dwarf galaxy. This list is based on a list originally published by M. Mateo in 1998, and augmented by discoveries of Local Group galaxies made between 1998–2005.

Table G.2 Nearby Galaxies in the Messier Catalog[a,b]

Galaxy Name (M / NGC)[c]	RA h	RA m	Dec °	Dec ′	RV_{hel}[d]	RV_{gal}[e]	Type[f]	Nickname
M31 / NGC224	00	43	+41	16	−300 ± 4	−122	Spiral	Andromeda
M32 / NGC221	00	43	+40	52	−145 ± 2	32	Elliptical	
M33 / NGC598	01	34	+30	40	−179 ± 3	−44	Spiral	Triangulum
M49 / NGC4472	12	30	+08	00	997 ± 7	929	Elliptical/Lenticular/Seyfert	
M51 / NGC5194	13	30	+47	12	463 ± 3	550	Spiral/Interacting	Whirlpool
M58 / NGC4579	12	38	+11	49	1,519 ± 6	1,468	Spiral/Seyfert	
M59 / NGC4621	12	42	+11	39	410 ± 6	361	Elliptical	
M60 / NGC4649	12	44	+11	33	1,117 ± 6	1,068	Elliptical	
M61 / NGC4303	12	22	+04	28	1,566 ± 2	1,483	Spiral/Seyfert	
M63 / NGC5055	13	16	+42	02	504 ± 4	570	Spiral	Sunflower
M64 / NGC4826	12	57	+21	41	408 ± 4	400	Spiral/Seyfert	Black Eye
M65 / NGC3623	11	19	+13	06	807 ± 3	723	Spiral	
M66 / NGC3627	11	20	+12	59	727 ± 3	643	Spiral/Seyfert	
M74 / NGC628	01	37	+15	47	657 ± 1	754	Spiral	
M77 / NGC1068	02	43	−00	01	1,137 ± 3	1,146	Spiral/Seyfert	
M81 / NGC3031	09	56	+69	04	−34 ± 4	73	Spiral/Seyfert	
M82 / NGC3034	09	56	+69	41	203 ± 4	312	Irregular/Starburst	
M83 / NGC5236	13	37	−29	52	516 ± 4	385	Spiral/Starburst	
M84 / NGC4374	12	25	+12	53	1,060 ± 6	1,005	Elliptical	
M85 / NGC4382	12	25	+18	11	729 ± 2	692	Spiral	
M86 / NGC4406	12	26	+12	57	−244 ± 5	−298	Elliptical/Lenticular	
M87 / NGC4486	12	30	+12	23	1,307 ± 7	1,254	Elliptical/Central Dominant/Seyfert	Virgo A
M88 / NGC4501	12	32	+14	25	2,281 ± 3	2,235	Spiral/Seyfert	
M89 / NGC4552	12	36	+12	33	340 ± 4	290	Elliptical	
M90 / NGC4569	12	37	+13	10	−235 ± 4	−282	Spiral/Seyfert	
M91 / NGC4548	12	35	+14	30	486 ± 4	442	Spiral/Seyfert	
M94 / NGC4736	12	51	+41	07	308 ± 1	360	Spiral	
M95 / NGC3351	10	44	+11	42	778 ± 4	677	Spiral/Starburst	
M96 / NGC3368	10	47	+11	49	897 ± 4	797	Spiral/Seyfert	
M98 / NGC4192	12	14	+14	54	−142 ± 4	−195	Spiral/Seyfert	
M99 / NGC4254	12	19	+14	25	2,407 ± 3	2,354	Spiral	
M100 / NGC4321	12	23	+15	49	1,571 ± 1	1,525	Spiral	
M101 / NGC5457	14	03	+54	21	241 ± 2	360	Spiral	
M104 / NGC4594	12	40	−11	37	1,024 ± 5	904	Spiral/Seyfert	Sombrero
M105 / NGC3379	10	48	+12	35	911 ± 2	814	Elliptical	
M106 / NGC4258	12	19	+47	18	448 ± 3	507	Spiral/Seyfert	
M108 / NGC3556	11	09	+55	57	695 ± 3	765	Spiral	
M109 / NGC3992	11	55	+53	39	1,048 ± 4	1,121	Spiral	
M110 / NGC205	00	38	+41	25	−241 ± 3	−61	Elliptical	

[a] Galaxies identified in the catalog published by Charles Messier in 1781; these galaxies are relatively easy to observe with small telescopes.

[b] Data obtained from NED: NASA/IPAC Extragalactic Database (http://ned.ipac.caltech.edu). The original Messier list of galaxies was obtained from SED, and the list data were updated to 2001 and M 102 was dropped.

[c] The galaxies are identified by the Messier number (M followed by a number) and by their NGC numbers, which come from the *New General Catalog* published in 1888.

[d] Radial velocity in km/s, with respect to the Sun (heliocentric). Positive values mean motion away from the Sun, and negative values are toward the Sun.

[e] Radial velocity in km/s, with respect to the Milky Way Galaxy, calculated from the RV_{hel} values with a correction for the Sun's motion around the galactic center.

[f] Galaxies are first listed by their primary type (spiral, elliptical, or irregular) and then by any other special categories that apply (see Chapter 15).

Table G.3 Nearby, X-ray Bright Clusters of Galaxies

Cluster Name	Redshift	Distance[a] (billions of ly)	Temperature of Intracluster Medium (millions of K)	Average Orbital Velocity of Galaxies[b] (km/sec)	Cluster Mass[c] (10^{15} M$_{Sun}$)
Abell 2142	0.0907	1.26	101. ± 2	1,132 ± 110	1.6
Abell 2029	0.0766	1.07	100. ± 3	1,164 ± 98	1.5
Abell 401	0.0737	1.03	95.2 ± 5	1,152 ± 86	1.4
Coma	0.0233	0.32	95.1 ± 1	821 ± 49	1.4
Abell 754	0.0539	0.75	93.3 ± 3	662 ± 77	1.4
Abell 2256	0.0589	0.82	87.0 ± 2	1,348 ± 86	1.4
Abell 399	0.0718	1.00	81.7 ± 7	1,116 ± 89	1.1
Abell 3571	0.0395	0.55	81.1 ± 3	1,045 ± 109	1.1
Abell 478	0.0882	1.23	78.9 ± 2	904 ± 281	1.1
Abell 3667	0.0566	0.79	78.5 ± 6	971 ± 62	1.1
Abell 3266	0.0599	0.84	78.2 ± 5	1,107 ± 82	1.1
Abell 1651a	0.0846	1.18	73.1 ± 6	685 ± 129	0.96
Abell 85	0.0560	0.78	70.9 ± 2	969 ± 95	0.92
Abell 119	0.0438	0.61	65.6 ± 5	679 ± 106	0.81
Abell 3558	0.0480	0.67	65.3 ± 2	977 ± 39	0.81
Abell 1795	0.0632	0.88	62.9 ± 2	834 ± 85	0.77
Abell 2199	0.0314	0.44	52.7 ± 1	801 ± 92	0.59
Abell 2147	0.0353	0.49	51.1 ± 4	821 ± 68	0.56
Abell 3562	0.0478	0.67	45.7 ± 8	736 ± 49	0.48
Abell 496	0.0325	0.45	45.3 ± 1	687 ± 89	0.47
Centaurus	0.0103	0.14	42.2 ± 1	863 ± 34	0.42
Abell 1367	0.0213	0.30	41.3 ± 2	822 ± 69	0.41
Hydra	0.0126	0.18	38.0 ± 1	610 ± 52	0.36
C0336	0.0349	0.49	37.4 ± 1	650 ± 170	0.35
Virgo	0.0038	0.05	25.7 ± 0.5	632 ± 41	0.20

Note: This table lists the 25 brightest clusters of galaxies in the X-ray sky from a catalog by J. P. Henry (2000).

[a] Cluster distances were computed using a value for Hubble's constant of 21.5 km/sec/million light-years.

[b] The average orbital velocities given in this column are the velocity component along our line of sight. This velocity should be multiplied by the square root of 2 to get the average orbital velocity used in Mathematical Insights 22.2 and 22.3.

[c] This column gives each cluster's mass within the largest radius at which the intracluster medium can be in gravitational equilibrium. Because our estimates of that radius depend on Hubble's constant, these masses are inversely proportional to Hubble's constant, which we have assumed to be 21.5 km/sec/million light-years.

H Selected Astronomical Web Sites

The Web contains a vast amount of astronomical information. For all your astronomical Web surfing, the best starting point is the Web site for this textbook:

MasteringAstronomy™
www.masteringastronomy.com

The following are some other sites that may be of particular use. In case any of the links change, you can always find live links to these sites, and many more, on the MasteringAstronomy Web site.

Key Mission Sites

The following table lists the Web pages for major current astronomy missions.

Site	Description	Web Address
NASA's Office of Space Science Missions Page	**Direct links to all past, present, and planned NASA space science missions**	**http://spacescience.nasa.gov/ missions**
Cassini/Huygens	Mission arrived at Saturn in 2004	http://saturn.jpl.nasa.gov/index.cfm
Chandra X-Ray Observatory	Latest discoveries, educational activities, and other information from the Chandra X-Ray Observatory	http://chandra.harvard.edu
Far Ultraviolet Spectroscopic Explorer (FUSE)	Ultraviolet observatory in space	http://fuse.pha.jhu.edu
Galileo	Mission orbiting Jupiter	http://www.jpl.nasa.gov/galileo
Hubble Space Telescope	Latest discoveries, educational activities, and other information from the Hubble Space Telescope	http://hubble.stsci.edu
Mars Exploration Program	Information on current and planned Mars missions	http://mars.jpl.nasa.gov
Microwave Anisotropy Probe (MAP)	Mission to study the cosmic microwave background	http://map.gsfc.nasa.gov
Spitzer Space Telescope	Infrared observatory launched in 2003	http://spitzer.caltech.edu
Stratospheric Observatory for Infrared Astronomy (SOFIA)	Airborne observatory scheduled to begin flights in 2002	http://sofia.arc.nasa.gov

Key Observatory Sites

The following table lists the Web pages leading to major ground-based observatories.

Site	Description	Web Address
World's Largest Optical Telescopes	**Direct links to most of the world's major optical observatories**	**http://www.seds.org/billa/ bigeyes.html**
Arecibo Observatory (Puerto Rico)	World's largest single-dish radio telescope	http://www.naic.edu
Cerro Tololo Inter-American Observatory	Links to major observatories on site in Cerro Tololo, Chile	http://www.ctio.noao.edu
European Southern Observatory	Links to European telescope projects in Chile, including the Very Large Telescope	http://www.eso.org
Mauna Kea Observatories	Links to major observatories in Hawaii, including Keck, Gemini, Subaru, CFHT, and others	http://www.ifa.hawaii.edu/mko
Mt. Palomar Observatory	Powerful telescope near San Diego	http://www.astro.caltech.edu/ palomarpublic
National Optical Astronomy Observatory	Home page for United States national observatories in Arizona, Hawaii, and Chile	http://www.noao.edu
National Radio Astronomy Observatory	Home page for United States national radio observatories, including the Very Large Array (VLA)	http://www.nrao.edu

More Astronomical Web Sites

The following Web sites are some of the authors' favorites among many other non-commercial resources for astronomy.

Site	Description	Web Address
MasteringAstronomy	**Don't forget to start here for all your astronomical Web surfing.**	**http://www.masteringastronomy.com**
American Association of Variable Star Observers (AAVSO)	One of the largest organizations of amateur astronomers in the world. Check this site if you are interested in serious amateur astronomy.	http://www.aavso.org
Astronomical Society of the Pacific	An organization for both professional astronomers and the general public, devoted largely to astronomy education.	http://www.astrosociety.org
Astronomy Picture of the Day	An archive of beautiful pictures, updated daily.	http://antwrp.gsfc.nasa.gov/apod
AstroWeb	Listing of major resources for astronomy on the Web.	http://www.stsci.edu/astroweb/astronomy.html
Canadian Space Agency	Home page for Canada's space program.	http://www.space.gc.ca
European Space Agency (ESA)	Home page for this international agency.	http://www.esa.int
The Extrasolar Planets Encyclopedia	Information about the search for and discoveries of extrasolar planets.	http://cfa-www.harvard.edu/planets
NASA Home Page	Learn almost anything you want about NASA.	http://www.nasa.gov
NASA Science News	Read the latest news from NASA; has option to subscribe to e-mail notices of news releases.	http://science.nasa.gov
The Nine Planets (University of Arizona)	A multimedia tour of the solar system.	http://www.nineplanets.org
The Planetary Society	Has more than 100,000 members who are interested in planetary exploration and the search for life in the universe.	http://planetary.org
The SETI Institute	Devoted to the search for other civilizations.	http://www.seti.org
Voyage Scale Model Solar System	Take a virtual tour of the Voyage Scale Model Solar System.	http://www.voyageonline.org

I The 88 Constellations

Constellation Names (English Equivalent in Parentheses)

Andromeda (The Chained Princess)
Antlia (The Air Pump)
Apus (The Bird of Paradise)
Aquarius (The Water Bearer)
Aquila (The Eagle)
Ara (The Altar)
Aries (The Ram)
Auriga (The Charioteer)
Boötes (The Herdsman)
Caelum (The Chisel)

Camelopardalis (The Giraffe)
Cancer (The Crab)
Canes Venatici (The Hunting Dogs)
Canis Major (The Great Dog)
Canis Minor (The Little Dog)
Capricornus (The Sea Goat)
Carina (The Keel)
Cassiopeia (The Queen)
Centaurus (The Centaur)
Cepheus (The King)

Cetus (The Whale)
Chamaeleon (The Chameleon)
Circinus (The Drawing Compass)
Columba (The Dove)
Coma Berenices (Berenice's Hair)
Corona Australis (The Southern Crown)
Corona Borealis (The Northern Crown)
Corvus (The Crow)
Crater (The Cup)
Crux (The Southern Cross)

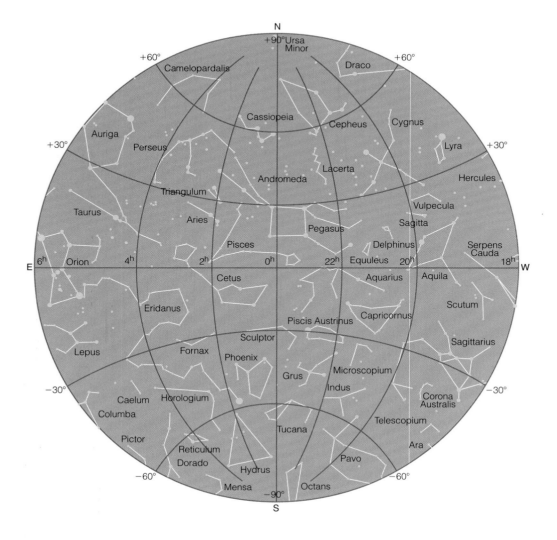

Cygnus (The Swan)
Delphinus (The Dolphin)
Dorado (The Goldfish)
Draco (The Dragon)
Equuleus (The Little Horse)
Eridanus (The River)
Fornax (The Furnace)
Gemini (The Twins)
Grus (The Crane)
Hercules
Horologium (The Clock)
Hydra (The Sea Serpent)
Hydrus (The Water Snake)
Indus (The Indian)
Lacerta (The Lizard)
Leo (The Lion)
Leo Minor (The Little Lion)
Lepus (The Hare)
Libra (The Scales)
Lupus (The Wolf)
Lynx (The Lynx)
Lyra (The Lyre)
Mensa (The Table)
Microscopium (The Microscope)

Monoceros (The Unicorn)
Musca (The Fly)
Norma (The Level)
Octans (The Octant)
Ophiuchus (The Serpent Bearer)
Orion (The Hunter)
Pavo (The Peacock)
Pegasus (The Winged Horse)
Perseus (The Hero)
Phoenix (The Phoenix)
Pictor (The Painter's Easel)
Pisces (The Fish)
Piscis Austrinus (The Southern Fish)
Puppis (The Stern)
Pyxis (The Compass)
Reticulum (The Reticle)
Sagitta (The Arrow)
Sagittarius (The Archer)
Scorpius (The Scorpion)
Sculptor (The Sculptor)
Scutum (The Shield)
Serpens (The Serpent)
Sextans (The Sextant)
Taurus (The Bull)

Telescopium (The Telescope)
Triangulum (The Triangle)
Triangulum Australe (Southern Triangle)
Tucana (The Toucan)
Ursa Major (The Great Bear)
Ursa Minor (The Little Bear)
Vela (The Sail)
Virgo (The Virgin)
Volans (The Flying Fish)
Vulpecula (The Fox)

Constellation Locations

Each of the charts on these pages shows half of the celestial sphere in projection, so you can use them to learn the approximate locations of the constellations. The grid lines are marked by right ascension and declination.

J Star Charts

How to use the star charts:

Check the times and dates under each chart to find the best one for you. Take it outdoors within an hour or so of the time listed for your date. Bring a dim flashlight to help you read it.

On each chart, the round outside edge represents the horizon all around you. Compass directions around the horizon are marked in yellow. Turn the chart around so the edge marked with the direction you're facing (for example, north, southeast) is down. The stars above this horizon now match the stars you are facing. Ignore the rest until you turn to look in a different direction.

The center of the chart represents the sky overhead, so a star plotted on the chart halfway from the edge to the center can be found in the sky halfway from the horizon to straight up.

The charts are drawn for 40°N latitude (for example, Denver, New York, Madrid). If you live far south of there, stars in the southern part of your sky will appear higher than on the chart and stars in the north will be lower. If you live far north of there, the reverse is true.

Jan–March

© Sky Publishing Corp.

© 1999 *Sky & Telescope*

Use this chart January, February, and March.

Early January — 1 A.M.　　　　Early February — 11 P.M.　　　　Early March — 9 P.M.
Late January — Midnight　　　Late February — 10 P.M.　　　　Late March — Dusk

© 1999 *Sky & Telescope*

Use this chart April, May, and June.

Early April — 3 A.M.*   Early May — 1 A.M.*   Early June — 11 P.M.*
Late April — 2 A.M.*   Late May — Midnight*   Late June — Dusk

*Daylight Saving Time

July–Sept.
© Sky Publishing Corp.

Use this chart July, August, and September.

Early July — 1 A.M.*   Early August — 11 P.M.*   Early September — 9 P.M.*
Late July — Midnight*   Late August — 10 P.M.*   Late September — Dusk

*Daylight Saving Time

Use this chart October, November, and December.

Early October — 1 A.M.*
Late October — Midnight*

Early November — 10 P.M.
Late November — 9 P.M.

Early December — 8 P.M.
Late December — 7 P.M.

*Daylight Saving Time

Glossary

21-cm line A spectral line from atomic hydrogen with wavelength 21 cm (in the radio portion of the spectrum).

absolute magnitude A measure of an object's luminosity; defined to be the apparent magnitude the object would have if it were located exactly 10 parsecs away.

absolute zero The coldest possible temperature, which is 0 K.

absorption (of light) The process by which matter absorbs radiative energy.

absorption-line spectrum A spectrum that contains absorption lines.

accelerating universe The possible fate of our universe in which a repulsive force (*see* cosmological constant) causes the expansion of the universe to accelerate with time. Its galaxies will recede from one another increasingly faster, and it will become cold and dark more quickly than a coasting universe.

acceleration The rate at which an object's velocity changes. Its standard units are m/s^2.

acceleration of gravity The acceleration of a falling object. On Earth, the acceleration of gravity, designated by *g*, is $9.8 \ m/s^2$.

accretion The process by which small objects gather together to make larger objects.

accretion disk A rapidly rotating disk of material that gradually falls inward as it orbits a starlike object (e.g., white dwarf, neutron star, or black hole).

active galactic nuclei The unusually luminous centers of some galaxies, thought to be powered by accretion onto supermassive black holes. Quasars are the brightest type of active galactic nuclei; radio galaxies also contain active galactic nuclei.

active galaxy A term sometimes used to describe a galaxy that contains an *active galactic nucleus*.

adaptive optics A technique in which telescope mirrors flex rapidly to compensate for the bending of starlight caused by atmospheric turbulence.

albedo Describes the fraction of sunlight reflected by a surface; albedo = 0 means no reflection at all (a perfectly black surface); albedo = 1 means all light is reflected (a perfectly white surface).

Algol Paradox A paradox concerning the binary star Algol, which contains a sub-giant star that is less massive than its main-sequence companion.

altitude (above horizon) The angular distance between the horizon and an object in the sky.

amino acids The building blocks of proteins.

analemma The figure-8 path traced by the Sun over the course of a year when viewed at the same place and the same time each day; represents the discrepancies between apparent and mean solar time.

Andromeda Galaxy (M 31; the Great Galaxy in Andromeda) The nearest large spiral galaxy to the Milky Way.

angular momentum Momentum attributable to rotation or revolution. The angular momentum of an object moving in a circle of radius *r* is the product $m \times v \times r$.

angular resolution (of a telescope) The smallest angular separation that two pointlike objects can have and still be seen as distinct points of light (rather than as a single point of light).

angular size (or **angular distance**) A measure of the angle formed by extending imaginary lines outward from our eyes to span an object (or between two objects).

annihilation *See* matter–antimatter annihilation

annular solar eclipse A solar eclipse during which the Moon is directly in front of the Sun but its angular size is not large enough to fully block the Sun; thus, a ring (or *annulus*) of sunlight is still visible around the Moon's disk.

Antarctic Circle The circle on the Earth with latitude 66.5°S.

antielectron *See* positron

antimatter Refers to any particle with the same mass as a particle of ordinary matter but whose other basic properties, such as electrical charge, are precisely opposite.

aphelion The point at which an object orbiting the Sun is farthest from the Sun.

apogee The point at which an object orbiting the Earth is farthest from the Earth.

apparent brightness The amount of light reaching us *per unit area* from a luminous object; often measured in units of watts/m^2.

apparent magnitude A measure of the apparent brightness of an object in the sky, based on the ancient system developed by Hipparchus.

apparent retrograde motion Refers to the apparent motion of a planet, as viewed from Earth, during the period of a few weeks or months when it moves westward relative to the stars in our sky.

apparent solar time Time measured by the actual position of the Sun in your local sky; defined so that noon is when the Sun is *on* the meridian.

arcminutes (or **minutes of arc**) One arcminute is 1/60 of 1°.

arcseconds (or **seconds of arc**) One arcsecond is 1/60 of an arcminute, or 1/3,600 of 1°.

Arctic Circle The circle on the Earth with latitude 66.5°N.

asteroid A relatively small and rocky object that orbits a star; asteroids are sometimes called *minor planets* because they are similar to planets but smaller.

asteroid belt The region of our solar system between the orbits of Mars and Jupiter in which asteroids are heavily concentrated.

astrobiology The study of life on Earth and beyond; emphasizes research into questions of the origin of life, the conditions under which life can survive, and the search for life beyond Earth.

astrometric technique The detection of extrasolar planets through the side-to-side motion of a star caused by gravitational tugs from the planet.

astronomical unit (AU) The average distance (semimajor axis) of the Earth from the Sun, which is about 150 million km.

atmosphere A layer of gas that surrounds a planet or moon, usually very thin compared to the size of the object.

atmospheric pressure The surface pressure resulting from the overlying weight of an atmosphere.

atmospheric structure The layering of a planetary atmosphere due to variations in temperature with altitude. For example, Earth's atmospheric structure from the ground up consists of the troposphere, stratosphere, thermosphere, and exosphere.

atomic mass number The combined number of protons and neutrons in an atom.

atomic number The number of protons in an atom.

atoms Consist of a nucleus made from protons and neutrons surrounded by a cloud of electrons.

aurora Dancing lights in the sky caused by charged particles entering our atmosphere; called the *aurora borealis* in the Northern Hemisphere and the *aurora australis* in the Southern Hemisphere.

autumnal equinox *See* fall equinox

axis tilt (of a planet in our solar system) The amount by which a planet's axis is tilted with respect to a line perpendicular to the ecliptic plane.

azimuth (usually called *direction* in this book) Direction around the horizon from due north, measured clockwise in degrees. E.g., the azimuth of due north is 0°, due east is 90°, due south is 180°, and due west is 270°.

bar The standard unit of pressure, approximately equal to the Earth's atmospheric pressure at sea level.

baryonic matter Refers to ordinary matter made from atoms (because the nuclei of atoms contain protons and neutrons, which are both baryons).

baryons Particles, including protons and neutrons, that are made from three quarks.

basalt A type of volcanic rock that makes a low-viscosity lava when molten.

belts (on a jovian planet) Dark bands of sinking air that encircle a jovian planet at a particular set of latitudes.

Big Bang The event that gave birth to the universe.

Big-Bang theory The scientific theory of the universe's earliest moments, stating that all the matter in our observable universe came into being at a single moment in time as an extremely hot, dense mixture of subatomic particles and radiation.

Big Crunch If gravity ever reverses the universal expansion, the universe will someday begin to collapse and presumably end in a Big Crunch.

binary star system A star system that contains two stars.

biosphere Refers to the "layer" of life on Earth.

BL Lac objects The name given to a class of active galactic nuclei that probably represent the centers of radio galaxies whose jets happen to be pointed directly at us.

blackbody radiation *See* thermal radiation

black hole A bottomless pit in spacetime. Nothing can escape from within a black hole, and we can never again detect or observe an object that falls into a black hole.

black smokers Structures around seafloor volcanic vents that support a wide variety of life.

blowout Ejection of the hot, gaseous contents of a superbubble when it grows so large that it bursts out of the cooler layer of gas filling the galaxy's disk.

blueshift A Doppler shift in which spectral features are shifted to shorter wavelengths, caused when an object is moving toward the observer.

bosons Particles, such as photons, to which the exclusion principle does not apply.

bound orbits Orbits on which an object travels repeatedly around another object; bound orbits are elliptical in shape.

brown dwarf An object too small to become an ordinary star because electron degeneracy pressure halts its gravitational collapse before fusion becomes self-sustaining; brown dwarfs have mass less than $0.08 M_{Sun}$.

bubble (interstellar) The surface of a bubble is an expanding shell of hot, ionized gas driven by stellar winds or supernovae; inside the bubble, the gas is very hot and has very low density.

bulge (of a spiral galaxy) The central portion of a spiral galaxy that is roughly spherical (or football shaped) and bulges above and below the plane of the galactic disk.

Cambrian explosion The dramatic diversification of life on Earth that occurred between about 540 and 500 million years ago.

carbon stars Stars whose atmospheres are especially carbon-rich, thought to be near the ends of their lives; carbon stars are the primary sources of carbon in the universe.

carbonate rock A carbon-rich rock, such as limestone, that forms underwater from chemical reactions between sediments and carbon dioxide. On Earth, most of the outgassed carbon dioxide currently resides in carbonate rocks.

carbon dioxide cycle The process that cycles carbon dioxide between the Earth's atmosphere and surface rocks.

Cassini division A large, dark gap in Saturn's rings, visible through small telescopes on Earth.

CCD (charge coupled device) A type of electronic light detector that has largely replaced photographic film in astronomical research.

celestial coordinates The coordinates of right ascension and declination that fix an object's position on the celestial sphere.

celestial equator (CE) The extension of the Earth's equator onto the celestial sphere.

celestial navigation Navigation on the surface of the Earth accomplished by observations of the Sun and stars.

celestial sphere The imaginary sphere on which objects in the sky appear to reside when observed from Earth.

Celsius (temperature scale) The temperature scale commonly used in daily activity internationally. Defined so that, on Earth's surface, water freezes at 0°C and boils at 100°C.

center of mass (or orbiting objects) The point at which two or more orbiting objects would balance if they were somehow connected; it is the point around which the orbiting objects actually orbit.

central dominant galaxy A giant elliptical galaxy found at the center of a dense cluster of galaxies, apparently formed by the merger of several individual galaxies.

Cepheid *See* Cepheid variable

Cepheid variable A particularly luminous type of pulsating variable star that follows a period–luminosity relation and hence is very useful for measuring cosmic distances.

Chandrasekhar limit *See* white dwarf limit

charged particle belts Zones in which ions and electrons accumulate and encircle a planet.

chemical enrichment The process by which the abundance of heavy elements (heavier than helium) in the interstellar medium gradually increases over time as these elements are produced by stars and released into space.

chondrites Another name for primitive meteorites, named for the round chodrules within them. Achondrites, meaning "without chondrules", is another name for processed meteorites.

chromosphere The layer of the Sun's atmosphere below the corona; most of the Sun's ultraviolet light is emitted from this region, in which the temperature is about 10,000 K.

circulation cells (also called *Hadley cells*) Large-scale cells (similar to convection cells) in a planet's atmosphere that transport heat between the equator and the poles.

circumpolar star A star that always remains above the horizon for a particular latitude.

climate Describes the long-term average of weather.

close binary A binary star system in which the two stars are very close together.

closed universe The universe is closed if its average density is greater than the critical density, in which case spacetime must curve back on itself to the point where its overall shape is analogous to that of the surface of a sphere. In the absence of a repulsive force (*see* cosmological constant), a closed universe would someday stop expanding and begin to contract.

cluster of galaxies A collection of a few dozen or more galaxies bound together by gravity; smaller collections of galaxies are simply called *groups*.

cluster of stars A group of anywhere from several hundred to a million or so stars; star clusters come in two types—open clusters and globular clusters.

CNO cycle The cycle of reactions by which intermediate- and high-mass stars fuse hydrogen into helium.

coasting universe The possible fate of our universe in which the mass density of the universe is *smaller* than the critical density, so that the collective gravity of all matter cannot halt the expansion. In the absence of a repulsive force (*see* cosmological constant), such a universe would keep expanding forever with little change in its rate of expansion.

coma (of a comet) The dusty atmosphere of a comet created by sublimation of ices in the nucleus when the comet is near the Sun.

comet A relatively small, icy object that orbits a star.

comparative planetology The study of the solar system by examining and understanding the similarities and differences among worlds.

compound (chemical) A substance made from molecules consisting of two or more atoms with different atomic numbers.

condensates Solid or liquid particles that condense from a cloud of gas.

condensation The formation of solid or liquid particles from a cloud of gas.

conduction (of energy) The process by which thermal energy is transferred by direct contact from warm material to cooler material.

conjunction (of a planet with the Sun) When a planet and the Sun line up in the sky.

conservation of angular momentum (law of) The principle that, in the absence of net torque (twisting force), the total angular momentum of a system remains constant.

conservation of energy (law of) The principle that energy (including mass-energy) can be neither created nor destroyed, but can only change from one form to another.

conservation of momentum (law of) The principle that, in the absence of net force, the total momentum of a system remains constant.

constellation A region of the sky; 88 official constellations cover the celestial sphere.

continental crust On Earth, the thick, low-density crust of rocks such as granite that is created near subduction zones.

convection The energy transport process in which warm material expands and rises, while cooler material contracts and falls.

convection cell An individual small region of convecting material.

convection zone (of a star) A region in which energy is transported outward by convection.

Copernican revolution The dramatic change, initiated by Copernicus, that occurred when we learned that Earth is a planet orbiting the Sun rather than the center of the universe.

core (of a planet) The dense central region of a planet that has undergone differentiation.

core (of a star) The central region of a star, in which nuclear fusion can occur.

Coriolis effect Causes air or objects moving on a rotating planet to deviate from straight-line trajectories.

corona (solar) The tenuous uppermost layer of the Sun's atmosphere; most of the Sun's X rays are emitted from this region, in which the temperature is about 1 million K.

coronal holes Regions of the corona that barely show up in X-ray images because they are nearly devoid of hot coronal gas.

coronal mass ejections Bursts of charged particles from the Sun's corona that travel outward into space.

cosmic microwave background The remnant radiation from the Big Bang, which we detect using radio telescopes sensitive to microwaves (which are short-wavelength radio waves).

cosmic rays Particles such as electrons, protons, and atomic nuclei that zip through interstellar space at close to the speed of light.

cosmological constant The name given to a term in Einstein's equations of general relativity. If it is not zero, then it represents a repulsive force or a type of energy (sometimes called *dark energy* or *quintessence*) that might cause the expansion of the universe to accelerate with time.

cosmological horizon The boundary of our observable universe, which is where the look-back time is equal to the age of the universe. Beyond this boundary in spacetime, we cannot see anything at all.

Cosmological Principle The idea that matter is distributed uniformly throughout the universe on very large scales, meaning that the universe has neither a center nor an edge.

cosmological redshift Refers to the redshifts we see from distant galaxies, caused by the fact that expansion of the universe stretches all the photons within it to longer, redder wavelengths.

cosmology The study of the overall structure and evolution of the universe.

cosmos An alternate name for the *universe*.

critical density The precise average density for the entire universe that marks the dividing line between a recollapsing universe and one that will expand forever.

critical universe The possible fate of our universe in which the mass density of the universe *equals* the critical density. The universe will never collapse, but in the absence of a repulsive force it will expand more and more slowly as time progresses. *See* cosmological constant

crust (of a planet) The low-density surface layer of a planet that has undergone differentiation.

curvature of spacetime A change in the geometry of space produced in the vicinity of a massive object and that is responsible for the force we call gravity. The overall geometry of the universe may also be curved, depending on its overall mass-energy content.

cycles per second Units of frequency for a wave; describes the number of peaks (or troughs) of a wave that pass by a given point each second. Equivalent to *hertz*.

dark energy Name sometimes given to energy that could be causing the expansion of the universe to accelerate. *See* cosmological constant.

dark matter Matter that we infer to exist from its gravitational effects but from which we have not detected any light; dark matter apparently dominates the total mass of the universe.

daylight saving time Standard time plus 1 hour, so that the Sun appears on the meridian around 1 P.M. rather than around noon.

declination (dec) Analogous to latitude, but on the celestial sphere; it is the angular north-south distance between the celestial equator and a location on the celestial sphere.

degeneracy pressure A type of pressure unrelated to an object's temperature, which arises when electrons (electron degeneracy pressure) or neutrons (neutron degeneracy pressure) are packed so tightly that the exclusion and uncertainty principles come into play.

degenerate object An object in which degeneracy pressure is the primary pressure pushing back against gravity, such as a brown dwarf, white dwarf, or neutron star.

deuterium A form of hydrogen in which the nucleus contains a proton and a neutron, rather than only a proton (as is the case for most hydrogen nuclei).

differential rotation Describes the rotation of an object in which the equator rotates at a different rate than the poles.

differentiation The process in which gravity separates materials according to density, with high-density materials sinking and low-density materials rising.

diffraction grating A finely etched surface that can split light into a spectrum.

diffraction limit The angular resolution that a telescope could achieve if it were limited only by the interference of light waves; it is smaller (i.e., better angular resolution) for larger telescopes.

dimension (mathematical) Describes the number of independent directions in which movement is possible; e.g., the surface of the Earth is two-dimensional because only two independent directions of motion are possible (north-south and east-west).

direction (in local sky) One of the two coordinates (the other is altitude) needed to pinpoint an object in the local sky. It is the direction, such as north, south, east, or west, in which you must face to see the object. *See also* azimuth

disk component (of a galaxy) The portion of a spiral galaxy that looks like a disk and contains an interstellar medium with cool gas and dust; stars of many ages are found in the disk component.

disk population Refers to stars that orbit within the disk of a spiral galaxy. Sometimes called Population I.

DNA (deoxyribonucleic acid) The molecule that represents the genetic material of life on Earth.

Doppler effect (shift) The effect that shifts the wavelengths of spectral features in objects that are moving toward or away from the observer.

Doppler technique the detection of extra-solar planets through the motion of a star towards and away from the observer caused by gravitational tugs from the planet.

double-shell burning star A star that is fusing helium into carbon in a shell around an inert carbon core and is fusing hydrogen into helium in a shell at the top of the helium layer.

down quark One of the two quark types (the other is the up quark) found in ordinary protons and neutrons. Has a charge of $-\frac{1}{3}$.

Drake equation An equation that lays out the factors that play a role in determining the number of communicating civilizations in our galaxy.

dust (or **dust grains**) Tiny solid flecks of material; in astronomy, we often discuss interplanetary dust (found within a star system) or interstellar dust (found between the stars in a galaxy). *See also* interstellar dust grains

dust tail (of a comet) One of two tails seen when a comet passes near the Sun (the other is the plasma tail); composed of small solid particles pushed away from the Sun by the radiation pressure of sunlight.

dwarf elliptical galaxy A small elliptical galaxy with less than about a billion stars.

dwarf galaxies Relatively small galaxies, consisting of less than about ten billion stars.

Earth-orbiters (spacecraft) Spacecraft designed to study the Earth or the universe from Earth orbit.

eccentricity A measure of how much an ellipse deviates from a perfect circle; defined as the center-to-focus distance divided by the length of the semimajor axis.

eclipse Occurs when one astronomical object casts a shadow on another or crosses our line of sight to the other object.

eclipse seasons Periods during which lunar and solar eclipses can occur because the nodes of the Moon's orbit are nearly aligned with the Earth and Sun.

eclipsing binary A binary star system in which the two stars happen to be orbiting in the plane of our line of sight, so that each star will periodically eclipse the other.

ecliptic The Sun's apparent annual path among the constellations.

ecliptic plane The plane of the Earth's orbit around the Sun.

ejecta (from an impact) Debris ejected by the blast of an impact.

electrical charge A measure of how a particle responds to the electromagnetic force.

electromagnetic field An abstract concept used to describe how a charged particle would affect other charged particles at a distance.

electromagnetic force One of the four fundamental forces; it is the force that dominates atomic and molecular interactions.

electromagnetic radiation Another name for light of all types, from radio waves through gamma rays.

electromagnetic spectrum The complete spectrum of light, including radio waves, infrared, visible light, ultraviolet light, X rays, and gamma rays.

electromagnetic wave A synonym for light, which consists of waves of electric and magnetic fields.

electron degeneracy pressure Degeneracy pressure exerted by electrons, as in brown dwarfs and white dwarfs.

electrons Fundamental particles with negative electric charge; the distribution of electrons in an atom gives the atom its size.

electron-volt (eV) A unit of energy equivalent to 1.60×10^{-19} joule.

electroweak era The era of the universe during which only three forces operated (gravity, strong force, and electroweak force), lasting from 10^{-38} second to 10^{-10} second after the Big Bang.

electroweak force The force that exists at high energies when the electromagnetic force and the weak force exist as a single force.

element (chemical) A substance made from individual atoms of a particular atomic number.

ellipse A type of oval that happens to be the shape of bound orbits. An ellipse can be drawn by moving a pencil along a string whose ends are tied to two tacks; the locations of the tacks are the foci (singular, focus) of the ellipse.

elliptical galaxies Galaxies that appear rounded in shape, often longer in one direction, like a football. They have no disks and contain very little cool gas and dust compared to spiral galaxies, though they often contain very hot, ionized gas.

elongation (greatest) For Mercury or Venus, the point at which it appears farthest from the Sun in our sky.

emission (of light) The process by which matter emits energy in the form of light.

emission-line spectrum A spectrum that contains emission lines.

emission nebula Another name for an ionization nebula. *See also* ionization nebula

energy Broadly speaking, energy is what can make matter move. The three basic types of energy are kinetic, potential, and radiative.

equation of time Describes the discrepancies between apparent and mean solar time.

equinox See *fall equinox* and *spring equinox*.

equivalence principle The fundamental starting point for general relativity, which states that the effects of gravity are exactly equivalent to the effects of acceleration.

era of atoms The era of the universe lasting from about 500,000 years to about 1 billion years after the Big Bang, during which it was cool enough for neutral atoms to form.

era of galaxies The present era of the universe, which began with the formation of galaxies when the universe was about 1 billion years old.

era of nuclei The era of the universe lasting from about 3 minutes to about 500,000 years after the Big Bang, during which matter in the universe was fully ionized and opaque to light. The cosmic background radiation was released at the end of this era.

era of nucleosynthesis The era of the universe lasting from about 0.001 second to about 3 minutes after the Big Bang, by the end of which virtually all of the neutrons and about one-seventh of the protons in the universe had fused into helium.

erosion The wearing down or building up of geological features by wind, water, ice, and other phenomena of planetary weather.

eruption The process of releasing hot lava on the planet's surface.

escape velocity The speed necessary for an object to completely escape the gravity of a large body such as a moon, planet, or star.

evaporation The process by which atoms or molecules escape into the gas phase from a liquid.

event Any particular point along a worldline represents a particular event; all observers will agree on the reality of an event but may disagree about its time and location.

event horizon The boundary that marks the "point of no return" between a black hole and the outside universe; events that occur within the event horizon can have no influence on our observable universe.

evolution (biological) The gradual change in populations of living organisms responsible for transforming life on Earth from its primitive origins to the great diversity of life today.

exchange particle According to the standard model of physics, each of the four fundamental forces is transmitted by the transfer of particular types of exchange particles.

excited state (of an atom) Any arrangement of electrons in an atom that has more energy than the ground state.

exclusion principle The law of quantum mechanics that states that two fermions cannot occupy the same quantum state at the same time.

exosphere The hot, outer layer of an atmosphere, where the atmosphere "fades away" to space.

expansion (of universe) The idea that the space between galaxies or clusters of galaxies is growing with time.

exposure time The amount of time for which light is collected to make a single image.

extrasolar planet A planet orbiting a star other than our Sun.

Fahrenheit (temperature scale) The temperature scale commonly used in daily activity in the United States. Defined so that, on Earth's surface, water freezes at 32°F and boils at 212°F.

fall equinox (autumnal equinox) Refers both to the point in Virgo on the celestial sphere where the ecliptic crosses the celestial equator and to the moment in time when the Sun appears at that point each year (around September 21).

false-color image An image displayed in colors that are *not* the true, visible-light colors of an object.

fault (geological) A place where rocks slip sideways relative to one another.

feedback processes Processes in which a small change in some property (such as temperature) which lead to changes in other properties which either amplify or diminish the original small change.

fermions Particles, such as electrons, neutrons, and protons, that obey the exclusion principle.

Fermi's paradox The question posed by Enrico Fermi about extraterrestrial intelligence— "So where is everybody?"—which asks why we have not observed other civilizations even though simple arguments would suggest that some ought to have spread throughout the galaxy by now.

field An abstract concept used to describe how a particle would interact with a force. For example, the idea of a *gravitational field* describes how a particle would react to the local strength of gravity, and the idea of an *electromagnetic field* describes how a charged particle would respond to forces from other charged particles.

filter (for light) A material that transmits only particular wavelengths of light.

fireball A particularly bright meteor.

fission The process by which one atomic nucleus breaks into two smaller nuclei. It releases energy if the two smaller nuclei together are less massive than the original nucleus.

flare star A small, spectral type M star that displays particularly strong flares on its surface.

flat (or **Euclidean**) **geometry** Refers to any case in which the rules of geometry for a flat plane hold, such as that the shortest distance between two points is a straight line.

flat universe A universe in which the overall geometry of spacetime is flat (Euclidean), as would be the case if the density of the universe is equal to the critical density.

flybys (spacecraft) Spacecraft that fly past a target object (such as a planet), usually just once, as opposed to entering a bound orbit of the object.

focal plane The place where an image created by a lens or mirror is in focus.

foci Plural of *focus*.

focus (of a lens or mirror) The point at which rays of light that were initially parallel (such as light from a distant star) converge.

focus (of an ellipse) One of two special points within an ellipse that lie along the major axis; these are the points around which we could stretch a pencil and string to draw an ellipse. When one object orbits a second object, the second object lies at one focus of the orbit.

force Anything that can cause a change in momentum.

formation properties (of planets) In this book, for the purpose of understanding geological processes, planets are defined to be born with four formation properties: size (mass and radius), distance from the Sun, composition, and rotation rate.

fossil Any relic of an organism that lived and died long ago.

frame of reference (in relativity) Two (or more) objects share the same frame of reference if they are *not* moving relative to each other.

free-fall Refers to conditions in which an object is falling without resistance; objects are weightless when in free-fall.

free-float frame A frame of reference in which all objects are weightless and hence float freely.

frequency Describes the rate at which peaks of a wave pass by a point; measured in units of 1/s, often called *cycles per second* or *hertz*.

frost line The boundary in the solar nebula beyond which ices could condense; only metals and rocks could condense within the frost line.

fundamental forces There are four known fundamental forces in nature: gravity, the electromagnetic force, the strong force, and the weak force.

fundamental particles Subatomic particles that cannot be divided into anything smaller.

fusion The process by which two atomic nuclei fuse together to make a single, more massive nucleus. It releases energy if the final nucleus is less massive than the two nuclei that went into the reaction.

galactic cannibalism The term sometimes used to describe the process by which large galaxies merge with other galaxies in collisions. *Central dominant galaxies* are products of galactic cannibalism.

galactic disk (of a spiral galaxy) *See* disk component

galactic fountain Refers to a model for the cycling of gas in the Milky Way Galaxy in which fountains of hot, ionized gas rise from the disk into the halo and then cool and form clouds as they sink back into the disk.

galactic wind A wind of low-density but extremely hot gas flowing out from a starburst galaxy, created by the combined energy of many supernovae.

galaxy A huge collection of anywhere from a few hundred million to more than a trillion stars, all bound together by gravity.

galaxy cluster *See* cluster of galaxies

galaxy evolution The formation and development of galaxies.

Galilean moons The four moons of Jupiter that were discovered by Galileo: Io, Europa, Ganymede, and Callisto.

gamma-ray burst A sudden burst of gamma rays from deep space; such bursts apparently come from distant galaxies, but their precise mechanism is unknown.

gamma rays Light with very short wavelengths (and hence high frequencies)—shorter than those of X rays.

gap moons Tiny moons located within a gap in a planet's ring system. The gravity of a gap moon helps clear the gap.

gas phase The phase of matter in which atoms or molecules can move essentially independently of one another.

gas pressure Describes the force (per unit area) pushing on any object due to surrounding gas. *See also* pressure

general theory of relativity Einstein's generalization of his special theory of relativity so that the general theory also applies when we consider effects of gravity or acceleration.

genetic code The "language" that living cells use to read the instructions chemically encoded in DNA.

geocentric model Any of the ancient Greek models that were used to predict planetary positions under the assumption that Earth lay in the center of the universe.

geocentric universe (ancient belief in) The idea that the Earth is the center of the entire universe.

geological activity Processes which change a planet's surface long after formation, such as volcanism, tectonics and erosion.

geological processes The four basic geological processes are impact cratering, volcanism, tectonics, and erosion.

geological time scale The time scale used by scientists to describe major eras in Earth's past.

geology The study of surface features (on a moon, planet, or asteroid) and the processes that create them.

geostationary satellite A satellite that appears to stay stationary in the sky as viewed from Earth's surface, because it orbits in the same time it takes Earth to rotate and orbits in Earth's equatorial plane.

geosynchronous satellite A satellite that orbits Earth in the same time it takes Earth to rotate, or one sidereal day.

giant impact A collision between a forming planet and a very large planetesimal, such as is thought to have formed our Moon.

giant molecular cloud A very large cloud of cold, dense interstellar gas, typically containing up to a million solar masses worth of material. *See also* molecular clouds

giants (luminosity class III) Stars that appear just below the supergiants on the H–R diagram because they are somewhat smaller in radius and lower in luminosity.

global positioning system (GPS) A system of navigation by satellites orbiting the Earth.

global warming An expected increase in Earth's global average temperature caused by human input of carbon dioxide and other greenhouse gases into the atmosphere.

global wind patterns (or **global circulation**) Wind patterns that remain fixed on a global scale, determined by the combination of surface heating and the planet's rotation.

globular cluster A spherically shaped cluster of up to a million or more stars; globular clusters are found primarily in the halos of galaxies and contain only very old stars.

gluons The exchange particles for the strong force.

grand unified theory (GUT) A theory that unifies three of the four fundamental forces—the strong force, the weak force, and the electromagnetic force (but not gravity)—in a single model.

granulation (on the Sun) The bubbling pattern visible in the photosphere, produced by the underlying convection.

gravitation (law of) *See* universal law of gravitation

gravitational constant The experimentally measured constant G that appears in the law of universal gravitation;

$$G = 6.67 \times 10^{-11} \frac{\text{m}^3}{\text{kg} \times \text{s}^2}.$$

gravitational contraction The process in which gravity causes an object to contract, thereby converting gravitational potential energy into thermal energy.

gravitational encounter Occurs when two (or more) objects pass near enough so that each can feel the effects of the other's gravity and can therefore exchange energy.

gravitational equilibrium Describes a state of balance in which the force of gravity pulling inward is precisely counteracted by pressure pushing outward.

gravitational lensing The magnification or distortion (into arcs, rings, or multiple images) of an image caused by light bending through a gravitational field, as predicted by Einstein's general theory of relativity.

gravitational redshift A redshift caused by the fact that time runs slow in gravitational fields.

gravitational time dilation The slowing of time that occurs in a gravitational field, as predicted by Einstein's general theory of relativity.

gravitational waves Predicted by Einstein's general theory of relativity, these waves travel at the speed of light and transmit distortions of space through the universe. Although not yet observed directly, we have strong indirect evidence that they exist.

gravitationally bound system Any system of objects, such as a star system or a galaxy, that is held together by gravity.

gravitons The exchange particles for the force of gravity.

gravity One of the four fundamental forces; it is the force that dominates on large scales.

grazing incidence (in telescopes) Reflections in which light grazes a mirror surface and is deflected at a small angle; commonly used to focus high-energy ultraviolet light and X rays.

great circle A circle on the surface of a sphere whose center is at the center of the sphere.

greatest elongation *See* elongation (greatest).

Great Red Spot A large, high-pressure storm on Jupiter.

greenhouse effect The process by which greenhouse gases in an atmosphere make a planet's surface temperature warmer than it would be in the absence of an atmosphere.

greenhouse gases Gases, such as carbon dioxide, water vapor, and methane, that are particularly good absorbers of infrared light but are transparent to visible light.

Gregorian calendar Our modern calendar, introduced by Pope Gregory in 1582.

ground state (of an atom) The lowest possible energy state of the electrons in an atom.

group (of galaxies) A few to a few dozen galaxies bound together by gravity. *See also* cluster of galaxies

GUT era The era of the universe during which only two forces operated (gravity and the grand-unified-theory or GUT force), lasting from 10^{-43} second to 10^{-38} second after the Big Bang.

GUT force The proposed force that exists at very high energies when the strong force, the weak force, and the electromagnetic force (but not gravity) all act as one.

H II region Another name for an ionization nebula. *See* ionization nebula

habitable world A world with environmental conditions under which life could *potentially* arise or survive.

habitable zone The region around a star in which planets could potentially have surface temperatures at which liquid water could exist.

Hadley cells *See* circulation cells

half-life The time it takes for half of the nuclei in a given quantity of a radioactive substance to decay.

halo (of a galaxy) The spherical region surrounding the disk of a spiral galaxy.

Hawking radiation Radiation predicted to arise from the evaporation of black holes.

heavy bombardment The period in the first few hundred million years after the solar system formed during which the tail end of planetary accretion created most of the craters found on ancient planetary surfaces.

heavy elements In astronomy, *heavy elements* generally refers to all elements *except* hydrogen and helium.

helium-burning star A star that is currently fusing helium into carbon in its core.

helium-capture reactions Fusion reactions that fuse a helium nucleus into some other nucleus; such reactions can fuse carbon into oxygen, oxygen into neon, neon into magnesium, and so on.

helium flash The event that marks the sudden onset of helium fusion in the previously inert helium core of a low-mass star.

helium fusion The fusion of three helium nuclei into one carbon nucleus; also called the *triple-alpha reaction.*

hertz (Hz) The standard unit of frequency for light waves; equivalent to units of 1/s.

Hertzsprung–Russell (H–R) diagram A graph plotting individual stars as points, with stellar luminosity on the vertical axis and spectral type (or surface temperature) on the horizontal axis.

high-mass stars Stars born with masses above about $8 M_{Sun}$; these stars will end their lives by exploding as supernovae.

horizon A boundary that divides what we can see from what we cannot see.

horizontal branch The horizontal line of stars that represents helium-burning stars on an H–R diagram for a cluster of stars.

horoscope A predictive chart made by an astrologer; in scientific studies, horoscopes have never been found to have any validity as predictive tools.

hot spot (geological) A place within a plate of the lithosphere where a localized plume of hot mantle material rises.

hour angle (HA) The angle or time (measured in hours) since an object was last on the meridian in the local sky. Defined to be 0 hours for objects that *are* on the meridian.

Hubble's constant A number that expresses the current rate of expansion of the universe; designated H_0, it is usually stated in units of km/s/Mpc. The reciprocal of Hubble's constant is the age the universe would be *if* the expansion rate had never changed.

Hubble's law Mathematically expresses the idea that more distant galaxies move away from us faster; its formula is $v = H_0 \times d$, where v is a galaxy's speed away from us, d is its distance, and H_0 is Hubble's constant.

hydrogen compounds Compounds that contain hydrogen and were common in the solar nebula, such as water (H_2O), ammonia (NH_3), and methane (CH_4).

hydrogen-shell burning Hydrogen fusion that occurs in a shell surrounding a stellar core.

hydrosphere Refers to the "layer" of water on the Earth consisting of oceans, lakes, rivers, ice caps, and other liquid water and ice.

hydrostatic equilibrium *See* gravitational equilibrium

hyperbola The precise mathematical shape of one type of unbound orbit (the other is a parabola) allowed under the force of gravity; at great distances from the attracting object, a hyperbolic path looks like a straight line.

hypernova A term sometimes used to describe a supernova (explosion) of a star so massive that it leaves a black hole behind.

hyperspace Any space with more than three dimensions.

hypothesis A tentative model proposed to explain some set of observed facts, but which has not yet been rigorously tested and confirmed.

ice ages Periods of global cooling during which the polar caps, glaciers and snow cover extend closer to the equator.

ices (in solar system theory) Materials that are solid only at low temperatures, such as the hydrogen compounds water, ammonia, and methane.

Ideal gas law The law relating the pressure, temperature, and number density of particles in an ideal gas.

image A picture of an object made by focusing light.

imaging (in astronomical research) The process of obtaining pictures of astronomical objects.

impact The collision of a small body (such as an asteroid or comet) with a larger object (such as a planet or moon).

impact basin A very large impact crater often filled by a lava flow.

impact crater A bowl-shaped depression left by the impact of an object that strikes a planetary surface (as opposed to burning up in the atmosphere).

impact cratering The excavation of bowl-shaped depressions (*impact craters*) by asteroids or comets striking a planet's surface.

impactor The object responsible for an impact.

inflation (of the universe) A sudden and dramatic expansion of the universe thought to have occurred at the end of the GUT era.

infrared light Light with wavelengths that fall in the portion of the electromagnetic spectrum between radio waves and visible light.

inner solar system Generally considered to encompass the region of our solar system out to about the orbit of Mars.

intensity (of light) A measure of the amount of energy coming from light of specific wavelength in the spectrum of an object.

interferometry A telescopic technique in which two or more telescopes are used in tandem to produce much better angular resolution than the telescopes could achieve individually.

intermediate-mass stars Stars born with masses between about $2 - 8 M_{Sun}$; these stars end their lives by ejecting a planetary nebula and becoming a white dwarf.

interstellar cloud A cloud of gas and dust between the stars.

interstellar dust grains Tiny solid flecks of carbon and silicon minerals found in cool interstellar clouds; they resemble particles of smoke and form in the winds of red giant stars.

interstellar medium Refers to gas and dust that fills the space between stars in a galaxy.

interstellar ramjet A hypothesized type of spaceship that uses a giant scoop to sweep up interstellar gas for use in a nuclear fusion engine.

interstellar reddening The change in the color of starlight as it passed through dusty gas. The light appears redder because dust grains absorb and scatter blue light more effectively than red light.

intracluster medium Hot, X-ray-emitting gas found between the galaxies within a cluster of galaxies.

inverse square law Any quantity that decreases with the square of the distance between two objects is said to follow an inverse square law.

inverse square law for light Law stating that the surface brightness of an object equals its luminosity divided by its distance squared, so that surface brightness depends on the inverse square of distance.

inversion (atmospheric) A local weather condition in which air is colder near the surface than higher up in the troposphere— the opposite of the usual condition, in which the troposphere is warmer at the bottom.

Io torus A donut-shaped charged-particle belt around Jupiter that approximately traces Io's orbit.

ionization The process of stripping an electron from an atom.

ionization nebula A colorful, wispy cloud of gas that glows because neighboring hot stars irradiate it with ultraviolet photons that can ionize hydrogen atoms.

ionosphere A portion of the thermosphere in which ions are particularly common (due to ionization by X rays from the Sun).

ions Atoms with a positive or negative electrical charge.

irregular galaxies Galaxies that look neither spiral nor elliptical.

isotopes Each different isotope of an element has the *same* number of protons but a *different* number of neutrons.

jets High-speed streams of gas ejected from an object into space.

joule The international unit of energy, equivalent to about 1/4,000 of a Calorie.

jovian nebulae The clouds of gas that swirled around the jovian planets, from which the moons formed.

jovian planets Giant gaseous planets similar in overall composition to Jupiter.

Julian calendar The calendar introduced in 46 B.C. by Julius Caesar and used until it was replaced by the Gregorian calendar.

Kelvin (temperature scale) The most commonly used temperature scale in science, defined such that absolute zero is 0 K and water freezes at 273.15 K.

Kepler's first law States that the orbit of each planet about the Sun is an ellipse with the Sun at one focus.

Kepler's laws of planetary motion Three laws discovered by Kepler that describe the motion of the planets around the Sun.

Kepler's second law States that, as a planet moves around its orbit, it sweeps out equal areas in equal times. This tells us that a planet moves faster when it is closer to the Sun (near perihelion) than when it is farther from the Sun (near aphelion) in its orbit.

Kepler's third law States that the square of a planet's orbital period is proportional to the cube of its average distance from the Sun (semimajor axis), which tells us that more distant planets move more slowly in their orbits. In its original form, written $p^2 = a^3$. *See also* Newton's version of Kepler's third law

kinetic energy Energy of motion, given by the formula $\frac{1}{2}mv^2$.

Kirchhoff's laws A set of rules that summarizes the conditions under which objects produce thermal, absorption line, or emission line spectra. In brief: (1) An opaque object produces thermal radiation. (2) An absorption line spectrum occurs when thermal radiation passes through a thin gas that is cooler than the object emitting the thermal radiation. (3) An emission line spectrum occurs when we view a cloud of gas that is warmer than any background source of light.

Kirkwood gaps On a plot of asteroid semimajor axes, regions with few asteroids as a result of orbital resonances with Jupiter.

K-T event (impact) The collision of an asteroid or comet 65 million years ago which caused the mass extinction best known for wiping out the dinosaurs. K-T stands for the initials of the geological layers above and below the event.

Kuiper belt The comet-rich region of our solar system that spans distances of about 30–100 AU from the Sun; Kuiper belt comets have orbits that lie fairly close to the plane of planetary orbits and travel around the Sun in the same direction as the planets.

Large Magellanic Cloud One of two small, irregular galaxies (the other is the Small Magellanic Cloud) located about 150,000 light-years away; it probably orbits the Milky Way Galaxy.

large-scale structure (of the universe) Generally refers to structure of the universe on size scales larger than that of clusters of galaxies.

latitude The angular north-south distance between the Earth's equator and a location on the Earth's surface.

leap year A calendar year with 366 rather than 365 days; our current calendar (the Gregorian calendar) has a leap year every 4 years (by adding February 29) except in century years that are not divisible by 400.

length contraction Refers to the effect in which you observe lengths to be shortened in reference frames moving relative to you.

lenticular galaxies Galaxies that look lens-shaped when seen edge-on, resembling spiral galaxies without arms. They tend to have less cool gas than normal spiral galaxies but more gas than elliptical galaxies.

leptons Fermions *not* made from quarks, such as electrons and neutrinos.

life track A track drawn on an H–R diagram to represent the changes in a star's surface temperature and luminosity during its life; also called an *evolutionary track*.

light-collecting area (of a telescope) The area of the primary mirror or lens that collects light in a telescope.

light curve A graph of an object's intensity against time.

light gases (in solar system theory) Refers to hydrogen and helium, which never condense under solar nebula conditions.

light pollution Human-made light that hinders astronomical observations.

light-year The distance that light can travel in 1 year, which is 9.46 trillion km.

liquid phase The phase of matter in which atoms or molecules are held together but move relatively freely.

lithosphere The relatively rigid outer layer of a planet; generally encompasses the crust and the uppermost portion of the mantle.

Local Bubble (interstellar) The bubble of hot gas in which our Sun and other nearby stars apparently reside. *See also* bubble (interstellar)

Local Group The group of more than 30 galaxies to which the Milky Way Galaxy belongs.

local sidereal time (LST) Sidereal time for a particular location, defined according to the position of the spring equinox in the local sky. More formally, the local sidereal time at any moment is defined to be the hour angle of the spring equinox.

local sky The sky as viewed from a particular location on Earth (or another solid object). Objects in the local sky are pinpointed by the coordinates of *altitude* and *direction* (or *azimuth*).

local solar neighborhood The portion of the Milky Way Galaxy that is located relatively close (within a few hundred to a couple thousand light-years) to our Sun.

Local Supercluster The supercluster of galaxies to which the Local Group belongs.

longitude The angular east-west distance between the prime meridian (which passes through Greenwich) and a location on the Earth's surface.

lookback time Refers to the amount of time since the light we see from a distant object was emitted. I.e., if an object has a lookback time of 400 million years, we are seeing it as it looked 400 million years ago.

low-mass stars Stars born with masses less than about $2M_{Sun}$; these stars end their lives by ejecting a planetary nebula and becoming a white dwarf.

luminosity The total power output of an object, usually measured in watts or in units of solar luminosities ($L_{Sun} = 3.8 \times 10^{26}$ watts).

luminosity class Describes the region of the H–R diagram in which a star falls. Luminosity class I represents supergiants, III represents giants, and V represents main-sequence stars; luminosity classes II and IV are intermediate to the others.

luminosity–distance formula The formula that relates apparent brightness, luminosity, and distance:

$$\text{apparent brightness} = \frac{\text{luminosity}}{4\pi \times (\text{distance})^2}$$

lunar eclipse Occurs when the Moon passes through the Earth's shadow, which can occur only at full moon; may be total, partial, or penumbral.

lunar maria The regions of the Moon that look smooth from Earth and actually are impact basins.

lunar month *See* synodic month

lunar phase Describes the appearance of the Moon as seen from Earth.

MACHOs Stands for *massive compact halo objects* and represents one possible form of dark matter in which the dark objects are relatively large, like planets or brown dwarfs.

magma Underground molten rock.

magnetic braking The process by which a star's rotation slows as its magnetic field transfers its angular momentum to the surrounding nebula.

magnetic field Describes the region surrounding a magnet in which it can affect other magnets or charged particles in its vicinity.

magnetic-field lines Lines that represent how the needles on a series of compasses would point if they were laid out in a magnetic field.

magnetosphere The region surrounding a planet in which charged particles are trapped by the planet's magnetic field.

magnitude system A system of describing stellar brightness by using numbers, called *magnitudes,* based on an ancient Greek way of describing the brightnesses of stars in the sky. This system uses *apparent magnitude* to describe a star's apparent brightness and *absolute magnitude* to describe a star's luminosity.

main sequence (luminosity class V) The prominent line of points running from the upper left to the lower right on an H–R diagram; main-sequence stars shine by fusing hydrogen in their cores.

main-sequence fitting A method for measuring the distance to a cluster of stars by comparing the apparent brightness of the cluster's main sequence with the standard main sequence.

main-sequence lifetime The length of time for which a star of a particular mass can shine by fusing hydrogen into helium in its core.

main-sequence stars Stars whose temperature and luminosity place them on the main sequence of the H-R diagram. Main-sequence stars are all releasing energy by fusing hydrogen into helium in their cores.

main-sequence turnoff A method for measuring the age of a cluster of stars from the point on its H–R diagram where its stars turn off from the main sequence; the age of the cluster is equal to the main-sequence lifetime of stars at the main-sequence turnoff point.

mantle (of a planet) The rocky layer that lies between a planet's core and crust.

Martian meteorite This term is used to describe meteorites found on Earth that are thought to have originated on Mars.

mass A measure of the amount of matter in an object.

mass-energy The potential energy of mass, which has an amount $E = mc^2$.

mass exchange (in close binary star systems) The process in which tidal forces cause matter to spill from one star to a companion star in a close binary system.

mass extinction An event in which a large fraction of the species living on Earth go extinct, such as the event in which the dinosaurs died out about 65 million years ago.

mass increase (in relativity) Refers to the effect in which an object moving past you seems to have a mass greater than its rest mass.

mass-to-light ratio The mass of an object divided by its luminosity, usually stated in units of solar masses per solar luminosity. Objects with high mass-to-light ratios must contain substantial quantities of dark matter.

massive-star supernova A supernova that occurs when a massive star dies, initiated by the catastrophic collapse of its iron core; often called a Type II supernova.

matter–antimatter annihilation Occurs when a particle of matter and a particle of antimatter meet and convert all of their mass-energy to photons.

mean solar time Time measured by the average position of the Sun in your local sky over the course of the year.

meridian A half-circle extending from your horizon (altitude 0°) due south, through your zenith, to your horizon due north.

metallic hydrogen Hydrogen that is so compressed that the hydrogen atoms all share electrons and thereby take on properties of metals, such as conducting electricity. Occurs only under very high pressure conditions, such as that found deep within Jupiter.

metals (in solar system theory) Elements, such as nickel, iron, and aluminum, that condense at fairly high temperatures.

meteor A flash of light caused when a particle from space burns up in our atmosphere.

meteor shower A period during which many more meteors than usual can be seen.

meteorite A rock from space that lands on Earth.

Metonic cycle The 19-year period, discovered by the Babylonian astronomer Meton, over which the lunar phases occur on the same dates.

microwaves Light with wavelengths in the range of micrometers to millimeters. Microwaves are generally considered to be a subset of the radio wave portion of the electromagnetic spectrum.

mid-ocean ridges (on Earth) Long ridges of undersea volcanoes, along which mantle material erupts onto the ocean floor and pushes apart the existing seafloor on either side. These ridges are essentially the source of new seafloor crust, which then makes its way along the ocean bottom for millions of years before returning to the mantle at a subduction zone.

Milankovitch cycles The cyclical changes in Earth's axis tilt and orbit which can change the climate and cause ice ages.

Milky Way Used both as the name of our galaxy and to refer to the band of light we see in the sky when we look into the plane of the Milky Way Galaxy.

millisecond pulsars Pulsars with rotation periods of a few thousandths of a second.

minor planets An alternate name for *asteroids.*

model (scientific) A representation of some aspect of nature that can be used to explain and predict real phenomena without invoking myth, magic, or the supernatural.

molecular bands The tightly bunched lines in an object's spectrum that are produced by molecules.

molecular cloud cores The densest regions of molecular clouds, which usually go on to form stars.

molecular clouds Cool, dense interstellar clouds in which the low temperatures allow hydrogen atoms to pair up into hydrogen molecules (H_2).

molecular dissociation The process by which a molecule splits into its component atoms.

molecule Technically the smallest unit of a chemical element or compound; in this text, the term refers only to combinations of two or more atoms held together by chemical bonds.

momentum The product of an object's mass and velocity.

moon An object that orbits a planet.

mutations Errors in the copying process when a living cell replicates itself.

natural selection The process by which mutations that make an organism better able to survive get passed on to future generations.

neap tides The lower-than-average tides on Earth that occur at first- and third-quarter moon, when the tidal forces from the Sun and Moon oppose one another.

nebula A cloud of gas in space, usually one that is glowing.

nebular capture The process by which icy planetesimals capture hydrogen and helium gas to form jovian planets.

nebular theory The detailed theory that describes how our solar system formed from a cloud of interstellar gas and dust.

net force The overall force to which an object responds; the net force is equal to the rate of change in the object's momentum, or equivalently to the object's mass $\times$ acceleration.

neutrino A type of fundamental particle that has extremely low mass and responds only to the weak force; neutrinos are leptons and come in three types—electron neutrinos, mu neutrinos, and tau neutrinos.

neutron degeneracy pressure Degeneracy pressure exerted by neutrons, as in neutron stars.

neutron star The compact corpse of a high-mass star left over after a supernova; typically contains a mass comparable to the mass of the Sun in a volume just a few kilometers in radius.

neutrons Particles with no electrical charge found in atomic nuclei, built from three quarks.

newton The standard unit of force in the metric system:

$$1 \text{ newton} = 1\frac{\text{kg} \times \text{m}}{\text{s}^2}$$

Newton's first law of motion States that, in the absence of a net force, an object moves with constant velocity.

Newton's laws of motion Three basic laws that describe how objects respond to forces.

Newton's second law of motion States how a net force affects an object's motion. Specifically: force = rate of change in momentum, or force = mass × acceleration.

Newton's third law of motion States that, for any force, there is always an equal and opposite reaction force.

Newton's universal law of gravitation *See* universal law of gravitation

Newton's version of Kepler's third law This generalization of Kepler's third law can be used to calculate the masses of orbiting objects from measurements of orbital period and distance. Usually written as:

$$p^2 = \frac{4\pi^2}{G(M_1 + M_2)}a^3$$

nodes (of Moon's orbit) The two points in the Moon's orbit where it crosses the ecliptic plane.

nonbaryonic matter Refers to exotic matter that is not part of the normal composition of atoms, such as neutrinos or the hypothetical WIMPs.

nonscience As defined in this book, non-science is any way of searching for knowledge that makes no claim to follow the scientific method, such as seeking knowledge through intuition, tradition, or faith.

north celestial pole (NCP) The point on the celestial sphere directly above the Earth's North Pole.

nova The dramatic brightening of a star that lasts for a few weeks and then subsides; occurs when a burst of hydrogen fusion ignites in a shell on the surface of an accreting white dwarf in a binary star system.

nuclear fission The process in which a larger nucleus splits into two (or more) smaller particles.

nuclear fusion The process in which two (or more) smaller nuclei slam together and make one larger nucleus.

nucleus (of an atom) The compact center of an atom made from protons and neutrons.

nucleus (of a comet) The solid portion of a comet, and the only portion that exists when the comet is far from the Sun.

observable universe The portion of the entire universe that, at least in principle, can be seen from Earth.

Occam's razor A principle often used in science, holding that scientists should prefer the simpler of two models that agree equally well with observations. Named after the medieval scholar William of Occam (1285–1349).

Olbers' paradox Asks the question of how the night sky can be dark if the universe is infinite and full of stars.

Oort cloud A huge, spherical region centered on the Sun, extending perhaps halfway to the nearest stars, in which trillions of comets orbit the Sun with random inclinations, orbital directions, and eccentricities.

opacity A measure of how much light a material absorbs compared to how much it transmits; materials with higher opacity absorb more light.

opaque (material) Describes a material that absorbs light.

open cluster A cluster of up to several thousand stars; open clusters are found only in the disks of galaxies and often contain young stars.

open universe The universe is open if its average density is less than the critical density, in which case spacetime has an overall shape analogous to the surface of a saddle.

opposition The point at which a planet appears opposite the Sun in our sky.

optical quality Describes the ability of a lens, mirror, or telescope to obtain clear and properly focused images.

orbital energy The sum of an orbiting object's kinetic and gravitational potential energies.

orbital resonance Describes any situation in which one object's orbital period is a simple ratio of another object's period, such as 1/2, 1/4, or 5/3. In such cases, the two objects periodically line up with each other, and the extra gravitational attractions at these times can affect the objects' orbits.

orbital velocity law This law (a variation on Newton's version of Kepler's third law) allows us to use a star's orbital speed and distance from the galactic center to determine the total mass of the galaxy contained *within* the star's orbit. Mathematically, it is written:

$$M_r = \frac{r \times v^2}{G}$$

where M_r is the mass contained within the star's orbit, r is the star's distance from the galactic center, v is the star's orbital velocity, and G is the gravitational constant.

orbiters (of other worlds) Spacecraft that go into orbit of another world for long-term study.

outer solar system Generally considered to encompass the region of our solar system beginning at about the orbit of Jupiter.

outgassing The process of releasing gases from a planetary interior, usually through volcanic eruptions.

oxidation Refers to chemical reactions, often with the surface of a planet, that remove oxygen from the atmosphere.

ozone The molecule O_3, which is a particularly good absorber of ultraviolet light.

ozone depletion Refers to the declining levels of atmospheric ozone found worldwide on Earth, especially in Antarctica, in recent years.

ozone hole A place where the concentration of ozone in the stratosphere is dramatically lower than is the norm.

pair production The process in which a concentration of energy spontaneously turns into a particle and its antiparticle.

parabola The precise mathematical shape of a special type of unbound orbit allowed under the force of gravity; if an object in a parabolic orbit loses only a tiny amount of energy, it will become bound.

paradigm (in science) Refers to general patterns of thought that tend to shape scientific beliefs during a particular time period.

paradox A situation that, at least at first, seems to violate common sense or contradict itself. Resolving paradoxes often leads to deeper understanding.

parallax The apparent shifting of an object against the background, due to viewing it from different positions. *See also* stellar parallax

parallax angle Half of a star's annual back-and-forth shift due to stellar parallax; related to the star's distance according to the formula

$$\text{distance in parsecs} = \frac{1}{p}$$

where p is the parallax angle in arcseconds.

parsec (pc) Approximately equal to 3.26 light-years; it is the distance to an object with a parallax angle of 1 arcsecond.

partial lunar eclipse A lunar eclipse in which the Moon becomes only partially covered by the Earth's umbral shadow.

partial solar eclipse A solar eclipse during which the Sun becomes only partially blocked by the disk of the Moon.

particle accelerator A machine designed to accelerate subatomic particles to high speeds in order to create new particles or to test fundamental theories of physics.

particle era The era of the universe lasting from 10^{-10} second to 0.001 second after the Big Bang, during which subatomic particles were continually created and destroyed and ending when matter annihilated antimatter.

peculiar velocity (of a galaxy) The component of a galaxy's velocity relative to the Milky Way that deviates from the velocity expected by Hubble's law.

penumbra The lighter, outlying regions of a shadow.

penumbral (lunar) eclipse A lunar eclipse in which the Moon passes only within the Earth's penumbral shadow and does not fall within the umbra.

perigee The point at which an object orbiting the Earth is nearest to the Earth.

perihelion The point at which an object orbiting the Sun is closest to the Sun.

period–luminosity relation The relation that describes how the luminosity of a Cepheid variable star is related to the period between peaks in its brightness; the longer the period, the more luminous the star.

phase (of matter) Describes the way in which atoms or molecules are held together; the common phases are solid, liquid, and gas.

photon An individual particle of light, characterized by a wavelength and a frequency.

photosphere The visible surface of the Sun, where the temperature averages just under 6,000 K.

pixel An individual "picture element" on a CCD.

Planck era The era of the universe prior to the Planck time.

Planck time The time when the universe was 10^{-43} second old, before which random energy fluctuations were so large that our current theories are powerless to describe what might have been happening.

Planck's constant A universal constant, abbreviated h, with value $h = 6.626 \times 10^{-43}$ joule $\times$ s.

planet An object that orbits a star and that, while much smaller than a star, is relatively large in size; there is no "official" minimum size for a planet, but the nine planets in our solar system all are at least 2,000 km in diameter.

planetary geology The extension of the study of Earth's surface and interior to apply to other solid bodies in the solar system such as terrestrial planets and jovian planet moons.

planetary nebula The glowing cloud of gas ejected from a low-mass star at the end of its life.

planetesimals The building blocks of planets, formed by accretion in the solar nebula.

plasma A gas consisting of ions and electrons.

plasma tail (of a comet) One of two tails seen when a comet passes near the Sun (the other is the dust tail); composed of ionized gas blown away from the Sun by the solar wind.

plate tectonics The geological process in which plates are moved around by stresses in a planet's mantle.

plates (on a planet) Pieces of a lithosphere that apparently float upon the denser mantle below.

polarization (of light) Light is polarized when the electric and magnetic fields of light waves are all aligned in some particular way.

Population I *See* disk population

Population II *See* spheroidal population

positron The antimatter equivalent of an electron. It is identical to an electron in virtually all respects, except it has a positive rather than a negative electrical charge.

potential energy Energy stored for later conversion into kinetic energy; includes gravitational potential energy, electrical potential energy, and chemical potential energy.

power The rate of energy usage, usually measured in watts (1 watt = 1 joule/s).

precession The gradual wobble of the axis of a rotating object around a vertical line.

precipitation condensed atmospheric gases which fall to the surface in the form of rain, snow or hail.

pressure Describes the force (per unit area) pushing on an object. In astronomy, we are generally interested in pressure applied by surrounding gas (or plasma). Ordinarily, such pressure is related to the temperature of the gas (*see* thermal pressure). In objects such as white dwarfs and neutron stars, pressure may arise from a quantum effect (*see* degeneracy pressure). Light can also exert pressure. (*See* radiation pressure.)

primary mirror The large, light-collecting mirror of a reflecting telescope.

prime focus (of a reflecting telescope) The first point at which light focuses after bouncing off the primary mirror; located in front of the primary mirror.

prime meridian The meridian of longitude that passes through Greenwich, England, defined to be longitude 0°.

primitive meteorites Meteorites that formed at the same time as the solar system itself, about 4.6 billion years ago. Primitive meteorites from the inner asteroid belt are usually stony, and those from the outer belt are usually carbon-rich.

processed meteorites Meteorites that apparently once were part of a larger object that "processed" the original material of the solar nebula into another form. Processed meteorites can be rocky if chipped from the surface or mantle, or metallic if blasted from the core.

proper motion The motion of a object in the plane of the sky, perpendicular to our line of sight.

protogalactic cloud A huge, collapsing cloud of intergalactic gas from which an individual galaxy formed.

proton–proton chain The chain of reactions by which low-mass stars (including the Sun) fuse hydrogen into helium.

protons Particles found in atomic nuclei with positive electrical charge, built from three quarks.

protoplanetary disk A disk of material surrounding a young star (or protostar) that may eventually form planets.

protostar A forming star that has not yet reached the point where sustained fusion can occur in its core.

protostellar disk A disk of material surrounding a protostar; essentially the same as a protoplanetary disk, but may not necessarily lead to planet formation.

protostellar wind The relatively strong wind from a protostar.

protosun The central object in the forming solar system that eventually became the Sun.

pseudoscience Something that purports to be science or may appear to be scientific but that does not adhere to the testing and verification requirements of the scientific method.

Ptolemaic model The geocentric model of the universe developed by Ptolemy in about 150 A.D.

pulsar A neutron star from which we see rapid pulses of radiation as it rotates.

pulsating variable stars Stars that alternately grow brighter and dimmer as their outer layers expand and contract in size.

quantum mechanics The branch of physics that deals with the very small, including molecules, atoms, and fundamental particles.

quantum state Refers to the complete description of the state of a subatomic particle, including its location, momentum, orbital angular momentum, and spin, to the extent allowed by the uncertainty principle.

quantum tunneling The process in which, thanks to the uncertainty principle, an electron or other subatomic particle appears on the other side of a barrier that it does not have the energy to overcome in a normal way.

quarks The building blocks of protons and neutrons, quarks are one of the two basic types of fermions (leptons are the other).

quasar The brightest type of active galactic nucleus.

radar mapping Imaging of a planet by bouncing radar waves off its surface, especially important for Venus and Titan where thick clouds mask the surface.

radar ranging A method of measuring distances within the solar system by bouncing radio waves off planets.

radial motion The component of an object's motion directed toward or away from us.

radial velocity The portion of any object's total velocity that is directed toward or away from us. This part of the velocity is the only part that we can measure with the Doppler effect.

radiation pressure Pressure exerted by photons of light.

radiation zone (of a star) A region of the interior in which energy is transported primarily by radiative diffusion.

radiative diffusion The process by which photons gradually migrate from a hot region (such as the solar core) to a cooler region (such as the solar surface).

radiative energy Energy carried by light; the energy of a photon is Planck's constant times its frequency, or $h \times f$.

radio galaxy A galaxy that emits unusually large quantities of radio waves; thought to contain an active galactic nucleus powered by a supermassive black hole.

radio lobes The huge regions of radio emission found on either side of radio galaxies. The lobes apparently contain plasma ejected by powerful jets from the galactic center.

radio waves Light with very long wavelengths (and hence low frequencies)—longer than those of infrared light.

radiometric dating The process of determining the age of a rock (i.e., the time since it solidified) by comparing the present amount of a radioactive substance to the amount of its decay product.

radioactive decay The spontaneous change of an atom, in which its nucleus breaks apart or a proton turns into an electron, thereby changing into a different element. It also releases heat to a planet's interior.

radioactive element (or **radioactive isotope**) A substance whose nucleus tends to fall apart spontaneously.

recession velocity (of a galaxy) The speed at which a distant galaxy is moving away from us due to the expansion of the universe.

recollapsing universe The possible fate of our universe in which the collective gravity of all its matter eventually halts and reverses the expansion. The galaxies will come crashing back together, and the universe will end in a fiery Big Crunch.

red giant A giant star that is red in color.

red-giant winds The relatively dense but slow winds from red giant stars.

redshift (Doppler) A Doppler shift in which spectral features are shifted to longer wavelengths, caused when an object is moving away from the observer.

reference frame (frame of reference) Two people (or objects) share the same reference frame if they are *not* moving relative to one another.

reflecting telescope A telescope that uses mirrors to focus light.

reflection (of light) The process by which matter changes the direction of light.

reflection nebula A nebula that we see as a result of starlight reflected from interstellar dust grains. Reflection nebulae tend to have blue and black tints.

refracting telescope A telescope that uses lenses to focus light.

resonance *See* orbital resonance

rest wavelength The wavelength of a spectral feature in the absence of any Doppler shift or gravitational redshift.

retrograde motion Motion that is backward compared to the norm; e.g., we see Mars in apparent retrograde motion during the periods of time when it moves westward, rather than the more common eastward, relative to the stars.

revolution The orbital motion of one object around another.

right ascension (RA) Analogous to longitude, but on the celestial sphere; it is the angular east-west distance between the vernal equinox and a location on the celestial sphere.

rings (planetary) Consist of numerous small particles orbiting a planet within its Roche zone.

Roche tidal zone The region within two to three planetary radii (of any planet) in which the tidal forces tugging an object apart become comparable to the gravitational forces holding it together; planetary rings are always found within the Roche tidal zone.

rocks (in solar system theory) Material common on the surface of the Earth, such as silicon-based minerals, that are solid at temperatures and pressures found on Earth but typically melt or vaporize at temperatures of 500–1,300 K.

rotation The spinning of an object around its axis.

rotation curve A graph that plots rotational (or orbital) velocity against distance from the center for any object or set of objects.

runaway greenhouse effect A positive feedback cycle in which heating caused by the greenhouse effect causes more greenhouse gases to enter the atmosphere, which further enhances the greenhouse effect.

saddle-shaped (or **hyperbolic**) **geometry** Refers to any case in which the rules of geometry for a saddle-shaped surface hold, such as that two lines that begin parallel eventually diverge.

Sagittarius Dwarf A small, dwarf elliptical galaxy that is currently passing through the disk of the Milky Way Galaxy.

saros cycle The period over which the basic pattern of eclipses repeats, which is about 18 years $11\frac{1}{3}$ days.

satellite Any object orbiting another object.

scattered light Light that is reflected into random directions.

Schwarzschild radius A measure of the size of the event horizon of a black hole.

science The search for knowledge that can be used to explain or predict natural phenomena in a way that can be confirmed by rigorous observations or experiments.

scientific method An organized approach to explaining observed facts through science.

scientific theory A model of some aspect of nature that has been rigorously tested and has passed all tests to date.

seafloor crust On Earth, the thin, dense crust of basalt created by seafloor spreading.

seafloor spreading On Earth, the creation of new seafloor crust at mid-ocean ridges.

secondary mirror A small mirror in a reflecting telescope, used to reflect light gathered by the primary mirror toward an eyepiece or instrument.

sedimentary rock A rock that formed from sediments created and deposited by erosional processes.

seismic waves Earthquake-induced vibrations that propagate through a planet.

semimajor axis Half the distance across the long axis of an ellipse; in this text, it is usually referred to as the *average* distance of an orbiting object, abbreviated *a* in the formula for Kepler's third law.

search for extraterrestrial intelligence (SETI) The name given to observing projects designed to search for signs of intelligent life beyond Earth.

Seyfert galaxies The name given to a class of galaxies found relatively nearby and that have nuclei much like those of quasars, except that they are less luminous.

shepherd moons Tiny moons within a planet's ring system that help force particles into a narrow ring. A variation on *gap moons*.

shield volcano A shallow-sloped volcano made from the flow of low-viscosity basaltic lava.

shock wave A wave of pressure generated by gas moving faster than the speed of sound.

sidereal day The time of 23 hours 56 minutes 4.09 seconds between successive appearances of any particular star on the meridian; essentially the true rotation period of the Earth.

sidereal month About $27\frac{1}{4}$ days, the time required for the Moon to orbit the Earth once (as measured against the stars).

sidereal period (of a planet) A planet's actual orbital period around the Sun.

sidereal time Time measured according to the position of stars in the sky rather than the position of the Sun in the sky. *See also* local sidereal time

sidereal year The time required for the Earth to complete exactly one orbit as measured against the stars; about 20 minutes longer than the tropical year on which our calendar is based.

silicate rock A silicon-rich rock.

singularity The place at the center of a black hole where, in principle, gravity crushes all matter to an infinitely tiny and dense point.

Small Magellanic Cloud One of two small, irregular galaxies (the other is the Large Magellanic Cloud) located about 150,000 light-years away; it probably orbits the Milky Way Galaxy.

snowball Earth Name given to a hypothesis suggesting that, some 600–700 million years ago, the Earth experienced a period in which it became cold enough for glaciers to exist worldwide, even in equatorial regions.

solar activity Refers to short-lived phenomena on the Sun, including the emergence and disappearance of individual sunspots, prominences, and flares; sometimes called *solar weather*.

solar circle The Sun's orbital path around the galaxy, which has a radius of about 28,000 light-years.

solar day Twenty-four hours, which is the average time between appearances of the Sun on the meridian.

solar eclipse Occurs when the Moon's shadow falls on the Earth, which can occur only at new moon; may be total, partial, or annular.

solar flares Huge and sudden releases of energy on the solar surface, probably caused when energy stored in magnetic fields is suddenly released.

solar luminosity The luminosity of the Sun, which is approximately 4×10^{26} watts.

solar maximum The time during each sunspot cycle at which the number of sunspots is the greatest.

solar minimum The time during each sunspot cycle at which the number of sunspots is the smallest.

solar nebula The piece of interstellar cloud from which our own solar system formed.

solar neutrino problem Refers to the disagreement between the predicted and observed number of neutrinos coming from the Sun.

solar prominences Vaulted loops of hot gas that rise above the Sun's surface and follow magnetic-field lines.

solar sail A large, highly reflective (and thin, to minimize mass) piece of material that can "sail" through space using pressure exerted by sunlight.

solar system (or **star system**) Consists of a star (sometimes more than one star) and all the objects that orbit it.

solar thermostat The regulation of the Sun's core temperature that comes about because pressure and gravity must balance within the Sun.

solar wind A stream of charged particles ejected from the Sun.

solid phase The phase of matter in which atoms or molecules are held rigidly in place.

solstice See *summer solstice* and *winter solstice*.

sound wave A wave of alternately rising and falling pressure.

south celestial pole (SCP) The point on the celestial sphere directly above the Earth's South Pole.

spacetime The inseparable, four-dimensional combination of space and time.

spacetime diagram A graph that plots a spatial dimension on one axis and time on another axis.

special theory of relativity Einstein's theory that describes the effects of the facts that all motion is relative and that everyone always measures the same speed of light.

spectral lines Bright or dark lines that appear in an object's spectrum, which we can see when we pass the object's light through a prismlike device that spreads out the light like a rainbow.

spectral resolution Describes the degree of detail that can be seen in a spectrum; the higher the spectral resolution, the more detail we can see.

spectral type A way of classifying a star by the lines that appear in its spectrum; it is related to surface temperature. The basic spectral types are designated by a letter (OBAFGKM, with O for the hottest stars and M for the coolest) and are subdivided with numbers from 0 through 9.

spectrograph An instrument used to record spectra.

spectroscopic binary A binary star system whose binary nature is revealed because we detect the spectral lines of one or both stars alternately becoming blueshifted and redshifted as the stars orbit each other.

spectroscopy (in astronomical research) The process of obtaining spectra from astronomical objects.

spectrum (of light) See *electromagnetic spectrum*

speed The rate at which an object moves. Its units are distance divided by time, such as m/s or km/hr.

speed of light The speed at which light travels, which is about 300,000 km/s.

spherical geometry Refers to any case in which the rules of geometry for the surface of a sphere hold, such as that lines that begin parallel eventually meet.

spheroidal component (of a galaxy) The portion of any galaxy that is spherical (or football-like) in shape and contains very little cool gas; generally contains only very old stars. Elliptical galaxies have only a spheroidal component, while spiral galaxies also have a disk component.

spheroidal galaxy Another name for an elliptical galaxy.

spheroidal population Refers to stars that orbit within the spheroidal component of a galaxy. Thus, elliptical galaxies have only a spheroidal population (they lack a disk population), while spiral galaxies have spheroidal population stars in their bulges and halos. Sometimes called Population II.

spin (quantum) See *spin angular momentum*

spin angular momentum Often simply called *spin*, it refers to the inherent angular momentum of a fundamental particle.

spiral arms The bright, prominent arms, usually in a spiral pattern, found in most spiral galaxies.

spiral density waves Gravitationally driven waves of enhanced density that move through a spiral galaxy and are responsible for maintaining its spiral arms.

spiral galaxies Galaxies that look like flat, white disks with yellowish bulges at their centers. The disks are filled with cool gas and dust, interspersed with hotter ionized gas, and usually display beautiful spiral arms.

spreading centers (geological) Places where hot mantle material rises upward between plates and then spreads sideways creating new seafloor crust.

spring equinox (vernal equinox) Refers both to the point in Pisces on the celestial sphere where the ecliptic crosses the celestial equator and to the moment in time when the Sun appears at that point each year (around March 21).

spring tides The higher-than-average tides on Earth that occur at new and full moon, when the tidal forces from the Sun and Moon both act along the same line.

standard candle An object for which we have some means of knowing its true luminosity, so that we can use its apparent brightness to determine its distance with the luminosity–distance formula.

standard model (of physics) The current theoretical model that describes the fundamental particles and forces in nature.

standard time Time measured according to the internationally recognized time zones.

star A large, glowing ball of gas that generates energy through nuclear fusion in its core. The term *star* is sometimes applied to objects that are in the process of becoming true stars (e.g., protostars) and to the remains of stars that have died (e.g., neutron stars).

star cluster See *cluster of stars*

starburst galaxy A galaxy in which stars are forming at an unusually high rate.

star-gas-star cycle The process of galactic recycling in which stars expel gas into space where it mixes with the interstellar medium and eventually forms new stars.

star system See *solar system*.

state (quantum) See *quantum state*

steady state theory A now-discredited theory that held that the universe had no beginning and looks about the same at all times.

Stefan–Boltzmann constant constant that appears in the laws of thermal radiation, with value

$$\sigma = 5.7 \times 10^{-8} \frac{\text{watt}}{m^2 \times \text{Kelvin}^4}.$$

stellar evolution The formation and development of stars.

stellar parallax The apparent shift in the position of a nearby star (relative to distant objects) that occurs as we view the star from different positions in the Earth's orbit of the Sun each year.

stellar wind A stream of charged particles ejected from the surface of a star.

stratosphere An intermediate-altitude layer of the atmosphere that is warmed by the absorption of ultraviolet light from the Sun.

stratovolcano A steep-sided volcano made from viscous lavas that can't flow very far before solidifying.

string theory New ideas, not yet well-tested, that attempt to explain all of physics in a much simpler way than current theories.

stromatolites Large bacterial "colonies."

strong force One of the four fundamental forces; it is the force that holds atomic nuclei together.

subduction (of tectonic plates) The process in which one plate slides under another.

subduction zones Places where one plate slides under another.

subgiant A star that is between being a main-sequence star and being a giant; subgiants have inert helium cores and hydrogen-burning shells.

sublimation The process by which atoms or molecules escape into the gas phase from a solid.

summer solstice Refers both to the point on the celestial sphere where the ecliptic is farthest north of the celestial equator and to the moment in time when the Sun appears at that point each year (around June 21).

sunspot cycle The period of about 11 years over which the number of sunspots on the Sun rises and falls.

sunspots Blotches on the surface of the Sun that appear darker than surrounding regions.

superbubble Essentially a giant interstellar bubble, formed when the shock waves of many individual bubbles merge to form a single, giant shock wave.

supercluster Superclusters consist of many clusters of galaxies, groups of galaxies, and individual galaxies and are the largest known structures in the universe.

supergiants (luminosity class I) The very large and very bright stars that appear at the top of an H–R diagram.

supermassive black hole Giant black hole, with a mass millions to billions of times that of our Sun, thought to reside in the centers of many galaxies and to power active galactic nuclei.

supernova The explosion of a star.

Supernova 1987A A supernova witnessed on Earth in 1987; it was the nearest supernova seen in nearly 400 years and helped astronomers refine theories of supernovae.

supernova remnant A glowing, expanding cloud of debris from a supernova explosion.

synchronous rotation Describes the rotation of an object that always shows the same face to an object that it is orbiting because its rotation period and orbital period are equal.

synchrotron radiation A type of radio emission that occurs when electrons moving at nearly the speed of light spiral around magnetic field lines.

synodic month (or **lunar month**) The time required for a complete cycle of lunar phases, which averages about $29\frac{1}{2}$ days.

synodic period (of a planet) The time between successive alignments of a planet and the Sun in our sky; measured from opposition to opposition for a planet beyond Earth's orbit, or from superior conjunction to superior conjunction for Mercury and Venus.

tangential motion The component of an object's motion directed across our line of sight.

tangential velocity The portion of any object's total velocity that is directed across (perpendicular to) our line-of-sight. This part of the velocity cannot be measured with the Doppler effect. It can be measured only by observing the object's gradual motion across our sky.

tectonics The disruption of a planet's surface by internal stresses.

temperature A measure of the average kinetic energy of particles in a substance.

terrestrial planets Rocky planets similar in overall composition to Earth.

theories of relativity (special and general) Einstein's theories that describe the nature of space, time, and gravity.

thermal emitter An object that produces a thermal radiation spectrum; sometimes called a "blackbody."

thermal energy Represents the collective kinetic energy, as measured by temperature, of the many individual particles moving within a substance.

thermal escape The process in which atoms or molecules in a planet's exosphere move fast enough to escape into space.

thermal pressure The ordinary pressure in a gas arising from motions of particles that can be attributed to the object's temperature.

thermal pulses The predicted upward spikes in the rate of helium fusion, occurring every few thousand years, that occur near the end of a low-mass star's life.

thermal radiation The spectrum of radiation produced by an opaque object that depends only on the object's temperature; sometimes called "blackbody radiation."

thermosphere A high, hot X-ray-absorbing layer of an atmosphere, just below the exosphere.

tidal force A force that is caused when the gravity pulling on one side of an object is larger than that on the other side, causing the object to stretch.

tidal friction Friction within an object that is caused by a tidal force.

tidal heating A source of internal heating created by tidal friction. It is particularly important for satellites with eccentric orbits such as Io and Europa.

time dilation Refers to the effect in which you observe time running slower in reference frames moving relative to you.

timing (in astronomical research) The process of tracking how the light intensity from an astronomical object varies with time.

theory of evolution The theory, first advanced by Charles Darwin, that explains *how* evolution occurs through the process of *natural selection*.

theory (in science) *See* scientific theory.

torque A twisting force that can cause a change in an object's angular momentum.

total apparent brightness *See* apparent brightness. We sometimes say "total apparent brightness" to distinguish it from wavelength-specific measures such as the apparent brightness measured in visible light.

total luminosity *See* luminosity. We sometimes say "total luminosity" to distinguish it from wavelength-specific measures such as the luminosity emitted in visible light or the X-ray luminosity.

total lunar eclipse A lunar eclipse in which the Moon becomes fully covered by the Earth's umbral shadow.

total solar eclipse A solar eclipse during which the Sun becomes fully blocked by the disk of the Moon.

totality (eclipse) The portion of either a total lunar eclipse during which the Moon is fully within the Earth's umbral shadow or a total solar eclipse during which the Sun's disk is fully blocked by the Moon.

transit A transit is an event in which a planet passes in front of a star (or the Sun) as seen from Earth. Only Mercury and Venus can be seen in transit of our Sun. The search for transits of extrasolar planets is an important planet detection strategy.

transmission (of light) The process in which light passes through matter without being absorbed.

transparent (material) Describes a material that transmits light.

triple-alpha reaction *See* helium fusion

Trojan asteroids Asteroids found within two stable zones that share Jupiter's orbit but lie 60° ahead of and behind Jupiter.

tropic of Cancer The circle on the Earth with latitude 23.5°N. It is the northernmost latitude at which the Sun ever passes directly overhead (at noon on the summer solstice).

tropic of Capricorn The circle on the Earth with latitude 23.5°S. It is the southernmost latitude at which the Sun ever passes directly overhead (at noon on the winter solstice).

tropical year The time from one spring equinox to the next, on which our calendar is based.

troposphere The lowest atmospheric layer, in which convection and weather occur.

Tully–Fisher relation A relationship among spiral galaxies showing that the faster a spiral galaxy's rotation speed, the more luminous it is; it is important because it allows us to determine the distance to a spiral galaxy once we measure its rotation rate and apply the luminosity–distance formula.

turbulence Rapid and random motion.

ultraviolet light Light with wavelengths that fall in the portion of the electromagnetic spectrum between visible light and X rays.

umbra The dark central region of a shadow.

unbound orbits Orbits on which an object comes in toward a large body only once, never to return; unbound orbits may be parabolic or hyperbolic in shape.

uncertainty principle The law of quantum mechanics that states that we can never know both a particle's position and its momentum, or both its energy and the time it has the energy, with absolute precision.

universal law of gravitation The law expressing the force of gravity (F_g) between two objects, given by the formula

$$F_g = G\frac{M_1 M_2}{d^2}$$

$$\left(G = 6.67 \times 10^{-11} \frac{m^3}{\text{kg} \times s^2}\right).$$

universal time (UT) Standard time in Greenwich (or anywhere on the prime meridian).

universe The sum total of all matter and energy.

up quark One of the two quark types (the other is the down quark) found in ordinary protons and neutrons. Has a charge of $+\frac{2}{3}$.

velocity The combination of speed and direction of motion; it can be stated as a speed in a particular direction, such as 100 km/hr due north.

vernal equinox *See* spring equinox

virtual particles Particles that "pop" in and out of existence so rapidly that, according to the uncertainty principle, they cannot be directly detected.

viscosity Describes the "thickness" of a liquid in terms of how rapidly it flows; low-viscosity liquids flow quickly (e.g., water), while high-viscosity liquids flow slowly (e.g., molasses).

visible light The light our eyes can see, ranging in wavelength from about 400 to 700 nm.

visual binary A binary star system in which we can resolve both stars through a telescope.

voids Huge volumes of space between superclusters that appear to contain very little matter.

volatiles Refers to substances, such as water, carbon dioxide, and methane, that are usually found as gases, liquids, or surface ices on the terrestrial worlds.

volcanic plains Vast, relatively smooth areas created by the eruption of very runny lava.

volcanism The eruption of molten rock, or lava, from a planet's interior onto its surface.

watt The standard unit of power in science: 1 watt = 1 joule/s.

wavelength The distance between adjacent peaks (or troughs) of a wave.

weak bosons The exchange particles for the weak force.

weak force One of the four fundamental forces; it is the force that mediates nuclear reactions; also the only force besides gravity felt by weakly interacting particles.

weakly interacting particles Particles, such as neutrinos and WIMPs, that respond only to the weak force and gravity; that is, they do not feel the strong force or the electromagnetic force.

weather Describes the ever-varying combination of winds, clouds, temperature, and pressure in a planet's troposphere.

weight The net force that an object applies to its surroundings; in the case of a stationary body on the surface of the Earth, weight = mass × acceleration of gravity.

weightless A weight of zero, as occurs during free-fall.

white-dwarf limit (also called the *Chandrasekhar limit*) The maximum possible mass for a white dwarf, which is about $1.4M_{Sun}$.

white dwarf supernova A supernova that occurs when an accreting white dwarf reaches the white-dwarf limit, ignites runaway carbon fusion, and explodes like a bomb; often called a *Type Ia supernova*.

white dwarfs The hot, compact corpses of low-mass stars, typically with a mass similar to the Sun compressed to a volume the size of the Earth.

WIMPs Stands for *weakly interacting massive particles* and represents a possible form of dark matter consisting of subatomic particles that are dark because they do not respond to the electromagnetic force.

winter solstice Refers both to the point on the celestial sphere where the ecliptic is farthest south of the celestial equator and to the moment in time when the Sun appears at that point each year (around December 21).

worldline A line that represents an object on a spacetime diagram.

wormholes The name given to hypothetical tunnels through hyperspace that might connect two distant places in our universe.

X rays Light with wavelengths that fall in the portion of the electromagnetic spectrum between ultraviolet light and gamma rays.

X-ray binary A binary star system that emits substantial amounts of X rays, thought to be from an accretion disk around a neutron star or black hole.

X-ray bursts Burst of X-rays coming from sudden ignition of fusion on the surface of an accreting neutron star in an X-ray binary system.

X-ray burster An object that emits a burst of X rays every few hours to every few days; each burst lasts a few seconds and is thought to be caused by helium fusion on the surface of an accreting neutron star in a binary system.

Zeeman effect The splitting of spectral lines by a magnetic field.

zenith The point directly overhead, which has an altitude of 90°.

zodiac The constellations on the celestial sphere through which the ecliptic passes.

zones (on a jovian planet) Bright bands of rising air that encircle a jovian planet at a particular set of latitudes.

Credits and Acknowledgments

(Magellan Mission) **9.14a** Gene Ahrens/Bruce Coleman Inc. **9.14b** Joachim Messerschmidt/Bruce Coleman Inc. **9.14c** Craig Aurness/CORBIS Bettmann Lunar and Planetary Institute **9.20** NASA **9.21a** NASA **9.21b** NASA **9.22a** NASA/Jet Propulsion Laboratory **9.22b** NASA/Jet Propulsion Laboratory **9.22c** Mark Robinson/Northwestern University **9.23** NASA and Mark Robinson **9.24** Lowell Observatory Photographs **9.25** NASA/Mars Global Surveyor/NSSDC **9.26** NASA/USGS **9.27** NASA/USGS **9.28** NASA/JPL **9.29a** USGS **9.29b** NASA/JPL Maline Space Science Systems **9.29c** R. P. Irwin III and G. A. Franz, National Air and Space Museum, Smithsonian Institution **9.29d** National Air and Space Museum/Smithsonian Institution **9.30** NASA/JPL **9.31** NASA/JPL **9.32** NASA, the Magellan project. Additional processing by John Weiss **9.33** NASA/JPL **9.34** NASA courtesy of NSSDC **9.35** National Geophysical Data Center **9.39** NASA/JPL **9.40** NASA/JPL **9.41** NASA/JPL **9.42** NASA/JPL **9.43a** Kurt Stier/CORBIS **9.43b** Associated Press, AP **9.44** (clockwise from top) Philip Rosenberg; Philip Rosenberg; University of Hawaii; HURL/Chave and Malahoff

CHAPTER 10 Opener: *NASA*

10.1a1 (Mercury-globe) NASA/Courtesy NSSDC **10.1a2** (Mercury-surface) Don Davis **10.1b1** (Venus-globe) NASA/JPL **10.1b2** (Venus-surface) From the Voyage scale model solar system, developed by Challenger Center for Space Science Education, the Smithsonian Institution, and NASA. ©2001 David P. Anderson, Southern Methodist University **10.1c1** (Earth-globe) NASA **10.1c2** (Earth-surface) Barrie Rokeach **10.1d1** (Moon-globe) Akira Fujii **10.1d2** (Moon-surface) Artis Planetarium/The Netherlands **10.1e1** (Mars-globe) NASA with enhancement by Calvin Hamilton **10.1e2** (Mars-surface) NASA/JPL/Caltech **10.2** NASA **10.5** NASA **10.9** Planetary Visions Limited **10.11b** Doug Duncan, University of Colorado **10.12** Planetary Visions Limited **10.13** Planetary Visions Limited **10.15a** Planetary Visions Limited **10.15b** GOES-10 satellite image. Colorization by ARC Science Simulations, ©1998. **10.16** Planetary Visions Limited **10.22a** M. Mendillo, J. Baumgartner, and J. Wilson of Boston University **10.22b** D. Potter, T. Morgan, and R. Killen **10.24** NASA, P. James (Univ. Toledo), T. Clancy (Space Science Inst.), S. Lee (Univ. Colorado), NSSDC **10.25** NASA/JPL and Malin Space Science Systems **10.26** J. Bell (Cornell), M. Wolff (Space Science Inst.), Hubble Heritage Team (STScI/AURA), and NASA **10.27** NASA/JPL/Malin Space Science Systems **10.28** NASA/JPL **10.30** NASA, courtesy NSSDC

CHAPTER 11 Opener: *Joe Bergeron*

11.01 NASA **11.09a** STScI and R. Beebe and A. Simon (NMSU) **11.09b** NASA Infrared Telescope Facility and Glenn Orton **11.11a** NASA/JPL **11.11b** NASA/JPL **11.11c** Lawrence Sromovsky, University of Wisconsin-Madison/ W. M. Keck Observatory **11.11d** NASA/JPL **11.12a** John Spencer/Lowell Observatory **11.12b** J. Clarke (Univ. of Michigan) and NASA **11.14** Courtesy Tim Parker/JPL **11.15** NASA/JPL **11.16** NASA/JPL/Galileo **11.17** PIRL/Lunar & Planetary Laboratory (University of Arizona) and NASA **11.18** University of Arizona and NASA **11.20a** NASA/JPL/Arizona State University **11.20b** DLR/NASA/JPL **11.20c** NASA/JPL/Arizona State University **11.22** Joe Bergeron **11.23** NASA/JPL/Galileo **11.24** Courtesy of Calvin Hamilton

11.25, 11.26a, 11.27, 11.31b, 11.32, 11.33 Cassini Imaging Team and NASA/JPL/Space Science Institute **11.26b** NASA/ESA NASA/JPL/Space Science Institute **11.29a** NASA/USGS **11.29b** NASA/JPL **11.30a** Jody Swann, Tammy Becker, and Alfred McEwen of U.S. Geological Survey in Flagstaff, Arizona **11.30b** NASA/JPL **11.31a** S. Larson (Univ. of Arizona/LPL) **11.31b** NASA/JPL **11.31c** William K. Hartmann **11.32** NASA/JPL **11.33** NASA/JPL **11.34a** NASA/Jet Propulsion Laboratory **11.34b** NASA/Jet Propulsion Laboratory **11.34c** Erich Karkoschka (Univ. of Arizona/LPL) and NASA **11.34d** Paul Schenk, Lunar and Planetary Institute, Houston

CHAPTER 12 Opener: *StockTrek/Photodisc/Getty Images*

12.1 Jeffery Bennett **12.2a** NASA/Jet Propulsion Laboratory **12.2b** NASA/Jet Propulsion Laboratory **12.2c** NASA/Goddard Institute for Space Studies **12.2d** NASA/Goddard Institute for Space Studies **12.3** Laird M. Close (Univ. of Arizona) and Wm. J. Merline (Southwest Research Institute, Boulder, Colorado **12.4, 12.5** Laird M. Close (Univ. of Arizona) and Wm. J. Merline (Southwest Research Institute, Boulder, Colorado and Data provided by Dave Tholen (Univ. of Hawaii) **12.6** Ivan and Colby Navarro/NASA **12.7** Jonathan Blair/Corbis/Bettmann **12.8** Meteorite specimens courtesy of Robert Haag Meteorites **12.9a** Peter Ceravolo **12.9b** Tony & Daphne Hallas **12.10** LASCO, SOHO Consortium, NRI, ESA, and NASA **12.11** NASA/JPL **12.12a** Jeffery Bennett **12.12b** European Space Agency **12.13b** Vic and Jen Winters/Icstars Astronomy **12.13c** ©Vic and Jen Winters/Icstars Astronomy **12.16** Courtesy NASA / JPL-Caltech **12.18** Eliot Young/Southwest Research Institute **12.19a** Hal Weaver and T.E. Smith (STScI) and NASA **12.19b** Courtesy of Paul Schenk (Lunar and Planetary Institute) **12.20a** MSSO, ANU/Science Library/Photo Researchers, Inc. **12.20b** HST Jupiter Imaging Science Team **12.20c** Jeffery Bennett **12.21** Kirk Johnson/Denver Museum of Natural History **12.22** Virgil L. Sharpton **12.23** Jeffery Bennett **12.24** Sovfoto/Eastfoto **12.26** Joe Bergeron

CHAPTER 13 Opener: *European Southern University*

p. 386 Stock Montage, Inc./Historical Pictures Collection **13.9** NASA/JPL **13.13** Dr. Mark Garlick **13.14** Geoffrey Bryden, JPL **13.15** NASA/Ames **13.16** NASA/JPL

CHAPTER S2 Opener: *CORBIS/Bettmann*

CHAPTER S3 Opener: *NASA, Andrew Fruchter and ERO Team (Sylvia Baggett/STScI, Richard Hook/ST-ECF, Zoltan Levay/STScI*

S3.20a ESA, NASA **S3.20b** HST/MIRLIN/Lindsay King (Univ. of Manchester)

CHAPTER S4 Opener: *Fermilab Visual Media Service, U.S. Department of Energy*

S4.1 Fermilab Visual Media Service **S4.6** Harold and Esther Edgerton Foundation, 1998. Palm Press, Inc.

CHAPTER 14 Opener: *Solar and Heliospheric Observatory (SOHO). SOHO is a project of international cooperation between ESA and NASA*

14.1 Corel Corporation **14.04** National Optical Astronomy Observatories/NSO, Sacramento

Peak **14.10b** Ambio-Royal Swedish Academy of Sciences **14.11** National Optical Astronomy Observatory **14.13** Brookhaven National Laboratory **14.14** Lawrence Berkeley National Laboratory **14.15a** Ambio-Royal Swedish Academy of Sciences **14.15b** National Solar Observatory **14.17b** NASA/Jet Propulsion Laboratory **14.18** Courtesy of SOHO. SOHO is a project of international corporation between ESA and NASA **14.19** Image from TRACE (Transition Region and Coronal Explorer), a mission of the Stanford-Lockheed Institute for Space Research (a joint program of the Lockheed-Martin Advanced Technology Center's Solar and Astrophysics laboratory and Stanford's Solar Observatories Group), ands part of the NASA Small Explorer program **14.20** Courtesy of B. Haisch and G. Slater (Lockheed Palo Alto Research Laboratory) **14.21** NASA/Goddard Space Flight Center

CHAPTER 15 Opener: *Hubble Heritage Team (AURA/STScI/NASA)*

15.4 Hubble Heritage Team (AURA/STScI/NASA) **Table 15.1 (Spectrum)** Roger Bell and Mike Briley, University of Maryland **15.5, page 510** Harvard College Observatory **15.6** NASA/Jet Propulsion Laboratory **15.9 (center)** Lowell Observatory **15.15** Anglo-Australian Observatory/David Malin **15.16** NASA/Jet Propulsion Laboratory

CHAPTER 16 Opener: *Michael Sherick/NASA*

16.01 Anglo-Australian Observatory/David Malin **16.03** Chris Brunt & Jerran Ontkean (DRAO, NRC Canada)/NASA **16.04** "Reprinted, with permission, from the Annual Review of Astronomy and Astrophysics, Volume 42 ©2004 by Annual Reviews www.annualreviews.org" **16.05** European Southern Observatory **16.06** Credit: NASA/JPL-Caltech/N. Evans (Univ. of Texas at Austin)/DSS **16.07** IPAC (Infrared Processing and Analysis Center) and Caltech/JPL **16.09** Matthew Bate, university of Exeter, UK **16.10** Bo Reipurth (University of Hawaii)/NASA and The Hubble Heritage Team (STScI /AURA) **16.11** Tom Abel/Laboratory for Computational Astrophysics University of California, San Diego **16.12** C. R. O'Dell (Vanderbilt University)/NASA and The Hubble Heritage Team (STScI /AURA) **16.13a** Harvard-Smithsonian Center for Astrophysics **16.13b** Jeffery Bennett **16.14a** NASA/Jet Propulsion Laboratory **16.14b** NASA/Jet Propulsion Laboratory **16.17** Davy Kirkpatrick, Caltech IPAC **16.18** European Southern Observatory **16.19** Don F. Figer (UCLA) and NASA

CHAPTER 17 Opener: *European Southern Observatory*

17.1 Rebecca Elson and Richard Sword, Cambridge UK, and NASA (Original WFPC2 image courtesy J. Westphal, Caltech) **17.7a** Nordic Optical Telescope, La Palma **17.7b** NASA, Andrew Fruchter and the ERO Team (Sylvia Baggett/STScI, Richard Hook/ST-ECF, Zoltan Levay/STScI) **17.7c** NASA and the Hubble Heritage Team (STScI/AURA) **17.7d** R. Sahai, J. Trauger (JPL), the WFPC2 Science Team, and NASA **17.17** European Southern Observatory **17.18a** Anglo-Australian Observatory/David Malin **17.18b** Anglo-Australian Observatory/David Malin **17.19a** P. Challis (Harvard-Smithsonian Center for Astrophysics) **17.19b** R. McCray (University of Colorado), D. Burrows and S. Park (Pennsylvania State University), and R. Manchester

(Australia Telescope National Facility) **17.20** Joe Bergeron **17.21** Jeffery Bennett

CHAPTER 18 Opener: *NASA, McGill and V. Kaspi at al.*

18.1 NASA/SAO/CXO **18.3** Joe Bergeron **18.4b** M. Shara, B. Williams, and D. Zurek (STScI); R. Gilmozzi (ESO): D. Prialnik (Tel Aviv Univ.) **18.6** NASA, McGill and V. Kaspi at al. **18.8** European Southern Observatory **18.9b** Joe Bergeron **Page 585** (*Bell*) Copyright is held by CWP/ Regents of the Univ. of California: www.physics .ucla.edu~cwp/Phase2/Burnell_Jocelyn_Bell@ 841234567.html; (*Oppenheimer*) CORBIS; (*others*) Courtesy of Astronomical Institute of Bonn University **18.17** Anglo-Australian Observatory, photography by David Malin **18.18** S. Kulkarni, J. Bloom, P. Price, Caltech -NRAO GRB Collaboration

CHAPTER 19 Opener: *Dr. Axel Mellinger*

19.1a Joe Bergeron **19.1b** Joe Bergeron **19.2** Jeffery Bennett **19.3a** Anglo-Australian Observatory/David Malin **19.3b** Paul A. Scowen, Research Professional **19.3c** John Bally **19.3d** NASA/Jet Propulsion Laboratory **19.3e** NASA Earth Observing System **19.3f** Anglo-Australian Observatory/David Malin **19.4** NASA/Jet Propulsion Laboratory **19.5** Anglo-Australian Observatory/David Malin **19.6** NASA/Jet Propulsion Laboratory **19.7a** Tony & Daphne Hallas **19.7b** Jeff Hester **19.8a** J. Keuhane, B. Koralesky, M. Anderson, L. Rudnick (Univ. of Minnesota) and R. Perley (NRAO) **19.8b** NASA/CXO/SAO/ Rutgers and J. Hughes **19.9b** Simulation by D. Strickland and I. Stevens **19.10** John Bally **19.11** Jeff Hester and Paul Scowen (Arizona State University) and NASA **19.12** NASA/Jet Propulsion Laboratory **19.13** Anglo-Australian Observatory/David Malin **19.15** Anglo-Australian Observatory/David Malin **19.16** Anglo-Australian Observatory/David Malin **19.17** NASA ,ESA , S. Beckwith (STScI), and The Hubble Heritage Team (STScI /AURA) **19.18** NASA/Jet Propulsion Laboratory **19.19** NASA/Jet Propulsion Laboratory **19.20** Jeffery Bennett **19.21** Jeffery Bennett **19.21d** European Southern Observatory **19.23** NASA/Jet Propulsion Laboratory

CHAPTER 20 Opener: *NASA, H. Ford (JHU), G. Illingworth (UCO-Lick), M. Clampin (STScI), G. Hartig (STScI), the ACS Science Team, and ESA*

20.1 Sky Publishing Corporation **20.2a** Anglo-Australian Observatory, CCD image by Steve Lee and David Malin **20.2b** Hubble Heritage Team (AURA/STScI/NASA) **20.2c** IAC photo from plates taken with the Isaac Newton telescope, photography by David Malin **20.3** NASA and The Hubble Heritage Team (STScI /AURA) **20.4** P. Knezek (WIYN), NASA, ESA, and The Hubble Heritage Team (STScI /AURA) **20.5** M. Carollo (Swiss Federal Institute of Technology, Zurich) **20.6a** Anglo-Australian Observatory/David Malin **20.6b** Anglo-Australian Observatory, photography David Malin **20.7** NASA/JPL **20.8(SBc)** National Optical Astronomy Observatories **20.8(SBb)** National Optical Astronomy Observatories **20.8 (SBa)** National Optical Astronomy Observatories **20.8(SBO)** Jeffery Bennett **20.8(EO)** Jeffery Bennett **20.8 (E5)** Jeffery Bennett **20.8 (SO)** Jeffery Bennett **20.8 (Sa)** National Optical Astronomy Observatories **20.8 (Sb)** National Optical Astronomy Observatories **20.8 (Sc)** National Optical Astronomy Observatories **20.9** NASA/Jet Propulsion Laboratory **20.10** NASA/Jet Propulsion Laboratory **20.13** High-Z Supernova Search Team, HST, NASA **20.14** NASA and A. Riess (STScI) **20.17** The Observatories of the Carnegie Institution of Washington **20.18** The Huntington Library, Art Collections, and Botanical Gardens

CHAPTER 21 Opener: *Subaru Telescope, National Astronomical Observatory of Japan, ©2000, NAOJ*

21.1 Robert Williams and the HDF Team (STScI) and NASA **21.2** Springel et al. (Virgo Consortium)/Max Planck Institute for Astrophysics **21.3** Quade Paul **21.4** Hyron Spinard, Univ. of California at Berkeley et al. **21.5a** NASA/Jet Propulsion Laboratory/ Brad Whitmore (STScI) **21.5b** W.C. Keel (Univ. of Alabama) **21.6** Robert Williams and the HDF Team (STScI) and NASA **21.7** Neg./Transparency no. [___]. (Photo by Frank Summers. Courtesy Dept. of Library Services, American Museum of Natural History. **21.8a** Anglo-Australian Observatory, photography David Malin **21.8b** Courtesy of Dr. Michael J. West/University of Hawaii **21.9** R. Thompson, M. Rieke, G. Schneider (Univ. of Arizona); N. Scoville (Caltech); and NASA **21.10a** NASA/ SAO/G. Fabbiano et al. **21.10b** Copyright © Subaru Telescope, National Astronomical Observatory of Japan. All rights reserved **21.11** NASA/ UCSB/C. Martinetal and NOAO/KPNO/C. Martin **21.12** Hubble heritage Team (AURA/STScI/ NASA) **21.14** John Bahcall/Institute for Advanced Study and NASA **21.15** Dr. William

Keel/NASA **21.17a,b** Alan Bridle (NRAO/AUI) **21.17c** NRAO/AUI **21.18** Space Telescope Science Institute/ John Biretta **21.19** NASA/Jet Propulsion Laboratory **21.20** Joe Bergeron **21.22** H. Ford (STScI/Johns Hopkins Univ.); L. Dressel, R. Harms, A. Kochar (Applied Research Corp.); Z. Tsvetanov, A Davidsen, G. Kriss (Johns Hopkins Univ.); R. Bohlin, G. Hartig (STScI); B. Margon (Univ. of Washington-Seattle); and NASA **21.24** Emiko-Rose Koike/fiVth.com

CHAPTER 22 Opener: *A. Fruchter, the ERO Team (STScI, ST.ECF)*

22.6 Courtesy of Caltec **22.7** NASA Earth Observing System **22.8** W. N. Colley and E. Turner (Princeton University), J. A. Tyson (Bell Labs, Lucent Technologies **22.10** A. Fruchter, the ERO Team (STScI, ST-ECF) **22.11(inset)** Lawrence Livermore National Laboratory/ Charles Alcock **22.12** Michael Strauss, Princeton University **22.13a** Harvard-Smithsonian Center for Astrophysics **22.13b** Emiko-Rose Koike/fiVth.com **22.14** Andrey Kravtsov/ Kavli Institute for Cosmological Physics, dept. of Astronomy and Astrophysics, University of Chicago

CHAPTER 23 Opener: E. Bunn/University of Richmond

23.2 Joe Bergeron **23.5** Roger Ressmeyer/ Corbis/Bettmann **23.8** E. Bunn/University of Richmond **23.18a** John Kieffer/Peter Arnold, Inc. **23.18b** Joel Gordon Photography

CHAPTER 24 Opener: *Seth Shostak*

24.2 Jeff Greenberg/Visuals Unlimited **24.3** James L. Amos/Photo Researchers, Inc. **24.5a** Stanley M. Awramik/Biological Photo Service **24.5b** Stanley M. Awramik/Biological Photo Service **24.5c** Stanley M. Awramik/Biological Photo Service **24.6** Nih R. Feldman/Visuals Unlimited **24.8a** Woods Hole Oceanographic Institution **24.8b** George Steinmetz Photography **24.9** Roger Ressmeyer/Corbis/Bettmann **24.10a** Jeffery Bennett **24.10b** F. M. Menger and Kurt Gabrielson **24.12** T. E. Adams/Visuals Unlimited **24.13** NASA/Jet Propulsion Laboratory **24.14a** NASA/Johnson Space Center **24.14b** NASA/Jet Propulsion Laboratory **24.14c** Jeffery Bennett **24.17** Yeshe Fenner **24.18** Seth Shostak **24.20** Seth Shostak **24.21** Jeffery Bennett **24.22** NASA/Jet Propulsion Laboratory **24.23** NASA/ Jet Propulsion Laboratory **24.24** NASA/Jet Propulsion Laboratory

Index